# Student Solutions Manual, Volume 1

# Calculus

## NINTH EDITION

## Ron Larson
The Pennsylvania State University
The Behrend College

## Bruce Edwards
University of Florida

Prepared by

## Bruce Edwards
University of Florida

BROOKS/COLE
CENGAGE Learning

Australia • Brazil • Japan • Korea • Mexico • Singapore • Spain • United Kingdom • United States

**BROOKS/COLE**
CENGAGE Learning™

ISBN-13: 978-0-547-21309-5
ISBN-10: 0-547-21309-3

**Brooks/Cole**
10 Davis Drive
Belmont, CA 94002-3098
USA

Cengage Learning is a leading provider of customized learning solutions with office locations around the globe, including Singapore, the United Kingdom, Australia, Mexico, Brazil, and Japan. Locate your local office at: **www.cengage.com/international**

Cengage Learning products are represented in Canada by Nelson Education, Ltd.

To learn more about Brooks/Cole, visit
**www.cengage.com/brookscole**

Purchase any of our products at your local college store or at our preferred online store
**www.ichapters.com**

2 3 4 5 6 7 12 11 10

# C H A P T E R   P
# Preparation for Calculus

# C H A P T E R   P
## Preparation for Calculus

## Section P.1   Graphs and Models

**1.** $y = -\frac{3}{2}x + 3$

$x$-intercept: $(2, 0)$

$y$-intercept: $(0, 3)$

Matches graph (b).

**3.** $y = 3 - x^2$

$x$-intercepts: $\left(\sqrt{3}, 0\right), \left(-\sqrt{3}, 0\right)$

$y$-intercept: $(0, 3)$

Matches graph (a).

**5.** $y = \frac{1}{2}x + 2$

| $x$ | $-4$ | $-2$ | 0 | 2 | 4 |
|---|---|---|---|---|---|
| $y$ | 0 | 1 | 2 | 3 | 4 |

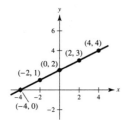

**7.** $y = 4 - x^2$

| $x$ | $-3$ | $-2$ | 0 | 2 | 3 |
|---|---|---|---|---|---|
| $y$ | $-5$ | 0 | 4 | 0 | $-5$ |

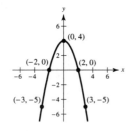

**9.** $y = |x + 2|$

| $x$ | $-5$ | $-4$ | $-3$ | $-2$ | $-1$ | 0 | 1 |
|---|---|---|---|---|---|---|---|
| $y$ | 3 | 2 | 1 | 0 | 1 | 2 | 3 |

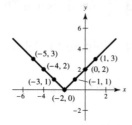

**11.** $y = \sqrt{x} - 6$

| $x$ | 0 | 1 | 4 | 9 | 16 |
|---|---|---|---|---|---|
| $y$ | $-6$ | $-5$ | $-4$ | $-3$ | $-2$ |

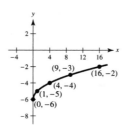

**13.** $y = \dfrac{3}{x}$

| $x$ | $-3$ | $-2$ | $-1$ | 0 | 1 | 2 | 3 |
|---|---|---|---|---|---|---|---|
| $y$ | $-1$ | $-\frac{3}{2}$ | $-3$ | Undef. | 3 | $\frac{3}{2}$ | 1 |

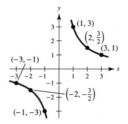

**15.**

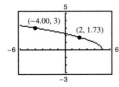

$$
\begin{aligned}
&\text{Xmin} = -5 \\
&\text{Xmax} = 4 \\
&\text{Xscl} = 1 \\
&\text{Ymin} = -5 \\
&\text{Ymax} = 8 \\
&\text{Yscl} = 1
\end{aligned}
$$

Note that $y = -3$ when $x = 0$ and $y = 0$ when $x = -1$.

**17.** $y = \sqrt{5 - x}$

(a) $(2, y) = (2, 1.73)$   $\left(y = \sqrt{5 - 2} = \sqrt{3} \approx 1.73\right)$

(b) $(x, 3) = (-4, 3)$   $\left(3 = \sqrt{5 - (-4)}\right)$

**19.** $y = 2x - 5$

$y$-intercept: $y = 2(0) - 5 = -5;\ (0, -5)$

$x$-intercept: $0 = 2x - 5$

$\qquad\qquad\quad 5 = 2x$

$\qquad\qquad\quad x = \frac{5}{2};\ \left(\frac{5}{2}, 0\right)$

**21.** $y = x^2 + x - 2$

$y$-intercept: $y = 0^2 + 0 - 2$

$\qquad\qquad\quad y = -2;\ (0, -2)$

$x$-intercepts: $0 = x^2 + x - 2$

$\qquad\qquad\quad 0 = (x + 2)(x - 1)$

$\qquad\qquad\quad x = -2, 1;\ (-2, 0),\ (1, 0)$

**23.** $y = x\sqrt{16 - x^2}$

$y$-intercept: $y = 0\sqrt{16 - 0^2} = 0;\ (0, 0)$

$x$-intercepts: $0 = x\sqrt{16 - x^2}$

$\qquad\qquad\quad 0 = x\sqrt{(4 - x)(4 + x)}$

$\qquad\qquad\quad x = 0, 4, -4;\ (0, 0),\ (4, 0),\ (-4, 0)$

**25.** $y = \dfrac{2 - \sqrt{x}}{5x}$

$y$-intercept: None. $x$ cannot equal 0.

$x$-intercept: $0 = \dfrac{2 - \sqrt{x}}{5x}$

$\qquad\qquad\quad 0 = 2 - \sqrt{x}$

$\qquad\qquad\quad x = 4;\ (4, 0)$

**27.** $x^2y - x^2 + 4y = 0$

$y$-intercept: $0^2(y) - 0^2 + 4y = 0$

$\qquad\qquad\qquad\qquad y = 0;\ (0, 0)$

$x$-intercept: $x^2(0) - x^2 + 4(0) = 0$

$\qquad\qquad\qquad\qquad x = 0;\ (0, 0)$

**29.** Symmetric with respect to the $y$-axis because

$y = (-x)^2 - 6 = x^2 - 6.$

**31.** Symmetric with respect to the $x$-axis because

$(-y)^2 = y^2 = x^3 - 8x.$

**33.** Symmetric with respect to the origin because

$(-x)(-y) = xy = 4.$

**35.** $y = 4 - \sqrt{x + 3}$

No symmetry with respect to either axis or the origin.

**37.** Symmetric with respect to the origin because

$$-y = \dfrac{-x}{(-x)^2 + 1}$$

$$y = \dfrac{x}{x^2 + 1}.$$

**39.** $y = \left|x^3 + x\right|$ is symmetric with respect to the $y$-axis

because $y = \left|(-x)^3 + (-x)\right| = \left|-(x^3 + x)\right| = \left|x^3 + x\right|.$

**41.** $y = 2 - 3x$

Intercepts: $(0, 2),\ \left(\frac{2}{3}, 0\right)$

Symmetry: None

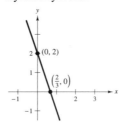

**43.** $y = \frac{1}{2}x - 4$

Intercepts: $(8, 0),\ (0, -4)$

Symmetry: none

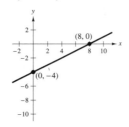

**45.** $y = 9 - x^2$

Intercepts: $(0, 9), (3, 0), (-3, 0)$

Symmetry: $y$-axis

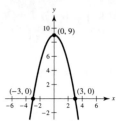

**47.** $y = (x + 3)^2$

Intercepts: $(-3, 0), (0, 9)$

Symmetry: none

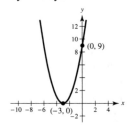

**49.** $y = x^3 + 2$

Intercepts: $\left(-\sqrt[3]{2}, 0\right), (0, 2)$

Symmetry: none

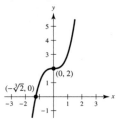

**51.** $y = x\sqrt{x + 5}$

Intercepts: $(0, 0), (-5, 0)$

Symmetry: none

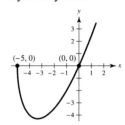

**53.** $x = y^3$

Intercept: $(0, 0)$

Symmetry: origin

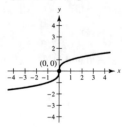

**55.** $y = \dfrac{8}{x}$

Intercepts: none

Symmetry: origin

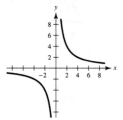

**57.** $y = 6 - |x|$

Intercepts: $(0, 6), (-6, 0), (6, 0)$

Symmetry: $y$-axis

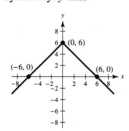

**59.** $y^2 - x = 9$

$$y^2 = x + 9$$

$$y = \pm\sqrt{x + 9}$$

Intercepts: $(0, 3), (0, -3), (-9, 0)$

Symmetry: $x$-axis

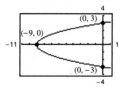

**61.** $x + 3y^2 = 6$

$$3y^2 = 6 - x$$

$$y = \pm\sqrt{\frac{6 - x}{3}}$$

Intercepts: $(6, 0), \left(0, \sqrt{2}\right), \left(0, -\sqrt{2}\right)$

Symmetry: $x$-axis

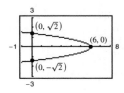

**63.** $x + y = 8 \Rightarrow y = 8 - x$

$4x - y = 7 \Rightarrow y = 4x - 7$

$$8 - x = 4x - 7$$

$$15 = 5x$$

$$3 = x$$

The corresponding $y$-value is $y = 5$.

Point of intersection: $(3, 5)$

**65.** $x^2 + y = 6 \Rightarrow y = 6 - x^2$

$x + y = 4 \Rightarrow y = 4 - x$

$$6 - x^2 = 4 - x$$

$$0 = x^2 - x - 2$$

$$0 = (x - 2)(x + 1)$$

$$x = 2, -1$$

The corresponding y-values are $y = 2$ (for $x = 2$) and $y = 5$ (for $x = -1$).

Points of intersection: $(2, 2), (-1, 5)$

**67.** $x^2 + y^2 = 5 \Rightarrow y^2 = 5 - x^2$

$x - y = 1 \Rightarrow y = x - 1$

$$5 - x^2 = (x - 1)^2$$

$$5 - x^2 = x^2 - 2x + 1$$

$$0 = 2x^2 - 2x - 4 = 2(x + 1)(x - 2)$$

$$x = -1 \text{ or } x = 2$$

The corresponding $y$-values are $y = -2$ (for $x = -1$) and $y = 1$ (for $x = 2$).

Points of intersection: $(-1, -2), (2, 1)$

**69.** $y = x^3$

$$y = x$$

$$x^3 = x$$

$$x^3 - x = 0$$

$$x(x + 1)(x - 1) = 0$$

$$x = 0, x = -1, \text{ or } x = 1$$

The corresponding $y$-values are

$y = 0$ (for $x = 0$), $y = -1$ (for $x = -1$), and $y = 1$ (for $x = 1$).

Points of intersection: $(0, 0), (-1, -1), (1, 1)$

**71.** Analytically,

$$y = x^3 - 2x^2 + x - 1$$

$$y = -x^2 + 3x - 1$$

$$x^3 - 2x^2 + x - 1 = -x^2 + 3x - 1$$

$$x^3 - x^2 - 2x = 0$$

$$x(x - 2)(x + 1) = 0$$

$$x = -1, 0, 2.$$

Points of intersection: $(-1, -5), (0, -1), (2, 1)$

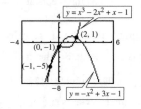

**73.** $y = \sqrt{x + 6}$

$$y = \sqrt{-x^2 - 4x}$$

Points of intersection: $(-2, 2), \left(-3, \sqrt{3}\right) \approx (-3, 1.732)$

Analytically,     $\sqrt{x + 6} = \sqrt{-x^2 - 4x}$

$$x + 6 = -x^2 - 4x$$

$$x^2 + 5x + 6 = 0$$

$$(x + 3)(x + 2) = 0$$

$$x = -3, -2.$$

**75.** (a) Using a graphing utility, you obtain

$$y = -0.027t^2 + 5.73t + 26.9.$$

(b)

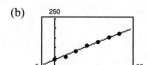

The model is a good fit for the data.

(c) For 2010, $t = 40$ and $y = 212.9$.

**77.**
$$C = R$$

$$5.5\sqrt{x} + 10{,}000 = 3.29x$$

$$\left(5.5\sqrt{x}\right)^2 = (3.29x - 10{,}000)^2$$

$$30.25x = 10.8241x^2 - 65{,}800x + 100{,}000{,}000$$

$$0 = 10.8241x^2 - 65{,}830.25x + 100{,}000{,}000 \quad \text{Use the Quadratic Formula.}$$

$$x \approx 3133 \text{ units}$$

The other root, $x \approx 2949$, does not satisfy the equation $R = C$.

This problem can also be solved by using a graphing utility and finding the intersection of the graphs of $C$ and $R$.

**79.** Answers may vary. *Sample answer*:

$y = (x + 4)(x - 3)(x - 8)$ has intercepts at $x = -4$, $x = 3$, and $x = 8$.

**81.** (a) If $(x, y)$ is on the graph, then so is $(-x, y)$ by $y$-axis symmetry. Because $(-x, y)$ is on the graph, then so is $(-x, -y)$ by $x$-axis symmetry. So, the graph is symmetric with respect to the origin. The converse is not true. For example, $y = x^3$ has origin symmetry but is not symmetric with respect to either the $x$-axis or the $y$-axis.

(b) Assume that the graph has $x$-axis and origin symmetry. If $(x, y)$ is on the graph, so is $(x, -y)$ by $x$-axis symmetry. Because $(x, -y)$ is on the graph, then so is $\left(-x, -(-y)\right) = (-x, y)$ by origin symmetry. Therefore, the graph is symmetric with respect to the $y$-axis. The argument is similar for $y$-axis and origin symmetry.

**83.** False. $x$-axis symmetry means that if $(-4, -5)$ is on the graph, then $(-4, 5)$ is also on the graph. For example, $(4, -5)$ is not on the graph of $x = y^2 - 29$, whereas $(-4, -5)$ is on the graph.

**85.** True. The $x$-intercepts are $\left(\dfrac{-b \pm \sqrt{b^2 - 4ac}}{2a}, 0\right)$.

**87.**

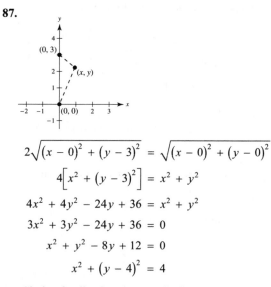

$$2\sqrt{(x - 0)^2 + (y - 3)^2} = \sqrt{(x - 0)^2 + (y - 0)^2}$$

$$4\left[x^2 + (y - 3)^2\right] = x^2 + y^2$$

$$4x^2 + 4y^2 - 24y + 36 = x^2 + y^2$$

$$3x^2 + 3y^2 - 24y + 36 = 0$$

$$x^2 + y^2 - 8y + 12 = 0$$

$$x^2 + (y - 4)^2 = 4$$

Circle of radius 2 and center $(0, 4)$.

# Section P.2   Linear Models and Rates of Change

**1.** $m = 1$

**3.** $m = 0$

**5.** $m = -12$

**7.**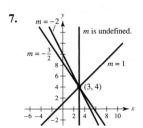

**9.** $m = \dfrac{2 - (-4)}{5 - 3} = \dfrac{6}{2} = 3$

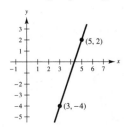

**11.** $m = \dfrac{1 - 6}{4 - 4} = \dfrac{-5}{0}$, undefined.

The line is vertical

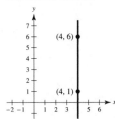

**13.** $m = \dfrac{\dfrac{2}{3} - \dfrac{1}{6}}{-\dfrac{1}{2} - \left(-\dfrac{3}{4}\right)} = \dfrac{\dfrac{1}{2}}{\dfrac{1}{4}} = 2$

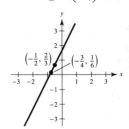

**15.** Because the slope is 0, the line is horizontal and its equation is $y = 2$. Therefore, three additional points are $(0, 2)$, $(1, 2)$, $(5, 2)$.

**17.** The equation of this line is

$$y - 7 = -3(x - 1)$$
$$y = -3x + 10.$$

Therefore, three additional points are $(0, 10)$, $(2, 4)$, and $(3, 1)$.

**19.** (a)  Slope $= \dfrac{\Delta y}{\Delta x} = \dfrac{1}{3}$

(b)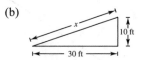

By the Pythagorean Theorem,

$$x^2 = 30^2 + 10^2 = 1000$$
$$x = 10\sqrt{10} \approx 31.623 \text{ feet.}$$

**21.** (a)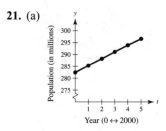

(b)  The slopes are:

$$\dfrac{285.3 - 282.4}{1 - 0} = 2.9$$

$$\dfrac{288.2 - 285.3}{2 - 1} = 2.9$$

$$\dfrac{291.1 - 288.2}{3 - 2} = 2.9$$

$$\dfrac{293.9 - 291.1}{4 - 3} = 2.8$$

$$\dfrac{296.6 - 293.9}{5 - 4} = 2.7$$

The population increased least rapidly from 2004 to 2005.

**23.**  $y = 4x - 3$

the slope is $m = 4$ and the $y$-intercept is $(0, -3)$.

**25.**  $x + 5y = 20$

$$y = -\tfrac{1}{5}x + 4$$

Therefore, the slope is $m = -\tfrac{1}{5}$ and the $y$-intercept is $(0, 4)$.

**27.**  $x = 4$

The line is vertical. Therefore, the slope is undefined and there is no $y$-intercept.

**29.** $y = \frac{3}{4}x + 3$

$4y = 3x + 12$

$0 = 3x - 4y + 12$

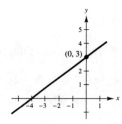

**31.** $y = \frac{2}{3}x$

$3y = 2x$

$0 = 2x - 3y$

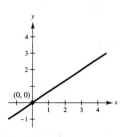

**33.** $y + 2 = 3(x - 3)$

$y + 2 = 3x - 9$

$y = 3x - 11$

$0 = 3x - y - 11$

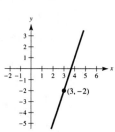

**35.** $m = \frac{8 - 0}{4 - 0} = 2$

$y - 0 = 2(x - 0)$

$y = 2x$

$0 = 2x - y$

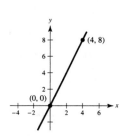

**37.** $m = \frac{1 - (-3)}{2 - 0} = 2$

$y - 1 = 2(x - 2)$

$y - 1 = 2x - 4$

$0 = 2x - y - 3$

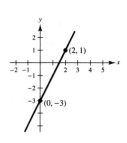

**39.** $m = \frac{8 - 0}{2 - 5} = -\frac{8}{3}$

$y - 0 = -\frac{8}{3}(x - 5)$

$y = -\frac{8}{3}x + \frac{40}{3}$

$8x + 3y - 40 = 0$

**41.** $m = \frac{8 - 3}{6 - 6} = \frac{5}{0}$, undefined

The line is horizontal.

$x = 6$

$x - 6 = 0$

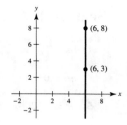

**43.** $m = \dfrac{\frac{7}{2} - \frac{3}{4}}{\frac{1}{2} - 0} = \dfrac{\frac{11}{4}}{\frac{1}{2}} = \frac{11}{2}$

$y - \frac{3}{4} = \frac{11}{2}(x - 0)$

$y = \frac{11}{2}x + \frac{3}{4}$

$0 = 22x - 4y + 3$

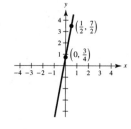

**45.** $x = 3$

$x - 3 = 0$

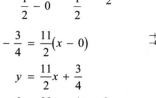

**47.** $\frac{x}{2} + \frac{y}{3} = 1$

$3x + 2y - 6 = 0$

**49.** $\frac{x}{a} + \frac{y}{a} = 1$

$\frac{1}{a} + \frac{2}{a} = 1$

$\frac{3}{a} = 1$

$a = 3 \Rightarrow x + y = 3$

$x + y - 3 = 0$

**51.** $y = -3$

**53.** $y = -2x + 1$

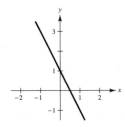

**55.** $y - 2 = \frac{3}{2}(x - 1)$

$y = \frac{3}{2}x + \frac{1}{2}$

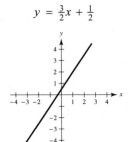

**57.** $2x - y - 3 = 0$

$y = 2x - 3$

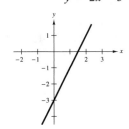

**59.** (a)

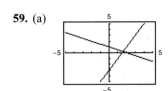

The lines do not appear perpendicular.

(b)

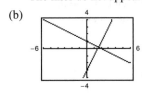

The lines appear perpendicular.

The lines are perpendicular because their slopes 2 and $-\frac{1}{2}$ are negative reciprocals of each other. You must use a square setting in order for perpendicular lines to appear perpendicular. Answers depend on calculator used.

**61.** The given line is vertical.

(a)  $x = -7$, or $x + 7 = 0$

(b)  $y = -2$, or $y + 2 = 0$

**63.** $4x - 2y = 3$

$y = 2x - \frac{3}{2}$

$m = 2$

(a)  $y - 1 = 2(x - 2)$

$y - 1 = 2x - 4$

$0 = 2x - y - 3$

(b)      $y - 1 = -\frac{1}{2}(x - 2)$

$2y - 2 = -x + 2$

$x + 2y - 4 = 0$

**65.** $5x - 3y = 0$

$y = \frac{5}{3}x$

$m = \frac{5}{3}$

(a)     $y - \frac{7}{8} = \frac{5}{3}\left(x - \frac{3}{4}\right)$

$24y - 21 = 40x - 30$

$0 = 40x - 24y - 9$

(b)         $y - \frac{7}{8} = -\frac{3}{5}\left(x - \frac{3}{4}\right)$

$40y - 35 = -24x + 18$

$24x + 40y - 53 = 0$

**67.** The slope is 250. $V = 1850$ when $t = 8$.

$V = 250(t - 8) + 1850 = 250t - 150$.

**69.** The slope is $-1600$. $V = 17{,}200$ when $t = 8$.

$V = -1600(t - 8) + 17{,}200 = -1600t + 30{,}000$

**71.**

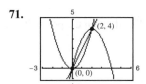

You can use the graphing utility to determine that the points of intersection are $(0, 0)$ and $(2, 4)$. Analytically,

$x^2 = 4x - x^2$

$2x^2 - 4x = 0$

$2x(x - 2) = 0$

$x = 0 \Rightarrow y = 0 \Rightarrow (0, 0)$

$x = 2 \Rightarrow y = 4 \Rightarrow (2, 4)$.

The slope of the line joining $(0, 0)$ and $(2, 4)$ is $m = (4 - 0)/(2 - 0) = 2$. So, an equation of the line is

$y - 0 = 2(x - 0)$

$y = 2x$.

**73.** $m_1 = \dfrac{1 - 0}{-2 - (-1)} = -1$

$m_2 = \dfrac{-2 - 0}{2 - (-1)} = -\dfrac{2}{3}$

$m_1 \neq m_2$

The points are not collinear.

**75.** Equations of perpendicular bisectors:

$y - \dfrac{c}{2} = \dfrac{a - b}{c}\left(x - \dfrac{a + b}{2}\right)$

$y - \dfrac{c}{2} = \dfrac{a + b}{-c}\left(x - \dfrac{b - a}{2}\right)$

Setting the right-hand sides of the two equations equal and solving for $x$ yields $x = 0$.

Letting $x = 0$ in either equation gives the point of intersection:

$\left(0, \dfrac{-a^2 + b^2 + c^2}{2c}\right)$.

This point lies on the third perpendicular bisector, $x = 0$.

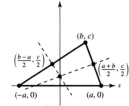

**77.** Equations of altitudes:

$y = \dfrac{a - b}{c}(x + a)$

$x = b$

$y = -\dfrac{a + b}{c}(x - a)$

Solving simultaneously, the point of intersection is

$\left(b, \dfrac{a^2 - b^2}{c}\right)$.

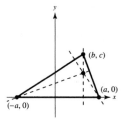

**79.** Find the equation of the line through the points $(0, 32)$ and $(100, 212)$.

$m = \dfrac{180}{100} = \dfrac{9}{5}$

$F - 32 = \dfrac{9}{5}(C - 0)$

$F = \dfrac{9}{5}C + 32$

or

$C = \dfrac{1}{9}(5F - 160)$

$5F - 9C - 160 = 0$

For $F = 72°$, $C \approx 22.2°$.

**81.** (a) $W_1 = 0.75x + 14.50$

$W_2 = 1.30x + 11.20$

(b)

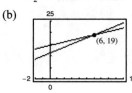

Using a graphing utility, the point of intersection is $(6, 19)$

Analytically,

$W_1 = W_2$

$0.75x + 14.50 = 1.30x + 11.20$

$3.3 = 0.55x$

$6 = x$

$y = 1.30(6) + 11.20 = 19.$

(c) When six units are produced, the wage for both options is $19.00 per hour. Choose option 1 if you think you will produce less than six units per hour, and choose option 2 if you think you will produce more than six.

**83.** (a) Two points are $(50, 780)$ and $(47, 825)$. The slope is

$m = \dfrac{825 - 780}{47 - 50} = \dfrac{45}{-3} = -15.$

$p - 780 = -15(x - 50)$

$p = -15x + 750 + 780 = -15x + 1530$

or

$x = \dfrac{1}{15}(1530 - p)$

(b)

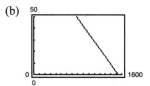

If $p = 855$, then $x = 45$ units.

(c) If $p = 795$, then $x = \dfrac{1}{15}(1530 - 795) = 49$ units.

**85.** The tangent line is perpendicular to the line joining the point $(5, 12)$ and the center $(0, 0)$.

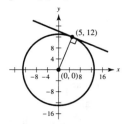

Slope of the line joining $(5, 12)$ and $(0, 0)$ is $\dfrac{12}{5}$.

The equation of the tangent line is

$y - 12 = \dfrac{-5}{12}(x - 5)$

$y = \dfrac{-5}{12}x + \dfrac{169}{12}$

$5x + 12y - 169 = 0.$

**87.** $4x + 3y - 10 = 0 \Rightarrow d = \dfrac{|4(0) + 3(0) - 10|}{\sqrt{4^2 + 3^2}}$

$$= \dfrac{10}{5} = 2$$

**89.** $x - y - 2 = 0 \Rightarrow d = \dfrac{|1(-2) + (-1)(1) - 2|}{\sqrt{1^2 + 1^2}}$

$$= \dfrac{5}{\sqrt{2}} = \dfrac{5\sqrt{2}}{2}$$

**91.** A point on the line $x + y = 1$ is $(0, 1)$. The distance from the point $(0, 1)$ to $x + y - 5 = 0$ is

$$d = \dfrac{|1(0) + 1(1) - 5|}{\sqrt{1^2 + 1^2}} = \dfrac{|1 - 5|}{\sqrt{2}} = \dfrac{4}{\sqrt{2}} = 2\sqrt{2}.$$

**93.** If $A = 0$, then $By + C = 0$ is the horizontal line $y = -C/B$. The distance to $(x_1, y_1)$ is

$$d = \left| y_1 - \left( \dfrac{-C}{B} \right) \right| = \dfrac{|By_1 + C|}{|B|} = \dfrac{|Ax_1 + By_1 + C|}{\sqrt{A^2 + B^2}}.$$

If $B = 0$, then $Ax + C = 0$ is the vertical line $x = -C/A$. The distance to $(x_1, y_1)$ is

$$d = \left| x_1 - \left( \dfrac{-C}{A} \right) \right| = \dfrac{|Ax_1 + C|}{|A|} = \dfrac{|Ax_1 + By_1 + C|}{\sqrt{A^2 + B^2}}.$$

(Note that A and B cannot both be zero.) The slope of the line $Ax + By + C = 0$ is $-A/B$.

The equation of the line through $(x_1, y_1)$ perpendicular to $Ax + By + C = 0$ is:

$$y - y_1 = \dfrac{B}{A}(x - x_1)$$
$$Ay - Ay_1 = Bx - Bx_1$$
$$Bx_1 - Ay_1 = Bx - Ay$$

The point of intersection of these two lines is:

$$Ax + By = -C \qquad \Rightarrow A^2x + ABy = -AC \qquad (1)$$
$$Bx - Ay = Bx_1 - Ay_1 \Rightarrow \underline{B^2x - ABy = B^2x_1 - ABy_1} \qquad (2)$$
$$(A^2 + B^2)x = -AC + B^2x_1 - ABy_1 \quad \left(\text{By adding equations (1) and (2)}\right)$$
$$x = \dfrac{-AC + B^2x_1 - ABy_1}{A^2 + B^2}$$

$$Ax + By = -C \qquad \Rightarrow \quad ABx + B^2y = -BC \qquad (3)$$
$$Bx - Ay = Bx_1 - Ay_1 \Rightarrow \underline{-ABx + A^2y = -ABx_1 + A^2y_1} \qquad (4)$$
$$(A^2 + B^2)y = -BC - ABx_1 + A^2y_1 \quad \left(\text{By adding equations (3) and (4)}\right)$$
$$y = \dfrac{-BC - ABx_1 + A^2y_1}{A^2 + B^2}$$

$$\left( \dfrac{-AC + B^2x_1 - ABy_1}{A^2 + B^2}, \dfrac{-BC - ABx_1 + A^2y_1}{A^2 + B^2} \right) \text{ point of intersection}$$

The distance between $(x_1, y_1)$ and this point gives you the distance between $(x_1, y_1)$ and the line $Ax + By + C = 0$.

$$d = \sqrt{\left[ \dfrac{-AC + B^2x_1 - ABy_1}{A^2 + B^2} - x_1 \right]^2 + \left[ \dfrac{-BC - ABx_1 + A^2y_1}{A^2 + B^2} - y_1 \right]^2}$$

$$= \sqrt{\left[ \dfrac{-AC - ABy_1 - A^2x_1}{A^2 + B^2} \right]^2 + \left[ \dfrac{-BC - ABx_1 - B^2y_1}{A^2 + B^2} \right]^2}$$

$$= \sqrt{\left[ \dfrac{-A(C + By_1 + Ax_1)}{A^2 + B^2} \right]^2 + \left[ \dfrac{-B(C + Ax_1 + By_1)}{A^2 + B^2} \right]^2} = \sqrt{\dfrac{(A^2 + B^2)(C + Ax_1 + By_1)^2}{(A^2 + B^2)^2}} = \dfrac{|Ax_1 + By_1 + C|}{\sqrt{A^2 + B^2}}$$

**95.** For simplicity, let the vertices of the rhombus be $(0, 0)$, $(a, 0)$, $(b, c)$, and $(a + b, c)$, as shown in the figure. The slopes of the diagonals are then $m_1 = \dfrac{c}{a + b}$ and

$m_2 = \dfrac{c}{b - a}$. Because the sides of the rhombus are equal, $a^2 = b^2 + c^2$, and you have

$$m_1 m_2 = \frac{c}{a + b} \cdot \frac{c}{b - a} = \frac{c^2}{b^2 - a^2} = \frac{c^2}{-c^2} = -1.$$

Therefore, the diagonals are perpendicular.

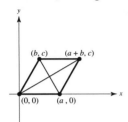

**97.** Consider the figure below in which the four points are collinear. Because the triangles are similar, the result immediately follows.

$$\frac{y_2^* - y_1^*}{x_2^* - x_1^*} = \frac{y_2 - y_1}{x_2 - x_1}$$

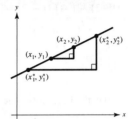

**99.** True.

$$ax + by = c_1 \Rightarrow y = -\frac{a}{b}x + \frac{c_1}{b} \Rightarrow m_1 = -\frac{a}{b}$$

$$bx - ay = c_2 \Rightarrow y = \frac{b}{a}x - \frac{c_2}{a} \Rightarrow m_2 = \frac{b}{a}$$

$$m_2 = -\frac{1}{m_1}$$

## Section P.3  Functions and Their Graphs

**1.** (a) Domain of $f$: $-4 \le x \le 4 \Rightarrow [-4, 4]$

Range of $f$: $-3 \le y \le 5 \Rightarrow [-3, 5]$

Domain of $g$: $-3 \le x \le 3 \Rightarrow [-3, 3]$

Range of $g$: $-4 \le y \le 4 \Rightarrow [-4, 4]$

(b) $f(-2) = -1$

$g(3) = -4$

(c) $f(x) = g(x)$ for $x = -1$

(d) $f(x) = 2$ for $x = 1$

(e) $g(x) = 0$ for $x = -1, 1$ and $2$

**3.** (a) $f(0) = 7(0) - 4 = -4$

(b) $f(-3) = 7(-3) - 4 = -25$

(c) $f(b) = 7(b) - 4 = 7b - 4$

(d) $f(x - 1) = 7(x - 1) - 4 = 7x - 11$

**5.** (a) $g(0) = 5 - 0^2 = 5$

(b) $g(\sqrt{5}) = 5 - (\sqrt{5})^2 = 5 - 5 = 0$

(c) $g(-2) = 5 - (-2)^2 = 5 - 4 = 1$

(d) $g(t - 1) = 5 - (t - 1)^2 = 5 - (t^2 - 2t + 1)$
$$= 4 + 2t - t^2$$

**7.** (a) $f(0) = \cos(2(0)) = \cos 0 = 1$

(b) $f\left(-\dfrac{\pi}{4}\right) = \cos\left(2\left(-\dfrac{\pi}{4}\right)\right) = \cos\left(-\dfrac{\pi}{2}\right) = 0$

(c) $f\left(\dfrac{\pi}{3}\right) = \cos\left(2\left(\dfrac{\pi}{3}\right)\right) = \cos\dfrac{2\pi}{3} = -\dfrac{1}{2}$

**9.** $\dfrac{f(x + \Delta x) - f(x)}{\Delta x} = \dfrac{(x + \Delta x)^3 - x^3}{\Delta x} = \dfrac{x^3 + 3x^2\Delta x + 3x^2(\Delta x)^2 + (\Delta x)^3 - x^3}{\Delta x} = 3x^2 + 3x\Delta x + (\Delta x)^2,\ \Delta x \ne 0$

**11.** $\dfrac{f(x) - f(2)}{x - 2} = \dfrac{\left(1/\sqrt{x - 1} - 1\right)}{x - 2}$

$$= \frac{1 - \sqrt{x - 1}}{(x - 2)\sqrt{x - 1}} \cdot \frac{1 + \sqrt{x - 1}}{1 + \sqrt{x - 1}} = \frac{2 - x}{(x - 2)\sqrt{x - 1}\left(1 + \sqrt{x - 1}\right)} = \frac{-1}{\sqrt{x - 1}\left(1 + \sqrt{x - 1}\right)},\ x \ne 2$$

**13.** $f(x) = 4x^2$

Domain: $(-\infty, \infty)$

Range: $[0, \infty)$

**15.** $g(x) = \sqrt{6x}$

Domain: $6x \geq 0$

$$x \geq 0 \Rightarrow [0, \infty)$$

Range: $[0, \infty)$

**17.** $f(t) = \sec\dfrac{\pi t}{4}$

$$\dfrac{\pi t}{4} \neq \dfrac{(2n + 1)\pi}{2} \Rightarrow t \neq 4n + 2$$

Domain: all $t \neq 4n + 2$, $n$ an integer

Range: $(-\infty, -1] \cup [1, \infty)$

**19.** $f(x) = \dfrac{3}{x}$

Domain: all $x \neq 0 \Rightarrow (-\infty, 0) \cup (0, \infty)$

Range: $(-\infty, 0) \cup (0, \infty)$

**21.** $f(x) = \sqrt{x} + \sqrt{1 - x}$

$x \geq 0$   and   $1 - x \geq 0$

$x \geq 0$   and   $x \leq 1$

Domain: $0 \leq x \leq 1 \Rightarrow [0, 1]$

**23.** $g(x) = \dfrac{2}{1 - \cos x}$

$1 - \cos x \neq 0$

$\cos x \neq 1$

Domain: all $x \neq 2n\pi$, $n$ an integer

**25.** $f(x) = \dfrac{1}{|x + 3|}$

$|x + 3| \neq 0$

$x + 3 \neq 0$

Domain: all $x \neq -3$

Domain: $(-\infty, -3) \cup (-3, \infty)$

**27.** $f(x) = \begin{cases} 2x + 1, & x < 0 \\ 2x + 2, & x \geq 0 \end{cases}$

(a) $f(-1) = 2(-1) + 1 = -1$

(b) $f(0) = 2(0) + 2 = 2$

(c) $f(2) = 2(2) + 2 = 6$

(d) $f(t^2 + 1) = 2(t^2 + 1) + 2 = 2t^2 + 4$

(**Note:** $t^2 + 1 \geq 0$ for all $t$)

Domain: $(-\infty, \infty)$

Range: $(-\infty, 1) \cup [2, \infty)$

**29.** $f(x) = \begin{cases} |x| + 1, & x < 1 \\ -x + 1, & x \geq 1 \end{cases}$

(a) $f(-3) = |-3| + 1 = 4$

(b) $f(1) = -1 + 1 = 0$

(c) $f(3) = -3 + 1 = -2$

(d) $f(b^2 + 1) = -(b^2 + 1) + 1 = -b^2$

Domain: $(-\infty, \infty)$

Range: $(-\infty, 0] \cup [1, \infty)$

**31.** $f(x) = 4 - x$

Domain: $(-\infty, \infty)$

Range: $(-\infty, \infty)$

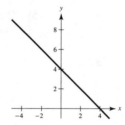

**33.** $h(x) = \sqrt{x - 6}$

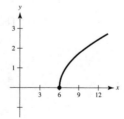

Domain: $x - 6 \geq 0$

$$x \geq 6 \Rightarrow [6, \infty)$$

Range: $[0, \infty)$

**35.** $f(x) = \sqrt{9 - x^2}$

Domain: $[-3, 3]$

Range: $[0, 3]$

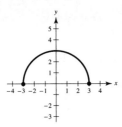

**37.** $g(t) = 3 \sin \pi t$

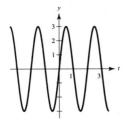

Domain: $(-\infty, \infty)$

Range: $[-3, 3]$

**39.** The student travels $\dfrac{2 - 0}{4 - 0} = \dfrac{1}{2}$ mi/min during the first 4 minutes. The student is stationary for the next 2 minutes. Finally, the student travels $\dfrac{6 - 2}{10 - 6} = 1$ mi/min during the final 4 minutes.

**41.** $x - y^2 = 0 \Rightarrow y = \pm\sqrt{x}$

$y$ is not a function of $x$. Some vertical lines intersect the graph twice.

**43.** $y$ is a function of $x$. Vertical lines intersect the graph at most once.

**45.** $x^2 + y^2 = 16 \Rightarrow y = \pm\sqrt{16 - x^2}$

$y$ is not a function of $x$ because there are two values of $y$ for some $x$.

**47.** $y^2 = x^2 - 1 \Rightarrow y = \pm\sqrt{x^2 - 1}$

$y$ is not a function of $x$ because there are two values of $y$ for some $x$.

**49.** $y = f(x + 5)$ is a horizontal shift 5 units to the left. Matches d.

**51.** $y = -f(-x) - 2$ is a reflection in the $y$-axis, a reflection in the $x$-axis, and a vertical shift downward 2 units. Matches c.

**53.** $y = f(x + 6) + 2$ is a horizontal shift to the left 6 units, and a vertical shift upward 2 units. Matches e.

**55.** (a)  the graph is shifted 3 units to the left.

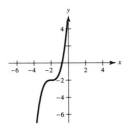

(b)  The graph is shifted 1 unit to the right.

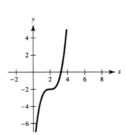

(c)  The graph is shifted 2 units upward.

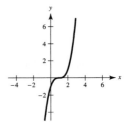

(d)  The graph is shifted 4 units downward.

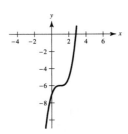

(e)  The graph is stretched vertically by a factor of 3.

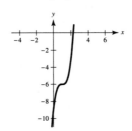

(f )  The graph is stretched vertically by a factor of $\frac{1}{4}$.

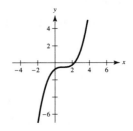

**57.** (a) $y = \sqrt{x} + 2$

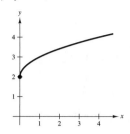

Vertical shift 2 units upward

(b) $y = -\sqrt{x}$

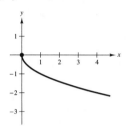

Reflection about the $x$-axis

(c) $y = \sqrt{x - 2}$

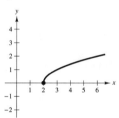

Horizontal shift 2 units to the right

**59.** (a) $f(g(1)) = f(0) = 0$

(b) $g(f(1)) = g(1) = 0$

(c) $g(f(0)) = g(0) = -1$

(d) $f(g(-4)) = f(15) = \sqrt{15}$

(e) $f(g(x)) = f(x^2 - 1) = \sqrt{x^2 - 1}$

(f) $g(f(x)) = g(\sqrt{x}) = (\sqrt{x})^2 - 1 = x - 1, (x \geq 0)$

**67.** $F(x) = \sqrt{2x - 2}$

Let $h(x) = 2x$, $g(x) = x - 2$ and $f(x) = \sqrt{x}$.

Then, $(f \circ g \circ h)(x) = f(g(2x)) = f((2x) - 2) = \sqrt{(2x) - 2} = \sqrt{2x - 2} = F(x)$.

[Other answers possible]

**69.** $f(-x) = (-x)^2(4 - (-x)^2) = x^2(4 - x^2) = f(x)$

Even

**61.** $f(x) = x^2$, $g(x) = \sqrt{x}$

$(f \circ g)(x) = f(g(x))$

$= f(\sqrt{x}) = (\sqrt{x})^2 = x, \; x \geq 0$

Domain: $[0, \infty)$

$(g \circ f)(x) = g(f(x)) = g(x^2) = \sqrt{x^2} = |x|$

Domain: $(-\infty, \infty)$

No. Their domains are different. $(f \circ g) = (g \circ f)$ for $x \geq 0$.

**63.** $f(x) = \dfrac{3}{x}$, $g(x) = x^2 - 1$

$(f \circ g)(x) = f(g(x)) = f(x^2 - 1) = \dfrac{3}{x^2 - 1}$

Domain: all $x \neq \pm 1 \Rightarrow (-\infty, -1) \cup (-1, 1) \cup (1, \infty)$

$(g \circ f)(x) = g(f(x))$

$= g\left(\dfrac{3}{x}\right) = \left(\dfrac{3}{x}\right)^2 - 1 = \dfrac{9}{x^2} - 1 = \dfrac{9 - x^2}{x^2}$

Domain: all $x \neq 0 \Rightarrow (-\infty, 0) \cup (0, \infty)$

No, $f \circ g \neq g \circ f$.

**65.** (a) $(f \circ g)(3) = f(g(3)) = f(-1) = 4$

(b) $g(f(2)) = g(1) = -2$

(c) $g(f(5)) = g(-5)$, which is undefined

(d) $(f \circ g)(-3) = f(g(-3)) = f(-2) = 3$

(e) $(g \circ f)(-1) = g(f(-1)) = g(4) = 2$

(f) $f(g(-1)) = f(-4)$, which is undefined

**71.** $f(-x) = (-x)\cos(-x) = -x \cos x = -f(x)$

Odd

**73.** (a) If $f$ is even, then $\left(\frac{3}{2}, 4\right)$ is on the graph.

    (b) If $f$ is odd, then $\left(\frac{3}{2}, -4\right)$ is on the graph.

**75.** $f$ is even because the graph is symmetric about the $y$-axis. $g$ is neither even nor odd. $h$ is odd because the graph is symmetric about the origin.

**77.** Slope $= \dfrac{4 - (-6)}{-2 - 0} = \dfrac{10}{-2} = -5$

$$y - 4 = -5(x - (-2))$$
$$y - 4 = -5x - 10$$
$$y = -5x - 6$$

For the line segment, you must restrict the domain.

$f(x) = -5x - 6,$

$-2 \le x \le 0$

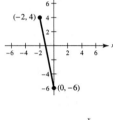

**79.** $x + y^2 = 0$

$$y^2 = -x$$
$$y = -\sqrt{-x}$$
$$f(x) = -\sqrt{-x}, \ x \le 0$$

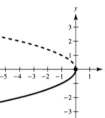

**81.** Answers will vary. *Sample answer*: Speed begins and ends at 0. The speed might be constant in the middle:

**83.** Answers will vary. *Sample answer*: In general , as the price decreases, the store will sell more.

**85.**      $y = \sqrt{c - x^2}$

$$y^2 = c - x^2$$
$$x^2 + y^2 = c, \text{ a circle.}$$

For the domain to be $[-5, 5]$, $c = 25$.

**87.** (a) $T(4) = 16°, \ T(15) \approx 23°$

    (b) If $H(t) = T(t - 1)$, then the changes in temperature will occur 1 hour later.

    (c) If $H(t) = T(t) - 1$, then the overall temperature would be 1 degree lower.

**89.** (a)

    (b) $A(20) \approx 384$ acres/farm

**91.** $f(x) = |x| + |x - 2|$

    If $x < 0$, then $f(x) = -x - (x - 2) = -2x + 2$.

    If $0 \le x < 2$, then $f(x) = x - (x - 2) = 2$.

    If $x \ge 2$, then $f(x) = x + (x - 2) = 2x - 2$.

    So,

$$f(x) = \begin{cases} -2x + 2, & x \le 0 \\ 2, & 0 < x < 2. \\ 2x - 2, & x \ge 2 \end{cases}$$

**93.** $f(-x) = a_{2n+1}(-x)^{2n+1} + \cdots + a_3(-x)^3 + a_1(-x)$

$$= -\left[a_{2n+1}x^{2n+1} + \cdots + a_3x^3 + a_1x\right]$$
$$= -f(x)$$

Odd

**95.** Let $F(x) = f(x)g(x)$ where $f$ and $g$ are even. Then $F(-x) = f(-x)g(-x) = f(x)g(x) = F(x)$.

So, $F(x)$ is even. Let $F(x) = f(x)g(x)$ where $f$ and $g$ are odd. Then

$$F(-x) = f(-x)g(-x) = \left[-f(x)\right]\left[-g(x)\right] = f(x)g(x) = F(x).$$

So, $F(x)$ is even.

**97.** (a)  $V = x(24 - 2x)^2$

Domain:  $0 < x < 12$

(b)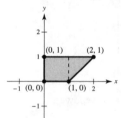

Maximum volume occurs at  $x = 4$. So, the dimensions of the box would be  $4 \times 16 \times 16$ cm.

(c)

| $x$ | Length and width | Volume |
|---|---|---|
| 1 | $24 - 2(1)$ | $1\left[24 - 2(1)\right]^2 = 484$ |
| 2 | $24 - 2(2)$ | $2\left[24 - 2(2)\right]^2 = 800$ |
| 3 | $24 - 2(3)$ | $3\left[24 - 2(3)\right]^2 = 972$ |
| 4 | $24 - 2(4)$ | $4\left[24 - 2(4)\right]^2 = 1024$ |
| 5 | $24 - 2(5)$ | $5\left[24 - 2(5)\right]^2 = 980$ |
| 6 | $24 - 2(6)$ | $6\left[24 - 2(6)\right]^2 = 864$ |

The dimensions of the box that yield a maximum volume appear to be  $4 \times 16 \times 16$ cm.

**99.** False. If  $f(x) = x^2$, then  $f(-3) = f(3) = 9$, but  $-3 \neq 3$.

**101.** True. The function is even.

**103.** First consider the portion of R in the first quadrant:  $x \geq 0,\ 0 \leq y \leq 1$ and  $x - y \leq 1$; shown below.

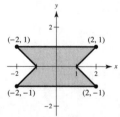

The area of this region is  $1 + \frac{1}{2} = \frac{3}{2}$.

By symmetry, you obtain the entire region R:

The area of R is  $4\left(\frac{3}{2}\right) = 6$.

# Section P.4   Fitting Models to Data

**1.** Trigonometric function

**3.** No relationship

**5.** (a), (b)

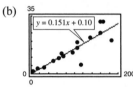

Yes. The cancer mortality increases linearly with increased exposure to the carcinogenic substance.

(c) If $x = 3$, then $y \approx 136$.

**9.** (a) Using a graphing utility,

$y = 0.151x + 0.10$

The correlation coefficient is $r \approx 0.880$.

(b)

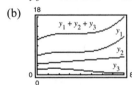

(c) Greater per capita energy consumption by a country tends to correspond to greater per capita gross national product. The four countries that differ most from the linear model are Venezuela, South Korea, Hong Kong and United Kingdom.

(d) Using a graphing utility,

$y = 0.155x + 0.22$ and $r \approx 0.984$.

**11.** (a) $y_1 = 0.04040t^3 - 0.3695t^2 + 1.123t + 5.88$

$y_2 = 0.264t + 3.35$

$y_3 = 0.01439t^3 - 0.1886t^2 + 0.476t + 1.59$

(b)

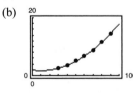

For year 12, $y_1 + y_2 + y_3 \approx 47.5$ cents/mile.

**13.** (a) $t = 0.002D^2 - 0.04D + 1.9$

(b)

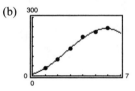

(c) According to the model, the times required to attain speeds of less than 20 miles per hour are all about the same.

(d) Adding $(0, 0)$ to the data produces

$t = 0.002D^2 + 0.02D + 0.1$

(e) No. From the graph in part (b), you can see that the model from part (a) follows the data more closely than the model from part (d).

**7.** (a) $d = 0.066F$

(b)

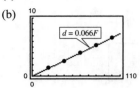

The model fits the data well.

(c) If $F = 55$, then $d \approx 0.066(55) = 3.63$ cm.

**15.** (a) $y = -1.806x^3 + 14.58x^2 + 16.4x + 10$

(b)

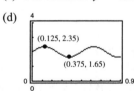

(c) If $x = 4.5$, $y \approx 214$ horsepower.

**17.** (a) Yes, $y$ is a function of $t$. At each time $t$, there is one and only one displacement $y$.

(b) The amplitude is approximately

$(2.35 - 1.65)/2 = 0.35$.

The period is approximately

$2(0.375 - 0.125) = 0.5$.

(c) One model is $y = 0.35 \sin(4\pi t) + 2$.

(d)

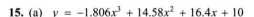

The model appears to fit the data.

**19.** Answers will vary.

**21.** Yes, $A_1 \leq A_2$. To see this, consider the two triangles of areas $A_1$ and $A_2$:

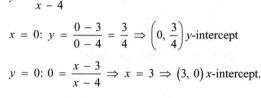

For $i = 1, 2$, the angles satisfy $\alpha i + \beta i + \gamma i = \pi$. At least one of $\alpha_1 \leq \alpha_2$, $\beta_1 \leq \beta_2$, $\gamma_1 \leq \gamma_2$ must hold.

Assume $\alpha_1 \leq \alpha_2$. Because $\alpha_2 \leq \pi/2$ (acute triangle), and the sine function increases on $[0, \pi/2]$, you have

$$A_1 = \tfrac{1}{2}b_1c_1 \sin \alpha_1 \leq \tfrac{1}{2} b_2c_2 \sin \alpha_1$$
$$\leq \tfrac{1}{2}b_2c_2 \sin \alpha_2 = A_2$$

# Review Exercises for Chapter P

**1.** $y = 5x - 8$

$x = 0$: $y = 5(0) - 8 = -8 \Rightarrow (0, -8)$ $y$-intercept

$y = 0$: $0 = 5x - 8 \Rightarrow x = \tfrac{8}{5} \Rightarrow \left(\tfrac{8}{5}, 0\right)$ $x$-intercept

**3.** $y = \dfrac{x - 3}{x - 4}$

$x = 0$: $y = \dfrac{0 - 3}{0 - 4} = \dfrac{3}{4} \Rightarrow \left(0, \dfrac{3}{4}\right)$ $y$-intercept

$y = 0$: $0 = \dfrac{x - 3}{x - 4} \Rightarrow x = 3 \Rightarrow (3, 0)$ $x$-intercept.

**5.** Symmetric with respect to $y$-axis because

$$(-x)^2 y - (-x)^2 + 4y = 0$$
$$x^2 y - x^2 + 4y = 0.$$

**7.** $y = -\tfrac{1}{2}x + \tfrac{3}{2}$

Slope: $-\tfrac{1}{2}$

$y$-intercept: $\tfrac{3}{2}$

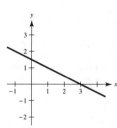

**9.** $-\tfrac{1}{3}x + \tfrac{5}{6}y = 1$

$-\tfrac{2}{5}x + y = \tfrac{6}{5}$

$y = \tfrac{2}{5}x + \tfrac{6}{5}$

Slope: $\tfrac{2}{5}$

$y$-intercept: $\tfrac{6}{5}$

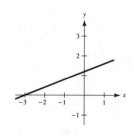

**11.** $y = 9 - 8x - x^2 = -(x - 1)(x + 9)$

$y$-intercept: $(0, 9)$

$x$-intercepts: $(1, 0)$, $(-9, 0)$

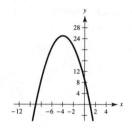

**13.** $y = 2\sqrt{4 - x}$

Domain: $(-\infty, 4]$

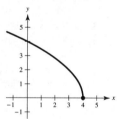

**15.** $y = 4x^2 - 25$

| Xmin = −5 |
| Xmax = 5 |
| Xscl = 1 |
| Ymin = −30 |
| Ymax = 10 |
| Yscl = 5 |

**17.** $5x + 3y = -1 \Rightarrow y = \frac{1}{3}(-5x - 1)$

$x - y = -5 \Rightarrow y = x + 5$

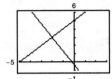

Using a graphing utility, the lines intersect at $(-2, 3)$. Analytically,

$$\frac{1}{3}(-5x - 1) = x + 5$$

$$-5x - 1 = 3x + 15$$

$$-16 = 8x$$

$$-2 = x.$$

For $x = -2$, $y = x + 5 = -2 + 5 = 3$.

**19.** Answers will vary. *Sample answer*:

You need factors $(x + 4)$ and $(x - 4)$.

Multiply by $x$ to obtain origin symmetry.

$$y = x(x + 4)(x - 4) = x^3 - 16x$$

**21.**

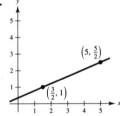

$$\text{Slope} = \frac{\left(\frac{5}{2}\right) - 1}{5 - \left(\frac{3}{2}\right)} = \frac{\frac{3}{2}}{\frac{7}{2}} = \frac{3}{7}$$

**23.** $\dfrac{t - 5}{0 - (-8)} = \dfrac{-1 - 5}{2 - (-8)}$

$$\frac{t - 5}{8} = \frac{-6}{10}$$

$$\frac{t - 5}{8} = -\frac{3}{5}$$

$$5t - 25 = -24$$

$$5t = 1$$

$$t = \frac{1}{5}$$

**25.** $y - (-5) = \frac{7}{4}(x - 3)$

$$y + 5 = \frac{7}{4}x - \frac{21}{4}$$

$$4y + 20 = 7x - 21$$

$$0 = 7x - 4y - 41$$

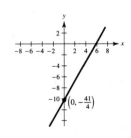

**27.** $y - 0 = -\frac{2}{3}(x - (-3))$

$$y = -\frac{2}{3}x - 2$$

$$2x + 3y + 6 = 0$$

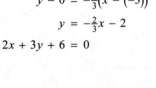

**29.** (a) $y - 5 = \frac{7}{16}(x + 3)$

$$16y - 80 = 7x + 21$$

$$0 = 7x - 16y + 101$$

(b) $5x - 3y = 3$ has slope $\frac{5}{3}$.

$$y - 5 = \frac{5}{3}(x + 3)$$

$$3y - 15 = 5x + 15$$

$$0 = 5x - 3y + 30$$

(c) $m = \dfrac{5 - 0}{-3 - 0} = -\dfrac{5}{3}$

$$y - 5 = -\frac{5}{3}(x + 3)$$

$$3y - 15 = -5x - 15$$

$$5x + 3y = 0$$

(d) Slope is undefined so the line is vertical.

$$x = -3$$

$$x + 3 = 0$$

**31.** The slope is $-850$.

$V = -850t + 12,500$.

$V(3) = -850(3) + 12,500 = \$9950$

**33.** $x - y^2 = 6$

$$y = \pm\sqrt{x - 6}$$

Not a function because there are two values of $y$ for some $x$.

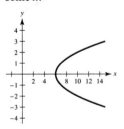

**35.** $y = \dfrac{|x - 2|}{x - 2}$

$y$ is a function of $x$ because there is one value of $y$ for each $x$.

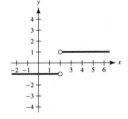

**37.** $f(x) = \dfrac{1}{x}$

(a) $f(0)$ does not exist.

(b) $\dfrac{f(1 + \Delta x) - f(1)}{\Delta x} = \dfrac{\dfrac{1}{1 + \Delta x} - \dfrac{1}{1}}{\Delta x} = \dfrac{1 - 1 - \Delta x}{(1 + \Delta x)\Delta x}$

$\qquad = \dfrac{-1}{1 + \Delta x}, \ \Delta x \neq -1, 0$

**39.** (a) Domain: $36 - x^2 \geq 0 \Rightarrow -6 \leq x \leq 6$ or $[-6, 6]$

Range: $[0, 6]$

(b) Domain: all $x \neq 5$ or $(-\infty, 5) \cup (5, \infty)$

Range: all $y \neq 0$ or $(-\infty, 0) \cup (0, \infty)$

(c) Domain: all $x$ or $(-\infty, \infty)$

Range: all $y$ or $(-\infty, \infty)$

**41.** (a) $f(x) = x^3 + c, \ c = -2, 0, 2$

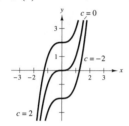

(b) $f(x) = (x - c)^3, \ c = -2, 0, 2$

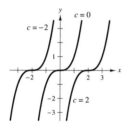

(c) $f(x) = (x - 2)^3 + c, \ c = -2, 0, 2$

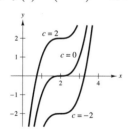

(d) $f(x) = cx^3, \ c = -2, 0, 2$

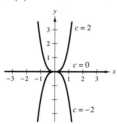

**43.** (a) Odd powers: $f(x) = x, \ g(x) = x^3, \ h(x) = x^5$

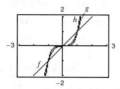

The graphs of $f$, $g$, and $h$ all rise to the right and fall to the left. As the degree increases, the graph rises and falls more steeply. All three graphs pass through the points $(0, 0)$, $(1, 1)$, and $(-1, -1)$ and are symmetric with respect to the origin.

Even powers: $f(x) = x^2, \ g(x) = x^4, \ h(x) = x^6$

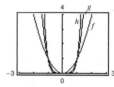

The graphs of $f$, $g$, and $h$ all rise to the left and to the right. As the degree increases, the graph rises more steeply. All three graphs pass through the points $(0, 0)$, $(1, 1)$, and $(-1, 1)$ and are symmetric with respect to the $y$-axis.

All of the graphs, even and odd, pass through the origin. As the powers increase, the graphs become flatter in the interval $-1 < x < 1$.

(b) $y = x^7$ will look like $h(x) = x^5$, but rise and fall even more steeply. $y = x^8$ will look like $h(x) = x^6$, but rise even more steeply.

**45.** (a)

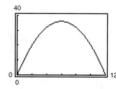

$2x + 2y = 24$

$y = 12 - x$

$A = xy = x(12 - x)$

(b) Domain: $0 < x < 12$ or $(0, 12)$

(c) Maximum area is $A = 36$ in.$^2$. In general, the maximum area is attained when the rectangle is a square. In this case, $x = 6$.

**47.** (a) 3 (cubic), negative leading coefficient

(b) 4 (quartic), positive leading coefficient

(c) 2 (quadratic), negative leading coefficient

(d) 5, positive leading coefficient

**49.** (a) Yes, $y$ is a function of $t$. At each time $t$, there is and only one displacement $y$.

(b) The amplitude is approximately $\left(0.25 - (-0.25)\right)/2 = 0.25$.

The period is approximately 1.1.

(c) One model is $y = \dfrac{1}{4}\cos\left(\dfrac{2\pi}{1.1}t\right) \approx \dfrac{1}{4}\cos(5.7t)$

(d)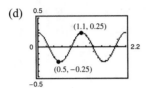

The model appears to fit the data.

# Problem Solving for Chapter P

**1.** (a)
$$x^2 - 6x + y^2 - 8y = 0$$
$$\left(x^2 - 6x + 9\right) + \left(y^2 - 8y + 16\right) = 9 + 16$$
$$(x - 3)^2 + (y - 4)^2 = 25$$

Center: $(3, 4)$; Radius: 5

(b) Slope of line from $(0, 0)$ to $(3, 4)$ is $\dfrac{4}{3}$.

Slope of tangent line is $-\dfrac{3}{4}$.

So, $y - 0 = -\dfrac{3}{4}(x - 0) \Rightarrow y = -\dfrac{3}{4}x$  Tangent line

(c) Slope of line from $(6, 0)$ to $(3, 4)$ is $\dfrac{4 - 0}{3 - 6} = -\dfrac{4}{3}$.

Slope of tangent line is $\dfrac{3}{4}$.

So, $y - 0 = \dfrac{3}{4}(x - 6) \Rightarrow y = \dfrac{3}{4}x - \dfrac{9}{2}$  Tangent line

(d)
$$-\frac{3}{4}x = \frac{3}{4}x - \frac{9}{2}$$
$$\frac{3}{2}x = \frac{9}{2}$$
$$x = 3$$

Intersection: $\left(3, -\dfrac{9}{4}\right)$

**3.** $H(x) = \begin{cases} 1, & x \geq 0 \\ 0, & x < 0 \end{cases}$

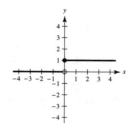

**(a)** $H(x) - 2 = \begin{cases} -1, & x \geq 0 \\ -2, & x < 0 \end{cases}$

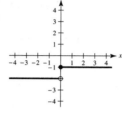

**(b)** $H(x - 2) = \begin{cases} 1, & x \geq 2 \\ 0, & x < 2 \end{cases}$

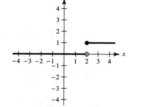

**(c)** $-H(x) = \begin{cases} -1, & x \geq 0 \\ 0, & x < 0 \end{cases}$

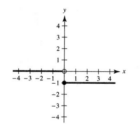

**(d)** $H(-x) = \begin{cases} 1, & x \leq 0 \\ 0, & x > 0 \end{cases}$

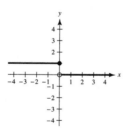

**(e)** $\frac{1}{2}H(x) = \begin{cases} \frac{1}{2}, & x \geq 0 \\ 0, & x < 0 \end{cases}$

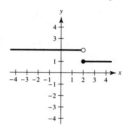

**(f)** $-H(x - 2) + 2 = \begin{cases} 1, & x \geq 2 \\ 2, & x < 2 \end{cases}$

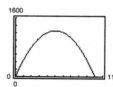

**5. (a)** $x + 2y = 100 \Rightarrow y = \dfrac{100 - x}{2}$

$A(x) = xy = x\left(\dfrac{100 - x}{2}\right) = -\dfrac{x^2}{2} + 50x$

Domain: $0 < x < 100$ or $(0, 100)$

**(b)**

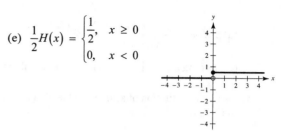

Maximum of 1250 m$^2$ at $x = 50$ m, $y = 25$ m.

**(c)** $A(x) = -\frac{1}{2}(x^2 - 100x)$

$= -\frac{1}{2}(x^2 - 100x + 2500) + 1250$

$= -\frac{1}{2}(x - 50)^2 + 1250$

$A(50) = 1250$ m$^2$ is the maximum.

$x = 50$ m, $y = 25$ m

**7.** The length of the trip in the water is $\sqrt{2^2 + x^2}$, and the length of the trip over land is $\sqrt{1 + (3 - x)^2}$.

So, the total time is $T = \dfrac{\sqrt{4 + x^2}}{2} + \dfrac{\sqrt{1 + (3 - x)^2}}{4}$ hours.

**9. (a)** Slope $= \dfrac{9 - 4}{3 - 2} = 5$. Slope of tangent line is less than 5.

**(b)** Slope $= \dfrac{4 - 1}{2 - 1} = 3$. Slope of tangent line is greater than 3.

**(c)** Slope $= \dfrac{4.41 - 4}{2.1 - 2} = 4.1$. Slope of tangent line is less than 4.1.

(d)  Slope $= \dfrac{f(2+h)-f(2)}{(2+h)-2} = \dfrac{(2+h)^2-4}{h} = \dfrac{4h+h^2}{h} = 4+h, \; h \neq 0$

(e)  Letting $h$ get closer and closer to 0, the slope approaches 4. So, the slope at $(2, 4)$ is 4.

**11.** Using the definition of absolute value, you can rewrite the equation.

$$y + |y| = x + |x|$$

$$\begin{cases} 2y, & y > 0 \\ 0, & y \leq 0 \end{cases} = \begin{cases} 2x, & x > 0 \\ 0, & x \leq 0 \end{cases}.$$

For $x > 0$ and $y > 0$, you have $2y = 2x \Rightarrow y = x$.

For any $x \leq 0$, y is any $y \leq 0$. So, the graph of $y + |y| = x + |x|$ is as follows.

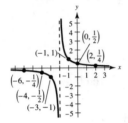

**13.** (a)  $\dfrac{I}{x^2+y^2} = \dfrac{kI}{(x-4)^2+y^2}$

$$(x-4)^2 + y^2 = k(x^2 + y^2)$$

$$(k-1)x^2 + 8x + (k-1)y^2 = 16$$

If $k = 1$, then $x = 2$ is a vertical line. Assume $k \neq 1$.

$$x^2 + \dfrac{8x}{k-1} + y^2 = \dfrac{16}{k-1}$$

$$x^2 + \dfrac{8x}{k-1} + \dfrac{16}{(k-1)^2} + y^2 = \dfrac{16}{k-1} + \dfrac{16}{(k-1)^2}$$

$$\left(x + \dfrac{4}{k-1}\right)^2 + y^2 = \dfrac{16k}{(k-1)^2}, \; \text{Circle}$$

(b)  If $k = 3$, $(x+2)^2 + y^2 = 12$

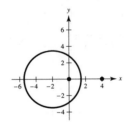

(c)  As $k$ becomes very large, $\dfrac{4}{k-1} \to 0$ and

$$\dfrac{16k}{(k-1)^2} \to 0.$$

The center of the circle gets closer to $(0, 0)$, and its radius approaches 0.

**15.**  $f(x) = y = \dfrac{1}{1-x}$

(a)  Domain: all $x \neq 1$ or $(-\infty, 1) \cup (1, \infty)$

Range: all $y \neq 0$ or $(-\infty, 0) \cup (0, \infty)$

(b)  $f(f(x)) = f\left(\dfrac{1}{1-x}\right) = \dfrac{1}{1-\left(\dfrac{1}{1-x}\right)}$

$$= \dfrac{1}{\dfrac{1-x-1}{1-x}} = \dfrac{1-x}{-x} = \dfrac{x-1}{x}$$

Domain: all $x \neq 0, 1$ or $(-\infty, 0) \cup (0, 1) \cup (1, \infty)$

(c)  $f(f(f(x))) = f\left(\dfrac{x-1}{x}\right) = \dfrac{1}{1-\left(\dfrac{x-1}{x}\right)} = \dfrac{1}{\dfrac{1}{x}} = x$

Domain: all $x \neq 0, 1$ or $(-\infty, 0) \cup (0, 1) \cup (1, \infty)$

(d)  The graph is not a line. it has holes at $(0, 0)$ and $(1, 1)$.

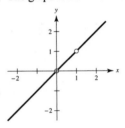

# CHAPTER 1
# Limits and Their Properties

# C H A P T E R   1
# Limits and Their Properties

## Section 1.1   A Preview of Calculus

**1.** Precalculus: $(20 \text{ ft/sec})(15 \text{ sec}) = 300 \text{ ft}$

**3.** Calculus required: Slope of the tangent line at $x = 2$ is the rate of change, and equals about 0.16.

**5.** (a) Precalculus: Area $= \frac{1}{2}bh = \frac{1}{2}(5)(4) = 10$ sq. units

 (b) Calculus required: Area $= bh \approx 2(2.5) = 5$ sq. units

**7.** $f(x) = 6x - x^2$

(a)

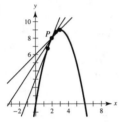

(b) slope $= m = \dfrac{(6x - x^2) - 8}{x - 2} = \dfrac{(x - 2)(4 - x)}{x - 2} = (4 - x), \ x \neq 2$

 For $x = 3, \ m = 4 - 3 = 1$

 For $x = 2.5, \ m = 4 - 2.5 = 1.5 = \dfrac{3}{2}$

 For $x = 1.5, \ m = 4 - 1.5 = 2.5 = \dfrac{5}{2}$

(c) At $P(2, 8)$, the slope is 2. You can improve your approximation by considering values of $x$ close to 2.

**9.** (a) Area $\approx 5 + \frac{5}{2} + \frac{5}{3} + \frac{5}{4} \approx 10.417$

 Area $\approx \frac{1}{2}\left(5 + \frac{5}{1.5} + \frac{5}{2} + \frac{5}{2.5} + \frac{5}{3} + \frac{5}{3.5} + \frac{5}{4} + \frac{5}{4.5}\right) \approx 9.145$

(b) You could improve the approximation by using more rectangles.

**11.** (a) $D_1 = \sqrt{(5 - 1)^2 + (1 - 5)^2} = \sqrt{16 + 16} \approx 5.66$

(b) $D_2 = \sqrt{1 + \left(\frac{5}{2}\right)^2} + \sqrt{1 + \left(\frac{5}{2} - \frac{5}{3}\right)^2} + \sqrt{1 + \left(\frac{5}{3} - \frac{5}{4}\right)^2} + \sqrt{1 + \left(\frac{5}{4} - 1\right)^2}$

 $\approx 2.693 + 1.302 + 1.083 + 1.031 \approx 6.11$

(c) Increase the number of line segments.

## Section 1.2   Finding Limits Graphically and Numerically

**1.**

| $x$ | 3.9 | 3.99 | 3.999 | 4.001 | 4.01 | 4.1 |
|-----|------|-------|--------|--------|-------|-------|
| $f(x)$ | 0.2041 | 0.2004 | 0.2000 | 0.2000 | 0.1996 | 0.1961 |

$\displaystyle\lim_{x \to 4} \frac{x - 4}{x^2 - 3x - 4} \approx 0.2000 \qquad \left(\text{Actual limit is } \frac{1}{5}.\right)$

**3.**

| $x$ | −0.1 | −0.01 | −0.001 | 0.001 | 0.01 | 0.1 |
|---|---|---|---|---|---|---|
| $f(x)$ | 0.2050 | 0.2042 | 0.2041 | 0.2041 | 0.2040 | 0.2033 |

$$\lim_{x \to 0} \frac{\sqrt{x + 6} - \sqrt{6}}{x} \approx 0.2041 \quad \left( \text{Actual limit is } \frac{1}{2\sqrt{6}}. \right)$$

**5.**

| $x$ | 2.9 | 2.99 | 2.999 | 3.001 | 3.01 | 3.1 |
|---|---|---|---|---|---|---|
| $f(x)$ | −0.0641 | −0.0627 | −0.0625 | −0.0625 | −0.0623 | −0.0610 |

$$\lim_{x \to 3} \frac{[1/(x + 1)] - (1/4)}{x - 3} \approx -0.0625 \quad \left( \text{Actual limit is } -\frac{1}{16}. \right)$$

**7.**

| $x$ | −0.1 | −0.01 | −0.001 | 0.001 | 0.01 | 0.1 |
|---|---|---|---|---|---|---|
| $f(x)$ | 0.9983 | 0.99998 | 1.0000 | 1.0000 | 0.99998 | 0.9983 |

$$\lim_{x \to 0} \frac{\sin x}{x} \approx 1.0000 \quad \left( \text{Actual limit is 1.} \right) \left( \text{Make sure you use radian mode.} \right)$$

**9.**

| $x$ | 0.9 | 0.99 | 0.999 | 1.001 | 1.01 | 1.1 |
|---|---|---|---|---|---|---|
| $f(x)$ | 0.2564 | 0.2506 | 0.2501 | 0.2499 | 0.2494 | 0.2439 |

$$\lim_{x \to 1} \frac{x - 2}{x^2 + x - 6} \approx 0.2500 \quad \left( \text{Actual limit is } \frac{1}{4}. \right)$$

**11.**

| $x$ | 0.9 | 0.99 | 0.999 | 1.001 | 1.01 | 1.1 |
|---|---|---|---|---|---|---|
| $f(x)$ | 0.7340 | 0.6733 | 0.6673 | 0.6660 | 0.6600 | 0.6015 |

$$\lim_{x \to 1} \frac{x^4 - 1}{x^6 - 1} \approx 0.6666 \quad \left( \text{Actual limit is } \frac{2}{3}. \right)$$

**13.**

| $x$ | −0.1 | −0.01 | −0.001 | 0.001 | 0.01 | 0.1 |
|---|---|---|---|---|---|---|
| $f(x)$ | 1.9867 | 1.9999 | 2.0000 | 2.0000 | 1.9999 | 1.9867 |

$$\lim_{x \to 0} \frac{\sin 2x}{x} \approx 2.0000 \quad \left( \text{Actual limit is 2.} \right) \left( \text{Make sure you use radian mode.} \right)$$

**15.** $\lim\limits_{x \to 3} (4 - x) = 1$

**17.** $\lim\limits_{x \to 2} f(x) = \lim\limits_{x \to 2} (4 - x) = 2$

**19.** $\lim\limits_{x \to 2} \dfrac{|x - 2|}{x - 2}$ does not exist.

For values of $x$ to the left of 2, $\dfrac{|x - 2|}{(x - 2)} = -1$, whereas

for values of $x$ to the right of 2, $\dfrac{|x - 2|}{(x - 2)} = 1$.

**21.** $\lim\limits_{x \to 1} \sin \pi x = 0$

**23.** $\lim\limits_{x \to 0} \cos(1/x)$ does not exist because the function

oscillates between −1 and 1 as $x$ approaches 0.

**25.** (a)  $f(1)$ exists. The black dot at $(1, 2)$ indicates that
   $f(1) = 2.$

   (b)  $\lim_{x \to 1} f(x)$ does not exist. As x approaches 1 from the
   left, $f(x)$ approaches 3.5, whereas as $x$ approaches
   1 from the right, $f(x)$ approaches 1.

   (c)  $f(4)$ does not exist. The hollow circle at
   $(4, 2)$ indicates that $f$ is not defined at 4.

   (d)  $\lim_{x \to 4} f(x)$ exists. As x approaches 4, $f(x)$ approaches
   2: $\lim_{x \to 4} f(x) = 2.$

**27.**  $\lim_{x \to c} f(x)$ exists for all  $c \neq -3.$

**29.**

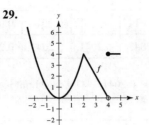

$\lim_{x \to c} f(x)$ exists for all values of $c \neq 4.$

**31.** One possible answer is

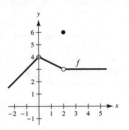

**33.**  $C(t) = 9.99 - 0.79 \left[\!\left[ -(t - 1) \right]\!\right]$

(a)

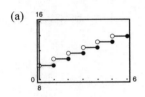

(b)

| $t$ | 3 | 3.3 | 3.4 | 3.5 | 3.6 | 3.7 | 4 |
|---|---|---|---|---|---|---|---|
| $C$ | 11.57 | 12.36 | 12.36 | 12.36 | 12.36 | 12.36 | 12.36 |

$\lim_{t \to 3.5} C(t) = 12.36$

(c)

| $t$ | 2 | 2.5 | 2.9 | 3 | 3.1 | 3.5 | 4 |
|---|---|---|---|---|---|---|---|
| $C$ | 10.78 | 11.57 | 11.57 | 11.57 | 12.36 | 12.36 | 12.36 |

The $\lim_{t \to 3} C(t)$ does not exist because the values of $C$ approach different values
as $t$ approaches 3 from both sides.

**35.** You need $\left| f(x) - 3 \right| = \left| (x + 1) - 3 \right| = \left| x - 2 \right| < 0.4.$ So, take $\delta = 0.4.$
If $0 < \left| x - 2 \right| < 0.4$, then $\left| x - 2 \right| = \left| (x + 1) - 3 \right| = \left| f(x) - 3 \right| < 0.4$, as desired.

**37.** You need to find $\delta$ such that $0 < |x - 1| < \delta$ implies

$$|f(x) - 1| = \left|\frac{1}{x} - 1\right| < 0.1. \text{ That is,}$$

$$-0.1 < \frac{1}{x} - 1 < 0.1$$

$$1 - 0.1 < \frac{1}{x} < 1 + 0.1$$

$$\frac{9}{10} < \frac{1}{x} < \frac{11}{10}$$

$$\frac{10}{9} > x > \frac{10}{11}$$

$$\frac{10}{9} - 1 > x - 1 > \frac{10}{11} - 1$$

$$\frac{1}{9} > x - 1 > -\frac{1}{11}.$$

So take $\delta = \dfrac{1}{11}$. Then $0 < |x - 1| < \delta$ implies

$$-\frac{1}{11} < x - 1 < \frac{1}{11}$$

$$-\frac{1}{11} < x - 1 < \frac{1}{9}.$$

Using the first series of equivalent inequalities, you obtain

$$|f(x) - 1| = \left|\frac{1}{x} - 1\right| < 0.1.$$

**39.** $\lim\limits_{x \to 2}(3x + 2) = 8 = L$

$$|(3x + 2) - 8| < 0.01$$

$$|3x - 6| < 0.01$$

$$3|x - 2| < 0.01$$

$$0 < |x - 2| < \tfrac{0.01}{3} \approx 0.0033 = \delta$$

So, if $0 < |x - 2| < \delta = \tfrac{0.01}{3}$, you have

$$3|x - 2| < 0.01$$

$$|3x - 6| < 0.01$$

$$|(3x + 2) - 8| < 0.01$$

$$|f(x) - L| < 0.01.$$

**41.** $\lim\limits_{x \to 2}(x^2 - 3) = 1 = L$

$$|(x^2 - 3) - 1| < 0.01$$

$$|x^2 - 4| < 0.01$$

$$|(x + 2)(x - 2)| < 0.01$$

$$|x + 2||x - 2| < 0.01$$

$$|x - 2| < \frac{0.01}{|x + 2|}$$

If you assume $1 < x < 3$, then $\delta = 0.01/5 = 0.002$.

So, if $0 < |x - 2| < \delta = 0.002$, you have

$$|x - 2| < 0.002 = \frac{1}{5}(0.01) < \frac{1}{|x + 2|}(0.01)$$

$$|x + 2||x - 2| < 0.01$$

$$|x^2 - 4| < 0.01$$

$$|(x^2 - 3) - 1| < 0.01$$

$$|f(x) - L| < 0.01.$$

**43.** $\lim\limits_{x \to 4}(x + 2) = 6$

Given $\varepsilon > 0$:

$$|(x + 2) - 6| < \varepsilon$$

$$|x - 4| < \varepsilon = \delta$$

So, let $\delta = \varepsilon$. So, if $0 < |x - 4| < \delta = \varepsilon$, you have

$$|x - 4| < \varepsilon$$

$$|(x + 2) - 6| < \varepsilon$$

$$|f(x) - L| < \varepsilon.$$

**45.** $\lim\limits_{x \to -4}\left(\tfrac{1}{2}x - 1\right) = \tfrac{1}{2}(-4) - 1 = -3$

Given $\varepsilon > 0$:

$$\left|\left(\tfrac{1}{2}x - 1\right) - (-3)\right| < \varepsilon$$

$$\left|\tfrac{1}{2}x + 2\right| < \varepsilon$$

$$\tfrac{1}{2}|x - (-4)| < \varepsilon$$

$$|x - (-4)| < 2\varepsilon$$

So, let $\delta = 2\varepsilon$.

So, if $0 < |x - (-4)| < \delta = 2\varepsilon$, you have

$$|x - (-4)| < 2\varepsilon$$

$$\left|\tfrac{1}{2}x + 2\right| < \varepsilon$$

$$\left|\left(\tfrac{1}{2}x - 1\right) + 3\right| < \varepsilon$$

$$|f(x) - L| < \varepsilon.$$

**47.** $\lim\limits_{x \to 6} 3 = 3$

Given $\varepsilon > 0$:

$$|3 - 3| < \varepsilon$$
$$0 < \varepsilon$$

So, any $\delta > 0$ will work.

So, for any $\delta > 0$, you have

$$|3 - 3| < \varepsilon$$
$$|f(x) - L| < \varepsilon.$$

**49.** $\lim\limits_{x \to 0} \sqrt[3]{x} = 0$

Given $\varepsilon > 0$: $\left|\sqrt[3]{x} - 0\right| < \varepsilon$

$$\left|\sqrt[3]{x}\right| < \varepsilon$$
$$|x| < \varepsilon^3 = \delta$$

So, let $\delta = \varepsilon^3$.

So, for $0|x - 0|\delta = \varepsilon^3$, you have

$$|x| < \varepsilon^3$$
$$\left|\sqrt[3]{x}\right| < \varepsilon$$
$$\left|\sqrt[3]{x} - 0\right| < \varepsilon$$
$$|f(x) - L| < \varepsilon.$$

**51.** $\lim\limits_{x \to -5} |x - 5| = |(-5) - 5| = |-10| = 10$

Given $\varepsilon > 0$: $\big||x - 5| - 10\big| < \varepsilon$

$$|-(x - 5) - 10| < \varepsilon \quad (x - 5 < 0)$$
$$|-x - 5| < \varepsilon$$
$$|x - (-5)| < \varepsilon$$

So, let $\delta = \varepsilon$.

So for $|x - (-5)| < \delta = \varepsilon$, you have

$$|-(x + 5)| < \varepsilon$$
$$|-(x - 5) - 10| < \varepsilon$$
$$\big||x - 5| - 10\big| < \varepsilon \quad \text{(because } x - 5 < 0\text{)}$$
$$|f(x) - L| < \varepsilon.$$

**53.** $\lim\limits_{x \to 1} (x^2 + 1) = 2$

Given $\varepsilon > 0$:

$$\left|(x^2 + 1) - 2\right| < \varepsilon$$
$$\left|x^2 - 1\right| < \varepsilon$$
$$|(x + 1)(x - 1)| < \varepsilon$$
$$|x - 1| < \frac{\varepsilon}{|x + 1|}$$

If you assume $0 < x < 2$, then $\delta = \varepsilon/3$.

So for $0 < |x - 1| < \delta = \dfrac{\varepsilon}{3}$, you have

$$|x - 1| < \frac{1}{3}\varepsilon < \frac{1}{|x + 1|}\varepsilon$$
$$\left|x^2 - 1\right| < \varepsilon$$
$$\left|(x^2 + 1) - 2\right| < \varepsilon$$
$$|f(x) - 2| < \varepsilon.$$

**55.** $\lim\limits_{x \to \pi} f(x) = \lim\limits_{x \to \pi} 4 = 4$

**57.** $f(x) = \dfrac{\sqrt{x + 5} - 3}{x - 4}$

$$\lim\limits_{x \to 4} f(x) = \frac{1}{6}$$

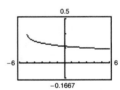

The domain is $[-5, 4) \cup (4, \infty)$. The graphing utility

does not show the hole at $\left(4, \dfrac{1}{6}\right)$.

**59.** $f(x) = \dfrac{x - 9}{\sqrt{x} - 3}$

$$\lim\limits_{x \to 9} f(x) = 6$$

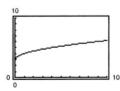

The domain is all $x \geq 0$ except $x = 9$. The graphing
utility does not show the hole at $(9, 6)$.

**61.** $\lim\limits_{x \to 8} f(x) = 25$ means that the values of $f$ approach 25 as $x$ gets closer and closer to 8.

**63.** (i) The values of $f$ approach different numbers as $x$ approaches $c$ from different sides of $c$:

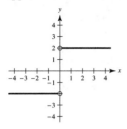

(ii) The values of $f$ increase without bound as $x$ approaches $c$:

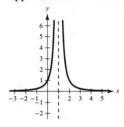

(iii) The values of $f$ oscillate between two fixed numbers as $x$ approaches $c$:

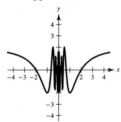

**65.** (a)  $C = 2\pi r$

$$r = \frac{C}{2\pi} = \frac{6}{2\pi} = \frac{3}{\pi} \approx 0.9549 \text{ cm}$$

(b) When $C = 5.5$: $r = \dfrac{5.5}{2\pi} \approx 0.87535$ cm

When $C = 6.5$: $r = \dfrac{6.5}{2\pi} \approx 1.03451$ cm

So $0.87535 < r < 1.03451$.

(c)  $\lim\limits_{x \to 3/\pi} (2\pi r) = 6$; $\varepsilon = 0.5$; $\delta \approx 0.0796$

**67.** $f(x) = (1 + x)^{1/x}$

$\lim\limits_{x \to 0}(1 + x)^{1/x} = e \approx 2.71828$

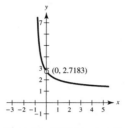

| $x$ | $f(x)$ | $x$ | $f(x)$ |
|---|---|---|---|
| $-0.1$ | 2.867972 | 0.1 | 2.593742 |
| $-0.01$ | 2.731999 | 0.01 | 2.704814 |
| $-0.001$ | 2.719642 | 0.001 | 2.716942 |
| $-0.0001$ | 2.718418 | 0.0001 | 2.718146 |
| $-0.00001$ | 2.718295 | 0.00001 | 2.718268 |
| $-0.000001$ | 2.718283 | 0.000001 | 2.718280 |

**69.**

Using the zoom and trace feature, $\delta = 0.001$. So $(2 - \delta, 2 + \delta) = (1.999, 2.001)$.

**Note:** $\dfrac{x^2 - 4}{x - 2} = x + 2$ for $x \neq 2$.

**71.** False. The existence or nonexistence of $f(x)$ at $x = c$ has no bearing on the existence of the limit of $f(x)$ as $x \to c$.

**73.** False. Let

$$f(x) = \begin{cases} x - 4, & x \neq 2 \\ 0, & x = 2 \end{cases}$$

$f(2) = 0$

$\lim\limits_{x \to 2} f(x) = \lim\limits_{x \to 2}(x - 4) = 2 \neq 0$

**75.** $f(x) = \sqrt{x}$

$\lim\limits_{x \to 0.25} \sqrt{x} = 0.5$ is true.

As $x$ approaches $0.25 = \frac{1}{4}$ from either side,

$f(x) = \sqrt{x}$ approaches $\frac{1}{2} = 0.5$.

**77.** Using a graphing utility, you see that

$$\lim_{x \to 0} \frac{\sin x}{x} = 1$$

$$\lim_{x \to 0} \frac{\sin 2x}{x} = 2, \text{ etc.}$$

So, $\lim\limits_{x \to 0} \dfrac{\sin nx}{x} = n$.

**79.** If $\lim\limits_{x \to c} f(x) = L_1$ and $\lim\limits_{x \to c} f(x) = L_2$, then for every $\varepsilon > 0$, there exists $\delta_1 > 0$ and $\delta_2 > 0$ such that

$|x - c| < \delta_1 \Rightarrow |f(x) - L_1| < \varepsilon$ and $|x - c| < \delta_2 \Rightarrow |f(x) - L_2| < \varepsilon$. Let $\delta$ equal the smaller of $\delta_1$ and $\delta_2$.

Then for $|x - c| < \delta$, you have $|L_1 - L_2| = |L_1 - f(x) + f(x) - L_2| \le |L_1 - f(x)| + |f(x) - L_2| < \varepsilon + \varepsilon$.

Therefore, $|L_1 - L_2| < 2\varepsilon$. Since $\varepsilon > 0$ is arbitrary, it follows that $L_1 = L_2$.

**81.** $\lim\limits_{x \to c}\left[ f(x) - L \right] = 0$ means that for every $\varepsilon > 0$ there

exists $\delta > 0$ such that if

$$0 < |x - c| < \delta,$$

then

$$\left| \left( f(x) - L \right) - 0 \right| < \varepsilon.$$

This means the same as $|f(x) - L| < \varepsilon$ when

$$0 < |x - c| < \delta.$$

So, $\lim\limits_{x \to c} f(x) = L.$

**83.** Answers will vary.

**85.** The radius $OP$ has a length equal to the altitude $z$ of the

triangle plus $\dfrac{h}{2}$. So, $z = 1 - \dfrac{h}{2}$.

Area triangle $= \dfrac{1}{2}b\left(1 - \dfrac{h}{2}\right)$

Area rectangle $= bh$

Because these are equal,

$$\frac{1}{2}b\left(1 - \frac{h}{2}\right) = bh$$

$$1 - \frac{h}{2} = 2h$$

$$\frac{5}{2}h = 1$$

$$h = \frac{2}{5}.$$

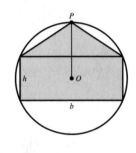

# Section 1.3 Evaluating Limits Analytically

**1.**

(a) $\lim\limits_{x \to 4} h(x) = 0$

(b) $\lim\limits_{x \to -1} h(x) = -5$

**3.**

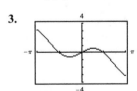

$f(x) = x \cos x$

(a) $\lim\limits_{x \to 0} f(x) = 0$

(b) $\lim\limits_{x \to \pi/3} f(x) \approx 0.524$

$$\left( = \frac{\pi}{6} \right)$$

**5.** $\lim\limits_{x \to 2} x^3 = 2^3 = 8$

**7.** $\lim\limits_{x \to 0} (2x - 1) = 2(0) - 1 = -1$

**9.** $\lim\limits_{x \to -3} (x^2 + 3x) = (-3)^2 + 3(-3) = 9 - 9 = 0$

**11.** $\lim\limits_{x \to -3} (2x^2 + 4x + 1) = 2(-3)^2 + 4(-3) + 1$
$$= 18 - 12 + 1 = 7$$

**13.** $\lim\limits_{x \to 3} \sqrt{x + 1} = \sqrt{3 + 1} = 2$

**15.** $\lim\limits_{x \to -4} (x + 3)^2 = (-4 + 3)^2 = 1$

**17.** $\lim\limits_{x \to 2} \dfrac{1}{x} = \dfrac{1}{2}$

**19.** $\lim\limits_{x \to 1} \dfrac{x}{x^2 + 4} = \dfrac{1}{1^2 + 4} = \dfrac{1}{5}$

**21.** $\lim\limits_{x \to 7} \dfrac{3x}{\sqrt{x+2}} = \dfrac{3(7)}{\sqrt{7+2}} = \dfrac{21}{3} = 7$

**23.** (a) $\lim\limits_{x \to 1} f(x) = 5 - 1 = 4$

   (b) $\lim\limits_{x \to 4} g(x) = 4^3 = 64$

   (c) $\lim\limits_{x \to 1} g(f(x)) = g(f(1)) = g(4) = 64$

**25.** (a) $\lim\limits_{x \to 1} f(x) = 4 - 1 = 3$

   (b) $\lim\limits_{x \to 3} g(x) = \sqrt{3+1} = 2$

   (c) $\lim\limits_{x \to 1} g(f(x)) = g(3) = 2$

**37.** (a) $\lim\limits_{x \to c} [5g(x)] = 5 \lim\limits_{x \to c} g(x) = 5(2) = 10$

   (b) $\lim\limits_{x \to c} [f(x) + g(x)] = \lim\limits_{x \to c} f(x) + \lim\limits_{x \to c} g(x) = 3 + 2 = 5$

   (c) $\lim\limits_{x \to c} [f(x)g(x)] = \left[\lim\limits_{x \to c} f(x)\right]\left[\lim\limits_{x \to c} g(x)\right] = (3)(2) = 6$

   (d) $\lim\limits_{x \to c} \dfrac{f(x)}{g(x)} = \dfrac{\lim\limits_{x \to c} f(x)}{\lim\limits_{x \to c} g(x)} = \dfrac{3}{2}$

**39.** (a) $\lim\limits_{x \to c} [f(x)]^3 = \left[\lim\limits_{x \to c} f(x)\right]^3 = (4)^3 = 64$

   (b) $\lim\limits_{x \to c} \sqrt{f(x)} = \sqrt{\lim\limits_{x \to c} f(x)} = \sqrt{4} = 2$

   (c) $\lim\limits_{x \to c} [3f(x)] = 3 \lim\limits_{x \to c} f(x) = 3(4) = 12$

   (d) $\lim\limits_{x \to c} [f(x)]^{3/2} = \left[\lim\limits_{x \to c} f(x)\right]^{3/2} = (4)^{3/2} = 8$

**41.** $f(x) = x - 1$ and $g(x) = \dfrac{x^2 - x}{x}$ agree except at $x = 0$.

   (a) $\lim\limits_{x \to 0} g(x) = \lim\limits_{x \to 0} f(x) = 0 - 1 = -1$

   (b) $\lim\limits_{x \to -1} g(x) = \lim\limits_{x \to -1} f(x) = -1 - 1 = -2$

**43.** $f(x) = x(x + 1)$ and $g(x) = \dfrac{x^3 - x}{x - 1}$ agree except at $x = 1$.

   (a) $\lim\limits_{x \to 1} g(x) = \lim\limits_{x \to 1} f(x) = 2$

   (b) $\lim\limits_{x \to -1} g(x) = \lim\limits_{x \to -1} f(x) = 0$

**45.** $f(x) = \dfrac{x^2 - 1}{x + 1}$ and $g(x) = x - 1$ agree except at $x = -1$.

   $\lim\limits_{x \to -1} f(x) = \lim\limits_{x \to -1} g(x) = -2$

**27.** $\lim\limits_{x \to \pi/2} \sin x = \sin \dfrac{\pi}{2} = 1$

**29.** $\lim\limits_{x \to 1} \cos \dfrac{\pi x}{3} = \cos \dfrac{\pi}{3} = \dfrac{1}{2}$

**31.** $\lim\limits_{x \to 0} \sec 2x = \sec 0 = 1$

**33.** $\lim\limits_{x \to 5\pi/6} \sin x = \sin \dfrac{5\pi}{6} = \dfrac{1}{2}$

**35.** $\lim\limits_{x \to 3} \tan\left(\dfrac{\pi x}{4}\right) = \tan \dfrac{3\pi}{4} = -1$

**47.** $f(x) = \dfrac{x^3 - 8}{x - 2}$ and $g(x) = x^2 + 2x + 4$ agree except at $x = 2$.

   $\lim\limits_{x \to 2} f(x) = \lim\limits_{x \to 2} g(x) = 12$

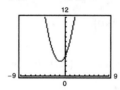

**49.** $\lim\limits_{x \to 0} \dfrac{x}{x^2 - x} = \lim\limits_{x \to 0} \dfrac{x}{x(x - 1)} = \lim\limits_{x \to 0} \dfrac{1}{x - 1} = -1$

**51.** $\lim\limits_{x \to 4} \dfrac{x - 4}{x^2 - 16} = \lim\limits_{x \to 4} \dfrac{x - 4}{(x + 4)(x - 4)}$

   $= \lim\limits_{x \to 4} \dfrac{1}{x + 4} = \dfrac{1}{8}$

**53.** $\lim\limits_{x \to -3} \dfrac{x^2 + x - 6}{x^2 - 9} = \lim\limits_{x \to -3} \dfrac{(x + 3)(x - 2)}{(x + 3)(x - 3)}$

   $= \lim\limits_{x \to -3} \dfrac{x - 2}{x - 3} = \dfrac{-5}{-6} = \dfrac{5}{6}$

**55.** $\displaystyle\lim_{x\to 4}\frac{\sqrt{x+5}-3}{x-4} = \lim_{x\to 4}\frac{\sqrt{x+5}-3}{x-4}\cdot\frac{\sqrt{x+5}+3}{\sqrt{x+5}+3}$

$\displaystyle = \lim_{x\to 4}\frac{(x+5)-9}{(x-4)\left(\sqrt{x+5}+3\right)} = \lim_{x\to 4}\frac{1}{\sqrt{x+5}+3} = \frac{1}{\sqrt{9}+3} = \frac{1}{6}$

**57.** $\displaystyle\lim_{x\to 0}\frac{\sqrt{x+5}-\sqrt{5}}{x} = \lim_{x\to 0}\frac{\sqrt{x+5}-\sqrt{5}}{x}\cdot\frac{\sqrt{x+5}+\sqrt{5}}{\sqrt{x+5}+\sqrt{5}}$

$\displaystyle = \lim_{x\to 0}\frac{(x+5)-5}{x\left(\sqrt{x+5}+\sqrt{5}\right)} = \lim_{x\to 0}\frac{1}{\sqrt{x+5}+\sqrt{5}} = \frac{1}{2\sqrt{5}} = \frac{\sqrt{5}}{10}$

**59.** $\displaystyle\lim_{x\to 0}\frac{\dfrac{1}{3+x}-\dfrac{1}{3}}{x} = \lim_{x\to 0}\frac{3-(3+x)}{(3+x)3(x)} = \lim_{x\to 0}\frac{-x}{(3+x)(3)(x)} = \lim_{x\to 0}\frac{-1}{(3+x)3} = -\frac{1}{9}$

**61.** $\displaystyle\lim_{\Delta x\to 0}\frac{2(x+\Delta x)-2x}{\Delta x} = \lim_{\Delta x\to 0}\frac{2x+2\Delta x-2x}{\Delta x} = \lim_{\Delta x\to 0}2 = 2$

**63.** $\displaystyle\lim_{\Delta x\to 0}\frac{(x+\Delta x)^2-2(x+\Delta x)+1-\left(x^2-2x+1\right)}{\Delta x} = \lim_{\Delta x\to 0}\frac{x^2+2x\Delta x+(\Delta x)^2-2x-2\Delta x+1-x^2+2x-1}{\Delta x}$

$\displaystyle = \lim_{\Delta x\to 0}(2x+\Delta x-2) = 2x-2$

**65.** $\displaystyle\lim_{x\to 0}\frac{\sin x}{5x} = \lim_{x\to 0}\left[\left(\frac{\sin x}{x}\right)\left(\frac{1}{5}\right)\right] = (1)\left(\frac{1}{5}\right) = \frac{1}{5}$

**67.** $\displaystyle\lim_{x\to 0}\frac{\sin x(1-\cos x)}{x^2} = \lim_{x\to 0}\left[\frac{\sin x}{x}\cdot\frac{1-\cos x}{x}\right]$

$\displaystyle = (1)(0) = 0$

**69.** $\displaystyle\lim_{x\to 0}\frac{\sin^2 x}{x} = \lim_{x\to 0}\left[\frac{\sin x}{x}\sin x\right] = (1)\sin 0 = 0$

**71.** $\displaystyle\lim_{h\to 0}\frac{(1-\cos h)^2}{h} = \lim_{h\to 0}\left[\frac{1-\cos h}{h}(1-\cos h)\right]$

$\displaystyle = (0)(0) = 0$

**73.** $\displaystyle\lim_{x\to \pi/2}\frac{\cos x}{\cot x} = \lim_{x\to \pi/2}\sin x = 1$

**75.** $\displaystyle\lim_{t\to 0}\frac{\sin 3t}{2t} = \lim_{t\to 0}\left(\frac{\sin 3t}{3t}\right)\left(\frac{3}{2}\right) = (1)\left(\frac{3}{2}\right) = \frac{3}{2}$

**77.** $f(x) = \dfrac{\sqrt{x+2}-\sqrt{2}}{x}$

| $x$ | $-0.1$ | $-0.01$ | $-0.001$ | 0 | 0.001 | 0.01 | 0.1 |
|---|---|---|---|---|---|---|---|
| $f(x)$ | 0.358 | 0.354 | 0.354 | ? | 0.354 | 0.353 | 0.349 |

It appears that the limit is 0.354.

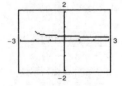

The graph has a hole at $x = 0$.

Analytically, $\displaystyle\lim_{x\to 0}\frac{\sqrt{x+2}-\sqrt{2}}{x} = \lim_{x\to 0}\frac{\sqrt{x+2}-\sqrt{2}}{x}\cdot\frac{\sqrt{x+2}+\sqrt{2}}{\sqrt{x+2}+\sqrt{2}}$

$\displaystyle = \lim_{x\to 0}\frac{x+2-2}{x\left(\sqrt{x+2}+\sqrt{2}\right)} = \lim_{x\to 0}\frac{1}{\sqrt{x+2}+\sqrt{2}} = \frac{1}{2\sqrt{2}} = \frac{\sqrt{2}}{4} \approx 0.354.$

**79.** $f(x) = \dfrac{\dfrac{1}{2+x} - \dfrac{1}{2}}{x}$

| $x$ | $-0.1$ | $-0.01$ | $-0.001$ | 0 | 0.001 | 0.01 | 0.1 |
|------|--------|---------|----------|---|-------|------|-----|
| $f(x)$ | $-0.263$ | $-0.251$ | $-0.250$ | ? | $-0.250$ | $-0.249$ | $-0.238$ |

It appears that the limit is $-0.250$.

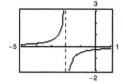

The graph has a hole at $x = 0$.

Analytically, $\lim\limits_{x \to 0} \dfrac{\dfrac{1}{2+x} - \dfrac{1}{2}}{x} = \lim\limits_{x \to 0} \dfrac{2 - (2+x)}{2(2+x)} \cdot \dfrac{1}{x} = \lim\limits_{x \to 0} \dfrac{-x}{2(2+x)} \cdot \dfrac{1}{x} = \lim\limits_{x \to 0} \dfrac{-1}{2(2+x)} = -\dfrac{1}{4}.$

**81.** $f(t) = \dfrac{\sin 3t}{t}$

| $t$ | $-0.1$ | $-0.01$ | $-0.001$ | 0 | 0.001 | 0.01 | 0.1 |
|------|--------|---------|----------|---|-------|------|-----|
| $f(t)$ | 2.96 | 2.9996 | 3 | ? | 3 | 2.9996 | 2.96 |

It appears that the limit is 3.

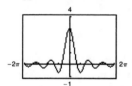

The graph has a hole at $t = 0$.

Analytically, $\lim\limits_{t \to 0} \dfrac{\sin 3t}{t} = \lim\limits_{t \to 0} 3\left(\dfrac{\sin 3t}{3t}\right) = 3(1) = 3.$

**83.** $f(x) = \dfrac{\sin x^2}{x}$

| $x$ | $-0.1$ | $-0.01$ | $-0.001$ | 0 | 0.001 | 0.01 | 0.1 |
|------|--------|---------|----------|---|-------|------|-----|
| $f(x)$ | $-0.099998$ | $-0.01$ | $-0.001$ | ? | 0.001 | 0.01 | 0.099998 |

It appears that the limit is 0.

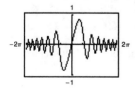

The graph has a hole at $x = 0$.

Analytically, $\lim\limits_{x \to 0} \dfrac{\sin x^2}{x} = \lim\limits_{x \to 0} x\left(\dfrac{\sin x^2}{x}\right) = 0(1) = 0.$

**85.** $\lim\limits_{\Delta x \to 0} \dfrac{f(x + \Delta x) - f(x)}{\Delta x} = \lim\limits_{\Delta x \to 0} \dfrac{3(x + \Delta x) - 2 - (3x - 2)}{\Delta x} = \lim\limits_{\Delta x \to 0} \dfrac{3x + 30x - 2 - 3x + 2}{\Delta x} = \lim\limits_{\Delta x \to 0} \dfrac{3\Delta x}{\Delta x} = 3$

**87.** $\lim\limits_{\Delta x \to 0} \dfrac{f(x + \Delta x) - f(x)}{\Delta x} = \lim\limits_{\Delta x \to 0} \dfrac{\dfrac{1}{x + \Delta x + 3} - \dfrac{1}{x + 3}}{\Delta x}$

$= \lim\limits_{\Delta x \to 0} \dfrac{x + 3 - (x + \Delta x + 3)}{(x + \Delta x + 3)(x + 3)} \cdot \dfrac{1}{\Delta x}$

$= \lim\limits_{\Delta x \to 0} \dfrac{-\Delta x}{(x + \Delta x + 3)(x + 3)\Delta x}$

$= \lim\limits_{\Delta x \to 0} \dfrac{-1}{(x + \Delta x + 3)(x + 3)} = \dfrac{-1}{(x + 3)^2}$

**89.** $\lim\limits_{x \to 0}(4 - x^2) \le \lim\limits_{x \to 0} f(x) \le \lim\limits_{x \to 0}(4 + x^2)$

$4 \le \lim\limits_{x \to 0} f(x) \le 4$

Therefore, $\lim\limits_{x \to 0} f(x) = 4$.

**91.** $f(x) = x \cos x$

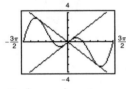

$\lim\limits_{x \to 0}(x \cos x) = 0$

**93.** $f(x) = |x| \sin x$

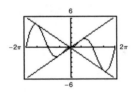

$\lim\limits_{x \to 0} |x| \sin x = 0$

**95.** $f(x) = x \sin \dfrac{1}{x}$

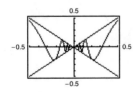

$\lim\limits_{x \to 0}\left( x \sin \dfrac{1}{x} \right) = 0$

**97.** You say that two functions $f$ and $g$ agree at all but one point (on an open interval) if $f(x) = g(x)$ for all $x$ in the interval except for $x = c$, where $c$ is in the interval.

**99.** An indeterminant form is obtained when evaluating a limit using direct substitution produces a meaningless fractional expression such as $0/0$. That is,

$$\lim\limits_{x \to c} \dfrac{f(x)}{g(x)}$$

for which $\lim\limits_{x \to c} f(x) = \lim\limits_{x \to c} g(x) = 0$

**101.** $f(x) = x,\ g(x) = \sin x,\ h(x) = \dfrac{\sin x}{x}$

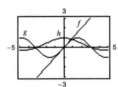

When the $x$-values are "close to" 0 the magnitude of $f$ is approximately equal to the magnitude of $g$. So, $|g|/|f| \approx 1$ when $x$ is "close to" 0.

**103.** $s(t) = -16t^2 + 500$

$\lim\limits_{t \to 2} \dfrac{s(2) - s(t)}{2 - t} = \lim\limits_{t \to 2} \dfrac{-16(2)^2 + 500 - (-16t^2 + 500)}{2 - t}$

$= \lim\limits_{t \to 2} \dfrac{436 + 16t^2 - 500}{2 - t}$

$= \lim\limits_{t \to 2} \dfrac{16(t^2 - 4)}{2 - t}$

$= \lim\limits_{t \to 2} \dfrac{16(t - 2)(t + 2)}{2 - t}$

$= \lim\limits_{t \to 2} -16(t + 2) = -64 \text{ ft/sec}$

The wrench is falling at about 64 feet/second.

**105.** $s(t) = -4.9t^2 + 200$

$$\lim_{t \to 3}\frac{s(3) - s(t)}{3 - t} = \lim_{t \to 3}\frac{-4.9(3)^2 + 200 - \left(-4.9t^2 + 200\right)}{3 - t} = \lim_{t \to 3}\frac{4.9\left(t^2 - 9\right)}{3 - t}$$

$$= \lim_{t \to 3}\frac{4.9(t - 3)(t + 3)}{3 - t} = \lim_{t \to 3}\left[-4.9(t + 3)\right] = -29.4 \text{ m/sec}$$

The object is falling about 29.4 m/sec.

**107.** Let $f(x) = 1/x$ and $g(x) = -1/x$. $\lim\limits_{x \to 0} f(x)$ and

$\lim\limits_{x \to 0} g(x)$ do not exist. However,

$$\lim_{x \to 0}\left[f(x) + g(x)\right] = \lim_{x \to 0}\left[\frac{1}{x} + \left(-\frac{1}{x}\right)\right] = \lim_{x \to 0}[0] = 0$$

and therefore does not exist.

**109.** Given $f(x) = b$, show that for every $\varepsilon > 0$ there exists

a $\delta > 0$ such that $\left|f(x) - b\right| < \varepsilon$ whenever

$\left|x - c\right| < \delta$. Because $\left|f(x) - b\right| = \left|b - b\right| = 0 < \varepsilon$

for every $\varepsilon > 0$, any value of $\delta > 0$ will work.

**111.** If $b = 0$, the property is true because both sides are

equal to 0. If $b \neq 0$, let $\varepsilon > 0$ be given. Because

$\lim\limits_{x \to c} f(x) = L$, there exists $\delta > 0$ such that

$\left|f(x) - L\right| < \varepsilon/|b|$ whenever $0 < \left|x - c\right| < \delta$.

So, whenever $0 < \left|x - c\right| < \delta$, we have

$$\left|b\right|\left|f(x) - L\right| < \varepsilon \quad \text{or} \quad \left|bf(x) - bL\right| < \varepsilon$$

which implies that $\lim\limits_{x \to c}\left[bf(x)\right] = bL$.

**113.**
$$-M\left|f(x)\right| \le f(x)g(x) \le M\left|f(x)\right|$$
$$\lim_{x \to c}\left(-M\left|f(x)\right|\right) \le \lim_{x \to c} f(x)g(x) \le \lim_{x \to c}\left(M\left|f(x)\right|\right)$$
$$-M(0) \le \lim_{x \to c} f(x)g(x) \le M(0)$$
$$0 \le \lim_{x \to c} f(x)g(x) \le 0$$

Therefore, $\lim\limits_{x \to c} f(x)g(x) = 0$.

**115.** Let

$$f(x) = \begin{cases} 4, & \text{if } x \ge 0 \\ -4, & \text{if } x < 0 \end{cases}$$

$$\lim_{x \to 0}\left|f(x)\right| = \lim_{x \to 0} 4 = 4.$$

$\lim\limits_{x \to 0} f(x)$ does not exist because for

$x < 0$, $f(x) = -4$ and for $x \ge 0$, $f(x) = 4$.

**117.** The limit does not exist

because the function

approaches 1 from the right

side of 0 and approaches

$-1$ from the left side of 0.

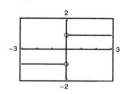

**119.** True.

**121.** False. The limit does not

exist because $f(x)$

approaches 3 from the left

side of 2 and approaches 0

from the right side of 2.

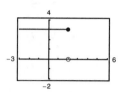

**123.** $\lim\limits_{x \to 0}\dfrac{1 - \cos x}{x} = \lim\limits_{x \to 0}\dfrac{1 - \cos x}{x} \cdot \dfrac{1 + \cos x}{1 + \cos x}$

$$= \lim_{x \to 0}\frac{1 - \cos^2 x}{x(1 + \cos x)} = \lim_{x \to 0}\frac{\sin^2 x}{x(1 + \cos x)}$$

$$= \lim_{x \to 0}\frac{\sin x}{x} \cdot \frac{\sin x}{1 + \cos x}$$

$$= \left[\lim_{x \to 0}\frac{\sin x}{x}\right]\left[\lim_{x \to 0}\frac{\sin x}{1 + \cos x}\right]$$

$$= (1)(0) = 0$$

**125.** $f(x) = \dfrac{\sec x - 1}{x^2}$

(a) The domain of $f$ is all $x \neq 0$, $\pi/2 + n\pi$.

(b)

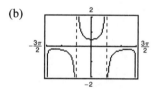

The domain is not obvious. The hole at $x = 0$ is not apparent.

(c) $\lim\limits_{x \to 0} f(x) = \dfrac{1}{2}$

(d) $\dfrac{\sec x - 1}{x^2} = \dfrac{\sec x - 1}{x^2} \cdot \dfrac{\sec x + 1}{\sec x + 1} = \dfrac{\sec^2 x - 1}{x^2(\sec x + 1)}$

$$= \frac{\tan^2 x}{x^2(\sec x + 1)} = \frac{1}{\cos^2 x}\left(\frac{\sin^2 x}{x^2}\right)\frac{1}{\sec x + 1}$$

So, $\lim\limits_{x \to 0}\dfrac{\sec x - 1}{x^2} = \lim\limits_{x \to 0}\dfrac{1}{\cos^2 x}\left(\dfrac{\sin^2 x}{x^2}\right)\dfrac{1}{\sec x + 1}$

$$= 1(1)\left(\frac{1}{2}\right) = \frac{1}{2}.$$

**127.** The graphing utility was set in degree mode, instead of *radian* mode.

## Section 1.4 Continuity and One-Sided Limits

**1.** (a) $\lim\limits_{x \to 4^+} f(x) = 3$

(b) $\lim\limits_{x \to 4^-} f(x) = 3$

(c) $\lim\limits_{x \to 4} f(x) = 3$

The function is continuous at $x = 4$ and is continuous on $(-\infty, \infty)$.

**3.** (a) $\lim\limits_{x \to 3^+} f(x) = 0$

(b) $\lim\limits_{x \to 3^-} f(x) = 0$

(c) $\lim\limits_{x \to 3} f(x) = 0$

The function is NOT continuous at $x = 3$.

**5.** (a) $\lim\limits_{x \to 2^+} f(x) = -3$

(b) $\lim\limits_{x \to 2^-} f(x) = 3$

(c) $\lim\limits_{x \to 2} f(x)$ does not exist

The function is NOT continuous at $x = 2$.

**15.** $\lim\limits_{\Delta x \to 0^-} \dfrac{\dfrac{1}{x + \Delta x} - \dfrac{1}{x}}{\Delta x} = \lim\limits_{\Delta x \to 0^-} \dfrac{x - (x + \Delta x)}{x(x + \Delta x)} \cdot \dfrac{1}{\Delta x} = \lim\limits_{\Delta x \to 0^-} \dfrac{-\Delta x}{x(x + \Delta x)} \cdot \dfrac{1}{\Delta x}$

$$= \lim\limits_{\Delta x \to 0^-} \dfrac{-1}{x(x + \Delta x)}$$

$$= \dfrac{-1}{x(x + 0)} = -\dfrac{1}{x^2}$$

**17.** $\lim\limits_{x \to 3^-} f(x) = \lim\limits_{x \to 3^-} \dfrac{x + 2}{2} = \dfrac{5}{2}$

**19.** $\lim\limits_{x \to 1^+} f(x) = \lim\limits_{x \to 1^+} (x + 1) = 2$

$\lim\limits_{x \to 1^-} f(x) = \lim\limits_{x \to 1^-} (x^3 + 1) = 2$

$\lim\limits_{x \to 1} f(x) = 2$

**21.** $\lim\limits_{x \to \pi} \cot x$ does not exist because

$\lim\limits_{x \to \pi^+} \cot x$ and $\lim\limits_{x \to \pi^-} \cot x$ do not exist.

**23.** $\lim\limits_{x \to 4^-} (5[\![x]\!] - 7) = 5(3) - 7 = 8$

$([\![x]\!] = 3 \text{ for } 3 \le x < 4)$

**7.** $\lim\limits_{x \to 8^+} \dfrac{1}{x + 8} = \dfrac{1}{8 + 8} = \dfrac{1}{16}$

**9.** $\lim\limits_{x \to 5^+} \dfrac{x - 5}{x^2 - 25} = \lim\limits_{x \to 5^+} \dfrac{1}{x + 5} = \dfrac{1}{10}$

**11.** $\lim\limits_{x \to -3^-} \dfrac{x}{\sqrt{x^2 - 9}}$ does not exist because

$\dfrac{x}{\sqrt{x^2 - 9}}$ decreases without bound as $x \to -3^-$.

**13.** $\lim\limits_{x \to 0^-} \dfrac{|x|}{x} = \lim\limits_{x \to 0^-} \dfrac{-x}{x} = -1$

**25.** $\lim\limits_{x \to 3} (2 - [\![-x]\!])$ does not exist because

$\lim\limits_{x \to 3^-} (2 - [\![-x]\!]) = 2 - (-3) = 5$

and

$\lim\limits_{x \to 3^+} (2 - [\![-x]\!]) = 2 - (-4) = 6.$

**27.** $f(x) = \dfrac{1}{x^2 - 4}$

has discontinuities at $x = -2$ and $x = 2$ because $f(-2)$ and $f(2)$ are not defined.

**29.** $f(x) = \dfrac{[\![x]\!]}{2} + x$

has discontinuities at each integer $k$ because $\lim\limits_{x \to k^-} f(x) \ne \lim\limits_{x \to k^+} f(x).$

**31.** $g(x) = \sqrt{49 - x^2}$ is continuous on $[-7, 7]$.

**33.** $\lim\limits_{x \to 0^-} f(x) = 3 = \lim\limits_{x \to 0^+} f(x)$. $f$ is continuous on $[-1, 4]$.

**35.** $f(x) = \dfrac{6}{x}$ has a nonremovable discontinuity at $x = 0$.

**37.** $f(x) = x^2 - 9$ is continuous for all real $x$.

**39.** $f(x) = \dfrac{1}{4 - x^2} = \dfrac{1}{(2 - x)(2 + x)}$ has nonremovable

discontinuities at $x = \pm 2$ because $\lim\limits_{x \to 2} f(x)$ and

$\lim\limits_{x \to -2} f(x)$ do not exist.

**41.** $f(x) = 3x - \cos x$ is continuous for all real $x$.

**43.** $f(x) = \dfrac{x}{x^2 - x}$ is not continuous at $x = 0, 1$. Because

$\dfrac{x}{x^2 - x} = \dfrac{1}{x - 1}$ for $x \neq 0$, $x = 0$ is a removable

discontinuity, whereas $x = 1$ is a nonremovable
discontinuity.

**45.** $f(x) = \dfrac{x}{x^2 + 1}$ is continuous for all real $x$.

**47.** $f(x) = \dfrac{x + 2}{(x + 2)(x - 5)}$

has a nonremovable discontinuity at $x = 5$ because
$\lim\limits_{x \to 5} f(x)$ does not exist, and has a removable

discontinuity at $x = -2$ because

$\lim\limits_{x \to -2} f(x) = \lim\limits_{x \to -2} \dfrac{1}{x - 5} = -\dfrac{1}{7}$.

**49.** $f(x) = \dfrac{|x + 7|}{x + 7}$

has a nonremovable discontinuity at $x = -7$ because
$\lim\limits_{x \to -7} f(x)$ does not exist.

**51.** $f(x) = \begin{cases} x, & x \leq 1 \\ x^2, & x > 1 \end{cases}$

has a **possible** discontinuity at $x = 1$.

 1. $f(1) = 1$

 2. $\left. \begin{aligned} \lim\limits_{x \to 1^-} f(x) &= \lim\limits_{x \to 1^-} x = 1 \\ \lim\limits_{x \to 1^+} f(x) &= \lim\limits_{x \to 1^+} x^2 = 1 \end{aligned} \right\} \lim\limits_{x \to 1} f(x) = 1$

 3. $f(-1) = \lim\limits_{x \to 1} f(x)$

$f$ is continuous at $x = 1$, therefore, $f$ is continuous for all
real $x$.

**53.** $f(x) = \begin{cases} \dfrac{x}{2} + 1, & x \leq 2 \\ 3 - x, & x > 2 \end{cases}$

has a **possible** discontinuity at $x = 2$.

 1. $f(2) = \dfrac{2}{2} + 1 = 2$

 2. $\left. \begin{aligned} \lim\limits_{x \to 2^-} f(x) &= \lim\limits_{x \to 2^-} \left( \dfrac{x}{2} + 1 \right) = 2 \\ \lim\limits_{x \to 2^+} f(x) &= \lim\limits_{x \to 2^+} (3 - x) = 1 \end{aligned} \right\} \lim\limits_{x \to 2} f(x)$ does not exist.

Therefore, $f$ has a nonremovable discontinuity at $x = 2$.

**55.** $f(x) = \begin{cases} \tan\dfrac{\pi x}{4}, & |x| < 1 \\ x, & |x| \geq 1 \end{cases}$

$= \begin{cases} \tan\dfrac{\pi x}{4}, & -1 < x < 1 \\ x, & x \leq -1 \text{ or } x \geq 1 \end{cases}$

has **possible** discontinuities at $x = -1$, $x = 1$.

 1. $f(-1) = -1$ $\qquad\qquad$ $f(1) = 1$

 2. $\lim\limits_{x \to -1} f(x) = -1$ $\qquad$ $\lim\limits_{x \to 1} f(x) = 1$

 3. $f(-1) = \lim\limits_{x \to -1} f(x)$ $\qquad$ $f(1) = \lim\limits_{x \to 1} f(x)$

$f$ is continuous at $x = \pm 1$, therefore, $f$ is continuous for all real $x$.

**57.** $f(x) = \csc 2x$ has nonremovable discontinuities at integer multiples of $\pi/2$.

**59.** $f(x) = [\![x - 8]\!]$ has nonremovable discontinuities at each integer $k$.

**61.** $\lim_{x \to 0^+} f(x) = 0$

$\lim_{x \to 0^-} f(x) = 0$

$f$ is not continuous at $x = -2$.

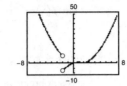

**63.** $f(1) = 3$

Find $a$ so that $\lim_{x \to 1^-} (ax - 4) = 3$

$a(1) - 4 = 3$

$a = 7$.

**65.** $f(2) = 8$

Find $a$ so that $\lim_{x \to 2^+} ax^2 = 8 \Rightarrow a = \frac{8}{2^2} = 2$.

**67.** Find $a$ and $b$ such that $\lim_{x \to -1^+} (ax + b) = -a + b = 2$ and $\lim_{x \to 3^-} (ax + b) = 3a + b = -2$.

$$a - b = -2$$
$$\underline{(+)3a + b = -2}$$
$$4a \quad\quad = -4$$
$$a = -1$$
$$b = 2 + (-1) = 1$$

$$f(x) = \begin{cases} 2, & x \le -1 \\ -x + 1, & -1 < x < 3 \\ -2, & x \ge 3 \end{cases}$$

**69.** $f(g(x)) = (x - 1)^2$

Continuous for all real $x$.

**71.** $f(g(x)) = \dfrac{1}{(x^2 + 5) - 6} = \dfrac{1}{x^2 - 1}$

Nonremovable discontinuities at $x = \pm 1$

**73.** $y = [\![x]\!] - x$

Nonremovable discontinuity at each integer

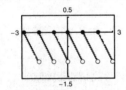

**75.** $g(x) = \begin{cases} x^2 - 3x, & x > 4 \\ 2x - 5, & x \le 4 \end{cases}$

There is a nonremovable discontinuity at $x = 4$.

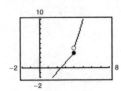

**77.** $f(x) = \dfrac{x}{x^2 + x + 2}$

Continuous on $(-\infty, \infty)$

**79.** $f(x) = \sec\dfrac{\pi x}{4}$

Continuous on:

$\dots, (-6, -2), (-2, 2), (2, 6), (6, 10), \dots$

**81.** $f(x) = \dfrac{\sin x}{x}$

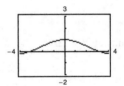

The graph **appears** to be continuous on the interval $[-4, 4]$. Because $f(0)$ is not defined, you know that $f$ has a discontinuity at $x = 0$. This discontinuity is removable so it does not show up on the graph.

**83.** $f(x) = \frac{1}{12}x^4 - x^3 + 4$ is continuous on the interval $[1, 2]$. $f(1) = \frac{37}{12}$ and $f(2) = -\frac{8}{3}$. By the Intermediate Value Theorem, there exists a number $c$ in $[1, 2]$ such that $f(c) = 0$.

**85.** $f(x) = x^2 - 2 - \cos x$ is continuous on $[0, \pi]$.

$f(0) = -3$ and $f(\pi) = \pi^2 - 1 \approx 8.87 > 0$. By the Intermediate Value Theorem, $f(c) = 0$ for at least one value of $c$ between 0 and $\pi$.

**87.** $f(x) = x^3 + x - 1$

$f(x)$ is continuous on $[0, 1]$.

$f(0) = -1$ and $f(1) = 1$

By the Intermediate Value Theorem, $f(c) = 0$ for at least one value of $c$ between 0 and 1. Using a graphing utility to zoom in on the graph of $f(x)$, you find that $x \approx 0.68$. Using the root feature, you find that $x \approx 0.6823$.

**89.** $g(t) = 2 \cos t - 3t$

$g$ is continuous on $[0, 1]$.

$g(0) = 2 > 0$ and $g(1) \approx -1.9 < 0$.

By the Intermediate Value Theorem, $g(c) = 0$ for at least one value of $c$ between 0 and 1. Using a graphing utility to zoom in on the graph of $g(t)$, you find that $t \approx 0.56$. Using the root feature, you find that $t \approx 0.5636$.

**91.** $f(x) = x^2 + x - 1$

$f$ is continuous on $[0, 5]$.

$f(0) = -1$ and $f(5) = 29$

$$-1 < 11 < 29$$

The Intermediate Value Theorem applies.

$$x^2 + x - 1 = 11$$
$$x^2 + x - 12 = 0$$
$$(x + 4)(x - 3) = 0$$
$$x = -4 \text{ or } x = 3$$

$c = 3$ ($x = -4$ is not in the interval.)

So, $f(3) = 11$.

**93.** $f(x) = x^3 - x^2 + x - 2$

$f$ is continuous on $[0, 3]$.

$f(0) = -2$ and $f(3) = 19$

$$-2 < 4 < 19$$

The Intermediate Value Theorem applies.

$$x^3 - x^2 + x - 2 = 4$$
$$x^3 - x^2 + x - 6 = 0$$
$$(x - 2)(x^2 + x + 3) = 0$$
$$x = 2$$

$(x^2 + x + 3$ has no real solution.)

$$c = 2$$

So, $f(2) = 4$.

**95.** (a) The limit does not exist at $x = c$.

(b) The function is not defined at $x = c$.

(c) The limit exists at $x = c$, but it is not equal to the value of the function at $x = c$.

(d) The limit does not exist at $x = c$.

**97.** If $f$ and $g$ are continuous for all real $x$, then so is $f + g$ (Theorem 1.11, part 2). However, $f/g$ might not be continuous if $g(x) = 0$. For example, let $f(x) = x$ and $g(x) = x^2 - 1$. Then $f$ and $g$ are continuous for all real $x$, but $f/g$ is not continuous at $x = \pm 1$.

**99.** True

1. $f(c) = L$ is defined.

2. $\lim_{x \to c} f(x) = L$ exists.

3. $f(c) = \lim_{x \to c} f(x)$

All of the conditions for continuity are met.

**101.** False. A rational function can be written as $P(x)/Q(x)$ where $P$ and $Q$ are polynomials of degree $m$ and $n$, respectively. It can have, at most, $n$ discontinuities.

**103.** $\lim_{t \to 4^-} f(t) \approx 28$

$\lim_{t \to 4^+} f(t) \approx 56$

At the end of day 3, the amount of chlorine in the pool has decreased to about 28 oz. At the beginning of day 4, more chlorine was added, and the amount is now about 56 oz.

**105.** $C(t) = \begin{cases} 0.40, & 0 < t \le 10 \\ 0.40 + 0.05[\![t - 9]\!], & t > 10, t \text{ not an integer} \\ 0.40 + 0.05(t - 10), & t > 10, t \text{ an integer} \end{cases}$

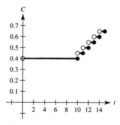

There is a nonremovable discontinuity at each integer greater than or equal to 10.

Note: You could also express $C$ as

$$C(t) = \begin{cases} 0.40, & 0 < t \le 10 \\ 0.40 - 0.05[\![10 - t]\!], & t > 10 \end{cases}$$

**107.** Let $s(t)$ be the position function for the run up to the campsite. $s(0) = 0$ ($t = 0$ corresponds to 8:00 A.M.,

$s(20) = k$ (distance to campsite)). Let $r(t)$ be the position function for the run back down the mountain:

$r(0) = k, r(10) = 0$. Let $f(t) = s(t) - r(t)$.

When $t = 0$ (8:00 A.M.), $f(0) = s(0) - r(0) = 0 - k < 0$.

When $t = 10$ (8:00 A.M.), $f(10) = s(10) - r(10) > 0$.

Because $f(0) < 0$ and $f(10) > 0$, then there must be a value $t$ in the interval $[0, 10]$ such that $f(t) = 0$.

If $f(t) = 0$, then $s(t) - r(t) = 0$, which gives us $s(t) = r(t)$. Therefore, at some time $t$, where $0 \le t \le 10$,

the position functions for the run up and the run down are equal.

**109.** Suppose there exists $x_1$ in $[a, b]$ such that $f(x_1) > 0$ and there exists $x_2$ in $[a, b]$ such that $f(x_2) < 0$.

Then by the Intermediate Value Theorem, $f(x)$ must equal zero for some value of $x$ in $[x_1, x_2]$ (or $[x_2, x_1]$ if $x_2 < x_1$).

So, $f$ would have a zero in $[a, b]$, which is a contradiction. Therefore, $f(x) > 0$ for all $x$ in $[a, b]$ or $f(x) < 0$ for

all $x$ in $[a, b]$.

**111.** If $x = 0$, then $f(0) = 0$ and $\lim\limits_{x \to 0} f(x) = 0$. So, $f$ is

continuous at $x = 0$.

If $x \neq 0$, then $\lim\limits_{t \to x} f(t) = 0$ for $x$ rational, whereas

$\lim\limits_{t \to x} f(t) = \lim\limits_{t \to x} kt = kx \neq 0$ for $x$ irrational. So, $f$ is

not continuous for all $x \neq 0$.

**113.** (a)

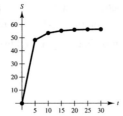

(b) There appears to be a limiting speed and a possible
cause is air resistance.

**115.** $f(x) = \begin{cases} 1 - x^2, & x \le c \\ x, & x > c \end{cases}$

$f$ is continuous for $x < c$ and for $x > c$. At $x = c$, you need $1 - c^2 = c$.

Solving $c^2 + c - 1$, you obtain $c = \dfrac{-1 \pm \sqrt{1 + 4}}{2} = \dfrac{-1 \pm \sqrt{5}}{2}$.

**117.** $f(x) = \dfrac{\sqrt{x + c^2} - c}{x}, c > 0$

Domain: $x + c^2 \ge 0 \Rightarrow x \ge -c^2$ and $x \neq 0, [-c^2, 0) \cup (0, \infty)$

$$\lim_{x \to 0} \frac{\sqrt{x + c^2} - c}{x} = \lim_{x \to 0} \frac{\sqrt{x + c^2} - c}{x} \cdot \frac{\sqrt{x + c^2} + c}{\sqrt{x + c^2} + c} = \lim_{x \to 0} \frac{(x + c^2) - c^2}{x\left[\sqrt{x + c^2} + c\right]} = \lim_{x \to 0} \frac{1}{\sqrt{x + c^2} + c} = \frac{1}{2c}$$

Define $f(0) = 1/(2c)$ to make $f$ continuous at $x = 0$.

**119.** $h(x) = x[\![x]\!]$

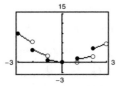

$h$ has nonremovable discontinuities at $x = \pm 1, \pm 2, \pm 3, \ldots$.

**121.** The statement is true.

If $y \geq 0$ and $y \leq 1$, then $y(y - 1) \leq 0 \leq x^2$, as desired. So assume $y > 1$. There are now two cases.

Case 1: If $x \leq y - \frac{1}{2}$, then $2x + 1 \leq 2y$ and

$$y(y - 1) = y(y + 1) - 2y$$
$$\leq (x + 1)^2 - 2y$$
$$= x^2 + 2x + 1 - 2y$$
$$\leq x^2 + 2y - 2y$$
$$= x^2$$

Case 2: If $x \geq y - \frac{1}{2}$

$$x^2 \geq \left(y - \frac{1}{2}\right)^2$$
$$= y^2 - y + \frac{1}{4}$$
$$> y^2 - y$$
$$= y(y - 1)$$

In both cases, $y(y - 1) \leq x^2$.

# Section 1.5   Infinite Limits

**1.** $f(x) = \dfrac{1}{x - 4}$

As $x$ approaches 4 from the left, $x - 4$ is a small negative number. So,

$$\lim_{x \to 4^-} f(x) = -\infty$$

As $x$ approaches 4 from the right, $x - 4$ is a small positive number. So,

$$\lim_{x \to 4^+} f(x) = \infty$$

**3.** $f(x) = \dfrac{1}{(x - 4)^2}$

As $x$ approaches 4 from the left or right, $(x - 4)^2$ is a small positive number. So,

$$\lim_{x \to 4^+} f(x) = \lim_{x \to 4^-} f(x) = \infty.$$

**5.** $\displaystyle\lim_{x \to -2^+} 2\left|\dfrac{x}{x^2 - 4}\right| = \infty$

$\displaystyle\lim_{x \to -2^-} 2\left|\dfrac{x}{x^2 - 4}\right| = \infty$

**7.** $\displaystyle\lim_{x \to -2^+} \tan \dfrac{\pi x}{4} = -\infty$

$\displaystyle\lim_{x \to -2^-} \tan \dfrac{\pi x}{4} = \infty$

**9.** $f(x) = \dfrac{1}{x^2 - 9}$

| $x$ | $-3.5$ | $-3.1$ | $-3.01$ | $-3.001$ | $-2.999$ | $-2.99$ | $-2.9$ | $-2.5$ |
|---|---|---|---|---|---|---|---|---|
| $f(x)$ | 0.308 | 1.639 | 16.64 | 166.6 | $-166.7$ | $-16.69$ | $-1.695$ | $-0.364$ |

$\displaystyle\lim_{x \to -3^-} f(x) = \infty$

$\displaystyle\lim_{x \to -3^+} f(x) = -\infty$

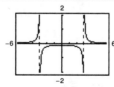

**11.** $f(x) = \dfrac{x^2}{x^2 - 9}$

| $x$ | $-3.5$ | $-3.1$ | $-3.01$ | $-3.001$ | $-2.999$ | $-2.99$ | $-2.9$ | $-2.5$ |
|---|---|---|---|---|---|---|---|---|
| $f(x)$ | 3.769 | 15.75 | 150.8 | 1501 | $-1499$ | $-149.3$ | $-14.25$ | $-2.273$ |

$\displaystyle\lim_{x \to -3^-} f(x) = \infty$

$\displaystyle\lim_{x \to -3^+} f(x) = -\infty$

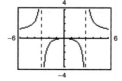

**13.** $\lim\limits_{x\to 0^+}\dfrac{1}{x^2}=\infty=\lim\limits_{x\to 0^-}\dfrac{1}{x^2}$

Therefore, $x=0$ is a vertical asymptote.

**15.** $\lim\limits_{x\to -2^-}\dfrac{x^2}{x^2-4}=\infty$ and $\lim\limits_{x\to -2^+}\dfrac{x^2}{x^2-4}=-\infty$

Therefore, $x=-2$ is a vertical asymptote.

$\lim\limits_{x\to 2^-}\dfrac{x^2}{x^2-4}=-\infty$ and $\lim\limits_{x\to 2^+}\dfrac{x^2}{x^2-4}=\infty$

Therefore, $x=2$ is a vertical asymptote.

**17.** No vertical asymptote because the denominator is never zero.

**19.** $\lim\limits_{x\to 2^+}\dfrac{x^2-2}{(x-2)(x+1)}=\infty$

$\lim\limits_{x\to 2^-}\dfrac{x^2-2}{(x-2)(x+1)}=-\infty$

Therefore, $x=2$ is a vertical asymptote.

$\lim\limits_{x\to -1^+}\dfrac{x^2-2}{(x-2)(x+1)}=\infty$

$\lim\limits_{x\to -1^-}\dfrac{x^2-2}{(x-2)(x+1)}=-\infty$

Therefore, $x=-1$ is a vertical asymptote.

**21.** $\lim\limits_{t\to 0^+}\left(1-\dfrac{4}{t^2}\right)=-\infty=\lim\limits_{t\to 0^-}\left(1-\dfrac{4}{t^2}\right)$

Therefore, $t=0$ is a vertical asymptote.

**23.** $f(x)=\dfrac{3}{x^2+x-2}=\dfrac{3}{(x+2)(x-1)}$

Vertical asymptotes at $x=-2$ and $x=1$.

**25.** $f(x)=\dfrac{x^3+1}{x+1}=\dfrac{(x+1)(x^2-x+1)}{x+1}$

has no vertical asymptote because

$\lim\limits_{x\to -1}f(x)=\lim\limits_{x\to -1}(x^2-x+1)=3.$

The graph has a hole at $x=-1$.

**27.** $f(x)=\dfrac{(x-5)(x+3)}{(x-5)(x^2+1)}=\dfrac{x+3}{x^2+1},\ x\neq 5$

No vertical asymptote. The graph has a hole at $x=5$.

**29.** $f(x)=\tan\pi x=\dfrac{\sin\pi x}{\cos\pi x}$ has vertical asymptotes at

$x=\dfrac{2n+1}{2},\ n$ any integer.

**31.** $s(t)=\dfrac{t}{\sin t}$ has vertical asymptotes at $t=n\pi,\ n$ a

nonzero integer. There is no vertical asymptote at $t=0$ since

$\lim\limits_{t\to 0}\dfrac{t}{\sin t}=1.$

**33.** $\lim\limits_{x\to -1}\dfrac{x^2-1}{x+1}=\lim\limits_{x\to -1}(x-1)=-2$

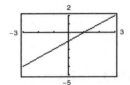

Removable discontinuity at $x=-1$

**35.** $\lim\limits_{x\to -1^+}\dfrac{x^2+1}{x+1}=\infty$

$\lim\limits_{x\to -1^-}\dfrac{x^2+1}{x+1}=-\infty$

Vertical asymptote at $x=-1$

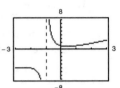

**37.** $\lim\limits_{x\to -1^+}\dfrac{1}{x+1}=\infty$

**39.** $\lim\limits_{x\to 2^+}\dfrac{x}{x-2}=\infty$

**41.** $\lim\limits_{x\to 1^+}\dfrac{x^2}{(x-1)^2}=\infty$

**43.** $\lim\limits_{x\to -3^-}\dfrac{x+3}{(x^2+x-6)}=\lim\limits_{x\to -3^-}\dfrac{x+3}{(x+3)(x-2)}$

$=\lim\limits_{x\to -3^-}\dfrac{1}{x-2}=-\dfrac{1}{5}$

**45.** $\lim\limits_{x\to 1}\dfrac{x-1}{(x^2+1)(x-1)}=\lim\limits_{x\to 1}\dfrac{1}{x^2+1}=\dfrac{1}{2}$

**47.** $\lim\limits_{x\to 0^-}\left(1+\dfrac{1}{x}\right)=-\infty$

**49.** $\lim\limits_{x\to 0^+}\dfrac{2}{\sin x}=\infty$

**51.** $\lim\limits_{x\to \pi}\dfrac{\sqrt{x}}{\csc x}=\lim\limits_{x\to \pi}\left(\sqrt{x}\sin x\right)=0$

**53.** $\lim\limits_{x\to (1/2)^-}x\sec(\pi x)=\infty$ and $\lim\limits_{x\to (1/2)^+}x\sec(\pi x)=-\infty.$

Therefore, $\lim\limits_{x\to (1/2)}x\sec(\pi x)$ does not exist.

**55.** $f(x) = \dfrac{x^2 + x + 1}{x^3 - 1} = \dfrac{x^2 + x + 1}{(x-1)(x^2 + x + 1)}$

$\displaystyle\lim_{x \to 1^+} f(x) = \lim_{x \to 1^+} \dfrac{1}{x - 1} = \infty$

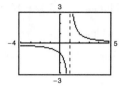

**57.** $f(x) = \dfrac{1}{x^2 - 25}$

$\displaystyle\lim_{x \to 5^-} f(x) = -\infty$

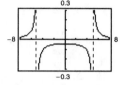

**59.** A limit in which $f(x)$ increases or decreases without bound as $x$ approaches $c$ is called an infinite limit. $\infty$ is not a number. Rather, the symbol

$\displaystyle\lim_{x \to c} f(x) = \infty$

says how the limit fails to exist.

**61.** One answer is

$f(x) = \dfrac{x - 3}{(x - 6)(x + 2)} = \dfrac{x - 3}{x^2 - 4x - 12}.$

**63.**

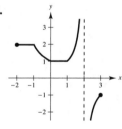

**65.** $m = \dfrac{m_0}{\sqrt{1 - (v^2/c^2)}}$

$\displaystyle\lim_{v \to c^-} m = \lim_{v \to c^-} \dfrac{m_0}{\sqrt{1 - (v^2/c^2)}} = \infty$

**67. (a)** $r = 50\pi \sec^2 \dfrac{\pi}{6} = \dfrac{200\pi}{3}$ ft/sec

**(b)** $r = 50\pi \sec^2 \dfrac{\pi}{3} = 200\pi$ ft/sec

**(c)** $\displaystyle\lim_{\theta \to (\pi/2)^-} \left[ 50\pi \sec^2 \theta \right] = \infty$

**69. (a)**  Average speed $= \dfrac{\text{Total distance}}{\text{Total time}}$

$50 = \dfrac{2d}{(d/x) + (d/y)}$

$50 = \dfrac{2xy}{y + x}$

$50y + 50x = 2xy$

$50x = 2xy - 50y$

$50x = 2y(x - 25)$

$\dfrac{25x}{x - 25} = y$

Domain: $x > 25$

**(b)**

| $x$ | 30 | 40 | 50 | 60 |
|---|---|---|---|---|
| $y$ | 150 | 66.667 | 50 | 42.857 |

**(c)** $\displaystyle\lim_{x \to 25^+} \dfrac{25x}{\sqrt{x - 25}} = \infty$

As $x$ gets close to 25 mi/h, $y$ becomes larger and larger.

**71. (a)** $A = \dfrac{1}{2}bh - \dfrac{1}{2}r^2\theta = \dfrac{1}{2}(10)(10 \tan \theta) - \dfrac{1}{2}(10)^2 \theta = 50 \tan \theta - 50\theta$

Domain: $\left(0, \dfrac{\pi}{2}\right)$

**(b)**

| $\theta$ | 0.3 | 0.6 | 0.9 | 1.2 | 1.5 |
|---|---|---|---|---|---|
| $f(\theta)$ | 0.47 | 4.21 | 18.0 | 68.6 | 630.1 |

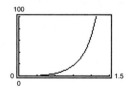

**(c)** $\displaystyle\lim_{\theta \to \pi/2^-} A = \infty$

**73.** False. For instance, let

$$f(x) = \frac{x^2 - 1}{x - 1} \text{ or}$$

$$g(x) = \frac{x}{x^2 + 1}.$$

**75.** False. The graphs of
$y = \tan x, \ y = \cot x, \ y = \sec x$ and $y = \csc x$ have vertical asymptotes.

**77.** Let $f(x) = \dfrac{1}{x^2}$ and $g(x) = \dfrac{1}{x^4}$, and $c = 0$.

$$\lim_{x \to 0} \frac{1}{x^2} = \infty \text{ and } \lim_{x \to 0} \frac{1}{x^4} = \infty, \text{ but}$$

$$\lim_{x \to 0} \left( \frac{1}{x^2} - \frac{1}{x^4} \right) = \lim_{x \to 0} \left( \frac{x^2 - 1}{x^4} \right) = -\infty \neq 0.$$

**79.** Given $\displaystyle\lim_{x \to c} f(x) = \infty$, let $g(x) = 1$. Then

$$\lim_{x \to c} \frac{g(x)}{f(x)} = 0 \text{ by Theorem 1.15.}$$

**81.** $f(x) = \dfrac{1}{x - 3}$ is defined for all $x > 3$. Let $M > 0$ be given. You need $\delta > 0$ such that

$$f(x) = \frac{1}{x - 3} > M \text{ whenever } 3 < x < 3 + \delta.$$

Equivalently, $x - 3 < \dfrac{1}{M}$ whenever

$$|x - 3| < \delta, x > 3.$$

So take $\delta = \dfrac{1}{M}$. Then for $x > 3$ and

$$|x - 3| < \delta, \frac{1}{x - 3} > \frac{1}{8} = M \text{ and so } f(x) > M.$$

# Review Exercises for Chapter 1

**1.** Calculus required. Using a graphing utility, you can estimate the length to be 8.3. Or, the length is slightly longer than the distance between the two points, approximately 8.25.

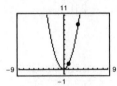

**3.** $f(x) = \dfrac{\dfrac{4}{x + 2} - 2}{x}$

| $x$ | $-0.1$ | $-0.01$ | $-0.001$ | $0.001$ | $0.01$ | $0.1$ |
|---|---|---|---|---|---|---|
| $f(x)$ | $-1.0526$ | $-1.0050$ | $-1.0005$ | $-0.9995$ | $-0.9950$ | $-0.9524$ |

$$\lim_{x \to 0} f(x) \approx -1.0$$

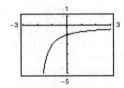

**5.** $\displaystyle\lim_{x \to 1}(x + 4) = 1 + 4 = 5$

Let $\varepsilon > 0$ be given. Choose $\delta = \varepsilon$. Then for $0 < |x - 1| < \delta = \varepsilon$, you have

$$|x - 1| < \varepsilon$$
$$|(x + 4) - 5| < \varepsilon$$
$$|f(x) - L| < \varepsilon.$$

**7.** $\lim\limits_{x \to 2}\left(1 - x^2\right) = 1 - 2^2 = -3$

Let $\varepsilon > 0$ be given. You need

$$\left|1 - x^2 - (-3)\right| < \varepsilon \Rightarrow \left|x^2 - 4\right| = |x - 2||x + 2| < \varepsilon \Rightarrow |x - 2| < \frac{1}{|x + 2|}\varepsilon$$

Assuming $1 < x < 3$, you can choose $\delta = \dfrac{\varepsilon}{5}$.

So, for $0 < |x - 2| < \delta = \dfrac{\varepsilon}{5}$, you have

$$|x - 2| < \frac{\varepsilon}{5} < \frac{\varepsilon}{|x + 2|}$$
$$|x - 2||x + 2| < \varepsilon$$
$$\left|x^2 - 4\right| < \varepsilon$$
$$\left|4 - x^2\right| < \varepsilon$$
$$\left|\left(1 - x^2\right) - (-3)\right| < \varepsilon$$
$$|f(x) - L| < \varepsilon.$$

**9.** $h(x) = \dfrac{4x - x^2}{x} = \dfrac{x(4 - x)}{x} = 4 - x, \; x \neq 0$

(a) $\lim\limits_{x \to 0} h(x) = 4 - 0 = 4$

(b) $\lim\limits_{x \to -1} h(x) = 4 - (-1) = 5$

**11.** $\lim\limits_{x \to 6}(x - 2)^2 = (6 - 2)^2 = 16$

**13.** $\lim\limits_{t \to 4}\sqrt{t + 2} = \sqrt{4 + 2} = \sqrt{6} = 2.45$

**15.** $\lim\limits_{t \to -2}\dfrac{t + 2}{t^2 - 4} = \lim\limits_{t \to -2}\dfrac{1}{t - 2} = -\dfrac{1}{4}$

**17.** $\lim\limits_{x \to 4}\dfrac{\sqrt{x - 3} - 1}{x - 4} = \lim\limits_{x \to 4}\dfrac{\sqrt{x - 3} - 1}{x - 4} \cdot \dfrac{\sqrt{x - 3} + 1}{\sqrt{x - 3} + 1}$

$\qquad = \lim\limits_{x \to 4}\dfrac{(x - 3) - 1}{(x - 4)\left(\sqrt{x - 3} + 1\right)}$

$\qquad = \lim\limits_{x \to 4}\dfrac{1}{\sqrt{x - 3} + 1} = \dfrac{1}{2}$

**19.** $\lim\limits_{x \to 0}\dfrac{\left[1/(x + 1)\right] - 1}{x} = \lim\limits_{x \to 0}\dfrac{1 - (x + 1)}{x(x + 1)}$

$\qquad = \lim\limits_{x \to 0}\dfrac{-1}{x + 1} = -1$

**21.** $\lim\limits_{x \to -5}\dfrac{x^3 + 125}{x + 5} = \lim\limits_{x \to -5}\dfrac{(x + 5)\left(x^2 - 5x + 25\right)}{x + 5}$

$\qquad = \lim\limits_{x \to -5}\left(x^2 - 5x + 25\right) = 75$

**23.** $\lim\limits_{x \to 0}\dfrac{1 - \cos x}{\sin x} = \lim\limits_{x \to 0}\left(\dfrac{x}{\sin x}\right)\left(\dfrac{1 - \cos x}{x}\right) = (1)(0) = 0$

**25.** $\lim\limits_{\Delta x \to 0}\dfrac{\sin\left[(\pi/6) + \Delta x\right] - (1/2)}{\Delta x} = \lim\limits_{\Delta x \to 0}\dfrac{\sin(\pi/6)\cos \Delta x + \cos(\pi/6)\sin \Delta x - (1/2)}{\Delta x}$

$\qquad = \lim\limits_{\Delta x \to 0}\dfrac{1}{2} \cdot \dfrac{(\cos \Delta x - 1)}{\Delta x} + \lim\limits_{\Delta x \to 0}\dfrac{\sqrt{3}}{2} \cdot \dfrac{\sin \Delta x}{\Delta x} = 0 + \dfrac{\sqrt{3}}{2}(1) = \dfrac{\sqrt{3}}{2}$

**27.** $\lim\limits_{x \to c}\left[f(x) \cdot g(x)\right] = \left(-\frac{3}{4}\right)\left(\frac{2}{3}\right) = -\frac{1}{2}$

**29.** $\lim\limits_{x \to c}\left[f(x) + 2g(x)\right] = -\frac{3}{4} + 2\left(\frac{2}{3}\right) = \frac{7}{12}$

**31.** $f(x) = \dfrac{\sqrt{2x+1} - \sqrt{3}}{x-1}$

(a)

| $x$ | 1.1 | 1.01 | 1.001 | 1.0001 |
|-----|------|------|-------|--------|
| $f(x)$ | 0.5680 | 0.5764 | 0.5773 | 0.5773 |

$\displaystyle\lim_{x\to 1^+} \dfrac{\sqrt{2x+1} - \sqrt{3}}{x-1} \approx 0.577$     $\left(\text{Actual limit is } \sqrt{3}/3.\right)$

(b)

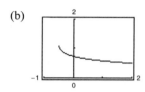

The graph has a hole at $x = 1$.

$\displaystyle\lim_{x\to 1^+} f(x) \approx 0.5774.$

(c) $\displaystyle\lim_{x\to 1^+} \dfrac{\sqrt{2x+1} - \sqrt{3}}{x-1} = \lim_{x\to 1^+} \dfrac{\sqrt{2x+1} - \sqrt{3}}{x-1} \cdot \dfrac{\sqrt{2x+1} + \sqrt{3}}{\sqrt{2x+1} + \sqrt{3}} = \lim_{x\to 1^+} \dfrac{(2x+1) - 3}{(x-1)\left(\sqrt{2x+1} + \sqrt{3}\right)}$

$= \displaystyle\lim_{x\to 1^+} \dfrac{2}{\sqrt{2x+1} + \sqrt{3}} = \dfrac{2}{2\sqrt{3}} = \dfrac{1}{\sqrt{3}} = \dfrac{\sqrt{3}}{3}$

**33.** $v = \displaystyle\lim_{t\to 4} \dfrac{s(4) - s(t)}{4-t}$

$= \displaystyle\lim_{t\to 4} \dfrac{\left[-4.9(16) + 250\right] - \left[-4.9t^2 + 250\right]}{4-t}$

$= \displaystyle\lim_{t\to 4} \dfrac{4.9\left(t^2 - 16\right)}{4-t}$

$= \displaystyle\lim_{t\to 4} \dfrac{4.9(t-4)(t+4)}{4-t}$

$= \displaystyle\lim_{t\to 4}\left[-4.9(t+4)\right] = -39.2 \text{ m/sec}$

The object is falling at about 39.2 m/sec.

**35.** $\displaystyle\lim_{x\to 3^-} \dfrac{|x-3|}{x-3} = \lim_{x\to 3^-} \dfrac{-(x-3)}{x-3} = -1$

**37.** $\displaystyle\lim_{x\to 2} f(x) = 0$

**39.** $\displaystyle\lim_{t\to 1} h(t)$ does not exist because $\displaystyle\lim_{t\to 1^-} h(t) = 1 + 1 = 2$

and $\displaystyle\lim_{t\to 1^+} h(t) = \tfrac{1}{2}(1+1) = 1.$

**41.** $f(x) = -3x^2 + 7$

Continuous on $(-\infty, \infty)$

**43.** $f(x) = [\![x + 3]\!]$

$\displaystyle\lim_{x\to k^+} [\![x+3]\!] = k + 3$ where $k$ is an integer.

$\displaystyle\lim_{x\to k^-} [\![x+3]\!] = k + 2$ where $k$ is an integer.

Nonremovable discontinuity at each integer $k$

Continuous on $(k, k+1)$ for all integers $k$

**45.** $f(x) = \dfrac{3x^2 - x - 2}{x-1} = \dfrac{(3x+2)(x-1)}{x-1}$

$\displaystyle\lim_{x\to 1} f(x) = \lim_{x\to 1}(3x+2) = 5$

Removable discontinuity at $x = 1$

Continuous on $(-\infty, 1) \cup (1, \infty)$

**47.** $f(x) = \dfrac{1}{(x-2)^2}$

$\displaystyle\lim_{x\to 2} \dfrac{1}{(x-2)^2} = \infty$

Nonremovable discontinuity at $x = 2$

Continuous on $(-\infty, 2) \cup (2, \infty)$

**49.** $f(x) = \dfrac{3}{x+1}$

$\lim\limits_{x \to 1^-} f(x) = -\infty$

$\lim\limits_{x \to 1^+} f(x) = \infty$

Nonremovable discontinuity at $x = -1$

Continuous on $(-\infty, -1) \cup (-1, \infty)$

**51.** $f(x) = \csc \dfrac{\pi x}{2}$

Nonremovable discontinuities at each even integer.

Continuous on

$(2k, 2k+2)$

for all integers $k$.

**53.** $f(2) = 5$

Find $c$ so that $\lim\limits_{x \to 2^+} (cx + 6) = 5$.

$c(2) + 6 = 5$

$\qquad 2c = -1$

$\qquad\quad c = -\dfrac{1}{2}$

**55.** $f$ is continuous on $[1, 2]$. $f(1) = -1 < 0$ and

$f(2) = 13 > 0$. Therefore by the Intermediate Value

Theorem, there is at least one value $c$ in $(1, 2)$ such that

$2c^3 - 3 = 0$.

**57.** $f(x) = \dfrac{x^2 - 4}{|x - 2|} = (x + 2)\left[\dfrac{x - 2}{|x - 2|}\right]$

(a) $\lim\limits_{x \to 2^-} f(x) = -4$

(b) $\lim\limits_{x \to 2^+} f(x) = 4$

(c) $\lim\limits_{x \to 2} f(x)$ does not exist.

**59.** $g(x) = 1 + \dfrac{2}{x}$

Vertical asymptote at $x = 0$

**61.** $f(x) = \dfrac{8}{(x - 10)^2}$

Vertical asymptote at $x = 10$

**63.** $\lim\limits_{x \to -2^-} \dfrac{2x^2 + x + 1}{x + 2} = -\infty$

**65.** $\lim\limits_{x \to -1^+} \dfrac{x + 1}{x^3 + 1} = \lim\limits_{x \to -1^+} \dfrac{1}{x^2 - x + 1} = \dfrac{1}{3}$

**67.** $\lim\limits_{x \to 1^-} \dfrac{x^2 + 2x + 1}{x - 1} = -\infty$

**69.** $\lim\limits_{x \to 0^+} \left(x - \dfrac{1}{x^3}\right) = -\infty$

**71.** $\lim\limits_{x \to 0^+} \dfrac{\sin 4x}{5x} = \lim\limits_{x \to 0^+} \left[\dfrac{4}{5}\left(\dfrac{\sin 4x}{4x}\right)\right] = \dfrac{4}{5}$

**73.** $\lim\limits_{x \to 0^+} \dfrac{\csc 2x}{x} = \lim\limits_{x \to 0^+} \dfrac{1}{x \sin 2x} = \infty$

**75.** $C = \dfrac{80{,}000p}{100 - p}, \, 0 \le p < 100$

(a) $C(15) \approx \$14{,}117.65$

(b) $C(50) = \$80.000$

(c) $C(90) = \$720{,}000$

(d) $\lim\limits_{p \to 100^-} \dfrac{80{,}000p}{100 - p} = \infty$

# Problem Solving for Chapter 1

**1.** (a)  Perimeter $\Delta PAO = \sqrt{x^2 + (y - 1)^2} + \sqrt{x^2 + y^2} + 1$

$\qquad\qquad\qquad\quad = \sqrt{x^2 + (x^2 - 1)^2} + \sqrt{x^2 + x^4} + 1$

$\qquad$ Perimeter $\Delta PBO = \sqrt{(x - 1)^2 + y^2} + \sqrt{x^2 + y^2} + 1$

$\qquad\qquad\qquad\quad = \sqrt{(x - 1)^2 + x^4} + \sqrt{x^2 + x^4} + 1$

(b)  $r(x) = \dfrac{\sqrt{x^2 + (x^2 - 1)^2} + \sqrt{x^2 + x^4} + 1}{\sqrt{(x - 1)^2 + x^4} + \sqrt{x^2 + x^4} + 1}$

| $x$ | 4 | 2 | 1 | 0.1 | 0.01 |
|---|---|---|---|---|---|
| Perimeter $\Delta PAO$ | 33.02 | 9.08 | 3.41 | 2.10 | 2.01 |
| Perimeter $\Delta PBO$ | 33.77 | 9.60 | 3.41 | 2.00 | 2.00 |
| $r(x)$ | 0.98 | 0.95 | 1 | 1.05 | 1.005 |

(c)  $\displaystyle\lim_{x \to 0^+} r(x) = \dfrac{1 + 0 + 1}{1 + 0 + 1} = \dfrac{2}{2} = 1$

**3.** (a)  There are 6 triangles, each with a central angle of $60° = \pi/3$. So,

$$\text{Area hexagon} = 6\left[\frac{1}{2}bh\right] = 6\left[\frac{1}{2}(1)\sin\frac{\pi}{3}\right] = \frac{3\sqrt{3}}{2} \approx 2.598.$$

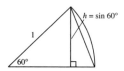

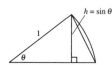

$$\text{Error} = \text{Area (Circle)} - \text{Area (Hexagon)} = \pi - \frac{3\sqrt{3}}{2} \approx 0.5435$$

(b)  There are $n$ triangles, each with central angle of $\theta = 2\pi/n$. So,

$$A_n = n\left[\frac{1}{2}bh\right] = n\left[\frac{1}{2}(1)\sin\frac{2\pi}{n}\right] = \frac{n\sin(2\pi/n)}{2}.$$

(c)

| $n$ | 6 | 12 | 24 | 48 | 96 |
|---|---|---|---|---|---|
| $A_n$ | 2.598 | 3 | 3.106 | 3.133 | 3.139 |

(d)  As $n$ gets larger and larger, $2\pi/n$ approaches 0. Letting $x = 2\pi/n$, $A_n = \dfrac{\sin(2\pi/n)}{2/n} = \dfrac{\sin(2\pi/n)}{(2\pi/n)}\pi = \dfrac{\sin x}{x}\pi$

which approaches $(1)\pi = \pi$.

**5.** (a)  Slope $= -\dfrac{12}{5}$

(b)  Slope of tangent line is $\dfrac{5}{12}$.

$$y + 12 = \frac{5}{12}(x - 5)$$

$$y = \frac{5}{12}x - \frac{169}{12} \quad \text{Tangent line}$$

(c)  $Q = (x, y) = \left(x, -\sqrt{169 - x^2}\right)$

$$m_x = \frac{-\sqrt{169 - x^2} + 12}{x - 5}$$

(d)  $\displaystyle\lim_{x \to 5} m_x = \lim_{x \to 5}\frac{12 - \sqrt{169 - x^2}}{x - 5} \cdot \frac{12 + \sqrt{169 - x^2}}{12 + \sqrt{169 - x^2}}$

$$= \lim_{x \to 5}\frac{144 - (169 - x^2)}{(x - 5)(12 + \sqrt{169 - x^2})}$$

$$= \lim_{x \to 5}\frac{x^2 - 25}{(x - 5)(12 + \sqrt{169 - x^2})}$$

$$= \lim_{x \to 5}\frac{(x + 5)}{12 + \sqrt{169 - x^2}} = \frac{10}{12 + 12} = \frac{5}{12}$$

This is the same slope as part (b).

**7.** (a) $3 + x^{1/3} \geq 0$

$$x^{1/3} \geq -3$$

$$x \geq -27$$

Domain: $x \geq -27,\ x \neq 1$ or $[-27, 1) \cup (1, \infty)$

(b)
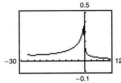

(c) $\displaystyle \lim_{x \to -27^+} f(x) = \frac{\sqrt{3 + (-27)^{1/3}} - 2}{-27 - 1} = \frac{-2}{-28} = \frac{1}{14} \approx 0.0714$

(d) $\displaystyle \lim_{x \to 1} f(x) = \lim_{x \to 1} \frac{\sqrt{3 + x^{1/3}} - 2}{x - 1} \cdot \frac{\sqrt{3 + x^{1/3}} + 2}{\sqrt{3 + x^{1/3}} + 2} = \lim_{x \to 1} \frac{3 + x^{1/3} - 4}{(x - 1)\left(\sqrt{3 + x^{1/3}} + 2\right)}$

$\displaystyle = \lim_{x \to 1} \frac{x^{1/3} - 1}{\left(x^{1/3} - 1\right)\left(x^{2/3} + x^{1/3} + 1\right)\left(\sqrt{3 + x^{1/3}} + 2\right)} = \lim_{x \to 1} \frac{1}{\left(x^{2/3} + x^{1/3} + 1\right)\left(\sqrt{3 + x^{1/3}} + 2\right)}$

$\displaystyle = \frac{1}{(1 + 1 + 1)(2 + 2)} = \frac{1}{12}$

**9.** (a) $\displaystyle \lim_{x \to 2} f(x) = 3$: $g_1, g_4$

(b) $f$ continuous at 2: $g_1$

(c) $\displaystyle \lim_{x \to 2^-} f(x) = 3$: $g_1, g_3, g_4$

**11.**

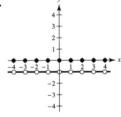

(a) $\quad f(1) = [\![1]\!] + [\![-1]\!] = 1 + (-1) = 0$

$f(0) = 0$

$f\left(\tfrac{1}{2}\right) = 0 + (-1) = -1$

$f(-2.7) = -3 + 2 = -1$

(b) $\displaystyle \lim_{x \to 1^-} f(x) = -1$

$\displaystyle \lim_{x \to 1^+} f(x) = -1$

$\displaystyle \lim_{x \to 1/2} f(x) = -1$

(c) $f$ is continuous for all real numbers except
$x = 0, \pm 1, \pm 2, \pm 3, \dots$

**13.** (a)

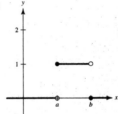

(b) (i) $\displaystyle \lim_{x \to a^+} P_{a,b}(x) = 1$

(ii) $\displaystyle \lim_{x \to a^-} P_{a,b}(x) = 0$

(iii) $\displaystyle \lim_{x \to b^+} P_{a,b}(x) = 0$

(iv) $\displaystyle \lim_{x \to b^-} P_{a,b}(x) = 1$

(c) $P_{a,b}$ is continuous for all positive real numbers
except $x = a, b$.

(d) The area under the graph of $U$, and above the $x$-axis,
is 1.

# CHAPTER 2
# Differentiation

# CHAPTER 2
## Differentiation

### Section 2.1   The Derivative and the Tangent Line Problem

**1.** (a) At $(x_1, y_1)$, slope $= 0$.

At $(x_2, y_2)$, slope $= \frac{5}{2}$.

(b) At $(x_1, y_1)$, slope $= -\frac{5}{2}$.

At $(x_2, y_2)$, slope $= 2$.

**3.** (a), (b)

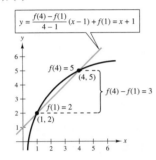

(c) $y = \dfrac{f(4) - f(1)}{4 - 1}(x - 1) + f(1)$

$= \dfrac{3}{3}(x - 1) + 2$

$= 1(x - 1) + 2$

$= x + 1$

**5.** $f(x) = 3 - 5x$ is a line. Slope $= -5$

**7.** Slope at $(2, -5) = \lim\limits_{\Delta x \to 0} \dfrac{g(2 + \Delta x) - g(2)}{\Delta x}$

$= \lim\limits_{\Delta x \to 0} \dfrac{(2 + \Delta x)^2 - 9 - (-5)}{\Delta x}$

$= \lim\limits_{\Delta x \to 0} \dfrac{4 + 4(\Delta x) + (\Delta x)^2 - 4}{\Delta x}$

$= \lim\limits_{\Delta x \to 0} (4 + \Delta x) = 4$

**17.** $f(x) = x^2 + x - 3$

$f'(x) = \lim\limits_{\Delta x \to 0} \dfrac{f(x + \Delta x) - f(x)}{\Delta x}$

$= \lim\limits_{\Delta x \to 0} \dfrac{(x + \Delta x)^2 + (x + \Delta x) - 3 - (x^2 + x - 3)}{\Delta x}$

$= \lim\limits_{\Delta x \to 0} \dfrac{x^2 + 2x(\Delta x) + (\Delta x)^2 + x + \Delta x - 3 - x^2 - x + 3}{\Delta x}$

$= \lim\limits_{\Delta x \to 0} \dfrac{2x(\Delta x) + (\Delta x)^2 + \Delta x}{\Delta x}$

$= \lim\limits_{\Delta x \to 0} (2x + \Delta x + 1) = 2x + 1$

**9.** Slope at $(0, 0) = \lim\limits_{\Delta t \to 0} \dfrac{f(0 + \Delta t) - f(0)}{\Delta t}$

$= \lim\limits_{\Delta t \to 0} \dfrac{3(\Delta t) - (\Delta t)^2 - 0}{\Delta t}$

$= \lim\limits_{\Delta t \to 0} (3 - \Delta t) = 3$

**11.** $f(x) = 7$

$f'(x) = \lim\limits_{\Delta x \to 0} \dfrac{f(x + \Delta x) - f(x)}{\Delta x}$

$= \lim\limits_{\Delta x \to 0} \dfrac{7 - 7}{\Delta x}$

$= \lim\limits_{\Delta x \to 0} 0 = 0$

**13.** $f(x) = -10x$

$f'(x) = \lim\limits_{\Delta x \to 0} \dfrac{f(x + \Delta x) - f(x)}{\Delta x}$

$= \lim\limits_{\Delta x \to 0} \dfrac{-10(x + \Delta x) - (-10x)}{\Delta x}$

$= \lim\limits_{\Delta x \to 0} \dfrac{-10\Delta x}{\Delta x}$

$= \lim\limits_{\Delta x \to 0} (-10) = -10$

**15.** $h(s) = 3 + \dfrac{2}{3}s$

$h'(s) = \lim\limits_{\Delta s \to 0} \dfrac{h(s + \Delta s) - h(s)}{\Delta s}$

$= \lim\limits_{\Delta s \to 0} \dfrac{3 + \dfrac{2}{3}(s + \Delta s) - \left(3 + \dfrac{2}{3}s\right)}{\Delta s}$

$= \lim\limits_{\Delta s \to 0} \dfrac{\dfrac{2}{3}\Delta s}{\Delta s} = \dfrac{2}{3}$

**19.** $f(x) = x^3 - 12x$

$$f'(x) = \lim_{\Delta x \to 0} \frac{f(x + \Delta x) - f(x)}{\Delta x}$$

$$= \lim_{\Delta x \to 0} \frac{\left[(x + \Delta x)^3 - 12(x + \Delta x)\right] - \left[x^3 - 12x\right]}{\Delta x}$$

$$= \lim_{\Delta x \to 0} \frac{x^3 + 3x^2 \Delta x + 3x(\Delta x)^2 + (\Delta x)^3 - 12x - 12\,\Delta x - x^3 + 12x}{\Delta x}$$

$$= \lim_{\Delta x \to 0} \frac{3x^2 \Delta x + 3x(\Delta x)^2 + (\Delta x)^3 - 12\,\Delta x}{\Delta x}$$

$$= \lim_{\Delta x \to 0} \left(3x^2 + 3x\,\Delta x + (\Delta x)^2 - 12\right) = 3x^2 - 12$$

**21.** $f(x) = \dfrac{1}{x - 1}$

$$f'(x) = \lim_{\Delta x \to 0} \frac{f(x + \Delta x) - f(x)}{\Delta x}$$

$$= \lim_{\Delta x \to 0} \frac{\dfrac{1}{x + \Delta x - 1} - \dfrac{1}{x - 1}}{\Delta x}$$

$$= \lim_{\Delta x \to 0} \frac{(x - 1) - (x + \Delta x - 1)}{\Delta x(x + \Delta x - 1)(x - 1)}$$

$$= \lim_{\Delta x \to 0} \frac{-\Delta x}{\Delta x(x + \Delta x - 1)(x - 1)}$$

$$= \lim_{\Delta x \to 0} \frac{-1}{(x + \Delta x - 1)(x - 1)} = -\frac{1}{(x - 1)^2}$$

**23.** $f(x) = \sqrt{x + 4}$

$$f'(x) = \lim_{\Delta x \to 0} \frac{f(x + \Delta x) - f(x)}{\Delta x}$$

$$= \lim_{\Delta x \to 0} \frac{\sqrt{x + \Delta x + 4} - \sqrt{x + 4}}{\Delta x} \cdot \left(\frac{\sqrt{x + \Delta x + 4} + \sqrt{x + 4}}{\sqrt{x + \Delta x + 4} + \sqrt{x + 4}}\right)$$

$$= \lim_{\Delta x \to 0} \frac{(x + \Delta x + 4) - (x + 4)}{\Delta x\left[\sqrt{x + \Delta x + 4} + \sqrt{x + 4}\right]}$$

$$= \lim_{\Delta x \to 0} \frac{1}{\sqrt{x + \Delta x + 4} + \sqrt{x + 4}} = \frac{1}{\sqrt{x + 4} + \sqrt{x + 4}} = \frac{1}{2\sqrt{x + 4}}$$

**25.** (a)   $f(x) = x^2 + 3$

$$f'(x) = \lim_{\Delta x \to 0} \frac{f(x + \Delta x) - f(x)}{\Delta x}$$

$$= \lim_{\Delta x \to 0} \frac{\left[(x + \Delta x)^2 + 3\right] - \left[x^2 + 3\right]}{\Delta x}$$

$$= \lim_{\Delta x \to 0} \frac{2x\,\Delta x + (\Delta x)^2}{\Delta x}$$

$$= \lim_{\Delta x \to 0} (2x + \Delta x) = 2x$$

At $(1, 4)$, the slope of the tangent line is $m = 2(1) = 2$. The equation of the tangent line is

$$y - 4 = 2(x - 1)$$
$$y - 4 = 2x - 2$$
$$y = 2x + 2.$$

(b)

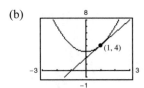

(c) Graphing utility confirms $\dfrac{dy}{dx} = 2$ at $(1, 4)$.

**27.** (a)  $f(x) = x^3$

$$f'(x) = \lim_{\Delta x \to 0} \frac{f(x + \Delta x) - f(x)}{\Delta x}$$

$$= \lim_{\Delta x \to 0} \frac{(x + \Delta x)^3 - x^3}{\Delta x}$$

$$= \lim_{\Delta x \to 0} \frac{3x^2 \Delta x + 3x(\Delta x)^2 + (\Delta x)^3}{\Delta x}$$

$$= \lim_{\Delta x \to 0} \left(3x^2 + 3x\,\Delta x + (\Delta x)^2\right) = 3x^2$$

At $(2, 8)$, the slope of the tangent is $m = 3(2)^2 = 12$. The equation of the tangent line is

$$y - 8 = 12(x - 2)$$
$$y = 12x - 16.$$

(b)

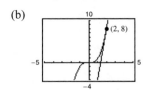

(c) Graphing utility confirms $\dfrac{dy}{dx} = 12$ at $(2, 8)$.

**29.** (a)  $f(x) = \sqrt{x}$

$$f'(x) = \lim_{\Delta x \to 0} \frac{f(x + \Delta x) - f(x)}{\Delta x}$$

$$= \lim_{\Delta x \to 0} \frac{\sqrt{x + \Delta x} - \sqrt{x}}{\Delta x} \cdot \frac{\sqrt{x + \Delta x} + \sqrt{x}}{\sqrt{x + \Delta x} + \sqrt{x}}$$

$$= \lim_{\Delta x \to 0} \frac{(x + \Delta x) - x}{\Delta x\left(\sqrt{x + \Delta x} + \sqrt{x}\right)}$$

$$= \lim_{\Delta x \to 0} \frac{1}{\sqrt{x + \Delta x} + \sqrt{x}} = \frac{1}{2\sqrt{x}}$$

At $(1, 1)$, the slope of the tangent line is $m = \dfrac{1}{2\sqrt{1}} = \dfrac{1}{2}$.

The equation of the tangent line is

$$y - 1 = \frac{1}{2}(x - 1)$$
$$y = \frac{1}{2}x + \frac{1}{2}.$$

(b)

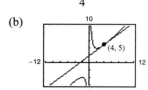

(c) Graphing utility confirms $\dfrac{dy}{dx} = \dfrac{1}{2}$ at $(1, 1)$.

**31.** (a) $f(x) = x + \dfrac{4}{x}$

$f'(x) = \lim\limits_{\Delta x \to 0} \dfrac{f(x + \Delta x) - f(x)}{\Delta x}$

$= \lim\limits_{\Delta x \to 0} \dfrac{(x + \Delta x) + \dfrac{4}{x + \Delta x} - \left(x + \dfrac{4}{x}\right)}{\Delta x}$

$= \lim\limits_{\Delta x \to 0} \dfrac{x(x + \Delta x)(x + \Delta x) + 4x - x^2(x + \Delta x) - 4(x + \Delta x)}{x(\Delta x)(x + \Delta x)}$

$= \lim\limits_{\Delta x \to 0} \dfrac{x^3 + 2x^2(\Delta x) + x(\Delta x)^2 - x^3 - x^2(\Delta x) - 4(\Delta x)}{x(\Delta x)(x + \Delta x)}$

$= \lim\limits_{\Delta x \to 0} \dfrac{x^2(\Delta x) + x(\Delta x)^2 - 4(\Delta x)}{x(\Delta x)(x + \Delta x)}$

$= \lim\limits_{\Delta x \to 0} \dfrac{x^2 + x(\Delta x) - 4}{x(x + \Delta x)} = \dfrac{x^2 - 4}{x^2} = 1 - \dfrac{4}{x^2}$

At $(4, 5)$, the slope of the tangent line is $m = 1 - \dfrac{4}{16} = \dfrac{3}{4}$.

The equation of the tangent line is

$y - 5 = \dfrac{3}{4}(x - 4)$

$y = \dfrac{3}{4}x + 2.$

(b)

(c) Graphing utility confirms $\dfrac{dy}{dx} = \dfrac{3}{4}$ at $(4, 5)$.

**33.** Using the limit definition of derivative, $f'(x) = 2x.$ Because the slope of the given line is 2, you have

$2x = 2$

$x = 1$

At the point $(1, 1)$ the tangent line is parallel to $2x - y + 1 = 0.$ The equation of this line is

$y - 1 = 2(x - 1)$

$y = 2x - 1.$

**35.** From Exercise 27 we know that $f'(x) = 3x^2.$ Because the slope of the given line is 3, you have

$3x^2 = 3$

$x = \pm 1.$

Therefore, at the points $(1, 1)$ and $(-1, -1)$ the tangent lines are parallel to $3x - y + 1 = 0.$ These lines have equations

$y - 1 = 3(x - 1)$ and $y + 1 = 3(x + 1)$

$y = 3x - 2 \qquad\qquad y = 3x + 2.$

**37.** Using the limit definition of derivative,

$$f'(x) = \frac{-1}{2x\sqrt{x}}.$$

Because the slope of the given line is $-\frac{1}{2}$, you have

$$-\frac{1}{2x\sqrt{x}} = -\frac{1}{2}$$

$$x = 1.$$

Therefore, at the point $(1, 1)$ the tangent line is parallel to $x + 2y - 6 = 0$. The equation of this line is

$$y - 1 = -\frac{1}{2}(x - 1)$$

$$y - 1 = -\frac{1}{2}x + \frac{1}{2}$$

$$y = -\frac{1}{2}x + \frac{3}{2}.$$

**39.** $f(x) = x \Rightarrow f'(x) = 1$       Matches (b).

**41.** $f(x) = \sqrt{x} \Rightarrow f'(x)$       Matches (a).

(decreasing slope as $x \to \infty$)

**43.** $g(4) = 5$ because the tangent line passes through $(4, 5)$.

$$g'(4) = \frac{5 - 0}{4 - 7} = -\frac{5}{3}$$

**45.** The slope of the graph of $f$ is 1 for all $x$-values.

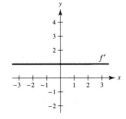

**47.** The slope of the graph of $f$ is negative for $x < 4$, positive for $x > 4$, and 0 at $x = 4$.

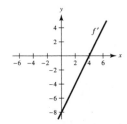

**49.** The slope of the graph of $f$ is negative for $x < 0$ and positive for $x > 0$. The slope is undefined at $x = 0$.

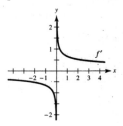

**51.** Answers will vary.

*Sample answer:* $y = -x$

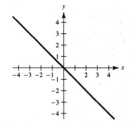

**53.** $f(x) = 5 - 3x$ and $c = 1$

**55.** $f(x) = -x^2$ and $c = 6$

**57.** $f(0) = 2$ and $f'(x) = -3, -\infty < x < \infty$

$$f(x) = -3x + 2$$

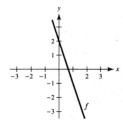

**59.** $f(0) = 0; f'(0) = 0; f'(x) > 0$ if $x \neq 0$

Answers will vary: *Sample answer:* $f(x) = x^3$

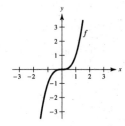

**61.** Let $(x_0, y_0)$ be a point of tangency on the graph of $f$. By the limit definition for the derivative, $f'(x) = 4 - 2x$. The slope of the line through $(2, 5)$ and $(x_0, y_0)$ equals the derivative of $f$ at $x_0$:

$$\frac{5 - y_0}{2 - x_0} = 4 - 2x_0$$

$$5 - y_0 = (2 - x_0)(4 - 2x_0)$$

$$5 - (4x_0 - x_0^2) = 8 - 8x_0 + 2x_0^2$$

$$0 = x_0^2 - 4x_0 + 3$$

$$0 = (x_0 - 1)(x_0 - 3) \Rightarrow x_0 = 1, 3$$

Therefore, the points of tangency are $(1, 3)$ and $(3, 3)$, and the corresponding slopes are 2 and $-2$. The equations of the tangent lines are:

$$y - 5 = 2(x - 2) \qquad y - 5 = -2(x - 2)$$
$$y = 2x + 1 \qquad\qquad y = -2x + 9$$

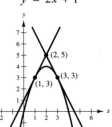

**63. (a)** $f(x) = x^2$

$$f'(x) = \lim_{\Delta x \to 0} \frac{f(x + \Delta x) - f(x)}{\Delta x}$$

$$= \lim_{\Delta x \to 0} \frac{(x + \Delta x)^2 - x^2}{\Delta x}$$

$$= \lim_{\Delta x \to 0} \frac{x^2 + 2x(\Delta x) + (\Delta x)^2 - x^2}{\Delta x}$$

$$= \lim_{\Delta x \to 0} \frac{\Delta x(2x + \Delta x)}{\Delta x}$$

$$= \lim_{\Delta x \to 0} (2x + \Delta x) = 2x$$

At $x = -1$, $f'(-1) = -2$ and the tangent line is

$$y - 1 = -2(x + 1) \qquad \text{or} \qquad y = -2x - 1.$$

At $x = 0$, $f'(0) = 0$ and the tangent line is $y = 0$.

At $x = 1$, $f'(1) = 2$ and the tangent line is

$$y = 2x - 1.$$

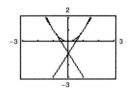

For this function, the slopes of the tangent lines are always distinct for different values of $x$.

**(b)** $g'(x) = \lim_{\Delta x \to 0} \frac{g(x + \Delta x) - g(x)}{\Delta x}$

$$= \lim_{\Delta x \to 0} \frac{(x + \Delta x)^3 - x^3}{\Delta x}$$

$$= \lim_{\Delta x \to 0} \frac{x^3 + 3x^2(\Delta x) + 3x(\Delta x)^2 + (\Delta x)^3 - x^3}{\Delta x}$$

$$= \lim_{\Delta x \to 0} \frac{\Delta x\left(3x^2 + 3x(\Delta x) + (\Delta x)^2\right)}{\Delta x}$$

$$= \lim_{\Delta x \to 0} \left(3x^2 + 3x(\Delta x) + (\Delta x)^2\right) = 3x^2$$

At $x = -1$, $g'(-1) = 3$ and the tangent line is

$$y + 1 = 3(x + 1) \qquad \text{or} \qquad y = 3x + 2.$$

At $x = 0$, $g'(0) = 0$ and the tangent line is $y = 0$.

At $x = 1$, $g'(1) = 3$ and the tangent line is

$$y - 1 = 3(x - 1) \qquad \text{or} \qquad y = 3x - 2.$$

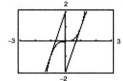    For this function, the slopes of the tangent lines are sometimes the same.

**65.** $f(x) = \frac{1}{2}x^2$

**(a)**

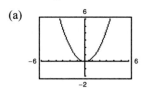

$$f'(0) = 0, \ f'(1/2) = 1/2, \ f'(1) = 1, \ f'(2) = 2$$

**(b)** By symmetry:

$$f'(-1/2) = -1/2, \ f'(-1) = -1, \ f'(-2) = -2$$

**(c)**

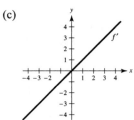

**(d)** $f'(x) = \lim_{\Delta x \to 0} \frac{f(x + \Delta x) - f(x)}{\Delta x}$

$$= \lim_{\Delta x \to 0} \frac{\frac{1}{2}(x + \Delta x)^2 - \frac{1}{2}x^2}{\Delta x}$$

$$= \lim_{\Delta x \to 0} \frac{\frac{1}{2}\left(x^2 + 2x(\Delta x) + (\Delta x)^2\right) - \frac{1}{2}x^2}{\Delta x}$$

$$= \lim_{\Delta x \to 0} \left(x + \frac{\Delta x}{2}\right) = x$$

**67.** $g(x) = \dfrac{f(x + 0.01) - f(x)}{0.01}$

$= \left[ 2(x + 0.01) - (x + 0.01)^2 - 2x + x^2 \right] 100$

$= 2 - 2x - 0.01$

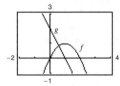

The graph of $g(x)$ is approximately the graph of

$f'(x) = 2 - 2x$.

**69.** $f(2) = 2(4 - 2) = 4, f(2.1) = 2.1(4 - 2.1) = 3.99$

$f'(2) \approx \dfrac{3.99 - 4}{2.1 - 2} = -0.1 \quad \left[ \text{Exact: } f'(2) = 0 \right]$

**71.** $f(x) = \dfrac{1}{\sqrt{x}}$ and $f'(x) = \dfrac{-1}{2x^{3/2}}$.

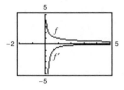

As $x \to \infty$, $f$ is nearly horizontal and thus $f' \approx 0$.

**73.** $f(x) = x^2 - 5, c = 3$

$f'(3) = \lim_{x \to 3} \dfrac{f(x) - f(3)}{x - 3}$

$= \lim_{x \to 3} \dfrac{x^2 - 5 - (9 - 5)}{x - 3}$

$= \lim_{x \to 3} \dfrac{(x - 3)(x + 3)}{x - 3}$

$= \lim_{x \to 3} (x + 3) = 6$

**75.** $f(x) = x^3 + 2x^2 + 1, c = -2$

$f'(-2) = \lim_{x \to -2} \dfrac{f(x) - f(-2)}{x + 2}$

$= \lim_{x \to -2} \dfrac{(x^3 + 2x^2 + 1) - 1}{x + 2}$

$= \lim_{x \to -2} \dfrac{x^2(x + 2)}{x + 2} = \lim_{x \to -2} x^2 = 4$

**77.** $g(x) = \sqrt{|x|}, c = 0$

$g'(0) = \lim_{x \to 0} \dfrac{g(x) - g(0)}{x - 0} = \lim_{x \to 0} \dfrac{\sqrt{|x|}}{x}$. Does not exist.

As $x \to 0^-$, $\dfrac{\sqrt{|x|}}{x} = \dfrac{-1}{\sqrt{|x|}} \to -\infty$.

As $x \to 0^+$, $\dfrac{\sqrt{|x|}}{x} = \dfrac{1}{\sqrt{x}} \to \infty$.

Therefore $g(x)$ is not differentiable at $x = 0$.

**79.** $f(x) = (x - 6)^{2/3}, c = 6$

$f'(6) = \lim_{x \to 6} \dfrac{f(x) - f(6)}{x - 6}$

$= \lim_{x \to 6} \dfrac{(x - 6)^{2/3} - 0}{x - 6} = \lim_{x \to 6} \dfrac{1}{(x - 6)^{1/3}}$.

Does not exist.

Therefore $f(x)$ is not differentiable at $x = 6$.

**81.** $h(x) = |x + 7|, c = -7$

$h'(-7) = \lim_{x \to -7} \dfrac{h(x) - h(-7)}{x - (-7)}$

$= \lim_{x \to -7} \dfrac{|x + 7| - 0}{x + 7} = \lim_{x \to -7} \dfrac{|x + 7|}{x + 7}$.

Does not exist.

Therefore $h(x)$ is not differentiable at $x = -7$.

**83.** $f(x)$ is differentiable everywhere except at
$x = 3$. (Discontinuity)

**85.** $f(x)$ is differentiable everywhere except at
$x = -4$. (Sharp turn in the graph)

**87.** $f(x)$ is differentiable on the interval $(1, \infty)$. (At
$x = 1$ the tangent line is vertical.)

**89.** $f(x) = |x - 5|$ is differentiable everywhere except at
$x = -5$. There is a sharp corner at $x = 5$.

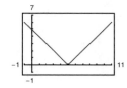

**91.** $f(x) = x^{2/5}$ is differentiable for all $x \neq 0$. There is a sharp corner at $x = 0$.

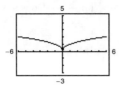

**93.** $f(x) = |x - 1|$

The derivative from the left is

$$\lim_{x \to 1^-} \frac{f(x) - f(1)}{x - 1} = \lim_{x \to 1^-} \frac{|x - 1| - 0}{x - 1} = -1.$$

The derivative from the right is

$$\lim_{x \to 1^+} \frac{f(x) - f(1)}{x - 1} = \lim_{x \to 1^+} \frac{|x - 1| - 0}{x - 1} = 1.$$

The one-sided limits are not equal. Therefore, $f$ is not differentiable at $x = 1$.

**95.** $f(x) = \begin{cases} (x - 1)^3, & x \leq 1 \\ (x - 1)^2, & x > 1 \end{cases}$

The derivative from the left is

$$\lim_{x \to 1^-} \frac{f(x) - f(1)}{x - 1} = \lim_{x \to 1^-} \frac{(x - 1)^3 - 0}{x - 1}$$
$$= \lim_{x \to 1^-} (x - 1)^2 = 0.$$

The derivative from the right is

$$\lim_{x \to 1^+} \frac{f(x) - f(1)}{x - 1} = \lim_{x \to 1^+} \frac{(x - 1)^2 - 0}{x - 1}$$
$$= \lim_{x \to 1^+} (x - 1) = 0.$$

The one-sided limits are equal. Therefore, $f$ is differentiable at $x = 1$. $\left( f'(1) = 0 \right)$

**97.** Note that $f$ is continuous at $x = 2$.

$$f(x) = \begin{cases} x^2 + 1, & x \leq 2 \\ 4x - 3, & x > 2 \end{cases}$$

The derivative from the left is $\displaystyle\lim_{x \to 2^-} \frac{f(x) - f(2)}{x - 2} = \lim_{x \to 2^-} \frac{(x^2 + 1) - 5}{x - 2} = \lim_{x \to 2^-} (x + 2) = 4.$

The derivative from the right is $\displaystyle\lim_{x \to 2^+} \frac{f(x) - f(2)}{x - 2} = \lim_{x \to 2^+} \frac{(4x - 3) - 5}{x - 2} = \lim_{x \to 2^+} 4 = 4.$

The one-sided limits are equal. Therefore, $f$ is differentiable at $x = 2$. $\left( f'(2) = 4 \right)$

**99.** (a) The distance from $(3, 1)$ to the line $mx - y + 4 = 0$ is $d = \dfrac{\left| Ax_1 + By_1 + C \right|}{\sqrt{A^2 + B^2}} = \dfrac{\left| m(3) - 1(1) + 4 \right|}{\sqrt{m^2 + 1}} = \dfrac{\left| 3m + 3 \right|}{\sqrt{m^2 + 1}}$

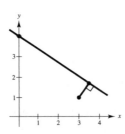

(b)

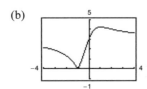

The function $d$ is not differentiable at $m = -1$. This corresponds to the line $y = -x + 4$, which passes through the point $(3, 1)$.

**101.** False. The slope is $\displaystyle \lim_{\Delta x \to 0} \frac{f(2 + \Delta x) - f(2)}{\Delta x}$.

**103.** False. If the derivative from the left of a point does not equal the derivative from the right of a point, then the derivative does not exist at that point. For example, if $f(x) = |x|$, then the derivative from the left at $x = 0$ is $-1$. and the derivative from the right at $x = 0$ is $1$. At $x = 0$, the derivative does not exist.

**105.** $f(x) = \begin{cases} x \sin(1/x), & x \neq 0 \\ 0, & x = 0 \end{cases}$

Using the Squeeze Theorem, you have $-|x| \leq x \sin(1/x) \leq |x|$, $x \neq 0$. So, $\displaystyle \lim_{x \to 0} x \sin(1/x) = 0 = f(0)$ and

$f$ is continuous at $x = 0$. Using the alternative form of the derivative, you have

$$\lim_{x \to 0} \frac{f(x) - f(0)}{x - 0} = \lim_{x \to 0} \frac{x \sin(1/x) - 0}{x - 0} = \lim_{x \to 0} \left( \sin \frac{1}{x} \right).$$

Because this limit does not exist ($\sin(1/x)$ oscillates between $-1$ and $1$), the function is not differentiable at $x = 0$.

$g(x) = \begin{cases} x^2 \sin(1/x), & x \neq 0 \\ 0, & x = 0 \end{cases}$

Using the Squeeze Theorem again, you have $-x^2 \leq x^2 \sin(1/x) \leq x^2$, $x \neq 0$. So, $\displaystyle \lim_{x \to 0} x^2 \sin(1/x) = 0 = g(0)$

and $g$ is continuous at $x = 0$. Using the alternative form of the derivative again, you have

$$\lim_{x \to 0} \frac{g(x) - g(0)}{x - 0} = \lim_{x \to 0} \frac{x^2 \sin(1/x) - 0}{x - 0} = \lim_{x \to 0} x \sin \frac{1}{x} = 0.$$

Therefore, $g$ is differentiable at $x = 0$, $g'(0) = 0$.

## Section 2.2   Basic Differentiation Rules and Rates of Change

**1. (a)** $\quad y = x^{1/2}$

$\qquad y' = \frac{1}{2} x^{-1/2}$

$\qquad y'(1) = \frac{1}{2}$

**(b)** $\quad y = x^3$

$\qquad y' = 3x^2$

$\qquad y'(1) = 3$

**3.** $y = 12$

$\quad y' = 0$

**5.** $y = x^7$

$\quad y' = 7x^6$

**7.** $y = \dfrac{1}{x^5} = x^{-5}$

$\quad y' = -5x^{-6} = -\dfrac{5}{x^6}$

**9.** $y = \sqrt[5]{x} = x^{1/5}$

$\quad y' = \dfrac{1}{5} x^{-4/5} = \dfrac{1}{5x^{4/5}}$

**11.** $f(x) = x + 11$

$\quad f'(x) = 1$

**13.** $f(t) = -2t^2 + 3t - 6$

$\quad f'(t) = -4t + 3$

**15.** $g(x) = x^2 + 4x^3$

$\quad g'(x) = 2x + 12x^2$

**17.** $s(t) = t^3 + 5t^2 - 3t + 8$

$\quad s'(t) = 3t^2 + 10t - 3$

**19.** $y = \dfrac{\pi}{2} \sin \theta - \cos \theta$

$\quad y' = \dfrac{\pi}{2} \cos \theta + \sin \theta$

**21.** $y = x^2 - \frac{1}{2} \cos x$

$\quad y' = 2x + \frac{1}{2} \sin x$

**23.** $y = \dfrac{1}{x} - 3 \sin x$

$\quad y' = -\dfrac{1}{x^2} - 3 \cos x$

|     | *Function*          | *Rewrite*              | *Differentiate*             | *Simplify*                |
|-----|---------------------|------------------------|-----------------------------|---------------------------|
| **25.** | $y = \dfrac{5}{2x^2}$ | $y = \dfrac{5}{2}x^{-2}$ | $y' = -5x^{-3}$ | $y' = -\dfrac{5}{x^3}$ |
| **27.** | $y = \dfrac{6}{(5x)^3}$ | $y = \dfrac{6}{125}x^{-3}$ | $y' = -\dfrac{18}{125}x^{-4}$ | $y' = -\dfrac{18}{125x^4}$ |
| **29.** | $y = \dfrac{\sqrt{x}}{x}$ | $y = x^{-1/2}$ | $y' = -\dfrac{1}{2}x^{-3/2}$ | $y' = -\dfrac{1}{2x^{3/2}}$ |

**31.** $f(x) = \dfrac{8}{x^2} = 8x^{-2}, \ (2, 2)$

$\qquad f'(x) = -16x^{-3} = -\dfrac{16}{x^3}$

$\qquad f'(2) = -2$

**33.** $f(x) = -\dfrac{1}{2} + \dfrac{7}{5}x^3, \ \left(0, -\dfrac{1}{2}\right)$

$\qquad f'(x) = \dfrac{21}{5}x^2$

$\qquad f'(0) = 0$

**35.** $\quad y = (4x + 1)^2, \ (0, 1)$

$\qquad = 16x^2 + 8x + 1$

$\qquad y' = 32x + 8$

$\qquad y'(0) = 8$

**37.** $f(\theta) = 4 \sin \theta - \theta, \ (0, 0)$

$\qquad f'(\theta) = 4 \cos \theta - 1$

$\qquad f'(0) = 4(1) - 1 = 3$

**39.** $f(x) = x^2 + 5 - 3x^{-2}$

$\qquad f'(x) = 2x + 6x^{-3} = 2x + \dfrac{6}{x^3}$

**41.** $g(t) = t^2 - \dfrac{4}{t^3} = t^2 - 4t^{-3}$

$\qquad g'(t) = 2t + 12t^{-4} = 2t + \dfrac{12}{t^4}$

**43.** $f(x) = \dfrac{4x^3 + 3x^2}{x} = 4x^2 + 3x$

$\qquad f'(x) = 8x + 3$

**45.** $f(x) = \dfrac{x^3 - 3x^2 + 4}{x^2} = x - 3 + 4x^{-2}$

$\qquad f'(x) = 1 - \dfrac{8}{x^3} = \dfrac{x^3 - 8}{x^3}$

**47.** $y = x(x^2 + 1) = x^3 + x$

$\qquad y' = 3x^2 + 1$

**49.** $f(x) = \sqrt{x} - 6\sqrt[3]{x} = x^{1/2} - 6x^{1/3}$

$\qquad f'(x) = \dfrac{1}{2}x^{-1/2} - 2x^{-2/3} = \dfrac{1}{2\sqrt{x}} - \dfrac{2}{x^{2/3}}$

**51.** $h(s) = s^{4/5} - s^{2/3}$

$\qquad h'(s) = \dfrac{4}{5}s^{-1/5} - \dfrac{2}{3}s^{-1/3} = \dfrac{4}{5s^{1/5}} - \dfrac{2}{3s^{1/3}}$

**53.** $f(x) = 6\sqrt{x} + 5 \cos x = 6x^{1/2} + 5 \cos x$

$\qquad f'(x) = 3x^{-1/2} - 5 \sin x = \dfrac{3}{\sqrt{x}} - 5 \sin x$

**55. (a)** $y = x^4 - 3x^2 + 2$

$\qquad y' = 4x^3 - 6x$

$\qquad$ At $(1, 0)$: $y' = 4(1)^3 - 6(1) = -2$

$\qquad$ Tangent line: $\qquad y - 0 = -2(x - 1)$

$\qquad\qquad\qquad\qquad\qquad 2x + y - 2 = 0$

**(b)**
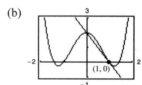

**57. (a)** $f(x) = \dfrac{2}{\sqrt[4]{x^3}} = 2x^{-3/4}$

$\qquad f'(x) = -\dfrac{3}{2}x^{-7/4} = -\dfrac{3}{2x^{7/4}}$

$\qquad$ At $(1, 2)$: $f'(1) = -\dfrac{3}{2}$

$\qquad$ Tangent line: $\qquad y - 2 = -\dfrac{3}{2}(x - 1)$

$\qquad\qquad\qquad\qquad\qquad y = -\dfrac{3}{2}x + \dfrac{7}{2}$

$\qquad\qquad\qquad\qquad 3x + 2y - 7 = 0$

**(b)**
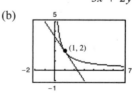

**59.**  $y = x^4 - 2x^2 + 3$

$\quad y' = 4x^3 - 4x$

$\quad\quad = 4x(x^2 - 1)$

$\quad\quad = 4x(x - 1)(x + 1)$

$\quad y' = 0 \Rightarrow x = 0, \pm 1$

Horizontal tangents: $(0, 3), (1, 2), (-1, 2)$

**61.**  $y = \dfrac{1}{x^2} = x^{-2}$

$\quad y' = -2x^{-3} = -\dfrac{2}{x^3}$ cannot equal zero.

Therefore, there are no horizontal tangents.

**63.**  $y = x + \sin x, 0 \le x < 2\pi$

$\quad y' = 1 + \cos x = 0$

$\quad \cos x = -1 \Rightarrow x = \pi$

At $x = \pi$: $y = \pi$

Horizontal tangent: $(\pi, \pi)$

**65.**  $x^2 - kx = 5x - 4$  Equate functions.

$\quad 2x - k = 5$  Equate derivatives.

So, $k = 2x - 5$ and

$\quad x^2 - (2x - 5)x = 5x - 4 \Rightarrow -x^2 = -4 \Rightarrow x = \pm 2.$

For $x = 2, k = -1$ and for $x = -2, k = -9.$

**67.**  $\dfrac{k}{x} = -\dfrac{3}{4}x + 3$  Equate functions.

$\quad -\dfrac{k}{x^2} = -\dfrac{3}{4}$  Equate derivatives.

So, $k = \dfrac{3}{4}x^2$ and

$\dfrac{\frac{3}{4}x^2}{x} = -\dfrac{3}{4}x + 3 \Rightarrow \dfrac{3}{4}x = -\dfrac{3}{4}x + 3$

$\quad\quad\quad\quad \Rightarrow \dfrac{3}{2}x = 3 \Rightarrow x = 2 \Rightarrow k = 3.$

**69.**  $kx^3 = x + 1$  Equate equations.

$\quad 3kx^2 = 1$  Equate derivatives.

So, $k = \dfrac{1}{3x^2}$ and

$\quad \left(\dfrac{1}{3x^2}\right)x^3 = x + 1$

$\quad\quad\quad \dfrac{1}{3}x = x + 1$

$\quad\quad\quad\quad x = -\dfrac{3}{2}, k = \dfrac{4}{27}.$

**71.** The graph of a function $f$ such that $f' > 0$ for all $x$ and the rate of change of the function is decreasing (i.e., $f'' < 0$) would, in general, look like the graph below.

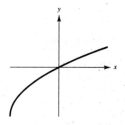

**73.**  $g(x) = f(x) + 6 \Rightarrow g'(x) = f'(x)$

**75.**

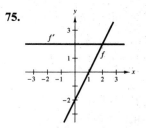

If $f$ is linear then its derivative is a constant function.

$\quad f(x) = ax + b$

$\quad f'(x) = a$

**77.** Let $(x_1, y_1)$ and $(x_2, y_2)$ be the points of tangency on $y = x^2$ and $y = -x^2 + 6x - 5$, respectively.

The derivatives of these functions are:

$y' = 2x \Rightarrow m = 2x_1$ and $y' = -2x + 6 \Rightarrow m = -2x_2 + 6$

$m = 2x_1 = -2x_2 + 6$

$\quad x_1 = -x_2 + 3$

Because $y_1 = x_1^2$ and $y_2 = -x_2^2 + 6x_2 - 5$:

$m = \dfrac{y_2 - y_1}{x_2 - x_1} = \dfrac{\left(-x_2^2 + 6x_2 - 5\right) - \left(x_1^2\right)}{x_2 - x_1} = -2x_2 + 6$

$\dfrac{\left(-x_2^2 + 6x_2 - 5\right) - \left(-x_2 + 3\right)^2}{x_2 - \left(-x_2 + 3\right)} = -2x_2 + 6$

$\left(-x_2^2 + 6x_2 - 5\right) - \left(x_2^2 - 6x_2 + 9\right) = \left(-2x_2 + 6\right)\left(2x_2 - 3\right)$

$-2x_2^2 + 12x_2 - 14 = -4x_2^2 + 18x_2 - 18$

$2x_2^2 - 6x_2 + 4 = 0$

$2\left(x_2 - 2\right)\left(x_2 - 1\right) = 0$

$x_2 = 1 \text{ or } 2$

$x_2 = 1 \Rightarrow y_2 = 0, \, x_1 = 2 \text{ and } y_1 = 4$

So, the tangent line through $(1, 0)$ and $(2, 4)$ is

$y - 0 = \left(\dfrac{4 - 0}{2 - 1}\right)(x - 1) \Rightarrow y = 4x - 4.$

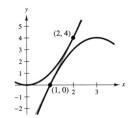

$x_2 = 2 \Rightarrow y_2 = 3, \, x_1 = 1 \text{ and } y_1 = 1$

So, the tangent line through $(2, 3)$ and $(1, 1)$ is

$y - 1 = \left(\dfrac{3 - 1}{2 - 1}\right)(x - 1) \Rightarrow y = 2x - 1.$

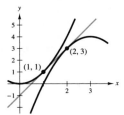

**79.** $f(x) = 3x + \sin x + 2$

$f'(x) = 3 + \cos x$

Because $\left|\cos x\right| \leq 1$, $f'(x) \neq 0$ for all $x$ and $f$ does not have a horizontal tangent line.

**81.** $f(x) = \sqrt{x}, \, (-4, 0)$

$f'(x) = \dfrac{1}{2}x^{-1/2} = \dfrac{1}{2\sqrt{x}}$

$\dfrac{1}{2\sqrt{x}} = \dfrac{0 - y}{-4 - x}$

$4 + x = 2\sqrt{x}\,y$

$4 + x = 2\sqrt{x}\sqrt{x}$

$4 + x = 2x$

$\quad x = 4, \, y = 2$

The point $(4, 2)$ is on the graph of $f$.

Tangent line: $y - 2 = \dfrac{0 - 2}{-4 - 4}(x - 4)$

$4y - 8 = x - 4$

$0 = x - 4y + 4$

**83.** $f'(1)$ appears to be close to $-1$.

$$f'(1) = -1$$

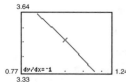

**85.** (a) One possible secant is between $(3.9, 7.7019)$ and $(4, 8)$:

$$y - 8 = \frac{8 - 7.7019}{4 - 3.9}(x - 4)$$

$$y - 8 = 2.981(x - 4)$$

$$y = S(x) = 2.981x - 3.924$$

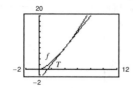

(b) $f'(x) = \frac{3}{2}x^{1/2} \Rightarrow f'(4) = \frac{3}{2}(2) = 3$

$$T(x) = 3(x - 4) + 8 = 3x - 4$$

The slope (and equation) of the secant line approaches that of the tangent line at $(4, 8)$ as you choose points closer and closer to $(4, 8)$.

(c) As you move further away from $(4, 8)$, the accuracy of the approximation $T$ gets worse.

(d)

| $\Delta x$ | $-3$ | $-2$ | $-1$ | $-0.5$ | $-0.1$ | $0$ | $0.1$ | $0.5$ | $1$ | $2$ | $3$ |
|---|---|---|---|---|---|---|---|---|---|---|---|
| $f(4 + \Delta x)$ | 1 | 2.828 | 5.196 | 6.548 | 7.702 | 8 | 8.302 | 9.546 | 11.180 | 14.697 | 18.520 |
| $T(4 + \Delta x)$ | $-1$ | 2 | 5 | 6.5 | 7.7 | 8 | 8.3 | 9.5 | 11 | 14 | 17 |

**87.** False. Let $f(x) = x$ and $g(x) = x + 1$. Then
$f'(x) = g'(x) = x$, but $f(x) \neq g(x)$.

**89.** False. If $y = \pi^2$, then $dy/dx = 0$. ( $\pi^2$ is a constant.)

**91.** True. If $g(x) = 3f(x)$, then $g'(x) = 3f'(x)$.

**93.** $f(t) = 4t + 5$,   $[1, 2]$

$f'(t) = 4$. So, $f'(1) = f'(2) = 4$.

Instantaneous rate of change is the constant 4. Average rate of change:

$$\frac{f(2) - f(1)}{2 - 1} = \frac{13 - 9}{1} = 4$$

(These are the same because $f$ is a line of slope 4.)

**95.** $f(x) = -\dfrac{1}{x}$,   $[1, 2]$

$$f'(x) = \frac{1}{x^2}$$

Instantaneous rate of change:

$$(1, -1) \Rightarrow f'(1) = 1$$

$$\left(2, -\frac{1}{2}\right) \Rightarrow f'(2) = \frac{1}{4}$$

Average rate of change:

$$\frac{f(2) - f(1)}{2 - 1} = \frac{(-1/2) - (-1)}{2 - 1} = \frac{1}{2}$$

**97.** (a) $s(t) = -16t^2 + 1362$

$v(t) = -32t$

(b) $\dfrac{s(2) - s(1)}{2 - 1} = 1298 - 1346 = -48$ ft/sec

(c) $v(t) = s'(t) = -32t$

When $t = 1$: $v(1) = -32$ ft/sec

When $t = 2$: $v(2) = -64$ ft/sec

(d) $-16t^2 + 1362 = 0$

$$t^2 = \frac{1362}{16} \Rightarrow t = \frac{\sqrt{1362}}{4} \approx 9.226 \text{ sec}$$

(e) $v\left(\dfrac{\sqrt{1362}}{4}\right) = -32\left(\dfrac{\sqrt{1362}}{4}\right)$

$$= -8\sqrt{1362} \approx -295.242 \text{ ft/sec}$$

**99.** $s(t) = -4.9t^2 + v_0 t + s_0$

$= -4.9t^2 + 120t$

$v(t) = -9.8t + 120$

$v(5) = -9.8(5) + 120 = 71$ m/sec

$v(10) = -9.8(10) + 120 = 22$ m/sec

**101.** From $(0, 0)$ to $(4, 2)$, $s(t) = \frac{1}{2}t \Rightarrow v(t) = \frac{1}{2}$ mi/min.

$v(t) = \frac{1}{2}(60) = 30$ mi/h for $0 < t < 4$

Similarly, $v(t) = 0$ for $4 < t < 6$. Finally, from $(6, 2)$ to $(10, 6)$,

$s(t) = t - 4 \Rightarrow v(t) = 1$ mi/min. $= 60$ mi/h.

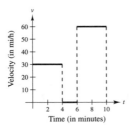

(The velocity has been converted to miles per hour.)

**103.** $v = 40$ mi/h $= \frac{2}{3}$ mi/min

$\left(\frac{2}{3} \text{ mi/min}\right)(6 \text{ min}) = 4$ mi

$v = 0$ mi/h $= 0$ mi/min

$(0 \text{ mi/min})(2 \text{ min}) = 0$ mi

$v = 60$ mi/h $= 1$ mi/min

$(1 \text{ mi/min})(2 \text{ min}) = 2$ mi

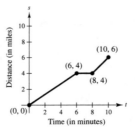

**105.** (a) Using a graphing utility,

$R(v) = 0.417v - 0.02$.

(b) Using a graphing utility,

$B(v) = 0.0056v^2 + 0.001v + 0.04$.

(c) $T(v) = R(v) + B(v) = 0.0056v^2 + 0.418v + 0.02$

(d)

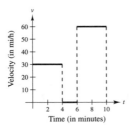

(e) $\dfrac{dT}{dv} = 0.0112v + 0.418$

For $v = 40$, $T'(40) \approx 0.866$

For $v = 80$, $T'(80) \approx 1.314$

For $v = 100$, $T'(100) \approx 1.538$

(f) For increasing speeds, the total stopping distance increases.

**107.** $V = s^3, \dfrac{dV}{ds} = 3s^2$

When $s = 6$ cm, $\dfrac{dV}{ds} = 108$ cm$^3$ per cm change in $s$.

**109.** $s(t) = -\dfrac{1}{2}at^2 + c$ and $s'(t) = -at$

Average velocity: $\dfrac{s(t_0 + \Delta t) - s(t_0 - \Delta t)}{(t_0 + \Delta t) - (t_0 - \Delta t)} = \dfrac{\left[-(1/2)a(t_0 + \Delta t)^2 + c\right] - \left[-(1/2)a(t_0 - \Delta t)^2 + c\right]}{2\Delta t}$

$= \dfrac{-(1/2)a\left(t_0^2 + 2t_0\Delta t + (\Delta t)^2\right) + (1/2)a\left(t_0^2 - 2t_0\Delta t + (\Delta t)^2\right)}{2\,\Delta t}$

$= \dfrac{-2at_0\,\Delta t}{2\,\Delta t} = -at_0 = s'(t_0)$ \qquad instantaneous velocity at $t = t_0$

**111.** $N = f(p)$

   (a) $f'(2.979)$ is the rate of change of the number of gallons of gasoline sold when the price is $2.979 per gallon.

   (b) $f'(2.979)$ is usually negative. As prices go up, sales go down.

**113.** $y = ax^2 + bx + c$

Because the parabola passes through $(0, 1)$ and $(1, 0)$, you have:

$(0, 1)$: $1 = a(0)^2 + b(0) + c \Rightarrow c = 1$

$(1, 0)$: $0 = a(1)^2 + b(1) + 1 \Rightarrow b = -a - 1$

So, $y = ax^2 + (-a - 1)x + 1$. From the tangent line $y = x - 1$, you know that the derivative is 1 at the point $(1, 0)$.

$y' = 2ax + (-a - 1)$

$1 = 2a(1) + (-a - 1)$

$1 = a - 1$

$a = 2$

$b = -a - 1 = -3$

Therefore, $y = 2x^2 - 3x + 1$.

**117.** $f(x) = \begin{cases} ax^3, & x \le 2 \\ x^2 + b, & x > 2 \end{cases}$

$f$ must be continuous at $x = 2$ to be differentiable at $x = 2$.

$\left. \begin{aligned} \lim_{x \to 2^-} f(x) &= \lim_{x \to 2^-} ax^3 = 8a \\ \lim_{x \to 2^+} f(x) &= \lim_{x \to 2^+} (x^2 + b) = 4 + b \end{aligned} \right\} \quad \begin{aligned} 8a &= 4 + b \\ 8a - 4 &= b \end{aligned}$

$f'(x) = \begin{cases} 3ax^2, & x < 2 \\ 2x, & x > 2 \end{cases}$

For $f$ to be differentiable at $x = 2$, the left derivative must equal the right derivative.

$3a(2)^2 = 2(2)$

$12a = 4$

$a = \frac{1}{3}$

$b = 8a - 4 = -\frac{4}{3}$

**119.** $f_1(x) = |\sin x|$ is differentiable for all $x \ne n\pi$, $n$ an integer.

$f_2(x) = \sin|x|$ is differentiable for all $x \ne 0$.

You can verify this by graphing $f_1$ and $f_2$ and observing the locations of the sharp turns.

**115.** $y = x^3 - 9x$

$y' = 3x^2 - 9$

Tangent lines through $(1, -9)$:

$$y + 9 = (3x^2 - 9)(x - 1)$$

$$(x^3 - 9x) + 9 = 3x^3 - 3x^2 - 9x + 9$$

$$0 = 2x^3 - 3x^2 = x^2(2x - 3)$$

$$x = 0 \text{ or } x = \frac{3}{2}$$

The points of tangency are $(0, 0)$ and $\left(\frac{3}{2}, -\frac{81}{8}\right)$.

At $(0, 0)$, the slope is $y'(0) = -9$. At $\left(\frac{3}{2}, -\frac{81}{8}\right)$, the slope is $y'\left(\frac{3}{2}\right) = -\frac{9}{4}$.

Tangent Lines:

$y - 0 = -9(x - 0)$ and $\quad y + \frac{81}{8} = -\frac{9}{4}\left(x - \frac{3}{2}\right)$

$\qquad y = -9x \qquad\qquad\qquad y = -\frac{9}{4}x - \frac{27}{4}$

$9x + y = 0 \qquad\qquad 9x + 4y + 27 = 0$

## Section 2.3   Product and Quotient Rules and Higher-Order Derivatives

**1.** $g(x) = (x^2 + 3)(x^2 - 4x)$

$g'(x) = (x^2 + 3)(2x - 4) + (x^2 - 4x)(2x)$

$\qquad = 2x^3 - 4x^2 + 6x - 12 + 2x^3 - 8x^2$

$\qquad = 4x^3 - 12x^2 + 6x - 12$

$\qquad = 2(2x^3 - 6x^2 + 3x - 6)$

**3.** $h(t) = \sqrt{t}(1 - t^2) = t^{1/2}(1 - t^2)$

$h'(t) = t^{1/2}(-2t) + (1 - t^2)\dfrac{1}{2}t^{-1/2}$

$\qquad = -2t^{3/2} + \dfrac{1}{2t^{1/2}} - \dfrac{1}{2}t^{3/2}$

$\qquad = -\dfrac{5}{2}t^{3/2} + \dfrac{1}{2t^{1/2}}$

$\qquad = \dfrac{1 - 5t^2}{2t^{1/2}} = \dfrac{1 - 5t^2}{2\sqrt{t}}$

**5.** $f(x) = x^3 \cos x$

$f'(x) = x^3(-\sin x) + \cos x(3x^2)$

$\qquad = 3x^2 \cos x - x^3 \sin x$

$\qquad = x^2(3 \cos x - x \sin x)$

**7.** $f(x) = \dfrac{x}{x^2 + 1}$

$f'(x) = \dfrac{(x^2 + 1)(1) - x(2x)}{(x^2 + 1)^2} = \dfrac{1 - x^2}{(x^2 + 1)^2}$

**9.** $h(x) = \dfrac{\sqrt{x}}{x^3 + 1} = \dfrac{x^{1/2}}{x^3 + 1}$

$h'(x) = \dfrac{(x^3 + 1)\dfrac{1}{2}x^{-1/2} - x^{1/2}(3x^2)}{(x^3 + 1)^2}$

$\qquad = \dfrac{x^3 + 1 - 6x^3}{2x^{1/2}(x^3 + 1)^2}$

$\qquad = \dfrac{1 - 5x^3}{2\sqrt{x}(x^3 + 1)^2}$

**11.** $g(x) = \dfrac{\sin x}{x^2}$

$g'(x) = \dfrac{x^2(\cos x) - \sin x(2x)}{(x^2)^2} = \dfrac{x \cos x - 2 \sin x}{x^3}$

**13.** $f(x) = (x^3 + 4x)(3x^2 + 2x - 5)$

$f'(x) = (x^3 + 4x)(6x + 2) + (3x^2 + 2x - 5)(3x^2 + 4)$

$\qquad = 6x^4 + 24x^2 + 2x^3 + 8x + 9x^4 + 6x^3 - 15x^2 + 12x^2 + 8x - 20$

$\qquad = 15x^4 + 8x^3 + 21x^2 + 16x - 20$

$f'(0) = -20$

**15.** $f(x) = \dfrac{x^2 - 4}{x - 3}$

$f'(x) = \dfrac{(x - 3)(2x) - (x^2 - 4)(1)}{(x - 3)^2}$

$\qquad = \dfrac{2x^2 - 6x - x^2 + 4}{(x - 3)^2}$

$\qquad = \dfrac{x^2 - 6x + 4}{(x - 3)^2}$

$f'(1) = \dfrac{1 - 6 + 4}{(1 - 3)^2} = -\dfrac{1}{4}$

**17.** $f(x) = x \cos x$

$f'(x) = (x)(-\sin x) + (\cos x)(1) = \cos x - x \sin x$

$f'\left(\dfrac{\pi}{4}\right) = \dfrac{\sqrt{2}}{2} - \dfrac{\pi}{4}\left(\dfrac{\sqrt{2}}{2}\right) = \dfrac{\sqrt{2}}{8}(4 - \pi)$

| *Function* | *Rewrite* | *Differentiate* | *Simplify* |
|---|---|---|---|

**19.** $y = \dfrac{x^2 + 3x}{7}$    $y = \dfrac{1}{7}x^2 + \dfrac{3}{7}x$    $y' = \dfrac{2}{7}x + \dfrac{3}{7}$    $y' = \dfrac{2x + 3}{7}$

**21.** $y = \dfrac{6}{7x^2}$    $y = \dfrac{6}{7}x^{-2}$    $y' = -\dfrac{12}{7}x^{-3}$    $y' = -\dfrac{12}{7x^3}$

**23.** $y = \dfrac{4x^{3/2}}{x}$    $y = 4x^{1/2}, \ x > 0$    $y' = 2x^{-1/2}$    $y' = \dfrac{2}{\sqrt{x}}, \ x > 0$

**25.** $f(x) = \dfrac{4 - 3x - x^2}{x^2 - 1}$

$f'(x) = \dfrac{(x^2 - 1)(-3 - 2x) - (4 - 3x - x^2)(2x)}{(x^2 - 1)^2}$

$\quad = \dfrac{-3x^2 + 3 - 2x^3 + 2x - 8x + 6x^2 + 2x^3}{(x^2 - 1)^2}$

$\quad = \dfrac{3x^2 - 6x + 3}{(x^2 - 1)^2} = \dfrac{3(x^2 - 2x + 1)}{(x^2 - 1)^2}$

$\quad = \dfrac{3(x - 1)^2}{(x - 1)^2(x + 1)^2} = \dfrac{3}{(x + 1)^2}, \ x \neq 1$

**27.** $f(x) = x\left(1 - \dfrac{4}{x + 3}\right) = x - \dfrac{4x}{x + 3}$

$f'(x) = 1 - \dfrac{(x + 3)4 - 4x(1)}{(x + 3)^2}$

$\quad = \dfrac{(x^2 + 6x + 9) - 12}{(x + 3)^2}$

$\quad = \dfrac{x^2 + 6x - 3}{(x + 3)^2}$

**29.** $f(x) = \dfrac{3x - 1}{\sqrt{x}} = 3x^{1/2} - x^{-1/2}$

$f'(x) = \dfrac{3}{2}x^{-1/2} + \dfrac{1}{2}x^{-3/2}$

$\quad = \dfrac{3x + 1}{2x^{3/2}}$

**Alternate solution:**

$f(x) = \dfrac{3x - 1}{\sqrt{x}} = \dfrac{3x - 1}{x^{1/2}}$

$f'(x) = \dfrac{x^{1/2}(3) - (3x - 1)\left(\dfrac{1}{2}\right)(x^{-1/2})}{x}$

$\quad = \dfrac{\dfrac{1}{2}x^{-1/2}(3x + 1)}{x}$

$\quad = \dfrac{3x + 1}{2x^{3/2}}$

**31.** $h(s) = (s^3 - 2)^2 = s^6 - 4s^3 + 4$

$h'(s) = 6s^5 - 12s^2 = 6s^2(s^3 - 2)$

**33.** $f(x) = \dfrac{2 - (1/x)}{x - 3} = \dfrac{2x - 1}{x(x - 3)} = \dfrac{2x - 1}{x^2 - 3x}$

$f'(x) = \dfrac{(x^2 - 3x)2 - (2x - 1)(2x - 3)}{(x^2 - 3x)^2} = \dfrac{2x^2 - 6x - 4x^2 + 8x - 3}{(x^2 - 3x)^2}$

$\quad = \dfrac{-2x^2 + 2x - 3}{(x^2 - 3x)^2} = -\dfrac{2x^2 - 2x + 3}{x^2(x - 3)^2}$

**35.** $f(x) = (2x^3 + 5x)(x - 3)(x + 2)$

$f'(x) = (6x^2 + 5)(x - 3)(x + 2) + (2x^3 + 5x)(1)(x + 2) + (2x^3 + 5x)(x - 3)(1)$

$\quad = (6x^2 + 5)(x^2 - x - 6) + (2x^3 + 5x)(x + 2) + (2x^3 + 5x)(x - 3)$

$\quad = (6x^4 + 5x^2 - 6x^3 - 5x - 36x^2 - 30) + (2x^4 + 4x^3 + 5x^2 + 10x) + (2x^4 + 5x^2 - 6x^3 - 15x)$

$\quad = 10x^4 - 8x^3 - 21x^2 - 10x - 30$

**Note:** You could simplify first:

$f(x) = (2x^3 + 5x)(x^2 - x - 6)$

**37.** $f(x) = \dfrac{x^2 + c^2}{x^2 - c^2}$

$f'(x) = \dfrac{\left(x^2 - c^2\right)(2x) - \left(x^2 + c^2\right)(2x)}{\left(x^2 - c^2\right)^2}$

$\quad = -\dfrac{4xc^2}{\left(x^2 - c^2\right)^2}$

**39.** $f(t) = t^2 \sin t$

$f'(t) = t^2 \cos t + 2t \sin t = t(t \cos t + 2 \sin t)$

**41.** $f(t) = \dfrac{\cos t}{t}$

$f'(t) = \dfrac{-t \sin t - \cos t}{t^2} = -\dfrac{t \sin t + \cos t}{t^2}$

**43.** $f(x) = -x + \tan x$

$f'(x) = -1 + \sec^2 x = \tan^2 x$

**45.** $g(t) = \sqrt[4]{t} + 6 \csc t = t^{1/4} + 6 \csc t$

$g'(t) = \dfrac{1}{4} t^{-3/4} - 6 \csc t \cot t$

$\quad = \dfrac{1}{4t^{3/4}} - 6 \csc t \cot t$

**47.** $y = \dfrac{3(1 - \sin x)}{2 \cos x} = \dfrac{3 - 3 \sin x}{2 \cos x}$

$y' = \dfrac{(-3 \cos x)(2 \cos x) - (3 - 3 \sin x)(-2 \sin x)}{(2 \cos x)^2}$

$\quad = \dfrac{-6 \cos^2 x + 6 \sin x - 6 \sin^2 x}{4 \cos^2 x}$

$\quad = \dfrac{3}{2}\left(-1 + \tan x \sec x - \tan^2 x\right)$

$\quad = \dfrac{3}{2} \sec x(\tan x - \sec x)$

**49.** $y = -\csc x - \sin x$

$y' = \csc x \cot x - \cos x$

$\quad = \dfrac{\cos x}{\sin^2 x} - \cos x$

$\quad = \cos x\left(\csc^2 x - 1\right)$

$\quad = \cos x \cot^2 x$

**51.** $f(x) = x^2 \tan x$

$f'(x) = x^2 \sec^2 x + 2x \tan x$

$\quad = x\left(x \sec^2 x + 2 \tan x\right)$

**53.** $y = 2x \sin x + x^2 \cos x$

$y' = 2x \cos x + 2 \sin x + x^2(-\sin x) + 2x \cos x$

$\quad = 4x \cos x + \left(2 - x^2\right) \sin x$

**55.** $g(x) = \left(\dfrac{x + 1}{x + 2}\right)(2x - 5)$

$g'(x) = \left(\dfrac{x + 1}{x + 2}\right)(2) + (2x - 5)\left[\dfrac{(x + 2)(1) - (x + 1)(1)}{(x + 2)^2}\right]$

$\quad = \dfrac{2x^2 + 8x - 1}{(x + 2)^2}$

(Form of answer may vary.)

**57.** $g(\theta) = \dfrac{\theta}{1 - \sin \theta}$

$g'(\theta) = \dfrac{1 - \sin \theta + \theta \cos \theta}{(1 - \sin \theta)^2}$

(Form of answer may vary.)

**59.**   $y = \dfrac{1 + \csc x}{1 - \csc x}$

$y' = \dfrac{(1 - \csc x)(-\csc x \cot x) - (1 + \csc x)(\csc x \cot x)}{(1 - \csc x)^2} = \dfrac{-2 \csc x \cot x}{(1 - \csc x)^2}$

$y'\left(\dfrac{\pi}{6}\right) = \dfrac{-2(2)\left(\sqrt{3}\right)}{(1 - 2)^2} = -4\sqrt{3}$

**61.** $h(t) = \dfrac{\sec t}{t}$

$h'(t) = \dfrac{t(\sec t \tan t) - (\sec t)(1)}{t^2} = \dfrac{\sec t(t \tan t - 1)}{t^2}$

$h'(\pi) = \dfrac{\sec \pi(\pi \tan \pi - 1)}{\pi^2} = \dfrac{1}{\pi^2}$

**63. (a)** $f(x) = (x^3 + 4x - 1)(x - 2), \quad (1, -4)$

$f'(x) = (x^3 + 4x - 1)(1) + (x - 2)(3x^2 + 4)$

$\quad = x^3 + 4x - 1 + 3x^3 - 6x^2 + 4x - 8$

$\quad = 4x^3 - 6x^2 + 8x - 9$

$f'(1) = -3;$ Slope at $(1, -4)$

Tangent line: $y + 4 = -3(x - 1) \Rightarrow y = -3x - 1$

**(b)**

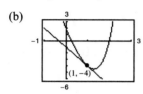

**(c)** Graphing utility confirms $\dfrac{dy}{dx} = -3$ at $(1, -4)$.

**65. (a)** $f(x) = \dfrac{x}{x + 4}, \quad (-5, 5)$

$f'(x) = \dfrac{(x + 4)(1) - x(1)}{(x + 4)^2} = \dfrac{4}{(x + 4)^2}$

$f'(-5) = \dfrac{4}{(-5 + 4)^2} = 4; \quad$ Slope at $(-5, 5)$

Tangent line: $y - 5 = 4(x + 5) \Rightarrow y = 4x + 25$

**(b)**

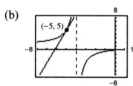

**(c)** Graphing utility confirms $\dfrac{dy}{dx} = 4$ at $(-5, 5)$.

**67. (a)** $f(x) = \tan x, \quad \left(\dfrac{\pi}{4}, 1\right)$

$f'(x) = \sec^2 x$

$f'\left(\dfrac{\pi}{4}\right) = 2; \quad$ Slope at $\left(\dfrac{\pi}{4}, 1\right)$

Tangent line: $\qquad y - 1 = 2\left(x - \dfrac{\pi}{4}\right)$

$\qquad\qquad\qquad y - 1 = 2x - \dfrac{\pi}{2}$

$\qquad\qquad 4x - 2y - \pi + 2 = 0$

**(b)**

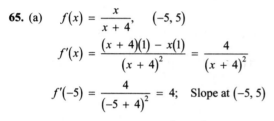

**(c)** Graphing utility confirms $\dfrac{dy}{dx} = 2$ at $\left(\dfrac{\pi}{4}, 1\right)$.

**69.** $f(x) = \dfrac{8}{x^2 + 4}; \quad (2, 1)$

$f'(x) = \dfrac{(x^2 + 4)(0) - 8(2x)}{(x^2 + 4)^2} = \dfrac{-16x}{(x^2 + 4)^2}$

$f'(2) = \dfrac{-16(2)}{(4 + 4)^2} = -\dfrac{1}{2}$

$y - 1 = -\dfrac{1}{2}(x - 2)$

$y = -\dfrac{1}{2}x + 2$

$2y + x - 4 = 0$

**71.** $f(x) = \dfrac{16x}{x^2 + 16}; \quad \left(-2, -\dfrac{8}{5}\right)$

$f'(x) = \dfrac{(x^2 + 16)(16) - 16x(2x)}{(x^2 + 16)^2} = \dfrac{256 - 16x^2}{(x^2 + 16)^2}$

$f'(-2) = \dfrac{256 - 16(4)}{20^2} = \dfrac{12}{25}$

$y + \dfrac{8}{5} = \dfrac{12}{25}(x + 2)$

$y = \dfrac{12}{25}x - \dfrac{16}{25}$

$25y - 12x + 16 = 0$

**73.** $f(x) = \dfrac{2x - 1}{x^2} = 2x^{-1} - x^{-2}$

$f'(x) = -2x^{-2} + 2x^{-3} = \dfrac{2(-x + 1)}{x^3}$

$f'(x) = 0$ when $x = 1$, and $f(1) = 1$.

Horizontal tangent at $(1, 1)$.

**75.** $f(x) = \dfrac{x^2}{x - 1}$

$f'(x) = \dfrac{(x - 1)(2x) - x^2(1)}{(x - 1)^2}$

$\quad = \dfrac{x^2 - 2x}{(x - 1)^2} = \dfrac{x(x - 2)}{(x - 1)^2}$

$f'(x) = 0$ when $x = 0$ or $x = 2$.

Horizontal tangents are at $(0, 0)$ and $(2, 4)$.

**77.** $f(x) = \dfrac{x+1}{x-1}$

$f'(x) = \dfrac{(x-1)-(x+1)}{(x-1)^2} = \dfrac{-2}{(x-1)^2}$

$2y + x = 6 \Rightarrow y = -\dfrac{1}{2}x + 3$; Slope: $-\dfrac{1}{2}$

$\dfrac{-2}{(x-1)^2} = -\dfrac{1}{2}$

$(x-1)^2 = 4$

$x - 1 = \pm 2$

$x = -1, 3;\ f(-1) = 0,\ f(3) = 2$

$y - 0 = -\dfrac{1}{2}(x+1) \Rightarrow y = -\dfrac{1}{2}x - \dfrac{1}{2}$

$y - 2 = -\dfrac{1}{2}(x-3) \Rightarrow y = -\dfrac{1}{2}x + \dfrac{7}{2}$

**79.** $f'(x) = \dfrac{(x+2)3 - 3x(1)}{(x+2)^2} = \dfrac{6}{(x+2)^2}$

$g'(x) = \dfrac{(x+2)5 - (5x+4)(1)}{(x+2)^2} = \dfrac{6}{(x+2)^2}$

$g(x) = \dfrac{5x+4}{(x+2)} = \dfrac{3x}{(x+2)} + \dfrac{2x+4}{(x+2)} = f(x) + 2$

$f$ and $g$ differ by a constant.

**81.** (a) $p'(x) = f'(x)g(x) + f(x)g'(x)$

$p'(1) = f'(1)g(1) + f(1)g'(1) = 1(4) + 6\left(-\dfrac{1}{2}\right) = 1$

(b) $q'(x) = \dfrac{g(x)f'(x) - f(x)g'(x)}{g(x)^2}$

$q'(4) = \dfrac{3(-1) - 7(0)}{3^2} = -\dfrac{1}{3}$

**83.** Area $= A(t) = (6t+5)\sqrt{t} = 6t^{3/2} + 5t^{1/2}$

$A'(t) = 9t^{1/2} + \dfrac{5}{2}t^{-1/2} = \dfrac{18t+5}{2\sqrt{t}}$ cm$^2$/sec

**85.** $C = 100\left(\dfrac{200}{x^2} + \dfrac{x}{x+30}\right),\ 1 \le x$

$\dfrac{dC}{dx} = 100\left(-\dfrac{400}{x^3} + \dfrac{30}{(x+30)^2}\right)$

(a) When $x = 10$: $\dfrac{dC}{dx} = -\$38.13$ thousand/100 components

(b) When $x = 15$: $\dfrac{dC}{dx} = -\$10.37$ thousand/100 components

(c) When $x = 20$: $\dfrac{dC}{dx} = -\$3.80$ thousand/100 components

As the order size increases, the cost per item decreases.

**87.** $P(t) = 500\left[1 + \dfrac{4t}{50 + t^2}\right]$

$P'(t) = 500\left[\dfrac{(50+t^2)(4) - (4t)(2t)}{(50+t^2)^2}\right] = 500\left[\dfrac{200 - 4t^2}{(50+t^2)^2}\right] = 2000\left[\dfrac{50 - t^2}{(50+t^2)^2}\right]$

$P'(2) \approx 31.55$ bacteria/h

**89.** (a) $\sec x = \dfrac{1}{\cos x}$

$\dfrac{d}{dx}[\sec x] = \dfrac{d}{dx}\left[\dfrac{1}{\cos x}\right] = \dfrac{(\cos x)(0) - (1)(-\sin x)}{(\cos x)^2} = \dfrac{\sin x}{\cos x \cos x} = \dfrac{1}{\cos x} \cdot \dfrac{\sin x}{\cos x} = \sec x \tan x$

(b)    $\csc x = \dfrac{1}{\sin x}$

$$\frac{d}{dx}[\csc x] = \frac{d}{dx}\left[\frac{1}{\sin x}\right] = \frac{(\sin x)(0) - (1)(\cos x)}{(\sin x)^2} = -\frac{\cos x}{\sin x \sin x} = -\frac{1}{\sin x} \cdot \frac{\cos x}{\sin x} = -\csc x \cot x$$

(c)    $\cot x = \dfrac{\cos x}{\sin x}$

$$\frac{d}{dx}[\cot x] = \frac{d}{dx}\left[\frac{\cos x}{\sin x}\right] = \frac{\sin x(-\sin x) - (\cos x)(\cos x)}{(\sin x)^2} = -\frac{\sin^2 x + \cos^2 x}{\sin^2 x} = -\frac{1}{\sin^2 x} = -\csc^2 x$$

**91.** (a)  Using a graphing utility,

$q(t) = -0.0546t^3 + 2.529t^2 - 36.89t + 186.6$

$v(t) = 0.0796t^3 - 2.162t^2 + 15.32t + 5.9.$

(b)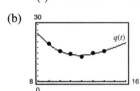

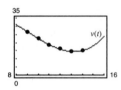

(c)  $A = \dfrac{v(t)}{q(t)} = \dfrac{0.0796t^3 - 2.162t^2 + 15.32t + 5.9}{-0.0546t^3 + 2.529t^2 - 36.89t + 186.6}$

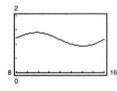

A represents the average value (in billions of dollars) per one million personal computers.

(d)  $A'(t)$ represents the rate of change of the average value per one million personal computers for the given year $t$.

**93.**  $f(x) = x^4 + 2x^3 - 3x^2 - x$

$f'(x) = 4x^3 + 6x^2 - 6x - 1$

$f''(x) = 12x^2 + 12x - 6$

**95.**  $f(x) = 4x^{3/2}$

$f'(x) = 6x^{1/2}$

$f''(x) = 3x^{-1/2} = \dfrac{3}{\sqrt{x}}$

**97.**  $f(x) = \dfrac{x}{x-1}$

$f'(x) = \dfrac{(x-1)(1) - x(1)}{(x-1)^2} = \dfrac{-1}{(x-1)^2}$

$f''(x) = \dfrac{2}{(x-1)^3}$

**99.**  $f(x) = x \sin x$

$f'(x) = x \cos x + \sin x$

$f''(x) = x(-\sin x) + \cos x + \cos x$

$\qquad = -x \sin x + 2 \cos x$

**101.**  $f'(x) = x^2$

$f''(x) = 2x$

**103.**  $f'''(x) = 2\sqrt{x}$

$f^{(4)}(x) = \dfrac{1}{2}(2)x^{-1/2} = \dfrac{1}{\sqrt{x}}$

**105.**  $f(x) = 2g(x) + h(x)$

$f'(x) = 2g'(x) + h'(x)$

$f'(2) = 2g'(2) + h'(2)$

$\qquad = 2(-2) + 4$

$\qquad = 0$

**107.**  $f(x) = \dfrac{g(x)}{h(x)}$

$f'(x) = \dfrac{h(x)g'(x) - g(x)h'(x)}{\left[h(x)\right]^2}$

$f'(2) = \dfrac{h(2)g'(2) - g(2)h'(2)}{\left[h(2)\right]^2}$

$\qquad = \dfrac{(-1)(-2) - (3)(4)}{(-1)^2}$

$\qquad = -10$

**109.** The graph of a differentiable function $f$ such that $f(2) = 0$, $f' < 0$ for $-\infty < x < 2$, and $f' > 0$ for $2 < x < \infty$ would, in general, look like the graph below.

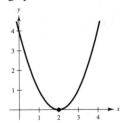

One such function is $f(x) = (x - 2)^2$.

**111.**

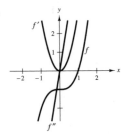

It appears that $f$ is cubic, so $f'$ would be quadratic and $f''$ would be linear.

**113.**

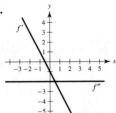

**115.**

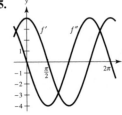

**117.** $v(t) = 36 - t^2, 0 \le t \le 6$

$a(t) = v'(t) = -2t$

$v(3) = 27$ m/sec

$a(3) = -6$ m/sec$^2$

The speed of the object is decreasing.

**119.** $s(t) = -8.25t^2 + 66t$

$v(t) = s'(t) = 16.50t + 66$

$a(t) = v'(t) = -16.50$

| $t$(sec) | 0 | 1 | 2 | 3 | 4 |
|---|---|---|---|---|---|
| $s(t)$ (ft) | 0 | 57.75 | 99 | 123.75 | 132 |
| $v(t) = s'(t)$ (ft/sec) | 66 | 49.5 | 33 | 16.5 | 0 |
| $a(t) = v'(t)$ (ft/sec$^2$) | −16.5 | −16.5 | −16.5 | −16.5 | −16.5 |

Average velocity on:

$[0, 1]$ is $\dfrac{57.75 - 0}{1 - 0} = 57.75$

$[1, 2]$ is $\dfrac{99 - 57.75}{2 - 1} = 41.25$

$[2, 3]$ is $\dfrac{123.75 - 99}{3 - 2} = 24.75$

$[3, 4]$ is $\dfrac{132 - 123.75}{4 - 3} = 8.25$

**121.** $f(x) = x^n$

$f^{(n)}(x) = n(n - 1)(n - 2) \cdots (2)(1) = n!$

**Note:** $n! = n(n - 1) \cdots 3 \cdot 2 \cdot 1$ (read "$n$ factorial")

**123.** $f(x) = g(x)h(x)$

(a)  $f'(x) = g(x)h'(x) + h(x)g'(x)$

$f''(x) = g(x)h''(x) + g'(x)h'(x) + h(x)g''(x) + h'(x)g'(x)$

$\quad = g(x)h''(x) + 2g'(x)h'(x) + h(x)g''(x)$

$f'''(x) = g(x)h'''(x) + g'(x)h''(x) + 2g'(x)h''(x) + 2g''(x)h'(x) + h(x)g'''(x) + h'(x)g''(x)$

$\quad = g(x)h'''(x) + 3g'(x)h''(x) + 3g''(x)h'(x) + g'''(x)h(x)$

$f^{(4)}(x) = g(x)h^{(4)}(x) + g'(x)h'''(x) + 3g'(x)h'''(x) + 3g''(x)h''(x) + 3g''(x)h''(x) + 3g'''(x)h'(x)$

$\quad\quad + g'''(x)h'(x) + g^{(4)}(x)h(x)$

$\quad = g(x)h^{(4)}(x) + 4g'(x)h'''(x) + 6g''(x)h''(x) + 4g'''(x)h'(x) + g^{(4)}(x)h(x)$

(b)  $f^{(n)}(x) = g(x)h^{(n)}(x) + \dfrac{n(n-1)(n-2)\cdots(2)(1)}{1\big[(n-1)(n-2)\cdots(2)(1)\big]}g'(x)h^{(n-1)}(x) + \dfrac{n(n-1)(n-2)\cdots(2)(1)}{(2)(1)\big[(n-2)(n-3)\cdots(2)(1)\big]}g''(x)h^{(n-2)}(x)$

$\quad\quad + \dfrac{n(n-1)(n-2)\cdots(2)(1)}{(3)(2)(1)\big[(n-3)(n-4)\cdots(2)(1)\big]}g'''(x)h^{(n-3)}(x) + \cdots$

$\quad\quad + \dfrac{n(n-1)(n-2)\cdots(2)(1)}{\big[(n-1)(n-2)\cdots(2)(1)\big](1)}g^{(n-1)}(x)h'(x) + g^{(n)}(x)h(x)$

$\quad = g(x)h^{(n)}(x) + \dfrac{n!}{1!(n-1)!}g'(x)h^{(n-1)}(x) + \dfrac{n!}{2!(n-2)!}g''(x)h^{(n-2)}(x) + \cdots$

$\quad\quad + \dfrac{n!}{(n-1)!1!}g^{(n-1)}(x)h'(x) + g^{(n)}(x)h(x)$

**Note:** $n! = n(n-1)\cdots 3 \cdot 2 \cdot 1$ (read "$n$ factorial")

**125.**  $f(x) = x^n \sin x$

$f'(x) = x^n \cos x + nx^{n-1}\sin x$

When $n = 1$: $f'(x) = x\cos x + \sin x$

When $n = 2$: $f'(x) = x^2 \cos x + 2\sin x$

When $n = 3$: $f'(x) = x^3 \cos x + 3x^2 \sin x$

When $n = 4$: $f'(x) = x^4 \cos x + 4x^3 \sin x$

For general $n$, $f'(x) = x^n \cos x + nx^{n-1}\sin x$.

**127.**  $y = \dfrac{1}{x}$, $y' = -\dfrac{1}{x^2}$, $y'' = \dfrac{2}{x^3}$

$x^3 y'' + 2x^2 y' = x^3\left[\dfrac{2}{x^3}\right] + 2x^2\left[-\dfrac{1}{x^2}\right] = 2 - 2 = 0$

**129.**  $y = 2\sin x + 3$

$y' = 2\cos x$

$y'' = -2\sin x$

$y'' + y = -2\sin x + (2\sin x + 3) = 3$

**131.** False. If $y = f(x)g(x)$, then

$\dfrac{dy}{dx} = f(x)g'(x) + g(x)f'(x)$.

**133.** True

$h'(c) = f(c)g'(c) + g(c)f'(c)$

$\quad = f(c)(0) + g(c)(0)$

$\quad = 0$

**135.** True

**137.**  $f(x) = ax^2 + bx + c$

$f'(x) = 2ax + b$

$x$-intercept at $(1, 0)$: $0 = a + b + c$

$(2, 7)$ on graph: $7 = 4a + 2b + c$

Slope 10 at $(2, 7)$: $10 = 4a + b$

Subtracting the third equation from the second, $-3 = b + c$. Subtracting this equation from the first, $3 = a$. Then, $10 = 4(3) + b \Rightarrow b = -2$. Finally, $-3 = (-2) + c \Rightarrow c = -1$.

$f(x) = 3x^2 - 2x - 1$

**139.** $f(x) = x|x| = \begin{cases} x^2, & \text{if } x \geq 0 \\ -x^2, & \text{if } x < 0 \end{cases}$

$f'(x) = \begin{cases} 2x, & \text{if } x \geq 0 \\ -2x, & \text{if } x < 0 \end{cases} = 2|x|$

$f''(x) = \begin{cases} 2, & \text{if } x > 0 \\ -2, & \text{if } x < 0 \end{cases}$

$f''(0)$ does not exist since the left and right derivatives are not equal.

**141.** $\dfrac{d}{dx}\big[f(x)g(x)h(x)\big] = \dfrac{d}{dx}\big[(f(x)g(x))h(x)\big]$

$= \dfrac{d}{dx}\big[f(x)g(x)\big]h(x) + f(x)g(x)h'(x)$

$= \big[f(x)g'(x) + g(x)f'(x)\big]h(x) + f(x)g(x)h'(x)$

$= f'(x)g(x)h(x) + f(x)g'(x)h(x) + f(x)g(x)h'(x)$

## Section 2.4   The Chain Rule

| $y = f(g(x))$ | $u = g(x)$ | $y = f(u)$ |
|---|---|---|
| **1.** $y = (5x - 8)^4$ | $u = 5x - 8$ | $y = u^4$ |
| **3.** $y = \sqrt{x^3 - 7}$ | $u = x^3 - 7$ | $y = \sqrt{u}$ |
| **5.** $y = \csc^3 x$ | $u = \csc x$ | $y = u^3$ |

**7.** $y = (4x - 1)^3$

$y' = 3(4x - 1)^2(4) = 12(4x - 1)^2$

**9.** $g(x) = 3(4 - 9x)^4$

$g'(x) = 12(4 - 9x)^3(-9) = -108(4 - 9x)^3$

**11.** $f(t) = \sqrt{5 - t} = (5 - t)^{1/2}$

$f'(t) = \dfrac{1}{2}(5 - t)^{-1/2}(-1) = \dfrac{-1}{2\sqrt{5 - t}}$

**13.** $y = \sqrt[3]{6x^2 + 1} = (6x^2 + 1)^{1/3}$

$y' = \dfrac{1}{3}(6x^2 + 1)^{-2/3}(12x) = \dfrac{4x}{(6x^2 + 1)^{2/3}} = \dfrac{4x}{\sqrt[3]{(6x^2 + 1)^2}}$

**15.** $y = 2\sqrt[4]{9 - x^2} = 2(9 - x^2)^{1/4}$

$y' = 2\left(\dfrac{1}{4}\right)(9 - x^2)^{-3/4}(-2x)$

$= \dfrac{-x}{(9 - x^2)^{3/4}} = \dfrac{-x}{\sqrt[4]{(9 - x^2)^3}}$

**17.** $y = (x - 2)^{-1}$

$y' = -1(x - 2)^{-2}(1) = \dfrac{-1}{(x - 2)^2}$

**19.** $f(t) = (t - 3)^{-2}$

$f'(t) = -2(t - 3)^{-3}(1) = \dfrac{-2}{(t - 3)^3}$

**21.** $y = (x + 2)^{-1/2}$

$\dfrac{dy}{dx} = -\dfrac{1}{2}(x + 2)^{-3/2} = -\dfrac{1}{2(x + 2)^{3/2}} = -\dfrac{1}{2\sqrt{(x + 2)^3}}$

**23.** $f(x) = x^2(x - 2)^4$

$f'(x) = x^2\big[4(x - 2)^3(1)\big] + (x - 2)^4(2x)$

$= 2x(x - 2)^3\big[2x + (x - 2)\big]$

$= 2x(x - 2)^3(3x - 2)$

**25.** $y = x\sqrt{1 - x^2} = x(1 - x^2)^{1/2}$

$y' = x\left[\dfrac{1}{2}(1 - x^2)^{-1/2}(-2x)\right] + (1 - x^2)^{1/2}(1)$

$= -x^2(1 - x^2)^{-1/2} + (1 - x^2)^{1/2}$

$= (1 - x^2)^{-1/2}\big[-x^2 + (1 - x^2)\big]$

$= \dfrac{1 - 2x^2}{\sqrt{1 - x^2}}$

**27.**  $y = \dfrac{x}{\sqrt{x^2 + 1}} = \dfrac{x}{\left(x^2 + 1\right)^{-1/2}}$

$y' = \dfrac{\left(x^2 + 1\right)^{1/2}(1) - x\left(\dfrac{1}{2}\right)\left(x^2 + 1\right)^{-1/2}(2x)}{\left[\left(x^2 + 1\right)^{1/2}\right]^2}$

$\quad = \dfrac{\left(x^2 + 1\right)^{1/2} - x^2\left(x^2 + 1\right)^{-1/2}}{x^2 + 1}$

$\quad = \dfrac{\left(x^2 + 1\right)^{-1/2}\left[x^2 + 1 - x^2\right]}{x^2 + 1}$

$\quad = \dfrac{1}{\left(x^2 + 1\right)^{3/2}} = \dfrac{1}{\sqrt{\left(x^2 + 1\right)^2}}$

**29.**  $g(x) = \left(\dfrac{x + 5}{x^2 + 2}\right)^2$

$g'(x) = 2\left(\dfrac{x + 5}{x^2 + 2}\right)\left(\dfrac{\left(x^2 + 2\right) - (x + 5)(2x)}{\left(x^2 + 2\right)^2}\right)$

$\quad = \dfrac{2(x + 5)\left(2 - 10x - x^2\right)}{\left(x^2 + 2\right)^3}$

$\quad = \dfrac{-2(x + 5)\left(x^2 + 10x - 2\right)}{\left(x^2 + 2\right)^3}$

**31.**  $f(v) = \left(\dfrac{1 - 2v}{1 + v}\right)^3$

$f'(v) = 3\left(\dfrac{1 - 2v}{1 + v}\right)^2\left(\dfrac{(1 + v)(-2) - (1 - 2v)}{\left(1 + v\right)^2}\right)$

$\quad = \dfrac{-9\left(1 - 2v\right)^2}{\left(1 + v\right)^4}$

**33.**  $f(x) = \left(\left(x^2 + 3\right)^5 + x\right)^2$

$f'(x) = 2\left(\left(x^2 + 3\right)^5 + x\right)\left(5\left(x^2 + 3\right)^4(2x) + 1\right)$

$\quad = 2\left[10x\left(x^2 + 3\right)^9 + \left(x^2 + 3\right)^5 + 10x^2\left(x^2 + 3\right)^4 + x\right] = 20x\left(x^2 + 3\right)^9 + 2\left(x^2 + 3\right)^5 + 20x^2\left(x^2 + 3\right)^4 + 2x$

**35.**  $f(x) = \sqrt{2 + \sqrt{2 + \sqrt{x}}}$

$\quad = \left[2 + \left(2 + x^{1/2}\right)^{1/2}\right]^{1/2}$

$f'(x) = \dfrac{1}{2}\left[2 + \left(2 + x^{1/2}\right)^{1/2}\right]^{-1/2}\left[\dfrac{1}{2}\left(2 + x^{1/2}\right)^{-1/2}\left(\dfrac{1}{2}x^{-1/2}\right)\right] = \dfrac{1}{8\sqrt{x}\sqrt{2 + \sqrt{x}}\sqrt{2 + \sqrt{2 + \sqrt{x}}}}$

**37.**  $y = \dfrac{\sqrt{x} + 1}{x^2 + 1}$

$y' = \dfrac{1 - 3x^2 - 4x^{3/2}}{2\sqrt{x}\left(x^2 + 1\right)^2}$

The zero of $y'$ corresponds to the point on the graph of $y$ where the tangent line is horizontal.

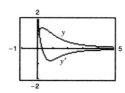

**41.**  $y = \dfrac{\cos \pi x + 1}{x}$

$\dfrac{dy}{dx} = \dfrac{-\pi x \sin \pi x - \cos \pi x - 1}{x^2}$

$\quad = -\dfrac{\pi x \sin \pi x + \cos \pi x + 1}{x^2}$

The zeros of $y'$ correspond to the points on the graph of $y$ where the tangent lines are horizontal.

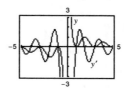

**39.**  $y = \sqrt{\dfrac{x + 1}{x}}$

$y' = -\dfrac{\sqrt{(x + 1)/x}}{2x(x + 1)}$

$y'$ has no zeros.

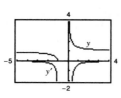

**43.** (a) $\quad y = \sin x$

$\quad\quad\quad y' = \cos x$

$\quad\quad\quad y'(0) = 1$

$\quad\quad\quad$ 1 cycle in $[0, 2\pi]$

$\quad$ (b) $\quad y = \sin 2x$

$\quad\quad\quad y' = 2 \cos 2x$

$\quad\quad\quad y'(0) = 2$

$\quad\quad\quad$ 2 cycles in $[0, 2\pi]$

$\quad\quad\quad$ The slope of sin $ax$ at the origin is $a$.

**45.** $\quad y = \cos 4x$

$\quad \dfrac{dy}{dx} = -4 \sin 4x$

**47.** $g(x) = 5 \tan 3x$

$\quad g'(x) = 15 \sec^2 3x$

**49.** $y = \sin(\pi x)^2 = \sin(\pi^2 x^2)$

$\quad y' = \cos(\pi^2 x^2)[2\pi^2 x] = 2\pi^2 x \cos(\pi^2 x^2)$

$\quad\quad = 2\pi^2 x \cos(\pi x)^2$

**51.** $h(x) = \sin 2x \cos 2x$

$\quad h'(x) = \sin 2x(-2 \sin 2x) + \cos 2x(2 \cos 2x)$

$\quad\quad = 2 \cos^2 2x - 2 \sin^2 2x$

$\quad\quad = 2 \cos 4x$

$\quad$ **Alternate solution:** $h(x) = \frac{1}{2} \sin 4x$

$\quad\quad\quad\quad\quad\quad\quad\quad h'(x) = \frac{1}{2} \cos 4x(4) = 2 \cos 4x$

**53.** $f(x) = \dfrac{\cot x}{\sin x} = \dfrac{\cos x}{\sin^2 x}$

$\quad f'(x) = \dfrac{\sin^2 x(-\sin x) - \cos x(2 \sin x \cos x)}{\sin^4 x}$

$\quad\quad = \dfrac{-\sin^2 x - 2 \cos^2 x}{\sin^3 x} = \dfrac{-1 - \cos^2 x}{\sin^3 x}$

**55.** $y = 4 \sec^2 x$

$\quad y' = 8 \sec x \cdot \sec x \tan x = 8 \sec^2 x \tan x$

**57.** $f(\theta) = \tan^2 5\theta = (\tan 5\theta)^2$

$\quad f'(\theta) = 2(\tan 5\theta)(\sec^2 5\theta)5 = 10 \tan 5\theta \sec^2 5\theta$

**59.** $f(\theta) = \frac{1}{4} \sin^2 2\theta = \frac{1}{4}(\sin 2\theta)^2$

$\quad f'(\theta) = 2(\frac{1}{4})(\sin 2\theta)(\cos 2\theta)(2)$

$\quad\quad = \sin 2\theta \cos 2\theta = \frac{1}{2} \sin 4\theta$

**61.** $f(t) = 3 \sec^2(\pi t - 1)$

$\quad f'(t) = 6 \sec(\pi t - 1) \sec(\pi t - 1) \tan(\pi t - 1)(\pi)$

$\quad\quad = 6\pi \sec^2(\pi t - 1) \tan(\pi t - 1) = \dfrac{6\pi \sin(\pi t - 1)}{\cos^3(\pi t - 1)}$

**63.** $y = \sqrt{x} + \frac{1}{4} \sin(2x)^2 = \sqrt{x} + \frac{1}{4} \sin(4x^2)$

$\quad \dfrac{dy}{dx} = \frac{1}{2} x^{-1/2} + \frac{1}{4} \cos(4x^2)(8x) = \dfrac{1}{2\sqrt{x}} + 2x \cos(2x)^2$

**65.** $y = \sin(\tan 2x)$

$\quad y' = \cos(\tan 2x)(\sec^2 2x)(2) = 2 \cos(\tan 2x) \sec^2 2x$

**67.** $s(t) = (t^2 + 6t - 2)^{1/2}, \quad (3, 5)$

$\quad s'(t) = \frac{1}{2}(t^2 + 6t - 2)^{-1/2}(2t + 6)$

$\quad\quad = \dfrac{t + 3}{\sqrt{t^2 + 6t - 2}}$

$\quad s'(3) = \dfrac{6}{5}$

**69.** $f(x) = \dfrac{5}{x^3 - 2} = 5(x^3 - 2)^{-1}, \quad \left(-2, -\frac{1}{2}\right)$

$\quad f'(x) = -5(x^3 - 2)^{-2}(3x^2) = \dfrac{-15x^2}{(x^3 - 2)^2}$

$\quad f'(-2) = -\dfrac{60}{100} = -\dfrac{3}{5}$

**71.** $f(t) = \dfrac{3t + 2}{t - 1}, \quad (0, -2)$

$\quad f'(t) = \dfrac{(t - 1)(3) - (3t + 2)(1)}{(t - 1)^2} = \dfrac{-5}{(t - 1)^2}$

$\quad f'(0) = -5$

**73.** $\quad y = 26 - \sec^3 4x, \quad (0, 25)$

$\quad\quad y' = -3 \sec^2 4x \sec 4x \tan 4x \cdot 4$

$\quad\quad\quad = -12 \sec^3 4x \tan 4x$

$\quad y'(0) = 0$

**75.** (a) $f(x) = (2x^2 - 7)^{1/2}$, $(4, 5)$

$$f'(x) = \frac{1}{2}(2x^2 - 7)^{-1/2}(4x) = \frac{2x}{\sqrt{2x^2 - 7}}$$

$$f'(4) = \frac{8}{5}$$

Tangent line:

$$y - 5 = \frac{8}{5}(x - 4) \Rightarrow 8x - 5y - 7 = 0$$

(b)

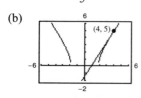

**77.** (a) $y = (4x^3 + 3)^2$, $(-1, 1)$

$$y' = 2(4x^3 + 3)(12x^2) = 24x^2(4x^3 + 3)$$

$$y'(-1) = -24$$

Tangent line:
$$y - 1 = -24(x + 1) \Rightarrow 24x + y + 23 = 0$$

(b)

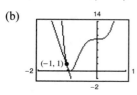

**79.** (a) $f(x) = \sin 2x$, $(\pi, 0)$

$$f'(x) = 2 \cos 2x$$

$$f'(\pi) = 2$$

Tangent line:
$$y = 2(x - \pi) \Rightarrow 2x - y - 2\pi = 0$$

(b)

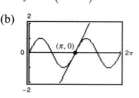

**81.** (a) $f(x) = \tan^2 x$, $\left(\frac{\pi}{4}, 1\right)$

$$f'(x) = 2 \tan x \sec^2 x$$

$$f'\left(\frac{\pi}{4}\right) = 2(1)(2) = 4$$

Tangent line:

$$y - 1 = 4\left(x - \frac{\pi}{4}\right) \Rightarrow 4x - y + (1 - \pi) = 0$$

(b)

**83.** (a) $g(t) = \frac{3t^2}{\sqrt{t^2 + 2t - 1}}$, $\left(\frac{1}{2}, \frac{3}{2}\right)$

$$g'(t) = \frac{3t(t^2 + 3t - 2)}{(t^2 + 2t - 1)^{3/2}}$$

$$g'\left(\frac{1}{2}\right) = -3$$

(b) $y - \frac{3}{2} = -3\left(x - \frac{1}{2}\right) \Rightarrow 3x + y - 3 = 0$

(c)

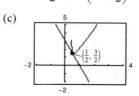

**85.** (a) $s(t) = \frac{(4 - 2t)\sqrt{1 + t}}{3}$, $\left(0, \frac{4}{3}\right)$

$$s'(t) = \frac{-2\sqrt{1 + t}}{3} + \frac{2 - t}{3\sqrt{1 + t}}$$

$$s'(0) = 0$$

(b) $y - \frac{4}{3} = 0(x - 0)$

$$y = \frac{4}{3}$$

(c)

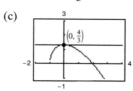

**87.** $f(x) = \sqrt{25 - x^2}$, $(3, 4)$

$$f'(x) = \frac{-x}{\sqrt{25 - x^2}}$$

$$f'(3) = -\frac{3}{4}$$

Tangent line:

$$y - 4 = -\frac{3}{4}(x - 3) \Rightarrow 3x + 4y - 25 = 0$$

**89.**
$$f(x) = 2\cos x + \sin 2x, \quad 0 < x < 2\pi$$
$$f'(x) = -2\sin x + 2\cos 2x$$
$$= -2\sin x + 2 - 4\sin^2 x = 0$$
$$2\sin^2 x + \sin x - 1 = 0$$
$$(\sin x + 1)(2\sin x - 1) = 0$$
$$\sin x = -1 \Rightarrow x = \frac{3\pi}{2}$$
$$\sin x = \frac{1}{2} \Rightarrow x = \frac{\pi}{6}, \frac{5\pi}{6}$$

Horizontal tangents at $x = \dfrac{\pi}{6}, \dfrac{3\pi}{2}, \dfrac{5\pi}{6}$

Horizontal tangent at the points $\left(\dfrac{\pi}{6}, \dfrac{3\sqrt{3}}{2}\right)$, $\left(\dfrac{3\pi}{2}, 0\right)$, and $\left(\dfrac{5\pi}{6}, -\dfrac{3\sqrt{3}}{2}\right)$

**91.** $f(x) = 5(2 - 7x)^4$

$f'(x) = 20(2 - 7x)^3(-7) = -140(2 - 7x)^3$

$f''(x) = -420(2 - 7x)^2(-7) = 2940(2 - 7x)^2$

**93.** $f(x) = \dfrac{1}{x - 6} = (x - 6)^{-1}$

$f'(x) = -(x - 6)^{-2}$

$f''(x) = 2(x - 6)^{-3} = \dfrac{2}{(x - 6)^3}$

**95.** $f(x) = \sin x^2$

$f'(x) = 2x\cos x^2$

$f''(x) = 2x\left[2x(-\sin x^2)\right] + 2\cos x^2$

$\phantom{f''(x)} = 2(\cos x^2 - 2x^2 \sin x^2)$

**97.** $h(x) = \frac{1}{9}(3x + 1)^3, \quad \left(1, \frac{64}{9}\right)$

$h'(x) = \frac{1}{9}3(3x + 1)^2(3) = (3x + 1)^2$

$h''(x) = 2(3x + 1)(3) = 18x + 6$

$h''(1) = 24$

**99.** $f(x) = \cos x^2, \quad (0, 1)$

$f'(x) = -\sin(x^2)(2x) = -2x\sin(x^2)$

$f''(x) = -2x\cos(x^2)(2x) - 2\sin(x^2)$

$\phantom{f''(x)} = -4x^2\cos(x^2) - 2\sin(x^2)$

$f''(0) = 0$

**101.**

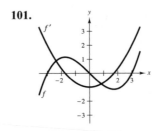

The zeros of $f'$ correspond to the points where the graph of $f$ has horizontal tangents.

**103.**

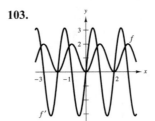

The zeros of $f'$ correspond to the points where the graph of $f$ has horizontal tangents.

**105.** $g(x) = f(3x)$

$g'(x) = f'(3x)(3) \Rightarrow g'(x) = 3f'(3x)$

**107.** (a) $g(x) = f(x) - 2 \Rightarrow g'(x) = f'(x)$

(b) $h(x) = 2f(x) \Rightarrow h'(x) = 2f'(x)$

(c) $r(x) = f(-3x) \Rightarrow r'(x) = f'(-3x)(-3) = -3f'(-3x)$

So, you need to know $f'(-3x)$.

$r'(0) = -3f'(0) = (-3)\left(-\frac{1}{3}\right) = 1$

$r'(-1) = -3f'(3) = (-3)(-4) = 12$

(d) $s(x) = f(x + 2) \Rightarrow s'(x) = f'(x + 2)$

So, you need to know $f'(x + 2)$.

$s'(-2) = f'(0) = -\frac{1}{3}$, etc.

| $x$ | $-2$ | $-1$ | $0$ | $1$ | $2$ | $3$ |
|---|---|---|---|---|---|---|
| $f'(x)$ | $4$ | $\frac{2}{3}$ | $-\frac{1}{3}$ | $-1$ | $-2$ | $-4$ |
| $g'(x)$ | $4$ | $\frac{2}{3}$ | $-\frac{1}{3}$ | $-1$ | $-2$ | $-4$ |
| $h'(x)$ | $8$ | $\frac{4}{3}$ | $-\frac{2}{3}$ | $-2$ | $-4$ | $-8$ |
| $r'(x)$ | | $12$ | $1$ | | | |
| $s'(x)$ | $-\frac{1}{3}$ | $-1$ | $-2$ | $-4$ | | |

**109.** (a) $h(x) = f(g(x)), g(1) = 4, g'(1) = -\frac{1}{2}, f'(4) = -1$

$h'(x) = f'(g(x))g'(x)$

$h'(1) = f'(g(1))g'(1) = f'(4)g'(1) = (-1)\left(-\frac{1}{2}\right) = \frac{1}{2}$

(b) $s(x) = g(f(x)), f(5) = 6, f'(5) = -1, g'(6)$ does not exist.

$s'(x) = g'(f(x))f'(x)$

$s'(5) = g'(f(5))f'(5) = g'(6)(-1)$

$s'(5)$ does not exist because $g$ is not differentiable at 6.

**111.** (a) $F = 132{,}400(331 - v)^{-1}$

$F' = (-1)(132{,}400)(331 - v)^{-2}(-1) = \dfrac{132{,}400}{(331 - v)^2}$

When $v = 30$, $F' \approx 1.461$.

(b) $F = 132{,}400(331 + v)^{-1}$

$F' = (-1)(132{,}400)(331 + v)^{-2}(-1) = \dfrac{-132{,}400}{(331 + v)^2}$

When $v = 30$, $F' \approx -1.016$.

**113.** $\theta = 0.2 \cos 8t$

The maximum angular displacement is $\theta = 0.2$ (because $-1 \le \cos 8t \le 1$).

$\dfrac{d\theta}{dt} = 0.2[-8 \sin 8t] = -1.6 \sin 8t$

When $t = 3$, $d\theta/dt = -1.6 \sin 24 \approx 1.4489$ rad/sec.

**115.** $S = C(R^2 - r^2)$

$\dfrac{dS}{dt} = C\left(2R\dfrac{dR}{dt} - 2r\dfrac{dr}{dt}\right)$

Because $r$ is constant, you have $dr/dt = 0$ and

$\dfrac{dS}{dt} = (1.76 \times 10^5)(2)(1.2 \times 10^{-2})(10^{-5})$

$= 4.224 \times 10^{-2} = 0.04224$ cm/sec.

**117.** (a) $x = -1.637t^3 + 19.31t^2 - 0.5t - 1$

(b) $C = 60x + 1350$

$= 60(-1.637t^3 + 19.31t^2 - 0.5t - 1) + 1350$

$\dfrac{dC}{dt} = 60(-4.911t^2 + 38.62t - 0.5)$

$= -294.66t^2 + 2317.2t - 30$

(c) The function $dC/dt$ is quadratic, not linear. The cost function levels off at the end of the day, perhaps due to fatigue.

**119.** (a) Yes, if $f(x + p) = f(x)$ for all $x$, then $f'(x + p) = f'(x)$, which shows that $f'$ is periodic as well.

(b) Yes, if $g(x) = f(2x)$, then $g'(x) = 2f'(2x)$. Because $f'$ is periodic, so is $g'$.

**121.** (a) $g(x) = \sin^2 x + \cos^2 x = 1 \Rightarrow g'(x) = 0$

$g'(x) = 2 \sin x \cos x + 2 \cos x(-\sin x) = 0$

(b) $\tan^2 x + 1 = \sec^2 x$

$g(x) + 1 = f(x)$

Taking derivatives of both sides, $g'(x) = f'(x)$.

Equivalently,

$f'(x) = 2 \sec x \cdot \sec x \tan x = 2 \sec^2 x \tan x$ and

$g'(x) = 2 \tan x \cdot \sec^2 x = 2 \sec^2 x \tan x$, which are the same.

**123.** $|u| = \sqrt{u^2}$

$$\frac{d}{dx}[|u|] = \frac{d}{dx}\left[\sqrt{u^2}\right] = \frac{1}{2}(u^2)^{-1/2}(2uu')$$

$$= \frac{uu'}{\sqrt{u^2}} = u'\frac{u}{|u|}, \quad u \neq 0$$

**125.** $f(x) = |x^2 - 9|$

$$f'(x) = 2x\left(\frac{x^2 - 9}{|x^2 - 9|}\right), \quad x \neq \pm 3$$

**127.** $f(x) = |\sin x|$

$$f'(x) = \cos x\left(\frac{\sin x}{|\sin x|}\right), x \neq k\pi$$

**129.** (a) $f(x) = \sec x$ $\qquad\qquad$ $f(\pi/6) = \dfrac{2}{\sqrt{3}}$

$\qquad$ $f'(x) = \sec x \tan x$ $\qquad\qquad$ $f'(\pi/6) = \dfrac{2}{3}$

$\qquad$ $f''(x) = \sec x(\sec^2 x) + \tan x(\sec x \tan x)$ $\qquad$ $f''(\pi/6) = \dfrac{10\sqrt{3}}{9}$

$\qquad\qquad$ $= \sec^3 x + \sec x \tan^2 x$

$\qquad$ $P_1(x) = \dfrac{2}{3}(x - \pi/6) + \dfrac{2}{\sqrt{3}}$

$\qquad$ $P_2(x) = \dfrac{1}{2} \cdot \left(\dfrac{10}{3\sqrt{3}}\right)\left(x - \dfrac{\pi}{6}\right)^2 + \dfrac{2}{3}\left(x - \dfrac{\pi}{6}\right) + \dfrac{2}{\sqrt{3}}$

$\qquad\qquad$ $= \left(\dfrac{5}{3\sqrt{3}}\right)\left(x - \dfrac{\pi}{6}\right)^2 + \dfrac{2}{3}\left(x - \dfrac{\pi}{6}\right) + \dfrac{2}{\sqrt{3}}$

(b)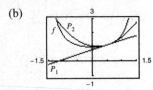

(c) $P_2$ is a better approximation than $P_1$.

(d) The accuracy worsens as you move away from $x = \pi/6$.

**131.** False. If $f(x) = \sin^2 2x$, then $f'(x) = 2(\sin 2x)(2\cos 2x)$.

**133.**
$$f(x) = a_1 \sin x + a_2 \sin 2x + \cdots + a_n \sin nx$$
$$f'(x) = a_1 \cos x + 2a_2 \cos 2x + \cdots + na_n \cos nx$$
$$f'(0) = a_1 + 2a_2 + \cdots + na_n$$

$$|a_1 + 2a_2 + \cdots + na_n| = |f'(0)| = \lim_{x\to 0}\left|\frac{f(x) - f(0)}{x - 0}\right| = \lim_{x\to 0}\left|\frac{f(x)}{\sin x}\right| \cdot \left|\frac{\sin x}{x}\right| = \lim_{x\to 0}\left|\frac{f(x)}{\sin x}\right| \leq 1$$

# Section 2.5 Implicit Differentiation

**1.** $x^2 + y^2 = 9$

$\qquad$ $2x + 2yy' = 0$

$\qquad\qquad$ $y' = -\dfrac{x}{y}$

**3.** $\qquad$ $x^{1/2} + y^{1/2} = 16$

$\qquad$ $\dfrac{1}{2}x^{-1/2} + \dfrac{1}{2}y^{-1/2}y' = 0$

$\qquad\qquad$ $y' = -\dfrac{x^{-1/2}}{y^{-1/2}}$

$\qquad\qquad$ $= -\sqrt{\dfrac{y}{x}}$

**5.** 
$$x^3 - xy + y^2 = 7$$
$$3x^2 - xy' - y + 2yy' = 0$$
$$(2y - x)y' = y - 3x^2$$
$$y' = \frac{y - 3x^2}{2y - x}$$

**7.** 
$$x^3y^3 - y - x = 0$$
$$3x^3y^2y' + 3x^2y^3 - y' - 1 = 0$$
$$(3x^3y^2 - 1)y' = 1 - 3x^2y^3$$
$$y' = \frac{1 - 3x^2y^3}{3x^3y^2 - 1}$$

**9.** 
$$x^3 - 3x^2y + 2xy^2 = 12$$
$$3x^2 - 3x^2y' - 6xy + 4xyy' + 2y^2 = 0$$
$$(4xy - 3x^2)y' = 6xy - 3x^2 - 2y^2$$
$$y' = \frac{6xy - 3x^2 - 2y^2}{4xy - 3x^2}$$

**11.** 
$$\sin x + 2\cos 2y = 1$$
$$\cos x - 4(\sin 2y)y' = 0$$
$$y' = \frac{\cos x}{4 \sin 2y}$$

**13.** 
$$\sin x = x(1 + \tan y)$$
$$\cos x = x(\sec^2 y)y' + (1 + \tan y)(1)$$
$$y' = \frac{\cos x - \tan y - 1}{x \sec^2 y}$$

**15.** 
$$y = \sin xy$$
$$y' = [xy' + y]\cos(xy)$$
$$y' - x\cos(xy)y' = y\cos(xy)$$
$$y' = \frac{y\cos(xy)}{1 - x\cos(xy)}$$

**17.** (a) $x^2 + y^2 = 64$
$$y^2 = 64 - x^2$$
$$y = \pm\sqrt{64 - x^2}$$

(b)

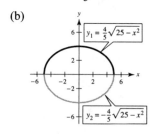

(c) Explicitly:
$$\frac{dy}{dx} = \pm\frac{1}{2}(64 - x^2)^{-1/2}(-2x) = \frac{\mp x}{\sqrt{64 - x^2}}$$
$$= \frac{-x}{\pm\sqrt{64 - x^2}} = -\frac{x}{y}$$

(d) Implicitly: $2x + 2yy' = 0$
$$y' = -\frac{x}{y}$$

**19.** (a) $16y^2 = 400 - 25x^2$
$$y^2 = \frac{1}{25}(400 - 16x^2) = \frac{16}{25}(25 - x^2)$$
$$y = \pm\frac{4}{5}\sqrt{25 - x^2}$$

(b)

(c) Explicitly:
$$\frac{dy}{dx} = \pm\frac{4}{10}(25 - x^2)^{-1/2}(-2x)$$
$$= \mp\frac{4x}{5\sqrt{25 - x^2}} = \frac{-4x}{5(5/4)y} = -\frac{16x}{25y}$$

(d) Implicitly: $32x + 50yy' = 0$
$$y' = -\frac{16x}{25y}$$

**21.** 
$$xy = 6$$
$$xy' + y(1) = 0$$
$$xy' = -y$$
$$y' = -\frac{y}{x}$$

At $(-6, -1)$: $y' = -\frac{1}{6}$

**23.** 
$$y^2 = \frac{x^2 - 49}{x^2 + 49}$$
$$2yy' = \frac{(x^2 + 49)(2x) - (x^2 - 49)(2x)}{(x^2 + 4)^2}$$
$$2yy' = \frac{196x}{(x^2 + 49)^2}$$
$$y' = \frac{98x}{y(x^2 + 49)^2}$$

At $(7, 0)$: $y'$ is undefined.

**25.** $x^{2/3} + y^{2/3} = 5$

$\dfrac{2}{3}x^{-1/3} + \dfrac{2}{3}y^{-1/3}y' = 0$

$y' = \dfrac{-x^{-1/3}}{y^{-1/3}} = -\sqrt[3]{\dfrac{y}{x}}$

At $(8, 1)$: $y' = -\dfrac{1}{2}$

**27.** $\tan(x + y) = x$

$(1 + y')\sec^2(x + y) = 1$

$y' = \dfrac{1 - \sec^2(x + y)}{\sec^2(x + y)}$

$= \dfrac{-\tan^2(x + y)}{\tan^2(x + y) + 1}$

$= -\sin^2(x + y)$

$= -\dfrac{x^2}{x^2 + 1}$

At $(0, 0)$: $y' = 0$

**29.** $(x^2 + 4)y = 8$

$(x^2 + 4)y' + y(2x) = 0$

$y' = \dfrac{-2xy}{x^2 + 4} = \dfrac{-2x\left[8/(x^2 + 4)\right]}{x^2 + 4}$

$= \dfrac{-16x}{(x^2 + 4)^2}$

At $(2, 1)$: $y' = \dfrac{-32}{64} = -\dfrac{1}{2}$

$\left(\text{Or, you could just solve for } y\text{: } y = \dfrac{8}{x^2 + 4}\right)$

**31.** $(x^2 + y^2)^2 = 4x^2y$

$2(x^2 + y^2)(2x + 2yy') = 4x^2y' + y(8x)$

$4x^3 + 4x^2yy' + 4xy^2 + 4y^3y' = 4x^2y' + 8xy$

$4x^2yy' + 4y^3y' - 4x^2y' = 8xy - 4x^3 - 4xy^2$

$4y'(x^2y + y^3 - x^2) = 4(2xy - x^3 - xy^2)$

$y' = \dfrac{2xy - x^3 - xy^2}{x^2y + y^3 - x^2}$

At $(1, 1)$: $y' = 0$

**33.** $(y - 3)^2 = 4(x - 5)$, $(6, 1)$

$2(y - 3)y' = 4$

$y' = \dfrac{2}{y - 3}$

At $(6, 1)$, $y' = \dfrac{2}{1 - 3} = -1$

Tangent line: $y - 1 = -1(x - 6)$

$y = -x + 7$

**35.** $xy = 1$, $(1, 1)$

$xy' + y = 0$

$y' = \dfrac{-y}{x}$

At $(1, 1)$: $y' = -1$

Tangent line: $y - 1 = -1(x - 1)$

$y = -x + 2$

**37.** $x^2y^2 - 9x^2 - 4y^2 = 0$, $\left(-4, 2\sqrt{3}\right)$

$x^2 2yy' + 2xy^2 - 18x - 8yy' = 0$

$y' = \dfrac{18x - 2xy^2}{2x^2y - 8y}$

At $\left(-4, 2\sqrt{3}\right)$: $y' = \dfrac{18(-4) - 2(-4)(12)}{2(16)\left(2\sqrt{3}\right) - 16\sqrt{3}}$

$= \dfrac{24}{48\sqrt{3}} = \dfrac{1}{2\sqrt{3}} = \dfrac{\sqrt{3}}{6}$

Tangent line: $y - 2\sqrt{3} = \dfrac{\sqrt{3}}{6}(x + 4)$

$y = \dfrac{\sqrt{3}}{6}x + \dfrac{8}{3}\sqrt{3}$

**39.** $3(x^2 + y^2)^2 = 100(x^2 - y^2)$, $(4, 2)$

$6(x^2 + y^2)(2x + 2yy') = 100(2x - 2yy')$

At $(4, 2)$:

$6(16 + 4)(8 + 4y') = 100(8 - 4y')$

$960 + 480y' = 800 - 400y'$

$880y' = -160$

$y' = -\dfrac{2}{11}$

Tangent line: $y - 2 = -\dfrac{2}{11}(x - 4)$

$11y + 2x - 30 = 0$

$y = -\dfrac{2}{11}x + \dfrac{30}{11}$

**41. (a)** $\dfrac{x^2}{2} + \dfrac{y^2}{8} = 1, \quad (1, 2)$

$$x + \dfrac{yy'}{4} = 0$$

$$y' = -\dfrac{4x}{y}$$

At $(1, 2)$: $y' = -2$

Tangent line: $y - 2 = -2(x - 1)$

$$y = -2x + 4$$

**(b)** $\dfrac{x^2}{a^2} + \dfrac{y^2}{b^2} = 1 \Rightarrow \dfrac{2x}{a^2} + \dfrac{2yy'}{b^2} = 0 \Rightarrow y' = \dfrac{-b^2 x}{a^2 y}$

$$y - y_0 = \dfrac{-b^2 x_0}{a^2 y_0}(x - x_0), \text{ Tangent line at } (x_0, y_0)$$

$$\dfrac{y_0 y}{b^2} - \dfrac{y_0^2}{b^2} = \dfrac{-x_0 x}{a^2} + \dfrac{x_0^2}{a^2}$$

Because $\dfrac{x_0^2}{a^2} + \dfrac{y_0^2}{b^2} = 1$, you have $\dfrac{y_0 y}{b^2} + \dfrac{x_0 x}{a^2} = 1$.

**Note:** From part (a),

$$\dfrac{1(x)}{2} + \dfrac{2(y)}{8} = 1 \Rightarrow \dfrac{1}{4}y = -\dfrac{1}{2}x + 1 \Rightarrow y = -2x + 4,$$

Tangent line.

**43.** $\tan y = x$

$y' \sec^2 y = 1$

$$y' = \dfrac{1}{\sec^2 y} = \cos^2 y, \quad -\dfrac{\pi}{2} < y < \dfrac{\pi}{2}$$

$\sec^2 y = 1 + \tan^2 y = 1 + x^2$

$$y' = \dfrac{1}{1 + x^2}$$

**45.** $x^2 + y^2 = 4$

$2x + 2yy' = 0$

$$y' = \dfrac{-x}{y}$$

$$y'' = \dfrac{y(-1) + xy'}{y^2} = \dfrac{-y + x(-x/y)}{y^2}$$

$$= \dfrac{-y^2 - x^2}{y^3} = -\dfrac{4}{y^3}$$

**47.** $x^2 - y^2 = 36$

$2x - 2yy' = 0$

$$y' = \dfrac{x}{y}$$

$x - yy' = 0$

$1 - yy'' - (y')^2 = 0$

$$1 - yy'' - \left(\dfrac{x}{y}\right)^2 = 0$$

$y^2 - y^3 y'' = x^2$

$$y'' = \dfrac{y^2 - x^2}{y^3} = -\dfrac{36}{y^3}$$

**49.** $y^2 = x^3$

$2yy' = 3x^2$

$$y' = \dfrac{3x^2}{2y} = \dfrac{3x^2}{2y} \cdot \dfrac{xy}{xy} = \dfrac{3y}{2x} \cdot \dfrac{x^3}{y^2} = \dfrac{3y}{2x}$$

$$y'' = \dfrac{2x(3y') - 3y(2)}{4x^2}$$

$$= \dfrac{2x[3 \cdot (3y/2x)] - 6y}{4x^2} = \dfrac{3y}{4x^2} = \dfrac{3x}{4y}$$

**51.** $\sqrt{x} + \sqrt{y} = 5$

$$\dfrac{1}{2}x^{-1/2} + \dfrac{1}{2}y^{-1/2}y' = 0$$

$$y' = \dfrac{-\sqrt{y}}{\sqrt{x}}$$

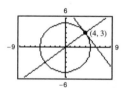

At $(9, 4)$: $y' = -\dfrac{2}{3}$

Tangent line: $\quad y - 4 = -\dfrac{2}{3}(x - 9)$

$$2x + 3y - 30 = 0$$

**53.** $x^2 + y^2 = 25$

$2x + 2yy' = 0$

$$y' = \dfrac{-x}{y}$$

At $(4, 3)$:

Tangent line:

$$y - 3 = \dfrac{-4}{3}(x - 4) \Rightarrow 4x + 3y - 25 = 0$$

Normal line: $y - 3 = \dfrac{3}{4}(x - 4) \Rightarrow 3x - 4y = 0$

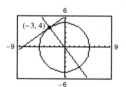

At $(-3, 4)$:

Tangent line:

$$y - 4 = \dfrac{3}{4}(x + 3) \Rightarrow 3x - 4y + 25 = 0$$

Normal line: $y - 4 = \dfrac{-4}{3}(x + 3) \Rightarrow 4x + 3y = 0$

**55.** $x^2 + y^2 = r^2$

$2x + 2yy' = 0$

$y' = \dfrac{-x}{y}$ = slope of tangent line

$\dfrac{y}{x}$ = slope of normal line

Let $(x_0, y_0)$ be a point on the circle. If $x_0 = 0$, then the tangent line is horizontal, the normal line is vertical and, hence, passes through the origin. If $x_0 \neq 0$, then the equation of the normal line is

$y - y_0 = \dfrac{y_0}{x_0}(x - x_0)$

$y = \dfrac{y_0}{x_0}x$

which passes through the origin.

**57.** $25x^2 + 16y^2 + 200x - 160y + 400 = 0$

$50x + 32yy' + 200 - 160y' = 0$

$y' = \dfrac{200 + 50x}{160 - 32y}$

Horizontal tangents occur when $x = -4$:

$25(16) + 16y^2 + 200(-4) - 160y + 400 = 0$

$y(y - 10) = 0 \Rightarrow y = 0, 10$

Horizontal tangents: $(-4, 0), (-4, 10)$

Vertical tangents occur when $y = 5$:

$25x^2 + 400 + 200x - 800 + 400 = 0$

$25x(x + 8) = 0 \Rightarrow x = 0, -8$

Vertical tangents: $(0, 5), (-8, 5)$

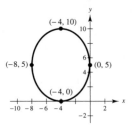

**59.** Find the points of intersection by letting $y^2 = 4x$ in the equation $2x^2 + y^2 = 6$.

$2x^2 + 4x = 6$ and $(x + 3)(x - 1) = 0$

The curves intersect at $(1, \pm 2)$.

| *Ellipse*: | *Parabola*: |
|---|---|
| $4x + 2yy' = 0$ | $2yy' = 4$ |
| $y' = -\dfrac{2x}{y}$ | $y' = \dfrac{2}{y}$ |

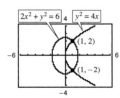

At $(1, 2)$, the slopes are:

$y' = -1 \qquad\qquad y' = 1$

At $(1, -2)$, the slopes are:

$y' = 1 \qquad\qquad y' = -1$

Tangents are perpendicular.

**61.** $y = -x$ and $x = \sin y$

Point of intersection: $(0, 0)$

| $y = -x$: | $x = \sin y$: |
|---|---|
| $y' = -1$ | $1 = y'\cos y$ |
| | $y' = \sec y$ |

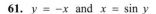

At $(0, 0)$, the slopes are:

$y' = -1 \qquad\qquad y' = 1$

Tangents are perpendicular.

**63.**     $xy = C$          $x^2 - y^2 = K$

$xy' + y = 0$      $2x - 2yy' = 0$

$$y' = -\frac{y}{x} \qquad\qquad y' = \frac{x}{y}$$

At any point of intersection $(x, y)$ the product of the slopes is $(-y/x)(x/y) = -1$.

The curves are orthogonal.

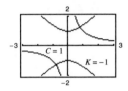 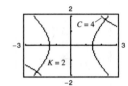

**65.** $2y^2 - 3x^4 = 0$

(a)   $4yy' - 12x^3 = 0$

$$4yy' = 12x^3$$

$$y' = \frac{12x^3}{4y} = \frac{3x^3}{y}$$

(b)   $4y\dfrac{dy}{dt} - 12x^3\dfrac{dx}{dt} = 0$

$$y\frac{dy}{dt} = 3x^3\frac{dx}{dt}$$

**67.** $\cos \pi y - 3 \sin \pi x = 1$

(a)   $-\pi \sin \pi y (y') - 3\pi \cos \pi x = 0$

$$y' = \frac{-3 \cos \pi x}{\sin \pi y}$$

(b)   $-\pi \sin \pi y \left(\dfrac{dy}{dt}\right) - 3\pi \cos \pi x \left(\dfrac{dx}{dt}\right) = 0$

$$-\sin \pi y \left(\frac{dy}{dt}\right) = 3 \cos \pi x \left(\frac{dx}{dt}\right)$$

**69.** Answers will vary. *Sample answer:* In the explicit form of a function, the variable is explicitly written as a function of $x$. In an implicit equation, the function is only implied by an equation. An example of an implicit function is $x^2 + xy = 5$. In explicit form it would be $y = \left(5 - x^2\right)\big/x$.

**71.**

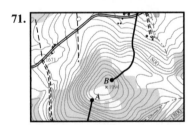

Use starting point $B$.

**73. (a)** $x^4 = 4(4x^2 - y^2)$

$4y^2 = 16x^2 - x^4$

$y^2 = 4x^2 - \dfrac{1}{4}x^4$

$y = \pm\sqrt{4x^2 - \dfrac{1}{4}x^4}$

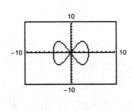

**(b)** $y = 3 \Rightarrow 9 = 4x^2 - \dfrac{1}{4}x^4$

$36 = 16x^2 - x^4$

$x^4 - 16x^2 + 36 = 0$

$x^2 = \dfrac{16 \pm \sqrt{256 - 144}}{2} = 8 \pm \sqrt{28}$

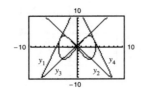

Note that $x^2 = 8 \pm \sqrt{28} = 8 \pm 2\sqrt{7} = \left(1 \pm \sqrt{7}\right)^2$. So, there are four values of $x$:

$-1 - \sqrt{7}, 1 - \sqrt{7}, -1 + \sqrt{7}, 1 + \sqrt{7}$

To find the slope, $2yy' = 8x - x^3 \Rightarrow y' = \dfrac{x(8 - x^2)}{2(3)}$.

For $x = -1 - \sqrt{7}$, $y' = \dfrac{1}{3}\left(\sqrt{7} + 7\right)$, and the line is

$y_1 = \dfrac{1}{3}\left(\sqrt{7} + 7\right)\left(x + 1 + \sqrt{7}\right) + 3 = \dfrac{1}{3}\left[\left(\sqrt{7} + 7\right)x + 8\sqrt{7} + 23\right].$

For $x = 1 - \sqrt{7}$, $y' = \dfrac{1}{3}\left(\sqrt{7} - 7\right)$, and the line is

$y_2 = \dfrac{1}{3}\left(\sqrt{7} - 7\right)\left(x - 1 + \sqrt{7}\right) + 3 = \dfrac{1}{3}\left[\left(\sqrt{7} - 7\right)x + 23 - 8\sqrt{7}\right].$

For $x = -1 + \sqrt{7}$, $y' = -\dfrac{1}{3}\left(\sqrt{7} - 7\right)$, and the line is

$y_3 = -\dfrac{1}{3}\left(\sqrt{7} - 7\right)\left(x + 1 - \sqrt{7}\right) + 3 = -\dfrac{1}{3}\left[\left(\sqrt{7} - 7\right)x - \left(23 - 8\sqrt{7}\right)\right].$

For $x = 1 + \sqrt{7}$, $y' = -\dfrac{1}{3}\left(\sqrt{7} + 7\right)$, and the line is

$y_4 = -\dfrac{1}{3}\left(\sqrt{7} + 7\right)\left(x - 1 - \sqrt{7}\right) + 3 = -\dfrac{1}{3}\left[\left(\sqrt{7} + 7\right)x - \left(8\sqrt{7} + 23\right)\right].$

**(c)** Equating $y_3$ and $y_4$:

$-\dfrac{1}{3}\left(\sqrt{7} - 7\right)\left(x + 1 - \sqrt{7}\right) + 3 = -\dfrac{1}{3}\left(\sqrt{7} + 7\right)\left(x - 1 - \sqrt{7}\right) + 3$

$\left(\sqrt{7} - 7\right)\left(x + 1 - \sqrt{7}\right) = \left(\sqrt{7} + 7\right)\left(x - 1 - \sqrt{7}\right)$

$\sqrt{7}x + \sqrt{7} - 7 - 7x - 7 + 7\sqrt{7} = \sqrt{7}x - \sqrt{7} - 7 + 7x - 7 - 7\sqrt{7}$

$16\sqrt{7} = 14x$

$x = \dfrac{8\sqrt{7}}{7}$

If $x = \dfrac{8\sqrt{7}}{7}$, then $y = 5$ and the lines intersect at $\left(\dfrac{8\sqrt{7}}{7}, 5\right)$.

**75.**    $\sqrt{x} + \sqrt{y} = \sqrt{c}$

$$\frac{1}{2\sqrt{x}} + \frac{1}{2\sqrt{y}}\frac{dy}{dx} = 0$$

$$\frac{dy}{dx} = -\frac{\sqrt{y}}{\sqrt{x}}$$

Tangent line at $(x_0, y_0)$: $y - y_0 = -\dfrac{\sqrt{y_0}}{\sqrt{x_0}}(x - x_0)$

$x$-intercept: $\left(x_0 + \sqrt{x_0}\sqrt{y_0},\, 0\right)$

$y$-intercept: $\left(0,\, y_0 + \sqrt{x_0}\sqrt{y_0}\right)$

Sum of intercepts:

$$\left(x_0 + \sqrt{x_0}\sqrt{y_0}\right) + \left(y_0 + \sqrt{x_0}\sqrt{y_0}\right) = x_0 + 2\sqrt{x_0}\sqrt{y_0} + y_0 = \left(\sqrt{x_0} + \sqrt{y_0}\right)^2 = \left(\sqrt{c}\right)^2 = c$$

**77.**    $x^2 + y^2 = 100$, slope $= \dfrac{3}{4}$

$$2x + 2yy' = 0$$

$$y' = -\frac{x}{y} = \frac{3}{4} \Rightarrow y = -\frac{4}{3}x$$

$$x^2 + \left(\frac{16}{9}x^2\right) = 100$$

$$\frac{25}{9}x^2 = 100$$

$$x = \pm 6$$

Points: $(6, -8)$ and $(-6, 8)$

**79.**    $\dfrac{x^2}{4} + \dfrac{y^2}{9} = 1$,   $(4, 0)$

$$\frac{2x}{4} + \frac{2yy'}{9} = 0$$

$$y' = \frac{-9x}{4y}$$

$$\frac{-9x}{4y} = \frac{y - 0}{x - 4}$$

$$-9x(x - 4) = 4y^2$$

But, $9x^2 + 4y^2 = 36 \Rightarrow 4y^2 = 36 - 9x^2$. So,

$-9x^2 + 36x = 4y^2 = 36 - 9x^2 \Rightarrow x = 1$.

Points on ellipse: $\left(1, \pm\dfrac{3}{2}\sqrt{3}\right)$

At $\left(1, \dfrac{3}{2}\sqrt{3}\right)$: $y' = \dfrac{-9x}{4y} = \dfrac{-9}{4\left[(3/2)\sqrt{3}\right]} = -\dfrac{\sqrt{3}}{2}$

At $\left(1, -\dfrac{3}{2}\sqrt{3}\right)$: $y' = \dfrac{\sqrt{3}}{2}$

Tangent lines: $y = -\dfrac{\sqrt{3}}{2}(x - 4) = -\dfrac{\sqrt{3}}{2}x + 2\sqrt{3}$

$$y = \frac{\sqrt{3}}{2}(x - 4) = \frac{\sqrt{3}}{2}x - 2\sqrt{3}$$

**81. (a)**    $\dfrac{x^2}{32} + \dfrac{y^2}{8} = 1$

$$\frac{2x}{32} + \frac{2yy'}{8} = 0 \Rightarrow y' = \frac{-x}{4y}$$

At $(4, 2)$: $y' = \dfrac{-4}{4(2)} = -\dfrac{1}{2}$

Slope of normal line is 2.

$$y - 2 = 2(x - 4)$$

$$y = 2x - 6$$

(b)

(c)  $\dfrac{x^2}{32} + \dfrac{(2x-6)^2}{8} = 1$

$x^2 + 4(4x^2 - 24x + 36) = 32$

$17x^2 - 96x + 112 = 0$

$(17x - 28)(x - 4) = 0 \Rightarrow x = 4, \dfrac{28}{17}$

Second point: $\left(\dfrac{28}{17}, -\dfrac{46}{17}\right)$

# Section 2.6   Related Rates

**1.**  $y = \sqrt{x}$

$\dfrac{dy}{dt} = \left(\dfrac{1}{2\sqrt{x}}\right)\dfrac{dx}{dt}$

$\dfrac{dx}{dt} = 2\sqrt{x}\,\dfrac{dy}{dt}$

(a) When $x = 4$ and $dx/dt = 3$,

$\dfrac{dy}{dt} = \dfrac{1}{2\sqrt{4}}(3) = \dfrac{3}{4}.$

(b) When $x = 25$ and $dy/dt = 2$,

$\dfrac{dx}{dt} = 2\sqrt{25}(2) = 20.$

**3.**  $xy = 4$

$x\dfrac{dy}{dt} + y\dfrac{dx}{dt} = 0$

$\dfrac{dy}{dt} = \left(-\dfrac{y}{x}\right)\dfrac{dx}{dt}$

$\dfrac{dx}{dt} = \left(-\dfrac{x}{y}\right)\dfrac{dy}{dt}$

(a) When $x = 8$, $y = 1/2$, and $dx/dt = 10$,

$\dfrac{dy}{dt} = -\dfrac{1/2}{8}(10) = -\dfrac{5}{8}.$

(b) When $x = 1$, $y = 4$, and $dy/dt = -6$,

$\dfrac{dx}{dt} = -\dfrac{1}{4}(-6) = \dfrac{3}{2}.$

**5.**  $y = 2x^2 + 1$

$\dfrac{dx}{dt} = 2$

$\dfrac{dy}{dt} = 4x\dfrac{dx}{dt}$

(a) When $x = -1$,

$\dfrac{dy}{dt} = 4(-1)(2) = -8$ cm/sec.

(b) When $x = 0$,

$\dfrac{dy}{dt} = 4(0)(2) = 0$ cm/sec.

(c) When $x = 1$,

$\dfrac{dy}{dt} = 4(1)(2) = 8$ cm/sec.

**7.**  $y = \tan x$

$\dfrac{dx}{dt} = 2$

$\dfrac{dy}{dt} = \sec^2 x\dfrac{dx}{dt}$

(a) When $x = -\pi/3$,

$\dfrac{dy}{dt} = (2)^2(2) = 8$ cm/sec.

(b) When $x = -\pi/4$,

$\dfrac{dy}{dt} = \left(\sqrt{2}\right)^2(2) = 4$ cm/sec.

(c) When $x = 0$,

$\dfrac{dy}{dt} = (1)^2(2) = 2$ cm/sec.

**9.** Yes, $y$ changes at a constant rate.

$$\frac{dy}{dt} = a \cdot \frac{dx}{dt}$$

No, the rate $dy/dt$ is a multiple of $dx/dt$.

**11.** $D = \sqrt{x^2 + y^2} = \sqrt{x^2 + (x^2 + 1)^2} = \sqrt{x^4 + 3x^2 + 1}$

$$\frac{dx}{dt} = 2$$

$$\frac{dD}{dt} = \frac{1}{2}(x^4 + 3x^2 + 1)^{-1/2}(4x^3 + 6x)\frac{dx}{dt}$$

$$= \frac{2x^3 + 3x}{\sqrt{x^4 + 3x^2 + 1}}\frac{dx}{dt}$$

$$= \frac{4x^3 + 6x}{\sqrt{x^4 + 3x^2 + 1}}$$

**13.** $A = \pi r^2$

$$\frac{dr}{dt} = 4$$

$$\frac{dA}{dt} = 2\pi r \frac{dr}{dt}$$

(a) When $r = 8$, $\dfrac{dA}{dt} = 2\pi(8)(4) = 64\pi$ cm²/min.

(b) When $r = 32$, $\dfrac{dA}{dt} = 2\pi(32)(4) = 256\pi$ cm²/min.

**15.** (a) $\sin\dfrac{\theta}{2} = \dfrac{(1/2)b}{s} \Rightarrow b = 2s\sin\dfrac{\theta}{2}$

$$\cos\frac{\theta}{2} = \frac{h}{s} \Rightarrow h = s\cos\frac{\theta}{2}$$

$$A = \frac{1}{2}bh = \frac{1}{2}\left(2s\sin\frac{\theta}{2}\right)\left(s\cos\frac{\theta}{2}\right)$$

$$= \frac{s^2}{2}\left(2\sin\frac{\theta}{2}\cos\frac{\theta}{2}\right) = \frac{s^2}{2}\sin\theta$$

(b) $\dfrac{dA}{dt} = \dfrac{s^2}{2}\cos\theta\dfrac{d\theta}{dt}$ where $\dfrac{d\theta}{dt} = \dfrac{1}{2}$ rad/min.

When $\theta = \dfrac{\pi}{6}$, $\dfrac{dA}{dt} = \dfrac{s^2}{2}\left(\dfrac{\sqrt{3}}{2}\right)\left(\dfrac{1}{2}\right) = \dfrac{\sqrt{3}s^2}{8}$.

When $\theta = \dfrac{\pi}{3}$, $\dfrac{dA}{dt} = \dfrac{s^2}{2}\left(\dfrac{1}{2}\right)\left(\dfrac{1}{2}\right) = \dfrac{s^2}{8}$.

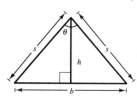

(c) If $s$ and $\dfrac{d\theta}{dt}$ is constant, $\dfrac{dA}{dt}$ is proportional to $\cos\theta$.

**17.** $V = \dfrac{4}{3}\pi r^3$, $\dfrac{dV}{dt} = 800$

$$\frac{dV}{dt} = 4\pi r^2 \frac{dr}{dt}$$

$$\frac{dr}{dt} = \frac{1}{4\pi r^2}\left(\frac{dV}{dt}\right) = \frac{1}{4\pi r^2}(800)$$

(a) When $r = 30$,

$$\frac{dr}{dt} = \frac{1}{4\pi(30)^2}(800) = \frac{2}{9\pi} \text{ cm/min.}$$

(b) When $r = 60$,

$$\frac{dr}{dt} = \frac{1}{4\pi(60)^2}(800) = \frac{1}{18\pi} \text{ cm/min.}$$

**19.** $s = 6x^2$

$$\frac{dx}{dt} = 6$$

$$\frac{ds}{dt} = 12x\frac{dx}{dt}$$

(a) When $x = 2$,

$$\frac{ds}{dt} = 12(2)(6) = 144 \text{ cm}^2/\text{sec.}$$

(b) When $x = 10$,

$$\frac{ds}{dt} = 12(10)(6) = 720 \text{ cm}^2/\text{sec.}$$

**21.** $V = \dfrac{1}{3}\pi r^2 h = \dfrac{1}{3}\pi\left(\dfrac{9}{4}h^2\right)h$   [because $2r = 3h$]

$$= \frac{3\pi}{4}h^3$$

$$\frac{dV}{dt} = 10$$

$$\frac{dV}{dt} = \frac{9\pi}{4}h^2\frac{dh}{dt} \Rightarrow \frac{dh}{dt} = \frac{4(dV/dt)}{9\pi h^2}$$

When $h = 15$,

$$\frac{dh}{dt} = \frac{4(10)}{9\pi(15)^2} = \frac{8}{405\pi} \text{ ft/min.}$$

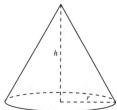

**23.**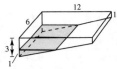

(a) Total volume of pool $= \frac{1}{2}(2)(12)(6) + (1)(6)(12) = 144 \text{ m}^3$

Volume of 1 m of water $= \frac{1}{2}(1)(6)(6) = 18 \text{ m}^3$    (see similar triangle diagram)

% pool filled $= \frac{18}{144}(100\%) = 12.5\%$

(b) Because for $0 \le h \le 2$, $b = 6h$, you have

$$V = \frac{1}{2}bh(6) = 3bh = 3(6h)h = 18h^2$$

$$\frac{dV}{dt} = 36h\frac{dh}{dt} = \frac{1}{4} \Rightarrow \frac{dh}{dt} = \frac{1}{144h} = \frac{1}{144(1)} = \frac{1}{144} \text{ m/min.}$$

**25.**    $x^2 + y^2 = 25^2$

$$2x\frac{dx}{dt} + 2y\frac{dy}{dt} = 0$$

$$\frac{dy}{dt} = \frac{-x}{y} \cdot \frac{dx}{dt} = \frac{-2x}{y} \quad \text{because } \frac{dx}{dt} = 2.$$

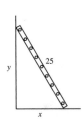

(a) When $x = 7$, $y = \sqrt{576} = 24$, $\frac{dy}{dt} = \frac{-2(7)}{24} = -\frac{7}{12} \text{ ft/sec.}$

When $x = 15$, $y = \sqrt{400} = 20$, $\frac{dy}{dt} = \frac{-2(15)}{20} = -\frac{3}{2} \text{ ft/sec.}$

When $x = 24$, $y = 7$, $\frac{dy}{dt} = \frac{-2(24)}{7} = -\frac{48}{7} \text{ ft/sec.}$

(b)    $A = \frac{1}{2}xy$

$$\frac{dA}{dt} = \frac{1}{2}\left(x\frac{dy}{dt} + y\frac{dx}{dt}\right)$$

From part (a) you have $x = 7$, $y = 24$, $\frac{dx}{dt} = 2$, and $\frac{dy}{dt} = -\frac{7}{12}$. So,

$$\frac{dA}{dt} = \frac{1}{2}\left[7\left(-\frac{7}{12}\right) + 24(2)\right] = \frac{527}{24} \text{ ft}^2/\text{sec.}$$

(c)    $\tan \theta = \frac{x}{y}$

$$\sec^2\theta \frac{d\theta}{dt} = \frac{1}{y} \cdot \frac{dx}{dt} - \frac{x}{y^2} \cdot \frac{dy}{dt}$$

$$\frac{d\theta}{dt} = \cos^2\theta\left[\frac{1}{y} \cdot \frac{dx}{dt} - \frac{x}{y^2} \cdot \frac{dy}{dt}\right]$$

Using $x = 7$, $y = 24$, $\frac{dx}{dt} = 2$, $\frac{dy}{dt} = -\frac{7}{12}$ and $\cos \theta = \frac{24}{25}$, you have

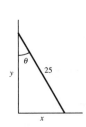

$$\frac{d\theta}{dt} = \left(\frac{24}{25}\right)^2\left[\frac{1}{24}(2) - \frac{7}{(24)^2}\left(-\frac{7}{12}\right)\right] = \frac{1}{12} \text{ rad/sec.}$$

**27.** When $y = 6$, $x = \sqrt{12^2 - 6^2} = 6\sqrt{3}$, and $s = \sqrt{x^2 + (12 - y)^2} = \sqrt{108 + 36} = 12$.

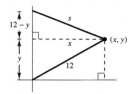

$$x^2 + (12 - y)^2 = s^2$$

$$2x\frac{dx}{dt} + 2(12 - y)(-1)\frac{dy}{dt} = 2s\frac{ds}{dt}$$

$$x\frac{dx}{dt} + (y - 12)\frac{dy}{dt} = s\frac{ds}{dt}$$

Also, $x^2 + y^2 = 12^2$.

$$2x\frac{dx}{dt} + 2y\frac{dy}{dt} = 0 \Rightarrow \frac{dy}{dt} = \frac{-x}{y}\frac{dx}{dt}$$

So, $x\frac{dx}{dt} + (y - 12)\left(\frac{-x}{y}\frac{dx}{dt}\right) = s\frac{ds}{dt}$.

$$\frac{dx}{dt}\left[x - x + \frac{12x}{y}\right] = s\frac{ds}{dt} \Rightarrow \frac{dx}{dt} = \frac{sy}{12x}\cdot\frac{ds}{dt} = \frac{(12)(6)}{(12)(6\sqrt{3})}(-0.2) = \frac{-1}{5\sqrt{3}} = \frac{-\sqrt{3}}{15}\text{ m/sec (horizontal)}$$

$$\frac{dy}{dt} = \frac{-x}{y}\frac{dx}{dt} = \frac{-6\sqrt{3}}{6}\cdot\frac{(-\sqrt{3})}{15} = \frac{1}{5}\text{ m/sec (vertical)}$$

**29. (a)**

$$s^2 = x^2 + y^2$$

$$\frac{dx}{dt} = -450$$

$$\frac{dy}{dt} = -600$$

$$2s\frac{ds}{dt} = 2x\frac{dx}{dt} + 2y\frac{dy}{dt}$$

$$\frac{ds}{dt} = \frac{x(dx/dt) + y(dy/dt)}{s}$$

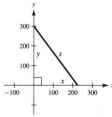

When $x = 225$ and $y = 300$, $s = 375$ and

$$\frac{ds}{dt} = \frac{225(-450) + 300(-600)}{375} = -750\text{ mi/h}.$$

**(b)** $t = \frac{375}{750} = \frac{1}{2}\text{ h} = 30\text{ min}$

**31.**

$$s^2 = 90^2 + x^2$$

$$x = 30$$

$$\frac{dx}{dt} = -25$$

$$2s\frac{ds}{dt} = 2x\frac{dx}{dt} \Rightarrow \frac{ds}{dt} = \frac{x}{s}\cdot\frac{dx}{dt}$$

When $x = 20$, $s = \sqrt{90^2 + 20^2} = 10\sqrt{85}$,

$$\frac{ds}{dt} = \frac{20}{10\sqrt{85}}(-25) = \frac{-50}{\sqrt{85}} \approx -5.42\text{ ft/sec.}$$

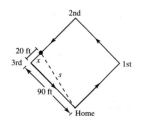

**33. (a)** $\dfrac{15}{6} = \dfrac{y}{y-x} \Rightarrow 15y - 15x = 6y$

$$y = \frac{5}{3}x$$

$$\frac{dx}{dt} = 5$$

$$\frac{dy}{dt} = \frac{5}{3} \cdot \frac{dx}{dt} = \frac{5}{3}(5) = \frac{25}{3} \text{ ft/sec}$$

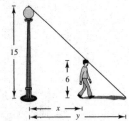

**(b)** $\dfrac{d(y-x)}{dt} = \dfrac{dy}{dt} - \dfrac{dx}{dt} = \dfrac{25}{3} - 5 = \dfrac{10}{3} \text{ ft/sec}$

**35.** $x(t) = \dfrac{1}{2} \sin \dfrac{\pi t}{6}, \ x^2 + y^2 = 1$

**(a)** Period: $\dfrac{2\pi}{\pi/6} = 12$ seconds

**(b)** When $x = \dfrac{1}{2}, \ y = \sqrt{1^2 - \left(\dfrac{1}{2}\right)^2} = \dfrac{\sqrt{3}}{2}$ m.

Lowest point: $\left(0, \dfrac{\sqrt{3}}{2}\right)$

**(c)** When

$$x = \frac{1}{4}, \quad y = \sqrt{1 - \left(\frac{1}{4}\right)^2} = \frac{\sqrt{15}}{4} \text{ and } t = 1:$$

$$\frac{dx}{dt} = \frac{1}{2}\left(\frac{\pi}{6}\right)\cos\frac{\pi t}{6} = \frac{\pi}{12}\cos\frac{\pi t}{6}$$

$$x^2 + y^2 = 1$$

$$2x\frac{dx}{dt} + 2y\frac{dy}{dt} = 0 \Rightarrow \frac{dy}{dt} = \frac{-x}{y}\frac{dx}{dt}$$

So, $\dfrac{dy}{dt} = -\dfrac{1/4}{\sqrt{15}/4} \cdot \dfrac{\pi}{12}\cos\left(\dfrac{\pi}{6}\right)$

$$= \frac{-\pi}{\sqrt{15}}\left(\frac{1}{12}\right)\frac{\sqrt{3}}{2} = \frac{-\pi}{24}\frac{1}{\sqrt{5}} = \frac{-\sqrt{5}\pi}{120}.$$

Speed $= \left|\dfrac{-\sqrt{5}\pi}{120}\right| = \dfrac{\sqrt{5}\pi}{120}$ m/sec

**37.** Because the evaporation rate is proportional to the surface area, $dV/dt = k(4\pi r^2)$. However, because $V = (4/3)\pi r^3$, you have

$$\frac{dV}{dt} = 4\pi r^2 \frac{dr}{dt}.$$

Therefore, $k(4\pi r^2) = 4\pi r^2 \dfrac{dr}{dt} \Rightarrow k = \dfrac{dr}{dt}.$

**39.** $\dfrac{1}{R} = \dfrac{1}{R_1} + \dfrac{1}{R_2}$

$$\frac{dR_1}{dt} = 1$$

$$\frac{dR_2}{dt} = 1.5$$

$$\frac{1}{R^2} \cdot \frac{dR}{dt} = \frac{1}{R_1^2} \cdot \frac{dR_1}{dt} + \frac{1}{R_2^2} \cdot \frac{dR_2}{dt}$$

When $R_1 = 50$ and $R_2 = 75$:

$R = 30$

$$\frac{dR}{dt} = (30)^2\left[\frac{1}{(50)^2}(1) + \frac{1}{(75)^2}(1.5)\right] = 0.6 \text{ ohm/sec}$$

**41.** $rg \tan \theta = v^2$

$32r \tan \theta = v^2, \quad r$ is a constant.

$$32r \sec^2\theta \frac{d\theta}{dt} = 2v\frac{dv}{dt}$$

$$\frac{dv}{dt} = \frac{16r}{v}\sec^2\theta\frac{d\theta}{dt}$$

Likewise, $\dfrac{d\theta}{dt} = \dfrac{v}{16r}\cos^2\theta\dfrac{dv}{dt}.$

**43.** $\sin\theta = \dfrac{10}{x}$

$$\frac{dx}{dt} = (-1) \text{ ft/sec}$$

$$\cos\theta\left(\frac{d\theta}{dt}\right) = \frac{-10}{x^2} \cdot \frac{dx}{dt}$$

$$\frac{d\theta}{dt} = \frac{-10}{x^2}\frac{dx}{dt}(\sec\theta)$$

$$= \frac{-10}{25^2}(-1)\frac{25}{\sqrt{25^2 - 10^2}}$$

$$= \frac{10}{25}\frac{1}{5\sqrt{21}} = \frac{2}{25\sqrt{21}}$$

$$= \frac{2\sqrt{21}}{525} \approx 0.017 \text{ rad/sec}$$

**45.** $\tan \theta = \dfrac{x}{50}$

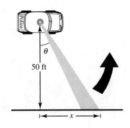

$\dfrac{d\theta}{dt} = 30(2\pi) = 60\pi$ rad/min $= \pi$ rad/sec

$\sec^2\theta\left(\dfrac{d\theta}{dt}\right) = \dfrac{1}{50}\left(\dfrac{dx}{dt}\right)$

$\dfrac{dx}{dt} = 50 \sec^2\theta\left(\dfrac{d\theta}{dt}\right)$

(a) When $\theta = 30°$, $\dfrac{dx}{dt} = \dfrac{200\pi}{3}$ ft/sec.

(b) When $\theta = 60°$, $\dfrac{dx}{dt} = 200\pi$ ft/sec.

(c) When $\theta = 70°$, $\dfrac{dx}{dt} \approx 427.43\pi$ ft/sec.

**47.** $\sin 18° = \dfrac{x}{y}$

$0 = -\dfrac{x}{y^2} \cdot \dfrac{dy}{dt} + \dfrac{1}{y} \cdot \dfrac{dx}{dt}$

$\dfrac{dx}{dt} = \dfrac{x}{y} \cdot \dfrac{dy}{dt} = (\sin 18°)(275) \approx 84.9797$ mi/hr

**49.** (a) $dy/dt = 3(dx/dt)$ means that $y$ changes three times as fast as $x$ changes.

(b) $y$ changes slowly when $x \approx 0$ or $x \approx L$. $y$ changes more rapidly when $x$ is near the middle of the interval

**51.** $L^2 = 144 + x^2$; acceleration of the boat $= \dfrac{d^2x}{dt^2}$

First derivative:   $2L\dfrac{dL}{dt} = 2x\dfrac{dx}{dt}$

$L\dfrac{dL}{dt} = x\dfrac{dx}{dt}$

Second derivative:   $L\dfrac{d^2L}{dt^2} + \dfrac{dL}{dt} \cdot \dfrac{dL}{dt} = x\dfrac{d^2x}{dt^2} + \dfrac{dx}{dt} \cdot \dfrac{dx}{dt}$

$\dfrac{d^2x}{dt^2} = \left(\dfrac{1}{x}\right)\left[L\dfrac{d^2L}{dt^2} + \left(\dfrac{dL}{dt}\right)^2 - \left(\dfrac{dx}{dt}\right)^2\right]$

When $L = 13$, $x = 5$, $\dfrac{dx}{dt} = -10.4$, and $\dfrac{dL}{dt} = -4$ (see Exercise 28). Because $\dfrac{dL}{dt}$ is constant, $\dfrac{d^2L}{dt^2} = 0$.

$\dfrac{d^2x}{dt^2} = \dfrac{1}{5}\left[13(0) + (-4)^2 - (-10.4)^2\right]$

$= \dfrac{1}{5}[16 - 108.16] = \dfrac{1}{5}[-92.16] = -18.432$ ft/sec$^2$

**53.** $y(t) = -4.9t^2 + 20$

$$\frac{dy}{dt} = -9.8t$$

$$y(1) = -4.9 + 20 = 15.1$$

$$y'(1) = -9.8$$

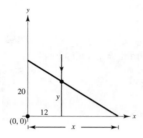

By similar triangles: $\dfrac{20}{x} = \dfrac{y}{x-12}$

$$20x - 240 = xy$$

When $y = 15.1$:  $20x - 240 = x(15.1)$

$$(20 - 15.1)x = 240$$

$$x = \frac{240}{4.9}$$

$$20x - 240 = xy$$

$$20\frac{dx}{dt} = x\frac{dy}{dt} + y\frac{dx}{dt}$$

$$\frac{dx}{dt} = \frac{x}{20-y}\frac{dy}{dt}$$

At $t = 1$, $\dfrac{dx}{dt} = \dfrac{240/4.9}{20 - 15.1}(-9.8) \approx -97.96$ m/sec.

# Review Exercises for Chapter 2

**1.** $f(x) = x^2 - 4x + 5$

$$f'(x) = \lim_{\Delta x \to 0} \frac{f(x + \Delta x) - f(x)}{\Delta x}$$

$$= \lim_{\Delta x \to 0} \frac{\left[(x + \Delta x)^2 - 4(x + \Delta x) + 5\right] - \left[x^2 - 4x + 5\right]}{\Delta x}$$

$$= \lim_{\Delta x \to 0} \frac{\left(x^2 + 2x(\Delta x) + (\Delta x)^2 - 4x - 4(\Delta x) + 5\right) - \left(x^2 - 4x + 5\right)}{\Delta x}$$

$$= \lim_{\Delta x \to 0} \frac{2x(\Delta x) + (\Delta x)^2 - 4(\Delta x)}{\Delta x} = \lim_{\Delta x \to 0} (2x + \Delta x - 4) = 2x - 4$$

**3.** $f(x) = \dfrac{x+1}{x-1}$

$$f'(x) = \lim_{\Delta x \to 0} \frac{f(x + \Delta x) - f(x)}{\Delta x} = \lim_{\Delta x \to 0} \frac{\dfrac{x + \Delta x + 1}{x + \Delta x - 1} - \dfrac{x+1}{x-1}}{\Delta x}$$

$$= \lim_{\Delta x \to 0} \frac{(x + \Delta x + 1)(x - 1) - (x + \Delta x - 1)(x + 1)}{\Delta x(x + \Delta x - 1)(x - 1)}$$

$$= \lim_{\Delta x \to 0} \frac{\left(x^2 + x\,\Delta x + x - x - \Delta x - 1\right) - \left(x^2 + x\,\Delta x - x + x + \Delta x - 1\right)}{\Delta x(x + \Delta x - 1)(x - 1)}$$

$$= \lim_{\Delta x \to 0} \frac{-2\,\Delta x}{\Delta x(x + \Delta x - 1)(x - 1)} = \lim_{\Delta x \to 0} \frac{-2}{(x + \Delta x - 1)(x - 1)} = \frac{-2}{(x-1)^2}$$

**5.** $f$ is differentiable for all $x \neq 3$.

**7.** $f(x) = 4 - |x - 2|$

    (a) Continuous at $x = 2$

    (b) Not differentiable at $x = 2$ because of the sharp turn in the graph. Also, the derivatives from the left and right are not equal.

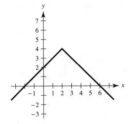

**9.** Using the limit definition, you obtain $g'(x) = \frac{4}{3}x - \frac{1}{6}$. At $x = -1$,

$$g'(-1) = -\frac{4}{3} - \frac{1}{6} = -\frac{3}{2}.$$

**11.** (a) Using the limit definition, $f'(x) = 3x^2$. At $x = -1$, $f'(-1) = 3$. The tangent line is

$$y - (-2) = 3(x - (-1))$$

$$y = 3x + 1.$$

    (b)

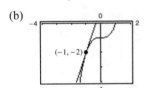

**13.** $g'(2) = \lim_{x \to 2} \dfrac{g(x) - g(2)}{x - 2}$

$\quad = \lim_{x \to 2} \dfrac{x^2(x - 1) - 4}{x - 2}$

$\quad = \lim_{x \to 2} \dfrac{x^3 - x^2 - 4}{x - 2}$

$\quad = \lim_{x \to 2} \dfrac{(x - 2)(x^2 + x + 2)}{x - 2}$

$\quad = \lim_{x \to 2} (x^2 + x + 2) = 8$

**15.** $y = 25$

$\quad y' = 0$

**17.** $f(x) = x^8$

$\quad f'(x) = 8x^7$

**19.** $h(t) = 13t^4$

$\quad h'(t) = 52t^3$

**21.** $f(x) = x^3 - 11x^2$

$\quad f'(x) = 3x^2 - 22x$

**23.** $h(x) = 6\sqrt{x} + 3\sqrt[3]{x} = 6x^{1/2} + 3x^{1/3}$

$\quad h'(x) = 3x^{-1/2} + x^{-2/3} = \dfrac{3}{\sqrt{x}} + \dfrac{1}{\sqrt[3]{x^2}}$

**25.** $g(t) = \dfrac{2}{3}t^{-2}$

$\quad g'(t) = \dfrac{-4}{3}t^{-3} = -\dfrac{4}{3t^3}$

**27.** $f(\theta) = 4\theta - 5 \sin \theta$

$\quad f'(\theta) = 4 - 5 \cos \theta$

**29.** $f(\theta) = 3 \cos \theta - \dfrac{\sin \theta}{4}$

$\quad f'(\theta) = -3 \sin \theta - \dfrac{\cos \theta}{4}$

**31.**

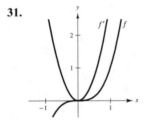

$f' > 0$ where the slopes of tangent lines to the graph of $f$ are positive.

**33.** $\quad F = 200\sqrt{T}$

$\quad F'(t) = \dfrac{100}{\sqrt{T}}$

    (a) When $T = 4$, $F'(4) = 50$ vibrations/sec/lb.

    (b) When $T = 9$, $F'(9) = 33\frac{1}{3}$ vibrations/sec/lb.

**35.** $\quad s(t) = -16t^2 + s_0$

$\quad s(9.2) = -16(9.2)^2 + s_0 = 0$

$\quad\quad s_0 = 1354.24$

The building is approximately 1354 feet high (or 415 m).

**37.** (a)

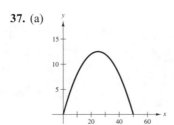

Total horizontal distance: 50

(b) $0 = x - 0.02x^2$

$0 = x\left(1 - \dfrac{x}{50}\right)$ implies $x = 50$.

(c) Ball reaches maximum height when $x = 25$.

(d)  $y = x - 0.02x^2$

$y' = 1 - 0.04x$

$y'(0) = 1$

$y'(10) = 0.6$

$y'(25) = 0$

$y'(30) = -0.2$

$y'(50) = -1$

(e) $y'(25) = 0$

**39.** $x(t) = t^2 - 3t + 2 = (t - 2)(t - 1)$

(a) $v(t) = x'(t) = 2t - 3$

(b) $v(t) < 0$ for $t < \dfrac{3}{2}$

(c) $v(t) = 0$ for $t = \dfrac{3}{2}$

$x = \left(\dfrac{3}{2} - 2\right)\left(\dfrac{3}{2} - 1\right) = \left(-\dfrac{1}{2}\right)\left(\dfrac{1}{2}\right) = -\dfrac{1}{4}$

(d) $x(t) = 0$ for $t = 1, 2$

$\left|v(1)\right| = \left|2(1) - 3\right| = 1$

$\left|v(2)\right| = \left|2(2) - 3\right| = 1$

The speed is 1 when the position is 0.

**41.** $f(x) = (5x^2 + 8)(x^2 - 4x - 6)$

$f'(x) = (5x^2 + 8)(2x - 4) + (x^2 - 4x - 6)(10x)$

$\quad = 10x^3 + 16x - 20x^2 - 32 + 10x^3 - 40x^2 - 60x$

$\quad = 20x^3 - 60x^2 - 44x - 32$

$\quad = 4(5x^3 - 15x^2 - 11x - 8)$

**43.** $h(x) = \sqrt{x}\,\sin x = x^{1/2}\sin x$

$h'(x) = \dfrac{1}{2\sqrt{x}}\sin x + \sqrt{x}\cos x$

**45.** $f(x) = \dfrac{x^2 + x - 1}{x^2 - 1}$

$f'(x) = \dfrac{(x^2 - 1)(2x + 1) - (x^2 + x - 1)(2x)}{(x^2 - 1)^2}$

$\quad = \dfrac{-(x^2 + 1)}{(x^2 - 1)^2}$

**47.** $f(x) = (9 - 4x^2)^{-1}$

$f'(x) = -(9 - 4x^2)^{-2}(-8x) = \dfrac{8x}{(9 - 4x^2)^2}$

**49.** $y = \dfrac{x^4}{\cos x}$

$y' = \dfrac{(\cos x)\,4x^3 - x^4(-\sin x)}{\cos^2 x}$

$\quad = \dfrac{4x^3 \cos x + x^4 \sin x}{\cos^2 x}$

**51.** $y = 3x^2 \sec x$

$y' = 3x^2 \sec x \tan x + 6x \sec x$

**53.** $y = x \cos x - \sin x$

$y' = -x \sin x + \cos x - \cos x = -x \sin x$

**55.** $f(x) = \dfrac{2x^3 - 1}{x^2} = 2x - x^{-2}, \quad (1, 1)$

$f'(x) = 2 + 2x^{-3}$

$f'(1) = 4$

Tangent line: $y - 1 = 4(x - 1)$

$\qquad\qquad\qquad y = 4x - 3$

**57.** $f(x) = -x \tan x, \quad (0, 0)$

$f'(x) = -x \sec^2 x - \tan x$

$f'(0) = 0$

Tangent line: $y - 0 = 0(x - 0)$

$\qquad\qquad\qquad y = 0$

**59.** $v(t) = 36 - t^2, \quad 0 \le t \le 6$

$a(t) = v'(t) = -2t$

$v(4) = 36 - 16 = 20$ m/sec

$a(4) = -8$ m/sec$^2$

**61.** $g(t) = -8t^3 - 5t + 12$

$g'(t) = -24t^2 - 5$

$g''(t) = -48t$

**63.** $f(x) = 15x^{5/2}$

$f'(x) = \frac{75}{2}x^{3/2}$

$f''(x) = \frac{225}{4}x^{1/2} = \frac{225}{4}\sqrt{x}$

**65.** $f(\theta) = 3\tan\theta$

$f'(\theta) = 3\sec^2\theta$

$f''(\theta) = 6\sec\theta(\sec\theta\tan\theta) = 6\sec^2\theta\tan\theta$

**67.** $y = 2\sin x + 3\cos x$

$y' = 2\cos x - 3\sin x$

$y'' = -2\sin x - 3\cos x$

$y'' + y = -(2\sin x + 3\cos x) + (2\sin x + 3\cos x)$

$\qquad = 0$

**69.** $h(x) = \left(\dfrac{x+5}{x^2+3}\right)^2$

$h'(x) = 2\left(\dfrac{x+5}{x^2+3}\right)\left(\dfrac{(x^2+3)(1)-(x+5)(2x)}{(x^2+3)^2}\right)$

$\qquad = \dfrac{2(x+5)(-x^2-10x+3)}{(x^2+3)^3}$

**71.** $f(s) = (s^2-1)^{5/2}(s^3+5)$

$f'(s) = (s^2-1)^{5/2}(3s^2) + (s^3+5)(\tfrac{5}{2})(s^2-1)^{3/2}(2s)$

$\qquad = s(s^2-1)^{3/2}[3s(s^2-1) + 5(s^3+5)]$

$\qquad = s(s^2-1)^{3/2}(8s^3 - 3s + 25)$

**73.** $y = 5\cos(9x+1)$

$y' = -5\sin(9x+1)(9) = -45\sin(9x+1)$

**75.** $y = \dfrac{x}{2} - \dfrac{\sin 2x}{4}$

$y' = \dfrac{1}{2} - \dfrac{1}{4}\cos 2x(2) = \dfrac{1}{2}(1-\cos 2x) = \sin^2 x$

**77.** $y = \dfrac{2}{3}\sin^{3/2}x - \dfrac{2}{7}\sin^{7/2}x$

$y' = \sin^{1/2}x\cos x - \sin^{5/2}x\cos x$

$\qquad = (\cos x)\sqrt{\sin x}(1-\sin^2 x)$

$\qquad = \cos^3 x\sqrt{\sin x}$

**79.** $y = \dfrac{\sin\pi x}{x+2}$

$y' = \dfrac{(x+2)\pi\cos\pi x - \sin\pi x}{(x+2)^2}$

**81.** $f(x) = \sqrt{1-x^3}$

$f'(x) = \dfrac{1}{2}(1-x^3)^{-1/2}(-3x^2) = \dfrac{-3x^2}{2\sqrt{1-x^3}}$

$f'(-2) = \dfrac{-12}{2(3)} = -2$

**83.** $\quad y = \dfrac{1}{2}\csc 2x$

$y' = -\csc 2x\cot 2x$

$y'\left(\dfrac{\pi}{4}\right) = 0$

**85.** $g(x) = 2x(x+1)^{-1/2}$

$g'(x) = \dfrac{x+2}{(x+1)^{3/2}}$

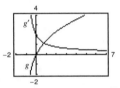

$g'$ does not equal zero for any value of $x$ in the domain. The graph of $g$ has no horizontal tangent lines.

**87.** $f(t) = \sqrt{t+1}\sqrt[3]{t+1}$

$f(t) = (t+1)^{1/2}(t+1)^{1/3} = (t+1)^{5/6}$

$f'(t) = \dfrac{5}{6(t+1)^{1/6}}$

$f'$ does not equal zero for any $x$ in the domain. The graph of $f$ has no horizontal tangent lines.

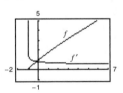

**89.** $f(t) = t^2(t-1)^5, \ (2,4)$

(a)  $f'(t) = t(t-1)^4(7t-2)$

$\quad f'(2) = 24$

(b)  $y - 4 = 24(t-2)$

$\quad y = 24t - 44$

(c)

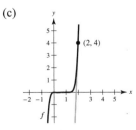

**91.** $f(x) = \tan \sqrt{1 - x}, \quad \left( -2, \tan \sqrt{3} \right)$

(a) $\quad f'(x) = \dfrac{-1}{2\sqrt{1 - x} \, \cos^2 \sqrt{1 - x}}$

$\quad f'(-2) = \dfrac{-\sqrt{3}}{6 \cos^2 \sqrt{3}} \approx -11.1983$

(b) $\quad y - \tan\sqrt{3} = \dfrac{-\sqrt{3}}{6 \cos^2 \sqrt{3}} (x + 2)$

$\quad y = -\dfrac{(x + 2)\sqrt{3}}{6 \cos^2 \sqrt{3}} + \tan\sqrt{3}$

(c)

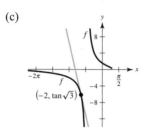

**93.** $y = 7x^2 + \cos 2x$

$y' = 14x - 2 \sin 2x$

$y'' = 14 - 4 \cos 2x$

**95.** $f(x) = \cot x$

$f'(x) = -\csc^2 x$

$f''(x) = -2 \csc x(-\csc x \cdot \cot x)$

$\qquad = 2 \csc^2 x \cot x$

**97.** $f(t) = \dfrac{4t^2}{(1 - t)^2}$

$f'(t) = \dfrac{8t}{(1 - t)^3}$

$f''(t) = \dfrac{8(2t + 1)}{(1 - t)^4}$

**99.** $g(\theta) = \tan 3\theta - \sin(\theta - 1)$

$g'(\theta) = 3 \sec^2 3\theta - \cos(\theta - 1)$

$g''(\theta) = 18 \sec^2 3\theta \tan 3\theta + \sin(\theta - 1)$

**101.** $T = \dfrac{700}{t^2 + 4t + 10}$

$T = 700\left(t^2 + 4t + 10\right)^{-1}$

$T' = \dfrac{-1400(t + 2)}{\left(t^2 + 4t + 10\right)^2}$

(a) When $t = 1$,

$T' = \dfrac{-1400(1 + 2)}{(1 + 4 + 10)^2} \approx -18.667$ deg/h.

(b) When $t = 3$,

$T' = \dfrac{-1400(3 + 2)}{(9 + 12 + 10)^2} \approx -7.284$ deg/h.

(c) When $t = 5$,

$T' = \dfrac{-1400(5 + 2)}{(25 + 20 + 10)^2} \approx -3.240$ deg/h.

(d) When $t = 10$,

$T' = \dfrac{-1400(10 + 2)}{(100 + 40 + 10)^2} \approx -0.747$ deg/h.

**103.** $\qquad x^2 + 3xy + y^3 = 10$

$2x + 3xy' + 3y + 3y^2 y' = 0$

$\qquad 3\left(x + y^2\right)y' = -(2x + 3y)$

$\qquad\qquad y' = \dfrac{-(2x + 3y)}{3\left(x + y^2\right)}$

**105.** $\qquad\qquad \sqrt{xy} = x - 4y$

$\dfrac{\sqrt{x}}{2\sqrt{y}} y' + \dfrac{\sqrt{y}}{2\sqrt{x}} = 1 - 4y'$

$\qquad xy' + y = 2\sqrt{xy} - 8\sqrt{xy}\, y'$

$\qquad x + 8\sqrt{xy}\, y' = 2\sqrt{xy} - y$

$\qquad\qquad y' = \dfrac{2\sqrt{xy} - y}{x + 8\sqrt{xy}} = \dfrac{2(x - 4y) - y}{x + 8(x - 4y)}$

$\qquad\qquad\quad = \dfrac{2x - 9y}{9x - 32y}$

**107.** $\qquad\qquad x \sin y = y \cos x$

$\left(x \cos y\right)y' + \sin y = -y \sin x + y' \cos x$

$y'\left(x \cos y - \cos x\right) = -y \sin x - \sin y$

$\qquad\qquad y' = \dfrac{y \sin x + \sin y}{\cos x - x \cos y}$

**109.** $x^2 + y^2 = 10$

$2x + 2yy' = 0$

$y' = \dfrac{-x}{y}$

At $(3, 1)$, $y' = -3$

Tangent line: $y - 1 = -3(x - 3) \Rightarrow 3x + y - 10 = 0$

Normal line: $y - 1 = \dfrac{1}{3}(x - 3) \Rightarrow x - 3y = 0$

**111.** $y = \sqrt{x}$

$\dfrac{dy}{dt} = 2$ units/sec

$\dfrac{dy}{dt} = \dfrac{1}{2\sqrt{x}} \dfrac{dx}{dt} \Rightarrow \dfrac{dx}{dt} = 2\sqrt{x}\dfrac{dy}{dt} = 4\sqrt{x}$

(a) When $x = \dfrac{1}{2}$, $\dfrac{dx}{dt} = 2\sqrt{2}$ units/sec.

(b) When $x = 1$, $\dfrac{dx}{dt} = 4$ units/sec.

(c) When $x = 4$, $\dfrac{dx}{dt} = 8$ units/sec.

**113.** $\dfrac{s}{h} = \dfrac{1/2}{2}$

$s = \dfrac{1}{4}h$

$\dfrac{dV}{dt} = 1$

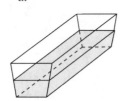

Width of water at depth $h$:

$w = 2 + 2s = 2 + 2\left(\dfrac{1}{4}h\right) = \dfrac{4 + h}{2}$

$V = \dfrac{5}{2}\left(2 + \dfrac{4 + h}{2}\right)h = \dfrac{5}{4}(8 + h)h$

$\dfrac{dV}{dt} = \dfrac{5}{2}(4 + h)\dfrac{dh}{dt}$

$\dfrac{dh}{dt} = \dfrac{2(dV/dt)}{5(4 + h)}$

When $h = 1$, $\dfrac{dh}{dt} = \dfrac{2}{25}$ m/min.

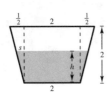

**115.** $s(t) = 60 - 4.9t^2$

$s'(t) = -9.8t$

$s = 35 = 60 - 4.9t^2$

$4.9t^2 = 25$

$t = \dfrac{5}{\sqrt{4.9}}$

$\tan 30 = \dfrac{1}{\sqrt{3}} = \dfrac{s(t)}{x(t)}$

$x(t) = \sqrt{3}s(t)$

$\dfrac{dx}{dt} = \sqrt{3}\dfrac{ds}{dt} = \sqrt{3}(-9.8)\dfrac{5}{\sqrt{4.9}} \approx -38.34$ m/sec

## Problem Solving for Chapter 2

**1.** (a)  $x^2 + (y - r)^2 = r^2$, Circle

$$x^2 = y, \text{ Parabola}$$

Substituting:

$$(y - r)^2 = r^2 - y$$
$$y^2 - 2ry + r^2 = r^2 - y$$
$$y^2 - 2ry + y = 0$$
$$y(y - 2r + 1) = 0$$

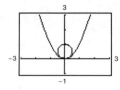

Because you want only one solution, let $1 - 2r = 0 \Rightarrow r = \dfrac{1}{2}$. Graph $y = x^2$ and $x^2 + \left(y - \dfrac{1}{2}\right)^2 = \dfrac{1}{4}$.

(b) Let $(x, y)$ be a point of tangency:

$$x^2 + (y - b)^2 = 1 \Rightarrow 2x + 2(y - b)y' = 0 \Rightarrow y' = \frac{x}{b - y}, \text{ Circle}$$

$$y = x^2 \Rightarrow y' = 2x, \text{ Parabola}$$

Equating:

$$2x = \frac{x}{b - y}$$
$$2(b - y) = 1$$
$$b - y = \frac{1}{2} \Rightarrow b = y + \frac{1}{2}$$

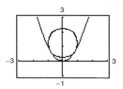

Also, $x^2 + (y - b)^2 = 1$ and $y = x^2$ imply:

$$y + (y - b)^2 = 1 \Rightarrow y + \left[y - \left(y + \frac{1}{2}\right)\right]^2 = 1 \Rightarrow y + \frac{1}{4} = 1 \Rightarrow y = \frac{3}{4} \text{ and } b = \frac{5}{4}$$

Center: $\left(0, \dfrac{5}{4}\right)$

Graph $y = x^2$ and $x^2 + \left(y - \dfrac{5}{4}\right)^2 = 1$.

**3.** (a)  $f(x) = \cos x$          $P_1(x) = a_0 + a_1 x$

       $f(0) = 1$              $P_1(0) = a_0 \Rightarrow a_0 = 1$

       $f'(0) = 0$            $P_1'(0) = a_1 \Rightarrow a_1 = 0$

       $P_1(x) = 1$

(b)  $f(x) = \cos x$          $P_2(x) = a_0 + a_1 x + a_2 x^2$

       $f(0) = 1$              $P_2(0) = a_0 \Rightarrow a_0 = 1$

       $f'(0) = 0$            $P_2'(0) = a_1 \Rightarrow a_1 = 0$

       $f''(0) = -1$         $P_2''(0) = 2a_2 \Rightarrow a_2 = -\frac{1}{2}$

       $P_2(x) = 1 - \frac{1}{2}x^2$

(c)

| $x$ | −1.0 | −0.1 | −0.001 | 0 | 0.001 | 0.1 | 1.0 |
|---|---|---|---|---|---|---|---|
| $\cos x$ | 0.5403 | 0.9950 | $\approx 1$ | 1 | $\approx 1$ | 0.9950 | 0.5403 |
| $P_2(x)$ | 0.5 | 0.9950 | $\approx 1$ | 1 | $\approx 1$ | 0.9950 | 0.5 |

$P_2(x)$ is a good approximation of $f(x) = \cos x$ when x is near 0.

(d)  $f(x) = \sin x$       $P_3(x) = a_0 + a_1 x + a_2 x^2 + a_3 x^3$

   $f(0) = 0$       $P_3(0) = a_0 \Rightarrow a_0 = 0$

   $f'(0) = 1$       $P_3'(0) = a_1 \Rightarrow a_1 = 1$

   $f''(0) = 0$       $P_3''(0) = 2a_2 \Rightarrow a_2 = 0$

   $f'''(0) = -1$       $P_3'''(0) = 6a_3 \Rightarrow a_3 = -\frac{1}{6}$

   $P_3(x) = x - \frac{1}{6}x^3$

**5.** Let  $p(x) = Ax^3 + Bx^2 + Cx + D$

   $p'(x) = 3Ax^2 + 2Bx + C.$

At $(1, 1)$:

   $A + B + C + D = 1$    Equation 1

   $3A + 2B + C \quad = 14$    Equation 2

At $(-1, -3)$:

   $A + B - C + D = -3$    Equation 3

   $3A + 2B + C \quad = -2$    Equation 4

Adding Equations 1 and 3: $2B + 2D = -2$

Subtracting Equations 1 and 3: $2A + 2C = 4$

Adding Equations 2 and 4: $6A + 2C = 12$

Subtracting Equations 2 and 4: $4B = 16$

So, $B = 4$ and $D = \frac{1}{2}(-2 - 2B) = -5.$ Subtracting

$2A + 2C = 4$ and $6A + 2C = 12,$ you obtain

$4A = 8 \Rightarrow A = 2.$ Finally, $C = \frac{1}{2}(4 - 2A) = 0.$

So, $p(x) = 2x^3 + 4x^2 - 5.$

**7.** (a)    $x^4 = a^2 x^2 - a^2 y^2$

   $a^2 y^2 = a^2 x^2 - x^4$

   $y = \dfrac{\pm\sqrt{a^2 x^2 - x^4}}{a}$

   Graph: $y_1 = \dfrac{\sqrt{a^2 x^2 - x^4}}{a}$ and

   $y_2 = -\dfrac{\sqrt{a^2 x^2 - x^4}}{a}.$

(b)
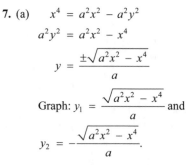

$(\pm a, 0)$ are the $x$-intercepts, along with $(0, 0)$.

(c) Differentiating implicitly:

   $4x^3 = 2a^2 x - 2a^2 yy'$

   $y' = \dfrac{2a^2 x - 4x^3}{2a^2 y}$

   $\quad = \dfrac{x(a^2 - 2x^2)}{a^2 y} = 0 \Rightarrow 2x^2 = a^2 \Rightarrow x = \dfrac{\pm a}{\sqrt{2}}$

   $\left(\dfrac{a^2}{2}\right)^2 = a^2\left(\dfrac{a^2}{2}\right) - a^2 y^2$

   $\dfrac{a^4}{4} = \dfrac{a^4}{2} - a^2 y^2$

   $a^2 y^2 = \dfrac{a^4}{4}$

   $y^2 = \dfrac{a^2}{4}$

   $y = \pm\dfrac{a}{2}$

   Four points: $\left(\dfrac{a}{\sqrt{2}}, \dfrac{a}{2}\right), \left(\dfrac{a}{\sqrt{2}}, -\dfrac{a}{2}\right), \left(-\dfrac{a}{\sqrt{2}}, \dfrac{a}{2}\right),$

   $\left(\dfrac{-a}{\sqrt{2}}, -\dfrac{a}{2}\right)$

**9.** (a)

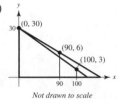

*Not drawn to scale*

Line determined by $(0, 30)$ and $(90, 6)$:

$$y - 30 = \frac{30 - 6}{0 - 90}(x - 0)$$

$$= -\frac{24}{90}x = -\frac{4}{15}x \Rightarrow y = -\frac{4}{15}x + 30$$

When $x = 100$:

$$y = -\frac{4}{15}(100) + 30 = \frac{10}{3} > 3$$

As you can see from the figure, the shadow determined by the man extends beyond the shadow determined by the child.

(b)

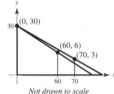

*Not drawn to scale*

Line determined by $(0, 30)$ and $(60, 6)$:

$$y - 30 = \frac{30 - 6}{0 - 60}(x - 0) = -\frac{2}{5}x \Rightarrow y = -\frac{2}{5}x + 30$$

When $x = 70$: $y = -\frac{2}{5}(70) + 30 = 2 < 3$

As you can see from the figure, the shadow determined by the child extends beyond the shadow determined by the man.

(c)  Need $(0, 30), (d, 6), (d + 10, 3)$ collinear.

$$\frac{30 - 6}{0 - d} = \frac{6 - 3}{d - (d + 10)} \Rightarrow \frac{24}{d} = \frac{3}{10} \Rightarrow d = 80 \text{ feet}$$

(d)  Let $y$ be the distance from the base of the street light to the tip of the shadow. You know that $dx/dt = -5$.

For $x > 80$, the shadow is determined by the man.

$$\frac{y}{30} = \frac{y - x}{6} \Rightarrow y = \frac{5}{4}x \text{ and } \frac{dy}{dt} = \frac{5}{4}\frac{dx}{dt} = \frac{-25}{4}$$

For $x < 80$, the shadow is determined by the child.

$$\frac{y}{30} = \frac{y - x - 10}{3} \Rightarrow y = \frac{10}{9}x + \frac{100}{9} \text{ and}$$

$$\frac{dy}{dt} = \frac{10}{9}\frac{dx}{dt} = -\frac{50}{9}$$

Therefore:

$$\frac{dy}{dt} = \begin{cases} -\dfrac{25}{4}, & x > 80 \\[2mm] -\dfrac{50}{9}, & 0 < x < 80 \end{cases}$$

$dy/dt$ is not continuous at $x = 80$.

**ALTERNATE SOLUTION for parts (a) and (b):**

(a)  As before, the line determined by the man's shadow is

$$y_m = -\frac{4}{15}x + 30$$

The line determined by the child's shadow is obtained by finding the line through $(0, 30)$ and $(100, 3)$:

$$y - 30 = \frac{30 - 3}{0 - 100}(x - 0) \Rightarrow y_c = -\frac{27}{100}x + 30$$

By setting $y_m = y_c = 0$, you can determine how far the shadows extend:

Man: $y_m = 0 \Rightarrow \dfrac{4}{15}x = 30 \Rightarrow x = 112.5 = 112\frac{1}{2}$

Child: $y_c = 0 \Rightarrow \dfrac{27}{100}x = 30 \Rightarrow x = 111.\overline{11} = 111\frac{1}{9}$

The man's shadow is $112\frac{1}{2} - 111\frac{1}{9} = 1\dfrac{7}{18}$ ft beyond the child's shadow.

(b)  As before, the line determined by the man's shadow is

$$y_m = -\frac{2}{5}x + 30$$

For the child's shadow,

$$y - 30 = \frac{30 - 3}{0 - 70}(x - 0) \Rightarrow y_c = -\frac{27}{70}x + 30$$

Man: $y_m = 0 \Rightarrow \dfrac{2}{5}x = 30 \Rightarrow x = 75$

Child: $y_c = 0 \Rightarrow \dfrac{27}{70}x = 30 \Rightarrow x = \dfrac{700}{9} = 77\frac{7}{9}$

So the child's shadow is $77\frac{7}{9} - 75 = 2\frac{7}{9}$ ft beyond the man's shadow.

**11.** $L'(x) = \lim\limits_{\Delta x \to 0} \dfrac{L(x + \Delta x) - L(x)}{\Delta x} = \lim\limits_{\Delta x \to 0} \dfrac{L(x) + L(\Delta x) - L(x)}{\Delta x} = \lim\limits_{\Delta x \to 0} \dfrac{L(\Delta x)}{\Delta x}$

Also, $L'(0) = \lim\limits_{\Delta x \to 0} \dfrac{L(\Delta x) - L(0)}{\Delta x}$. But, $L(0) = 0$ because $L(0) = L(0 + 0) = L(0) + L(0) \Rightarrow L(0) = 0.$

So, $L'(x) = L'(0)$ for all $x$. The graph of $L$ is a line through the origin of slope $L'(0)$.

**13. (a)**

| $z$ (degrees) | 0.1 | 0.01 | 0.0001 |
|---|---|---|---|
| $\dfrac{\sin z}{z}$ | 0.0174524 | 0.0174533 | 0.0174533 |

**(b)** $\lim\limits_{z \to 0} \dfrac{\sin z}{z} \approx 0.0174533$

In fact, $\lim\limits_{z \to 0} \dfrac{\sin z}{z} = \dfrac{\pi}{180}$.

**(c)** $\dfrac{d}{dz}(\sin z) = \lim\limits_{\Delta z \to 0} \dfrac{\sin (z + \Delta z) - \sin z}{\Delta z}$

$= \lim\limits_{\Delta z \to 0} \dfrac{\sin z \cdot \cos \Delta z + \sin \Delta z \cdot \cos z - \sin z}{\Delta z}$

$= \lim\limits_{\Delta z \to 0} \left[ \sin z \left( \dfrac{\cos \Delta z - 1}{\Delta z} \right) \right] + \lim\limits_{\Delta z \to 0} \left[ \cos z \left( \dfrac{\sin \Delta z}{\Delta z} \right) \right]$

$= (\sin z)(0) + (\cos z)\left( \dfrac{\pi}{180} \right) = \dfrac{\pi}{180} \cos z$

**(d)** $S(90) = \sin\left( \dfrac{\pi}{180} 90 \right) = \sin\dfrac{\pi}{2} = 1$

$C(180) = \cos\left( \dfrac{\pi}{180} 180 \right) = -1$

$\dfrac{d}{dz} S(z) = \dfrac{d}{dz}\sin(cz) = c \cdot \cos(cz) = \dfrac{\pi}{180} C(z)$

**(e)** The formulas for the derivatives are more complicated in degrees.

**15.** $j(t) = a'(t)$

**(a)** $j(t)$ is the rate of change of acceleration.

**(b)** $s(t) = -8.25t^2 + 66t$

$v(t) = -16.5t + 66$

$a(t) = -16.5$

$a'(t) = j(t) = 0$

The acceleration is constant, so $j(t) = 0$.

**(c)** $a$ is position.

$b$ is acceleration.

$c$ is jerk.

$d$ is velocity.

# C H A P T E R   3
# Applications of Differentiation

# CHAPTER 3
# Applications of Differentiation

## Section 3.1   Extrema on an Interval

**1.** $f(x) = \dfrac{x^2}{x^2 + 4}$

$f'(x) = \dfrac{(x^2 + 4)(2x) - (x^2)(2x)}{(x^2 + 4)^2} = \dfrac{8x}{(x^2 + 4)^2}$

$f'(0) = 0$

**3.** $f(x) = x + \dfrac{4}{x^2} = x + 4x^{-2}$

$f'(x) = 1 - 8x^{-3} = 1 - \dfrac{8}{x^3}$

$f'(2) = 0$

**5.** $f(x) = (x + 2)^{2/3}$

$f'(x) = \frac{2}{3}(x + 2)^{-1/3}$

$f'(-2)$ is undefined.

**7.** Critical number: $x = 2$

   $x = 2$: absolute maximum (and relative maximum)

**9.** Critical numbers: $x = 1, 2, 3$

   $x = 1, 3$: absolute maxima (and relative maxima)

   $x = 2$: absolute minimum (and relative minimum)

**11.** $f(x) = x^3 - 3x^2$

$f'(x) = 3x^2 - 6x = 3x(x - 2)$

Critical numbers: $x = 0, 2$

**13.** $g(t) = t\sqrt{4 - t}, \ t < 3$

$g'(t) = t\left[\frac{1}{2}(4 - t)^{-1/2}(-1)\right] + (4 - t)^{1/2}$

$\quad = \frac{1}{2}(4 - t)^{-1/2}\left[-t + 2(4 - t)\right]$

$\quad = \dfrac{8 - 3t}{2\sqrt{4 - t}}$

Critical number: $t = \dfrac{8}{3}$

**15.** $h(x) = \sin^2 x + \cos x, \ 0 < x < 2\pi$

$h'(x) = 2 \sin x \cos x - \sin x = \sin x(2 \cos x - 1)$

Critical numbers in $(0, 2\pi)$: $x = \dfrac{\pi}{3}, \pi, \dfrac{5\pi}{3}$

**17.** $f(x) = 3 - x, \quad [-1, 2]$

$f'(x) = -1 \Rightarrow$ no critical numbers

Left endpoint: $(-1, 4)$ Maximum

Right endpoint: $(2, 1)$ Minimum

**19.** $g(x) = x^2 - 2x, \quad [0, 4]$

$g'(x) = 2x - 2 = 2(x - 1)$

Critical number: $x = 1$

Left endpoint: $(0, 0)$

Critical number: $(1, -1)$ Minimum

Right endpoint: $(4, 8)$ Maximum

**21.** $f(x) = x^3 - \frac{3}{2}x^2, \quad [-1, 2]$

$f'(x) = 3x^2 - 3x = 3x(x - 1)$

Left endpoint: $\left(-1, -\frac{5}{2}\right)$ Minimum

Right endpoint: $(2, 2)$ Maximum

Critical number: $(0, 0)$

Critical number: $\left(1, -\frac{1}{2}\right)$

**23.** $f(x) = 3x^{2/3} - 2x, \quad [-1, 1]$

$f'(x) = 2x^{-1/3} - 2 = \dfrac{2\left(1 - \sqrt[3]{x}\right)}{\sqrt[3]{x}}$

Left endpoint: $(-1, 5)$ Maximum

Critical number: $(0, 0)$ Minimum

Right endpoint: $(1, 1)$

**25.** $g(t) = \dfrac{t^2}{t^2 + 3}, \quad [-1, 1]$

$g'(t) = \dfrac{6t}{(t^2 + 3)^2}$

Left endpoint: $\left(-1, \dfrac{1}{4}\right)$ Maximum

Critical number: $(0, 0)$ Minimum

Right endpoint: $\left(1, \dfrac{1}{4}\right)$ Maximum

**27.** $h(s) = \dfrac{1}{s-2}$, $[0, 1]$

$h'(s) = \dfrac{-1}{(s-2)^2}$

Left endpoint: $\left(0, -\dfrac{1}{2}\right)$ Maximum

Right endpoint: $(1, -1)$ Minimum

**29.** $y = 3 - |t - 3|$, $[-1, 5]$

For $x < 3$, $y = 3 + (t - 3) = t$

and $y' = 1 \neq 0$ on $[-1, 3)$

For $x > 3$, $y = 3 - (t - 3) = 6 - t$

and $y' = -1 \neq 0$ on $(3, 5]$

So, $x = 3$ is the only critical number.

Left endpoint: $(-1, -1)$ Minimum

Right endpoint: $(5, 1)$

Critical number: $(3, 3)$ Maximum

**31.** $f(x) = [\![x]\!]$, $[-2, 2]$

From the graph of $f$, you see that the maximum value of $f$ is 2 for $x = 2$, and the minimum value is $-2$ for $-2 \leq x < -1$.

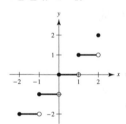

**33.** $f(x) = \cos \pi x$, $\left[0, \dfrac{1}{6}\right]$

$f'(x) = -\pi \sin \pi x$

Left endpoint: $(0, 1)$ Maximum

Right endpoint: $\left(\dfrac{1}{6}, \dfrac{\sqrt{3}}{2}\right)$ Minimum

**35.** $y = 3 \cos x$, $[0, 2\pi]$

$y' = -3 \sin x$

Critical number in $(0, 2\pi)$: $x = \pi$

Left endpoint: $(0, 3)$ Maximum

Critical number: $(\pi, -3)$ Minimum

Right endpoint: $(2\pi, 3)$ Maximum

**37.** $f(x) = 2x - 3$

(a) Minimum: $(0, -3)$

  Maximum: $(2, 1)$

(b) Minimum: $(0, -3)$

(c) Maximum: $(2, 1)$

(d) No extrema

**39.** $f(x) = x^2 - 2x$

(a) Minimum: $(1, -1)$

  Maximum: $(-1, 3)$

(b) Maximum: $(3, 3)$

(c) Minimum: $(1, -1)$

(d) Minimum: $(1, -1)$

**41.** $f(x) = \begin{cases} 2x + 2, & 0 \leq x \leq 1 \\ 4x^2, & 1 < x \leq 3 \end{cases}$

Left endpoint: $(0, 2)$ Minimum

Right endpoint: $(3, 36)$ Maximum

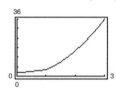

**43.** $f(x) = \dfrac{3}{x - 1}$, $(1, 4]$

Right endpoint: $(4, 1)$ Minimum

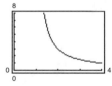

**45.** $f(x) = x^4 - 2x^3 + x + 1$, $[-1, 3]$

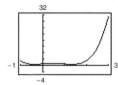

$f'(x) = 4x^3 - 6x^2 + 1 = (2x - 1)(2x^2 - 2x - 1) = 0$

$x = \dfrac{1}{2}, \dfrac{1 \pm \sqrt{3}}{2} \approx 0.5, -0.366, 1.366$

Right endpoint: $(3, 31)$ Maximum

Critical points: $\left(\dfrac{1 \pm \sqrt{3}}{2}, \dfrac{3}{4}\right)$ Minima

**47.** (a)

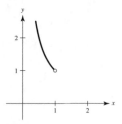

Minimum: $(0.4398, -1.0613)$

(b)
$$f(x) = 3.2x^5 + 5x^3 - 3.5x, \quad [0, 1]$$
$$f'(x) = 16x^4 + 15x^2 - 3.5$$
$$16x^4 + 15x^2 - 3.5 = 0$$

$$x^2 = \frac{-15 \pm \sqrt{(15)^2 - 4(16)(-3.5)}}{2(16)} = \frac{-15 \pm \sqrt{449}}{32}$$

$$x = \sqrt{\frac{-15 + \sqrt{449}}{32}} \approx 0.4398$$

Left endpoint: $(0, 0)$

Critical point: $(0.4398, -1.0613)$ Minimum

Right endpoint: $(1, 4.7)$ Maximum

**49.** $f(x) = (1 + x^3)^{1/2}, \quad [0, 2]$

$$f'(x) = \frac{3}{2}x^2(1 + x^3)^{-1/2}$$

$$f''(x) = \frac{3}{4}(x^4 + 4x)(1 + x^3)^{-3/2}$$

$$f'''(x) = -\frac{3}{8}(x^6 + 20x^3 - 8)(1 + x^3)^{-5/2}$$

Setting $f''' = 0$, you have $x^6 + 20x^3 - 8 = 0$.

$$x^3 = \frac{-20 \pm \sqrt{400 - 4(1)(-8)}}{2}$$

$$x = \sqrt[3]{-10 \pm \sqrt{108}} = \sqrt{3} - 1$$

In the interval $[0, 2]$, choose

$$x = \sqrt[3]{-10 \pm \sqrt{108}} = \sqrt{3} - 1 \approx 0.732.$$

$$\left| f''\left(\sqrt[3]{-10 + \sqrt{108}}\right) \right| \approx 1.47 \text{ is the maximum value.}$$

**51.** $f(x) = (x + 1)^{2/3}, \quad [0, 2]$

$$f'(x) = \frac{2}{3}(x + 1)^{-1/3}$$

$$f''(x) = -\frac{2}{9}(x + 1)^{-4/3}$$

$$f'''(x) = \frac{8}{27}(x + 1)^{-7/3}$$

$$f^{(4)}(x) = -\frac{56}{81}(x + 1)^{-10/3}$$

$$f^{(5)}(x) = \frac{560}{243}(x + 1)^{-13/3}$$

$$\left| f^{(4)}(0) \right| = \frac{56}{81} \text{ is the maximum value.}$$

**53.** Answers will vary. *Sample answer*:

$$y = \frac{1}{x} \text{ on the interval } (0, 1)$$

There is no maximum or minimum value.

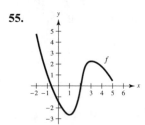

**55.**

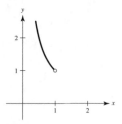

**57.** (a) Yes

(b) No

**59.** (a) No

(b) Yes

**61.** $P = VI - RI^2 = 12I - 0.5I^2, 0 \le I \le 15$

$P = 0$ when $I = 0$.

$P = 67.5$ when $I = 15$.

$P' = 12 - I = 0$

Critical number: $I = 12$ amps

When $I = 12$ amps, $P = 72$, the maximum output.

No, a 20-amp fuse would not increase the power output. $P$ is decreasing for $I > 12$.

**63.**
$$S = 6hs + \frac{3s^2}{2}\left(\frac{\sqrt{3} - \cos\theta}{\sin\theta}\right), \frac{\pi}{6} \le \theta \le \frac{\pi}{2}$$

$$\frac{dS}{d\theta} = \frac{3s^2}{2}\left(-\sqrt{3}\csc\theta\cot\theta + \csc^2\theta\right)$$

$$= \frac{3s^2}{2}\csc\theta\left(-\sqrt{3}\cot\theta + \csc\theta\right) = 0$$

$$\csc\theta = \sqrt{3}\cot\theta$$

$$\sec\theta = \sqrt{3}$$

$$\theta = \text{arcsec}\sqrt{3} \approx 0.9553 \text{ radians}$$

$$S\left(\frac{\pi}{6}\right) = 6hs + \frac{3s^2}{2}\left(\sqrt{3}\right)$$

$$S\left(\frac{\pi}{6}\right) = 6hs + \frac{3s^2}{2}\left(\sqrt{3}\right)$$

$$S\left(\text{arcsec}\sqrt{3}\right) = 6hs + \frac{3s^2}{2}\left(\sqrt{2}\right)$$

$S$ is minimum when $\theta = \text{arcsec}\sqrt{3} \approx 0.9553$ radian.

**65.** True. See Exercise 25.

**67.** True

**69.** If $f$ has a maximum value at $x = c$, then $f(c) \ge f(x)$ for all $x$ in $I$. So, $-f(c) \le -f(x)$ for all $x$ in $I$. So, $-f$ has a minimum value at $x = c$.

**71.** First do an example: Let $a = 4$ and $f(x) = 4$. Then $R$ is the square $0 \le x \le 4, 0 \le y \le 4$. Its area and perimeter are both $k = 16$.

Claim that all real numbers $a > 2$ work. On the one hand, if $a > 2$ is given, then let $f(x) = 2a/(a - 2)$. Then the rectangle

$$R = \left\{(x, y): 0 \le x \le a, 0 \le y \le \frac{2a}{a - 2}\right\}$$

has $k = \frac{2a^2}{a - 2}$:

$$\text{Area} = a\left(\frac{2a}{a - 2}\right) = \frac{2a^2}{a - 2}$$

$$\text{Perimeter} = 2a + 2\left(\frac{2a}{a - 2}\right)$$

$$= \frac{2a(a - 2) + 2(2a)}{a - 2} = \frac{2a^2}{a - 2}.$$

To see that $a$ must be greater than 2, consider

$$R = \{(x, y): 0 \le x \le a, 0 \le y \le f(x)\}.$$

$f$ attains its maximum value on $[0, a]$ at some point $P(x_0, y_0)$, as indicated in the figure.

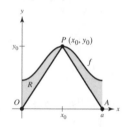

Draw segments $\overline{OP}$ and $\overline{PA}$. The region $R$ is bounded by the rectangle $0 \le x \le a, 0 \le y \le y_0$, so $\text{area}(R) = k \le ay_0$. Furthermore, from the figure, $y_0 < \overline{OP}$ and $y_0 < \overline{PA}$. So, $k = \text{Perimeter}(R) > \overline{OP} + \overline{PA} > 2y_0$. Combining, $2y_0 < k \le ay_0 \Rightarrow a > 2$.

## Section 3.2 Rolle's Theorem and the Mean Value Theorem

**1.** $f(x) = \left|\frac{1}{x}\right|$

$f(-1) = f(1) = 1$. But, $f$ is not continuous on $[-1, 1]$.

**3.** Rolle's Theorem does not apply to $f(x) = 1 - |x - 1|$ over $[0, 2]$ because $f$ is not differentiable at $x = 1$.

**5.** $f(x) = x^2 - x - 2 = (x - 2)(x + 1)$

$x$-intercepts: $(-1, 0), (2, 0)$

$f'(x) = 2x - 1 = 0$ at $x = \frac{1}{2}$.

**7.** $f(x) = x\sqrt{x + 4}$

$x$-intercepts: $(-4, 0), (0, 0)$

$$f'(x) = x\frac{1}{2}(x + 4)^{-1/2} + (x + 4)^{1/2}$$

$$= (x + 4)^{-1/2}\left(\frac{x}{2} + (x + 4)\right)$$

$$f'(x) = \left(\frac{3}{2}x + 4\right)(x + 4)^{-1/2} = 0 \text{ at } x = -\frac{8}{3}$$

**9.** $f(x) = x^2 + 2x - 3 = (x + 3)(x - 1)$

$f(-3) = f(1) = 0$

$f'(x) = 2x + 2 = 0$   at   $x = -1$

$c = -1$   and   $f'(-1) = 0$.

**11.** $f(x) = -x^2 + 3x$,   $[0, 3]$

$f(0) = f(3) = 0$

$f$ is continuous on $[0, 3]$ and differentiable on $(0, 3)$.
Rolle's Theorem applies.

$f'(x) = -2x + 3$

$-2x + 3 = 0 \Rightarrow x = \frac{3}{2}$

$c$-value: $\frac{3}{2}$

**13.** $f(x) = (x - 1)(x - 2)(x - 3)$, $[1, 3]$

$f(1) = f(3) = 0$

$f$ is continuous on $[1, 3]$. $f$ is differentiable on $(1, 3)$.
Rolle's Theorem applies.

$$f(x) = x^3 - 6x^2 + 11x - 6$$
$$f'(x) = 3x^2 - 12x + 11$$

$3x^2 - 12x + 11 = 0 \Rightarrow x = \dfrac{6 \pm \sqrt{3}}{3}$

$c$-values: $\dfrac{6 - \sqrt{3}}{3}, \dfrac{6 + \sqrt{3}}{3}$

**15.** $f(x) = x^{2/3} - 1$, $[-8, 8]$

$f(-8) = f(8) = 3$

$f$ is continuous on $[-8, 8]$. $f$ is not differentiable on $(-8, 8)$ because $f'(0)$ does not exist. Rolle's Theorem does not apply.

**17.** $f(x) = \dfrac{x^2 - 2x - 3}{x + 2}$, $[-1, 3]$

$f(-1) = f(3) = 0$

$f$ is continuous on $[-1, 3]$. (**Note:** the discontinuity, $x = -2$, is not in the interval.) $f$ is differentiable on $(-1, 3)$. Rolle's Theorem applies.

$$f'(x) = \frac{(x + 2)(2x - 2) - (x^2 - 2x - 3)(1)}{(x + 2)^2} = 0$$

$$\frac{x^2 + 4x - 1}{(x + 2)^2} = 0$$

$$x = \frac{-4 \pm 2\sqrt{5}}{2} = -2 \pm \sqrt{5}$$

$c$-value: $-2 + \sqrt{5}$

**19.** $f(x) = \sin x$, $[0, 2\pi]$

$f(0) = f(2\pi) = 0$

$f$ is continuous on $[0, 2\pi]$. $f$ is differentiable on $(0, 2\pi)$. Rolle's Theorem applies.

$f'(x) = \cos x$

$c$-values: $\dfrac{\pi}{2}, \dfrac{3\pi}{2}$

**21.** $f(x) = \dfrac{6x}{\pi} - 4 \sin^2 x$, $\left[0, \dfrac{\pi}{6}\right]$

$f(0) = f\left(\dfrac{\pi}{6}\right) = 0$

$f$ is continuous on $[0, \pi/6]$. $f$ is differentiable on $(0, \pi/6)$. Rolle's Theorem applies.

$$f'(x) = \frac{6}{\pi} - 8 \sin x \cos x = 0$$

$$\frac{6}{\pi} = 8 \sin x \cos x$$

$$\frac{6}{\pi} = 4 \sin 2x$$

$$\frac{3}{2\pi} = \sin 2x$$

$$\frac{1}{2} \arcsin\left(\frac{3}{2\pi}\right) = x$$

$$x \approx 0.2489$$

$c$-value: $0.2489$

**23.** $f(x) = \tan x, [0, \pi]$

$f(0) = f(\pi) = 0$

$f$ is not continuous on $[0, \pi]$ because $f(\pi/2)$ does not exist. Rolle's Theorem does not apply.

**25.** $f(x) = |x| - 1, [-1, 1]$

$f(-1) = f(1) = 0$

$f$ is continuous on $[-1, 1]$. $f$ is not differentiable on $(-1, 1)$ because $f'(0)$ does not exist. Rolle's Theorem does not apply.

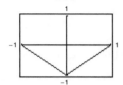

**27.** $f(x) = x - \tan \pi x, \left[-\frac{1}{4}, \frac{1}{4}\right]$

$f\left(-\frac{1}{4}\right) = -\frac{1}{4} + 1 = \frac{3}{4}$

$f\left(\frac{1}{4}\right) = \frac{1}{4} - 1 = -\frac{3}{4}$

Rolle's Theorem does not apply

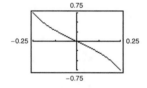

**29.** $f(t) = -16t^2 + 48t + 6$

(a) $f(1) = f(2) = 38$

(b) $v = f'(t)$ must be 0 at some time in $(1, 2)$.

$f'(t) = -32t + 48 = 0$

$t = \frac{3}{2}$ sec

**31.**

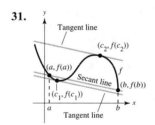

**33.** $f$ is not continuous on the interval $[0, 6]$. ($f$ is not continuous at $x = 2$.)

**35.** $f(x) = \dfrac{1}{x - 3}, [0, 6]$

$f$ has a discontinuity at $x = 3$.

**37.** $f(x) = -x^2 + 5$

(a) Slope $= \dfrac{1 - 4}{2 + 1} = -1$

Secant line: $\quad y - 4 = -(x + 1)$

$y = -x + 3$

$x + y - 3 = 0$

(b) $f'(x) = -2x = -1 \quad \Rightarrow \quad x = c = \dfrac{1}{2}$

(c) $f(c) = f\left(\dfrac{1}{2}\right) = -\dfrac{1}{4} + 5 = \dfrac{19}{4}$

Tangent line: $\quad y - \dfrac{19}{4} = -\left(x - \dfrac{1}{2}\right)$

$4y - 19 = -4x + 2$

$4x + 4y - 21 = 0$

(d)

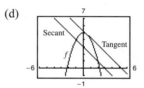

**39.** $f(x) = x^2$ is continuous on $[-2, 1]$ and differentiable on $(-2, 1)$.

$\dfrac{f(1) - f(-2)}{1 - (-2)} = \dfrac{1 - 4}{3} = -1$

$f'(x) = 2x = -1$

$x = -\dfrac{1}{2}$

$c = -\dfrac{1}{2}$

**41.** $f(x) = x^3 + 2x$ is continuous on $[-1, 1]$ and differentiable on $(-1, 1)$.

$\dfrac{f(1) - f(-1)}{1 - (-1)} = \dfrac{3 - (-3)}{2} = 3$

$f'(x) = 3x^2 + 2 = 3$

$3x^2 = 1$

$x = \pm\dfrac{1}{\sqrt{3}}$

$c = \pm\dfrac{\sqrt{3}}{3}$

**43.** $f(x) = x^{2/3}$ is continuous on $[0, 1]$ and differentiable on $(0, 1)$.

$$\frac{f(1) - f(0)}{1 - 0} = 1$$

$$f'(x) = \frac{2}{3}x^{-1/3} = 1$$

$$x = \left(\frac{2}{3}\right)^3 = \frac{8}{27}$$

$$c = \frac{8}{27}$$

**45.** $f(x) = |2x + 1|$ is not differentiable at $x = -1/2$. The Mean Value Theorem does not apply.

**47.** $f(x) = \sin x$ is continuous on $[0, \pi]$ and differentiable on $(0, \pi)$.

$$\frac{f(\pi) - f(0)}{\pi - 0} = \frac{0 - 0}{\pi} = 0$$

$$f'(x) = \cos x = 0$$

$$x = \pi/2$$

$$c = \frac{\pi}{2}$$

**49.** $f(x) = \dfrac{x}{x + 1}, \left[-\dfrac{1}{2}, 2\right]$

(a)–(c)

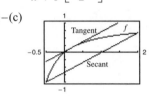

(b) Secant line:

$$\text{slope} = \frac{f(2) - f(-1/2)}{2 - (-1/2)} = \frac{2/3 - (-1)}{5/2} = \frac{2}{3}$$

$$y - \frac{2}{3} = \frac{2}{3}(x - 2)$$

$$y = \frac{2}{3}(x - 1)$$

(c) $f'(x) = \dfrac{1}{(x + 1)^2} = \dfrac{2}{3}$

$$(x + 1)^2 = \frac{3}{2}$$

$$x = -1 \pm \sqrt{\frac{3}{2}} = -1 \pm \frac{\sqrt{6}}{2}$$

In the interval $[-1/2, 2]$: $c = -1 + \left(\sqrt{6}/2\right)$

$$f(c) = \frac{-1 + \left(\sqrt{6}/2\right)}{\left[-1 + \left(\sqrt{6}/2\right)\right] + 1} = \frac{-2 + \sqrt{6}}{\sqrt{6}} = \frac{-2}{\sqrt{6}} + 1$$

Tangent line: $y - 1 + \dfrac{2}{\sqrt{6}} = \dfrac{2}{3}\left(x - \dfrac{\sqrt{6}}{2} + 1\right)$

$$y - 1 + \frac{\sqrt{6}}{3} = \frac{2}{3}x - \frac{\sqrt{6}}{3} + \frac{2}{3}$$

$$y = \frac{1}{3}\left(2x + 5 - 2\sqrt{6}\right)$$

**51.** $f(x) = \sqrt{x}, [1, 9]$

(a)–(c)

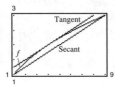

(b) Secant line:

$$\text{slope} = \frac{f(9) - f(1)}{9 - 1} = \frac{3 - 1}{8} = \frac{1}{4}$$

$$y - 1 = \frac{1}{4}(x - 1)$$

$$y = \frac{1}{4}x + \frac{3}{4}$$

(c) $f'(x) = \dfrac{1}{2\sqrt{x}} = \dfrac{1}{4}$

$$x = c = 4$$

$$f(4) = 2$$

Tangent line: $y - 2 = \dfrac{1}{4}(x - 4)$

$$y = \frac{1}{4}x + 1$$

**53.** $s(t) = -4.9t^2 + 300$

(a) $v_{\text{avg}} = \dfrac{s(3) - s(0)}{3 - 0} = \dfrac{255.9 - 300}{3} = -14.7 \text{ m/sec}$

(b) $s(t)$ is continuous on $[0, 3]$ and differentiable on $(0, 3)$. Therefore, the Mean Value Theorem applies.

$$v(t) = s'(t) = -9.8t = -14.7 \text{ m/sec}$$

$$t = \frac{-14.7}{-9.8} = 1.5 \text{ sec}$$

**55.** No. Let $f(x) = x^2$ on $[-1, 2]$.

$$f'(x) = 2x$$

$f'(0) = 0$ and zero is in the interval $(-1, 2)$ but $f(-1) \neq f(2)$.

**57.** $f(x) = \begin{cases} 0, & x = 0 \\ 1 - x, & 0 < x \leq 1 \end{cases}$

No, this does not contradict Rolle's Theorem. $f$ is not continuous on $[0, 1]$.

**59.** Let $S(t)$ be the position function of the plane. If $t = 0$ corresponds to 2 P.M., $S(0) = 0$, $S(5.5) = 2500$ and the Mean Value Theorem says that there exists a time $t_0$, $0 < t_0 < 5.5$, such that

$$S'(t_0) = v(t_0) = \frac{2500 - 0}{5.5 - 0} \approx 454.54.$$

Applying the Intermediate Value Theorem to the velocity function on the intervals $[0, t_0]$ and $[t_0, 5.5]$, you see that there are at least two times during the flight when the speed was 400 miles per hour. $(0 < 400 < 454.54)$

**61.** Let $S(t)$ be the difference in the positions of the 2 bicyclists, $S(t) = S_1(t) - S_2(t)$. Because $S(0) = S(2.25) = 0$, there must exist a time $t_0 \in (0, 2.25)$ such that $S'(t_0) = v(t_0) = 0$.

At this time, $v_1(t_0) = v_2(t_0)$.

**63.** $f(x) = 3\cos^2\left(\dfrac{\pi x}{2}\right)$, $f'(x) = 6\cos\left(\dfrac{\pi x}{2}\right)\left(-\sin\left(\dfrac{\pi x}{2}\right)\right)\left(\dfrac{\pi}{2}\right) = -3\pi\cos\left(\dfrac{\pi x}{2}\right)\sin\left(\dfrac{\pi x}{2}\right)$

(a)

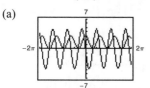

(b) $f$ and $f'$ are both continuous on the entire real line.

(c) Because $f(-1) = f(1) = 0$, Rolle's Theorem applies on $[-1, 1]$. Because $f(1) = 0$ and $f(2) = 3$, Rolle's Theorem does not apply on $[1, 2]$.

(d) $\displaystyle\lim_{x \to 3^-} f'(x) = 0$

$\displaystyle\lim_{x \to 3^+} f'(x) = 0$

**65.** $f$ is continuous on $[-5, 5]$ and does not satisfy the conditions of the Mean Value Theorem. $\Rightarrow$ $f$ is not differentiable on $(-5, 5)$. Example: $f(x) = |x|$

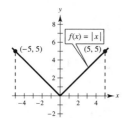

**67.** $f(x) = x^5 + x^3 + x + 1$

$f$ is differentiable for all $x$.

$f(-1) = -2$ and $f(0) = 1$, so the Intermediate Value Theorem implies that $f$ has at least one zero $c$ in $[-1, 0]$, $f(c) = 0$.

Suppose $f$ had 2 zeros, $f(c_1) = f(c_2) = 0$. Then Rolle's Theorem would guarantee the existence of a number $a$ such that

$$f'(a) = f(c_2) - f(c_1) = 0.$$

But, $f'(x) = 5x^4 + 3x^2 + 1 > 0$ for all $x$. So, $f$ has exactly one real solution.

**69.** $f(x) = 3x + 1 - \sin x$

$f$ is differentiable for all $x$.

$f(-\pi) = -3\pi + 1 < 0$ and $f(0) = 1 > 0$, so the Intermediate Value Theorem implies that $f$ has at least one zero $c$ in $[-\pi, 0]$, $f(c) = 0$.

Suppose $f$ had 2 zeros, $f(c_1) = f(c_2) = 0$. Then Rolle's Theorem would guarantee the existence of a number $a$ such that

$$f'(a) = f(c_2) - f(c_1) = 0.$$

But $f'(x) = 3 - \cos x > 0$ for all $x$. So, $f(x) = 0$ has exactly one real solution.

**71.** $f$ continuous at $x = 0$: $1 = b$

$f$ continuous at $x = 1$: $a + 1 = 5 + c$

$f$ differentiable at $x = 1$: $a = 2 + 4 = 6$. So, $c = 2$.

$$f(x) = \begin{cases} 1, & x = 0 \\ 6x + 1, & 0 < x \le 1 \\ x^2 + 4x + 2, & 1 < x \le 3 \end{cases}$$

$$= \begin{cases} 6x + 1, & 0 \le x \le 1 \\ x^2 + 4x + 2, & 1 < x \le 3 \end{cases}$$

**73.** $f'(x) = 0$

$f(x) = c$

$f(2) = 5$

So, $f(x) = 5$.

**75.** $f'(x) = 2x$

$f(x) = x^2 + c$

$f(1) = 0 \Rightarrow 0 = 1 + c \Rightarrow c = -1$

So, $f(x) = x^2 - 1$.

**77.** False. $f(x) = 1/x$ has a discontinuity at $x = 0$.

**79.** True. A polynomial is continuous and differentiable everywhere.

**81.** Suppose that $p(x) = x^{2n+1} + ax + b$ has two real roots $x_1$ and $x_2$. Then by Rolle's Theorem, because $p(x_1) = p(x_2) = 0$, there exists $c$ in $(x_1, x_2)$ such that $p'(c) = 0$. But $p'(x) = (2n + 1)x^{2n} + a \neq 0$, because $n > 0, a > 0$. Therefore, $p(x)$ cannot have two real roots.

**83.** If $p(x) = Ax^2 + Bx + C$, then

$$p'(x) = 2Ax + B = \frac{f(b) - f(a)}{b - a} = \frac{\left(Ab^2 + Bb + C\right) - \left(Aa^2 + Ba + C\right)}{b - a}$$

$$= \frac{A\left(b^2 - a^2\right) + B(b - a)}{b - a}$$

$$= \frac{(b - a)\left[A(b + a) + B\right]}{b - a}$$

$$= A(b + a) + B.$$

So, $2Ax = A(b + a)$ and $x = (b + a)/2$ which is the midpoint of $[a, b]$.

**85.** Suppose $f(x)$ has two fixed points $c_1$ and $c_2$. Then, by the Mean Value Theorem, there exists $c$ such that

$$f'(c) = \frac{f(c_2) - f(c_1)}{c_2 - c_1} = \frac{c_2 - c_1}{c_2 - c_1} = 1.$$

This contradicts the fact that $f'(x) < 1$ for all $x$.

**87.** Let $f(x) = \cos x$. $f$ is continuous and differentiable for all real numbers. By the Mean Value Theorem, for any interval $[a, b]$, there exists $c$ in $(a, b)$ such that

$$\frac{f(b) - f(a)}{b - a} = f'(c)$$

$$\frac{\cos b - \cos a}{b - a} = -\sin c$$

$$\cos b - \cos a = (-\sin c)(b - a)$$

$$|\cos b - \cos a| = |-\sin c||b - a|$$

$$|\cos b - \cos a| \leq |b - a| \text{ since } |-\sin c| \leq 1.$$

**89.** Let $0 < a < b$. $f(x) = \sqrt{x}$ satisfies the hypotheses of the Mean Value Theorem on $[a, b]$. Hence, there exists $c$ in $(a, b)$ such that

$$f'(c) = \frac{1}{2\sqrt{c}} = \frac{f(b) - f(a)}{b - a} = \frac{\sqrt{b} - \sqrt{a}}{b - a}.$$

So $\sqrt{b} - \sqrt{a} = (b - a)\dfrac{1}{2\sqrt{c}} < \dfrac{b - a}{2\sqrt{a}}$.

## Section 3.3   Increasing and Decreasing Functions and the First Derivative Test

**1.** (a) Increasing: $(0, 6)$ and $(8, 9)$. Largest: $(0, 6)$

(b) Decreasing: $(6, 8)$ and $(9, 10)$. Largest: $(6, 8)$

**3.** $f(x) = x^2 - 6x + 8$

From the graph, $f$ is decreasing on $(-\infty, 3)$ and increasing on $(3, \infty)$.

Analytically, $f'(x) = 2x - 6$.

Critical number: $x = 3$

| Test Intervals: | $-\infty < x < 3$ | $3 < x < \infty$ |
|---|---|---|
| Sign of $f'(x)$: | $f' < 0$ | $f' > 0$ |
| Conclusion: | Decreasing | Increasing |

**5.** $y = \dfrac{x^3}{4} - 3x$

From the graph, $y$ is increasing on $(-\infty, -2)$ and $(2, \infty)$, and decreasing on $(-2, 2)$.

Analytically, $y' = \dfrac{3x^2}{4} - 3 = \dfrac{3}{4}(x^2 - 4) = \dfrac{3}{4}(x - 2)(x + 2)$

Critical numbers: $x = \pm 2$

| Test Intervals: | $-\infty < x < -2$ | $-2 < x < 2$ | $2 < x < \infty$ |
|---|---|---|---|
| Sign of $y'$: | $y' > 0$ | $y' < 0$ | $y' > 0$ |
| Conclusion: | Increasing | Decreasing | Increasing |

**7.** $f(x) = \dfrac{1}{(x + 1)^2}$

From the graph, $f$ is increasing on $(-\infty, -1)$ and decreasing on $(-1, \infty)$.

Analytically, $f'(x) = \dfrac{-2}{(x + 1)^3}$.

No critical numbers. Discontinuity: $x = -1$

| Test Intervals: | $-\infty < x < -1$ | $-1 < x < \infty$ |
|---|---|---|
| Sign of $f'(x)$: | $f' > 0$ | $f' < 0$ |
| Conclusion: | Increasing | Decreasing |

**9.** $g(x) = x^2 - 2x - 8$

$g'(x) = 2x - 2$

Critical number: $x = 1$

| Test Intervals: | $-\infty < x < 1$ | $1 < x < \infty$ |
|---|---|---|
| Sign of $g'(x)$: | $g' < 0$ | $g' > 0$ |
| Conclusion: | Decreasing | Increasing |

Increasing on: $(1, \infty)$

Decreasing on: $(-\infty, 1)$

**11.** $y = x\sqrt{16 - x^2}$   Domain: $[-4, 4]$

$$y' = \frac{-2(x^2 - 8)}{\sqrt{16 - x^2}} = \frac{-2}{\sqrt{16 - x^2}}(x - 2\sqrt{2})(x + 2\sqrt{2})$$

Critical numbers: $x = \pm 2\sqrt{2}$

| Test intervals: | $-4 < x < -2\sqrt{2}$ | $-2\sqrt{2} < x < 2\sqrt{2}$ | $2\sqrt{2} < x < 4$ |
|---|---|---|---|
| Sign of $y'$: | $y' < 0$ | $y' > 0$ | $y' < 0$ |
| Conclusion: | Decreasing | Increasing | Decreasing |

Increasing on: $\left(-2\sqrt{2}, 2\sqrt{2}\right)$

Decreasing on: $\left(-4, -2\sqrt{2}\right), \left(2\sqrt{2}, 4\right)$

**13.** $f(x) = \sin x - 1,\quad 0 < x < 2\pi$

$f'(x) = \cos x$

Critical numbers: $x = \dfrac{\pi}{2}, \dfrac{3\pi}{2}$

| Test intervals: | $0 < x < \dfrac{\pi}{2}$ | $\dfrac{\pi}{2} < x < \dfrac{3\pi}{2}$ | $\dfrac{3\pi}{2} < x < 2\pi$ |
|---|---|---|---|
| Sign of $f'(x)$: | $f' > 0$ | $f' < 0$ | $f' > 0$ |
| Conclusion: | Increasing | Decreasing | Increasing |

Increasing on: $\left(0, \dfrac{\pi}{2}\right), \left(\dfrac{3\pi}{2}, 2\pi\right)$

Decreasing on: $\left(\dfrac{\pi}{2}, \dfrac{3\pi}{2}\right)$

**15.** $y = x - 2\cos x,\quad 0 < x < 2\pi$

$y' = 1 + 2\sin x$

$y' = 0$: $\sin x = -\dfrac{1}{2}$

Critical numbers: $x = \dfrac{7\pi}{6}, \dfrac{11\pi}{6}$

| Test intervals: | $0 < x < \dfrac{7\pi}{6}$ | $\dfrac{7\pi}{6} < x < \dfrac{11\pi}{6}$ | $\dfrac{11\pi}{6} < x < 2\pi$ |
|---|---|---|---|
| Sign of $y'$: | $y' > 0$ | $y' < 0$ | $y' > 0$ |
| Conclusion: | Increasing | Decreasing | Increasing |

Increasing on: $\left(0, \dfrac{7\pi}{6}\right), \left(\dfrac{11\pi}{6}, 2\pi\right)$

Decreasing on: $\left(\dfrac{7\pi}{6}, \dfrac{11\pi}{6}\right)$

**17.** (a)  $f(x) = x^2 - 4x$

$f'(x) = 2x - 4$

Critical number: $x = 2$

(b)

| Test intervals: | $-\infty < x < 2$ | $2 < x < \infty$ |
|---|---|---|
| Sign of $f'$: | $f' < 0$ | $f' > 0$ |
| Conclusion: | Decreasing | Increasing |

Decreasing on: $(-\infty, 2)$

Increasing on: $(2, \infty)$

(c)  Relative minimum: $(2, -4)$

**19.** (a)  $f(x) = -2x^2 + 4x + 3$

$f'(x) = -4x + 4 = 0$

Critical number: $x = 1$

(b)

| Test intervals: | $-\infty < x < 1$ | $1 < x < \infty$ |
|---|---|---|
| Sign of $f'(x)$: | $f' > 0$ | $f' < 0$ |
| Conclusion: | Increasing | Decreasing |

Increasing on: $(-\infty, 1)$

Decreasing on: $(1, \infty)$

(c)  Relative maximum: $(1, 5)$

**21.** (a)  $f(x) = 2x^3 + 3x^2 - 12x$

$f'(x) = 6x^2 + 6x - 12 = 6(x + 2)(x - 1) = 0$

Critical numbers: $x = -2, 1$

(b)

| Test intervals: | $-\infty < x < -2$ | $-2 < x < 1$ | $1 < x < \infty$ |
|---|---|---|---|
| Sign of $f'(x)$: | $f' > 0$ | $f' < 0$ | $f' > 0$ |
| Conclusion: | Increasing | Decreasing | Increasing |

Increasing on: $(-\infty, -2), (1, \infty)$

Decreasing on: $(-2, 1)$

(c)  Relative maximum: $(-2, 20)$

Relative minimum: $(1, -7)$

**23.** (a) $f(x) = (x - 1)^2(x + 3) = x^3 + x^2 - 5x + 3$

$f'(x) = 3x^2 + 2x - 5 = (x - 1)(3x + 5)$

Critical numbers: $x = 1, -\frac{5}{3}$

(b)

| Test intervals: | $-\infty < x < -\frac{5}{3}$ | $-5/3 < x < 1$ | $1 < x < \infty$ |
|---|---|---|---|
| Sign of $f'$: | $f' > 0$ | $f' < 0$ | $f' > 0$ |
| Conclusion: | Increasing | Decreasing | Increasing |

Increasing on: $\left(-\infty, -\frac{5}{3}\right)$ and $(1, \infty)$

Decreasing on: $\left(-\frac{5}{3}, 1\right)$

(c) Relative maximum: $\left(-\frac{5}{3}, \frac{256}{27}\right)$

Relative minimum: $(1, 0)$

**25.** (a) $f(x) = \dfrac{x^5 - 5x}{5}$

$f'(x) = x^4 - 1$

Critical numbers: $x = -1, 1$

(b)

| Test intervals: | $-\infty < x < -1$ | $-1 < x < 1$ | $1 < x < \infty$ |
|---|---|---|---|
| Sign of $f'(x)$: | $f' > 0$ | $f' < 0$ | $f' > 0$ |
| Conclusion: | Increasing | Decreasing | Increasing |

Increasing on: $(-\infty, -1), (1, \infty)$

Decreasing on: $(-1, 1)$

(c) Relative maximum: $\left(-1, \dfrac{4}{5}\right)$

Relative minimum: $\left(1, -\dfrac{4}{5}\right)$

**27.** (a) $f(x) = x^{1/3} + 1$

$f'(x) = \dfrac{1}{3}x^{-2/3} = \dfrac{1}{3x^{2/3}}$

Critical number: $x = 0$

(b)

| Test intervals: | $-\infty < x < 0$ | $0 < x < \infty$ |
|---|---|---|
| Sign of $f'(x)$: | $f' > 0$ | $f' > 0$ |
| Conclusion: | Increasing | Increasing |

Increasing on: $(-\infty, \infty)$

(c) No relative extrema

**29.** (a) $f(x) = (x + 2)^{2/3}$

$f'(x) = \dfrac{2}{3}(x + 2)^{-1/3} = \dfrac{2}{3(x + 2)^{1/3}}$

Critical number: $x = -2$

(b)

| Test intervals: | $-\infty < x < -2$ | $-2 < x < \infty$ |
|---|---|---|
| Sign of $f'$: | $f' < 0$ | $f' > 0$ |
| Conclusion: | Decreasing | Increasing |

Decreasing on: $(-\infty, -2)$

Increasing on: $(-2, \infty)$

(c) Relative minimum: $(-2, 0)$

**31. (a)** $f(x) = 5 - |x - 5|$

$$f'(x) = -\frac{x - 5}{|x - 5|} = \begin{cases} 1, & x < 5 \\ -1, & x > 5 \end{cases}$$

Critical number: $x = 5$

**(b)**

| Test intervals: | $-\infty < x < 5$ | $5 < x < \infty$ |
|---|---|---|
| Sign of $f'(x)$: | $f' > 0$ | $f' < 0$ |
| Conclusion: | Increasing | Decreasing |

Increasing on: $(-\infty, 5)$

Decreasing on: $(5, \infty)$

**(c)** Relative maximum: $(5, 5)$

**33. (a)** $f(x) = 2x + \dfrac{1}{x}$

$$f'(x) = 2 - \frac{1}{x^2} = \frac{2x^2 - 1}{x^2}$$

Critical numbers: $x = \pm\dfrac{\sqrt{2}}{2}$

Discontinuity: $x = 0$

**(b)**

| Test intervals: | $-\infty < x < -\dfrac{\sqrt{2}}{2}$ | $-\dfrac{\sqrt{2}}{2} < x < 0$ | $0 < x < \dfrac{\sqrt{2}}{2}$ | $\dfrac{\sqrt{2}}{2} < x < \infty$ |
|---|---|---|---|---|
| Sign of $f'$: | $f' > 0$ | $f' < 0$ | $f' < 0$ | $f' > 0$ |
| Conclusion: | Increasing | Decreasing | Decreasing | Increasing |

Increasing on: $\left(-\infty, -\dfrac{\sqrt{2}}{2}\right)$   and   $\left(\dfrac{\sqrt{2}}{2}, \infty\right)$

Decreasing on: $\left(-\dfrac{\sqrt{2}}{2}, 0\right)$   and   $\left(0, \dfrac{\sqrt{2}}{2}\right)$

**(c)** Relative maximum: $\left(-\dfrac{\sqrt{2}}{2}, -2\sqrt{2}\right)$

Relative minimum: $\left(\dfrac{\sqrt{2}}{2}, 2\sqrt{2}\right)$

**35.** (a) $f(x) = \dfrac{x^2}{x^2 - 9}$

$$f'(x) = \frac{(x^2 - 9)(2x) - (x^2)(2x)}{(x^2 - 9)^2} = \frac{-18x}{(x^2 - 9)^2}$$

Critical number: $x = 0$

Discontinuities: $x = -3, 3$

(b)

| Test intervals: | $-\infty < x < -3$ | $-3 < x < 0$ | $0 < x < 3$ | $3 < x < \infty$ |
|---|---|---|---|---|
| Sign of $f'(x)$: | $f' > 0$ | $f' > 0$ | $f' < 0$ | $f' < 0$ |
| Conclusion: | Increasing | Increasing | Decreasing | Decreasing |

Increasing on: $(-\infty, -3), (-3, 0)$

Decreasing on: $(0, 3), (3, \infty)$

(c) Relative maximum: $(0, 0)$

**37.** (a) $f(x) = \dfrac{x^2 - 2x + 1}{x + 1}$

$$f'(x) = \frac{(x + 1)(2x - 2) - (x^2 - 2x + 1)(1)}{(x + 1)^2} = \frac{x^2 + 2x - 3}{(x + 1)^2} = \frac{(x + 3)(x - 1)}{(x + 1)^2}$$

Critical numbers: $x = -3, 1$

Discontinuity: $x = -1$

(b)

| Test intervals: | $-\infty < x < -3$ | $-3 < x < -1$ | $-1 < x < 1$ | $1 < x < \infty$ |
|---|---|---|---|---|
| Sign of $f'(x)$: | $f' > 0$ | $f' < 0$ | $f' < 0$ | $f' > 0$ |
| Conclusion: | Increasing | Decreasing | Decreasing | Increasing |

Increasing on: $(-\infty, -3), (1, \infty)$

Decreasing on: $(-3, -1), (-1, 1)$

(c) Relative maximum: $(-3, -8)$

Relative minimum: $(1, 0)$

**39.** (a) $f(x) = \begin{cases} 4 - x^2, & x \le 0 \\ -2x, & x > 0 \end{cases}$

$$f'(x) = \begin{cases} -2x, & x < 0 \\ -2, & x > 0 \end{cases}$$

Critical number: $x = 0$

(b)

| Test intervals: | $-\infty < x < 0$ | $0 < x < \infty$ |
|---|---|---|
| Sign of $f'$: | $f' > 0$ | $f' < 0$ |
| Conclusion: | Increasing | Decreasing |

Increasing on: $(-\infty, 0)$

Decreasing on: $(0, \infty)$

(c) No relative extrema. (Note: $(0, 4)$ is an absolute maximum)

**41. (a)** $f(x) = \begin{cases} 3x + 1, & x \le 1 \\ 5 - x^2, & x > 1 \end{cases}$

$f'(x) = \begin{cases} 3, & x < 1 \\ -2x, & x > 1 \end{cases}$

Critical number: $x = 1$

**(b)**

| Test intervals: | $-\infty < x < 1$ | $1 < x < \infty$ |
|---|---|---|
| Sign of $f'$: | $f' > 0$ | $f' < 0$ |
| Conclusion: | Increasing | Decreasing |

Increasing on: $(-\infty, 1)$

Decreasing on: $(1, \infty)$

**(c)** Relative maximum: $(1, 4)$

**43. (a)** $f(x) = \dfrac{x}{2} + \cos x, \, 0 < x < 2\pi$

$f'(x) = \dfrac{1}{2} - \sin x = 0$

Critical numbers: $x = \dfrac{\pi}{6}, \dfrac{5\pi}{6}$

| Test intervals: | $0 < x < \dfrac{\pi}{6}$ | $\dfrac{\pi}{6} < x < \dfrac{5\pi}{6}$ | $\dfrac{5\pi}{6} < x < 2\pi$ |
|---|---|---|---|
| Sign of $f'(x)$: | $f' > 0$ | $f' < 0$ | $f' > 0$ |
| Conclusion: | Increasing | Decreasing | Increasing |

Increasing on: $\left(0, \dfrac{\pi}{6}\right), \left(\dfrac{5\pi}{6}, 2\pi\right)$

Decreasing on: $\left(\dfrac{\pi}{6}, \dfrac{5\pi}{6}\right)$

**(b)** Relative maximum: $\left(\dfrac{\pi}{6}, \dfrac{\pi + 6\sqrt{3}}{12}\right)$

Relative minimum: $\left(\dfrac{5\pi}{6}, \dfrac{5\pi - 6\sqrt{3}}{12}\right)$

**(c)**

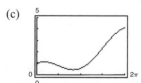

**45.** (a) $f(x) = \sin x + \cos x, \quad 0 < x < 2\pi$

$f'(x) = \cos x - \sin x = 0 \Rightarrow \sin x = \cos x$

Critical numbers: $x = \dfrac{\pi}{4}, \dfrac{5\pi}{4}$

| Test intervals: | $0 < x < \dfrac{\pi}{4}$ | $\dfrac{\pi}{4} < x < \dfrac{5\pi}{4}$ | $\dfrac{5\pi}{4} < x < 2\pi$ |
|---|---|---|---|
| Sign of $f'(x)$: | $f' > 0$ | $f' < 0$ | $f' > 0$ |
| Conclusion: | Increasing | Decreasing | Increasing |

Increasing on: $\left(0, \dfrac{\pi}{4}\right), \left(\dfrac{5\pi}{4}, 2\pi\right)$

Decreasing on: $\left(\dfrac{\pi}{4}, \dfrac{5\pi}{4}\right)$

(b) Relative maximum: $\left(\dfrac{\pi}{4}, \sqrt{2}\right)$

Relative minimum: $\left(\dfrac{5\pi}{4}, -\sqrt{2}\right)$

(c)

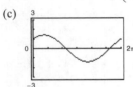

**47.** (a) $f(x) = \cos^2(2x), \quad 0 < x < 2\pi$

$f'(x) = -4\cos 2x \sin 2x = 0 \Rightarrow \cos 2x = 0 \text{ or } \sin 2x = 0$

Critical numbers: $x = \dfrac{\pi}{4}, \dfrac{3\pi}{4}, \dfrac{5\pi}{4}, \dfrac{7\pi}{4}, \dfrac{\pi}{2}, \pi, \dfrac{3\pi}{2}$

| Test intervals: | $0 < x < \dfrac{\pi}{4}$ | $\dfrac{\pi}{4} < x < \dfrac{\pi}{2}$ | $\dfrac{\pi}{2} < x < \dfrac{3\pi}{4}$ | $\dfrac{3\pi}{4} < x < \pi$ |
|---|---|---|---|---|
| Sign of $f'(x)$: | $f' < 0$ | $f' > 0$ | $f' < 0$ | $f' > 0$ |
| Conclusion: | Decreasing | Increasing | Decreasing | Increasing |

| Test intervals: | $\pi < x < \dfrac{5\pi}{4}$ | $\dfrac{5\pi}{4} < x < \dfrac{3\pi}{2}$ | $\dfrac{3\pi}{2} < x < \dfrac{7\pi}{4}$ | $\dfrac{7\pi}{4} < x < 2\pi$ |
|---|---|---|---|---|
| Sign of $f'(x)$: | $f' < 0$ | $f' > 0$ | $f' < 0$ | $f' > 0$ |
| Conclusion: | Decreasing | Increasing | Decreasing | Increasing |

Increasing on: $\left(\dfrac{\pi}{4}, \dfrac{\pi}{2}\right), \left(\dfrac{3\pi}{4}, \pi\right), \left(\dfrac{5\pi}{4}, \dfrac{3\pi}{2}\right), \left(\dfrac{7\pi}{4}, 2\pi\right)$

Decreasing on: $\left(0, \dfrac{\pi}{4}\right), \left(\dfrac{\pi}{2}, \dfrac{3\pi}{4}\right), \left(\pi, \dfrac{5\pi}{4}\right), \left(\dfrac{3\pi}{2}, \dfrac{7\pi}{4}\right)$

(b) Relative maxima: $\left(\dfrac{\pi}{2}, 1\right), (\pi, 1), \left(\dfrac{3\pi}{2}, 1\right)$

Relative minima: $\left(\dfrac{\pi}{4}, 0\right), \left(\dfrac{3\pi}{4}, 0\right), \left(\dfrac{5\pi}{4}, 0\right), \left(\dfrac{7\pi}{4}, 0\right)$

(c)

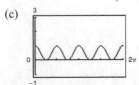

**49. (a)** $f(x) = \sin^2 x + \sin x, \quad 0 < x < 2\pi$

$f'(x) = 2 \sin x \cos x + \cos x = \cos x(2 \sin x + 1) = 0$

Critical numbers: $x = \dfrac{\pi}{2}, \dfrac{7\pi}{6}, \dfrac{3\pi}{2}, \dfrac{11\pi}{6}$

| Test intervals: | $0 < x < \dfrac{\pi}{2}$ | $\dfrac{\pi}{2} < x < \dfrac{7\pi}{6}$ | $\dfrac{7\pi}{6} < x < \dfrac{3\pi}{2}$ | $\dfrac{3\pi}{2} < x < \dfrac{11\pi}{6}$ | $\dfrac{11\pi}{6} < x < 2\pi$ |
|---|---|---|---|---|---|
| Sign of $f'(x)$: | $f' > 0$ | $f' < 0$ | $f' > 0$ | $f' < 0$ | $f' > 0$ |
| Conclusion: | Increasing | Decreasing | Increasing | Decreasing | Increasing |

Increasing on: $\left(0, \dfrac{\pi}{2}\right), \left(\dfrac{7\pi}{6}, \dfrac{3\pi}{2}\right), \left(\dfrac{11\pi}{6}, 2\pi\right)$

Decreasing on: $\left(\dfrac{\pi}{2}, \dfrac{7\pi}{6}\right), \left(\dfrac{3\pi}{2}, \dfrac{11\pi}{6}\right)$

**(b)** Relative minima: $\left(\dfrac{7\pi}{6}, -\dfrac{1}{4}\right), \left(\dfrac{11\pi}{6}, -\dfrac{1}{4}\right)$

Relative maxima: $\left(\dfrac{\pi}{2}, 2\right), \left(\dfrac{3\pi}{2}, 0\right)$

**(c)**

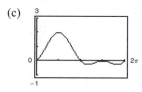

**51.** $f(x) = 2x\sqrt{9 - x^2}, [-3, 3]$

**(a)** $f'(x) = \dfrac{2(9 - 2x^2)}{\sqrt{9 - x^2}}$

**(b)**

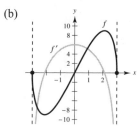

**(c)** $\dfrac{2(9 - 2x^2)}{\sqrt{9 - x^2}} = 0$

Critical numbers: $x = \pm\dfrac{3}{\sqrt{2}} = \pm\dfrac{3\sqrt{2}}{2}$

**(d)** Intervals:

| $\left(-3, -\dfrac{3\sqrt{2}}{2}\right)$ | $\left(-\dfrac{3\sqrt{2}}{2}, \dfrac{3\sqrt{2}}{2}\right)$ | $\left(\dfrac{3\sqrt{2}}{2}, 3\right)$ |
|---|---|---|
| $f'(x) < 0$ | $f'(x) > 0$ | $f'(x) < 0$ |
| Decreasing | Increasing | Decreasing |

$f$ is increasing when $f'$ is positive and decreasing when $f'$ is negative.

**53.** $f(t) = t^2 \sin t, [0, 2\pi]$

**(a)** $f'(t) = t^2 \cos t + 2t \sin t = t(t \cos t + 2 \sin t)$

**(b)**

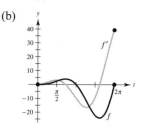

**(c)** $t(t \cos t + 2 \sin t) = 0$

$t = 0 \text{ or } t = -2 \tan t$

$t \cot t = -2$

$t \approx 2.2889, 5.0870 \text{ (graphing utility)}$

Critical numbers: $t = 2.2889, 5.0870$

**(d)** Intervals:

| $(0, 2.2889)$ | $(2.2889, 5.0870)$ | $(5.0870, 2\pi)$ |
|---|---|---|
| $f'(t) > 0$ | $f'(t) < 0$ | $f'(t) > 0$ |
| Increasing | Decreasing | Increasing |

$f$ is increasing when $f'$ is positive and decreasing when $f'$ is negative.

**55. (a)** $f(x) = -3 \sin \dfrac{x}{3}, [0, 6\pi]$

$f'(x) = -\cos \dfrac{x}{3}$

**(b)**

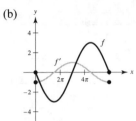

**(c)** Critical numbers: $x = \dfrac{3\pi}{2}, \dfrac{9\pi}{2}$

**(d)** Intervals:

$$\left(0, \dfrac{3\pi}{2}\right) \qquad \left(\dfrac{3\pi}{2}, \dfrac{9\pi}{2}\right) \qquad \left(\dfrac{9\pi}{2}, 6\pi\right)$$

$\quad f' < 0 \qquad\quad f' > 0 \qquad\quad f' < 0$

Decreasing   Increasing   Decreasing

$f$ is increasing when $f'$ is positive and decreasing when $f'$ is negative.

**57.** $f(x) = \dfrac{x^5 - 4x^3 + 3x}{x^2 - 1} = \dfrac{(x^2 - 1)(x^3 - 3x)}{x^2 - 1} = x^3 - 3x, \; x \neq \pm 1$

$f(x) = g(x) = x^3 - 3x$ for all $x \neq \pm 1$.

$f'(x) = 3x^2 - 3 = 3(x^2 - 1), \; x \neq \pm 1 \Rightarrow f'(x) \neq 0$

$f$ symmetric about origin

zeros of $f$: $(0, 0), \left(\pm\sqrt{3}, 0\right)$

$g(x)$ is continuous on $(-\infty, \infty)$ and $f(x)$ has holes at $(-1, 2)$ and $(1, -2)$.

**59.** $f(x) = c$ is constant $\Rightarrow f'(x) = 0$.

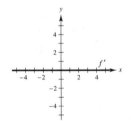

**61.** $f$ is quadratic $\Rightarrow f'$ is a line.

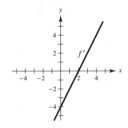

**63.** $f$ has positive, but decreasing slope

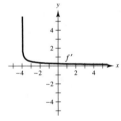

**65. (a)** $f$ increasing on $(2, \infty)$ because $f' > 0$ on $(2, \infty)$

$\quad f$ decreasing on $(-\infty, 2)$ because $f' < 0$ on $(-\infty, 2)$

**(b)** $f$ has a relative minimum at $x = 2$.

**67. (a)** $f$ increasing on $(-\infty, -1)$ and $(0, 1)$ because

$\quad f' > 0$

$\quad f$ decreasing on $(-1, 0)$ and $(1, \infty)$ because

$\quad f' < 0$

**(b)** $f$ has a relative maximum at $x = -1$ and $x = 1$.

$\quad f$ has a relative minimum at $x = 0$.

**69.** (a)  $f' = 0$ at $x = -1, 1, 2$

Critical numbers: $x = -1, 1, 2$

(b)  $x = 1$ is a relative maximum because $f'$ changes from positive to negative.

$x = 2$ is a relative minimum because $f'$ changes from negative to positive. $x = -1$ is not a relative extremum.

**In Exercises 71–75,**  $f'(x) > 0$ **on** $(-\infty, -4)$, $f'(x) < 0$ **on** $(-4, 6)$ **and** $f'(x) > 0$ **on** $(6, \infty)$.

**71.**  $g(x) = f(x) + 5$

$g'(x) = f'(x)$

$g'(0) = f'(0) < 0$

**73.**  $g(x) = -f(x)$

$g'(x) = -f'(x)$

$g'(-6) = -f'(-6) < 0$

**75.**  $g(x) = f(x - 10)$

$g'(x) = f'(x - 10)$

$g'(0) = f'(-10) > 0$

**77.**  $f'(x) \begin{cases} > 0, & x < 4 \Rightarrow f \text{ is increasing on } (-\infty, 4). \\ \text{undefined,} & x = 4 \\ < 0, & x > 4 \Rightarrow f \text{ is decreasing on } (4 \, \infty). \end{cases}$

Two possibilities for $f(x)$ are given below.

(a)

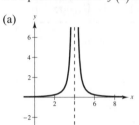

(b)

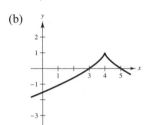

**In Exercise 79, answers will vary.**

*Sample answers:*

**79.** (a)

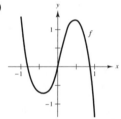

(b) The critical numbers are in intervals $(-0.50, -0.25)$ and $(0.25, 0.50)$ because the sign of $f'$ changes in these intervals.

$f$ is decreasing on approximately $(-1, -0.40)$, $(0.48, 1)$, and increasing on $(-0.40, 0.48)$.

(c) Relative minimum when $x \approx -0.40$: $(-0.40, 0.75)$

Relative maximum when $x \approx 0.48$: $(0.48, 1.25)$

**81.**  $s(t) = 4.9(\sin \theta)t^2$

(a)  $s'(t) = 4.9(\sin \theta)(2t) = 9.8(\sin \theta)t$

speed $= |s'(t)| = |9.8(\sin \theta)t|$

(b)

| $\theta$ | 0 | $\dfrac{\pi}{4}$ | $\dfrac{\pi}{3}$ | $\dfrac{\pi}{2}$ | $\dfrac{2\pi}{3}$ | $\dfrac{3\pi}{4}$ | $\pi$ |
|---|---|---|---|---|---|---|---|
| $|s'(t)|$ | 0 | $4.9\sqrt{2}t$ | $4.9\sqrt{3}t$ | $9.8t$ | $4.9\sqrt{3}t$ | $4.9\sqrt{2}t$ | 0 |

The speed is maximum for $\theta = \dfrac{\pi}{2}$.

**83.** $f(x) = x, g(x) = \sin x, 0 < x < \pi$

(a)

| $x$ | 0.5 | 1 | 1.5 | 2 | 2.5 | 3 |
|---|---|---|---|---|---|---|
| $f(x)$ | 0.5 | 1 | 1.5 | 2 | 2.5 | 3 |
| $g(x)$ | 0.479 | 0.841 | 0.997 | 0.909 | 0.598 | 0.141 |

$f(x)$ seems greater than $g(x)$ on $(0, \pi)$.

(b)

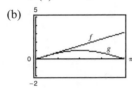

$x > \sin x$ on $(0, \pi)$ so, $f(x) > g(x)$.

(c) Let $h(x) = f(x) - g(x) = x - \sin x$

$h'(x) = 1 - \cos x > 0$ on $(0, \pi)$.

Therefore, $h(x)$ is increasing on $(0, \pi)$. Because $h(0) = 0$ and $h'(x) > 0$ on $(0, \pi)$,

$$h(x) > 0$$
$$x - \sin x > 0$$
$$x > \sin x$$
$$f(x) > g(x) \text{ on } (0, \pi)$$

**85.** $v = k(R - r)r^2 = k(Rr^2 - r^3)$

$v' = k(2Rr - 3r^2)$

$\quad = kr(2R - 3r) = 0$

$r = 0 \text{ or } \frac{2}{3}R$

Maximum when $r = \frac{2}{3}R$.

**87.** $R = \sqrt{0.001T^4 - 4T + 100}$

(a) $R' = \dfrac{0.004T^3 - 4}{2\sqrt{0.001T^4 - 4T + 100}} = 0$

Critical number: $T = 10°$

Minimum resistance: $R \approx 8.3666$ ohms

(b)

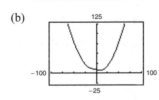

The minimum resistance is approximately $R \approx 8.37$ ohms at $T = 10°$.

**89.** (a) $s(t) = 6t - t^2, t \geq 0$

$v(t) = 6 - 2t$

(b) $v(t) = 0$ when $t = 3$.

Moving in positive direction for $0 < t < 3$ because $v(t) > 0$ on $0 < t < 3$.

(c) Moving in negative direction when $t > 3$.

(d) The particle changes direction at $t = 3$.

**91.** (a) $s(t) = t^3 - 5t^2 + 4t, t \geq 0$

$v(t) = 3t^2 - 10t + 4$

(b) $v(t) = 0$ for $t = \dfrac{10 \pm \sqrt{100 - 48}}{6} = \dfrac{5 \pm \sqrt{13}}{3}$

Particle is moving in a positive direction on

$\left(0, \dfrac{5 - \sqrt{13}}{3}\right) \approx (0, 0.4648)$ and

$\left(\dfrac{5 + \sqrt{13}}{3}, \infty\right) \approx (2.8685, \infty)$ because $v > 0$ on

these intervals.

(c) Particle is moving in a negative direction on

$\left(\dfrac{5 - \sqrt{13}}{3}, \dfrac{5 + \sqrt{13}}{3}\right) \approx (0.4648, 2.8685)$

(d) The particle changes direction at $t = \dfrac{5 \pm \sqrt{13}}{3}$.

**93.** Answers will vary.

**95.** (a) Use a cubic polynomial

$$f(x) = a_3x^3 + a_2x^2 + a_1x + a_0$$

(b) $f'(x) = 3a_3x^2 + 2a_2x + a_1$.

$$f(0) = 0: \quad a_3(0)^3 + a_2(0)^2 + a_1(0) + a_0 = 0 \Rightarrow \qquad a_0 = 0$$

$$f'(0) = 0: \qquad 3a_3(0)^2 + 2a_2(0) + a_1 = 0 \Rightarrow \qquad a_1 = 0$$

$$f(2) = 2: \quad a_3(2)^3 + a_2(2)^2 + a_1(2) + a_0 = 2 \Rightarrow \quad 8a_3 + 4a_2 = 2$$

$$f'(2) = 0: \qquad 3a_3(2)^2 + 2a_2(2) + a_1 = 0 \Rightarrow \quad 12a_3 + 4a_2 = 0$$

(c) The solution is $a_0 = a_1 = 0, a_2 = \frac{3}{2}, a_3 = -\frac{1}{2}$:

$$f(x) = -\frac{1}{2}x^3 + \frac{3}{2}x^2.$$

(d)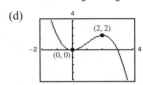

**97.** (a) Use a fourth degree polynomial

$$f(x) = a_4x^4 + a_3x^3 + a_2x^2 + a_1x + a_0.$$

(b) $f'(x) = 4a_4x^3 + 3a_3x^2 + 2a_2x + a_1$

$$f(0) = 0: \quad a_4(0)^4 + a_3(0)^3 + a_2(0)^2 + a_1(0) + a_0 = 0 \Rightarrow \qquad a_0 = 0$$

$$f'(0) = 0: \qquad 4a_4(0)^3 + 3a_3(0)^2 + 2a_2(0) + a_1 = 0 \Rightarrow \qquad a_1 = 0$$

$$f(4) = 0: \quad a_4(4)^4 + a_3(4)^3 + a_2(4)^2 + a_1(4) + a_0 = 0 \Rightarrow \quad 256a_4 + 64a_3 + 16a_2 = 0$$

$$f'(4) = 0: \qquad 4a_4(4)^3 + 3a_3(4)^2 + 2a_2(4) + a_1 = 0 \Rightarrow \quad 256a_4 + 48a_3 + 8a_2 = 0$$

$$f(2) = 4: \quad a_4(2)^4 + a_3(2)^3 + a_2(2)^2 + a_1(2) + a_0 = 4 \Rightarrow \quad 16a_4 + 8a_3 + 4a_2 = 4$$

$$f'(2) = 0: \qquad 4a_4(2)^3 + 3a_3(2)^2 + 2a_2(2) + a_1 = 0 \Rightarrow \quad 32a_4 + 12a_3 + 4a_2 = 0$$

(c) The solution is $a_0 = a_1 = 0, a_2 = 4, a_3 = -2, \quad a_4 = \frac{1}{4}$.

$$f(x) = \frac{1}{4}x^4 - 2x^3 + 4x^2$$

(d)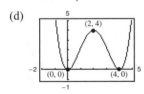

**99.** True.

Let $h(x) = f(x) + g(x)$ where $f$ and $g$ are increasing.

Then $h'(x) = f'(x) + g'(x) > 0$ because

$f'(x) > 0$ and $g'(x) > 0$.

**101.** False.

Let $f(x) = x^3$, then $f'(x) = 3x^2$ and $f$ only has one

critical number. Or, let $f(x) = x^3 + 3x + 1$, then

$f'(x) = 3(x^2 + 1)$ has no critical numbers.

**103.** False. For example, $f(x) = x^3$ does not have a relative

extrema at the critical number $x = 0$.

**105.** Suppose $f'(x)$ changes from positive to negative at $c$.

Then there exists $a$ and $b$ in $I$ such that $f'(x) > 0$ for all

$x$ in $(a, c)$ and $f'(x) < 0$ for all $x$ in $(c, b)$. By Theorem

3.5, $f$ is increasing on $(a, c)$ and decreasing on $(c, b)$.

Therefore, $f(c)$ is a maximum of $f$ on $(a, b)$ and so, a

relative maximum of $f$.

**107.** Let $x_1$ and $x_2$ be two positive real numbers, $0 < x_1 < x_2$. Then

$$\frac{1}{x_1} > \frac{1}{x_2}$$

$$f(x_1) > f(x_2)$$

So, $f$ is decreasing on $(0, \infty)$.

# Section 3.4    Concavity and the Second Derivative Test

**1.** The graph of $f$ is increasing and concave upwards:
$f' > 0, \ f'' > 0$

**3.** The graph of $f$ is decreasing and concave downward:
$f' < 0, \ f'' < 0$

**5.** $y = x^2 - x - 2$

$y' = 2x - 1$

$y'' = 2$

Concave upward: $(-\infty, \infty)$

**7.** $g(x) = 3x^2 - x^3$

$g'(x) = 6x - 3x^2$

$g''(x) = 6 - 6x$

Concave upward: $(-\infty, 1)$

Concave downward: $(1, \infty)$

**9.** $f(x) = -x^3 + 6x^2 - 9x - 1$

$f'(x) = -3x^2 + 12x - 9$

$f''(x) = -6x + 12 = -6(x - 2)$

Concave upward: $(-\infty, 2)$

Concave downward: $(2, \infty)$

**11.** $f(x) = \dfrac{24}{x^2 + 12}$

$f' = \dfrac{-48x}{\left(x^2 + 12\right)^2}$

$f'' = \dfrac{-144\left(4 - x^2\right)}{\left(x^2 + 12\right)^3}$

Concave upward: $(-\infty, -2), (2, \infty)$

Concave downward: $(-2, 2)$

**13.** $f(x) = \dfrac{x^2 + 1}{x^2 - 1}$

$f' = \dfrac{-4x}{\left(x^2 - 1\right)^2}$

$f'' = \dfrac{4\left(3x^2 + 1\right)}{\left(x^2 - 1\right)^3}$

Concave upward: $(-\infty, -1), (1, \infty)$

Concave downward: $(-1, 1)$

**15.** $g(x) = \dfrac{x^2 + 4}{4 - x^2}$

$g'(x) = \dfrac{16x}{\left(4 - x^2\right)^2}$

$g''(x) = \dfrac{16\left(3x^2 + 4\right)}{\left(4 - x^2\right)^3} = \dfrac{16\left(3x^2 + 4\right)}{(2 - x)^3(2 + x)^3}$

Concave upward: $(-2, 2)$

Concave downward: $(-\infty, -2), (2, \infty)$

**17.** $y = 2x - \tan x, \left(-\dfrac{\pi}{2}, \dfrac{\pi}{2}\right)$

$y' = 2 - \sec^2 x$

$y'' = -2 \sec^2 x \tan x$

Concave upward: $\left(-\dfrac{\pi}{2}, 0\right)$

Concave downward: $\left(0, \dfrac{\pi}{2}\right)$

**19.** $f(x) = \frac{1}{2}x^4 + 2x^3$

$f'(x) = 2x^3 + 6x^2$

$f''(x) = 6x^2 + 12x = 6x(x + 2)$

$f''(x) = 0$ when $x = 0, -2$

Concave upward: $(-\infty, -2), (0, \infty)$

Concave downward: $(-2, 0)$

Points of inflection: $(-2, -8)$ and $(0, 0)$

**21.** $f(x) = x^3 - 6x^2 + 12x$

$f'(x) = 3x^2 - 12x + 12$

$f''(x) = 6(x - 2) = 0$ when $x = 2$.

Concave upward: $(2, \infty)$

Concave downward: $(-\infty, 2)$

Point of inflection: $(2, 8)$

**23.** $f(x) = \dfrac{1}{4}x^4 - 2x^2$

$f'(x) = x^3 - 4x$

$f''(x) = 3x^2 - 4$

$f''(x) = 3x^2 - 4 = 0$ when $x = \pm\dfrac{2}{\sqrt{3}}$.

| Test interval: | $-\infty < x < -\dfrac{2}{\sqrt{3}}$ | $-\dfrac{2}{\sqrt{3}} < x < \dfrac{2}{\sqrt{3}}$ | $\dfrac{2}{\sqrt{3}} < x < \infty$ |
|---|---|---|---|
| Sign of $f''(x)$: | $f''(x) > 0$ | $f''(x) < 0$ | $f''(x) > 0$ |
| Conclusion: | Concave upward | Concave downward | Concave upward |

Points of inflection: $\left(\pm\dfrac{2}{\sqrt{3}}, -\dfrac{20}{9}\right)$

**25.** $f(x) = x(x - 4)^3$

$f'(x) = x\left[3(x - 4)^2\right] + (x - 4)^3 = (x - 4)^2(4x - 4)$

$f''(x) = 4(x - 1)\left[2(x - 4)\right] + 4(x - 4)^2 = 4(x - 4)\left[2(x - 1) + (x - 4)\right] = 4(x - 4)(3x - 6) = 12(x - 4)(x - 2)$

$f''(x) = 12(x - 4)(x - 2) = 0$ when $x = 2, 4$.

| Test interval: | $-\infty < x < 2$ | $2 < x < 4$ | $4 < x < \infty$ |
|---|---|---|---|
| Sign of $f''(x)$: | $f''(x) > 0$ | $f''(x) < 0$ | $f''(x) > 0$ |
| Conclusion: | Concave upward | Concave downward | Concave upward |

Points of inflection: $(2, -16), (4, 0)$

**27.** $f(x) = x\sqrt{x + 3}$, Domain: $[-3, \infty)$

$f'(x) = x\left(\dfrac{1}{2}\right)(x + 3)^{-1/2} + \sqrt{x + 3} = \dfrac{3(x + 2)}{2\sqrt{x + 3}}$

$f''(x) = \dfrac{6\sqrt{x + 3} - 3(x + 2)(x + 3)^{-1/2}}{4(x + 3)}$

$= \dfrac{3(x + 4)}{4(x + 3)^{3/2}}$

$f''(x) > 0$ on the entire domain of $f$ (except for $x = -3$, for which $f''(x)$ is undefined). There are no points of inflection.

Concave upward: $(-3, \infty)$

**29.** $f(x) = \dfrac{4}{x^2 + 1}$

$f'(x) = \dfrac{-8x}{\left(x^2 + 1\right)^2}$

$f''(x) = \dfrac{8\left(3x^2 - 1\right)}{\left(x^2 + 1\right)^3}$

$f''(x) = 0$ for $x = \pm\dfrac{\sqrt{3}}{3}$

Concave upward: $\left(-\infty, -\dfrac{\sqrt{3}}{3}\right), \left(\dfrac{\sqrt{3}}{3}, \infty\right)$

Concave downward: $\left(-\dfrac{\sqrt{3}}{3}, \dfrac{\sqrt{3}}{3}\right)$

Points of inflection: $\left(-\dfrac{\sqrt{3}}{3}, 3\right)$ and $\left(\dfrac{\sqrt{3}}{3}, 3\right)$

**31.** $f(x) = \sin\dfrac{x}{2}, 0 \le x \le 4\pi$

$f'(x) = \dfrac{1}{2}\cos\left(\dfrac{x}{2}\right)$

$f''(x) = -\dfrac{1}{4}\sin\left(\dfrac{x}{2}\right)$

$f''(x) = 0$ when $x = 0, 2\pi, 4\pi$.

Point of inflection: $(2\pi, 0)$

| Test interval: | $0 < x < 2\pi$ | $2\pi < x < 4\pi$ |
|---|---|---|
| Sign of $f''(x)$: | $f'' < 0$ | $f'' > 0$ |
| Conclusion: | Concave downward | Concave upward |

**33.** $f(x) = \sec\left(x - \dfrac{\pi}{2}\right), 0 < x < 4\pi$

$f'(x) = \sec\left(x - \dfrac{\pi}{2}\right)\tan\left(x - \dfrac{\pi}{2}\right)$

$f''(x) = \sec^3\left(x - \dfrac{\pi}{2}\right) + \sec\left(x - \dfrac{\pi}{2}\right)\tan^2\left(x - \dfrac{\pi}{2}\right) \ne 0$ for any $x$ in the domain of $f$.

Concave upward: $(0, \pi), (2\pi, 3\pi)$

Concave downward: $(\pi, 2\pi), (3\pi, 4\pi)$

No point of inflection

**35.** $f(x) = 2\sin x + \sin 2x, 0 \le x \le 2\pi$

$f'(x) = 2\cos x + 2\cos 2x$

$f''(x) = -2\sin x - 4\sin 2x = -2\sin x(1 + 4\cos x)$

$f''(x) = 0$ when $x = 0, 1.823, \pi, 4.460$.

| Test interval: | $0 < x < 1.823$ | $1.823 < x < \pi$ | $\pi < x < 4.460$ | $4.460 < x < 2\pi$ |
|---|---|---|---|---|
| Sign of $f''(x)$: | $f'' < 0$ | $f'' > 0$ | $f'' < 0$ | $f'' > 0$ |
| Conclusion: | Concave downward | Concave upward | Concave downward | Concave upward |

Points of inflection: $(1.823, 1.452), (\pi, 0), (4.46, -1.452)$

**37.** $f(x) = (x - 5)^2$

$f'(x) = 2(x - 5)$

$f''(x) = 2$

Critical number: $x = 5$

$f''(5) > 0$

Therefore, $(5, 0)$ is a relative minimum.

**39.** $f(x) = 6x - x^2$

$f'(x) = 6 - 2x$

$f''(x) = -2$

Critical number: $x = 3$

$f''(3) < 0$

Therefore, $(3, 9)$ is a relative maximum.

**41.** $f(x) = x^3 - 3x^2 + 3$

$f'(x) = 3x^2 - 6x = 3x(x - 2)$

$f''(x) = 6x - 6 = 6(x - 1)$

Critical numbers: $x = 0, x = 2$

$f''(0) = -6 < 0$

Therefore, $(0, 3)$ is a relative maximum.

$f''(2) = 6 > 0$

Therefore, $(2, -1)$ is a relative minimum.

**43.** $f(x) = x^4 - 4x^3 + 2$

$f'(x) = 4x^3 - 12x^2 = 4x^2(x - 3)$

$f''(x) = 12x^2 - 24x = 12x(x - 2)$

Critical numbers: $x = 0, x = 3$

However, $f''(0) = 0$, so you must use the First Derivative

Test. $f'(x) < 0$ on the intervals $(-\infty, 0)$ and $(0, 3)$; so,

$(0, 2)$ is not an extremum. $f''(3) > 0$ so $(3, -25)$ is a

relative minimum.

**45.** $g(x) = x^2(6 - x)^3$

$g'(x) = x(x - 6)^2(12 - 5x)$

$g''(x) = 4(6 - x)(5x^2 - 24x + 18)$

Critical numbers: $x = 0, \frac{12}{5}, 6$

$g''(0) = 432 > 0$

Therefore, $(0, 0)$ is a relative minimum.

$g''\left(\frac{12}{5}\right) = -155.52 < 0$

Therefore, $\left(\frac{12}{5}, 268.7\right)$ is a relative maximum.

$g''(6) = 0$

Test fails. By the First Derivative Test, $(6, 0)$ is not an extremum.

**47.** $f(x) = x^{2/3} - 3$

$f'(x) = \dfrac{2}{3x^{1/3}}$

$f''(x) = -\dfrac{2}{9x^{4/3}}$

Critical number: $x = 0$

However, $f''(0)$ is undefined, so you must use the First Derivative Test. Because $f'(x) < 0$ on $(-\infty, 0)$ and $f'(x) > 0$ on $(0, \infty)$, $(0, -3)$ is a relative minimum.

**49.** $f(x) = x + \dfrac{4}{x}$

$f'(x) = 1 - \dfrac{4}{x^2} = \dfrac{x^2 - 4}{x^2}$

$f''(x) = \dfrac{8}{x^3}$

Critical numbers: $x = \pm 2$

$f''(-2) < 0$

Therefore, $(-2, -4)$ is a relative maximum.

$f''(2) > 0$

Therefore, $(2, 4)$ is a relative minimum.

**51.** $f(x) = \cos x - x, 0 \le x \le 4\pi$

$f'(x) = -\sin x - 1 \le 0$

Therefore, $f$ is non-increasing and there are no relative extrema.

**53.** $f(x) = 0.2x^2(x - 3)^3, [-1, 4]$

(a) $f'(x) = 0.2x(5x - 6)(x - 3)^2$

$f''(x) = (x - 3)(4x^2 - 9.6x + 3.6)$

$\quad\quad = 0.4(x - 3)(10x^2 - 24x + 9)$

(b) $f''(0) < 0 \Rightarrow (0, 0)$ is a relative maximum.

$f''\left(\frac{6}{5}\right) > 0 \Rightarrow (1.2, -1.6796)$ is a relative minimum.

Points of inflection:

$(3, 0), (0.4652, -0.7048), (1.9348, -0.9049)$

(c)

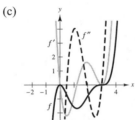

$f$ is increasing when $f' > 0$ and decreasing when $f' < 0$. $f$ is concave upward when $f'' > 0$ and concave downward when $f'' < 0$.

**55.** $f(x) = \sin x - \dfrac{1}{3}\sin 3x + \dfrac{1}{5}\sin 5x, \quad [0, \pi]$

(a) $f'(x) = \cos x - \cos 3x + \cos 5x$

$f'(x) = 0$ when $x = \dfrac{\pi}{6}, x = \dfrac{\pi}{2}, x = \dfrac{5\pi}{6}$.

$f''(x) = -\sin x + 3\sin 3x - 5\sin 5x$

$f''(x) = 0$ when $x = \dfrac{\pi}{6}, x = \dfrac{5\pi}{6}$,

$x \approx 1.1731, x \approx 1.9685$

(b) $f''\left(\dfrac{\pi}{2}\right) < 0 \Rightarrow \left(\dfrac{\pi}{2}, 1.53333\right)$ is a relative maximum.

Points of inflection: $\left(\dfrac{\pi}{6}, 0.2667\right), (1.1731, 0.9638),$

$(1.9685, 0.9637), \left(\dfrac{5\pi}{6}, 0.2667\right)$

**Note:** $(0, 0)$ and $(\pi, 0)$ are not points of inflection because they are endpoints.

(c)

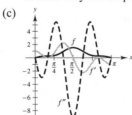

The graph of $f$ is increasing when $f' > 0$ and decreasing when $f' < 0$. $f$ is concave upward when $f'' > 0$ and concave downward when $f'' < 0$.

**57. (a)**

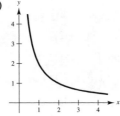

$f' < 0$ means $f$ decreasing

$f'$ increasing means concave upward

**(b)**

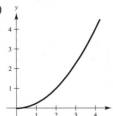

$f' > 0$ means $f$ increasing

$f'$ increasing means concave upward

**59.** Answers will vary. *Sample answer*:

Let $f(x) = x^4$.

$f''(x) = 12x^2$

$f''(0) = 0$, but $(0, 0)$ is not a point of inflection.

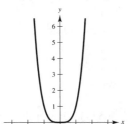

**61.**

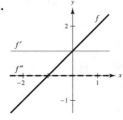

**63.**

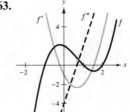

**65.**

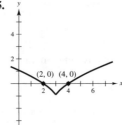

**67.**

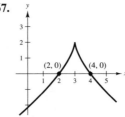

**69.**

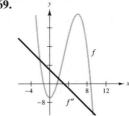

$f''$ is linear.

$f'$ is quadratic.

$f$ is cubic.

$f$ concave upward on $(-\infty, 3)$, downward on $(3, \infty)$.

**71. (a)**

| $n = 1$: | $n = 2$: | $n = 3$: | $n = 4$: |
|---|---|---|---|
| $f(x) = x - 2$ | $f(x) = (x - 2)^2$ | $f(x) = (x - 2)^3$ | $f(x) = (x - 2)^4$ |
| $f'(x) = 1$ | $f'(x) = 2(x - 2)$ | $f'(x) = 3(x - 2)^2$ | $f'(x) = 4(x - 2)^3$ |
| $f''(x) = 0$ | $f''(x) = 2$ | $f''(x) = 6(x - 2)$ | $f''(x) = 12(x - 2)^2$ |
| No point of inflection | No point of inflection | Point of inflection: $(2, 0)$ | No point of inflection |
| | Relative minimum: $(2, 0)$ | | Relative minimum: $(2, 0)$ |

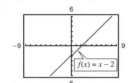
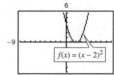

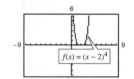

**Conclusion:** If $n \geq 3$ and $n$ is odd, then $(2, 0)$ is point of inflection. If $n \geq 2$ and $n$ is even, then $(2, 0)$ is a relative minimum.

**(b)** Let $f(x) = (x - 2)^n$, $f'(x) = n(x - 2)^{n-1}$, $f''(x) = n(n - 1)(x - 2)^{n-2}$.

For $n \geq 3$ and odd, $n - 2$ is also odd and the concavity changes at $x = 2$.

For $n \geq 4$ and even, $n - 2$ is also even and the concavity does not change at $x = 2$.

So, $x = 2$ is point of inflection if and only if $n \geq 3$ is odd.

**73.** $f(x) = ax^3 + bx^2 + cx + d$

Relative maximum: $(3, 3)$

Relative minimum: $(5, 1)$

Point of inflection: $(4, 2)$

$f'(x) = 3ax^2 + 2bx + c,\ f''(x) = 6ax + 2b$

$\left.\begin{array}{l} f(3) = 27a + 9b + 3c + d = 3 \\ f(5) = 125a + 25b + 5c + d = 1 \end{array}\right\} 98a + 16b + 2c = -2 \Rightarrow 49a + 8b + c = -1$

$f'(3) = 27a + 6b + c = 0,\ f''(4) = 24a + 2b = 0$

$$
\begin{array}{ll}
49a + 8b + c = -1 & 24a + 2b = \phantom{-}0 \\
\underline{27a + 6b + c = \phantom{-}0} & \underline{22a + 2b = -1} \\
22a + 2b \phantom{+ c} = -1 & \phantom{22}2a \phantom{+ 2b} = \phantom{-}1
\end{array}
$$

$a = \frac{1}{2}, b = -6, c = \frac{45}{2}, d = -24$

$f(x) = \frac{1}{2}x^3 - 6x^2 + \frac{45}{2}x - 24$

**75.** $f(x) = ax^3 + bx^2 + cx + d$

Maximum: $(-4, 1)$

Minimum: $(0, 0)$

**(a)** $f'(x) = 3ax^2 + 2bx + c,\qquad f''(x) = 6ax + 2b$

$$
\begin{array}{l}
f(0) = 0 \Rightarrow d = 0 \\
f(-4) = 1 \Rightarrow \phantom{-}{-64a} + 16b - 4c = 1 \\
f'(-4) = 0 \Rightarrow \phantom{-}48a - \phantom{1}8b + \phantom{4}c = 0 \\
f'(0) = 0 \Rightarrow \phantom{48a - 8b +}\ c = 0
\end{array}
$$

Solving this system yields $a = \frac{1}{32}$ and $b = 6a = \frac{3}{16}$.

$f(x) = \frac{1}{32}x^3 + \frac{3}{16}x^2$

**(b)** The plane would be descending at the greatest rate at the point of inflection.

$f''(x) = 6ax + 2b = \frac{3}{16}x + \frac{3}{8} = 0 \Rightarrow x = -2$.

Two miles from touchdown.

**77.** $D = 2x^4 - 5Lx^3 + 3L^2x^2$

$D' = 8x^3 - 15Lx^2 + 6L^2x = x(8x^2 - 15Lx + 6L^2) = 0$

$x = 0$ or $x = \dfrac{15L \pm \sqrt{33}L}{16} = \left(\dfrac{15 \pm \sqrt{33}}{16}\right)L$

By the Second Derivative Test, the deflection is maximum when

$x = \left(\dfrac{15 - \sqrt{33}}{16}\right)L \approx 0.578L.$

**79.** $C = 0.5x^2 + 15x + 5000$

$\overline{C} = \dfrac{C}{x} = 0.5x + 15 + \dfrac{5000}{x}$

$\overline{C} = $ average cost per unit

$\dfrac{d\overline{C}}{dx} = 0.5 - \dfrac{5000}{x^2} = 0$ when $x = 100$

By the First Derivative Test, $\overline{C}$ is minimized when $x = 100$ units.

**81.** $S = \dfrac{5000t^2}{8 + t^2}, 0 \le t \le 3$

(a)

| $t$ | 0.5 | 1 | 1.5 | 2 | 2.5 | 3 |
|-----|-----|-----|-----|-----|-----|-----|
| $S$ | 151.5 | 555.6 | 1097.6 | 1666.7 | 2193.0 | 2647.1 |

Increasing at greatest rate when $1.5 < t < 2$

(b)

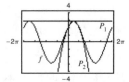

Increasing at greatest rate when $t \approx 1.5$.

(c) $S = \dfrac{5000t^2}{8 + t^2}$

$S'(t) = \dfrac{80{,}000t}{(8 + t^2)^2}$

$S''(t) = \dfrac{80{,}000(8 - 3t^2)}{(8 + t^2)^3}$

$S''(t) = 0$ for $t = \pm\sqrt{\dfrac{8}{3}}$. So, $t = \dfrac{2\sqrt{6}}{3} \approx 1.633$ yrs.

**83.** $f(x) = 2(\sin x + \cos x), \quad f\left(\dfrac{\pi}{4}\right) = 2\sqrt{2}$

$f'(x) = 2(\cos x - \sin x), \quad f'\left(\dfrac{\pi}{4}\right) = 0$

$f''(x) = 2(-\sin x - \cos x), \quad f''\left(\dfrac{\pi}{4}\right) = -2\sqrt{2}$

$P_1(x) = 2\sqrt{2} + 0\left(x - \dfrac{\pi}{4}\right) = 2\sqrt{2}$

$P_1'(x) = 0$

$P_2(x) = 2\sqrt{2} + 0\left(x - \dfrac{\pi}{4}\right) + \dfrac{1}{2}(-2\sqrt{2})\left(x - \dfrac{\pi}{4}\right)^2 = 2\sqrt{2} - \sqrt{2}\left(x - \dfrac{\pi}{4}\right)^2$

$P_2'(x) = -2\sqrt{2}\left(x - \dfrac{\pi}{4}\right)$

$P_2''(x) = -2\sqrt{2}$

The values of $f$, $P_1$, $P_2$, and their first derivatives are equal at $x = \pi/4$. The values of the second derivatives of $f$ and $P_2$ are equal at $x = \pi/4$. The approximations worsen as you move away from $x = \pi/4$.

**85.** $f(x) = \sqrt{1 - x},$ $\qquad f(0) = 1$

$f'(x) = -\dfrac{1}{2\sqrt{1 - x}},$ $\qquad f'(0) = -\dfrac{1}{2}$

$f''(x) = -\dfrac{1}{4(1 - x)^{3/2}},$ $\qquad f''(0) = -\dfrac{1}{4}$

$P_1(x) = 1 + \left(-\dfrac{1}{2}\right)(x - 0) = 1 - \dfrac{x}{2}$

$P_1'(x) = -\dfrac{1}{2}$

$P_2(x) = 1 + \left(-\dfrac{1}{2}\right)(x - 0) + \dfrac{1}{2}\left(-\dfrac{1}{4}\right)(x - 0)^2 = 1 - \dfrac{x}{2} - \dfrac{x^2}{8}$

$P_2'(x) = -\dfrac{1}{2} - \dfrac{x}{4}$

$P_2''(x) = -\dfrac{1}{4}$

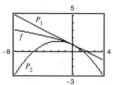

The values of $f$, $P_1$, $P_2$, and their first derivatives are equal at $x = 0$. The values of the second derivatives of $f$ and $P_2$ are equal at $x = 0$. The approximations worsen as you move away from $x = 0$.

**87.** $f(x) = x \sin\left(\dfrac{1}{x}\right)$

$f'(x) = x\left[-\dfrac{1}{x^2}\cos\left(\dfrac{1}{x}\right)\right] + \sin\left(\dfrac{1}{x}\right) = -\dfrac{1}{x}\cos\left(\dfrac{1}{x}\right) + \sin\left(\dfrac{1}{x}\right)$

$f''(x) = -\dfrac{1}{x}\left[\dfrac{1}{x^2}\sin\left(\dfrac{1}{x}\right)\right] + \dfrac{1}{x^2}\cos\left(\dfrac{1}{x}\right) - \dfrac{1}{x^2}\cos\left(\dfrac{1}{x}\right) = -\dfrac{1}{x^3}\sin\left(\dfrac{1}{x}\right) = 0$

$x = \dfrac{1}{\pi}$

Point of inflection: $\left(\dfrac{1}{\pi}, 0\right)$

When $x > 1/\pi$, $f'' < 0$, so the graph is concave downward.

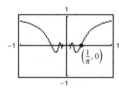

**89.** Assume the zero of $f$ are all real. Then express the function as $f(x) = a(x - r_1)(x - r_2)(x - r_3)$ where $r_1$, $r_2$, and $r_3$ are the distinct zeros of $f$. From the Product Rule for a function involving three factors, we have

$f'(x) = a\left[(x - r_1)(x - r_2) + (x - r_1)(x - r_3) + (x - r_2)(x - r_3)\right]$

$f''(x) = a\left[(x - r_1) + (x - r_2) + (x - r_1) + (x - r_3) + (x - r_2) + (x - r_3)\right] = a\left[6x - 2(r_1 + r_2 + r_3)\right].$

Consequently, $f''(x) = 0$ if $x = \dfrac{2(r_1 + r_2 + r_3)}{6} = \dfrac{r_1 + r_2 + r_3}{3} = $ (Average of $r_1$, $r_2$, and $r_3$).

**91.** True. Let $y = ax^3 + bx^2 + cx + d, a \neq 0$. Then $y'' = 6ax + 2b = 0$ when $x = -(b/3a)$, and the concavity changes at this point.

**93.** False. Concavity is determined by $f''$. For example, let $f(x) = x$ and $c = 2$. $f'(c) = f'(2) > 0$, but $f$ is not concave upward at $c = 2$.

**95.** $f$ and $g$ are concave upward on $(a, b)$ implies that $f'$ and $g'$ are increasing on $(a, b)$, and $f'' > 0$ and $g'' > 0$. So, $(f + g)'' > 0 \Rightarrow f + g$ is concave upward on $(a, b)$ by Theorem 3.7

## Section 3.5 Limits at Infinity

**1.** $f(x) = \dfrac{2x^2}{x^2 + 2}$

No vertical asymptotes

Horizontal asymptote: $y = 2$

Matches (f )

**3.** $f(x) = \dfrac{x}{x^2 + 2}$

No vertical asymptotes

Horizontal asymptote: $y = 0$

$f(1) < 1$

Matches (d)

**5.** $f(x) = \dfrac{4 \sin x}{x^2 + 1}$

No vertical asymptotes

Horizontal asymptote: $y = 0$

$f(1) > 1$

Matches (b)

**7.** $f(x) = \dfrac{4x + 3}{2x - 1}$

| $x$ | $10^0$ | $10^1$ | $10^2$ | $10^3$ | $10^4$ | $10^5$ | $10^6$ |
|-----|--------|--------|--------|--------|--------|--------|--------|
| $f(x)$ | 7 | 2.26 | 2.025 | 2.0025 | 2.0003 | 2 | 2 |

$\lim\limits_{x \to \infty} f(x) = 2$

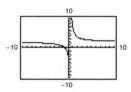

**9.** $f(x) = \dfrac{-6x}{\sqrt{4x^2 + 5}}$

| $x$ | $10^0$ | $10^1$ | $10^2$ | $10^3$ | $10^4$ | $10^5$ | $10^6$ |
|-----|--------|--------|--------|--------|--------|--------|--------|
| $f(x)$ | $-2$ | $-2.98$ | $-2.9998$ | $-3$ | $-3$ | $-3$ | $-3$ |

$\lim\limits_{x \to \infty} f(x) = -3$

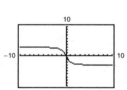

**11.** $f(x) = 5 - \dfrac{1}{x^2 + 1}$

| $x$ | $10^0$ | $10^1$ | $10^2$ | $10^3$ | $10^4$ | $10^5$ | $10^6$ |
|-----|--------|--------|--------|--------|--------|--------|--------|
| $f(x)$ | 4.5 | 4.99 | 4.9999 | 4.999999 | 5 | 5 | 5 |

$\lim\limits_{x \to \infty} f(x) = 5$

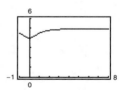

**13.** (a) $h(x) = \dfrac{f(x)}{x^2} = \dfrac{5x^3 - 3x^2 + 10x}{x^2} = 5x - 3 + \dfrac{10}{x}$

$\lim\limits_{x \to \infty} h(x) = \infty$  (Limit does not exist)

(b) $h(x) = \dfrac{f(x)}{x^3} = \dfrac{5x^3 - 3x^2 + 10x}{x^3} = 5 - \dfrac{3}{x} + \dfrac{10}{x^2}$

$\lim\limits_{x \to \infty} h(x) = 5$

(c) $h(x) = \dfrac{f(x)}{x^4} = \dfrac{5x^3 - 3x^2 + 10x}{x^4} = \dfrac{5}{x} - \dfrac{3}{x^2} + \dfrac{10}{x^3}$

$\lim\limits_{x \to \infty} h(x) = 0$

**15.** (a) $\lim_{x \to \infty} \dfrac{x^2 + 2}{x^3 - 1} = 0$

    (b) $\lim_{x \to \infty} \dfrac{x^2 + 2}{x^2 - 1} = 1$

    (c) $\lim_{x \to \infty} \dfrac{x^2 + 2}{x - 1} = \infty$   (Limit does not exist)

**17.** (a) $\lim_{x \to \infty} \dfrac{5 - 2x^{3/2}}{3x^2 - 4} = 0$

    (b) $\lim_{x \to \infty} \dfrac{5 - 2x^{3/2}}{3x^{3/2} - 4} = -\dfrac{2}{3}$

    (c) $\lim_{x \to \infty} \dfrac{5 - 2x^{3/2}}{3x - 4} = -\infty$   (Limit does not exist)

**19.** $\lim_{x \to \infty} \left( 4 + \dfrac{3}{x} \right) = 4 + 0 = 4$

**21.** $\lim_{x \to \infty} \dfrac{2x - 1}{3x + 2} = \lim_{x \to \infty} \dfrac{2 - (1/x)}{3 + (2/x)} = \dfrac{2 - 0}{3 + 0} = \dfrac{2}{3}$

**23.** $\lim_{x \to \infty} \dfrac{x}{x^2 - 1} = \lim_{x \to \infty} \dfrac{1/x}{1 - \left( 1/x^2 \right)} = \dfrac{0}{1} = 0$

**25.** $\lim_{x \to -\infty} \dfrac{5x^2}{x + 3} = \lim_{x \to -\infty} \dfrac{5x}{1 + (3/x)} = -\infty$

Limit does not exist

**27.** $\lim_{x \to -\infty} \dfrac{x}{\sqrt{x^2 - x}} = \lim_{x \to -\infty} \dfrac{1}{\left( \dfrac{\sqrt{x^2 - x}}{-\sqrt{x^2}} \right)} = \lim_{x \to -\infty} \dfrac{-1}{\sqrt{1 - (1/x)}} = -1, \left( \text{for } x < 0 \text{ we have } x = -\sqrt{x^2} \right)$

**29.** $\lim_{x \to -\infty} \dfrac{2x + 1}{\sqrt{x^2 - x}} = \lim_{x \to -\infty} \dfrac{2 + \dfrac{1}{x}}{\left( \dfrac{\sqrt{x^2 - x}}{-\sqrt{x^2}} \right)} = \lim_{x \to -\infty} \dfrac{-2 - \left( \dfrac{1}{x} \right)}{\sqrt{1 - \dfrac{1}{x}}} = -2, \left( \text{for } x < 0, x = -\sqrt{x^2} \right)$

**31.** $\lim_{x \to \infty} \dfrac{\sqrt{x^2 - 1}}{2x - 1} = \lim_{x \to \infty} \dfrac{\sqrt{x^2 - 1}/\sqrt{x^2}}{2 - 1/x}$

              $= \lim_{x \to \infty} \dfrac{\sqrt{1 - 1/x^2}}{2 - 1/x} = \dfrac{1}{2}$

**33.** $\lim_{x \to \infty} \dfrac{x + 1}{\left( x^2 + 1 \right)^{1/3}} = \lim_{x \to \infty} \dfrac{x + 1}{\left( x^2 + 1 \right)^{1/3}} \left( \dfrac{1/x^{2/3}}{1/\left( x^2 \right)^{1/3}} \right)$

              $= \lim_{x \to \infty} \dfrac{x^{1/3} + 1/x^{2/3}}{\left( 1 + 1/x^2 \right)^{1/3}} = \infty$

          (Limit does not exist)

**35.** $\lim_{x \to \infty} \dfrac{1}{2x + \sin x} = 0$

**37.** Because $(-1/x) \le (\sin 2x)/x \le (1/x)$ for all $x \ne 0$, you have by the Squeeze Theorem,

$$\lim_{x \to \infty} -\dfrac{1}{x} \le \lim_{x \to \infty} \dfrac{\sin 2x}{x} \le \lim_{x \to \infty} \dfrac{1}{x}$$

$$0 \le \lim_{x \to \infty} \dfrac{\sin 2x}{x} \le 0.$$

Therefore, $\lim_{x \to \infty} \dfrac{\sin 2x}{x} = 0.$

**39.** $f(x) = \dfrac{|x|}{x + 1}$

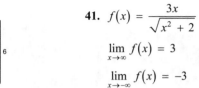

$\displaystyle\lim_{x \to \infty} \dfrac{|x|}{x + 1} = 1$

$\displaystyle\lim_{x \to -\infty} \dfrac{|x|}{x + 1} = -1$

Therefore, $y = 1$ and $y = -1$ are both horizontal asymptotes.

**41.** $f(x) = \dfrac{3x}{\sqrt{x^2 + 2}}$

$\displaystyle\lim_{x \to \infty} f(x) = 3$

$\displaystyle\lim_{x \to -\infty} f(x) = -3$

Therefore, $y = 3$ and $y = -3$ are both horizontal asymptotes.

**43.** $\displaystyle\lim_{x \to \infty} x \sin \dfrac{1}{x} = \lim_{t \to 0^+} \dfrac{\sin t}{t} = 1$

(Let $x = 1/t$.)

**45.** $\displaystyle\lim_{x \to -\infty}\left(x + \sqrt{x^2 + 3}\right) = \lim_{x \to -\infty}\left[\left(x + \sqrt{x^2 + 3}\right) \cdot \dfrac{x - \sqrt{x^2 + 3}}{x - \sqrt{x^2 + 3}}\right] = \lim_{x \to -\infty} \dfrac{-3}{x - \sqrt{x^2 + 3}} = 0$

**47.** $\displaystyle\lim_{x \to -\infty}\left(3x + \sqrt{9x^2 - x}\right) = \lim_{x \to -\infty}\left[\left(3x + \sqrt{9x^2 - x}\right) \cdot \dfrac{3x - \sqrt{9x^2 - x}}{3x - \sqrt{9x^2 - x}}\right]$

$= \displaystyle\lim_{x \to -\infty} \dfrac{x}{3x - \sqrt{9x^2 - x}}$

$= \displaystyle\lim_{x \to -\infty} \dfrac{1}{3 - \dfrac{\sqrt{9x^2 - x}}{-\sqrt{x^2}}}$    $\left(\text{for } x < 0 \text{ you have } x = -\sqrt{x^2}\right)$

$= \displaystyle\lim_{x \to -\infty} \dfrac{1}{3 + \sqrt{9 - (1/x)}} = \dfrac{1}{6}$

**49.**

| $x$ | $10^0$ | $10^1$ | $10^2$ | $10^3$ | $10^4$ | $10^5$ | $10^6$ |
|---|---|---|---|---|---|---|---|
| $f(x)$ | 1 | 0.513 | 0.501 | 0.500 | 0.500 | 0.500 | 0.500 |

$\displaystyle\lim_{x \to \infty}\left(x - \sqrt{x(x - 1)}\right) = \lim_{x \to \infty} \dfrac{x - \sqrt{x^2 - x}}{1} \cdot \dfrac{x + \sqrt{x^2 - x}}{x + \sqrt{x^2 - x}} = \lim_{x \to \infty} \dfrac{x}{x + \sqrt{x^2 - x}} = \lim_{x \to \infty} \dfrac{1}{1 + \sqrt{1 - (1/x)}} = \dfrac{1}{2}$

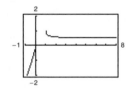

**51.**

| $x$ | $10^0$ | $10^1$ | $10^2$ | $10^3$ | $10^4$ | $10^5$ | $10^6$ |
|---|---|---|---|---|---|---|---|
| $f(x)$ | 0.479 | 0.500 | 0.500 | 0.500 | 0.500 | 0.500 | 0.500 |

Let $x = 1/t$.

$\displaystyle\lim_{x \to \infty} x \sin\left(\dfrac{1}{2x}\right) = \lim_{t \to 0^+} \dfrac{\sin(t/2)}{t} = \lim_{t \to 0^+} \dfrac{1}{2} \dfrac{\sin(t/2)}{t/2} = \dfrac{1}{2}$

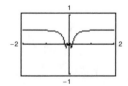

**53.** $\lim_{x \to \infty} f(x) = 4$ means that $f(x)$ approaches 4 as $x$ becomes large.

**55.** $x = 2$ is a critical number.

$f'(x) < 0$ for $x < 2$.

$f'(x) > 0$ for $x > 2$.

$\lim_{x \to -\infty} f(x) = \lim_{x \to \infty} f(x) = 6$

For example, let $f(x) = \dfrac{-6}{0.1(x - 2)^2 + 1} + 6$.

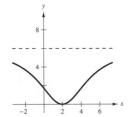

**57.** (a) The function is even: $\lim_{x \to -\infty} f(x) = 5$

(b) The function is odd: $\lim_{x \to -\infty} f(x) = -5$

**59.** $y = \dfrac{x}{1 - x}$

Intercept: $(0, 0)$

Symmetry: none

Horizontal asymptote: $y = -1$

Vertical asymptote: $x = 1$

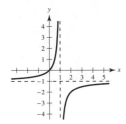

**61.** $y = \dfrac{x + 1}{x^2 - 4}$

Intercepts: $(0, -1/4), (-1, 0)$

Symmetry: none

Horizontal asymptote: $y = 0$

Vertical asymptotes: $x = \pm 2$

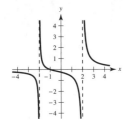

**63.** $y = \dfrac{x^2}{x^2 + 16}$

Intercept: $(0, 0)$

Symmetry: $y$-axis

Horizontal asymptote: $y = 1$

$y' = \dfrac{32x}{(x^2 + 16)^2}$

Relative minimum: $(0, 0)$

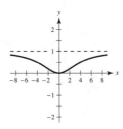

**65.** $y = \dfrac{2x^2}{x^2 - 4}$

Intercept: $(0, 0)$

Symmetry: $y$-axis

Horizontal asymptote: $y = 2$

Vertical asymptotes: $x = \pm 2$

Relative maximum: $(0, 0)$

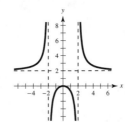

**67.** $xy^2 = 9$

Domain: $x > 0$

Intercepts: none

Symmetry: $x$-axis

$y = \pm \dfrac{3}{\sqrt{x}}$

Horizontal asymptote: $y = 0$

Vertical asymptote: $x = 0$

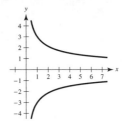

**69.** $y = \dfrac{3x}{1-x}$

Intercept: $(0, 0)$

Symmetry: none

Horizontal asymptote: $y = -3$

Vertical asymptote: $x = 1$

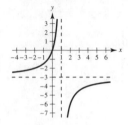

**71.** $y = 2 - \dfrac{3}{x^2}$

Intercepts: $\left( \pm\sqrt{\dfrac{3}{2}}, 0 \right)$

Symmetry: $y$-axis

Horizontal asymptote: $y = 2$ because

$$\lim_{x \to -\infty}\left( 2 - \dfrac{3}{x^2} \right) = 2 = \lim_{x \to \infty}\left( 2 - \dfrac{3}{x^2} \right).$$

Discontinuity: $x = 0$ (Vertical asymptote)

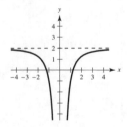

**73.** $y = 3 + \dfrac{2}{x}$

Intercept:

$$y = 0 = 3 + \dfrac{2}{x} \Rightarrow \dfrac{2}{x} = -3 \Rightarrow x = -\dfrac{2}{3}; \left( -\dfrac{2}{3}, 0 \right)$$

Symmetry: none

Horizontal asymptote: $y = 3$

Vertical asymptote: $x = 0$

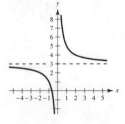

**75.** $y = \dfrac{x^3}{\sqrt{x^2 - 4}}$

Domain: $(-\infty, -2), (2, \infty)$

Intercepts: none

Symmetry: origin

Horizontal asymptote: none

Vertical asymptotes: $x = \pm 2$ (discontinuities)

**77.** $f(x) = 9 - \dfrac{5}{x^2}$

Domain: all $x \neq 0$

$f'(x) = \dfrac{10}{x^3} \Rightarrow$ No relative extrema

$f''(x) = -\dfrac{30}{x^4} \Rightarrow$ No points of inflection

Vertical asymptote: $x = 0$

Horizontal asymptote: $y = 9$

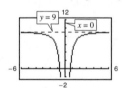

**79.** $f(x) = \dfrac{x-2}{x^2 - 4x + 3} = \dfrac{x-2}{(x-1)(x-3)}$

$f'(x) = \dfrac{(x^2 - 4x + 3) - (x-2)(2x-4)}{(x^2 - 4x + 3)^2} = \dfrac{-x^2 + 4x - 5}{(x^2 - 4x + 3)^2} \neq 0$

$f''(x) = \dfrac{(x^2 - 4x + 3)^2(-2x+4) - (-x^2 + 4x - 5)(2)(x^2 - 4x + 3)(2x-4)}{(x^2 - 4x + 3)^4}$

$\quad = \dfrac{2(x^3 - 6x^2 + 15x - 14)}{(x^2 - 4x + 3)^3} = \dfrac{2(x-2)(x^2 - 4x + 7)}{(x^2 - 4x + 3)^3} = 0$ when $x = 2.$

Because $f''(x) > 0$ on $(1, 2)$ and $f''(x) < 0$ on $(2, 3)$, then $(2, 0)$ is a point of inflection.

Vertical asymptotes: $x = 1, x = 3$

Horizontal asymptote: $y = 0$

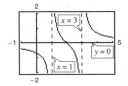

**81.** $f(x) = \dfrac{3x}{\sqrt{4x^2 + 1}}$

$f'(x) = \dfrac{3}{(4x^2 + 1)^{3/2}} \Rightarrow$ No relative extrema

$f''(x) = \dfrac{-36x}{(4x^2 + 1)^{5/2}} = 0$ when $x = 0.$

Point of inflection: $(0, 0)$

Horizontal asymptotes: $y = \pm\dfrac{3}{2}$

No vertical asymptotes

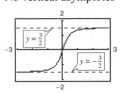

**83.** $g(x) = \sin\left(\dfrac{x}{x-2}\right), 3 < x < \infty$

$g'(x) = \dfrac{-2\cos\left(\dfrac{x}{x-2}\right)}{(x-2)^2}$

Horizontal asymptote:

$y = \sin(1)$

Relative maximum:

$\dfrac{x}{x-2} = \dfrac{\pi}{2} \Rightarrow x = \dfrac{2\pi}{\pi - 2} \approx 5.5039$

No vertical asymptotes

**85.** $f(x) = \dfrac{x^3 - 3x^2 + 2}{x(x-3)}, g(x) = x + \dfrac{2}{x(x-3)}$

(a)

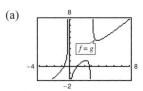

(b) $f(x) = \dfrac{x^3 - 3x^2 + 2}{x(x-3)}$

$\quad = \dfrac{x^2(x-3)}{x(x-3)} + \dfrac{2}{x(x-3)}$

$\quad = x + \dfrac{2}{x(x-3)} = g(x)$

(c)

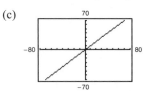

The graph appears as the slant asymptote $y = x.$

**87.** $\displaystyle\lim_{v_1/v_2 \to \infty} 100\left[1 - \dfrac{1}{(v_1/v_2)^c}\right] = 100[1 - 0] = 100\%$

**89.** $\displaystyle\lim_{t \to \infty} N(t) = \infty$

$\displaystyle\lim_{t \to \infty} E(t) = c$

**91.** $y = \dfrac{3.351t^2 + 42.461t - 543.730}{t^2}$

(a)

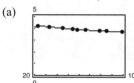

(b) Yes. $\lim\limits_{t \to \infty} y = 3.351$

**93.** (a) $T_1(t) = -0.003t^2 + 0.68t + 26.6$

(b)

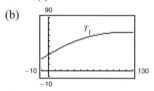

(c)

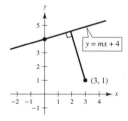

$T_2 = \dfrac{1451 + 86t}{58 + t}$

(d) $T_1(0) \approx 26.6°$

$T_2(0) \approx 25.0°$

(e) $\lim\limits_{t \to \infty} T_2 = \dfrac{86}{1} = 86$

(f) No. The limiting temperature is $86°$.

$T_1$ has no horizontal asymptote.

**95.** line: $mx - y + 4 = 0$

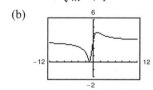

(a) $d = \dfrac{|Ax_1 + By_1 + C|}{\sqrt{A^2 + B^2}} = \dfrac{|m(3) - 1(1) + 4|}{\sqrt{m^2 + 1}}$

$= \dfrac{|3m + 3|}{\sqrt{m^2 + 1}}$

(b)

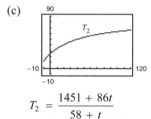

(c) $\lim\limits_{m \to \infty} d(m) = 3 = \lim\limits_{m \to -\infty} d(m)$

The line approaches the vertical line $x = 0$. So, the distance from $(3, 1)$ approaches 3.

**97.** $f(x) = \dfrac{2x^2}{x^2 + 2}$

(a) $\lim\limits_{x \to \infty} f(x) = 2 = L$

(b) $f(x_1) + \varepsilon = \dfrac{2x_1^2}{x_1^2 + 2} + \varepsilon = 2$

$2x_1^2 + \varepsilon x_1^2 + 2\varepsilon = 2x_1^2 + 4$

$x_1^2 \varepsilon = 4 - 2\varepsilon$

$x_1 = \sqrt{\dfrac{4 - 2\varepsilon}{\varepsilon}}$

$x_2 = -x_1$ by symmetry

(c) Let $M = \sqrt{\dfrac{4 - 2\varepsilon}{\varepsilon}} > 0$. For $x > M$:

$x > \sqrt{\dfrac{4 - 2\varepsilon}{\varepsilon}}$

$x^2 \varepsilon > 4 - 2\varepsilon$

$2x^2 + x^2 \varepsilon + 2\varepsilon > 2x^2 + 4$

$\dfrac{2x^2}{x^2 + 2} + \varepsilon > 2$

$\left| \dfrac{2x^2}{x^2 + 2} - 2 \right| > |-\varepsilon| = \varepsilon$

$|f(x) - L| > \varepsilon$

(d) Similarly, $N = -\sqrt{\dfrac{4 - 2\varepsilon}{\varepsilon}}$.

**99.** $\lim\limits_{x \to \infty} \dfrac{3x}{\sqrt{x^2 + 3}} = 3$

$f(x_1) + \varepsilon = \dfrac{3x_1}{\sqrt{x_1^2 + 3}} + \varepsilon = 3$

$3x_1 = (3 - \varepsilon)\sqrt{x_1^2 + 3}$

$9x_1^2 = (3 - \varepsilon)^2(x_1^2 + 3)$

$9x_1^2 - (3 - \varepsilon)^2 x_1^2 = 3(3 - \varepsilon)^2$

$x_1^2(9 - 9 + 6\varepsilon - \varepsilon^2) = 3(3 - \varepsilon)^2$

$x_1^2 = \dfrac{3(3 - \varepsilon)^2}{6\varepsilon - \varepsilon^2}$

$x_1 = (3 - \varepsilon)\sqrt{\dfrac{3}{6\varepsilon - \varepsilon^2}}$

Let $M = x_1 = (3 - \varepsilon)\sqrt{\dfrac{3}{6\varepsilon - \varepsilon^2}}$

(a) When $\varepsilon = 0.5$:

$M = (3 - 0.5)\sqrt{\dfrac{3}{6(0.5) - (0.5)^2}} = \dfrac{5\sqrt{33}}{11}$

(b) When $\varepsilon = 0.1$:

$M = (3 - 0.1)\sqrt{\dfrac{3}{6(0.1) - (0.1)^2}} = \dfrac{29\sqrt{177}}{59}$

**101.** $\lim\limits_{x \to \infty} \dfrac{1}{x^2} = 0$. Let $\varepsilon > 0$ be given. You need

$M > 0$ such that

$$\left| f(x) - L \right| = \left| \dfrac{1}{x^2} - 0 \right| = \dfrac{1}{x^2} < \varepsilon \text{ whenever } x > M.$$

$$x^2 > \dfrac{1}{\varepsilon} \Rightarrow x > \dfrac{1}{\sqrt{\varepsilon}}$$

Let $M = \dfrac{1}{\sqrt{\varepsilon}}$.

For $x > M$, you have

$$x > \dfrac{1}{\sqrt{\varepsilon}} \Rightarrow x^2 > \dfrac{1}{\varepsilon} \Rightarrow \dfrac{1}{x^2} < \varepsilon \Rightarrow \left| f(x) - L \right| < \varepsilon.$$

**103.** $\lim\limits_{x \to -\infty} \dfrac{1}{x^3} = 0$. Let $\varepsilon > 0$. You need $N < 0$ such that

$$\left| f(x) - L \right| = \left| \dfrac{1}{x^3} - 0 \right| = \dfrac{-1}{x^3} < \varepsilon \text{ whenever } x < N.$$

$$\dfrac{-1}{x^3} < \varepsilon \Rightarrow -x^3 > \dfrac{1}{\varepsilon} \Rightarrow x < \dfrac{-1}{\varepsilon^{1/3}}$$

Let $N = \dfrac{-1}{\sqrt[3]{\varepsilon}}$.

For $x < N < \dfrac{-1}{\sqrt[3]{\varepsilon}}$,

$$\dfrac{1}{x} > -\sqrt[3]{\varepsilon}$$

$$-\dfrac{1}{x} < \sqrt[3]{\varepsilon}$$

$$-\dfrac{1}{x^3} < \varepsilon$$

$$\Rightarrow \left| f(x) - L \right| < \varepsilon.$$

**105.** $\lim\limits_{x \to \infty} \dfrac{p(x)}{q(x)} = \lim\limits_{x \to \infty} \dfrac{a_n x^n + \cdots + a_1 x + a_0}{b_m x^m + \cdots + b_1 x + b_0}$

Divide $p(x)$ and $q(x)$ by $x^m$.

**Case 1:** If $n < m$: $\lim\limits_{x \to \infty} \dfrac{p(x)}{q(x)} = \lim\limits_{x \to \infty} \dfrac{\dfrac{a_n}{x^{m-n}} + \cdots + \dfrac{a_1}{x^{m-1}} + \dfrac{a_0}{x^m}}{b_m + \cdots + \dfrac{b_1}{x^{m-1}} + \dfrac{b_0}{x^m}} = \dfrac{0 + \cdots + 0 + 0}{b_m + \cdots + 0 + 0} = \dfrac{0}{b_m} = 0.$

**Case 2:** If $m = n$: $\lim\limits_{x \to \infty} \dfrac{p(x)}{q(x)} = \lim\limits_{x \to \infty} \dfrac{a_n + \cdots + \dfrac{a_1}{x^{m-1}} + \dfrac{a_0}{x^m}}{b_m + \cdots + \dfrac{b_1}{x^{m-1}} + \dfrac{b_0}{x^m}} = \dfrac{a_n + \cdots + 0 + 0}{b_m + \cdots + 0 + 0} = \dfrac{a_n}{b_m}.$

**Case 3:** If $n > m$: $\lim\limits_{x \to \infty} \dfrac{p(x)}{q(x)} = \lim\limits_{x \to \infty} \dfrac{a_n x^{n-m} + \cdots + \dfrac{a_1}{x^{m-1}} + \dfrac{a_0}{x^m}}{b_m + \cdots + \dfrac{b_1}{x^{m-1}} + \dfrac{b_0}{x^m}} = \dfrac{\pm\infty + \cdots + 0}{b_m + \cdots + 0} = \pm\infty.$

**107.** False. Let $f(x) = \dfrac{2x}{\sqrt{x^2 + 2}}$. (See Exercise 54.)

# Section 3.6   A Summary of Curve Sketching

**1.** $f$ has constant negative slope. Matches (d)

**3.** The slope is periodic, and zero at $x = 0$. Matches (a)

**5.** $y = \dfrac{1}{x - 2} - 3$

$y' = -\dfrac{1}{(x - 2)^2} < 0$ when $x \neq 2$.

$y'' = \dfrac{2}{(x - 2)^3}$

No relative extrema,
no points of inflection

Intercepts: $\left( \dfrac{7}{3}, 0 \right), \left( 0, -\dfrac{7}{2} \right)$

Vertical asymptote: $x = 2$

Horizontal asymptote: $y = -3$

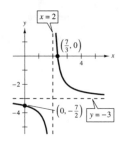

**7.** $y = \dfrac{x^2}{x^2 + 3}$

$y' = \dfrac{6x}{\left(x^2 + 3\right)^2} = 0$ when $x = 0$.

$y'' = \dfrac{18\left(1 - x^2\right)}{\left(x^2 + 3\right)^3} = 0$ when $x = \pm 1$.

Horizontal asymptote: $y = 1$

|  | $y$ | $y'$ | $y''$ | Conclusion |
|---|---|---|---|---|
| $-\infty < x < -1$ |  | $-$ | $-$ | Decreasing, concave down |
| $x = -1$ | $\dfrac{1}{4}$ | $-$ | $0$ | Point of inflection |
| $-1 < x < 0$ |  | $-$ | $+$ | Decreasing, concave up |
| $x = 0$ | $0$ | $0$ | $+$ | Relative minimum |
| $0 < x < 1$ |  | $+$ | $+$ | Increasing, concave up |
| $x = 1$ | $\dfrac{1}{4}$ | $+$ | $0$ | Point of inflection |
| $1 < x < \infty$ |  | $+$ | $-$ | Increasing, concave down |

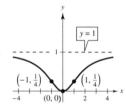

**9.** $y = \dfrac{3x}{x^2 - 1}$

$y' = \dfrac{-3\left(x^2 + 1\right)}{\left(x^2 - 1\right)^2} < 0$ if $x \neq \pm 1$

$y'' = \dfrac{6x\left(x^2 + 3\right)}{\left(x^2 - 1\right)^3}$

Inflection point: $(0, 0)$

Intercept: $(0, 0)$

Symmetry with respect to origin

Vertical asymptotes: $x = \pm 1$

Horizontal asymptote: $y = 0$

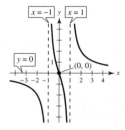

**11.** $g(x) = x - \dfrac{8}{x^2} = \dfrac{x^3 - 8}{x^2}$

$g'(x) = \dfrac{x^3 + 16}{x^3} = 0$ if $x = \sqrt[3]{-16}$

$g''(x) = -48/x^4 < 0$ if $x \neq 0$

Therefore $\left(\sqrt[3]{-16}, -3\sqrt[3]{2}\right) \approx (-2.52, -3.78)$ is a relative maximum.

Intercept: $(2, 0)$

Vertical asymptote: $x = 0$

Slant asymptote: $y = x$

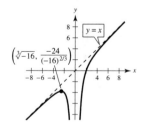

**13.**  $f(x) = \dfrac{x^2 + 1}{x} = x + \dfrac{1}{x}$

$f'(x) = 1 - \dfrac{1}{x^2} = 0$ when $x = \pm 1$.

$f''(x) = \dfrac{2}{x^3} \neq 0$

Relative maximum: $(-1, -2)$

Relative minimum: $(1, 2)$

Vertical asymptote: $x = 0$

Slant asymptote: $y = x$

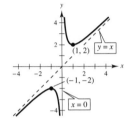

**15.**  $y = \dfrac{x^2 - 6x + 12}{x - 4} = x - 2 + \dfrac{4}{x - 4}$

$y' = 1 - \dfrac{4}{(x - 4)^2}$

$\quad = \dfrac{(x - 2)(x - 6)}{(x - 4)^2} = 0$ when $x = 2, 6$.

$y'' = \dfrac{8}{(x - 4)^3}$

$y'' < 0$ when $x = 2$.

Therefore, $(2, -2)$ is a relative maximum.

$y'' > 0$ when $x = 6$.

Therefore, $(6, 6)$ is a relative minimum.

Vertical asymptote: $x = 4$

Slant asymptote: $y = x - 2$

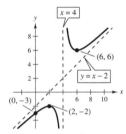

**17.**  $y = x\sqrt{4 - x}$,

Domain: $(-\infty, 4]$

$y' = \dfrac{8 - 3x}{2\sqrt{4 - x}} = 0$ when $x = \dfrac{8}{3}$ and undefined when $x = 4$.

$y'' = \dfrac{3x - 16}{4(4 - x)^{3/2}} = 0$ when $x = \dfrac{16}{3}$ and undefined when $x = 4$.

**Note:** $x = \dfrac{16}{3}$ is not in the domain.

|  | $y$ | $y'$ | $y''$ | Conclusion |
|---|---|---|---|---|
| $-\infty < x < \dfrac{8}{3}$ |  | $+$ | $-$ | Increasing, concave down |
| $x = \dfrac{8}{3}$ | $\dfrac{16}{3\sqrt{3}}$ | $0$ | $-$ | Relative maximum |
| $\dfrac{8}{3} < x < 4$ |  | $-$ | $-$ | Decreasing, concave down |
| $x = 4$ | $0$ | Undefined | Undefined | Endpoint |

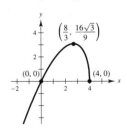

**19.** $h(x) = x\sqrt{4 - x^2}$

Domain. $-2 \le x \le 2$

$h'(x) = \dfrac{4 - 2x^2}{\sqrt{4 - x^2}} = 0$ when $x = \pm\sqrt{2}$

$h''(x) = \dfrac{2x(x^2 - 6)}{(4 - x^2)^{3/2}} = 0$ when $x = 0$

Relative maximum: $\left(\sqrt{2}, 2\right)$

Relative minimum: $\left(-\sqrt{2}, -2\right)$

Intercepts $(-2, 0), (0, 0), (2, 0)$

Symmetric with respect to origin.

Point of inflection: $(0, 0)$

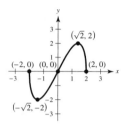

**21.** $y = 3x^{2/3} - 2x$

$y' = 2x^{-1/3} - 2 = \dfrac{2\left(1 - x^{1/3}\right)}{x^{1/3}}$

$= 0$ when $x = 1$ and undefined when $x = 0$.

$y'' = \dfrac{-2}{3x^{4/3}} < 0$ when $x \ne 0$.

|  | $y$ | $y'$ | $y''$ | Conclusion |
|---|---|---|---|---|
| $-\infty < x < 0$ |  | $-$ | $-$ | Decreasing, concave down |
| $x = 0$ | 0 | Undefined | Undefined | Relative minimum |
| $0 < x < 1$ |  | $+$ | $-$ | Increasing, concave down |
| $x = 1$ | 1 | 0 | $-$ | Relative maximum |
| $1 < x < \infty$ |  | $-$ | $-$ | Decreasing, concave down |

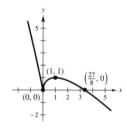

**23.** $y = x^3 - 3x^2 + 3$

$y' = 3x^2 - 6x = 3x(x - 2) = 0$ when $x = 0, x = 2$.

$y'' = 6x - 6 = 6(x - 1) = 0$ when $x = 1$.

|  | $y$ | $y'$ | $y''$ | Conclusion |
|---|---|---|---|---|
| $-\infty < x < 0$ |  | $+$ | $-$ | Increasing, concave down |
| $x = 0$ | 3 | 0 | $-$ | Relative maximum |
| $0 < x < 1$ |  | $-$ | $-$ | Decreasing, concave down |
| $x = 1$ | 1 | $-$ | 0 | Point of inflection |
| $1 < x < 2$ |  | $-$ | $+$ | Decreasing, concave up |
| $x = 2$ | $-1$ | 0 | $+$ | Relative minimum |
| $2 < x < \infty$ |  | $+$ | $+$ | Increasing, concave up |

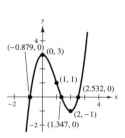

**25.** $y = 2 - x - x^3$

$y' = -1 - 3x^2$

No critical numbers

$y'' = -6x = 0$ when $x = 0$.

| | $y$ | $y'$ | $y''$ | Conclusion |
|---|---|---|---|---|
| $-\infty < x < 0$ | | $-$ | $+$ | Decreasing, concave up |
| $x = 0$ | 2 | $-$ | 0 | Point of inflection |
| $0 < x < \infty$ | | $-$ | $-$ | Decreasing, concave down |

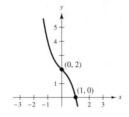

**27.** $y = 3x^4 + 4x^3$

$y' = 12x^3 + 12x^2 = 12x^2(x + 1) = 0$ when $x = 0, x = -1$.

$y'' = 36x^2 + 24x = 12x(3x + 2) = 0$ when $x = 0, x = -\frac{2}{3}$.

| | $y$ | $y'$ | $y''$ | Conclusion |
|---|---|---|---|---|
| $-\infty < x < -1$ | | $-$ | $+$ | Decreasing, concave up |
| $x = -1$ | $-1$ | 0 | $+$ | Relative minimum |
| $-1 < x < -\frac{2}{3}$ | | $+$ | $+$ | Increasing, concave up |
| $x = -\frac{2}{3}$ | $-\frac{16}{27}$ | $+$ | 0 | Point of inflection |
| $-\frac{2}{3} < x < 0$ | | $+$ | $-$ | Increasing, concave down |
| $x = 0$ | 0 | 0 | 0 | Point of inflection |
| $0 < x < \infty$ | | $+$ | $+$ | Increasing, concave up |

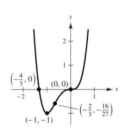

**29.** $y = x^5 - 5x$

$y' = 5x^4 - 5 = 5(x^4 - 1) = 0$ when $x = \pm 1$.

$y'' = 20x^3 = 0$ when $x = 0$.

| | $y$ | $y'$ | $y''$ | Conclusion |
|---|---|---|---|---|
| $-\infty < x < -1$ | | $+$ | $-$ | Increasing, concave down |
| $x = -1$ | 4 | 0 | $-$ | Relative maximum |
| $-1 < x < 0$ | | $-$ | $-$ | Decreasing, concave down |
| $x = 0$ | 0 | $-$ | 0 | Point of inflection |
| $0 < x < 1$ | | $-$ | $+$ | Decreasing, concave up |
| $x = 1$ | $-4$ | 0 | $+$ | Relative minimum |
| $1 < x < \infty$ | | $+$ | $+$ | Increasing, concave up |

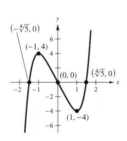

**31.** $y = |2x - 3|$

$y' = \dfrac{2(2x - 3)}{|2x - 3|}$ undefined at $x = \dfrac{3}{2}$.

$y'' = 0$

| | $y$ | $y'$ | Conclusion |
|---|---|---|---|
| $-\infty < x < \dfrac{3}{2}$ | | $-$ | Decreasing |
| $x = \dfrac{3}{2}$ | | 0  Undefined | Relative minimum |
| $\dfrac{3}{2} < x < \infty$ | | $+$ | Increasing |

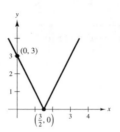

**33.** $f(x) = \dfrac{20x}{x^2 + 1} - \dfrac{1}{x} = \dfrac{19x^2 - 1}{x(x^2 + 1)}$

$f'(x) = \dfrac{-(19x^4 - 22x^2 - 1)}{x^2(x^2 + 1)^2} = 0$ for $x \approx \pm 1.10$

$f''(x) = \dfrac{2(19x^6 - 63x^9 - 3x^2 - 1)}{x^3(x^2 + 1)^3} = 0$ for $x \approx \pm 1.84$

Vertical asymptote: $x = 0$

Horizontal asymptote: $y = 0$

Minimum: $(-1.10, -9.05)$

Maximum: $(1.10, 9.05)$

Points of inflection:

$(-1.84, -7.86), (1.84, 7.86)$

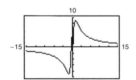

**35.** $f(x) = \dfrac{-2x}{\sqrt{x^2 + 7}}$

$f'(x) = \dfrac{-14}{(x^2 + 7)^{3/2}} < 0$

$f''(x) = \dfrac{42x}{(x^2 + 7)^{5/2}} = 0$ at $x = 0$

Horizontal asymptotes: $y = \pm 2$

Point of inflection: $(0, 0)$

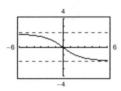

**37.** $f(x) = 2x - 4 \sin x, \ 0 \le x \le 2\pi$

$f'(x) = 2 - 4 \cos x$

$f''(x) = 4 \sin x$

$f'(x) = 0 \Rightarrow \cos x = \dfrac{1}{2} \Rightarrow x = \dfrac{\pi}{3}, \dfrac{5\pi}{3}$

$f''(x) = 0 \Rightarrow x = 0, \pi, 2\pi$

Relative minimum: $\left( \dfrac{\pi}{3}, \dfrac{2\pi}{3} - 2\sqrt{3} \right)$

Relative maximum: $\left( \dfrac{5\pi}{3}, \dfrac{10\pi}{3} + 2\sqrt{3} \right)$

Points of inflection: $(0, 0), (\pi, 2\pi), (2\pi, 4\pi)$

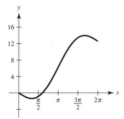

**39.** $y = \sin x - \dfrac{1}{18} \sin 3x, \, 0 \le x \le 2\pi$

$y' = \cos x - \dfrac{1}{6} \cos 3x$

$= \cos x - \dfrac{1}{6}[\cos 2x \cos x - \sin 2x \sin x]$

$= \cos x - \dfrac{1}{6}\left[\left(1 - 2\sin^2 x\right)\cos x - 2\sin^2 x \cos x\right]$

$= \cos x\left[1 - \dfrac{1}{6}\left(1 - 2\sin^2 x - 2\sin^2 x\right)\right] = \cos x\left[\dfrac{5}{6} + \dfrac{2}{3}\sin^2 x\right]$

$y' = 0$:        $\cos x = 0 \Rightarrow x = \pi/2, 3\pi/2$

$\dfrac{5}{6} + \dfrac{2}{3}\sin^2 x = 0 \Rightarrow \sin^2 x = -5/4$, impossible

$y'' = -\sin x + \dfrac{1}{2}\sin 3x = 0 \Rightarrow 2 \sin x = \sin 3x$

$\qquad\qquad\qquad\qquad = \sin 2x \cos x + \cos 2x \sin x$

$\qquad\qquad\qquad\qquad = 2 \sin x \cos^2 x + \left(2\cos^2 x - 1\right)\sin x$

$\qquad\qquad\qquad\qquad = \sin x\left(2\cos^2 x + 2\cos^2 x - 1\right)$

$\qquad\qquad\qquad\qquad = \sin x\left(4\cos^2 x - 1\right)$

$\sin x = 0 \Rightarrow x = 0, \pi, 2\pi$

$2 = 4\cos^2 x - 1 \Rightarrow \cos x = \pm\sqrt{3}/2 \Rightarrow x = \dfrac{\pi}{6}, \dfrac{5\pi}{6}, \dfrac{7\pi}{6}, \dfrac{11\pi}{6}$

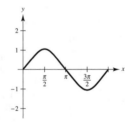

Relative maximum: $\left(\dfrac{\pi}{2}, \dfrac{19}{18}\right)$

Relative minimum: $\left(\dfrac{3\pi}{2}, -\dfrac{19}{18}\right)$

Points of inflection: $\left(\dfrac{\pi}{6}, \dfrac{4}{9}\right), \left(\dfrac{5\pi}{6}, \dfrac{4}{9}\right), (\pi, 0), \left(\dfrac{7\pi}{6}, -\dfrac{4}{9}\right), \left(\dfrac{11\pi}{6}, -\dfrac{4}{9}\right)$

**41.** $y = 2x - \tan x, \, -\dfrac{\pi}{2} < x < \dfrac{\pi}{2}$

$y' = 2 - \sec^2 x = 0$ when $x = \pm\dfrac{\pi}{4}$.

$y'' = -2\sec^2 x \tan x = 0$ when $x = 0$.

Relative maximum: $\left(\dfrac{\pi}{4}, \dfrac{\pi}{2} - 1\right)$

Relative minimum: $\left(-\dfrac{\pi}{4}, 1 - \dfrac{\pi}{2}\right)$

Point of inflection: $(0, 0)$

Vertical asymptotes: $x = \pm\dfrac{\pi}{2}$

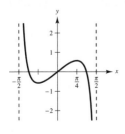

**43.** $y = 2(\csc x + \sec x), \, 0 < x < \dfrac{\pi}{2}$

$y' = 2(\sec x \tan x - \csc x \cot x) = 0 \Rightarrow x = \dfrac{\pi}{4}$

Relative minimum: $\left(\dfrac{\pi}{4}, 4\sqrt{2}\right)$

Vertical asymptotes: $x = 0, \dfrac{\pi}{2}$

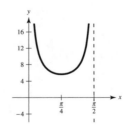

**45.** $g(x) = x \tan x,\ -\dfrac{3\pi}{2} < x < \dfrac{3\pi}{2}$

$g'(x) = \dfrac{x + \sin x \cos x}{\cos^2 x} = 0$ when $x = 0$.

$g''(x) = \dfrac{2(\cos x + x \sin x)}{\cos^3 x}$

Vertical asymptotes: $x = -\dfrac{3\pi}{2}, -\dfrac{\pi}{2}, \dfrac{\pi}{2}, \dfrac{3\pi}{2}$

Intercepts: $(-\pi, 0), (0, 0), (\pi, 0)$

Symmetric with respect to $y$-axis.

Increasing on $\left(0, \dfrac{\pi}{2}\right)$ and $\left(\dfrac{\pi}{2}, \dfrac{3\pi}{2}\right)$

Points of inflection: $(\pm 2.80, -1)$

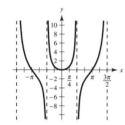

**47.** Because the slope is negative, the function is decreasing on $(2, 8)$, and so $f(3) > f(5)$.

**49.** $f$ is cubic.

$f'$ is quadratic.

$f''$ is linear.

The zeros of $f'$ correspond to the points where the graph of $f$ has horizontal tangents. The zero of $f''$ corresponds to the point where the graph of $f'$ has a horizontal tangent.

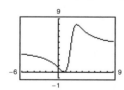

**51.** $f(x) = \dfrac{4(x - 1)^2}{x^2 - 4x + 5}$

Vertical asymptote: none

Horizontal asymptote: $y = 4$

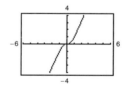

The graph crosses the horizontal asymptote $y = 4$. If a function has a vertical asymptote at $x = c$, the graph would not cross it because $f(c)$ is undefined.

**53.** $h(x) = \dfrac{\sin 2x}{x}$

Vertical asymptote: none

Horizontal asymptote: $y = 0$

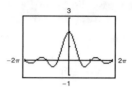

Yes, it is possible for a graph to cross its horizontal asymptote.

It is not possible to cross a vertical asymptote because the function is not continuous there.

**55.** $h(x) = \dfrac{6 - 2x}{3 - x}$

$= \dfrac{2(3 - x)}{3 - x} = \begin{cases} 2, & \text{if } x \neq 3 \\ \text{Undefined,} & \text{if } x = 3 \end{cases}$

The rational function is not reduced to lowest terms.

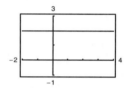

There is a hole at $(3, 2)$.

**57.** $f(x) = -\dfrac{x^2 - 3x - 1}{x - 2} = -x + 1 + \dfrac{3}{x - 2}$

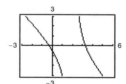

The graph appears to approach the slant asymptote $y = -x + 1$.

**59.** $f(x) = \dfrac{2x^3}{x^2 + 1} = 2x - \dfrac{2x}{x^2 + 1}$

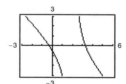

The graph appears to approach the slant asymptote $y = 2x$.

**61.**

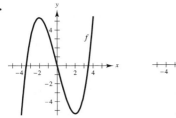 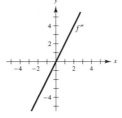

(or any vertical translation of $f$)

**63.**

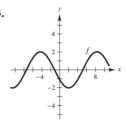 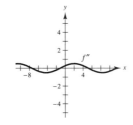

(or any vertical translation of $f$)

**65.** $f(x) = \dfrac{\cos^2 \pi x}{\sqrt{x^2 + 1}}$, $(0, 4)$

(a)

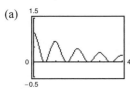

On $(0, 4)$ there seem to be 7 critical numbers: 0.5, 1.0, 1.5, 2.0, 2.5, 3.0, 3.5

(b) $f'(x) = \dfrac{-\cos \pi x \left( x \cos \pi x + 2\pi \left( x^2 + 1 \right) \sin \pi x \right)}{\left( x^2 + 1 \right)^{3/2}} = 0$

Critical numbers $\approx \dfrac{1}{2}, 0.97, \dfrac{3}{2}, 1.98, \dfrac{5}{2}, 2.98, \dfrac{7}{2}$.

The critical numbers where maxima occur appear to be integers in part (a), but approximating them using $f'$ shows that they are not integers.

**67.** Vertical asymptote: $x = 3$

Horizontal asymptote: $y = 0$

$y = \dfrac{1}{x - 3}$

**69.** Vertical asymptote: $x = 3$

Slant asymptote: $y = 3x + 2$

$y = 3x + 2 + \dfrac{1}{x - 3} = \dfrac{3x^2 - 7x - 5}{x - 3}$

**71.** (a) $f'(x) = 0$ for $x = -2$ and $x = 2$

   $f'$ is negative for $-2 < x < 2$ (decreasing function).

   $f'$ is positive for $x > 2$ and $x < -2$ (increasing function).

(b) $f''(x) = 0$ at $x = 0$ (Point of inflection).

   $f''$ is positive for $x > 0$ (Concave upward).

   $f''$ is negative for $x < 0$ (Concave downward).

(c) $f'$ is increasing on $(0, \infty)$. $\left( f'' > 0 \right)$

(d) $f'(x)$ is minimum at $x = 0$. The rate of change of $f$ at $x = 0$ is less than the rate of change of $f$ for all other values of $x$.

**73.** $f(x) = \dfrac{ax}{\left( x - b \right)^2}$

Answers will vary. *Sample answer*: The graph has a vertical asymptote at $x = b$. If $a$ and $b$ are both positive, or both negative, then the graph of $f$ approaches $\infty$ as $x$ approaches $b$, and the graph has a minimum at $x = -b$. If $a$ and $b$ have opposite signs, then the graph of $f$ approaches $-\infty$ as $x$ approaches $b$, and the graph has a maximum at $x = -b$.

**75.** $f(x) = \dfrac{2x^n}{x^4 + 1}$

(a) For $n$ even, $f$ is symmetric about the $y$-axis. For $n$ odd, $f$ is symmetric about the origin.

(b) The $x$-axis will be the horizontal asymptote if the degree of the numerator is less than 4. That is, $n = 0, 1, 2, 3$.

(c) $n = 4$ gives $y = 2$ as the horizontal asymptote.

(d) There is a slant asymptote $y = 2x$ if $n = 5$:

$\dfrac{2x^5}{x^4 + 1} = 2x - \dfrac{2x}{x^4 + 1}$.

(e)

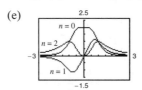

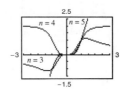

| $n$ | 0 | 1 | 2 | 3 | 4 | 5 |
|-----|---|---|---|---|---|---|
| $M$ | 1 | 2 | 3 | 2 | 1 | 0 |
| $N$ | 2 | 3 | 4 | 5 | 2 | 3 |

**77.** (a)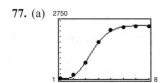

(b) When $t = 10$, $N = 2434$ bacteria.

(c) $N$ is greatest $(\approx 2518)$ at $t \approx 7.2$.

(d) $N'(t)$ is greatest when $t \approx 3.2$.

   (Find the $t$-value of the point of inflection.)

(e) $\displaystyle \lim_{t \to \infty} N(t) = \frac{13{,}250}{7} \approx 1893$ bacteria

**79.** $y = \sqrt{x^2 + 6x} = \sqrt{(x+3)^2 - 9}$

$y \to x + 3$ as $x \to \infty$, and $y \to -x - 3$ as $x \to -\infty$.

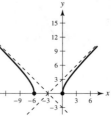

## Section 3.7   Optimization Problems

**1.** (a)

| First Number, $x$ | Second Number | Product, $P$ |
|---|---|---|
| 10 | $110 - 10$ | $10(110 - 10) = 1000$ |
| 20 | $110 - 20$ | $20(110 - 20) = 1800$ |
| 30 | $110 - 30$ | $30(110 - 30) = 2400$ |
| 40 | $110 - 40$ | $40(110 - 40) = 2800$ |
| 50 | $110 - 50$ | $50(110 - 50) = 3000$ |
| 60 | $110 - 60$ | $60(110 - 60) = 3000$ |

(b)

| First Number, $x$ | Second Number | Product, $P$ |
|---|---|---|
| 10 | $110 - 10$ | $10(110 - 10) = 1000$ |
| 20 | $110 - 20$ | $20(110 - 20) = 1800$ |
| 30 | $110 - 30$ | $30(110 - 30) = 2400$ |
| 40 | $110 - 40$ | $40(110 - 40) = 2800$ |
| 50 | $110 - 50$ | $50(110 - 50) = 3000$ |
| 60 | $110 - 60$ | $60(110 - 60) = 3000$ |
| 70 | $110 - 70$ | $70(110 - 70) = 2800$ |
| 80 | $110 - 80$ | $80(110 - 80) = 2400$ |
| 90 | $110 - 90$ | $90(110 - 90) = 1800$ |
| 100 | $110 - 100$ | $100(110 - 100) = 1000$ |

The maximum is attained near $x = 50$ and $60$.

(c) $P = x(110 - x) = 110x - x^2$

(d)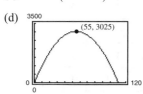

The solution appears to be $x = 55$.

(e)  $\dfrac{dP}{dx} = 110 - 2x = 0$ when $x = 55$.

$\dfrac{d^2P}{dx^2} = -2 < 0$

$P$ is a maximum when $x = 110 - x = 55$. The two numbers are 55 and 55.

**3.** Let $x$ and $y$ be two positive numbers such that $x + y = S$.

$P = xy = x(S - x) = Sx - x^2$

$\dfrac{dP}{dx} = S - 2x = 0$ when $x = \dfrac{S}{2}$.

$\dfrac{d^2P}{dx^2} = -2 < 0$ when $x = \dfrac{S}{2}$.

$P$ is a maximum when $x = y = S/2$.

**5.** Let $x$ and $y$ be two positive numbers such that $xy = 147$.

$S = x + 3y = \dfrac{147}{y} + 3y$

$\dfrac{dS}{dy} = 3 - \dfrac{147}{y^2} = 0$ when $y = 7$.

$\dfrac{d^2S}{dy^2} = \dfrac{294}{y^3} > 0$ when $y = 7$.

$S$ is minimum when $y = 7$ and $x = 21$.

**7.** Let $x$ and $y$ be two positive numbers such that $x + 2y = 108$.

$P = xy = y(108 - 2y) = 108y - 2y^2$

$\dfrac{dP}{dy} = 108 - 4y = 0$ when $y = 27$.

$\dfrac{d^2P}{dy^2} = -4 < 0$ when $y = 27$.

$P$ is a maximum when $x = 54$ and $y = 27$.

**9.** Let $x$ be the length and $y$ the width of the rectangle.

$2x + 2y = 80$

$y = 40 - x$

$A = xy = x(40 - x) = 40x - x^2$

$\dfrac{dA}{dx} = 40 - 2x = 0$ when $x = 20$.

$\dfrac{d^2A}{dx^2} = -2 < 0$ when $x = 20$.

$A$ is maximum when $x = y = 20$ m.

**11.** Let $x$ be the length and $y$ the width of the rectangle.

$xy = 32$

$y = \dfrac{32}{x}$

$P = 2x + 2y = 2x + 2\left(\dfrac{32}{x}\right) = 2x + \dfrac{64}{x}$

$\dfrac{dP}{dx} = 2 - \dfrac{64}{x^2} = 0$ when $x = 4\sqrt{2}$.

$\dfrac{d^2P}{dx^2} = \dfrac{128}{x^3} > 0$ when $x = 4\sqrt{2}$.

$P$ is minimum when $x = y = 4\sqrt{2}$ ft.

**13.** $d = \sqrt{(x-2)^2 + \left[x^2 - (1/2)\right]^2} = \sqrt{x^4 - 4x + (17/4)}$

Because $d$ is smallest when the expression inside the radical is smallest, you need only find the critical numbers of

$f(x) = x^4 - 4x + \dfrac{17}{4}$.

$f'(x) = 4x^3 - 4 = 0$

$x = 1$

By the First Derivative Test, the point nearest to $\left(2, \frac{1}{2}\right)$ is $(1, 1)$.

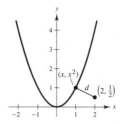

**15.** $d = \sqrt{(x-4)^2 + \left(\sqrt{x} - 0\right)^2} = \sqrt{x^2 - 7x + 16}$

Because $d$ is smallest when the expression inside the radical is smallest, you need only find the critical numbers of

$f(x) = x^2 - 7x + 16$.

$f'(x) = 2x - 7 = 0$

$x = \dfrac{7}{2}$

By the First Derivative Test, the point nearest to $(4, 0)$ is $\left(7/2, \sqrt{7/2}\right)$.

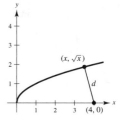

**17.** $xy = 30 \Rightarrow y = \dfrac{30}{x}$

$A = (x + 2)\left(\dfrac{30}{x} + 2\right)$ (see figure)

$\dfrac{dA}{dx} = (x + 2)\left(\dfrac{-30}{x^2}\right) + \left(\dfrac{30}{x} + 2\right)$

$\quad = \dfrac{2(x^2 - 30)}{x^2} = 0$ when $x = \sqrt{30}$.

$y = \dfrac{30}{\sqrt{30}} = \sqrt{30}$

By the First Derivative Test, the dimensions $(x + 2)$ by $(y + 2)$ are $\left(2 + \sqrt{30}\right)$ by $\left(2 + \sqrt{30}\right)$ (approximately 7.477 by 7.477). These dimensions yield a minimum area.

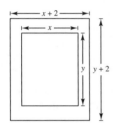

**19.** $\dfrac{dQ}{dx} = kx(Q_0 - x) = kQ_0x - kx^2$

$\dfrac{d^2Q}{dx^2} = kQ_0 - 2kx$

$\quad = k(Q_0 - 2x) = 0$ when $x = \dfrac{Q_0}{2}$.

$\dfrac{d^3Q}{dx^3} = -2k < 0$ when $x = \dfrac{Q_0}{2}$.

$dQ/dx$ is maximum when $x = Q_0/2$.

**21.** $xy = 245,000$ (see figure)

$S = x + 2y$

$\quad = \left(x + \dfrac{490,000}{x}\right)$ where $S$ is the length

$\qquad\qquad\qquad\qquad$ of fence needed.

$\dfrac{dS}{dx} = 1 - \dfrac{490,000}{x^2} = 0$ when $x = 700$.

$\dfrac{d^2S}{dx^2} = \dfrac{980,000}{x^3} > 0$ when $x = 700$.

$S$ is a minimum when $x = 700$ m and $y = 350$ m.

**23.** (a)  $A = 4(\text{area of side}) + 2(\text{area of Top})$

(1) $A = 4(3)(11) + 2(3)(3) = 150$ in$^2$.

(2) $A = 4(5)(5) + 2(5)(5) = 150$ in$^2$.

(3) $A = 4(3.25)(6) + 2(6)(6) = 150$ in$^2$.

(b)  $V = (\text{length})(\text{width})(\text{height})$

(1) $V = (3)(3)(11) = 99$ in$^3$.

(2) $V = (5)(5)(5) = 125$ in$^3$.

(3) $V = (6)(6)(3.25) = 117$ in$^3$.

(c)  $S = 4xy + 2x^2 = 150 \Rightarrow y = \dfrac{150 - 2x^2}{4x}$

$V = x^2y = x^2\left(\dfrac{150 - 2x^2}{4x}\right) = \dfrac{75}{2}x - \dfrac{1}{2}x^3$

$V' = \dfrac{75}{2} - \dfrac{3}{2}x^2 = 0 \Rightarrow x = \pm 5$

By the First Derivative Test, $x = 5$ yields the maximum volume. Dimensions: $5 \times 5 \times 5$ in. (A cube!)

**25.**   $16 = 2y + x + \pi\left(\dfrac{x}{2}\right)$

$32 = 4y + 2x + \pi x$

$y = \dfrac{32 - 2x - \pi x}{4}$

$A = xy + \dfrac{\pi}{2}\left(\dfrac{x}{2}\right)^2 = \left(\dfrac{32 - 2x - \pi x}{4}\right)x + \dfrac{\pi x^2}{8} = 8x - \dfrac{1}{2}x^2 - \dfrac{\pi}{4}x^2 + \dfrac{\pi}{8}x^2$

$\dfrac{dA}{dx} = 8 - x - \dfrac{\pi}{2}x + \dfrac{\pi}{4}x = 8 - x\left(1 + \dfrac{\pi}{4}\right) = 0$ when $x = \dfrac{8}{1 + (\pi/4)} = \dfrac{32}{4 + \pi}$.

$\dfrac{d^2A}{dx^2} = -\left(1 + \dfrac{\pi}{4}\right) < 0$ when $x = \dfrac{32}{4 + \pi}$.

$y = \dfrac{32 - 2\left[32/(4 + \pi)\right] - \pi\left[32/(4 + \pi)\right]}{4} = \dfrac{16}{4 + \pi}$

The area is maximum when $y = \dfrac{16}{4 + \pi}$ ft and $x = \dfrac{32}{4 + \pi}$ ft.

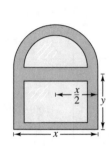

**27. (a)**   $\dfrac{y - 2}{0 - 1} = \dfrac{0 - 2}{x - 1}$

$y = 2 + \dfrac{2}{x - 1}$

$L = \sqrt{x^2 + y^2} = \sqrt{x^2 + \left(2 + \dfrac{2}{x - 1}\right)^2} = \sqrt{x^2 + 4 + \dfrac{8}{x - 1} + \dfrac{4}{(x - 1)^2}},\quad x > 1$

**(b)**

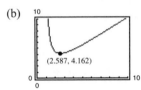

(2.587, 4.162)

$L$ is minimum when $x \approx 2.587$ and $L \approx 4.162$.

**(c)**   Area $= A(x) = \dfrac{1}{2}xy = \dfrac{1}{2}x\left(2 + \dfrac{2}{x - 1}\right) = x + \dfrac{x}{x - 1}$

$A'(x) = 1 + \dfrac{(x - 1) - x}{(x - 1)^2} = 1 - \dfrac{1}{(x - 1)^2} = 0$

$(x - 1)^2 = 1$

$x - 1 = \pm 1$

$x = 0, 2 \left(\text{select } x = 2\right)$

They $y = 4$ and $A = 4$.

Vertices: $(0, 0), (2, 0), (0, 4)$

**29.**   $A = 2xy = 2x\sqrt{25 - x^2}$ (see figure)

$\dfrac{dA}{dx} = 2x\left(\dfrac{1}{2}\right)\left(\dfrac{-2x}{\sqrt{25 - x^2}}\right) + 2\sqrt{25 - x^2} = 2\left(\dfrac{25 - 2x^2}{\sqrt{25 - x^2}}\right) = 0$

when $x = y = \dfrac{5\sqrt{2}}{2} \approx 3.54$.

By the First Derivative Test, the inscribed rectangle of

maximum area has vertices $\left(\pm\dfrac{5\sqrt{2}}{2}, 0\right), \left(\pm\dfrac{5\sqrt{2}}{2}, \dfrac{5\sqrt{2}}{2}\right)$.

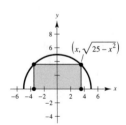

Width: $\dfrac{5\sqrt{2}}{2}$; Length: $5\sqrt{2}$

**31.** (a) $P = 2x + 2\pi r = 2x + 2\pi\left(\dfrac{y}{2}\right) = 2x + \pi y = 200 \Rightarrow y = \dfrac{200 - 2x}{\pi} = \dfrac{2}{\pi}(100 - x)$

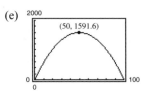

(b)

| Length, $x$ | Width, $y$ | Area, $xy$ |
|---|---|---|
| 10 | $\dfrac{2}{\pi}(100 - 10)$ | $(10)\dfrac{2}{\pi}(100 - 10) \approx 573$ |
| 20 | $\dfrac{2}{\pi}(100 - 20)$ | $(20)\dfrac{2}{\pi}(100 - 20) \approx 1019$ |
| 30 | $\dfrac{2}{\pi}(100 - 30)$ | $(30)\dfrac{2}{\pi}(100 - 30) \approx 1337$ |
| 40 | $\dfrac{2}{\pi}(100 - 40)$ | $(40)\dfrac{2}{\pi}(100 - 40) \approx 1528$ |
| 50 | $\dfrac{2}{\pi}(100 - 50)$ | $(50)\dfrac{2}{\pi}(100 - 50) \approx 1592$ |
| 60 | $\dfrac{2}{\pi}(100 - 60)$ | $(60)\dfrac{2}{\pi}(100 - 60) \approx 1528$ |

The maximum area of the rectangle is approximately 1592 m$^2$.

(c) $A = xy = x\dfrac{2}{\pi}(100 - x) = \dfrac{2}{\pi}\left(100x - x^2\right)$

(d) $A' = \dfrac{2}{\pi}(100 - 2x)$. $A' = 0$ when $x = 50$.

Maximum value is approximately 1592 when length $= 50$ m and width $= \dfrac{100}{\pi}$.

(e)

Maximum area is approximately

1591.55 m$^2$ $(x = 50$ m$)$.

**33.** Let $x$ be the sides of the square ends and $y$ the length of the package.

$P = 4x + y = 108 \Rightarrow y = 108 - 4x$

$V = x^2 y = x^2(108 - 4x) = 108x^2 - 4x^3$

$\dfrac{dV}{dx} = 216x - 12x^2 = 12x(18 - x) = 0$ when $x = 18$.

$\dfrac{d^2V}{dx^2} = 216 - 24x = -216 < 0$ when $x = 18$.

The volume is maximum when $x = 18$ in. and $y = 108 - 4(18) = 36$ in.

**35.**  $V = \dfrac{1}{3}\pi x^2 h = \dfrac{1}{3}\pi x^2 \left( r + \sqrt{r^2 - x^2} \right)$ (see figure)

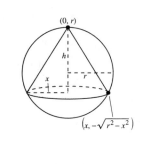

$\dfrac{dV}{dx} = \dfrac{1}{3}\pi\left[ \dfrac{-x^3}{\sqrt{r^2 - x^2}} + 2x\left( r + \sqrt{r^2 - x^2} \right) \right] = \dfrac{\pi x}{3\sqrt{r^2 - x^2}}\left( 2r^2 + 2r\sqrt{r^2 - x^2} - 3x^2 \right) = 0$

$2r^2 + 2r\sqrt{r^2 - x^2} - 3x^2 = 0$

$\qquad 2r\sqrt{r^2 - x^2} = 3x^2 - 2r^2$

$\qquad 4r^2\left( r^2 - x^2 \right) = 9x^4 - 12x^2 r^2 + 4r^4$

$\qquad 0 = 9x^4 - 8x^2 r^2 = x^2\left( 9x^2 - 8r^2 \right)$

$\qquad\qquad x = 0, \dfrac{2\sqrt{2}r}{3}$

By the First Derivative Test, the volume is a maximum when $x = \dfrac{2\sqrt{2}r}{3}$ and $h = r + \sqrt{r^2 - x^2} = \dfrac{4r}{3}$.

Thus, the maximum volume is $V = \dfrac{1}{3}\pi\left( \dfrac{8r^2}{9} \right)\left( \dfrac{4r}{3} \right) = \dfrac{32\pi r^3}{81}$ cubic units.

**37.** No. The volume will change because the shape of the container changes when squeezed.

**39.**  $V = 14 = \dfrac{4}{3}\pi r^3 + \pi r^2 h$

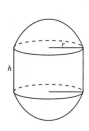

$\quad h = \dfrac{14 - (4/3)\pi r^3}{\pi r^2} = \dfrac{14}{\pi r^2} - \dfrac{4}{3}r$

$\quad S = 4\pi r^2 + 2\pi r h = 4\pi r^2 + 2\pi r\left( \dfrac{14}{\pi r^2} - \dfrac{4}{3}r \right) = 4\pi r^2 + \dfrac{28}{r} - \dfrac{8}{3}\pi r^2 = \dfrac{4}{3}\pi r^2 + \dfrac{28}{r}$

$\dfrac{dS}{dr} = \dfrac{8}{3}\pi r - \dfrac{28}{r^2} = 0$ when $r = \sqrt[3]{\dfrac{21}{2\pi}} \approx 1.495$ cm.

$\dfrac{d^2 S}{dr^2} = \dfrac{8}{3}\pi + \dfrac{56}{r^3} > 0$ when $r = \sqrt[3]{\dfrac{21}{2\pi}}$.

The surface area is minimum when $r = \sqrt[3]{\dfrac{21}{2\pi}}$ cm and $h = 0$. The resulting solid is a sphere of radius $r \approx 1.495$ cm.

**41.** Let $x$ be the length of a side of the square and $y$ the length of a side of the triangle.

$\qquad 4x + 3y = 10$

$\qquad\qquad A = x^2 + \dfrac{1}{2}y\left( \dfrac{\sqrt{3}}{2}y \right) = \dfrac{(10 - 3y)^2}{16} + \dfrac{\sqrt{3}}{4}y^2$

$\qquad\qquad \dfrac{dA}{dy} = \dfrac{1}{8}(10 - 3y)(-3) + \dfrac{\sqrt{3}}{2}y = 0$

$\qquad -30 + 9y + 4\sqrt{3}y = 0$

$\qquad\qquad y = \dfrac{30}{9 + 4\sqrt{3}}$

$\qquad\qquad \dfrac{d^2 A}{dy^2} = \dfrac{9 + 4\sqrt{3}}{8} > 0$

$A$ is minimum when $y = \dfrac{30}{9 + 4\sqrt{3}}$ and $x = \dfrac{10\sqrt{3}}{9 + 4\sqrt{3}}$.

**43.** Let $S$ be the strength and $k$ the constant of proportionality. Given $h^2 + w^2 = 20^2$, $h^2 = 20^2 - w^2$,

$$S = kwh^2$$

$$S = kw(400 - w^2) = k(400w - w^3)$$

$$\frac{dS}{dw} = k(400 - 3w^2) = 0 \text{ when } w = \frac{20\sqrt{3}}{3} \text{ in. and } h = \frac{20\sqrt{6}}{3} \text{ in.}$$

$$\frac{d^2S}{dw^2} = -6kw < 0 \text{ when } w = \frac{20\sqrt{3}}{3}.$$

These values yield a maximum.

**45.**     $R = \frac{v_0^2}{g} \sin 2\theta$

$$\frac{dR}{d\theta} = \frac{2v_0^2}{g} \cos 2\theta = 0 \text{ when } \theta = \frac{\pi}{4}, \frac{3\pi}{4}.$$

$$\frac{d^2R}{d\theta^2} = -\frac{4v_0^2}{g} \sin 2\theta < 0 \text{ when } \theta = \frac{\pi}{4}.$$

By the Second Derivative Test, $R$ is maximum when $\theta = \pi/4$.

**47.**   $\sin \alpha = \dfrac{h}{s} \Rightarrow s = \dfrac{h}{\sin \alpha}, 0 < \alpha < \dfrac{\pi}{2}$

$$\tan \alpha = \frac{h}{2} \Rightarrow h = 2\tan \alpha \Rightarrow s = \frac{2\tan \alpha}{\sin \alpha} = 2\sec \alpha$$

$$I = \frac{k \sin \alpha}{s^2} = \frac{k \sin \alpha}{4 \sec^2 \alpha} = \frac{k}{4} \sin \alpha \cos^2 \alpha$$

$$\frac{dI}{d\alpha} = \frac{k}{4}\Big[\sin \alpha(-2 \sin \alpha \cos \alpha) + \cos^2 \alpha(\cos \alpha)\Big]$$

$$= \frac{k}{4} \cos \alpha \Big[\cos^2 \alpha - 2 \sin^2 \alpha\Big]$$

$$= \frac{k}{4} \cos \alpha \Big[1 - 3 \sin^2 \alpha\Big]$$

$$= 0 \text{ when } \alpha = \frac{\pi}{2}, \frac{3\pi}{2}, \text{ or when } \sin \alpha = \pm\frac{1}{\sqrt{3}}.$$

Because $\alpha$ is acute, you have

$$\sin \alpha = \frac{1}{\sqrt{3}} \Rightarrow h = 2\tan \alpha = 2\left(\frac{1}{\sqrt{2}}\right) = \sqrt{2} \text{ ft.}$$

Because $(d^2I)/(d\alpha^2) = (k/4) \sin \alpha(9 \sin^2\alpha - 7) < 0$ when $\sin \alpha = 1/\sqrt{3}$, this yields a maximum.

**49.**
$$S = \sqrt{x^2 + 4},\, L = \sqrt{1 + (3 - x)^2}$$

$$\text{Time} = T = \frac{\sqrt{x^2 + 4}}{2} + \frac{\sqrt{x^2 - 6x + 10}}{4}$$

$$\frac{dT}{dx} = \frac{x}{2\sqrt{x^2 + 4}} + \frac{x - 3}{4\sqrt{x^2 - 6x + 10}} = 0$$

$$\frac{x^2}{x^2 + 4} = \frac{9 - 6x + x^2}{4(x^2 - 6x + 10)}$$

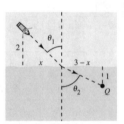

$$x^4 - 6x^3 + 9x^2 + 8x - 12 = 0$$

You need to find the roots of this equation in the interval $[0, 3]$. By using a computer or graphing utility you can determine that this equation has only one root in this interval $(x = 1)$. Testing at this value and at the endpoints, you see that $x = 1$ yields the minimum time. So, the man should row to a point 1 mile from the nearest point on the coast.

**51.**    $$T = \frac{\sqrt{x^2 + 4}}{v_1} + \frac{\sqrt{x^2 - 6x + 10}}{v_2}$$

$$\frac{dT}{dx} = \frac{x}{v_1\sqrt{x^2 + 4}} + \frac{x - 3}{v_2\sqrt{x^2 - 6x + 10}} = 0$$

Because

$$\frac{x}{\sqrt{x^2 + 4}} = \sin \theta_1 \text{ and } \frac{x - 3}{\sqrt{x^2 - 6x + 10}} = -\sin \theta_2$$

you have

$$\frac{\sin \theta_1}{v_1} - \frac{\sin \theta_2}{v_2} = 0 \Rightarrow \frac{\sin \theta_1}{v_1} = \frac{\sin \theta_2}{v_2}.$$

Because

$$\frac{d^2T}{dx^2} = \frac{4}{v_1(x^2 + 4)^{3/2}} + \frac{1}{v_2(x^2 - 6x + 10)^{3/2}} > 0$$

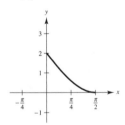

this condition yields a minimum time.

**53.**  $f(x) = 2 - 2\sin x$

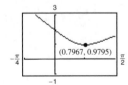

(a)  Distance from origin to $y$-intercept is 2.

Distance from origin to $x$-intercept is $\pi/2 \approx 1.57$.

(b)  $d = \sqrt{x^2 + y^2} = \sqrt{x^2 + (2 - 2\sin x)^2}$

(0.7967, 0.9795)

Minimum distance $= 0.9795$ at $x = 0.7967$.

(c) Let $f(x) = d^2(x) = x^2 + (2 - 2 \sin x)^2$.

$$f'(x) = 2x + 2(2 - 2 \sin x)(-2 \cos x)$$

Setting $f'(x) = 0$, you obtain $x \approx 0.7967$, which corresponds to $d = 0.9795$.

**55.** $F \cos \theta = k(W - F \sin \theta)$

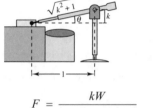

$$F = \frac{kW}{\cos \theta + k \sin \theta}$$

$$\frac{dF}{d\theta} = \frac{-kW(k \cos \theta - \sin \theta)}{(\cos \theta + k \sin \theta)^2} = 0$$

$$k \cos \theta = \sin \theta \Rightarrow k = \tan \theta \Rightarrow \theta = \arctan k$$

Because $\cos \theta + k \sin \theta = \dfrac{1}{\sqrt{k^2 + 1}} + \dfrac{k^2}{\sqrt{k^2 + 1}} = \sqrt{k^2 + 1}$,

the minimum force is $F = \dfrac{kW}{\cos \theta + k \sin \theta} = \dfrac{kW}{\sqrt{k^2 + 1}}$.

**57.** (a)

| Base 1 | Base 2 | Altitude | Area |
|---|---|---|---|
| 8 | $8 + 16 \cos 10°$ | $8 \sin 10°$ | $\approx 22.1$ |
| 8 | $8 + 16 \cos 20°$ | $8 \sin 20°$ | $\approx 42.5$ |
| 8 | $8 + 16 \cos 30°$ | $8 \sin 30°$ | $\approx 59.7$ |
| 8 | $8 + 16 \cos 40°$ | $8 \sin 40°$ | $\approx 72.7$ |
| 8 | $8 + 16 \cos 50°$ | $8 \sin 50°$ | $\approx 80.5$ |
| 8 | $8 + 16 \cos 60°$ | $8 \sin 60°$ | $\approx 83.1$ |

(b)

| Base 1 | Base 2 | Altitude | Area |
|---|---|---|---|
| 8 | $8 + 16 \cos 10°$ | $8 \sin 10°$ | $\approx 22.1$ |
| 8 | $8 + 16 \cos 20°$ | $8 \sin 20°$ | $\approx 42.5$ |
| 8 | $8 + 16 \cos 30°$ | $8 \sin 30°$ | $\approx 59.7$ |
| 8 | $8 + 16 \cos 40°$ | $8 \sin 40°$ | $\approx 72.7$ |
| 8 | $8 + 16 \cos 50°$ | $8 \sin 50°$ | $\approx 80.5$ |
| 8 | $8 + 16 \cos 60°$ | $8 \sin 60°$ | $\approx 83.1$ |
| 8 | $8 + 16 \cos 70°$ | $8 \sin 70°$ | $\approx 80.7$ |
| 8 | $8 + 16 \cos 80°$ | $8 \sin 80°$ | $\approx 74.0$ |
| 8 | $8 + 16 \cos 90°$ | $8 \sin 90°$ | $\approx 64.0$ |

The maximum cross-sectional area is approximately 83.1 ft$^2$.

(c)  $A = (a + b)\dfrac{h}{2}$

$$= \left[8 + (8 + 16 \cos \theta)\right]\dfrac{8 \sin \theta}{2} = 64(1 + \cos \theta) \sin \theta, 0° < \theta < 90°$$

(d)  $\dfrac{dA}{d\theta} = 64(1 + \cos \theta) \cos \theta + (-64 \sin \theta) \sin \theta$

$$= 64\left(\cos \theta + \cos^2 \theta - \sin^2 \theta\right)$$

$$= 64\left(2 \cos^2 \theta + \cos \theta - 1\right)$$

$$= 64(2 \cos \theta - 1)(\cos \theta + 1) = 0 \text{ when } \theta = 60°, 180°, 300°.$$

The maximum occurs when $\theta = 60°$.

(e)

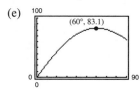

**59.**  $C = 100\left(\dfrac{200}{x^2} + \dfrac{x}{x + 30}\right), 1 \le x$

$C' = 100\left(-\dfrac{400}{x^3} + \dfrac{30}{(x + 30)^2}\right)$

Approximation: $x \approx 40.45$ hundred units, or 4045 units

**61.**  $S_1 = (4m - 1)^2 + (5m - 6)^2 + (10m - 3)^2$

$\dfrac{dS_1}{dm} = 2(4m - 1)(4) + 2(5m - 6)(5) + 2(10m - 3)(10)$

$= 282m - 128 = 0 \text{ when } m = \dfrac{64}{141}.$

Line:  $y = \dfrac{64}{141}x$

$S = \left|4\left(\dfrac{64}{141}\right) - 1\right| + \left|5\left(\dfrac{64}{141}\right) - 6\right| + \left|10\left(\dfrac{64}{141}\right) - 3\right|$

$= \left|\dfrac{256}{141} - 1\right| + \left|\dfrac{320}{141} - 6\right| + \left|\dfrac{640}{141} - 3\right| = \dfrac{858}{141} \approx 6.1 \text{ mi}$

**63.**  $S_3 = \dfrac{|4m - 1|}{\sqrt{m^2 + 1}} + \dfrac{|5m - 6|}{\sqrt{m^2 + 1}} + \dfrac{|10m - 3|}{\sqrt{m^2 + 1}}$

Using a graphing utility, you can see that the minimum occurs when $x \approx 0.3$.

Line:  $y \approx 0.3x$

$S_3 = \dfrac{|4(0.3) - 1| + |5(0.3) - 6| + |10(0.3) - 3|}{\sqrt{(0.3)^2 + 1}} \approx 4.5 \text{ mi}.$

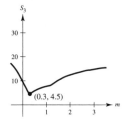

**65.**  $f(x) = x^3 - 3x; x^4 + 36 \le 13x^2$

$x^4 - 13x^2 + 36 = \left(x^2 - 9\right)\left(x^2 - 4\right)$

$$= (x - 3)(x - 2)(x + 2)(x + 3) \le 0$$

So, $-3 \le x \le -2$ or $2 \le x \le 3$.

$f'(x) = 3x^2 - 3 = 3(x + 1)(x - 1)$

$f$ is increasing on $(-\infty, -1)$ and $(1, \infty)$.

So, $f$ is increasing on $[-3, -2]$ and $[2, 3]$.

$f(-2) = -2, f(3) = 18$. The maximum value of $f$ is 18.

## Section 3.8 Newton's Method

**The following solutions may vary depending on the software or calculator used, and on rounding.**

**1.** $f(x) = x^2 - 5$

$f'(x) = 2x$

$x_1 = 2.2$

| $n$ | $x_n$ | $f(x_n)$ | $f'(x_n)$ | $\dfrac{f(x_n)}{f'(x_n)}$ | $x_n - \dfrac{f(x_n)}{f'(x_n)}$ |
|---|---|---|---|---|---|
| 1 | 2.2000 | −0.1600 | 4.4000 | −0.0364 | 2.2364 |
| 2 | 2.2364 | 0.0013 | 4.4727 | 0.0003 | 2.2361 |

**3.** $f(x) = \cos x$

$f'(x) = -\sin x$

$x_1 = 1.6$

| $n$ | $x_n$ | $f(x_n)$ | $f'(x_n)$ | $\dfrac{f(x_n)}{f'(x_n)}$ | $x_n - \dfrac{f(x_n)}{f'(x_n)}$ |
|---|---|---|---|---|---|
| 1 | 1.6000 | −0.0292 | −0.9996 | 0.0292 | 1.5708 |
| 2 | 1.5708 | 0.0000 | −1.0000 | 0.0000 | 1.5708 |

**5.** $f(x) = x^3 + 4$

$f'(x) = 3x^2$

$x_1 = -2$

| $n$ | $x_n$ | $f(x_n)$ | $f'(x_n)$ | $\dfrac{f(x_n)}{f'(x_n)}$ | $x_n - \dfrac{f(x_n)}{f'(x_n)}$ |
|---|---|---|---|---|---|
| 1 | −2.0000 | −4.0000 | 12.0000 | −0.3333 | −1.6667 |
| 2 | −1.6667 | −0.6296 | 8.3333 | −0.0756 | −1.5911 |
| 3 | −1.5911 | −0.0281 | 7.5949 | −0.0037 | −1.5874 |
| 4 | −1.5874 | −0.0000 | 7.5596 | 0.0000 | −1.5874 |

Approximation of the zero of $f$ is −1.587.

**7.** $f(x) = x^3 + x - 1$

$f'(x) = 3x^2 + 1$

| $n$ | $x_n$ | $f(x_n)$ | $f'(x_n)$ | $\dfrac{f(x_n)}{f'(x_n)}$ | $x_n - \dfrac{f(x_n)}{f'(x_n)}$ |
|---|---|---|---|---|---|
| 1 | 0.5000 | −0.3750 | 1.7500 | −0.2143 | 0.7143 |
| 2 | 0.7143 | 0.0788 | 2.5307 | 0.0311 | 0.6832 |
| 3 | 0.6832 | 0.0021 | 2.4003 | 0.0009 | 0.6823 |

Approximation of the zero of $f$ is 0.682.

**9.** $f(x) = 5\sqrt{x - 1} - 2x$

$f'(x) = \dfrac{5}{2\sqrt{x - 1}} - 2$

From the graph you see that these are two zeros. Begin with $x = 1.2$.

| $n$ | $x_n$ | $f(x_n)$ | $f'(x_n)$ | $\dfrac{f(x_n)}{f'(x_n)}$ | $x_n - \dfrac{f(x_n)}{f'(x_n)}$ |
|---|---|---|---|---|---|
| 1 | 1.2000 | −0.1639 | 3.5902 | −0.0457 | 1.2457 |
| 2 | 1.2457 | −0.0131 | 3.0440 | −0.0043 | 1.2500 |
| 3 | 1.2500 | −0.0001 | 3.0003 | −0.0003 | 1.2500 |

Approximation of the zero of $f$ is 1.250.

Similarly, the other zero is approximately 5.000.

(**Note:** These answers are exact)

**11.** $f(x) = x^3 - 3.9x^2 + 4.79x - 1.881$

$f'(x) = 3x^2 - 7.8x + 4.79$

| $n$ | $x_n$ | $f(x_n)$ | $f'(x_n)$ | $\dfrac{f(x_n)}{f'(x_n)}$ | $x_n - \dfrac{f(x_n)}{f'(x_n)}$ |
|---|---|---|---|---|---|
| 1 | 0.5000 | −0.3360 | 1.6400 | −0.2049 | 0.7049 |
| 2 | 0.7049 | −0.0921 | 0.7824 | −0.1177 | 0.8226 |
| 3 | 0.8226 | −0.0231 | 0.4037 | −0.0573 | 0.8799 |
| 4 | 0.8799 | −0.0045 | 0.2495 | −0.0181 | 0.8980 |
| 5 | 0.8980 | −0.0004 | 0.2048 | −0.0020 | 0.9000 |
| 6 | 0.9000 | 0.0000 | 0.2000 | 0.0000 | 0.9000 |

Approximation of the zero of $f$ is 0.900.

| $n$ | $x_n$ | $f(x_n)$ | $f'(x_n)$ | $\dfrac{f(x_n)}{f'(x_n)}$ | $x_n - \dfrac{f(x_n)}{f'(x_n)}$ |
|---|---|---|---|---|---|
| 1 | 1.1 | 0.0000 | −0.1600 | −0.0000 | 1.1000 |

Approximation of the zero of $f$ is 1.100.

| $n$ | $x_n$ | $f(x_n)$ | $f'(x_n)$ | $\dfrac{f(x_n)}{f'(x_n)}$ | $x_n - \dfrac{f(x_n)}{f'(x_n)}$ |
|---|---|---|---|---|---|
| 1 | 1.9 | 0.0000 | 0.8000 | 0.0000 | 1.9000 |

Approximation of the zero of $f$ is 1.900.

**13.**  $f(x) = 1 - x + \sin x$

$f'(x) = -1 + \cos x$

$x_1 = 2$

| $n$ | $x_n$ | $f(x_n)$ | $f'(x_n)$ | $\dfrac{f(x_n)}{f'(x_n)}$ | $x_n - \dfrac{f(x_n)}{f'(x_n)}$ |
|---|---|---|---|---|---|
| 1 | 2.0000 | −0.0907 | −1.4161 | 0.0640 | 1.9360 |
| 2 | 1.9360 | −0.0019 | −1.3571 | 0.0014 | 1.9346 |
| 3 | 1.9346 | 0.0000 | −1.3558 | 0.0000 | 1.9346 |

Approximate zero:  $x \approx 1.935$

**15.**  $h(x) = f(x) - g(x) = 2x + 1 - \sqrt{x + 4}$

$h'(x) = 2 - \dfrac{1}{2\sqrt{x + 4}}$

| $n$ | $x_n$ | $h(x_n)$ | $h'(x_n)$ | $\dfrac{h(x_n)}{h'(x_n)}$ | $x_n - \dfrac{h(x_n)}{h'(x_n)}$ |
|---|---|---|---|---|---|
| 1 | 0.6000 | 0.0552 | 1.7669 | 0.0313 | 0.5687 |
| 2 | 0.5687 | 0.0000 | 1.7661 | 0.0000 | 0.5687 |

Point of intersection of the graphs of $f$ and $g$ occurs when  $x \approx 0.569$.

**17.**  $h(x) = f(x) - g(x) = x - \tan x$

$h'(x) = 1 - \sec^2 x$

| $n$ | $x_n$ | $h(x_n)$ | $h'(x_n)$ | $\dfrac{h(x_n)}{h'(x_n)}$ | $x_n - \dfrac{h(x_n)}{h'(x_n)}$ |
|---|---|---|---|---|---|
| 1 | 4.5000 | −0.1373 | −21.5048 | 0.0064 | 4.4936 |
| 2 | 4.4936 | −0.0039 | −20.2271 | 0.0002 | 4.4934 |

Point of intersection of the graphs of $f$ and $g$ occurs when  $x \approx 4.493$.

**Note:**  $f(x) = x$ and  $g(x) = \tan x$ intersect infinitely often.

**19.** (a)  $f(x) = x^2 - a,\, a > 0$

$f'(x) = 2x$

$x_{n+1} = x_n - \dfrac{f(x_n)}{f'(x_n)}$

$= x_n - \dfrac{x_n^2 - a}{2x_n}$

$= \dfrac{1}{2}\left(x_n + \dfrac{a}{x_n}\right)$

(b) $\sqrt{5}$: $x_{n+1} = \frac{1}{2}\left(x_n + \frac{5}{x_n}\right)$, $x_1 = 2$

| $n$ | 1 | 2 | 3 | 4 |
|-----|---|------|--------|--------|
| $x_n$ | 2 | 2.25 | 2.2361 | 2.2361 |

For example, given $x_1 = 2$,

$$x_2 = \frac{1}{2}\left(2 + \frac{5}{2}\right) = \frac{9}{4} = 2.25.$$

$\sqrt{5} \approx 2.236$

$\sqrt{7}$: $x_{n+1} = \frac{1}{2}\left(x_n + \frac{7}{x_n}\right)$, $x_1 = 2$

| $n$ | 1 | 2 | 3 | 4 | 5 |
|-----|---|------|--------|--------|--------|
| $x_n$ | 2 | 2.75 | 2.6477 | 2.6458 | 2.6458 |

$\sqrt{7} \approx 2.646$

**21.** $y = 2x^3 - 6x^2 + 6x - 1 = f(x)$

$y' = 6x^2 - 12x + 6 = f'(x)$

$x_1 = 1$

$f'(x) = 0$; therefore, the method fails.

| $n$ | $x_n$ | $f(x_n)$ | $f'(x_n)$ |
|-----|-------|----------|-----------|
| 1 | 1 | 1 | 0 |

**23.** $y = -x^3 + 6x^2 - 10x + 6 = f(x)$

$y' = -3x^2 + 12x - 10 = f'(x)$

$x_1 = 2$

$x_2 = 1$

$x_3 = 2$

$x_4 = 1$     and so on.

Fails to converge

**25.** Let $g(x) = f(x) - x = \cos x - x$

$g'(x) = -\sin x - 1.$

| $n$ | $x_n$ | $g(x_n)$ | $g'(x_n)$ | $\dfrac{g(x_n)}{g'(x_n)}$ | $x_n - \dfrac{g(x_n)}{g'(x_n)}$ |
|-----|--------|----------|-----------|---------------------------|----------------------------------|
| 1 | 1.0000 | −0.4597 | −1.8415 | 0.2496 | 0.7504 |
| 2 | 0.7504 | −0.0190 | −1.6819 | 0.0113 | 0.7391 |
| 3 | 0.7391 | 0.0000 | −1.6736 | 0.0000 | 0.7391 |

The fixed point is approximately 0.74.

**27.** $f(x) = \dfrac{1}{x} - a = 0$

$$f'(x) = -\dfrac{1}{x^2}$$

$$x_{n+1} = x_n - \dfrac{(1/x_n) - a}{-1/x_n^2} = x_n + x_n^2\left(\dfrac{1}{x_n} - a\right) = x_n + x_n - x_n^2 a = 2x_n - x_n^2 a = x_n(2 - ax_n)$$

**29.** $f(x) = x^3 - 3x^2 + 3,\ f'(x) = 3x^2 - 6x$

(a)

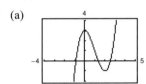

(b) $x_1 = 1$

$$x_2 = x_1 - \dfrac{f(x_1)}{f'(x_1)} \approx 1.333$$

Continuing, the zero is 1.347.

(c) $x_1 = \dfrac{1}{4}$

$$x_2 = x_1 - \dfrac{f(x_1)}{f'(x_1)} \approx 2.405$$

Continuing, the zero is 2.532.

(d)

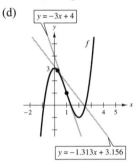

The $x$-intercept of $y = -3x + 4$ is $\dfrac{4}{3}$. The $x$-intercept of $y = 1.313x + 3.156$ is approximately 2.405.

The $x$-intercepts correspond to the values resulting from the first iteration of Newton's Method.

(e) If the initial guess $x_1$ is not "close to" the desired zero of the function, the $x$-intercept of the tangent line may approximate another zero of the function.

**31.** Answers will vary. See page 229.

If $f$ is a function continuous on $[a, b]$ and differentiable on $(a, b)$ where $c \in [a, b]$ and $f(c) = 0$, Newton's Method uses tangent lines to approximate $c$ such that $f(c) = 0$.

First, estimate an initial $x_1$ close to $c$ (see graph).

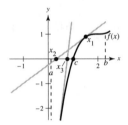

Then determine $x_2$ by $x_2 = x_1 - \dfrac{f(x_1)}{f'(x_1)}$.

Calculate a third estimate by $x_3 = x_2 - \dfrac{f(x_2)}{f'(x_2)}$.

Continue this process until $|x_n - x_{n+1}|$ is within the desired accuracy.

Let $x_{n+1}$ be the final approximation of $c$.

**33.** $f(x) = x \cos x, (0, \pi)$

$f'(x) = -x \sin x + \cos x = 0$

Letting $F(x) = f'(x)$, you can use Newton's Method as follows.

$\left[ F'(x) = -2 \sin x + x \cos x \right]$

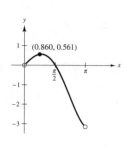

| $n$ | $x_n$ | $F(x_n)$ | $F'(x_n)$ | $\dfrac{F(x_n)}{F'(x_n)}$ | $x_n - \dfrac{F(x_n)}{F'(x_n)}$ |
|---|---|---|---|---|---|
| 1 | 0.9000 | −0.0834 | −2.1261 | 0.0392 | 0.8608 |
| 2 | 0.8608 | −0.0010 | −2.0778 | 0.0005 | 0.8603 |

Approximation to the critical number: 0.860

**35.** $y = f(x) = 4 - x^2, (1, 0)$

$d = \sqrt{(x-1)^2 + (y-0)^2} = \sqrt{(x-1)^2 + (4-x^2)^2} = \sqrt{x^4 - 7x^2 - 2x + 17}$

$d$ is minimized when $D = x^4 - 7x^2 - 2x + 17$ is a minimum.

$g(x) = D' = 4x^3 - 14x - 2$

$g'(x) = 12x^2 - 14$

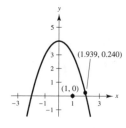

| $n$ | $x_n$ | $g(x_n)$ | $g'(x_n)$ | $\dfrac{g(x_n)}{g'(x_n)}$ | $x_n - \dfrac{g(x_n)}{g'(x_n)}$ |
|---|---|---|---|---|---|
| 1 | 2.0000 | 2.0000 | 34.0000 | 0.0588 | 1.9412 |
| 2 | 1.9412 | 0.0830 | 31.2191 | 0.0027 | 1.9385 |
| 3 | 1.9385 | −0.0012 | 31.0934 | 0.0000 | 1.9385 |

$x \approx 1.939$

Point closest to $(1, 0)$ is $\approx (1.939, 0.240)$.

**37.**

Minimize: $T = \dfrac{\text{Distance rowed}}{\text{Rate rowed}} + \dfrac{\text{Distance walked}}{\text{Rate walked}}$

$T = \dfrac{\sqrt{x^2 + 4}}{3} + \dfrac{\sqrt{x^2 - 6x + 10}}{4}$

$T' = \dfrac{x}{3\sqrt{x^2 + 4}} + \dfrac{x - 3}{4\sqrt{x^2 - 6x + 10}} = 0$

$4x\sqrt{x^2 - 6x + 10} = -3(x - 3)\sqrt{x^2 + 4}$

$16x^2(x^2 - 6x + 10) = 9(x - 3)^2(x^2 + 4)$

$7x^4 - 42x^3 + 43x^2 + 216x - 324 = 0$

Let $f(x) = 7x^4 - 42x^3 + 43x^2 + 216x - 324$ and $f'(x) = 28x^3 - 126x^2 + 86x + 216$.

Because $f(1) = -100$ and $f(2) = 56$, the solution is in the interval $(1, 2)$.

| $n$ | $x_n$ | $f(x_n)$ | $f'(x_n)$ | $\dfrac{f(x_n)}{f'(x_n)}$ | $x_n - \dfrac{f(x_n)}{f'(x_n)}$ |
|---|---|---|---|---|---|
| 1 | 1.7000 | 19.5887 | 135.6240 | 0.1444 | 1.5556 |
| 2 | 1.5556 | −1.0480 | 150.2780 | −0.0070 | 1.5626 |
| 3 | 1.5626 | 0.0014 | 49.5591 | 0.0000 | 1.5626 |

Approximation: $x \approx 1.563$ mi

**39.** $225 = 0.602x^3 - 41.44x^2 + 922.8x - 6330, 14 \le x \le 27.$

$0.602x^3 - 41.44x^2 + 922.8x - 6555 = 0$

Using Newton's Method with $x_1 = 15$, you get 15.1 years

Using Newton's Method with $x_1 = 25$, you get 26.8 years

**41.** False. Let $f(x) = (x^2 - 1)/(x - 1). x = 1$ is a discontinuity. It is not a zero of $f(x)$. This statement would be true if $f(x) = p(x)/q(x)$ was given in **reduced** form.

**43.** True

**45.** $f(x) = -\sin x$

$f'(x) = -\cos x$

Let $(x_0, y_1) = (x_0, -\sin(x_0))$ be a point on the graph of $f$. If $(x_0, y_0)$ is a point of tangency, then

$$-\cos(x_0) = \frac{y_0 - 0}{x_0 - 0} = \frac{y_0}{x_0} = \frac{-\sin(x_0)}{x_0}.$$

So, $x_0 = \tan(x_0)$.

$x_0 \approx 4.4934$

Slope $= -\cos(x_0) \approx 0.217$

You can verify this answer by graphing $y_1 = -\sin x$ and the tangent line $y_2 = 0.217x$.

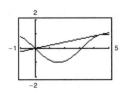

## Section 3.9    Differentials

**1.** $f(x) = x^2$

$f'(x) = 2x$

Tangent line at $(2, 4)$: $y - f(2) = f'(2)(x - 2)$

$y - 4 = 4(x - 2)$

$y = 4x - 4$

| $x$ | 1.9 | 1.99 | 2 | 2.01 | 2.1 |
|---|---|---|---|---|---|
| $f(x) = x^2$ | 3.6100 | 3.9601 | 4 | 4.0401 | 4.4100 |
| $T(x) = 4x - 4$ | 3.6000 | 3.9600 | 4 | 4.0400 | 4.4000 |

**3.** $f(x) = x^5$

$f'(x) = 5x^4$

Tangent line at $(2, 32)$:

$y - f(2) = f'(2)(x - 2)$

$y - 32 = 80(x - 2)$

$y = 80x - 128$

| $x$ | 1.9 | 1.99 | 2 | 2.01 | 2.1 |
|---|---|---|---|---|---|
| $f(x) = x^5$ | 24.7610 | 31.2080 | 32 | 32.8080 | 40.8410 |
| $T(x) = 80x - 128$ | 24.0000 | 31.2000 | 32 | 32.8000 | 40.0000 |

**5.** $f(x) = \sin x$

$f'(x) = \cos x$

Tangent line at $(2, \sin 2)$:

$y - f(2) = f'(2)(x - 2)$

$y - \sin 2 = (\cos 2)(x - 2)$

$\qquad y = (\cos 2)(x - 2) + \sin 2$

| $x$ | 1.9 | 1.99 | 2 | 2.01 | 2.1 |
|---|---|---|---|---|---|
| $f(x) = \sin x$ | 0.9463 | 0.9134 | 0.9093 | 0.9051 | 0.8632 |
| $T(x) = (\cos 2)(x - 2) + \sin 2$ | 0.9509 | 0.9135 | 0.9093 | 0.9051 | 0.8677 |

**7.** $y = f(x) = x^3,\ f'(x) = 3x^2,\ x = 1,\ \Delta x = dx = 0.1$

$\Delta y = f(x + \Delta x) - f(x) \qquad\qquad dy = f'(x)\,dx$

$\quad = f(1.1) - f(1) \qquad\qquad\qquad = f'(1)(0.1)$

$\quad = 0.331 \qquad\qquad\qquad\qquad\quad = 3(0.1)$

$\qquad\qquad\qquad\qquad\qquad\qquad\qquad = 0.3$

**9.** $y = f(x) = x^4 + 1,\ f'(x) = 4x^3,\ x = -1,\ \Delta x = dx = 0.01$

$\Delta y = f(x + \Delta x) - f(x) \qquad\qquad dy = f'(x)\,dx$

$\quad = f(-0.99) - f(-1) \qquad\qquad\quad = f'(-1)(0.01)$

$\quad = \left[(-0.99)^4 + 1\right] - \left[(-1)^4 + 1\right] \approx -0.0394 \qquad = (-4)(0.01) = -0.04$

**11.** $y = 3x^2 - 4$

$dy = 6x\,dx$

**13.** $y = \dfrac{x + 1}{2x - 1}$

$dy = -\dfrac{3}{(2x - 1)^2}\,dx$

**15.** $y = x\sqrt{1 - x^2}$

$dy = \left(x\dfrac{-x}{\sqrt{1 - x^2}} + \sqrt{1 - x^2}\right)dx = \dfrac{1 - 2x^2}{\sqrt{1 - x^2}}\,dx$

**17.** $y = 3x - \sin^2 x$

$dy = (3 - 2\sin x \cos x)\,dx = (3 - \sin 2x)\,dx$

**19.** $y = \dfrac{1}{3}\cos\left(\dfrac{6\pi x - 1}{2}\right)$

$dy = -\pi \sin\left(\dfrac{6\pi x - 1}{2}\right)dx$

**21.** (a) $f(1.9) = f(2 - 0.1) \approx f(2) + f'(2)(-0.1)$

$\qquad\qquad\qquad\qquad\qquad \approx 1 + (1)(-0.1) = 0.9$

$\quad$ (b) $f(2.04) = f(2 + 0.04) \approx f(2) + f'(2)(0.04)$

$\qquad\qquad\qquad\qquad\qquad \approx 1 + (1)(0.04) = 1.04$

**23.** (a) $f(1.9) = f(2 - 0.1) \approx f(2) + f'(2)(-0.1)$

$\qquad\qquad\qquad\qquad\qquad \approx 1 + \left(-\tfrac{1}{2}\right)(-0.1) = 1.05$

$\quad$ (b) $f(2.04) = f(2 + 0.04) \approx f(2) + f'(2)(0.04)$

$\qquad\qquad\qquad\qquad\qquad \approx 1 + \left(-\tfrac{1}{2}\right)(0.04) = 0.98$

**25.** (a) $g(2.93) = g(3 - 0.07) \approx g(3) + g'(3)(-0.07)$

$\qquad\qquad\qquad\qquad\qquad \approx 8 + \left(-\tfrac{1}{2}\right)(-0.07) = 8.035$

$\quad$ (b) $g(3.1) = g(3 + 0.1) \approx g(3) + g'(3)(0.1)$

$\qquad\qquad\qquad\qquad\qquad \approx 8 + \left(-\tfrac{1}{2}\right)(0.1) = 7.95$

**27.** $A = x^2$

$\quad x = 10$

$\quad \Delta x = dx = \pm\dfrac{1}{32}$

$\quad dA = 2x\,dx$

$\quad \Delta A \approx dA = 2(10)\left(\pm\tfrac{1}{32}\right)$

$\qquad\quad = \pm\tfrac{5}{8}\ \text{in.}^2$

**29.** $A = \pi r^2$

$\quad r = 16$

$\quad \Delta r = dr = \pm\frac{1}{4}$

$\quad \Delta A \approx dA = 2\pi r\, dr = \pi(32)\left(\pm\frac{1}{4}\right) = \pm 8\pi \text{ in.}^2$

**31. (a)** $\quad x = 12 \text{ cm}$

$\quad\quad \Delta x = dx = \pm 0.05 \text{ cm}$

$\quad\quad A = x^2$

$\quad\quad \Delta A \approx dA = 2x\, dx = 2(12)(\pm 0.05) = \pm 1.2 \text{ cm}^2$

$\quad$ Percentage error:

$$\frac{dA}{A} = \frac{1.2}{(12)^2} = 0.00833\ldots = \frac{5}{6}\%$$

**(b)** $\dfrac{dA}{A} = \dfrac{2x\, dx}{x^2} = \dfrac{2\, dx}{x} \le 0.025$

$\quad\quad \dfrac{dx}{x} \le \dfrac{0.025}{2} = 0.0125 = 1.25\%$

**33.** $r = 8 \text{ in.}$

$\quad \Delta r = dr = \pm 0.02 \text{ in.}$

**(a)** $\quad V = \dfrac{4}{3}\pi r^3$

$\quad\quad dV = 4\pi r^2 dr = 4\pi(8)^2(\pm 0.02) = \pm 5.12\pi \text{ in.}^3$

**(b)** $\quad S = 4\pi r^2$

$\quad\quad dS = 8\pi r\, dr = 8\pi(8)(\pm 0.02) = \pm 1.28\pi \text{ in.}^2$

**(c)** Relative error: $\dfrac{dV}{V} = \dfrac{4\pi r^2 dr}{(4/3)\pi r^3} = \dfrac{3\, dr}{r}$

$$= \frac{3}{8}(0.02) = 0.0075 = 0.75\%$$

$\quad$ Relative error: $\dfrac{dS}{S} = \dfrac{8\pi r\, dr}{4\pi r^2} = \dfrac{2\, dr}{r}$

$$= \frac{2(0.02)}{8} = 0.005\ldots = \frac{1}{2}\%$$

**35.** $V = \pi r^2 h = 40\pi r^2, r = 5 \text{ cm}, h = 40 \text{ cm}, dr = 0.2 \text{ cm}$

$\quad \Delta V \approx dV = 80\pi r\, dr = 80\pi(5)(0.2) = 80\pi \text{ cm}^3$

**37. (a)** $\quad T = 2\pi\sqrt{L/g}$

$$dT = \frac{\pi}{g\sqrt{L/g}}dL$$

$\quad$ Relative error:

$$\frac{dT}{T} = \frac{(\pi\, dL)/(g\sqrt{L/g})}{2\pi\sqrt{L/g}}$$

$$= \frac{dL}{2L}$$

$$= \frac{1}{2}\left(\text{relative error in } L\right)$$

$$= \frac{1}{2}(0.005) = 0.0025$$

$\quad$ Percentage error: $\dfrac{dT}{T}(100) = 0.25\% = \dfrac{1}{4}\%$

**(b)** $(0.0025)(3600)(24) = 216 \text{ sec} = 3.6 \text{ min}$

**39.** $\quad \theta = 26°45' = 26.75°$

$\quad d\theta = \pm 15' = \pm 0.25°$

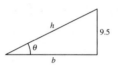

**(a)** $\quad h = 9.5 \csc \theta$

$\quad\quad dh = -9.5 \csc \theta \cot \theta\, d\theta$

$\quad\quad \dfrac{dh}{h} = -\cot \theta\, d\theta$

$\quad\quad \left|\dfrac{dh}{h}\right| = (\cot 26.75°)(0.25°)$

$\quad$ Converting to radians,

$\quad\quad (\cot 0.4669)(0.0044) \approx 0.0087$

$$= 0.87\% \text{ (in radians)}.$$

**(b)** $\left|\dfrac{dh}{h}\right| = \cot \theta\, d\theta \le 0.02$

$\quad\quad \dfrac{d\theta}{\theta} \le \dfrac{0.02}{\theta(\cot \theta)} = \dfrac{0.02 \tan \theta}{\theta}$

$\quad\quad \dfrac{d\theta}{\theta} \le \dfrac{0.02 \tan 26.75°}{26.75°} \approx \dfrac{0.02 \tan 0.4669}{0.4669}$

$$\approx 0.0216$$

$$= 2.16\% \text{ (in radians)}$$

**41.** $R = \dfrac{v_0^2}{32}(\sin 2\theta)$

$v_0 = 2500$ ft/sec

$\theta$ changes from $10°$ to $11°$.

$dR = \dfrac{v_0^2}{32}(2)(\cos 2\theta)d\theta = \dfrac{(2500)^2}{16}\cos 2\theta\, d\theta$

$\theta = 10\left(\dfrac{\pi}{180}\right)$

$d\theta = (11 - 10)\dfrac{\pi}{180}$

$\Delta R \approx dR$

$\qquad = \dfrac{(2500)^2}{16}\cos\left(\dfrac{20\pi}{180}\right)\left(\dfrac{\pi}{180}\right)$

$\qquad \approx 6407$ ft

**43.** Let $f(x) = \sqrt{x}$, $x = 100$, $dx = -0.6$.

$f(x + \Delta x) \approx f(x) + f'(x)\, dx$

$\qquad = \sqrt{x} + \dfrac{1}{2\sqrt{x}}dx$

$f(x + \Delta x) = \sqrt{99.4}$

$\qquad \approx \sqrt{100} + \dfrac{1}{2\sqrt{100}}(-0.6) = 9.97$

Using a calculator: $\sqrt{99.4} \approx 9.96995$

**45.** Let $f(x) = \sqrt[4]{x}$, $x = 625$, $dx = -1$.

$f(x + \Delta x) \approx f(x) + f'(x)\, dx = \sqrt[4]{x} + \dfrac{1}{4\sqrt[4]{x^3}}dx$

$f(x + \Delta x) = \sqrt[4]{624} \approx \sqrt[4]{625} + \dfrac{1}{4\left(\sqrt[4]{625}\right)^3}(-1)$

$\qquad = 5 - \dfrac{1}{500} = 4.998$

Using a calculator, $\sqrt[4]{624} \approx 4.9980$.

# Review Exercises for Chapter 3

**1.** A number $c$ in the domain of $f$ is a critical number if $f'(c) = 0$ or $f'$ is undefined at $c$.

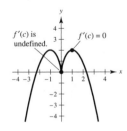

**47.** $f(x) = \sqrt{x + 4}$

$f'(x) = \dfrac{1}{2\sqrt{x + 4}}$

At $(0, 2)$, $f(0) = 2$, $f'(0) = \dfrac{1}{4}$

Tangent line: $y - 2 = \dfrac{1}{4}(x - 0)$

$\qquad\qquad y = \dfrac{1}{4}x + 2$

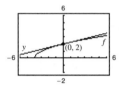

**49.** In general, when $\Delta x \to 0$, $dy$ approaches $\Delta y$.

**51.** (a) Let $f(x) = \sqrt{x}$, $x = 4$, $dx = 0.02$,

$f'(x) = 1/\left(2\sqrt{x}\right)$.

Then

$f(4.02) \approx f(4) + f'(4)\, dx$

$\sqrt{4.02} \approx \sqrt{4} + \dfrac{1}{2\sqrt{4}}(0.02) = 2 + \dfrac{1}{4}(0.02)$.

(b) Let

$f(x) = \tan x$, $x = 0$, $dx = 0.05$, $f'(x) = \sec^2 x$.

Then

$f(0.05) \approx f(0) + f'(0)\, dx$

$\tan 0.05 \approx \tan 0 + \sec^2 0(0.05) = 0 + 1(0.05)$.

**53.** True

**55.** True

**3.** $f(x) = x^2 + 5x$,  $[-4, 0]$

$f'(x) = 2x + 5 = 0$ when $x = -5/2$

Critical number: $x = -5/2$

Left endpoint: $(-4, -4)$

Critical number: $(-5/2, -25/4)$    Minimum

Right endpoint: $(0, 0)$    Maximum

**5.** $g(x) = 2x + 5 \cos x, [0, 2\pi]$

$g'(x) = 2 - 5 \sin x = 0$ when $\sin x = \frac{2}{5}$.

Critical numbers: $x \approx 0.41, x \approx 2.73$

Left endpoint: $(0, 5)$

Critical number: $(0.41, 5.41)$

Critical number: $(2.73, 0.88)$    Minimum

Right endpoint: $(2\pi, 17.57)$    Maximum

**7.** No, Rolle's Theorem cannot be applied.

$f(0) = -7 \neq 25 = f(4)$

**9.** No. $f(x) = \dfrac{x^2}{1 - x^2}$ is not continuous on $[-2, 2]$. $f(-1)$

is not defined.

**11.** $f(x) = 3 - |x - 4|$

(a)

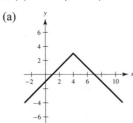

$f(1) = f(7) = 0$

(b) $f$ is not differentiable at $x = 4$.

**13.** $f(x) = x^{2/3}, 1 \leq x \leq 8$

$f'(x) = \dfrac{2}{3} x^{-1/3}$

$\dfrac{f(b) - f(a)}{b - a} = \dfrac{4 - 1}{8 - 1} = \dfrac{3}{7}$

$f'(c) = \dfrac{2}{3} c^{-1/3} = \dfrac{3}{7}$

$c = \left(\dfrac{14}{9}\right)^3 = \dfrac{2744}{729} \approx 3.764$

**23.** $f(x) = (x - 1)^2 (x - 3)$

$f'(x) = (x - 1)^2 (1) + (x - 3)(2)(x - 1) = (x - 1)(3x - 7)$

Critical numbers: $x = 1$ and $x = \frac{7}{3}$

| Intervals: | $-\infty < x < 1$ | $1 < x < \frac{7}{3}$ | $\frac{7}{3} < x < \infty$ |
|---|---|---|---|
| Sign of $f'(x)$: | $f'(x) > 0$ | $f'(x) < 0$ | $f'(x) > 0$ |
| Conclusion: | Increasing | Decreasing | Increasing |

**15.** The Mean Value Theorem cannot be applied. $f$ is not differentiable at $x = 5$ in $[2, 6]$.

**17.** $f(x) = x - \cos x, -\dfrac{\pi}{2} \leq x \leq \dfrac{\pi}{2}$

$f'(x) = 1 + \sin x$

$\dfrac{f(b) - f(a)}{b - a} = \dfrac{(\pi/2) - (-\pi/2)}{(\pi/2) - (-\pi/2)} = 1$

$f'(c) = 1 + \sin c = 1$

$c = 0$

**19.** $f(x) = Ax^2 + Bx + C$

$f'(x) = 2Ax + B$

$\dfrac{f(x_2) - f(x_1)}{x_2 - x_1} = \dfrac{A(x_2^2 - x_1^2) + B(x_2 - x_1)}{x_2 - x_1}$

$= A(x_1 + x_2) + B$

$f'(c) = 2Ac + B = A(x_1 + x_2) + B$

$2Ac = A(x_1 + x_2)$

$c = \dfrac{x_1 + x_2}{2} = $ Midpoint of $[x_1, x_2]$

**21.** $f(x) = x^2 + 3x - 12$

$f'(x) = 2x + 3$

Critical number: $x = -\dfrac{3}{2}$

| Intervals: | $-\infty < x < -\frac{3}{2}$ | $-\frac{3}{2} < x < \infty$ |
|---|---|---|
| Sign of $f'(x)$: | $f'(x) < 0$ | $f'(x) > 0$ |
| Conclusion: | Decreasing | Increasing |

**25.** $h(x) = \sqrt{x}(x - 3) = x^{3/2} - 3x^{1/2}$

Domain: $(0, \infty)$

$$h'(x) = \frac{3}{2}x^{3/2} - \frac{3}{2}x^{-1/2} = \frac{3}{2}x^{-1/2}(x - 1) = \frac{3(x - 1)}{2\sqrt{x}}$$

Critical number: $x = 1$

| Intervals: | $0 < x < 1$ | $1 < x < \infty$ |
|---|---|---|
| Sign of $h'(x)$: | $h'(x) < 0$ | $h'(x) > 0$ |
| Conclusion: | Decreasing | Increasing |

**27.** $f(x) = 4x^3 - 5x$

$$f'(x) = 12x^2 - 5 = 0 \text{ when } x = \pm\sqrt{\frac{5}{12}} = \pm\frac{\sqrt{15}}{6}$$

| Intervals: | $-\infty < x < \dfrac{\sqrt{15}}{6}$ | $-\dfrac{\sqrt{15}}{6} < x < \dfrac{\sqrt{15}}{6}$ | $\dfrac{\sqrt{15}}{6} < x < \infty$ |
|---|---|---|---|
| Sign of $f'(x)$: | $f'(x) > 0$ | $f'(x) < 0$ | $f'(x) > 0$ |
| Conclusion: | Increasing | Decreasing | Increasing |

Relative maximum: $\left(-\dfrac{\sqrt{15}}{6}, \dfrac{5\sqrt{15}}{9}\right)$

Relative minimum: $\left(\dfrac{\sqrt{15}}{6}, -\dfrac{5\sqrt{15}}{9}\right)$

**29.** $h(t) = \frac{1}{4}t^4 - 8t$

$h'(t) = t^3 - 8 = 0$ when $t = 2$.

Relative minimum: $(2, -12)$

| Test Intervals: | $-\infty < t < 2$ | $2 < t < \infty$ |
|---|---|---|
| Sign of $h'(t)$: | $h'(t) < 0$ | $h'(t) > 0$ |
| Conclusion: | Decreasing | Increasing |

**31.** $y = \dfrac{1}{3}\cos(12t) - \dfrac{1}{4}\sin(12t)$

$v = y' = -4\sin(12t) - 3\cos(12t)$

(a) When $t = \dfrac{\pi}{8}$, $y = \dfrac{1}{4}$ in. and $v = y' = 4$ in./sec.

(b) $y' = -4\sin(12t) - 3\cos(12t) = 0$ when $\dfrac{\sin(12t)}{\cos(12t)} = -\dfrac{3}{4} \Rightarrow \tan(12t) = -\dfrac{3}{4}$.

Therefore, $\sin(12t) = -\dfrac{3}{5}$ and $\cos(12t) = \dfrac{4}{5}$. The maximum displacement is

$$y = \left(\frac{1}{3}\right)\left(\frac{4}{5}\right) - \frac{1}{4}\left(-\frac{3}{5}\right) = \frac{5}{12} \text{ in.}$$

(c) Period: $\dfrac{2\pi}{12} = \dfrac{\pi}{6}$

Frequency: $\dfrac{1}{\pi/6} = \dfrac{6}{\pi}$

**33.** $f(x) = x^3 - 9x^2$

$f'(x) = 3x^2 - 18x$

$f''(x) = 6x - 18 = 0$ when $x = 3$

| Intervals: | $-\infty < x < 3$ | $3 < x < \infty$ |
|---|---|---|
| Sign of $f''(x)$: | $f''(x) < 0$ | $f''(x) > 0$ |
| Conclusion: | Concave downward | Concave upward |

Point of inflection: $(3, -54)$

**35.** $f(x) = x + \cos x, 0 \le x \le 2\pi$

$f'(x) = 1 - \sin x$

$f''(x) = -\cos x = 0$ when $x = \dfrac{\pi}{2}, \dfrac{3\pi}{2}$.

Points of inflection: $\left(\dfrac{\pi}{2}, \dfrac{\pi}{2}\right), \left(\dfrac{3\pi}{2}, \dfrac{3\pi}{2}\right)$

| Test Intervals: | $0 < x < \dfrac{\pi}{2}$ | $\dfrac{\pi}{2} < x < \dfrac{3\pi}{2}$ | $\dfrac{3\pi}{2} < x < 2\pi$ |
|---|---|---|---|
| Sign of $f''(x)$: | $f''(x) < 0$ | $f''(x) > 0$ | $f''(x) < 0$ |
| Conclusion: | Concave downward | Concave upward | Concave downward |

**37.** $f(x) = (x + 9)^2$

$f'(x) = 2(x + 9) = 0 \Rightarrow x = -9$

$f''(x) = 2 > 0 \Rightarrow (-9, 0)$ is a relative minimum.

**39.** $g(x) = 2x^2(1 - x^2)$

$g'(x) = -4x(2x^2 - 1) = 0 \Rightarrow x = 0, \pm\dfrac{1}{\sqrt{2}}$

$g''(x) = 4 - 24x^2$

$g''(0) = 4 > 0 \quad (0, 0)$ is a relative minimum.

$g''\left(\pm\dfrac{1}{\sqrt{2}}\right) = -8 < 0 \left(\pm\dfrac{1}{\sqrt{2}}, \dfrac{1}{2}\right)$ are relative maxima.

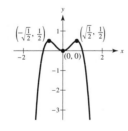

**41.**

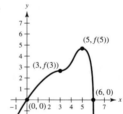

**43.** The first derivative is positive and the second derivative is negative. The graph is increasing and is concave down.

**45.** (a) $D = 0.00430t^4 - 0.2856t^3 + 5.833t^2 - 26.85t + 87.1, \quad 0 \le t \le 35$

(b)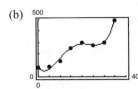

(c) Maximum occurs at $t = 35 \quad (2005)$

Minimum occurs at $t \approx 2.9 \quad (1972)$

(d) $D'(t)$ is greatest at $t = 35 \quad (2005)$

**47.** $\lim\limits_{x \to \infty}\left(8 + \dfrac{1}{x}\right) = 8 + 0 = 8$

**49.** $\lim\limits_{x \to \infty}\dfrac{2x^2}{3x^2 + 5} = \lim\limits_{x \to \infty}\dfrac{2}{3 + 5/x^2} = \dfrac{2}{3}$

**51.** $\lim\limits_{x \to -\infty}\dfrac{3x^2}{x + 5} = -\infty$

**53.** $\lim\limits_{x \to \infty}\dfrac{5 \cos x}{x} = 0$, because $|5 \cos x| \le 5$.

**55.** $\lim\limits_{x \to -\infty}\dfrac{6x}{x + \cos x} = 6$

**57.** $f(x) = \dfrac{3}{x} - 2$

Discontinuity: $x = 0$

$\lim\limits_{x \to \infty}\left(\dfrac{3}{x} - 2\right) = -2$

Vertical asymptote: $x = 0$

Horizontal asymptote: $y = -2$

**59.** $h(x) = \dfrac{2x + 3}{x - 4}$

Discontinuity: $x = 4$

$\lim\limits_{x \to \infty}\dfrac{2x + 3}{x - 4} = \lim\limits_{x \to \infty}\dfrac{2 + (3/x)}{1 - (4/x)} = 2$

Vertical asymptote: $x = 4$

Horizontal asymptote: $y = 2$

**61.** $f(x) = x^3 + \dfrac{243}{x}$

Relative minimum: $(3, 108)$

Relative maximum: $(-3, -108)$

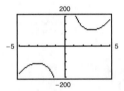

Vertical asymptote: $x = 0$

**63.** $f(x) = \dfrac{x - 1}{1 + 3x^2}$

Relative minimum: $(-0.155, -1.077)$

Relative maximum: $(2.155, 0.077)$

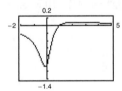

Horizontal asymptote: $y = 0$

**65.** $f(x) = 4x - x^2 = x(4 - x)$

Domain: $(-\infty, \infty)$; Range: $(-\infty, 4]$

$f'(x) = 4 - 2x = 0$ when $x = 2$.

$f''(x) = -2$

Therefore, $(2, 4)$ is a relative maximum.

Intercepts: $(0, 0), (4, 0)$

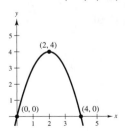

**67.** $f(x) = x\sqrt{16 - x^2}$

Domain: $[-4, 4]$; Range: $[-8, 8]$

$f'(x) = \dfrac{16 - 2x^2}{\sqrt{16 - x^2}} = 0$ when $x = \pm 2\sqrt{2}$ and undefined when $x = \pm 4$.

$f''(x) = \dfrac{2x(x^2 - 24)}{(16 - x^2)^{3/2}}$

$f''(-2\sqrt{2}) > 0$

Therefore, $(-2\sqrt{2}, -8)$ is a relative minimum.

$f''(2\sqrt{2}) < 0$

Therefore, $(2\sqrt{2}, 8)$ is a relative maximum.

Point of inflection: $(0, 0)$

Intercepts: $(-4, 0), (0, 0), (4, 0)$

Symmetry with respect to origin

**69.** $f(x) = (x - 1)^3(x - 3)^2$

Domain: $(-\infty, \infty)$; Range: $(-\infty, \infty)$

$f'(x) = (x - 1)^2(x - 3)(5x - 11) = 0$ when $x = 1, \dfrac{11}{5}, 3$.

$f''(x) = 4(x - 1)(5x^2 - 22x + 23) = 0$ when

$x = 1, \dfrac{11 \pm \sqrt{6}}{5}$.

$f''(3) > 0$

Therefore, $(3, 0)$ is a relative minimum.

$f''\left(\dfrac{11}{5}\right) < 0$

Therefore, $\left(\dfrac{11}{5}, \dfrac{3456}{3125}\right)$ is a relative maximum.

Points of inflection:

$(1, 0), \left(\dfrac{11 - \sqrt{6}}{5}, 0.60\right), \left(\dfrac{11 + \sqrt{6}}{5}, 0.46\right)$

Intercepts: $(0, -9), (1, 0), (3, 0)$

**71.** $f(x) = x^{1/3}(x + 3)^{2/3}$

Domain: $(-\infty, \infty)$; Range: $(-\infty, \infty)$

$f'(x) = \dfrac{x + 1}{(x + 3)^{1/3} x^{2/3}} = 0$ when $x = -1$ and undefined when $x = -3, 0$.

$f''(x) = \dfrac{-2}{x^{5/3}(x + 3)^{4/3}}$ is undefined when $x = 0, -3$.

By the First Derivative Test $(-3, 0)$ is a relative maximum and $\left(-1, -\sqrt[3]{4}\right)$ is a relative minimum. $(0, 0)$ is a point of inflection.

Intercepts: $(-3, 0), (0, 0)$

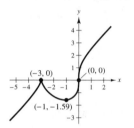

**73.** $f(x) = \dfrac{5 - 3x}{x - 2}$

$f'(x) = \dfrac{1}{(x - 2)^2} > 0$ for all $x \neq 2$

$f''(x) = \dfrac{-2}{(x - 2)^3}$

Concave upward on $(-\infty, 2)$

Concave downward on $(2, \infty)$

Vertical asymptote: $x = 2$

Horizontal asymptote: $y = -3$

Intercepts: $\left(\dfrac{5}{3}, 0\right), \left(0, -\dfrac{5}{2}\right)$

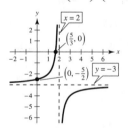

**75.** $f(x) = \dfrac{4}{1 + x^2}$

Domain: $(-\infty, \infty)$; Range: $(0, 4]$

$f'(x) = \dfrac{-8x}{\left(1 + x^2\right)^2} = 0$ when $x = 0$.

$f''(x) = \dfrac{-8\left(1 - 3x^2\right)}{\left(1 + x^2\right)^3} = 0$ when $x = \pm\dfrac{\sqrt{3}}{3}$.

$f''(0) < 0$

Therefore, $(0, 4)$ is a relative maximum.

Points of inflection: $\left(\pm\sqrt{3}/3, 3\right)$

Intercept: $(0, 4)$

Symmetric to the $y$-axis

Horizontal asymptote: $y = 0$

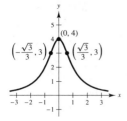

**77.** $f(x) = x^3 + x + \dfrac{4}{x}$

Domain: $(-\infty, 0), (0, \infty)$; Range: $(-\infty, -6], [6, \infty)$

$f'(x) = 3x^2 + 1 - \dfrac{4}{x^2}$

$\qquad = \dfrac{3x^4 + x^2 - 4}{x^2} = \dfrac{\left(3x^2 + 4\right)\left(x^2 - 1\right)}{x^2} = 0$

when $x = \pm1$.

$f''(x) = 6x + \dfrac{8}{x^3} = \dfrac{6x^4 + 8}{x^3} \neq 0$

$f''(-1) < 0$

Therefore, $(-1, -6)$ is a relative maximum.

$f''(1) > 0$

Therefore, $(1, 6)$ is a relative minimum.

Vertical asymptote: $x = 0$

Symmetric with respect to origin

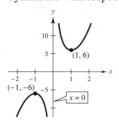

**79.** $f(x) = \left|x^2 - 9\right|$

Domain: $(-\infty, \infty)$; Range: $[0, \infty)$

$f'(x) = \dfrac{2x\left(x^2 - 9\right)}{\left|x^2 - 9\right|} = 0$ when $x = 0$ and is undefined

when $x = \pm3$.

$f''(x) = \dfrac{2\left(x^2 - 9\right)}{\left|x^2 - 9\right|}$ is undefined at $x = \pm3$.

$f''(0) < 0$

Therefore, $(0, 9)$
is a relative maximum.

Relative minima: $(\pm3, 0)$

Intercepts: $(\pm3, 0), (0, 9)$

Symmetric to the $y$-axis

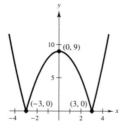

**81.** $f(x) = x + \cos x$

Domain: $[0, 2\pi]$; Range: $[1, 1 + 2\pi]$

$f'(x) = 1 - \sin x \geq 0$, $f$ is increasing.

$f''(x) = -\cos x = 0$ when $x = \dfrac{\pi}{2}, \dfrac{3\pi}{2}$.

Points of inflection: $\left(\dfrac{\pi}{2}, \dfrac{\pi}{2}\right), \left(\dfrac{3\pi}{2}, \dfrac{3\pi}{2}\right)$

Intercept: $(0, 1)$

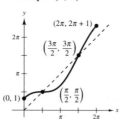

**83.** $x^2 + 4y^2 - 2x - 16y + 13 = 0$

(a) $(x^2 - 2x + 1) + 4(y^2 - 4y + 4) = -13 + 1 + 16$

$(x - 1)^2 + 4(y - 2)^2 = 4$

$\dfrac{(x - 1)^2}{4} + \dfrac{(y - 2)^2}{1} = 1$

The graph is an ellipse:

Maximum: $(1, 3)$

Minimum: $(1, 1)$

(b) $x^2 + 4y^2 - 2x - 16y + 13 = 0$

$2x + 8y\dfrac{dy}{dx} - 2 - 16\dfrac{dy}{dx} = 0$

$\dfrac{dy}{dx}(8y - 16) = 2 - 2x$

$\dfrac{dy}{dx} = \dfrac{2 - 2x}{8y - 16} = \dfrac{1 - x}{4y - 8}$

The critical numbers are $x = 1$ and $y = 2$. These correspond to the points $(1, 1)$, $(1, 3)$, $(2, -1)$, and $(2, 3)$. So, the maximum is $(1, 3)$ and the minimum is $(1, 1)$.

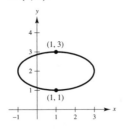

**85.** Let $t = 0$ at noon.

$L = d^2 = (100 - 12t)^2 + (-10t)^2$

$= 10{,}000 - 2400t + 244t^2$

$\dfrac{dL}{dt} = -2400 + 488t = 0$ when $t = \dfrac{300}{61} \approx 4.92$ hr.

Ship $A$ at $(40.98, 0)$; Ship $B$ at $(0, -49.18)$

$d^2 = 10{,}000 - 2400t + 244t^2$

$\approx 4098.36$ when $t \approx 4.92 \approx 4{:}55$ P.M.

$d \approx 64$ km

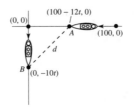

**87.** You have points $(0, y)$, $(x, 0)$, and $(1, 8)$. So,

$m = \dfrac{y - 8}{0 - 1} = \dfrac{0 - 8}{x - 1}$ or $y = \dfrac{8x}{x - 1}$.

Let $f(x) = L^2 = x^2 + \left(\dfrac{8x}{x - 1}\right)^2$.

$f'(x) = 2x + 128\left(\dfrac{x}{x - 1}\right)\left[\dfrac{(x - 1) - x}{(x - 1)^2}\right] = 0$

$x - \dfrac{64x}{(x - 1)^3} = 0$

$x\left[(x - 1)^3 - 64\right] = 0$ when $x = 0, 5$ (minimum).

Vertices of triangle: $(0, 0)$, $(5, 0)$, $(0, 10)$

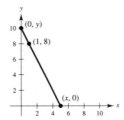

**89.** $A = (\text{Average of bases})(\text{Height}) = \left(\dfrac{x + s}{2}\right)\dfrac{\sqrt{3s^2 + 2sx - x^2}}{2}$ (see figure)

$\dfrac{dA}{dx} = \dfrac{1}{4}\left[\dfrac{(s - x)(s + x)}{\sqrt{3s^2 + 2sx - x^2}} + \sqrt{3s^2 + 2sx - x^2}\right] = \dfrac{2(2s - x)(s + x)}{4\sqrt{3s^2 + 2sx - x^2}} = 0$ when $x = 2s$.

$A$ is a maximum when $x = 2s$.

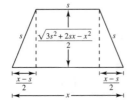

**91.** You can form a right triangle with vertices $(0, 0)$, $(x, 0)$ and $(0, y)$. Assume that the hypotenuse of length $L$ passes through $(4, 6)$.

$$m = \frac{y - 6}{0 - 4} = \frac{6 - 0}{4 - x} \text{ or } y = \frac{6x}{x - 4}$$

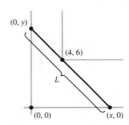

Let $f(x) = L^2 = x^2 + y^2 = x^2 + \left(\frac{6x}{x - 4}\right)^2$.

$$f'(x) = 2x + 72\left(\frac{x}{x - 4}\right)\left[\frac{-4}{(x - 4)^2}\right] = 0$$

$$x\left[(x - 4)^3 - 144\right] = 0 \text{ when } x = 0 \text{ or } x = 4 + \sqrt[3]{144}.$$

$$L \approx 14.05 \text{ ft}$$

**93.**   $\csc \theta = \dfrac{L_1}{6}$ or $L_1 = 6 \csc \theta$   (see figure)

$$\sec \theta = \frac{L_2}{9} \text{ or } L_2 = 9 \sec \theta$$

$$L = L_1 + L_2 = 6 \csc \theta + 9 \sec \theta$$

$$\frac{dL}{d\theta} = -6 \csc \theta \cot \theta + 9 \sec \theta \tan \theta = 0$$

$$\tan^3 \theta = \frac{2}{3} \Rightarrow \tan \theta = \frac{\sqrt[3]{2}}{\sqrt[3]{3}}$$

$$\sec \theta = \sqrt{1 + \tan^2 \theta} = \sqrt{1 + \left(\frac{2}{3}\right)^{2/3}} = \frac{\sqrt{3^{2/3} + 2^{2/3}}}{3^{1/3}}$$

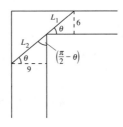

$$\csc \theta = \frac{\sec \theta}{\tan \theta} = \frac{\sqrt{3^{2/3} + 2^{2/3}}}{2^{1/3}}$$

$$L = 6\frac{\left(3^{2/3} + 2^{2/3}\right)^{1/2}}{2^{1/3}} + 9\frac{\left(3^{2/3} + 2^{2/3}\right)^{1/2}}{3^{1/3}}$$

$$= 3\left(3^{2/3} + 2^{2/3}\right)^{3/2} \text{ ft} \approx 21.07 \text{ ft}$$

(Compare to Exercise 92 using $a = 9$ and $b = 6$.)

**95.**   Total cost $= (\text{Cost per hour})(\text{Number of hours})$

$$T = \left(\frac{v^2}{600} + 5\right)\left(\frac{110}{v}\right) = \frac{11v}{60} + \frac{550}{v}$$

$$\frac{dT}{dv} = \frac{11}{60} - \frac{550}{v^2} = \frac{11v^2 - 33,000}{60v^2} = 0 \text{ when } v = \sqrt{3000} = 10\sqrt{30} \approx 54.77 \text{ mi/h}$$

$$\frac{d^2T}{dv^2} = \frac{1100}{v^3} > 0 \text{ when } v = 10\sqrt{30} \text{ so this value yields a minimum.}$$

**97.** $f(x) = x^3 - 3x - 1$

From the graph you can see that $f(x)$ has three real zeros.

$f'(x) = 3x^2 - 3$

| $n$ | $x_n$ | $f(x_n)$ | $f'(x_n)$ | $\dfrac{f(x_n)}{f'(x_n)}$ | $x_n - \dfrac{f(x_n)}{f'(x_n)}$ |
|---|---|---|---|---|---|
| 1 | −1.5000 | 0.1250 | 3.7500 | 0.0333 | −1.5333 |
| 2 | −1.5333 | −0.0049 | 4.0530 | −0.0012 | −1.5321 |

| $n$ | $x_n$ | $f(x_n)$ | $f'(x_n)$ | $\dfrac{f(x_n)}{f'(x_n)}$ | $x_n - \dfrac{f(x_n)}{f'(x_n)}$ |
|---|---|---|---|---|---|
| 1 | −0.5000 | 0.3750 | −2.2500 | −0.1667 | −0.3333 |
| 2 | −0.3333 | −0.0371 | −2.6667 | 0.0139 | −0.3472 |
| 3 | −0.3472 | −0.0003 | −2.6384 | 0.0001 | −0.3473 |

| $n$ | $x_n$ | $f(x_n)$ | $f'(x_n)$ | $\dfrac{f(x_n)}{f'(x_n)}$ | $x_n - \dfrac{f(x_n)}{f'(x_n)}$ |
|---|---|---|---|---|---|
| 1 | 1.9000 | 0.1590 | 7.8300 | 0.0203 | 1.8797 |
| 2 | 1.8797 | 0.0024 | 7.5998 | 0.0003 | 1.8794 |

The three real zeros of $f(x)$ are $x \approx -1.532$, $x \approx -0.347$, and $x \approx 1.879$.

**99.** Find the zeros of $f(x) = x^4 - x - 3$.

$f'(x) = 4x^3 - 1$

From the graph you can see that $f(x)$ has two real zeros.

$f$ changes sign in $[-2, -1]$.

| $n$ | $x_n$ | $f(x_n)$ | $f'(x_n)$ | $\dfrac{f(x_n)}{f'(x_n)}$ | $x_n - \dfrac{f(x_n)}{f'(x_n)}$ |
|---|---|---|---|---|---|
| 1 | −1.2000 | 0.2736 | −7.9120 | −0.0346 | −1.1654 |
| 2 | −1.1654 | 0.0100 | −7.3312 | −0.0014 | −1.1640 |

On the interval $[-2, -1]$: $x \approx -1.164$.

$f$ changes sign in $[1, 2]$.

| $n$ | $x_n$ | $f(x_n)$ | $f'(x_n)$ | $\dfrac{f(x_n)}{f'(x_n)}$ | $x_n - \dfrac{f(x_n)}{f'(x_n)}$ |
|---|---|---|---|---|---|
| 1 | 1.5000 | 0.5625 | 12.5000 | 0.0450 | 1.4550 |
| 2 | 1.4550 | 0.0268 | 11.3211 | 0.0024 | 1.4526 |
| 3 | 1.4526 | −0.0003 | 11.2602 | 0.0000 | 1.4526 |

On the interval $[1, 2]$: $x \approx 1.453$.

**101.** $y = x(1 - \cos x) = x - x\cos x$

$\dfrac{dy}{dx} = 1 + x\sin x - \cos x$

$dy = (1 + x\sin x - \cos x)dx$

**103.** $S = 4\pi r^2 \quad dr = \Delta r = \pm0.025$

$dS = 8\pi r\, dr = 8\pi(9)(\pm0.025) = \pm1.8\pi \text{ cm}^2$

$\dfrac{dS}{S}(100) = \dfrac{8\pi r\, dr}{4\pi r^2}(100) = \dfrac{2\, dr}{r}(100) = \dfrac{2(\pm0.025)}{9}(100) \approx \pm0.56\%$

$V = \dfrac{4}{3}\pi r^3$

$dV = 4\pi r^2\, dr = 4\pi(9)^2(\pm0.025) = \pm8.1\pi \text{ cm}^3$

$\dfrac{dV}{V}(100) = \dfrac{4\pi r^2\, dr}{(4/3)\pi r^3}(100) = \dfrac{3\, dr}{r}(100) = \dfrac{3(\pm0.025)}{9}(100) \approx \pm0.83\%$

# Problem Solving for Chapter 3

**1.** $p(x) = x^4 + ax^2 + 1$

   (a) $p'(x) = 4x^3 + 2ax = 2x(2x^2 + a)$

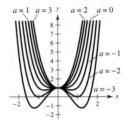

      $p''(x) = 12x^2 + 2a$

      For $a \geq 0$, there is one relative minimum at $(0, 1)$.

   (b) For $a < 0$, there is a relative maximum at $(0, 1)$.

   (c) For $a < 0$, there are two relative minima at $x = \pm\sqrt{-\dfrac{a}{2}}$.

   (d) If $a < 0$, there are three critical points; if $a > 0$, there is only one critical point.

**3.** $f(x) = \dfrac{c}{x} + x^2$

$f'(x) = -\dfrac{c}{x^2} + 2x = 0 \Rightarrow \dfrac{c}{x^2} = 2x \Rightarrow x^3 = \dfrac{c}{2} \Rightarrow x = \sqrt[3]{\dfrac{c}{2}}$

$f''(x) = \dfrac{2c}{x^3} + 2$

If $c = 0$, $f(x) = x^2$ has a relative minimum, but no relative maximum.

If $c > 0$, $x = \sqrt[3]{\dfrac{c}{2}}$ is a relative minimum, because $f''\left(\sqrt[3]{\dfrac{c}{2}}\right) > 0$.

If $c < 0$, $x = \sqrt[3]{\dfrac{c}{2}}$ is a relative minimum, too.

Answer: All $c$.

**5.** Assume $y_1 < d < y_2$. Let $g(x) = f(x) - d(x - a)$. $g$ is continuous on $[a, b]$ and therefore

has a minimum $(c, g(c))$ on $[a, b]$. The point $c$ cannot be an endpoint of $[a, b]$ because

$g'(a) = f'(a) - d = y_1 - d < 0$

$g'(b) = f'(b) - d = y_2 - d > 0$.

So, $a < c < b$ and $g'(c) = 0 \Rightarrow f'(c) = d$.

**7.** Set $\dfrac{f(b) - f(a) - f'(a)(b - a)}{(b - a)^2} = k$.

Define $F(x) = f(x) - f(a) - f'(a)(x - a) - k(x - a)^2$.

$F(a) = 0, F(b) = f(b) - f(a) - f'(a)(b - a) - k(b - a)^2 = 0$

$F$ is continuous on $[a, b]$ and differentiable on $(a, b)$.

There exists $c_1, a < c_1 < b$, satisfying $F'(c_1) = 0$.

$F'(x) = f'(x) - f'(a) - 2k(x - a)$ satisfies the hypothesis of Rolle's Theorem on $[a, c_1]$:

$F'(a) = 0, F'(c_1) = 0$.

There exists $c_2, a < c_2 < c_1$ satisfying $F''(c_2) = 0$.

Finally, $F''(x) = f''(x) - 2k$ and $F''(c_2) = 0$ implies that

$$k = \frac{f''(c_2)}{2}.$$

So, $k = \dfrac{f(b) - f(a) - f'(a)(b - a)}{(b - a)^2} = \dfrac{f''(c_2)}{2} \Rightarrow f(b) = f(a) + f'(a)(b - a) + \dfrac{1}{2}f''(c_2)(b - a)^2$.

**9.**

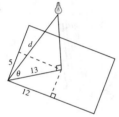

$d = \sqrt{13^2 + x^2}$, $\sin \theta = \dfrac{x}{d}$.

Let $A$ be the amount of illumination at one of the corners, as indicated in the figure. Then

$$A = \frac{kI}{\left(13^2 + x^2\right)}\sin \theta = \frac{kIx}{\left(13^2 + x^2\right)^{3/2}}$$

$$A'(x) = kI\frac{\left(x^2 + 169\right)^{3/2}(1) - x\left(\dfrac{3}{2}\right)\left(x^2 + 169\right)^{1/2}(2x)}{\left(169 + x^2\right)^3} = 0$$

$$\Rightarrow \left(x^2 + 169\right)^{3/2} = 3x^2\left(x^2 + 169\right)^{1/2}$$

$$x^2 + 169 = 3x^2$$

$$2x^2 = 169$$

$$x = \frac{13}{\sqrt{2}} \approx 9.19 \text{ ft}$$

By the First Derivative Test, this is a maximum.

**11.** Let $T$ be the intersection of $PQ$ and $RS$. Let $MN$ be the perpendicular to $SQ$ and $PR$ passing through $T$.
Let $TM = x$ and $TN = b - x$.

$$\frac{SN}{b - x} = \frac{MR}{x} \Rightarrow SN = \frac{b - x}{x} MR$$

$$\frac{NQ}{b - x} = \frac{PM}{x} \Rightarrow NQ = \frac{b - x}{x} PM$$

$$SQ = \frac{b - x}{x}(MR + PM) = \frac{b - x}{x} d$$

$$A(x) = \text{Area} = \frac{1}{2}dx + \frac{1}{2}\left(\frac{b - x}{x}d\right)(b - x) = \frac{1}{2}d\left[x + \frac{(b - x)^2}{x}\right] = \frac{1}{2}d\left[\frac{2x^2 - 2bx + b^2}{x}\right]$$

$$A'(x) = \frac{1}{2}d\left[\frac{x(4x - 2b) - (2x^2 - 2bx + b^2)}{x^2}\right]$$

$$A'(x) = 0 \Rightarrow 4x^2 - 2xb = 2x^2 - 2bx + b^2$$

$$2x^2 = b^2$$

$$x = \frac{b}{\sqrt{2}}$$

So, you have $SQ = \frac{b - x}{x}d = \frac{b - (b/\sqrt{2})}{b/\sqrt{2}}d = (\sqrt{2} - 1)d$.

Using the Second Derivative Test, this is a minimum. There is no maximum.

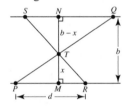

**13.** (a) Let $M > 0$ be given. Take $N = \sqrt{M}$. Then whenever $x > N = \sqrt{M}$, you have

$$f(x) = x^2 > M.$$

(b) Let $\varepsilon > 0$ be given. Let $M = \sqrt{\frac{1}{\varepsilon}}$. Then whenever $x > M = \sqrt{\frac{1}{\varepsilon}}$, you have $x^2 > \frac{1}{\varepsilon} \Rightarrow \frac{1}{x^2} < \varepsilon \Rightarrow \left|\frac{1}{x^2} - 0\right| < \varepsilon$.

(c) Let $\varepsilon > 0$ be given. There exists $N > 0$ such that $\left|f(x) - L\right| < \varepsilon$ whenever $x > N$.

Let $\delta = \frac{1}{N}$. Let $x = \frac{1}{y}$.

If $0 < y < \delta = \frac{1}{N}$, then $\frac{1}{x} < \frac{1}{N} \Rightarrow x > N$ and $\left|f(x) - L\right| = \left|f\left(\frac{1}{y}\right) - L\right| < \varepsilon$.

**15.** (a)

| $x$ | 0 | 0.5 | 1 | 2 |
|---|---|---|---|---|
| $\sqrt{1 + x}$ | 1 | 1.2247 | 1.4142 | 1.7321 |
| $\frac{1}{2}x + 1$ | 1 | 1.25 | 1.5 | 2 |

(b) Let $f(x) = \sqrt{1 + x}$. Using the Mean Value Theorem on the interval $[0, x]$, there exists $c, 0 < c < x$, satisfying

$$f'(c) = \frac{1}{2\sqrt{1 + c}} = \frac{f(x) - f(0)}{x - 0} = \frac{\sqrt{1 + x} - 1}{x}.$$

So $\sqrt{1 + x} = \frac{x}{2\sqrt{1 + c}} + 1 < \frac{x}{2} + 1 \left(\text{because } \sqrt{1 + c} > 1\right)$.

**17.** (a) $s = \dfrac{v\dfrac{km}{h}\left(1000\dfrac{m}{km}\right)}{\left(3600\dfrac{sec}{h}\right)} = \dfrac{5}{18}v$

| $v$ | 20 | 40 | 60 | 80 | 100 |
|---|---|---|---|---|---|
| $s$ | 5.56 | 11.11 | 16.67 | 22.22 | 27.78 |
| $d$ | 5.1 | 13.7 | 27.2 | 44.2 | 66.4 |

$d(s) = 0.071s^2 + 0.389s + 0.727$

(b) The distance between the back of the first vehicle and the front of the second vehicle is $d(s)$, the safe stopping distance.

The first vehicle passes the given point in $5.5/s$ seconds, and the second vehicle takes $d(s)/s$ more seconds. So,

$$T = \frac{d(s)}{s} + \frac{5.5}{s}.$$

(c)
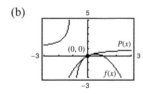

$$T = \frac{1}{s}\left(0.071s^2 + 0.389s + 0.727\right) + \frac{5.5}{s}$$

The minimum is attained when $s \approx 9.365$ m/sec.

(d) $T(s) = 0.071s + 0.389 + \dfrac{6.227}{s}$

$T'(s) = 0.071 - \dfrac{6.227}{s^2} \Rightarrow s^2 = \dfrac{6.227}{0.071} \Rightarrow s \approx 9.365$ m/sec

$T(9.365) \approx 1.719$ seconds

$9.365$ m/sec $\cdot \dfrac{3600}{1000} = 33.7$ km/h

(e) $d(9.365) = 10.597$ m

**19.** (a) $f(x) = \dfrac{x}{x+1}, f'(x) = \dfrac{1}{(x+1)^2}, f''(x) = \dfrac{-2}{(x+1)^3}$

$P(0) = f(0)$: $c_0 = 0$

$P'(0) = f'(0)$: $c_1 = 1$

$P''(0) = f''(0)$: $2c_2 = -2 \Rightarrow c_2 = -1$

$P(x) = x - x^2$

(b)

# CHAPTER 4
# Integration

# C H A P T E R  4
# Integration

## Section 4.1   Antiderivatives and Indefinite Integration

**1.** $\dfrac{d}{dx}\left(\dfrac{2}{x^3} + C\right) = \dfrac{d}{dx}\left(2x^{-3} + C\right) = -6x^{-4} = \dfrac{-6}{x^4}$

**3.** $\dfrac{d}{dx}\left(\dfrac{1}{3}x^3 - 16x + C\right) = x^2 - 16 = (x - 4)(x + 4)$

**5.** $\dfrac{dy}{dt} = 9t^2$

  $y = 3t^3 + C$

  **Check:** $\dfrac{d}{dt}\left[3t^3 + C\right] = 9t^2$

**7.** $\dfrac{dy}{dx} = x^{3/2}$

  $y = \dfrac{2}{5}x^{5/2} + C$

  **Check:** $\dfrac{d}{dx}\left[\dfrac{2}{5}x^{5/2} + C\right] = x^{3/2}$

| | *Given* | *Rewrite* | *Integrate* | *Simplify* |
|---|---|---|---|---|
| **9.** | $\displaystyle\int \sqrt[3]{x}\ dx$ | $\displaystyle\int x^{1/3}\ dx$ | $\dfrac{x^{4/3}}{4/3} + C$ | $\dfrac{3}{4}x^{4/3} + C$ |
| **11.** | $\displaystyle\int \dfrac{1}{x\sqrt{x}}\ dx$ | $\displaystyle\int x^{-3/2}\ dx$ | $\dfrac{x^{-1/2}}{-1/2} + C$ | $-\dfrac{2}{\sqrt{x}} + C$ |
| **13.** | $\displaystyle\int \dfrac{1}{2x^3}\ dx$ | $\dfrac{1}{2}\displaystyle\int x^{-3}\ dx$ | $\dfrac{1}{2}\left(\dfrac{x^{-2}}{-2}\right) + C$ | $-\dfrac{1}{4x^2} + C$ |

**15.** $\displaystyle\int (x + 7)\ dx = \dfrac{x^2}{2} + 7x + C$

  **Check:** $\dfrac{d}{dx}\left[\dfrac{x^2}{2} + 7x + C\right] = x + 7$

**17.** $\displaystyle\int (2x - 3x^2)\ dx = x^2 - x^3 + C$

  **Check:** $\dfrac{d}{dx}\left[x^2 - x^3 + C\right] = 2x - 3x^2$

**19.** $\displaystyle\int (x^5 + 1)\ dx = \dfrac{x^6}{6} + x + C$

  **Check:** $\dfrac{d}{dx}\left(\dfrac{x^6}{6} + x + C\right) = x^5 + 1$

**21.** $\displaystyle\int (x^{3/2} + 2x + 1)\ dx = \dfrac{2}{5}x^{5/2} + x^2 + x + C$

  **Check:** $\dfrac{d}{dx}\left(\dfrac{2}{5}x^{5/2} + x^2 + x + C\right) = x^{3/2} + 2x + 1$

**23.** $\displaystyle\int \sqrt[3]{x^2}\ dx = \displaystyle\int x^{2/3}\ dx = \dfrac{x^{5/3}}{5/3} + C = \dfrac{3}{5}x^{5/3} + C$

  **Check:** $\dfrac{d}{dx}\left(\dfrac{3}{5}x^{5/3} + C\right) = x^{2/3} = \sqrt[3]{x^2}$

**25.** $\displaystyle\int \dfrac{1}{x^5}\ dx = \displaystyle\int x^{-5}\ dx = \dfrac{x^{-4}}{-4} + C = \dfrac{-1}{4x^4} + C$

  **Check:** $\dfrac{d}{dx}\left(\dfrac{-1}{4x^4} + C\right) = \dfrac{d}{dx}\left(-\dfrac{1}{4}x^{-4} + C\right)$

  $= -\dfrac{1}{4}(-4x^{-5}) = \dfrac{1}{x^5}$

**27.** $\displaystyle\int \dfrac{x + 6}{\sqrt{x}}\ dx = \displaystyle\int (x^{1/2} + 6x^{-1/2})\ dx$

  $= \dfrac{x^{3/2}}{3/2} + 6\dfrac{x^{1/2}}{1/2} + C$

  $= \dfrac{2}{3}x^{3/2} + 12x^{1/2} + C$

  $= \dfrac{2}{3}x^{1/2}(x + 18) + C$

  **Check:** $\dfrac{d}{dx}\left(\dfrac{2}{3}x^{3/2} + 12x^{1/2} + C\right)$

  $= \dfrac{2}{3}\left(\dfrac{3}{2}x^{1/2}\right) + 12\left(\dfrac{1}{2}x^{-1/2}\right)$

  $= x^{1/2} + 6x^{-1/2} = \dfrac{x + 6}{\sqrt{x}}$

**29.** $\int (x + 1)(3x - 2)\,dx = \int (3x^2 + x - 2)\,dx$

$$= x^3 + \frac{1}{2}x^2 - 2x + C$$

**Check:** $\dfrac{d}{dx}\left( x^3 + \dfrac{1}{2}x^2 - 2x + C \right) = 3x^2 + x - 2$

$$= (x + 1)(3x - 2)$$

**31.** $\int y^2 \sqrt{y}\,dy = \int y^{5/2}\,dy = \dfrac{2}{7}y^{7/2} + C$

**Check:** $\dfrac{d}{dy}\left( \dfrac{2}{7}y^{7/2} + C \right) = y^{5/2} = y^2\sqrt{y}$

**33.** $\int dx = \int 1\,dx = x + C$

**Check:** $\dfrac{d}{dx}(x + C) = 1$

**35.** $\int (5 \cos x + 4 \sin x)\,dx = 5 \sin x - 4 \cos x + C$

**Check:**

$\dfrac{d}{dx}(5 \sin x - 4 \cos x + C) = 5 \cos x + 4 \sin x$

**37.** $\int (1 - \csc t \cot t)\,dt = t + \csc t + C$

**Check:** $\dfrac{d}{dt}(t + \csc t + C) = 1 - \csc t \cot t$

**39.** $\int (\sec^2 \theta - \sin \theta)\,d\theta = \tan \theta + \cos \theta + C$

**Check:** $\dfrac{d}{d\theta}(\tan \theta + \cos \theta + C) = \sec^2 \theta - \sin \theta$

**41.** $\int (\tan^2 y + 1)\,dy = \int \sec^2 y\,dy = \tan y + C$

**Check:** $\dfrac{d}{dy}(\tan y + C) = \sec^2 y = \tan^2 y + 1$

**43.** $\int \dfrac{\cos x}{1 - \cos^2 x}\,dx = \int \dfrac{\cos x}{\sin^2 x}\,dx = \int \left( \dfrac{1}{\sin x} \right)\left( \dfrac{\cos x}{\sin x} \right)\,dx$

$$= \int \csc x \cot x\,dx = -\csc x + C$$

**Check:** $\dfrac{d}{dx}[-\csc x + C] = \csc x \cot x + \dfrac{1}{\sin x} \cdot \dfrac{\cos x}{\sin x}$

$$= \dfrac{\cos x}{1 - \cos^2 x}$$

**45.** $f'(x) = 4$

$f(x) = 4x + C$

Answers will vary.

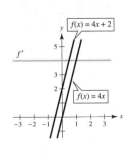

**47.** $f'(x) = 2 - x^2$

$f(x) = 2x - \dfrac{x^3}{3} + C$

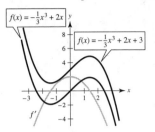

Answers will vary.

**49.** $\dfrac{dy}{dx} = 2x - 1,\ (1, 1)$

$y = \int (2x - 1)\,dx = x^2 - x + C$

$1 = (1)^2 - (1) + C \Rightarrow C = 1$

$y = x^2 - x + 1$

**51.** (a)  Answers will vary. *Sample answer.*

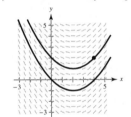

(b) $\dfrac{dy}{dx} = \dfrac{1}{2}x - 1,\ (4, 2)$

$y = \dfrac{x^2}{4} - x + C$

$2 = \dfrac{4^2}{4} - 4 + C$

$2 = C$

$y = \dfrac{x^2}{4} - x + 2$

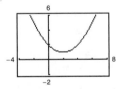

**53.** (a) Answers will vary. *Sample answer:*

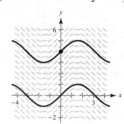

(b) $\dfrac{dy}{dx} = \cos x, \ (0, 4)$

$y = \displaystyle\int \cos x \, dx = \sin x + C$

$4 = \sin(0) + C \Rightarrow C = 4$

$y = \sin x + 4$

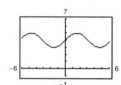

**55.** (a)

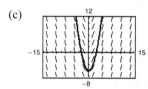

(b) $\dfrac{dy}{dx} = 2x, \ (-2, -2)$

$y = \displaystyle\int 2x \, dx = x^2 + C$

$-2 = (-2)^2 + C = 4 + C \Rightarrow C = -6$

$y = x^2 - 6$

(c)

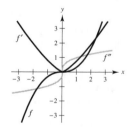

**57.** $f'(x) = 6x, \ f(0) = 8$

$f(x) = \displaystyle\int 6x \, dx = 3x^2 + C$

$f(0) = 8 = 3(0)^2 + C \Rightarrow C = 8$

$f(x) = 3x^2 + 8$

**59.** $h'(t) = 8t^3 + 5, \ h(1) = -4$

$h(t) = \displaystyle\int (8t^3 + 5) \, dt = 2t^4 + 5t + C$

$h(1) = -4 = 2 + 5 + C \Rightarrow C = -11$

$h(t) = 2t^4 + 5t - 11$

**61.** $f''(x) = 2$

$f'(2) = 5$

$f(2) = 10$

$f'(x) = \displaystyle\int 2 \, dx = 2x + C_1$

$f'(2) = 4 + C_1 = 5 \Rightarrow C_1 = 1$

$f'(x) = 2x + 1$

$f(x) = \displaystyle\int (2x + 1) \, dx = x^2 + x + C_2$

$f(2) = 6 + C_2 = 10 \Rightarrow C_2 = 4$

$f(x) = x^2 + x + 4$

**63.** $f''(x) = x^{-3/2}$

$f'(4) = 2$

$f(0) = 0$

$f'(x) = \displaystyle\int x^{-3/2} \, dx = -2x^{-1/2} + C_1 = -\dfrac{2}{\sqrt{x}} + C_1$

$f'(4) = -\dfrac{2}{2} + C_1 = 2 \Rightarrow C_1 = 3$

$f'(x) = -\dfrac{2}{\sqrt{x}} + 3$

$f(x) = \displaystyle\int \left(-2x^{-1/2} + 3\right) dx = -4x^{1/2} + 3x + C_2$

$f(0) = 0 + 0 + C_2 = 0 \Rightarrow C_2 = 0$

$f(x) = -4x^{1/2} + 3x = -4\sqrt{x} + 3x$

**65.** (a) $h(t) = \displaystyle\int (1.5t + 5) \, dt = 0.75t^2 + 5t + C$

$h(0) = 0 + 0 + C = 12 \Rightarrow C = 12$

$h(t) = 0.75t^2 + 5t + 12$

(b) $h(6) = 0.75(6)^2 + 5(6) + 12 = 69$ cm

**67.** They are the same. In both cases you are finding a function $F(x)$ such that $F'(x) = f(x)$.

**69.** Because $f''$ is negative on $(-\infty, 0)$, $f'$ is decreasing on $(-\infty, 0)$. Because $f''$ is positive on $(0, \infty)$, $f'$ is increasing on $(0, \infty)$. $f'$ has a relative minimum at $(0, 0)$. Because $f'$ is positive on $(-\infty, \infty)$, $f$ is increasing on $(-\infty, \infty)$.

**71.** $a(t) = -32 \text{ ft/sec}^2$

$v(t) = \int -32 \, dt = -32t + C_1$

$v(0) = 60 = C_1$

$s(t) = \int (-32t + 60) \, dt = -16t^2 + 60t + C_2$

$s(0) = 6 = C_2$

$s(t) = -16t^2 + 60t + 6$, Position function

The ball reaches its maximum height when

$v(t) = -32t + 60 = 0$

$32t = 60$

$t = \frac{15}{8}$ seconds.

$s\left(\frac{15}{8}\right) = -16\left(\frac{15}{8}\right)^2 + 60\left(\frac{15}{8}\right) + 6 = 62.25$ feet

**73.** From Exercise 72:

$s(t) = -16t^2 + v_0 t$

$s'(t) = -32t + v_0 = 0$ when $t = \dfrac{v_0}{32}$ = time to reach

maximum height.

$s\left(\dfrac{v_0}{32}\right) = -16\left(\dfrac{v_0}{32}\right)^2 + v_0\left(\dfrac{v_0}{32}\right) = 550$

$-\dfrac{v_0^2}{64} + \dfrac{v_0^2}{32} = 550$

$v_0^2 = 35,200$

$v_0 \approx 187.617 \text{ ft/sec}$

**75.** $a(t) = -9.8$

$v(t) = \int -9.8 \, dt = -9.8t + C_1$

$v(0) = v_0 = C_1 \Rightarrow v(t) = -9.8t + v_0$

$f(t) = \int (-9.8t + v_0) \, dt = -4.9t^2 + v_0 t + C_2$

$f(0) = s_0 = C_2 \Rightarrow f(t) = -4.9t^2 + v_0 t + s_0$

**77.** From Exercise 75, $f(t) = -4.9t^2 + 10t + 2$.

$v(t) = -9.8t + 10 = 0$ (Maximum height when $v = 0$.)

$9.8t = 10$

$t = \dfrac{10}{9.8}$

$f\left(\dfrac{10}{9.8}\right) \approx 7.1 \text{ m}$

**79.** $a = -1.6$

$v(t) = \int -1.6 \, dt = -1.6t + v_0 = -1.6t,$

because the stone was dropped, $v_0 = 0$.

$s(t) = \int (-1.6t) \, dt = -0.8t^2 + s_0$

$s(20) = 0 \Rightarrow -0.8(20)^2 + s_0 = 0$

$s_0 = 320$

So, the height of the cliff is 320 meters.

$v(t) = -1.6t$

$v(20) = -32 \text{ m/sec}$

**81.** $x(t) = t^3 - 6t^2 + 9t - 2, \quad 0 \le t \le 5$

(a) $v(t) = x'(t) = 3t^2 - 12t + 9$

$\qquad = 3(t^2 - 4t + 3) = 3(t - 1)(t - 3)$

$\quad a(t) = v'(t) = 6t - 12 = 6(t - 2)$

(b) $v(t) > 0$ when $0 < t < 1$ or $3 < t < 5$.

(c) $a(t) = 6(t - 2) = 0$ when $t = 2$.

$\quad v(2) = 3(1)(-1) = -3$

**83.** $v(t) = \dfrac{1}{\sqrt{t}} = t^{-1/2} \quad t > 0$

$x(t) = \int v(t) \, dt = 2t^{1/2} + C$

$x(1) = 4 = 2(1) + C \Rightarrow C = 2$

Position function: $x(t) = 2t^{1/2} + 2$

Acceleration function: $a(t) = v'(t) = -\dfrac{1}{2}t^{-3/2} = \dfrac{-1}{2t^{3/2}}$

**85.** (a) $v(0) = 25 \text{ km/h} = 25 \cdot \dfrac{1000}{3600} = \dfrac{250}{36} \text{ m/sec}$

$\quad v(13) = 80 \text{ km/h} = 80 \cdot \dfrac{1000}{3600} = \dfrac{800}{36} \text{ m/sec}$

$\quad a(t) = a \text{ (constant acceleration)}$

$\quad v(t) = at + C$

$\quad v(0) = \dfrac{250}{36} \Rightarrow v(t) = at + \dfrac{250}{36}$

$\quad v(13) = \dfrac{800}{36} = 13a + \dfrac{250}{36}$

$\quad \dfrac{550}{36} = 13a$

$\quad a = \dfrac{550}{468} = \dfrac{275}{234} \approx 1.175 \text{ m/sec}^2$

(b) $s(t) = a\dfrac{t^2}{2} + \dfrac{250}{36}t \quad (s(0) = 0)$

$\quad s(13) = \dfrac{275}{234}\dfrac{(13)^2}{2} + \dfrac{250}{36}(13) \approx 189.58 \text{ m}$

**87.** Truck: $v(t) = 30$

$s(t) = 30t$ $\left(\text{Let } s(0) = 0.\right)$

Automobile: $a(t) = 6$

$v(t) = 6t$ $\left(\text{Let } v(0) = 0.\right)$

$s(t) = 3t^2$ $\left(\text{Let } s(0) = 0.\right)$

At the point where the automobile overtakes the truck:

$30t = 3t^2$

$0 = 3t^2 - 30t$

$0 = 3t(t - 10)$ when $t = 10$ sec.

(a) $s(10) = 3(10)^2 = 300$ ft

(b) $v(10) = 6(10) = 60$ ft/sec $\approx 41$ mi/h

**89.** Let the planes be located 10 and 17 miles away from the airport, as indicated in the figure.

$v_A(t) = k_A t - 150$

$s_A(t) = \frac{1}{2}k_A t^2 - 150t + 10$

$v_B = k_B t - 250$

$s_B = \frac{1}{2}k_B t^2 - 250t + 17$

Airport $\quad \leftarrow A \quad \leftarrow B$

$0 \qquad\qquad 10 \qquad 17$

(a) When airplane A lands at time $t_A$ you have

$v_A(t_A) = k_A t_A - 150 = -100 \Rightarrow k_A = \dfrac{50}{t_A}$

$s_A(t_A) = \dfrac{1}{2}k_A t_A^2 - 150t_A + 10 = 0$

$\dfrac{1}{2}\left(\dfrac{50}{t_A}\right)t_A^2 - 150t_A = -10$

$125t_A = 10$

$t_A = \dfrac{10}{125}h.$

$k_A = \dfrac{50}{t_A} = 50\left(\dfrac{125}{10}\right) = 625 \Rightarrow s_A(t) = \dfrac{625}{2}t^2 - 150t + 10$

Similarly, when airplane B lands at time $t_B$ you have

$v_B(t_B) = k_B t_B - 250 = -115 \Rightarrow k_B = \dfrac{135}{t_B}$

$s_B(t_B) = \dfrac{1}{2}k_B t_B^2 - 250t_B + 17 = 0$

$\dfrac{1}{2}\left(\dfrac{135}{t_B}\right)t_B^2 - 250t_B = -17$

$\dfrac{365}{2}t_B = 17$

$t_B = \dfrac{34}{365}h.$

$k_B = \dfrac{135}{t_B} = 135\left(\dfrac{365}{34}\right) = \dfrac{49{,}275}{34} \Rightarrow s_B(t) = \dfrac{49{,}275}{68}t^2 - 250t + 17$

(b)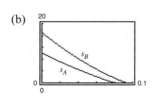

(c) $d = s_B(t) - s_A(t) = \dfrac{28{,}025}{68}t^2 - 100t + 7$

Yes, $d < 3$ for $t > 0.0505$ h.

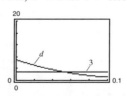

**91.** True

**93.** True

**95.** False. $f$ has an infinite number of antiderivatives, each differing by a constant.

**97.** $f'(x) = \begin{cases} -1, & 0 \le x < 2 \\ 2, & 2 < x < 3 \\ 0, & 3 < x \le 4 \end{cases}$

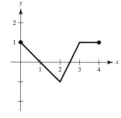

$f(x) = \begin{cases} -x + C_1, & 0 \le x < 2 \\ 2x + C_2, & 2 < x < 3 \\ C_3, & 3 < x \le 4 \end{cases}$

$f(0) = 1 \Rightarrow C_1 = 1$

$f$ continuous at $x = 2 \Rightarrow -2 + 1 = 4 + C_2 \Rightarrow C_2 = -5$

$f$ continuous at $x = 3 \Rightarrow 6 - 5 = C_3 = 1$

$f(x) = \begin{cases} -x + 1, & 0 \le x < 2 \\ 2x - 5, & 2 \le x < 3 \\ 1, & 3 \le x \le 4 \end{cases}$

**99.** $\dfrac{d}{dx}\Big[\big[s(x)\big]^2 + \big[c(x)\big]^2\Big] = 2s(x)s'(x) + 2c(x)c'(x)$

$\qquad\qquad\qquad\qquad\qquad = 2s(x)c(x) - 2c(x)s(x) = 0$

So, $\big[s(x)\big]^2 + \big[c(x)\big]^2 = k$ for some constant $k$. Because, $s(0) = 0$ and $c(0) = 1$, $k = 1$.

Therefore, $\big[s(x)\big]^2 + \big[c(x)\big]^2 = 1$.

[Note that $s(x) = \sin x$ and $c(x) = \cos x$ satisfy these properties.]

## Section 4.2    Area

**1.** $\displaystyle\sum_{i=1}^{6}(3i + 2) = 3\sum_{i=1}^{6}i + \sum_{i=1}^{6}2 = 3(1 + 2 + 3 + 4 + 5 + 6) + 12 = 75$

**3.** $\displaystyle\sum_{k=0}^{4}\frac{1}{k^2 + 1} = 1 + \frac{1}{2} + \frac{1}{5} + \frac{1}{10} + \frac{1}{17} = \frac{158}{85}$

**15.** $\displaystyle\sum_{i=1}^{12}7 = 7(12) = 84$

**5.** $\displaystyle\sum_{k=1}^{4}c = c + c + c + c = 4c$

**17.** $\displaystyle\sum_{i=1}^{24}4i = 4\sum_{i=1}^{24}i = 4\left[\frac{24(25)}{2}\right] = 1200$

**7.** $\displaystyle\sum_{i=1}^{11}\frac{1}{5i}$

**19.** $\displaystyle\sum_{i=1}^{20}(i - 1)^2 = \sum_{i=1}^{19}i^2 = \left[\frac{19(20)(39)}{6}\right] = 2470$

**9.** $\displaystyle\sum_{j=1}^{6}\left[7\left(\frac{j}{6}\right) + 5\right]$

**21.** $\displaystyle\sum_{i=1}^{15}i(i - 1)^2 = \sum_{i=1}^{15}i^3 - 2\sum_{i=1}^{15}i^2 + \sum_{i=1}^{15}i$

**11.** $\dfrac{2}{n}\displaystyle\sum_{i=1}^{n}\left[\left(\frac{2i}{n}\right)^3 - \left(\frac{2i}{n}\right)\right]$

$\qquad = \dfrac{15^2(16)^2}{4} - 2\dfrac{15(16)(31)}{6} + \dfrac{15(16)}{2}$

$\qquad = 14{,}400 - 2480 + 120 = 12{,}040$

**13.** $\dfrac{3}{n}\displaystyle\sum_{i=1}^{n}\left[2\left(1 + \frac{3i}{n}\right)^2\right]$

**23.** $\displaystyle\sum_{i=1}^{20} \left(i^2 + 3\right) = \frac{20(20 + 1)\left(2(20) + 1\right)}{6} + 3(20) = \frac{(20)(21)(41)}{6} + 60 = 2930$

**25.** (a)

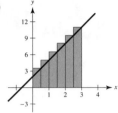

The width $\Delta x$ of each rectangle is $\frac{1}{2}$. The heights are given by the right endpoints.

Area $\approx \frac{1}{2}\left[\left(3\left(\frac{1}{2}\right) + 2\right) + \left(3(1) + 2\right) + \left(3\left(\frac{3}{2}\right) + 2\right) + \left(3(2) + 2\right) + \left(3\left(\frac{5}{2}\right) + 2\right) + \left(3(3) + 2\right)\right] = \frac{87}{4} = 21.75$

(b)

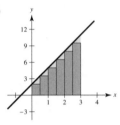

Using left endpoints,

Area $\approx \frac{1}{2}\left[\left(3(0) + 2\right) + \left(3\left(\frac{1}{2}\right) + 2\right) + \left(3(1) + 2\right) + \left(3\left(\frac{3}{2}\right) + 2\right) + \left(3(2) + 2\right) + \left(3\left(\frac{5}{2}\right) + 2\right)\right] = \frac{69}{4} = 17.25$

**27.**

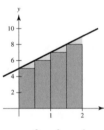 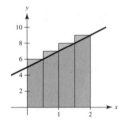

$\Delta x = \dfrac{2 - 0}{4} = \dfrac{1}{2}$

Left endpoints: Area $\approx \frac{1}{2}\left[5 + 6 + 7 + 8\right] = \frac{26}{2} = 13$

Right endpoints: Area $\approx \frac{1}{2}\left[6 + 7 + 8 + 9\right] = \frac{30}{2} = 15$

$13 < \text{Area} < 15$

**29.**

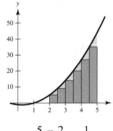 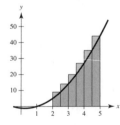

$\Delta x = \dfrac{5 - 2}{6} = \dfrac{1}{2}$

Left endpoints: Area $\approx \frac{1}{2}\left[5 + 9 + 14 + 20 + 27 + 35\right] = 55$

Right endpoints: Area $\approx \frac{1}{2}\left[9 + 14 + 20 + 27 + 35 + 44\right] = \frac{149}{2} = 74.5$

$55 < \text{Area} < 74.5$

**31.**

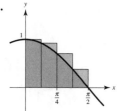

$$\Delta x = \frac{\frac{\pi}{2} - 0}{4} = \frac{\pi}{8}$$

Left endpoints: Area $\approx \dfrac{\pi}{8}\left[\cos(0) + \cos\left(\dfrac{\pi}{8}\right) + \cos\left(\dfrac{\pi}{4}\right) + \cos\left(\dfrac{3\pi}{8}\right)\right] \approx 1.1835$

Right endpoints: Area $\approx \dfrac{\pi}{8}\left[\cos\left(\dfrac{\pi}{8}\right) + \cos\left(\dfrac{\pi}{4}\right) + \cos\left(\dfrac{3\pi}{8}\right) + \cos\left(\dfrac{\pi}{2}\right)\right] \approx 0.7908$

$0.7908 < \text{Area} < 1.1835$

**33.** $S = \left[3 + 4 + \frac{9}{2} + 5\right](1) = \frac{33}{2} = 16.5$

$s = \left[1 + 3 + 4 + \frac{9}{2}\right](1) = \frac{25}{2} = 12.5$

**35.** $S = \left[3 + 3 + 5\right](1) = 11$

$s = \left[2 + 2 + 3\right](1) = 7$

**37.** $\displaystyle\lim_{n\to\infty}\left[\left(\frac{81}{n^4}\right)\frac{n^2(n+1)^2}{4}\right] = \frac{81}{4}\lim_{n\to\infty}\left[\frac{n^4 + 2n^3 + n^2}{n^4}\right] = \frac{81}{4}(1) = \frac{81}{4}$

**39.** $\displaystyle\lim_{n\to\infty}\left[\left(\frac{18}{n^2}\right)\frac{n(n+1)}{2}\right] = \frac{18}{2}\lim_{n\to\infty}\left[\frac{n^2 + n}{n^2}\right] = \frac{18}{2}(1) = 9$

**41.** $S(4) = \sqrt{\frac{1}{4}}\left(\frac{1}{4}\right) + \sqrt{\frac{1}{2}}\left(\frac{1}{4}\right) + \sqrt{\frac{3}{4}}\left(\frac{1}{4}\right) + \sqrt{1}\left(\frac{1}{4}\right) = \dfrac{1 + \sqrt{2} + \sqrt{3} + 2}{8} \approx 0.768$

$s(4) = 0\left(\frac{1}{4}\right) + \sqrt{\frac{1}{4}}\left(\frac{1}{4}\right) + \sqrt{\frac{1}{2}}\left(\frac{1}{4}\right) + \sqrt{\frac{3}{4}}\left(\frac{1}{4}\right) = \dfrac{1 + \sqrt{2} + \sqrt{3}}{8} \approx 0.518$

**43.** $S(5) = 1\left(\frac{1}{5}\right) + \frac{1}{6/5}\left(\frac{1}{5}\right) + \frac{1}{7/5}\left(\frac{1}{5}\right) + \frac{1}{8/5}\left(\frac{1}{5}\right) + \frac{1}{9/5}\left(\frac{1}{5}\right) = \frac{1}{5} + \frac{1}{6} + \frac{1}{7} + \frac{1}{8} + \frac{1}{9} \approx 0.746$

$s(5) = \frac{1}{6/5}\left(\frac{1}{5}\right) + \frac{1}{7/5}\left(\frac{1}{5}\right) + \frac{1}{8/5}\left(\frac{1}{5}\right) + \frac{1}{9/5}\left(\frac{1}{5}\right) + \frac{1}{2}\left(\frac{1}{5}\right) = \frac{1}{6} + \frac{1}{7} + \frac{1}{8} + \frac{1}{9} + \frac{1}{10} \approx 0.646$

**45.** $\displaystyle\sum_{i=1}^{n}\frac{2i+1}{n^2} = \frac{1}{n^2}\sum_{i=1}^{n}(2i+1) = \frac{1}{n^2}\left[2\frac{n(n+1)}{2} + n\right] = \frac{n+2}{n} = 1 + \frac{2}{n} = S(n)$

$S(10) = \dfrac{12}{10} = 1.2$

$S(100) = 1.02$

$S(1000) = 1.002$

$S(10{,}000) = 1.0002$

**47.** $\displaystyle\sum_{k=1}^{n}\frac{6k(k-1)}{n^3} = \frac{6}{n^3}\sum_{k=1}^{n}(k^2-k) = \frac{6}{n^3}\left[\frac{n(n+1)(2n+1)}{6} - \frac{n(n+1)}{2}\right]$

$\displaystyle\qquad\qquad = \frac{6}{n^2}\left[\frac{2n^2+3n+1-3n-3}{6}\right] = \frac{1}{n^2}\left[2n^2-2\right] = 2 - \frac{2}{n^2} = S(n)$

$S(10) = 1.98$

$S(100) = 1.9998$

$S(1000) = 1.999998$

$S(10,000) = 1.99999998$

**49.** $\displaystyle\lim_{n\to\infty}\sum_{i=1}^{n}\left(\frac{24i}{n^2}\right) = \lim_{n\to\infty}\frac{24}{n^2}\sum_{i=1}^{n}i = \lim_{n\to\infty}\frac{24}{n^2}\left(\frac{n(n+1)}{2}\right) = \lim_{n\to\infty}\left[12\left(\frac{n^2+n}{n^2}\right)\right] = 12\lim_{n\to\infty}\left(1+\frac{1}{n}\right) = 12$

**51.** $\displaystyle\lim_{n\to\infty}\sum_{i=1}^{n}\frac{1}{n^3}(i-1)^2 = \lim_{n\to\infty}\frac{1}{n^3}\sum_{i=1}^{n-1}i^2 = \lim_{n\to\infty}\frac{1}{n^3}\left[\frac{(n-1)(n)(2n-1)}{6}\right]$

$\displaystyle\qquad\qquad = \lim_{n\to\infty}\frac{1}{6}\left[\frac{2n^3-3n^2+n}{n^3}\right] = \lim_{n\to\infty}\left[\frac{1}{6}\left(\frac{2-(3/n)+(1/n^2)}{1}\right)\right] = \frac{1}{3}$

**53.** $\displaystyle\lim_{n\to\infty}\sum_{i=1}^{n}\left(1+\frac{i}{n}\right)\left(\frac{2}{n}\right) = 2\lim_{n\to\infty}\frac{1}{n}\left[\sum_{i=1}^{n}1 + \frac{1}{n}\sum_{i=1}^{n}i\right] = 2\lim_{n\to\infty}\frac{1}{n}\left[n + \frac{1}{n}\left(\frac{n(n+1)}{2}\right)\right] = 2\lim_{n\to\infty}\left[1+\frac{n^2+n}{2n^2}\right] = 2\left(1+\frac{1}{2}\right) = 3$

**55.** (a)

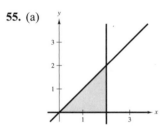

(b)  $\Delta x = \dfrac{2-0}{n} = \dfrac{2}{n}$

Endpoints:  $0 < 1\left(\dfrac{2}{n}\right) < 2\left(\dfrac{2}{n}\right) < \ldots < (n-1)\left(\dfrac{2}{n}\right) < n\left(\dfrac{2}{n}\right) = 2$

(c)  Because $y = x$ is increasing, $f(m_i) = f(x_{i-1})$ on $[x_{i-1}, x_i]$.

$\displaystyle s(n) = \sum_{i=1}^{n}f(x_{i-1})\Delta x = \sum_{i=1}^{n}f\left(\frac{2i-2}{n}\right)\left(\frac{2}{n}\right) = \sum_{i=1}^{n}\left[(i-1)\left(\frac{2}{n}\right)\right]\left(\frac{2}{n}\right)$

(d)  $f(M_i) = f(x_i)$ on $[x_{i-1}, x_i]$

$\displaystyle S(n) = \sum_{i=1}^{n}f(x_i)\,\Delta x = \sum_{i=1}^{n}f\left(\frac{2i}{n}\right)\frac{2}{n} = \sum_{i=1}^{n}\left[i\left(\frac{2}{n}\right)\right]\left(\frac{2}{n}\right)$

(e)

| $x$ | 5 | 10 | 50 | 100 |
|---|---|---|---|---|
| $s(n)$ | 1.6 | 1.8 | 1.96 | 1.98 |
| $S(n)$ | 2.4 | 2.2 | 2.04 | 2.02 |

(f)  $\displaystyle\lim_{n\to\infty}\sum_{i=1}^{n}\left[(i-1)\left(\frac{2}{n}\right)\right]\left(\frac{2}{n}\right) = \lim_{n\to\infty}\frac{4}{n^2}\sum_{i=1}^{n}(i-1) = \lim_{n\to\infty}\frac{4}{n^2}\left[\frac{n(n+1)}{2}-n\right] = \lim_{n\to\infty}\left[\frac{2(n+1)}{n}-\frac{4}{n}\right] = 2$

$\displaystyle\qquad\lim_{n\to\infty}\sum_{i=1}^{n}\left[i\left(\frac{2}{n}\right)\right]\left(\frac{2}{n}\right) = \lim_{n\to\infty}\frac{4}{n^2}\sum_{i=1}^{n}i = \lim_{n\to\infty}\left(\frac{4}{n^2}\right)\frac{n(n+1)}{2} = \lim_{n\to\infty}\frac{2(n+1)}{n} = 2$

**57.** $y = -4x + 5$ on $[0, 1]$. $\left(\textbf{Note: } \Delta x = \dfrac{1}{n}\right)$

$$s(n) = \sum_{i=1}^{n} f\left(\frac{i}{n}\right)\left(\frac{1}{n}\right) = \sum_{i=1}^{n}\left[-4\left(\frac{i}{n}\right) + 5\right]\left(\frac{1}{n}\right)$$

$$= -\frac{4}{n^2}\sum_{i=1}^{n} i + 5$$

$$= -\frac{4}{n^2}\frac{n(n+1)}{2} + 5$$

$$= -2\left(1 + \frac{1}{n}\right) + 5$$

Area $= \lim_{n\to\infty} s(n) = 3$

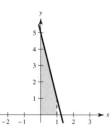

**59.** $y = x^2 + 2$ on $[0, 1]$. $\left(\textbf{Note: } \Delta x = \dfrac{1}{n}\right)$

$$S(n) = \sum_{i=1}^{n} f\left(\frac{i}{n}\right)\left(\frac{1}{n}\right)$$

$$= \sum_{i=1}^{n}\left[\left(\frac{i}{n}\right)^2 + 2\right]\left(\frac{1}{n}\right)$$

$$= \left[\frac{1}{n^3}\sum_{i=1}^{n} i^2\right] + 2$$

$$= \frac{n(n+1)(2n+1)}{6n^3} + 2 = \frac{1}{6}\left(2 + \frac{3}{n} + \frac{1}{n^2}\right) + 2$$

Area $= \lim_{n\to\infty} S(n) = \dfrac{7}{3}$

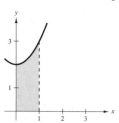

**61.** $y = 25 - x^2$ on $[1, 4]$. $\left(\textbf{Note: } \Delta x = \dfrac{3}{n}\right)$

$$s(n) = \sum_{i=1}^{n} f\left(1 + \frac{3i}{n}\right)\left(\frac{3}{n}\right) = \sum_{i=1}^{n}\left[25 - \left(1 + \frac{3i}{n}\right)^2\right]\left(\frac{3}{n}\right)$$

$$= \frac{3}{n}\sum_{i=1}^{n}\left[24 - \frac{9i^2}{n^2} - \frac{6i}{n}\right]$$

$$= \frac{3}{n}\left[24n - \frac{9}{n^2}\frac{n(n+1)(2n+1)}{6} - \frac{6}{n}\frac{n(n+1)}{2}\right]$$

$$= 72 - \frac{9}{2n^2}(n+1)(2n+1) - \frac{9}{n}(n+1)$$

Area $= \lim_{n\to\infty} s(n) = 72 - 9 - 9 = 54$

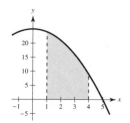

**63.** $y = 27 - x^3$ on $[1, 3]$. $\left(\textbf{Note: } \Delta x = \dfrac{3-1}{n} = \dfrac{2}{n}\right)$

$$s(n) = \sum_{i=1}^{n} f\left(1 + \frac{2i}{n}\right)\left(\frac{2}{n}\right) = \sum_{i=1}^{n}\left[27 - \left(1 + \frac{2i}{n}\right)^3\right]\left(\frac{2}{n}\right)$$

$$= \frac{2}{n}\sum_{i=1}^{n}\left[26 - \frac{8i^3}{n^3} - \frac{12i^2}{n^2} - \frac{6i}{n}\right]$$

$$= \frac{2}{n}\left[26n - \frac{8}{n^3}\frac{n^2(n+1)^2}{4} - \frac{12}{n^2}\frac{n(n+1)(2n+1)}{6} - \frac{6}{n}\frac{n(n+1)}{2}\right]$$

$$= 52 - \frac{4}{n^2}(n+1)^2 - \frac{4}{n^2}(n+1)(2n+1) - \frac{6n+1}{n}$$

Area $= \lim_{n\to\infty} s(n) = 52 - 4 - 8 - 6 = 34$

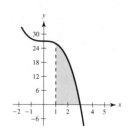

**65.** $y = x^2 - x^3$ on $[-1, 1]$. $\left(\textbf{Note: } \Delta x = \dfrac{1 - (-1)}{n} = \dfrac{2}{n}\right)$

Because $y$ both increases and decreases on $[-1, 1]$, $T(n)$ is neither an upper nor a lower sum.

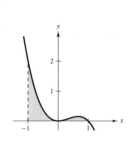

$$T(n) = \sum_{i=1}^{n} f\left(-1 + \frac{2i}{n}\right)\left(\frac{2}{n}\right) = \sum_{i=1}^{n}\left[\left(-1 + \frac{2i}{n}\right)^2 - \left(-1 + \frac{2i}{n}\right)^3\right]\left(\frac{2}{n}\right)$$

$$= \sum_{i=1}^{n}\left[\left(1 - \frac{4i}{n} + \frac{4i^2}{n^2}\right) - \left(-1 + \frac{6i}{n} - \frac{12i^2}{n^2} + \frac{8i^3}{n^3}\right)\right]\left(\frac{2}{n}\right)$$

$$= \sum_{i=1}^{n}\left[2 - \frac{10i}{n} + \frac{16i^2}{n^2} - \frac{8i^3}{n^3}\right]\left(\frac{2}{n}\right) = \frac{4}{n}\sum_{i=1}^{n}1 - \frac{20}{n^2}\sum_{i=1}^{n}i + \frac{32}{n^3}\sum_{i=1}^{n}i^2 - \frac{16}{n^4}\sum_{i=1}^{n}i^3$$

$$= \frac{4}{n}(n) - \frac{20}{n^2}\cdot\frac{n(n + 1)}{2} + \frac{32}{n^3}\cdot\frac{n(n + 1)(2n + 1)}{6} - \frac{16}{n^4}\cdot\frac{n^2(n + 1)^2}{4}$$

$$= 4 - 10\left(1 + \frac{1}{n}\right) + \frac{16}{3}\left(2 + \frac{3}{n} + \frac{1}{n^2}\right) - 4\left(1 + \frac{2}{n} + \frac{1}{n^2}\right)$$

$$\text{Area} = \lim_{n\to\infty} T(n) = 4 - 10 + \frac{32}{3} - 4 = \frac{2}{3}$$

**67.** $f(y) = 4y,\ 0 \le y \le 2$ $\left(\textbf{Note: } \Delta y = \dfrac{2 - 0}{n} = \dfrac{2}{n}\right)$

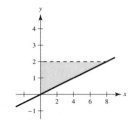

$$S(n) = \sum_{i=1}^{n} f(m_i)\Delta y = \sum_{i=1}^{n} f\left(\frac{2i}{n}\right)\left(\frac{2}{n}\right) = \sum_{i=1}^{n} 4\left(\frac{2i}{n}\right)\left(\frac{2}{n}\right) = \frac{16}{n^2}\sum_{i=1}^{n} i$$

$$= \left(\frac{16}{n^2}\right)\cdot\frac{n(n + 1)}{2} = \frac{8(n + 1)}{n} = 8 + \frac{8}{n}$$

$$\text{Area} = \lim_{n\to\infty} S(n) = \lim_{n\to\infty}\left(8 + \frac{8}{n}\right) = 8$$

**69.** $f(y) = y^2,\ 0 \le y \le 5$ $\left(\textbf{Note: } \Delta y = \dfrac{5 - 0}{n} = \dfrac{5}{n}\right)$

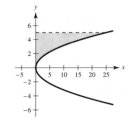

$$S(n) = \sum_{i=1}^{n} f\left(\frac{5i}{n}\right)\left(\frac{5}{n}\right)$$

$$= \sum_{i=1}^{n}\left(\frac{5i}{n}\right)^2\left(\frac{5}{n}\right)$$

$$= \frac{125}{n^3}\sum_{i=1}^{n} i^2$$

$$= \frac{125}{n^3}\cdot\frac{n(n + 1)(2n + 1)}{6}$$

$$= \frac{125}{n^2}\left(\frac{2n^2 + 3n + 1}{6}\right) = \frac{125}{3} + \frac{125}{2n} + \frac{125}{6n^2}$$

$$\text{Area } \lim_{n\to\infty} S(n) = \lim_{n\to\infty}\left(\frac{125}{3} + \frac{125}{2n} + \frac{125}{6n^2}\right) = \frac{125}{3}$$

**71.** $g(y) = 4y^2 - y^3$, $1 \le y \le 3$. $\left(\textbf{Note: } \Delta y = \dfrac{3-1}{n} = \dfrac{2}{n}\right)$

$$S(n) = \sum_{i=1}^{n} g\left(1 + \frac{2i}{n}\right)\left(\frac{2}{n}\right)$$

$$= \sum_{i=1}^{n} \left[4\left(1 + \frac{2i}{n}\right)^2 - \left(1 + \frac{2i}{n}\right)^3\right]\frac{2}{n}$$

$$= \frac{2}{n}\sum_{n=1}^{n} 4\left[1 + \frac{4i}{n} + \frac{4i^2}{n^2}\right] - \left[1 + \frac{6i}{n} + \frac{12i^2}{n^2} + \frac{8i^3}{n^3}\right]$$

$$= \frac{2}{n}\sum_{i=1}^{n} \left[3 + \frac{10i}{n} + \frac{4i^2}{n^2} - \frac{8i^3}{n^3}\right]$$

$$= \frac{2}{n}\left[3n + \frac{10}{n}\frac{n(n+1)}{2} + \frac{4}{n^2}\frac{n(n+1)(2n+1)}{6} - \frac{8}{n^2}\frac{n^2(n+1)^2}{4}\right]$$

Area $= \lim_{n\to\infty} S(n) = 6 + 10 + \dfrac{8}{3} - 4 = \dfrac{44}{3}$

**73.** $f(x) = x^2 + 3$, $0 \le x \le 2$, $n = 4$

Let $c_i = \dfrac{x_i + x_{i-1}}{2}$.

$\Delta x = \dfrac{1}{2}$, $c_1 = \dfrac{1}{4}$, $c_2 = \dfrac{3}{4}$, $c_3 = \dfrac{5}{4}$, $c_4 = \dfrac{7}{4}$

Area $\approx \sum_{i=1}^{n} f(c_i)\Delta x = \sum_{i=1}^{4}\left[c_i^2 + 3\right]\left(\dfrac{1}{2}\right) = \dfrac{1}{2}\left[\left(\dfrac{1}{16} + 3\right) + \left(\dfrac{9}{16} + 3\right) + \left(\dfrac{25}{16} + 3\right) + \left(\dfrac{49}{16} + 3\right)\right] = \dfrac{69}{8}$

**75.** $f(x) = \tan x$, $0 \le x \le \dfrac{\pi}{4}$, $n = 4$

Let $c_i = \dfrac{x_i + x_{i-1}}{2}$.

$\Delta x = \dfrac{\pi}{16}$, $c_1 = \dfrac{\pi}{32}$, $c_2 = \dfrac{3\pi}{32}$, $c_3 = \dfrac{5\pi}{32}$, $c_4 = \dfrac{7\pi}{32}$

Area $\approx \sum_{i=1}^{n} f(c_i)\Delta x = \sum_{i=1}^{4}(\tan c_i)\left(\dfrac{\pi}{16}\right) = \dfrac{\pi}{16}\left(\tan\dfrac{\pi}{32} + \tan\dfrac{3\pi}{32} + \tan\dfrac{5\pi}{32} + \tan\dfrac{7\pi}{32}\right) \approx 0.345$

**77.** $f(x) = \sqrt{x}$ on $[0, 4]$.

| $n$ | 4 | 8 | 12 | 16 | 20 |
|---|---|---|---|---|---|
| Approximate area | 5.3838 | 5.3523 | 5.3439 | 5.3403 | 5.3384 |

(Exact value is $16/3$.)

**79.** $f(x) = \tan\left(\dfrac{\pi x}{8}\right)$ on $[1, 3]$.

| $n$ | 4 | 8 | 12 | 16 | 20 |
|---|---|---|---|---|---|
| Approximate area | 2.2223 | 2.2387 | 2.2418 | 2.2430 | 2.2435 |

**81.**

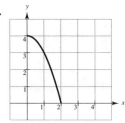

(b)   $A \approx 6$ square units

**83.** You can use the line $y = x$ bounded by $x = a$ and $x = b$. The sum of the areas of these inscribed rectangles is the lower sum.

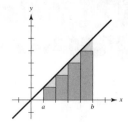

The sum of the areas of these circumscribed rectangles is the upper sum.

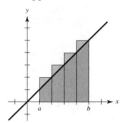

You can see that the rectangles do not contain all of the area in the first graph and the rectangles in the second graph cover more than the area of the region. The exact value of the area lies between these two sums.

**85.** (a)

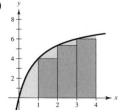

Lower sum: $s(4) = 0 + 4 + 5\frac{1}{3} + 6 = 15\frac{1}{3} = \frac{46}{3} \approx 15.333$

(b)

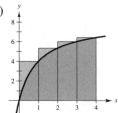

Upper sum: $S(4) = 4 + 5\frac{1}{3} + 6 + 6\frac{2}{5} = 21\frac{11}{15} = \frac{326}{15} \approx 21.733$

(c)

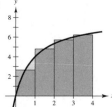

Midpoint Rule: $M(4) = 2\frac{2}{3} + 4\frac{4}{5} + 5\frac{5}{7} + 6\frac{2}{9} = \frac{6112}{315} \approx 19.403$

(d)   In each case, $\Delta x = 4/n$. The lower sum uses left end-points, $(i - 1)(4/n)$. The upper sum uses right endpoints, $(i)(4/n)$. The Midpoint Rule uses midpoints, $(i - \frac{1}{2})(4/n)$.

(e)

| N | 4 | 8 | 20 | 100 | 200 |
|---|---|---|---|---|---|
| $s(n)$ | 15.333 | 17.368 | 18.459 | 18.995 | 19.06 |
| $S(n)$ | 21.733 | 20.568 | 19.739 | 19.251 | 19.188 |
| $M(n)$ | 19.403 | 19.201 | 19.137 | 19.125 | 19.125 |

(f)  $s(n)$ increases because the lower sum approaches the exact value as $n$ increases. $S(n)$ decreases because the upper sum
approaches the exact value as $n$ increases. Because of the shape of the graph, the lower sum is always smaller than the
exact value, whereas the upper sum is always larger.

**87.** True.  (Theorem 4.2 (2))

**89.** Suppose there are $n$ rows and $n + 1$ columns in the figure. The stars on the left total $1 + 2 + \cdots + n$, as do the stars on the
right. There are $n(n + 1)$ stars in total, so

$$2[1 + 2 + \cdots + n] = n(n + 1)$$
$$1 + 2 + \cdots + n = \tfrac{1}{2}(n)(n + 1).$$

**91.** (a)  $y = \left(-4.09 \times 10^{-5}\right)x^3 + 0.016x^2 - 2.67x + 452.9$

(b)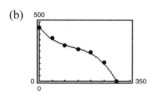

(c)  Using the integration capability of a graphing utility, you obtain $A \approx 76{,}897.5 \text{ ft}^2$.

**93.** (a)  $\displaystyle\sum_{i=1}^{n} 2i = n(n + 1)$

The formula is true for $n = 1$: $2 = 1(1 + 1) = 2$.

Assume that the formula is true for $n = k$: $\displaystyle\sum_{i=1}^{k} 2i = k(k + 1)$.

Then you have $\displaystyle\sum_{i=1}^{k+1} 2i = \sum_{i=1}^{k} 2i + 2(k + 1) = k(k + 1) + 2(k + 1) = (k + 1)(k + 2)$

which shows that the formula is true for $n = k + 1$.

(b)  $\displaystyle\sum_{i=1}^{n} i^3 = \dfrac{n^2(n + 1)^2}{4}$

The formula is true for $n = 1$ because $1^3 = \dfrac{1^2(1 + 1)^2}{4} = \dfrac{4}{4} = 1$.

Assume that the formula is true for $n = k$: $\displaystyle\sum_{i=1}^{k} i^3 = \dfrac{k^2(k + 1)^2}{4}$.

Then you have $\displaystyle\sum_{i=1}^{k+1} i^3 = \sum_{i=1}^{k} i^3 + (k + 1)^3 = \dfrac{k^2(k + 1)^2}{4} + (k + 1)^3 = \dfrac{(k + 1)^2}{4}\left[k^2 + 4(k + 1)\right] = \dfrac{(k + 1)^2}{4}(k + 2)^2$

which shows that the formula is true for $n = k + 1$.

## Section 4.3   Riemann Sums and Definite Integrals

**1.** $f(x) = \sqrt{x}$, $y = 0$, $x = 0$, $x = 3$, $c_i = \dfrac{3i^2}{n^2}$

$$\Delta x_i = \frac{3i^2}{n^2} - \frac{3(i-1)^2}{n^2} = \frac{3}{n^2}(2i-1)$$

$$\lim_{n\to\infty}\sum_{i=1}^{n} f(c_i)\Delta x_i = \lim_{n\to\infty}\sum_{i=1}^{n}\sqrt{\frac{3i^2}{n^2}}\,\frac{3}{n^2}(2i-1)$$

$$= \lim_{n\to\infty}\frac{3\sqrt{3}}{n^3}\sum_{i=1}^{n}\left(2i^2 - i\right)$$

$$= \lim_{n\to\infty}\frac{3\sqrt{3}}{n^3}\left[2\frac{n(n+1)(2n+1)}{6} - \frac{n(n+1)}{2}\right]$$

$$= \lim_{n\to\infty}3\sqrt{3}\left[\frac{(n+1)(2n+1)}{3n^2} - \frac{n+1}{2n^2}\right]$$

$$= 3\sqrt{3}\left[\frac{2}{3} - 0\right] = 2\sqrt{3} \approx 3.464$$

**3.** $y = 8$ on $[2, 6]$. $\left(\textbf{Note: } \Delta x = \dfrac{6-2}{n} = \dfrac{4}{n}, \|\Delta\| \to 0 \text{ as } n \to \infty\right)$

$$\sum_{i=1}^{n} f(c_i)\,\Delta x_i = \sum_{i=1}^{n} f\left(2 + \frac{4i}{n}\right)\left(\frac{4}{n}\right) = \sum_{i=1}^{n} 8\left(\frac{4}{n}\right) = \sum_{i=1}^{n}\frac{32}{n} = \frac{1}{n}\sum_{i=1}^{n}32 = \frac{1}{n}(32n) = 32$$

$$\int_{2}^{6} 8\,dx = \lim_{n\to\infty} 32 = 32$$

**5.** $y = x^3$ on $[-1, 1]$. $\left(\textbf{Note:} \Delta x = \dfrac{1-(-1)}{n} = \dfrac{2}{n}, \|\Delta\| \to 0 \text{ as } n \to \infty\right)$

$$\sum_{i=1}^{n} f(c_i)\Delta x_i = \sum_{i=1}^{n} f\left(-1 + \frac{2i}{n}\right)\left(\frac{2}{n}\right)$$

$$= \sum_{i=1}^{n}\left(-1 + \frac{2i}{n}\right)^3\left(\frac{2}{n}\right)$$

$$= \sum_{i=1}^{n}\left[-1 + \frac{6i}{n} - \frac{12i^2}{n^2} + \frac{8i^3}{n^3}\right]\left(\frac{2}{n}\right)$$

$$= -2 + \frac{12}{n^2}\sum_{i=1}^{n} i - \frac{24}{n^3}\sum_{i=1}^{n} i^2 + \frac{16}{n^4}\sum_{i=1}^{n} i^3$$

$$= -2 + 6\left(1 + \frac{1}{n}\right) - 4\left(2 + \frac{3}{n} + \frac{1}{n^2}\right) + 4\left(1 + \frac{2}{n} + \frac{1}{n^2}\right) = \frac{2}{n}$$

$$\int_{-1}^{1} x^3\,dx = \lim_{n\to\infty}\frac{2}{n} = 0$$

**7.** $y = x^2 + 1$ on $[1, 2]$. $\left(\textbf{Note: } \Delta x = \dfrac{2 - 1}{n} = \dfrac{1}{n}, \|\Delta\| \to 0 \text{ as } n \to \infty\right)$

$$\sum_{i=1}^{n} f(c_i)\,\Delta x_i = \sum_{i=1}^{n} f\left(1 + \frac{i}{n}\right)\left(\frac{1}{n}\right)$$

$$= \sum_{i=1}^{n} \left[\left(1 + \frac{i}{n}\right)^2 + 1\right]\left(\frac{1}{n}\right)$$

$$= \sum_{i=1}^{n} \left[1 + \frac{2i}{n} + \frac{i^2}{n^2} + 1\right]\left(\frac{1}{n}\right)$$

$$= 2 + \frac{2}{n^2}\sum_{i=1}^{n} i + \frac{1}{n^3}\sum_{i=1}^{n} i^2 = 2 + \left(1 + \frac{1}{n}\right) + \frac{1}{6}\left(2 + \frac{3}{n} + \frac{1}{n^2}\right) = \frac{10}{3} + \frac{3}{2n} + \frac{1}{6n^2}$$

$$\int_{1}^{2} \left(x^2 + 1\right) dx = \lim_{n \to \infty}\left(\frac{10}{3} + \frac{3}{2n} + \frac{1}{6n^2}\right) = \frac{10}{3}$$

**9.** $\displaystyle\lim_{\|\Delta\| \to 0} \sum_{i=1}^{n} (3c_i + 10)\,\Delta x_i = \int_{-1}^{5} (3x + 10)\,dx$

on the interval $[-1, 5]$.

**11.** $\displaystyle\lim_{\|\Delta\| \to 0} \sum_{i=1}^{n} \sqrt{c_i^2 + 4}\,\Delta x_i = \int_{0}^{3} \sqrt{x^2 + 4}\,dx$

on the interval $[0, 3]$.

**13.** $\displaystyle\int_{0}^{4} 5\,dx$

**15.** $\displaystyle\int_{-4}^{4} \left(4 - |x|\right) dx$

**17.** $\displaystyle\int_{-5}^{5} \left(25 - x^2\right) dx$

**19.** $\displaystyle\int_{0}^{\pi/2} \cos x\,dx$

**21.** $\displaystyle\int_{0}^{2} y^3\,dy$

**23.** Rectangle

$A = bh = 3(4)$

$A = \displaystyle\int_{0}^{3} 4\,dx = 12$

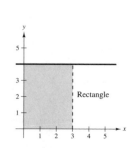

**25.** Triangle

$A = \frac{1}{2}bh = \frac{1}{2}(4)(4) = 8$

$A = \displaystyle\int_{0}^{4} x\,dx = 8$

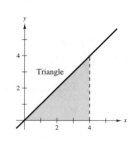

**27.** Trapezoid

$A = \dfrac{b_1 + b_2}{2}h = \left(\dfrac{4 + 10}{2}\right)2 = 14$

$A = \displaystyle\int_{0}^{2} (3x + 4)\,dx = 14$

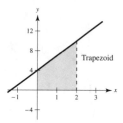

**29.** Triangle

$A = \frac{1}{2}bh = \frac{1}{2}(2)(1) = 1$

$A = \displaystyle\int_{-1}^{1} \left(1 - |x|\right) dx = 1$

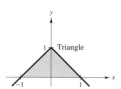

**31.** Semicircle

$A = \dfrac{1}{2}\pi r^2 = \dfrac{1}{2}\pi(7)^2 = \dfrac{49\pi}{2}$

$A = \displaystyle\int_{-7}^{7} \sqrt{49 - x^2}\,dx = \dfrac{49\pi}{2}$

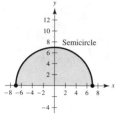

In Exercises 33–39, $\int_2^4 x^3\,dx = 60$, $\int_2^4 x\,dx = 6$,

$\int_2^4 dx = 2$

**33.** $\int_4^2 x\,dx = -\int_2^4 x\,dx = -6$

**35.** $\int_2^4 8x\,dx = 8\int_2^4 x\,dx = 8(6) = 48$

**37.** $\int_2^4 (x-9)\,dx = \int_2^4 x\,dx - 9\int_2^4 dx = 6 - 9(2) = -12$

**39.** $\int_2^4 \left(\frac{1}{2}x^3 - 3x + 2\right)dx = \frac{1}{2}\int_2^4 x^3\,dx - 3\int_2^4 x\,dx + 2\int_2^4 dx$

$\qquad = \frac{1}{2}(60) - 3(6) + 2(2) = 16$

**41. (a)** $\int_0^7 f(x)\,dx = \int_0^5 f(x)\,dx + \int_5^7 f(x)\,dx = 10 + 3 = 13$

**(b)** $\int_5^0 f(x)\,dx = -\int_0^5 f(x)\,dx = -10$

**(c)** $\int_5^5 f(x)\,dx = 0$

**(d)** $\int_0^5 3f(x)\,dx = 3\int_0^5 f(x)\,dx = 3(10) = 30$

**43. (a)** $\int_2^6 \left[f(x) + g(x)\right]dx = \int_2^6 f(x)\,dx + \int_2^6 g(x)\,dx$

$\qquad = 10 + (-2) = 8$

**(b)** $\int_2^6 \left[g(x) - f(x)\right]dx = \int_2^6 g(x)\,dx - \int_2^6 f(x)\,dx$

$\qquad = -2 - 10 = -12$

**(c)** $\int_2^6 2g(x)\,dx = 2\int_2^6 g(x)\,dx = 2(-2) = -4$

**(d)** $\int_2^6 3f(x)\,dx = 3\int_2^6 f(x)\,dx = 3(10) = 30$

**45.** Lower estimate: $[24 + 12 - 4 - 20 - 36](2) = -48$

Upper estimate: $[32 + 24 + 12 - 4 - 20](2) = 88$

**47. (a)** Quarter circle below $x$-axis:

$-\frac{1}{4}\pi r^2 = -\frac{1}{4}\pi(2)^2 = -\pi$

**(b)** Triangle: $\frac{1}{2}bh = \frac{1}{2}(4)(2) = 4$

**(c)** Triangle + Semicircle below $x$-axis:

$-\frac{1}{2}(2)(1) - \frac{1}{2}\pi(2)^2 = -(1 + 2\pi)$

**(d)** Sum of parts (b) and (c): $4 - (1 + 2\pi) = 3 - 2\pi$

**(e)** Sum of absolute values of (b) and (c):

$4 + (1 + 2\pi) = 5 + 2\pi$

**(f)** Answers to (d) plus

$2(10) = 20$: $(3 - 2\pi) + 20 = 23 - 2\pi$

**49. (a)** $\int_0^5 \left[f(x) + 2\right]dx = \int_0^5 f(x)\,dx + \int_0^5 2\,dx$

$\qquad = 4 + 10 = 14$

**(b)** $\int_{-2}^3 f(x+2)\,dx = \int_0^5 f(x)\,dx = 4$ (Let $u = x + 2$.)

**(c)** $\int_{-5}^5 f(x)\,dx = 2\int_0^5 f(x)\,dx = 2(4) = 8$ ($f$ even)

**(d)** $\int_{-5}^5 f(x)\,dx = 0$ ($f$ odd)

**51.** $f(x) = \begin{cases} 6, & x > 6 \\ -\frac{1}{2}x + 9, & x \le 6 \end{cases}$

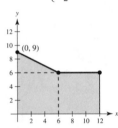

$\int_0^{12} f(x)\,dx = 6(6) + \frac{1}{2}6(3) + 6(6) = 36 + 9 + 36 = 81$

**53.** The left endpoint approximation will be greater than the actual area so,

$\sum_{i=1}^n f(x_i)\Delta x > \int_1^5 f(x)\,dx.$

**55.** $f(x) = \dfrac{1}{x - 4}$

is not integrable on the interval $[3, 5]$ because $f$ has a discontinuity at $x = 4$.

**57.**

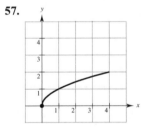

**(a)** $A \approx 5$ square units

**59.**

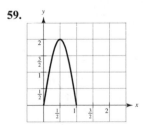

**(d)** $\int_0^1 2\sin \pi x\,dx \approx \frac{1}{2}(1)(2) \approx 1$

**61.** $\int_0^3 x\sqrt{3-x}\,dx$

| $N$ | 4 | 8 | 12 | 16 | 20 |
|---|---|---|---|---|---|
| $L(n)$ | 3.6830 | 3.9956 | 4.0707 | 4.1016 | 4.1177 |
| $M(n)$ | 4.3082 | 4.2076 | 4.1838 | 4.1740 | 4.1690 |
| $R(n)$ | 3.6830 | 3.9956 | 4.0707 | 4.1016 | 4.1177 |

**63.** $\int_0^{\pi/2} \sin^2 x\,dx$

| $n$ | 4 | 8 | 12 | 16 | 20 |
|---|---|---|---|---|---|
| $L(n)$ | 0.5890 | 0.6872 | 0.7199 | 0.7363 | 0.7461 |
| $M(n)$ | 0.7854 | 0.7854 | 0.7854 | 0.7854 | 0.7854 |
| $R(n)$ | 0.9817 | 0.8836 | 0.8508 | 0.8345 | 0.8247 |

**65.** True

**67.** True

**69.** False

$$\int_0^2 (-x)\,dx = -2$$

**71.** $f(x) = x^2 + 3x,\ [0, 8]$

$x_0 = 0,\ x_1 = 1,\ x_2 = 3,\ x_3 = 7,\ x_4 = 8$

$\Delta x_1 = 1,\ \Delta x_2 = 2,\ \Delta x_3 = 4,\ \Delta x_4 = 1$

$c_1 = 1,\ c_2 = 2,\ c_3 = 5,\ c_4 = 8$

$$\sum_{i=1}^{4} f(c_i)\Delta x = f(1)\Delta x_1 + f(2)\Delta x_2 + f(5)\Delta x_3 + f(8)\Delta x_4$$

$$= (4)(1) + (10)(2) + (40)(4) + (88)(1) = 272$$

**73.** $\Delta x = \dfrac{b-a}{n},\ c_i = a + i(\Delta x) = a + i\left(\dfrac{b-a}{n}\right)$

$$\int_0^b x\,dx = \lim_{\|\Delta\|\to 0} \sum_{i=1}^{n} f(c_i)\Delta x$$

$$= \lim_{n\to\infty} \sum_{i=1}^{n} \left[a + i\left(\frac{b-a}{n}\right)\right]\left(\frac{b-a}{n}\right)$$

$$= \lim_{n\to\infty} \left[\left(\frac{b-a}{n}\right)\sum_{i=1}^{n} a + \left(\frac{b-a}{n}\right)^2 \sum_{i=1}^{n} i\right]$$

$$= \lim_{n\to\infty} \left[\frac{b-a}{n}(an) + \left(\frac{b-a}{n}\right)^2 \frac{n(n+1)}{2}\right]$$

$$= \lim_{n\to\infty} \left[a(b-a) + \frac{(b-a)^2}{n}\cdot\frac{n+1}{2}\right]$$

$$= a(b-a) + \frac{(b-a)^2}{2}$$

$$= (b-a)\left[a + \frac{b-a}{2}\right]$$

$$= \frac{(b-a)(a+b)}{2} = \frac{b^2-a^2}{2}$$

**75.** $f(x) = \begin{cases} 1, & x \text{ is rational} \\ 0, & x \text{ is irrational} \end{cases}$

is not integrable on the interval $[0, 1]$. As

$\|\Delta\| \to 0$, $f(c_i) = 1$ or $f(c_i) = 0$ in each subinterval

because there are an infinite number of both rational and irrational numbers in any interval, no matter how small.

**77.** The function $f$ is nonnegative between $x = -1$ and $x = 1$.

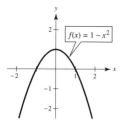

So,

$$\int_a^b (1 - x^2)\,dx$$

is a maximum for $a = -1$ and $b = 1$.

**79.** Let $f(x) = x^2$, $0 \le x \le 1$, and $\Delta x_i = 1/n$. The appropriate Riemann Sum is

$$\sum_{i=1}^{n} f(c_i)\Delta x_i = \sum_{i=1}^{n} \left(\frac{i}{n}\right)^2 \frac{1}{n} = \frac{1}{n^3}\sum_{i=1}^{n} i^2.$$

$$\lim_{n\to\infty} \frac{1}{n^3}\left[1^2 + 2^2 + 3^2 + \cdots + n^2\right] = \lim_{n\to\infty} \frac{1}{n^3}\cdot\frac{n(2n+1)(n+1)}{6} = \lim_{n\to\infty} \frac{2n^2 + 3n + 1}{6n^2} = \lim_{n\to\infty}\left(\frac{1}{3} + \frac{1}{2n} + \frac{1}{6n^2}\right) = \frac{1}{3}$$

## Section 4.4 The Fundamental Theorem of Calculus

**1.** $f(x) = \dfrac{4}{x^2 + 1}$

$\displaystyle\int_0^\pi \dfrac{4}{x^2 + 1}\,dx$ is positive.

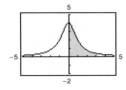

**3.** $f(x) = x\sqrt{x^2 + 1}$

$\displaystyle\int_{-2}^2 x\sqrt{x^2 + 1}\,dx = 0$

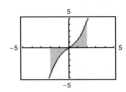

**5.** $\displaystyle\int_0^2 6x\,dx = \left[3x^2\right]_0^2 = 12 - 0 = 12$

**7.** $\displaystyle\int_{-1}^0 (2x - 1)\,dx = \left[x^2 - x\right]_{-1}^0$

$$= 0 - \left((-1)^2 - (-1)\right) = -(1 + 1) = -2$$

**9.** $\displaystyle\int_{-1}^1 (t^2 - 2)\,dt = \left[\dfrac{t^3}{3} - 2t\right]_{-1}^1$

$$= \left(\dfrac{1}{3} - 2\right) - \left(-\dfrac{1}{3} + 2\right) = -\dfrac{10}{3}$$

**11.** $\displaystyle\int_0^1 (2t - 1)^2\,dt = \int_0^1 \left(4t^2 - 4t + 1\right)\,dt$

$$= \left[\tfrac{4}{3}t^3 - 2t^2 + t\right]_0^1$$

$$= \tfrac{4}{3} - 2 + 1 = \tfrac{1}{3}$$

**13.** $\displaystyle\int_1^2 \left(\dfrac{3}{x^2} - 1\right)dx = \left[-\dfrac{3}{x} - x\right]_1^2 = \left(-\dfrac{3}{2} - 2\right) - (-3 - 1) = \dfrac{1}{2}$

**15.** $\displaystyle\int_1^4 \dfrac{u - 2}{\sqrt{u}}\,du = \int_1^4 \left(u^{1/2} - 2u^{-1/2}\right)du = \left[\dfrac{2}{3}u^{3/2} - 4u^{1/2}\right]_1^4 = \left[\dfrac{2}{3}\left(\sqrt{4}\right)^3 - 4\sqrt{4}\right] - \left[\dfrac{2}{3} - 4\right] = \dfrac{2}{3}$

**17.** $\displaystyle\int_{-1}^1 \left(\sqrt[3]{t} - 2\right)dt = \left[\tfrac{3}{4}t^{4/3} - 2t\right]_{-1}^1 = \left(\tfrac{3}{4} - 2\right) - \left(\tfrac{3}{4} + 2\right) = -4$

**19.** $\displaystyle\int_0^1 \dfrac{x - \sqrt{x}}{3}\,dx = \dfrac{1}{3}\int_0^1 \left(x - x^{1/2}\right)dx = \dfrac{1}{3}\left[\dfrac{x^2}{2} - \dfrac{2}{3}x^{3/2}\right]_0^1 = \dfrac{1}{3}\left(\dfrac{1}{2} - \dfrac{2}{3}\right) = -\dfrac{1}{18}$

**21.** $\displaystyle\int_{-1}^0 \left(t^{1/3} - t^{2/3}\right)dt = \left[\tfrac{3}{4}t^{4/3} - \tfrac{3}{5}t^{5/3}\right]_{-1}^0 = 0 - \left(\tfrac{3}{4} + \tfrac{3}{5}\right) = -\dfrac{27}{20}$

**23.** $\displaystyle\int_0^5 |2x - 5|\,dx = \int_0^{5/2} (5 - 2x)\,dx + \int_{5/2}^5 (2x - 5)\,dx$ $\left(\text{split up the integral at the zero } x = \dfrac{5}{2}\right)$

$$= \left[5x - x^2\right]_0^{5/2} + \left[x^2 - 5x\right]_{5/2}^5 = \left(\dfrac{25}{2} - \dfrac{25}{4}\right) - 0 + (25 - 25) - \left(\dfrac{25}{4} - \dfrac{25}{2}\right) = 2\left(\dfrac{25}{2} - \dfrac{25}{4}\right) = \dfrac{25}{2}$$

**Note:** By Symmetry, $\displaystyle\int_0^5 |2x - 5|\,dx = 2\int_{5/2}^5 (2x - 5)\,dx$.

**25.** $\displaystyle\int_0^4 |x^2 - 9|\,dx = \int_0^3 (9 - x^2)\,dx + \int_3^4 (x^2 - 9)\,dx$ (split up integral at the zero $x = 3$)

$$= \left[9x - \dfrac{x^3}{3}\right]_0^3 + \left[\dfrac{x^3}{3} - 9x\right]_3^4 = (27 - 9) + \left(\dfrac{64}{3} - 36\right) - (9 - 27) = \dfrac{64}{3}$$

**27.** $\displaystyle\int_0^\pi (1 + \sin x)\,dx = \left[x - \cos x\right]_0^\pi = (\pi + 1) - (0 - 1) = 2 + \pi$

**29.** $\int_0^{\pi/4} \dfrac{1 - \sin^2 \theta}{\cos^2 \theta} \, d\theta = \int_0^{\pi/4} d\theta = \left[ \theta \right]_0^{\pi/4} = \dfrac{\pi}{4}$

**31.** $\int_{-\pi/6}^{\pi/6} \sec^2 x \, dx = \left[ \tan x \right]_{-\pi/6}^{\pi/6} = \dfrac{\sqrt{3}}{3} - \left( -\dfrac{\sqrt{3}}{3} \right) = \dfrac{2\sqrt{3}}{3}$

**33.** $\int_{-\pi/3}^{\pi/3} 4 \sec \theta \tan \theta \, d\theta = \left[ 4 \sec \theta \right]_{-\pi/3}^{\pi/3} = 4(2) - 4(2) = 0$

**35.** $A = \int_0^1 \left( x - x^2 \right) dx = \left[ \dfrac{x^2}{2} - \dfrac{x^3}{3} \right]_0^1 = \dfrac{1}{6}$

**37.** $A = \int_0^{\pi/2} \cos x \, dx = \left[ \sin x \right]_0^{\pi/2} = 1$

**39.** Because $y > 0$ on $[0, 2]$,

   $\text{Area} = \int_0^2 \left( 5x^2 + 2 \right) dx = \left[ \dfrac{5}{3}x^3 + 2x \right]_0^2 = \dfrac{40}{3} + 4 = \dfrac{52}{3}.$

**41.** Because $y > 0$ on $[0, 8]$,

   $\text{Area} = \int_0^8 \left( 1 + x^{1/3} \right) dx = \left[ x + \dfrac{3}{4}x^{4/3} \right]_0^8 = 8 + \dfrac{3}{4}(16) = 20.$

**43.** Because $y > 0$ on $[0, 4]$,

   $\text{Area} = \int_0^4 \left( -x^2 + 4x \right) dx = \left[ -\dfrac{x^3}{3} + 2x^2 \right]_0^4 = -\dfrac{64}{3} + 32 = \dfrac{32}{3}.$

**45.** $\int_0^3 x^3 \, dx = \left[ \dfrac{x^4}{4} \right]_0^3 = \dfrac{81}{4}$

   $f(c)(3 - 0) = \dfrac{81}{4}$

   $f(c) = \dfrac{27}{4}$

   $c^3 = \dfrac{27}{4}$

   $c = \dfrac{3}{\sqrt[3]{4}} = \dfrac{3}{2}\sqrt[3]{2} \approx 1.8899$

**47.** $\int_4^9 \sqrt{x} \, dx = \left[ \dfrac{2}{3}x^{3/2} \right]_4^9 = \dfrac{2}{3}(27 - 8) = \dfrac{38}{3}$

   $f(c)(9 - 4) = \dfrac{38}{3}$

   $f(c) = \dfrac{38}{15}$

   $\sqrt{c} = \dfrac{38}{15}$

   $c = \dfrac{1444}{225} \approx 6.4178$

**49.** $\int_{-\pi/4}^{\pi/4} 2 \sec^2 x \, dx = \left[ 2 \tan x \right]_{-\pi/4}^{\pi/4} = 2(1) - 2(-1) = 4$

   $f(c)\left[ \dfrac{\pi}{4} - \left( -\dfrac{\pi}{4} \right) \right] = 4$

   $2 \sec^2 c = \dfrac{8}{\pi}$

   $\sec^2 c = \dfrac{4}{\pi}$

   $\sec c = \pm \dfrac{2}{\sqrt{\pi}}$

   $c = \pm \text{arcsec}\left( \dfrac{2}{\sqrt{\pi}} \right)$

   $= \pm \text{arccos}\dfrac{\sqrt{\pi}}{2} \approx \pm 0.4817$

**51.** $\dfrac{1}{3 - (-3)} \displaystyle\int_{-3}^{3} \left(9 - x^2\right) dx = \dfrac{1}{6}\left[9x - \dfrac{1}{3}x^3\right]_{-3}^{3}$

$\qquad\qquad\qquad\qquad = \dfrac{1}{6}\left[(27 - 9) - (-27 + 9)\right]$

$\qquad\qquad\qquad\qquad = 6$

Average value $= 6$

$9 - x^2 = 6$ when $x^2 = 9 - 6$ or $x = \pm\sqrt{3} \approx \pm 1.7321$.

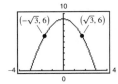

**53.** $\dfrac{1}{1 - 0} \displaystyle\int_{0}^{1} x^3 \, dx = \left[\dfrac{x^4}{4}\right]_{0}^{1} = \dfrac{1}{4}$

Average value $= \dfrac{1}{4}$

$x^3 = \dfrac{1}{4}$

$x = \sqrt[3]{\dfrac{1}{4}} = \dfrac{\sqrt[3]{2}}{2} \approx 0.6300$

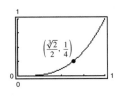

**55.** $\dfrac{1}{\pi - 0} \displaystyle\int_{0}^{\pi} \sin x \, dx = \left[-\dfrac{1}{\pi}\cos x\right]_{0}^{\pi} = \dfrac{2}{\pi}$

Average value $= \dfrac{2}{\pi}$

$\sin x = \dfrac{2}{\pi}$

$x \approx 0.690, \ 2.451$

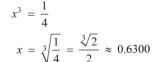

**57.** The distance traveled is $\displaystyle\int_{0}^{8} v(t)\, dt$. The area under the curve from $0 \le t \le 8$ is approximately $(18 \text{ squares})(30) \approx 540$ ft.

**59.** (a) $\displaystyle\int_{1}^{7} f(x)\, dx = $ Sum of the areas

$\qquad = A_1 + A_2 + A_3 + A_4$

$\qquad = \dfrac{1}{2}(3 + 1) + \dfrac{1}{2}(1 + 2) + \dfrac{1}{2}(2 + 1) + (3)(1)$

$\qquad = 8$

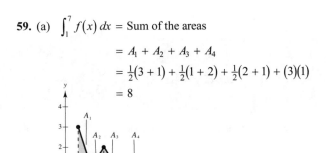

(b) Average value $= \dfrac{\displaystyle\int_{1}^{7} f(x)\, dx}{7 - 1} = \dfrac{8}{6} = \dfrac{4}{3}$

(c) $A = 8 + (6)(2) = 20$

Average value $= \dfrac{20}{6} = \dfrac{10}{3}$

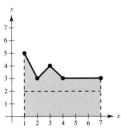

**61.** (a) $F(x) = k \sec^2 x$

$\qquad F(0) = k = 500$

$\qquad F(x) = 500 \sec^2 x$

(b) $\dfrac{1}{\pi/3 - 0} \displaystyle\int_{0}^{\pi/3} 500 \sec^2 x \, dx = \dfrac{1500}{\pi}\left[\tan x\right]_{0}^{\pi/3}$

$\qquad\qquad\qquad\qquad = \dfrac{1500}{\pi}\left(\sqrt{3} - 0\right)$

$\qquad\qquad\qquad\qquad \approx 826.99 \text{ newtons}$

$\qquad\qquad\qquad\qquad \approx 827 \text{ newtons}$

**63.** $\dfrac{1}{5 - 0} \displaystyle\int_{0}^{5} \left(0.1729t + 0.1522t^2 - 0.0374t^3\right) dt \approx \dfrac{1}{5}\left[0.08645t^2 + 0.05073t^3 - 0.00935t^4\right]_{0}^{5} \approx 0.5318 \text{ liter}$

**65.** (a) $v = -0.00086t^3 + 0.0782t^2 - 0.208t + 0.10$

(b)

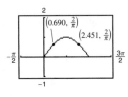

(c) $\displaystyle\int_{0}^{60} v(t)\, dt = \left[\dfrac{-0.00086t^4}{4} + \dfrac{0.0782t^3}{3} - \dfrac{0.208t^2}{2} + 0.10t\right]_{0}^{60} \approx 2476 \text{ meters}$

**67.** $F(x) = \int_0^x (4t - 7)\,dt = \left[2t^2 - 7t\right]_0^x = 2x^2 - 7x$

$F(2) = 2(2^2) - 7(2) = -6$

$F(5) = 2(5^2) - 7(5) = 15$

$F(8) = 2(8^2) - 7(8) = 72$

**69.** $F(x) = \int_1^x \dfrac{20}{v^2}\,dv = \int_1^x 20v^{-2}\,dv = -\dfrac{20}{v}\Big]_1^x$

$= -\dfrac{20}{x} + 20 = 20\left(1 - \dfrac{1}{x}\right)$

$F(2) = 20\left(\tfrac{1}{2}\right) = 10$

$F(5) = 20\left(\tfrac{4}{5}\right) = 16$

$F(8) = 20\left(\tfrac{7}{8}\right) = \tfrac{35}{2}$

**71.** $F(x) = \int_1^x \cos\theta\,d\theta = \sin\theta\Big]_1^x = \sin x - \sin 1$

$F(2) = \sin 2 - \sin 1 \approx 0.0678$

$F(5) = \sin 5 - \sin 1 \approx -1.8004$

$F(8) = \sin 8 - \sin 1 \approx 0.1479$

**73.** $g(x) = \int_0^x f(t)\,dt$

(a) $g(0) = \int_0^0 f(t)\,dt = 0$

$g(2) = \int_0^2 f(t)\,dt \approx 4 + 2 + 1 = 7$

$g(4) = \int_0^4 f(t)\,dt \approx 7 + 2 = 9$

$g(6) = \int_0^6 f(t)\,dt \approx 9 + (-1) = 8$

$g(8) = \int_0^8 f(t)\,dt \approx 8 - 3 = 5$

(b) $g$ increasing on $(0, 4)$ and decreasing on $(4, 8)$

(c) $g$ is a maximum of 9 at $x = 4$.

(d) 

**75.** (a) $\int_0^x (t + 2)\,dt = \left[\dfrac{t^2}{2} + 2t\right]_0^x = \dfrac{1}{2}x^2 + 2x$

(b) $\dfrac{d}{dx}\left[\dfrac{1}{2}x^2 + 2x\right] = x + 2$

**77.** (a) $\int_8^x \sqrt[3]{t}\,dt = \left[\dfrac{3}{4}t^{4/3}\right]_8^x = \dfrac{3}{4}\left(x^{4/3} - 16\right) = \dfrac{3}{4}x^{4/3} - 12$

(b) $\dfrac{d}{dx}\left[\dfrac{3}{4}x^{4/3} - 12\right] = x^{1/3} = \sqrt[3]{x}$

**79.** (a) $\int_{x/4}^x \sec^2 t\,dt = \left[\tan t\right]_{x/4}^x = \tan x - 1$

(b) $\dfrac{d}{dx}[\tan x - 1] = \sec^2 x$

**81.** $F(x) = \int_{-2}^x (t^2 - 2t)\,dt$

$F'(x) = x^2 - 2x$

**83.** $F(x) = \int_{-1}^x \sqrt{t^4 + 1}\,dt$

$F'(x) = \sqrt{x^4 + 1}$

**85.** $F(x) = \int_0^x t\cos t\,dt$

$F'(x) = x\cos x$

**87.** $F(x) = \int_x^{x+2} (4t + 1)\,dt$

$= \left[2t^2 + t\right]_x^{x+2}$

$= \left[2(x + 2)^2 + (x + 2)\right] - \left[2x^2 + x\right]$

$= 8x + 10$

$F'(x) = 8$

**Alternate solution:**

$F(x) = \int_x^{x+2} (4t + 1)\,dt$

$= \int_x^0 (4t + 1)\,dt + \int_0^{x+2} (4t + 1)\,dt$

$= -\int_0^x (4t + 1)\,dt + \int_0^{x+2} (4t + 1)\,dt$

$F'(x) = -(4x + 1) + 4(x + 2) + 1 = 8$

**89.** $F(x) = \int_0^{\sin x} \sqrt{t}\,dt = \left[\dfrac{2}{3}t^{3/2}\right]_0^{\sin x} = \dfrac{2}{3}(\sin x)^{3/2}$

$F'(x) = (\sin x)^{1/2}\cos x = \cos x\sqrt{\sin x}$

**Alternate solution:**

$F(x) = \int_0^{\sin x} \sqrt{t}\,dt$

$F'(x) = \sqrt{\sin x}\,\dfrac{d}{dx}(\sin x) = \sqrt{\sin x}\,(\cos x)$

**91.** $F(x) = \int_0^{x^3} \sin t^2\,dt$

$F'(x) = \sin(x^3)^2 \cdot 3x^2 = 3x^2 \sin x^6$

**93.** $g(x) = \int_0^x f(t)\,dt$

$g(0) = 0, g(1) \approx \frac{1}{2}, g(2) \approx 1, g(3) \approx \frac{1}{2}, g(4) = 0$

$g$ has a relative maximum at $x = 2$.

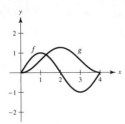

**95.** (a) $C(x) = 5000\left(25 + 3\int_0^x t^{1/4}dt\right) = 5000\left(25 + 3\left[\frac{4}{5}t^{5/4}\right]_0^x\right) = 5000\left(25 + \frac{12}{5}x^{5/4}\right) = 1000\left(125 + 12x^{5/4}\right)$

(b) $C(1) = 1000\left(125 + 12(1)\right) = \$137,000$

$C(5) = 1000\left(125 + 12(5)^{5/4}\right) \approx \$214,721$

$C(10) = 1000\left(125 + 12(10)^{5/4}\right) \approx \$338,394$

**97.** (a) $v(t) = 5t - 7, \ 0 \le t \le 3$

Displacement $= \int_0^3 (5t - 7)\,dt = \left[\frac{5t^2}{2} - 7t\right]_0^3 = \frac{45}{2} - 21 = \frac{3}{2}$ ft to the right

(b) Total distance traveled $= \int_0^3 |5t - 7|\,dt$

$= \int_0^{7/5} (7 - 5t)\,dt + \int_{7/5}^3 (5t - 7)\,dt$

$= \left[7t - \frac{5t^2}{2}\right]_0^{7/5} + \left[\frac{5t^2}{2} - 7t\right]_{7/5}^3$

$= 7\left(\frac{7}{5}\right) - \frac{5}{2}\left(\frac{7}{2}\right)^2 + \left(\frac{5}{2}(9) - 21\right) - \left(\frac{5}{2}\left(\frac{7}{5}\right)^2 - 7\left(\frac{7}{5}\right)\right)$

$= \frac{49}{5} - \frac{49}{10} + \frac{45}{2} - 21 - \frac{49}{10} + \frac{49}{5} = \frac{113}{10}$ ft

**99.** (a) $v(t) = t^3 - 10t^2 + 27t - 18 = (t - 1)(t - 3)(t - 6), \ 1 \le t \le 7$

Displacement $= \int_1^7 \left(t^3 - 10t^2 + 27t - 18\right)\,dt$

$= \left[\frac{t^4}{4} - \frac{10t^3}{3} + \frac{27t^2}{2} - 18t\right]_1^7$

$= \left[\frac{7^4}{4} - \frac{10(7^3)}{3} + \frac{27(7^2)}{2} - 18(7)\right] - \left[\frac{1}{4} - \frac{10}{3} + \frac{27}{2} - 18\right] = -\frac{91}{12} - \left(-\frac{91}{12}\right) = 0$

(b) Total distance traveled $= \int_1^7 |v(t)|\,dt$

$= \int_1^3 \left(t^3 - 10t^2 + 27t - 18\right)\,dt - \int_3^6 \left(t^3 - 10t^2 + 27t - 18\right)\,dt + \int_6^7 \left(t^3 - 10t^2 + 27t - 18\right)\,dt$

Evaluating each of these integrals, you obtain

Total distance $= \frac{16}{3} - \left(-\frac{63}{4}\right) + \frac{125}{12} = \frac{63}{2}$ ft

**101.** (a)  $v(t) = \dfrac{1}{\sqrt{t}},\ 1 \le t \le 4$

  Because $v(t) > 0,$

  Displacement = Total Distance

  Displacement $= \displaystyle\int_{1}^{4} t^{-1/2}\, dt = \left[2t^{1/2}\right]_{1}^{4} = 4 - 2 = 2$ ft to the right

  (b)  Total distance = 2 ft

**103.**  $x(t) = t^3 - 6t^2 + 9t - 2$

  $x'(t) = 3t^2 - 12t + 9 = 3\left(t^2 - 4t + 3\right) = 3(t - 3)(t - 1)$

  Total distance $= \displaystyle\int_{0}^{5} \left|x'(t)\right|\, dt$

  $\phantom{Total distance}= \displaystyle\int_{0}^{5} 3\left|(t - 3)(t - 1)\right|\, dt$

  $\phantom{Total distance}= 3\displaystyle\int_{0}^{1} \left(t^2 - 4t + 3\right) dt - 3\int_{1}^{3}\left(t^2 - 4t + 3\right) dt + 3\int_{3}^{5}\left(t^2 - 4t + 3\right) dt = 4 + 4 + 20 = 28$ units

**105.** Let $c(t)$ be the amount of water that is flowing out of the tank. Then $c'(t) = 500 - 5t$ L/min is the rate of flow.

  $\displaystyle\int_{0}^{18} c'(t)\,dt = \int_{0}^{18}\left(500 - 5t\right) dt = \left[500t - \frac{5t^2}{2}\right]_{0}^{18} = 9000 - 810 = 8190$ L

**107.** The function $f(x) = x^{-2}$ is not continuous on $[-1, 1].$

  $\displaystyle\int_{-1}^{1} x^{-2}\, dx = \int_{-1}^{0} x^{-2}\, dx + \int_{0}^{1} x^{-2}\, dx$

  Each of these integrals is infinite. $f(x) = x^{-2}$ has a nonremovable discontinuity at $x = 0.$

**109.** The function $f(x) = \sec^2 x$ is not continuous on $\left[\dfrac{\pi}{4}, \dfrac{3\pi}{4}\right].$

  $\displaystyle\int_{\pi/4}^{3\pi/4} \sec^2 x\, dx = \int_{\pi/4}^{\pi/2} \sec^2 x\, dx + \int_{\pi/2}^{3\pi/4} \sec^2 x\, dx$

  Each of these integrals is infinite. $f(x) = \sec^2 x$ has a nonremovable discontinuity at $x = \dfrac{\pi}{2}$

**111.**  $P = \dfrac{2}{\pi} \displaystyle\int_{0}^{\pi/2} \sin\theta\, d\theta = \left[-\frac{2}{\pi}\cos\theta\right]_{0}^{\pi/2} = -\frac{2}{\pi}(0 - 1) = \frac{2}{\pi} \approx 63.7\%$

**113.** True

**115.**  $f(x) = \displaystyle\int_{0}^{1/x} \frac{1}{t^2 + 1}\, dt + \int_{0}^{x} \frac{1}{t^2 + 1}\, dt$

  By the Second Fundamental Theorem of Calculus, you have $f'(x) = \dfrac{1}{\left(1/x\right)^2 + 1}\left(-\dfrac{1}{x^2}\right) + \dfrac{1}{x^2 + 1} = -\dfrac{1}{1 + x^2} + \dfrac{1}{x^2 + 1} = 0.$

  Because $f'(x) = 0,$ $f(x)$ must be constant.

**117.** $G(x) = \int_0^x \left[ s \int_0^s f(t)\, dt \right] ds$

(a) $G(0) = \int_0^0 \left[ s \int_0^s f(t)\, dt \right] ds = 0$

(b) Let $F(s) = s \int_0^s f(t)\, dt$.

$G(x) = \int_0^x F(s)\, ds$

$G'(x) = F(x) = x \int_0^x f(t)\, dt$

$G'(0) = 0 \int_0^0 f(t)\, dt = 0$

(c) $G''(x) = x \cdot f(x) + \int_0^x f(t)\, dt$

(d) $G''(0) = 0 \cdot f(0) + \int_0^0 f(t)\, dt = 0$

## Section 4.5   Integration by Substitution

| $\int f(g(x))g'(x)\, dx$ | $u = g(x)$ | $du = g'(x)\, dx$ |
|---|---|---|
| **1.** $\int (8x^2 + 1)^2 (16x)\, dx$ | $8x^2 + 1$ | $16x\, dx$ |
| **3.** $\int \dfrac{x}{\sqrt{x^2 + 1}}\, dx$ | $x^2 + 1$ | $2x\, dx$ |
| **5.** $\int \tan^2 x \sec^2 x\, dx$ | $\tan x$ | $\sec^2 x\, dx$ |

**7.** $\int \sqrt{x}(6 - x)\, dx = \int \left( 6x^{1/2} - x^{3/2} \right) dx$

Substitution not needed

**9.** $\int x\sqrt[3]{1 + x^2}\, dx = \int \left( 1 + x^2 \right)^{1/3} x\, dx$

Substitution is necessary $\left( u = 1 + x^2 \right)$

**11.** $\int (1 + 6x)^4 (6)\, dx = \dfrac{(1 + 6x)^5}{5} + C$

**Check:** $\dfrac{d}{dx}\left[ \dfrac{(1 + 6x)^5}{5} + C \right] = 6(1 + 6x)^4$

**13.** $\int \sqrt{25 - x^2}(-2x)\, dx = \dfrac{(25 - x^2)^{3/2}}{3/2} + C = \dfrac{2}{3}(25 - x^2)^{3/2} + C$

**Check:** $\dfrac{d}{dx}\left[ \dfrac{2}{3}(25 - x^2)^{3/2} + C \right] = \dfrac{2}{3}\left( \dfrac{3}{2} \right)(25 - x^2)^{1/2}(-2x) = \sqrt{25 - x^2}(-2x)$

**15.** $\int x^3(x^4 + 3)^2\, dx = \dfrac{1}{4}\int (x^4 + 3)^2 (4x^3)\, dx = \dfrac{1}{4}\dfrac{(x^4 + 3)^3}{3} + C = \dfrac{(x^4 + 3)^3}{12} + C$

**Check:** $\dfrac{d}{dx}\left[ \dfrac{(x^4 + 3)^3}{12} + C \right] = \dfrac{3(x^4 + 3)^2}{12}(4x^3) = (x^4 + 3)^2(x^3)$

**17.** $\int x^2 (x^3 - 1)^4 \, dx = \frac{1}{3} \int (x^3 - 1)^4 (3x^2) \, dx = \frac{1}{3} \left[ \frac{(x^3 - 1)^5}{5} \right] + C = \frac{(x^3 - 1)^5}{15} + C$

**Check:** $\frac{d}{dx} \left[ \frac{(x^3 - 1)^5}{15} + C \right] = \frac{5(x^3 - 1)^4 (3x^2)}{15} = x^2 (x^3 - 1)^4$

**19.** $\int t\sqrt{t^2 + 2} \, dt = \frac{1}{2} \int (t^2 + 2)^{1/2} (2t) \, dt = \frac{1}{2} \frac{(t^2 + 2)^{3/2}}{3/2} + C = \frac{(t^2 + 2)^{3/2}}{3} + C$

**Check:** $\frac{d}{dt} \left[ \frac{(t^2 + 2)^{3/2}}{3} + C \right] = \frac{3/2(t^2 + 2)^{1/2}(2t)}{3} = (t^2 + 2)^{1/2} t$

**21.** $\int 5x(1 - x^2)^{1/3} \, dx = -\frac{5}{2} \int (1 - x^2)^{1/3} (-2x) \, dx = -\frac{5}{2} \cdot \frac{(1 - x^2)^{4/3}}{4/3} + C = -\frac{15}{8}(1 - x^2)^{4/3} + C$

**Check:** $\frac{d}{dx} \left[ -\frac{15}{8}(1 - x^2)^{4/3} + C \right] = -\frac{15}{8} \cdot \frac{4}{3}(1 - x^2)^{1/3}(-2x) = 5x(1 - x^2)^{1/3} = 5x\sqrt[3]{1 - x^2}$

**23.** $\int \frac{x}{(1 - x^2)^3} \, dx = -\frac{1}{2} \int (1 - x^2)^{-3}(-2x) \, dx = -\frac{1}{2} \frac{(1 - x^2)^{-2}}{-2} + C = \frac{1}{4(1 - x^2)^2} + C$

**Check:** $\frac{d}{dx} \left[ \frac{1}{4(1 - x^2)^2} + C \right] = \frac{1}{4}(-2)(1 - x^2)^{-3}(-2x) = \frac{x}{(1 - x^2)^3}$

**25.** $\int \frac{x^2}{(1 + x^3)^2} \, dx = \frac{1}{3} \int (1 + x^3)^{-2}(3x^2) \, dx = \frac{1}{3} \left[ \frac{(1 + x^3)^{-1}}{-1} \right] + C = -\frac{1}{3(1 + x^3)} + C$

**Check:** $\frac{d}{dx} \left[ -\frac{1}{3(1 + x^3)} + C \right] = -\frac{1}{3}(-1)(1 + x^3)^{-2}(3x^2) = \frac{x^2}{(1 + x^3)^2}$

**27.** $\int \frac{x}{\sqrt{1 - x^2}} \, dx = -\frac{1}{2} \int (1 - x^2)^{-1/2}(-2x) \, dx = -\frac{1}{2} \frac{(1 - x^2)^{1/2}}{1/2} + C = -\sqrt{1 - x^2} + C$

**Check:** $\frac{d}{dx} \left[ -(1 - x^2)^{1/2} + C \right] = -\frac{1}{2}(1 - x^2)^{-1/2}(-2x) = \frac{x}{\sqrt{1 - x^2}}$

**29.** $\int \left(1 + \frac{1}{t}\right)^3 \left(\frac{1}{t^2}\right) \, dt = -\int \left(1 + \frac{1}{t}\right)^3 \left(-\frac{1}{t^2}\right) \, dt = -\frac{\left[1 + \left(\frac{1}{t}\right)\right]^4}{4} + C$

**Check:** $\frac{d}{dt} \left[ -\frac{[1 + (1/t)]^4}{4} + C \right] = -\frac{1}{4}(4)\left(1 + \frac{1}{t}\right)^3 \left(-\frac{1}{t^2}\right) = \frac{1}{t^2}\left(1 + \frac{1}{t}\right)^3$

**31.** $\displaystyle \int \frac{1}{\sqrt{2x}}\, dx = \frac{1}{2}\int (2x)^{-1/2}\, 2\, dx = \frac{1}{2}\left[\frac{(2x)^{1/2}}{1/2}\right] + C = \sqrt{2x} + C$

**Alternate Solution:** $\displaystyle \int \frac{1}{\sqrt{2x}}\, dx = \frac{1}{\sqrt{2}}\int x^{-1/2}\, dx = \frac{1}{\sqrt{2}}\frac{x^{1/2}}{(1/2)} + C = \sqrt{2x} + C$

**Check:** $\displaystyle \frac{d}{dx}\left[\sqrt{2x} + C\right] = \frac{1}{2}(2x)^{-1/2}(2) = \frac{1}{\sqrt{2x}}$

**33.** $\displaystyle \int \frac{x^2 + 5x - 8}{\sqrt{x}}\, dx = \int \left(x^{3/2} + 5x^{1/2} - 8x^{-1/2}\right) dx = \frac{2}{5}x^{5/2} + \frac{10}{3}x^{3/2} - 16x^{1/2} + C = \frac{1}{15}\sqrt{x}\left[6x^2 + 50x - 240\right] + C$

**Check:** $\displaystyle \frac{d}{dx}\left[\frac{2}{5}x^{5/2} + \frac{10}{3}x^{3/2} - 16x^{1/2} + C\right] = x^{3/2} + 5x^{1/2} - 8x^{-1/2} = \frac{x^2 + 5x - 8}{\sqrt{x}}$

**35.** $\displaystyle \int t^2\left(t - \frac{8}{t}\right) dt = \int \left(t^3 - 8t\right) dt = \frac{t^4}{4} - 4t^2 + C$

**Check:** $\displaystyle \frac{d}{dt}\left[\frac{t^4}{4} - 4t^2 + C\right] = t^3 - 8t = t^2\left(t - \frac{8}{t}\right)$

**37.** $\displaystyle \int (9 - y)\sqrt{y}\, dy = \int \left(9y^{1/2} - y^{3/2}\right) dy = 9\left(\frac{2}{3}y^{3/2}\right) - \frac{2}{5}y^{5/2} + C = \frac{2}{5}y^{3/2}(15 - y) + C$

**Check:** $\displaystyle \frac{d}{dy}\left[\frac{2}{5}y^{3/2}(15 - y) + C\right] = \frac{d}{dy}\left[6y^{3/2} - \frac{2}{5}y^{5/2} + C\right] = 9y^{1/2} - y^{3/2} = (9 - y)\sqrt{y}$

**39.** $\displaystyle y = \int \left[4x + \frac{4x}{\sqrt{16 - x^2}}\right] dx = 4\int x\, dx - 2\int \left(16 - x^2\right)^{-1/2}(-2x)\, dx = 4\left(\frac{x^2}{2}\right) - 2\left[\frac{\left(16 - x^2\right)^{1/2}}{1/2}\right] + C = 2x^2 - 4\sqrt{16 - x^2} + C$

**41.** $\displaystyle y = \int \frac{x + 1}{\left(x^2 + 2x - 3\right)^2}\, dx = \frac{1}{2}\int \left(x^2 + 2x - 3\right)^{-2}(2x + 2)\, dx = \frac{1}{2}\left[\frac{\left(x^2 + 2x - 3\right)^{-1}}{-1}\right] + C = -\frac{1}{2\left(x^2 + 2x - 3\right)} + C$

**43.** (a) Answers will vary. *Sample answer:*

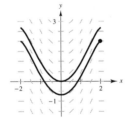

(b) $\displaystyle \frac{dy}{dx} = x\sqrt{4 - x^2},\ (2, 2)$

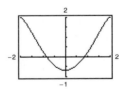

$\displaystyle y = \int x\sqrt{4 - x^2}\, dx = -\frac{1}{2}\int \left(4 - x^2\right)^{1/2}(-2x\, dx)$

$\displaystyle = -\frac{1}{2}\cdot\frac{2}{3}\left(4 - x^2\right)^{3/2} + C = -\frac{1}{3}\left(4 - x^2\right)^{3/2} + C$

$\displaystyle (2, 2):\ 2 = -\frac{1}{3}\left(4 - 2^2\right)^{3/2} + C \Rightarrow C = 2$

$\displaystyle y = -\frac{1}{3}\left(4 - x^2\right)^{3/2} + 2$

**45.** (a)  Answers will vary. *Sample answer:*

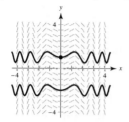

(b) $\dfrac{dy}{dx} = x \cos x^2,\ (0, 1)$

$$y = \int x \cos x^2\ dx = \frac{1}{2}\int \cos(x^2)2x\ dx = \frac{1}{2}\sin(x^2) + C$$

$$(0, 1):\ 1 = \frac{1}{2}\sin(0) + C \Rightarrow C = 1$$

$$y = \frac{1}{2}\sin(x^2) + 1$$

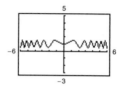

**47.** $\displaystyle\int \pi \sin \pi x\ dx = -\cos \pi x + C$

**49.** $\displaystyle\int \sin 4x\ dx = \frac{1}{4}\int (\sin 4x)(4)\ dx = -\frac{1}{4}\cos 4x + C$

**51.** $\displaystyle\int \frac{1}{\theta^2}\cos\frac{1}{\theta}\ d\theta = -\int \cos\frac{1}{\theta}\left(-\frac{1}{\theta^2}\right) d\theta = -\sin\frac{1}{\theta} + C$

**53.** $\displaystyle\int \sin 2x \cos 2x\ dx = \frac{1}{2}\int (\sin 2x)(2\cos 2x)\ dx = \frac{1}{2}\frac{(\sin 2x)^2}{2} + C = \frac{1}{4}\sin^2 2x + C\ \ \text{OR}$

$\displaystyle\int \sin 2x \cos 2x\ dx = -\frac{1}{2}\int (\cos 2x)(-2\sin 2x)\ dx = -\frac{1}{2}\frac{(\cos 2x)^2}{2} + C_1 = -\frac{1}{4}\cos^2 2x + C_1\ \ \text{OR}$

$\displaystyle\int \sin 2x \cos 2x\ dx = \frac{1}{2}\int 2\sin 2x \cos 2x\ dx = \frac{1}{2}\int \sin 4x\ dx = -\frac{1}{8}\cos 4x + C_2$

**55.** $\displaystyle\int \tan^4 x \sec^2 x\ dx = \frac{\tan^5 x}{5} + C = \frac{1}{5}\tan^5 x + C$

**57.** $\displaystyle\int \frac{\csc^2 x}{\cot^3 x}\ dx = -\int (\cot x)^{-3}\left(-\csc^2 x\right) dx$

$$= -\frac{(\cot x)^{-2}}{-2} + C = \frac{1}{2\cot^2 x} + C = \frac{1}{2}\tan^2 x + C = \frac{1}{2}\left(\sec^2 x - 1\right) + C = \frac{1}{2}\sec^2 x + C_1$$

**59.** $\displaystyle\int \cot^2 x\ dx = \int \left(\csc^2 x - 1\right) dx = -\cot x - x + C$

**61.** $f(x) = \displaystyle\int -\sin\frac{x}{2}\ dx = 2\cos\frac{x}{2} + C$

Because $f(0) = 6 = 2\cos\left(\dfrac{0}{2}\right) + C,\ C = 4.$ So,

$$f(x) = 2\cos\frac{x}{2} + 4.$$

**63.** $f'(x) = 2\sin 4x,\ \left(\dfrac{\pi}{4}, \dfrac{-1}{2}\right)$

$$f(x) = \frac{-1}{2}\cos 4x + C$$

$$f\left(\frac{\pi}{4}\right) = \frac{-1}{2}\cos\left(4\left(\frac{\pi}{4}\right)\right) + C = \frac{-1}{2}$$

$$-\frac{1}{2}(-1) + C = \frac{-1}{2}$$

$$C = -1$$

$$f(x) = -\frac{1}{2}\cos 4x - 1$$

**65.** $f'(x) = 2x(4x^2 - 10)^2$, $(2, 10)$

$$f(x) = \frac{(4x^2 - 10)^3}{12} + C = \frac{2(2x^2 - 5)^3}{3} + C$$

$$f(2) = \frac{2(8 - 5)^3}{3} + C = 18 + C = 10 \Rightarrow C = -8$$

$$f(x) = \frac{2}{3}(2x^2 - 5)^3 - 8$$

**67.** $u = x + 6$, $x = u - 6$, $dx = du$

$$\int x\sqrt{x + 6}\, dx = \int (u - 6)\sqrt{u}\, du$$

$$= \int (u^{3/2} - 6u^{1/2})\, du$$

$$= \frac{2}{5}u^{5/2} - 4u^{3/2} + C$$

$$= \frac{2u^{3/2}}{5}(u - 10) + C$$

$$= \frac{2}{5}(x + 6)^{3/2}\big[(x + 6) - 10\big] + C$$

$$= \frac{2}{5}(x + 6)^{3/2}(x - 4) + C$$

**69.** $u = 1 - x$, $x = 1 - u$, $dx = -du$

$$\int x^2\sqrt{1 - x}\, dx = -\int (1 - u)^2\sqrt{u}\, du$$

$$= -\int (u^{1/2} - 2u^{3/2} + u^{5/2})\, du$$

$$= -\left(\frac{2}{3}u^{3/2} - \frac{4}{5}u^{5/2} + \frac{2}{7}u^{7/2}\right) + C$$

$$= -\frac{2u^{3/2}}{105}(35 - 42u + 15u^2) + C$$

$$= -\frac{2}{105}(1 - x)^{3/2}\big[35 - 42(1 - x) + 15(1 - x)^2\big] + C$$

$$= -\frac{2}{105}(1 - x)^{3/2}(15x^2 + 12x + 8) + C$$

**71.** $u = 2x - 1$, $x = \frac{1}{2}(u + 1)$, $dx = \frac{1}{2}\, du$

$$\int \frac{x^2 - 1}{\sqrt{2x - 1}}\, dx = \int \frac{\big[(1/2)(u + 1)\big]^2 - 1}{\sqrt{u}}\frac{1}{2}\, du$$

$$= \frac{1}{8}\int u^{-1/2}\big[(u^2 + 2u + 1) - 4\big]\, du$$

$$= \frac{1}{8}\int (u^{3/2} + 2u^{1/2} - 3u^{-1/2})\, du$$

$$= \frac{1}{8}\left(\frac{2}{5}u^{5/2} + \frac{4}{3}u^{3/2} - 6u^{1/2}\right) + C$$

$$= \frac{u^{1/2}}{60}(3u^2 + 10u - 45) + C$$

$$= \frac{\sqrt{2x - 1}}{60}\big[3(2x - 1)^2 + 10(2x - 1) - 45\big] + C$$

$$= \frac{1}{60}\sqrt{2x - 1}(12x^2 + 8x - 52) + C$$

$$= \frac{1}{15}\sqrt{2x - 1}(3x^2 + 2x - 13) + C$$

**73.** $u = x + 1, \ x = u - 1, \ dx = du$

$$\int \frac{-x}{(x + 1) - \sqrt{x + 1}} \, dx = \int \frac{-(u - 1)}{u - \sqrt{u}} \, du$$

$$= -\int \frac{(\sqrt{u} + 1)(\sqrt{u} - 1)}{\sqrt{u}(\sqrt{u} - 1)} \, du$$

$$= -\int \left(1 + u^{-1/2}\right) du$$

$$= -\left(u + 2u^{1/2}\right) + C$$

$$= -u - 2\sqrt{u} + C$$

$$= -(x + 1) - 2\sqrt{x + 1} + C$$

$$= -x - 2\sqrt{x + 1} - 1 + C$$

$$= -\left(x + 2\sqrt{x + 1}\right) + C_1$$

where $C_1 = -1 + C$.

**75.** Let $u = x^2 + 1, \ du = 2x \, dx$.

$$\int_{-1}^{1} x\left(x^2 + 1\right)^3 dx = \frac{1}{2} \int_{-1}^{1} \left(x^2 + 1\right)^3 (2x) \, dx = \left[\frac{1}{8}\left(x^2 + 1\right)^4\right]_{-1}^{1} = 0$$

**77.** Let $u = x^3 + 1, du = 3x^2 \, dx$.

$$\int_{1}^{2} 2x^2 \sqrt{x^3 + 1} \, dx = 2 \cdot \frac{1}{3} \int_{1}^{2} \left(x^3 + 1\right)^{1/2} (3x^2) \, dx = \left[\frac{\left(x^3 + 1\right)^{3/2}}{3/2}\right]_{1}^{2} = \frac{4}{9}\left[\left(x^3 + 1\right)^{3/2}\right]_{1}^{2} = \frac{4}{9}\left[27 - 2\sqrt{2}\right] = 12 - \frac{8}{9}\sqrt{2}$$

**79.** Let $u = 2x + 1, \ du = 2 \, dx$.

$$\int_{0}^{4} \frac{1}{\sqrt{2x + 1}} \, dx = \frac{1}{2} \int_{0}^{4} (2x + 1)^{-1/2} (2) \, dx = \left[\sqrt{2x + 1}\right]_{0}^{4} = \sqrt{9} - \sqrt{1} = 2$$

**81.** Let $u = 1 + \sqrt{x}, du = \dfrac{1}{2\sqrt{x}} \, dx$.

$$\int_{1}^{9} \frac{1}{\sqrt{x}\left(1 + \sqrt{x}\right)^2} \, dx = 2 \int_{1}^{9} \left(1 + \sqrt{x}\right)^{-2} \left(\frac{1}{2\sqrt{x}}\right) dx = \left[-\frac{2}{1 + \sqrt{x}}\right]_{1}^{9} = -\frac{1}{2} + 1 = \frac{1}{2}$$

**83.** $u = 2 - x, \ x = 2 - u, \ dx = -du$

When $x = 1, u = 1$. When $x = 2, u = 0$.

$$\int_{1}^{2} (x - 1)\sqrt{2 - x} \, dx = \int_{1}^{0} -[(2 - u) - 1]\sqrt{u} \, du = \int_{1}^{0} \left(u^{3/2} - u^{1/2}\right) du = \left[\frac{2}{5}u^{5/2} - \frac{2}{3}u^{3/2}\right]_{1}^{0} = -\left[\frac{2}{5} - \frac{2}{3}\right] = \frac{4}{15}$$

**85.** $\displaystyle\int_{0}^{\pi/2} \cos\left(\frac{2}{3}x\right) dx = \left[\frac{3}{2}\sin\left(\frac{2}{3}x\right)\right]_{0}^{\pi/2} = \frac{3}{2}\left(\frac{\sqrt{3}}{2}\right) = \frac{3\sqrt{3}}{4}$

**87.** $\dfrac{dy}{dx} = 18x^2\left(2x^3 + 1\right)^2, (0, 4)$

$$y = 3\int \left(2x^3 + 1\right)^2 \left(6x^2\right) dx \quad \left(u = 2x^3 + 1\right)$$

$$y = 3\frac{\left(2x^3 + 1\right)^3}{3} + C = \left(2x^3 + 1\right)^3 + C$$

$$4 = 1^3 + C \Rightarrow C = 3$$

$$y = \left(2x^3 + 1\right)^3 + 3$$

**89.** $\dfrac{dy}{dx} = \dfrac{2x}{\sqrt{2x^2 - 1}}, (5, 4)$

$$y = \frac{1}{2}\int \left(2x^2 - 1\right)^{-1/2} (4x\,dx) \quad \left(u = 2x^2 - 1\right)$$

$$y = \frac{1}{2}\frac{\left(2x^2 - 1\right)^{1/2}}{1/2} + C = \sqrt{2x^2 - 1} + C$$

$$4 = \sqrt{49} + C = 7 + C \Rightarrow C = -3$$

$$y = \sqrt{2x^2 - 1} - 3$$

**91.** $u = x + 1, \ x = u - 1, \ dx = du$

When $x = 0, u = 1$. When $x = 7, u = 8$.

$$\text{Area} = \int_0^7 x\sqrt[3]{x + 1}\, dx = \int_1^8 (u - 1)\sqrt[3]{u}\, du$$

$$= \int_1^8 \left(u^{4/3} - u^{1/3}\right) du$$

$$= \left[\frac{3}{7}u^{7/3} - \frac{3}{4}u^{4/3}\right]_1^8$$

$$= \left(\frac{384}{7} - 12\right) - \left(\frac{3}{7} - \frac{3}{4}\right)$$

$$= \frac{1209}{28}$$

**93.** $\text{Area} = \int_0^\pi (2\sin x + \sin 2x)\, dx$

$$= -\left[2\cos x + \frac{1}{2}\cos 2x\right]_0^\pi = 4$$

**95.** $\text{Area} = \int_{\pi/2}^{2\pi/3} \sec^2\left(\frac{x}{2}\right) dx$

$$= 2\int_{\pi/2}^{2\pi/3} \sec^2\left(\frac{x}{2}\right)\left(\frac{1}{2}\right) dx$$

$$= \left[2\tan\left(\frac{x}{2}\right)\right]_{\pi/2}^{2\pi/3} = 2\left(\sqrt{3} - 1\right)$$

**97.** $\displaystyle\int_0^6 \dfrac{x}{\sqrt{4x + 1}}\, dx \approx 4.667 = \dfrac{14}{3}$

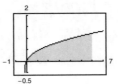

**99.** $\displaystyle\int_3^7 x\sqrt{x - 3}\, dx \approx 28.8 = \dfrac{144}{5}$

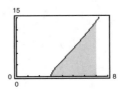

**101.** $\displaystyle\int_1^4 \left(\theta + \sin\left(\dfrac{\theta}{4}\right)\right) d\theta \approx 9.2144$

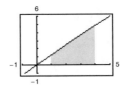

**103.** $f(x) = x^2\left(x^2 + 1\right)$ is even.

$$\int_{-2}^2 x^2\left(x^2 + 1\right) dx = 2\int_0^2 \left(x^4 + x^2\right) dx = 2\left[\frac{x^5}{5} + \frac{x^3}{3}\right]_0^2$$

$$= 2\left[\frac{32}{5} + \frac{8}{3}\right] = \frac{272}{15}$$

**105.** $f(x) = \sin^2 x \cos x$ is even.

$$\int_{-\pi/2}^{\pi/2} \sin^2 x \cos x\, dx = \int_0^{\pi/2} \sin^2 x (\cos x)\, dx$$

$$= 2\left[\frac{\sin^3 x}{3}\right]_0^{\pi/2} = \frac{2}{3}$$

**107.** $\displaystyle\int_0^4 x^2\, dx = \left[\dfrac{x^3}{3}\right]_0^4 = \dfrac{64}{3}$; the function $x^2$ is an even function.

(a) $\displaystyle\int_{-4}^0 x^2\, dx = \int_0^4 x^2\, dx = \dfrac{64}{3}$

(b) $\displaystyle\int_{-4}^4 x^2\, dx = 2\int_0^4 x^2\, dx = \dfrac{128}{3}$

(c) $\displaystyle\int_0^4 \left(-x^2\right) dx = -\int_0^4 x^2\, dx = -\dfrac{64}{3}$

(d) $\displaystyle\int_{-4}^0 3x^2\, dx = 3\int_0^4 x^2\, dx = 64$

**109.** $\int_{-3}^{3} \left(x^3 + 4x^2 - 2x - 6\right) dx = \int_{-3}^{3} \left(x^3 - 3x\right) dx + \int_{-3}^{3} \left(4x^2 - 6\right) dx = 0 + 2\int_{0}^{3} \left(4x^2 - 6\right) dx = 2\left[\frac{4}{3}x^3 - 6x\right]_{0}^{3} = 36$

**111.** If $u = 5 - x^2$, then $du = -2x\, dx$ and $\int x\left(5 - x^2\right)^3 dx = -\frac{1}{2}\int \left(5 - x^2\right)^3 (-2x)\, dx = -\frac{1}{2}\int u^3\, du.$

**113.** Let $u = 2x$, $du = 2\, dx$

When $x = 0$, $u = 0$, and when $x = 4$, $u = 8$. So, $\int_{0}^{4} f(2x)\, dx = \int_{0}^{8} f(u)\frac{1}{2}\, du = \frac{1}{2}\int_{0}^{8} f(u)\, du = \frac{1}{2}(32) = 16$

**115.**  $\dfrac{dQ}{dt} = k(100 - t)^2$

$Q(t) = \int k(100 - t)^2\, dt = -\dfrac{k}{3}(100 - t)^3 + C$

$Q(100) = C = 0$

$Q(t) = -\dfrac{k}{3}(100 - t)^3$

$Q(0) = -\dfrac{k}{3}(100)^3 = 2{,}000{,}000 \Rightarrow k = -6$

So, $Q(t) = 2(100 - t)^3$. When

$t = 50$, $Q(50) = \$250{,}000$.

**117.** (a) $R(t) = 2.876 + 2.202 \sin(0.576t + 0.847)$

$R'(t) = 2.202 \cos(0.576t + 0.847)(0.576)$

$\qquad = 1.268352 \cos(0.576t + 0.847)$

$R'(t) = 0$ when

$0.576t + 0.847 = \dfrac{\pi}{2}, \dfrac{3\pi}{2}, \dfrac{5\pi}{2}, \dfrac{7\pi}{2}, \dots .$

So, $t = 1.257, 6.711$ (others are outside $[0, 12]$).

By the First Derivative Test,

$(6.7, 0.7)$ is a relative minimum (July)

$(1.3, 5.1)$ is a relative maximum (February)

(b) $\int_{0}^{12} R(t)\, dt \approx 36.68$ inches

(c) $\frac{1}{3}\int_{9}^{12} R(t)\, dt \approx 3.99$ inches

**119.** (a)

Maximum flow: $R \approx 61.713$ at $t = 9.36$.

$[(18.861, 61.178)$ is a relative maximum.$]$

(b) Volume $= \int_{0}^{24} R(t)\, dt \approx 1272$ thousand gallons

**121.** $u = 1 - x$, $x = 1 - u$, $dx = -du$

When $x = a$, $u = 1 - a$. When $x = b$, $u = 1 - b$.

$P_{a,b} = \int_{a}^{b} \dfrac{15}{4}x\sqrt{1 - x}\, dx = \dfrac{15}{4}\int_{1-a}^{1-b} -(1 - u)\sqrt{u}\, du$

$= \dfrac{15}{4}\int_{1-a}^{1-b} \left(u^{3/2} - u^{1/2}\right) du = \dfrac{15}{4}\left[\dfrac{2}{5}u^{5/2} - \dfrac{2}{3}u^{3/2}\right]_{1-a}^{1-b} = \dfrac{15}{4}\left[\dfrac{2u^{3/2}}{15}(3u - 5)\right]_{1-a}^{1-b} = \left[-\dfrac{(1 - x)^{3/2}}{2}(3x + 2)\right]_{a}^{b}$

(a) $P_{0.50, 0.75} = \left[-\dfrac{(1 - x)^{3/2}}{2}(3x + 2)\right]_{0.50}^{0.75} = 0.353 = 35.3\%$

(b) $P_{0,b} = \left[ -\frac{(1-x)^{3/2}}{2}(3x+2) \right]_{0}^{b} = -\frac{(1-b)^{3/2}}{2}(3b+2) + 1 = 0.5$

$$(1-b)^{3/2}(3b+2) = 1$$

$$b \approx 0.586 = 58.6\%$$

**123.** (a) $C = 0.1 \int_{8}^{20} \left[ 12 \sin \frac{\pi(t-8)}{12} \right] dt = \left[ -\frac{14.4}{\pi} \cos \frac{\pi(t-8)}{12} \right]_{8}^{20} = \frac{-14.4}{\pi}(-1-1) \approx \$9.17$

(b) $C = 0.1 \int_{10}^{18} \left[ 12 \sin \frac{\pi(t-8)}{12} - 6 \right] dt = \left[ -\frac{14.4}{\pi} \cos \frac{\pi(t-8)}{12} - 0.6t \right]_{10}^{18}$

$$= \left[ \frac{-14.4}{\pi}\left( \frac{-\sqrt{3}}{2} \right) - 10.8 \right] - \left[ \frac{-14.4}{\pi}\left( \frac{\sqrt{3}}{2} \right) - 6 \right] \approx \$3.14$$

Savings $\approx 9.17 - 3.14 = \$6.03$.

**125.** (a)

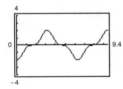

(b) $g$ is nonnegative because the graph of $f$ is positive at the beginning, and generally has more positive sections than negative ones.

(c) The points on $g$ that correspond to the extrema of $f$ are points of inflection of $g$.

(d) No, some zeros of $f$, like $x = \pi/2$, do not correspond to an extrema of $g$. The graph of $g$ continues to increase after $x = \pi/2$ because $f$ remains above the $x$-axis.

(e) The graph of $h$ is that of $g$ shifted 2 units downward.

$$g(t) = \int_{0}^{t} f(x)\, dx = \int_{0}^{\pi/2} f(x)\, dx + \int_{\pi/2}^{t} f(x)\, dx = 2 + h(t).$$

**127.** (a) Let $u = 1 - x,\ du = -dx,\ x = 1 - u$

$$x = 0 \Rightarrow u = 1,\ x = 1 \Rightarrow u = 0$$

$$\int_{0}^{1} x^2(1-x)^5\, dx = \int_{1}^{0} (1-u)^2 u^5(-du)$$

$$= \int_{0}^{1} u^5(1-u)^2\, du$$

$$= \int_{0}^{1} x^5(1-x)^2\, dx$$

(b) Let $u = 1 - x,\ du = -dx,\ x = 1 - u$

$$x = 0 \Rightarrow u = 1,\ x = 1 \Rightarrow u = 0$$

$$\int_{0}^{1} x^a(1-x)^b\, dx = \int_{1}^{0} (1-u)^a u^b(-du)$$

$$= \int_{0}^{1} u^b(1-u)^a\, du$$

$$= \int_{0}^{1} x^b(1-x)^a\, dx$$

**129.** False

$$\int (2x+1)^2\, dx = \frac{1}{2}\int (2x+1)^2 2\, dx = \frac{1}{6}(2x+1)^3 + C$$

**131.** True

$$\int_{-10}^{10} \left(ax^3 + bx^2 + cx + d\right) dx = \int_{-10}^{10} \left(ax^3 + cx\right) dx + \int_{-10}^{10} \left(bx^2 + d\right) dx = 0 + 2\int_{0}^{10} \left(bx^2 + d\right) dx$$

                                                     Odd                      Even

**133.** True

$$4\int \sin x \cos x \, dx = 2 \int \sin 2x \, dx = -\cos 2x + C$$

**135.** Let $u = cx, du = c \, dx$:

$$c\int_{a}^{b} f(cx) \, dx = c\int_{ca}^{cb} f(u) \, \frac{du}{c}$$

$$= \int_{ca}^{cb} f(u) \, du$$

$$= \int_{ca}^{cb} f(x) \, dx$$

**137.** Because $f$ is odd, $f(-x) = -f(x)$. Then

$$\int_{-a}^{a} f(x) \, dx = \int_{-a}^{0} f(x) \, dx + \int_{0}^{a} f(x) \, dx$$

$$= -\int_{0}^{-a} f(x) \, dx + \int_{0}^{a} f(x) \, dx.$$

Let $x = -u,\ dx = -du$ in the first integral.

When $x = 0,\ u = 0$. When $x = -a, u = a$.

$$\int_{-a}^{1} f(x) \, dx = -\int_{0}^{a} f(-u)(-du) + \int_{0}^{a} f(x) \, dx$$

$$= -\int_{0}^{a} f(u) \, du + \int_{0}^{a} f(x) \, dx = 0$$

**139.** Let $f(x) = a_0 + a_1 x + a_2 x^2 + \cdots + a_n x^n$.

$$\int_{0}^{1} f(x) \, dx = \left[ a_0 x + a_1 \frac{x^2}{2} + a_2 \frac{x^3}{3} + \cdots + a_n \frac{x^{n+1}}{n+1} \right]_{0}^{1}$$

$$= a_0 + \frac{a_1}{2} + \frac{a_2}{3} + \cdots + \frac{a_n}{n+1} = 0 \, (\text{Given})$$

By the Mean Value Theorem for Integrals, there exists $c$ in $[0, 1]$ such that

$$\int_{0}^{1} f(x) \, dx = f(c)(1 - 0)$$

$$0 = f(c).$$

So the equation has at least one real zero.

## Section 4.6   Numerical Integration

**1.** Exact: $\displaystyle\int_{0}^{2} x^2 \, dx = \left[\tfrac{1}{3}x^3\right]_{0}^{2} = \tfrac{8}{3} \approx 2.6667$

    Trapezoidal: $\displaystyle\int_{0}^{2} x^2 \, dx \approx \tfrac{1}{4}\left[0 + 2\left(\tfrac{1}{2}\right)^2 + 2(1)^2 + 2\left(\tfrac{3}{2}\right)^2 + (2)^2\right] = \tfrac{11}{4} = 2.7500$

    Simpson's: $\displaystyle\int_{0}^{2} x^2 \, dx \approx \tfrac{1}{6}\left[0 + 4\left(\tfrac{1}{2}\right)^2 + 2(1)^2 + 4\left(\tfrac{3}{2}\right)^2 + (2)^2\right] = \tfrac{8}{3} \approx 2.6667$

**3.** Exact: $\displaystyle\int_{0}^{2} x^3 \, dx = \left[\frac{x^4}{4}\right]_{0}^{2} = 4.0000$

    Trapezoidal: $\displaystyle\int_{0}^{2} x^3 \, dx \approx \frac{1}{4}\left[0 + 2\left(\frac{1}{2}\right)^3 + 2(1)^3 + 2\left(\frac{3}{2}\right)^3 + (2)^3\right] = \frac{17}{4} = 4.2500$

    Simpson's: $\displaystyle\int_{0}^{2} x^3 \, dx \approx \frac{1}{6}\left[0 + 4\left(\frac{1}{2}\right)^3 + 2(1)^3 + 4\left(\frac{3}{2}\right)^3 + (2)^3\right] = \frac{24}{6} = 4.0000$

**5.** Exact: $\int_1^3 x^3 \, dx = \left[ \dfrac{x^4}{4} \right]_1^3 = \dfrac{81}{4} - \dfrac{1}{4} = 20$

Trapezoidal: $\int_1^3 x^3 \, dx \approx \dfrac{1}{6}\left[ 1 + 2\left(\dfrac{4}{3}\right)^3 + 2\left(\dfrac{5}{3}\right)^3 + 2(2)^3 + 2\left(\dfrac{7}{3}\right)^3 + 2\left(\dfrac{8}{3}\right)^3 + 27 \right] \approx 20.2222$

Simpson's: $\int_1^3 x^3 \, dx \approx \dfrac{1}{9}\left[ 1 + 4\left(\dfrac{4}{3}\right)^3 + 2\left(\dfrac{5}{3}\right)^3 + 4(2)^3 + 2\left(\dfrac{7}{3}\right)^3 + 4\left(\dfrac{8}{3}\right)^3 + 27 \right] = 20.0000$

**7.** Exact: $\int_4^9 \sqrt{x} \, dx = \left[ \dfrac{2}{3}x^{3/2} \right]_4^9 = 18 - \dfrac{16}{3} = \dfrac{38}{3} \approx 12.6667$

Trapezoidal: $\int_4^9 \sqrt{x} \, dx \approx \dfrac{5}{16}\left[ 2 + 2\sqrt{\dfrac{37}{8}} + 2\sqrt{\dfrac{21}{4}} + 2\sqrt{\dfrac{47}{8}} + 2\sqrt{\dfrac{26}{4}} + 2\sqrt{\dfrac{57}{8}} + 2\sqrt{\dfrac{31}{4}} + 2\sqrt{\dfrac{67}{8}} + 3 \right] \approx 12.6640$

Simpson's: $\int_4^9 \sqrt{x} \, dx \approx \dfrac{5}{24}\left[ 2 + 4\sqrt{\dfrac{37}{8}} + \sqrt{21} + 4\sqrt{\dfrac{47}{8}} + \sqrt{26} + 4\sqrt{\dfrac{57}{8}} + \sqrt{31} + 4\sqrt{\dfrac{67}{8}} + 3 \right] \approx 12.6667$

**9.** Exact: $\int_0^1 \dfrac{2}{(x+2)^2} \, dx = \left[ \dfrac{-2}{(x+2)} \right]_0^1 = \dfrac{-2}{3} + \dfrac{2}{2} = \dfrac{1}{3}$

Trapezoidal: $\int_0^1 \dfrac{2}{(x+2)^2} \, dx \approx \dfrac{1}{8}\left[ \dfrac{1}{2} + 2\left(\dfrac{2}{((1/4)+2)^2}\right) + 2\left(\dfrac{2}{((1/2)+2)^2}\right) + 2\left(\dfrac{2}{((3/4)+2)^2}\right) + \dfrac{2}{9} \right]$

$= \dfrac{1}{8}\left[ \dfrac{1}{2} + 2\left(\dfrac{32}{81}\right) + 2\left(\dfrac{8}{25}\right) + 2\left(\dfrac{32}{121}\right) + \dfrac{2}{9} \right] \approx 0.3352$

Simpson's: $\int_0^1 \dfrac{2}{(x+2)^2} \, dx \approx \dfrac{1}{12}\left[ \dfrac{1}{2} + 4\left(\dfrac{2}{((1/4)+2)^2}\right) + 2\left(\dfrac{2}{((1/2)+2)^2}\right) + 4\left(\dfrac{2}{((3/4)+2)^2}\right) + \dfrac{2}{9} \right]$

$= \dfrac{1}{12}\left[ \dfrac{1}{2} + 4\left(\dfrac{32}{81}\right) + 2\left(\dfrac{8}{25}\right) + 4\left(\dfrac{32}{121}\right) + \dfrac{2}{9} \right] \approx 0.3334$

**11.** Trapezoidal: $\int_0^2 \sqrt{1+x^3} \, dx \approx \dfrac{1}{4}\left[ 1 + 2\sqrt{1+\left(\dfrac{1}{8}\right)} + 2\sqrt{2} + 2\sqrt{1+\left(\dfrac{27}{8}\right)} + 3 \right] \approx 3.283$

Simpson's: $\int_0^2 \sqrt{1+x^3} \, dx \approx \dfrac{1}{6}\left[ 1 + 4\sqrt{1+\left(\dfrac{1}{8}\right)} + 2\sqrt{2} + 4\sqrt{1+\left(\dfrac{27}{8}\right)} + 3 \right] \approx 3.240$

Graphing utility: 3.241

**13.** $\int_0^1 \sqrt{x}\sqrt{1-x} \, dx = \int_0^1 \sqrt{x(1-x)} \, dx$

Trapezoidal: $\int_0^1 \sqrt{x(1-x)} \, dx \approx \dfrac{1}{8}\left[ 0 + 2\sqrt{\dfrac{1}{4}\left(1-\dfrac{1}{4}\right)} + 2\sqrt{\dfrac{1}{2}\left(1-\dfrac{1}{2}\right)} + 2\sqrt{\dfrac{3}{4}\left(1-\dfrac{3}{4}\right)} \right] \approx 0.342$

Simpson's: $\int_0^1 \sqrt{x(1-x)} \, dx \approx \dfrac{1}{12}\left[ 0 + 4\sqrt{\dfrac{1}{4}\left(1-\dfrac{1}{4}\right)} + 2\sqrt{\dfrac{1}{2}\left(1-\dfrac{1}{2}\right)} + 4\sqrt{\dfrac{3}{4}\left(1-\dfrac{3}{4}\right)} \right] \approx 0.372$

Graphing utility: 0.393

**15.** Trapezoidal: $\int_0^{\sqrt{\pi/2}} \sin(x^2) \, dx \approx \dfrac{\sqrt{\pi/2}}{8}\left[ \sin 0 + 2\sin\left(\dfrac{\sqrt{\pi/2}}{4}\right)^2 + 2\sin\left(\dfrac{\sqrt{\pi/2}}{2}\right)^2 + 2\sin\left(\dfrac{3\sqrt{\pi/2}}{4}\right)^2 + \sin\left(\sqrt{\dfrac{\pi}{2}}\right)^2 \right] \approx 0.550$

Simpson's: $\int_0^{\sqrt{\pi/2}} \sin(x^2) \, dx \approx \dfrac{\sqrt{\pi/2}}{12}\left[ \sin 0 + 4\sin\left(\dfrac{\sqrt{\pi/2}}{4}\right)^2 + 2\sin\left(\dfrac{\sqrt{\pi/2}}{2}\right)^2 + 4\sin\left(\dfrac{3\sqrt{\pi/2}}{4}\right)^2 + \sin\left(\sqrt{\dfrac{\pi}{2}}\right)^2 \right] \approx 0.548$

Graphing utility: 0.549

**17.** Trapezoidal: $\int_{3}^{3.1} \cos x^2 \, dx \approx \frac{0.1}{8}\left[ \cos(3)^2 + 2\cos(3.025)^2 + 2\cos(3.05)^2 + 2\cos(3.075)^2 + \cos(3.1)^2 \right] \approx -0.098$

Simpson's: $\int_{3}^{3.1} \cos x^2 \, dx \approx \frac{0.1}{12}\left[ \cos(3)^2 + 4\cos(3.025)^2 + 2\cos(3.05)^2 + 4\cos(3.075)^2 + \cos(3.1)^2 \right] \approx -0.098$

Graphing utility: $-0.098$

**19.** Trapezoidal: $\int_{0}^{\pi/4} x \tan x \, dx \approx \frac{\pi}{32}\left[ 0 + 2\left(\frac{\pi}{16}\right)\tan\left(\frac{\pi}{16}\right) + 2\left(\frac{2\pi}{16}\right)\tan\left(\frac{2\pi}{16}\right) + 2\left(\frac{3\pi}{16}\right)\tan\left(\frac{3\pi}{16}\right) + \frac{\pi}{4} \right] \approx 0.194$

Simpson's: $\int_{0}^{\pi/4} x \tan x \, dx \approx \frac{\pi}{48}\left[ 0 + 4\left(\frac{\pi}{16}\right)\tan\left(\frac{\pi}{16}\right) + 2\left(\frac{2\pi}{16}\right)\tan\left(\frac{2\pi}{16}\right) + 4\left(\frac{3\pi}{16}\right)\tan\left(\frac{3\pi}{16}\right) + \frac{\pi}{4} \right] \approx 0.186$

Graphing utility: $0.186$

**21.** Trapezoidal: Linear polynomials

Simpson's: Quadratic polynomials

**23.**  $f(x) = 2x^3$

$f'(x) = 6x^2$

$f''(x) = 12x$

$f'''(x) = 12$

$f^{(4)}(x) = 0$

(a) Trapezoidal: Error $\leq \dfrac{(3-1)^3}{12(4^2)}(36) = 1.5$ because

$\left| f''(x) \right|$ is maximum in $[1, 3]$ when $x = 3$.

(b) Simpson's: Error $\leq \dfrac{(3-1)^5}{180(4^4)}(0) = 0$ because

$f^{(4)}(x) = 0.$

**25.**  $f(x) = \dfrac{1}{x+1}$

$f'(x) = \dfrac{-1}{(x+1)^2}$

$f''(x) = \dfrac{2}{(x+1)^3}$

$f'''(x) = \dfrac{-6}{(x+1)^4}$

$f^{(4)}(x) = \dfrac{24}{(x+1)^5}$

(a) Trapezoidal: Error $\leq \dfrac{(1-0)^2}{12(4^2)}(2) = \dfrac{1}{96} \approx 0.01$

because $f''(x)$ is maximum in $[0, 1]$ when $x = 0$.

(b) Simpson's: Error $\leq \dfrac{(1-0)^5}{180(4^4)}(24) = \dfrac{1}{1920} \approx 0.0005$

because $f^{(4)}(x)$ is maximum in $[0, 1]$ when $x = 0$.

**27.**  $f(x) = \cos x$

$f'(x) = -\sin x$

$f''(x) = -\cos x$

$f'''(x) = \sin x$

$f^{(4)}(x) = \cos x$

(a) Trapezoidal: Error $\leq \dfrac{(\pi-0)^3}{12(4^2)}(1) = \dfrac{\pi^3}{192} \approx 0.1615$

because $\left| f''(x) \right|$ is at most 1 on $[0, \pi]$.

(b) Simpson's:

Error $\leq \dfrac{(\pi-0)^5}{180(4^4)}(1) = \dfrac{\pi^5}{46,080} \approx 0.006641$

because $\left| f^{(4)}(x) \right|$ is at most 1 on $[0, \pi]$.

**29.**  $f(x) = x^{-1}, \quad 1 \leq x \leq 3$

$f'(x) = -x^{-2}$

$f''(x) = 2x^{-3}$

$f'''(x) = -6x^{-4}$

$f^{(4)}(x) = 24x^{-5}$

(a) Maximum of $\left| f''(x) \right| = \left| 2x^{-3} \right|$ is 2.

Trapezoidal:

Error $\leq \dfrac{2^3}{12n^2}(2) \leq 0.00001$, $n^2 \geq 133{,}333.33$,

$n \geq 365.15$ Let $n = 366$.

(b) Maximum of $\left| f^{(4)}(x) \right| = \left| 24x^{-5} \right|$ is 24.

Simpson's: Error $\leq \dfrac{2^5}{180n^4}(24) \leq 0.00001$,

$n^4 \geq 426{,}666.67$, $n \geq 25.56$ Let $n = 26$.

**31.** $f(x) = (x + 2)^{1/2}, \quad 0 \le x \le 2$

$$f'(x) = \frac{1}{2}(x + 2)^{-1/2}$$

$$f''(x) = -\frac{1}{4}(x + 2)^{-3/2}$$

$$f'''(x) = \frac{3}{8}(x + 2)^{-5/2}$$

$$f^{(4)}(x) = \frac{-15}{16}(x + 2)^{-7/2}$$

(a) Maximum of $\left| f''(x) \right| = \left| \dfrac{-1}{4(x + 2)^{3/2}} \right|$ is

$$\frac{\sqrt{2}}{16} \approx 0.0884.$$

Trapezoidal:

$$\text{Error} \le \frac{(2 - 0)^3}{12n^2}\left(\frac{\sqrt{2}}{16}\right) \le 0.00001$$

$$n^2 \ge \frac{8\sqrt{2}}{12(16)}10^5 = \frac{\sqrt{2}}{24}10^5$$

$$n \ge 76.8. \text{ Let } n = 77.$$

(b) Maximum of $\left| f^{(4)}(x) \right| = \left| \dfrac{-15}{16(x + 2)^{7/2}} \right|$ is

$$\frac{15\sqrt{2}}{256} \approx 0.0829.$$

Simpson's:

$$\text{Error} \le \frac{2^5}{180n^4}\left(\frac{15\sqrt{2}}{256}\right) \le 0.00001$$

$$n^4 \ge \frac{32(15)\sqrt{2}}{180(256)}10^5$$

$$= \frac{\sqrt{2}}{96}10^5$$

$$n \ge 6.2. \text{ Let } n = 8 \text{ (even)}.$$

**33.** $f(x) = \cos(\pi x), \quad 0 \le x \le 1$

$$f'(x) = -\pi \sin(\pi x)$$

$$f''(x) = -\pi^2 \cos(\pi x)$$

$$f'''(x) = \pi^3 \sin(\pi x)$$

$$f^{(4)}(x) = \pi^4 \cos(\pi x)$$

(a) Maximum of $\left| f''(x) \right| = \left| -\pi^2 \cos(\pi x) \right|$ is $\pi^2$.

Trapezoidal:

$$\text{Error} \le \frac{(1 - 0)^3}{12n^2}\pi^2 \le 0.00001$$

$$n^2 \ge \frac{\pi^2}{12}10^5$$

$$n \ge 286.8. \text{ Let } n = 287$$

(b) Maximum of $\left| f^{(4)}(x) \right| = \left| \pi^4 \cos(\pi x) \right|$ is $\pi^4$.

Simpson's: Error $\le \dfrac{1}{180n^4}\pi^4 \le 0.00001$

$$n^4 \ge \frac{\pi^4}{180}10^5$$

$$n \ge 15.3. \text{ Let } n = 16.$$

**35.** $f(x) = \sqrt{1 + x}$

(a) $f''(x) = -\dfrac{1}{4(1 + x)^{3/2}}$ in $[0, 2]$.

$\left| f''(x) \right|$ is maximum when $x = 0$ and $\left| f''(0) \right| = \dfrac{1}{4}$.

Trapezoidal: Error $\le \dfrac{8}{12n^2}\left(\dfrac{1}{4}\right) \le 0.00001$,

$n^2 \ge 16,666.67, \ n \ge 129.10$; let $n = 130$.

(b) $f^{(4)}(x) = -\dfrac{15}{16(1 + x)^{7/2}}$ in $[0, 2]$

$\left| f^{(4)}(x) \right|$ is maximum when $x = 0$ and

$\left| f^{(4)}(0) \right| = \dfrac{15}{16}$.

Simpson's: Error $\le \dfrac{32}{180n^4}\left(\dfrac{15}{16}\right) \le 0.00001$,

$n^4 \ge 16,666.67, \ n \ge 11.36$; let $n = 12$.

**37.** $f(x) = \tan(x^2)$

(a) $f''(x) = 2\sec^2(x^2)\left[1 + 4x^2 \tan(x^2)\right]$ in $[0, 1]$.

$\left| f''(x) \right|$ is maximum when $x = 1$ and $\left| f''(1) \right| \approx 49.5305$.

Trapezoidal: Error $\le \dfrac{(1 - 0)^3}{12n^2}(49.5305) \le 0.00001, \ n^2 \ge 412,754.17, \ n \ge 642.46$; let $n = 643$.

(b)   $f^{(4)}(x) = 8\sec^2(x^2)\left[12x^2 + (3 + 32x^4)\tan(x^2) + 36x^2\tan^2(x^2) + 48x^4\tan^3(x^2)\right]$ in $[0, 1]$

$\left|f^{(4)}(x)\right|$ is maximum when $x = 1$ and $\left|f^{(4)}(1)\right| \approx 9184.4734$.

Simpson's: Error $\leq \dfrac{(1 - 0)^5}{180n^4}(9184.4734) \leq 0.00001$, $n^4 \geq 5{,}102{,}485.22$, $n \geq 47.53$; let $n = 48$.

**39.**   $n = 4$, $b - a = 4 - 0 = 4$

(a)   $\displaystyle\int_0^4 f(x)\,dx \approx \frac{4}{8}\left[3 + 2(7) + 2(9) + 2(7) + 0\right] = \frac{1}{2}(49) = \frac{49}{2} = 24.5$

(b)   $\displaystyle\int_0^4 f(x)\,dx \approx \frac{4}{12}\left[3 + 4(7) + 2(9) + 4(7) + 0\right] = \frac{77}{3} \approx 25.67$

**41.** The program will vary depending upon the computer or programmable calculator that you use.

**43.**   $f(x) = \sqrt{1 - x^2}$ on $[0, 1]$.

| $n$ | $L(n)$ | $M(n)$ | $R(n)$ | $T(n)$ | $S(n)$ |
|-----|--------|--------|--------|--------|--------|
| 4 | 0.8739 | 0.7960 | 0.6239 | 0.7489 | 0.7709 |
| 8 | 0.8350 | 0.7892 | 0.7100 | 0.7725 | 0.7803 |
| 10 | 0.8261 | 0.7881 | 0.7261 | 0.7761 | 0.7818 |
| 12 | 0.8200 | 0.7875 | 0.7367 | 0.7783 | 0.7826 |
| 16 | 0.8121 | 0.7867 | 0.7496 | 0.7808 | 0.7836 |
| 20 | 0.8071 | 0.7864 | 0.7571 | 0.7821 | 0.7841 |

**45.**   $A = \displaystyle\int_0^{\pi/2} \sqrt{x}\cos x\,dx$

Simpson's Rule: $n = 14$

$\displaystyle\int_0^{\pi/2} \sqrt{x}\cos x\,dx \approx \frac{\pi}{84}\left[\sqrt{0}\cos 0 + 4\sqrt{\frac{\pi}{28}}\cos\frac{\pi}{28} + 2\sqrt{\frac{\pi}{14}}\cos\frac{\pi}{14} + 4\sqrt{\frac{3\pi}{28}}\cos\frac{3\pi}{28} + \cdots + \sqrt{\frac{\pi}{2}}\cos\frac{\pi}{2}\right] \approx 0.701$

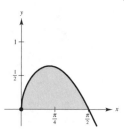

**47.** Simpson's Rule: $n = 8$

$8\sqrt{3}\displaystyle\int_0^{\pi/2}\sqrt{1 - \frac{2}{3}\sin^2\theta}\,d\theta \approx \frac{\sqrt{3}\pi}{6}\left[\sqrt{1 - \frac{2}{3}\sin^2 0} + 4\sqrt{1 - \frac{2}{3}\sin^2\frac{\pi}{16}} + 2\sqrt{1 - \frac{2}{3}\sin^2\frac{\pi}{8}} + \cdots + \sqrt{1 - \frac{2}{3}\sin^2\frac{\pi}{2}}\right]$

$\approx 17.476$

**49.** (a) Trapezoidal:

$\displaystyle\int_0^2 f(x)\,dx \approx \frac{2}{2(8)}\left[4.32 + 2(4.36) + 2(4.58) + 2(5.79) + 2(6.14) + 2(7.25) + 2(7.64) + 2(8.08) + 8.14\right] \approx 12.518$

Simpson's:

$\displaystyle\int_0^2 f(x)\,dx \approx \frac{2}{3(8)}\left[4.32 + 4(4.36) + 2(4.58) + 4(5.79) + 2(6.14) + 4(7.25) + 2(7.64) + 4(8.08) + 8.14\right] \approx 12.592$

(b) Using a graphing utility,

$y = -1.37266x^3 + 4.0092x^2 - 0.620x + 4.28$. Integrating, $\int_0^2 y \, dx \approx 12.521$.

**51.** Simpson's Rule: $n = 6$

$$\pi = 4\int_0^1 \frac{1}{1+x^2} \, dx \approx \frac{4}{3(6)}\left[1 + \frac{4}{1+(1/6)^2} + \frac{2}{1+(2/6)^2} + \frac{4}{1+(3/6)^2} + \frac{2}{1+(4/6)^2} + \frac{4}{1+(5/6)^2} + \frac{1}{2}\right] \approx 3.14159$$

**53.** Area $\approx \dfrac{120}{2(12)}\Big[75 + 2(81) + 2(84) + 2(76) + 2(67) + 2(68) + 2(69) + 2(72) + 2(68) + 2(56) + 2(42) + 2(23) + 0\Big]$

$= 7435 \text{ m}^2$

**55.** $\int_0^t \sin\sqrt{x} \, dx = 2, \; n = 10$

By trial and error, you obtain $t \approx 2.477$.

# Review Exercises for Chapter 4

**1.**

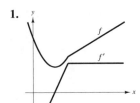

**3.** $\int\left(4x^2 + x + 3\right) dx = \frac{4}{3}x^3 + \frac{1}{2}x^2 + 3x + C$

**5.** $\int \dfrac{x^4 + 8}{x^3} \, dx = \int\left(x + 8x^{-3}\right) dx = \dfrac{1}{2}x^2 - \dfrac{4}{x^2} + C$

**7.** $\int(2x - 9\sin x) \, dx = x^2 + 9\cos x + C$

**9.** $f'(x) = -6x, \; (-1, 1)$

$f(x) = \int -6x \, dx = -3x^2 + C$

When $x = 1$:

$f(1) = -3(1)^2 + C = -2$

$C = 1$

$f(x) = 1 - 3x^2$

**11.** (a) Answers will vary. *Sample answer:*

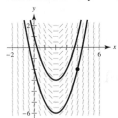

(b) $\dfrac{dy}{dx} = 2x - 4, \; (4, -2)$

$y = \int(2x - 4) \, dx = x^2 - 4x + C$

$-2 = 16 - 16 + C \Rightarrow C = -2$

$y = x^2 - 4x - 2$

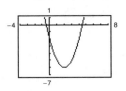

**13.** $a(t) = a$

$v(t) = \int a \, dt = at + C_1$

$v(0) = 0 + C_1 = 0$ when $C_1 = 0$.

$v(t) = at$

$s(t) = \int at \, dt = \dfrac{a}{2}t^2 + C_2$

$s(0) = 0 + C_2 = 0$ when $C_2 = 0$.

$s(t) = \dfrac{a}{2}t^2$

$s(30) = \dfrac{a}{2}(30)^2 = 3600$ or

$$a = \frac{2(3600)}{(30)^2} = 8 \text{ ft/sec}^2.$$

$v(30) = 8(30) = 240 \text{ ft/sec}$

**15.** $a(t) = -32$

$v(t) = -32t + 96$

$s(t) = -16t^2 + 96t$

(a) $v(t) = -32t + 96 = 0$ when $t = 3$ sec.

$s(3) = -144 + 288 = 144$ ft

(b) $v(t) = -32t + 96 = \frac{96}{2}$ when $t = \frac{3}{2}$ sec.

(c) $s\left(\frac{3}{2}\right) = -16\left(\frac{9}{4}\right) + 96\left(\frac{3}{2}\right) = 108$ ft

**17.** $\displaystyle\sum_{i=1}^{10} \frac{1}{3i} = \frac{1}{3(1)} + \frac{1}{3(2)} + \cdots + \frac{1}{3(10)}$

**19.** $\displaystyle\sum_{i=1}^{20} 2i = 2\left(\frac{20(21)}{2}\right) = 420$

**21.** $\displaystyle\sum_{i=1}^{20} (i+1)^2 = \sum_{i=1}^{20} (i^2 + 2i + 1)$

$= \frac{20(21)(41)}{6} = 2\frac{20(21)}{2} + 20$

$= 2870 + 420 + 20 = 3310$

**23.** (a) $\displaystyle\sum_{i=1}^{10} (2i - 1)$

(b) $\displaystyle\sum_{i=1}^{n} i^3$

(c) $\displaystyle\sum_{i=1}^{10} (4i + 2)$

**25.** $y = \dfrac{10}{x^2 + 1}, \ \Delta x = \dfrac{1}{2}, \ n = 4$

$S(n) = S(4) = \dfrac{1}{2}\left[\dfrac{10}{1} + \dfrac{10}{(1/2)^2 + 1} + \dfrac{10}{(1)^2 + 1} + \dfrac{10}{(3/2)^2 + 1}\right] \approx 13.0385$

$s(n) = s(4) = \dfrac{1}{2}\left[\dfrac{10}{(1/2)^2 + 1} + \dfrac{10}{1 + 1} + \dfrac{10}{(3/2)^2 + 1} + \dfrac{10}{2^2 + 1}\right] \approx 9.0385$

$9.0385 < \text{Area of Region} < 13.0385$

**27.** $y = 8 - 2x, \ \Delta x = \dfrac{3}{n}$, right endpoints

$\text{Area} = \displaystyle\lim_{n \to \infty} \sum_{i=1}^{n} f(ci)\Delta x$

$= \displaystyle\lim_{n \to \infty} \sum_{i=1}^{n} \left(8 - 2\left(\frac{3i}{n}\right)\right)\frac{3}{n}$

$= \displaystyle\lim_{n \to \infty} \frac{3}{n} \sum_{i=1}^{n} \left(8 - \frac{6i}{n}\right)$

$= \displaystyle\lim_{n \to \infty} \frac{3}{n}\left[8n - \frac{6}{n}\frac{n(n+1)}{2}\right]$

$= \displaystyle\lim_{n \to \infty}\left[24 - 9\frac{n+1}{n}\right] = 24 - 9 = 15$

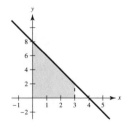

**29.** $y = 5 - x^2, \ \Delta x = \dfrac{3}{n}$

$\text{Area} = \displaystyle\lim_{n \to \infty} \sum_{i=1}^{n} f(c_i)\Delta x$

$= \displaystyle\lim_{n \to \infty} \sum_{i=1}^{n}\left[5 - \left(-2 + \frac{3i}{n}\right)^2\right]\left(\frac{3}{n}\right)$

$= \displaystyle\lim_{n \to \infty} \frac{3}{n}\sum_{i=1}^{n}\left[1 + \frac{12i}{n} - \frac{9i^2}{n^2}\right]$

$= \displaystyle\lim_{n \to \infty} \frac{3}{n}\left[n + \frac{12}{n}\frac{n(n+1)}{2} - \frac{9}{n^2}\frac{n(n+1)(2n+1)}{6}\right]$

$= \displaystyle\lim_{n \to \infty}\left[3 + 18\frac{n+1}{n} - \frac{9}{2}\frac{(n+1)(2n+1)}{n^2}\right]$

$= 3 + 18 - 9 = 12$

**31.** $x = 5y - y^2, \; 2 \le y \le 5, \; \Delta y = \dfrac{3}{n}$

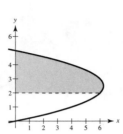

$$\text{Area} = \lim_{n \to \infty} \sum_{i=1}^{n} \left[ 5\left(2 + \frac{3i}{n}\right) - \left(2 + \frac{3i}{n}\right)^2 \right]\left(\frac{3}{n}\right)$$

$$= \lim_{n \to \infty} \frac{3}{n} \sum_{i=1}^{n} \left[ 10 + \frac{15i}{n} - 4 - 12\frac{i}{n} - \frac{9i^2}{n^2} \right]$$

$$= \lim_{n \to \infty} \frac{3}{n} \sum_{i=1}^{n} \left[ 6 + \frac{3i}{n} - \frac{9i^2}{n^2} \right]$$

$$= \lim_{n \to \infty} \frac{3}{n} \left[ 6n + \frac{3}{n}\frac{n(n+1)}{2} - \frac{9}{n^2}\frac{n(n+1)(2n+1)}{6} \right]$$

$$= \left[ 18 + \frac{9}{2} - 9 \right] = \frac{27}{2}$$

**33.** $\displaystyle \lim_{\|\Delta\| \to 0} \sum_{i=1}^{n} (2c_i - 3)\Delta x_i = \int_4^6 (2x - 3)\, dx$

**35.** $\displaystyle \int_{-4}^{0} (2x + 8)\, dx$

**37.**

$$\int_0^5 (5 - |x - 5|)\, dx = \frac{1}{2}(5)(5) = \frac{25}{2}$$
(triangle)

**39.** (a) $\displaystyle \int_4^8 [f(x) + g(x)]\, dx = \int_4^8 f(x)\, dx + \int_4^8 g(x)\, dx = 12 + 5 = 17$

    (b) $\displaystyle \int_4^8 [f(x) - g(x)]\, dx = \int_4^8 f(x)\, dx - \int_4^8 g(x)\, dx = 12 - 5 = 7$

    (c) $\displaystyle \int_4^8 [2f(x) - 3g(x)]\, dx = 2\int_4^8 f(x)\, dx - 3\int_4^8 g(x)\, dx = 2(12) - 3(5) = 9$

    (d) $\displaystyle \int_4^8 7f(x)\, dx = 7\int_4^8 f(x)\, dx = 7(12) = 84$

**41.** $\displaystyle \int_0^8 (3 + x)\, dx = \left[ 3x + \frac{x^2}{2} \right]_0^8 = 24 + \frac{64}{2} = 56$

**43.** $\displaystyle \int_{-1}^{1} (4t^3 - 2t)\, dt = \left[ t^4 - t^2 \right]_{-1}^{1} = 0$

**45.** $\displaystyle \int_4^9 x\sqrt{x}\, dx = \int_4^9 x^{3/2}\, dx = \left[ \frac{2}{5}x^{5/2} \right]_4^9 = \frac{2}{5}\left[ \left(\sqrt{9}\right)^5 - \left(\sqrt{4}\right)^5 \right] = \frac{2}{5}(243 - 32) = \frac{422}{5}$

**47.** $\displaystyle \int_0^{3\pi/4} \sin\theta\, d\theta = \left[ -\cos\theta \right]_0^{3\pi/4} = -\left( -\frac{\sqrt{2}}{2} \right) + 1 = 1 + \frac{\sqrt{2}}{2} = \frac{\sqrt{2} + 2}{2}$

**49.** $\int_2^4 (3x - 4)\,dx = \left[\dfrac{3x^2}{2} - 4x\right]_2^4$

$= (24 - 16) - (6 - 8) = 10$

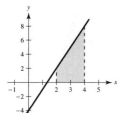

**51.** $\int_3^4 (x^2 - 9)\,dx = \left[\dfrac{x^3}{3} - 9x\right]_3^4$

$= \left(\dfrac{64}{3} - 36\right) - (9 - 27)$

$= \dfrac{64}{3} - \dfrac{54}{3} = \dfrac{10}{3}$

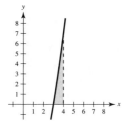

**53.** $\int_0^1 (x - x^3)\,dx = \left[\dfrac{x^2}{2} - \dfrac{x^4}{4}\right]_0^1 = \dfrac{1}{2} - \dfrac{1}{4} = \dfrac{1}{4}$

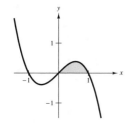

**55.** Area $= \int_0^2 \sin x\,dx$

$= [-\cos x]_0^2$

$= -\cos(2) + 1$

$= 1 - \cos(2)$

$\approx 1.416$

**57.** Area $= \int_1^9 \dfrac{4}{\sqrt{x}}\,dx = \left[\dfrac{4x^{1/2}}{(1/2)}\right]_1^9 = 8(3 - 1) = 16$

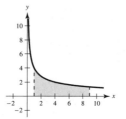

**59.** Average value: $\dfrac{1}{9 - 4}\int_4^9 \dfrac{1}{\sqrt{x}}\,dx = \left[\dfrac{1}{5}2\sqrt{x}\right]_4^9$

$= \dfrac{2}{5}(3 - 2) = \dfrac{2}{5}$

$\dfrac{2}{5} = \dfrac{1}{\sqrt{x}}$

$\sqrt{x} = \dfrac{5}{2}$

$x = \dfrac{25}{4}$

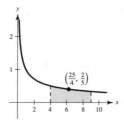

**61.** $F'(x) = x^2\sqrt{1 + x^3}$

**63.** $F'(x) = x^2 + 3x + 2$

**65.** $\int (x^2 - 3)^3\,dx = \int (27 - 27x^2 + 9x^4 - x^6)\,dx$

$= 27x - 9x^3 + \dfrac{9}{5}x^5 - \dfrac{x^7}{7} + C$

**67.** $u = x^3 + 3,\ du = 3x^2\,dx$

$\int \dfrac{x^2}{\sqrt{x^3 + 3}}\,dx = \int (x^3 + 3)^{-1/2}x^2\,dx$

$= \dfrac{1}{3}\int (x^3 + 3)^{-1/2}3x^2\,dx$

$= \dfrac{2}{3}(x^3 + 3)^{1/2} + C$

**69.** $u = 1 - 3x^2, \, du = -6x \, dx$

$$\int x(1 - 3x^2)^4 \, dx = -\frac{1}{6} \int (1 - 3x^2)^4 (-6x \, dx)$$
$$= -\frac{1}{30}(1 - 3x^2)^5 + C$$
$$= \frac{1}{30}(3x^2 - 1)^5 + C$$

**71.** $\int \sin^3 x \cos x \, dx = \frac{1}{4}\sin^4 x + C$

**73.** $\int \dfrac{\cos \theta}{\sqrt{1 - \sin \theta}} \, d\theta = -\int (1 - \sin \theta)^{-1/2}(-\cos \theta) \, d\theta$

$$= -2(1 - \sin \theta)^{1/2} + C$$
$$= -2\sqrt{1 - \sin \theta} + C$$

**75.** $\int (1 + \sec \pi x)^2 \sec \pi x \tan \pi x \, dx = \dfrac{1}{\pi} \int (1 + \sec \pi x)^2 (\pi \sec \pi x \tan \pi x) \, dx = \dfrac{1}{3\pi}(1 + \sec \pi x)^3 + C$

**77.** $\displaystyle\int_{-2}^{1} x(x^2 - 6) \, dx = \int_{-2}^{1} (x^3 - 6x) \, dx = \left[\dfrac{x^4}{4} - 3x^2\right]_{-2}^{1}$

$$= \left(\frac{1}{4} - 3\right) - (4 - 12) = \frac{21}{4}$$

**79.** $\displaystyle\int_{0}^{3} \dfrac{1}{\sqrt{1 + x}} \, dx = \int_{0}^{3} (1 + x)^{-1/2} \, dx = \left[2(1 + x)^{1/2}\right]_{0}^{3} = 4 - 2 = 2$

**81.** $u = 1 - y, \, y = 1 - u, \, dy = -du$

When $y = 0, u = 1$. When $y = 1, u = 0$.

$$2\pi \int_{0}^{1} (y + 1)\sqrt{1 - y} \, dy = 2\pi \int_{1}^{0} -\big[(1 - u) + 1\big]\sqrt{u} \, du = 2\pi \int_{1}^{0} \left(u^{3/2} - 2u^{1/2}\right) du = 2\pi \left[\frac{2}{5}u^{5/2} - \frac{4}{3}u^{3/2}\right]_{1}^{0} = \frac{28\pi}{15}$$

**83.** $\displaystyle\int_{0}^{\pi} \cos\left(\dfrac{x}{2}\right) dx = 2\int_{0}^{\pi} \cos\left(\dfrac{x}{2}\right)\dfrac{1}{2} \, dx = \left[2\sin\left(\dfrac{x}{2}\right)\right]_{0}^{\pi} = 2$

**85.** (a)  Answers will vary. *Sample answer:*

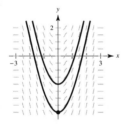

(b) $\dfrac{dy}{dx} = x\sqrt{9 - x^2}, \, (0, -4)$

$$y = \int (9 - x^2)^{1/2} x \, dx = \frac{-1}{2}\frac{(9 - x^2)^{3/2}}{3/2} + C = -\frac{1}{3}(9 - x^2)^{3/2} + C$$

$$-4 = -\frac{1}{3}(9 - 0)^{3/2} + C = -\frac{1}{3}(27) + C \Rightarrow C = 5$$

$$y = -\frac{1}{3}(9 - x^2)^{3/2} + 5$$

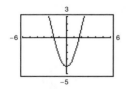

**87.** $\int_1^9 x(x-1)^{1/3}\,dx.$ Let $u = x - 1,\ du = dx.$

$$A = \int_0^8 (u+1)u^{1/3}\,du$$

$$= \int_0^8 \left(u^{4/3} + u^{1/3}\right)du$$

$$= \left[\frac{3u^{7/3}}{7} + \frac{3u^{4/3}}{4}\right]_0^8$$

$$= \frac{3}{7}(128) + \frac{3}{4}(16) = \frac{468}{7}$$

**89.** (a) $\int_0^{12}\left[2.880 + 2.125\sin(0.578t + 0.745)\right]dt = \left[2.880t - \dfrac{2.125}{0.578}\cos(0.578t + 0.745)\right]_0^{12} \approx 36.63$ in.

(b) $\frac{1}{2}\int_8^{10} R(t)\,dt \approx 2.22$ in.

**91.** Trapezoidal Rule $(n = 4)$: $\int_2^3 \dfrac{2}{1+x^2}\,dx$

$$\approx \frac{1}{8}\left[\frac{2}{1+2^2} + 2\left(\frac{2}{1+(9/4)^2}\right) + 2\left(\frac{2}{1+(5/2)^2}\right) + 2\left(\frac{2}{1+(11/4)^2}\right) + \frac{2}{1+3^2}\right] \approx 0.285$$

Simpson's Rule $(n = 4)$: $\int_2^3 \dfrac{2}{1+x^2}\,dx$

$$\approx \frac{1}{12}\left[\frac{2}{1+2^2} + 4\left(\frac{2}{1+(9/4)^2}\right) + 2\left(\frac{2}{1+(5/2)^2}\right) + 4\left(\frac{2}{1+(11/4)^2}\right) + \frac{2}{1+3^2}\right] \approx 0.284$$

Graphing utility: 0.284

**93.** Trapezoidal Rule: $(n = 4)$: $\int_0^{\pi/2} \sqrt{x}\cos x\,dx \approx 0.637$

Simpson's Rule $(n = 4)$: 0.685

Graphing Utility: 0.704

# Problem Solving for Chapter 4

**1.** (a) $L(1) = \int_1^1 \dfrac{1}{t}\,dt = 0$

(b) $L'(x) = \dfrac{1}{x}$ by the Second Fundamental Theorem of Calculus.

$L'(1) = 1$

(c) $L(x) = 1 = \int_1^x \dfrac{1}{t}\,dt$ for $x \approx 2.718$

$$\int_1^{2.718} \frac{1}{t}\,dt = 0.999896$$

(**Note:** The exact value of $x$ is $e$, the base of the natural logarithm function.)

(d) First show that $\int_{1}^{x_1} \dfrac{1}{t}\, dt = \int_{1/x_1}^{1} \dfrac{1}{t}\, dt.$

To see this, let $u = \dfrac{t}{x_1}$ and $du = \dfrac{1}{x_1}\, dt.$

Then $\int_{1}^{x_1} \dfrac{1}{t}\, dt = \int_{1/x_1}^{1} \dfrac{1}{ux_1}(x_1\, du) = \int_{1/x_1}^{1} \dfrac{1}{u}\, du = \int_{1/x_1}^{1} \dfrac{1}{t}\, dt.$

Now, $L(x_1\, x_2) = \int_{1}^{x_1 x_2} \dfrac{1}{t}\, dt = \int_{1/x_1}^{x_2} \dfrac{1}{u}\, du \left( \text{using } u = \dfrac{t}{x_1} \right)$

$\qquad\qquad = \int_{1/x_1}^{1} \dfrac{1}{u}\, du + \int_{1}^{x_2} \dfrac{1}{u}\, du$

$\qquad\qquad = \int_{1}^{x_1} \dfrac{1}{u}\, du + \int_{1}^{x_2} \dfrac{1}{u}\, du$

$\qquad\qquad = L(x_1) + L(x_2).$

**3.** $y = x^4 - 4x^3 + 4x^2,\ [0, 2],\ c_i = \dfrac{2i}{n}$

(a) $\Delta x = \dfrac{2}{n},\ f(x) = x^4 - 4x^3 + 4x^2$

$A = \displaystyle\lim_{n\to\infty}\sum_{i=1}^{n} f(c_i)\Delta x = \lim_{n\to\infty}\sum_{i=1}^{n}\left[\left(\dfrac{2i}{n}\right)^4 - 4\left(\dfrac{2i}{n}\right)^3 + 4\left(\dfrac{2i}{n}\right)^2\right]\dfrac{2}{n} = \lim_{n\to\infty}$

(b) $\left[\dfrac{32}{n^5}\displaystyle\sum_{i=1}^{n} i^4 - \dfrac{64}{n^4}\sum_{i=1}^{n} i^3 + \dfrac{32}{n^3}\sum_{i=1}^{n} i^2\right] = \left[\dfrac{32 \cdot n(n+1)(2n+1)(3n^2+3n-1)}{n^5 \cdot 30} - \dfrac{64 \cdot n^2(n+1)^2}{n^4 \cdot 4} + \dfrac{32 \cdot n(n+1)(2n+1)}{n^3 \cdot 6}\right]$

$\qquad\qquad = \left[\dfrac{16(n+1)(6n^3+9n^2+n-1)}{15n^4} - \dfrac{16(n+1)^2}{n^2} + \dfrac{16(n+1)(2n+1)}{3n^2}\right]$

$\qquad\qquad = \left[\dfrac{16(n+1)(n^3-n^2+n-1)}{15n^4}\right] = \dfrac{16n^4-16}{15n^4}$

(c) $A = \displaystyle\lim_{n\to\infty}\left[\dfrac{16n^4-16}{15n^4}\right] = \dfrac{16}{15}$

**5.** $S(x) = \displaystyle\int_{0}^{x} \sin\left(\dfrac{\pi t^2}{2}\right) dt$

(a)

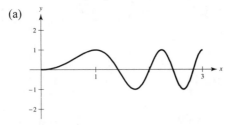

(b)

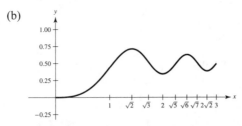

The zeros of $y = \sin\dfrac{\pi x^2}{2}$ correspond to the relative extrema of $S(x)$.

(c) $S'(x) = \sin\dfrac{\pi x^2}{2} = 0 \Rightarrow \dfrac{\pi x^2}{2} = n\pi \Rightarrow x^2 = 2n \Rightarrow x = \sqrt{2n}$, $n$ integer

Relative maxima at $x = \sqrt{2} \approx 1.4142$ and $x = \sqrt{6} \approx 2.4495$

Relative minima at $x = 2$ and $x = 2\sqrt{2} \approx 2.8284$

(d) $S''(x) = \cos\left(\dfrac{\pi x^2}{2}\right)(\pi x) = 0 \Rightarrow \dfrac{\pi x^2}{2} = \dfrac{\pi}{2} + n\pi \Rightarrow x^2 = 1 + 2n \Rightarrow x = \sqrt{1 + 2n}$, $n$ integer

Points of inflection at $x = 1, \sqrt{3}, \sqrt{5}$, and $\sqrt{7}$.

**7.** (a)

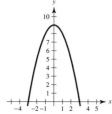

Area $= \displaystyle\int_{-3}^{3}\left(9 - x^2\right)dx = 2\int_{0}^{3}\left(9 - x^2\right)dx = 2\left[9x - \dfrac{x^3}{3}\right]_{0}^{3} = 2[27 - 9] = 36$

(b) Base $= 6$, height $= 9$, Area $= \dfrac{2}{3}bh = \dfrac{2}{3}(6)(9) = 36$

(c) Let the parabola be given by $y = b^2 - a^2x^2$, $a, b > 0$.

Area $= 2\displaystyle\int_{0}^{b/a}\left(b^2 - a^2x^2\right)dx$

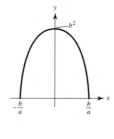

$= 2\left[b^2x - a^2\dfrac{x^3}{3}\right]_{0}^{b/a}$

$= 2\left[b^2\left(\dfrac{b}{a}\right) - \dfrac{a^2}{3}\left(\dfrac{b}{a}\right)^3\right]$

$= 2\left[\dfrac{b^3}{a} - \dfrac{1}{3}\dfrac{b^3}{a}\right] = \dfrac{4}{3}\dfrac{b^3}{a}$

Base $= \dfrac{2b}{a}$, height $= b^2$

Archimedes' Formula: Area $= \dfrac{2}{3}\left(\dfrac{2b}{a}\right)\left(b^2\right) = \dfrac{4}{3}\dfrac{b^3}{a}$

**9.** (a)

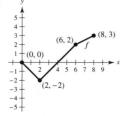

(b)

| $x$ | 0 | 1 | 2 | 3 | 4 | 5 | 6 | 7 | 8 |
|---|---|---|---|---|---|---|---|---|---|
| $F(x)$ | 0 | $-\dfrac{1}{2}$ | $-2$ | $-\dfrac{7}{2}$ | $-4$ | $-\dfrac{7}{2}$ | $-2$ | $\dfrac{1}{4}$ | 3 |

(c) $f(x) = \begin{cases} -x, & 0 \le x < 2 \\ x - 4, & 2 \le x < 6 \\ \frac{1}{2}x - 1, & 6 \le x \le 8 \end{cases}$

$F(x) = \int_0^x f(t)\, dt = \begin{cases} (-x^2/2), & 0 \le x < 2 \\ (x^2/2) - 4x + 4, & 2 \le x < 6 \\ (1/4)x^2 - x - 5, & 6 \le x \le 8 \end{cases}$

$F'(x) = f(x)$. $F$ is decreasing on $(0, 4)$ and increasing on $(4, 8)$. Therefore, the minimum is $-4$ at $x = 4$, and the maximum is 3 at $x = 8$.

(d) $F''(x) = f'(x) = \begin{cases} -1, & 0 < x < 2 \\ 1, & 2 < x < 6 \\ \frac{1}{2}, & 6 < x < 8 \end{cases}$

$x = 2$ is a point of inflection, whereas $x = 6$ is not.

**11.** $\int_0^x f(t)(x - t)\, dt = \int_0^x xf(t)\, dt - \int_0^x tf(t)\, dt = x\int_0^x f(t)\, dt - \int_0^x tf(t)\, dt$

So, $\dfrac{d}{dx}\int_0^x f(t)(x - t)\, dt = xf(x) + \int_0^x f(t)\, dt - xf(x) = \int_0^x f(t)\, dt$

Differentiating the other integral,

$\dfrac{d}{dx}\int_0^x \left(\int_0^x f(v)\, dv\right) dt = \int_0^x f(v)\, dv.$

So, the two original integrals have equal derivatives,

$\int_0^x f(t)(x - t)\, dt = \int_0^x \left(\int_0^t f(v)\, dv\right) dt + C.$

Letting $x = 0$, you see that $C = 0$.

**13.** Consider $\int_0^1 \sqrt{x}\, dx = \frac{2}{3}x^{3/2}\Big]_0^1 = \frac{2}{3}$. The corresponding Riemann Sum using right-hand endpoints is

$S(n) = \dfrac{1}{n}\left[\sqrt{\dfrac{1}{n}} + \sqrt{\dfrac{2}{n}} + \cdots + \sqrt{\dfrac{n}{n}}\right] = \dfrac{1}{n^{3/2}}\left[\sqrt{1} + \sqrt{2} + \cdots + \sqrt{n}\right]$. So, $\displaystyle\lim_{n \to \infty} \dfrac{\sqrt{1} + \sqrt{2} + \cdots + \sqrt{n}}{n^{3/2}} = \dfrac{2}{3}$.

**15.** By Theorem 4.8, $0 < f(x) \le M \Rightarrow \int_a^b f(x)\, dx \le \int_a^b M\, dx = M(b - a)$.

Similarly, $m \le f(x) \Rightarrow m(b - a) = \int_a^b m\, dx \le \int_a^b f(x)\, dx$.

So, $m(b - a) \le \int_a^b f(x)\, dx \le M(b - a)$. On the interval $[0, 1]$, $1 \le \sqrt{1 + x^4} \le \sqrt{2}$ and $b - a = 1$.

So, $1 \le \int_0^1 \sqrt{1 + x^4}\, dx \le \sqrt{2}$. $\left(\textbf{Note: } \int_0^1 \sqrt{1 + x^4}\, dx \approx 1.0894\right)$

**17.** (a) $(1 + i)^3 = 1 + 3i + 3i^2 + i^3 \Rightarrow (1 + i)^3 - i^3 = 3i^2 + 3i + 1$

(b) $\qquad 3i^2 + 3i + 1 = (i + 1)^3 - i^3$

$\displaystyle\sum_{i=1}^n (3i^2 + 3i + 1) = \sum_{i=1}^n \left[(i + 1)^3 - i^3\right]$

$\qquad = (2^3 - 1^3) + (3^3 - 2^3) + \cdots + \left[((n + 1)^3 - n^3)\right] = (n + 1)^3 - 1$

So, $(n + 1)^3 = \displaystyle\sum_{i=1}^n (3i^2 + 3i + 1) + 1$.

(c) $(n + 1)^3 - 1 = \sum_{i=1}^{n}(3i^2 + 3i + 1) = \sum_{i=1}^{n} 3i^2 + \dfrac{3(n)(n + 1)}{2} + n$

$\Rightarrow \sum_{i=1}^{n} 3i^2 = n^3 + 3n^2 + 3n - \dfrac{3n(n + 1)}{2} - n$

$= \dfrac{2n^3 + 6n^2 + 6n - 3n^2 - 3n - 2n}{2}$

$= \dfrac{2n^3 + 3n^2 + n}{2}$

$= \dfrac{n(n + 1)(2n + 1)}{2}$

$\Rightarrow \sum_{i=1}^{n} i^2 = \dfrac{n(n + 1)(2n + 1)}{6}$

**19.** (a) $R < I < T < L$

(b) $S(4) = \dfrac{4 - 0}{3(4)}\big[f(0) + 4f(1) + 2f(2) + 4f(3) + f(4)\big] \approx \dfrac{1}{3}\Big[4 + 4(2) + 2(1) + 4\Big(\dfrac{1}{2}\Big) + \dfrac{1}{4}\Big] \approx 5.417$

**21.** The graph of $y = x^2 - 16$ is a parabola:

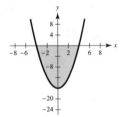

The integral $\displaystyle\int_{a}^{b}\left(x^2 - 16\right) dx$ will be a minimum when $a = -4$ and $b = 4$, as indicated in the figure.

# CHAPTER 5
# Logarithmic, Exponential, and Other Transcendental Functions

# CHAPTER 5
# Logarithmic, Exponential, and Other Transcendental Functions

## Section 5.1   The Natural Logarithmic Function: Differentiation

**1.** Simpson's Rule: $n = 10$

| $x$ | 0.5 | 1.5 | 2 | 2.5 | 3 | 3.5 | 4 |
|---|---|---|---|---|---|---|---|
| $\int_1^x \frac{1}{t}\,dt$ | −0.6932 | 0.4055 | 0.6932 | 0.9163 | 1.0987 | 1.2529 | 1.3865 |

**Note:** $\int_1^{0.5} \frac{1}{t}\,dt = -\int_{0.5}^1 \frac{1}{t}\,dt$

**3.** (a)  $\ln 45 \approx 3.8067$

(b)  $\int_1^{45} \frac{1}{t}\,dt \approx 3.8067$

**5.** (a)  $\ln 0.8 \approx -0.2231$

(b)  $\int_1^{0.8} \frac{1}{t}\,dt \approx -0.2231$

**7.** $f(x) = \ln x + 1$

Vertical shift 1 unit upward

Matches (b)

**9.** $f(x) = \ln(x - 1)$

Horizontal shift 1 unit to the right

Matches (a)

**11.** $f(x) = 3 \ln x$

Domain: $x > 0$

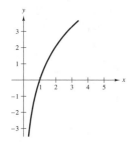

**13.** $f(x) = \ln 2x$

Domain: $x > 0$

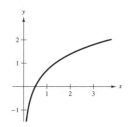

**15.** $f(x) = \ln(x - 1)$

Domain: $x > 1$

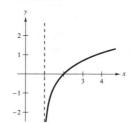

**17.** $h(x) = \ln(x + 2)$

Domain: $x > -2$

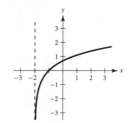

**19.** (a)  $\ln 6 = \ln 2 + \ln 3 \approx 1.7917$

(b)  $\ln \frac{2}{3} = \ln 2 - \ln 3 \approx -0.4055$

(c)  $\ln 81 = \ln 3^4 = 4 \ln 3 \approx 4.3944$

(d)  $\ln \sqrt{3} = \ln 3^{1/2} = \frac{1}{2} \ln 3 \approx 0.5493$

**21.** $\ln \frac{x}{4} = \ln x - \ln 4$

**23.** $\ln \frac{xy}{z} = \ln x + \ln y - \ln z$

**25.** $\ln\left(x\sqrt{x^2 + 5}\right) = \ln x + \ln\left(x^2 + 5\right)^{1/2}$

$$= \ln x + \frac{1}{2} \ln\left(x^2 + 5\right)$$

**27.** $\ln\sqrt{\dfrac{x-1}{x}} = \ln\left(\dfrac{x-1}{x}\right)^{1/2} = \dfrac{1}{2}\ln\left(\dfrac{x-1}{x}\right)$

$\qquad = \dfrac{1}{2}\Big[\ln(x-1) - \ln x\Big]$

$\qquad = \dfrac{1}{2}\ln(x-1) - \dfrac{1}{2}\ln x$

**29.** $\ln z(z-1)^2 = \ln z + \ln(z-1)^2 = \ln z + 2\ln(z-1)$

**31.** $\ln(x-2) - \ln(x+2) = \ln\dfrac{x-2}{x+2}$

**33.** $\dfrac{1}{3}\Big[2\ln(x+3) + \ln x - \ln(x^2-1)\Big] = \dfrac{1}{3}\ln\dfrac{x(x+3)^2}{x^2-1}$

$\qquad = \ln\sqrt[3]{\dfrac{x(x+3)^2}{x^2-1}}$

**35.** $2\ln 3 - \dfrac{1}{2}\ln(x^2+1) = \ln 9 - \ln\sqrt{x^2+1} = \ln\dfrac{9}{\sqrt{x^2+1}}$

**37.** (a)

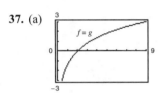

(b) $f(x) = \ln\dfrac{x^2}{4} = \ln x^2 - \ln 4 = 2\ln x - \ln 4 = g(x)$

because $x > 0$.

**39.** $\displaystyle\lim_{x\to 3^+} \ln(x-3) = -\infty$

**41.** $\displaystyle\lim_{x\to 2^-} \ln\Big[x^2(3-x)\Big] = \ln 4 \approx 1.3863$

**43.** $y = \ln x^3 = 3\ln x$

$y' = \dfrac{3}{x}$

Slope at $(1, 0)$ is $\dfrac{3}{1} = 3$.

Tangent line:

$y - 0 = 3(x-1)$

$\qquad y = 3x - 3$

**45.** $y = \ln x^4 = 4\ln x$

$y' = \dfrac{4}{x}$

Slope at $(1, 0)$ is 4.

Tangent line:

$y - 0 = 4(x-1)$

$\qquad y = 4x - 4$

**47.** $f(x) = \ln(3x)$

$f'(x) = \dfrac{1}{3x}(3) = \dfrac{1}{x}$

**49.** $g(x) = \ln x^2 = 2\ln x$

$g'(x) = \dfrac{2}{x}$

**51.** $y = (\ln x)^4$

$\dfrac{dy}{dx} = 4(\ln x)^3\left(\dfrac{1}{x}\right) = \dfrac{4(\ln x)^3}{x}$

**53.** $y = \ln(t+1)^2 = 2\ln(t+1)$

$y' = 2\dfrac{1}{t+1} = \dfrac{2}{t+1}$

**55.** $y = \ln\Big[x\sqrt{x^2-1}\Big] = \ln x + \dfrac{1}{2}\ln(x^2-1)$

$\dfrac{dy}{dx} = \dfrac{1}{x} + \dfrac{1}{2}\left(\dfrac{2x}{x^2-1}\right) = \dfrac{2x^2-1}{x(x^2-1)}$

**57.** $f(x) = \ln\dfrac{x}{x^2+1} = \ln x - \ln(x^2+1)$

$f'(x) = \dfrac{1}{x} - \dfrac{2x}{x^2+1} = \dfrac{1-x^2}{x(x^2+1)}$

**59.** $g(t) = \dfrac{\ln t}{t^2}$

$g'(t) = \dfrac{t^2(1/t) - 2t\ln t}{t^4} = \dfrac{1-2\ln t}{t^3}$

**61.** $y = \ln(\ln x^2)$

$\dfrac{dy}{dx} = \dfrac{1}{\ln x^2}\dfrac{d}{dx}(\ln x^2) = \dfrac{(2x/x^2)}{\ln x^2} = \dfrac{2}{x\ln x^2} = \dfrac{1}{x\ln x}$

**63.** $y = \ln\sqrt{\dfrac{x+1}{x-1}} = \dfrac{1}{2}\Big[\ln(x+1) - \ln(x-1)\Big]$

$\dfrac{dy}{dx} = \dfrac{1}{2}\left[\dfrac{1}{x+1} - \dfrac{1}{x-1}\right] = \dfrac{1}{1-x^2}$

**65.** $f(x) = \ln\dfrac{\sqrt{4+x^2}}{x} = \dfrac{1}{2}\ln(4+x^2) - \ln x$

$f'(x) = \dfrac{x}{4+x^2} - \dfrac{1}{x} = \dfrac{-4}{x(x^2+4)}$

**67.**  $y = \dfrac{-\sqrt{x^2 + 1}}{x} + \ln\left(x + \sqrt{x^2 + 1}\right)$

$\dfrac{dy}{dx} = \dfrac{-x\left(x/\sqrt{x^2 + 1}\right) + \sqrt{x^2 + 1}}{x^2} + \left(\dfrac{1}{x + \sqrt{x^2 + 1}}\right)\left(1 + \dfrac{x}{\sqrt{x^2 + 1}}\right)$

$= \dfrac{1}{x^2\sqrt{x^2 + 1}} + \left(\dfrac{1}{x + \sqrt{x^2 + 1}}\right)\left(\dfrac{\sqrt{x^2 + 1} + x}{\sqrt{x^2 + 1}}\right) = \dfrac{1}{x^2\sqrt{x^2 + 1}} + \dfrac{1}{\sqrt{x^2 + 1}} = \dfrac{1 + x^2}{x^2\sqrt{x^2 + 1}} = \dfrac{\sqrt{x^2 + 1}}{x^2}$

**69.**  $y = \ln|\sin x|$

$\dfrac{dy}{dx} = \dfrac{\cos x}{\sin x} = \cot x$

**71.**  $y = \ln\left|\dfrac{\cos x}{\cos x - 1}\right| = \ln|\cos x| - \ln|\cos x - 1|$

$\dfrac{dy}{dx} = \dfrac{-\sin x}{\cos x} - \dfrac{-\sin x}{\cos x - 1} = -\tan x + \dfrac{\sin x}{\cos x - 1}$

**73.**  $y = \ln\left|\dfrac{-1 + \sin x}{2 + \sin x}\right| = \ln|-1 + \sin x| - \ln|2 + \sin x|$

$\dfrac{dy}{dx} = \dfrac{\cos x}{-1 + \sin x} - \dfrac{\cos x}{2 + \sin x} = \dfrac{3\cos x}{(\sin x - 1)(\sin x + 2)}$

**75.**  $f(x) = \displaystyle\int_{2}^{\ln(2x)} (t + 1)\, dt$

$f'(x) = \left[\ln(2x) + 1\right]\left(\dfrac{1}{x}\right) = \dfrac{\ln(2x) + 1}{x}$

**Alternate solution:**

$f(x) = \displaystyle\int_{2}^{\ln(2x)} (t + 1)\, dt$

$= \left[\dfrac{t^2}{2} + t\right]_{2}^{\ln 2x} = \left[\dfrac{[\ln(2x)]^2}{2} + \ln(2x)\right] - [2 + 2]$

$f'(x) = \dfrac{1}{2}\, 2\ln(2x)\dfrac{1}{x} + \dfrac{1}{2x}(2) = \dfrac{\ln(2x) + 1}{x}$

**77. (a)**   $y = 3x^2 - \ln x, \quad (1, 3)$

$\dfrac{dy}{dx} = 6x - \dfrac{1}{x}$

When $x = 1$, $\dfrac{dy}{dx} = 5$.

Tangent line:  $y - 3 = 5(x - 1)$

$y = 5x - 2$

$0 = 5x - y - 2$

**(b)**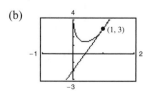

**79. (a)**   $f(x) = \ln\sqrt{1 + \sin^2 x}$

$= \dfrac{1}{2}\ln(1 + \sin^2 x), \quad \left(\dfrac{\pi}{4}, \ln\sqrt{\dfrac{3}{2}}\right)$

$f'(x) = \dfrac{2\sin x\cos x}{2(1 + \sin^2 x)} = \dfrac{\sin x\cos x}{1 + \sin^2 x}$

$f'\left(\dfrac{\pi}{4}\right) = \dfrac{\left(\sqrt{2}/2\right)\left(\sqrt{2}/2\right)}{(3/2)} = \dfrac{1}{3}$

Tangent line:  $y - \ln\sqrt{\dfrac{3}{2}} = \dfrac{1}{3}\left(x - \dfrac{\pi}{4}\right)$

$y = \dfrac{1}{3}x + \dfrac{1}{2}\ln\left(\dfrac{3}{2}\right) - \dfrac{\pi}{12}$

**(b)**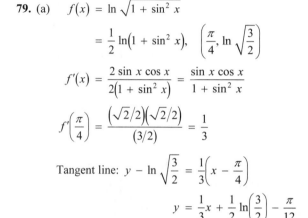

**81. (a)**   $f(x) = x^3 \ln x, \quad (1, 0)$

$f'(x) = 3x^2 \ln x + x^2$

$f'(1) = 1$

Tangent line:  $y - 0 = 1(x - 1)$

$y = x - 1$

**(b)**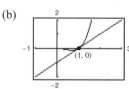

**83.**   $x^2 - 3\ln y + y^2 = 10$

$2x - \dfrac{3}{y}\dfrac{dy}{dx} + 2y\dfrac{dy}{dx} = 0$

$2x = \dfrac{dy}{dx}\left(\dfrac{3}{y} - 2y\right)$

$\dfrac{dy}{dx} = \dfrac{2x}{(3/y) - 2y} = \dfrac{2xy}{3 - 2y^2}$

**85.** $4x^3 + \ln y^2 + 2y = 2x$

$$12x^2 + \frac{2}{y}y' + 2y' = 2$$

$$\left(\frac{2}{y} + 2\right)y' = 2 - 12x^2$$

$$y' = \frac{2 - 12x^2}{2/y + 2}$$

$$y' = \frac{y - 6yx^2}{1 + y} = \frac{y(1 - 6x^2)}{1 + y}$$

**87.** $\qquad x + y - 1 = \ln(x^2 + y^2), \quad (1, 0)$

$$1 + y' = \frac{2x + 2yy'}{x^2 + y^2}$$

$$x^2 + y^2 + (x^2 + y^2)y' = 2x + 2yy'$$

At $(1, 0)$: $1 + y' = 2$

$$y' = 1$$

Tangent line: $y = x - 1$

**89.** $y = 2(\ln x) + 3$

$$y' = \frac{2}{x}$$

$$y'' = -\frac{2}{x^2}$$

$$xy'' + y' = x\left(-\frac{2}{x^2}\right) + \frac{2}{x} = 0$$

**91.** $y = \frac{x^2}{2} - \ln x$

Domain: $x > 0$

$$y' = x - \frac{1}{x}$$

$$= \frac{(x + 1)(x - 1)}{x}$$

$$= 0 \text{ when } x = 1.$$

$$y'' = 1 + \frac{1}{x^2} > 0$$

Relative minimum: $\left(1, \frac{1}{2}\right)$

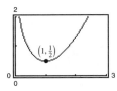

**93.** $y = x \ln x$

Domain: $x > 0$

$$y' = x\left(\frac{1}{x}\right) + \ln x = 1 + \ln x = 0 \text{ when } x = e^{-1}.$$

$$y'' = \frac{1}{x} > 0$$

Relative minimum: $\left(e^{-1}, -e^{-1}\right)$

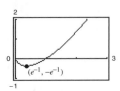

**95.** $y = \frac{x}{\ln x}$

Domain: $0 < x < 1, x > 1$

$$y' = \frac{(\ln x)(1) - (x)(1/x)}{(\ln x)^2} = \frac{\ln x - 1}{(\ln x)^2} = 0 \text{ when } x = e.$$

$$y'' = \frac{(\ln x)^2(1/x) - (\ln x - 1)(2/x)\ln x}{(\ln x)^4}$$

$$= \frac{2 - \ln x}{x(\ln x)^3} = 0 \text{ when } x = e^2.$$

Relative minimum: $(e, e)$

Point of inflection: $\left(e^2, \frac{e^2}{2}\right)$

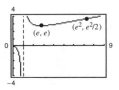

**97.**

$$f(x) = \ln x, \qquad\qquad\qquad f(1) = 0$$

$$f'(x) = \frac{1}{x}, \qquad\qquad\qquad f'(1) = 1$$

$$f''(x) = -\frac{1}{x^2}, \qquad\qquad\qquad f''(1) = -1$$

$$P_1(x) = f(1) + f'(1)(x-1) = x-1, \qquad P_1(1) = 0$$

$$P_2(x) = f(1) + f'(1)(x-1) + \frac{1}{2}f''(1)(x-1)^2$$

$$\qquad = (x-1) - \frac{1}{2}(x-1)^2, \qquad\qquad P_2(1) = 0$$

$$P_1'(x) = 1, \qquad\qquad\qquad P_1'(1) = 1$$

$$P_2'(x) = 1 - (x-1) = 2-x, \qquad\qquad P_2'(1) = 1$$

$$P_2''(x) = -1, \qquad\qquad\qquad P_2''(1) = -1$$

The values of $f$, $P_1$, $P_2$, and their first derivatives agree at $x = 1$. The values of the second derivatives of $f$ and $P_2$ agree at $x = 1$.

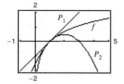

**99.** Find $x$ such that $\ln x = -x$.

$$f(x) = \ln x + x = 0$$

$$f'(x) = \frac{1}{x} + 1$$

$$x_{n+1} = x_n - \frac{f(x_n)}{f'(x_n)} = x_n\left[\frac{1 - \ln x_n}{1 + x_n}\right]$$

| $n$ | 1 | 2 | 3 |
|---|---|---|---|
| $x_n$ | 0.5 | 0.5644 | 0.5671 |
| $f(x_n)$ | −0.1931 | −0.0076 | −0.0001 |

Approximate root: $x \approx 0.567$

**101.**

$$y = x\sqrt{x^2+1}$$

$$\ln y = \ln x + \frac{1}{2}\ln(x^2+1)$$

$$\frac{1}{y}\left(\frac{dy}{dx}\right) = \frac{1}{x} + \frac{x}{x^2+1}$$

$$\frac{dy}{dx} = y\left[\frac{2x^2+1}{x(x^2+1)}\right] = \frac{2x^2+1}{\sqrt{x^2+1}}$$

**103.**

$$y = \frac{x^2\sqrt{3x-2}}{(x+1)^2}$$

$$\ln y = 2\ln x + \frac{1}{2}\ln(3x-2) - 2\ln(x+1)$$

$$\frac{1}{y}\left(\frac{dy}{dx}\right) = \frac{2}{x} + \frac{3}{2(3x-2)} - \frac{2}{x+1}$$

$$\frac{dy}{dx} = y\left[\frac{3x^2+15x-8}{2x(3x-2)(x+1)}\right] = \frac{3x^3+15x^2-8x}{2(x+1)^3\sqrt{3x-2}}$$

**105.**

$$y = \frac{x(x-1)^{3/2}}{\sqrt{x+1}}$$

$$\ln y = \ln x + \frac{3}{2}\ln(x-1) - \frac{1}{2}\ln(x+1)$$

$$\frac{1}{y}\left(\frac{dy}{dx}\right) = \frac{1}{x} + \frac{3}{2}\left(\frac{1}{x-1}\right) - \frac{1}{2}\left(\frac{1}{x+1}\right)$$

$$\frac{dy}{dx} = \frac{y}{2}\left[\frac{2}{x} + \frac{3}{x-1} - \frac{1}{x+1}\right]$$

$$\qquad = \frac{y}{2}\left[\frac{4x^2+4x-2}{x(x^2-1)}\right] = \frac{(2x^2+2x-1)\sqrt{x-1}}{(x+1)^{3/2}}$$

**107.** The domain of the natural logarithmic function is $(0, \infty)$ and the range is $(-\infty, \infty)$. The function is continuous, increasing, and one-to-one, and its graph is concave downward. In addition, if $a$ and $b$ are positive numbers and $n$ is rational, then $\ln(1) = 0$,

$$\ln(a \cdot b) = \ln a + \ln b,\ \ln(a^n) = n\ln a,\ \text{and}$$

$$\ln(a/b) = \ln a - \ln b.$$

**109.** $g(x) = \ln f(x), \quad f(x) > 0$

$$g'(x) = \frac{f'(x)}{f(x)}$$

(a) Yes. If the graph of $g$ is increasing, then $g'(x) > 0$. Because $f(x) > 0$, you know that $f'(x) = g'(x)f(x)$ and so, $f'(x) > 0$. Therefore, the graph of $f$ is increasing.

(b) No. Let $f(x) = x^2 + 1$ (positive and concave up).

$g(x) = \ln(x^2 + 1)$ is not concave up.

**111.** False

$$\ln x + \ln 25 = \ln(25x) \neq \ln(x + 25)$$

**113.** False; $\pi$ is a constant.

$$\frac{d}{dx}[\ln \pi] = 0$$

**115.** $t = 13.375 \ln\left(\dfrac{x}{x - 1250}\right), \quad x > 1250$

(a)
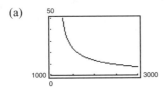

(b) When $x = 1398.43$: $t \approx 30$ years

Total amount paid $= (1398.43)(30)(12) = \$503,434.80$

(c) When $x = 1611.19$: $t \approx 20$ years

Total amount paid $= (1611.19)(20)(12) = \$386,685.60$

(d) $\dfrac{dt}{dx} = \dfrac{d}{dx}\Big[13.375\big(\ln x - \ln(x - 1250)\big)\Big] = 13.375\left[\dfrac{1}{x} - \dfrac{1}{x - 1250}\right] = \dfrac{-16718.75}{x(x - 1250)}$

When $x = 1398.43$: $\dfrac{dt}{dx} \approx -0.0805$

When $x = 1611.19$: $\dfrac{dt}{dx} \approx -0.0287$

(e) The benefits include a shorter term, and a lower total amount paid.

**117.** (a)
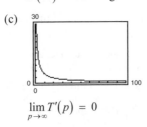

(b) $T'(p) = \dfrac{34.96}{p} + \dfrac{3.955}{\sqrt{p}}$

$T'(10) \approx 4.75$ deg/lb/in.$^2$

$T'(70) \approx 0.97$ deg/lb/in.$^2$

(c)
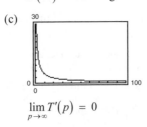

$\displaystyle\lim_{p \to \infty} T'(p) = 0$

Answers will vary. *Sample answer*: As the pounds per square inch approach infinity, the temperature will not change.

**119.** $y = 10 \ln \left( \dfrac{10 + \sqrt{100 - x^2}}{x} \right) - \sqrt{100 - x^2} = 10 \left[ \ln \left( 10 + \sqrt{100 - x^2} \right) - \ln x \right] - \sqrt{100 - x^2}$

(a)

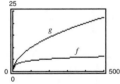

(b) $\dfrac{dy}{dx} = 10 \left[ \dfrac{-x}{\sqrt{100 - x^2} \left( 10 + \sqrt{100 - x^2} \right)} - \dfrac{1}{x} \right] + \dfrac{x}{\sqrt{100 - x^2}}$

$= \dfrac{x}{\sqrt{100 - x^2}} \left[ \dfrac{-10}{10 + \sqrt{100 - x^2}} \right] - \dfrac{10}{x} + \dfrac{x}{\sqrt{100 - x^2}}$

$= \dfrac{x}{\sqrt{100 - x^2}} \left[ \dfrac{-10}{10 + \sqrt{100 - x^2}} + 1 \right] - \dfrac{10}{x}$

$= \dfrac{x}{\sqrt{100 - x^2}} \left[ \dfrac{\sqrt{100 - x^2}}{10 + \sqrt{100 - x^2}} \right] - \dfrac{10}{x}$

$= \dfrac{x}{10 + \sqrt{100 - x^2}} - \dfrac{10}{x}$

$= \dfrac{x \left( 10 - \sqrt{100 - x^2} \right)}{x^2} - \dfrac{10}{x} = -\dfrac{\sqrt{100 - x^2}}{x}$

When $x = 5$, $dy/dx = -\sqrt{3}$.

When $x = 9$, $dy/dx = -\sqrt{19}/9$.

(c) $\lim\limits_{x \to 10^-} \dfrac{dy}{dx} = 0$

**121.** (a) $f(x) = \ln x$, $g(x) = \sqrt{x}$

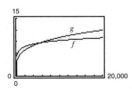

$f'(x) = \dfrac{1}{x}$, $g'(x) = \dfrac{1}{2\sqrt{x}}$

For $x > 4$, $g'(x) > f'(x)$. $g$ is increasing at a faster rate than $f$ for "large" values of $x$.

(b) $f(x) = \ln x$, $g(x) = \sqrt[4]{x}$

$f'(x) = \dfrac{1}{x}$, $g'(x) = \dfrac{1}{4\sqrt[4]{x^3}}$

For $x > 256$, $g'(x) > f'(x)$. $g$ is increasing at a faster rate than $f$ for "large" values of $x$. $f(x) = \ln x$ increases very slowly for "large" values of $x$.

## Section 5.2    The Natural Logarithmic Function: Integration

**1.** $\displaystyle\int \frac{5}{x}\,dx = 5\int \frac{1}{x}\,dx = 5\ln|x| + C$

**3.** $u = x + 1,\, du = dx$

$\displaystyle\int \frac{1}{x + 1}\,dx = \ln|x + 1| + C$

**5.** $u = 2x + 5,\, du = 2\,dx$

$\displaystyle\int \frac{1}{2x + 5}\,dx = \frac{1}{2}\int \frac{1}{2x + 5}(2)\,dx$

$\displaystyle\qquad\qquad\qquad = \frac{1}{2}\ln|2x + 5| + C$

**7.** $u = x^2 - 3,\, du = 2x\,dx$

$\displaystyle\int \frac{x}{x^2 - 3}\,dx = \frac{1}{2}\int \frac{1}{x^2 - 3}(2x)\,dx$

$\displaystyle\qquad\qquad\qquad = \frac{1}{2}\ln|x^2 - 3| + C$

**9.** $u = x^4 + 3x,\, du = \left(4x^3 + 3\right)dx$

$\displaystyle\int \frac{4x^3 + 3}{x^4 + 3x}\,dx = \int \frac{1}{x^4 + 3x}\left(4x^3 + 3\right)dx$

$\displaystyle\qquad\qquad\qquad\quad = \ln|x^4 + 3x| + C$

**11.** $\displaystyle\int \frac{x^2 - 4}{x}\,dx = \int \left(x - \frac{4}{x}\right)dx$

$\displaystyle\qquad\qquad\quad = \frac{x^2}{2} - 4\ln|x| + C$

$\displaystyle\qquad\qquad\quad = \frac{x^2}{2} - \ln\left(x^4\right) + C$

**13.** $u = x^3 + 3x^2 + 9x,\, du = 3\left(x^2 + 2x + 3\right)dx$

$\displaystyle\int \frac{x^2 + 2x + 3}{x^3 + 3x^2 + 9x}\,dx = \frac{1}{3}\int \frac{3\left(x^2 + 2x + 3\right)}{x^3 + 3x^2 + 9x}\,dx$

$\displaystyle\qquad\qquad\qquad\qquad\quad = \frac{1}{3}\ln|x^3 + 3x^2 + 9x| + C$

**15.** $\displaystyle\int \frac{x^2 - 3x + 2}{x + 1}\,dx = \int \left(x - 4 + \frac{6}{x + 1}\right)dx$

$\displaystyle\qquad\qquad\qquad\qquad = \frac{x^2}{2} - 4x + 6\ln|x + 1| + C$

**17.** $\displaystyle\int \frac{x^3 - 3x^2 + 5}{x - 3}\,dx = \int \left(x^2 + \frac{5}{x - 3}\right)dx$

$\displaystyle\qquad\qquad\qquad\qquad = \frac{x^3}{3} + 5\ln|x - 3| + C$

**19.** $\displaystyle\int \frac{x^4 + x - 4}{x^2 + 2}\,dx = \int \left(x^2 - 2 + \frac{x}{x^2 + 2}\right)dx$

$\displaystyle\qquad\qquad\qquad\qquad = \frac{x^3}{3} - 2x + \frac{1}{2}\ln\left(x^2 + 2\right) + C$

$\displaystyle\qquad\qquad\qquad\qquad = \frac{x^3}{3} - 2x + \ln\sqrt{x^2 + 2} + C$

**21.** $u = \ln x,\, du = \frac{1}{x}dx$

$\displaystyle\int \frac{(\ln x)^2}{x}\,dx = \frac{1}{3}(\ln x)^3 + C$

**23.** $u = x + 1,\, du = dx$

$\displaystyle\int \frac{1}{\sqrt{x + 1}}\,dx = \int (x + 1)^{-1/2}\,dx$

$\displaystyle\qquad\qquad\qquad = 2(x + 1)^{1/2} + C$

$\displaystyle\qquad\qquad\qquad = 2\sqrt{x + 1} + C$

**25.** $\displaystyle\int \frac{2x}{(x - 1)^2}\,dx = \int \frac{2x - 2 + 2}{(x - 1)^2}\,dx$

$\displaystyle\qquad\qquad\qquad = \int \frac{2(x - 1)}{(x - 1)^2}\,dx + 2\int \frac{1}{(x - 1)^2}\,dx$

$\displaystyle\qquad\qquad\qquad = 2\int \frac{1}{x - 1}\,dx + 2\int \frac{1}{(x - 1)^2}\,dx$

$\displaystyle\qquad\qquad\qquad = 2\ln|x - 1| - \frac{2}{(x - 1)} + C$

**27.** $u = 1 + \sqrt{2x},\, du = \frac{1}{\sqrt{2x}}\,dx \Rightarrow (u - 1)\,du = dx$

$\displaystyle\int \frac{1}{1 + \sqrt{2x}}\,dx = \int \frac{(u - 1)}{u}\,du = \int \left(1 - \frac{1}{u}\right)du$

$\displaystyle\qquad\qquad\qquad = u - \ln|u| + C_1$

$\displaystyle\qquad\qquad\qquad = \left(1 + \sqrt{2x}\right) - \ln\left|1 + \sqrt{2x}\right| + C_1$

$\displaystyle\qquad\qquad\qquad = \sqrt{2x} - \ln\left(1 + \sqrt{2x}\right) + C$

where $C = C_1 + 1$.

**29.** $u = \sqrt{x} - 3, \; du = \dfrac{1}{2\sqrt{x}} \, dx \Rightarrow 2(u + 3) \, du = dx$

$$\int \frac{\sqrt{x}}{\sqrt{x} - 3} \, dx = 2 \int \frac{(u + 3)^2}{u} \, du$$

$$= 2 \int \frac{u^2 + 6u + 9}{u} \, du = 2 \int \left( u + 6 + \frac{9}{u} \right) du$$

$$= 2 \left[ \frac{u^2}{2} + 6u + 9 \ln|u| \right] + C_1$$

$$= u^2 + 12u + 18 \ln|u| + C_1$$

$$= \left( \sqrt{x} - 3 \right)^2 + 12 \left( \sqrt{x} - 3 \right) + 18 \ln \left| \sqrt{x} - 3 \right| + C_1$$

$$= x + 6\sqrt{x} + 18 \ln \left| \sqrt{x} - 3 \right| + C$$

where $C = C_1 - 27$.

**31.** $\displaystyle \int \cot\left( \frac{\theta}{3} \right) d\theta = 3 \int \cot\left( \frac{\theta}{3} \right)\left( \frac{1}{3} \right) d\theta$

$$= 3 \ln \left| \sin \frac{\theta}{3} \right| + C$$

**33.** $\displaystyle \int \csc 2x \, dx = \frac{1}{2} \int (\csc 2x)(2) \, dx$

$$= -\frac{1}{2} \ln |\csc 2x + \cot 2x| + C$$

**35.** $\displaystyle \int (\cos 3\theta - 1) \, d\theta = \frac{1}{3} \int \cos 3\theta(3) \, d\theta - \int d\theta$

$$= \frac{1}{3} \sin 3\theta - \theta + C$$

**37.** $u = 1 + \sin t, \; du = \cos t \, dt$

$$\int \frac{\cos t}{1 + \sin t} \, dt = \ln |1 + \sin t| + C$$

**39.** $u = \sec x - 1, \; du = \sec x \tan x \, dx$

$$\int \frac{\sec x \tan x}{\sec x - 1} \, dx = \ln |\sec x - 1| + C$$

**41.** $\displaystyle y = \int \frac{4}{x} \, dx = 4 \ln |x| + C$

$\quad (1, 2): \; 2 = 4 \ln |1| + C \Rightarrow C = 2$

$\quad\quad y = 4 \ln |x| + 2$

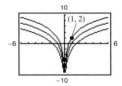

**43.** $\displaystyle y = \int \frac{3}{2 - x} \, dx = -3 \int \frac{1}{x - 2} \, dx = -3 \ln |x - 2| + C$

$\quad (1, 0): \; 0 = -3 \ln |1 - 2| + C \Rightarrow C = 0$

$\quad\quad y = -3 \ln |x - 2|$

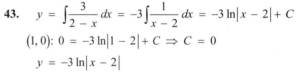

**45.** $\displaystyle s = \int \tan(2\theta) \, d\theta$

$$= \frac{1}{2} \int \tan(2\theta)(2 \, d\theta)$$

$$= -\frac{1}{2} \ln |\cos 2\theta| + C$$

$\quad (0, 2): \; 2 = -\frac{1}{2} \ln |\cos(0)| + C \Rightarrow C = 2$

$\quad\quad s = -\frac{1}{2} \ln |\cos 2\theta| + 2$

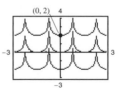

**47.** $f''(x) = \dfrac{2}{x^2} = 2x^{-2}, \quad x > 0$

$\quad f'(x) = \dfrac{-2}{x} + C$

$\quad f'(1) = 1 = -2 + C \Rightarrow C = 3$

$\quad f'(x) = \dfrac{-2}{x} + 3$

$\quad f(x) = -2 \ln x + 3x + C_1$

$\quad f(1) = 1 = -2(0) + 3 + C_1 \Rightarrow C_1 = -2$

$\quad f(x) = -2 \ln x + 3x - 2$

**49.** $\dfrac{dy}{dx} = \dfrac{1}{x + 2}, \ (0, 1)$

(a)

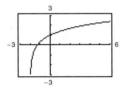

(b) $\quad y = \displaystyle\int \dfrac{1}{x + 2}\, dx = \ln|x + 2| + C$

$y(0) = 1 \Rightarrow 1 = \ln 2 + C \Rightarrow C = 1 - \ln 2$

So, $y = \ln|x + 2| + 1 - \ln 2 = \ln\left(\dfrac{x + 2}{2}\right) + 1.$

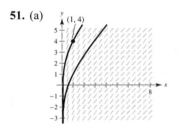

**51.** (a)

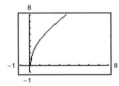

(b) $\dfrac{dy}{dx} = 1 + \dfrac{1}{x}, \ (1, 4)$

$y = x + \ln x + C$

$4 = 1 + 0 + C \Rightarrow C = 3$

$y = x + \ln x + 3$

**53.** $\displaystyle\int_0^4 \dfrac{5}{3x + 1}\, dx = \left[\dfrac{5}{3}\ln|3x + 1|\right]_0^4 = \dfrac{5}{3}\ln 13 \approx 4.275$

**55.** $u = 1 + \ln x, \ du = \dfrac{1}{x}\, dx$

$\displaystyle\int_1^e \dfrac{(1 + \ln x)^2}{x}\, dx = \left[\dfrac{1}{3}(1 + \ln x)^3\right]_1^e = \dfrac{7}{3}$

**57.** $\displaystyle\int_0^2 \dfrac{x^2 - 2}{x + 1}\, dx = \int_0^2 \left(x - 1 - \dfrac{1}{x + 1}\right) dx$

$= \left[\dfrac{1}{2}x^2 - x - \ln|x + 1|\right]_0^2 = -\ln 3$

$\approx -1.099$

**59.** $\displaystyle\int_1^2 \dfrac{1 - \cos\theta}{\theta - \sin\theta}\, d\theta = \left[\ln|\theta - \sin\theta|\right]_1^2$

$= \ln\left|\dfrac{2 - \sin 2}{1 - \sin 1}\right| \approx 1.929$

**61.** $\displaystyle\int \dfrac{1}{1 + \sqrt{x}}\, dx = 2\sqrt{x} - 2\ln\left(1 + \sqrt{x}\right) + C$

**63.** $\displaystyle\int \dfrac{\sqrt{x}}{x - 1}\, dx = \ln\left(\dfrac{\sqrt{x} - 1}{\sqrt{x} + 1}\right) + 2\sqrt{x} + C$

**65.** $\displaystyle\int_{\pi/4}^{\pi/2} \left(\csc x - \sin x\right) dx = \ln\left(\sqrt{2} + 1\right) - \dfrac{\sqrt{2}}{2} \approx 0.174$

**Note: In Exercises 67 and 69, you can use the Second Fundamental Theorem of Calculus or integrate the function.**

**67.** $F(x) = \displaystyle\int_1^x \dfrac{1}{t}\, dt$

$F'(x) = \dfrac{1}{x}$

**69.** $F(x) = \displaystyle\int_1^{3x} \dfrac{1}{t}\, dt$

$F'(x) = \dfrac{1}{3x}(3) = \dfrac{1}{x}$

(by Second Fundamental Theorem of Calculus)

**Alternate Solution:**

$F(x) = \displaystyle\int_1^{3x} \dfrac{1}{t}\, dt = \left[\ln|t|\right]_1^{3x} = \ln|3x|$

$F'(x) = \dfrac{1}{3x}(3) = \dfrac{1}{x}$

**71.**

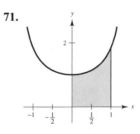

$A \approx 1.25$; Matches (d)

**73.** $A = \displaystyle\int_1^3 \dfrac{6}{x}\, dx = \left[6\ln|x|\right]_1^3 = 6\ln 3$

**75.** $A = \displaystyle\int_0^{\pi/4} \tan x \, dx = -\ln|\cos x|\Big]_0^{\pi/4} = -\ln\dfrac{\sqrt{2}}{2} + 0 = \ln\sqrt{2} = \dfrac{\ln 2}{2}$

**77.** $A = \displaystyle\int_1^4 \dfrac{x^2 + 4}{x}\,dx = \int_1^4 \left(x + \dfrac{4}{x}\right)dx = \left[\dfrac{x^2}{2} + 4\ln x\right]_1^4 = (8 + 4\ln 4) - \dfrac{1}{2} = \dfrac{15}{2} + 8\ln 2 \approx 13.045$

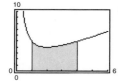

**79.** $\displaystyle\int_0^2 2\sec\dfrac{\pi x}{6}\,dx = \dfrac{12}{\pi}\int_0^2 \sec\left(\dfrac{\pi x}{6}\right)\dfrac{\pi}{6}\,dx = \dfrac{12}{\pi}\left[\ln\left|\sec\dfrac{\pi x}{6} + \tan\dfrac{\pi x}{6}\right|\right]_0^2$

$\qquad = \dfrac{12}{\pi}\left(\ln\left|\sec\dfrac{\pi}{3} + \tan\dfrac{\pi}{3}\right| - \ln|1 + 0|\right) = \dfrac{12}{\pi}\ln\left(2 + \sqrt{3}\right) \approx 5.03041$

**81.** $f(x) = \dfrac{12}{x},\, b - a = 5 - 1 = 4,\, n = 4$

Trapezoid: $\dfrac{4}{2(4)}\big[f(1) + 2f(2) + 2f(3) + 2f(4) + f(5)\big] = \dfrac{1}{2}[12 + 12 + 8 + 6 + 2.4] = 20.2$

Simpson: $\dfrac{4}{3(4)}\big[f(1) + 4f(2) + 2f(3) + 4f(4) + f(5)\big] = \dfrac{1}{3}[12 + 24 + 8 + 12 + 2.4] \approx 19.4667$

Calculator: $\displaystyle\int_1^5 \dfrac{12}{x}\,dx \approx 19.3133$

Exact: $12\ln 5$

**83.** $f(x) = \ln x,\, b - a = 6 - 2 = 4,\, n = 4$

Trapezoid: $\dfrac{4}{2(4)}\big[f(2) + 2f(3) + 2f(4) + 2f(5) + f(6)\big] = \dfrac{1}{2}[0.6931 + 2.1972 + 2.7726 + 3.2189 + 1.7918] \approx 5.3368$

Simpson: $\dfrac{4}{3(4)}\big[f(2) + 4f(3) + 2f(4) + 4f(5) + f(6)\big] \approx 5.3632$

Calculator: $\displaystyle\int_2^6 \ln x \, dx \approx 5.3643$

**85.** Power Rule

**87.** Substitution: $\left(u = x^2 + 4\right)$ and Log Rule

**89.** $\displaystyle\int_1^x \dfrac{3}{t}\,dt = \int_{1/4}^x \dfrac{1}{t}\,dt$

$\big[3\ln|t|\big]_1^x = \big[\ln|t|\big]_{1/4}^x$

$3\ln x = \ln x - \ln\left(\dfrac{1}{4}\right)$

$2\ln x = -\ln\left(\dfrac{1}{4}\right) = \ln 4$

$\ln x = \dfrac{1}{2}\ln 4 = \ln 2$

$x = 2$

**91.** $\int \cot u \, du = \int \dfrac{\cos u}{\sin u} \, du = \ln|\sin u| + C$

**Alternate solution:**

$\dfrac{d}{du}\Big[\ln|\sin u| + C\Big] = \dfrac{1}{\sin u} \cos u + C = \cot u + C$

**93.** $-\ln|\cos x| + C = \ln\left|\dfrac{1}{\cos x}\right| + C = \ln|\sec x| + C$

**95.** $\ln|\sec x + \tan x| + C = \ln\left|\dfrac{(\sec x + \tan x)(\sec x - \tan x)}{(\sec x - \tan x)}\right| + C$

$= \ln\left|\dfrac{\sec^2 x - \tan^2 x}{\sec x - \tan x}\right| + C = \ln\left|\dfrac{1}{\sec x - \tan x}\right| + C = -\ln|\sec x - \tan x| + C$

**97.** Average value $= \dfrac{1}{4-2}\displaystyle\int_2^4 \dfrac{8}{x^2} \, dx$

$= 4\displaystyle\int_2^4 x^{-2} \, dx$

$= \left[-4\dfrac{1}{x}\right]_2^4 = -4\left(\dfrac{1}{4} - \dfrac{1}{2}\right) = 1$

**99.** Average value $= \dfrac{1}{e-1}\displaystyle\int_1^e \dfrac{2\ln x}{x} \, dx$

$= \dfrac{2}{e-1}\left[\dfrac{(\ln x)^2}{2}\right]_1^e$

$= \dfrac{1}{e-1}(1-0) = \dfrac{1}{e-1} \approx 0.582$

**101.** $P(t) = \displaystyle\int \dfrac{3000}{1+0.25t} \, dt = (3000)(4)\displaystyle\int \dfrac{0.25}{1+0.25t} \, dt$

$= 12{,}000\ln|1 + 0.25t| + C$

$P(0) = 12{,}000\ln|1 + 0.25(0)| + C = 1000$

$C = 1000$

$P(t) = 12{,}000\ln|1 + 0.25t| + 1000$

$= 1000\big[12\ln|1 + 0.25t| + 1\big]$

$P(3) = 1000\big[12(\ln 1.75) + 1\big] \approx 7715$

**103.** $\dfrac{1}{50-40}\displaystyle\int_{40}^{50} \dfrac{90{,}000}{400+3x} \, dx = \Big[3000\ln|400 + 3x|\Big]_{40}^{50}$

$\approx \$168.27$

**105.** False

$\dfrac{1}{2}(\ln x) = \ln\left(x^{1/2}\right) \neq (\ln x)^{1/2}$

**107.** True

$\displaystyle\int \dfrac{1}{x} \, dx = \ln|x| + C_1 = \ln|x| + \ln|C| = \ln|Cx|, \; C \neq 0$

**109. (a)** $2x^2 - y^2 = 8$

$y^2 = 2x^2 - 8$

$y_1 = \sqrt{2x^2 - 8}$

$y_2 = -\sqrt{2x^2 - 8}$

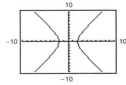

**(b)** $y^2 = e^{-\int(1/x)\,dx} = e^{-\ln x + C} = e^{\ln(1/x)}\left(e^C\right) = \dfrac{1}{x}k$

Let $k = 4$ and graph

$y^2 = \dfrac{4}{x}. \; \left(y_1 = \dfrac{2}{\sqrt{x}}, \; y_2 = \dfrac{2}{\sqrt{x}}\right)$

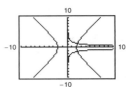

**(c)** In part (a): $2x^2 - y^2 = 8$

$4x - 2yy' = 0$

$y' = \dfrac{2x}{y}$

In part (b): $y^2 = \dfrac{4}{x} = 4x^{-1}$

$2yy' = \dfrac{-4}{x^2}$

$y' = \dfrac{-2}{yx^2} = \dfrac{-2y}{y^2x^2} = \dfrac{-2y}{4x} = \dfrac{-y}{2x}$

Using a graphing utility the graphs intersect at $(2.214, 1.344)$. The slopes are 3.295 and $-0.304 = (-1)/3.295$, respectively.

**111.** Let $f(t) = \ln t$ on $[x, y]$,  $0 < x < y$.

By the Mean Value Theorem,

$$\frac{f(y) - f(x)}{y - x} = f'(c), \quad x < c < y,$$

$$\frac{\ln y - \ln x}{y - x} = \frac{1}{c}.$$

Because $0 < x < c < y, \dfrac{1}{x} > \dfrac{1}{c} > \dfrac{1}{y}$. So,

$$\frac{1}{y} < \frac{\ln y - \ln x}{y - x} < \frac{1}{x}.$$

# Section 5.3   Inverse Functions

**1.** (a)    $f(x) = 5x + 1$

$$g(x) = \frac{x - 1}{5}$$

$$f(g(x)) = f\left(\frac{x - 1}{5}\right) = 5\left(\frac{x - 1}{5}\right) + 1 = x$$

$$g(f(x)) = g(5x + 1) = \frac{(5x - 1) - 1}{5} = x$$

(b)

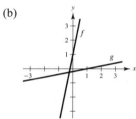

**3.** (a)    $f(x) = x^3$

$$g(x) = \sqrt[3]{x}$$

$$f(g(x)) = f(\sqrt[3]{x}) = (\sqrt[3]{x})^3 = x$$

$$g(f(x)) = g(x^3) = \sqrt[3]{x^3} = x$$

(b)

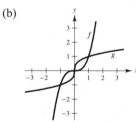

**5.** (a)    $f(x) = \sqrt{x - 4}$

$$g(x) = x^2 + 4, \quad x \geq 0$$

$$f(g(x)) = f(x^2 + 4)$$

$$= \sqrt{(x^2 + 4) - 4} = \sqrt{x^2} = x$$

$$g(f(x)) = g(\sqrt{x - 4})$$

$$= (\sqrt{x - 4})^2 + 4 = x - 4 + 4 = x$$

(b)

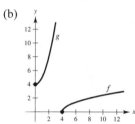

**7.** (a)    $f(x) = \dfrac{1}{x}$

$$g(x) = \frac{1}{x}$$

$$f(g(x)) = \frac{1}{1/x} = x$$

$$g(f(x)) = \frac{1}{1/x} = x$$

(b)

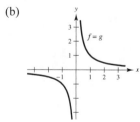

**9.** Matches (c)

**11.** Matches (a)

**13.** $f(x) = \frac{3}{4}x + 6$

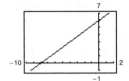

One-to-one; has an inverse

**15.** $f(\theta) = \sin \theta$

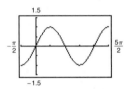

Not one-to-one; does not have an inverse

**17.** $h(s) = \dfrac{1}{s - 2} - 3$

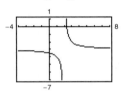

One-to-one; has an inverse

**19.** $f(x) = \ln x$

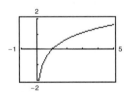

One-to-one; has an inverse

**21.** $g(x) = (x + 5)^3$

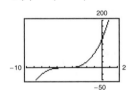

One-to-one; has an inverse

**23.** (a)  $f(x) = 2x - 3 = y$

$$x = \frac{y + 3}{2}$$

$$y = \frac{x + 3}{2}$$

$$f^{-1}(x) = \frac{x + 3}{2}$$

(b)

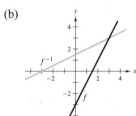

(c) The graphs of $f$ and $f^{-1}$ are reflections of each other across the line $y = x$.

(d) Domain of $f$:   all real numbers

Range of $f$:   all real numbers

Domain of $f^{-1}$:   all real numbers

Range of $f^{-1}$:   all real numbers

**25.** (a)  $f(x) = x^5 = y$

$$x = \sqrt[5]{y}$$

$$y = \sqrt[5]{x}$$

$$f^{-1}(x) = \sqrt[5]{x} = x^{1/5}$$

(b)

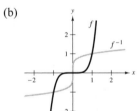

(c) The graphs of $f$ and $f^{-1}$ are reflections of each other across the line $y = x$.

(d) Domain of $f$:   all real numbers

Range of $f$:   all real numbers

Domain of $f^{-1}$:   all real numbers

Range of $f^{-1}$:   all real numbers

**27.** (a) $f(x) = \sqrt{x} = y$

$$x = y^2$$
$$y = x^2$$
$$f^{-1}(x) = x^2, \quad x \geq 0$$

(b)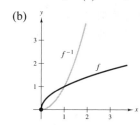

(c) The graphs of $f$ and $f^{-1}$ are reflections of each other across the line $y = x$.

(d) Domain of $f$:      $x \geq 0$

Range of $f$:      $y \geq 0$

Domain of $f^{-1}$:  $x \geq 0$

Range of $f^{-1}$:  $y \geq 0$

**29.** (a) $f(x) = \sqrt{4 - x^2} = y, \quad 0 \leq x \leq 2$

$$4 - x^2 = y^2$$
$$x^2 = 4 - y^2$$
$$x = \sqrt{4 - y^2}$$
$$y = \sqrt{4 - x^2}$$
$$f^{-1}(x) = \sqrt{4 - x^2}, \quad 0 \leq x \leq 2$$

(b)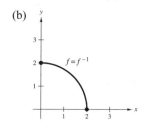

(c) The graphs of $f$ and $f^{-1}$ are reflections of each other across the line $y = x$. In fact, the graphs are identical.

(d) Domain of $f$:      $0 \leq x \leq 2$

Range of $f$:      $0 \leq y \leq 2$

Domain of $f^{-1}$:  $0 \leq x \leq 2$

Range of $f^{-1}$:  $0 \leq y \leq 2$

**31.** (a) $f(x) = \sqrt[3]{x - 1} = y$

$$x - 1 = y^3$$
$$x = y^3 + 1$$
$$y = x^3 + 1$$
$$f^{-1}(x) = x^3 + 1$$

(b)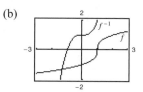

(c) The graphs of $f$ and $f^{-1}$ are reflections of each other across the line $y = x$

(d) Domain of $f$:      all real numbers

Range of $f$:      all real numbers

Domain of $f^{-1}$:  all real numbers

Range of $f^{-1}$:  all real numbers

**33.** (a) $f(x) = x^{2/3} = y, \quad x \geq 0$

$$x = y^{3/2}$$
$$y = x^{3/2}$$
$$f^{-1}(x) = x^{3/2}, \quad x \geq 0$$

(b)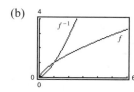

(c) The graphs of $f$ and $f^{-1}$ are reflections of each other across the line $y = x$

(d) Domain of $f$:      $x \geq 0$

Range of $f$:      $y \geq 0$

Domain of $f^{-1}$:  $x \geq 0$

Range of $f^{-1}$:  $y \geq 0$

**35.** (a) $f(x) = \dfrac{x}{\sqrt{x^2 + 7}} = y$

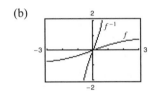

$$x = y\sqrt{x^2 + 7}$$

$$x^2 = y^2(x^2 + 7) = y^2 x^2 + 7y^2$$

$$x^2(1 - y^2) = 7y^2$$

$$x = \frac{\sqrt{7}\,y}{\sqrt{1 - y^2}}$$

$$y = \frac{\sqrt{7}\,x}{\sqrt{1 - x^2}}$$

$$f^{-1}(x) = \frac{\sqrt{7}\,x}{\sqrt{1 - x^2}}, \quad -1 < x < 1$$

(b)

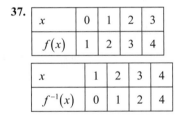

(c) The graphs of $f$ and $f^{-1}$ are reflections of each other in the line $y = x$

(d) Domain of $f$:   all real numbers

   Range of $f$:     $-1 < y < 1$

   Domain of $f^{-1}$:  $-1 < x < 1$

   Range of $f^{-1}$:   all real numbers

**37.**

| $x$    | 0 | 1 | 2 | 3 |
|--------|---|---|---|---|
| $f(x)$ | 1 | 2 | 3 | 4 |

| $x$         | 1 | 2 | 3 | 4 |
|-------------|---|---|---|---|
| $f^{-1}(x)$ | 0 | 1 | 2 | 4 |

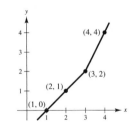

**39.** (a) Let $x$ be the number of pounds of the commodity costing 1.25 per pound. Because there are 50 pounds total, the amount of the second commodity is $50 - x$. The total cost is

$$y = 1.25x + 1.60(50 - x)$$

$$= -0.35x + 80, \quad 0 \le x \le 50.$$

(b) Find the inverse of the original function.

$$y = -0.35x + 80$$

$$0.35x = 80 - y$$

$$x = \frac{100}{35}(80 - y)$$

Inverse: $y = \frac{100}{35}(80 - x) = \frac{20}{7}(80 - x)$

$x$ represents cost and $y$ represents pounds.

(c) Domain of inverse is $62.5 \le x \le 80$.

(d) If $x = 73$ in the inverse function,

$$y = \frac{100}{35}(80 - 73) = \frac{100}{5} = 20 \text{ pounds.}$$

**41.** $f(x) = 2 - x - x^3$

$f'(x) = -1 - 3x^2 < 0$ for all $x$

$f$ is decreasing on $(-\infty, \infty)$. Therefore, $f$ is strictly monotonic and has an inverse.

**43.** $f(x) = \dfrac{x^4}{4} - 2x^2$

$f'(x) = x^3 - 4x = 0$ when $x = 0, 2, -2$

$f$ is not strictly monotonic on $(-\infty, \infty)$. Therefore, $f$ does not have an inverse.

**45.** $f(x) = \ln(x - 3), \quad x > 3$

$f'(x) = \dfrac{1}{x - 3} > 0$ for $x > 3$

$f$ is increasing on $(3, \infty)$. Therefore, $f$ is strictly monotonic and has an inverse.

**47.** $f(x) = (x - 4)^2$ on $[4, \infty)$

$f'(x) = 2(x - 4) > 0$ on $[4, \infty)$

$f$ is increasing on $[4, \infty)$. Therefore, $f$ is strictly monotonic and has an inverse.

**49.** $f(x) = \dfrac{4}{x^2}$ on $(0, \infty)$

$f'(x) = -\dfrac{8}{x^3} < 0$ on $(0, \infty)$

$f$ is decreasing on $(0, \infty)$. Therefore, $f$ is strictly monotonic and has an inverse.

**51.** $f(x) = \cos x$ on $[0, \pi]$

$f'(x) = -\sin x < 0$ on $(0, \pi)$

$f$ is decreasing on $[0, \pi]$. Therefore, $f$ is strictly monotonic and has an inverse.

**53.** $f(x) = \dfrac{x}{x^2 - 4} = y$ on $(-2, 2)$

$x^2 y - 4y = x$

$x^2 y - x - 4y = 0$

$(a = y, b = -1, c = -4y)$

$x = \dfrac{1 \pm \sqrt{1 - 4(y)(-4y)}}{2y} = \dfrac{1 \pm \sqrt{1 + 16y^2}}{2y}$

$y = f^{-1}(x) = \begin{cases} \left(1 - \sqrt{1 + 16x^2}\right)\big/2x, & \text{if } x \neq 0 \\ 0, & \text{if } x = 0 \end{cases}$

Domain of $f^{-1}$: all $x$

Range of $f^{-1}$: $-2 < y < 2$

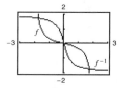

The graphs of $f$ and $f^{-1}$ are reflections of each other across the line $y = x$.

**55.** (a), (b)

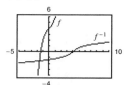

(c) Yes, $f$ is one-to-one and has an inverse. The inverse relation is an inverse function.

**57.** (a), (b)

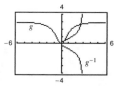

(c) $g$ is not one-to-one and does not have an inverse. The inverse relation is not an inverse function.

**59.** $f(x) = \sqrt{x - 2}$, Domain: $x \geq 2$

$f'(x) = \dfrac{1}{2\sqrt{x - 2}} > 0$ for $x > 2$

$f$ is one-to-one; has an inverse

$\sqrt{x - 2} = y$

$x - 2 = y^2$

$x = y^2 + 2$

$y = x^2 + 2$

$f^{-1}(x) = x^2 + 2, \quad x \geq 0$

**61.** $f(x) = |x - 2|, \quad x \leq 2$

$= -(x - 2)$

$= 2 - x$

$f$ is one-to-one; has an inverse

$2 - x = y$

$2 - y = x$

$f^{-1}(x) = 2 - x, \quad x \geq 0$

**63.** $f(x) = (x - 3)^2$ is one-to-one for $x \geq 3$.

$(x - 3)^2 = y$

$x - 3 = \sqrt{y}$

$x = \sqrt{y} + 3$

$y = \sqrt{x} + 3$

$f^{-1}(x) = \sqrt{x} + 3, \quad x \geq 0$

(Answer is not unique.)

**65.** $f(x) = |x + 3|$ is one-to-one for $x \geq -3$.

$x + 3 = y$

$x = y - 3$

$y = x - 3$

$f^{-1}(x) = x - 3, \quad x \geq 0$

(Answer is not unique.)

**67.** Yes, the volume is an increasing function, and therefore one-to-one. The inverse function gives the time $t$ corresponding to the volume $V$.

**69.** No, $C(t)$ is not one-to-one because long distance costs are step functions. A call lasting 2.1 minutes costs the same as one lasting 2.2 minutes.

**71.** $f(x) = x^3 - 1, \quad a = 26$

$f'(x) = 3x^2$

$f$ is monotonic (increasing) on $(-\infty, \infty)$ therefore $f$ has an inverse.

$$f(3) = 26 \Rightarrow f^{-1}(26) = 3$$

$$\left(f^{-1}\right)'(26) = \frac{1}{f'\left(f^{-1}(26)\right)} = \frac{1}{f'(3)} = \frac{1}{3(3^2)} = \frac{1}{27}$$

**73.** $f(x) = x^3 + 2x - 1, \quad a = 2$

$f'(x) = 3x^2 + 2 > 0$

$f$ is monotonic (increasing) on $(-\infty, \infty)$ therefore $f$ has an inverse.

$$f(1) = 2 \Rightarrow f^{-1}(2) = 1$$

$$\left(f^{-1}\right)'(2) = \frac{1}{f'\left(f^{-1}(2)\right)} = \frac{1}{f'(1)} = \frac{1}{3(1^2) + 2} = \frac{1}{5}$$

**75.** $f(x) = \sin x, \quad a = 1/2, -\dfrac{\pi}{2} \le x \le \dfrac{\pi}{2}$

$f'(x) = \cos x > 0$ on $\left(-\dfrac{\pi}{2}, \dfrac{\pi}{2}\right)$

$f$ is monotonic (increasing) on $\left[-\dfrac{\pi}{2}, \dfrac{\pi}{2}\right]$ therefore $f$ has an inverse.

$$f\left(\frac{\pi}{6}\right) = \sin \frac{\pi}{6} = \frac{1}{2} \Rightarrow f^{-1}\left(\frac{1}{2}\right) = \frac{\pi}{6}$$

$$\left(f^{-1}\right)'\left(\frac{1}{2}\right) = \frac{1}{f'\left(f^{-1}\left(\frac{1}{2}\right)\right)}$$

$$= \frac{1}{f'\left(\frac{\pi}{6}\right)} = \frac{1}{\cos\left(\frac{\pi}{6}\right)} = \frac{2}{\sqrt{3}} = \frac{2\sqrt{3}}{3}$$

**77.** $f(x) = \dfrac{x + 6}{x - 2}, \quad x > 0, a = 3$

$$f'(x) = \frac{(x - 2)(1) - (x + 6)(1)}{(x - 2)^2}$$

$$= \frac{-8}{(x - 2)^2} < 0 \text{ on } (2, \infty)$$

$f$ is monotonic (decreasing) on $(2, \infty)$ therefore $f$ has an inverse.

$$f(6) = 3 \Rightarrow f^{-1}(3) = 6$$

$$\left(f^{-1}\right)'(3) = \frac{1}{f'\left(f^{-1}(3)\right)} = \frac{1}{f'(6)} = \frac{1}{-8/(6 - 2)^2} = -2$$

**79.** $f(x) = x^3 - \dfrac{4}{x}, \quad a = 6, x > 0$

$$f'(x) = 3x^2 + \frac{4}{x^2} > 0$$

$f$ is monotonic (increasing) on $(0, \infty)$ therefore $f$ has an inverse.

$$f(2) = 6 \Rightarrow f^{-1}(6) = 2$$

$$\left(f^{-1}\right)'(6) = \frac{1}{f'\left(f^{-1}(6)\right)} = \frac{1}{f'(2)} = \frac{1}{3(2^2) + 4/2^2} = \frac{1}{13}$$

**81.** (a)  Domain $f = $ Domain $f^{-1} = (-\infty, \infty)$

(b)  Range $f = $ Range $f^{-1} = (-\infty, \infty)$

(c)

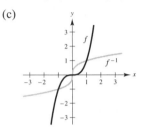

(d)  $f(x) = x^3, \quad \left(\dfrac{1}{2}, \dfrac{1}{8}\right)$

$f'(x) = 3x^2$

$$f'\left(\frac{1}{2}\right) = \frac{3}{4}$$

$$f^{-1}(x) = \sqrt[3]{x}, \quad \left(\frac{1}{8}, \frac{1}{2}\right)$$

$$\left(f^{-1}\right)'(x) = \frac{1}{3\sqrt[3]{x}}$$

$$\left(f^{-1}\right)'\left(\frac{1}{8}\right) = \frac{4}{3}$$

**83.** (a)  Domain $f = [4, \infty)$, Domain $f^{-1} = [0, \infty)$

(b)  Range $f = [0, \infty)$, Range $f^{-1} = [4, \infty)$

(c)

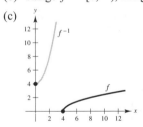

(d)  $f(x) = \sqrt{x - 4}, \quad (5, 1)$

$$f'(x) = \frac{1}{2\sqrt{x - 4}}$$

$$f'(5) = \frac{1}{2}$$

$$f^{-1}(x) = x^2 + 4, \quad (1, 5)$$

$$\left(f^{-1}\right)'(x) = 2x$$

$$\left(f^{-1}\right)'(1) = 2$$

**85.** $x = y^3 - 7y^2 + 2$

$$1 = 3y^2 \frac{dy}{dx} - 14y \frac{dy}{dx}$$

$$\frac{dy}{dx} = \frac{1}{3y^2 - 14y}$$

At $(-4, 1)$, $\dfrac{dy}{dx} = \dfrac{1}{3 - 14} = \dfrac{-1}{11}$.

**Alternate Solution:**

Let $f(x) = x^3 - 7x^2 + 2$. Then

$f'(x) = 3x^2 - 14x$ and $f'(1) = -11$. So,

$$\frac{dy}{dx} = \frac{1}{-11} = \frac{-1}{11}.$$

**In Exercises 87 and 89, use the following.**

$f(x) = \frac{1}{8}x - 3$ **and** $g(x) = x^3$

$f^{-1}(x) = 8(x + 3)$ **and** $g^{-1}(x) = \sqrt[3]{x}$

**87.** $\left(f^{-1} \circ g^{-1}\right)(1) = f^{-1}\left(g^{-1}(1)\right) = f^{-1}(1) = 32$

**89.** $\left(f^{-1} \circ f^{-1}\right)(6) = f^{-1}\left(f^{-1}(6)\right) = f^{-1}(72) = 600$

**In Exercises 91 and 93, use the following.**

$f(x) = x + 4$ **and** $g(x) = 2x - 5$

$f^{-1}(x) = x - 4$ **and** $g^{-1}(x) = \dfrac{x + 5}{2}$

**91.** $\left(g^{-1} \circ f^{-1}\right)(x) = g^{-1}\left(f^{-1}(x)\right)$

$$= g^{-1}(x - 4)$$

$$= \frac{(x - 4) + 5}{2} = \frac{x + 1}{2}$$

**93.** $(f \circ g)(x) = f\left(g(x)\right)$

$$= f(2x - 5)$$

$$= (2x - 5) + 4 = 2x - 1$$

So, $(f \circ g)^{-1}(x) = \dfrac{x + 1}{2}$.

**Note:** $(f \circ g)^{-1} = g^{-1} \circ f^{-1}$

**95.** Let $y = f(x)$ be one-to-one. Solve for $x$ as a function of $y$. Interchange $x$ and $y$ to get $y = f^{-1}(x)$. Let the domain of $f^{-1}$ be the range of $f$. Verify that $f\left(f^{-1}(x)\right) = x$ and $f^{-1}\left(f(x)\right) = x$.

Example:

$f(x) = x^3$; $y = x^3$; $x = \sqrt[3]{y}$; $y = \sqrt[3]{x}$;

$f^{-1}(x) = \sqrt[3]{x}$

**97.** $f$ is not one-to-one because many different $x$-values yield the same $y$-value.

Example: $f(0) = f(\pi) = 0$

Not continuous at $\dfrac{(2n - 1)\pi}{2}$, where $n$ is an integer.

**99.** $f(x) = k\left(2 - x - x^3\right)$ is one-to-one. Because

$f^{-1}(3) = -2$,

$f(-2) = 3 = k\left(2 - (-2) - (-2)^3\right) = 12k \Rightarrow k = \frac{1}{4}$.

**101.** False. Let $f(x) = x^2$.

**103.** True

**105.** (a) $f(x) = 2x^3 + 3x^2 - 36x$

$f$ does not pass the horizontal line test.

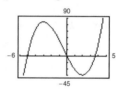

(b) $f'(x) = 6x^2 + 6x - 36$

$$= 6\left(x^2 + x - 6\right) = 6(x + 3)(x - 2)$$

$f'(x) = 0$ at $x = 2, -3$

On the interval $(-2, 2)$, $f$ is one-to-one, so, $c = 2$.

**107.** If $f$ has an inverse, then $f$ and $f^{-1}$ are both one-to-one.

Let $\left(f^{-1}\right)^{-1}(x) = y$ then $x = f^{-1}(y)$ and $f(x) = y$. So,

$\left(f^{-1}\right)^{-1} = f$.

**109.** If $f$ has an inverse and $f(x_1) = f(x_2)$, then

$f^{-1}\left(f(x_1)\right) = f^{-1}\left(f(x_2)\right) \Rightarrow x_1 = x_2$. Therefore, $f$ is one-to-one. If $f(x)$ is one-to-one, then for every value $b$ in the range, there corresponds exactly one value $a$ in the domain. Define $g(x)$ such that the domain of $g$ equals the range of $f$ and $g(b) = a$. By the reflexive property of inverses, $g = f^{-1}$.

**111.** From Theorem 5.9, you have:

$$g'(x) = \frac{1}{f'(g(x))}$$

$$g''(x) = \frac{f'(g(x))(0) - f''(g(x))g'(x)}{\left[f'(g(x))\right]^2}$$

$$= -\frac{f''(g(x)) \cdot \left[1/(f'(g(x)))\right]}{\left[f'(g(x))\right]^2} = -\frac{f''(g(x))}{\left[f'(g(x))\right]^3}$$

If $f$ is increasing and concave down, then $f' > 0$ and $f'' < 0$ which implies that $g$ is increasing and concave up.

**115.** $f(x) = \dfrac{ax + b}{cx + d}$

(a)  Assume $bc - ad \neq 0$ and $f(x_1) = f(x_2)$. Then

$$\frac{ax_1 + b}{cx_1 + d} = \frac{ax_2 + b}{cx_2 + d}$$

$$acx_1x_2 + bcx_2 + adx_1 + bd = acx_1x_2 + adx_2 + bcx_1 + bd$$

$$(ad - bc)x_1 = (ad - bc)x_2$$

$$x_1 = x_2 \qquad (\text{because } ad - bc \neq 0)$$

So, $f$ is one-to-one.

Now assume $f$ is one-to-one. Suppose, on the contrary, that $ad = bc$. If $d = 0$, then either $b = 0$ or $c = 0$. In both cases, $f$ is not one-to-one. Similarly, if $b = 0$, then $a = 0$ or $d = 0$ and $f$ is not one-to-one. So consider

$$f(x) = \frac{ax + b}{cx + d} = \frac{adx + bd}{bcx + bd} \cdot \frac{b}{d} = \frac{bcx + bd}{bcx + bd} \cdot \frac{b}{d} = \frac{b}{d},$$

which is not one-to-one.

**Alternate Solution:**

$$f(x) = \frac{ax + b}{cx + d} \implies f'(x) = \frac{ad - bc}{(cx + d)^2}$$

$f$ is monotonic (and therefore one-to-one) if and only if $ad - bc \neq 0$.

(b) $\qquad\qquad y = \dfrac{ax + b}{cx + d}$

$$cyx + dy = ax + b$$

$$(cy - a)x = b - dy$$

$$x = \frac{b - dy}{cy - a}$$

$$f^{-1}(x) = y = \frac{b - dx}{cx - a}, \quad bc - ad \neq 0$$

(c) $\dfrac{ax + b}{cx + d} = \dfrac{b - dx}{cx - a}$

$$acx^2 + bcx - a^2x - ab = bcx - cdx^2 + bd - d^2x$$

$$(ac + cd)x^2 + (d^2 - a^2)x - bd - ab = 0$$

$$c(a + d)x^2 + (d - a)(d + a)x - b(a + d) = 0$$

So, $f = f^{-1}$ if $a = -d$, or if $c = b = 0$ and $a = d$.

**113.** $f(x) = \displaystyle\int_2^x \sqrt{1 + t^2}\, dt,\ f(2) = 0$

$$f'(x) = \sqrt{1 + x^2}$$

$f'(x) > 0$ for all $x \implies f$ increasing on $(-\infty, \infty) \implies f$ is one-to-one.

$$\left(f^{-1}\right)'(0) = \frac{1}{f'(2)} = \frac{1}{\sqrt{5}} = \frac{\sqrt{5}}{5}$$

# Section 5.4   Exponential Functions: Differentiation and Integration

**1.** $e^{\ln x} = 4$

   $x = 4$

**3.** $e^x = 12$

   $x = \ln 12 \approx 2.485$

**5.** $9 - 2e^x = 7$

   $2e^x = 2$

   $e^x = 1$

   $x = 0$

**7.** $50e^{-x} = 30$

   $e^{-x} = \frac{3}{5}$

   $-x = \ln\left(\frac{3}{5}\right)$

   $x = \ln\left(\frac{5}{3}\right)$

   $\approx 0.511$

**9.** $\dfrac{800}{100 - e^{x/2}} = 50$

   $\dfrac{800}{50} = 100 - e^{x/2}$

   $84 = e^{x/2}$

   $\ln 84 = \dfrac{x}{2}$

   $x = 2 \ln 84 \approx 8.862$

**11.** $\ln x = 2$

   $x = e^2 \approx 7.389$

**13.** $\ln(x - 3) = 2$

   $x - 3 = e^2$

   $x = 3 + e^2 \approx 10.389$

**15.** $\ln \sqrt{x + 2} = 1$

   $\sqrt{x + 2} = e^1 = e$

   $x + 2 = e^2$

   $x = e^2 - 2 \approx 5.389$

**17.** $y = e^{-x}$

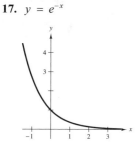

**19.** $y = e^x + 2$

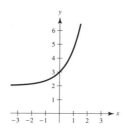

**21.** $y = e^{-x^2}$

Symmetric with respect to the $y$-axis

Horizontal asymptote: $y = 0$

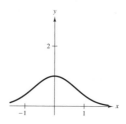

**23.** (a)

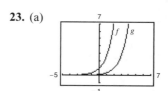

Horizontal shift 2 units to the right

(b)

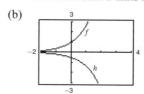

A reflection in the $x$-axis and a vertical shrink

(c)

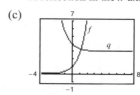

Vertical shift 3 units upward and a reflection in the $y$-axis

**25.** $y = Ce^{ax}$

Horizontal asymptote: $y = 0$

Matches (c)

**27.** $y = C\left(1 - e^{-ax}\right)$

Vertical shift $C$ units

Reflection in both the $x$- and $y$-axes

Matches (a)

**29.** $f(x) = e^{2x}$

$g(x) = \ln\sqrt{x} = \dfrac{1}{2}\ln x$

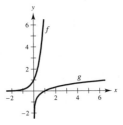

**31.** $f(x) = e^x - 1$

$g(x) = \ln(x + 1)$

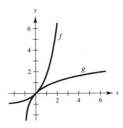

**33.**

As $x \to \infty$, the graph of $f$ approaches the graph of $g$.

$\lim\limits_{x\to\infty}\left(1 + \dfrac{0.5}{x}\right)^x = e^{0.5}$

**35.** $\left(1 + \dfrac{1}{1,000,000}\right)^{1,000,000} \approx 2.718280469$

$e \approx 2.718281828$

$e > \left(1 + \dfrac{1}{1,000,000}\right)^{1,000,000}$

**37. (a)**

$y = e^{3x}$

$y' = 3e^{3x}$

$y'(0) = 3$

Tangent line: $y - 1 = 3(x - 0)$

$y = 3x + 1$

**(b)**

$y = e^{-3x}$

$y' = -3e^{-3x}$

$y'(0) = -3$

Tangent line: $y - 1 = -3(x - 0)$

$y = -3x + 1$

**39.** $f(x) = e^{2x}$

$f'(x) = 2e^{2x}$

**41.** $y = e^{\sqrt{x}}$

$\dfrac{dy}{dx} = \dfrac{e^{\sqrt{x}}}{2\sqrt{x}}$

**43.** $y = e^{x-4}$

$y' = e^{x-4}$

**45.** $y = e^x \ln x$

$y' = e^x\left(\dfrac{1}{x}\right) + e^x \ln x = e^x\left(\dfrac{1}{x} + \ln x\right)$

**47.** $y = x^3 e^x$

$y' = x^3 e^x + 3x^2\left(e^x\right) = x^2 e^x(x + 3) = e^x\left(x^3 + 3x^2\right)$

**49.** $g(t) = \left(e^{-t} + e^t\right)^3$

$g'(t) = 3\left(e^{-t} + e^t\right)^2\left(e^t - e^{-t}\right)$

**51.** $y = \ln\left(1 + e^{2x}\right)$

$\dfrac{dy}{dx} = \dfrac{2e^{2x}}{1 + e^{2x}}$

**53.** $y = \dfrac{2}{e^x + e^{-x}} = 2\left(e^x + e^{-x}\right)^{-1}$

$\dfrac{dy}{dx} = -2\left(e^x + e^{-x}\right)^{-2}\left(e^x - e^{-x}\right) = \dfrac{-2\left(e^x - e^{-x}\right)}{\left(e^x + e^{-x}\right)^2}$

**55.** $y = \dfrac{e^x + 1}{e^x - 1}$

$y' = \dfrac{\left(e^x - 1\right)e^x - \left(e^x + 1\right)e^x}{\left(e^x - 1\right)^2} = \dfrac{-2e^x}{\left(e^x - 1\right)^2}$

**57.**  $y = e^x(\sin x + \cos x)$

$\dfrac{dy}{dx} = e^x(\cos x - \sin x) + (\sin x + \cos x)(e^x)$

$\qquad = e^x(2 \cos x) = 2e^x \cos x$

**59.**  $F(x) = \displaystyle\int_{\pi}^{\ln x} \cos e^t \, dt$

$F'(x) = \cos(e^{\ln x}) \cdot \dfrac{1}{x} = \dfrac{\cos(x)}{x}$

**61.**  $f(x) = e^{1-x}, \quad (1, 1)$

$f'(x) = -e^{1-x}, \ f'(1) = -1$

Tangent line: $y - 1 = -1(x - 1)$

$\qquad\qquad\quad y = -x + 2$

**63.**  $y = \ln(e^{x^2}) = x^2, \quad (-2, 4)$

$\quad y' = 2x$

$y'(-2) = -4$

Tangent line: $y - 4 = -4(x + 2)$

$\qquad\qquad\quad y = -4x - 4$

**65.**  $y = x^2 e^x - 2xe^x + 2e^x, \quad (1, e)$

$\quad y' = x^2 e^x + 2xe^x - 2xe^x - 2e^x + 2e^x = x^2 e^x$

$y'(1) = e$

Tangent line: $y - e = e(x - 1)$

$\qquad\qquad\quad y = ex$

**67.**  $f(x) = e^{-x} \ln x, \quad (1, 0)$

$f'(x) = e^{-x}\left(\dfrac{1}{x}\right) - e^{-x} \ln x = e^{-x}\left(\dfrac{1}{x} - \ln x\right)$

$f'(1) = e^{-1}$

Tangent line: $y - 0 = e^{-1}(x - 1)$

$\qquad\qquad\quad y = \dfrac{1}{e}x - \dfrac{1}{e}$

**69.**  $xe^y - 10x + 3y = 0$

$xe^y \dfrac{dy}{dx} + e^y - 10 + 3\dfrac{dy}{dx} = 0$

$\dfrac{dy}{dx}(xe^y + 3) = 10 - e^y$

$\dfrac{dy}{dx} = \dfrac{10 - e^y}{xe^y + 3}$

**71.**  $xe^y + ye^x = 1, \quad (0, 1)$

$xe^y y' + e^y + ye^x + y'e^x = 0$

At $(0, 1)$: $e + 1 + y' = 0$

$\qquad\qquad\quad y' = -e - 1$

Tangent line: $y - 1 = (-e - 1)(x - 0)$

$\qquad\qquad\quad y = (-e - 1)x + 1$

**73.**  $f(x) = (3 + 2x)e^{-3x}$

$f'(x) = (3 + 2x)(-3e^{-3x}) + 2e^{-3x} = (-7 - 6x)e^{-3x}$

$f''(x) = (-7 - 6x)(-3e^{-3x}) - 6e^{-3x} = 3(6x + 5)e^{-3x}$

**75.**  $\quad y = 4e^{-x}$

$\quad y' = -4e^{-x}$

$\quad y'' = 4e^{-x}$

$y'' - y = 4e^{-x} - 4e^{-x} = 0$

**77.**  $y = e^x\left(\cos \sqrt{2}x + \sin \sqrt{2}x\right)$

$y' = e^x\left(-\sqrt{2} \sin \sqrt{2}x + \sqrt{2} \cos \sqrt{2}x\right) + e^x\left(\cos \sqrt{2}x + \sin \sqrt{2}x\right)$

$\quad = e^x\left[\left(1 + \sqrt{2}\right) \cos \sqrt{2}x + \left(1 - \sqrt{2}\right) \sin \sqrt{2}x\right]$

$y'' = e^x\left[-\left(\sqrt{2} + 2\right) \sin \sqrt{2}x + \left(\sqrt{2} - 2\right) \cos \sqrt{2}x\right] + e^x\left[\left(1 + \sqrt{2}\right) \cos \sqrt{2}x + \left(1 - \sqrt{2}\right) \sin \sqrt{2}x\right]$

$\quad = e^x\left[\left(-1 - 2\sqrt{2}\right) \sin \sqrt{2}x + \left(-1 + 2\sqrt{2}\right) \cos \sqrt{2}x\right]$

$-2y' + 3y = -2e^x\left[\left(1 + \sqrt{2}\right) \cos \sqrt{2}x + \left(1 - \sqrt{2}\right) \sin \sqrt{2}x\right] + 3e^x\left[\cos \sqrt{2}x + \sin \sqrt{2}x\right]$

$\quad = e^x\left[\left(1 - 2\sqrt{2}\right) \cos \sqrt{2}x + \left(1 + 2\sqrt{2}\right) \sin \sqrt{2}x\right] = -y''$

Therefore, $-2y' + 3y = -y'' \Rightarrow y'' - 2y' + 3y = 0$.

**79.** $f(x) = \dfrac{e^x + e^{-x}}{2}$

$f'(x) = \dfrac{e^x - e^{-x}}{2} = 0$ when $x = 0$.

$f''(x) = \dfrac{e^x + e^{-x}}{2} > 0$

Relative minimum: $(0, 1)$

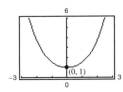

**81.** $g(x) = \dfrac{1}{\sqrt{2\pi}} e^{-(x-2)^2/2}$

$g'(x) = \dfrac{-1}{\sqrt{2\pi}}(x - 2)e^{-(x-2)^2/2}$

$g''(x) = \dfrac{1}{\sqrt{2\pi}}(x - 1)(x - 3)e^{-(x-2)^2/2}$

Relative maximum:

$\left(2, \dfrac{1}{\sqrt{2\pi}}\right) \approx (2, 0.399)$

Points of inflection:

$\left(1, \dfrac{1}{\sqrt{2\pi}}e^{-1/2}\right), \left(3, \dfrac{1}{\sqrt{2\pi}}e^{-1/2}\right) \approx (1, 0.242), (3, 0.242)$

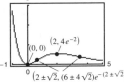

**83.** $f(x) = x^2 e^{-x}$

$f'(x) = -x^2 e^{-x} + 2xe^{-x}$

$\quad = xe^{-x}(2 - x) = 0$ when $x = 0, 2$.

$f''(x) = -e^{-x}(2x - x^2) + e^{-x}(2 - 2x)$

$\quad = e^{-x}(x^2 - 4x + 2) = 0$ when $x = 2 \pm \sqrt{2}$.

Relative minimum: $(0, 0)$

Relative maximum: $(2, 4e^{-2})$

$x = 2 \pm \sqrt{2}$

$y = \left(2 \pm \sqrt{2}\right)^2 e^{-\left(2 \pm \sqrt{2}\right)}$

$\quad = \left(6 \pm 4\sqrt{2}\right)e^{-\left(2 \pm \sqrt{2}\right)}$

Points of inflection:

$\left(2 \pm \sqrt{2}, \left(6 \pm 4\sqrt{2}\right)e^{-\left(2 \pm \sqrt{2}\right)}\right)$

$\approx (3.414, 0.384), (0.586, 0.191)$

**85.** $g(t) = 1 + (2 + t)e^{-t}$

$g'(t) = -(1 + t)e^{-t}$

$g''(t) = te^{-t}$

Relative maximum: $(-1, 1 + e) \approx (-1, 3.718)$

Point of inflection: $(0, 3)$

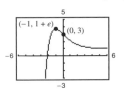

**87.** $A = (\text{base})(\text{height}) = 2xe^{-x^2}$

$\dfrac{dA}{dx} = -4x^2 e^{-x^2} + 2e^{-x^2}$

$\quad = 2e^{-x^2}\left(1 - 2x^2\right) = 0$ when $x = \dfrac{\sqrt{2}}{2}$.

$A = \sqrt{2}e^{-1/2}$

**89.** $f(x) = e^{2x}$

$f'(x) = 2e^{2x}$

Let $(x, y) = (x, e^{2x})$ be the point on the graph where the tangent line passes through the origin. Equating slopes,

$2e^{2x} = \dfrac{e^{2x} - 0}{x - 0}$

$2 = \dfrac{1}{x}$

$x = \dfrac{1}{2}, \quad y = e, \quad y' = 2e.$

Point: $\left(\dfrac{1}{2}, e\right)$

Tangent line: $y - e = 2e\left(x - \dfrac{1}{2}\right)$

$\qquad\qquad\qquad y = 2ex$

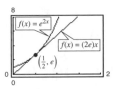

**91.** $V = 15{,}000e^{-0.6286t}, \quad 0 \le t \le 10$

(a)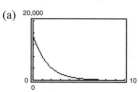

(b) $\dfrac{dV}{dt} = -9429e^{-0.6286t}$

   When $t = 1$, $\dfrac{dV}{dt} \approx -5028.84$.

   When $t = 5$, $\dfrac{dV}{dt} \approx -406.89$.

(c)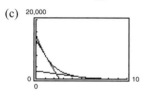

**93.**

| $h$ | 0 | 5 | 10 | 15 | 20 |
|-----|---|---|----|----|----|
| $P$ | 10,332 | 5583 | 2376 | 1240 | 517 |
| $\ln P$ | 9.243 | 8.627 | 7.773 | 7.123 | 6.248 |

(a)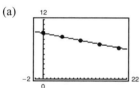

   $y = -0.1499h + 9.3018$ is the regression line for data $(h, \ln P)$.

(b) $\ln P = ah + b$

   $P = e^{ah + b} = e^{b}e^{ah}$

   $P = Ce^{ah}, C = e^{b}$

   $a = -0.1499$ and $C = e^{9.3018} = 10{,}957.7$.

   So, $P = 10{,}957.7e^{-0.1499h}$

(c)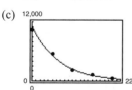

(d) $\dfrac{dP}{dh} = (10{,}957.71)(-0.1499)e^{-0.1499h}$

   $= -1642.56e^{-0.1499h}$

   For $h = 5$, $\dfrac{dP}{dh} = -776.3$.

   For $h = 18$, $\dfrac{dP}{dh} \approx -110.6$.

**95.** $f(x) = e^{x}$      $f(0) = 1$

     $f'(x) = e^{x}$      $f'(0) = 1$

     $f''(x) = e^{x}$      $f''(0) = 1$

     $P_1(x) = 1 + 1(x - 0) = 1 + x$

     $P_2(x) = 1 + 1(x - 0) = \dfrac{1}{2}(1)(x - 0)^2 = 1 + x + \dfrac{x^2}{2}$

The values of $f$, $P_1$, and $P_2$ and their first derivatives agree at $x = 0$.

**97.** $n = 12. 12! = 12 \cdot 11 \cdot 10 \cdots 3 \cdot 2 \cdot 1 = 479{,}001{,}600$

Stirlings Formula:

$12! \approx \left(\dfrac{12}{e}\right)^{12} \sqrt{2\pi(12)} \approx 475{,}687{,}487$

**99.** Let $u = 5x$, $du = 5\,dx$.

$\displaystyle\int e^{5x}(5)\,dx = e^{5x} + C$

**101.** Let $u = 2x - 1$, $du = 2\,dx$.

$\displaystyle\int e^{2x-1}\,dx = \dfrac{1}{2}\int e^{2x-1}(2)\,dx = \dfrac{1}{2}e^{2x-1} + C$

**103.** Let $u = x^3$, $du = 3x^2\,dx$.

$\displaystyle\int x^2 e^{x^3}\,dx = \dfrac{1}{3}\int e^{x^3}(3x^2)\,dx = \dfrac{1}{3}e^{x^3} + C$

**105.** Let $u = \sqrt{x}$, $du = \dfrac{1}{2\sqrt{x}}\,dx$.

$\displaystyle\int \dfrac{e^{\sqrt{x}}}{\sqrt{x}}\,dx = 2\int e^{\sqrt{x}}\left(\dfrac{1}{2\sqrt{x}}\right)dx = 2e^{\sqrt{x}} + C$

**107.** Let $u = 1 + e^{-x}$, $du = -e^{-x}\,dx$.

$\displaystyle\int \dfrac{e^{-x}}{1 + e^{-x}}\,dx = -\int \dfrac{-e^{-x}}{1 + e^{-x}}\,dx = -\ln(1 + e^{-x}) + C$

    $= \ln\left(\dfrac{e^x}{e^x + 1}\right) + C$

    $= x - \ln(e^x + 1) + C$

**109.** Let $u = 1 - e^{x}$, $du = -e^{x}\,dx$.

$\displaystyle\int e^{x}\sqrt{1 - e^{x}}\,dx = -\int (1 - e^{x})^{1/2}(-e^{x})\,dx$

    $= -\dfrac{2}{3}(1 - e^{x})^{3/2} + C$

**111.** Let $u = e^x - e^{-x}$, $du = \left(e^x + e^{-x}\right) dx$.

$$\int \frac{e^x + e^{-x}}{e^x - e^{-x}} \, dx = \ln\left|e^x - e^{-x}\right| + C$$

**113.** $\int \dfrac{5 - e^x}{e^{2x}} \, dx = \int 5e^{-2x} \, dx - \int e^{-x} \, dx$

$$= -\frac{5}{2}e^{-2x} + e^{-x} + C$$

**115.** $\int e^{-x} \tan\left(e^{-x}\right) dx = -\int \left[\tan\left(e^{-x}\right)\right]\left(-e^{-x}\right) dx$

$$= \ln\left|\cos\left(e^{-x}\right)\right| + C$$

**117.** $\displaystyle\int_0^1 e^{-2x} \, dx = -\frac{1}{2}\int_0^1 e^{-2x}(-2) \, dx = \left[-\frac{1}{2}e^{-2x}\right]_0^1$

$$= \frac{1}{2}\left(1 - e^{-2}\right) = \frac{e^2 - 1}{2e^2}$$

**119.** $\displaystyle\int_0^1 xe^{-x^2} \, dx = -\frac{1}{2}\int_0^1 e^{-x^2}(-2x) \, dx$

$$= -\frac{1}{2}\left[e^{-x^2}\right]_0^1$$

$$= -\frac{1}{2}\left[e^{-1} - 1\right]$$

$$= \frac{1 - (1/e)}{2} = \frac{e - 1}{2e}$$

**121.** Let $u = \dfrac{3}{x}$, $du = -\dfrac{3}{x^2} \, dx$.

$$\int_1^3 \frac{e^{3/x}}{x^2} \, dx = -\frac{1}{3}\int_1^3 e^{3/x}\left(-\frac{3}{x^2}\right) dx$$

$$= \left[-\frac{1}{3}e^{3/x}\right]_1^3 = \frac{e}{3}\left(e^2 - 1\right)$$

**123.** Let $u = 1 + e^{2x}$, $du = 2e^{2x} \, dx$.

$$\int_0^3 \frac{2e^{2x}}{1 + e^{2x}} \, dx = \left[\ln\left(1 + e^{2x}\right)\right]_0^3$$

$$= \ln\left(1 + e^6\right) - \ln 2 = \ln\left(\frac{1 + e^6}{2}\right)$$

**125.** Let $u = \sin \pi x$, $du = \pi \cos \pi x \, dx$.

$$\int_0^{\pi/2} e^{\sin \pi x} \cos \pi x \, dx = \frac{1}{\pi}\int_0^{\pi/2} e^{\sin \pi x}(\pi \cos \pi x) \, dx$$

$$= \frac{1}{\pi}\left[e^{\sin \pi x}\right]_0^{\pi/2}$$

$$= \frac{1}{\pi}\left[e^{\sin(\pi^2/2)} - 1\right]$$

**127.** Let $u = ax^2$, $du = 2ax \, dx$. (Assume $a \neq 0$.)

$$y = \int xe^{ax^2} \, dx$$

$$= \frac{1}{2a}\int e^{ax^2}(2ax) \, dx = \frac{1}{2a}e^{ax^2} + C$$

**129.** $f'(x) = \int \frac{1}{2}\left(e^x + e^{-x}\right)dx = \frac{1}{2}\left(e^x - e^{-x}\right) + C_1$

$f'(0) = C_1 = 0$

$f(x) = \int \frac{1}{2}\left(e^x - e^{-x}\right)dx = \frac{1}{2}\left(e^x + e^{-x}\right) + C_2$

$f(0) = 1 + C_2 = 1 \Rightarrow C_2 = 0$

$f(x) = \frac{1}{2}\left(e^x + e^{-x}\right)$

**131.** (a)

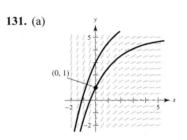

(b) $\dfrac{dy}{dx} = 2e^{-x/2}$, $(0, 1)$

$$y = \int 2e^{-x/2} \, dx = -4\int e^{-x/2}\left(-\frac{1}{2} \, dx\right)$$

$$= -4e^{-x/2} + C$$

$(0, 1)$: $1 = -4e^0 + C = -4 + C \Rightarrow C = 5$

$y = -4e^{-x/2} + 5$

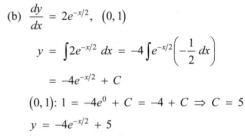

**133.** $\displaystyle\int_0^5 e^x dx = \left[e^x\right]_0^5 = e^5 - 1 \approx 147.413$

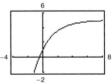

**135.** $\displaystyle\int_0^{\sqrt{6}} xe^{-x^2/4} dx = \left[-2e^{-x^2/4}\right]_0^{\sqrt{6}}$

$$= -2e^{-3/2} + 2 \approx 1.554$$

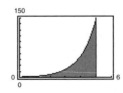

**137.** $\int_0^4 \sqrt{x} e^x dx$, $n = 12$

Midpoint Rule:    92.1898

Trapezoidal Rule:  93.8371

Simpson's Rule:   92.7385

Graphing utility:  92.7437

**139.** $0.0665 \int_{48}^{60} e^{-0.0139(t-48)^2} dt$

Graphing utility: $0.4772 = 47.72\%$

**141.** $x(t) = Ae^{kt} + Be^{-kt}$; $A, B, k > 0$

(a) $x'(t) = Ake^{kt} - Bke^{-kt} = 0$

$$Ae^{kt} = Be^{-kt}$$

$$e^{2kt} = \frac{B}{A}$$

$$2kt = \ln\left(\frac{B}{A}\right)$$

$$t = \frac{1}{2k}\ln\left(\frac{B}{A}\right)$$

By the first Derivative Test, this is a minimum.

(b) $x''(t) = Ak^2 e^{kt} + Bk^2 e^{-kt}$

$$= k^2\left(Ae^{kt} + Be^{-kt}\right) = k^2 x(t)$$

$k^2$ is the constant of proportionality.

**143.** $f(x) = e^x$. Domain is $(-\infty, \infty)$ and range is $(0, \infty)$. $f$ is continuous, increasing, one-to-one, and concave upwards on its entire domain.

$$\lim_{x \to -\infty} e^x = 0 \text{ and } \lim_{x \to \infty} e^x = \infty.$$

**145.** (a) Log Rule: $\left(u = e^x + 1\right)$

(b) Substitution: $\left(u = x^2\right)$

**147.** $\int_0^x e^t \, dt \geq \int_0^x 1 \, dt$

$$\left[e^t\right]_0^x \geq \left[t\right]_0^x$$

$$e^x - 1 \geq x \Rightarrow e^x \geq 1 + x \text{ for } x \geq 0$$

**149.** $e^{-x} = x \Rightarrow f(x) = x - e^{-x}$

$$f'(x) = 1 + e^{-x}$$

$$x_{n+1} = x_n - \frac{f(x_n)}{f'(x_n)} = x_n - \frac{x_n - e^{-x_n}}{1 + e^{-x_n}}$$

$$x_1 = 1$$

$$x_2 = x_1 - \frac{f(x_1)}{f'(x_1)} \approx 0.5379$$

$$x_3 = x_2 - \frac{f(x_2)}{f'(x_2)} \approx 0.5670$$

$$x_4 = x_3 - \frac{f(x_3)}{f'(x_3)} \approx 0.5671$$

Approximate the root of $f$ to be $x \approx 0.567$.

**151.** $y = \dfrac{L}{1 + ae^{-x/b}}$, $\quad a > 0, b > 0, L > 0$

$$y' = \frac{-L\left(-\dfrac{a}{b}e^{-x/b}\right)}{\left(1 + ae^{-x/b}\right)^2} = \frac{\dfrac{aL}{b}e^{-x/b}}{\left(1 + ae^{-x/b}\right)^2}$$

$$y'' = \frac{\left(1 + ae^{-x/b}\right)^2\left(\dfrac{-aL}{b^2}e^{-x/b}\right) - \left(\dfrac{aL}{b}e^{-x/b}\right)2\left(1 + ae^{-x/b}\right)\left(\dfrac{-a}{b}e^{-x/b}\right)}{\left(1 + ae^{-x/b}\right)^4}$$

$$= \frac{\left(1 + ae^{-x/b}\right)\left(\dfrac{-aL}{b^2}e^{-x/b}\right) + 2\left(\dfrac{aL}{b}e^{-x/b}\right)\left(\dfrac{a}{b}e^{-x/b}\right)}{\left(1 + ae^{-x/b}\right)^3} = \frac{Lae^{-x/b}\left[ae^{-x/b} - 1\right]}{\left(1 + ae^{-x/b}\right)^3 b^2}$$

$y'' = 0$ if $ae^{-x/b} = 1 \Rightarrow \dfrac{-x}{b} = \ln\left(\dfrac{1}{a}\right) \Rightarrow x = b \ln a$

$$y(b \ln a) = \frac{L}{1 + ae^{-(b \ln a)/b}} = \frac{L}{1 + a(1/a)} = \frac{L}{2}$$

Therefore, the $y$-coordinate of the inflection point is $L/2$.

## Section 5.5 Bases Other than *e* and Applications

**1.** $\log_2 \frac{1}{8} = \log_2 2^{-3} = -3$

**3.** $\log_7 1 = 0$

**5.** (a) $\quad 2^3 = 8$

$\qquad \log_2 8 = 3$

(b) $\quad 3^{-1} = \frac{1}{3}$

$\qquad \log_3 \frac{1}{3} = -1$

**7.** (a) $\log_{10} 0.01 = -2$

$\qquad 10^{-2} = 0.01$

(b) $\log_{0.5} 8 = -3$

$\qquad 0.5^{-3} = 8$

$\qquad \left(\frac{1}{2}\right)^{-3} = 8$

**9.** $y = 3^x$

| $x$ | $-2$ | $-1$ | 0 | 1 | 2 |
|-----|------|------|---|---|---|
| $y$ | $\frac{1}{9}$ | $\frac{1}{3}$ | 1 | 3 | 9 |

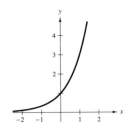

**11.** $y = \left(\frac{1}{3}\right)^x = 3^{-x}$

| $x$ | $-2$ | $-1$ | 0 | 1 | 2 |
|-----|------|------|---|---|---|
| $y$ | 9 | 3 | 1 | $\frac{1}{3}$ | $\frac{1}{9}$ |

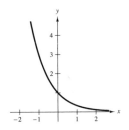

**13.** $h(x) = 5^{x-2}$

| $x$ | $-1$ | 0 | 1 | 2 | 3 |
|-----|------|---|---|---|---|
| $y$ | $\frac{1}{125}$ | $\frac{1}{25}$ | $\frac{1}{5}$ | 1 | 5 |

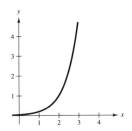

**15.** $f(x) = 3^x$

Increasing function that passes through $(0, 1)$ and $(1, 3)$. Matches (d).

**17.** $f(x) = 3^x - 1$

Increasing function that passes through $(0, 0)$ and $(1, 2)$. Matches (b).

**19.** (a) $\log_{10} 1000 = x$

$\qquad 10^x = 1000$

$\qquad x = 3$

(b) $\log_{10} 0.1 = x$

$\qquad 10^x = 0.1$

$\qquad x = -1$

**21.** (a) $\log_3 x = -1$

$\qquad 3^{-1} = x$

$\qquad x = \frac{1}{3}$

(b) $\log_2 x = -4$

$\qquad 2^{-4} = x$

$\qquad x = \frac{1}{16}$

**23.** (a) $\qquad x^2 - x = \log_5 25$

$\qquad x^2 - x = \log_5 5^2 = 2$

$\qquad x^2 - x - 2 = 0$

$\qquad (x + 1)(x - 2) = 0$

$\qquad\qquad x = -1 \quad \text{OR} \quad x = 2$

(b) $3x + 5 = \log_2 64$

$\qquad 3x + 5 = \log_2 2^6 = 6$

$\qquad 3x = 1$

$\qquad x = \frac{1}{3}$

**25.**   $3^{2x} = 75$

$2x \ln 3 = \ln 75$

$x = \left(\dfrac{1}{2}\right)\dfrac{\ln 75}{\ln 3} \approx 1.965$

**27.**   $2^{3-z} = 625$

$(3 - z) \ln 2 = \ln 625$

$3 - z = \dfrac{\ln 625}{\ln 2}$

$z = 3 - \dfrac{\ln 625}{\ln 2} \approx -6.288$

**29.**   $\left(1 + \dfrac{0.09}{12}\right)^{12t} = 3$

$12t \ln\left(1 + \dfrac{0.09}{12}\right) = \ln 3$

$t = \left(\dfrac{1}{12}\right)\dfrac{\ln 3}{\ln\left(1 + \dfrac{0.09}{12}\right)} \approx 12.253$

**31.**   $\log_2(x - 1) = 5$

$x - 1 = 2^5 = 32$

$x = 33$

**33.**   $\log_3 x^2 = 4.5$

$x^2 = 3^{4.5}$

$x = \pm\sqrt{3^{4.5}} \approx \pm 11.845$

**35.**   $g(x) = 6\left(2^{1-x}\right) - 25$

Zero: $x \approx -1.059$

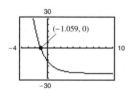

**37.**   $h(s) = 32 \log_{10}(s - 2) + 15$

Zero: $s \approx 2.340$

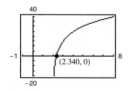

**39.**   $f(x) = 4^x$

$g(x) = \log_4 x$

| $x$    | $-2$           | $-1$          | $0$ | $\frac{1}{2}$ | $1$ |
|--------|----------------|---------------|-----|---------------|-----|
| $f(x)$ | $\frac{1}{16}$ | $\frac{1}{4}$ | $1$ | $2$           | $4$ |

| $x$    | $\frac{1}{16}$ | $\frac{1}{4}$ | $1$ | $2$           | $4$ |
|--------|----------------|---------------|-----|---------------|-----|
| $g(x)$ | $-2$           | $-1$          | $0$ | $\frac{1}{2}$ | $1$ |

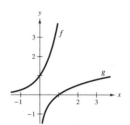

**41.**   $f(x) = 4^x$

$f'(x) = (\ln 4)4^x$

**43.**   $y = 5^{-4x}$

$y' = -4(\ln 5)5^{-4x} = \dfrac{-4 \ln 5}{625^x}$

**45.**   $f(x) = x9^x$

$f'(x) = x(\ln 9)9^x + 9^x = 9^x(1 + x \ln 9)$

**47.**   $g(t) = t^2 2^t$

$g'(t) = t^2(\ln 2)2^t + (2t)2^t$

$= t2^t(t \ln 2 + 2)$

$= 2^t t(2 + t \ln 2)$

**49.**   $h(\theta) = 2^{-\theta} \cos \pi\theta$

$h'(\theta) = 2^{-\theta}(-\pi \sin \pi\theta) - (\ln 2)2^{-\theta} \cos \pi\theta$

$= -2^{-\theta}\left[(\ln 2) \cos \pi\theta + \pi \sin \pi\theta\right]$

**51.**   $y = \log_4(5x + 1)$

$y' = \dfrac{1}{(5x + 1) \ln 4}(5) = \dfrac{5}{\ln 4(5x + 1)}$

**53.**   $h(t) = \log_5(4 - t)^2 = 2 \log_5(4 - t)$

$h'(t) = 2\dfrac{-1}{\ln(5)(4 - t)} = \dfrac{2}{(t - 4) \ln 5}$

**55.** $y = \log_5 \sqrt{x^2 - 1} = \frac{1}{2} \log_5(x^2 - 1)$

$\dfrac{dy}{dx} = \dfrac{1}{2} \cdot \dfrac{2x}{(x^2 - 1)\ln 5} = \dfrac{x}{(x^2 - 1)\ln 5}$

**57.** $f(x) = \log_2 \dfrac{x^2}{x - 1} = 2\log_2 x - \log_2(x - 1)$

$f'(x) = \dfrac{2}{x \ln 2} - \dfrac{1}{(x - 1)\ln 2} = \dfrac{x - 2}{(\ln 2)x(x - 1)}$

**59.** $h(x) = \log_3 \dfrac{x\sqrt{x - 1}}{2}$

$= \log_3 x + \dfrac{1}{2}\log_3(x - 1) - \log_3 2$

$h'(x) = \dfrac{1}{x \ln 3} + \dfrac{1}{2} \cdot \dfrac{1}{(x - 1)\ln 3} - 0$

$= \dfrac{1}{\ln 3}\left[\dfrac{1}{x} + \dfrac{1}{2(x - 1)}\right] = \dfrac{1}{\ln 3}\left[\dfrac{3x - 2}{2x(x - 1)}\right]$

**61.** $g(t) = \dfrac{10 \log_4 t}{t} = \dfrac{10}{\ln 4}\left(\dfrac{\ln t}{t}\right)$

$g'(t) = \dfrac{10}{\ln 4}\left[\dfrac{t(1/t) - \ln t}{t^2}\right]$

$= \dfrac{10}{t^2 \ln 4}[1 - \ln t] = \dfrac{5}{t^2 \ln 2}(1 - \ln t)$

**63.** $y = 2^{-x}, \quad (-1, 2)$

$y' = -2^{-x}\ln(2)$

At $(-1, 2)$, $y' = -2\ln(2)$.

Tangent line: $y - 2 = -2\ln(2)(x + 1)$

$\qquad\qquad\qquad y = -2x \ln 2 + 2 - 2\ln 2$

**65.** $y = \log_3 x, \quad (27, 3)$

$y' = \dfrac{1}{x \ln 3}$

At $(27, 3)$, $y' = \dfrac{1}{27 \ln 3}$.

Tangent line: $y - 3 = \dfrac{1}{27 \ln 3}(x - 27)$

$\qquad\qquad\qquad y = \dfrac{1}{27 \ln 3}x + 3 - \dfrac{1}{\ln 3}$

**67.** $y = x^{2/x}$

$\ln y = \dfrac{2}{x}\ln x$

$\dfrac{1}{y}\left(\dfrac{dy}{dx}\right) = \dfrac{2}{x}\left(\dfrac{1}{x}\right) + \ln x\left(-\dfrac{2}{x^2}\right) = \dfrac{2}{x^2}(1 - \ln x)$

$\dfrac{dy}{dx} = \dfrac{2y}{x^2}(1 - \ln x) = 2x^{(2/x)-2}(1 - \ln x)$

**69.** $y = (x - 2)^{x+1}$

$\ln y = (x + 1)\ln(x - 2)$

$\dfrac{1}{y}\left(\dfrac{dy}{dx}\right) = (x + 1)\left(\dfrac{1}{x - 2}\right) + \ln(x - 2)$

$\dfrac{dy}{dx} = y\left[\dfrac{x + 1}{x - 2} + \ln(x - 2)\right]$

$= (x - 2)^{x+1}\left[\dfrac{x + 1}{x - 2} + \ln(x - 2)\right]$

**71.** $y = x^{\sin x}, \quad \left(\dfrac{\pi}{2}, \dfrac{\pi}{2}\right)$

$\ln y = \sin x \ln x$

$\dfrac{y'}{y} = \dfrac{\sin x}{x} + \cos x \ln x$

At $\left(\dfrac{\pi}{2}, \dfrac{\pi}{2}\right)$: $\dfrac{y'}{(\pi/2)} = \dfrac{1}{(\pi/2)} + 0$

$\qquad\qquad\qquad y' = 1$

Tangent line: $y - \dfrac{\pi}{2} = 1\left(x - \dfrac{\pi}{2}\right)$

$\qquad\qquad\qquad y = x$

**73.** $y = (\ln x)^{\cos x}, \quad (e, 1)$

$\ln y = \cos x \cdot \ln(\ln x)$

$\dfrac{y'}{y} = \cos x \cdot \dfrac{1}{x \ln x} - \sin x \cdot \ln(\ln x)$

At $(e, 1)$, $y' = \cos(e)\dfrac{1}{e} - 0$.

Tangent line: $y - 1 = \dfrac{\cos e}{e}(x - e)$

$\qquad\qquad\qquad y = \dfrac{\cos e}{e}x + 1 - \cos e$

**75.** $\displaystyle\int 3^x \, dx = \dfrac{3^x}{\ln 3} + C$

**77.** $\displaystyle\int (x^2 + 2^{-x}) \, dx = \dfrac{x^3}{3} + \dfrac{-1}{\ln 2}2^{-x} + C$

$= \dfrac{1}{3}x^3 - \dfrac{2^{-x}}{\ln 2} + C$

**79.** $\displaystyle\int x(5^{-x^2}) \, dx = -\dfrac{1}{2}\int 5^{-x^2}(-2x) \, dx$

$= -\left(\dfrac{1}{2}\right)\dfrac{5^{-x^2}}{\ln 5} + C = \dfrac{-1}{2\ln 5}(5^{-x^2}) + C$

**81.** $\int \dfrac{3^{2x}}{1 + 3^{2x}}\, dx$, $u = 1 + 3^{2x}$, $du = 2(\ln 3)3^{2x}\, dx$

$$\dfrac{1}{2 \ln 3} \int \dfrac{(2 \ln 3)3^{2x}}{1 + 3^{2x}}\, dx = \dfrac{1}{2 \ln 3} \ln\!\left(1 + 3^{2x}\right) + C$$

**83.** $\displaystyle\int_{-1}^{2} 2^x\, dx = \left[\dfrac{2^x}{\ln 2}\right]_{-1}^{2} = \dfrac{1}{\ln 2}\left[4 - \dfrac{1}{2}\right] = \dfrac{7}{2 \ln 2} = \dfrac{7}{\ln 4}$

**85.** $\displaystyle\int_{0}^{1}\left(5^x - 3^x\right) dx = \left[\dfrac{5^x}{\ln 5} - \dfrac{3^x}{\ln 3}\right]_{0}^{1}$

$$= \left(\dfrac{5}{\ln 5} - \dfrac{3}{\ln 3}\right) - \left(\dfrac{1}{\ln 5} - \dfrac{1}{\ln 3}\right)$$

$$= \dfrac{4}{\ln 5} - \dfrac{2}{\ln 3}$$

**87.** Area $= \displaystyle\int_{0}^{3} 3^x\, dx = \left[\dfrac{1}{\ln 3} 3^x\right]_{0}^{3} = \dfrac{1}{\ln 3}(27 - 1) = \dfrac{26}{\ln 3}$

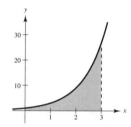

**89.** $\dfrac{dy}{dx} = 0.4^{x/3}$, $\left(0, \dfrac{1}{2}\right)$

(a)

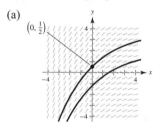

(b) $y = \displaystyle\int 0.4^{x/3}\, dx = 3 \int 0.4^{x/3}\left(\dfrac{1}{3}\, dx\right) = \dfrac{3}{\ln 0.4} 0.4^{x/3} + C$

$$\dfrac{1}{2} = \dfrac{3}{\ln 0.4} + C \Rightarrow C = \dfrac{1}{2} - \dfrac{3}{\ln 0.4}$$

$$y = \dfrac{3}{\ln 0.4}(0.4)^{x/3} + \dfrac{1}{2} - \dfrac{3}{\ln 0.4}$$

$$= \dfrac{3}{\ln 0.4}\left(0.4^{x/3} - 1\right) + \dfrac{1}{2} = \dfrac{3\left(1 - 0.4^{x/3}\right)}{\ln 2.5} + \dfrac{1}{2}$$

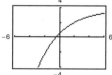

**91.** $f(x) = \log_{10} x$

(a) Domain: $x > 0$

(b) $y = \log_{10} x$

$$10^y = x$$

$$f^{-1}(x) = 10^x$$

(c) $\log_{10} 1000 = \log_{10} 10^3 = 3$

$\log_{10} 10{,}000 = \log_{10} 10^4 = 4$

If $1000 \le x \le 10{,}000$, then $3 \le f(x) \le 4$.

(d) If $f(x) < 0$, then $0 < x < 1$.

(e) $f(x) + 1 = \log_{10} x + \log_{10} 10 = \log_{10}(10x)$

$x$ must have been increased by a factor of 10.

(f) $\log_{10}\!\left(\dfrac{x_1}{x_2}\right) = \log_{10} x_1 - \log_{10} x_2$

$$= 3n - n = 2n$$

So, $x_1/x_2 = 10^{2n} = 100^n$.

**93.** (a) $y = x^a$

$y' = ax^{a-1}$

(b) $y = a^x$

$y' = (\ln a)a^x$

(c) $y = x^x$

$\ln y = x \ln x$

$\dfrac{1}{y}y' = x \cdot \dfrac{1}{x} + (1) \ln x$

$y' = y(1 + \ln x)$

$y' = x^x(1 + \ln x)$

(d) $y = a^a$

$y' = 0$

**95.** $C(t) = P(1.05)^t$

(a) $C(10) = 24.95(1.05)^{10}$

$\approx \$40.64$

(b) $\dfrac{dC}{dt} = P(\ln 1.05)(1.05)^t$

When $t = 1$, $\dfrac{dC}{dt} \approx 0.051P$.

When $t = 8$, $\dfrac{dC}{dt} \approx 0.072P$.

(c) $\dfrac{dC}{dt} = (\ln 1.05)\left[P(1.05)^t\right] = (\ln 1.05)C(t)$

The constant of proportionality is $\ln 1.05$.

**97.** $P = \$1000, r = 3\frac{1}{2}\% = 0.035, t = 10$

$$A = 1000\left(1 + \frac{0.035}{n}\right)^{10n}$$

$$A = 1000e^{(0.035)(10)} = 1419.07$$

| $n$ | 1 | 2 | 4 | 12 | 365 | Continuous |
|---|---|---|---|---|---|---|
| $A$ | 1410.60 | 1414.78 | 1416.91 | 1418.34 | 1419.04 | 1419.07 |

**99.** $P = \$1000, r = 5\% = 0.05, t = 30$

$$A = 1000\left(1 + \frac{0.05}{n}\right)^{30n}$$

$$A = 1000e^{(0.05)30} = 4481.69$$

| $n$ | 1 | 2 | 4 | 12 | 365 | Continuous |
|---|---|---|---|---|---|---|
| $A$ | 4321.94 | 4399.79 | 4440.21 | 4467.74 | 4481.23 | 4481.69 |

**101.** $100,000 = Pe^{0.05t} \Rightarrow P = 100,000e^{-0.05t}$

| $t$ | 1 | 10 | 20 | 30 | 40 | 50 |
|---|---|---|---|---|---|---|
| $P$ | 95,122.94 | 60,653.07 | 36,787.94 | 22,313.02 | 13,533.53 | 8208.50 |

**103.** $100,000 = P\left(1 + \frac{0.05}{12}\right)^{12t} \Rightarrow P = 100,000\left(1 + \frac{0.05}{12}\right)^{-12t}$

| $t$ | 1 | 10 | 20 | 30 | 40 | 50 |
|---|---|---|---|---|---|---|
| $P$ | 95,132.82 | 60,716.10 | 36,864.45 | 22,382.66 | 13,589.88 | 8251.24 |

**105.** (a) $A = 20,000\left(1 + \frac{0.06}{365}\right)^{(365)(8)} \approx \$32,320.21$

    (b) $A = \$30,000$

    (c) $A = 8000\left(1 + \frac{0.06}{365}\right)^{(365)(8)} + 20,000\left(1 + \frac{0.06}{365}\right)^{(365)(4)} \approx \$12,928.09 + 25,424.48 = \$38,352.57$

    (d) $A = 9000\left[\left(1 + \frac{0.06}{365}\right)^{(365)(8)} + \left(1 + \frac{0.06}{365}\right)^{(365)(4)} + 1\right] \approx \$34,985.11$

    Take option (c).

**107.** (a) $\lim\limits_{t \to \infty} 6.7e^{(-48.1)/t} = 6.7e^0 = 6.7$ million ft$^3$/acre

    (b) $\quad V' = \frac{322.27}{t^2}e^{-(48.1)/t}$

        $V'(20) \approx 0.073$ million ft$^3$/acre/yr

        $V'(60) \approx 0.040$ million ft$^3$/acre/yr

**109.** $y = \dfrac{300}{3 + 17e^{-0.0625x}}$

(a)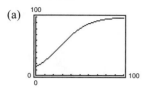

(b) If $x = 2$ (2000 egg masses), $y \approx 16.67 \approx 16.7\%$.

(c) If $y = 66.67\%$, then $x \approx 38.8$ or 38,800 egg masses.

(d) $y = 300\left(3 + 17e^{-0.0625x}\right)^{-1}$

$y' = \dfrac{318.75e^{-0.0625x}}{\left(3 + 17e^{-0.0625x}\right)^2}$

$y'' = \dfrac{19.921875e^{-0.0625x}\left(17e^{-0.0625x} - 3\right)}{\left(3 + 17e^{-0.0625x}\right)^3}$

$17e^{-0.0625x} - 3 = 0 \Rightarrow$

$x \approx 27.75$ or 27,750 egg masses.

**111.** (a) $B = 4.7539(6.7744)^d = 4.7539e^{1.9132d}$

(b)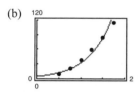

(c) $B'(d) = 9.0952e^{1.9132d}$

$B'(0.8) \approx 42.03$ tons/inch

$B'(1.5) \approx 160.38$ tons/inch

**113.** (a) $\displaystyle\int_0^4 f(t)\,dt \approx 5.67$

$\displaystyle\int_0^4 g(t)\,dt \approx 5.67$

$\displaystyle\int_0^4 h(t)\,dt \approx 5.67$

(b)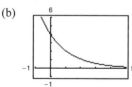

(c) The functions appear to be equal:
$f(t) = g(t) = h(t)$. Analytically,

$f(t) = 4\left(\dfrac{3}{8}\right)^{2t/3} = 4\left[\left(\dfrac{3}{8}\right)^{2/3}\right]^t = 4\left(\dfrac{9^{1/3}}{4}\right)^t = g(t)$

$h(t) = 4e^{-0.653886t} = 4\left[e^{-0.653886}\right]^t \approx 4(0.52002)^t$

$g(t) = 4\left(\dfrac{9^{1/3}}{4}\right)^t \approx 4(0.52002)^t$

No. The definite integrals over a given interval may be equal when the functions are not equal.

**115.**

| $t$ | 0 | 1 | 2 | 3 | 4 |
|---|---|---|---|---|---|
| $y$ | 1200 | 720 | 432 | 259.20 | 155.52 |

$y = C\left(k^t\right)$

When $t = 0$, $y = 1200 \Rightarrow C = 1200$.

$y = 1200\left(k^t\right)$

$\dfrac{720}{1200} = 0.6, \dfrac{432}{720} = 0.6, \dfrac{259.20}{432} = 0.6, \dfrac{155.52}{259.20} = 0.6$

Let $k = 0.6$.

$y = 1200(0.6)^t$

**117.** $\quad 5^{1/\ln 5} = x$

$\ln\left[5^{1/\ln 5}\right] = \ln x$

$\dfrac{1}{\ln 5}[\ln 5] = \ln x$

$1 = \ln x$

$x = e$

**119.** $\quad 9^{1/\ln 3} = x$

$\ln\left[9^{1/\ln 3}\right] = \ln x$

$\dfrac{1}{\ln 3}\ln\left(3^2\right) = \ln x$

$\dfrac{1}{\ln 3}\,2\ln 3 = \ln x$

$\ln x = 2$

$x = e^2$

**121.** False. $e$ is an irrational number.

**123.** True

$f(g(x)) = 2 + e^{\ln(x-2)} = 2 + x - 2 = x$

$g(f(x)) = \ln(2 + e^x - 2) = \ln e^x = x$

**125.** True

$\dfrac{d}{dx}\left[e^x\right] = e^x$ and $\dfrac{d}{dx}\left[e^{-x}\right] = -e^{-x}$

$e^x = e^{-x}$ when $x = 0$.

$\left(e^0\right)\left(-e^{-0}\right) = -1$

**127.** (a) $\left(2^3\right)^2 = 8^2 = 64$

$2^{\left(3^2\right)} = 2^9 = 512$

(b) In general, $f(x) = \left(x^x\right)^x = x^{\left(x^2\right)}$ and $g(x) = x^{\left(x^x\right)}$ are not the same.

For example, $f(3) = 3^9 = 19683$, whereas $g(3) = 3^{27}$. Note that when $x = 2$, $f(2) = g(2) = 16$.

(c) (i) $\quad y = f(x) = \left(x^x\right)^x = x^{\left(x^2\right)}$

$\ln y = x^2 \ln x$

$\dfrac{y'}{y} = 2x \ln x + x^2\left(\dfrac{1}{x}\right)$

$y' = x^{\left(x^2\right)}(x)(2 \ln x + 1)$

$\quad = x^{x^2 + 1}(2 \ln x + 1)$

(ii) $\quad y = g(x) = x^{\left(x^x\right)}$

$\ln y = x^x \ln x$

(**Note:** $\dfrac{d}{dx}\left[x^x\right] = x^x(1 + \ln x)$ from Example 5)

$\dfrac{y'}{y} = x^x(1 + \ln x)\ln x + x^x\dfrac{1}{x}$

$y' = x^{\left(x^x\right)} \cdot x^x\left[(1 + \ln x)\ln x + \dfrac{1}{x}\right]$

$\quad = x^{\left(x^x\right) + x - 1}\left[x(\ln x)^2 + x \ln x + 1\right]$

**129.**
$$\dfrac{dy}{dt} = \dfrac{8}{25}y\left(\dfrac{5}{4} - y\right), \quad y(0) = 1$$

$$\dfrac{dy}{y\left[(5/4) - y\right]} = \dfrac{8}{25}\,dt$$

$$\dfrac{4}{5}\int\left(\dfrac{1}{y} + \dfrac{1}{(5/4) - y}\right)dy = \int\dfrac{8}{25}\,dt$$

$$\ln y - \ln\left(\dfrac{5}{4} - y\right) = \dfrac{2}{5}t + C$$

$$\ln\left(\dfrac{y}{(5/4) - y}\right) = \dfrac{2}{5}t + C$$

$$\dfrac{y}{(5/4) - y} = e^{(2/5)t + C} = C_1 e^{(2/5)t}$$

When $t = 0$, $y = 1 \Rightarrow C_1 = 4 \Rightarrow 4e^{(2/5)t} = \dfrac{y}{(5/4) - y}$

$$4e^{(2/5)t}\left(\dfrac{5}{4} - y\right) = y$$

$$5e^{(2/5)t} = 4e^{(2/5)t}y + y = \left(4e^{(2/5)t} + 1\right)y$$

$$y = \dfrac{5e^{(2/5)t}}{4e^{(2/5)t} + 1} = \dfrac{5}{4 + e^{-0.4t}} = \dfrac{1.25}{1 + 0.25e^{-0.4t}}$$

**131.** (a)
$$y^x = x^y$$
$$x \ln y = y \ln x$$
$$x\frac{y'}{y} + \ln y = \frac{y}{x} + y' \ln x$$
$$y'\left[\frac{x}{y} - \ln x\right] = \frac{y}{x} - \ln y$$
$$y' = \frac{(y/x) - \ln y}{(x/y) - \ln x}$$
$$y' = \frac{y^2 - xy \ln y}{x^2 - xy \ln x}$$

(b) (i)   At $(c, c)$: $y' = \dfrac{c^2 - c^2 \ln c}{c^2 - c^2 \ln c} = 1, (c \neq 0, e)$

   (ii)   At $(2, 4)$: $y' = \dfrac{16 - 8 \ln 4}{4 - 8 \ln 2} = \dfrac{4 - 4 \ln 2}{1 - 2 \ln 2} \approx -3.1774$

   (iii)  At $(4, 2)$: $y' = \dfrac{4 - 8 \ln 2}{16 - 8 \ln 4} = \dfrac{1 - 2 \ln 2}{4 - 4 \ln 2} \approx -0.3147$

(c)  $y'$ is undefined for
$$x^2 = xy \ln x$$
$$x = y \ln x = \ln x^y$$
$$e^x = x^y.$$
At $(e, e)$, $y'$ is undefined.

**133.** Let $f(x) = \dfrac{\ln x}{x}, \quad x > 0.$

$f'(x) = \dfrac{1 - \ln x}{x^2} < 0$ for $x > e \Rightarrow f$ is decreasing for $x \geq e.$ So, for $e \leq x < y$:

$$f(x) > f(y)$$
$$\frac{\ln x}{x} > \frac{\ln y}{y}$$
$$(xy)\frac{\ln x}{x} > (xy)\frac{\ln y}{y}$$
$$\ln x^y > \ln y^x$$
$$x^y > y^x$$

For $n \geq 8, e < \sqrt{n} < \sqrt{n + 1}, \left(\sqrt{8} \approx 2.828\right)$ and so letting $x = \sqrt{n}, y = \sqrt{n + 1},$ you have

$$\left(\sqrt{n}\right)^{\sqrt{n+1}} > \left(\sqrt{n + 1}\right)^{\sqrt{n}}.$$

**Note:** $\sqrt{8}^{\sqrt{9}} \approx 22.6$ and $\sqrt{9}^{\sqrt{8}} \approx 22.4.$

**Note:** This same argument shows $e^\pi > \pi^e.$

## Section 5.6 Inverse Trigonometric Functions: Differentiation

**1.** $y = \arcsin x$

(a)

| $x$ | $-1$ | $-0.8$ | $-0.6$ | $-0.4$ | $-0.2$ | 0 | 0.2 | 0.4 | 0.6 | 0.8 | 1 |
|-----|------|--------|--------|--------|--------|---|-----|-----|-----|-----|---|
| $y$ | $-1.571$ | $-0.927$ | $-0.644$ | $-0.412$ | $-0.201$ | 0 | 0.201 | 0.412 | 0.644 | 0.927 | 1.571 |

(b)

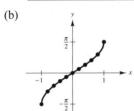

(c)

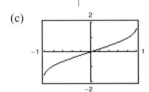

(d) Symmetric about origin: $\arcsin(-x) = -\arcsin x$

Intercept: $(0, 0)$

**3.** $y = \arccos x$

$\left(-\dfrac{\sqrt{2}}{2}, \dfrac{3\pi}{4}\right)$ because $\cos\left(\dfrac{3\pi}{4}\right) = -\dfrac{\sqrt{2}}{2}$

$\left(\dfrac{1}{2}, \dfrac{\pi}{3}\right)$ because $\cos\left(\dfrac{\pi}{3}\right) = \dfrac{1}{2}$

$\left(\dfrac{\sqrt{3}}{2}, \dfrac{\pi}{6}\right)$ because $\cos\left(\dfrac{\pi}{6}\right) = \dfrac{\sqrt{3}}{2}$

**5.** $\arcsin \dfrac{1}{2} = \dfrac{\pi}{6}$

**7.** $\arccos \dfrac{1}{2} = \dfrac{\pi}{3}$

**9.** $\arctan \dfrac{\sqrt{3}}{3} = \dfrac{\pi}{6}$

**11.** $\text{arccsc}\left(-\sqrt{2}\right) = -\dfrac{\pi}{4}$

**13.** $\arccos(-0.8) \approx 2.50$

**15.** $\text{arcsec}(1.269) = \arccos\left(\dfrac{1}{1.269}\right) \approx 0.66$

**17.** (a) $\sin\left(\arctan \dfrac{3}{4}\right) = \dfrac{3}{5}$

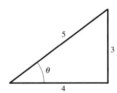

(b) $\sec\left(\arcsin \dfrac{4}{5}\right) = \dfrac{5}{3}$

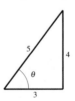

**19.** (a) $\cot\left[\arcsin\left(-\dfrac{1}{2}\right)\right] = \cot\left(-\dfrac{\pi}{6}\right) = -\sqrt{3}$

(b) $\csc\left[\arctan\left(-\dfrac{5}{12}\right)\right] = -\dfrac{13}{5}$

**In Exercises 21–25, use the triangle.**

**21.**   $y = \arccos x$

$\cos y = x$

**23.**   $\tan y = \dfrac{\sqrt{1 - x^2}}{x}$

**25.**   $\sec y = \dfrac{1}{x}$

**27.**   $y = \cos(\arcsin 2x)$

$\theta = \arcsin 2x$

$y = \cos \theta = \sqrt{1 - 4x^2}$

**29.**   $y = \sin(\text{arcsec } x)$

$\theta = \text{arcsec } x, \, 0 \le \theta \le \pi, \, \theta \ne \dfrac{\pi}{2}$

$y = \sin \theta = \dfrac{\sqrt{x^2 - 1}}{|x|}$

The absolute value bars on $x$ are necessary because of the restriction $0 \le \theta \le \pi, \, \theta \ne \pi/2$, and $\sin \theta$ for this domain must always be nonnegative.

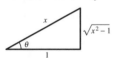

**31.**   $y = \tan\left(\text{arcsec } \dfrac{x}{3}\right)$

$\theta = \text{arcsec } \dfrac{x}{3}$

$y = \tan \theta = \dfrac{\sqrt{x^2 - 9}}{3}$

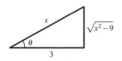

**33.**   $y = \csc\left(\arctan \dfrac{x}{\sqrt{2}}\right)$

$\theta = \arctan \dfrac{x}{\sqrt{2}}$

$y = \csc \theta = \dfrac{\sqrt{x^2 + 2}}{x}$

**35.** (a)

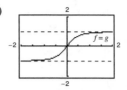

(b)   Let   $y = \arctan(2x)$

$\tan y = 2x$

$\sin y = \sin(\arctan(2x))$

$\phantom{\sin y} = \dfrac{2x}{\sqrt{1 + 4x^2}}.$

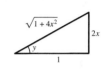

(c)   Asymptotes:  $y = \pm 1$

**37.**   $\arcsin(3x - \pi) = \dfrac{1}{2}$

$3x - \pi = \sin\left(\dfrac{1}{2}\right)$

$x = \dfrac{1}{3}\left[\sin\left(\dfrac{1}{2}\right) + \pi\right] \approx 1.207$

**39.**   $\arcsin \sqrt{2x} = \arccos \sqrt{x}$

$\sqrt{2x} = \sin(\arccos \sqrt{x})$

$\sqrt{2x} = \sqrt{1 - x}, \quad 0 \le x \le 1$

$2x = 1 - x$

$3x = 1$

$x = \dfrac{1}{3}$

**41.** (a)   $\text{arccsc } x = \arcsin \dfrac{1}{x}, \quad |x| \ge 1$

Let  $y = \text{arccsc } x$. Then for

$-\dfrac{\pi}{2} \le y < 0$ and $0 < y \le \dfrac{\pi}{2}$,

$\csc y = x \Rightarrow \sin y = 1/x$. So,  $y = \arcsin(1/x)$.

Therefore,  $\text{arccsc } x = \arcsin(1/x)$.

(b)   $\arctan x + \arctan \dfrac{1}{x} = \dfrac{\pi}{2}, \quad x > 0$

Let  $y = \arctan x + \arctan(1/x)$. Then,

$\tan y = \dfrac{\tan(\arctan x) + \tan[\arctan(1/x)]}{1 - \tan(\arctan x) \tan[\arctan(1/x)]}$

$\phantom{\tan y} = \dfrac{x + (1/x)}{1 - x(1/x)}$

$\phantom{\tan y} = \dfrac{x + (1/x)}{0} \text{ (which is undefined)}.$

So,  $y = \pi/2$. Therefore,

$\arctan x + \arctan(1/x) = \pi/2$.

**43.** $f(x) = 2 \arcsin(x - 1)$

$$f'(x) = \frac{2}{\sqrt{1 - (x - 1)^2}} = \frac{2}{\sqrt{2x - x^2}}$$

**45.** $g(x) = 3 \arccos \dfrac{x}{2}$

$$g'(x) = \frac{-3(1/2)}{\sqrt{1 - (x^2/4)}} = \frac{-3}{\sqrt{4 - x^2}}$$

**47.** $f(x) = \arctan(e^x)$

$$f'(x) = \frac{1}{1 + (e^x)^2} e^x = \frac{e^x}{1 + e^{2x}}$$

**49.** $g(x) = \dfrac{\arcsin 3x}{x}$

$$g'(x) = \frac{x\left(3/\sqrt{1 - 9x^2}\right) - \arcsin 3x}{x^2}$$

$$= \frac{3x - \sqrt{1 - 9x^2}\,\arcsin 3x}{x^2\sqrt{1 - 9x^2}}$$

**51.** $h(t) = \sin(\arccos t) = \sqrt{1 - t^2}$

$$h'(t) = \frac{1}{2}(1 - t^2)^{-1/2}(-2t)$$

$$= \frac{-t}{\sqrt{1 - t^2}}$$

**53.** $y = 2x \arccos x - 2\sqrt{1 - x^2}$

$$y' = 2 \arccos x - 2x\frac{1}{\sqrt{1 - x^2}} - 2\left(\frac{1}{2}\right)(1 - x^2)^{-1/2}(-2x)$$

$$= 2 \arccos x - \frac{2x}{\sqrt{1 - x^2}} + \frac{2x}{\sqrt{1 - x^2}} = 2 \arccos x$$

**55.** $y = \dfrac{1}{2}\left(\dfrac{1}{2}\ln \dfrac{x + 1}{x - 1} + \arctan x\right) = \dfrac{1}{4}\left[\ln(x + 1) - \ln(x - 1)\right] + \dfrac{1}{2}\arctan x$

$$\frac{dy}{dx} = \frac{1}{4}\left(\frac{1}{x + 1} - \frac{1}{x - 1}\right) + \frac{1/2}{1 + x^2} = \frac{1}{1 - x^4}$$

**57.** $y = x \arcsin x + \sqrt{1 - x^2}$

$$\frac{dy}{dx} = x\left(\frac{1}{\sqrt{1 - x^2}}\right) + \arcsin x - \frac{x}{\sqrt{1 - x^2}} = \arcsin x$$

**59.** $y = 8 \arcsin \dfrac{x}{4} - \dfrac{x\sqrt{16 - x^2}}{2}$

$$y' = 2\frac{1}{\sqrt{1 - (x/4)^2}} - \frac{\sqrt{16 - x^2}}{2} - \frac{x}{4}(16 - x^2)^{-1/2}(-2x)$$

$$= \frac{8}{\sqrt{16 - x^2}} - \frac{\sqrt{16 - x^2}}{2} + \frac{x^2}{2\sqrt{16 - x^2}} = \frac{16 - (16 - x^2) + x^2}{2\sqrt{16 - x^2}} = \frac{x^2}{\sqrt{16 - x^2}}$$

**61.** $y = \arctan x + \dfrac{x}{1 + x^2}$

$$y' = \frac{1}{1 + x^2} + \frac{(1 + x^2) - x(2x)}{(1 + x^2)^2}$$

$$= \frac{(1 + x^2) + (1 - x^2)}{(1 + x^2)^2}$$

$$= \frac{2}{(1 + x^2)^2}$$

**63.** $y = 2 \arcsin x, \quad \left(\dfrac{1}{2}, \dfrac{\pi}{3}\right)$

$$y' = \frac{2}{\sqrt{1 - x^2}}$$

At $\left(\dfrac{1}{2}, \dfrac{\pi}{3}\right)$, $y' = \dfrac{2}{\sqrt{1 - (1/4)}} = \dfrac{4}{\sqrt{3}}$.

Tangent line: $y - \dfrac{\pi}{3} = \dfrac{4}{\sqrt{3}}\left(x - \dfrac{1}{2}\right)$

$$y = \frac{4}{\sqrt{3}}x + \frac{\pi}{3} - \frac{2}{\sqrt{3}}$$

$$y = \frac{4\sqrt{3}}{3}x + \frac{\pi}{3} - \frac{2\sqrt{3}}{3}$$

**65.**  $y = \arctan\left(\dfrac{x}{2}\right), \quad \left(2, \dfrac{\pi}{4}\right)$

$y' = \dfrac{1}{1 + (x^2/4)}\left(\dfrac{1}{2}\right) = \dfrac{2}{4 + x^2}$

At $\left(2, \dfrac{\pi}{4}\right)$, $y' = \dfrac{2}{4 + 4} = \dfrac{1}{4}$.

Tangent line: $y - \dfrac{\pi}{4} = \dfrac{1}{4}(x - 2)$

$\qquad\qquad\quad y = \dfrac{1}{4}x + \dfrac{\pi}{4} - \dfrac{1}{2}$

**67.**  $y = 4x \arccos(x - 1), \quad (1, 2\pi)$

$y' = 4x\dfrac{-1}{\sqrt{1 - (x - 1)^2}} + 4\arccos(x - 1)$

At $(1, 2\pi)$, $y' = -4 + 2\pi$.

Tangent line: $y - 2\pi = (2\pi - 4)(x - 1)$

$\qquad\qquad\quad y = (2\pi - 4)x + 4$

**69.**  $f(x) = \arctan x, \quad a = 0$

$f(0) = 0$

$f'(x) = \dfrac{1}{1 + x^2}, \quad f'(0) = 1$

$f''(x) = \dfrac{-2x}{\left(1 + x^2\right)^2}, f''(0) = 0$

$P_1(x) = f(0) + f'(0)x = x$

$P_2(x) = f(0) + f'(0)x + \dfrac{1}{2}f''(0)x^2 = x$

**71.**  $f(x) = \arcsin x, \quad a = \dfrac{1}{2}$

$f'(x) = \dfrac{1}{\sqrt{1 - x^2}}$

$f''(x) = \dfrac{x}{\left(1 - x^2\right)^{3/2}}$

$P_1(x) = f\left(\dfrac{1}{2}\right) + f'\left(\dfrac{1}{2}\right)\left(x - \dfrac{1}{2}\right) = \dfrac{\pi}{6} + \dfrac{2\sqrt{3}}{3}\left(x - \dfrac{1}{2}\right)$

$P_2(x) = f\left(\dfrac{1}{2}\right) + f'\left(\dfrac{1}{2}\right)\left(x - \dfrac{1}{2}\right) + \dfrac{1}{2}f''\left(\dfrac{1}{2}\right)\left(x - \dfrac{1}{2}\right)^2$

$\qquad = \dfrac{\pi}{6} + \dfrac{2\sqrt{3}}{3}\left(x - \dfrac{1}{2}\right) + \dfrac{2\sqrt{3}}{9}\left(x - \dfrac{1}{2}\right)^2$

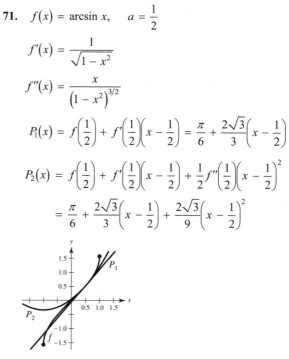

**73.**  $f(x) = \operatorname{arcsec} x - x$

$f'(x) = \dfrac{1}{|x|\sqrt{x^2 - 1}} - 1 = 0 \text{ when } |x|\sqrt{x^2 - 1} = 1$

$x^2\left(x^2 - 1\right) = 1$

$x^4 - x^2 - 1 = 0 \text{ when } x^2 = \dfrac{1 + \sqrt{5}}{2} \text{ or}$

$\qquad\qquad x = \pm\sqrt{\dfrac{1 + \sqrt{5}}{2}} = \pm 1.272$

Relative maximum: $(1.272, -0.606)$

Relative minimum: $(-1.272, 3.747)$

**75.**  $f(x) = \arctan x - \arctan(x - 4)$

$f'(x) = \dfrac{1}{1 + x^2} - \dfrac{1}{1 + (x - 4)^2} = 0$

$\qquad\qquad\qquad 1 + x^2 = 1 + (x - 4)^2$

$\qquad\qquad\qquad\quad 0 = -8x + 16$

$\qquad\qquad\qquad\quad x = 2$

By the First Derivative Test, $(2, 2.214)$ is a relative maximum.

**77.** $f(x) = \arcsin(x - 1)$

$$f'(x) = \frac{1}{\sqrt{1 - (x - 1)^2}} = \frac{1}{\sqrt{2x - x^2}}$$

$$f''(x) = \frac{x - 1}{(2x - x^2)^{3/2}}$$

Maximum: $\left(2, \dfrac{\pi}{2}\right)$

Minimum: $\left(0, -\dfrac{\pi}{2}\right)$

Point of inflection: $(1, 0)$

Domain: $[0, 2]$

Range: $\left[-\dfrac{\pi}{2}, \dfrac{\pi}{2}\right]$

The graph of $f$ is $y = \arcsin x$ shifted 1 unit to the right.

**79.** $f(x) = \text{arcsec } 2x$

$$f'(x) = \frac{1}{|x|\sqrt{4x^2 - 1}}$$

Domain: $\left(-\infty, -\dfrac{1}{2}\right] \cup \left[\dfrac{1}{2}, \infty\right)$

Range: $\left[0, \dfrac{\pi}{2}\right) \cup \left(\dfrac{\pi}{2}, \pi\right]$

Maximum: $\left(-\dfrac{1}{2}, \pi\right)$   Minimum: $\left(\dfrac{1}{2}, 0\right)$

Horizontal asymptote: $y = \dfrac{\pi}{2}$

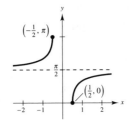

**81.** $\qquad x^2 + x \arctan y = y - 1, \qquad \left(-\dfrac{\pi}{4}, 1\right)$

$$2x + \arctan y + \frac{x}{1 + y^2}y' = y'$$

$$\left(1 - \frac{x}{1 + y^2}\right)y' = 2x + \arctan y$$

$$y' = \frac{2x + \arctan y}{1 - \dfrac{x}{1 + y^2}}$$

At $\left(-\dfrac{\pi}{4}, 1\right)$: $y' = \dfrac{-\dfrac{\pi}{2} + \dfrac{\pi}{4}}{1 - \dfrac{-\pi/4}{2}} = \dfrac{-\dfrac{\pi}{2}}{2 + \dfrac{\pi}{4}} = \dfrac{-2\pi}{8 + \pi}$

Tangent line: $y - 1 = \dfrac{-2\pi}{8 + \pi}\left(x + \dfrac{\pi}{4}\right)$

$$y = \frac{-2\pi}{8 + \pi}x + 1 - \frac{\pi^2}{16 + 2\pi}$$

**83.** $\qquad \arcsin x + \arcsin y = \dfrac{\pi}{2}, \qquad \left(\dfrac{\sqrt{2}}{2}, \dfrac{\sqrt{2}}{2}\right)$

$$\frac{1}{\sqrt{1 - x^2}} + \frac{1}{\sqrt{1 - y^2}}y' = 0$$

$$\frac{1}{\sqrt{1 - y^2}}y' = \frac{-1}{\sqrt{1 - x^2}}$$

At $\left(\dfrac{\sqrt{2}}{2}, \dfrac{\sqrt{2}}{2}\right)$: $y' = -1$

Tangent line: $y - \dfrac{\sqrt{2}}{2} = -1\left(x - \dfrac{\sqrt{2}}{2}\right)$

$$y = -x + \sqrt{2}$$

**85.** The trigonometric functions are not one-to-one on $(-\infty, \infty)$, so their domains must be restricted to intervals on which they are one-to-one.

**87.** $\qquad y = \text{arccot } x \qquad 0 < y < \pi$

$\qquad x = \cot y$

$\tan y = \dfrac{1}{x}$

So, graph the function $y = \arctan(1/x)$ for $x > 0$ and $y = \arctan(1/x) + \pi$ for $x < 0$.

**89.** (a)  $\arcsin(\arcsin(0.5)) \approx 0.551$

$\arcsin(\arcsin(1.0))$ does not exist

(b) In order for $f(x) = \arcsin(\arcsin x)$ to be real, you

must have $-1 \le \arcsin x \le 1$.

Because $\arcsin x = 1 \Rightarrow \sin(1) = x$ and

$\arcsin x = -1 \Rightarrow \sin(-1) = -\sin 1 = x$, you have

$-\sin(1) \le x \le \sin(1)$

$-0.84147 \le x \le 0.84147$

**91.** False

$\arccos \dfrac{1}{2} = \dfrac{\pi}{3}$

because the range is $[0, \pi]$.

**93.** True

$\dfrac{d}{dx}[\arctan x] = \dfrac{1}{1 + x^2} > 0$ for all $x$.

**95.** True

$\dfrac{d}{dx}[\arctan(\tan x)] = \dfrac{\sec^2 x}{1 + \tan^2 x} = \dfrac{\sec^2 x}{\sec^2 x} = 1$

**97.** (a)  $\cot \theta = \dfrac{x}{5}$

$\theta = \text{arccot}\left(\dfrac{x}{5}\right)$

(b) $\dfrac{d\theta}{dt} = \dfrac{-1/5}{1 + (x/5)^2}\dfrac{dx}{dt} = \dfrac{-5}{x^2 + 25}\dfrac{dx}{dt}$

If $\dfrac{dx}{dt} = -400$ and $x = 10$, $\dfrac{d\theta}{dt} = 16$ rad/h.

If $\dfrac{dx}{dt} = -400$ and $x = 3$, $\dfrac{d\theta}{dt} \approx 58.824$ rad/h.

**99.** (a)  $h(t) = -16t^2 + 256$

$-16t^2 + 256 = 0$ when $t = 4$ sec

(b) $\tan \theta = \dfrac{h}{500} = \dfrac{-16t^2 + 256}{500}$

$\theta = \arctan\left[\dfrac{16}{500}(-t^2 + 16)\right]$

$\dfrac{d\theta}{dt} = \dfrac{-8t/125}{1 + \left[(4/125)(-t^2 + 16)\right]^2}$

$= \dfrac{-1000t}{15{,}625 + 16(16 - t^2)^2}$

When $t = 1$, $d\theta/dt \approx -0.0520$ rad/sec.

When $t = 2$, $d\theta/dt \approx -0.1116$ rad/sec.

**101.** $\tan \alpha = \dfrac{40}{x}$

$\tan(\alpha + \theta) = \dfrac{40 + 85}{x} = \dfrac{125}{x}$

$\tan(\alpha + \theta) = \dfrac{\tan \alpha + \tan \theta}{1 - \tan \alpha \tan \theta}$

$\dfrac{125}{x} = \dfrac{40/x + \tan \theta}{1 - \dfrac{40}{x}\tan \theta} = \dfrac{40 + x \tan \theta}{x - 40 \tan \theta}$

$125(x - 40 \tan \theta) = x(40 + x \tan \theta)$

$85x = (x^2 + 5000)\tan \theta$

$\theta = \arctan\left(\dfrac{85x}{x^2 + 5000}\right)$

$\dfrac{d\theta}{dx} = \dfrac{85(5000 - x^2)}{(x^2 + 1600)(x^2 + 15625)} = 0 \Rightarrow x = \sqrt{5000} = 50\sqrt{2}$

By the First Derivative Test, this is a maximum $x = 50\sqrt{2} \approx 70.71$ ft.

**103.** (a) $\tan(\arctan x + \arctan y) = \dfrac{\tan(\arctan x) + \tan(\arctan y)}{1 - \tan(\arctan x)\tan(\arctan y)} = \dfrac{x + y}{1 - xy}, \qquad xy \neq 1$

Therefore, $\arctan x + \arctan y = \arctan\left(\dfrac{x + y}{1 - xy}\right), \; xy \neq 1.$

(b) Let $x = \dfrac{1}{2}$ and $y = \dfrac{1}{3}.$

$\arctan\left(\dfrac{1}{2}\right) + \arctan\left(\dfrac{1}{3}\right) = \arctan\dfrac{(1/2) + (1/3)}{1 - \left[(1/2)\cdot(1/3)\right]} = \arctan\dfrac{5/6}{1 - (1/6)} = \arctan\dfrac{5/6}{5/6} = \arctan 1 = \dfrac{\pi}{4}$

**105.** $f(x) = kx + \sin x$

$f'(x) = k + \cos x \geq 0$ for $k \geq 1$

$f'(x) = k + \cos x \leq 0$ for $k \leq -1$

Therefore, $f(x) = kx + \sin x$ is strictly monotonic and has an inverse for $k \leq -1$ or $k \geq 1.$

**107.** (a) $f(x) = \arccos x + \arcsin x$

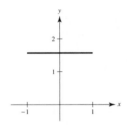

(b) The graph of $f$ is the constant function $y = \pi/2.$

(c) Let $\quad u = \arccos x \quad$ and $\quad v = \arcsin x$

$\qquad\; \cos u = x \qquad$ and $\quad \sin v = x.$

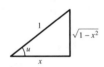

$\sin(u + v) = \sin u \cos v + \sin v \cos u$

$\qquad\qquad\quad = \sqrt{1 - x^2}\,\sqrt{1 - x^2} + x \cdot x$

$\qquad\qquad\quad = 1 - x^2 + x^2 = 1$

So, $u + v = \pi/2.$ Therefore,

$\arccos x + \arcsin x = \pi/2.$

**109.**

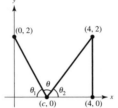

$\tan\theta_1 = \dfrac{2}{c}, \; \tan\theta_2 = \dfrac{2}{4 - c}, \quad 0 < c < 4$

To maximize $\theta$, minimize $f(c) = \theta_1 + \theta_2.$

$f(c) = \arctan\left(\dfrac{2}{c}\right) + \arctan\left(\dfrac{2}{4 - c}\right)$

$f'(c) = \dfrac{-2}{c^2 + 4} + \dfrac{2}{(4 - c)^2 + 4} = 0$

$\dfrac{1}{c^2 + 4} = \dfrac{1}{(4 - c)^2 + 4}$

$c^2 + 4 = c^2 - 8c + 16 + 4$

$8c = 16$

$c = 2$

By the First Derivative Test, $c = 2$ is a minimum. So, $(c, f(c)) = (2, \pi/2)$ is a relative maximum for the angle $\theta.$ Checking the endpoints:

$c = 0: \tan\theta = \dfrac{4}{2} = 2 \Rightarrow \theta \approx 1.107$

$c = 4: \tan\theta = \dfrac{4}{2} = 2 \Rightarrow \theta \approx 1.107$

$c = 2: \theta = \pi - \theta_1 - \theta_2 = \dfrac{\pi}{2} \approx 1.5708$

So, $(2, \pi/2)$ is the absolute maximum.

**111.** $f(x) = \sec x, \quad 0 \le x < \dfrac{\pi}{2}, \pi \le x < \dfrac{3\pi}{2}$

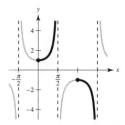

(a)  $y = \text{arcsec } x, \quad x \le -1 \quad \text{or} \quad x \ge 1$

$0 \le y < \dfrac{\pi}{2} \quad \text{or} \quad \pi \le y < \dfrac{3\pi}{2}$

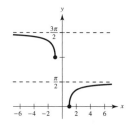

(b)        $y = \text{arcsec } x$

$x = \sec y$

$1 = \sec y \tan y \cdot y'$

$y' = \dfrac{1}{\sec y \tan y}$

$\quad = \dfrac{1}{x\sqrt{x^2 - 1}}$

$\tan^2 y + 1 = \sec^2 y$

$\tan y = \pm\sqrt{\sec^2 y - 1}$

On $0 \le y < \pi/2$ and $\pi \le y < 3\pi/2$, $\tan y \ge 0$.

## Section 5.7   Inverse Trigonometric Functions: Integration

**1.** $\displaystyle\int \dfrac{dx}{\sqrt{9 - x^2}} = \arcsin\left(\dfrac{x}{3}\right) + C$

**3.** $\displaystyle\int \dfrac{7}{16 + x^2}\, dx = \dfrac{7}{4} \arctan\left(\dfrac{x}{4}\right) + C$

**5.** $\displaystyle\int \dfrac{1}{x\sqrt{4x^2 - 1}}\, dx = \int \dfrac{2}{2x\sqrt{(2x)^2 - 1}}\, dx$

$\qquad\qquad\qquad = \text{arcsec}|2x| + C$

**7.** $\displaystyle\int \dfrac{1}{\sqrt{1 - (x + 1)^2}}\, dx = \arcsin(x + 1) + C$

**9.** Let $u = t^2$, $du = 2t\, dt$.

$\displaystyle\int \dfrac{t}{\sqrt{1 - t^4}}\, dt = \dfrac{1}{2}\int \dfrac{1}{\sqrt{1 - (t^2)^2}}(2t)\, dt$

$\qquad\qquad\qquad = \dfrac{1}{2}\arcsin t^2 + C$

**11.** $\displaystyle\int \dfrac{t}{t^4 + 25}\, dt = \dfrac{1}{2}\int \dfrac{1}{(t^2)^2 + 5^2}(2)\, dt$

$\qquad\qquad\qquad = \dfrac{1}{2}\dfrac{1}{5}\arctan\left(\dfrac{t^2}{5}\right) + C$

$\qquad\qquad\qquad = \dfrac{1}{10}\arctan\left(\dfrac{t^2}{5}\right) + C$

**13.** Let $u = e^{2x}$, $du = 2e^{2x} \, dx$.

$$\int \frac{e^{2x}}{4 + e^{4x}} \, dx = \frac{1}{2} \int \frac{2e^{2x}}{4 + \left(e^{2x}\right)^2} \, dx = \frac{1}{4} \arctan \frac{e^{2x}}{2} + C$$

**15.** $\int \frac{\sec^2 x}{\sqrt{25 - \tan^2 x}} \, dx = \int \frac{\sec^2 x}{\sqrt{5^2 - \left(\tan x\right)^2}} \, dx = \arcsin\left(\frac{\tan x}{5}\right) + C$

**17.** $\int \frac{x^3}{x^2 + 1} \, dx = \int \left[x - \frac{x}{x^2 + 1}\right] dx = \int x \, dx - \frac{1}{2} \int \frac{2x}{x^2 + 1} \, dx = \frac{1}{2}x^2 - \frac{1}{2}\ln\left(x^2 + 1\right) + C$ (Use long division.)

**19.** $\int \frac{1}{\sqrt{x}\sqrt{1 - x}} \, dx, u = \sqrt{x}, x = u^2, dx = 2u \, du$

$$\int \frac{1}{u\sqrt{1 - u^2}}(2u \, du) = 2 \int \frac{du}{\sqrt{1 - u^2}} = 2 \arcsin u + C = 2 \arcsin \sqrt{x} + C$$

**21.** $\int \frac{x - 3}{x^2 + 1} \, dx = \frac{1}{2} \int \frac{2x}{x^2 + 1} \, dx - 3 \int \frac{1}{x^2 + 1} \, dx = \frac{1}{2}\ln\left(x^2 + 1\right) - 3 \arctan x + C$

**23.** $\int \frac{x + 5}{\sqrt{9 - (x - 3)^2}} \, dx = \int \frac{(x - 3)}{\sqrt{9 - (x - 3)^2}} \, dx + \int \frac{8}{\sqrt{9 - (x - 3)^2}} \, dx$

$$= -\sqrt{9 - (x - 3)^2} + 8 \arcsin\left(\frac{x - 3}{3}\right) + C = -\sqrt{6x - x^2} + 8 \arcsin\left(\frac{x}{3} - 1\right) + C$$

**25.** Let $u = 3x$, $du = 3 \, dx$.

$$\int_0^{1/6} \frac{3}{\sqrt{1 - 9x^2}} \, dx = \int_0^{1/6} \frac{1}{\sqrt{1 - (3x)^2}}(3) \, dx$$

$$= \left[\arcsin(3x)\right]_0^{1/6} = \frac{\pi}{6}$$

**27.** Let $u = 2x$, $du = 2 \, dx$.

$$\int_0^{\sqrt{3}/2} \frac{1}{1 + 4x^2} \, dx = \frac{1}{2} \int_0^{\sqrt{3}/2} \frac{2}{1 + (2x)^2} \, dx$$

$$= \left[\frac{1}{2}\arctan(2x)\right]_0^{\sqrt{3}/2} = \frac{\pi}{6}$$

**29.** Let $u = 1 - x^2$, $du = -2x \, dx$.

$$\int_{-1/2}^0 \frac{x}{\sqrt{1 - x^2}} \, dx = -\frac{1}{2} \int_{-1/2}^0 \left(1 - x^2\right)^{-1/2}(-2x) \, dx$$

$$= \left[-\sqrt{1 - x^2}\right]_{-1/2}^0 = \frac{\sqrt{3} - 2}{2}$$

$$\approx -0.134$$

**31.** $\int_3^6 \frac{1}{25 + (x - 3)^2} \, dx = \left[\frac{1}{5}\arctan\left(\frac{x - 3}{5}\right)\right]_3^6$

$$= \frac{1}{5}\arctan(3/5) \approx 0.108$$

**33.** Let $u = e^x$, $du = e^x \, dx$

$$\int_0^{\ln 5} \frac{e^x}{1 + e^{2x}} \, dx = \left[\arctan\left(e^x\right)\right]_0^{\ln 5}$$

$$= \arctan 5 - \frac{\pi}{4} \approx 0.588$$

**35.** Let $u = \cos x$, $du = -\sin x \, dx$.

$$\int_{\pi/2}^\pi \frac{\sin x}{1 + \cos^2 x} \, dx = -\int_{\pi/2}^\pi \frac{-\sin x}{1 + \cos^2 x} \, dx$$

$$= \left[-\arctan(\cos x)\right]_{\pi/2}^\pi = \frac{\pi}{4}$$

**37.** Let $u = \arcsin x$, $du = \frac{1}{\sqrt{1 - x^2}} \, dx$.

$$\int_0^{1/\sqrt{2}} \frac{\arcsin x}{\sqrt{1 - x^2}} \, dx = \left[\frac{1}{2}\arcsin^2 x\right]_0^{1/\sqrt{2}}$$

$$= \frac{\pi^2}{32} \approx 0.308$$

**39.** $\int_0^2 \frac{dx}{x^2 - 2x + 2} = \int_0^2 \frac{1}{1 + (x - 1)^2} \, dx$

$$= \left[\arctan(x - 1)\right]_0^2 = \frac{\pi}{2}$$

**41.** $\displaystyle\int \frac{2x}{x^2 + 6x + 13}\,dx = \int \frac{2x + 6}{x^2 + 6x + 13}\,dx - 6\int \frac{1}{x^2 + 6x + 13}\,dx$

$\displaystyle = \int \frac{2x + 6}{x^2 + 6x + 13}\,dx - 6\int \frac{1}{4 + (x + 3)^2}\,dx = \ln\left|x^2 + 6x + 13\right| - 3\arctan\left(\frac{x + 3}{2}\right) + C$

**43.** $\displaystyle\int \frac{1}{\sqrt{-x^2 - 4x}}\,dx = \int \frac{1}{\sqrt{4 - (x + 2)^2}}\,dx = \arcsin\left(\frac{x + 2}{2}\right) + C$

**45.** Let $u = -x^2 - 4x$, $du = (-2x - 4)\,dx$.

$\displaystyle\int \frac{x + 2}{\sqrt{-x^2 - 4x}}\,dx = -\frac{1}{2}\int \left(-x^2 - 4x\right)^{-1/2}(-2x - 4)\,dx = -\sqrt{-x^2 - 4x} + C$

**47.** $\displaystyle\int_2^3 \frac{2x - 3}{\sqrt{4x - x^2}}\,dx = \int_2^3 \frac{2x - 4}{\sqrt{4x - x^2}}\,dx + \int_2^3 \frac{1}{\sqrt{4x - x^2}}\,dx$

$\displaystyle = -\int_2^3 \left(4x - x^2\right)^{-1/2}(4 - 2x)\,dx + \int_2^3 \frac{1}{\sqrt{4 - (x - 2)^2}}\,dx$

$\displaystyle = \left[-2\sqrt{4x - x^2} + \arcsin\left(\frac{x - 2}{2}\right)\right]_2^3 = 4 - 2\sqrt{3} + \frac{\pi}{6} \approx 1.059$

**49.** Let $u = x^2 + 1$, $du = 2x\,dx$.

$\displaystyle\int \frac{x}{x^4 + 2x^2 + 2}\,dx = \frac{1}{2}\int \frac{2x}{\left(x^2 + 1\right)^2 + 1}\,dx = \frac{1}{2}\arctan\left(x^2 + 1\right) + C$

**51.** Let $u = \sqrt{e^t - 3}$. Then $u^2 + 3 = e^t$, $2u\,du = e^t\,dt$, and $\dfrac{2u\,du}{u^2 + 3} = dt$.

$\displaystyle\int \sqrt{e^t - 3}\,dt = \int \frac{2u^2}{u^2 + 3}\,du = \int 2\,du - \int 6\frac{1}{u^2 + 3}\,du$

$\displaystyle = 2u - 2\sqrt{3}\arctan\frac{u}{\sqrt{3}} + C = 2\sqrt{e^t - 3} - 2\sqrt{3}\arctan\sqrt{\frac{e^t - 3}{3}} + C$

**53.** $\displaystyle\int_1^3 \frac{dx}{\sqrt{x}(1 + x)}$

Let $u = \sqrt{x}$, $u^2 = x$, $2u\,du = dx$, $1 + x = 1 + u^2$.

$\displaystyle\int_1^{\sqrt{3}} \frac{2u\,du}{u(1 + u^2)} = \int_1^{\sqrt{3}} \frac{2}{1 + u^2}\,du = \left[2\arctan(u)\right]_1^{\sqrt{3}} = 2\left(\frac{\pi}{3} - \frac{\pi}{4}\right) = \frac{\pi}{6}$

**55.** (a) $\displaystyle\int \frac{1}{\sqrt{1 - x^2}}\,dx = \arcsin x + C$, $\quad u = x$

(b) $\displaystyle\int \frac{x}{\sqrt{1 - x^2}}\,dx = -\sqrt{1 - x^2} + C$, $\quad u = 1 - x^2$

(c) $\displaystyle\int \frac{1}{x\sqrt{1 - x^2}}\,dx$ cannot be evaluated using the basic integration rules.

**57. (a)** $\int \sqrt{x-1}\,dx = \frac{2}{3}(x-1)^{3/2} + C, \quad u = x - 1$

**(b)** Let $u = \sqrt{x-1}$. Then $x = u^2 + 1$ and $dx = 2u\,du$.

$$\int x\sqrt{x-1}\,dx = \int (u^2 + 1)(u)(2u)\,du$$

$$= 2\int (u^4 + u^2)\,du$$

$$= 2\left(\frac{u^5}{5} + \frac{u^3}{3}\right) + C$$

$$= \frac{2}{15}u^3(3u^2 + 5) + C$$

$$= \frac{2}{15}(x-1)^{3/2}\left[3(x-1) + 5\right] + C$$

$$= \frac{2}{15}(x-1)^{3/2}(3x + 2) + C$$

**(c)** Let $u = \sqrt{x-1}$. Then $x = u^2 + 1$ and $dx = 2u\,du$.

$$\int \frac{x}{\sqrt{x-1}}\,dx = \int \frac{u^2 + 1}{u}(2u)\,du$$

$$= 2\int (u^2 + 1)\,du$$

$$= 2\left(\frac{u^3}{3} + u\right) + C$$

$$= \frac{2}{3}u(u^2 + 3) + C$$

$$= \frac{2}{3}\sqrt{x-1}(x + 2) + C$$

**Note:** In (b) and (c), substitution was necessary *before* the basic integration rules could be used.

**59.** No. This integral does not correspond to any of the basic differentiation rules.

**61.** $y' = \dfrac{1}{\sqrt{4 - x^2}}, \quad (0, \pi)$

$y = \int \dfrac{1}{\sqrt{4 - x^2}}\,dx = \arcsin\left(\dfrac{x}{2}\right) + C$

When $x = 0,\ y = \pi \Rightarrow C = \pi$

$y = \arcsin\left(\dfrac{x}{2}\right) + \pi$

**63. (a)**

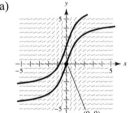

$(0, 0)$

**(b)** $\dfrac{dy}{dx} = \dfrac{3}{1 + x^2}, \quad (0, 0)$

$y = 3\int \dfrac{dx}{1 + x^2} = 3\arctan x + C$

$(0, 0): 0 = 3\arctan(0) + C \Rightarrow C = 0$

$y = 3\arctan x$

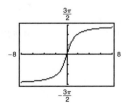

**65.** (a)

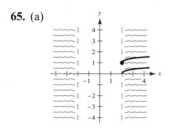

(b) $y' = \dfrac{1}{x\sqrt{x^2-4}}, \quad (2,1)$

$y = \displaystyle\int \dfrac{1}{x\sqrt{x^2-4}}\,dx = \dfrac{1}{2}\operatorname{arcsec}\dfrac{|x|}{2}+C$

$1 = \dfrac{1}{2}\operatorname{arcsec}(1)+C = C$

$y = \dfrac{1}{2}\operatorname{arcsec}\dfrac{x}{2}+1, \quad x \ge 2$

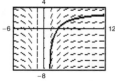

**67.** $\dfrac{dy}{dx} = \dfrac{10}{x\sqrt{x^2-1}}, \quad (3,0)$

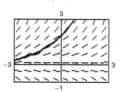

**69.** $\dfrac{dy}{dx} = \dfrac{2y}{\sqrt{16-x^2}}, \quad (0,2)$

**71.** Area $= \displaystyle\int_0^1 \dfrac{2}{\sqrt{4-x^2}}\,dx$

$= \left[2\arcsin\left(\dfrac{x}{2}\right)\right]_0^1$

$= 2\arcsin\left(\dfrac{1}{2}\right) - 2\arcsin(0) = 2\left(\dfrac{\pi}{6}\right) = \dfrac{\pi}{3}$

**73.** Area $= \displaystyle\int_1^3 \dfrac{1}{x^2-2x+5}\,dx = \int_1^3 \dfrac{1}{(x-1)^2+4}\,dx$

$= \left[\dfrac{1}{2}\arctan\left(\dfrac{x-1}{2}\right)\right]_1^3$

$= \dfrac{1}{2}\arctan(1) - \dfrac{1}{2}\arctan(0)$

$= \dfrac{\pi}{8}$

**75.** Area $= \displaystyle\int_{-\pi/2}^{\pi/2} \dfrac{3\cos x}{1+\sin^2 x}\,dx = 3\int_{-\pi/2}^{\pi/2}\dfrac{1}{1+\sin^2 x}(\cos x\,dx)$

$= \left[3\arctan(\sin x)\right]_{-\pi/2}^{\pi/2}$

$= 3\arctan(1) - 3\arctan(-1)$

$= \dfrac{3\pi}{4} + \dfrac{3\pi}{4} = \dfrac{3\pi}{2}$

**77.** (a) $\dfrac{d}{dx}\left[\ln x - \dfrac{1}{2}\ln(1+x^2) - \dfrac{\arctan x}{x}+C\right] = \dfrac{1}{x} - \dfrac{x}{1+x^2} - \left(\dfrac{x\left[1/(1+x^2)\right]-\arctan x}{x^2}\right)$

$= \dfrac{1+x^2-x^2}{x(1+x^2)} - \dfrac{1}{x(1+x^2)} + \dfrac{\arctan x}{x^2} = \dfrac{\arctan x}{x^2}$

So, $\displaystyle\int \dfrac{\arctan x}{x^2}\,dx = \ln x - \dfrac{1}{2}\ln(1+x^2) - \dfrac{\arctan x}{x}+C.$

(b) $A = \displaystyle\int_1^{\sqrt{3}} \dfrac{\arctan x}{x^2}\,dx$

$= \left[\ln x - \dfrac{1}{2}\ln(1+x^2) - \dfrac{\arctan x}{x}\right]_1^{\sqrt{3}}$

$= \left(\ln\sqrt{3} - \dfrac{1}{2}\ln(4) - \dfrac{\arctan\sqrt{3}}{\sqrt{3}}\right) - \left(\dfrac{-1}{2}\ln 2 - \arctan(1)\right)$

$= \dfrac{1}{2}\ln 3 - \dfrac{1}{2}\ln 2 - \dfrac{\pi\sqrt{3}}{9} + \dfrac{\pi}{4} = \ln\dfrac{\sqrt{6}}{2} + \dfrac{9\pi-4\pi\sqrt{3}}{36} \approx 0.3835$

**79.** (a)

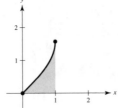

Shaded area is given by $\int_0^1 \arcsin x \, dx$.

(b) $\int_0^1 \arcsin x \, dx \approx 0.5708$

(c) Divide the rectangle into two regions.

$$\text{Area rectangle} = (\text{base})(\text{height}) = 1\left(\frac{\pi}{2}\right) = \frac{\pi}{2}$$

$$\text{Area rectangle} = \int_0^1 \arcsin x \, dx + \int_0^{\pi/2} \sin y \, dy$$

$$\frac{\pi}{2} = \int_0^1 \arcsin x \, dx + \left(-\cos y\right)\Big]_0^{\pi/2} = \int_0^1 \arcsin x \, dx + 1$$

So, $\int_0^1 \arcsin x \, dx = \dfrac{\pi}{2} - 1$, $(\approx 0.5708)$.

**81.** $F(x) = \dfrac{1}{2}\displaystyle\int_x^{x+2} \dfrac{2}{t^2 + 1}\, dt$

(a) $F(x)$ represents the average value of $f(x)$ over the interval $[x, x + 2]$. Maximum at $x = -1$, because the graph is greatest on $[-1, 1]$.

(b) $F(x) = \left[\arctan t\right]_x^{x+2} = \arctan(x + 2) - \arctan x$

$$F'(x) = \frac{1}{1 + (x + 2)^2} - \frac{1}{1 + x^2} = \frac{(1 + x^2) - (x^2 + 4x + 5)}{(x^2 + 1)(x^2 + 4x + 5)} = \frac{-4(x + 1)}{(x^2 + 1)(x^2 + 4x + 5)} = 0 \text{ when } x = -1.$$

**83.** False, $\displaystyle\int \dfrac{dx}{3x\sqrt{9x^2 - 16}} = \dfrac{1}{12}\operatorname{arcsec}\dfrac{|3x|}{4} + C$

**85.** True

$$\frac{d}{dx}\left[-\arccos\frac{x}{2} + C\right] = \frac{1/2}{\sqrt{1 - (x/2)^2}} = \frac{1}{\sqrt{4 - x^2}}$$

**87.** $\dfrac{d}{dx}\left[\arcsin\left(\dfrac{u}{a}\right) + C\right] = \dfrac{1}{\sqrt{1 - (u^2/a^2)}}\left(\dfrac{u'}{a}\right) = \dfrac{u'}{\sqrt{a^2 - u^2}}$

So, $\displaystyle\int \dfrac{du}{\sqrt{a^2 - u^2}} = \arcsin\left(\dfrac{u}{a}\right) + C.$

**89.** Assume $u > 0$.

$$\frac{d}{dx}\left[\frac{1}{a}\operatorname{arcsec}\frac{u}{a} + C\right] = \frac{1}{a}\left[\frac{u'/a}{(u/a)\sqrt{(u/a)^2 - 1}}\right] = \frac{1}{a}\left[\frac{u'}{u\sqrt{(u^2 - a^2)/a^2}}\right] = \frac{u'}{u\sqrt{u^2 - a^2}}.$$

The case $u < 0$ is handled in a similar manner.

So, $\displaystyle\int \dfrac{du}{u\sqrt{u^2 - a^2}} = \int \dfrac{u'}{u\sqrt{u^2 - a^2}}\, dx = \dfrac{1}{a}\operatorname{arcsec}\dfrac{|u|}{a} + C.$

**91.** (a) $v(t) = -32t + 500$

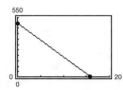

(b) $s(t) = \int v(t)\, dt = \int (-32t + 500)\, dt = -16t^2 + 500t + C$

$s(0) = -16(0) + 500(0) + C = 0 \Rightarrow C = 0$

$s(t) = -16t^2 + 500t$

hen the object reaches its maximum height, $v(t) = 0$.

$v(t) = -32t + 500 = 0$

$-32t = -500$

$t = 15.625$

$s(15.625) = -16(15.625)^2 + 500(15.625) = 3906.25$ ft $\left(\text{Maximum height}\right)$

(c) $\qquad \displaystyle \int \frac{1}{32 + kv^2}\, dv = -\int dt$

$\dfrac{1}{\sqrt{32k}} \arctan\!\left(\sqrt{\dfrac{k}{32}}\,v\right) = -t + C_1$

$\arctan\!\left(\sqrt{\dfrac{k}{32}}\,v\right) = -\sqrt{32k}\,t + C$

$\sqrt{\dfrac{k}{32}}\,v = \tan\!\left(C - \sqrt{32k}\,t\right)$

$v = \sqrt{\dfrac{32}{k}}\,\tan\!\left(C - \sqrt{32k}\,t\right)$

When $t = 0, v = 500, C = \arctan\!\left(500\sqrt{k/32}\right)$, and you have

$v(t) = \sqrt{\dfrac{32}{k}}\,\tan\!\left[\arctan\!\left(500\sqrt{\dfrac{k}{32}}\right) - \sqrt{32k}\,t\right]$.

(d) When $k = 0.001$:

$v(t) = \sqrt{32{,}000}\,\tan\!\left[\arctan\!\left(500\sqrt{0.00003125}\right) - \sqrt{0.032t}\,\right]$

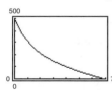

$v(t) = 0$ when $t_0 \approx 6.86$ sec.

(e) $h = \displaystyle \int_0^{6.86} \sqrt{32{,}000}\,\tan\!\left[\arctan\!\left(500\sqrt{0.00003125}\right) - \sqrt{0.032}\; t\right] dt$

Simpson's Rule: $n = 10$; $h \approx 1088$ feet

(f) Air resistance lowers the maximum height.

## Section 5.8 Hyperbolic Functions

**1.** (a) $\sinh 3 = \dfrac{e^3 - e^{-3}}{2} \approx 10.018$

(b) $\tanh(-2) = \dfrac{\sinh(-2)}{\cosh(-2)} = \dfrac{e^{-2} - e^2}{e^{-2} + e^2} \approx -0.964$

**3.** (a) $\operatorname{csch}(\ln 2) = \dfrac{2}{e^{\ln 2} - e^{-\ln 2}} = \dfrac{2}{2 - (1/2)} = \dfrac{4}{3}$

(b) $\coth(\ln 5) = \dfrac{\cosh(\ln 5)}{\sinh(\ln 5)} = \dfrac{e^{\ln 5} + e^{-\ln 5}}{e^{\ln 5} - e^{-\ln 5}}$

$= \dfrac{5 + (1/5)}{5 - (1/5)} = \dfrac{13}{12}$

**5.** (a) $\cosh^{-1} 2 = \ln\left(2 + \sqrt{3}\right) \approx 1.317$

(b) $\operatorname{sech}^{-1} \dfrac{2}{3} = \ln\left(\dfrac{1 + \sqrt{1 - (4/9)}}{2/3}\right) \approx 0.962$

**7.** $\sinh x + \cosh x = \dfrac{e^x - e^{-x}}{2} + \dfrac{e^x + e^{-x}}{2} = e^x$

**9.** $\tanh^2 x + \operatorname{sech}^2 x = \left(\dfrac{e^x - e^{-x}}{e^x + e^{-x}}\right)^2 + \left(\dfrac{2}{e^x + e^{-x}}\right)^2$

$= \dfrac{e^{2x} - 2 + e^{-2x} + 4}{\left(e^x + e^{-x}\right)^2}$

$= \dfrac{e^{2x} + 2 + e^{-2x}}{e^{2x} + 2 + e^{-2x}} = 1$

**11.** $\dfrac{1 + \cosh 2x}{2} = \dfrac{1 + \left(e^{2x} + e^{-2x}\right)/2}{2}$

$= \dfrac{e^{2x} + 2 + e^{-2x}}{4}$

$= \left(\dfrac{e^x + e^{-x}}{2}\right)^2 = \cosh^2 x$

**13.** $\sinh x \cosh y + \cosh x \sinh y = \left(\dfrac{e^x - e^{-x}}{2}\right)\left(\dfrac{e^y + e^{-y}}{2}\right) + \left(\dfrac{e^x + e^{-x}}{2}\right)\left(\dfrac{e^y - e^{-y}}{2}\right)$

$= \dfrac{1}{4}\left[e^{x+y} - e^{-x+y} + e^{x-y} - e^{-(x+y)} + e^{x+y} + e^{-x+y} - e^{x-y} - e^{-(x+y)}\right]$

$= \dfrac{1}{4}\left[2\left(e^{x+y} - e^{-(x+y)}\right)\right] = \dfrac{e^{(x+y)} - e^{-(x+y)}}{2} = \sinh(x + y)$

**15.** $3 \sinh x + 4 \sinh^3 x = \sinh x\left(3 + 4 \sinh^2 x\right) = \left(\dfrac{e^x - e^{-x}}{2}\right)\left[3 + 4\left(\dfrac{e^x - e^{-x}}{2}\right)^2\right]$

$= \left(\dfrac{e^x - e^{-x}}{2}\right)\left[3 + e^{2x} - 2 + e^{-2x}\right] = \dfrac{1}{2}\left(e^x - e^{-x}\right)\left(e^{2x} + e^{-2x} + 1\right)$

$= \dfrac{1}{2}\left[e^{3x} + e^{-x} + e^x - e^x - e^{-3x} - e^{-x}\right] = \dfrac{e^{3x} - e^{-3x}}{2} = \sinh(3x)$

**17.** $\sinh x = \dfrac{3}{2}$

$\cosh^2 x - \left(\dfrac{3}{2}\right)^2 = 1 \Rightarrow \cosh^2 x = \dfrac{13}{4} \Rightarrow \cosh x = \dfrac{\sqrt{13}}{2}$

$\tanh x = \dfrac{3/2}{\sqrt{13}/2} = \dfrac{3\sqrt{13}}{13}$

$\operatorname{csch} x = \dfrac{1}{3/2} = \dfrac{2}{3}$

$\operatorname{sech} x = \dfrac{1}{\sqrt{13}/2} = \dfrac{2\sqrt{13}}{13}$

$\coth x = \dfrac{1}{3/\sqrt{13}} = \dfrac{\sqrt{13}}{3}$

**19.** $f(x) = \sinh(3x)$

$f'(x) = 3\cosh(3x)$

**21.** $y = \operatorname{sech}\left(5x^2\right)$

$y' = -\operatorname{sech}\left(5x^2\right)\tanh\left(5x^2\right)(10x)$

$= -10x\operatorname{sech}\left(5x^2\right)\tanh\left(5x^2\right)$

**23.** $f(x) = \ln(\sinh x)$

$f'(x) = \dfrac{1}{\sinh x}(\cosh x) = \coth x$

**25.** $y = \ln\left(\tanh\dfrac{x}{2}\right)$

$y' = \dfrac{1/2}{\tanh(x/2)}\operatorname{sech}^2\left(\dfrac{x}{2}\right)$

$= \dfrac{1}{2\sinh(x/2)\cosh(x/2)} = \dfrac{1}{\sinh x} = \operatorname{csch} x$

**27.** $h(x) = \dfrac{1}{4}\sinh 2x - \dfrac{x}{2}$

$h'(x) = \dfrac{1}{2}\cosh(2x) - \dfrac{1}{2} = \dfrac{\cosh(2x) - 1}{2} = \sinh^2 x$

**29.** $f(t) = \arctan(\sinh t)$

$f'(t) = \dfrac{1}{1 + \sinh^2 t}(\cosh t) = \dfrac{\cosh t}{\cosh^2 t} = \operatorname{sech} t$

**31.** $y = \sinh(1 - x^2), \quad (1, 0)$

$y' = \cosh(1 - x^2)(-2x)$

$y'(1) = -2$

Tangent line: $y - 0 = -2(x - 1)$

$y = -2x + 2$

**33.** $y = (\cosh x - \sinh x)^2, \quad (0, 1)$

$y' = 2(\cosh x - \sinh x)(\sinh x - \cosh x)$

At $(0, 1)$, $y' = 2(1)(-1) = -2$.

Tangent line: $y - 1 = -2(x - 0)$

$y = -2x + 1$

**35.** $f(x) = \sin x \sinh x - \cos x \cosh x, \quad -4 \le x \le 4$

$f'(x) = \sin x \cosh x + \cos x \sinh x - \cos x \sinh x + \sin x \cosh x$

$\quad = 2\sin x \cosh x = 0$ when $x = 0, \pm\pi$.

Relative maxima: $(\pm\pi, \cosh \pi)$

Relative minimum: $(0, -1)$

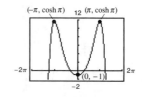

**37.**   $g(x) = x \operatorname{sech} x$

$g'(x) = \operatorname{sech} x - x \operatorname{sech} x \tanh x$

$\quad = \operatorname{sech} x(1 - x \tanh x) = 0$

$x \tanh x = 1$

Using a graphing utility, $x \approx \pm 1.1997$.

By the First Derivative Test, $(1.1997, 0.6627)$ is a relative maximum and $(-1.1997, -0.6627)$ is a relative minimum.

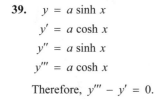

**39.**   $y = a \sinh x$

$y' = a \cosh x$

$y'' = a \sinh x$

$y''' = a \cosh x$

Therefore, $y''' - y' = 0$.

**41.** $f(x) = \tanh x, \qquad\qquad f(0) = 0$

$f'(x) = \operatorname{sech}^2 x, \qquad\quad f'(0) = 1$

$f''(x) = -2\operatorname{sech}^2 x \tanh x, \quad f''(0) = 0$

$P_1(x) = f(0) + f'(0)(x - 0) = x$

$P_2(x) = f(0) + f'(0)(x - 0) + \dfrac{1}{2}f''(0)(x - 0)^2 = x$

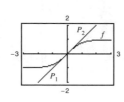

**43.** (a) $y = 10 + 15 \cosh \dfrac{x}{15}$, $-15 \le x \le 15$

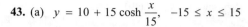
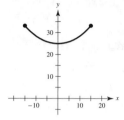

(b) At $x = \pm15$, $y = 10 + 15 \cosh(1) \approx 33.146$.

At $x = 0$, $y = 10 + 15 \cosh(0) = 25$.

(c) $y' = \sinh \dfrac{x}{15}$. At $x = 15$, $y' = \sinh(1) \approx 1.175$.

**45.** $\displaystyle\int \cosh 2x \, dx = \frac{1}{2} \int \cosh(2x)(2) \, dx$

$$= \frac{1}{2} \sinh 2x + C$$

**47.** Let $u = 1 - 2x$, $du = -2 \, dx$.

$\displaystyle\int \sinh(1 - 2x) \, dx = -\frac{1}{2} \int \sinh(1 - 2x)(-2) \, dx$

$$= -\frac{1}{2} \cosh(1 - 2x) + C$$

**49.** Let $u = \cosh(x - 1)$, $du = \sinh(x - 1) \, dx$.

$\displaystyle\int \cosh^2(x - 1) \sinh(x - 1) \, dx = \frac{1}{3} \cosh^3(x - 1) + C$

**51.** Let $u = \sinh x$, $du = \cosh x \, dx$.

$\displaystyle\int \frac{\cosh x}{\sinh x} \, dx = \ln|\sinh x| + C$

**53.** Let $u = \dfrac{x^2}{2}$, $du = x \, dx$.

$\displaystyle\int x \operatorname{csch}^2 \frac{x^2}{2} \, dx = \int \left( \operatorname{csch}^2 \frac{x^2}{2} \right) x \, dx = -\coth \frac{x^2}{2} + C$

**55.** Let $u = \dfrac{1}{x}$, $du = -\dfrac{1}{x^2} \, dx$.

$\displaystyle\int \frac{\operatorname{csch}(1/x) \coth(1/x)}{x^2} \, dx = -\int \operatorname{csch} \frac{1}{x} \coth \frac{1}{x} \left( -\frac{1}{x^2} \right) dx$

$$= \operatorname{csch} \frac{1}{x} + C$$

**57.** Let $u = x^2$, $du = 2x \, dx$.

$\displaystyle\int \frac{x}{x^4 + 1} \, dx = \frac{1}{2} \int \frac{2x}{\left( x^2 \right)^2 + 1} \, dx = \frac{1}{2} \arctan\left( x^2 \right) + C$

**59.** $\displaystyle\int_0^{\ln 2} \tanh x \, dx = \int_0^{\ln 2} \frac{\sinh x}{\cosh x} \, dx$, $(u = \cosh x)$

$$= \Big[ \ln(\cosh x) \Big]_0^{\ln 2}$$

$$= \ln(\cosh(\ln 2) - \ln(\cosh(0))$$

$$= \ln\left( \frac{5}{4} \right) - 0 = \ln\left( \frac{5}{4} \right)$$

**Note:** $\cosh(\ln 2) = \dfrac{e^{\ln 2} + e^{-\ln 2}}{2} = \dfrac{2 + (1/2)}{2} = \dfrac{5}{4}$

**61.** $\displaystyle\int_0^4 \frac{1}{25 - x^2} \, dx = \frac{1}{10} \int \frac{1}{5 - x} \, dx + \frac{1}{10} \int \frac{1}{5 + x} \, dx$

$$= \left[ \frac{1}{10} \ln \left| \frac{5 + x}{5 - x} \right| \right]_0^4 = \frac{1}{10} \ln 9 = \frac{1}{5} \ln 3$$

**63.** Let $u = 2x$, $du = 2 \, dx$.

$\displaystyle\int_0^{\sqrt{2}/4} \frac{2}{\sqrt{1 - 4x^2}} \, dx = \int_0^{\sqrt{2}/4} \frac{1}{\sqrt{1 - (2x)^2}} (2) \, dx$

$$= \Big[ \arcsin(2x) \Big]_0^{\sqrt{2}/4} = \frac{\pi}{4}$$

**65.** $y = \cosh^{-1}(3x)$

$$y' = \frac{3}{\sqrt{9x^2 - 1}}$$

**67.** $y = \tanh^{-1} \sqrt{x}$

$$y' = \frac{1}{1 - \left( \sqrt{x} \right)^2} \left( \frac{1}{2} x^{-1/2} \right)$$

$$= \frac{1}{2\sqrt{x}(1 - x)}$$

**69.** $y = \sinh^{-1}(\tan x)$

$$y' = \frac{1}{\sqrt{\tan^2 x + 1}} (\sec^2 x) = |\sec x|$$

**71.** $y = \left( \operatorname{csch}^{-1} x \right)^2$

$$y' = 2 \operatorname{csch}^{-1} x \left( \frac{-1}{|x| \sqrt{1 + x^2}} \right) = \frac{-2 \operatorname{csch}^{-1} x}{|x| \sqrt{1 + x^2}}$$

**73.** $y = 2x \sinh^{-1}(2x) - \sqrt{1 + 4x^2}$

$$y' = 2x \left( \frac{2}{\sqrt{1 + 4x^2}} \right) + 2 \sinh^{-1}(2x) - \frac{4x}{\sqrt{1 + 4x^2}}$$

$$= 2 \sinh^{-1}(2x)$$

**75.** Answers will vary.

**77.** The derivatives of $f(x) = \cosh x$ and
$f(x) = \operatorname{sech} x$ differ by a minus sign.

**79.** $\lim_{x \to \infty} \sinh x = \infty$

**81.** $\lim_{x \to \infty} \tanh x = 1$

**83.** $\lim_{x \to \infty} \operatorname{sech} x = 0$

**85.** $\lim_{x \to 0} \dfrac{\sinh x}{x} = \lim_{x \to 0} \dfrac{e^x - e^{-x}}{2x} = 1$

**87.** $\displaystyle \int \dfrac{1}{3 - 9x^2}\, dx = \dfrac{1}{3} \int \dfrac{1}{3 - (3x)^2}\,(3)\, dx$

$$= \dfrac{1}{3} \dfrac{1}{2\sqrt{3}} \ln \left| \dfrac{\sqrt{3} + 3x}{\sqrt{3} - 3x} \right| + C$$

$$= \dfrac{\sqrt{3}}{18} \ln \left| \dfrac{1 + \sqrt{3}x}{1 - \sqrt{3}x} \right| + C$$

**89.** $\displaystyle \int \dfrac{1}{\sqrt{1 + e^{2x}}}\, dx = \int \dfrac{e^x}{e^x \sqrt{1 + (e^x)^2}}\, dx$

$$= -\operatorname{csch}^{-1}(e^x) + C$$

$$= -\ln \left( \dfrac{1 + \sqrt{1 + e^{2x}}}{e^x} \right) + C$$

$$= \ln \left( \dfrac{e^x}{1 + \sqrt{1 + e^{2x}}} \right) + C$$

$$= \ln \left( \dfrac{-e^x + e^x \sqrt{1 + e^{2x}}}{e^{2x}} \right) + C$$

$$= \ln \left( \sqrt{1 + e^{2x}} - 1 \right) - x + C$$

**91.** Let $u = \sqrt{x}$, $du = \dfrac{1}{2\sqrt{x}}\, dx$.

$$\int \dfrac{1}{\sqrt{x}\sqrt{1 + x}}\, dx = 2 \int \dfrac{1}{\sqrt{1 + (\sqrt{x})^2}} \left( \dfrac{1}{2\sqrt{x}} \right) dx$$

$$= 2 \sinh^{-1} \sqrt{x} + C$$

$$= 2 \ln \left( \sqrt{x} + \sqrt{1 + x} \right) + C$$

**93.** $\displaystyle \int \dfrac{-1}{4x - x^2}\, dx = \int \dfrac{1}{(x - 2)^2 - 4}\, dx$

$$= \dfrac{1}{4} \ln \left| \dfrac{(x - 2) - 2}{(x - 2) + 2} \right|$$

$$= \dfrac{1}{4} \ln \left| \dfrac{x - 4}{x} \right| + C$$

**95.** $\displaystyle \int \dfrac{1}{1 - 4x - 2x^2}\, dx = \int \dfrac{1}{3 - 2(x + 1)^2}\, dx$

$$= \dfrac{1}{\sqrt{2}} \int \dfrac{\sqrt{2}}{(\sqrt{3})^2 - [\sqrt{2}(x + 1)]^2}\, dx$$

$$= \dfrac{1}{\sqrt{2}} \cdot \dfrac{1}{2\sqrt{3}} \ln \left| \dfrac{\sqrt{3} + \sqrt{2}(x + 1)}{\sqrt{3} - \sqrt{2}(x + 1)} \right| + C = \dfrac{1}{2\sqrt{6}} \ln \left| \dfrac{\sqrt{2}(x + 1) + \sqrt{3}}{\sqrt{2}(x + 1) - \sqrt{3}} \right| + C$$

**97.** $\displaystyle \int_3^7 \dfrac{1}{\sqrt{x^2 - 4}}\, dx = \left[ \ln \left( x + \sqrt{x^2 - 4} \right) \right]_3^7 = \ln \left( 7 + \sqrt{45} \right) - \ln \left( 3 + \sqrt{5} \right) = \ln \left( \dfrac{7 + \sqrt{45}}{3 + \sqrt{5}} \right) = \ln \left( \dfrac{\sqrt{5} + 3}{2} \right)$

**99.** $\int_{-1}^{1} \frac{1}{16 - 9x^2} \, dx = \frac{1}{3}\int_{-1}^{1} \frac{1}{4^2 - (3x)^2}(3) \, dx$

$$= \left[\frac{1}{3}\frac{1}{4}\frac{1}{2} \ln\left|\frac{4 + 3x}{4 - 3x}\right|\right]_{-1}^{1}$$

$$= \frac{1}{24}\left[\ln(7) - \ln\left(\frac{1}{7}\right)\right]$$

$$= \frac{1}{24}\left[\ln 7 - \ln 1 + \ln 7\right] = \frac{1}{12} \ln 7$$

**101.** Let $u = 4x - 1$, $du = 4 \, dx$.

$$y = \int \frac{1}{\sqrt{80 + 8x - 16x^2}} \, dx$$

$$= \frac{1}{4}\int \frac{4}{\sqrt{81 - (4x - 1)^2}} \, dx$$

$$= \frac{1}{4} \arcsin\left(\frac{4x - 1}{9}\right) + C$$

**103.** $y = \int \frac{x^3 - 21x}{5 + 4x - x^2} \, dx = \int\left(-x - 4 + \frac{20}{5 + 4x - x^2}\right) dx$

$$= \int(-x - 4) \, dx + 20\int \frac{1}{3^2 - (x - 2)^2} \, dx$$

$$= -\frac{x^2}{2} - 4x + \frac{20}{6} \ln\left|\frac{3 + (x - 2)}{3 - (x - 2)}\right| + C$$

$$= -\frac{x^2}{2} - 4x + \frac{10}{3} \ln\left|\frac{1 + x}{5 - x}\right| + C$$

$$= \frac{-x^2}{2} - 4x - \frac{10}{3} \ln\left|\frac{5 - x}{x + 1}\right| + C$$

**105.** $A = 2\int_{0}^{4} \text{sech} \frac{x}{2} \, dx$

$$= 2\int_{0}^{4} \frac{2}{e^{x/2} + e^{-x/2}} \, dx$$

$$= 4\int_{0}^{4} \frac{e^{x/2}}{\left(e^{x/2}\right)^2 + 1} \, dx$$

$$= \left[8 \arctan\left(e^{x/2}\right)\right]_{0}^{4} = 8 \arctan\left(e^2\right) - 2\pi \approx 5.207$$

**107.** $A = \int_{0}^{2} \frac{5x}{\sqrt{x^4 + 1}} \, dx$

$$= \frac{5}{2}\int_{0}^{2} \frac{2x}{\sqrt{\left(x^2\right)^2 + 1}} \, dx$$

$$= \left[\frac{5}{2} \ln\left(x^2 + \sqrt{x^4 + 1}\right)\right]_{0}^{2}$$

$$= \frac{5}{2} \ln\left(4 + \sqrt{17}\right) \approx 5.237$$

**109. (a)** $\int_{0}^{\sqrt{3}} \frac{dx}{\sqrt{x^2 + 1}} = \ln\left(x + \sqrt{x^2 + 1}\right)\Big]_{0}^{\sqrt{3}}$

$$= \ln\left(\sqrt{3} + 2\right) \approx 1.317$$

**(b)** $\int_{0}^{\sqrt{3}} \frac{dx}{\sqrt{x^2 + 1}} = \left[\sinh^{-1} x\right]_{0}^{\sqrt{3}}$

$$= \sinh^{-1}\left(\sqrt{3}\right) \approx 1.317$$

**111.** $\int \frac{3k}{16}\,dt = \int \frac{1}{x^2 - 12x + 32}\,dx$

$\dfrac{3kt}{16} = \int \dfrac{1}{(x-6)^2 - 4}\,dx = \dfrac{1}{2(2)}\ln\left|\dfrac{(x-6)-2}{(x-6)+2}\right| + C = \dfrac{1}{4}\ln\left|\dfrac{x-8}{x-4}\right| + C$

When $x = 0$:            When $x = 1$:                                When $x = 20$:

$t = 0$                        $t = 10$                            $\left(\dfrac{3}{16}\right)\left(\dfrac{2}{15}\right)\ln\left(\dfrac{7}{6}\right)(20) = \dfrac{1}{4}\ln\dfrac{x-8}{2x-8}$

$C = -\dfrac{1}{4}\ln(2)$        $\dfrac{30k}{16} = \dfrac{1}{4}\ln\left|\dfrac{-7}{-3}\right| - \dfrac{1}{4}\ln(2) = \dfrac{1}{4}\ln\left(\dfrac{7}{6}\right)$        $\ln\left(\dfrac{7}{6}\right)^2 = \ln\dfrac{x-8}{2x-8}$

$k = \dfrac{2}{15}\ln\left(\dfrac{7}{6}\right)$        $\dfrac{49}{36} = \dfrac{x-8}{2x-8}$

$62x = 104$

$x = \dfrac{104}{62} = \dfrac{52}{31} \approx 1.677 \text{ kg}$

**113.** $y = a\,\text{sech}^{-1}\dfrac{x}{a} - \sqrt{a^2 - x^2},\ a > 0$

$\dfrac{dy}{dx} = \dfrac{-1}{(x/a)\sqrt{1-(x^2/a^2)}} + \dfrac{x}{\sqrt{a^2-x^2}} = \dfrac{-a^2}{x\sqrt{a^2-x^2}} + \dfrac{x}{a^2-x^2} = \dfrac{x^2-a^2}{x\sqrt{a^2-x^2}} = \dfrac{-\sqrt{a^2-x^2}}{x}$

**115.** Let $u = \tanh^{-1}x,\ -1 < x < 1$

$\tanh u = x.$

$\dfrac{\sinh u}{\cosh u} = \dfrac{e^u - e^{-u}}{e^u + e^{-u}} = x$

$e^u - e^{-u} = xe^u + xe^{-u}$

$e^{2u} - 1 = xe^{2u} + x$

$e^{2u}(1 - x) = 1 + x$

$e^{2u} = \dfrac{1+x}{1-x}$

$2u = \ln\left(\dfrac{1+x}{1-x}\right)$

$u = \dfrac{1}{2}\ln\left(\dfrac{1+x}{1-x}\right),\ -1 < x < 1$

**117.** Let $y = \arcsin(\tanh x)$. Then,

$\sin y = \tanh x = \dfrac{e^x - e^{-x}}{e^x + e^{-x}}$ and

$\tan y = \dfrac{e^x - e^{-x}}{2} = \sinh x.$

So, $y = \arctan(\sinh x)$. Therefore,

$\arctan(\sinh x) = \arcsin(\tanh x).$

**119.** $y = \text{sech}^{-1}x$

$\text{sech}\,y = x$

$-(\text{sech}\,y)(\tanh y)y' = 1$

$y' = \dfrac{-1}{(\text{sech}\,y)(\tanh y)}$

$= \dfrac{-1}{(\text{sech}\,y)\sqrt{1 - \text{sech}^2 y}}$

$= \dfrac{-1}{x\sqrt{1-x^2}}$

**121.** $y = \sinh^{-1}x$

$\sinh y = x$

$(\cosh y)y' = 1$

$y' = \dfrac{1}{\cosh y} = \dfrac{1}{\sqrt{\sinh^2 y + 1}} = \dfrac{1}{\sqrt{x^2 + 1}}$

**123.** $y = \coth x = \dfrac{\cosh x}{\sinh x}$

$y' = \dfrac{\sinh^2 x - \cosh^2 x}{\sinh^2 x} = \dfrac{-1}{\sinh^2 x} = -\text{csch}^2 x$

**125.** $y = c \cosh \dfrac{x}{c}$

Let $P(x_1, y_1)$ be a point on the catenary.

$$y' = \sinh \dfrac{x}{c}$$

The slope at $P$ is $\sinh(x_1/c)$. The equation of line $L$ is $y - c = \dfrac{-1}{\sinh(x_1/c)}(x - 0)$.

When $y = 0$, $c = \dfrac{x}{\sinh(x_1/c)} \Rightarrow x = c \sinh\left(\dfrac{x_1}{c}\right)$. The length of $L$ is $\sqrt{c^2 \sinh^2\left(\dfrac{x_1}{c}\right) + c^2} = c \cdot \cosh\dfrac{x_1}{c} = y_1$,

the ordinate $y_1$ of the point $P$.

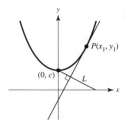

# Review Exercises for Chapter 5

**1.** $f(x) = \ln x - 3$

Vertical shift 3 units downward

Vertical asymptote: $x = 0$

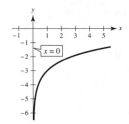

**3.** $\ln \sqrt[5]{\dfrac{4x^2 - 1}{4x^2 + 1}} = \dfrac{1}{5} \ln \dfrac{(2x - 1)(2x + 1)}{4x^2 + 1} = \dfrac{1}{5}\left[\ln(2x - 1) + \ln(2x + 1) - \ln(4x^2 + 1)\right]$

**5.** $\ln 3 + \dfrac{1}{3}\ln(4 - x^2) - \ln x = \ln 3 + \ln \sqrt[3]{4 - x^2} - \ln x = \ln\left(\dfrac{3\sqrt[3]{4 - x^2}}{x}\right)$

**7.** $\ln\sqrt{x + 1} = 2$

$\sqrt{x + 1} = e^2$

$x + 1 = e^4$

$x = e^4 - 1 \approx 53.598$

**9.** $g(x) = \ln\sqrt{2x} = \ln(2x)^{1/2} = \dfrac{1}{2}\ln 2x$

$g'(x) = \dfrac{1}{2}\dfrac{1}{2x}(2) = \dfrac{1}{2x}$

**11.** $f(x) = x\sqrt{\ln x}$

$f'(x) = \left(\dfrac{x}{2}\right)(\ln x)^{-1/2}\left(\dfrac{1}{x}\right) + \sqrt{\ln x}$

$\qquad = \dfrac{1}{2\sqrt{\ln x}} + \sqrt{\ln x} = \dfrac{1 + 2\ln x}{2\sqrt{\ln x}}$

**13.** $y = \dfrac{1}{b^2}\left[a + bx - a\ln(a + bx)\right]$

$\dfrac{dy}{dx} = \dfrac{1}{b^2}\left(b - \dfrac{ab}{a + bx}\right) = \dfrac{x}{a + bx}$

**15.**    $y = \ln(2 + x) + \dfrac{2}{2 + x},\quad (-1, 2)$

$y' = \dfrac{1}{2 + x} - \dfrac{2}{(2 + x)^2}$

$y'(-1) = 1 - 2 = -1$

Tangent line:  $y - 2 = -1(x + 1)$

$y = -x + 1$

**17.**  $u = 7x - 2,\, du = 7\, dx$

$\displaystyle\int \dfrac{1}{7x - 2}\, dx = \dfrac{1}{7}\int \dfrac{1}{7x - 2}(7)\, dx = \dfrac{1}{7}\ln|7x - 2| + C$

**19.**  $\displaystyle\int \dfrac{\sin x}{1 + \cos x}\, dx = -\int \dfrac{-\sin x}{1 + \cos x}\, dx = -\ln|1 + \cos x| + C$

**21.**  $\displaystyle\int_1^4 \dfrac{2x + 1}{2x}\, dx = \int_1^4 \left(1 + \dfrac{1}{2x}\right) dx$

$= \left[ x + \dfrac{1}{2}\ln|x| \right]_1^4$

$= 4 + \dfrac{1}{2}\ln 4 - 1 = 3 + \ln 2$

**23.**  $\displaystyle\int_0^{\pi/3} \sec\theta\, d\theta = \Big[ \ln|\sec\theta + \tan\theta| \Big]_0^{\pi/3} = \ln\!\left(2 + \sqrt{3}\right)$

**25.** (a)    $f(x) = \tfrac{1}{2}x - 3$

$y = \tfrac{1}{2}x - 3$

$2(y + 3) = x$

$2(x + 3) = y$

$f^{-1}(x) = 2x + 6$

(b)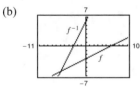

(c)  $f^{-1}(f(x)) = f^{-1}\!\left(\tfrac{1}{2}x - 3\right) = 2\!\left(\tfrac{1}{2}x - 3\right) + 6 = x$

$f\!\left(f^{-1}(x)\right) = f(2x + 6) = \tfrac{1}{2}(2x + 6) - 3 = x$

(d)  Domain $f$: all real numbers; Range $f$: all real numbers

Domain $f^{-1}$: all real numbers; Range $f^{-1}$: all real0020numbers

**27.** (a)    $f(x) = \sqrt{x + 1}$

$y = \sqrt{x + 1}$

$y^2 - 1 = x$

$x^2 - 1 = y$

$f^{-1}(x) = x^2 - 1,\ x \geq 0$

(b)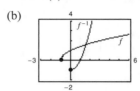

(c)  $f^{-1}(f(x)) = f^{-1}\!\left(\sqrt{x + 1}\right) = \sqrt{\left(x^2 - 1\right)^2} - 1 = x$

$f\!\left(f^{-1}(x)\right) = f\!\left(x^2 - 1\right) = \sqrt{\left(x^2 - 1\right) + 1}$

$= \sqrt{x^2} = x$ for $x \geq 0$.

(d)  Domain $f$: $x \geq -1$; Range $f$: $y \geq 0$

Domain $f^{-1}$: $x \geq 0$; Range $f^{-1}$: $y \geq -1$

**29.** (a)    $f(x) = \sqrt[3]{x + 1}$

$y = \sqrt[3]{x + 1}$

$y^3 - 1 = x$

$x^3 - 1 = y$

$f^{-1}(x) = x^3 - 1$

(b)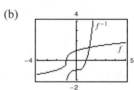

(c)  $f^{-1}(f(x)) = f^{-1}\!\left(\sqrt[3]{x + 1}\right)$

$= \left(\sqrt[3]{x + 1}\right)^3 - 1 = x$

$f\!\left(f^{-1}(x)\right) = f\!\left(x^3 - 1\right) = \sqrt[3]{\left(x^3 - 1\right) + 1} = x$

(d)  Domain $f$: all real numbers; Range $f$: all real numbers

Domain $f^{-1}$: all real numbers; Range $f^{-1}$: all real numbers

**31.**  $f(x) = x^3 + 2,\quad a = -1$

$f'(x) = 3x^2 > 0$

$f$ is monotonic (increasing) on $(-\infty, \infty)$ therefore $f$ has an inverse.

$f\!\left(-3^{1/3}\right) = -1 \Rightarrow f^{-1}(-1) = -3^{1/3}$

$f'\!\left(-3^{1/3}\right) = 3^{2/3}$

$\left(f^{-1}\right)'(-1) = \dfrac{1}{f'\!\left(f^{-1}(-1)\right)} = \dfrac{1}{f'\!\left(-3^{1/3}\right)} = \dfrac{1}{3\left(3^{2/3}\right)} = \dfrac{1}{3^{5/3}}$

**33.** $f(x) = \tan x, \quad a = \dfrac{\sqrt{3}}{3}, -\dfrac{\pi}{4} \le x \le \dfrac{\pi}{4}$

$f'(x) = \sec^2 x > 0 \text{ on } \left(-\dfrac{\pi}{4}, \dfrac{\pi}{4}\right)$

$f$ is monotonic (increasing) on $\left[-\dfrac{\pi}{4}, \dfrac{\pi}{4}\right]$ therefore $f$ has an inverse.

$f\left(\dfrac{\pi}{6}\right) = \dfrac{\sqrt{3}}{3} \Rightarrow f^{-1}\left(\dfrac{\sqrt{3}}{3}\right) = \dfrac{\pi}{6}$

$f'\left(\dfrac{\pi}{6}\right) = \dfrac{4}{3}$

$(f^{-1})'\left(\dfrac{\sqrt{3}}{3}\right) = \dfrac{1}{f'\left(f^{-1}\left(\dfrac{\sqrt{3}}{3}\right)\right)} = \dfrac{1}{f'\left(\dfrac{\pi}{6}\right)} = \dfrac{1}{\left(\dfrac{4}{3}\right)} = \dfrac{3}{4}$

**35.** (a) $f(x) = \ln \sqrt{x}$

$y = \ln \sqrt{x}$

$e^y = \sqrt{x}$

$e^{2y} = x$

$e^{2x} = y$

$f^{-1}(x) = e^{2x}$

(b)

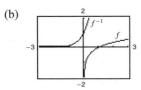

(c) $f^{-1}(f(x)) = f^{-1}\left(\ln \sqrt{x}\right) = e^{2\ln\sqrt{x}} = e^{\ln x} = x$

$f\left(f^{-1}(x)\right) = f(e^{2x}) = \ln\sqrt{e^{2x}} = \ln e^x = x$

(d) Domain $f$: $x > 0$; Range $f$: all real numbers

Domain $f^{-1}$: all real numbers; Range $f^{-1}$: $y > 0$

**37.** $f = e^{-x/2}$

**39.** $g(t) = t^2 e^t$

$g'(t) = t^2 e^t + 2te^t = te^t(t + 2)$

**41.** $y = \sqrt{e^{2x} + e^{-2x}}$

$y' = \dfrac{1}{2}\left(e^{2x} + e^{-2x}\right)^{-1/2}\left(2e^{2x} - 2e^{-2x}\right) = \dfrac{e^{2x} - e^{-2x}}{\sqrt{e^{2x} + e^{-2x}}}$

**43.** $g(x) = \dfrac{x^2}{e^x}$

$g'(x) = \dfrac{e^x(2x) - x^2 e^x}{e^{2x}} = \dfrac{x(2 - x)}{e^x}$

**45.** $f(x) = \ln\left(e^{-x^2}\right) = -x^2, \quad (2, -4)$

$f'(x) = -2x$

$f'(2) = -4$

Tangent line: $y + 4 = -4(x - 2)$

$y = -4x + 4$

**47.** $\qquad\qquad y(\ln x) + y^2 = 0$

$y\left(\dfrac{1}{x}\right) + (\ln x)\left(\dfrac{dy}{dx}\right) + 2y\left(\dfrac{dy}{dx}\right) = 0$

$(2y + \ln x)\dfrac{dy}{dx} = \dfrac{-y}{x}$

$\dfrac{dy}{dx} = \dfrac{-y}{x(2y + \ln x)}$

**49.** $\displaystyle\int_0^1 xe^{-3x^2}\,dx = -\dfrac{1}{6}\int_0^1 e^{-3x^2}(-6x\,dx)$

$= \left[-\dfrac{1}{6}e^{-3x^2}\right]_0^1$

$= -\dfrac{1}{6}\left[e^{-3} - 1\right]$

$= \dfrac{1}{6}\left(1 - \dfrac{1}{e^3}\right) \approx 0.158$

**51.** $\displaystyle\int\dfrac{e^{4x} - e^{2x} + 1}{e^x}\,dx = \int\left(e^{3x} - e^x + e^{-x}\right)dx$

$= \dfrac{1}{3}e^{3x} - e^x - e^{-x} + C$

$= \dfrac{e^{4x} - 3e^{2x} - 3}{3e^x} + C$

**53.** $\displaystyle\int xe^{1-x^2}\,dx = -\dfrac{1}{2}\int e^{1-x^2}(-2x)\,dx = -\dfrac{1}{2}e^{1-x^2} + C$

**55.** $\int_1^3 \dfrac{e^x}{e^x - 1}\, dx$

Let $u = e^x - 1$, $du = e^x dx$.

$$\int_1^3 \frac{e^x}{e^x - 1}\, dx = \left[\ln\left|e^x - 1\right|\right]_1^3$$

$$= \ln\!\left(e^3 - 1\right) - \ln\!\left(e - 1\right)$$

$$= \ln\!\left(\frac{e^3 - 1}{e - 1}\right)$$

$$= \ln\!\left(e^2 + e + 1\right) \approx 2.408$$

**57.**
$$y = e^x(a \cos 3x + b \sin 3x)$$

$$y' = e^x(-3a \sin 3x + 3b \cos 3x) + e^x(a \cos 3x + b \sin 3x)$$

$$= e^x\!\left[(-3a + b)\sin 3x + (a + 3b)\cos 3x\right]$$

$$y'' = e^x\!\left[3(-3a + b)\cos 3x - 3(a + 3b)\sin 3x\right] + e^x\!\left[(-3a + b)\sin 3x + (a + 3b)\cos 3x\right]$$

$$= e^x\!\left[(-6a - 8b)\sin 3x + (-8a + 6b)\cos 3x\right]$$

$$y'' - 2y' + 10y = e^x\!\left\{\left[(-6a - 8b) - 2(-3a + b) + 10b\right]\sin 3x + \left[(-8a + 6b) - 2(a + 3b) + 10a\right]\cos 3x\right\} = 0$$

**59.** Area $= \displaystyle\int_0^4 xe^{-x^2}\, dx = \left[-\tfrac{1}{2}e^{-x^2}\right]_0^4 = -\tfrac{1}{2}\!\left(e^{-16} - 1\right) \approx 0.500$

**61.** $y = 3^{x/2}$

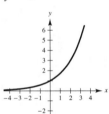

**63.** $y = \log_2(x - 1)$

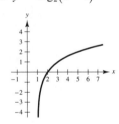

**65.** $f(x) = 3^{x-1}$

$f'(x) = 3^{x-1}\ln 3$

**67.** $y = x^{2x+1}$

$\ln y = (2x + 1)\ln x$

$\dfrac{y'}{y} = \dfrac{2x + 1}{x} + 2\ln x$

$y' = y\!\left(\dfrac{2x + 1}{x} + 2\ln x\right) = x^{2x+1}\!\left(\dfrac{2x + 1}{x} + 2\ln x\right)$

**69.** $g(x) = \log_3\sqrt{1 - x} = \dfrac{1}{2}\log_3(1 - x)$

$g'(x) = \left(\dfrac{1}{2}\right)\dfrac{-1}{(1 - x)\ln 3} = \dfrac{1}{2(x - 1)\ln 3}$

**71.** $\displaystyle\int (x + 1)5^{(x+1)^2}\, dx = \left(\dfrac{1}{2}\right)\dfrac{1}{\ln 5}5^{(x+1)^2} + C$

**73.** $t = 50\log_{10}\!\left(\dfrac{18{,}000}{18{,}000 - h}\right)$

(a) Domain: $0 \le h < 18{,}000$

(b)

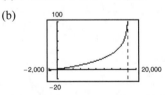

Vertical asymptote: $h = 18{,}000$

(c) $t = 50\log_{10}\!\left(\dfrac{18{,}000}{18{,}000 - h}\right)$

$$10^{t/50} = \frac{18{,}000}{18{,}000 - h}$$

$$18{,}000 - h = 18{,}000\!\left(10^{-t/50}\right)$$

$$h = 18{,}000\!\left(1 - 10^{-t/50}\right)$$

$\dfrac{dh}{dt} = 360\ln 10\!\left(\dfrac{1}{10}\right)^{t/50}$ is greatest when $t = 0$.

**75.** $f(x) = 2 \arctan(x + 3)$

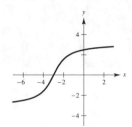

**77. (a)** Let $\theta = \arcsin \dfrac{1}{2}$

$$\sin \theta = \dfrac{1}{2}$$

$$\sin\left(\arcsin \dfrac{1}{2}\right) = \sin \theta = \dfrac{1}{2}.$$

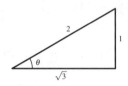

**(b)** Let $\theta = \arcsin \dfrac{1}{2}$

$$\sin \theta = \dfrac{1}{2}$$

$$\cos\left(\arcsin \dfrac{1}{2}\right) = \cos \theta = \dfrac{\sqrt{3}}{2}.$$

**79.** $y = \tan(\arcsin x) = \dfrac{x}{\sqrt{1 - x^2}}$

$$y' = \dfrac{\left(1 - x^2\right)^{1/2} + x^2\left(1 - x^2\right)^{-1/2}}{1 - x^2} = \left(1 - x^2\right)^{-3/2}$$

**81.** $y = x \operatorname{arcsec} x$

$$y' = \dfrac{x}{|x|\sqrt{x^2 - 1}} + \operatorname{arcsec} x$$

**83.** $y = x(\arcsin x)^2 - 2x + 2\sqrt{1 - x^2}\,\arcsin x$

$$y' = \dfrac{2x \arcsin x}{\sqrt{1 - x^2}} + (\arcsin x)^2 - 2 + \dfrac{2\sqrt{1 - x^2}}{\sqrt{1 - x^2}} - \dfrac{2x}{\sqrt{1 - x^2}}\arcsin x = (\arcsin x)^2$$

**85.** Let $u = e^{2x}$, $du = 2e^{2x}\,dx$.

$$\int \dfrac{1}{e^{2x} + e^{-2x}}\,dx = \int \dfrac{e^{2x}}{1 + e^{4x}}\,dx = \dfrac{1}{2}\int \dfrac{1}{1 + \left(e^{2x}\right)^2}\left(2e^{2x}\right)dx = \dfrac{1}{2}\arctan\left(e^{2x}\right) + C$$

**87.** Let $u = x^2$, $du = 2x\,dx$.

$$\int \dfrac{x}{\sqrt{1 - x^4}}\,dx = \dfrac{1}{2}\int \dfrac{1}{\sqrt{1 - \left(x^2\right)^2}}(2x)\,dx = \dfrac{1}{2}\arcsin x^2 + C$$

**89.** Let $u = \arctan\left(\dfrac{x}{2}\right)$, $du = \dfrac{2}{4 + x^2}\,dx$.

$$\int \dfrac{\arctan(x/2)}{4 + x^2}\,dx = \dfrac{1}{2}\int \left(\arctan \dfrac{x}{2}\right)\left(\dfrac{2}{4 + x^2}\right)dx = \dfrac{1}{4}\left(\arctan \dfrac{x}{2}\right)^2 + C$$

**91.** $A = \int_0^1 \dfrac{4 - x}{\sqrt{4 - x^2}}\, dx$

$= 4\int_0^1 \dfrac{1}{\sqrt{4 - x^2}}\, dx + \dfrac{1}{2}\int_0^1 (4 - x^2)^{-1/2}(-2x)\, dx$

$= \left[ 4\arcsin\left(\dfrac{x}{2}\right) + \sqrt{4 - x^2}\, \right]_0^1$

$= \left( 4\arcsin\left(\dfrac{1}{2}\right) + \sqrt{3}\, \right) - 2$

$= \dfrac{2\pi}{3} + \sqrt{3} - 2 \approx 1.8264$

**93.** $\displaystyle\int \dfrac{dy}{\sqrt{A^2 - y^2}} = \int \sqrt{\dfrac{k}{m}}\, dt$

$\arcsin\left(\dfrac{y}{A}\right) = \sqrt{\dfrac{k}{m}}\, t + C$

Because $y = 0$ when $t = 0$, you have $C = 0$. So,

$\sin\left(\sqrt{\dfrac{k}{m}}\, t\right) = \dfrac{y}{A}$

$y = A\sin\left(\sqrt{\dfrac{k}{m}}\, t\right).$

**95.** $y = x\tanh^{-1} 2x$

$y' = x\left(\dfrac{2}{1 - 4x^2}\right) + \tanh^{-1} 2x = \dfrac{2x}{1 - 4x^2} + \tanh^{-1} 2x$

**97.** Let $u = x^3,\ du = 3x^2\, dx$.

$\displaystyle\int x^2 (\operatorname{sech} x^3)^2\, dx = \dfrac{1}{3}\int (\operatorname{sech} x^3)^2 (3x^2)\, dx$

$= \dfrac{1}{3}\tanh x^3 + C$

# Problem Solving for Chapter 5

**1.** $\tan\theta_1 = \dfrac{3}{x}$

$\tan\theta_2 = \dfrac{6}{10 - x}$

Minimize $\theta_1 + \theta_2$:   $f(x) = \theta_1 + \theta_2 = \arctan\left(\dfrac{3}{x}\right) + \arctan\left(\dfrac{6}{10 - x}\right)$

$f'(x) = \dfrac{1}{1 + \dfrac{9}{x^2}}\left(\dfrac{-3}{x^2}\right) + \dfrac{1}{1 + \dfrac{36}{(10 - x)^2}}\left(\dfrac{6}{(10 - x)^2}\right) = 0$

$\dfrac{3}{x^2 + 9} = \dfrac{6}{(10 - x)^2 + 36}$

$(10 - x)^2 + 36 = 2(x^2 + 9)$

$100 - 20x + x^2 + 36 = 2x^2 + 18$

$x^2 + 20x - 118 = 0$

$x = \dfrac{-20 \pm \sqrt{20^2 - 4(-118)}}{2} = -10 \pm \sqrt{218}$

$a = -10 + \sqrt{218} \approx 4.7648$   $f(a) \approx 1.4153$

$\theta = \pi - (\theta_1 + \theta_2) \approx 1.7263$ or $98.9°$

Endpoints: $a = 0$:   $\theta \approx 1.0304$

$a = 10$:   $\theta \approx 1.2793$

Maximum is $1.7263$ at $a = -10 + \sqrt{218} \approx 4.7648.$

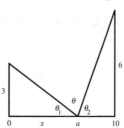

**3.** (a) $f(x) = \dfrac{\ln(x + 1)}{x}, \quad -1 \le x \le 1$

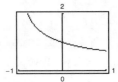

(b) $\lim\limits_{x \to 0} f(x) = 1$

(c) Let $g(x) = \ln x$, $g'(x) = 1/x$, and $g'(1) = 1$. From the definition of derivative

$$g'(1) = \lim_{x \to 0} \frac{g(1 + x) - g(1)}{x} = \lim_{x \to 0} \frac{\ln(1 + x)}{x}.$$

So, $\lim\limits_{x \to 0} f(x) = 1$.

**5.** $y = 0.5^x$ and $y = 1.2^x$ intersect $y = x$. $y = 2^x$ does not intersect $y = x$.

Suppose $y = x$ is tangent to $y = a^x$ at $(x, y)$.

$a^x = x \Rightarrow a = x^{1/x}$.

$y' = a^x \ln a = 1 \Rightarrow x \ln x^{1/x} = 1 \Rightarrow \ln x = 1 \Rightarrow x = e, a = e^{1/e}$

For $0 < a \le e^{1/e} \approx 1.445$, the curve $y = a^x$ intersects $y = x$.

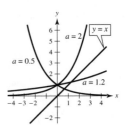

**7.** $f(x) = \ln x$ is continuous on $[1, e]$ and differentiable on $(1, e)$.

By the Mean Value Theorem, there exists $c \in (1, e)$ such that

$$f'(c) = \frac{f(e) - f(1)}{e - 1} = \frac{1 - 0}{e - 1}$$

$$\frac{1}{c} = \frac{1}{e - 1}$$

$$c = e - 1.$$

**9.** (a) $y = f(x) = \arcsin x$

$\sin y = x$

$$\text{Area } A = \int_{\pi/6}^{\pi/4} \sin y \cdot dy = \left[-\cos y\right]_{\pi/6}^{\pi/4} = -\frac{\sqrt{2}}{2} + \frac{\sqrt{3}}{2} = \frac{\sqrt{3} - \sqrt{2}}{2} \approx 0.1589$$

$$\text{Area } B = \left(\frac{1}{2}\right)\left(\frac{\pi}{6}\right) = \frac{\pi}{12} \approx 0.2618$$

(b) $\displaystyle\int_{1/2}^{\sqrt{2}/2} \arcsin x \, dx = \text{Area}(C) = \left(\frac{\pi}{4}\right)\left(\frac{\sqrt{2}}{2}\right) - A - B$

$$= \frac{\pi\sqrt{2}}{8} - \frac{\sqrt{3} - \sqrt{2}}{2} - \frac{\pi}{12} = \pi\left(\frac{\sqrt{2}}{8} - \frac{1}{12}\right) + \frac{\sqrt{2} - \sqrt{3}}{2} \approx 0.1346$$

(c) $\text{Area } A = \displaystyle\int_0^{\ln 3} e^y \, dy = \left[e^y\right]_0^{\ln 3} = 3 - 1 = 2$

$\text{Area } B = \displaystyle\int_1^3 \ln x \, dx = 3(\ln 3) - A = 3 \ln 3 - 2 = \ln 27 - 2 \approx 1.2958$

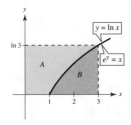

(d)   $\tan y = x$

$$\text{Area } A = \int_{\pi/4}^{\pi/3} \tan y \, dy = \left[ -\ln|\cos y| \right]_{\pi/4}^{\pi/3} = -\ln \frac{1}{2} + \ln \frac{\sqrt{2}}{2} = \ln \sqrt{2} = \frac{1}{2} \ln 2$$

$$\text{Area } C = \int_{1}^{\sqrt{3}} \arctan x \, dx = \left( \frac{\pi}{3} \right) (\sqrt{3}) - \frac{1}{2} \ln 2 - \left( \frac{\pi}{4} \right) (1) = \frac{\pi}{12} (4\sqrt{3} - 3) - \frac{1}{2} \ln 2 \approx 0.6818$$

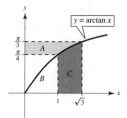

**11.**  $y = e^x$

$y' = e^x$

Tangent line: $y - b = e^a(x - a)$

$$y = e^a x - ae^a + b$$

If $y = 0$: $e^a x = ae^a - b$

$$bx = ab - b \quad (b = e^a)$$

$$x = a - 1$$

$$c = a - 1$$

So, $a - c = a - (a - 1) = 1$.

**13.** Let $u = 1 + \sqrt{x}, \sqrt{x} = u - 1$,

$x = u^2 - 2u + 1, dx = (2u - 2) \, du$.

$$\text{Area} = \int_{1}^{4} \frac{1}{\sqrt{x} + x} \, dx$$

$$= \int_{2}^{3} \frac{2u - 2}{(u - 1) + (u^2 - 2u + 1)} \, du$$

$$= \int_{2}^{3} \frac{2(u - 1)}{u^2 - u} \, du$$

$$= \int_{2}^{3} \frac{2}{u} \, du = \left[ 2 \ln|u| \right]_{2}^{3}$$

$$= 2 \ln 3 - 2 \ln 2$$

$$= 2 \ln \left( \frac{3}{2} \right) \approx 0.8109$$

**15. (a) (i)** $y = e^x$

$$y_1 = 1 + x$$

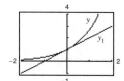

**(ii)** $y = e^x$

$$y_2 = 1 + x + \left( \frac{x^2}{2} \right)$$

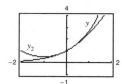

**(iii)** $y = e^x$

$$y_3 = 1 + x + \frac{x^2}{2} + \frac{x^3}{6}$$

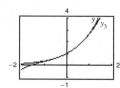

**(b)** $n^{\text{th}}$ term is $x^n/n!$ in polynomial:

$$y_4 = 1 + x + \frac{x^2}{2!} + \frac{x^3}{3!} + \frac{x^4}{4!}$$

**(c)** Conjecture: $e^x = 1 + x + \frac{x^2}{2!} + \frac{x^3}{3!} + \cdots$

# CHAPTER 6
# Differential Equations

# CHAPTER 6
# Differential Equations

## Section 6.1  Slope Fields and Euler's Method

**1.** Differential equation: $y' = 4y$

Solution: $y = Ce^{4x}$

Check: $y' = 4Ce^{4x} = 4y$

**3.** Differential equation: $y' = \dfrac{2xy}{x^2 - y^2}$

Solution: $x^2 + y^2 = Cy$

Check: $2x + 2yy' = Cy'$

$$y' = \frac{-2x}{(2y - C)}$$

$$y' = \frac{-2xy}{2y^2 - Cy} = \frac{-2xy}{2y^2 - \left(x^2 + y^2\right)} = \frac{-2xy}{y^2 - x^2} = \frac{2xy}{x^2 - y^2}$$

**5.** Differential Equation: $y'' + y = 0$

Solution:  $y = C_1 \sin x - C_2 \cos x$

$\qquad\quad y' = C_1 \cos x + C_2 \sin x$

$\qquad\quad y'' = -C_1 \sin x + C_2 \cos x$

Check: $y'' + y = \left(-C_1 \sin x + C_2 \cos x\right) + \left(C_1 \sin x - C_2 \cos x\right) = 0$

**7.** Differential Equation: $y'' + y = \tan x$

Solution:  $y = -\cos x \ln\left|\sec x + \tan x\right|$

$$y' = (-\cos x)\frac{1}{\sec x + \tan x}\left(\sec x \cdot \tan x + \sec^2 x\right) + \sin x \ln\left|\sec x + \tan x\right|$$

$$= \frac{(-\cos x)}{\sec x + \tan x}(\sec x)(\tan x + \sec x) + \sin x \ln\left|\sec x + \tan x\right| = -1 + \sin x \ln\left|\sec x + \tan x\right|$$

$$y'' = (\sin x)\frac{1}{\sec x + \tan x}\left(\sec x \cdot \tan x + \sec^2 x\right) + \cos x \ln\left|\sec x + \tan x\right|$$

$$= (\sin x)(\sec x) + \cos x \ln\left|\sec x + \tan x\right|$$

Check: $y'' + y = (\sin x)(\sec x) + \cos x \ln\left|\sec x + \tan x\right| - \cos x \ln\left|\sec x + \tan x\right| = \tan x.$

**9.**  $y = \sin x \cos x - \cos^2 x$

$y' = -\sin^2 x + \cos^2 x + 2 \cos x \sin x = -1 + 2 \cos^2 x + \sin 2x$

Differential Equation: $2y + y' = 2\left(\sin x \cos x - \cos^2 x\right) + \left(-1 + 2\cos^2 x + \sin 2x\right)$

$$= 2 \sin x \cos x - 1 + \sin 2x = 2 \sin 2x - 1$$

Initial condition $\left(\dfrac{\pi}{4}, 0\right)$: $\sin \dfrac{\pi}{4} \cos \dfrac{\pi}{4} - \cos^2 \dfrac{\pi}{4} = \dfrac{\sqrt{2}}{2} \cdot \dfrac{\sqrt{2}}{2} - \left(\dfrac{\sqrt{2}}{2}\right)^2 = 0$

**11.** $y = 4e^{-6x^2}$

$y' = 4e^{-6x^2}(-12x) = -48xe^{-6x^2}$

Differential equation: $y' = -12xy = -12x\left(4e^{-6x^2}\right) = -48xe^{-6x^2}$

Initial condition $(0, 4)$: $4e^0 = 4$

**In Exercises 13–19, the differential equation is $y^{(4)} - 16y = 0$.**

**13.**     $y = 3\cos x$

$y^{(4)} = 3\cos x$

$y^{(4)} - 16y = -45\cos x \neq 0,$

No

**15.**     $y = 3\cos 2x$

$y^{(4)} = 48\cos 2x$

$y^{(4)} - 16y = 48\cos 2x - 48\cos 2x = 0,$

Yes

**17.**     $y = e^{-2x}$

$y^{(4)} = 16e^{-2x}$

$y^{(4)} - 16y = 16e^{-2x} - 16e^{-2x} = 0,$

Yes

**19.**     $y = C_1e^{2x} + C_2e^{-2x} + C_3\sin 2x + C_4\cos 2x$

$y^{(4)} = 16C_1e^{2x} + 16C_2e^{-2x} + 16C_3\sin 2x + 16C_4\cos 2x$

$y^{(4)} - 16y = 0,$

Yes

**In Exercises 21–27, the differential equation is $xy' - 2y = x^3e^x$.**

**21.** $y = x^2,\ y' = 2x$

$xy' - 2y = x(2x) - 2\left(x^2\right) = 0 \neq x^3e^x,$

No

**23.** $y = x^2e^x,\ y' = x^2e^x + 2xe^x = e^x\left(x^2 + 2x\right)$

$xy' - 2y = x\left(e^x\left(x^2 + 2x\right)\right) - 2\left(x^2e^x\right) = x^3e^x,$

Yes

**25.** $y = \sin x,\ y' = \cos x$

$xy' - 2y = x(\cos x) - 2(\sin x) \neq x^3e^x,$

No

**27.** $y = \ln x,\ y' = \dfrac{1}{x}$

$xy' - 2y = x\left(\dfrac{1}{x}\right) - 2\ln x \neq x^3e^x,$

No

**29.** $y = Ce^{-x/2}$ passes through $(0, 3)$.

$3 = Ce^0 = C \Rightarrow C = 3$

Particular solution: $y = 3e^{-x/2}$

**31.** $y^2 = Cx^3$ passes through $(4, 4)$.

$16 = C(64) \Rightarrow C = \frac{1}{4}$

Particular solution: $y^2 = \frac{1}{4}x^3$ or $4y^2 = x^3$

**33.** Differential equation: $4yy' - x = 0$

General solution: $4y^2 - x^2 = C$

Particular solutions: $C = 0$, Two intersecting lines
$\qquad C = \pm 1, C = \pm 4$, Hyperbolas

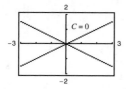
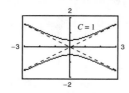
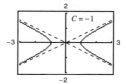

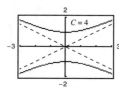
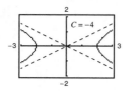

**35.** Differential equation: $y' + 2y = 0$

General solution: $y = Ce^{-2x}$

$y' + 2y = C(-2)e^{-2x} + 2\left(Ce^{-2x}\right) = 0$

Initial condition $(0, 3)$: $3 = Ce^0 = C$

Particular solution: $y = 3e^{-2x}$

**37.** Differential equation: $y'' + 9y = 0$

General solution: $y = C_1 \sin 3x + C_2 \cos 3x$

$y' = 3C_1 \cos 3x - 3C_2 \sin 3x,$

$y'' = -9C_1 \sin 3x - 9C_2 \cos 3x$

$y'' + 9y = \left(-9C_1 \sin 3x - 9C_2 \cos 3x\right) + 9\left(C_1 \sin 3x + C_2 \cos 3x\right) = 0$

Initial conditions $\left(\dfrac{\pi}{6}, 2\right)$ and $y' = 1$ when $x = \dfrac{\pi}{6}$:

$2 = C_1 \sin\left(\dfrac{\pi}{2}\right) + C_2 \cos\left(\dfrac{\pi}{2}\right) \Rightarrow C_1 = 2$

$y' = 3C_1 \cos 3x - 3C_2 \sin 3x$

$1 = 3C_1 \cos\left(\dfrac{\pi}{2}\right) - 3C_2 \sin\left(\dfrac{\pi}{2}\right) = -3C_2 \Rightarrow C_2 = -\dfrac{1}{3}$

Particular solution: $y = 2 \sin 3x - \dfrac{1}{3} \cos 3x$

**39.** Differential equation: $x^2y'' - 3xy' + 3y = 0$

General solution: $y = C_1x + C_2x^3$

$y' = C_1 + 3C_2x^2, \ y'' = 6C_2x$

$x^2y'' - 3xy' + 3y = x^2(6C_2x) - 3x(C_1 + 3C_2x^2) + 3(C_1x + C_2x^3) = 0$

Initial conditions $(2, 0)$ and $y' = 4$ when $x = 2$:

$0 = 2C_1 + 8C_2$

$y' = C_1 + 3C_2x^2$

$4 = C_1 + 12C_2$

$\left.\begin{array}{c} C_1 + 4C_2 = 0 \\ C_1 + 12C_2 = 4 \end{array}\right\} \ C_2 = \tfrac{1}{2}, \ C_1 = -2$

Particular solution: $y = -2x + \tfrac{1}{2}x^3$

**41.** $\dfrac{dy}{dx} = 6x^2$

$y = \displaystyle\int 6x^2 \, dx = 2x^3 + C$

**43.** $\dfrac{dy}{dx} = \dfrac{x}{1 + x^2}$

$y = \displaystyle\int \dfrac{x}{1 + x^2} \, dx = \dfrac{1}{2}\ln(1 + x^2) + C$

$(u = 1 + x^2, \, du = 2x \, dx)$

**45.** $\dfrac{dy}{dx} = \dfrac{x - 2}{x} = 1 - \dfrac{2}{x}$

$y = \displaystyle\int \left(1 - \dfrac{2}{x}\right) dx$

$= x - 2\ln|x| + C = x - \ln x^2 + C$

**47.** $\dfrac{dy}{dx} = \sin 2x$

$y = \displaystyle\int \sin 2x \, dx = -\dfrac{1}{2}\cos 2x + C$

$(u = 2x, \, du = 2 \, dx)$

**49.** $\dfrac{dy}{dx} = x\sqrt{x - 6}$

Let $u = \sqrt{x - 6}$, then $x = u^2 + 6$ and $dx = 2u \, du$.

$y = \displaystyle\int x\sqrt{x - 6} \, dx = \int (u^2 + 6)(u)(2u) \, du = 2\int (u^4 + 6u^2) \, du$

$\qquad = 2\left(\dfrac{u^5}{5} + 2u^3\right) + C = \dfrac{2}{5}(x - 6)^{5/2} + 4(x - 6)^{3/2} + C$

$\qquad = \dfrac{2}{5}(x - 6)^{3/2}(x - 6 + 10) + C = \dfrac{2}{5}(x - 6)^{3/2}(x + 4) + C$

**51.** $\dfrac{dy}{dx} = xe^{x^2}$

$y = \displaystyle\int xe^{x^2} \, dx = \dfrac{1}{2}e^{x^2} + C$

$(u = x^2, \, du = 2x \, dx)$

**53.**

| $x$ | $-4$ | $-2$ | 0 | 2 | 4 | 8 |
|---|---|---|---|---|---|---|
| $y$ | 2 | 0 | 4 | 4 | 6 | 8 |
| $dy/dx$ | $-4$ | Undef. | 0 | 1 | $\frac{4}{3}$ | 2 |

**55.**

| $x$ | $-4$ | $-2$ | 0 | 2 | 4 | 8 |
|---|---|---|---|---|---|---|
| $y$ | 2 | 0 | 4 | 4 | 6 | 8 |
| $dy/dx$ | $-2\sqrt{2}$ | $-2$ | 0 | 0 | $-2\sqrt{2}$ | $-8$ |

**57.** $\dfrac{dy}{dx} = \sin 2x$

For $x = 0, \dfrac{dy}{dx} = 0$. Matches (b).

**59.** $\dfrac{dy}{dx} = e^{-2x}$

As $x \to \infty, \dfrac{dy}{dx} \to 0$. Matches (d).

**61.** (a), (b)

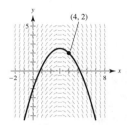

(4, 2)

(c) As $x \to \infty$, $y \to -\infty$

As $x \to -\infty$, $y \to -\infty$

**63.** (a), (b)

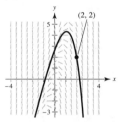

(2, 2)

(c) As $x \to \infty$, $y \to -\infty$

As $x \to -\infty$, $y \to -\infty$

**65.** (a)  $y' = \dfrac{1}{x}$, $(1, 0)$

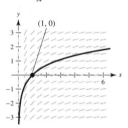

(1, 0)

As $x \to \infty$, $y \to \infty$

[Note: The solution is $y = \ln x$.]

(b)  $y' = \dfrac{1}{x}$, $(2, -1)$

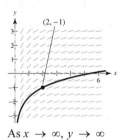

(2, −1)

As $x \to \infty$, $y \to \infty$

**67.** $\dfrac{dy}{dx} = 0.25y$, $y(0) = 4$

(a), (b)

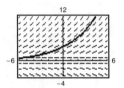

**69.** $\dfrac{dy}{dx} = 0.02y(10 - y)$, $y(0) = 2$

(a), (b)

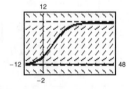

**71.** $\dfrac{dy}{dx} = 0.4y(3 - x)$, $y(0) = 1$

(a), (b)

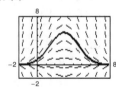

**73.** $y' = x + y$,   $y(0) = 2$,   $n = 10$,   $h = 0.1$

$y_1 = y_0 + hF(x_0, y_0) = 2 + (0.1)(0 + 2) = 2.2$

$y_2 = y_1 + hF(x_1, y_1) = 2.2 + (0.1)(0.1 + 2.2) = 2.43$, etc.

| $n$ | 0 | 1 | 2 | 3 | 4 | 5 | 6 | 7 | 8 | 9 | 10 |
|-----|---|-----|------|-------|-------|-------|-------|-------|-------|-------|------|
| $x_n$ | 0 | 0.1 | 0.2 | 0.3 | 0.4 | 0.5 | 0.6 | 0.7 | 0.8 | 0.9 | 1.0 |
| $y_n$ | 2 | 2.2 | 2.43 | 2.693 | 2.992 | 3.332 | 3.715 | 4.146 | 4.631 | 5.174 | 5.781 |

**75.** $y' = 3x - 2y,\quad y(0) = 3,\quad n = 10,\quad h = 0.05$

$y_1 = y_0 + hF(x_0, y_0) = 3 + (0.05)(3(0) - 2(3)) = 2.7$

$y_2 = y_1 + hF(x_1, y_1) = 2.7 + (0.05)(3(0.05) + 2(2.7)) = 2.4375,$ etc.

| $n$ | 0 | 1 | 2 | 3 | 4 | 5 | 6 | 7 | 8 | 9 | 10 |
|---|---|---|---|---|---|---|---|---|---|---|---|
| $x_n$ | 0 | 0.05 | 0.1 | 0.15 | 0.2 | 0.25 | 0.3 | 0.35 | 0.4 | 0.45 | 0.5 |
| $y_n$ | 3 | 2.7 | 2.438 | 2.209 | 2.010 | 1.839 | 1.693 | 1.569 | 1.464 | 1.378 | 1.308 |

**77.** $y' = e^{xy},\quad y(0) = 1,\quad n = 10,\quad h = 0.1$

$y_1 = y_0 + hF(x_0, y_0) = 1 + (0.1)e^{0(1)} = 1.1$

$y_2 = y_1 + hF(x_1, y_1) = 1.1 + (0.1)e^{(0.1)(1.1)} \approx 1.2116,$ etc.

| $n$ | 0 | 1 | 2 | 3 | 4 | 5 | 6 | 7 | 8 | 9 | 10 |
|---|---|---|---|---|---|---|---|---|---|---|---|
| $x_n$ | 0 | 0.1 | 0.2 | 0.3 | 0.4 | 0.5 | 0.6 | 0.7 | 0.8 | 0.9 | 1.0 |
| $y_n$ | 1 | 1.1 | 1.212 | 1.339 | 1.488 | 1.670 | 1.900 | 2.213 | 2.684 | 3.540 | 5.958 |

**79.** $\dfrac{dy}{dx} = y,\ y = 3e^x,\ (0, 3)$

| $x$ | 0 | 0.2 | 0.4 | 0.6 | 0.8 | 1 |
|---|---|---|---|---|---|---|
| $y(x)$ (exact) | 3 | 3.6642 | 4.4755 | 5.4664 | 6.6766 | 8.1548 |
| $y(x)\,(h = 0.2)$ | 3 | 3.6000 | 4.3200 | 5.1840 | 6.2208 | 7.4650 |
| $y(x)\,(h = 0.1)$ | 3 | 3.6300 | 4.3923 | 5.3147 | 6.4308 | 7.7812 |

**81.** $\dfrac{dy}{dx} = y + \cos x,\ y = \dfrac{1}{2}(\sin x - \cos x + e^x),\quad (0, 0)$

| $x$ | 0 | 0.2 | 0.4 | 0.6 | 0.8 | 1 |
|---|---|---|---|---|---|---|
| $y(x)$ (exact) | 0 | 0.2200 | 0.4801 | 0.7807 | 1.1231 | 1.5097 |
| $y(x)\,(h = 0.2)$ | 0 | 0.2000 | 0.4360 | 0.7074 | 1.0140 | 1.3561 |
| $y(x)\,(h = 0.1)$ | 0 | 0.2095 | 0.4568 | 0.7418 | 1.0649 | 1.4273 |

**83.** $\dfrac{dy}{dt} = -\dfrac{1}{2}(y - 72),\quad (0, 140),\ h = 0.1$

(a)

| $t$ | 0 | 1 | 2 | 3 |
|---|---|---|---|---|
| Euler | 140 | 112.7 | 96.4 | 86.6 |

(c) $\dfrac{dy}{dt} = -\dfrac{1}{2}(y - 72),\quad (0, 140),\ h = 0.05$

| $t$ | 0 | 1 | 2 | 3 |
|---|---|---|---|---|
| Euler | 140 | 112.98 | 96.7 | 86.9 |

The approximations are better using $h = 0.05$.

(b) $y = 72 + 68e^{-t/2}$   exact

| $t$ | 0 | 1 | 2 | 3 |
|---|---|---|---|---|
| Exact | 140 | 113.24 | 97.016 | 87.173 |

**85.** The general solution is a family of curves that satisfies the differential equation. A particular solution is one member of the family that satisfies given conditions.

**87.** Consider $y' = F(x, y)$, $y(x_0) = y_0$. Begin with a point $(x_0, y_0)$ that satisfies the initial condition, $y(x_0) = y_0$.

Then, using a step size of $h$, find the point $(x_1, y_1) = (x_0 + h, y_0 + hF(x_0, y_0))$.

Continue generating the sequence of points $(x_{n+1}, y_{n+1}) = (x_n + h, y_n + hF(x_n, y_n))$.

**89.** False. Consider Example 2. $y = x^3$ is a solution to $xy' - 3y = 0$, but $y = x^3 + 1$ is not a solution.

**91.** True

**93.** $\dfrac{dy}{dx} = -2y$, $y(0) = 4$, $y = 4e^{-2x}$

(a)

| $x$ | 0 | 0.2 | 0.4 | 0.6 | 0.8 | 1 |
|---|---|---|---|---|---|---|
| $y$ | 4 | 2.6813 | 1.7973 | 1.2048 | 0.8076 | 0.5413 |
| $y_1$ | 4 | 2.5600 | 1.6384 | 1.0486 | 0.6711 | 0.4295 |
| $y_2$ | 4 | 2.4000 | 1.4400 | 0.8640 | 0.5184 | 0.3110 |
| $e_1$ | 0 | 0.1213 | 0.1589 | 0.1562 | 0.1365 | 0.1118 |
| $e_2$ | 0 | 0.2813 | 0.3573 | 0.3408 | 0.2892 | 0.2303 |
| $r$ | | 0.4312 | 0.4447 | 0.4583 | 0.4720 | 0.4855 |

(b) If $h$ is halved, then the error is approximately halved $(r \approx 0.5)$.

(c) When $h = 0.05$, the errors will again be approximately halved.

**95.** (a) $L\dfrac{dI}{dt} + RI = E(t)$

$4\dfrac{dI}{dt} + 12I = 24$

$\dfrac{dI}{dt} = \dfrac{1}{4}(24 - 12I) = 6 - 3I$

(b) As $t \to \infty$, $I \to 2$. That is, $\lim\limits_{t \to \infty} I(t) = 2$. In fact, $I = 2$ is a solution to the differential equation.

**97.** $y = A \sin \omega t$

$y' = A\omega \cos \omega t$

$y'' = -A\omega^2 \sin \omega t$

$y'' + 16y = 0$

$-A\omega^2 \sin \omega t + 16A \sin \omega t = 0$

$A \sin \omega t \left[16 - \omega^2\right] = 0$

If $A \neq 0$, then $\omega = \pm 4$

**99.** Let the vertical line $x = k$ cut the graph of the solution $y = f(x)$ at $(k, t)$. The tangent line at $(k, t)$ is

$$y - t = f'(k)(x - k)$$

Because $y' + p(x)y = q(x)$, you have

$$y - t = \left[q(k) - p(k)t\right](x - k)$$

For any value of $t$, this line passes through the point $\left(k + \dfrac{1}{p(k)}, \dfrac{q(k)}{p(k)}\right)$.

To see this, note that

$$\frac{q(k)}{p(k)} - t \overset{?}{=} \left[q(k) - p(k)t\right]\left(k + \frac{1}{p(k)} - k\right)$$

$$\overset{?}{=} q(k)k - p(k)tk + \frac{q(k)}{p(k)} - t - kq(k) + p(k)kt = \frac{q(k)}{p(k)} - t.$$

## Section 6.2 Differential Equations: Growth and Decay

**1.** $\dfrac{dy}{dx} = x + 3$

$\quad y = \displaystyle\int (x + 3)\, dx = \dfrac{x^2}{2} + 3x + C$

**3.** $\qquad \dfrac{dy}{dx} = y + 3$

$\qquad \dfrac{dy}{y + 3} = dx$

$\quad \displaystyle\int \dfrac{1}{y + 3}\, dy = \int dx$

$\qquad \ln|y + 3| = x + C_1$

$\qquad\quad y + 3 = e^{x + C_1} = Ce^x$

$\qquad\qquad y = Ce^x - 3$

**5.** $\qquad y' = \dfrac{5x}{y}$

$\qquad\quad yy' = 5x$

$\quad \displaystyle\int yy'\, dx = \int 5x\, dx$

$\quad \displaystyle\int y\, dy = \int 5x\, dx$

$\qquad \dfrac{1}{2}y^2 = \dfrac{5}{2}x^2 + C_1$

$\quad y^2 - 5x^2 = C$

**7.** $\qquad y' = \sqrt{x}\,y$

$\qquad \dfrac{y'}{y} = \sqrt{x}$

$\quad \displaystyle\int \dfrac{y'}{y}\, dx = \int \sqrt{x}\, dx$

$\quad \displaystyle\int \dfrac{dy}{y} = \int \sqrt{x}\, dx$

$\qquad \ln|y| = \dfrac{2}{3}x^{3/2} + C_1$

$\qquad\quad y = e^{(2/3)x^{3/2} + C_1} = e^{C_1}e^{(2/3)x^{3/2}} = Ce^{\left(2x^{3/2}\right)/3}$

**9.** $\left(1 + x^2\right)y' - 2xy = 0$

$\qquad y' = \dfrac{2xy}{1 + x^2}$

$\qquad \dfrac{y'}{y} = \dfrac{2x}{1 + x^2}$

$\quad \displaystyle\int \dfrac{y'}{y}\, dx = \int \dfrac{2x}{1 + x^2}\, dx$

$\quad \displaystyle\int \dfrac{dy}{y} = \int \dfrac{2x}{1 + x^2}\, dx$

$\qquad \ln|y| = \ln\left(1 + x^2\right) + C_1$

$\qquad \ln|y| = \ln\left(1 + x^2\right) + \ln C$

$\qquad \ln|y| = \ln\left[C\left(1 + x^2\right)\right]$

$\qquad\quad y = C\left(1 + x^2\right)$

**11.**   $\dfrac{dQ}{dt} = \dfrac{k}{t^2}$

$\displaystyle\int \dfrac{dQ}{dt}\,dt = \int \dfrac{k}{t^2}\,dt$

$\displaystyle\int dQ = -\dfrac{k}{t} + C$

$Q = -\dfrac{k}{t} + C$

**13.**   $\dfrac{dN}{ds} = k(500 - s)$

$\displaystyle\int \dfrac{dN}{ds}\,ds = \int k(500 - s)\,ds$

$\displaystyle\int dN = -\dfrac{k}{2}(500 - s)^2 + C$

$N = -\dfrac{k}{2}(500 - s)^2 + C$

**15. (a)**

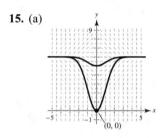

**(b)**   $\dfrac{dy}{dx} = x(6 - y), \quad (0, 0)$

$\dfrac{dy}{y - 6} = -x\,dx$

$\ln|y - 6| = \dfrac{-x^2}{2} + C$

$y - 6 = e^{-x^2/2 + C} = C_1 e^{-x^2/2}$

$y = 6 + C_1 e^{-x^2/2}$

$(0, 0): \ 0 = 6 + C_1 \ \Rightarrow \ C_1 = -6$

$y = 6 - 6e^{-x^2/2}$

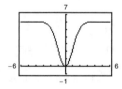

**17.**   $\dfrac{dy}{dt} = \dfrac{1}{2}t, \quad (0, 10)$

$\displaystyle\int dy = \int \dfrac{1}{2}t\,dt$

$y = \dfrac{1}{4}t^2 + C$

$10 = \dfrac{1}{4}(0)^2 + C \ \Rightarrow \ C = 10$

$y = \dfrac{1}{4}t^2 + 10$

**19.**   $\dfrac{dy}{dt} = -\dfrac{1}{2}y, \quad (0, 10)$

$\displaystyle\int \dfrac{dy}{y} = \int -\dfrac{1}{2}\,dt$

$\ln|y| = -\dfrac{1}{2}t + C_1$

$y = e^{-(t/2) + C_1} = e^{C_1} e^{-t/2} = Ce^{-t/2}$

$10 = Ce^0 \ \Rightarrow \ C = 10$

$y = 10e^{-t/2}$

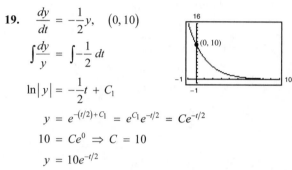

**21.**   $\dfrac{dy}{dx} = ky$

$y = Ce^{kx}$   (Theorem 6.1)

$(0, 6): \ 6 = Ce^{k(0)} = C$

$(4, 15): \ 15 = 6e^{k(4)} \ \Rightarrow \ k = \dfrac{1}{4}\ln\!\left(\dfrac{5}{2}\right)$

$y = 6e^{[1/4\,\ln(5/2)]x} \approx 6e^{0.2291x}$

When $x = 8$,

$\qquad y = 6e^{1/4\,\ln(5/2)(8)} = 6e^{\ln(5/2)^2} = 6\!\left(\dfrac{25}{4}\right) = \dfrac{75}{2}$

**23.**   $\dfrac{dV}{dt} = kV$

$V = Ce^{kt}$   (Theorem 6.1)

$(0, 20{,}000): \ C = 20{,}000$

$(4, 12{,}500): \ 12{,}500 = 20{,}000e^{4k} \ \Rightarrow \ k = \dfrac{1}{4}\ln\!\left(\dfrac{5}{8}\right)$

$V = 20{,}000e^{[1/4\,\ln(5/8)]t} \approx 20{,}000e^{-0.1175t}$

When $t = 6$, $V = 20{,}000e^{(1/4)\ln(5/8)(6)} = 20{,}000e^{\ln(5/8)^{3/2}}$

$$= 20{,}000\!\left(\dfrac{5}{8}\right)^{3/2} \approx 9882.118.$$

**25.** $y = Ce^{kt}, \quad \left(0, \frac{1}{2}\right), (5, 5)$

$$C = \frac{1}{2}$$

$$y = \frac{1}{2}e^{kt}$$

$$5 = \frac{1}{2}e^{5k}$$

$$k = \frac{\ln 10}{5}$$

$$y = \frac{1}{2}e^{[(\ln 10)/5]t} = \frac{1}{2}(10^{t/5}) \text{ or } y \approx \frac{1}{2}e^{0.4605t}$$

**27.** $y = Ce^{kt}, \quad (1, 5), (5, 2)$

$$5 = Ce^{k} \Rightarrow 10 = 2Ce^{k}$$

$$2 = Ce^{5k} \Rightarrow 10 = 5Ce^{k}$$

$$2Ce^{k} = 5Ce^{5k}$$

$$2e^{k} = 5e^{5k}$$

$$\frac{2}{5} = e^{4k}$$

$$k = \frac{1}{4}\ln\left(\frac{2}{5}\right) = \ln\left(\frac{2}{5}\right)^{1/4}$$

$$C = 5e^{-k} = 5e^{-1/4\ln(2/5)} = 5\left(\frac{2}{5}\right)^{-1/4} = 5\left(\frac{5}{2}\right)^{1/4}$$

$$y = 5\left(\frac{5}{2}\right)^{1/4}e^{[1/4\ln(2/5)]t} \approx 6.2872\,e^{-0.2291t}$$

**29.** In the model $y = Ce^{kt}$, $C$ represents the initial value of $y$ (when $t = 0$). $k$ is the proportionality constant.

**31.** $\dfrac{dy}{dx} = \dfrac{1}{2}xy$

$\dfrac{dy}{dx} > 0$ when $xy > 0$. Quadrants I and III.

**33.** Because the initial quantity is 20 grams,

$$y = 20e^{kt}.$$

Because the half-life is 1599 years,

$$10 = 20e^{k(1599)}$$

$$k = \frac{1}{1599}\ln\left(\frac{1}{2}\right).$$

So, $y = 20e^{[\ln(1/2)/1599]t}$.

When $t = 1000$, $y = 20e^{[\ln(1/2)/1599](1000)} \approx 12.96g$.

When $t = 10,000$, $y \approx 0.26g$.

**35.** Because the half-life is 1599 years,

$$\frac{1}{2} = 1e^{k(1599)}$$

$$k = \frac{1}{1599}\ln\left(\frac{1}{2}\right).$$

Because there are 0.1 gram after 10,000 years,

$$0.1 = Ce^{[\ln(1/2)/1599](10,000)}$$

$$C \approx 7.63.$$

So, the initial quantity is approximately 7.63 g.

When $t = 1000$, $y = 7.63e^{[\ln(1/2)/1599](1000)} \approx 4.95$ g.

**37.** Because the initial quantity is 5 grams, $C = 5$.

Because the half-life is 5715 years,

$$2.5 = 5e^{k(5715)}$$

$$k = \frac{1}{5715}\ln\left(\frac{1}{2}\right).$$

When $t = 1000$ years, $y = 5e^{[\ln(1/2)/5715](1000)} \approx 4.43$ g.

When $t = 10,000$ years, $y = 5e^{[\ln(1/2)/5715](10,000)}$

$$\approx 1.49 \text{ g.}$$

**39.** Because the half-life is 24,100 years,

$$\frac{1}{2} = 1e^{k(24,100)}$$

$$k = \frac{1}{24,100}\ln\left(\frac{1}{2}\right).$$

Because there are 2.1 grams after 1000 years,

$$2.1 = Ce^{[\ln(1/2)/24,100](1000)}$$

$$C \approx 2.161.$$

So, the initial quantity is approximately 2.161 g.

When $t = 10,000$, $y = 2.161e^{[\ln(1/2)/24,100](10,000)}$

$$\approx 1.62 \text{ g.}$$

**41.** $y = Ce^{kt}$

$$\frac{1}{2}C = Ce^{k(1599)}$$

$$k = \frac{1}{1599}\ln\left(\frac{1}{2}\right)$$

When $t = 100$, $y = Ce^{[\ln(1/2)/1599](100)} \approx 0.9576\,C$

Therefore, 95.76% remains after 100 years.

**43.** Because $A = 4000e^{0.06t}$, the time to double is given by

$$8000 = 4000e^{0.06t}$$

$$2 = e^{0.06t}$$

$$\ln 2 = 0.06t$$

$$t = \frac{\ln 2}{0.06} \approx 11.55 \text{ years.}$$

Amount after 10 years: $A = 4000e^{(0.06)(10)} \approx \$7288.48$

**45.** Because $A = 750e^{rt}$ and $A = 1500$ when $t = 7.75$, you have the following.

$$1500 = 750e^{7.75r}$$

$$r = \frac{\ln 2}{7.75} \approx 0.0894 = 8.94\%$$

Amount after 10 years: $A = 750e^{0.0894(10)} \approx \$1833.67$

**47.** Because $A = 500e^{rt}$ and $A = 1292.85$ when $t = 10$, you have the following.

$$1292.85 = 500e^{10r}$$

$$r = \frac{\ln(1292.85/500)}{10} \approx 0.0950 = 9.50\%$$

The time to double is given by

$$1000 = 500e^{0.0950t}$$

$$t = \frac{\ln 2}{0.095} \approx 7.30 \text{ years.}$$

**49.** $1{,}000{,}000 = P\left(1 + \dfrac{0.075}{12}\right)^{(12)(20)}$

$P = 1{,}000{,}000\left(1 + \dfrac{0.075}{12}\right)^{-240} \approx \$224{,}174.18$

**51.** $1{,}000{,}000 = P\left(1 + \dfrac{0.08}{12}\right)^{(12)(35)}$

$P = 1{,}000{,}000\left(1 + \dfrac{0.08}{12}\right)^{-420} = \$61{,}377.75$

**53.** (a) $2000 = 1000(1 + 0.07)^{t}$

$2 = 1.07^{t}$

$\ln 2 = t \ln 1.07$

$t = \dfrac{\ln 2}{\ln 1.07} \approx 10.24 \text{ years}$

(b) $2000 = 1000\left(1 + \dfrac{0.07}{12}\right)^{12t}$

$2 = \left(1 + \dfrac{0.007}{12}\right)^{12t}$

$\ln 2 = 12t \ln\left(1 + \dfrac{0.07}{12}\right)$

$t = \dfrac{\ln 2}{12 \ln\left(1 + (0.07/12)\right)} \approx 9.93 \text{ years}$

(c) $2000 = 1000\left(1 + \dfrac{0.07}{365}\right)^{365t}$

$2 = \left(1 + \dfrac{0.07}{365}\right)^{365t}$

$\ln 2 = 365t \ln\left(1 + \dfrac{0.07}{365}\right)$

$t = \dfrac{\ln 2}{365 \ln\left(1 + (0.07/365)\right)} \approx 9.90 \text{ years}$

(d) $2000 = 1000e^{(0.07)t}$

$2 = e^{0.07t}$

$\ln 2 = 0.07t$

$t = \dfrac{\ln 2}{0.07} \approx 9.90 \text{ years}$

**55.** (a) $2000 = 1000(1 + 0.085)^{t}$

$2 = 1.085^{t}$

$\ln 2 = t \ln 1.085$

$t = \dfrac{\ln 2}{\ln 1.085} \approx 8.50 \text{ years}$

(b) $2000 = 1000\left(1 + \dfrac{0.085}{12}\right)^{12t}$

$2 = \left(1 + \dfrac{0.085}{12}\right)^{12t}$

$\ln 2 = 12t \ln\left(1 + \dfrac{0.085}{12}\right)$

$t = \dfrac{1}{12}\dfrac{\ln 2}{\ln\left(1 + \dfrac{0.085}{12}\right)} \approx 8.18 \text{ years}$

(c) $2000 = 1000\left(1 + \dfrac{0.085}{365}\right)^{365t}$

$2 = \left(1 + \dfrac{0.085}{365}\right)^{365t}$

$\ln 2 = 365t \ln\left(1 + \dfrac{0.085}{365}\right)$

$t = \dfrac{1}{365}\dfrac{\ln 2}{\ln\left(1 + \dfrac{0.085}{365}\right)} \approx 8.16 \text{ years}$

(d) $2000 = 1000e^{0.085t}$

$2 = e^{0.085t}$

$\ln 2 = 0.085t$

$t = \dfrac{\ln 2}{0.085} \approx 8.15 \text{ years}$

**57.** (a) $P = Ce^{kt} = Ce^{-0.006t}$

$$P(7) = 2.3 = Ce^{-0.006(7)} \Rightarrow C \approx 2.40$$

$$P = 2.40e^{-0.006t}$$

(b) For $t = 15$, $P = 2.40e^{-0.006(15)} \approx 2.19$ million

(c) Because $k < 0$, the population is decreasing.

**59.** (a) $P = Ce^{kt} = Ce^{0.024t}$

$$P(7) = 6.7 = Ce^{0.024(7)} \Rightarrow C \approx 5.66$$

$$P = 5.66e^{0.024t}$$

(b) For $t = 15$, $P = 5.66e^{0.024(15)} \approx 8.11$ million

(c) Because $k > 0$, the population is increasing.

**61.** (a) $P = Ce^{kt} = Ce^{0.036t}$

$$P(7) = 30.3 = Ce^{0.036(7)} \Rightarrow C \approx 23.55$$

$$P = 23.55e^{0.036t}$$

(b) For $t = 15$, $P = 23.55e^{0.036(15)} \approx 40.41$ million

(c) Because $k > 0$, the population is increasing.

**63.** (a) $N = 100.1596(1.2455)^t$

(b) $N = 400$ when $t = 6.3$ hours (graphing utility)

Analytically,

$$400 = 100.1596(1.2455)^t$$

$$1.2455^t = \frac{400}{100.1596} = 3.9936$$

$$t \ln 1.2455 = \ln 3.9936$$

$$t = \frac{\ln 3.9936}{\ln 1.2455} \approx 6.3 \text{ hours.}$$

**65.** (a) $19 = 30\left(1 - e^{20k}\right)$

$$30e^{20k} = 11$$

$$k = \frac{\ln(11/30)}{20} \approx -0.0502$$

$$N \approx 30\left(1 - e^{-0.0502t}\right)$$

(b) $25 = 30\left(1 - e^{-0.0502t}\right)$

$$e^{-0.0502t} = \frac{1}{6}$$

$$t = \frac{-\ln 6}{-0.0502} \approx 36 \text{ days}$$

**67.** (a) $P_1 = Ce^{kt} = 181e^{kt}$

$$205 = 181e^{10k} \Rightarrow k = \frac{1}{10}\ln\left(\frac{205}{181}\right) \approx 0.01245$$

$$P_1 \approx 181e^{0.01245t} \approx 181(1.01253)^t$$

(b) Using a graphing utility, $P_2 \approx 182.3248(1.01091)^t$

(c)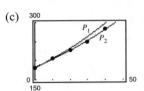

The model $P_2$ fits the data better.

(d) Using the model $P_2$,

$$320 = 182.3248(1.01091)^t$$

$$\frac{320}{182.3248} = (1.01091)^t$$

$$t = \frac{\ln(320/182.3248)}{\ln(1.01091)}$$

$$\approx 51.8 \text{ years, or 2011.}$$

**69.** $\beta(I) = 10\log_{10}\frac{I}{I_0}$, $I_0 = 10^{-16}$

(a) $\beta(10^{-14}) = 10\log_{10}\frac{10^{-14}}{10^{-16}} = 20$ decibels

(b) $\beta(10^{-9}) = 10\log_{10}\frac{10^{-9}}{10^{-16}} = 70$ decibels

(c) $\beta(10^{-6.5}) = 10\log_{10}\frac{10^{-6.5}}{10^{-16}} = 95$ decibels

(d) $\beta(10^{-4}) = 10\log_{10}\frac{10^{-4}}{10^{-16}} = 120$ decibels

**71.** $A(t) = V(t)e^{-0.10t}$

$$= 100,000e^{0.8\sqrt{t}}e^{-0.10t}$$

$$= 100,000e^{0.8\sqrt{t}-0.10t}$$

$$\frac{dA}{dt} = 100,000\left(\frac{0.4}{\sqrt{t}} - 0.10\right)e^{0.8\sqrt{t}-0.10t} = 0 \text{ when 16.}$$

The timber should be harvested in the year 2024, (2008 + 16). **Note:** You could also use a graphing utility to graph $A(t)$ and find the maximum of $A(t)$. Use the viewing rectangle $0 \le x \le 30$ and $0 \le y \le 600,000$.

**73.** Because $\dfrac{dy}{dt} = k(y - 80)$

$$\int \frac{1}{y - 80}\,dy = \int k\,dt$$

$\ln(y - 80) = kt + C.$

When $t = 0$, $y = 1500$. So, $C = \ln 1420$.

When $t = 1$, $y = 1120$. So,

$k(1) + \ln 1420 = \ln(1120 - 80)$

$\qquad k = \ln 1040 - \ln 1420 = \ln \dfrac{104}{142}.$

So, $y = 1420e^{\left[\ln(104/142)\right]t} + 80.$

When $t = 5$, $y \approx 379.2°\text{F}.$

**75.** False. If $y = Ce^{kt}$, $y' = Cke^{kt} \ne$ constant.

**77.** True

# Section 6.3   Separation of Variables and the Logistic Equation

**1.**  $\dfrac{dy}{dx} = \dfrac{x}{y}$

$\displaystyle\int y\,dy = \int x\,dx$

$\dfrac{y^2}{2} = \dfrac{x^2}{2} + C_1$

$y^2 - x^2 = C$

**3.**  $x^2 + 5y\dfrac{dy}{dx} = 0$

$5y\dfrac{dy}{dx} = -x^2$

$\displaystyle\int 5y\,dy = \int -x^2\,dx$

$\dfrac{5y^2}{2} = \dfrac{-x^3}{3} + C_1$

$15y^2 + 2x^3 = C$

**5.**  $\dfrac{dr}{ds} = 0.75\,r$

$\displaystyle\int \frac{dr}{r} = \int 0.75\,ds$

$\ln|r| = 0.75\,s + C_1$

$r = e^{0.75\,s + C_1}$

$r = Ce^{0.75\,s}$

**7.**  $(2 + x)y' = 3y$

$\displaystyle\int \frac{dy}{y} = \int \frac{3}{2 + x}\,dx$

$\ln|y| = 3\ln|2 + x| + \ln C = \ln\left| C(2 + x)^3 \right|$

$y = C(x + 2)^3$

**9.**  $yy' = 4\sin x$

$y\dfrac{dy}{dx} = 4\sin x$

$\displaystyle\int y\,dy = \int 4\sin x\,dx$

$\dfrac{y^2}{2} = -4\cos x + C_1$

$y^2 = C - 8\cos x$

**11.**  $\sqrt{1 - 4x^2}\,y' = x$

$dy = \dfrac{x}{\sqrt{1 - 4x^2}}\,dx$

$\displaystyle\int dy = \int \frac{x}{\sqrt{1 - 4x^2}}\,dx$

$\qquad = -\dfrac{1}{8}\int\left(1 - 4x^2\right)^{-1/2}(-8x\,dx)$

$y = -\dfrac{1}{4}\sqrt{1 - 4x^2} + C$

**13.**  $y\ln x - xy' = 0$

$\displaystyle\int \frac{dy}{y} = \int \frac{\ln x}{x}\,dx \quad \left( u = \ln x,\ du = \frac{dx}{x}\right)$

$\ln|y| = \dfrac{1}{2}(\ln x)^2 + C_1$

$y = e^{(1/2)(\ln x)^2 + C_1} = Ce^{(\ln x)^2/2}$

**15.** $yy' - 2e^x = 0$

$$y \frac{dy}{dx} = 2e^x$$

$$\int y \, dy = \int 2e^x \, dx$$

$$\frac{y^2}{2} = 2e^x + C$$

Initial condition $(0, 3)$: $\dfrac{9}{2} = 2 + C \Rightarrow C = \dfrac{5}{2}$

Particular solution: $\dfrac{y^2}{2} = 2e^x + \dfrac{5}{2}$

$$y^2 = 4e^x + 5$$

**17.** $y(x + 1) + y' = 0$

$$\int \frac{dy}{y} = -\int (x + 1) \, dx$$

$$\ln|y| = -\frac{(x + 1)^2}{2} + C_1$$

$$y = Ce^{-(x+1)^2/2}$$

Initial condition $(-2, 1)$: $1 = Ce^{-1/2}, \ C = e^{1/2}$

Particular solution: $y = e^{\left[1-(x+1)^2\right]/2} = e^{-\left(x^2+2x\right)/2}$

**19.** $y(1 + x^2)y' = x(1 + y^2)$

$$\frac{y}{1 + y^2} \, dy = \frac{x}{1 + x^2} \, dx$$

$$\frac{1}{2} \ln(1 + y^2) = \frac{1}{2} \ln(1 + x^2) + C_1$$

$$\ln(1 + y^2) = \ln(1 + x^2) + \ln C = \ln\left[C(1 + x^2)\right]$$

$$1 + y^2 = C(1 + x^2)$$

Initial condition $\left(0, \sqrt{3}\right)$: $1 + 3 = C \Rightarrow C = 4$

Particular solution: $1 + y^2 = 4(1 + x^2)$

$$y^2 = 3 + 4x^2$$

**21.** $\dfrac{du}{dv} = uv \sin v^2$

$$\int \frac{du}{u} = \int v \sin v^2 \, dv$$

$$\ln|u| = -\frac{1}{2} \cos v^2 + C_1$$

$$u = Ce^{-\left(\cos v^2\right)/2}$$

Initial condition: $u(0) = 1$: $C = \dfrac{1}{e^{-1/2}} = e^{1/2}$

Particular solution: $u = e^{\left(1-\cos v^2\right)/2}$

**23.** $dP - kP \, dt = 0$

$$\int \frac{dP}{P} = k \int dt$$

$$\ln|P| = kt + C_1$$

$$P = Ce^{kt}$$

Initial condition: $P(0) = P_0, \ P_0 = Ce^0 = C$

Particular solution: $P = P_0 e^{kt}$

**25.** $y' = \dfrac{dy}{dx} = \dfrac{x}{4y}$

$$\int 4y \, dy = \int x \, dx$$

$$2y^2 = \frac{x^2}{2} + C$$

Initial condition $(0, 2)$: $2(2^2) = 0 + C \Rightarrow C = 8$

Particular solution: $2y^2 = \dfrac{x^2}{2} + 8$

$$4y^2 - x^2 = 16$$

**27.** $y' = \dfrac{dy}{dx} = \dfrac{y}{2x}$

$$\int \frac{2}{y} \, dy = \int \frac{1}{x} \, dx$$

$$2 \ln|y| = \ln|x| + C_1 = \ln|x| + \ln C$$

$$y^2 = Cx$$

Initial condition $(9, 1)$: $1 = 9C \Rightarrow C = \dfrac{1}{9}$

Particular solution: $\qquad y^2 = \dfrac{1}{9}x$

$$9y^2 - x = 0$$

$$y = \frac{1}{3}\sqrt{x}$$

**29.** $m = \dfrac{dy}{dx} = \dfrac{0 - y}{(x + 2) - x} = -\dfrac{y}{2}$

$$\int \frac{dy}{y} = \int -\frac{1}{2} \, dx$$

$$\ln|y| = -\frac{1}{2}x + C_1$$

$$y = Ce^{-x/2}$$

**31.** $f(x, y) = x^3 - 4xy^2 + y^3$

$$f(tx, ty) = t^3x^3 - 4txt^2y^2 + t^3y^3$$

$$= t^3\left(x^3 - 4xy^2 + y^3\right)$$

Homogeneous of degree 3

**33.**  $f(x, y) = \dfrac{x^2 y^2}{\sqrt{x^2 + y^2}}$

$f(tx, ty) = \dfrac{t^4 x^2 y^2}{\sqrt{t^2 x^2 + t^2 y^2}} = t^3 \dfrac{x^2 y^2}{\sqrt{x^2 + y^2}}$

Homogeneous of degree 3

**35.**  $f(x, y) = 2 \ln xy$

$f(tx, ty) = 2 \ln[txty] = 2 \ln[t^2 xy] = 2(\ln t^2 + \ln xy)$

Not homogeneous

**37.**  $f(x, y) = 2 \ln \dfrac{x}{y}$

$f(tx, ty) = 2 \ln \dfrac{tx}{ty} = 2 \ln \dfrac{x}{y}$

Homogeneous of degree 0

**39.**  $y' = \dfrac{x + y}{2x}, \quad y = vx$

$v + x\dfrac{dv}{dx} = \dfrac{x + vx}{2x}$

$x\dfrac{dv}{dx} = \dfrac{1 + v}{2} - v = \dfrac{1 - v}{2}$

$2\displaystyle\int \dfrac{dv}{1 - v} = \int \dfrac{dx}{x}$

$-\ln(1 - v)^2 = \ln|x| + \ln C = \ln|Cx|$

$\dfrac{1}{(1 - v)^2} = |Cx|$

$\dfrac{1}{[1 - (y/x)]^2} = |Cx|$

$\dfrac{x^2}{(x - y)^2} = |Cx|$

$|x| = C(x - y)^2$

**41.**  $y' = \dfrac{x - y}{x + y}, \quad y = vx$

$v + x\dfrac{dv}{dx} = \dfrac{x - xv}{x + xv}$

$v\, dx + x\, dv = \dfrac{1 - v}{1 + v}\, dx$

$x\, dv = \left(\dfrac{1 - v}{1 + v} - v\right) dx = \dfrac{1 - 2v - v^2}{1 + v}\, dx$

$\displaystyle\int \dfrac{v + 1}{v^2 + 2v - 1}\, dv = -\int \dfrac{dx}{x}$

$\dfrac{1}{2} \ln|v^2 + 2v - 1| = -\ln|x| + \ln C_1 = \ln\left|\dfrac{C_1}{x}\right|$

$|v^2 + 2v - 1| = \dfrac{C}{x^2}$

$\left|\dfrac{y^2}{x^2} + 2\dfrac{y}{x} - 1\right| = \dfrac{C}{x^2}$

$|y^2 + 2xy - x^2| = C$

**43.**  $y' = \dfrac{xy}{x^2 - y^2}, \quad y = vx$

$v + x\dfrac{dv}{dx} = \dfrac{x^2 v}{x^2 - x^2 v^2}$

$v\, dx + x\, dv = \dfrac{v}{1 - v^2}\, dx$

$x\, dv = \left(\dfrac{v}{1 - v^2} - v\right) dx = \left(\dfrac{v^3}{1 - v^2}\right) dx$

$\displaystyle\int \dfrac{1 - v^2}{v^3}\, dv = \int \dfrac{dx}{x}$

$-\dfrac{1}{2v^2} - \ln|v| = \ln|x| + \ln C_1 = \ln|C_1 x|$

$\dfrac{-1}{2v^2} = \ln|C_1 xv|$

$\dfrac{-x^2}{2y^2} = \ln|C_1 y|$

$y = Ce^{-x^2/2y^2}$

**45.**  $x\, dy - (2xe^{-y/x} + y)\, dx = 0, \quad y = vx$

$x(v\, dx + x\, dv) - (2xe^{-v} + vx)\, dx = 0$

$\displaystyle\int e^v dv = \int \dfrac{2}{x}\, dx$

$e^v = \ln C_1 x^2$

$e^{y/x} = \ln C_1 + \ln x^2$

$e^{y/x} = C + \ln x^2$

Initial condition $(1, 0)$:  $1 = C$

Particular solution:  $e^{y/x} = 1 + \ln x^2$

**47.**  $\left(x \sec \dfrac{y}{x} + y\right) dx - x\, dy = 0, \quad y = vx$

$(x \sec v + xv)\, dx - x(v\, dx + x\, dv) = 0$

$(\sec v + v)\, dx = v\, dx + x\, dv$

$\displaystyle\int \cos v\, dv = \int \dfrac{dx}{x}$

$\sin v = \ln x + \ln C_1$

$x = Ce^{\sin v}$

$= Ce^{\sin(y/x)}$

Initial condition $(1, 0)$:  $1 = Ce^0 = C$

Particular solution:  $x = e^{\sin(y/x)}$

**49.** $\dfrac{dy}{dx} = x$

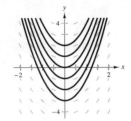

$$y = \int x\, dx = \frac{1}{2}x^2 + C$$

**51.** $\dfrac{dy}{dx} = 4 - y$

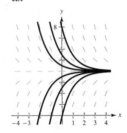

$$\int \frac{dy}{4 - y} = \int dx$$

$$\ln|4 - y| = -x + C_1$$

$$4 - y = e^{-x + C_1}$$

$$y = 4 + Ce^{-x}$$

**53.** (a) Euler's Method gives $y \approx 0.1602$ when $x = 1$.

(b) $\dfrac{dy}{dx} = -6xy$

$$\int \frac{dy}{y} = \int -6x$$

$$\ln|y| = -3x^2 + C_1$$

$$y = Ce^{-3x^2}$$

$$y(0) = 5 \Rightarrow C = 5$$

$$y = 5e^{-3x^2}$$

(c) At $x = 1$, $y = 5e^{-3(1)} \approx 0.2489$.

Error: $0.2489 - 0.1602 \approx 0.0887$

**63.** $\dfrac{dw}{dt} = k(1200 - w)$

$$\int \frac{dw}{1200 - w} = \int k\, dt$$

$$\ln|1200 - w| = -kt + C_1$$

$$1200 - w = e^{-kt + C_1} = Ce^{-kt}$$

$$w = 1200 - Ce^{-kt}$$

$$w(0) = 60 = 1200 - C \Rightarrow C = 1200 - 60 = 1140$$

$$w = 1200 - 1140e^{-kt}$$

**55.** (a) Euler's Method gives $y \approx 3.0318$ when $x = 2$.

(b) $\dfrac{dy}{dx} = \dfrac{2x + 12}{3y^2 - 4}$

$$\int (3y^2 - 4)dy = \int (2x + 12)\, dx$$

$$y^3 - 4y = x^2 + 12x + C$$

$$y(1) = 2: 2^3 - 4(2) = 1 + 12 + C \Rightarrow C = -13$$

$$y^3 - 4y = x^2 + 12x - 13$$

(c) At $x = 2$,

$$y^3 - 4y = 2^2 + 12(2) - 13 = 15$$

$$y^3 - 4y - 15 = 0$$

$$(y - 3)(y^2 + 3y + 5) = 0 \Rightarrow y = 3.$$

Error: $3.0318 - 3 = 0.0318$

**57.** $\dfrac{dy}{dt} = ky$, $\quad y = Ce^{kt}$

Initial amount: $y(0) = y_0 = C$

Half-life: $\dfrac{y_0}{2} = y_0 e^{k(1599)}$

$$k = \frac{1}{1599} \ln\left(\frac{1}{2}\right)$$

$$y = Ce^{[\ln(1/2)/1599]t}$$

When $t = 50$, $y = 0.9786C$ or $97.86\%$.

**59.** (a) $\dfrac{dy}{dx} = k(y - 4)$

(b) The direction field satisfies $(dy/dx) = 0$ along $y = 4$; but not along $y = 0$. Matches (a).

**61.** (a) $\dfrac{dy}{dx} = ky(y - 4)$

(b) The direction field satisfies $(dy/dx) = 0$ along $y = 0$ and $y = 4$. Matches (c).

(a)

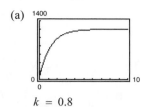

$k = 0.8$

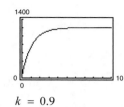

$k = 0.9$

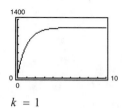

$k = 1$

(b)  $k = 0.8$:    $t = 1.31$ years

  $k = 0.9$:    $t = 1.16$ years

  $k = 1.0$:    $t = 1.05$ years

(c)  Maximum weight: 1200 pounds

  $\lim\limits_{x \to \infty} w = 1200$

**65.** Given family (circles):    $x^2 + y^2 = C$

$$2x + 2yy' = 0$$

$$y' = -\frac{x}{y}$$

Orthogonal trajectory (lines):    $y' = \dfrac{y}{x}$

$$\int \frac{dy}{y} = \int \frac{dx}{x}$$

$$\ln|y| = \ln|x| + \ln K$$

$$y = Kx$$

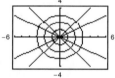

**67.** Given family (parabolas):  $x^2 = Cy$

$$2x = Cy'$$

$$y' = \frac{2x}{C} = \frac{2x}{x^2/y} = \frac{2y}{x}$$

Orthogonal trajectory (ellipses):    $y' = -\dfrac{x}{2y}$

$$2\int y \, dy = -\int x \, dx$$

$$y^2 = -\frac{x^2}{2} + K_1$$

$$x^2 + 2y^2 = K$$

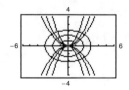

**69.** Given family:    $y^2 = Cx^3$

$$2yy' = 3Cx^2$$

$$y' = \frac{3Cx^2}{2y} = \frac{3x^2}{2y}\left(\frac{y^2}{x^3}\right) = \frac{3y}{2x}$$

Orthogonal trajectory (ellipses):    $y' = -\dfrac{2x}{3y}$

$$3\int y \, dy = -2\int x \, dx$$

$$\frac{3y^2}{2} = -x^2 + K_1$$

$$3y^2 + 2x^2 = K$$

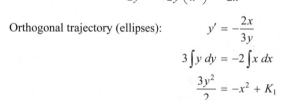

**71.**  $y = \dfrac{12}{1 + e^{-x}}$

Because  $y(0) = 6$, it matches (c) or (d).

Because (d) approaches its horizontal asymptote slower than (c), it matches (d).

**73.**  $y = \dfrac{12}{1 + \dfrac{1}{2}e^{-x}}$

Because  $y(0) = \dfrac{12}{\left(\dfrac{3}{2}\right)} = 8$, it matches (b).

**75.** $P(t) = \dfrac{2100}{1 + 29e^{-0.75t}}$

(a) $k = 0.75$

(b) $L = 2100$

(c) $P(0) = \dfrac{2100}{1 + 29} = 70$

(d) $1050 = \dfrac{2100}{1 + 29e^{-0.75t}}$

$1 + 29e^{-0.75t} = 2$

$e^{-0.75t} = \dfrac{1}{29}$

$-0.75t = \ln\left(\dfrac{1}{29}\right) = -\ln 29$

$t = \dfrac{\ln 29}{0.75} \approx 4.4897 \text{ yr}$

(e) $\dfrac{dP}{dt} = 0.75P\left(1 - \dfrac{P}{2100}\right), \quad P(0) = 70$

**77.** $\dfrac{dP}{dt} = 3P\left(1 - \dfrac{P}{100}\right)$

(a) $k = 3$

(b) $L = 100$

(c)

(d) $\dfrac{d^2P}{dt^2} = 3P'\left(1 - \dfrac{P}{100}\right) + 3P\left(\dfrac{-P'}{100}\right)$

$= 3\left[3P\left(1 - \dfrac{P}{100}\right)\right]\left(1 - \dfrac{P}{100}\right) - \dfrac{3P}{100}\left[3P\left(1 - \dfrac{P}{100}\right)\right] = 9P\left(1 - \dfrac{P}{100}\right)\left(1 - \dfrac{P}{100} - \dfrac{P}{100}\right) = 9P\left(1 - \dfrac{P}{100}\right)\left(1 - \dfrac{2P}{100}\right)$

$\dfrac{d^2P}{dt^2} = 0$ for $P = 50$, and by the first Derivative Test, this is a maximum. $\left(\text{Note: } P = 50 = \dfrac{L}{2} = \dfrac{100}{2}\right)$

**79.** $\dfrac{dy}{dt} = y\left(1 - \dfrac{y}{36}\right), \quad y(0) = 4$

$k = 1, L = 36$

$y = \dfrac{L}{1 + be^{-kt}} = \dfrac{36}{1 + be^{-t}}$

$(0, 4): 4 = \dfrac{36}{1 + b} \Rightarrow b = 8$

Solution: $y = \dfrac{36}{1 + 8e^{-t}}$

**81.** $\dfrac{dy}{dt} = \dfrac{4y}{5} - \dfrac{y^2}{150} = \dfrac{4}{5}y\left(1 - \dfrac{y}{120}\right), \quad y(0) = 8$

$k = \dfrac{4}{5} = 0.8, L = 120$

$y = \dfrac{L}{1 + be^{-kt}} = \dfrac{120}{1 + be^{-0.8t}}$

$(0, 8): 8 = \dfrac{120}{1 + b} \Rightarrow b = 14$

Solution: $y = \dfrac{120}{1 + 14e^{-0.8t}}$

**83.** (a) $P = \dfrac{L}{1 + be^{-kt}}$, $L = 200$, $P(0) = 25$

$$25 = \frac{200}{1 + b} \Rightarrow b = 7$$

$$39 = \frac{200}{1 + 7e^{-k(2)}}$$

$$1 + 7e^{-2k} = \frac{200}{39}$$

$$e^{-2k} = \frac{23}{39}$$

$$k = -\frac{1}{2}\ln\left(\frac{23}{39}\right) = \frac{1}{2}\ln\left(\frac{39}{23}\right) \approx 0.2640$$

$$P = \frac{200}{1 + 7e^{-0.2640t}}$$

(b) For $t = 5$, $P \approx 70$ panthers.

(c) $$100 = \frac{200}{1 + 7e^{-0.264t}}$$

$$1 + 7e^{-0.264t} = 2$$

$$-0.264t = \ln\left(\frac{1}{7}\right)$$

$$t \approx 7.37 \text{ years}$$

(d) $$\frac{dP}{dt} = kP\left(1 - \frac{P}{L}\right)$$

$$= 0.264P\left(1 - \frac{P}{200}\right), \quad P(0) = 25$$

Using Euler's Method, $P \approx 65.6$ when $t = 5$.

(e) $P$ is increasing most rapidly where $P = 200/2 = 100$, corresponds to $t \approx 7.37$ years.

**85.** A differential equation can be solved by separation of variables if it can be written in the form

$$M(x) + N(y)\frac{dy}{dx} = 0.$$

To solve a separable equation, rewrite as,

$$M(x)\,dx = -N(y)\,dy$$

and integrate both sides.

**87.** Two families of curves are mutually orthogonal if each curve in the first family intersects each curve in the second family at right angles.

**89.** $y = \dfrac{1}{1 + be^{-kt}}$

$$y' = \frac{-1}{\left(1 + be^{-kt}\right)^2}\left(-bke^{-kt}\right)$$

$$= \frac{k}{\left(1 + be^{-kt}\right)} \cdot \frac{be^{-kt}}{\left(1 + be^{-kt}\right)}$$

$$= \frac{k}{\left(1 + be^{-kt}\right)} \cdot \frac{1 + be^{-kt} - 1}{\left(1 + be^{-kt}\right)}$$

$$= \frac{k}{\left(1 + be^{-kt}\right)} \cdot \left(1 - \frac{1}{1 + be^{-kt}}\right) = ky(1 - y)$$

**91.** False. $\dfrac{dy}{dx} = \dfrac{x}{y}$ is separable, but $y = 0$ is not a solution.

**93.** False

$$f(tx, ty) = t^2x^2 - 4xyt^2 + 6t^2y^2 + 1$$

$$\neq t^2 f(x, y)$$

**95.** $fg' + gf' = f'g'$   Product Rule

$$(f - f')g' + gf' = 0$$

$$g' + \frac{f'}{f - f'}g = 0$$

Need $f - f' = e^{x^2} - 2xe^{x^2} = (1 - 2x)e^{x^2} \neq 0$, so avoid $x = \dfrac{1}{2}$.

$$\frac{g'}{g} = \frac{f'}{f' - f} = \frac{2xe^{x^2}}{(2x - 1)e^{x^2}} = 1 + \frac{1}{2x - 1}$$

$$\ln|g(x)| = x + \frac{1}{2}\ln|2x - 1| + C_1$$

$$g(x) = Ce^x|2x - 1|^{1/2}$$

So there exists $g$ and interval $(a, b)$, as long as

$$\frac{1}{2} \notin (a, b).$$

# Section 6.4   First-Order Linear Differential Equations

**1.** $x^3y' + xy = e^x + 1$

$$y' + \frac{1}{x^2}y = \frac{1}{x^3}(e^x + 1)$$

Linear

**3.** $y' - y\sin x = xy^2$

Not linear, because of the $xy^2$-term.

**5.** $\dfrac{dy}{dx} + \left(\dfrac{1}{x}\right)y = 6x + 2$

Integrating factor: $e^{\int(1/x)\,dx} = e^{\ln x} = x$

$$xy = \int x(6x + 2)\,dx = 2x^3 + x^2 + C$$

$$y = 2x^2 + x + \frac{C}{x}$$

**7.** $y' - y = 16$

Integrating factor: $e^{\int -1\, dx} = e^{-x}$

$e^{-x}y' - e^{-x}y = 16e^{-x}$

$ye^{-x} = \int 16e^{-x}\, dx = -16e^{-x} + C$

$y = -16 + Ce^{x}$

**9.** $(y + 1)\cos x\, dx = dy$

$y' = (y + 1)\cos x = y\cos x + \cos x$

$y' - (\cos x)y = \cos x$

Integrating factor: $e^{\int -\cos x\, dx} = e^{-\sin x}$

$y'e^{-\sin x} - (\cos x)e^{-\sin x}y = (\cos x)e^{-\sin x}$

$ye^{-\sin x} = \int (\cos x)e^{-\sin x}\, dx$

$\qquad = -e^{-\sin x} + C$

$\qquad y = -1 + Ce^{\sin x}$

**11.** $(x - 1)y' + y = x^2 - 1$

$y' + \left(\dfrac{1}{x - 1}\right)y = x + 1$

Integrating factor: $e^{\int [1/(x-1)]\, dx} = e^{\ln|x-1|} = x - 1$

$y(x - 1) = \int (x^2 - 1)\, dx = \dfrac{1}{3}x^3 - x + C_1$

$\qquad y = \dfrac{x^3 - 3x + C}{3(x - 1)}$

**13.** $y' - 3x^2y = e^{x^3}$

Integrating factor: $e^{-\int 3x^2\, dx} = e^{-x^3}$

$ye^{-x^3} = \int e^{x^3}e^{-x^3}\, dx = \int dx = x + C$

$\qquad y = (x + C)e^{x^3}$

**15.** (a) Answers will vary.

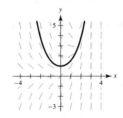

**(b)** $\dfrac{dy}{dx} = e^x - y$

$\dfrac{dy}{dx} + y = e^x$ $\qquad$ Integrating factor: $e^{\int dx} = e^x$

$e^xy' + e^xy = e^{2x}$

$(ye^x) = \int e^{2x}\, dx$

$ye^x = \dfrac{1}{2}e^{2x} + C$

$y(0) = 1 \Rightarrow 1 = \dfrac{1}{2} + C \Rightarrow C = \dfrac{1}{2}$

$ye^x = \dfrac{1}{2}e^{2x} + \dfrac{1}{2}$

$y = \dfrac{1}{2}e^x + \dfrac{1}{2}e^{-x} = \dfrac{1}{2}\left(e^x + e^{-x}\right)$

**(c)**

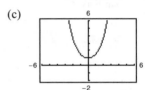

**17.** $y'\cos^2 x + y - 1 = 0$

$y' + (\sec^2 x)y = \sec^2 x$

Integrating factor: $e^{\int \sec^2 x\, dx} = e^{\tan x}$

$ye^{\tan x} = \int \sec^2 x\, e^{\tan x}\, dx = e^{\tan x} + C$

$\qquad y = 1 + Ce^{-\tan x}$

Initial condition: $y(0) = 5, C = 4$

Particular solution: $y = 1 + 4e^{-\tan x}$

**19.** $y' + y\tan x = \sec x + \cos x$

Integrating factor: $e^{\int \tan x\, dx} = e^{\ln|\sec x|} = \sec x$

$y\sec x = \int \sec x(\sec x + \cos x)\, dx = \tan x + x + C$

$\qquad y = \sin x + x\cos x + C\cos x$

Initial condition: $y(0) = 1, 1 = C$

Particular solution: $y = \sin x + (x + 1)\cos x$

**21.** $y' + \left(\dfrac{1}{x}\right)y = 0$

Integrating factor: $e^{\int (1/x)\,dx} = e^{\ln|x|} = x$

Separation of variables:

$$\dfrac{dy}{dx} = -\dfrac{y}{x}$$

$$\int \dfrac{1}{y}\,dy = \int -\dfrac{1}{x}\,dx$$

$$\ln y = -\ln x + \ln C$$

$$\ln xy = \ln C$$

$$xy = C$$

Initial condition: $y(2) = 2, C = 4$

Particular solution: $xy = 4$

**23.** $x\,dy = (x + y + 2)\,dx$

$$\dfrac{dy}{dx} = \dfrac{x + y + 2}{x} = \dfrac{y}{x} + 1 + \dfrac{2}{x}$$

$$\dfrac{dy}{dx} - \dfrac{1}{x}y = 1 + \dfrac{2}{x} \qquad \text{Linear}$$

$$u(x) = e^{\int -(1/x)\,dx} = \dfrac{1}{x}$$

$$y = x\int \left(1 + \dfrac{2}{x}\right)\dfrac{1}{x}\,dx = x\int \left(\dfrac{1}{x} + \dfrac{2}{x^2}\right)dx$$

$$= x\left[\ln|x| + \dfrac{-2}{x} + C\right]$$

$$= -2 + x\ln|x| + Cx$$

$$y(1) = 10 = -2 + C \Rightarrow C = 12$$

$$y = -2 + x\ln|x| + 12x$$

**25.** $\qquad \dfrac{dP}{dt} = kP + N, N$ constant

$$\dfrac{dP}{kP + N} = dt$$

$$\int \dfrac{1}{kP + N}\,dP = \int dt$$

$$\dfrac{1}{k}\ln(kP + N) = t + C_1$$

$$\ln(kP + N) = kt + C_2$$

$$kP + N = e^{kt + C_2}$$

$$P = \dfrac{C_3 e^{kt} - N}{k}$$

$$P = Ce^{kt} - \dfrac{N}{k}$$

When $t = 0: P = P_0$

$$P_0 = C - \dfrac{N}{k} \Rightarrow C = P_0 + \dfrac{N}{k}$$

$$P = \left(P_0 + \dfrac{N}{k}\right)e^{kt} - \dfrac{N}{k}$$

**27. (a)** $A = \dfrac{P}{r}\left(e^{rt} - 1\right)$

$$A = \dfrac{275{,}000}{0.06}\left(e^{0.08(10)} - 1\right) \approx \$4{,}212{,}796.94$$

**(b)** $A = \dfrac{550{,}000}{0.05}\left(e^{0.059(25)} - 1\right) \approx \$31{,}424{,}909.75$

**29. (a)** $\dfrac{dQ}{dt} = q - kQ, q$ constant

**(b)** $Q' + kQ = q$

Let $P(t) = k, Q(t) = q$, then the integrating factor is $u(t) = e^{kt}$.

$$Q = e^{-kt}\int qe^{kt}\,dt = e^{-kt}\left(\dfrac{q}{k}e^{kt} + C\right) = \dfrac{q}{k} + Ce^{-kt}$$

When $t = 0: Q = Q_0$

$$Q_0 = \dfrac{q}{k} + C \Rightarrow C = Q_0 - \dfrac{q}{k}$$

$$Q = \dfrac{q}{k} + \left(Q_0 - \dfrac{q}{k}\right)e^{-kt}$$

**(c)** $\lim\limits_{t \to \infty} Q = \dfrac{q}{k}$

**31.** Let $Q$ be the number of pounds of concentrate in the solution at any time $t$. Because the number of gallons of solution in the tank at any time $t$ is $v_0 + (r_1 - r_2)t$ and because the tank loses $r_2$ gallons of solution per minute, it must lose concentrate at the rate

$$\left[\dfrac{Q}{v_0 + (r_1 - r_2)t}\right]r_2.$$

The solution gains concentrate at the rate $r_1 q_1$. Therefore, the net rate of change is

$$\dfrac{dQ}{dt} = q_1 r_1 - \left[\dfrac{Q}{v_0 + (r_1 - r_2)t}\right]r_2$$

or

$$\dfrac{dQ}{dt} + \dfrac{r_2 Q}{v_0 + (r_1 - r_2)t} = q_1 r_1.$$

**33.** (a) $Q' + \dfrac{r^2 Q}{v_0 + (r_1 - r_2)t} = q_1 r_1$

$Q(0) = q_0, q_0 = 25, q_1 = 0, v_0 = 200,$

$r_1 = 10, r_2 = 10, Q' + \dfrac{1}{20}Q = 0$

$\displaystyle\int \dfrac{1}{Q}\,dQ = \int -\dfrac{1}{20}\,dt$

$\ln Q = -\dfrac{1}{20}t + \ln C_1$

$Q = Ce^{-(1/20)t}$

Initial condition: $Q(0) = 25, C = 25$

Particular solution: $Q = 25e^{-(1/20)t}$

(b) $\quad 15 = 25e^{-(1/20)t}$

$\ln\left(\dfrac{3}{5}\right) = -\dfrac{1}{20}t$

$t = -20\ln\left(\dfrac{3}{5}\right) \approx 10.2 \text{ min}$

(c) $\displaystyle\lim_{t\to\infty} 25e^{-(1/20)t} = 0$

**35.** (a) The volume of the solution in the tank is given by

$v_0 + (r_1 - r_2)t.$ Therefore, $100 + (5 - 3)t = 200$ or $t = 50$ minutes.

(b) $Q' + \dfrac{r_2 Q}{v_0 + (r_1 - r_2)t} = q_1 r_1$

$Q(0) = q_0, q_0 = 0, q_1 = 0.5, v_0 = 100, r_1 = 5, r_2 = 3, Q' + \dfrac{3}{100 + 2t}Q = 2.5$

Integrating factor: $e^{\int [3/(100+2t)]\,dt} = (50 + t)^{3/2}$

$Q(50 + t)^{3/2} = \displaystyle\int 2.5(50 + t)^{3/2}\,dt = (50 + t)^{5/2} + C$

$Q = (50 + t) + C(50 + t)^{-3/2}$

Initial condition: $Q(0) = 0, 0 = 50 + C(50^{-3/2}), C = -50^{5/2}$

Particular solution: $\quad Q = (50 + t) - 50^{-5/2}(50 + t)^{-3/2}$

$Q(50) = 100 - 50^{5/2}(100)^{-3/2} = 100 - \dfrac{25}{\sqrt{2}} \approx 82.32 \text{ lbs}$

(c) The volume of the solution is given by $v_0 + (r_1 - r_2)t = 100 + (5 - 3)t = 200 \Rightarrow t = 50$ minutes.

$Q' + \dfrac{r_2 Q}{v_0 + (r_1 - r_2)t} = q_1 r_1$

$Q(0) = q_0 = 0, q_1 = 1, v_0 = 100, r_1 = 5, r_2 = 3$

$Q' + \dfrac{3Q}{100 + 2t} = 5$

Integrating factor is $(50 + t)^{3/2}$, as in #43.

$Q(50 + t)^{3/2} = \displaystyle\int 5(50 + t)^{3/2}\,dt = 2(50 + t)^{5/2} + C$

$Q = 2(50 + t) + C(50 + t)^{-3/2}$

$Q(0) = 0: 0 = 100 + C(50)^{-3/2} \Rightarrow C = -100(50)^{3/2} = -2(50)^{5/2}$

$Q = 2(50 + t) - 2(50)^{5/2}(50 + t)^{-3/2}$

When $t = 50$, $Q = 200 - 2(50)^{5/2}(100)^{-3/2} = 200 - \dfrac{50}{\sqrt{2}} \approx 164.64 \text{ lbs}$   (double the answer to part (b))

**37.** From Example 6,

$$\frac{dv}{dt} + \frac{kv}{m} = g$$

$$v = \frac{mg}{k}\left(1 - e^{-kt/m}\right), \quad \text{Solution}$$

$g = -32$, $mg = -8$, $v(5) = -101$, $m = \dfrac{-8}{g} = \dfrac{1}{4}$ implies that $-101 = \dfrac{-8}{k}\left(1 - e^{-5k/(1/4)}\right)$.

Using a graphing utility, $k \approx 0.050165$, and $v = -159.47\left(1 - e^{-0.2007t}\right)$.

As $t \to \infty$, $v \to -159.47$ ft/sec. The graph of $v$ is shown below.

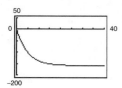

**39.** $L\dfrac{dI}{dt} + RI = E_0$, $I' + \dfrac{R}{L}I = \dfrac{E_0}{L}$

Integrating factor: $e^{\int (R/L)\,dt} = e^{Rt/L}$

$$I\,e^{Rt/L} = \int \frac{E_0}{L}e^{Rt/L}\,dt = \frac{E_0}{R}e^{Rt/L} + C$$

$$I = \frac{E_0}{R} + Ce^{-Rt/L}$$

**41.** $\dfrac{dy}{dx} + P(x)y = Q(x)$     Standard form

$u(x) = e^{\int P(x)\,dx}$     Integrating factor

**43.** $y' - 2x = 0$

$$\int dy = \int 2x\,dx$$

$$y = x^2 + C$$

Matches c.

**45.** $y' - 2xy = 0$

$$\int \frac{dy}{y} = \int 2x\,dx$$

$$\ln y = x^2 + C_1$$

$$y = Ce^{x^2}$$

Matches a.

**47.** $y' + 3x^2 y = x^2 y^3$

$n = 3$, $Q = x^2$, $P = 3x^2$

$$y^{-2}e^{\int(-2)3x^2\,dx} = \int (-2)x^2 e^{\int(-2)3x^2\,dx}\,dx$$

$$y^{-2}e^{-2x^3} = -\int 2x^2 e^{-2x^3}\,dx$$

$$y^{-2}e^{-2x^3} = \frac{1}{3}e^{-2x^3} + C$$

$$y^{-2} = \frac{1}{3} + Ce^{2x^3}$$

$$\frac{1}{y^2} + Ce^{2x^3} + \frac{1}{3}$$

**49.** $y' + \left(\dfrac{1}{x}\right)y = xy^2$

$n = 2$, $Q = x$, $P = x^{-1}$

$e^{\int -(1/x)\,dx} = e^{-\ln|x|} = x^{-1}$

$$y^{-1}x^{-1} = \int -x\left(x^{-1}\right)\,dx = -x + C$$

$$\frac{1}{y} = -x^2 + Cx$$

$$y = \frac{1}{Cx - x^2}$$

**51.** $xy' + y = xy^3$

$$y' + \frac{1}{x}y = y^3$$

$n = 3$, $Q = 1$, $P = \dfrac{1}{x}$,     $e^{\int \frac{-2}{x}\,dx} = e^{-2\ln x} = x^{-2}$

$$y^{-2}x^{-2} = \int -2x^{-2}\,dx + C = 2x^{-1} + C$$

$$y^{-2} = 2x + Cx^2$$

$$y^2 = \frac{1}{2x + Cx^2} \quad \text{or} \quad \frac{1}{y^2} = 2x + Cx^2$$

**53.** $y' - y = e^x \sqrt[3]{y}, n = \frac{1}{3}, Q = e^x, P = -1$

$e^{\int -(2/3)\,dx} = e^{-(2/3)x}$

$y^{2/3}e^{-(2/3)x} = \int \frac{2}{3}e^x e^{-(2/3)x}\,dx = \int \frac{2}{3}e^{(1/3)x}\,dx$

$y^{2/3}e^{-(2/3)x} = 2e^{(1/3)x} + C$

$y^{2/3} = 2e^x + Ce^{2x/3}$

**55. (a)**

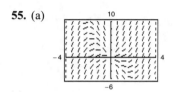

**(b)** $\dfrac{dy}{dx} - \dfrac{1}{x}y = x^2$

Integrating factor: $e^{-1/x\,dx} = e^{-\ln x} = \dfrac{1}{x}$

$\dfrac{1}{x}y' - \dfrac{1}{x^2}y = x$

$\left(\dfrac{1}{x}y\right) = \int x\,dx = \dfrac{x^2}{2} + C$

$y = \dfrac{x^3}{2} + Cx$

$(-2, 4): 4 = \dfrac{-8}{2} - 2C \Rightarrow C = -4 \Rightarrow y = \dfrac{x^3}{2} - 4x = \dfrac{1}{2}x\left(x^2 - 8\right)$

$(2, 8): 8 = \dfrac{8}{2} + 2C \Rightarrow C = 2 \Rightarrow y = \dfrac{x^3}{2} + 2x = \dfrac{1}{2}x\left(x^2 + 4\right)$

**(c)**

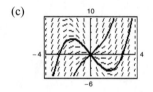

**57. (a)**

**(b)** $y' + (\cot x)y = 2$

Integrating factor: $e^{\int \cot x\,dx} = e^{\ln|\sin x|} = \sin x$

$y'\sin x + (\cos x)y = 2\sin x$

$y\sin x = \int 2\sin x\,dx = -2\cos x + C$

$y = -2\cot x + C\csc x$

$(1, 1): 1 = -2\cot 1 + C\csc 1 \Rightarrow C = \dfrac{1 + 2\cot 1}{\csc 1} = \sin 1 + 2\cos 1$

$y = -2\cot x + (\sin 1 + 2\cos 1)\csc x$

$(3, -1): -1 = -2\cot 3 + C\csc 3 \Rightarrow C = \dfrac{2\cot 3 - 1}{\csc 3} = 2\cos 3 - \sin 3$

$y = -2\cot x + (2\cos 3 - \sin 3)\csc x$

(c)

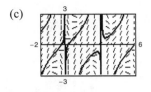

**59.** $e^{2x+y} \, dx - e^{x-y} \, dy = 0$

Separation of variables:

$$e^{2x}e^y \, dx = e^x e^{-y} \, dy$$

$$\int e^x \, dx = \int e^{-2y} \, dy$$

$$e^x = -\frac{1}{2}e^{-2y} + C_1$$

$$2e^x + e^{-2y} = C$$

**61.** $(y \cos x - \cos x) \, dx + dy = 0$

Separation of variables:

$$\int \cos x \, dx = \int \frac{-1}{y-1} \, dy$$

$$\sin x = -\ln(y-1) + \ln C$$

$$\ln(y-1) = -\sin x + \ln C$$

$$y = Ce^{-\sin x} + 1$$

**63.** $(3y^2 + 4xy) \, dx + (2xy + x^2) \, dy = 0$

Homogeneous: $y = vx, \, dy = v \, dx + x \, dv$

$$(3v^2x^2 + 4vx^2) \, dx + (2vx^2 + x^2)(v \, dx + x \, dv) = 0$$

$$\int \frac{5}{x} \, dx + \int \left(\frac{2v+1}{v^2+v}\right) dv = 0$$

$$\ln x^5 + \ln\left|v^2 + v\right| = \ln C$$

$$x^5(v^2 + v) = C$$

$$x^3y^2 + x^4y = C$$

**65.** $(2y - e^x) \, dx + x \, dy = 0$

Linear: $y' + \left(\frac{2}{x}\right)y = \frac{1}{x}e^x$

Integrating factor: $e^{\int (2/x) \, dx} = e^{\ln x^2} = x^2$

$$yx^2 = \int x^2 \frac{1}{x}e^x \, dx = e^x(x-1) + C$$

$$y = \frac{e^x}{x^2}(x-1) + \frac{C}{x^2}$$

**67.** $(x^2y^4 - 1) \, dx + x^3y^3 \, dy = 0$

$$y' + \left(\frac{1}{x}\right)y = x^{-3}y^{-3}$$

Bernoulli: $n = -3, \, Q = x^{-3}, \, P = x^{-1},$

$$e^{\int (4/x) \, dx} = e^{\ln x^4} = x^4$$

$$y^4x^4 = \int 4(x^{-3})(x^4) \, dx = 2x^2 + C$$

$$x^4y^4 - 2x^2 = C$$

**69.** $3(y - 4x^2) \, dx = -x \, dy$

$$x\frac{dy}{dx} = -3y + 12x^2$$

$$y' + \frac{3}{x}y = 12x$$

Integrating factor: $e^{\int (3/x) \, dx} = e^{3\ln x} = x^3$

$$y'x^3 + \frac{3}{x}x^3y = 12x(x^3) = 12x^4$$

$$yx^3 = \int 12x^4 \, dx = \frac{12}{5}x^5 + C$$

$$y = \frac{12}{5}x^2 + \frac{C}{x^3}$$

**71.** False. The equation contains $\sqrt{y}$.

# Review Exercises for Chapter 6

**1.** $y = x^3, \, y' = 3x^2$

$2xy' + 4y = 2x(3x^2) + 4(x^3) = 10x^3.$

Yes, it is a solution.

**3.** $\dfrac{dy}{dx} = 4x^2 + 7$

$$y = \int (4x^2 + 7) \, dx = \frac{4x^3}{3} + 7x + C$$

**5.** $\dfrac{dy}{dx} = \cos 2x$

$\quad y = \displaystyle\int \cos 2x\,dx = \dfrac{1}{2}\sin 2x + C$

**7.** $\dfrac{dy}{dx} = x\sqrt{x-5}$

$\quad y = \displaystyle\int x\sqrt{x-5}\,dx$

Let $u = x - 5,\, du = dx,\, x = u + 5$

$\quad y = \displaystyle\int (u+5)\sqrt{u}\,du$

$\quad\quad = \displaystyle\int \left(u^{3/2} + 5u^{1/2}\right)du$

$\quad\quad = \dfrac{2}{5}u^{5/2} + \dfrac{10}{3}u^{3/2} + C$

$\quad\quad = \dfrac{2}{5}(x-5)^{5/2} + \dfrac{10}{3}(x-5)^{3/2} + C$

$\quad\quad = \dfrac{1}{15}(x-5)^{3/2}\left[6(x-5) + 50\right] + C$

$\quad\quad = \dfrac{1}{15}(x-5)^{3/2}(6x+20) + C$

**9.** $\dfrac{dy}{dx} = e^{2-x}$

$\quad y = \displaystyle\int e^{2-x}\,dx = -e^{2-x} + C$

**11.** $\dfrac{dy}{dx} = 2x - y$

| $x$ | $-4$ | $-2$ | $0$ | $2$ | $4$ | $8$ |
|---|---|---|---|---|---|---|
| $y$ | $2$ | $0$ | $4$ | $4$ | $6$ | $8$ |
| $dy/dx$ | $-10$ | $-4$ | $-4$ | $0$ | $2$ | $8$ |

**13.** $y' = 3 - x, \quad (2, 1)$

(a) and (b)

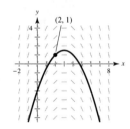

**15.** $y' = \dfrac{1}{4}x^2 - \dfrac{1}{3}x, \quad (0, 3)$

(a) and (b)

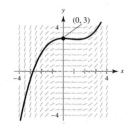

**17.** $y' = \dfrac{xy}{x^2 + 4}, \quad (0, 1)$

(a) and (b)

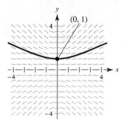

**19.** $\dfrac{dy}{dx} = 8 - x$

$\quad y = \displaystyle\int (8-x)\,dx = 8x - \dfrac{x^2}{2} + C$

**21.** $\qquad \dfrac{dy}{dx} = (3 + y)^2$

$\quad \displaystyle\int (3+y)^{-2}\,dy = \int dx$

$\quad\quad -(3+y)^{-1} = x + C$

$\quad\quad\quad 3 + y = \dfrac{-1}{x + C}$

$\quad\quad\quad\quad y = -3 - \dfrac{1}{x + C}$

**23.** $(2 + x)y' - xy = 0$

$\quad (2 + x)\dfrac{dy}{dx} = xy$

$\quad\quad \dfrac{1}{y}\,dy = \dfrac{x}{2+x}\,dx$

$\quad\quad \dfrac{1}{y}\,dy = \left(1 - \dfrac{2}{2+x}\right)dx$

$\quad\quad \ln|y| = x - 2\ln|2 + x| + C_1$

$\quad\quad\quad y = Ce^x(2+x)^{-2} = \dfrac{Ce^x}{(2+x)^2}$

**25.** $y = Ce^{kt}$

$\quad \left(0, \dfrac{3}{4}\right): \dfrac{3}{4} = C$

$\quad (5, 5): 5 = \dfrac{3}{4}e^{k(5)}$

$\quad\quad \dfrac{20}{3} = e^{5k}$

$\quad\quad k = \dfrac{1}{5}\ln\left(\dfrac{20}{3}\right)$

$\quad y = \dfrac{3}{4}e^{[\ln(20/3)/5]t} \approx \dfrac{3}{4}e^{0.379t}$

**27.** $y = Ce^{kt}$

$(0, 5): C = 5$

$\left(5, \dfrac{1}{6}\right): \dfrac{1}{6} = 5e^{5k}$

$k = \dfrac{1}{5} \ln\left(\dfrac{1}{30}\right) = \dfrac{-\ln 30}{5}$

$y = 5e^{[-t \ln 30]/5} \approx 5e^{-0.680t}$

**29.** $\dfrac{dP}{dh} = kp, \quad P(0) = 30$

$P(h) = 30e^{kh}$

$P(18{,}000) = 30e^{18{,}000k} = 15$

$k = \dfrac{\ln(1/2)}{18{,}000} = \dfrac{-\ln 2}{18{,}000}$

$P(h) = 30e^{-(h \ln 2)/18{,}000}$

$P(35{,}000) = 30e^{-(35{,}000 \ln 2)/18{,}000} \approx 7.79$ inches

**31.** $S = Ce^{k/t}$

(a) $\qquad S = 5$ when $t = 1$

$\qquad 5 = Ce^{k}$

$\qquad \lim_{t \to \infty} Ce^{k/t} = C = 30$

$\qquad 5 = 30e^{k}$

$\qquad k = \ln \dfrac{1}{6} \approx -1.7918$

$\qquad S = 30\left(\dfrac{1}{6}\right) \approx 30e^{-1.7918/t}$

(b) When $t = 5$, $S \approx 20.9646$ which is 20,965 units.

(c)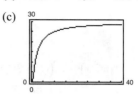

**33.** $\qquad P = Ce^{0.0185t}$

$\qquad 2C = Ce^{0.0185t}$

$\qquad 2 = e^{0.0185t}$

$\qquad \ln 2 = 0.0185t$

$\qquad t = \dfrac{\ln 2}{0.0185} \approx 37.5$ years

**35.** $\dfrac{dy}{dx} = \dfrac{x^5 + 7}{x} = x^4 + \dfrac{7}{x}$

$y = \int\left(x^4 + \dfrac{7}{x}\right) dx = \dfrac{x^5}{5} + 7 \ln|x| + C$

**37.** $y' - 16xy = 0$

$\dfrac{dy}{dx} = 16xy$

$\int \dfrac{1}{y} \, dy = \int 16x \, dx$

$\ln|y| = 8x^2 + C_1$

$e^{8x^2 + C_1} = y$

$y = Ce^{8x^2}$

**39.** $\dfrac{dy}{dx} = \dfrac{x^2 + y^2}{2xy}$ (homogeneous differential equation)

$(x^2 + y^2) \, dx - 2xy \, dy = 0$

Let $y = vx, \, dy = x \, dv + v \, dx.$

$(x^2 + v^2 x^2) \, dx - 2x(vx)(x \, dv + v \, dx) = 0$

$(x^2 + v^2 x^2 - 2x^2 v^2) \, dx - 2x^3 v \, dv = 0$

$(x^2 - x^2 v^2) \, dx = 2x^3 v \, dv$

$(1 - v^2) \, dx = 2xv \, dv$

$\int \dfrac{dx}{x} = \int \dfrac{2v}{1 - v^2} \, dv$

$\ln|x| = -\ln|1 - v^2| + C_1 = -\ln|1 - v^2| + \ln C$

$x = \dfrac{C}{1 - v^2} = \dfrac{C}{1 - (y/x)^2} = \dfrac{Cx^2}{x^2 - y^2}$

$1 = \dfrac{Cx}{x^2 - y^2} \quad \text{or} \quad C_1 = \dfrac{x}{x^2 - y^2}$

**41.** $y = C_1 x + C_2 x^3$

$y' = C_1 + 3C_2 x^2$

$y'' = 6C_2 x$

$x^2 y'' - 3xy' + 3y = x^2(6C_2 x) - 3x(C_1 + 3C_2 x^2) + 3(C_1 x + C_2 x^3)$

$\qquad\qquad = 6C_2 x^3 - 3C_1 x - 9C_2 x^3 + 3C_1 x + 3C_2 x^3 = 0$

$x = 2, y = 0: \quad 0 = 2C_1 + 8C_2 \Rightarrow C_1 = -4C_2$

$x = 2, y' = 4: \quad 4 = C_1 + 12C_2$

$\qquad\qquad 4 = (-4C_2) + 12C_2 = 8C_2 \Rightarrow C_2 = \frac{1}{2}, C_1 = -2$

$y = -2x + \frac{1}{2}x^3$

**43.** $\dfrac{dy}{dx} = \dfrac{-4x}{y}$

$\displaystyle\int y \, dy = \int -4x \, dx$

$\dfrac{y^2}{2} = -2x^2 + C_1$

$4x^2 + y^2 = C \qquad$ ellipses

**45.** $P(t) = \dfrac{5250}{1 + 34e^{-0.55t}}$

(a) $k = 0.55$

(b) $L = 5250$

(c) $P(0) = \dfrac{5250}{1 + 44} = 150$

(d) $\qquad 2625 = \dfrac{5250}{1 + 34e^{-0.55t}}$

$\qquad 1 + 34e^{-0.55t} = 2$

$\qquad e^{-0.55t} = \dfrac{1}{34}$

$\qquad t = \dfrac{-1}{0.55}\ln\left(\dfrac{1}{34}\right) \approx 6.41 \text{ yr}$

(e) $\dfrac{dP}{dt} = 0.55P\left(1 - \dfrac{P}{5250}\right)$

**47.** $\dfrac{dy}{dt} = y\left(1 - \dfrac{y}{80}\right), \quad (0, 8)$

$k = 1, L = 80$

$y = \dfrac{L}{1 + be^{-kt}} = \dfrac{80}{1 + be^{-t}}$

$y(0) = 8: \quad 8 = \dfrac{80}{1 + b} \Rightarrow b = 9$

Solution: $y = \dfrac{80}{1 + 9e^{-t}}$

**49.** (a) $L = 20{,}400, y(0) = 1200, y(1) = 2000$

$y = \dfrac{20{,}400}{1 + be^{-kt}}$

$y(0) = 1200 = \dfrac{20{,}400}{1 + b} \Rightarrow b = 16$

$y(1) = 2000 = \dfrac{20{,}400}{1 + 16e^{-k}}$

$16e^{-k} = \dfrac{46}{5}$

$k = -\ln\dfrac{23}{40} = \ln\dfrac{40}{23} \approx 0.553$

$y = \dfrac{20{,}400}{1 + 16e^{-0.553t}}$

(b) $y(8) \approx 17{,}118$ trout

(c) $10{,}000 = \dfrac{20{,}400}{1 + 16e^{-0.553t}} \Rightarrow t \approx 4.94 \text{ yr}$

**51.** $y' - y = 10$

$P(x) = -1, Q(x) = 10$

$u(x) = e^{\int -dx} = e^{-x}$

$y = \dfrac{1}{e^{-x}}\int 10e^{-x}\,dx$

$\quad = e^x(-10e^{-x} + C)$

$\quad = -10 + Ce^x$

**53.** $\qquad 4y' = e^{x/y} + y$

$y' - \dfrac{1}{4}y = \dfrac{1}{4}e^{x/4}$

$P(x) = -\dfrac{1}{4}, Q(x) = \dfrac{1}{4}e^{x/4}$

$u(x) = e^{\int -(1/4)\,dx} = e^{-(1/4)x}$

$y = \dfrac{1}{e^{-(1/4)x}}\int \dfrac{1}{4}e^{x/4}e^{-(1/4)x}\,dx$

$\quad = e^{(1/4)x}\left(\dfrac{1}{4}x + C\right) = \dfrac{1}{4}xe^{x/4} + Ce^{x/4}$

**55.** $(x - 2)y' + y = 1$

$$\frac{dy}{dx} + \frac{1}{x - 2}y = \frac{1}{x - 2}$$

$$P(x) = \frac{1}{x - 2}, Q(x) = \frac{1}{x - 2}$$

$$u(x) = e^{\int(1/x-2)\,dx} = e^{\ln|x-2|} = x - 2$$

$$y = \frac{1}{x - 2}\int\left(\frac{1}{x - 2}\right)(x - 2)\,dx = \frac{1}{x - 2}(x + C)$$

**57.** $(3y + \sin 2x)\,dx - dy = 0$

$$y' - 3y = \sin 2x$$

Integrating factor: $e^{\int -3\,dx} = e^{-3x}$

$$ye^{-3x} = \int e^{-3x} \sin 2x\,dx$$

$$= \frac{1}{13}e^{-3x}(-3 \sin 2x - 2 \cos 2x) + C$$

$$y = -\frac{1}{13}(3 \sin 2x + 2 \cos 2x) + Ce^{3x}$$

**59.** $y' + 5y = e^{5x}$

Integrating factor: $e^{\int 5\,dx} = e^{5x}$

$$ye^{5x} = \int e^{10x}\,dx = \frac{1}{10}e^{10x} + C$$

$$y = \frac{1}{10}e^{5x} + Ce^{-5x}$$

**61.** $y' + y = xy^2$     Bernoulli equation

$n = 2$, let $z = y^{1-2} = y^{-1}, z' = -y^{-2}y'.$

$$\left(-y^{-2}\right)y' + \left(-y^{-2}\right)y = -x$$

$$z' - z = -x \qquad \text{Linear equation}$$

$$u(x) = e^{\int -dx} = e^{-x}$$

$$z = \frac{1}{e^{-x}}\int -xe^{-x}\,dx = e^x\left[xe^{-x} + e^{-x} + C\right]$$

$$y^{-1} = x + 1 + Ce^x$$

$$y = \frac{1}{x + 1 + Ce^x}$$

**63.** $y' + \frac{1}{x}y = \frac{y^3}{x^2}$     Bernoulli equation

$n = 3$, let $z = y^{1-3} = y^{-2}, z' = -2y^{-3}y'.$

$$\left(-2y^{-3}\right)y' + \frac{1}{x}y\left(-2y^{-3}\right) = \frac{-2}{x^2}$$

$$z' - \frac{2}{x}z = \frac{-2}{x^2} \qquad \text{Linear equation}$$

$$u(x) = e^{\int -(2/x)\,dx} = e^{-2 \ln x} = x^{-2}$$

$$z = \frac{1}{x^{-2}}\int \frac{-2}{x^2}\left(x^{-2}\right)dx = x^2\left(\frac{2x^{-3}}{3} + C\right)$$

$$\frac{1}{y^2} = \frac{2}{3x} + Cx^2$$

**65.** Answers will vary. *Sample answer:* $\left(x^2 + 3y^2\right)dx - 2xy\,dy = 0$

Solution: Let $y = vx, dy = x\,dv + v\,dx.$

$$\left(x^2 + 3v^2x^2\right)dx - 2x(vx)(x\,dv + v\,dx) = 0$$

$$\left(x^2 + v^2x^2\right)dx - 2x^3v\,dv = 0$$

$$\left(1 + v^2\right)dx = 2xv\,dv$$

$$\int\frac{dx}{x} = \int\frac{2v}{1 + v^2}\,dv$$

$$\ln|x| = \ln\left|1 + v^2\right| + C_1$$

$$x = C\left(1 + v^2\right) = C\left(1 + \frac{y^2}{x^2}\right)$$

$$x^3 = C\left(x^2 + y^2\right)$$

**67.** Answers will vary.

*Sample answer:* $x^3y' + 2x^2y = 1$

$$y' + \frac{2}{x}y = \frac{1}{x^3}$$

$$u(x) = e^{\int(2/x)\,dx} = x^2$$

$$y = \frac{1}{x^2}\int\frac{1}{x^3}\left(x^2\right)dx = \frac{1}{x^2}\left[\ln|x| + C\right]$$

# Problem Solving for Chapter 6

**1. (a)**
$$\frac{dy}{dt} = y^{1.01}$$

$$\int y^{-1.01} \, dy = \int dt$$

$$\frac{y^{-0.01}}{-0.01} = t + C_1$$

$$\frac{1}{y^{0.01}} = -0.01t + C$$

$$y^{0.01} = \frac{1}{C - 0.01t}$$

$$y = \frac{1}{(C - 0.01t)^{100}}$$

$$y(0) = 1: 1 = \frac{1}{C^{100}} \Rightarrow C = 1$$

So, $y = \dfrac{1}{(1 - 0.01t)^{100}}$.

For $T = 100$, $\lim\limits_{t \to T^-} y = \infty$.

**(b)**
$$\int y^{-(1+\varepsilon)} \, dy = \int k \, dt$$

$$\frac{y^{-\varepsilon}}{-\varepsilon} = kt + C_1$$

$$y^{-\varepsilon} = -\varepsilon kt + C$$

$$y = \frac{1}{(C - \varepsilon kt)^{1/\varepsilon}}$$

$$y(0) = y_0 = \frac{1}{C^{1/\varepsilon}} \Rightarrow C^{1/\varepsilon} = \frac{1}{y_0} \Rightarrow C = \left(\frac{1}{y_0}\right)^{\varepsilon}$$

So, $y = \dfrac{1}{\left(\dfrac{1}{y_0^{\varepsilon}} - \varepsilon kt\right)^{1/\varepsilon}}$.

For $t \to \dfrac{1}{y_0^{\varepsilon} \varepsilon k}$, $y \to \infty$.

**3. (a)**
$$\frac{dS}{dt} = k_1 S(L - S)$$

$S = \dfrac{L}{1 + Ce^{-kt}}$ is a solution because

$$\frac{dS}{dt} = -L\left(1 + Ce^{-kt}\right)^{-2}\left(-Cke^{-kt}\right)$$

$$= \frac{LC\,ke^{-kt}}{\left(1 + Ce^{-kt}\right)^2}$$

$$= \left(\frac{k}{L}\right)\frac{L}{1 + Ce^{-kt}} \cdot \frac{C\,Le^{-kt}}{1 + Ce^{-kt}}$$

$$= \left(\frac{k}{L}\right)\frac{L}{1 + Ce^{-kt}} \cdot \left(L - \frac{L}{1 + Ce^{-kt}}\right)$$

$$= k_1 S(L - S), \text{ where } k_1 = \frac{k}{L}.$$

$L = 100$. Also, $S = 10$ when $t = 0 \Rightarrow C = 9$.

And, $S = 20$ when $t = 1 \Rightarrow k = -\ln\dfrac{4}{9}$.

Particular Solution: $S = \dfrac{100}{1 + 9e^{\ln(4/9)t}} = \dfrac{100}{1 + 9e^{-0.8109t}}$

**(b)**
$$\frac{dS}{dt} = k_1 S(100 - S)$$

$$\frac{d^2S}{dt^2} = k_1\left[S\left(-\frac{dS}{dt}\right) + (100 - S)\frac{dS}{dt}\right]$$

$$= k_1(100 - 2S)\frac{dS}{dt}$$

$$= 0 \text{ when } S = 50 \text{ or } \frac{dS}{dt} = 0.$$

Choosing $S = 50$, you have:

$$50 = \frac{100}{1 + 9e^{\ln(4/9)t}}$$

$$2 = 1 + 9e^{\ln(4/9)t}$$

$$\frac{\ln(1/9)}{\ln(4/9)} = t$$

$$t \approx 2.7 \text{ months}$$

(This is the point of inflection.)

**(c)**

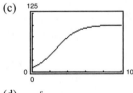

**(d)**

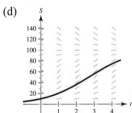

**(e)** Sales will decrease toward the line $S = L$.

**5.** Let $u = \dfrac{1}{2}k\left(t - \dfrac{\ln b}{k}\right)$.

$$1 + \tanh u = 1 + \frac{e^4 - e^{-u}}{e^u + e^{-u}} = \frac{2}{1 + e^{-2u}}$$

$$e^{-2u} = e^{-k(t - (\ln b/k))} = e^{\ln b}e^{-kt} = be^{-kt}$$

Finally,

$$\frac{1}{2}L\left[1 + \tanh\left(\frac{1}{2}k\left(t - \frac{\ln b}{k}\right)\right)\right] = \frac{L}{2}[1 + \tanh u]$$

$$= \frac{L}{2}\frac{2}{1 + be^{-kt}}$$

$$= \frac{L}{1 + be^{-kt}}.$$

Notice the graph of the logistics function is just a shift of the graph of the hyperbolic tangent. (See section 5.8.)

**7.** $k = \left(\dfrac{1}{12}\right)^2 \pi$

$g = 32$

$x^2 + (y - 6)^2 = 36$    Equation of tank

$\qquad x^2 = 36 - (y - 6)^2 = 12y - y^2$

Area of cross section: $A(h) = (12h - h^2)\pi$

$$A(h)\frac{dh}{dt} = -k\sqrt{2gh}$$

$$(12h - h^2)\pi\frac{dh}{dt} = -\frac{1}{144}\pi\sqrt{64h}$$

$$(12h - h^2)\frac{dh}{dt} = -\frac{1}{18}h^{1/2}$$

$$\int(18h^{3/2} - 216h^{1/2})\,dh = \int dt$$

$$\frac{36}{5}h^{5/2} - 144h^{3/2} = t + C$$

$$\frac{h^{3/2}}{5}(36h - 720) = t + C$$

When $h = 6, t = 0$ and $C = \dfrac{6^{3/2}}{5}(-504) \approx -1481.45.$

The tank is completely drained when

$h = 0 \Rightarrow t = 1481.45$ sec $\approx$ 24 min, 41 sec

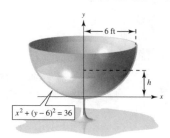

$x^2 + (y - 6)^2 = 36$

**9.** $A(h)\dfrac{dh}{dt} = -k\sqrt{2gh}$

$$\pi 64\frac{dh}{dt} = \frac{-\pi}{36}8\sqrt{h}$$

$$\int h^{-1/2}\,dh = \int\frac{-1}{288}\,dt$$

$$2\sqrt{h} = \frac{-t}{288} + C$$

$h = 20: 2\sqrt{20} = C = 4\sqrt{5}$

$$2\sqrt{h} = \frac{-t}{288} + 4\sqrt{5}$$

$h = 0 \Rightarrow t = 4\sqrt{5}(288)$

$\qquad\qquad \approx 2575.95$ sec $\approx$ 42 min, 56 sec

**11.** $\dfrac{ds}{dt} = 3.5 - 0.019s$

(a) $\displaystyle\int\frac{-ds}{3.5 - 0.019s} = -\int dt$

$$\frac{1}{0.019}\ln|3.5 - 0.019s| = -t + C_1$$

$$\ln|3.5 - 0.019s| = -0.019t + C_2$$

$$3.5 - 0.019s = C_3e^{-0.019t}$$

$$-0.019s = 3.5 - C_3e^{-0.019t}$$

$$s = 184.21 - Ce^{-0.019t}$$

(b)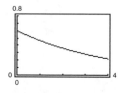

(c) As $t \to \infty$, $Ce^{-0.019t} \to 0$, and $s \to 184.21$.

**13.** From Exercises 12, you have $C = C_0e^{-Rt/V}$.

(a) For $V = 2$, $R = 0.5$, and $C_0 = 0.6$,

you have $C = 0.6e^{-0.25t}$

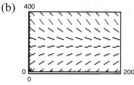

(b) For $V = 2$, $R = 1.5$, and $C_0 = 0.6$,

you have $C = 0.6e^{-0.75t}$.

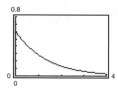

# CHAPTER 7
## Applications of Integration

# C H A P T E R   7
## Applications of Integration

### Section 7.1   Area of a Region Between Two Curves

**1.** $A = \int_0^6 \left[ 0 - \left( x^2 - 6x \right) \right] dx = -\int_0^6 \left( x^2 - 6x \right) dx$

**3.** $A = \int_0^3 \left[ \left( -x^2 + 2x + 3 \right) - \left( x^2 - 4x + 3 \right) \right] dx$

$\quad = \int_0^3 \left( -2x^2 + 6x \right) dx$

**5.** $A = 2\int_{-1}^0 3\left( x^3 - x \right) dx = 6\int_{-1}^0 \left( x^3 - x \right) dx$

$\quad$ or $-6\int_0^1 \left( x^3 - x \right) dx$

**7.** $\int_0^4 \left[ (x + 1) - \dfrac{x}{2} \right] dx$

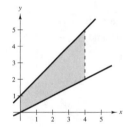

**9.** $\int_0^6 \left[ 4\left( 2^{-x/3} \right) - \dfrac{x}{6} \right] dx$

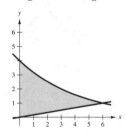

**11.** $\int_{-\pi/3}^{\pi/3} (2 - \sec x)\, dx$

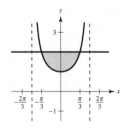

**13.** $\int_{-2}^1 \left[ (2 - y) - y^2 \right] dy$

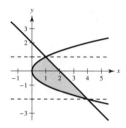

**15.** $f(x) = x + 1$

$\quad g(x) = (x - 1)^2$

$\quad A \approx 4$

$\quad$ Matches (d)

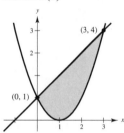

**17.** (a) $\qquad\qquad x = 4 - y^2$

$\qquad\qquad\qquad x = y - 2$

$\qquad\qquad 4 - y^2 = y - 2$

$\qquad\qquad y^2 + y - 6 = 0$

$\qquad\qquad (y + 3)(y - 2) = 0$

$\quad$ Intersection points: $(0, 2)$ and $(-5, -3)$

$\quad A = \int_{-5}^0 \left[ (x + 2) + \sqrt{4 - x} \right] dx + \int_0^4 2\sqrt{4 - x}\, dx$

$\quad = \dfrac{61}{6} + \dfrac{32}{3} = \dfrac{125}{6}$

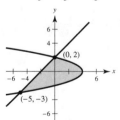

$\quad$ (b) $A = \int_{-3}^2 \left[ \left( 4 - y^2 \right) - (y - 2) \right] dy = \dfrac{125}{6}$

$\quad$ (c) The second method is simpler. Explanations will vary.

**19.**

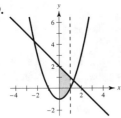

$$A = \int_0^1 \left[ (-x + 2) - \left( x^2 - 1 \right) \right] dx$$

$$= \int_0^1 \left( -x^2 - x + 3 \right) dx$$

$$= \left[ \frac{-x^3}{3} - \frac{x^2}{2} + 3x \right]_0^1$$

$$= \left( -\frac{1}{3} - \frac{1}{2} + 3 \right) - 0 = \frac{13}{6}$$

**21.** $A = \int_0^2 \left[ \left( \frac{1}{2}x^3 + 2 \right) - (x + 1) \right] dx$

$$= \int_0^2 \left( \frac{1}{2}x^3 - x + 1 \right) dx$$

$$= \left[ \frac{x^4}{8} - \frac{x^2}{2} + x \right]_0^2$$

$$= \left( \frac{16}{8} - \frac{4}{2} + 2 \right) - 0 = 2$$

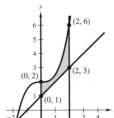

**23.** The points of intersection are given by:

$$x^2 - 4x = 0$$

$$x(x - 4) = 0 \quad \text{when } x = 0, 4$$

$$A = \int_0^4 \left[ g(x) - f(x) \right] dx$$

$$= -\int_0^4 \left( x^2 - 4x \right) dx = -\left[ \frac{x^3}{3} - 2x^2 \right]_0^4 = \frac{32}{3}$$

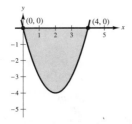

**25.** The points of intersection are given by:

$$x^2 + 2x = x + 2$$

$$x^2 + x - 2 = 0$$

$$(x + 2)(x - 1) = 0 \quad \text{when } x = -2, 1$$

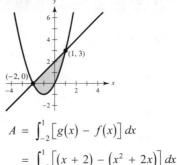

$$A = \int_{-2}^1 \left[ g(x) - f(x) \right] dx$$

$$= \int_{-2}^1 \left[ (x + 2) - \left( x^2 + 2x \right) \right] dx$$

$$= \left[ \frac{-x^3}{3} - \frac{x^2}{2} + 2x \right]_{-2}^1$$

$$= \left( -\frac{1}{3} - \frac{1}{2} + 2 \right) - \left( \frac{8}{3} - 2 - 4 \right) = \frac{9}{2}$$

**27.** The points of intersection are given by:

$$x = 2 - x \quad \text{and} \quad x = 0 \quad \text{and} \quad 2 - x = 0$$

$$x = 1 \qquad\qquad x = 0 \qquad\qquad x = 2$$

$$A = \int_0^1 \left[ (2 - y) - (y) \right] dy = \left[ 2y - y^2 \right]_0^1 = 1$$

Note that if you integrate
with respect to $x$, you need
two integrals. Also, note
that the region is a triangle.

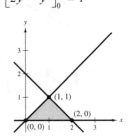

**29.** The points of intersection are given by:

$$\sqrt{x} + 3 = \frac{1}{2}x + 3$$

$$\sqrt{x} = \frac{1}{2}x$$

$$x = \frac{x^2}{4} \quad \text{when } x = 0, 4$$

$$A = \int_0^4 \left[ \left( \sqrt{x} + 3 \right) - \left( \frac{1}{2}x + 3 \right) \right] dx$$

$$= \left[ \frac{2}{3}x^{3/2} - \frac{x^2}{4} \right]_0^4 = \frac{16}{3} - 4 = \frac{4}{3}$$

**31.** The points of intersection are given by:

$$y^2 = y + 2$$

$$(y - 2)(y + 1) = 0 \quad \text{when } y = -1, 2$$

$$A = \int_{-1}^{2} \left[ g(y) - f(y) \right] dy$$

$$= \int_{-1}^{2} \left[ (y + 2) - y^2 \right] dy$$

$$= \left[ 2y + \frac{y^2}{2} - \frac{y^3}{3} \right]_{-1}^{2} = \frac{9}{2}$$

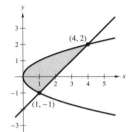

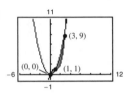

**33.** $A = \int_{-1}^{2} \left[ f(y) - g(y) \right] dy$

$$= \int_{-1}^{2} \left[ (y^2 + 1) - 0 \right] dy$$

$$= \left[ \frac{y^3}{3} + y \right]_{-1}^{2} = 6$$

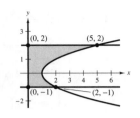

**35.** $y = \dfrac{10}{x} \Rightarrow x = \dfrac{10}{y}$

$$A = \int_{2}^{10} \frac{10}{y} \, dy$$

$$= \left[ 10 \ln y \right]_{2}^{10}$$

$$= 10(\ln 10 - \ln 2)$$

$$= 10 \ln 5 \approx 16.0944$$

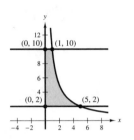

**37.** (a)

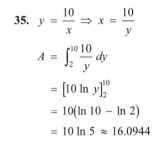

(b) The points of intersection are given by:

$$x^3 - 3x^2 + 3x = x^2$$

$$x(x - 1)(x - 3) = 0 \quad \text{when } x = 0, 1, 3$$

$$A = \int_{0}^{1} \left[ f(x) - g(x) \right] dx + \int_{1}^{3} \left[ g(x) - f(x) \right] dx$$

$$= \int_{0}^{1} \left[ (x^3 - 3x^2 + 3x) - x^2 \right] dx + \int_{1}^{3} \left[ x^2 - (x^3 - 3x^2 + 3x) \right] dx$$

$$= \int_{0}^{1} (x^3 - 4x^2 + 3x) \, dx + \int_{1}^{3} (-x^3 + 4x^2 - 3x) \, dx = \left[ \frac{x^4}{4} - \frac{4}{3}x^3 + \frac{3}{2}x^2 \right]_{0}^{1} + \left[ \frac{-x^4}{4} + \frac{4}{3}x^3 - \frac{3}{2}x^2 \right]_{1}^{3} = \frac{5}{12} + \frac{8}{3} = \frac{37}{12}$$

(c) Numerical approximation: $0.417 + 2.667 \approx 3.083$

**39.** (a)

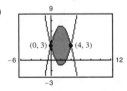

(b) The points of intersection are given by:

$$x^2 - 4x + 3 = 3 + 4x - x^2$$

$$2x(x - 4) = 0 \quad \text{when } x = 0, 4$$

$$A = \int_{0}^{4} \left[ (3 + 4x - x^2) - (x^2 - 4x + 3) \right] dx$$

$$= \int_{0}^{4} (-2x^2 + 8x) \, dx = \left[ -\frac{2x^3}{3} + 4x^2 \right]_{0}^{4} = \frac{64}{3}$$

(c) Numerical approximation: $21.333$

**41.** (a) $f(x) = x^4 - 4x^2$, $g(x) = x^2 - 4$

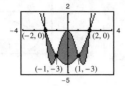

(b) The points of intersection are given by:

$$x^4 - 4x^2 = x^2 - 4$$

$$x^4 - 5x^2 + 4 = 0$$

$$(x^2 - 4)(x^2 - 1) = 0 \quad \text{when } x = \pm 2, \pm 1$$

By symmetry:

$$A = 2\int_0^1 \left[(x^4 - 4x^2) - (x^2 - 4)\right] dx + 2\int_1^2 \left[(x^2 - 4) - (x^4 - 4x^2)\right] dx$$

$$= 2\int_0^1 (x^4 - 5x^2 + 4) \, dx + 2\int_1^2 (-x^4 + 5x^2 - 4) \, dx$$

$$= 2\left[\frac{x^5}{5} - \frac{5x^3}{3} + 4x\right]_0^1 + 2\left[-\frac{x^5}{5} + \frac{5x^3}{3} - 4x\right]_1^2$$

$$= 2\left[\frac{1}{5} - \frac{5}{3} + 4\right] + 2\left[\left(-\frac{32}{5} + \frac{40}{3} - 8\right) - \left(-\frac{1}{5} + \frac{5}{3} - 4\right)\right] = 8$$

(c) Numerical approximation:

$$5.067 + 2.933 = 8.0$$

**43.** (a)

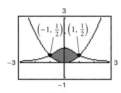

(b) The points of intersection are given by:

$$\frac{1}{1 + x^2} = \frac{x^2}{2}$$

$$x^4 + x^2 - 2 = 0$$

$$(x^2 + 2)(x^2 - 1) = 0$$

$$x = \pm 1$$

$$A = 2\int_0^1 \left[f(x) - g(x)\right] dx$$

$$= 2\int_0^1 \left[\frac{1}{1 + x^2} - \frac{x^2}{2}\right] dx$$

$$= 2\left[\arctan x - \frac{x^3}{6}\right]_0^1$$

$$= 2\left(\frac{\pi}{4} - \frac{1}{6}\right) = \frac{\pi}{2} - \frac{1}{3} \approx 1.237$$

(c) Numerical approximation: 1.237

**45.** (a)

(b) and (c) $\sqrt{1 + x^3} \le \frac{1}{2}x + 2$ on $[0, 2]$

You must use numerical integration because $y = \sqrt{1 + x^3}$ does not have an elementary antiderivative.

$$A = \int_0^2 \left[\frac{1}{2}x + 2 - \sqrt{1 + x^3}\right] dx \approx 1.759$$

**47.** $A = \int_0^{2\pi} \left[(2 - \cos x) - \cos x\right] dx$

$$= 2\int_0^{2\pi} (1 - \cos x) \, dx$$

$$= 2\left[x - \sin x\right]_0^{2\pi} = 4\pi \approx 12.566$$

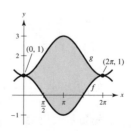

**49.**  $A = 2 \int_0^{\pi/3} \left[ f(x) - g(x) \right] dx$

$= 2 \int_0^{\pi/3} (2 \sin x - \tan x)\, dx$

$= 2 \left[ -2 \cos x + \ln|\cos x| \right]_0^{\pi/3} = 2(1 - \ln 2) \approx 0.614$

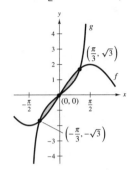

**51.**  $A = \int_0^1 \left[ xe^{-x^2} - 0 \right] dx$

$= \left[ -\frac{1}{2} e^{-x^2} \right]_0^1 = \frac{1}{2}\left( 1 - \frac{1}{e} \right) \approx 0.316$

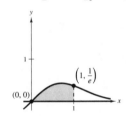

**53.** (a)

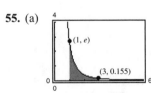

(b)  $A = \int_0^{\pi} (2 \sin x + \sin 2x)\, dx$

$= \left[ -2 \cos x - \tfrac{1}{2} \cos 2x \right]_0^{\pi}$

$= \left( 2 - \tfrac{1}{2} \right) - \left( -2 - \tfrac{1}{2} \right) = 4$

(c)  Numerical approximation: 4.0

**55.** (a)

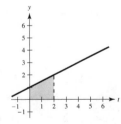

(b)  $A = \int_1^3 \frac{1}{x^2} e^{1/x}\, dx$

$= \left[ -e^{-1/x} \right]_1^3$

$= e - e^{1/3}$

(c)  Numerical approximation: 1.323

**57.** (a)

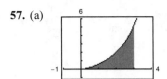

(b)  The integral

$$A = \int_0^3 \sqrt{\frac{x^3}{4 - x}}\, dx$$

does not have an elementary antiderivative.

(c)  $A \approx 4.7721$

**59.** (a)

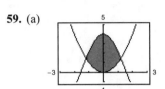

(b)  The intersection points are difficult to determine by hand.

(c)  Area $= \int_{-c}^{c} \left[ 4 \cos x - x^2 \right] dx \approx 6.3043$ where

$c \approx 1.201538$.

**61.**  $F(x) = \int_0^x \left( \frac{1}{2} t + 1 \right) dt = \left[ \frac{t^2}{4} + t \right]_0^x = \frac{x^2}{4} + x$

(a)  $F(0) = 0$

(b)  $F(2) = \frac{2^2}{4} + 2 = 3$

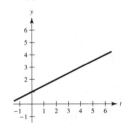

(c)  $F(6) = \frac{6^2}{4} + 6 = 15$

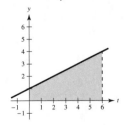

**63.** $F(\alpha) = \int_{-1}^{\alpha} \cos\frac{\pi\theta}{2}\,d\theta = \left[\frac{2}{\pi}\sin\frac{\pi\theta}{2}\right]_{-1}^{\alpha} = \frac{2}{\pi}\sin\frac{\pi\alpha}{2} + \frac{2}{\pi}$

(a) $F(-1) = 0$

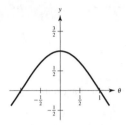

(b) $F(0) = \frac{2}{\pi} \approx 0.6366$

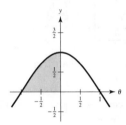

(c) $F\left(\frac{1}{2}\right) = \frac{2 + \sqrt{2}}{\pi} \approx 1.0868$

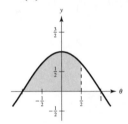

**65.** $A = \int_{2}^{4}\left[\left(\frac{9}{2}x - 12\right) - (x - 5)\right]dx + \int_{4}^{6}\left[\left(-\frac{5}{2}x + 16\right) - (x - 5)\right]dx$

$= \int_{2}^{4}\left(\frac{7}{2}x - 7\right)dx + \int_{4}^{6}\left(-\frac{7}{2}x + 21\right)dx$

$= \left[\frac{7}{4}x^2 - 7x\right]_{2}^{4} + \left[-\frac{7}{4}x^2 + 21x\right]_{4}^{6} = 7 + 7 = 14$

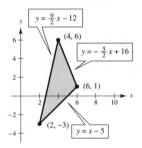

**67.** Left boundary line: $y = x + 2 \Leftrightarrow x = y - 2$

Right boundary line: $y = x - 2 \Leftrightarrow x = y + 2$

$A = \int_{-2}^{2}\left[(y + 2) - (y - 2)\right]dy$

$= \int_{-2}^{2} 4\,dy = [4y]_{-2}^{2} = 8 - (-8) = 16$

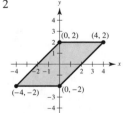

**69.** Answers will vary. *Sample answer*: If you let $\Delta x = 6$ and $n = 10, b - a = 10(6) = 60$.

(a) Area $\approx \dfrac{60}{2(10)}\big[0 + 2(14) + 2(14) + 2(12) + 2(12) + 2(15) + 2(20) + 2(23) + 2(25) + 2(26) + 0\big] = 3[322] = 966 \text{ ft}^2$

(b) Area $\approx \dfrac{60}{3(10)}\big[0 + 4(14) + 2(14) + 4(12) + 2(12) + 4(15) + 2(20) + 4(23) + 2(25) + 4(26) + 0\big] = 2[502] = 1004 \text{ ft}^2$

**71.** On the interval $\left[0, \dfrac{\pi}{4}\right]$, the points of intersection of $y_1 = \sin 2x$ and $y_2 = \cos 4x$ are given by:

$$\sin 2x = \cos 4x = 1 - 2\sin^2 2x$$

$$2\sin^2 2x + \sin 2x - 1 = 0$$

$$(2\sin 2x - 1)(\sin 2x + 1) = 0$$

$$\sin 2x = \frac{1}{2} \quad \text{when } x = \frac{\pi}{12}$$

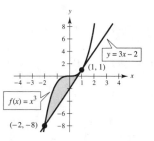

$$\int_0^{\pi/4}|\sin 2x - \cos 4x|\,dx = \int_0^{\pi/12}(\cos 4x - \sin 2x)\,dx + \int_{\pi/12}^{\pi/4}(\sin 2x - \cos 4x)\,dx$$

$$= \left[\frac{\sin 4x}{4} + \frac{\cos 2x}{2}\right]_0^{\pi/12} + \left[\frac{-\cos 2x}{2} - \frac{\sin 4x}{4}\right]_{\pi/12}^{\pi/4}$$

$$= \left[\left(\frac{\sqrt{3}}{8} + \frac{\sqrt{3}}{4}\right) - \frac{1}{2}\right] + \left[0 + \frac{\sqrt{3}}{4} + \frac{\sqrt{3}}{8}\right] = \frac{3\sqrt{3}}{4} - \frac{1}{2} \approx 0.7990$$

**73.** $f(x) = x^3$

$f'(x) = 3x^2$

At $(1, 1), f'(1) = 3$.

Tangent line: $y - 1 = 3(x - 1)$ or $y = 3x - 2$

The tangent line intersects $f(x) = x^3$ at $x = -2$.

$$A = \int_{-2}^1 \left[x^3 - (3x - 2)\right]dx = \left[\frac{x^4}{4} - \frac{3x^2}{2} + 2x\right]_{-2}^1 = \frac{27}{4}$$

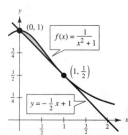

**75.** $f(x) = \dfrac{1}{x^2 + 1}$

$f'(x) = -\dfrac{2x}{\left(x^2 + 1\right)^2}$

At $\left(1, \dfrac{1}{2}\right), f'(1) = -\dfrac{1}{2}$.

Tangent line: $y - \dfrac{1}{2} = -\dfrac{1}{2}(x - 1)$ or $y = -\dfrac{1}{2}x + 1$

The tangent line intersects $f(x) = \dfrac{1}{x^2 + 1}$ at $x = 0$.

$$A = \int_0^1\left[\frac{1}{x^2 + 1} - \left(-\frac{1}{2}x + 1\right)\right]dx = \left[\arctan x + \frac{x^2}{4} - x\right]_0^1 = \frac{\pi - 3}{4} \approx 0.0354$$

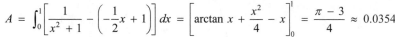

**77.** $x^4 - 2x^2 + 1 \le 1 - x^2$ on $[-1, 1]$

$$A = \int_{-1}^{1} \left[ \left( 1 - x^2 \right) - \left( x^4 - 2x^2 + 1 \right) \right] dx$$

$$= \int_{-1}^{1} \left( x^2 - x^4 \right) dx$$

$$= \left[ \frac{x^3}{3} - \frac{x^5}{5} \right]_{-1}^{1} = \frac{4}{15}$$

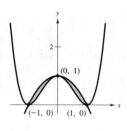

You can use a single integral because $x^4 - 2x^2 + 1 \le 1 - x^2$ on $[-1, 1]$.

**79.** Offer 2 is better because the accumulated salary (area under the curve) is larger.

**81.** (a) $\int_0^5 \left[ v_1(t) - v_2(t) \right] dt = 10$ means that Car 1 traveled
10 more meters than Car 2 on the interval $0 \le t \le 5$.

$\int_0^{10} \left[ v_1(t) - v_2(t) \right] dt = 30$ means that Car 1 traveled 30 more meters than Car 2 on the interval $0 \le t \le 10$.

$\int_{20}^{30} \left[ v_1(t) - v_2(t) \right] dt = -5$ means that Car 2

traveled 5 more meters than Car 1 on the interval $20 \le t \le 30$.

(b) No, it is not possible because you do not know the initial distance between the cars.

(c) At $t = 10$, Car 1 is ahead by 30 meters.

(d) At $t = 20$, Car 1 is ahead of Car 2 by 13 meters. From part (a), at $t = 30$, Car 1 is ahead by $13 - 5 = 8$ meters.

**83.**
$$A = \int_{-3}^{3} \left( 9 - x^2 \right) dx = 36$$

$$\int_{-\sqrt{9-b}}^{\sqrt{9-b}} \left[ \left( 9 - x^2 \right) - b \right] dx = 18$$

$$\int_{0}^{\sqrt{9-b}} \left[ \left( 9 - b \right) - x^2 \right] dx = 9$$

$$\left[ \left( 9 - b \right) x - \frac{x^3}{3} \right]_0^{\sqrt{9-b}} = 9$$

$$\frac{2}{3} \left( 9 - b \right)^{3/2} = 9$$

$$\left( 9 - b \right)^{3/2} = \frac{27}{2}$$

$$9 - b = \frac{9}{\sqrt[3]{4}}$$

$$b = 9 - \frac{9}{\sqrt[3]{4}} \approx 3.330$$

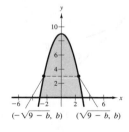

$(-\sqrt{9-b}, b)$     $(\sqrt{9-b}, b)$

**85.** Area of triangle $OAB$ is $\frac{1}{2}(4)(4) = 8$.

$$4 = \int_0^a \left( 4 - x \right) dx = \left[ 4x - \frac{x^2}{2} \right]_0^a = 4a - \frac{a^2}{2}$$

$$a^2 - 8a + 8 = 0$$

$$a = 4 \pm 2\sqrt{2}$$

Because $0 < a < 4$, select $a = 4 - 2\sqrt{2} \approx 1.172$.

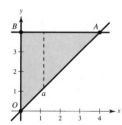

**87.** $\lim\limits_{\|\Delta\| \to 0} \sum\limits_{i=1}^{n} \left( x_i - x_i^2 \right) \Delta x$

where $x_i = \dfrac{i}{n}$ and $\Delta x = \dfrac{1}{n}$ is the same as

$$\int_0^1 \left( x - x^2 \right) dx = \left[ \frac{x^2}{2} - \frac{x^3}{3} \right]_0^1 = \frac{1}{6}.$$

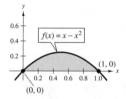

**89.** $f(x) = x^4 + 4x^3 + x + 7$

$f'(x) = 4x^3 + 12x^2 + 1$

$f''(x) = 12x^2 + 24x = 12x(x + 2)$

(a) Concavity changes when $x = 0, -2$.

Points of inflection: $(0, 7), (-2, -11)$

(b) Slope $= \frac{18}{2} = 9$

Tangent line: $y - 7 = 9(x - 0)$

$y = 9x + 7$ or $g(x) = 9x + 7$

(c) Intersection points:

$$x^4 + 4x^3 + x + 7 = 9x + 7$$

$$x^4 + 4x^3 - 8x = 0$$

$$x(x + 2)(x + \sqrt{5} + 1)(x - \sqrt{5} + 1) = 0$$

$$\left[ x = -\sqrt{5} - 1, -2, 0, \sqrt{5} - 1 \right]$$

$$\int_{-\sqrt{5}-1}^{-2} \left[ g(x) - f(x) \right] dx = \frac{16}{5}$$

$$\int_{-2}^{0} \left[ f(x) - g(x) \right] dx = \frac{32}{5}$$

$$\int_{0}^{\sqrt{5}-1} \left[ g(x) - f(x) \right] dx = \frac{16}{5}$$

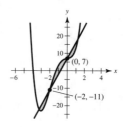

The area between the two inflection points is the sum of the areas between the other two regions.

**91.** $\int_{8}^{13} \left[ (7.21 + 0.58t) - (7.21 + 0.45t) \right] dt = \int_{8}^{13} 0.13t \, dt = \left[ \frac{0.13t^2}{2} \right]_{8}^{13} = \$6.825$ billion

**93.** (a) $y_1 = 0.0124x^2 - 0.385x + 7.85$

(b)

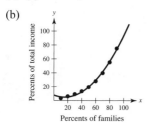

(c)

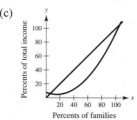

(d) Income inequality $= \int_{0}^{100} \left[ x - y_1 \right] dx \approx 2006.7$

**95.** The total area is 8 times the area of the shaded region to the right. A point $(x, y)$ is on the upper boundary of the region if

$$\sqrt{x^2 + y^2} = 2 - y$$
$$x^2 + y^2 = 4 - 4y + y^2$$
$$x^2 = 4 - 4y$$
$$4y = 4 - x^2$$
$$y = 1 - \frac{x^2}{4}.$$

Now determine where this curve intersects the line $y = x$.

$$x = 1 - \frac{x^2}{4}$$
$$x^2 + 4x - 4 = 0$$
$$x = \frac{-4 \pm \sqrt{16 + 16}}{2} = -2 \pm 2\sqrt{2} \Rightarrow x = -2 + 2\sqrt{2}$$

$$\text{Total area} = 8 \int_0^{-2+2\sqrt{2}} \left( 1 - \frac{x^2}{4} - x \right) dx = 8 \left[ x - \frac{x^3}{12} - \frac{x^2}{2} \right]_0^{-2+2\sqrt{2}} = \frac{16}{3}\left(4\sqrt{2} - 5\right) \approx 8(0.4379) = 3.503$$

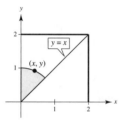

**97.** (a) $A = 2 \left[ \int_0^5 \left( 1 - \frac{1}{3}\sqrt{5 - x} \right) dx + \int_5^{5.5} (1 - 0) \, dx \right]$

$$= 2 \left( \left[ x + \frac{2}{9}(5 - x)^{3/2} \right]_0^5 + \left[ x \right]_5^{5.5} \right)$$

$$= 2 \left( 5 - \frac{10\sqrt{5}}{9} + 5.5 - 5 \right) \approx 6.031 \text{ m}^2$$

(b) $V = 2A \approx 2(6.031) \approx 12.062 \text{ m}^3$

(c) $5000 V \approx 5000(12.062) = 60,310$ pounds

**99.** True. The region has been shifted $C$ units upwards (if $C > 0$), or $C$ units downwards (if $C < 0$).

**101.** False. Let $f(x) = x$ and $g(x) = 2x - x^2$, $f$ and $g$ intersect at $(1, 1)$, the midpoint of $[0, 2]$, but

$$\int_a^b \left[ f(x) - g(x) \right] dx = \int_0^2 \left[ x - \left( 2x - x^2 \right) \right] dx = \frac{2}{3} \ne 0.$$

**103.** Line: $y = \frac{-3}{7\pi}x$

$$A = \int_0^{7\pi/6} \left[ \sin x + \frac{3x}{7\pi} \right] dx = \left[ -\cos x + \frac{3x^2}{14\pi} \right]_0^{7\pi/6}$$

$$= \frac{\sqrt{3}}{2} + \frac{7\pi}{24} + 1 \approx 2.7823$$

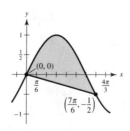

**105.** You want to find $c$ such that:

$$\int_0^b \left[ \left(2x - 3x^3\right) - c \right] dx = 0$$

$$\left[ x^2 - \tfrac{3}{4}x^4 - cx \right]_0^b = 0$$

$$b^2 - \tfrac{3}{4}b^4 - cb = 0$$

But, $c = 2b - 3b^3$ because $(b, c)$ is on the graph.

$$b^2 - \tfrac{3}{4}b^4 - \left(2b - 3b^3\right)b = 0$$

$$4 - 3b^2 - 8 + 12b^2 = 0$$

$$9b^2 = 4$$

$$b = \tfrac{2}{3}$$

$$c = \tfrac{4}{9}$$

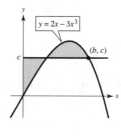

# Section 7.2   Volume: The Disk Method

**1.** $V = \pi \int_0^1 (-x + 1)^2 \, dx = \pi \int_0^1 \left(x^2 - 2x + 1\right) dx = \pi \left[ \dfrac{x^3}{3} - x^2 + x \right]_0^1 = \dfrac{\pi}{3}$

**3.** $V = \pi \int_1^4 \left(\sqrt{x}\right)^2 dx = \pi \int_1^4 x \, dx = \pi \left[ \dfrac{x^2}{2} \right]_1^4 = \dfrac{15\pi}{2}$

**5.** $V = \pi \int_0^1 \left[ \left(x^2\right)^2 - \left(x^5\right)^2 \right] dx = \pi \int_0^1 \left(x^4 - x^{10}\right) dx$

   $= \pi \left[ \dfrac{x^5}{5} - \dfrac{x^{11}}{11} \right]_0^1 = \pi \left( \dfrac{1}{5} - \dfrac{1}{11} \right) = \dfrac{6\pi}{55}$

**7.** $y = x^2 \Rightarrow x = \sqrt{y}$

   $V = \pi \int_0^4 \left(\sqrt{y}\right)^2 dy = \pi \int_0^4 y \, dy = \pi \left[ \dfrac{y^2}{2} \right]_0^4 = 8\pi$

**9.** $y = x^{2/3} \Rightarrow x = y^{3/2}$

   $V = \pi \int_0^1 \left(y^{3/2}\right)^2 dy = \pi \int_0^1 y^3 \, dy = \pi \left[ \dfrac{y^4}{4} \right]_0^1 = \dfrac{\pi}{4}$

**11.** $y = \sqrt{x}, \; y = 0, \; x = 3$

   (a)  $R(x) = \sqrt{x}, \; r(x) = 0$

   $V = \pi \int_0^3 \left(\sqrt{x}\right)^2 dx = \pi \int_0^3 x \, dx = \pi \left[ \dfrac{x^2}{2} \right]_0^3$

   $= \dfrac{9\pi}{2}$

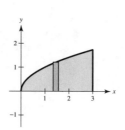

(b) $R(y) = 3, r(y) = y^2$

$$V = \pi \int_0^{\sqrt{3}} \left[ 3^2 - \left(y^2\right)^2 \right] dy = \pi \int_0^{\sqrt{3}} \left(9 - y^4\right) dy$$

$$= \pi \left[ 9y - \frac{y^5}{5} \right]_0^{\sqrt{3}} = \pi \left[ 9\sqrt{3} - \frac{9}{5}\sqrt{3} \right]$$

$$= \frac{36\sqrt{3}\pi}{5}$$

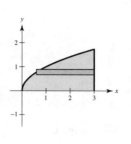

(c) $R(y) = 3 - y^2, r(y) = 0$

$$V = \pi \int_0^{\sqrt{3}} \left(3 - y^2\right)^2 dy = \pi \int_0^{\sqrt{3}} \left(9 - 6y^2 + y^4\right) dy$$

$$= \pi \left[ 9y - 2y^3 + \frac{y^5}{5} \right]_0^{\sqrt{3}} = \pi \left[ 9\sqrt{3} - 6\sqrt{3} + \frac{9\sqrt{3}}{5} \right]$$

$$= \frac{24\sqrt{3}\pi}{5}$$

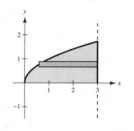

(d) $R(y) = 3 + \left(3 - y^2\right) = 6 - y^2, r(y) = 3$

$$V = \pi \int_0^{\sqrt{3}} \left[ \left(6 - y^2\right)^2 - 3^2 \right] dy = \pi \int_0^{\sqrt{3}} \left(y^4 - 12y^2 + 27\right) dy$$

$$= \pi \left[ \frac{y^5}{5} - 4y^3 + 27y \right]_0^{\sqrt{3}} = \pi \left[ \frac{9\sqrt{3}}{5} - 12\sqrt{3} + 27\sqrt{3} \right]$$

$$= \frac{84\sqrt{3}\pi}{5}$$

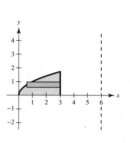

**13.** $y = x^2, y = 4x - x^2$ intersect at $(0, 0)$ and $(2, 4)$.

(a) $R(x) = 4x - x^2, r(x) = x^2$

$$V = \pi \int_0^2 \left[ \left(4x - x^2\right)^2 - x^4 \right] dx$$

$$= \pi \int_0^2 \left(16x^2 - 8x^3\right) dx = \pi \left[ \frac{16}{3}x^3 - 2x^4 \right]_0^2$$

$$= \frac{32\pi}{3}$$

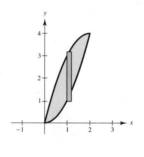

(b) $R(x) = 6 - x^2, r(x) = 6 - \left(4x - x^2\right)-$

$$V = \pi \int_0^2 \left[ \left(6 - x^2\right)^2 - \left(6 - 4x + x^2\right)^2 \right] dx$$

$$= 8\pi \int_0^2 \left(x^3 - 5x^2 + 6x\right) dx = 8\pi \left[ \frac{x^4}{4} - \frac{5}{3}x^3 + 3x^2 \right]_0^2$$

$$= \frac{64\pi}{3}$$

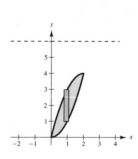

**15.** $R(x) = 4 - x, r(x) = 1$

$$V = \pi \int_0^3 \left[ (4 - x)^2 - (1)^2 \right] dx$$

$$= \pi \int_0^3 \left( x^2 - 8x + 15 \right) dx$$

$$= \pi \left[ \frac{x^3}{3} - 4x^2 + 15x \right]_0^3$$

$$= 18\pi$$

**17.** $R(x) = 4, r(x) = 4 - \dfrac{3}{1 + x}$

$$V = \pi \int_0^3 \left[ 4^2 - \left( 4 - \frac{3}{1 + x} \right)^2 \right] dx$$

$$= \pi \int_0^3 \left[ \frac{24}{1 + x} - \frac{9}{(1 + x)^2} \right] dx$$

$$= \pi \left[ 24 \ln |1 + x| + \frac{9}{1 + x} \right]_0^3$$

$$= \pi \left[ \left( 24 \ln 4 + \frac{9}{4} \right) - 9 \right]$$

$$= \left( 48 \ln 2 - \frac{27}{4} \right) \pi \approx 83.318$$

**19.** $R(y) = 5 - y, r(y) = 0$

$$V = \pi \int_0^4 (5 - y)^2 \, dy$$

$$= \pi \int_0^4 \left( 25 - 10y + y^2 \right) dy$$

$$= \pi \left[ 25y - 5y^2 + \frac{y^3}{3} \right]_0^4$$

$$= \pi \left[ 100 - 80 + \frac{64}{3} \right] = \frac{124\pi}{3}$$

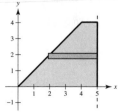

**21.** $R(y) = 5 - y^2, r(y) = 1$

$$V = \pi \int_{-2}^2 \left[ \left( 5 - y^2 \right)^2 - 1 \right] dy$$

$$= 2\pi \int_0^2 \left[ y^4 - 10y^2 + 24 \right] dy$$

$$= 2\pi \left[ \frac{y^5}{5} - \frac{10y^3}{3} + 24y \right]_0^2$$

$$= 2\pi \left[ \frac{32}{5} - \frac{80}{3} + 48 \right] = \frac{832\pi}{15}$$

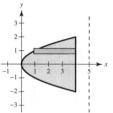

**23.** $R(x) = \dfrac{1}{\sqrt{x + 1}}, r(x) = 0$

$$V = \pi \int_0^4 \left( \frac{1}{\sqrt{x + 1}} \right)^2 dx$$

$$= \pi \int_0^4 \frac{1}{x + 1} dx = \pi \Big[ \ln |x + 1| \Big]_0^4 = \pi \ln 5$$

**25.** $R(x) = \dfrac{1}{x}, r(x) = 0$

$$V = \pi \int_1^3 \left( \frac{1}{x} \right)^2 dx$$

$$= \pi \left[ -\frac{1}{x} \right]_1^3$$

$$= \pi \left[ -\frac{1}{3} + 1 \right] = \frac{2}{3}\pi$$

**27.** $R(x) = e^{-x}, r(x) = 0$

$$V = \pi \int_0^1 \left(e^{-x}\right)^2 dx$$

$$= \pi \int_0^1 e^{-2x} dx$$

$$= \left[-\frac{\pi}{2} e^{-2x}\right]_0^1 = \frac{\pi}{2}\left(1 - e^{-2}\right) \approx 1.358$$

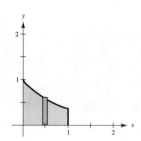

**29.**
$$x^2 + 1 = -x^2 + 2x + 5$$

$$2x^2 - 2x - 4 = 0$$

$$x^2 - x - 2 = 0$$

$$(x - 2)(x + 1) = 0$$

The curves intersect at $(-1, 2)$ and $(2, 5)$.

$$V = \pi \int_0^2 \left[\left(5 + 2x - x^2\right)^2 - \left(x^2 + 1\right)^2\right] dx + \pi \int_2^3 \left[\left(x^2 + 1\right)^2 - \left(5 + 2x - x^2\right)^2\right] dx$$

$$= \pi \int_0^2 \left(-4x^3 - 8x^2 + 20x + 24\right) dx + \pi \int_2^3 \left(4x^3 + 8x^2 - 20x - 24\right) dx$$

$$= \pi \left[-x^4 - \frac{8}{3}x^3 + 10x^2 + 24x\right]_0^2 + \pi \left[x^4 + \frac{8}{3}x^3 - 10x^2 - 24x\right]_2^3$$

$$= \pi \frac{152}{3} + \pi \frac{125}{3} = \frac{277\pi}{3}$$

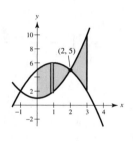

**31.** $y = 6 - 3x \Rightarrow x = \frac{1}{3}(6 - y)$

$$V = \pi \int_0^6 \left[\frac{1}{3}(6 - y)\right]^2 dy$$

$$= \frac{\pi}{9} \int_0^6 \left[36 - 12y + y^2\right] dy$$

$$= \frac{\pi}{9} \left[36y - 6y^2 + \frac{y^3}{3}\right]_0^6$$

$$= \frac{\pi}{9} \left[216 - 216 + \frac{216}{3}\right]$$

$$= 8\pi = \frac{1}{3}\pi r^2 h, \text{ Volume of cone}$$

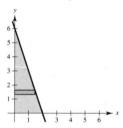

**33.** $V = \pi \int_0^\pi \left(\sin x\right)^2 dx$

$$= \pi \int_0^\pi \frac{1 - \cos 2x}{2} dx$$

$$= \frac{\pi}{2} \left[x - \frac{1}{2}\sin 2x\right]_0^\pi = \frac{\pi}{2}[\pi] = \frac{\pi^2}{2}$$

Numerical approximation: 4.9348

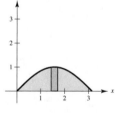

**35.** $V = \pi \int_1^2 \left(e^{x-1}\right)^2 dx$

$$= \pi \int_1^2 e^{2x-2} dx$$

$$= \frac{\pi}{2} e^{2x-2} \Big]_1^2$$

$$= \frac{\pi}{2}\left(e^2 - 1\right)$$

Numerical approximation: 10.0359

**37.** $V = \pi \int_0^2 \left[ e^{-x^2} \right]^2 dx \approx 1.9686$

**39.** $V = \pi \int_0^5 \left[ 2 \arctan(0.2x) \right]^2 dx$
$\approx 15.4115$

**41.** $V = \pi \int_0^1 y^2 \, dy = \pi \frac{y^3}{3} \Big]_0^1 = \frac{\pi}{3}$

**43.** $V = \pi \int_0^1 \left( x^2 - x^4 \right) dx$
$= \pi \left[ \frac{x^3}{3} - \frac{x^5}{5} \right]_0^1$
$= \pi \left( \frac{1}{3} - \frac{1}{5} \right)$
$= \frac{2\pi}{15}$

**45.** $V = \pi \int_0^1 (1 - y) \, dy$
$= \pi \left[ y - \frac{y^2}{2} \right]_0^1 = \pi \left( 1 - \frac{1}{2} \right) = \frac{\pi}{2}$

**47.** $V = \pi \int_0^1 \left( y - y^2 \right) dy$
$= \pi \left[ \frac{y^2}{2} - \frac{y^3}{3} \right]_0^1 = \pi \left( \frac{1}{2} - \frac{1}{3} \right) = \frac{\pi}{6}$

**49.** $A \approx 3$
Matches (a)

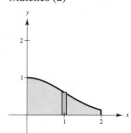

**51.** $\pi \int_0^{\pi/2} \sin^2 x \, dx$ represents the volume of the solid generated by revolving the region bounded by $y = \sin x$, $y = 0$, $x = 0$, $x = \pi/2$ about the $x$-axis.

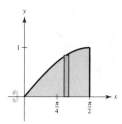

**53.**

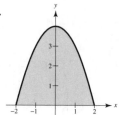

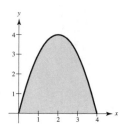

The volumes are the same because the solid has been translated horizontally. $\left( 4x - x^2 = 4 - (x - 2)^2 \right)$

**55.** (a) True. Answers will vary.
(b) False. Answers will vary.

**57.** $R(x) = \frac{1}{2}x, \, r(x) = 0$

$V = \pi \int_0^6 \frac{1}{4}x^2 \, dx = \left[ \frac{\pi}{12}x^3 \right]_0^6 = 18\pi$

**Note:** $V = \frac{1}{3}\pi r^2 h = \frac{1}{3}\pi \left( 3^2 \right) 6 = 18\pi$

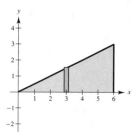

**59.** $R(x) = \sqrt{r^2 - x^2}, \, r(x) = 0$

$V = \pi \int_{-r}^r \left( r^2 - x^2 \right) dx$
$= 2\pi \int_0^r \left( r^2 - x^2 \right) dx$
$= 2\pi \left[ r^2 x - \frac{1}{3}x^3 \right]_0^r$
$= 2\pi \left( r^3 - \frac{1}{3}r^3 \right) = \frac{4}{3}\pi r^3$

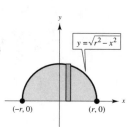

**61.** $x = r - \dfrac{r}{H}y = r\left(1 - \dfrac{y}{H}\right)$, $R(y) = r\left(1 - \dfrac{y}{H}\right)$, $r(y) = 0$

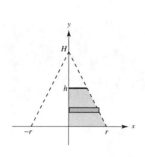

$$V = \pi \int_0^h \left[r\left(1 - \dfrac{y}{H}\right)\right]^2 dy = \pi r^2 \int_0^h \left(1 - \dfrac{2}{H}y + \dfrac{1}{H^2}y^2\right) dy$$

$$= \pi r^2 \left[y - \dfrac{1}{H}y^2 + \dfrac{1}{3H^2}y^3\right]_0^h$$

$$= \pi r^2 \left(h - \dfrac{h^2}{H} + \dfrac{h^3}{3H^2}\right) = \pi r^2 h\left(1 - \dfrac{h}{H} + \dfrac{h^2}{3H^2}\right)$$

**63.**

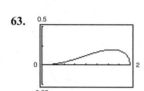

$$V = \pi \int_0^2 \left(\dfrac{1}{8}x^2\sqrt{2 - x}\right)^2 dx = \dfrac{\pi}{64}\int_0^2 x^4(2 - x)\, dx = \dfrac{\pi}{64}\left[\dfrac{2x^5}{5} - \dfrac{x^6}{6}\right]_0^2 = \dfrac{\pi}{30}\ \text{m}^3$$

**65.** (a)  $R(x) = \dfrac{3}{5}\sqrt{25 - x^2}$, $r(x) = 0$

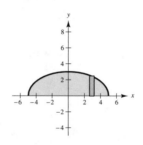

$$V = \dfrac{9\pi}{25}\int_{-5}^5 \left(25 - x^2\right) dx$$

$$= \dfrac{18\pi}{25}\int_0^5 \left(25 - x^2\right) dx$$

$$= \dfrac{18\pi}{25}\left[25x - \dfrac{x^3}{3}\right]_0^5 = 60\pi$$

(b)  $R(y) = \dfrac{5}{3}\sqrt{9 - y^2}$, $r(y) = 0$, $x \geq 0$

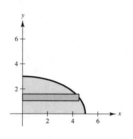

$$V = \dfrac{25\pi}{9}\int_0^3 \left(9 - y^2\right) dy$$

$$= \dfrac{25\pi}{9}\left[9y - \dfrac{y^3}{3}\right]_0^3 = 50\pi$$

**67.** (a)  First find where $y = b$ intersects the parabola:

$$b = 4 - \dfrac{x^2}{4}$$

$$x^2 = 16 - 4b = 4(4 - b)$$

$$x = 2\sqrt{4 - b}$$

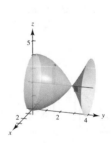

$$V = \int_0^{2\sqrt{4-b}} \pi\left[4 - \dfrac{x^2}{4} - b\right]^2 dx + \int_{2\sqrt{4-b}}^4 \pi\left[b - 4 + \dfrac{x^2}{4}\right]^2 dx$$

$$= \int_0^4 \pi\left[4 - \dfrac{x^2}{4} - b\right]^2 dx = \pi\int_0^4 \left[\dfrac{x^4}{16} - 2x^2 + \dfrac{bx^2}{2} + b^2 - 8b + 16\right] dx$$

$$= \pi\left[\dfrac{x^5}{80} - \dfrac{2x^3}{3} + \dfrac{bx^3}{6} + b^2 x - 8bx + 16x\right]_0^4$$

$$= \pi\left(\dfrac{64}{5} - \dfrac{128}{3} + \dfrac{32}{3}b + 4b^2 - 32b + 64\right) = \pi\left(4b^2 - \dfrac{64}{3}b + \dfrac{512}{15}\right)$$

(b) Graph of $V(b) = \pi \left( 4b^2 - \dfrac{64}{3}b + \dfrac{512}{15} \right)$

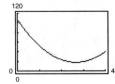

Minimum volume is 17.87 for $b = 2.67$.

(c) $V'(b) = \pi \left( 8b - \dfrac{64}{3} \right) = 0 \Rightarrow b = \dfrac{64/3}{8} = \dfrac{8}{3} = 2\dfrac{2}{3}$

$V''(b) = 8\pi > 0 \Rightarrow b = \dfrac{8}{3}$ is a relative minimum.

**69.** (a) $\pi \displaystyle\int_0^h r^2 \, dx$ (ii)

is the volume of a right circular
cylinder with radius $r$ and height $h$.

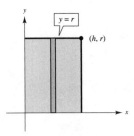

(b) $\pi \displaystyle\int_{-b}^{b} \left( a\sqrt{1 - \dfrac{x^2}{b^2}} \right)^2 dx$ (iv)

is the volume of an ellipsoid
with axes $2a$ and $2b$.

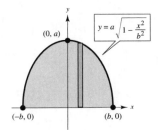

(c) $\pi \displaystyle\int_{-r}^{r} \left( \sqrt{r^2 - x^2} \right)^2 dx$ (iii)

is the volume of a sphere with radius $r$.

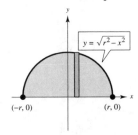

(d) $\pi \displaystyle\int_0^h \left( \dfrac{rx}{h} \right)^2 dx$ (i)

is the volume of a right circular
cone with the radius of the base
as $r$ and height $h$.

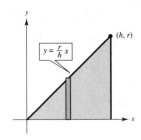

(e) $\pi \int_{-r}^{r} \left[ \left( R + \sqrt{r^2 - x^2} \right)^2 - \left( R - \sqrt{r^2 - x^2} \right)^2 \right] dx$  (v)

is the volume of a torus with the radius of its circular cross section as $r$ and the distance from the axis of the torus to the center of its cross section as $R$.

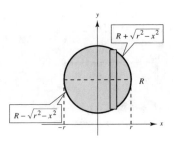

**71.**

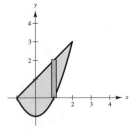

Base of cross section $= (x + 1) - (x^2 - 1) = 2 + x - x^2$

(a) $A(x) = b^2 = \left( 2 + x - x^2 \right)^2 = 4 + 4x - 3x^2 - 2x^3 + x^4$

$$V = \int_{-1}^{2} \left( 4 + 4x - 3x^2 - 2x^3 + x^4 \right) dx = \left[ 4x + 2x^2 - x^3 - \frac{1}{2}x^4 + \frac{1}{5}x^5 \right]_{-1}^{2} = \frac{81}{10}$$

$2 + x - x^2$

$2 + x - x^2$

(b) $A(x) = bh = \left( 2 + x - x^2 \right) 1$

$$V = \int_{-1}^{2} \left( 2 + x - x^2 \right) dx = \left[ 2x + \frac{x^2}{2} - \frac{x^3}{3} \right]_{-1}^{2} = \frac{9}{2}$$

$2 + x - x^2$

1

**73.** The cross sections are squares. By symmetry, you can set up an integral for an eighth of the volume and multiply by 8.

$A(y) = b^2 = \left( \sqrt{r^2 - y^2} \right)^2$

$V = 8 \int_{0}^{r} \left( r^2 - y^2 \right) dy$

$\quad = 8 \left[ r^2 y - \frac{1}{3} y^3 \right]_{0}^{r}$

$\quad = \frac{16}{3} r^3$

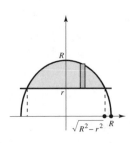

**75.** $V = \pi \int_{-\sqrt{R^2 - r^2}}^{\sqrt{R^2 - r^2}} \left[ \left( \sqrt{R^2 - x^2} \right)^2 - r^2 \right] dx$

$\quad = 2\pi \int_{0}^{\sqrt{R^2 - r^2}} \left( R^2 - r^2 - x^2 \right) dx$

$\quad = 2\pi \left[ \left( R^2 - r^2 \right) x - \frac{x^3}{3} \right]_{0}^{\sqrt{R^2 - r^2}}$

$\quad = 2\pi \left[ \left( R^2 - r^2 \right)^{3/2} - \frac{\left( R^2 - r^2 \right)^{3/2}}{3} \right] = \frac{4}{3} \pi \left( R^2 - r^2 \right)^{3/2}$

**77.** $y = \dfrac{8x}{9 + x^2}$

$$V = \pi \int_0^5 \left( \dfrac{8x}{9 + x^2} \right)^2 dx$$

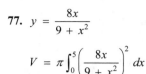

Using Simpson's Rule with $n = 10$, you obtain

$V \approx 19.7444.$

**79. (a)** Because the cross sections are isosceles right triangles:

$$A(x) = \frac{1}{2} bh = \frac{1}{2}\left(\sqrt{r^2 - y^2}\right)\left(\sqrt{r^2 - y^2}\right) = \frac{1}{2}\left(r^2 - y^2\right)$$

$$V = \frac{1}{2}\int_{-r}^r \left(r^2 - y^2\right) dy = \int_0^r \left(r^2 - y^2\right) dy = \left[r^2 y - \frac{y^3}{3}\right]_0^r = \frac{2}{3}r^3$$

**(b)** $A(x) = \dfrac{1}{2} bh = \dfrac{1}{2}\sqrt{r^2 - y^2}\left(\sqrt{r^2 - y^2} \tan \theta\right) = \dfrac{\tan \theta}{2}\left(r^2 - y^2\right)$

$$V = \frac{\tan \theta}{2}\int_{-r}^r \left(r^2 - y^2\right) dy = \tan \theta \int_0^r \left(r^2 - y^2\right) dy = \tan \theta \left[r^2 y - \frac{y^3}{3}\right]_0^r = \frac{2}{3}r^3 \tan \theta$$

As $\theta \to 90°, V \to \infty.$

## Section 7.3    Volume: The Shell Method

**1.** $p(x) = x, h(x) = x$

$$V = 2\pi \int_0^2 x(x)\, dx = \left[\frac{2\pi x^3}{3}\right]_0^2 = \frac{16\pi}{3}$$

**3.** $p(x) = x, h(x) = \sqrt{x}$

$$V = 2\pi \int_0^4 x\sqrt{x}\, dx = 2\pi \int_0^4 x^{3/2}\, dx = \left[\frac{4\pi}{5}x^{5/2}\right]_0^4 = \frac{128\pi}{5}$$

**5.** $p(x) = x, h(x) = x^2$

$$V = 2\pi \int_0^3 x\left(x^2\right) dx$$

$$= 2\pi \left[\frac{x^4}{4}\right]_0^3$$

$$= 2\pi \left(\frac{81}{4}\right) = \frac{81\pi}{2}$$

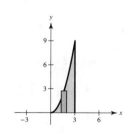

**7.** $p(x) = x, h(x) = \left(4x - x^2\right) - x^2 = 4x - 2x^2$

$$V = 2\pi \int_0^2 x\left(4x - 2x^2\right) dx$$

$$= 4\pi \int_0^2 \left(2x^2 - x^3\right) dx$$

$$= 4\pi \left[\frac{2}{3}x^3 - \frac{1}{4}x^4\right]_0^2 = \frac{16\pi}{3}$$

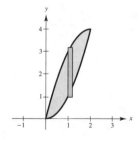

**9.** $p(x) = x$

$h(x) = 4 - \left(4x - x^2\right)$

$\qquad = x^2 - 4x + 4$

$V = 2\pi \int_0^2 x\left(x^2 - 4x + 4\right) dx$

$V = 2\pi \int_0^2 \left(x^3 - 4x^2 + 4x\right) dx$

$\qquad = 2\pi \left[\dfrac{x^4}{4} - \dfrac{4}{3}x^3 + 2x^2\right]_0^2$

$\qquad = \dfrac{8\pi}{3}$

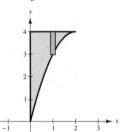

**11.** $p(x) = x, h(x) = \sqrt{x - 2}$

$V = 2\pi \int_2^4 x\sqrt{x - 2}\, dx$

Let $u = x - 2, x = u + 2, du = dx$.

When $x = 2, u = 0$.

When $x = 4, u = 2$.

$V = 2\pi \int_0^2 (u + 2)u^{1/2}\, du$

$\qquad = 2\pi \left[\dfrac{2}{5}u^{5/2} + \dfrac{4}{3}u^{3/2}\right]_0^2$

$\qquad = 2\pi \left[\dfrac{2}{5}(2)^{5/2} + \dfrac{4}{3}(2)^{3/2}\right]$

$\qquad = 2\pi\sqrt{2}\left[\dfrac{2}{5}(4) + \dfrac{4}{3}(2)\right] = \dfrac{128\sqrt{2}\pi}{15}$

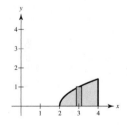

**13.** $p(x) = x, h(x) = \dfrac{1}{\sqrt{2\pi}}e^{-x^2/2}$

$V = 2\pi \int_0^1 x\left(\dfrac{1}{\sqrt{2\pi}}e^{-x^2/2}\right) dx$

$\qquad = \sqrt{2\pi} \int_0^1 e^{-x^2/2}x\, dx$

$\qquad = \left[-\sqrt{2\pi}e^{-x^2/2}\right]_0^1$

$\qquad = \sqrt{2\pi}\left(1 - \dfrac{1}{\sqrt{e}}\right)$

$\qquad \approx 0.986$

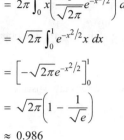

**15.** $p(y) = y, h(y) = 2 - y$

$V = 2\pi \int_0^2 y(2 - y)\, dy$

$\qquad = 2\pi \int_0^2 \left(2y - y^2\right) dy$

$\qquad = 2\pi \left[y^2 - \dfrac{y^3}{3}\right]_0^2 = \dfrac{8\pi}{3}$

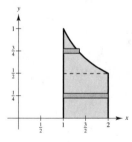

**17.** $p(y) = y$ and $h(y) = 1$ if $0 \le y < \dfrac{1}{2}$.

$p(y) = y$ and $h(y) = \dfrac{1}{y} - 1$ if $\dfrac{1}{2} \le y \le 1$.

$V = 2\pi \int_0^{1/2} y\, dy + 2\pi \int_{1/2}^1 (1 - y)\, dy$

$\qquad = 2\pi \left[\dfrac{y^2}{2}\right]_0^{1/2} + 2\pi \left[y - \dfrac{y^2}{2}\right]_{1/2}^1 = \dfrac{\pi}{4} + \dfrac{\pi}{4} = \dfrac{\pi}{2}$

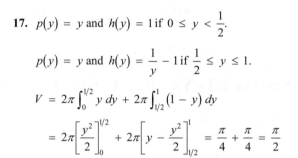

**19.** $p(y) = y, h(y) = \sqrt[3]{y}$

$V = 2\pi \int_0^8 y\sqrt[3]{y}\, dy$

$\qquad = 2\pi \int_0^8 y^{4/3}\, dy$

$\qquad = \left[2\pi\left(\dfrac{3}{7}\right)y^{7/3}\right]_0^8$

$\qquad = \dfrac{6\pi}{7}\left(2^7\right) = \dfrac{768\pi}{7}$

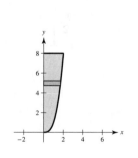

**21.** $p(y) = y, h(y) = (4 - y) - (y) = 4 - 2y$

$$V = 2\pi \int_0^2 y(4 - 2y)\, dy$$

$$= 2\pi \int_0^2 (4y - 2y^2)\, dy$$

$$= 2\pi \left[ 2y^2 - \frac{2}{3}y^3 \right]_0^2$$

$$= 2\pi \left( 8 - \frac{16}{3} \right) = \frac{16\pi}{3}$$

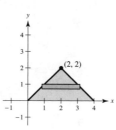

**23.** $p(x) = 5 - x, h(x) = 4x - x^2$

$$V = 2\pi \int_0^4 (5 - x)(4x - x^2)\, dx$$

$$= 2\pi \int_0^4 (x^3 - 9x^2 + 20x)\, dx$$

$$= 2\pi \left[ \frac{x^4}{4} - 3x^3 + 10x^2 \right]_0^4 = 64\pi$$

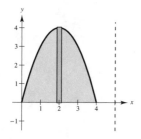

**25.** $p(x) = 4 - x, h(x) = 4x - x^2 - x^2 = 4x - 2x^2$

$$V = 2\pi \int_0^2 (4 - x)(4x - 2x^2)\, dx$$

$$= 2\pi(2) \int_0^2 (x^3 - 6x^2 + 8x)\, dx$$

$$= 4\pi \left[ \frac{x^4}{4} - 2x^3 + 4x^2 \right]_0^2 = 16\pi$$

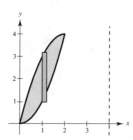

**27.** The shell method would be easier:

$$V = 2\pi \int_0^4 \left[ 4 - (y - 2)^2 \right] y\, dy$$

Using the disk method:

$$V = \pi \int_0^4 \left[ \left( 2 + \sqrt{4 - x} \right)^2 - \left( 2 - \sqrt{4 - x} \right)^2 \right] dx$$

$$\left[ \textbf{Note: } V = \frac{128\pi}{3} \right]$$

**29.** (a) **Disk**

$$R(x) = x^3, r(x) = 0$$

$$V = \pi \int_0^2 x^6\, dx = \pi \left[ \frac{x^7}{7} \right]_0^2 = \frac{128\pi}{7}$$

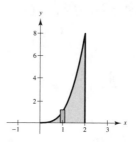

(b) **Shell**

$$p(x) = x, h(x) = x^3$$

$$V = 2\pi \int_0^2 x^4\, dx = 2\pi \left[ \frac{x^5}{5} \right]_0^2 = \frac{64\pi}{5}$$

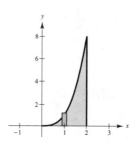

(c) **Shell**

$$p(x) = 4 - x, h(x) = x^3$$

$$V = 2\pi \int_0^2 (4 - x)x^3\, dx$$

$$= 2\pi \int_0^2 (4x^3 - x^4)\, dx$$

$$= 2\pi \left[ x^4 - \frac{1}{5}x^5 \right]_0^2 = \frac{96\pi}{5}$$

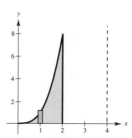

**31.** (a) **Shell**

$$p(y) = y, h(y) = \left(a^{1/2} - y^{1/2}\right)^2$$

$$V = 2\pi \int_0^a y\left(a - 2a^{1/2}y^{1/2} + y\right) dy$$

$$= 2\pi \int_0^a \left(ay - 2a^{1/2}y^{3/2} + y^2\right) dy$$

$$= 2\pi\left[\frac{a}{2}y^2 - \frac{4a^{1/2}}{5}y^{5/2} + \frac{y^3}{3}\right]_0^a$$

$$= 2\pi\left(\frac{a^3}{2} - \frac{4a^3}{5} + \frac{a^3}{3}\right) = \frac{\pi a^3}{15}$$

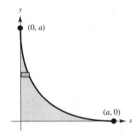

(b) Same as part (a) by symmetry

(c) **Shell**

$$p(x) = a - x, h(x) = \left(a^{1/2} - x^{1/2}\right)^2$$

$$V = 2\pi \int_0^a (a - x)\left(a^{1/2} - x^{1/2}\right)^2 dx$$

$$= 2\pi \int_0^a \left(a^2 - 2a^{3/2}x^{1/2} + 2a^{1/2}x^{3/2} - x^2\right) dx$$

$$= 2\pi\left[a^2 x - \frac{4}{3}a^{3/2}x^{3/2} + \frac{4}{5}a^{1/2}x^{5/2} - \frac{1}{3}x^3\right]_0^a$$

$$= \frac{4\pi a^3}{15}$$

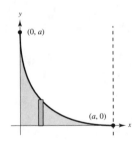

**33.** (a)

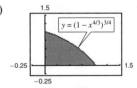

(b) $x^{4/3} + y^{4/3} = 1, x = 0, y = 0$

$$y = \left(1 - x^{4/3}\right)^{3/4}$$

$$V = 2\pi \int_0^1 x\left(1 - x^{4/3}\right)^{3/4} dx \approx 1.5056$$

**35.** (a)

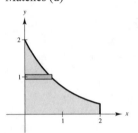

(b) $V = 2\pi \int_2^6 x\sqrt[3]{(x - 2)^2(x - 6)^2}\, dx \approx 187.249$

**37.** $y = 2e^{-x}, y = 0, x = 0, x = 2$

Volume $\approx 7.5$

Matches (d)

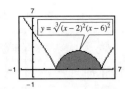

**39.**

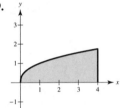

(a) Around $x$-axis: $V = \pi \int_0^4 \left(x^{2/5}\right)^2 dx = \left[\pi\frac{5}{9}x^{9/5}\right]_0^4$

$$= \frac{5}{9}\pi(4)^{9/5} \approx 6.7365\pi$$

(b) Around $y$-axis: $V = 2\pi \int_0^4 x\left(x^{2/5}\right) dx$

$$= \left[2\pi\frac{5}{12}x^{12/5}\right]_0^4 \approx 23.2147\pi$$

(c) Around $x = 4$:

$$V = 2\pi \int_0^4 (4 - x)x^{2/5}\, dx \approx 16.5819\pi$$

So, a < c < b.

**41.** $\pi \int_1^5 (x - 1)\, dx = \pi \int_1^5 \left(\sqrt{x-1}\right)^2 dx$

This integral represents the volume of the solid generated by revolving the region bounded by $y = \sqrt{x-1}$, $y = 0$, and $x = 5$ about the $x$-axis by using the disk method.

$2\pi \int_0^2 y\left[5 - \left(y^2 + 1\right)\right] dy$

represents this same volume by using the shell method.

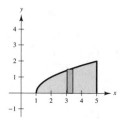

Disk method

**43.** Answers will vary.

(a) The rectangles would be vertical.

(b) The rectangles would be horizontal.

**45.** $p(x) = x,\ h(x) = 2 - \frac{1}{2}x^2$

$V = 2\pi \int_0^2 x\left(2 - \frac{1}{2}x^2\right) dx$

$\quad = 2\pi \int_0^2 \left(2x - \frac{1}{2}x^3\right) dx$

$\quad = 2\pi\left[x^2 - \frac{1}{8}x^4\right]_0^2 = 4\pi$  (total volume)

Now find $x_0$ such that:

$$\pi = 2\pi \int_0^{x_0} \left(2x - \frac{1}{2}x^3\right) dx$$

$$1 = 2\left[x^2 - \frac{1}{8}x^4\right]_0^{x_0}$$

$$1 = 2x_0^2 - \frac{1}{4}x_0^4$$

$x_0^4 - 8x_0^2 + 4 = 0$

$\qquad x_0^2 = 4 \pm 2\sqrt{3}$  (Quadratic Formula)

Take $x_0 = \sqrt{4 - 2\sqrt{3}} \approx 0.73205$, because the other root is too large.

Diameter: $2\sqrt{4 - 2\sqrt{3}} \approx 1.464$

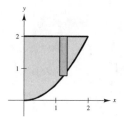

**47.** $V = 4\pi \int_{-1}^1 (2 - x)\sqrt{1 - x^2}\, dx$

$\quad = 8\pi \int_{-1}^1 \sqrt{1 - x^2}\, dx - 4\pi \int_{-1}^1 x\sqrt{1 - x^2}\, dx$

$\quad = 8\pi\left(\frac{\pi}{2}\right) + 2\pi \int_{-1}^1 x\left(1 - x^2\right)^{1/2}(-2)\, dx$

$\quad = 4\pi^2 + \left[2\pi\left(\frac{2}{3}\right)\left(1 - x^2\right)^{3/2}\right]_{-1}^1 = 4\pi^2$

**49.** $2\pi \int_0^2 x^3\, dx = 2\pi \int_0^2 x\left(x^2\right) dx$

(a) Plane region bounded by

$\qquad y = x^2,\ y = 0,\ x = 0,\ x = 2$

(b) Revolved about the $y$-axis

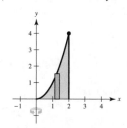

Other answers possible

**51.** $2\pi \int_0^6 (y + 2)\sqrt{6 - y}\, dy$

(a) Plane region bounded by

$\qquad x = \sqrt{6 - y},\ x = 0,\ y = 0$

(b) Revolved around line $y = -2$

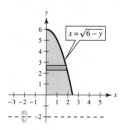

Other answers possible

**53. (a)** $\dfrac{d}{dx}[\sin x - x\cos x + C] = \cos x + x\sin x - \cos x$

$$= x\sin x$$

So, $\displaystyle\int x\sin x\,dx = \sin x - x\cos x + C.$

**(b) (i)** $p(x) = x,\ h(x) = \sin x$

$$V = 2\pi\int_0^{\pi/2} x\sin x\,dx$$

$$= 2\pi[\sin x - x\cos x]_0^{\pi/2}$$

$$= 2\pi[(1 - 0) - 0] = 2\pi$$

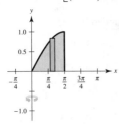

**(ii)** $p(x) = x,\ h(x) = 2\sin x - (-\sin x) = 3\sin x$

$$V = 2\pi\int_0^{\pi} x(3\sin x)\,dx$$

$$= 6\pi\int_0^{\pi} x\sin x\,dx$$

$$= 6\pi[\sin x - x\cos x]_0^{\pi}$$

$$= 6\pi(\pi) = 6\pi^2$$

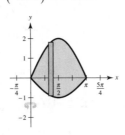

**55. Disk Method**

$$R(y) = \sqrt{r^2 - y^2}$$

$$r(y) = 0$$

$$V = \pi\int_{r-h}^{r}(r^2 - y^2)\,dy$$

$$= \pi\left[r^2 y - \frac{y^3}{3}\right]_{r-h}^{r}$$

$$= \frac{1}{3}\pi h^2(3r - h)$$

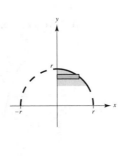

**57. (a)** Area of region $= \displaystyle\int_0^b [ab^n - ax^n]\,dx$

$$= \left[ab^n x - a\frac{x^{n+1}}{n+1}\right]_0^b$$

$$= ab^{n+1} - a\frac{b^{n+1}}{n+1}$$

$$= ab^{n+1}\left(1 - \frac{1}{n+1}\right)$$

$$= ab^{n+1}\left(\frac{n}{n+1}\right)$$

$$R_1(n) = \frac{ab^{n+1}[n/(n+1)]}{(ab^n)b} = \frac{n}{n+1}$$

**(b)** $\displaystyle\lim_{n\to\infty} R_1(n) = \lim_{n\to\infty}\frac{n}{n+1} = 1$

$$\lim_{n\to\infty}(ab^n)b = \infty$$

**(c) Disk Method:**

$$V = 2\pi\int_0^b x(ab^n - ax^n)\,dx$$

$$= 2\pi a\int_0^b (xb^n - x^{n+1})\,dx$$

$$= 2\pi a\left[\frac{b^n}{2}x^2 - \frac{x^{n+2}}{n+2}\right]_0^b$$

$$= 2\pi a\left[\frac{b^{n+2}}{2} - \frac{b^{n+2}}{n+2}\right] = \pi ab^{n+2}\left(\frac{n}{n+2}\right)$$

$$R_2(n) = \frac{\pi ab^{n+2}[n/(n+2)]}{(\pi b^2)(ab^n)} = \left(\frac{n}{n+2}\right)$$

**(d)** $\displaystyle\lim_{n\to\infty} R_2(n) = \lim_{n\to\infty}\left(\frac{n}{n+2}\right) = 1$

$$\lim_{n\to\infty}(\pi b^2)(ab^n) = \infty$$

**(e)** As $n\to\infty$, the graph approaches the line $x = b$.

**59. (a)** $V = 2\pi\displaystyle\int_0^4 xf(x)\,dx = \dfrac{2\pi(40)}{3(4)}\Big[0 + 4(10)(45) + 2(20)(40) + 4(30)(20) + 0\Big] = \dfrac{20\pi}{3}(5800) \approx 121{,}475\ \text{ft}^3$

**(b)** Top line: $y - 50 = \dfrac{40 - 50}{20 - 0}(x - 0) = -\dfrac{1}{2}x \Rightarrow y = -\dfrac{1}{2}x + 50$

Bottom line: $y - 40 = \dfrac{0 - 40}{40 - 20}(x - 20) = -2(x - 20) \Rightarrow y = -2x + 80$

$$V = 2\pi\int_0^{20} x\left(-\frac{1}{2}x + 50\right)dx + 2\pi\int_{20}^{40} x(-2x + 80)\,dx = 2\pi\int_0^{20}\left(-\frac{1}{2}x^2 + 50x\right)dx + 2\pi\int_{20}^{40}(-2x^2 + 80x)\,dx$$

$$= 2\pi\left[-\frac{x^3}{6} + 25x^2\right]_0^{20} + 2\pi\left[-\frac{2x^3}{3} + 40x^2\right]_{20}^{40} = 2\pi\left(\frac{26{,}000}{3}\right) + 2\pi\left(\frac{32{,}000}{3}\right) \approx 121{,}475\ \text{ft}^3$$

(Note that Simpson's Rule is exact for this problem.)

**61.** $V_1 = \pi \int_{1/4}^{c} \frac{1}{x^2}\, dx = \pi\left[-\frac{1}{x}\right]_{1/4}^{c} = \pi\left[-\frac{1}{c} + 4\right] = \frac{4c-1}{c}\pi$

$V_2 = \left[2\pi \int_{1/4}^{c} x\left(\frac{1}{x}\right) dx = 2\pi x\right]_{1/4}^{c} = 2\pi\left(c - \frac{1}{4}\right)$

$V_1 = V_2 \Rightarrow \frac{4c-1}{c}\pi = 2\pi\left(c - \frac{1}{4}\right)$

$\qquad\qquad 4c - 1 = 2c\left(c - \frac{1}{4}\right)$

$\qquad\qquad 4c^2 - 9c + 2 = 0$

$\qquad\qquad (4c-1)(c-2) = 0$

$\qquad\qquad\qquad c = 2 \left(c = \frac{1}{4} \text{ yields no volume.}\right)$

**63.** $y^2 = x(4-x)^2, \quad 0 \le x \le 4$

$y_1 = \sqrt{x(4-x)^2} = (4-x)\sqrt{x}$

$y_2 = -\sqrt{x(4-x)^2} = -(4-x)\sqrt{x}$

(a) $V = \pi \int_0^4 x(4-x)^2\, dx$

$\quad = \pi \int_0^4 \left(x^3 - 8x^2 + 16x\right) dx$

$\quad = \pi\left[\frac{x^4}{4} - \frac{8x^3}{3} + 8x^2\right]_0^4 = \frac{64\pi}{3}$

(b) $V = 4\pi \int_0^4 x(4-x)\sqrt{x}\, dx$

$\quad = 4\pi \int_0^4 \left(4x^{3/2} - x^{5/2}\right) dx$

$\quad = 4\pi\left[\frac{8}{5}x^{5/2} - \frac{2}{7}x^{7/2}\right]_0^4 = \frac{2048\pi}{35}$

(c) $V = 4\pi \int_0^4 (4-x)(4-x)\sqrt{x}\, dx$

$\quad = 4\pi \int_0^4 \left(16\sqrt{x} - 8x^{3/2} + x^{5/2}\right) dx$

$\quad = 4\pi\left[\frac{32}{3}x^{3/2} - \frac{16}{5}x^{5/2} + \frac{2}{7}x^{7/2}\right]_0^4 = \frac{8192\pi}{105}$

# Section 7.4  Arc Length and Surfaces of Revolution

**1.** $(0,0), (8,15)$

(a) $d = \sqrt{(8-0)^2 + (15-0)^2}$

$\quad = \sqrt{64 + 225}$

$\quad = \sqrt{289} = 17$

(b) $y = \frac{15}{8}x$

$\quad y' = \frac{15}{8}$

$\quad s = \int_0^8 \sqrt{1 + \left(\frac{15}{8}\right)^2}\, dx = \int_0^8 \frac{17}{8}\, dx = \left[\frac{17}{8}x\right]_0^8 = 17$

**3.** $y = \frac{2}{3}\left(x^2 + 1\right)^{3/2}$

$y' = \left(x^2 + 1\right)^{1/2}(2x), \quad 0 \le x \le 1$

$1 + (y')^2 = 1 + 4x^2\left(x^2 + 1\right)$

$\qquad\qquad = 4x^4 + 4x^2 + 1 = \left(2x^2 + 1\right)^2$

$s = \int_0^1 \sqrt{1 + (y')^2}\, dx$

$\quad = \int_0^1 \left(2x^2 + 1\right) dx = \left[\frac{2x^3}{3} + x\right]_0^1 = \frac{5}{3}$

**5.** $y = \frac{2}{3}x^{3/2} + 1$

$y' = x^{1/2}, \quad 0 \le x \le 1$

$s = \int_0^1 \sqrt{1 + x}\, dx$

$\quad = \left[\frac{2}{3}(1+x)^{3/2}\right]_0^1 = \frac{2}{3}\left(\sqrt{8} - 1\right) \approx 1.219$

**7.** $y = \dfrac{3}{2}x^{2/3}$

$y' = \dfrac{1}{x^{1/3}}, \quad 1 \le x \le 8$

$s = \displaystyle\int_1^8 \sqrt{1 + \left(\dfrac{1}{x^{1/3}}\right)^2}\, dx = \int_1^8 \sqrt{\dfrac{x^{2/3} + 1}{x^{2/3}}}\, dx$

$\phantom{s}= \dfrac{3}{2}\displaystyle\int_1^8 \sqrt{x^{2/3} + 1}\left(\dfrac{2}{3x^{1/3}}\right) dx = \dfrac{3}{2}\left[\dfrac{2}{3}\left(x^{2/3} + 1\right)^{3/2}\right]_1^8 = 5\sqrt{5} - 2\sqrt{2} \approx 8.352$

**9.** $y = \dfrac{x^5}{10} + \dfrac{1}{6x^3}, \quad 2 \le x \le 5$

$y' = \dfrac{x^4}{2} - \dfrac{1}{2x^4} = \dfrac{1}{2}\left(x^4 - \dfrac{1}{x^4}\right)$

$1 + (y')^2 = 1 + \dfrac{1}{4}\left(x^4 - \dfrac{1}{x^4}\right)^2 = 1 + \dfrac{1}{4}\left(x^8 - 2 + \dfrac{1}{x^8}\right)$

$\phantom{1 + (y')^2}= \dfrac{1}{4}\left(x^8 + 2 + \dfrac{1}{x^8}\right) = \dfrac{1}{4}\left(x^4 + \dfrac{1}{x^4}\right)^2$

$s = \displaystyle\int_2^5 \sqrt{1 + (y')^2}\, dx = \int_2^5 \dfrac{1}{2}\left(x^4 + \dfrac{1}{x^4}\right) dx$

$\phantom{s}= \dfrac{1}{2}\left[\dfrac{x^5}{5} - \dfrac{1}{3x^3}\right]_2^5 = \dfrac{1}{2}\left[\left(625 - \dfrac{1}{375}\right) - \left(\dfrac{32}{5} - \dfrac{1}{24}\right)\right] = \dfrac{618639}{2000} \approx 309.320$

**11.** $y = \ln(\sin x), \quad \left[\dfrac{\pi}{4}, \dfrac{3\pi}{4}\right]$

$y' = \dfrac{1}{\sin x}\cos x = \cot x$

$1 + (y')^2 = 1 + \cot^2 x = \csc^2 x$

$s = \displaystyle\int_{\pi/4}^{3\pi/4} \csc x\, dx$

$\phantom{s}= \Big[\ln|\csc x - \cot x|\Big]_{\pi/4}^{3\pi/4}$

$\phantom{s}= \ln\left(\sqrt{2} + 1\right) - \ln\left(\sqrt{2} - 1\right) \approx 1.763$

**13.** $y = \dfrac{1}{2}\left(e^x + e^{-x}\right)$

$y' = \dfrac{1}{2}\left(e^x - e^{-x}\right), \quad [0, 2]$

$1 + (y')^2 = \left[\dfrac{1}{2}\left(e^x + e^{-x}\right)\right]^2, \quad [0, 2]$

$s = \displaystyle\int_0^2 \sqrt{\left[\dfrac{1}{2}\left(e^x + e^{-x}\right)\right]^2}\, dx$

$\phantom{s}= \dfrac{1}{2}\displaystyle\int_0^2 \left(e^x + e^{-x}\right) dx$

$\phantom{s}= \dfrac{1}{2}\left[e^x - e^{-x}\right]_0^2 = \dfrac{1}{2}\left(e^2 - \dfrac{1}{e^2}\right) \approx 3.627$

**15.** $x = \dfrac{1}{3}\left(y^2 + 2\right)^{3/2}, \quad 0 \le y \le 4$

$\dfrac{dx}{dy} = y\left(y^2 + 2\right)^{1/2}$

$s = \displaystyle\int_0^4 \sqrt{1 + y^2\left(y^2 + 2\right)}\, dy$

$\phantom{s}= \displaystyle\int_0^4 \sqrt{y^4 + 2y^2 + 1}\, dy$

$\phantom{s}= \displaystyle\int_0^4 \left(y^2 + 1\right) dy$

$\phantom{s}= \left[\dfrac{y^3}{3} + y\right]_0^4 = \dfrac{64}{3} + 4 = \dfrac{76}{3}$

**17.** (a) $y = 4 - x^2, \quad 0 \le x \le 2$

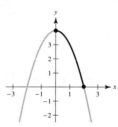

(b) $\qquad y' = -2x$

$\qquad 1 + (y')^2 = 1 + 4x^2$

$\qquad L = \displaystyle\int_0^2 \sqrt{1 + 4x^2}\, dx$

(c) $L \approx 4.647$

**19.** (a) $y = \dfrac{1}{x}$,   $1 \le x \le 3$

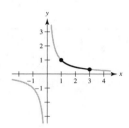

(b)      $y' = -\dfrac{1}{x^2}$

$$1 + \left(y'\right)^2 = 1 + \dfrac{1}{x^4}$$

$$L = \int_1^3 \sqrt{1 + \dfrac{1}{x^4}} \; dx$$

(c)  $L \approx 2.147$

**21.** (a) $y = \sin x$,   $0 \le x \le \pi$

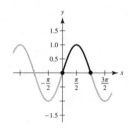

(b)      $y' = \cos x$

$$1 + \left(y'\right)^2 = 1 + \cos^2 x$$

$$L = \int_0^\pi \sqrt{1 + \cos^2 x} \; dx$$

(c)  $L \approx 3.820$

**23.** (a) $x = e^{-y}$,   $0 \le y \le 2$

$y = -\ln x$

$1 \ge x \ge e^{-2} \approx 0.135$

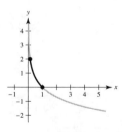

(b)      $y' = -\dfrac{1}{x}$

$$1 + \left(y'\right)^2 = 1 + \dfrac{1}{x^2}$$

$$L = \int_{e^{-2}}^1 \sqrt{1 + \dfrac{1}{x^2}} \; dx$$

(c)  $L \approx 2.221$

Alternatively, you can do all the computations with respect to $y$.

(a)  $x = e^{-y}$,   $0 \le y \le 2$

(b)      $\dfrac{dx}{dy} = -e^{-y}$

$$1 + \left(\dfrac{dx}{dy}\right)^2 = 1 + e^{-2y}$$

$$L = \int_0^2 \sqrt{1 + e^{-2y}} \; dy$$

(c)  $L \approx 2.221$

**25.** (a) $y = 2 \arctan x$,   $0 \le x \le 1$

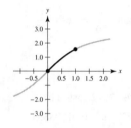

(b)  $y' = \dfrac{2}{1 + x^2}$

$$L = \int_0^1 \sqrt{1 + \dfrac{4}{\left(1 + x^2\right)^2}} \; dx$$

(c)  $L \approx 1.871$

**27.** $\displaystyle\int_0^2 \sqrt{1 + \left[\dfrac{d}{dx}\left(\dfrac{5}{x^2 + 1}\right)\right]^2} \; dx$

$s \approx 5$

Matches (b)

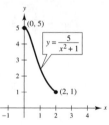

**29.** $y = x^3$, $[0, 4]$

(a) $d = \sqrt{(4-0)^2 + (64-0)^2} \approx 64.125$

(b) $d = \sqrt{(1-0)^2 + (1-0)^2} + \sqrt{(2-1)^2 + (8-1)^2} + \sqrt{(3-2)^2 + (27-8)^2} + \sqrt{(4-3)^2 + (64-27)^2}$

$\approx 64.525$

(c) $s = \int_0^4 \sqrt{1 + (3x^2)^2} \, dx = \int_0^4 \sqrt{1 + 9x^4} \, dx \approx 64.666$    (Simpson's Rule, $n = 10$)

(d) $64.672$

**31.** $y = 20 \cosh \dfrac{x}{20}$, $-20 \le x \le 20$

$y' = \sinh \dfrac{x}{20}$

$1 + (y')^2 = 1 + \sinh^2 \dfrac{x}{20} = \cosh^2 \dfrac{x}{20}$

$L = \int_{-20}^{20} \cosh \dfrac{x}{20} \, dx = 2 \int_0^{20} \cosh \dfrac{x}{20} \, dx = \left[ 2(20) \sinh \dfrac{x}{20} \right]_0^{20} = 40 \sinh(1) \approx 47.008 \text{ m}$

**33.** $y = 693.8597 - 68.7672 \cosh 0.0100333x$

$y' = -0.6899619478 \sinh 0.0100333x$

$s = \int_{-299.2239}^{299.2239} \sqrt{1 + (-0.6899619478 \sinh 0.0100333x)^2} \, dx \approx 1480$

(Use Simpson's Rule with $n = 100$ or a graphing utility.)

**35.** $y = \sqrt{9 - x^2}$

$y' = \dfrac{-x}{\sqrt{9 - x^2}}$

$1 + (y')^2 = \dfrac{9}{9 - x^2}$

$s = \int_0^2 \sqrt{\dfrac{9}{9 - x^2}} \, dx = \int_0^2 \dfrac{3}{\sqrt{9 - x^2}} \, dx$

$= \left[ 3 \arcsin \dfrac{x}{3} \right]_0^2 = 3 \left( \arcsin \dfrac{2}{3} - \arcsin 0 \right)$

$= 3 \arcsin \dfrac{2}{3} \approx 2.1892$

**37.** $y = \dfrac{x^3}{3}$

$y' = x^2$, $[0, 3]$

$S = 2\pi \int_0^3 \dfrac{x^3}{3} \sqrt{1 + x^4} \, dx$

$= \dfrac{\pi}{6} \int_0^3 (1 + x^4)^{1/2} (4x^3) \, dx$

$= \left[ \dfrac{\pi}{9} (1 + x^4)^{3/2} \right]_0^3$

$= \dfrac{\pi}{9} (82\sqrt{82} - 1) \approx 258.85$

**39.** $y = \dfrac{x^3}{6} + \dfrac{1}{2x}$

$y' = \dfrac{x^2}{2} - \dfrac{1}{2x^2}$

$1 + (y')^2 = \left( \dfrac{x^2}{2} + \dfrac{1}{2x^2} \right)^2$, $[1, 2]$

$S = 2\pi \int_1^2 \left( \dfrac{x^3}{6} + \dfrac{1}{2x} \right) \left( \dfrac{x^2}{2} + \dfrac{1}{2x^2} \right) dx$

$= 2\pi \int_1^2 \left( \dfrac{x^5}{12} + \dfrac{x}{3} + \dfrac{1}{4x^3} \right) dx$

$= 2\pi \left[ \dfrac{x^6}{72} + \dfrac{x^2}{6} - \dfrac{1}{8x^2} \right]_1^2 = \dfrac{47\pi}{16}$

**41.** $y = \sqrt{4 - x^2}$

$y' = \dfrac{1}{2}(4 - x^2)^{-1/2}(-2x) = \dfrac{-x}{\sqrt{4 - x^2}}$, $-1 \le x \le 1$

$1 + (y')^2 = 1 + \dfrac{x^2}{4 - x^2} = \dfrac{4}{4 - x^2}$

$S = 2\pi \int_{-1}^1 \sqrt{4 - x^2} \cdot \sqrt{\dfrac{4}{4 - x^2}} \, dx$

$= 4\pi \int_{-1}^1 dx = 4\pi [x]_{-1}^1 = 8\pi$

**43.** $y = \sqrt[3]{x} + 2$

$$y' = \frac{1}{3x^{2/3}}, \quad [1, 8]$$

$$S = 2\pi \int_1^8 x\sqrt{1 + \frac{1}{9x^{4/3}}} \, dx$$

$$= \frac{2\pi}{3} \int_1^8 x^{1/3}\sqrt{9x^{4/3} + 1} \, dx$$

$$= \frac{\pi}{18} \int_1^8 \left(9x^{4/3} + 1\right)^{1/2}\left(12x^{1/3}\right) dx$$

$$= \left[\frac{\pi}{27}\left(9x^{4/3} + 1\right)^{3/2}\right]_1^8$$

$$= \frac{\pi}{27}\left(145\sqrt{145} - 10\sqrt{10}\right) \approx 199.48$$

**45.** $\qquad y = 1 - \frac{x^2}{4}$

$$y' = -\frac{x}{2}, \quad 0 \le x \le 2$$

$$1 + \left(y'\right)^2 = 1 + \frac{x^2}{4} = \frac{4 + x^2}{4}$$

$$S = 2\pi \int_0^2 x\sqrt{\frac{4 + x^2}{4}} \, dx$$

$$= \pi \int_0^2 x\sqrt{4 + x^2} \, dx$$

$$= \frac{1}{2}\pi \int_0^2 \left(4 + x^2\right)^{1/2}(2x) \, dx$$

$$= \frac{1}{2}\pi \left[\frac{2}{3}\left(4 + x^2\right)^{3/2}\right]_0^2$$

$$= \frac{\pi}{3}\left(8^{3/2} - 4^{3/2}\right)$$

$$= \frac{\pi}{3}\left(16\sqrt{2} - 8\right) \approx 15.318$$

**47.** $y = \sin x$

$$y' = \cos x, \quad [0, \pi]$$

$$S = 2\pi \int_0^\pi \sin x\sqrt{1 + \cos^2 x} \, dx \approx 14.4236$$

**49.** A rectifiable curve is one that has a finite arc length.

**51.** The precalculus formula is the surface area formula for the lateral surface of the frustum of a right circular cone. The formula is $S = 2\pi r L$, where $r = \frac{1}{2}\left(r_1 + r_2\right)$, which is the average radius of the frustum, and $L$ is the length of a line segment on the frustum. The representative element is

$$2\pi f(d_i)\sqrt{\Delta x_i^2 + \Delta y_i^2} = 2\pi f(d_i)\sqrt{1 + \left(\frac{\Delta y_i}{\Delta x_i}\right)^2} \, \Delta x_i.$$

**53.** (a)

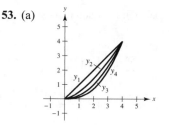

(b) $y_1, y_2, y_3, y_4$

(c) $y_1' = 1, \quad s_1 = \int_0^4 \sqrt{2} \, dx \approx 5.657$

$$y_2' = \frac{3}{4}x^{1/2}, \quad s_2 = \int_0^4 \sqrt{1 + \frac{9x}{16}} \, dx \approx 5.759$$

$$y_3' = \frac{1}{2}x, \quad s_3 = \int_0^4 \sqrt{1 + \frac{x^2}{4}} \, dx \approx 5.916$$

$$y_4' = \frac{5}{16}x^{3/2}, \quad s_4 = \int_0^4 \sqrt{1 + \frac{25}{256}x^3} \, dx \approx 6.063$$

**55.** $\qquad y = \frac{3x}{4}, \quad y' = \frac{3}{4}$

$$1 + \left(y'\right)^2 = 1 + \frac{9}{16} = 25/16$$

$$S = 2\pi \int_0^4 x\sqrt{\frac{25}{16}} \, dx = \frac{5\pi}{2}\left[\frac{x^2}{2}\right]_0^4 = 20\pi$$

**57.** $\qquad y = \sqrt{9 - x^2}$

$$y' = \frac{-x}{\sqrt{9 - x^2}}$$

$$\sqrt{1 + \left(y'\right)^2} = \frac{3}{\sqrt{9 - x^2}}$$

$$S = 2\pi \int_0^2 \frac{3x}{\sqrt{9 - x^2}} \, dx$$

$$= -3\pi \int_0^2 \frac{-2x}{\sqrt{9 - x^2}} \, dx$$

$$= \left[-6\pi\sqrt{9 - x^2}\right]_0^2$$

$$= 6\pi\left(3 - \sqrt{5}\right) \approx 14.40$$

See figure in Exercise 58.

**59.** (a) Approximate the volume by summing six disks of thickness 3 and circumference $C_i$ equal to the average of the given circumferences:

$$V \approx \sum_{i=1}^{6} \pi r_i^2(3) = \sum_{i=1}^{6} \pi \left(\frac{C_i}{2\pi}\right)^2 (3) = \frac{3}{4\pi} \sum_{i=1}^{6} C_i^2$$

$$= \frac{3}{4\pi}\left[\left(\frac{50 + 65.5}{2}\right)^2 + \left(\frac{65.5 + 70}{2}\right)^2 + \left(\frac{70 + 66}{2}\right)^2 + \left(\frac{66 + 58}{2}\right)^2 + \left(\frac{58 + 51}{2}\right)^2 + \left(\frac{51 + 48}{2}\right)^2\right]$$

$$= \frac{3}{4\pi}\left[57.75^2 + 67.75^2 + 68^2 + 62^2 + 54.5^2 + 49.5^2\right] = \frac{3}{4\pi}(21813.625) = 5207.62 \text{ in.}^3$$

(b) The lateral surface area of a frustum of a right circular cone is $\pi s(R + r)$. For the first frustum:

$$S_1 \approx \pi\left[3^2 + \left(\frac{65.5 - 50}{2\pi}\right)^2\right]^{1/2}\left[\frac{50}{2\pi} + \frac{65.5}{2\pi}\right]$$

$$= \left(\frac{50 + 65.5}{2}\right)\left[9 + \left(\frac{65.5 - 50}{2\pi}\right)^2\right]^{1/2}.$$

Adding the six frustums together:

$$S \approx \left(\frac{50 + 65.5}{2}\right)\left[9 + \left(\frac{15.5}{2\pi}\right)^2\right]^{1/2} + \left(\frac{65.5 + 70}{2}\right)\left[9 + \left(\frac{4.5}{2\pi}\right)^2\right]^{1/2}$$

$$+ \left(\frac{70 + 66}{2}\right)\left[9 + \left(\frac{4}{2\pi}\right)^2\right]^{1/2} + \left(\frac{66 + 58}{2}\right)\left[9 + \left(\frac{8}{2\pi}\right)^2\right]^{1/2}$$

$$+ \left(\frac{58 + 51}{2}\right)\left[9 + \left(\frac{7}{2\pi}\right)^2\right]^{1/2} + \left(\frac{51 + 48}{2}\right)\left[9 + \left(\frac{3}{2\pi}\right)^2\right]^{1/2}$$

$$\approx 224.30 + 208.96 + 208.54 + 202.06 + 174.41 + 150.37 = 1168.64$$

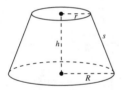

(c) $r = 0.00401y^3 - 0.1416y^2 + 1.232y + 7.943$

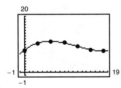

(d) $V = \int_0^{18} \pi r^2 \, dy \approx 5275.9 \text{ in.}^3$

$$S = \int_0^{18} 2\pi r(y)\sqrt{1 + r'(y)^2} \, dy \approx 1179.5 \text{ in.}^2$$

**61. (a)** $V = \pi \int_1^b \frac{1}{x^2}\, dx = \left[ -\frac{\pi}{x} \right]_1^b = \pi\left( 1 - \frac{1}{b} \right)$

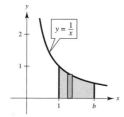

**(b)** $S = 2\pi \int_1^b \frac{1}{x} \sqrt{1 + \left( -\frac{1}{x^2} \right)^2}\, dx = 2\pi \int_1^b \frac{1}{x}\sqrt{1 + \frac{1}{x^4}}\, dx = 2\pi \int_1^b \frac{\sqrt{x^4 + 1}}{x^3}\, dx$

**(c)** $\lim\limits_{b \to \infty} V = \lim\limits_{b \to \infty} \pi\left( 1 - \frac{1}{b} \right) = \pi$

**(d)** Because $\dfrac{\sqrt{x^4 + 1}}{x^3} > \dfrac{\sqrt{x^4}}{x^3} = \dfrac{1}{x} > 0$ on $[1, b]$,

you have $\int_1^b \dfrac{\sqrt{x^4 + 1}}{x^3}\, dx > \int_1^b \dfrac{1}{x}\, dx = \left[ \ln x \right]_1^b = \ln b$ and $\lim\limits_{b \to \infty} \ln b \to \infty$.

So, $\lim\limits_{b \to \infty} 2\pi \int_1^b \dfrac{\sqrt{x^4 + 1}}{x^3}\, dx = \infty$.

**63. (a)** $\dfrac{x^2}{9} + \dfrac{y^2}{4} = 1$

Ellipse: $y_1 = 2\sqrt{1 - \dfrac{x^2}{9}}$

$y_2 = -2\sqrt{1 - \dfrac{x^2}{9}}$

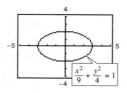

**(b)** $y = 2\sqrt{1 - \dfrac{x^2}{9}}, \quad 0 \le x \le 3$

$y' = 2\left( \dfrac{1}{2} \right)\left( 1 - \dfrac{x^2}{9} \right)^{-1/2}\left( \dfrac{-2x}{9} \right) = \dfrac{-2x}{9\sqrt{1 - \left( x^2/9 \right)}} = \dfrac{-2x}{3\sqrt{9 - x^2}}$

$L = \int_0^3 \sqrt{1 + \dfrac{4x^2}{81 - 9x^2}}\, dx$

**(c)** You cannot evaluate this definite integral, because the integrand is not defined at $x = 3$. Simpson's Rule will not work for the same reason. Also, the integrand does not have an elementary antiderivative.

**65.** $y = \dfrac{1}{3}\left( x^{3/2} - 3x^{1/2} + 2 \right)$

When $x = 0$, $y = \dfrac{2}{3}$. So, the fleeing object has traveled $\dfrac{2}{3}$ unit when it is caught.

$y' = \dfrac{1}{3}\left( \dfrac{3}{2}x^{1/2} - \dfrac{3}{2}x^{-1/2} \right) = \left( \dfrac{1}{2} \right)\dfrac{x - 1}{x^{1/2}}$

$1 + \left( y' \right)^2 = 1 + \dfrac{(x - 1)^2}{4x} = \dfrac{(x + 1)^2}{4x}$

$s = \int_0^1 \dfrac{x + 1}{2x^{1/2}}\, dx = \dfrac{1}{2}\int_0^1 \left( x^{1/2} + x^{-1/2} \right)\, dx = \dfrac{1}{2}\left[ \dfrac{2}{3}x^{3/2} + 2x^{1/2} \right]_0^1 = \dfrac{4}{3} = 2\left( \dfrac{2}{3} \right)$

The pursuer has traveled twice the distance that the fleeing object has traveled when it is caught.

**67.** $x^{2/3} + y^{2/3} = 4$

$$y^{2/3} = 4 - x^{2/3}$$

$$y = \left(4 - x^{2/3}\right)^{3/2}, \quad 0 \le x \le 8$$

$$y' = \frac{3}{2}\left(4 - x^{2/3}\right)^{1/2}\left(-\frac{2}{3}x^{-1/3}\right) = \frac{-\left(4 - x^{2/3}\right)^{1/2}}{x^{1/3}}$$

$$1 + \left(y'\right)^2 = 1 + \frac{4 - x^{2/3}}{x^{2/3}} = \frac{4}{x^{2/3}}$$

$$S = 2\pi \int_0^8 \left(4 - x^{2/3}\right)^{3/2}\sqrt{\frac{4}{x^{2/3}}}\,dx = 4\pi\int_0^8 \frac{\left(4 - x^{2/3}\right)^{3/2}}{x^{1/3}}\,dx = \left[-\frac{12\pi}{5}\left(4 - x^{2/3}\right)^{5/2}\right]_0^8 = \frac{384\pi}{5}$$

[Surface area of portion above the *x*-axis]

**69.** $y = kx^2, \ y' = 2kx$

$$1 + \left(y'\right)^2 = 1 + 4k^2x^2$$

$$h = kw^2 \Rightarrow k = \frac{h}{w^2} \Rightarrow 1 + \left(y'\right) = 1 + \frac{4h^2}{w^4}x^2$$

By symmetry, $C = 2\int_0^w \sqrt{1 + \frac{4h^2}{w^4}x^2}\,dx.$

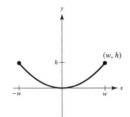

**71.**

$$y = f(x) = \cosh x$$

$$y' = \sinh x$$

$$1 + \left(y'\right)^2 = 1 + \sinh^2 x = \cosh^2 x$$

$$\text{Area} = \int_0^t \cosh x\,dx = \left[\sinh x\right]_0^t = \sinh t$$

$$\text{Arc length} = \int_0^t \sqrt{1 + \left(y'\right)^2}\,dx$$

$$= \int_0^t \cosh x\,dx = \sinh x\bigg]_0^t$$

$$= \sinh t.$$

Another curve with this property is $g(x) = 1$.

$$\text{Area} = \int_0^t dx = t$$

$$\text{Arc length} = t$$

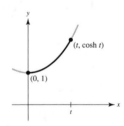

# Section 7.5 Work

**1.** $W = Fd = (100)(20) = 2000$ ft-lb

**3.** $W = Fd = (112)(8) = 896$ joules (Newton-meters)

**5.** $F(x) = kx$

$$5 = k(3)$$

$$k = \frac{5}{3}$$

$$F(x) = \frac{5}{3}x$$

$$W = \int_0^7 F(x)\,dx = \int_0^7 \frac{5}{3}x\,dx = \left[\frac{5}{6}x^2\right]_0^7 = \frac{245}{6}\ \text{in.-lb}$$

$$\approx 40.833\ \text{in.-lb} \approx 3.403\ \text{ft-lb}$$

**7.** $F(x) = kx$

$$250 = k(30) \Rightarrow k = \frac{25}{3}$$

$$W = \int_{20}^{50} F(x)\,dx$$

$$= \int_{20}^{50} \frac{25}{3}x\,dx = \frac{25x^2}{6}\bigg]_{20}^{50}$$

$$= 8750\ \text{n-cm}$$

$$= 87.5\ \text{joules or Nm}$$

**9.** $F(x) = kx$

$20 = k(9)$

$k = \frac{20}{9}$

$W = \int_0^{12} \frac{20}{9} x \, dx = \left[ \frac{10}{9} x^2 \right]_0^{12} = 160 \text{ in.-lb} = \frac{40}{3} \text{ ft-lb}$

**11.** $W = 18 = \int_0^{1/3} kx \, dx = \frac{kx^2}{2} \bigg]_0^{1/3} = \frac{k}{18} \Rightarrow k = 324$

$W = \int_{1/3}^{7/12} 324x \, dx = \left[ 162x^2 \right]_{1/3}^{7/12} = 37.125 \text{ ft-lb}$

$\left[ \textbf{Note: } 4 \text{ inches} = \frac{1}{3} \text{ foot} \right]$

**13.** Assume that Earth has a radius of 4000 miles.

$F(x) = \frac{k}{x^2}$

$5 = \frac{k}{(4000)^2}$

$k = 80{,}000{,}000$

$F(x) = \frac{80{,}000{,}000}{x^2}$

(a) $W = \int_{4000}^{4100} \frac{80{,}000{,}000}{x^2} \, dx = \left[ \frac{-80{,}000{,}000}{x} \right]_{4000}^{4100}$

$\approx 487.8 \text{ mi-tons} \approx 5.15 \times 10^9 \text{ ft-lb}$

(b) $W = \int_{4000}^{4300} \frac{80{,}000{,}000}{x^2} \, dx$

$\approx 1395.3 \text{ mi-ton} \approx 1.47 \times 10^{10} \text{ ft-ton}$

**15.** Assume that Earth has a radius of 4000 miles.

$F(x) = \frac{k}{x^2}$

$10 = \frac{k}{(4000)^2}$

$k = 160{,}000{,}000$

$F(x) = \frac{160{,}000{,}000}{x^2}$

(a) $W = \int_{4000}^{15{,}000} \frac{160{,}000{,}000}{x^2} \, dx = \left[ -\frac{160{,}000{,}000}{x} \right]_{4000}^{15{,}000} \approx -10{,}666.667 + 40{,}000$

$= 29{,}333.333 \text{ mi-ton}$

$\approx 2.93 \times 10^4 \text{ mi-ton}$

$\approx 3.10 \times 10^{11} \text{ ft-lb}$

(b) $W = \int_{4000}^{26{,}000} \frac{160{,}000{,}000}{x^2} \, dx = \left[ -\frac{160{,}000{,}000}{x} \right]_{4000}^{26{,}000} \approx -6{,}153.846 + 40{,}000$

$= 33{,}846.154 \text{ mi-ton}$

$\approx 3.38 \times 10^4 \text{ mi-ton}$

$\approx 3.57 \times 10^{11} \text{ ft-lb}$

**17.** Weight of each layer: $62.4(20) \Delta y$

Distance: $4 - y$

(a) $W = \int_2^4 62.4(20)(4 - y) \, dy = \left[ 4992y - 624y^2 \right]_2^4 = 2496 \text{ ft-lb}$

(b) $W = \int_0^4 62.4(20)(4 - y) \, dy = \left[ 4992y - 624y^2 \right]_0^4 = 9984 \text{ ft-lb}$

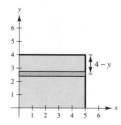

**19.** Volume of disk: $\pi(2)^2 \, \Delta y = 4\pi \, \Delta y$

Weight of disk of water: $9800(4\pi) \, \Delta y$

Distance the disk of water is moved: $5 - y$

$$W = \int_0^4 (5 - y)(9800)4\pi \, dy = 39{,}200\pi \int_0^4 (5 - y) \, dy$$

$$= 39{,}200\pi \left[ 5y - \frac{y^2}{2} \right]_0^4$$

$$= 39{,}200\pi(12) = 470{,}400\pi \text{ newton–meters}$$

**21.** Volume of disk: $\pi \left( \frac{2}{3}y \right)^2 \Delta y$

Weight of disk: $62.4\pi \left( \frac{2}{3}y \right)^2 \Delta y$

Distance: $6 - y$

$$W = \frac{4(62.4)\pi}{9} \int_0^6 (6 - y)y^2 \, dy$$

$$= \frac{4}{9}(62.4)\pi \left[ 2y^3 - \frac{1}{4}y^4 \right]_0^6$$

$$= 2995.2\pi \text{ ft-lb}$$

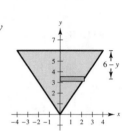

**23.** Volume of disk: $\pi \left( \sqrt{36 - y^2} \right)^2 \Delta y$

Weight of disk: $62.4\pi \left( 36 - y^2 \right) \Delta y$

Distance: $y$

$$W = 62.4\pi \int_0^6 y\left(36 - y^2\right) dy = 62.4\pi \int_0^6 \left(36y - y^3\right) dy$$

$$= 62.4\pi \left[ 18y^2 - \frac{1}{4}y^4 \right]_0^6 = 20{,}217.6\pi \text{ ft-lb}$$

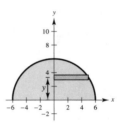

**25.** Volume of layer: $V = lwh = 4(2)\sqrt{(9/4) - y^2} \, \Delta y$

Weight of layer: $W = 42(8)\sqrt{(9/4) - y^2} \, \Delta y$

Distance: $\frac{13}{2} - y$

$$W = \int_{-1.5}^{1.5} 42(8)\sqrt{\frac{9}{4} - y^2}\left(\frac{13}{2} - y\right) dy$$

$$= 336\left[ \frac{13}{2} \int_{-1.5}^{1.5} \sqrt{\frac{9}{4} - y^2} \, dy - \int_{-1.5}^{1.5} \sqrt{\frac{9}{4} - y^2}\, y \, dy \right]$$

The second integral is zero because the integrand is odd and the limits of integration are symmetric to the origin. The first integral represents the area of a semicircle of radius

$\frac{3}{2}$. So, the work is

$$W = 336\left(\tfrac{13}{2}\right)\pi\left(\tfrac{3}{2}\right)^2\left(\tfrac{1}{2}\right) = 2457\pi \text{ ft-lb.}$$

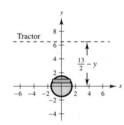

**27.** Weight of section of chain: $3 \, \Delta y$

Distance: $20 - y.$ $\quad \Delta W = (\text{force increment})(\text{distance}) = (3 \, \Delta y)(20 - y)$

$$W = \int_0^{20} (20 - y)3 \, dy = 3\left[ 20y - \frac{y^2}{2} \right]_0^{20} = 3\left[ 400 - \frac{400}{2} \right] = 600 \text{ ft-lb}$$

**29.** The lower 10 feet of fence are raised 10 feet with a constant force.

$$W_1 = 3(10)(10) = 300 \text{ ft-lb}$$

The top 10 feet are raised with a variable force.

Weight of section: $3 \Delta y$

Distance: $10 - y$

$$W_2 = \int_0^{10} 3(10 - y) \, dy = 3\left[10y - \frac{y^2}{2}\right]_0^{10} = 150 \text{ ft-lb}$$

$$W = W_1 + W_2 = 300 + 150 = 450 \text{ ft-lb}$$

**31.** Weight of section of chain: $3 \Delta y$

Distance: $15 - 2y$

$$W = 3\int_0^{7.5} (15 - 2y) \, dy = \left[-\frac{3}{4}(15 - 2y)^2\right]_0^{7.5}$$

$$= \frac{3}{4}(15)^2 = 168.75 \text{ ft-lb}$$

**33.** If an object is moved a distance $D$ in the direction of an applied constant force $F$, then the work $W$ done by the force is defined as force times distance, $W = FD$.

**35.** (a) requires more work. In part (b) no work is done because the books are not moved:

$W = $ force $\times$ distance

**37.** (a) $W = \int_0^9 6 \, dx = 54 \text{ ft-lb}$

(b) $W = \int_0^7 20 \, dx + \int_7^9 (-10x + 90) \, dx = 140 + 20$

$$= 160 \text{ ft-lb}$$

(c) $W = \int_0^9 \frac{1}{27}x^2 \, dx = \frac{x^3}{81}\Big]_0^9 = 9 \text{ ft-lb}$

(d) $W = \int_0^9 \sqrt{x} \, dx = \frac{2}{3}x^{3/2}\Big]_0^9 = \frac{2}{3}(27) = 18 \text{ ft-lb}$

**39.** $p = \dfrac{k}{V}$

$$1000 = \frac{k}{2}$$

$$k = 2000$$

$$W = \int_2^3 \frac{2000}{V} \, dV$$

$$= \left[2000 \ln|V|\right]_2^3 = 2000 \ln\left(\frac{3}{2}\right) \approx 810.93 \text{ ft-lb}$$

**41.** $F(x) = \dfrac{k}{(2 - x)^2}$

$$W = \int_{-2}^1 \frac{k}{(2 - x)^2} \, dx = \left[\frac{k}{2 - x}\right]_{-2}^1 = k\left(1 - \frac{1}{4}\right)$$

$$= \frac{3k}{4} \text{ (units of work)}$$

**43.** $W = \int_0^5 1000\left[1.8 - \ln(x + 1)\right] dx \approx 3249.44 \text{ ft-lb}$

**45.** $W = \int_0^5 100x\sqrt{125 - x^3} \, dx \approx 10{,}330.3 \text{ ft-lb}$

## Section 7.6    Moments, Centers of Mass, and Centroids

**1.** $\bar{x} = \dfrac{7(-5) + 3(0) + 5(3)}{7 + 3 + 5} = \dfrac{-20}{15} = -\dfrac{4}{3}$

**3.** $\bar{x} = \dfrac{1(7) + 1(8) + 1(12) + 1(15) + 1(18)}{1 + 1 + 1 + 1 + 1} = 12$

**5.** (a) Add 4 to each $x$-value because each point is translated 4 units to the right.

$$\bar{x} = \frac{1(7 + 4) + 1(8 + 4) + 1(12 + 4) + 1(15 + 4) + 1(18 + 4)}{1 + 1 + 1 + 1 + 1} = \frac{80}{5} = 16$$

**Note:** From Exercise 3, $12 + 4 = 16$.

(b) Subtract 2 from each $x$-value because each point is translated 2 units to the left.

$$\bar{x} = \frac{12(-6 - 2) + 1(-4 - 2) + 6(-2 - 2) + 3(0 - 2) + 11(8 - 2)}{12 + 1 + 6 + 3 + 11} = \frac{-66}{33} = -2$$

**Note:** From Exercise 4, $0 - 2 = -2$.

**7.** $48x = 72(L - x) = 72(10 - x)$

$\quad 48x = 720 - 72x$

$\quad 120x = 720$

$\qquad x = 6 \text{ ft}$

**9.** $\quad \overline{x} = \dfrac{5(2) + 1(-3) + 3(1)}{5 + 1 + 3} = \dfrac{10}{9}$

$\quad \overline{y} = \dfrac{5(2) + 1(1) + 3(-4)}{5 + 1 + 3} = -\dfrac{1}{9}$

$\quad (\overline{x}, \overline{y}) = \left(\dfrac{10}{9}, -\dfrac{1}{9}\right)$

**11.** $\quad \overline{x} = \dfrac{12(2) + 6(-1) + (9/2)(6) + 15(2)}{12 + 6 + (9/2) + 15} = \dfrac{75}{37.5} = 2$

$\quad \overline{y} = \dfrac{12(3) + 6(5) + (9/2)(8) + 15(-2)}{12 + 6 + (9/2) + 15} = \dfrac{72}{37.5} = \dfrac{48}{25}$

$\quad (\overline{x}, \overline{y}) = \left(2, \dfrac{48}{25}\right)$

**13.** $\quad m = \rho \displaystyle\int_0^2 \dfrac{x}{2}\, dx = \left[\rho \dfrac{x^2}{4}\right]_0^2 = \rho$

$\quad M_x = \rho \displaystyle\int_0^2 \dfrac{1}{2}\left(\dfrac{x}{2}\right)^2 dx$

$\qquad = \dfrac{\rho}{8}\left[\dfrac{x^3}{3}\right]_0^2 = \dfrac{\rho}{3}$

$\quad \overline{y} = \dfrac{M_x}{m} = \dfrac{\rho/3}{\rho} = \dfrac{1}{3}$

$\quad M_y = \rho \displaystyle\int_0^2 x\left(\dfrac{x}{2}\right) dx = \dfrac{\rho}{2}\left[\dfrac{x^3}{3}\right]_0^2 = \dfrac{4}{3}\rho$

$\quad \overline{x} = \dfrac{M_y}{m} = \dfrac{4/3\rho}{\rho} = \dfrac{4}{3}$

$\quad (\overline{x}, \overline{y}) = \left(\dfrac{4}{3}, \dfrac{1}{3}\right)$

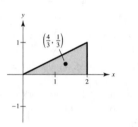

**15.** $\quad m = \rho \displaystyle\int_0^4 \sqrt{x}\, dx = \left[\dfrac{2\rho}{3}x^{3/2}\right]_0^4 = \dfrac{16\rho}{3}$

$\quad M_x = \rho \displaystyle\int_0^4 \dfrac{\sqrt{x}}{2}\left(\sqrt{x}\right) dx = \left[\rho \dfrac{x^2}{4}\right]_0^4 = 4\rho$

$\quad \overline{y} = \dfrac{M_x}{m} = 4\rho\left(\dfrac{3}{16\rho}\right) = \dfrac{3}{4}$

$\quad M_y = \rho \displaystyle\int_0^4 x\sqrt{x}\, dx = \left[\rho \dfrac{2}{5}x^{5/2}\right]_0^4 = \dfrac{64\rho}{5}$

$\quad \overline{x} = \dfrac{M_y}{m} = \dfrac{64\rho}{5}\left(\dfrac{3}{16\rho}\right) = \dfrac{12}{5}$

$\quad (\overline{x}, \overline{y}) = \left(\dfrac{12}{5}, \dfrac{3}{4}\right)$

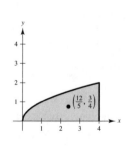

**17.**  $m = \rho \int_0^1 (x^2 - x^3)\, dx = \rho\left[\dfrac{x^3}{3} - \dfrac{x^4}{4}\right]_0^1 = \dfrac{\rho}{12}$

$M_x = \rho \int_0^1 \dfrac{(x^2 + x^3)}{2}(x^2 - x^3)\, dx = \dfrac{\rho}{2}\int_0^1 (x^4 - x^6)\, dx = \dfrac{\rho}{2}\left[\dfrac{x^5}{5} - \dfrac{x^7}{7}\right]_0^1 = \dfrac{\rho}{35}$

$\bar{y} = \dfrac{M_x}{m} = \dfrac{\rho}{35}\left(\dfrac{12}{\rho}\right) = \dfrac{12}{35}$

$M_y = \rho \int_0^1 x(x^2 - x^3)\, dx = \rho \int_0^1 (x^3 - x^4)\, dx = \rho\left[\dfrac{x^4}{4} - \dfrac{x^5}{5}\right]_0^1 = \dfrac{\rho}{20}$

$\bar{x} = \dfrac{M_y}{m} = \dfrac{\rho}{20}\left(\dfrac{12}{\rho}\right) = \dfrac{3}{5}$

$(\bar{x}, \bar{y}) = \left(\dfrac{3}{5}, \dfrac{12}{35}\right)$

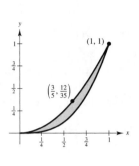

**19.**  $m = \rho \int_0^3 \left[(-x^2 + 4x + 2) - (x + 2)\right] dx = -\rho\left[\dfrac{x^3}{3} + \dfrac{3x^2}{2}\right]_0^3 = \dfrac{9\rho}{2}$

$M_x = \rho \int_0^3 \left[\dfrac{(-x^2 + 4x + 2) + (x + 2)}{2}\right]\left[(-x^2 + 4x + 2) - (x + 2)\right] dx$

$= \dfrac{\rho}{2}\int_0^3 (-x^2 + 5x + 4)(-x^2 + 3x)\, dx = \dfrac{\rho}{2}\int_0^3 (x^4 - 8x^3 + 11x^2 + 12x)\, dx$

$= \dfrac{\rho}{2}\left[\dfrac{x^5}{5} - 2x^4 + \dfrac{11x^3}{3} + 6x^2\right]_0^3 = \dfrac{99\rho}{5}$

$\bar{y} = \dfrac{M_x}{m} = \dfrac{99\rho}{5}\left(\dfrac{2}{9\rho}\right) = \dfrac{22}{5}$

$M_y = \rho \int_0^3 x\left[(-x^2 + 4x - 2) - (x + 2)\right] dx = \rho \int_0^3 (-x^3 + 3x^2)\, dx = \rho\left[-\dfrac{x^4}{4} + x^3\right]_0^3 = \dfrac{27\rho}{4}$

$\bar{x} = \dfrac{M_y}{m} = \dfrac{27\rho}{4}\left(\dfrac{2}{9\rho}\right) = \dfrac{3}{2}$

$(\bar{x}, \bar{y}) = \left(\dfrac{3}{2}, \dfrac{22}{5}\right)$

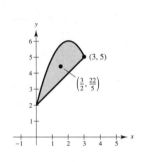

**21.**  $m = \rho \int_0^8 x^{2/3}\, dx = \rho\left[\dfrac{3}{5}x^{5/3}\right]_0^8 = \dfrac{96\rho}{5}$

$M_x = \rho \int_0^8 \dfrac{x^{2/3}}{2}(x^{2/3})\, dx = \dfrac{\rho}{2}\left[\dfrac{3}{7}x^{7/3}\right]_0^8 = \dfrac{192\rho}{7}$

$\bar{y} = \dfrac{M_x}{m} = \dfrac{192\rho}{7}\left(\dfrac{5}{96\rho}\right) = \dfrac{10}{7}$

$M_y = \rho \int_0^8 x(x^{2/3})\, dx = \rho\left[\dfrac{3}{8}x^{8/3}\right]_0^8 = 96\rho$

$\bar{x} = \dfrac{M_y}{m} = 96\rho\left(\dfrac{5}{96\rho}\right) = 5$

$(\bar{x}, \bar{y}) = \left(5, \dfrac{10}{7}\right)$

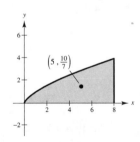

**23.**    $m = 2\rho \int_0^2 \left(4 - y^2\right) dy = 2\rho \left[4y - \dfrac{y^3}{3}\right]_0^2 = \dfrac{32\rho}{3}$

$M_y = 2\rho \int_0^2 \left(\dfrac{4 - y^2}{2}\right)\left(4 - y^2\right) dy = \rho \left[16y - \dfrac{8}{3}y^3 + \dfrac{y^5}{5}\right]_0^2 = \dfrac{256\rho}{15}$

$\overline{x} = \dfrac{M_y}{m} = \dfrac{256\rho}{15}\left(\dfrac{3}{32\rho}\right) = \dfrac{8}{5}$

By symmetry, $M_x$ and $\overline{y} = 0$.

$\left(\overline{x}, \overline{y}\right) = \left(\dfrac{8}{5}, 0\right)$

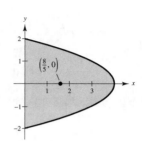

**25.**    $m = \rho \int_0^3 \left[\left(2y - y^2\right) - (-y)\right] dy = \rho \left[\dfrac{3y^2}{2} - \dfrac{y^3}{3}\right]_0^3 = \dfrac{9\rho}{2}$

$M_y = \rho \int_0^3 \dfrac{\left[\left(2y - y^2\right) + (-y)\right]}{2}\left[\left(2y - y^2\right) - (-y)\right] dy = \dfrac{\rho}{2} \int_0^3 \left(y - y^2\right)\left(3y - y^2\right) dy$

$= \dfrac{\rho}{2} \int_0^3 \left(y^4 - 4y^3 + 3y^2\right) dy = \dfrac{\rho}{2}\left[\dfrac{y^5}{5} - y^4 + y^3\right]_0^3 = -\dfrac{27\rho}{10}$

$\overline{x} = \dfrac{M_y}{m} = -\dfrac{27\rho}{10}\left(\dfrac{2}{9\rho}\right) = -\dfrac{3}{5}$

$M_x = \rho \int_0^3 y\left[\left(2y - y^2\right) - (-y)\right] dy = \rho \int_0^3 \left(3y^2 - y^3\right) dy = \rho \left[y^3 - \dfrac{y^4}{4}\right]_0^3 = \dfrac{27\rho}{4}$

$\overline{y} = \dfrac{M_x}{m} = \dfrac{27\rho}{4}\left(\dfrac{2}{9\rho}\right) = \dfrac{3}{2}$

$\left(\overline{x}, \overline{y}\right) = \left(-\dfrac{3}{5}, \dfrac{3}{2}\right)$

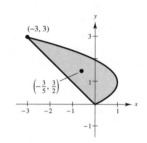

**27.**    $A = \int_0^2 \left(2x - x^2\right) dx = \left[x^2 - \dfrac{x^3}{3}\right]_0^2 = 4 - \dfrac{8}{3} = \dfrac{4}{3}$

$M_x = \dfrac{1}{2} \int_0^2 \left[(2x)^2 - \left(x^2\right)^2\right] dx = \dfrac{1}{2}\left[\dfrac{4x^3}{3} - \dfrac{x^5}{5}\right]_0^2 = \dfrac{1}{2}\left[\dfrac{32}{3} - \dfrac{32}{5}\right] = \dfrac{32}{15}$

$M_y = \int_0^2 x\left(2x - x^2\right) dx = \left[\dfrac{2x^3}{3} - \dfrac{x^4}{4}\right]_0^2 = \dfrac{16}{3} - 4 = \dfrac{4}{3}$

**29.**    $A = \int_0^3 (2x + 4) dx = \left[x^2 + 4x\right]_0^3 = 9 + 12 = 21$

$M_x = \dfrac{1}{2} \int_0^3 (2x + 4)^2 dx = \int_0^3 \left(2x^2 + 8x + 8\right) dx = \left[\dfrac{2x^3}{3} + 4x^2 + 8x\right]_0^3 = 18 + 36 + 24 = 78$

$M_y = \int_0^3 \left(2x^2 + 4x\right) dx = \left[\dfrac{2x^3}{3} + 2x^2\right]_0^3 = 18 + 18 = 36$

**31.**   $m = \rho \int_0^5 10x\sqrt{125 - x^3}\, dx \approx 1033.0\rho$

$M_x = \rho \int_0^5 \left( \dfrac{10x\sqrt{125 - x^3}}{2} \right)\left(10x\sqrt{125 - x^3}\right) dx = 50\rho \int_0^5 x^2\left(125 - x^3\right) dx = \dfrac{3{,}124{,}375\rho}{24} \approx 130{,}208\rho$

$M_y = \rho \int_0^5 10x^2\sqrt{125 - x^3}\, dx = -\dfrac{10\rho}{3} \int_0^5 \sqrt{125 - x^3}\left(-3x^2\right) dx = \dfrac{12{,}500\sqrt{5}\rho}{9} \approx 3105.6\rho$

$\bar{x} = \dfrac{M_y}{m} \approx 3.0$

$\bar{y} = \dfrac{M_x}{m} \approx 126.0$

Therefore, the centroid is $(3.0, 126.0)$.

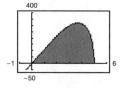

**33.**   $m = \rho \int_{-20}^{20} 5\sqrt[3]{400 - x^2}\, dx \approx 1239.76\rho$

$M_x = \rho \int_{-20}^{20} \dfrac{5\sqrt[3]{400 - x^2}}{2}\left(5\sqrt[3]{400 - x^2}\right) dx$

$\quad = \dfrac{25\rho}{2} \int_{-20}^{20} \left(400 - x^2\right)^{2/3} dx \approx 20064.27$

$\bar{y} = \dfrac{M_x}{m} \approx 16.18$

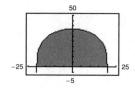

$\bar{x} = 0$ by symmetry. Therefore, the centroid is $(0, 16.2)$.

**35.**   $A = \dfrac{1}{2}(2a)c = ac$

$\dfrac{1}{A} = \dfrac{1}{ac}$

$\bar{x} = \left(\dfrac{1}{ac}\right)\dfrac{1}{2}\int_0^c \left[ \left(\dfrac{b - a}{c}y + a\right)^2 - \left(\dfrac{b + a}{c}y - a\right)^2 \right] dy$

$\quad = \dfrac{1}{2ac}\int_0^c \left[\dfrac{4ab}{c}y - \dfrac{4ab}{c^2}y^2\right] dy = \dfrac{1}{2ac}\left[\dfrac{2ab}{c}y^2 - \dfrac{4ab}{3c^2}y^3\right]_0^c = \dfrac{1}{2ac}\left(\dfrac{2}{3}abc\right) = \dfrac{b}{3}$

$\bar{y} = \dfrac{1}{ac}\int_0^c y\left[\left(\dfrac{b - a}{c}y + a\right) - \left(\dfrac{b + a}{c}y - a\right)\right] dy$

$\quad = \dfrac{1}{ac}\int_0^c y\left(-\dfrac{2a}{c}y + 2a\right) dy = \dfrac{2}{c}\int_0^c \left(y - \dfrac{y^2}{c}\right) dy = \dfrac{2}{c}\left[\dfrac{y^2}{2} - \dfrac{y^3}{3c}\right]_0^c = \dfrac{c}{3}$

$(\bar{x}, \bar{y}) = \left(\dfrac{b}{3}, \dfrac{c}{3}\right)$

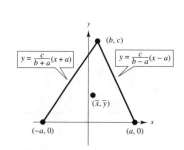

From elementary geometry, $(b/3, c/3)$ is the point of intersection of the medians.

**37.** $A = \dfrac{c}{2}(a + b)$

$\dfrac{1}{A} = \dfrac{2}{c(a + b)}$

$\bar{x} = \dfrac{2}{c(a + b)} \displaystyle\int_0^c x\left(\dfrac{b - a}{c}x + a\right) dx = \dfrac{2}{c(a + b)} \displaystyle\int_0^c \left(\dfrac{b - a}{c}x^2 + ax\right) dx = \dfrac{2}{c(a + b)}\left[\dfrac{b - a}{c}\dfrac{x^3}{3} + \dfrac{ax^2}{2}\right]_0^c$

$= \dfrac{2}{c(a + b)}\left[\dfrac{(b - a)c^2}{3} + \dfrac{ac^2}{2}\right] = \dfrac{2}{c(a + b)}\left[\dfrac{2bc^2 - 2ac^2 + 3ac^2}{6}\right] = \dfrac{c(2b + a)}{3(a + b)} = \dfrac{(a + 2b)c}{3(a + b)}$

$\bar{y} = \dfrac{2}{c(a + b)}\dfrac{1}{2}\displaystyle\int_0^c \left(\dfrac{b - a}{c}x + a\right)^2 dx = \dfrac{1}{c(a + b)}\displaystyle\int_0^c \left[\left(\dfrac{b - a}{c}\right)^2 x^2 + \dfrac{2a(b - a)}{c}x + a^2\right] dx$

$= \dfrac{1}{c(a + b)}\left[\left(\dfrac{b - a}{c}\right)^2 \dfrac{x^3}{3} + \dfrac{2a(b - a)}{c}\dfrac{x^2}{2} + a^2 x\right]_0^c = \dfrac{1}{c(a + b)}\left[\dfrac{(b - a)^2 c}{3} + ac(b - a) + a^2 c\right]$

$= \dfrac{1}{3c(a + b)}\left[(b^2 - 2ab + a^2)c + 3ac(b - a) + 3a^2 c\right]$

$= \dfrac{1}{3(a + b)}\left[b^2 - 2ab + a^2 + 3ab - 3a^2 + 3a^2\right] = \dfrac{a^2 + ab + b^2}{3(a + b)}$

So, $(\bar{x}, \bar{y}) = \left(\dfrac{(a + 2b)c}{3(a + b)}, \dfrac{a^2 + ab + b^2}{3(a + b)}\right)$.

The one line passes through $(0, a/2)$ and $(c, b/2)$. It's equation is $y = \dfrac{b - a}{2c}x + \dfrac{a}{2}$.

The other line passes through $(0, -b)$ and $(c, a + b)$. It's equation is $y = \dfrac{a + 2b}{c}x - b$.

$(\bar{x}, \bar{y})$ is the point of intersection of these two lines.

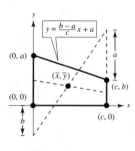

**39.** $\bar{x} = 0$ by symmetry.

$A = \dfrac{1}{2}\pi ab$

$\dfrac{1}{A} = \dfrac{2}{\pi ab}$

$\bar{y} = \dfrac{2}{\pi ab}\dfrac{1}{2}\displaystyle\int_{-a}^a \left(\dfrac{b}{a}\sqrt{a^2 - x^2}\right)^2 dx$

$= \dfrac{1}{\pi ab}\left(\dfrac{b^2}{a^2}\right)\left[a^2 x - \dfrac{x^3}{3}\right]_{-a}^a = \dfrac{b}{\pi a^3}\left(\dfrac{4a^3}{3}\right) = \dfrac{4b}{3\pi}$

$(\bar{x}, \bar{y}) = \left(0, \dfrac{4b}{3\pi}\right)$

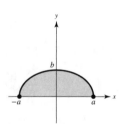

**41.** (a)

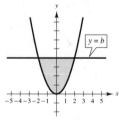

(b) $\bar{x} = 0$ by symmetry.

(c) $M_y = \displaystyle\int_{-\sqrt{b}}^{\sqrt{b}} x(b - x^2)\, dx = 0$ because $bx - x^3$ is odd.

(d) $\bar{y} > \dfrac{b}{2}$ because there is more area above $y = \dfrac{b}{2}$ than below.

(e) $M_x = \displaystyle\int_{-\sqrt{b}}^{\sqrt{b}} \dfrac{\left(b + x^2\right)\left(b - x^2\right)}{2}\,dx = \int_{-\sqrt{b}}^{\sqrt{b}} \dfrac{b^2 - x^4}{2}\,dx = \dfrac{1}{2}\left[b^2 x - \dfrac{x^5}{5}\right]_{-\sqrt{b}}^{\sqrt{b}} = b^2\sqrt{b} - \dfrac{b^2\sqrt{b}}{5} = \dfrac{4b^2\sqrt{b}}{5}$

$A = \displaystyle\int_{-\sqrt{b}}^{\sqrt{b}} \left(b - x^2\right)dx = \left[bx - \dfrac{x^3}{3}\right]_{-\sqrt{b}}^{\sqrt{b}} = \left(b\sqrt{b} - \dfrac{b\sqrt{b}}{3}\right)2 = 4\dfrac{b\sqrt{b}}{3}$

$\bar{y} = \dfrac{M_x}{A} = \dfrac{4b^2\sqrt{b}/5}{4b\sqrt{b}/3} = \dfrac{3}{5}b$

**43.** (a) $\bar{x} = 0$ by symmetry.

$A = 2\displaystyle\int_0^{40} f(x)\,dx = \dfrac{2(40)}{3(4)}\Big[30 + 4(29) + 2(26) + 4(20) + 0\Big] = \dfrac{20}{3}(278) = \dfrac{5560}{3}$

$M_x = \displaystyle\int_{-40}^{40} \dfrac{f(x)^2}{2}\,dx = \dfrac{40}{3(4)}\Big[30^2 + 4(29)^2 + 2(26)^2 + 4(20)^2 + 0\Big] = \dfrac{10}{3}(7216) = \dfrac{72{,}160}{3}$

$\bar{y} = \dfrac{M_x}{A} = \dfrac{72{,}160/3}{5560/3} = \dfrac{72{,}160}{5560} \approx 12.98$

$(\bar{x}, \bar{y}) = (0, 12.98)$

(b) $y = \left(-1.02 \times 10^{-5}\right)x^4 - 0.0019x^2 + 29.28$    (Use nine data points.)

(c) $\bar{y} = \dfrac{M_x}{A} \approx \dfrac{23{,}697.68}{1843.54} \approx 12.85$

$(\bar{x}, \bar{y}) = (0, 12.85)$

**45.** Centroids of the given regions: $(1, 0)$ and $(3, 0)$

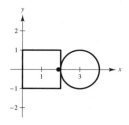

Area: $A = 4 + \pi$

$\bar{x} = \dfrac{4(1) + \pi(3)}{4 + \pi} = \dfrac{4 + 3\pi}{4 + \pi}$

$\bar{y} = \dfrac{4(0) + \pi(0)}{4 + \pi} = 0$

$(\bar{x}, \bar{y}) = \left(\dfrac{4 + 3\pi}{4 + \pi}, 0\right) \approx (1.88, 0)$

**47.** Centroids of the given regions: $\left(0, \dfrac{3}{2}\right)$, $(0, 5)$, and $\left(0, \dfrac{15}{2}\right)$

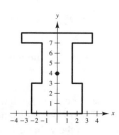

Area: $A = 15 + 12 + 7 = 34$

$\bar{x} = \dfrac{15(0) + 12(0) + 7(0)}{34} = 0$

$\bar{y} = \dfrac{15(3/2) + 12(5) + 7(15/2)}{34} = \dfrac{135}{34}$

$(\bar{x}, \bar{y}) = \left(0, \dfrac{135}{34}\right) \approx (0, 3.97)$

**49.** Centroids of the given regions: $(1, 0)$ and $(3, 0)$

Mass: $4 + 2\pi$

$$\bar{x} = \frac{4(1) + 2\pi(3)}{4 + 2\pi} = \frac{2 + 3\pi}{2 + \pi}$$

$$\bar{y} = 0$$

$$(\bar{x}, \bar{y}) = \left(\frac{2 + 3\pi}{2 + \pi}, 0\right) \approx (2.22, 0)$$

**51.** $r = 5$ is distance between center of circle and $y$-axis.

$A \approx \pi(4)^2 = 16\pi$ is area of circle. So,

$$V = 2\pi rA = 2\pi(5)(16\pi) = 160\pi^2 \approx 1579.14.$$

**53.** $A = \frac{1}{2}(4)(4) = 8$

$$\bar{y} = \left(\frac{1}{8}\right)\frac{1}{2}\int_0^4 (4 + x)(4 - x)\, dx = \frac{1}{16}\left[16x - \frac{x^3}{3}\right]_0^4 = \frac{8}{3}$$

$$r = \bar{y} = \frac{8}{3}$$

$$V = 2\pi rA = 2\pi\left(\frac{8}{3}\right)(8) = \frac{128\pi}{3} \approx 134.04$$

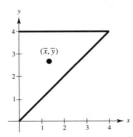

**55.** The center of mass $(\bar{x}, \bar{y})$ is $\bar{x} = M_y/m$ and $\bar{y} = M_x/m$, where:

1. $m = m_1 + m_2 + \cdots + m_n$ is the total mass of the system.

2. $M_y = m_1x_1 + m_2x_2 + \cdots + m_nx_n$ is the moment about the $y$-axis.

3. $M_x = m_1y_1 + m_2y_2 + \cdots + m_ny_n$ is the moment about the $x$-axis.

**57.** Let $R$ be a region in a plane and let $L$ be a line such that $L$ does not intersect the interior of $R$. If $r$ is the distance between the centroid of $R$ and $L$, then the volume $V$ of the solid of revolution formed by revolving $R$ about $L$ is $V = 2\pi rA$ where $A$ is the area of $R$.

**59.** The surface area of the sphere is $S = 4\pi r^2$. The arc length of $C$ is $s = \pi r$. The distance traveled by the centroid is

$$d = \frac{S}{s} = \frac{4\pi r^2}{\pi r} = 4r.$$

This distance is also the circumference of the circle of radius $y$.

$$d = 2\pi y$$

So, $2\pi y = 4r$ and you have $y = 2r/\pi$. Therefore, the centroid of the semicircle $y = \sqrt{r^2 - x^2}$ is $(0, 2r/\pi)$.

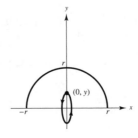

**61.**
$$A = \int_0^1 x^n\, dx = \left[\frac{x^{n+1}}{n+1}\right]_0^1 = \frac{1}{n+1}$$

$$m = \rho A = \frac{\rho}{n+1}$$

$$M_x = \frac{\rho}{2}\int_0^1 (x^n)^2\, dx = \left[\frac{\rho}{2} \cdot \frac{x^{2n+1}}{2n+1}\right]_0^1 = \frac{\rho}{2(2n+1)}$$

$$M_y = \rho\int_0^1 x(x^n)\, dx = \left[\rho \cdot \frac{x^{n+2}}{n+2}\right]_0^1 = \frac{\rho}{n+2}$$

$$\bar{x} = \frac{M_y}{m} = \frac{n+1}{n+2}$$

$$\bar{y} = \frac{M_x}{m} = \frac{n+1}{2(2n+1)} = \frac{n+1}{4n+2}$$

Centroid: $\left(\dfrac{n+1}{n+2}, \dfrac{n+1}{4n+2}\right)$

As $n \to \infty$, $(\bar{x}, \bar{y}) \to \left(1, \dfrac{1}{4}\right)$. The graph approaches the $x$-axis and the line $x = 1$ as $n \to \infty$.

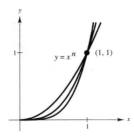

## Section 7.7  Fluid Pressure and Fluid Force

**1.** $F = PA = \left[62.4\,(8)\right]3 = 1497.6$ lb

**3.** $F = PA = \left[62.4\,(8)\right]10 = 4992$ lb

**5.** $F = 62.4(h + 2)(6) - (62.4)(h)(6)$

   $\phantom{F} = 62.4(2)(6) = 748.8$ lb

**7.** $h(y) = 3 - y$

   $L(y) = 4$

   $F = 62.4 \int_0^3 (3 - y)(4)\, dy$

   $\phantom{F} = 249.6 \int_0^3 (3 - y)\, dy$

   $\phantom{F} = 249.6 \left[3y - \dfrac{y^2}{2}\right]_0^3 = 1123.2$ lb

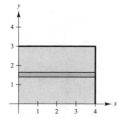

**9.** $h(y) = 3 - y$

   $L(y) = 2\left(\dfrac{y}{3} + 1\right)$

   $F = 2(62.4) \int_0^3 (3 - y)\left(\dfrac{y}{3} + 1\right) dy$

   $\phantom{F} = 124.8 \int_0^3 \left(3 - \dfrac{y^2}{3}\right) dy$

   $\phantom{F} = 124.8 \left[3y - \dfrac{y^3}{9}\right]_0^3 = 748.8$ lb

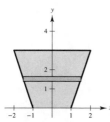

**11.** $h(y) = 4 - y$

   $L(y) = 2\sqrt{y}$

   $F = 2(62.4) \int_0^4 (4 - y)\sqrt{y}\, dy$

   $\phantom{F} = 124.8 \int_0^4 \left(4y^{1/2} - y^{3/2}\right) dy$

   $\phantom{F} = 124.8 \left[\dfrac{8y^{3/2}}{3} - \dfrac{2y^{5/2}}{5}\right]_0^4 = 1064.96$ lb

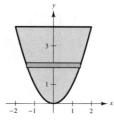

**13.** $h(y) = 4 - y$

   $L(y) = 2$

   $F = 9800 \int_0^2 2(4 - y)\, dy$

   $\phantom{F} = 9800 \left[8y - y^2\right]_0^2 = 117{,}600$ newtons

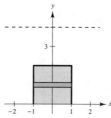

**15.** $h(y) = 12 - y$

   $L(y) = 6 - \dfrac{2y}{3}$

   $F = 9800 \int_0^9 (12 - y)\left(6 - \dfrac{2y}{3}\right) dy = 9800 \left[72y - 7y^2 + \dfrac{2y^3}{9}\right]_0^9 = 2{,}381{,}400$ newtons

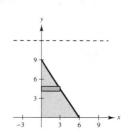

**17.** $h(y) = 2 - y$

$L(y) = 10$

$F = 140.7 \int_0^2 (2 - y)(10) \, dy$

$= 1407 \int_0^2 (2 - y) \, dy$

$= 1407 \left[ 2y - \dfrac{y^2}{2} \right]_0^2 = 2814$ lb

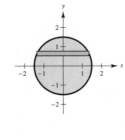

**21.** $h(y) = -y$

$L(y) = 2\left(\dfrac{1}{2}\right)\sqrt{9 - 4y^2}$

$F = 42 \int_{-3/2}^0 (-y)\sqrt{9 - 4y^2} \, dy$

$= \dfrac{42}{8} \int_{-3/2}^0 \left(9 - 4y^2\right)^{1/2} (-8y) \, dy$

$= \left[ \left(\dfrac{21}{4}\right)\left(\dfrac{2}{3}\right)\left(9 - 4y^2\right)^{3/2} \right]_{-3/2}^0 = 94.5$ lb

**19.** $h(y) = 4 - y$

$L(y) = 6$

$F = 140.7 \int_0^4 (4 - y)(6) \, dy$

$= 844.2 \int_0^4 (4 - y) \, dy$

$= 844.2 \left[ 4y - \dfrac{y^2}{2} \right]_0^4 = 6753.6$ lb

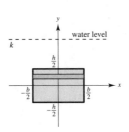

**23.** $h(y) = k - y$

$L(y) = 2\sqrt{r^2 - y^2}$

$F = w \int_{-r}^r (k - y)\sqrt{r^2 - y^2}\,(2) \, dy = w \left[ 2k \int_{-r}^r \sqrt{r^2 - y^2} \, dy + \int_{-r}^r \sqrt{r^2 - y^2}\,(-2y) \, dy \right]$

The second integral is zero because its integrand is odd and the limits of integration are symmetric to the origin. The first integral is the area of a semicircle with radius $r$.

$F = w \left[ (2k)\dfrac{\pi r^2}{2} + 0 \right] = wk\pi r^2$

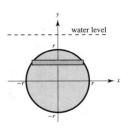

**25.** $h(y) = k - y$

$L(y) = b$

$F = w \int_{-h/2}^{h/2} (k - y)b \, dy$

$= wb \left[ ky - \dfrac{y^2}{2} \right]_{-h/2}^{h/2} = wb(hk) = wkhb$

**27.** From Exercise 25:

$F = 64(15)(1)(1) = 960$ lb

**29.** $h(y) = 4 - y$

$F = 62.4 \int_0^4 (4 - y)L(y) \, dy$

Using Simpson's Rule with $n = 8$ you have:

$F \approx 62.4\left(\dfrac{4-0}{3(8)}\right)\left[0 + 4(3.5)(3) + 2(3)(5) + 4(2.5)(8) + 2(2)(9) + 4(1.5)(10) + 2(1)(10.25) + 4(0.5)(10.5) + 0\right]$

$= 3010.8$ lb

**31.** $h(y) = 15 - y$

$$L(y) = 2\left(4^{2/3} - y^{2/3}\right)^{3/2}$$

$$F = 62.4 \int_0^4 2(15 - y)\left(4^{2/3} - y^{2/3}\right)^{3/2} dy$$

$$\approx 8213.04 \text{ lb}$$

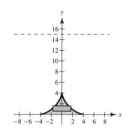

**33.** If the fluid force is one-half of 1123.2 lb, and the height of the water is $b$, then

$$h(y) = b - y$$

$$L(y) = 4$$

$$F = 62.4 \int_0^b (b - y)(4) \, dy = \frac{1}{2}(1123.2)$$

$$\int_0^b (b - y) \, dy = 2.25$$

$$\left[by - \frac{y^2}{2}\right]_0^b = 2.25$$

$$b^2 - \frac{b^2}{2} = 2.25$$

$$b^2 = 4.5 \Rightarrow b \approx 2.12 \text{ ft.}$$

The pressure increases with increasing depth.

**35.** You use horizontal representative rectangles because you are measuring total force against a region between two depths.

# Review Exercises for Chapter 7

**1.** $A = \int_1^5 \frac{1}{x^2} dx = \left[-\frac{1}{x}\right]_1^5 = \frac{4}{5}$

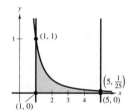

**3.** $A = \int_{-1}^1 \frac{1}{x^2 + 1} dx = \left[\arctan x\right]_{-1}^1 = \frac{\pi}{4} - \left(-\frac{\pi}{4}\right) = \frac{\pi}{2}$

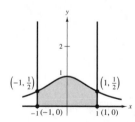

**5.** $A = 2\int_0^1 \left(x - x^3\right) dx = 2\left[\frac{1}{2}x^2 - \frac{1}{4}x^4\right]_0^1 = \frac{1}{2}$

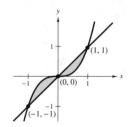

**7.** $A = \int_0^2 \left(e^2 - e^x\right) dx = \left[xe^2 - e^x\right]_0^2 = e^2 + 1$

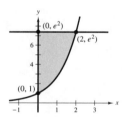

**9.** $A = \int_{\pi/4}^{5\pi/4} \left(\sin x - \cos x\right) dx$

$$= \left[-\cos x - \sin x\right]_{\pi/4}^{5\pi/4}$$

$$= \left(\frac{1}{\sqrt{2}} + \frac{1}{\sqrt{2}}\right) - \left(-\frac{1}{\sqrt{2}} - \frac{1}{\sqrt{2}}\right)$$

$$= \frac{4}{\sqrt{2}} = 2\sqrt{2}$$

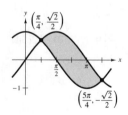

**11.** Points of intersection:

$$x^2 - 8x + 3 = 3 + 8x - x^2$$

$$2x^2 - 16x = 0 \quad \text{when} \quad x = 0, 8$$

$$A = \int_0^8 \left[ \left( 3 + 8x - x^2 \right) - \left( x^2 - 8x + 3 \right) \right] dx$$

$$= \int_0^8 \left( 16x - 2x^2 \right) dx$$

$$= \left[ 8x^2 - \tfrac{2}{3}x^3 \right]_0^8 = \tfrac{512}{3} \approx 170.667$$

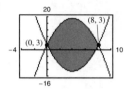

**13.** $y = \left( 1 - \sqrt{x} \right)^2$

$$A = \int_0^1 \left( 1 - \sqrt{x} \right)^2 dx$$

$$= \int_0^1 \left( 1 - 2x^{1/2} + x \right) dx$$

$$= \left[ x - \tfrac{4}{3}x^{3/2} + \tfrac{1}{2}x^2 \right]_0^1 = \tfrac{1}{6} \approx 0.1667$$

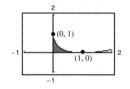

**15.** $x = y^2 - 2y \Rightarrow x + 1 = (y - 1)^2 \Rightarrow y = 1 \pm \sqrt{x + 1}$

$$A = \int_{-1}^0 \left[ \left( 1 + \sqrt{x + 1} \right) - \left( 1 - \sqrt{x + 1} \right) \right] dx$$

$$= \int_{-1}^0 2\sqrt{x + 1}\, dx$$

$$A = \int_0^2 \left[ 0 - \left( y^2 - 2y \right) \right] dy$$

$$= \int_0^2 \left( 2y - y^2 \right) dy = \left[ y^2 - \tfrac{1}{3}y^3 \right]_0^2 = \tfrac{4}{3}$$

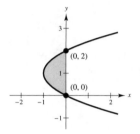

**17.** $A = \int_0^2 \left[ 1 - \left( 1 - \dfrac{x}{2} \right) \right] dx + \int_2^3 \left[ 1 - (x - 2) \right] dx$

$$= \int_0^2 \dfrac{x}{2}\, dx + \int_2^3 (3 - x)\, dx$$

$$y = 1 - \dfrac{x}{2} \Rightarrow x = 2 - 2y$$

$$y = x - 2 \Rightarrow x = y + 2, \, y = 1$$

$$A = \int_0^1 \left[ (y + 2) - (2 - 2y) \right] dy$$

$$= \int_0^1 3y\, dy = \left[ \dfrac{3}{2}y^2 \right]_0^1 = \dfrac{3}{2}$$

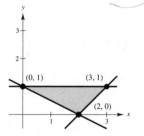

**19.** (a) Trapezoidal: Area $\approx \dfrac{160}{2(8)} \left[ 0 + 2(50) + 2(54) + 2(82) + 2(82) + 2(73) + 2(75) + 2(80) + 0 \right] = 9920$ ft$^2$

(b) Simpson's: Area $\approx \dfrac{160}{3(8)} \left[ 0 + 4(50) + 2(54) + 4(82) + 2(82) + 4(73) + 2(75) + 4(80) + 0 \right] = 10{,}413\tfrac{1}{3}$ ft$^2$

**21.** (a) **Disk**

$$V = \pi \int_0^3 x^2 \, dx = \left[ \frac{\pi x^3}{3} \right]_0^3 = 9\pi$$

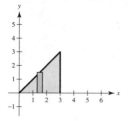

(b) **Shell**

$$V = 2\pi \int_0^3 x(x) \, dx = 2\pi \left[ \frac{x^3}{3} \right]_0^3 = 18\pi$$

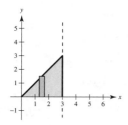

(c) **Shell**

$$V = 2\pi \int_0^3 (3 - x)x \, dx = 2\pi \left[ \frac{3x^2}{2} - \frac{x^3}{3} \right]_0^3 = 9\pi$$

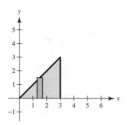

(d) **Shell**

$$V = 2\pi \int_0^3 (6 - x)x \, dx = 2\pi \left[ 3x^2 - \frac{x^3}{3} \right]_0^3 = 36\pi$$

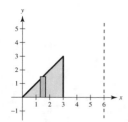

**23.** (a) **Shell**

$$V = 4\pi \int_0^4 x \left( \frac{3}{4} \right) \sqrt{16 - x^2} \, dx$$

$$= \left[ 3\pi \left( -\frac{1}{2} \right) \left( \frac{2}{3} \right) \left( 16 - x^2 \right)^{3/2} \right]_0^4 = 64\pi$$

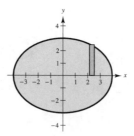

(b) **Disk**

$$V = 2\pi \int_0^4 \left[ \frac{3}{4} \sqrt{16 - x^2} \right]^2 \, dx$$

$$= \frac{9\pi}{8} \left[ 16x - \frac{x^3}{3} \right]_0^4 = 48\pi$$

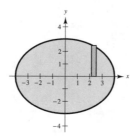

**25. Shell**

$$V = 2\pi \int_0^1 \frac{x}{x^4 + 1} \, dx = \pi \int_0^1 \frac{(2x)}{\left( x^2 \right)^2 + 1} \, dx$$

$$= \left[ \pi \arctan\left( x^2 \right) \right]_0^1 = \pi \left( \frac{\pi}{4} - 0 \right) = \frac{\pi^2}{4}$$

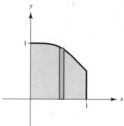

**27. Shell:** $V = 2\pi \int_2^6 \frac{x}{1 + \sqrt{x-2}}\, dx$

$u = \sqrt{x-2}$

$x = u^2 + 2$

$dx = 2u\, du$

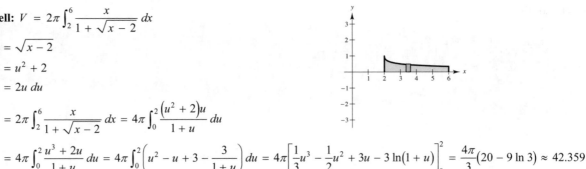

$V = 2\pi \int_2^6 \frac{x}{1 + \sqrt{x-2}}\, dx = 4\pi \int_0^2 \frac{(u^2 + 2)u}{1 + u}\, du$

$= 4\pi \int_0^2 \frac{u^3 + 2u}{1 + u}\, du = 4\pi \int_0^2 \left( u^2 - u + 3 - \frac{3}{1+u} \right) du = 4\pi \left[ \frac{1}{3}u^3 - \frac{1}{2}u^2 + 3u - 3\ln(1+u) \right]_0^2 = \frac{4\pi}{3}(20 - 9\ln 3) \approx 42.359$

**29. (a)** Because $y \le 0$, $A = -\int_{-1}^0 x\sqrt{x+1}\, dx$.

$u = x + 1$

$x = u - 1$

$dx = du$

$A = -\int_0^1 (u-1)\sqrt{u}\, du$

$= -\int_0^1 (u^{3/2} - u^{1/2})\, du$

$= -\left[ \frac{2}{5}u^{5/2} - \frac{2}{3}u^{3/2} \right]_0^1$

$= \frac{4}{15}$

**(b) Disk**

$V = \pi \int_{-1}^0 x^2(x+1)\, dx$

$= \pi \int_{-1}^0 (x^3 + x^2)\, dx$

$= \pi \left[ \frac{x^4}{4} + \frac{x^3}{3} \right]_{-1}^0$

$= \frac{\pi}{12}$

**(c) Shell**

$u = \sqrt{x+1}$

$x = u^2 - 1$

$dx = 2u\, du$

$V = 2\pi \int_{-1}^0 x^2 \sqrt{x+1}\, dx$

$= 4\pi \int_0^1 (u^2 - 1)^2 u^2\, du$

$= 4\pi \int_0^1 (u^6 - 2u^4 + u^2)\, du$

$= 4\pi \left[ \frac{1}{7}u^7 - \frac{2}{5}u^5 + \frac{1}{3}u^3 \right]_0^1 = \frac{32\pi}{105}$

**31.** From Exercise 23(a) you have: $V = 64\pi$ ft$^3$

$$\frac{1}{4}V = 16\pi$$

**Disk:** $\pi \int_{-3}^{y_0} \frac{16}{9}(9 - y^2)\, dy = 16\pi$

$\frac{1}{9} \int_{-3}^{y_0} (9 - y^2)\, dy = 1$

$\left[ 9y - \frac{1}{3}y^3 \right]_{-3}^{y_0} = 9$

$\left( 9y_0 - \frac{1}{3}y_0^3 \right) - (-27 + 9) = 9$

$y_0^3 - 27y_0 - 27 = 0$

By Newton's Method, $y_0 \approx -1.042$ and the depth of the gasoline is $3 - 1.042 = 1.958$ feet.

**33.**
$f(x) = \frac{4}{5}x^{5/4}$

$f'(x) = x^{1/4}$

$1 + [f'(x)]^2 = 1 + \sqrt{x}$

$u = 1 + \sqrt{x}$

$x = (u - 1)^2$

$dx = 2(u - 1)\, du$

$s = \int_0^4 \sqrt{1 + \sqrt{x}}\, dx = 2\int_1^3 \sqrt{u}(u - 1)\, du$

$= 2\int_1^3 (u^{3/2} - u^{1/2})\, du$

$= 2\left[ \frac{2}{5}u^{5/2} - \frac{2}{3}u^{3/2} \right]_1^3 = \frac{4}{15}\left[ u^{3/2}(3u - 5) \right]_1^3$

$= \frac{8}{15}(1 + 6\sqrt{3}) \approx 6.076$

**35.** $y = 300\cosh\left( \frac{x}{2000} \right) - 280,\ -2000 \le x \le 2000$

$y' = \frac{3}{20}\sinh\left( \frac{x}{2000} \right)$

$s = \int_{-2000}^{2000} \sqrt{1 + \left[ \frac{3}{20}\sinh\left( \frac{x}{2000} \right) \right]^2}\, dx$

$= \frac{1}{20} \int_{-2000}^{2000} \sqrt{400 + 9\sinh^2\left( \frac{x}{2000} \right)}\, dx$

$= 4018.2$ ft (by Simpson's Rule or graphing utility)

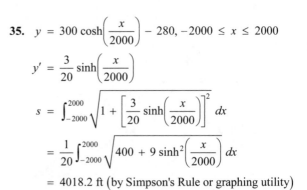

**37.** $y = \dfrac{3}{4}x$

$$y' = \dfrac{3}{4}$$

$$1 + (y')^2 = \dfrac{25}{16}$$

$$S = 2\pi \int_0^4 \left(\dfrac{3}{4}x\right)\sqrt{\dfrac{25}{16}}\, dx = \left[\left(\dfrac{15\pi}{8}\right)\dfrac{x^2}{2}\right]_0^4 = 15\pi$$

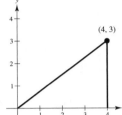

(4, 3)

**39.** $F = kx$

$5 = k(1)$

$F = 5x$

$$W = \int_0^5 5x\, dx = \dfrac{5x^2}{2}\Bigg]_0^5 = \dfrac{125}{2} \text{ in-lb} \approx 5.21 \text{ ft-lb}$$

**41.** Volume of disk: $\pi\left(\dfrac{1}{3}\right)^2 \Delta y \quad \left[\text{diameter} = \dfrac{2}{3} \text{ ft}\right]$

Weight of disk: $62.4\pi\left(\dfrac{1}{3}\right)^2 \Delta y$

Distance: $190 - y$

$$W = \dfrac{62.4\pi}{9}\int_0^{165}(190 - y)\, dy$$

$$= \dfrac{62.4\pi}{9}\left[190y - \dfrac{y^2}{2}\right]_0^{165}$$

$$= \dfrac{62.4\pi}{9}\left[\dfrac{35,475}{2}\right] = 122,980\pi \text{ ft-lb}$$

$$\approx 193.2 \text{ foot-tons}$$

**43.** Weight of section of chain: $4\,\Delta x$

Distance moved: $10 - x$

$$W = 4\int_0^{10}(10 - x)\, dx = 4\left[10x - \dfrac{x^2}{2}\right]_0^{10}$$

$$= 200 \text{ ft-lb}$$

**45.** $W = \displaystyle\int_a^b F(x)\, dx$

$$80 = \int_0^4 ax^2\, dx = \left[\dfrac{ax^3}{3}\right]_0^4 = \dfrac{64}{3}a$$

$$a = \dfrac{3(80)}{64} = \dfrac{15}{4} = 3.75$$

**47.** $A = \displaystyle\int_0^a \left(\sqrt{a} - \sqrt{x}\right)^2 dx = \int_0^a \left(a - 2\sqrt{a}x^{1/2} + x\right) dx = \left[ax - \dfrac{4}{3}\sqrt{a}x^{3/2} + \dfrac{1}{2}x^2\right]_0^a = \dfrac{a^2}{6}$

$$\dfrac{1}{A} = \dfrac{6}{a^2}$$

$$\bar{x} = \dfrac{6}{a^2}\int_0^a x\left(\sqrt{a} - \sqrt{x}\right)^2 dx = \dfrac{6}{a^2}\int_0^a \left(ax - 2\sqrt{a}x^{3/2} + x^2\right) dx = \dfrac{a}{5}$$

$$\bar{y} = \left(\dfrac{6}{a^2}\right)\dfrac{1}{2}\int_0^a \left(\sqrt{a} - \sqrt{x}\right)^4 dx$$

$$= \dfrac{3}{a^2}\int_0^a \left(a^2 - 4a^{3/2}x^{1/2} + 6ax - 4a^{1/2}x^{3/2} + x^2\right) dx$$

$$= \dfrac{3}{a^2}\left[a^2x - \dfrac{8}{3}a^{3/2}x^{3/2} + 3ax^2 - \dfrac{8}{5}a^{1/2}x^{5/2} + \dfrac{1}{3}x^3\right]_0^a = \dfrac{a}{5}$$

$$(\bar{x}, \bar{y}) = \left(\dfrac{a}{5}, \dfrac{a}{5}\right)$$

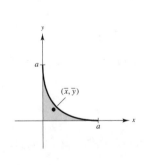

$(\bar{x}, \bar{y})$

**49.** By symmetry, $x = 0$.

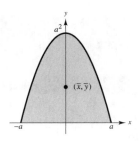

$$A = 2\int_0^1 \left(a^2 - x^2\right) dx = 2\left[a^2x - \frac{x^3}{3}\right]_0^a = \frac{4a^3}{3}$$

$$\frac{1}{A} = \frac{3}{4a^3}$$

$$\overline{y} = \left(\frac{3}{4a^3}\right)\frac{1}{2}\int_{-a}^a \left(a^2 - x^2\right)^2 dx$$

$$= \frac{6}{8a^3}\int_0^a \left(a^4 - 2a^2x^2 + x^4\right) dx$$

$$= \frac{6}{8a^3}\left[a^4x - \frac{2a^2}{3}x^3 + \frac{1}{5}x^5\right]_0^a$$

$$= \frac{6}{8a^3}\left(a^5 - \frac{2}{3}a^5 + \frac{1}{5}a^5\right) = \frac{2a^2}{5}$$

$$(\overline{x}, \overline{y}) = \left(0, \frac{2a^2}{5}\right)$$

**51.** $\overline{y} = 0$ by symmetry.

For the trapezoid:

$$m = \left[(4)(6) - (1)(6)\right]\rho = 18\rho$$

$$M_y = \rho\int_0^6 x\left[\left(\frac{1}{6}x + 1\right) - \left(-\frac{1}{6}x - 1\right)\right] dx = \rho\int_0^6 \left(\frac{1}{3}x^2 + 2x\right) dx = \rho\left[\frac{x^3}{9} + x^2\right]_0^6 = 60\rho$$

For the semicircle:

$$m = \left(\frac{1}{2}\right)(\pi)(2)^2 \rho = 2\pi\rho$$

$$M_y = \rho\int_6^8 x\left[\sqrt{4 - (x - 6)^2} - \left(-\sqrt{4 - (x - 6)^2}\right)\right] dx = 2\rho\int_6^8 x\sqrt{4 - (x - 6)^2}\, dx$$

Let $u = x - 6$, then $x = u + 6$ and $dx = du$. When $x = 6, u = 0$. When $x = 8, u = 2$.

$$M_y = 2\rho\int_0^2 (u + 6)\sqrt{4 - u^2}\, du = 2\rho\int_0^2 u\sqrt{4 - u^2}\, du + 12\rho\int_0^2 \sqrt{4 - u^2}\, du$$

$$= 2\rho\left[\left(-\frac{1}{2}\right)\left(\frac{2}{3}\right)\left(4 - u^2\right)^{3/2}\right]_0^2 + 12\rho\left[\frac{\pi(2)^2}{4}\right] = \frac{16\rho}{3} + 12\pi\rho = \frac{4\rho(4 + 9\pi)}{3}$$

So, you have:

$$\overline{x}(18\rho + 2\pi\rho) = 60\rho + \frac{4\rho(4 + 9\pi)}{3}$$

$$\overline{x} = \frac{180\rho + 4\rho(4 + 9\pi)}{3} \cdot \frac{1}{2\rho(9 + \pi)} = \frac{2(9\pi + 49)}{3(\pi + 9)}$$

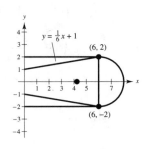

The centroid of the blade is $\left(\dfrac{2(9\pi + 49)}{3(\pi + 9)}, 0\right)$.

**53.** Let $D$ = surface of liquid; $\rho$ = weight per cubic volume.

$$F = \rho \int_c^d (D - y)\big[f(y) - g(y)\big]\, dy$$

$$= \rho \left[ \int_c^d D\big[f(y) - g(y)\big]\, dy - \int_c^d y\big[f(y) - g(y)\big]\, dy \right]$$

$$= \rho \left[ \int_c^d \big[f(y) - g(y)\big]\, dy \right] \left[ D - \frac{\int_c^d y\big[f(y) - g(y)\big]\, dy}{\int_c^d \big[f(y) - g(y)\big]\, dy} \right]$$

$$= \rho(\text{Area})(D - \bar{y})$$

$$= \rho(\text{Area})(\text{depth of centroid})$$

# Problem Solving for Chapter 7

**1.** $T = \dfrac{1}{2}c(c^2) = \dfrac{1}{2}c^3$

$$R = \int_0^c (cx - x^2)\, dx = \left[\frac{cx^2}{2} - \frac{x^3}{3}\right]_0^c = \frac{c^3}{2} - \frac{c^3}{3} = \frac{c^3}{6}$$

$$\lim_{c\to 0^+} \frac{T}{R} = \lim_{c\to 0^+} \frac{\dfrac{1}{2}c^3}{\dfrac{1}{6}c^3} = 3$$

**3.** $R = \displaystyle\int_0^1 x(1 - x)\, dx = \left[\frac{x^2}{2} - \frac{x^3}{3}\right]_0^1 = \frac{1}{2} - \frac{1}{3} = \frac{1}{6}$

Let $(c, mc)$ be the intersection of the line and the parabola.

Then, $mc = c(1 - c) \Rightarrow m = 1 - c$ or $c = 1 - m$.

$$\frac{1}{2}\left(\frac{1}{6}\right) = \int_0^{1-m} (x - x^2 - mx)\, dx$$

$$\frac{1}{12} = \left[\frac{x^2}{2} - \frac{x^3}{3} - m\frac{x^2}{2}\right]_0^{1-m}$$

$$= \frac{(1-m)^2}{2} - \frac{(1-m)^3}{3} - m\frac{(1-m)^2}{2}$$

$$1 = 6(1 - m)^2 - 4(1 - m)^3 - 6m(1 - m)^2$$

$$= (1 - m)^2\big(6 - 4(1 - m) - 6m\big)$$

$$= (1 - m)^2(2 - 2m)$$

$$\frac{1}{2} = (1 - m)^3$$

$$\left(\frac{1}{2}\right)^{1/3} = 1 - m$$

$$m = 1 - \left(\frac{1}{2}\right)^{1/3} \approx 0.2063$$

So, $y = 0.2063x$.

**5.** $8y^2 = x^2(1 - x^2)$

$$y = \pm\frac{|x|\sqrt{1 - x^2}}{2\sqrt{2}}$$

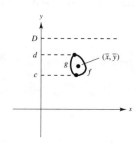

For $x > 0$, $y' = \dfrac{1 - 2x^2}{2\sqrt{2}\sqrt{1 - x^2}}$

$$S = 2(2\pi)\int_0^1 x\sqrt{1 + \left(\frac{1 - 2x^2}{2\sqrt{2}\sqrt{1 - x^2}}\right)^2}\, dx$$

$$= \frac{5\sqrt{2}\pi}{3}$$

**7.** By the Theorem of Pappus,

$$V = 2\pi r A$$

$$= 2\pi\left[d + \frac{1}{2}\sqrt{w^2 + l^2}\right] lw$$

**9.** $f'(x)^2 = e^x$

$$f'(x) = e^{x/2}$$

$$f(x) = 2e^{x/2} + C$$

$$f(0) = 0 \Rightarrow C = -2$$

$$f(x) = 2e^{x/2} - 2$$

**11.** Let $\rho_f$ be the density of the fluid and $\rho_0$ the density of the iceberg. The buoyant force is

$$F = \rho_f g \int_{-h}^{0} A(y)\, dy$$

where $A(y)$ is a typical cross section and $g$ is the acceleration due to gravity. The weight of the object is

$$W = \rho_0 g \int_{-h}^{L-h} A(y)\, dy.$$

$$F = W$$

$$\rho_f g \int_{-h}^{0} A(y)\, dy = \rho_0 g \int_{-h}^{L-h} A(y)\, dy$$

$$\frac{\rho_0}{\rho_f} = \frac{\text{submerged volume}}{\text{total volume}} = \frac{0.92 \times 10^3}{1.03 \times 10^3} = 0.893 \text{ or } 89.3\%$$

**13.** (a) $\bar{y} = 0$ by symmetry

$$M_y = 2 \int_{1}^{6} x \frac{1}{x^4}\, dx = 2 \int_{1}^{6} \frac{1}{x^3}\, dx = \frac{35}{36}$$

$$m = 2 \int_{1}^{6} \frac{1}{x^4}\, dx = \frac{215}{324}$$

$$\bar{x} = \frac{35/36}{215/324} = \frac{63}{43} \qquad (\bar{x}, \bar{y}) = \left(\frac{63}{43}, 0\right)$$

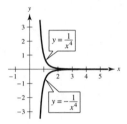

(b) $M_y = 2 \int_{1}^{b} \frac{1}{x^3}\, dx = \frac{b^2 - 1}{b^2}$

$$m = 2 \int_{1}^{b} \frac{1}{x^4}\, dx = \frac{2(b^3 - 1)}{3b^3}$$

$$\bar{x} = \frac{(b^2 - 1)/b^2}{2(b^3 - 1)/3b^3} = \frac{3b(b + 1)}{2(b^2 + b + 1)} \qquad (\bar{x}, \bar{y}) = \left(\frac{3b(b + 1)}{2(b^2 + b + 1)}, 0\right)$$

(c) $\displaystyle\lim_{b \to \infty} \bar{x} = \frac{3b(b + 1)}{2(b^2 + b + 1)} = \frac{3}{2} \qquad (\bar{x}, \bar{y}) = \left(\frac{3}{2}, 0\right)$

**15.** Point of equilibrium: $50 - 0.5x = 0.125x$

$$x = 80, \ p = 10$$

$$(P_0, x_0) = (10, 80)$$

Consumer surplus $= \int_{0}^{80} \left[(50 - 0.5x) - 10\right] dx = 1600$

Producer surplus $= \int_{0}^{80} (10 - 0.125x)\, dx = 400$

**17.** Use Exercise 25, Section 7.7, which gives $F = wkhb$ for a rectangle plate.

Wall at shallow end

From Exercise 25: $F = 62.4(2)(4)(20) = 9984$ lb

Wall at deep end

From Exercise 25: $F = 62.4(4)(8)(20) = 39{,}936$ lb

Side wall

From Exercise 25: $F_1 = 62.4(2)(4)(40) = 19{,}968$ lb

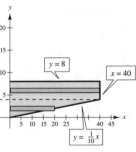

$$F_2 = 62.4 \int_{0}^{4} (8 - y)(10y)\, dy = 624 \int_{0}^{4} \left(8y - y^2\right) dy = 624\left[4y^2 - \frac{y^3}{3}\right]_{0}^{4} = 26{,}624 \text{ lb}$$

Total force: $F_1 + F_2 = 46{,}592$ lb

# CHAPTER 8
# Integration Techniques, L'Hôpital's Rule, and Improper Integrals

# C H A P T E R   8
# Integration Techniques, L'Hôpital's Rule, and Improper Integrals

## Section 8.1   Basic Integration Rules

**1.** (a) $\dfrac{d}{dx}\left[2\sqrt{x^2+1}+C\right] = 2\left(\dfrac{1}{2}\right)\left(x^2+1\right)^{-1/2}(2x) = \dfrac{2x}{\sqrt{x^2+1}}$

(b) $\dfrac{d}{dx}\left[\sqrt{x^2+1}+C\right] = \dfrac{1}{2}\left(x^2+1\right)^{-1/2}(2x) = \dfrac{x}{\sqrt{x^2+1}}$

(c) $\dfrac{d}{dx}\left[\dfrac{1}{2}\sqrt{x^2+1}+C\right] = \dfrac{1}{2}\left(\dfrac{1}{2}\right)\left(x^2+1\right)^{-1/2}(2x) = \dfrac{x}{2\sqrt{x^2+1}}$

(d) $\dfrac{d}{dx}\left[\ln\left(x^2+1\right)+C\right] = \dfrac{2x}{x^2+1}$

$\displaystyle\int \dfrac{x}{\sqrt{x^2+1}}\,dx$ matches (b).

**3.** (a) $\dfrac{d}{dx}\left[\ln\sqrt{x^2+1}+C\right] = \dfrac{1}{2}\left(\dfrac{2x}{x^2+1}\right) = \dfrac{x}{x^2+1}$

(b) $\dfrac{d}{dx}\left[\dfrac{2x}{\left(x^2+1\right)^2}+C\right] = \dfrac{\left(x^2+1\right)^2(2) - (2x)(2)\left(x^2+1\right)(2x)}{\left(x^2+1\right)^4} = \dfrac{2\left(1-3x^2\right)}{\left(x^2+1\right)^3}$

(c) $\dfrac{d}{dx}[\arctan x + C] = \dfrac{1}{1+x^2}$

(d) $\dfrac{d}{dx}\left[\ln\left(x^2+1\right)+C\right] = \dfrac{2x}{x^2+1}$

$\displaystyle\int \dfrac{1}{x^2+1}\,dx$ matches (c).

**5.** $\displaystyle\int (5x-3)^4\,dx$

$u = 5x - 3,\ du = 5\,dx,\ n = 4$

Use $\displaystyle\int u^n\,du$.

**7.** $\displaystyle\int \dfrac{1}{\sqrt{x}\left(1-2\sqrt{x}\right)}\,dx$

$u = 1 - 2\sqrt{x},\ du = -\dfrac{1}{\sqrt{x}}\,dx$

Use $\displaystyle\int \dfrac{du}{u}$.

**9.** $\displaystyle\int \dfrac{3}{\sqrt{1-t^2}}\,dt$

$u = t,\ du = dt,\ a = 1$

Use $\displaystyle\int \dfrac{du}{\sqrt{a^2-u^2}}$.

**11.** $\displaystyle\int t\sin t^2\,dt$

$u = t^2,\ du = 2t\,dt$

Use $\displaystyle\int \sin u\,du$.

**13.** $\displaystyle\int (\cos x)e^{\sin x}\,dx$

$u = \sin x,\ du = \cos x\,dx$

Use $\displaystyle\int e^u\,du$.

**15.** Let $u = x - 5,\ du = dx$.

$\displaystyle\int 14(x-5)^6\,dx = 14\int (x-5)^6\,dx = 2(x-5)^7 + C$

**17.** Let $u = z - 10,\ du = dz$.

$\displaystyle\int \dfrac{7}{(z-10)^7}\,dz = 7\int (z-10)^{-7}\,dz = -\dfrac{7}{6(z-10)^6} + C$

**19.** $\displaystyle\int \left[v + \dfrac{1}{(3v-1)^3}\right]dv = \int v\,dv + \dfrac{1}{3}\int (3v-1)^{-3}(3)\,dv$

$= \dfrac{1}{2}v^2 - \dfrac{1}{6(3v-1)^2} + C$

**21.** Let $u = -t^3 + 9t + 1$,

$$du = \left(-3t^2 + 9\right) dt = -3\left(t^2 - 3\right) dt.$$

$$\int \frac{t^2 - 3}{-t^3 + 9t + 1} dt = -\frac{1}{3} \int \frac{-3\left(t^2 - 3\right)}{-t^3 + 9t + 1} dt$$

$$= -\frac{1}{3} \ln\left| -t^3 + 9t + 1 \right| + C$$

**23.** $\int \frac{x^2}{x - 1} dx = \int (x + 1) dx + \int \frac{1}{x - 1} dx$

$$= \frac{1}{2}x^2 + x + \ln|x - 1| + C$$

**25.** Let $u = 1 + e^x$, $du = e^x dx$.

$$\int \frac{e^x}{1 + e^x} dx = \ln\left(1 + e^x\right) + C$$

**27.** $\int \left(5 + 4x^2\right)^2 dx = \int \left(25 + 40x^2 + 16x^4\right) dx$

$$= 25x + \frac{40}{3}x^3 + \frac{16}{5}x^5 + C$$

$$= \frac{x}{15}\left(48x^5 + 200x^3 + 375\right) + C$$

**29.** Let $u = 2\pi x^2$, $du = 4\pi x \, dx$.

$$\int x\left(\cos 2\pi x^2\right) dx = \frac{1}{4\pi} \int \left(\cos 2\pi x^2\right)(4\pi x) dx$$

$$= \frac{1}{4\pi} \sin 2\pi x^2 + C$$

**31.** Let $u = \pi x$, $du = \pi \, dx$.

$$\int \csc \pi x \cot \pi x \, dx = \frac{1}{\pi} \int (\csc \pi x)(\cot \pi x) \pi \, dx$$

$$= -\frac{1}{\pi} \csc \pi x + C$$

**33.** $\int e^{11x} dx = \frac{1}{11} \int e^{11x}(11) \, dx = \frac{1}{11} e^{11x} + C$

**35.** Let $u = 1 + e^x$, $du = e^x dx$.

$$\int \frac{2}{e^{-x} + 1} dx = 2\int \left(\frac{2}{e^{-x} + 1}\right)\left(\frac{e^x}{e^x}\right) dx$$

$$= 2\int \frac{e^x}{1 + e^x} dx = 2 \ln\left(1 + e^x\right) + C$$

**37.** $\int \frac{\ln x^2}{x} dx = 2 \int (\ln x)\frac{1}{x} dx$

$$= 2 \frac{(\ln x)^2}{2} + C = (\ln x)^2 + C$$

**39.** $\int \frac{1 + \sin x}{\cos x} dx = \int \frac{1 + \sin x}{\cos x} \cdot \frac{1 - \sin x}{1 - \sin x} dx$

$$= \int \frac{1 - \sin^2 x}{\cos x(1 - \sin x)} dx$$

$$= \int \frac{\cos^2 x}{\cos x(1 - \sin x)} dx$$

$$= -\int \frac{-\cos x}{1 - \sin x} dx$$

$$= -\ln(1 - \sin x) + C, \qquad (u = 1 - \sin x)$$

**Alternate Solution:**

$$\int \frac{1 + \sin x}{\cos x} dx = \int (\sec x + \tan x) dx$$

$$= \ln|\sec x + \tan x| + \ln|\sec x| + C$$

$$= \ln|\sec x(\sec x + \tan x)| + C$$

**41.**

$$\frac{1}{\cos \theta - 1} = \frac{1}{\cos \theta - 1} \cdot \frac{\cos \theta + 1}{\cos \theta + 1} = \frac{\cos \theta + 1}{\cos^2 \theta - 1}$$

$$= \frac{\cos \theta + 1}{-\sin^2 \theta} = -\csc \theta \cdot \cot \theta - \csc^2 \theta$$

$$\int \frac{1}{\cos \theta - 1} d\theta = \int \left(-\csc \theta \cot \theta - \csc^2 \theta\right) d\theta$$

$$= \csc \theta + \cot \theta + C$$

$$= \frac{1}{\sin \theta} + \frac{\cos \theta}{\sin \theta} + C$$

$$= \frac{1 + \cos \theta}{\sin \theta} + C$$

**43.** Let $u = 4t + 1$, $du = 4 \, dt$.

$$\int \frac{-1}{\sqrt{1 - (4t + 1)^2}} dt = -\frac{1}{4} \int \frac{4}{\sqrt{1 - (4t + 1)^2}} dt$$

$$= -\frac{1}{4} \arcsin(4t + 1) + C$$

**45.** Let $u = \cos\left(\frac{2}{t}\right)$, $du = \frac{2 \sin(2/t)}{t^2} dt$.

$$\int \frac{\tan(2/t)}{t^2} dt = \frac{1}{2} \int \frac{1}{\cos(2/t)} \left[\frac{2 \sin(2/t)}{t^2}\right] dt$$

$$= \frac{1}{2} \ln\left| \cos\left(\frac{2}{t}\right) \right| + C$$

**47.** Note: $10x - x^2 = 25 - (25 - 10x + x^2)$

$$= 25 - (5 - x)^2$$

$$\int \frac{6}{\sqrt{10x - x^2}} \, dx = 6 \int \frac{1}{\sqrt{25 - (5 - x)^2}} \, dx$$

$$= 6 \int \frac{-1}{\sqrt{5^2 - (5 - x)^2}} \, dx$$

$$= -6 \arcsin \frac{(5 - x)}{5} + C$$

$$= 6 \arcsin \left( \frac{x - 5}{5} \right) + C$$

**49.** $\int \frac{4}{4x^2 + 4x + 65} \, dx = \int \frac{1}{[x + (1/2)]^2 + 16} \, dx$

$$= \frac{1}{4} \arctan \left[ \frac{x + (1/2)}{4} \right] + C$$

$$= \frac{1}{4} \arctan \left( \frac{2x + 1}{8} \right) + C$$

**51.** $\int \frac{1}{\sqrt{1 - 4x - x^2}} \, dx = \int \frac{1}{\sqrt{5 - (x^2 + 4x + 4)}} \, dx$

$$= \int \frac{1}{\sqrt{5 - (x + 2)^2}} \, dx$$

$$= \arcsin \left( \frac{x + 2}{\sqrt{5}} \right) + C, \left( a = \sqrt{5} \right)$$

**53.** $\dfrac{ds}{dt} = \dfrac{t}{\sqrt{1 - t^4}}, \quad \left( 0, -\dfrac{1}{2} \right)$

(a)

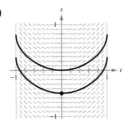

(b) $u = t^2, du = 2t \, dt$

$$\int \frac{t}{\sqrt{1 - t^4}} \, dt = \frac{1}{2} \int \frac{2t}{\sqrt{1 - (t^2)^2}} \, dt = \frac{1}{2} \arcsin t^2 + C$$

$$\left( 0, -\frac{1}{2} \right): -\frac{1}{2} = \frac{1}{2} \arcsin 0 + C \Rightarrow C = -\frac{1}{2}$$

$$s = \frac{1}{2} \arcsin t^2 - \frac{1}{2}$$

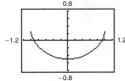

**55.** (a) $\dfrac{dy}{dx} = (\sec x + \tan x), \quad (0, 1)$

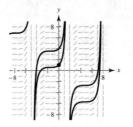

(b) $y = \int (\sec x + \tan x)^2 \, dx$

$$= \int (\sec^2 x + 2 \sec x \tan x + \tan^2 x) \, dx$$

$$= \int (\sec^2 x + 2 \sec x \tan x + (\sec^2 x - 1)) \, dx$$

$$= \int (2 \sec^2 x + 2 \sec x \tan x - 1) \, dx$$

$$= 2 \tan x + 2 \sec x - x + C$$

$$(0, 1): 1 = 0 + 2 - 0 + C \Rightarrow C = -1$$

$$y = 2 \tan x + 2 \sec x - x - 1$$

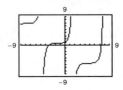

**57.**

$$y = 4e^{0.8x}$$

**59.** $\dfrac{dy}{dx} = (e^x + 5)^2 = e^{2x} + 10e^x + 25$

$$y = \int (e^{2x} + 10e^x + 25) \, dx$$

$$= \frac{1}{2} e^{2x} + 10e^x + 25x + C$$

**61.** $\dfrac{dr}{dt} = \dfrac{10e^t}{\sqrt{1 - e^{2t}}}$

$$r = \int \frac{10e^t}{1 - (e^t)^2} \, dt = 10 \arcsin (e^t) + C$$

**63.** $\dfrac{dy}{dx} = \dfrac{\sec^2 x}{4 + \tan^2 x}$

Let $u = \tan x, du = \sec^2 x \, dx$.

$$y = \int \frac{\sec^2 x}{4 + \tan^2 x} \, dx = \frac{1}{2} \arctan \left( \frac{\tan x}{2} \right) + C$$

**65.** Let $u = 2x$, $du = 2\,dx$.

$$\int_0^{\pi/4} \cos 2x\,dx = \frac{1}{2}\int_0^{\pi/4} \cos 2x(2)\,dx$$

$$= \left[\frac{1}{2}\sin 2x\right]_0^{\pi/4} = \frac{1}{2}$$

**67.** Let $u = -x^2$, $du = -2x\,dx$.

$$\int_0^1 xe^{-x^2}\,dx = -\frac{1}{2}\int_0^1 e^{-x^2}(-2x)\,dx = \left[-\frac{1}{2}e^{-x^2}\right]_0^1$$

$$= \frac{1}{2}\left(1 - e^{-1}\right) \approx 0.316$$

**69.** Let $u = x^2 + 36$, $du = 2x\,dx$.

$$\int_0^8 \frac{2x}{\sqrt{x^2 + 36}}\,dx = \int_0^8 \left(x^2 + 36\right)^{-1/2}(2x)\,dx$$

$$= 2\left[\left(x^2 + 36\right)^{1/2}\right]_0^8 = 8$$

**71.** Let $u = 3x$, $du = 3\,dx$.

$$\int_0^{2/\sqrt{3}} \frac{1}{4 + 9x^2}\,dx = \frac{1}{3}\int_0^{2/\sqrt{3}} \frac{3}{4 + (3x)^2}\,dx$$

$$= \left[\frac{1}{6}\arctan\left(\frac{3x}{2}\right)\right]_0^{2/\sqrt{3}}$$

$$= \frac{\pi}{18} \approx 0.175$$

**73.** $A = \displaystyle\int_0^{3/2} (-4x + 6)^{3/2}\,dx$

$$= -\frac{1}{4}\int_0^{3/2} (6 - 4x)^{3/2}(-4)\,dx$$

$$= -\frac{1}{4}\left[\frac{2}{5}(6 - 4x)^{5/2}\right]_0^{3/2}$$

$$= -\frac{1}{10}\left(0 - 6^{5/2}\right)$$

$$= \frac{18}{5}\sqrt{6} \approx 8.8182$$

**75.** $A = \displaystyle\int_0^5 \frac{3x + 2}{x^2 + 9}\,dx$

$$= \int_0^5 \frac{3x}{x^2 + 9}\,dx + \int_0^5 \frac{2}{x^2 + 9}\,dx$$

$$= \left[\frac{3}{2}\ln\left|x^2 + 9\right| + \frac{2}{3}\arctan\left(\frac{x}{3}\right)\right]_0^5$$

$$= \frac{3}{2}\ln(34) + \frac{2}{3}\arctan\left(\frac{5}{3}\right) - \frac{3}{2}\ln 9$$

$$= \frac{3}{2}\ln\left(\frac{34}{9}\right) + \frac{2}{3}\arctan\left(\frac{5}{3}\right)$$

$$\approx 2.6806$$

**77.** $y^2 = x^2\left(1 - x^2\right)$

$$y = \pm\sqrt{x^2\left(1 - x^2\right)}$$

$$A = 4\int_0^1 x\sqrt{1 - x^2}\,dx$$

$$= -2\int_0^1 \left(1 - x^2\right)^{1/2}(-2x)\,dx$$

$$= -\frac{4}{3}\left[\left(1 - x\right)^{3/2}\right]_0^1$$

$$= -\frac{4}{3}(0 - 1) = \frac{4}{3}$$

**79.** $\displaystyle\int \frac{1}{x^2 + 4x + 13}\,dx = \frac{1}{3}\arctan\left(\frac{x + 2}{3}\right) + C$

The antiderivatives are vertical translations of each other.

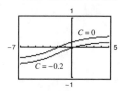

**81.** $\displaystyle\int \frac{1}{1 + \sin\theta}\,d\theta = \tan\theta - \sec\theta + C \quad \left(\text{or } \frac{-2}{1 + \tan(\theta/2)}\right)$

The antiderivatives are vertical translations of each other.

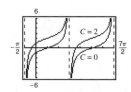

**83.** Power Rule: $\displaystyle\int u^n\,du = \frac{u^{n+1}}{n + 1} + C, \quad n \ne -1$

$$u = x^2 + 1, n = 3$$

**85.** Log Rule: $\displaystyle\int \frac{du}{u} = \ln|u| + C, \quad u = x^2 + 1$

**87.** $\sin x + \cos x = a \sin(x + b)$

$\sin x + \cos x = a \sin x \cos b + a \cos x \sin b$

$\sin x + \cos x = (a \cos b) \sin x + (a \sin b) \cos x$

Equate coefficients of like terms to obtain the following.

$1 = a \cos b$ and $1 = a \sin b$

So, $a = 1/\cos b$. Now, substitute for $a$ in $1 = a \sin b$.

$1 = \left(\dfrac{1}{\cos b}\right) \sin b$

$1 = \tan b \Rightarrow b = \dfrac{\pi}{4}$

Because $b = \dfrac{\pi}{4}, a = \dfrac{1}{\cos(\pi/4)} = \sqrt{2}$. So,

$\sin x + \cos x = \sqrt{2} \sin\left(x + \dfrac{\pi}{4}\right)$.

$\displaystyle\int \dfrac{dx}{\sin x + \cos x} = \int \dfrac{dx}{\sqrt{2} \sin(x + (\pi/4))}$

$\displaystyle = \dfrac{1}{\sqrt{2}} \int \csc\left(x + \dfrac{\pi}{4}\right) dx$

$\displaystyle = -\dfrac{1}{\sqrt{2}} \ln\left|\csc\left(x + \dfrac{\pi}{4}\right) + \cot\left(x + \dfrac{\pi}{4}\right)\right| + C$

**89.** $\displaystyle\int_0^{1/a} (x - ax^2)\, dx = \left[\dfrac{1}{2}x^2 - \dfrac{a}{3}x^3\right]_0^{1/a} = \dfrac{1}{6a^2}$

Let $\dfrac{1}{6a^2} = \dfrac{2}{3}, 12a^2 = 3, a = \dfrac{1}{2}$.

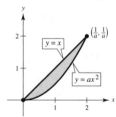

**91.** $f(x) = \dfrac{1}{5}(x^3 - 7x^2 + 10x)$

$\displaystyle\int_0^5 f(x)\, dx < 0$ because more area is below the $x$-axis than above.

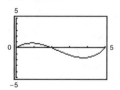

**93.** $\displaystyle\int_0^2 \dfrac{4x}{x^2 + 1}\, dx \approx 3$

Matches (a).

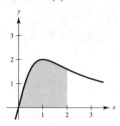

**95.** (a) $y = 2\pi x^2, \quad 0 \le x \le 2$

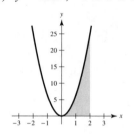

(b) $y = \sqrt{2}x, \quad 0 \le x \le 2$

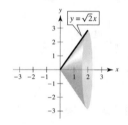

(c) $y = x, \quad 0 \le x \le 2$

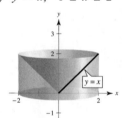

**97.** (a) **Shell Method:**

Let $u = -x^2$, $du = -2x\,dx$.

$$V = 2\pi \int_0^1 xe^{-x^2}\,dx$$

$$= -\pi \int_0^1 e^{-x^2}(-2x)\,dx$$

$$= \left[ -\pi e^{-x^2} \right]_0^1$$

$$= \pi\left(1 - e^{-1}\right) \approx 1.986$$

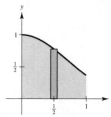

(b) **Shell Method:**

$$V = 2\pi \int_0^b xe^{-x^2}\,dx$$

$$= \left[ -\pi e^{-x^2} \right]_0^b$$

$$= \pi\left(1 - e^{-b^2}\right) = \frac{4}{3}$$

$$e^{-b^2} = \frac{3\pi - 4}{3\pi}$$

$$b = \sqrt{\ln\left(\frac{3\pi}{3\pi - 4}\right)} \approx 0.743$$

**99.** $y = f(x) = \ln(\sin x)$

$$f'(x) = \frac{\cos x}{\sin x}$$

$$s = \int_{\pi/4}^{\pi/2} \sqrt{1 + \frac{\cos^2 x}{\sin^2 x}}\,dx = \int_{\pi/4}^{\pi/2} \sqrt{\frac{\sin^2 x + \cos^2 x}{\sin^2 x}}\,dx$$

$$= \int_{\pi/4}^{\pi/2} \frac{1}{\sin x}\,dx = \int_{\pi/4}^{\pi/2} \csc x\,dx$$

$$= \left[ -\ln\left| \csc x + \cot x \right| \right]_{\pi/4}^{\pi/2}$$

$$= -\ln(1) + \ln\left(\sqrt{2} + 1\right)$$

$$= \ln\left(\sqrt{2} + 1\right) \approx 0.8814$$

**101.** $y = 2\sqrt{x}$

$$y' = \frac{1}{\sqrt{x}}$$

$$1 + (y')^2 = 1 + \frac{1}{x} = \frac{x + 1}{x}$$

$$S = 2\pi \int_0^9 2\sqrt{x}\sqrt{\frac{x + 1}{x}}\,dx$$

$$= 2\pi \int_0^9 2\sqrt{x + 1}\,dx$$

$$= \left[ 4\pi\left(\frac{2}{3}\right)(x + 1)^{3/2} \right]_0^9 = \frac{8\pi}{3}\left(10\sqrt{10} - 1\right) \approx 256.545$$

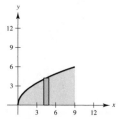

**103.** Average value $= \dfrac{1}{b - a}\displaystyle\int_a^b f(x)\,dx$

$$= \frac{1}{3 - (-3)}\int_{-3}^3 \frac{1}{1 + x^2}\,dx$$

$$= \frac{1}{6}\left[ \arctan(x) \right]_{-3}^3$$

$$= \frac{1}{6}\left[ \arctan(3) - \arctan(-3) \right]$$

$$= \frac{1}{3}\arctan(3) \approx 0.4163$$

**105.** $\qquad y = \tan(\pi x)$

$$y' = \pi \sec^2(\pi x)$$

$$1 + (y')^2 = 1 + \pi^2 \sec^4(\pi x)$$

$$s = \int_0^{1/4} \sqrt{1 + \pi^2 \sec^4(\pi x)}\,dx \approx 1.0320$$

**107.** (a) $\displaystyle\int \cos^3 x\,dx = \int\left(1 - \sin^2 x\right)\cos x\,dx = \sin x - \frac{\sin^3 x}{3} + C = \frac{1}{3}\sin x\left(\cos^2 x + 2\right) + C$

(b) $\displaystyle\int \cos^5 x\,dx = \int\left(1 - \sin^2 x\right)^2 \cos x\,dx = \int\left(1 - 2\sin^2 x + \sin^4 x\right)\cos x\,dx$

$$= \sin x - \frac{2}{3}\sin^3 x + \frac{\sin^5 x}{5} + C = \frac{1}{15}\sin x\left(3\cos^4 x + 4\cos^2 x + 8\right) + C$$

(c) $\int \cos^7 x\, dx = \int \left(1 - \sin^2 x\right)^3 \cos x\, dx$

$$= \int \left(1 - 3\sin^2 x + 3\sin^4 x - \sin^6 x\right) \cos x\, dx$$

$$= \sin x - \sin^3 x + \frac{3}{5}\sin^5 x - \frac{1}{7}\sin^7 x + C$$

$$= \frac{1}{35}\sin x\left(5\cos^6 x + 6\cos^4 x + 8\cos^2 x + 16\right) + C$$

(d) $\int \cos^{15} x\, dx = \int \left(1 - \sin^2 x\right)^7 \cos x\, dx$

You would expand $\left(1 - \sin^2 x\right)^7$.

**109.** Let $f(x) = \frac{1}{2}\left(x\sqrt{x^2 + 1} + \ln\left|x + \sqrt{x^2 + 1}\right|\right) + C.$

$$f'(x) = \frac{1}{2}\left(x\frac{1}{2}\left(x^2 + 1\right)^{-1/2}(2x) + \sqrt{x^2 + 1} + \frac{1}{x + \sqrt{x^2 + 1}}\left(1 + \frac{1}{2}\left(x^2 + 1\right)^{-1/2}(2x)\right)\right)$$

$$= \frac{1}{2}\left(\frac{x^2}{\sqrt{x^2 + 1}} + \sqrt{x^2 + 1} + \frac{1}{x + \sqrt{x^2 + 1}}\left(1 + \frac{x}{\sqrt{x^2 + 1}}\right)\right)$$

$$= \frac{1}{2}\left(\frac{x^2 + \left(x^2 + 1\right)}{\sqrt{x^2 + 1}} + \frac{1}{x + \sqrt{x^2 + 1}}\left(\frac{\sqrt{x^2 + 1} + x}{\sqrt{x^2 + 1}}\right)\right)$$

$$= \frac{1}{2}\left(\frac{2x^2 + 1}{\sqrt{x^2 + 1}} + \frac{1}{\sqrt{x^2 + 1}}\right) = \frac{1}{2}\left(\frac{2\left(x^2 + 1\right)}{\sqrt{x^2 + 1}}\right) = \sqrt{x^2 + 1}$$

So, $\int \sqrt{x^2 + 1}\, dx = \frac{1}{2}\left(x\sqrt{x^2 + 1} + \ln\left|x + \sqrt{x^2 + 1}\right|\right) + C.$

Let $g(x) = \frac{1}{2}\left(x\sqrt{x^2 + 1} + \operatorname{arcsinh}(x)\right).$

$$g'(x) = \frac{1}{2}\left(x\frac{1}{2}\left(x^2 + 1\right)^{-1/2}(2x) + \sqrt{x^2 + 1} + \frac{1}{\sqrt{x^2 + 1}}\right)$$

$$= \frac{1}{2}\left(\frac{x^2}{\sqrt{x^2 + 1}} + \sqrt{x^2 + 1} + \frac{1}{\sqrt{x^2 + 1}}\right)$$

$$= \frac{1}{2}\left(\frac{x^2 + \left(x^2 + 1\right) + 1}{\sqrt{x^2 + 1}}\right)$$

$$= \frac{1}{2}\left(\frac{2\left(x^2 + 1\right)}{\sqrt{x^2 + 1}}\right) = \sqrt{x^2 + 1}$$

So, $\int \sqrt{x^2 + 1}\, dx = \frac{1}{2}\left(x\sqrt{x^2 + 1} + \operatorname{arcsinh}(x)\right) + C.$

## Section 8.2   Integration by Parts

**1.** $\int xe^{2x}\, dx$

$u = x,\, dv = e^{2x}\, dx$

**3.** $\int (\ln x)^2\, dx$

$u = (\ln x)^2,\, dv = dx$

**5.** $\int x\sec^2 x\, dx$

$u = x,\, dv = \sec^2 x\, dx$

**7.** $dv = x^3 \, dx \Rightarrow v = \int x^3 \, dx = \dfrac{x^4}{4}$

$u = \ln x \Rightarrow du = \dfrac{1}{x} dx$

$\int x^3 \ln x \, dx = uv - \int v \, du$

$\qquad = (\ln x)\dfrac{x^4}{4} - \int \dfrac{x^4}{4}\dfrac{1}{x} dx$

$\qquad = \dfrac{x^4}{4} \ln x - \dfrac{1}{4}\int x^3 \, dx$

$\qquad = \dfrac{x^4}{4} \ln x - \dfrac{1}{16}x^4 + C$

$\qquad = \dfrac{1}{16}x^4(4 \ln x - 1) + C$

**9.** $dv = \sin 3x \, dx \Rightarrow v = \int \sin 3x \, dx = -\dfrac{1}{3}\cos 3x$

$u = x \qquad \Rightarrow du = dx$

$\int x \sin 3x \, dx = uv - \int v \, du$

$\qquad = x\left(-\dfrac{1}{3}\cos 3x\right) - \int -\dfrac{1}{3}\cos 3x \, dx$

$\qquad = -\dfrac{x}{3}\cos 3x + \dfrac{1}{9}\sin 3x + C$

**11.** $dv = e^{-4x} \, dx \Rightarrow v = \int e^{-4x} \, dx = -\dfrac{1}{4}e^{-4x}$

$u = x \qquad \Rightarrow du = dx$

$\int xe^{-4x} \, dx = x\left(-\dfrac{1}{4}e^{-4x}\right) - \int -\dfrac{1}{4}e^{-4x} \, dx$

$\qquad = -\dfrac{x}{4}e^{-4x} - \dfrac{1}{16}e^{-4x} + C$

$\qquad = -\dfrac{1}{16e^{4x}}(1 + 4x) + C$

**13.** Use integration by parts three times.

(1) $dv = e^x \, dx \Rightarrow v = \int e^x \, dx = e^x$

$\quad u = x^3 \quad \Rightarrow du = 3x^2 \, dx$

(2) $dv = e^x \, dx \Rightarrow v = \int e^x \, dx = e^x$

$\quad u = x^2 \quad \Rightarrow du = 2x \, dx$

(3) $dv = e^x \, dx \Rightarrow v = \int e^x \, dx = e^x$

$\quad u = x \quad \Rightarrow du = dx$

$\int x^3 e^x \, dx = x^3 e^x - 3\int x^2 e^x \, dx = x^3 e^x - 3x^2 e^x + 6\int xe^x \, dx$

$\qquad = x^3 e^x - 3x^2 e^x + 6xe^x - 6e^x + C = e^x(x^3 - 3x^2 + 6x - 6) + C$

**15.** $\int x^2 e^{x^3} \, dx = \dfrac{1}{3}\int e^{x^3}(3x^2) \, dx = \dfrac{1}{3}e^{x^3} + C$

**17.** $dv = t \, dt \quad \Rightarrow v = \int t \, dt = \dfrac{t^2}{2}$

$u = \ln(t + 1) \Rightarrow du = \dfrac{1}{t+1} dt$

$\int t \ln(t + 1) \, dt = \dfrac{t^2}{2}\ln(t + 1) - \dfrac{1}{2}\int \dfrac{t^2}{t+1} dt$

$\qquad = \dfrac{t^2}{2}\ln(t + 1) - \dfrac{1}{2}\int\left(t - 1 + \dfrac{1}{t+1}\right)dt$

$\qquad = \dfrac{t^2}{2}\ln(t + 1) - \dfrac{1}{2}\left[\dfrac{t^2}{2} - t + \ln(t + 1)\right] + C$

$\qquad = \dfrac{1}{4}\left[2(t^2 - 1)\ln|t + 1| - t^2 + 2t\right] + C$

**19.** Let $u = \ln x$, $du = \dfrac{1}{x} dx$.

$\int \dfrac{(\ln x)^2}{x} dx = \int (\ln x)^2\left(\dfrac{1}{x}\right)dx = \dfrac{(\ln x)^3}{3} + C$

**21.** $dv = \dfrac{1}{(2x + 1)^2} dx \Rightarrow v = \int (2x + 1)^{-2} \, dx$

$\qquad\qquad\qquad\qquad = -\dfrac{1}{2(2x + 1)}$

$u = xe^{2x} \qquad \Rightarrow du = (2xe^{2x} + e^{2x}) \, dx$

$\qquad\qquad\qquad\qquad = e^{2x}(2x + 1) \, dx$

$\int \dfrac{xe^{2x}}{(2x + 1)^2} dx = -\dfrac{xe^{2x}}{2(2x + 1)} + \int \dfrac{e^{2x}}{2} dx$

$\qquad = \dfrac{-xe^{2x}}{2(2x + 1)} + \dfrac{e^{2x}}{4} + C = \dfrac{e^{2x}}{4(2x + 1)} + C$

**23.** Use integration by parts twice.

(1) $dv = e^x \, dx \Rightarrow v = \int e^x \, dx = e^x$      (2) $dv = e^x \, dx \Rightarrow v = \int e^x \, dx = e^x$

    $u = x^2 \Rightarrow du = 2x \, dx$          $u = x \Rightarrow du = dx$

$\int (x^2 - 1) e^x \, dx = \int x^2 e^x \, dx - \int e^x \, dx = x^2 e^x - 2 \int x e^x \, dx - e^x$

$\qquad = x^2 e^x - 2\left( x e^x - \int e^x \, dx \right) - e^x = x^2 e^x - 2x e^x + e^x + C = (x - 1)^2 e^x + C$

**25.** $dv = \sqrt{x - 5} \, dx \Rightarrow v = \int (x - 5)^{1/2} \, dx = \frac{2}{3}(x - 5)^{3/2}$

    $u = x \Rightarrow du = dx$

$\int x \sqrt{x - 5} \, dx = x \frac{2}{3}(x - 5)^{3/2} - \int \frac{2}{3}(x - 5)^{3/2} \, dx$

$\qquad = \frac{2}{3} x (x - 5)^{3/2} - \frac{4}{15}(x - 5)^{5/2} + C$

$\qquad = \frac{2}{15}(x - 5)^{3/2} \left( 5x - 2(x - 5) \right) + C$

$\qquad = \frac{2}{15}(x - 5)^{3/2} (3x + 10) + C$

**27.** $dv = \cos x \, dx \Rightarrow v = \int \cos x \, dx = \sin x$

    $u = x \Rightarrow du = dx$

$\int x \cos x \, dx = x \sin x - \int \sin x \, dx = x \sin x + \cos x + C$

**29.** Use integration by parts three times.

(1) $u = x^3, du = 3x^2 \, dx, dv = \sin x \, dx, v = -\cos x$

$\int x^3 \sin dx = -x^3 \cos x + 3 \int x^2 \cos x \, dx$

(2) $u = x^2, du = 2x \, dx, dv = \cos x \, dx, v = \sin x$

$\int x^3 \sin x \, dx = -x^3 \cos x + 3\left( x^2 \sin x - 2\int x \sin x \, dx \right) = -x^3 \cos x + 3x^2 \sin x - 6\int x \sin x \, dx$

(3) $u = x, du = dx, dv = \sin x \, dx, v = -\cos x$

$\int x^3 \sin x \, dx = -x^3 \cos x + 3x^2 \sin x - 6\left( -x \cos x + \int \cos x \, dx \right)$

$\qquad = -x^3 \cos x + 3x^2 \sin x + 6x \cos x - 6 \sin x + C$

$\qquad = \left( 6x - x^3 \right) \cos x + \left( 3x^2 - 6 \right) \sin x + C$

**31.** $u = t, du = dt, dv = \csc t \cot t \, dt, v = -\csc t$

$\int t \csc t \cot t \, dt = -t \csc t + \int \csc t \, dt = -t \csc t - \ln|\csc t + \cot t| + C$

**33.** $dv = dx \Rightarrow v = \int dx = x$

    $u = \arctan x \Rightarrow du = \frac{1}{1 + x^2} \, dx$

$\int \arctan x \, dx = x \arctan x - \int \frac{x}{1 + x^2} \, dx$

$\qquad = x \arctan x - \frac{1}{2} \ln(1 + x^2) + C$

**35.** Use integration by parts twice.

(1) $dv = e^{2x} \, dx \implies v = \int e^{2x} \, dx = \frac{1}{2}e^{2x}$     (2) $dv = e^{2x} \, dx \implies v = \int e^{2x} \, dx = \frac{1}{2}e^{2x}$

    $u = \sin x \implies du = \cos x \, dx$             $u = \cos x \implies du = -\sin x \, dx$

$$\int e^{2x} \sin x \, dx = \frac{1}{2}e^{2x} \sin x - \frac{1}{2}\int e^{2x} \cos x \, dx = \frac{1}{2}e^{2x} \sin x - \frac{1}{2}\left(\frac{1}{2}e^{2x} \cos x + \frac{1}{2}\int e^{2x} \sin x \, dx\right)$$

$$\frac{5}{4}\int e^{2x} \sin x \, dx = \frac{1}{2}e^{2x} \sin x - \frac{1}{4}e^{2x} \cos x$$

$$\int e^{2x} \sin x \, dx = \frac{1}{5}e^{2x}(2 \sin x - \cos x) + C$$

**37.** Use integration by parts twice.

(1) $dv = e^{-x} \, dx \implies v = \int e^{-x} \, dx = -e^{-x}$

    $u = \cos 2x \implies du = -2 \sin 2x \, dx$

$$\int e^{-x} \cos 2x \, dx = \cos 2x\left(-e^{-x}\right) - \int \left(-e^{-x}\right)(-2 \sin 2x) \, dx = -e^{-x} \cos 2x - 2\int e^{-x} \sin 2x \, dx$$

(2) $dv = e^{-x} \, dx \implies v = \int e^{-x} \, dx = -e^{-x}$

    $u = \sin 2x \implies du = 2 \cos 2x \, dx$

$$\int e^{-x} \cos 2x \, dx = -e^{-x} \cos 2x - 2\left[\sin 2x\left(-e^{-x}\right) - \int \left(-e^{-x}\right)(2 \cos 2x) \, dx\right]$$

$$= -e^{-x} \cos 2x + 2e^{-x} \sin 2x - 4\int e^{-x} \cos 2x \, dx$$

$$(4 + 1)\int e^{-x} \cos 2x \, dx = -e^{-x} \cos 2x + 2e^{-x} \sin 2x$$

$$\int e^{-x} \cos 2x \, dx = \frac{1}{5}e^{-x}(2 \sin 2x - \cos 2x) + C$$

**39.** $y' = xe^{x^2}$

$$y = \int xe^{x^2} \, dx = \frac{1}{2}e^{x^2} + C$$

**41.** Use integration by parts twice.

(1) $dv = \dfrac{1}{\sqrt{3 + 5t}} \, dt \implies v = \int (3 + 5t)^{-1/2} \, dt = \dfrac{2}{5}(3 + 5t)^{1/2}$

    $u = t^2 \qquad\qquad\quad \implies du = 2t \, dt$

$$\int \frac{t^2}{\sqrt{3 + 5t}} \, dt = \frac{2t^2}{5}(3 + 5t)^{1/2} - \int \frac{2}{5}(3 + 5t)^{1/2} 2t \, dt$$

$$= \frac{2}{5}t^2(3 + 5t)^{1/2} - \frac{4}{5}\int t(3 + 5t)^{1/2} \, dt$$

(2) $dv = (3 + 5t)^{1/2} \, dt \implies v = \int (3 + 5t)^{1/2} \, dt = \dfrac{2}{15}(3 + 5t)^{3/2}$

    $u = t \qquad\qquad\quad \implies du = dt$

$$\int \frac{t^2}{\sqrt{3 + 5t}} \, dt = \frac{2}{5}t^2(3 + 5t)^{1/2} - \frac{4}{5}\left[\frac{2t}{15}(3 + 5t)^{3/2} - \int \frac{2}{15}(3 + 5t)^{3/2} \, dt\right]$$

$$= \frac{2}{5}t^2(3 + 5t)^{1/2} - \frac{8t}{75}(3 + 5t)^{3/2} + \frac{8}{75}\int (3 + 5t)^{3/2} \, dt$$

$$= \frac{2}{5}t^2(3 + 5t)^{1/2} - \frac{8t}{75}(3 + 5t)^{3/2} + \frac{16}{1875}(3 + 5t)^{5/2} + C$$

$$= \frac{2}{1875}\sqrt{3 + 5t}\left(375t^2 - 100t(3 + 5t) + 8(3 + 5t)^2\right) + C$$

$$= \frac{2}{625}\sqrt{3 + 5t}\left(25t^2 - 20t + 24\right) + C$$

**43.** $(\cos y)y' = 2x$

$$\int \cos y \, dy = \int 2x \, dx$$

$$\sin y = x^2 + C$$

**45. (a)**

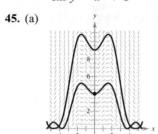

**(b)**      $\dfrac{dy}{dx} = x\sqrt{y} \cos x, \quad (0, 4)$

$$\int \frac{dy}{\sqrt{y}} = \int x \cos x \, dx$$

$$\int y^{-1/2} \, dy = \int x \cos x \, dx \qquad (u = x, du = dx, dv = \cos x \, dx, v = \sin x)$$

$$2y^{1/2} = x \sin x - \int \sin x \, dx = x \sin x + \cos x + C$$

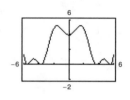

$$(0, 4): 2(4)^{1/2} = 0 + 1 + C \Rightarrow C = 3$$

$$2\sqrt{y} = x \sin x + \cos x + 3$$

**47.** $\dfrac{dy}{dx} = \dfrac{x}{y} e^{x/8}, \, y(0) = 2$

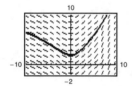

**49.** $u = x, du = dx, dv = e^{x/2} \, dx, v = 2e^{x/2}$

$$\int xe^{x/2} \, dx = 2xe^{x/2} - \int 2e^{x/2} \, dx = 2xe^{x/2} - 4e^{x/2} + C$$

So,

$$\int_0^3 xe^{x/2} \, dx = \left[ 2xe^{x/2} - 4e^{x/2} \right]_0^3 = \left( 6e^{3/2} - 4e^{3/2} \right) - (-4) = 4 + 2e^{3/2} \approx 12.963$$

**51.** $u = x, du = dx, dv = \cos 2x \, dx, v = \dfrac{1}{2} \sin 2x$

$$\int x \cos 2x \, dx = \frac{1}{2}x \sin 2x - \int \frac{1}{2} \sin 2x \, dx = \frac{1}{2}x \sin 2x + \frac{1}{4} \cos 2x + C$$

So,

$$\int_0^{\pi/4} x \cos 2x \, dx = \left[ \frac{1}{2}x \sin 2x + \frac{1}{4} \cos 2x \right]_0^{\pi/4}$$

$$= \left( \frac{\pi}{8}(1) + 0 \right) - \left( 0 + \frac{1}{4} \right)$$

$$= \frac{\pi}{8} - \frac{1}{4} \approx 0.143$$

**53.** $u = \arccos x, \; du = -\dfrac{1}{\sqrt{1-x^2}}\, dx, \; dv = dx, \; v = x$

$$\int \arccos x \, dx = x \arccos x + \int \frac{x}{\sqrt{1-x^2}}\, dx$$

$$= x \arccos x - \sqrt{1-x^2} + C$$

So,

$$\int_0^{1/2} \arccos x = \left[ x \arccos x - \sqrt{1-x^2} \right]_0^{1/2}$$

$$= \frac{1}{2} \arccos\left(\frac{1}{2}\right) - \sqrt{\frac{3}{4}} + 1$$

$$= \frac{\pi}{6} - \frac{\sqrt{3}}{2} + 1 \approx 0.658.$$

**55.** Use integration by parts twice.

(1)  $dv = e^x \, dx \;\Rightarrow\; v = \int e^x \, dx = e^x$    (2)  $dv = e^x \, dx \;\Rightarrow\; v = \int e^x \, dx = e^x$

   $u = \sin x \;\Rightarrow\; du = \cos x \, dx$        $u = \cos x \;\Rightarrow\; du = -\sin x \, dx$

$$\int e^x \sin x \, dx = e^x \sin x - \int e^x \cos x \, dx = e^x \sin x - e^x \cos x - \int e^x \sin x \, dx$$

$$2 \int e^x \sin x \, dx = e^x (\sin x - \cos x)$$

$$\int e^x \sin x \, dx = \frac{e^x}{2}(\sin x - \cos x) + C$$

So, $\displaystyle\int_0^1 e^x \sin x \, dx = \left[ \frac{e^x}{2}(\sin x - \cos x) \right]_0^1 = \frac{e}{2}(\sin 1 - \cos 1) + \frac{1}{2} = \frac{e(\sin 1 - \cos 1) + 1}{2} \approx 0.909.$

**57.** $u = \ln x, \; du = \dfrac{1}{x}\, dx, \; dv = \sqrt{x}\, dx, \; v = \dfrac{2}{3}x^{3/2}$

$$\int \sqrt{x} \ln x \, dx = \frac{2}{3}x^{3/2} \ln x - \int \frac{2}{3}x^{3/2}\frac{1}{x}\, dx = \frac{2}{3}x^{3/2} \ln x - \frac{2}{3}\int x^{1/2}\, dx = \frac{2}{3}x^{3/2} \ln x - \frac{4}{9}x^{3/2} + C$$

So, $\displaystyle\int_1^2 \sqrt{x} \ln x \, dx = \left[ \frac{2}{3}x^{3/2} \ln x - \frac{4}{9}x^{3/2} \right]_1^2 = \left( \frac{4}{3}\sqrt{2} \ln 2 - \frac{4}{9}2\sqrt{2} \right) - \left( 0 - \frac{4}{9} \right) = \frac{4}{3}\sqrt{2} \ln 2 - \frac{8}{9}\sqrt{2} + \frac{4}{9} \approx 0.494$

**59.** $dv = x \, dx, \; v = \dfrac{x^2}{2}, \; u = \operatorname{arcsec} x, \; du = \dfrac{1}{x\sqrt{x^2-1}}\, dx$

$$\int x \operatorname{arcsec} x \, dx = \frac{x^2}{2} \operatorname{arcsec} x - \int \frac{x^2/2}{x\sqrt{x^2-1}}\, dx = \frac{x^2}{2} \operatorname{arcsec} x - \frac{1}{4}\int \frac{2x}{\sqrt{x^2-1}}\, dx = \frac{x^2}{2} \operatorname{arcsec} x - \frac{1}{2}\sqrt{x^2-1} + C$$

So,

$$\int_2^4 x \operatorname{arcsec} x \, dx = \left[ \frac{x^2}{2} \operatorname{arcsec} x - \frac{1}{2}\sqrt{x^2-1} \right]_2^4 = \left( 8 \operatorname{arcsec} 4 - \frac{\sqrt{15}}{2} \right) - \left( \frac{2\pi}{3} - \frac{\sqrt{3}}{2} \right) = 8 \operatorname{arcsec} 4 - \frac{\sqrt{15}}{2} + \frac{\sqrt{3}}{2} - \frac{2\pi}{3} \approx 7.380.$$

**61.** $\int x^2 e^{2x}\, dx = x^2\left(\frac{1}{2}e^{2x}\right) - (2x)\left(\frac{1}{4}e^{2x}\right) + 2\left(\frac{1}{8}e^{2x}\right) + C$

$= \frac{1}{2}x^2 e^{2x} - \frac{1}{2}xe^{2x} + \frac{1}{4}e^{2x} + C$

$= \frac{1}{4}e^{2x}\left(2x^2 - 2x + 1\right) + C$

| Alternate signs | $u$ and its derivatives | $v'$ and its antiderivatives |
| --- | --- | --- |
| + | $x^2$ | $e^{2x}$ |
| − | $2x$ | $\frac{1}{2}e^{2x}$ |
| + | $2$ | $\frac{1}{4}e^{2x}$ |
| − | $0$ | $\frac{1}{8}e^{2x}$ |

**63.** $\int x^3 \sin x\, dx = x^3(-\cos x) - 3x^2(-\sin x) + 6x \cos x - 6 \sin x + C$

$= -x^3 \cos x + 3x^2 \sin x + 6x \cos x - 6 \sin x + C$

$= \left(3x^2 - 6\right)\sin x - \left(x^3 - 6x\right)\cos x + C$

| Alternate signs | $u$ and its derivatives | $v'$ and its antiderivatives |
| --- | --- | --- |
| + | $x^3$ | $\sin x$ |
| − | $3x^2$ | $-\cos x$ |
| + | $6x$ | $-\sin x$ |
| − | $6$ | $\cos x$ |
| + | $0$ | $\sin x$ |

**65.** $\int x \sec^2 x\, dx = x \tan x + \ln\left|\cos x\right| + C$

| Alternate signs | $u$ and its derivatives | $v'$ and its antiderivatives |
| --- | --- | --- |
| + | $x$ | $\sec^2 x$ |
| − | $1$ | $\tan x$ |
| + | $0$ | $-\ln\left|\cos x\right|$ |

**67.** $u = \sqrt{x} \Rightarrow u^2 = x \Rightarrow 2u\, du = dx$

$\int \sin \sqrt{x}\, dx = \int \sin u(2u\, du) = 2\int u \sin u\, du$

Integration by parts: $w = u,\ dw = du,\ dv = \sin u\, du,\ v = -\cos u$

$2\int u \sin u\, du = 2\left(-u \cos u + \int \cos u\, du\right) = 2(-u \cos u + \sin u) + C = 2\left(-\sqrt{x} \cos \sqrt{x} + \sin \sqrt{x}\right) + C$

**69.** Let $u = 4 - x,\ du = -dx,\ x = 4 - u.$

$\int_0^4 x\sqrt{4 - x}\, dx = \int_4^0 (4 - u)u^{1/2}(-du)$

$= \int_0^4 \left(4u^{1/2} - u^{3/2}\right) du$

$= \left[\frac{8}{3}u^{3/2} - \frac{2}{5}u^{5/2}\right]_0^4$

$= \frac{8}{3}(8) - \frac{2}{5}(32) = \frac{128}{15}$

**71.** $u = x^2, du = 2x\,dx$

$$\int x^5 e^{x^2}\,dx = \frac{1}{2}\int e^{x^2} x^4\,2x\,dx = \frac{1}{2}\int e^u u^2\,du$$

Integration by parts twice.

(1)  $w = u^2, dw = 2u\,du, dv = e^u\,du, v = e^u$

$$\frac{1}{2}\int e^u u^2\,du = \frac{1}{2}\left[u^2 e^u - \int 2u e^u\,du\right]$$

$$= \frac{1}{2}u^2 e^u - \int u e^u\,du$$

(2)  $w = u, dw = du, dv = e^u\,du, v = e^u$

$$\frac{1}{2}\int e^u u^2\,du = \frac{1}{2}u^2 e^u - \left(u e^u - \int e^u\,du\right)$$

$$= \frac{1}{2}u^2 e^u - u e^u + e^u + C$$

$$= \frac{1}{2}x^4 e^{x^2} - x^2 e^{x^2} + e^{x^2} + C$$

$$= \frac{e^{x^2}}{2}\left(x^4 - 2x^2 + 2\right) + C$$

**73.** Let $w = \ln x, dw = \dfrac{1}{x}\,dx, x = e^w, dx = e^w\,dw.$

$$\int \cos(\ln x)\,dx = \int \cos w\left(e^w\,dw\right)$$

Now use integration by parts twice.

$$\int (\cos w)e^w\,dw = (\cos w)e^w + \int (\sin w)e^w\,dw \qquad \left[u = \cos w, dv = e^w\,dw\right]$$

$$= (\cos w)e^w + \left[(\sin w)e^w - \int (\cos w)e^w\,dw\right] \quad \left[u = \sin w, dv = e^w\,dw\right]$$

$$2\int (\cos w)e^w\,dw = (\cos w)e^w + (\sin w)e^w$$

$$\int (\cos w)e^w\,dw = \frac{1}{2}e^w(\cos w + \sin w) + C$$

$$\int \cos(\ln x)\,dx = \frac{1}{2}x\left[\cos(\ln x) + \sin(\ln x)\right] + C$$

**75.** Integration by parts is based on the Product Rule.

**77.** In order for the integration by parts technique to be efficient, you want $dv$ to be the most complicated portion of the integrand and you want $u$ to be the portion of the integrand whose derivative is a function simpler than $u$. Suppose you let $u = \sin x$ and $dv = x\,dx$. Then $du = \cos x\,dx$ and $v = x^2/2$. So

$$\int x \sin x\,dx = uv - \int v\,du = \frac{x^2}{2}\sin x - \int \frac{x^2}{2}\cos x\,dx,$$

which is a more complicated integral than the original one.

**79.** (a)  $\int t^3 e^{-4t}\,dt = \dfrac{-e^{-4t}}{128}\left(32t^3 + 24t^2 + 12t + 3\right) + C$

(b)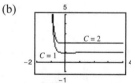

(c)  The graphs are vertical translations of each other.

**81. (a)**   $\displaystyle\int e^{-2x}\sin 3x\,dx = \frac{e^{-2x}}{13}(-2\sin 3x - 3\cos 3x) + C$

$\displaystyle\int_0^{\pi/2} e^{-2x}\sin 3x\,dx = \frac{1}{13}\left(2e^{-\pi} + 3\right) \approx 0.2374$

**(b)**

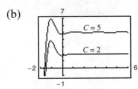

**(c)**   The graphs are vertical translations of each other.

**83. (a)**   $dv = \sqrt{2x - 3}\,dx \;\Rightarrow\; v = \int(2x-3)^{1/2}\,dx = \frac{1}{3}(2x-3)^{3/2}$

$u = 2x \qquad\qquad \Rightarrow\; du = 2\,dx$

$\displaystyle\int 2x\sqrt{2x-3}\,dx = \frac{2}{3}x(2x-3)^{3/2} - \frac{2}{3}\int(2x-3)^{3/2}\,dx$

$\displaystyle\qquad\qquad = \frac{2}{3}x(2x-3)^{3/2} - \frac{2}{15}(2x-3)^{5/2} + C$

$\displaystyle\qquad\qquad = \frac{2}{15}(2x-3)^{3/2}(3x+3) + C = \frac{2}{5}(2x-3)^{3/2}(x+1) + C$

**(b)**   $u = 2x - 3 \;\Rightarrow\; x = \dfrac{u+3}{2}$ and $dx = \dfrac{1}{2}\,du$

$\displaystyle\int 2x\sqrt{2x-3}\,dx = \int 2\left(\frac{u+3}{2}\right)u^{1/2}\left(\frac{1}{2}\right)du = \frac{1}{2}\int\left(u^{3/2} + 3u^{1/2}\right)du$

$\displaystyle\qquad\qquad = \frac{1}{2}\left[\frac{2}{5}u^{5/2} + 2u^{3/2}\right] + C$

$\displaystyle\qquad\qquad = \frac{1}{5}u^{3/2}(u+5) + C$

$\displaystyle\qquad\qquad = \frac{1}{5}(2x-3)^{3/2}\left[(2x-3) + 5\right] + C$

$\displaystyle\qquad\qquad = \frac{2}{5}(2x-3)^{3/2}(x+1) + C$

**85. (a)**   $dv = \dfrac{x}{\sqrt{4+x^2}}\,dx \;\Rightarrow\; v = \int(4+x^2)^{-1/2}x\,dx = \sqrt{4+x^2}$

$u = x^2 \qquad\qquad \Rightarrow\; du = 2x\,dx$

$\displaystyle\int\frac{x^3}{\sqrt{4+x^2}}\,dx = x^2\sqrt{4+x^2} - 2\int x\sqrt{4+x^2}\,dx$

$\displaystyle\qquad\qquad = x^2\sqrt{4+x^2} - \frac{2}{3}(4+x^2)^{3/2} + C = \frac{1}{3}\sqrt{4+x^2}(x^2 - 8) + C$

**(b)**   $u = 4 + x^2 \;\Rightarrow\; x^2 = u - 4$ and $2x\,dx = du \;\Rightarrow\; x\,dx = \dfrac{1}{2}du$

$\displaystyle\int\frac{x^3}{\sqrt{4+x^2}}\,dx = \int\frac{x^2}{\sqrt{4+x^2}}x\,dx = \int\frac{u-4}{\sqrt{u}}\frac{1}{2}\,du$

$\displaystyle\qquad\qquad = \frac{1}{2}\int\left(u^{1/2} - 4u^{-1/2}\right)du = \frac{1}{2}\left(\frac{2}{3}u^{3/2} - 8u^{1/2}\right) + C$

$\displaystyle\qquad\qquad = \frac{1}{3}u^{1/2}(u - 12) + C$

$\displaystyle\qquad\qquad = \frac{1}{3}\sqrt{4+x^2}\left[(4+x^2) - 12\right] + C = \frac{1}{3}\sqrt{4+x^2}(x^2 - 8) + C$

**87.** $n = 0$:  $\int \ln x \; dx = x(\ln x - 1) + C$

$\quad n = 1$:  $\int x \ln x \; dx = \dfrac{x^2}{4}(2 \ln x - 1) + C$

$\quad n = 2$:  $\int x^2 \ln x \; dx = \dfrac{x^3}{9}(3 \ln x - 1) + C$

$\quad n = 3$:  $\int x^3 \ln x \; dx = \dfrac{x^4}{16}(4 \ln x - 1) + C$

$\quad n = 4$:  $\int x^4 \ln x \; dx = \dfrac{x^5}{25}(5 \ln x - 1) + C$

In general, $\int x^n \ln x \; dx = \dfrac{x^{n+1}}{(n+1)^2}\big[(n+1)\ln x - 1\big] + C.$

**89.** $dv = \sin x \; dx \quad \Rightarrow \quad v = -\cos x$

$\quad u = x^n \qquad\quad \Rightarrow \quad du = nx^{n-1} \; dx$

$\quad \int x^n \sin x \; dx = -x^n \cos x + n \int x^{n-1} \cos x \; dx$

**91.** $dv = x^n \; dx \quad \Rightarrow \quad v = \dfrac{x^{n+1}}{n+1}$

$\quad u = \ln x \quad \Rightarrow \quad du = \dfrac{1}{x} dx$

$\quad \int x^n \ln x \; dx = \dfrac{x^{n+1}}{n+1} \ln x - \int \dfrac{x^n}{n+1} \; dx$

$\qquad\qquad\qquad = \dfrac{x^{n+1}}{n+1} \ln x - \dfrac{x^{n+1}}{(n+1)^2} + C$

$\qquad\qquad\qquad = \dfrac{x^{n+1}}{(n+1)^2}\big[(n+1)\ln x - 1\big] + C$

**93.** Use integration by parts twice.

(1) $dv = e^{ax} \; dx \quad \Rightarrow \quad v = \dfrac{1}{a} e^{ax}$ $\qquad\qquad$ (2) $dv = e^{ax} \; dx \quad \Rightarrow \quad v = \dfrac{1}{a} e^{ax}$

$\quad u = \sin bx \quad \Rightarrow \quad du = b \cos bx \; dx$ $\qquad\qquad\quad u = \cos bx \quad \Rightarrow \quad du = -b \sin bx \; dx$

$\int e^{ax} \sin bx \; dx = \dfrac{e^{ax} \sin bx}{a} - \dfrac{b}{a} \int e^{ax} \cos bx \; dx$

$\qquad\qquad\qquad = \dfrac{e^{ax} \sin bx}{a} - \dfrac{b}{a}\left( \dfrac{e^{ax} \cos bx}{a} + \dfrac{b}{a} \int e^{ax} \sin bx \; dx \right) = \dfrac{e^{ax} \sin bx}{a} - \dfrac{b}{a^2} e^{ax} \cos bx - \dfrac{b^2}{a^2} \int e^{ax} \sin bx \; dx$

Therefore, $\left( 1 + \dfrac{b^2}{a^2} \right) \int e^{ax} \sin bx \; dx = \dfrac{e^{ax}\left( a \sin bx - b \cos bx \right)}{a^2}$

$\qquad\qquad\qquad \int e^{ax} \sin bx \; dx = \dfrac{e^{ax}\left( a \sin bx - b \cos bx \right)}{a^2 + b^2} + C.$

**95.** $n = 5$   (Use formula in Exercise 91.)

$\quad \int x^5 \ln x \; dx = \dfrac{x^6}{6^2}(-1 + 6 \ln x) + C = \dfrac{x^6}{36}(-1 + 6 \ln x) + C$

**97.** $a = 2, b = 3$,    (Use formula in Exercise 94.)

$\quad \int e^{2x} \cos 3x \; dx = \dfrac{e^{2x}\left( 2 \cos 3x + 3 \sin 3x \right)}{13} + C$

**99.**

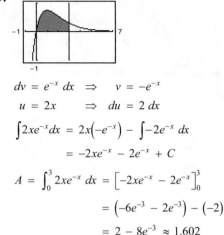

$$dv = e^{-x}\,dx \quad \Rightarrow \quad v = -e^{-x}$$

$$u = 2x \quad \Rightarrow \quad du = 2\,dx$$

$$\int 2xe^{-x}\,dx = 2x\left(-e^{-x}\right) - \int -2e^{-x}\,dx$$

$$= -2xe^{-x} - 2e^{-x} + C$$

$$A = \int_0^3 2xe^{-x}\,dx = \left[-2xe^{-x} - 2e^{-x}\right]_0^3$$

$$= \left(-6e^{-3} - 2e^{-3}\right) - \left(-2\right)$$

$$= 2 - 8e^{-3} \approx 1.602$$

**101.** $A = \int_0^1 e^{-x}\sin(\pi x)\,dx$

$$= \left[\frac{e^{-x}\left(-\sin \pi x - \pi\cos \pi x\right)}{1 + \pi^2}\right]_0^1$$

$$= \frac{1}{1 + \pi^2}\left(\frac{\pi}{e} + \pi\right)$$

$$= \frac{\pi}{1 + \pi^2}\left(\frac{1}{e} + 1\right)$$

$$\approx 0.395 \quad \text{(See Exercise 93.)}$$

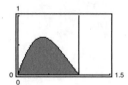

**103. (a)** $dv = dx \quad \Rightarrow \quad v = x$

$$u = \ln x \quad \Rightarrow \quad du = \frac{1}{x}\,dx$$

$$A = \int_1^e \ln x\,dx = \left[x\ln x - x\right]_1^e = 1 \quad \text{(Use integration by parts once.)}$$

**(b)** $R(x) = \ln x, r(x) = 0$

$$V = \pi\int_1^e \left(\ln x\right)^2 dx$$

$$= \pi\left[x\left(\ln x\right)^2 - 2x\ln x + 2x\right]_1^e \quad \text{(Use integration by parts twice, see Exercise 3.)}$$

$$= \pi\left(e - 2\right) \approx 2.257$$

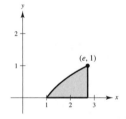

**(c)** $p(x) = x, h(x) = \ln x$

$$V = 2\pi\int_1^e x\ln x\,dx = 2\pi\left[\frac{x^2}{4}\left(-1 + 2\ln x\right)\right]_1^e$$

$$= \frac{\left(e^2 + 1\right)\pi}{2} \approx 13.177 \ \left(\text{See Exercise 91.}\right)$$

**(d)** $\quad \overline{x} = \dfrac{\displaystyle\int_1^e x\ln x\,dx}{1} = \dfrac{e^2 + 1}{4} \approx 2.097$

$$\overline{y} = \frac{\dfrac{1}{2}\displaystyle\int_1^e \left(\ln x\right)^2 dx}{1} = \frac{e - 2}{2} \approx 0.359$$

$$\left(\overline{x}, \overline{y}\right) = \left(\frac{e^2 + 1}{4}, \frac{e - 2}{2}\right) \approx \left(2.097, 0.359\right)$$

**105.** In Example 6, you showed that the centroid of an equivalent region was $(1, \pi/8)$. By symmetry, the centroid of this region is $(\pi/8, 1)$. You can also solve this problem directly.

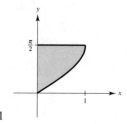

$$A = \int_0^1 \left(\frac{\pi}{2} - \arcsin x\right) dx = \left[\frac{\pi}{2}x - x \arcsin x - \sqrt{1 - x^2}\right]_0^1 \qquad \text{(Example 3)}$$

$$= \left(\frac{\pi}{2} - \frac{\pi}{2} - 0\right) - (-1) = 1$$

$$\bar{x} = \frac{M_y}{A} = \int_0^1 x\left(\frac{\pi}{2} - \arcsin x\right) dx = \frac{\pi}{8}, \quad \bar{y} = \frac{M_x}{A} = \int_0^1 \frac{(\pi/2) + \arcsin x}{2}\left(\frac{\pi}{2} - \arcsin x\right) dx = 1$$

**107.** Average value $= \dfrac{1}{\pi}\displaystyle\int_0^\pi e^{-4t}(\cos 2t + 5 \sin 2t)\, dt$

$$= \frac{1}{\pi}\left[e^{-4t}\left(\frac{-4\cos 2t + 2\sin 2t}{20}\right) + 5e^{-4t}\left(\frac{-4\sin 2t - 2\cos 2t}{20}\right)\right]_0^\pi \quad \text{(From Exercises 93 and 94)}$$

$$= \frac{7}{10\pi}\left(1 - e^{-4\pi}\right) \approx 0.223$$

**109.** $c(t) = 100{,}000 + 4000t, r = 5\%, t_1 = 10$

$$P = \int_0^{10} (100{,}000 + 4000t)e^{-0.05t}\, dt = 4000 \int_0^{10} (25 + t)e^{-0.05t}\, dt$$

Let $u = 25 + t$, $dv = e^{-0.05t}dt$, $du = dt$, $v = -\dfrac{100}{5}e^{-0.05t}$.

$$P = 4000\left\{\left[(25 + t)\left(-\frac{100}{5}e^{-0.05t}\right)\right]_0^{10} + \frac{100}{5}\int_0^{10} e^{-0.05t}dt\right\} = 4000\left\{\left[(25 + t)\left(-\frac{100}{5}e^{-0.05t}\right)\right]_0^{10} - \left[\frac{10{,}000}{25}e^{-0.05t}\right]_0^{10}\right\} \approx \$931{,}265$$

**111.** $\displaystyle\int_{-\pi}^{\pi} x \sin nx\, dx = \left[-\frac{x}{n}\cos nx + \frac{1}{n^2}\sin nx\right]_{-\pi}^{\pi} = -\frac{\pi}{n}\cos \pi n - \frac{\pi}{n}\cos(-\pi n) = -\frac{2\pi}{n}\cos \pi n = \begin{cases} -(2\pi/n), & \text{if } n \text{ is even} \\ (2\pi/n), & \text{if } n \text{ is odd} \end{cases}$

**113.** Let $u = x$, $dv = \sin\left(\dfrac{n\pi}{2}x\right) dx$, $du = dx$, $v = -\dfrac{2}{n\pi}\cos\left(\dfrac{n\pi}{2}x\right)$.

$$I_1 = \int_0^1 x \sin\left(\frac{n\pi}{2}x\right) dx = \left[\frac{-2x}{n\pi}\cos\left(\frac{n\pi}{2}x\right)\right]_0^1 + \frac{2}{n\pi}\int_0^1 \cos\left(\frac{n\pi}{2}x\right) dx$$

$$= -\frac{2}{n\pi}\cos\left(\frac{n\pi}{2}\right) + \left[\left(\frac{2}{n\pi}\right)^2 \sin\left(\frac{n\pi}{2}x\right)\right]_0^1$$

$$= -\frac{2}{n\pi}\cos\left(\frac{n\pi}{2}\right) + \left(\frac{2}{n\pi}\right)^2 \sin\left(\frac{n\pi}{2}\right)$$

Let $u = (-x + 2)$, $dv = \sin\left(\dfrac{n\pi}{2}x\right) dx$, $du = -dx$, $v = -\dfrac{2}{n\pi}\cos\left(\dfrac{n\pi}{2}x\right)$.

$$I_2 = \int_1^2 (-x + 2) \sin\left(\frac{n\pi}{2}x\right) dx = \left[\frac{-2(-x + 2)}{n\pi}\cos\left(\frac{n\pi}{2}x\right)\right]_1^2 - \frac{2}{n\pi}\int_1^2 \cos\left(\frac{n\pi}{2}x\right) dx$$

$$= \frac{2}{n\pi}\cos\left(\frac{n\pi}{2}\right) - \left[\left(\frac{2}{n\pi}\right)^2 \sin\left(\frac{n\pi}{2}x\right)\right]_1^2$$

$$= \frac{2}{n\pi}\cos\left(\frac{n\pi}{2}\right) + \left(\frac{2}{n\pi}\right)^2 \sin\left(\frac{n\pi}{2}\right)$$

$$h(I_1 + I_2) = b_n = h\left[\left(\frac{2}{n\pi}\right)^2 \sin\left(\frac{n\pi}{2}\right) + \left(\frac{2}{n\pi}\right)^2 \sin\left(\frac{n\pi}{2}\right)\right] = \frac{8h}{(n\pi)^2}\sin\left(\frac{n\pi}{2}\right)$$

**115. Shell Method:**

$$V = 2\pi \int_a^b xf(x)\,dx$$

$$dv = x\,dx \quad \Rightarrow \quad v = \frac{x^2}{2}$$

$$u = f(x) \quad \Rightarrow \quad du = f'(x)\,dx$$

$$V = 2\pi\left[\frac{x^2}{2}f(x) - \int \frac{x^2}{2}f'(x)\,dx\right]_a^b$$

$$= \pi\left[b^2 f(b) - a^2 f(a) - \int_a^b x^2 f'(x)\,dx\right]$$

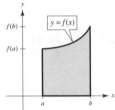

**Disk Method:**

$$V = \pi \int_0^{f(a)} \left(b^2 - a^2\right)dy + \pi \int_{f(a)}^{f(b)}\left[b^2 - \left[f^{-1}(y)\right]^2\right]dy$$

$$= \pi\left(b^2 - a^2\right)f(a) + \pi b^2\left(f(b) - f(a)\right)$$

$$\quad - \pi \int_{f(a)}^{f(b)}\left[f^{-1}(y)\right]^2 dy$$

$$= \pi\left[b^2 f(b) - a^2 f(a) - \int_{f(a)}^{f(b)}\left[f^{-1}(y)\right]^2 dy\right]$$

Because $x = f^{-1}(y)$, you have $f(x) = y$ and $f'(x)\,dx = dy$. When $y = f(a)$, $x = a$. When $y = f(b)$, $x = b$. So,

$$\int_{f(a)}^{f(b)}\left[f^{-1}(y)\right]^2 dy = \int_a^b x^2 f'(x)\,dx \text{ and the volumes}$$

are the same.

**117.** $f'(x) = 3x \sin(2x)$, $f(0) = 0$

(a) $f(x) = \int 3x \sin 2x\,dx$

$$= -\tfrac{3}{4}(2x \cos 2x - \sin 2x) + C$$

(Parts: $u = 3x$, $dv = \sin 2x\,dx$)

$$f(0) = 0 = -\tfrac{3}{4}(0) + C \Rightarrow C = 0$$

$$f(x) = -\tfrac{3}{4}(2x \cos 2x - \sin 2x)$$

(b)

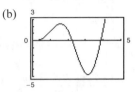

(c) Using $h = 0.05$, you obtain the points:

| $n$ | $x_n$ | $y_n$ |
|---|---|---|
| 0 | 0 | 0 |
| 1 | 0.05 | 0 |
| 2 | 0.10 | $7.4875 \times 10^{-4}$ |
| 3 | 0.15 | 0.0037 |
| 4 | 0.20 | 0.0104 |
| $\vdots$ | $\vdots$ | $\vdots$ |
| 80 | 4.0 | 1.3181 |

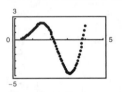

(d) Using $h = 0.1$, you obtain the points:

| $n$ | $x_n$ | $y_n$ |
|---|---|---|
| 0 | 0 | 0 |
| 1 | 0.1 | 0 |
| 2 | 0.2 | 0.0060 |
| 3 | 0.3 | 0.0293 |
| 4 | 0.4 | 0.0801 |
| $\vdots$ | $\vdots$ | $\vdots$ |
| 40 | 4.0 | 1.0210 |

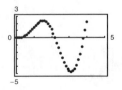

**119.** On $\left[0, \dfrac{\pi}{2}\right]$, $\sin x \le 1 \Rightarrow x \sin x \le x \Rightarrow \displaystyle\int_0^{\pi/2} x \sin x \, dx \le \int_0^{\pi/2} x \, dx$.

## Section 8.3   Trigonometric Integrals

**1.** $y = \sec x$

   $y' = \sec x \tan x = \sin x \sec^2 x$

   $\displaystyle\int \sin x \sec^2 x \, dx = \sec x + C$

   Matches (c)

**3.** $y = x - \tan x + \dfrac{1}{3}\tan^3 x$

   $y' = 1 - \sec^2 x + \tan^2 x\left(\sec^2 x\right)$

   $\quad = -\tan^2 x + \tan^2 x\left(1 + \tan^2 x\right)$

   $\quad = \tan^4 x$

   $\displaystyle\int \tan^4 x \, dx = x - \tan x + \dfrac{1}{3}\tan^3 x + C$

   Matches (d)

**5.** Let $u = \cos x$, $du = -\sin x \, dx$.

   $\displaystyle\int \cos^5 x \sin x \, dx = -\int \cos^5 x \left(-\sin x\right) dx = -\dfrac{\cos^6 x}{6} + C$

**7.** Let $u = \sin 2x$, $du = 2\cos 2x \, dx$.

   $\displaystyle\int \sin^7 2x \cos 2x \, dx = \dfrac{1}{2}\int \sin^7 2x(2\cos 2x) \, dx$

   $\qquad\qquad\qquad\quad = \dfrac{1}{2}\dfrac{\sin^8 2x}{8} + C$

   $\qquad\qquad\qquad\quad = \dfrac{1}{16}\sin^8 2x + C$

**9.** $\displaystyle\int \sin^3 x \cos^2 x \, dx = \int \left(1 - \cos^2 x\right)\cos^2 x \sin x \, dx$

   $\qquad\qquad\qquad\quad = \int \left(\cos^2 x - \cos^4 x\right)\sin x \, dx$

   $\qquad\qquad\qquad\quad = -\int \left(\cos^2 x - \cos^4 x\right)\left(-\sin x\right) dx$

   $\qquad\qquad\qquad\quad = -\dfrac{\cos^3 x}{3} + \dfrac{\cos^5 x}{5} + C$

**11.** $\displaystyle\int \sin^3 2\theta \sqrt{\cos 2\theta} \, d\theta = \int \left(1 - \cos^2 2\theta\right)\sqrt{\cos 2\theta} \sin 2\theta \, d\theta$

   $\qquad\qquad\qquad\qquad = \int \left[\left(\cos 2\theta\right)^{1/2} - \left(\cos 2\theta\right)^{5/2}\right] \sin 2\theta \, d\theta$

   $\qquad\qquad\qquad\qquad = -\dfrac{1}{2}\int \left[\left(\cos 2\theta\right)^{1/2} - \left(\cos 2\theta\right)^{5/2}\right]\left(-2\sin 2\theta\right) d\theta$

   $\qquad\qquad\qquad\qquad = -\dfrac{1}{2}\left[\dfrac{2}{3}\left(\cos 2\theta\right)^{3/2} - \dfrac{2}{7}\left(\cos 2\theta\right)^{7/2}\right] + C$

   $\qquad\qquad\qquad\qquad = -\dfrac{1}{3}\left(\cos 2\theta\right)^{3/2} + \dfrac{1}{7}\left(\cos 2\theta\right)^{7/2} + C$

**13.** $\int \cos^2 3x\, dx = \int \dfrac{1 + \cos 6x}{2}\, dx = \dfrac{1}{2}\left(x + \dfrac{1}{6}\sin 6x\right) + C = \dfrac{1}{12}(6x + \sin 6x) + C$

**15.** $\int \cos^4 3\alpha\, d\alpha = \int \left(\dfrac{1 + \cos 6\alpha}{2}\right)\left(\dfrac{1 + \cos 6\alpha}{2}\right) d\alpha$

$\qquad = \dfrac{1}{4}\int \left(1 + 2\cos 6\alpha + \cos^2 6\alpha\right) d\alpha$

$\qquad = \dfrac{1}{4}\int \left(1 + 2\cos 6\alpha + \dfrac{1 + \cos 12\alpha}{2}\right) d\alpha$

$\qquad = \dfrac{1}{4}\left(\dfrac{3}{2}\alpha + \dfrac{\sin 6\alpha}{3} + \dfrac{\sin 12\alpha}{24}\right) + C$

$\qquad = \dfrac{3}{8}\alpha + \dfrac{\sin 6\alpha}{12} + \dfrac{\sin 12\alpha}{96} + C$

**17.** Integration by parts:

$\qquad dv = \sin^2 x\, dx = \dfrac{1 - \cos 2x}{2} \Rightarrow v = \dfrac{x}{2} - \dfrac{\sin 2x}{4} = \dfrac{1}{4}(2x - \sin 2x)$

$\qquad u = x \Rightarrow du = dx$

$\qquad \int x \sin^2 x\, dx = \dfrac{1}{4}x(2x - \sin 2x) - \dfrac{1}{4}\int (2x - \sin 2x)\, dx$

$\qquad\qquad = \dfrac{1}{4}x(2x - \sin 2x) - \dfrac{1}{4}\left(x^2 + \dfrac{1}{2}\cos 2x\right) + C = \dfrac{1}{8}\left(2x^2 - 2x\sin 2x - \cos 2x\right) + C$

**19.** $\displaystyle\int_0^{\pi/2} \cos^7 x\, dx = \left(\dfrac{2}{3}\right)\left(\dfrac{4}{5}\right)\left(\dfrac{6}{7}\right) = \dfrac{16}{35}, \,(n = 7)$

**21.** $\displaystyle\int_0^{\pi/2} \cos^{10} x\, dx = \left(\dfrac{1}{2}\right)\left(\dfrac{3}{4}\right)\left(\dfrac{5}{6}\right)\left(\dfrac{7}{8}\right)\left(\dfrac{9}{10}\right)\left(\dfrac{\pi}{2}\right)$

$\qquad\qquad = \dfrac{63}{512}\pi, \,(n = 10)$

**23.** $\displaystyle\int_0^{\pi/2} \sin^6 x\, dx = \left(\dfrac{1}{2}\right)\left(\dfrac{3}{4}\right)\left(\dfrac{5}{6}\right)\dfrac{\pi}{2} = \dfrac{5\pi}{32}, \,(n = 6)$

**25.** $\int \sec 7x\, dx = \dfrac{1}{7}\int \sec(7x)\, 7\, dx$

$\qquad\qquad = \dfrac{1}{7}\ln\left|\sec 7x + \tan 7x\right| + C$

**27.** $\int \sec^4 5x\, dx = \int (1 + \tan^2 5x)\sec^2 5x\, dx = \dfrac{1}{5}\left(\tan 5x + \dfrac{\tan^3 5x}{3}\right) + C = \dfrac{\tan 5x}{15}(3 + \tan^2 5x) + C$

**29.** $dv = \sec^2 \pi x\, dx \quad \Rightarrow \quad v = \dfrac{1}{\pi}\tan \pi x$

$\qquad u = \sec \pi x \qquad \Rightarrow \quad du = \pi \sec \pi x \tan \pi x\, dx$

$\qquad \int \sec^3 \pi x\, dx = \dfrac{1}{\pi}\sec \pi x \tan \pi x - \int \sec \pi x \tan^2 \pi x\, dx = \dfrac{1}{\pi}\sec \pi x \tan \pi x - \int \sec \pi x\left(\sec^2 \pi x - 1\right) dx$

$\qquad 2\int \sec^3 \pi x\, dx = \dfrac{1}{\pi}\left(\sec \pi x \tan \pi x + \ln\left|\sec \pi x + \tan \pi x\right|\right) + C_1$

$\qquad \int \sec^3 \pi x\, dx = \dfrac{1}{2\pi}\left(\sec \pi x \tan \pi x + \ln\left|\sec \pi x + \tan \pi x\right|\right) + C$

**31.** $\displaystyle\int\tan^5\frac{x}{2}\,dx = \int\left(\sec^2\frac{x}{2}-1\right)\tan^3\frac{x}{2}\,dx$

$\displaystyle = \int\tan^3\frac{x}{2}\sec^2\frac{x}{2}\,dx - \int\tan^3\frac{x}{2}\,dx$

$\displaystyle = \frac{\tan^4\frac{x}{2}}{2} - \int\left(\sec^2\frac{x}{2}-1\right)\tan\frac{x}{2}\,dx$

$\displaystyle = \frac{1}{2}\tan^4\frac{x}{2} - \tan^2\frac{x}{2} - 2\ln\left|\cos\frac{x}{2}\right| + C$

**33.** $u = \tan x,\ du = \sec^2 x\,dx$

$\displaystyle\int\sec^2 x\tan x\,dx = \tfrac{1}{2}\tan^2 x + C$

or, $u = \sec x,\ du = \sec x\tan x\,dx$,

$\displaystyle\int\sec^2 x\tan x\,dx = \tfrac{1}{2}\sec^2 x + C$

**35.** $\displaystyle\int\tan^2 x\sec^4 x\,dx = \int\tan^2 x\left(\tan^2 x+1\right)\sec^2 x\,dx$

$\displaystyle = \int\tan^4 x\sec^2 x\,dx + \int\tan^2 x\sec^2 x\,dx$

$\displaystyle = \frac{\tan^5 x}{5} + \frac{\tan^3 x}{3} + C$

**37.** $\displaystyle\int\sec^6 4x\tan 4x\,dx = \frac{1}{4}\int\sec^5 4x\left(4\sec 4x\tan 4x\right)dx$

$\displaystyle = \frac{\sec^6 4x}{24} + C$

**39.** $\displaystyle\int\sec^5 x\tan^3 x\,dx = \int\sec^4 x\tan^2 x\left(\sec x\tan x\right)dx$

$\displaystyle = \int\sec^4 x\left(\sec^2 x-1\right)\left(\sec x\tan x\right)dx$

$\displaystyle = \int\left(\sec^6 x-\sec^4 x\right)\left(\sec x\tan x\right)dx$

$\displaystyle = \frac{\sec^7 x}{7} - \frac{\sec^5 x}{5} + C$

**41.** $\displaystyle\int\frac{\tan^2 x}{\sec x}\,dx = \int\frac{\left(\sec^2 x-1\right)}{\sec x}\,dx$

$\displaystyle = \int\left(\sec x - \cos x\right)dx$

$\displaystyle = \ln\left|\sec x + \tan x\right| - \sin x + C$

**43.** $\displaystyle r = \int\sin^4(\pi\theta)\,d\theta = \frac{1}{4}\int\left[1-\cos(2\pi\theta)\right]^2 d\theta$

$\displaystyle = \frac{1}{4}\int\left[1-2\cos(2\pi\theta)+\cos^2(2\pi\theta)\right]d\theta$

$\displaystyle = \frac{1}{4}\int\left[1-2\cos(2\pi\theta)+\frac{1+\cos(4\pi\theta)}{2}\right]d\theta$

$\displaystyle = \frac{1}{4}\left[\theta - \frac{1}{\pi}\sin(2\pi\theta) + \frac{\theta}{2} + \frac{1}{8\pi}\sin(4\pi\theta)\right] + C$

$\displaystyle = \frac{1}{32\pi}\left[12\pi\theta - 8\sin(2\pi\theta) + \sin(4\pi\theta)\right] + C$

**45.** $\displaystyle y = \int\tan^3 3x\sec 3x\,dx$

$\displaystyle = \int\left(\sec^2 3x-1\right)\sec 3x\tan 3x\,dx$

$\displaystyle = \frac{1}{3}\int\sec^2 3x\left(3\sec 3x\tan 3x\right)dx - \frac{1}{3}\int 3\sec 3x\tan 3x\,dx$

$\displaystyle = \frac{1}{9}\sec^3 3x - \frac{1}{3}\sec 3x + C$

**47. (a)**

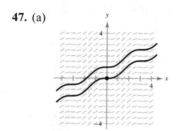

**(b)** $\displaystyle\frac{dy}{dx} = \sin^2 x,\quad (0,0)$

$\displaystyle y = \int\sin^2 x\,dx = \int\frac{1-\cos 2x}{2}\,dx$

$\displaystyle = \frac{1}{2}x - \frac{\sin 2x}{4} + C$

$\displaystyle (0,0):\ 0 = C,\ y = \frac{1}{2}x - \frac{\sin 2x}{4}$

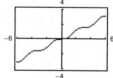

**49.** $\displaystyle\frac{dy}{dx} = \frac{3\sin x}{y},\ y(0) = 2$

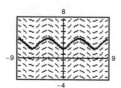

**51.** $\int \cos 2x \cos 6x \, dx = \dfrac{1}{2} \int \Big[ \cos((2-6)x) + \cos((2+6)x) \Big] dx$

$\qquad\qquad\qquad\quad = \dfrac{1}{2} \int \Big[ \cos(-4x) + \cos 8x \Big] dx$

$\qquad\qquad\qquad\quad = \dfrac{1}{2} \int (\cos 4x + \cos 8x) \, dx$

$\qquad\qquad\qquad\quad = \dfrac{1}{2} \left[ \dfrac{\sin 4x}{4} + \dfrac{\sin 8x}{8} \right] + C$

$\qquad\qquad\qquad\quad = \dfrac{\sin 4x}{8} + \dfrac{\sin 8x}{16} + C$

$\qquad\qquad\qquad\quad = \dfrac{1}{16}(2 \sin 4x + \sin 8x) + C$

**53.** $\int \sin 2x \cos 4x \, dx = \dfrac{1}{2} \int \Big[ \sin((2-4)x) + \sin((2+4)x) \Big] dx$

$\qquad\qquad\qquad\quad = \dfrac{1}{2} \int (\sin(-2x) + \sin 6x) \, dx$

$\qquad\qquad\qquad\quad = \dfrac{1}{2} \int (-\sin 2x + \sin 6x) \, dx$

$\qquad\qquad\qquad\quad = \dfrac{1}{2} \left[ \dfrac{\cos 2x}{2} - \dfrac{\cos 6x}{6} \right] + C$

$\qquad\qquad\qquad\quad = \dfrac{1}{4} \cos 2x - \dfrac{1}{12} \cos 6x + C$

$\qquad\qquad\qquad\quad = \dfrac{1}{12}(3 \cos 2x - \cos 6x) + C$

**55.** $\int \sin \theta \sin 3\theta \, d\theta = \tfrac{1}{2} \int (\cos 2\theta - \cos 4\theta) \, d\theta$

$\qquad\qquad\qquad\quad = \tfrac{1}{2} \left( \tfrac{1}{2} \sin 2\theta - \tfrac{1}{4} \sin 4\theta \right) + C$

$\qquad\qquad\qquad\quad = \tfrac{1}{8}(2 \sin 2\theta - \sin 4\theta) + C$

**57.** $\int \cot^3 2x \, dx = \int (\csc^2 2x - 1) \cot 2x \, dx$

$\qquad\qquad\quad = -\dfrac{1}{2} \int \cot 2x(-2 \csc^2 2x) \, dx - \dfrac{1}{2} \int \dfrac{2 \cos 2x}{\sin 2x} \, dx$

$\qquad\qquad\quad = -\dfrac{1}{4} \cot^2 2x - \dfrac{1}{2} \ln|\sin 2x| + C$

$\qquad\qquad\quad = \dfrac{1}{4} \Big( \ln|\csc^2 2x| - \cot^2 2x \Big) + C$

**59.** $\int \csc^4 2x \, dx = \int \csc^2 2x(1 + \cot^2 2x) \, dx$

$\qquad\qquad\quad = \int \csc^2 2x \, dx + \int \cot^2 2x \csc^2 2x \, dx$

$\qquad\qquad\quad = -\dfrac{1}{2} \cot 2x - \dfrac{\cot^3 2x}{6} + C$

**61.** $\int \dfrac{\cot^2 t}{\csc t} \, dt = \int \dfrac{\csc^2 t - 1}{\csc t} \, dt$

$\qquad\qquad\quad = \int (\csc t - \sin t) \, dt$

$\qquad\qquad\quad = \ln|\csc t - \cot t| + \cos t + C$

**63.** $\int \dfrac{1}{\sec x \tan x} \, dx = \int \dfrac{\cos^2 x}{\sin x} \, dx = \int \dfrac{1 - \sin^2 x}{\sin x} \, dx$

$\qquad\qquad\qquad\quad = \int (\csc x - \sin x) \, dx$

$\qquad\qquad\qquad\quad = \ln|\csc x - \cot x| + \cos x + C$

**65.** $\int \left(\tan^4 t - \sec^4 t\right) dt = \int \left(\tan^2 t + \sec^2 t\right)\left(\tan^2 t - \sec^2 t\right) dt,$     $\left(\tan^2 t - \sec^2 t = -1\right)$

$$= -\int \left(\tan^2 t + \sec^2 t\right) dt = -\int \left(2 \sec^2 t - 1\right) dt = -2 \tan t + t + C$$

**67.** $\int_{-\pi}^{\pi} \sin^2 x\, dx = 2 \int_0^{\pi} \frac{1 - \cos 2x}{2}\, dx$

$$= \left[ x - \frac{1}{2} \sin 2x \right]_0^{\pi} = \pi$$

**69.** $\int_0^{\pi/4} 6 \tan^3 x\, dx = 6 \int_0^{\pi/4} \left(\sec^2 x - 1\right) \tan x\, dx$

$$= 6 \int_0^{\pi/4} \left[ \tan x \sec^2 x - \tan x \right] dx$$

$$= 6 \left[ \frac{\tan^2 x}{2} + \ln|\cos x| \right]_0^{\pi/4}$$

$$= 6 \left[ \frac{1}{2} + \ln\left(\frac{\sqrt{2}}{2}\right) \right] = 6 \left( \frac{1}{2} - \ln \sqrt{2} \right)$$

$$= 3\left(1 - \ln 2\right)$$

**71.** Let $u = 1 + \sin t,\ du = \cos t\, dt.$

$$\int_0^{\pi/2} \frac{\cos t}{1 + \sin t}\, dt = \left[ \ln|1 + \sin t| \right]_0^{\pi/2} = \ln 2$$

**73.** $\int_{-\pi/2}^{\pi/2} 3 \cos^3 x\, dx = 3 \int_{-\pi/2}^{\pi/2} \left(1 - \sin^2 x\right) \cos x\, dx$

$$= 3 \left[ \sin x + \frac{\sin^3 x}{3} \right]_{-\pi/2}^{\pi/2}$$

$$= 3 \left[ \left(1 + \frac{1}{3}\right) - \left(-1 - \frac{1}{3}\right) \right] = 4$$

**75.** $\int \cos^4 \frac{x}{2}\, dx = \frac{1}{16}\left(6x + 8 \sin x + \sin 2x\right) + C$

$$= \frac{1}{8}\left( 4 \sin \frac{x}{2} \cos^3 \frac{x}{2} + 6 \sin \frac{x}{2} \cos \frac{x}{2} + 3x \right) + C$$

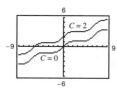

**77.** $\int \sec^5 \pi x\, dx = \frac{1}{4\pi}\left\{ \sec^3 \pi x \tan \pi x + \frac{3}{2}\left( \sec \pi x \tan \pi x + \ln|\sec \pi x + \tan \pi x| \right) \right\} + C$

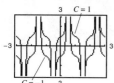

**79.** $\int \sec^5 \pi x \tan \pi x \, dx = \dfrac{1}{5\pi} \sec^5 \pi x + C$

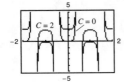

**81.** $\displaystyle\int_0^{\pi/4} \sin 3\theta \sin 4\theta \, d\theta$

$$= \left[ \dfrac{1}{2} \sin \theta - \dfrac{1}{14} \sin 7\theta \right]_0^{\pi/4} = \dfrac{2\sqrt{2}}{7}$$

**83.** $\displaystyle\int_0^{\pi/2} \sin^4 x \, dx = \dfrac{1}{4} \left[ \dfrac{3x}{2} - \sin 2x + \dfrac{1}{8} \sin 4x \right]_0^{\pi/2} = \dfrac{3\pi}{16}$

**85.** (a) Save one sine factor and convert the remaining factors to cosines. Then expand and integrate.

    (b) Save one cosine factor and convert the remaining factors to sines. Then expand and integrate.

    (c) Make repeated use of the power reducing formulas to convert the integrand to odd powers of the cosine. Then proceed as in part (b).

**89.** (a) Let $u = \tan 3x$, $du = 3 \sec^2 3x \, dx$.

$$\int \sec^4 3x \tan^3 3x \, dx = \int \sec^2 3x \tan^3 3x \sec^2 3x \, dx = \dfrac{1}{3} \int \left( \tan^2 3x + 1 \right) \tan^3 3x \left( 3 \sec^2 3x \right) dx$$

$$= \dfrac{1}{3} \int \left( \tan^5 3x + \tan^3 3x \right) \left( 3 \sec^2 3x \right) dx = \dfrac{\tan^6 3x}{18} + \dfrac{\tan^4 3x}{12} + C_1$$

Or let $u = \sec 3x$, $du = 3 \sec 3x \tan 3x \, dx$.

$$\int \sec^4 3x \tan^3 3x \, dx = \int \sec^3 3x \tan^2 3x \sec 3x \tan 3x \, dx$$

$$= \dfrac{1}{3} \int \sec^3 3x \left( \sec^2 3x - 1 \right) \left( 3 \sec 3x \tan 3x \right) dx = \dfrac{\sec^6 3x}{18} - \dfrac{\sec^4 3x}{12} + C$$

    (b)

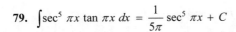

    (c) $\dfrac{\sec^6 3x}{18} - \dfrac{\sec^4 3x}{12} + C = \dfrac{\left( 1 + \tan^2 3x \right)^3}{18} - \dfrac{\left( 1 + \tan^2 3x \right)^2}{12} + C$

$$= \dfrac{1}{18} \tan^6 3x + \dfrac{1}{6} \tan^4 3x + \dfrac{1}{6} \tan^2 3x + \dfrac{1}{18} - \dfrac{1}{12} \tan^4 3x - \dfrac{1}{6} \tan^2 3x - \dfrac{1}{12} + C$$

$$= \dfrac{\tan^6 3x}{18} + \dfrac{\tan^4 3x}{12} + \left( \dfrac{1}{18} - \dfrac{1}{12} \right) + C$$

$$= \dfrac{\tan^6 3x}{18} + \dfrac{\tan^4 3x}{12} + C_2$$

**87.** (a) $\int \sin x \cos x \, dx = \dfrac{\sin^2 x}{2} + C$

    (b) $- \int \cos x \left( -\sin x \right) dx = -\dfrac{\cos^2 x}{2} + C$

    (c) $dv = \cos x \, dx \implies v = \sin x$

       $u = \sin x \implies du = \cos x \, dx$

$$\int \sin x \cos x \, dx = \sin^2 x - \int \sin x \cos x \, dx$$

$$2 \int \sin x \cos x \, dx = \sin^2 x$$

$$\int \sin x \cos x \, dx = \dfrac{\sin^2 x}{2} + C$$

    (Answers will vary)

    (d) $\int \sin x \cos x \, dx = \int \dfrac{1}{2} \sin 2x \, dx = -\dfrac{1}{4} \cos 2x + C$

    The answers all differ by a constant.

**91.** $A = \int_0^{\pi/2} \left(\sin x - \sin^3 x\right) dx$

$\quad = \int_0^{\pi/2} \sin x\, dx - \int_0^{\pi/2} \sin^3 x\, dx$

$\quad = \left[-\cos x\right]_0^{\pi/2} - \dfrac{2}{3}$ \quad (Wallis's Formula)

$\quad = 1 - \dfrac{2}{3} = \dfrac{1}{3}$

**93.** $A = \int_{-\pi/4}^{\pi/4} \left[\cos^2 x - \sin^2 x\right] dx$

$\quad = \int_{-\pi/4}^{\pi/4} \cos 2x\, dx$

$\quad = \left[\dfrac{\sin 2x}{2}\right]_{-\pi/4}^{\pi/4} = \dfrac{1}{2} + \dfrac{1}{2} = 1$

**95. Disks**

$\quad R(x) = \tan x,\, r(x) = 0$

$\quad V = 2\pi \int_0^{\pi/4} \tan^2 x\, dx$

$\quad\quad = 2\pi \int_0^{\pi/4} \left(\sec^2 x - 1\right) dx$

$\quad\quad = 2\pi \left[\tan x - x\right]_0^{\pi/4}$

$\quad\quad = 2\pi \left(1 - \dfrac{\pi}{4}\right) \approx 1.348$

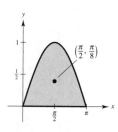

**97. (a)** $V = \pi \int_0^{\pi} \sin^2 x\, dx = \dfrac{\pi}{2} \int_0^{\pi} (1 - \cos 2x)\, dx = \dfrac{\pi}{2}\left[x - \dfrac{1}{2}\sin 2x\right]_0^{\pi} = \dfrac{\pi^2}{2}$

**(b)** $A = \int_0^{\pi} \sin x\, dx = \left[-\cos x\right]_0^{\pi} = 1 + 1 = 2$

Let $u = x,\, dv = \sin x\, dx,\, du = dx,\, v = -\cos x.$

$\bar{x} = \dfrac{1}{A}\int_0^{\pi} x \sin x\, dx = \dfrac{1}{2}\left[\left[-x\cos x\right]_0^{\pi} + \int_0^{\pi} \cos x\, dx\right] = \dfrac{1}{2}\left[-x\cos x + \sin x\right]_0^{\pi} = \dfrac{\pi}{2}$

$\bar{y} = \dfrac{1}{2A}\int_0^{\pi} \sin^2 x\, dx = \dfrac{1}{8}\int_0^{\pi} (1 - \cos 2x)\, dx = \dfrac{1}{8}\left[x - \dfrac{1}{2}\sin 2x\right]_0^{\pi} = \dfrac{\pi}{8}$

$\left(\bar{x}, \bar{y}\right) = \left(\dfrac{\pi}{2}, \dfrac{\pi}{8}\right)$

**99.** $dv = \sin x\, dx \Rightarrow v = -\cos x$

$u = \sin^{n-1} x \Rightarrow du = (n-1)\sin^{n-2} x \cos x\, dx$

$\int \sin^n x\, dx = -\sin^{n-1} x \cos x + (n-1)\int \sin^{n-2} x \cos^2 x\, dx = -\sin^{n-1} x \cos x + (n-1)\int \sin^{n-2} x\left(1 - \sin^2 x\right) dx$

$\quad = -\sin^{n-1} x \cos x + (n-1)\int \sin^{n-2} x\, dx - (n-1)\int \sin^n x\, dx$

Therefore, $n\int \sin^n x\, dx = -\sin^{n-1} x \cos x + (n-1)\int \sin^{n-2} x\, dx$

$\int \sin^n x\, dx = \dfrac{-\sin^{n-1} x \cos x}{n} + \dfrac{n-1}{n}\int \sin^{n-2} x\, dx.$

**101.** Let $u = \sin^{n-1} x,\, du = (n-1)\sin^{n-2} x \cos x\, dx,\, dv = \cos^m x \sin x\, dx,\, v = \dfrac{-\cos^{m+1} x}{m+1}.$

$\int \cos^m x \sin^n x\, dx = \dfrac{-\sin^{n-1} x \cos^{m+1} x}{m+1} + \dfrac{n-1}{m+1}\int \sin^{n-2} x \cos^{m+2} x\, dx$

$\quad = \dfrac{-\sin^{n-1} x \cos^{m+1} x}{m+1} + \dfrac{n-1}{m+1}\int \sin^{n-2} x \cos^m x\left(1 - \sin^2 x\right) dx$

$\quad = \dfrac{-\sin^{n-1} x \cos^{m+1} x}{m+1} + \dfrac{n-1}{m+1}\int \sin^{n-2} x \cos^m x\, dx - \dfrac{n-1}{m+1}\int \sin^n x \cos^m x\, dx$

$\dfrac{m+n}{m+1}\int \cos^m x \sin^n x\, dx = \dfrac{-\sin^{n-1} x \cos^{m+1} x}{m+1} + \dfrac{n-1}{m+1}\int \sin^{n-2} x \cos^m x\, dx$

$\int \cos^m x \sin^n x\, dx = \dfrac{-\cos^{m+1} x \sin^{n-1} x}{m+n} + \dfrac{n-1}{m+n}\int \cos^m x \sin^{n-2} x\, dx$

**103.** $\displaystyle\int \sin^5 x \, dx = -\frac{\sin^4 x \cos x}{5} + \frac{4}{5}\int \sin^3 x \, dx$

$$= -\frac{\sin^4 x \cos x}{5} + \frac{4}{5}\left(-\frac{\sin^2 x \cos x}{3} + \frac{2}{3}\int \sin x \, dx\right)$$

$$= -\frac{1}{5}\sin^4 x \cos x - \frac{4}{15}\sin^2 x \cos x - \frac{8}{15}\cos x + C$$

$$= -\frac{\cos x}{15}\left(3 \sin^4 x + 4 \sin^2 x + 8\right) + C$$

**105.** $\displaystyle\int \sec^4 \frac{2\pi x}{5} \, dx = \frac{5}{2\pi}\int \sec^4\left(\frac{2\pi x}{5}\right)\frac{2\pi}{5} \, dx$

$$= \frac{5}{2\pi}\left[\frac{1}{3}\sec^2\left(\frac{2\pi x}{5}\right)\tan\left(\frac{2\pi x}{5}\right) + \frac{2}{3}\int \sec^2\left(\frac{2\pi x}{5}\right)\frac{2\pi}{5} \, dx\right]$$

$$= \frac{5}{6\pi}\left[\sec^2\left(\frac{2\pi x}{5}\right)\tan\left(\frac{2\pi x}{5}\right) + 2 \tan\left(\frac{2\pi x}{5}\right)\right] + C$$

$$= \frac{5}{6\pi}\tan\left(\frac{2\pi x}{5}\right)\left[\sec^2\left(\frac{2\pi x}{5}\right) + 2\right] + C$$

**107.** $f(t) = a_0 + a_1 \cos\dfrac{\pi t}{6} + b_1 \sin\dfrac{\pi t}{6}$

$$a_0 = \frac{1}{12}\int_0^{12} f(t) \, dt, \quad a_1 = \frac{1}{6}\int_0^{12} f(t)\cos\frac{\pi t}{6} \, dt, \quad b_1 = \frac{1}{6}\int_0^{12} f(t)\sin\frac{\pi t}{6} \, dt$$

(a) $a_0 \approx \dfrac{1}{12} \cdot \dfrac{(12-0)}{3(12)}[33.5 + 4(35.4) + 2(44.7) + 4(55.6) + 2(67.4) + 4(76.2) + 2(80.4) + 4(79.0) + 2(72.0)$

$$+ 4(61.0) + 2(49.3) + 4(38.6) + 33.5]$$

$$\approx 57.72$$

$$a_1 \approx -23.36$$

$$b_1 \approx -2.75 \qquad \text{(Answers will vary.)}$$

$$H(t) \approx 57.72 - 23.36 \cos\left(\frac{\pi t}{6}\right) - 2.75 \sin\left(\frac{\pi t}{6}\right)$$

(b) $L(t) \approx 42.04 - 20.91 \cos\left(\dfrac{\pi t}{6}\right) - 4.33 \sin\left(\dfrac{\pi t}{6}\right)$

(c)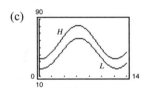

Temperature difference is greatest in the summer $(t \approx 4.9$ or end of May$)$.

**109.** $\displaystyle\int_{-\pi}^{\pi}\cos(mx)\cos(nx)\,dx = \frac{1}{2}\left[\frac{\sin(m+n)x}{m+n}+\frac{\sin(m-n)x}{m-n}\right]_{-\pi}^{\pi}=0, \quad (m\neq n)$

$\displaystyle\int_{-\pi}^{\pi}\sin(mx)\sin(nx)\,dx = \frac{1}{2}\int_{-\pi}^{\pi}\left[\cos(m-n)x-\cos(m+n)x\right]\,dx=\frac{1}{2}\left[\frac{\sin(m-n)x}{m-n}-\frac{\sin(m+n)x}{m+n}\right]_{-\pi}^{\pi}=0,\;(m\neq n)$

$\displaystyle\int_{-\pi}^{\pi}\sin(mx)\cos(nx)\,dx = \frac{1}{2}\int_{-\pi}^{\pi}\left[\sin(m+n)x+\sin(m-n)x\right]\,dx$

$\displaystyle\qquad = -\frac{1}{2}\left[\frac{\cos(m+n)x}{m+n}+\frac{\cos(m-n)x}{m-n}\right]_{-\pi}^{\pi}, \quad (m\neq n)$

$\displaystyle\qquad = -\frac{1}{2}\left[\left(\frac{\cos(m+n)\pi}{m+n}+\frac{\cos(m-n)\pi}{m-n}\right)-\left(\frac{\cos(m+n)(-\pi)}{m+n}+\frac{\cos(m-n)(-\pi)}{m-n}\right)\right]$

$\displaystyle\qquad = 0, \text{ because } \cos(-\theta)=\cos\theta.$

$\displaystyle\int_{-\pi}^{\pi}\sin(mx)\cos(mx)\,dx = \frac{1}{m}\left[\frac{\sin^2(mx)}{2}\right]_{-\pi}^{\pi}=0$

## Section 8.4   Trigonometric Substitution

**1.** Use $x = 3\tan\theta$

**3.** Use $x = 4\sin\theta$

**5.** Let $x = 4\sin\theta,\, dx = 4\cos\theta\,d\theta,\, \sqrt{16-x^2}=4\cos\theta.$

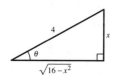

$\displaystyle\int\frac{1}{\left(16-x^2\right)^{3/2}}\,dx = \int\frac{4\cos\theta}{(4\cos\theta)^3}\,d\theta=\frac{1}{16}\int\sec^2\theta\,d\theta=\frac{1}{16}\tan\theta+C=\frac{1}{16}\frac{x}{\sqrt{16-x^2}}+C$

**7.** Same substitution as in Exercise 5

$\displaystyle\int\frac{\sqrt{16-x^2}}{x}\,dx = \int\frac{4\cos\theta}{4\sin\theta}\,4\cos\theta\,d\theta$

$\displaystyle\qquad = 4\int\frac{\cos^2\theta}{\sin\theta}\,d\theta$

$\displaystyle\qquad = 4\int\frac{1-\sin^2\theta}{\sin\theta}\,d\theta$

$\displaystyle\qquad = 4\int(\csc\theta-\sin\theta)\,d\theta$

$\displaystyle\qquad = -4\ln|\csc\theta+\cot\theta|+4\cos\theta+C$

$\displaystyle\qquad = -4\ln\left|\frac{4}{x}+\frac{\sqrt{16-x^2}}{x}\right|+4\frac{\sqrt{16-x^2}}{4}+C$

$\displaystyle\qquad = -4\ln\left|\frac{4+\sqrt{16-x^2}}{x}\right|+\sqrt{16-x^2}+C$

$\displaystyle\qquad = 4\ln\left|\frac{4-\sqrt{16-x^2}}{x}\right|+\sqrt{16-x^2}+C$

**9.** Let $x = 5 \sec \theta$, $dx = 5 \sec \theta \tan \theta \, d\theta$,

$\sqrt{x^2 - 25} = 5 \tan \theta$

$$\int \frac{1}{\sqrt{x^2 - 25}} \, dx = \int \frac{5 \sec \theta \tan \theta}{5 \tan \theta} \, d\theta$$

$$= \int \sec \theta \, d\theta$$

$$= \ln \left| \sec \theta + \tan \theta \right| + C$$

$$= \ln \left| \frac{x}{5} + \frac{\sqrt{x^2 - 25}}{5} \right| + C$$

$$= \ln \left| x + \sqrt{x^2 - 25} \right| + C$$

**11.** Same substitution as in Exercise 9

$$\int x^3 \sqrt{x^2 - 25} \, dx = \int (5 \sec \theta)^3 (5 \tan \theta)(5 \sec \theta \tan \theta) \, d\theta$$

$$= 3125 \int \sec^4 \theta \tan^2 \theta \, d\theta$$

$$= 3125 \int \left( 1 + \tan^2 \theta \right) \tan^2 \theta \sec^2 \theta \, d\theta$$

$$= 3125 \int \left( \tan^2 \theta + \tan^4 \theta \right) \sec^2 \theta \, d\theta$$

$$= 3125 \left[ \frac{\tan^3 \theta}{3} + \frac{\tan^5 \theta}{5} \right] + C$$

$$= 3125 \left[ \frac{\left( x^2 - 25 \right)^{3/2}}{125(3)} + \frac{\left( x^2 - 25 \right)^{5/2}}{5^5(5)} \right] + C$$

$$= \frac{1}{15} \left( x^2 - 25 \right)^{3/2} \left[ 125 + 3 \left( x^2 - 25 \right) \right] + C$$

$$= \frac{1}{15} \left( x^2 - 25 \right)^{3/2} \left( 50 + 3x^2 \right) + C$$

**13.** Let $x = \tan \theta$, $dx = \sec^2 \theta \, d\theta$, $\sqrt{1 + x^2} = \sec \theta$.

$$\int x \sqrt{1 + x^2} \, dx = \int \tan \theta (\sec \theta) \sec^2 \theta \, d\theta = \frac{\sec^3 \theta}{3} + C = \frac{1}{3} \left( 1 + x^2 \right)^{3/2} + C$$

**Note:** This integral could have been evaluated with the Power Rule.

**15.** Same substitution as in Exercise 13

$$\int \frac{1}{\left(1 + x^2\right)^2} \, dx = \int \frac{1}{\left(\sqrt{1 + x^2}\right)^4} \, dx = \int \frac{\sec^2 \theta \, d\theta}{\sec^4 \theta}$$

$$= \int \cos^2 \theta \, d\theta = \frac{1}{2} \int (1 + \cos 2\theta) \, d\theta$$

$$= \frac{1}{2}\left[\theta + \frac{\sin 2\theta}{2}\right]$$

$$= \frac{1}{2}[\theta + \sin \theta \cos \theta] + C$$

$$= \frac{1}{2}\left[\arctan x + \left(\frac{x}{\sqrt{1 + x^2}}\right)\left(\frac{1}{\sqrt{1 + x^2}}\right)\right] + C$$

$$= \frac{1}{2}\left(\arctan x + \frac{x}{1 + x^2}\right) + C$$

**17.** Let $u = 4x, a = 3, du = 4 \, dx$.

$$\int \sqrt{9 + 16x^2} \, dx = \frac{1}{4} \int \sqrt{(4x)^2 + 3^2} \, (4) \, dx$$

$$= \frac{1}{4} \cdot \frac{1}{2}\left[4x\sqrt{16x^2 + 9} + 9 \ln\left|4x + \sqrt{16x^2 + 9}\right|\right] + C$$

$$= \frac{1}{2}x\sqrt{16x^2 + 9} + \frac{9}{8} \ln\left|4x + \sqrt{16x^2 + 9}\right| + C$$

**19.** $\displaystyle\int \sqrt{25 - 4x^2} \, dx = \int 2\sqrt{\frac{25}{4} - x^2} \, dx, \quad a = \frac{5}{2}$

$$= 2\frac{1}{2}\left[\frac{25}{4} \arcsin\left(\frac{2x}{5}\right) + x\sqrt{\frac{25}{4} - x^2}\right] + C$$

$$= \frac{25}{4} \arcsin\left(\frac{2x}{5}\right) + \frac{x}{2}\sqrt{25 - 4x^2} + C$$

**21.** $\displaystyle\int \frac{x}{\sqrt{x^2 + 36}} \, dx = \frac{1}{2} \int (x^2 + 36)^{-1/2} (2x) \, dx$

$$= \frac{1}{2}\frac{(x^2 + 36)^{1/2}}{1/2} + C$$

$$= \sqrt{x^2 + 36} + C$$

**23.** $\displaystyle\int \frac{1}{\sqrt{16 - x^2}} \, dx = \arcsin\left(\frac{x}{4}\right) + C$

**25.** Let $x = 2 \sin \theta, dx = 2 \cos \theta \, d\theta$,

$$\sqrt{4 - x^2} = 2 \cos \theta.$$

$$\int \sqrt{16 - 4x^2} \, dx = 2 \int \sqrt{4 - x^2} \, dx$$

$$= 2 \int 2 \cos \theta (2 \cos \theta \, d\theta)$$

$$= 8 \int \cos^2 \theta \, d\theta$$

$$= 4 \int (1 + \cos 2\theta) \, d\theta$$

$$= 4\left(\theta + \frac{1}{2} \sin 2\theta\right) + C$$

$$= 4\theta + 4 \sin \theta \cos \theta + C$$

$$= 4 \arcsin\left(\frac{x}{2}\right) + x\sqrt{4 - x^2} + C$$

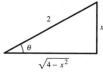

**27.** Let $x = 2 \sec \theta$, $dx = 2 \sec \theta \tan \theta \, d\theta$,

$\sqrt{x^2 - 4} = 2 \tan \theta$.

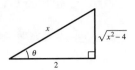

$$\int \frac{1}{\sqrt{x^2 - 4}} \, dx = \int \frac{2 \sec \theta \tan \theta}{2 \tan \theta} \, d\theta$$

$$= \int \sec \theta \, d\theta$$

$$= \ln|\sec \theta + \tan \theta| + C$$

$$= \ln \left| \frac{x}{2} + \frac{\sqrt{x^2 - 4}}{2} \right| + C$$

$$= \ln \left| x + \sqrt{x^2 - 4} \right| + C$$

**29.** Let $x = \sin \theta$, $dx = \cos \theta \, d\theta$, $\sqrt{1 - x^2} = \cos \theta$.

$$\int \frac{\sqrt{1 - x^2}}{x^4} \, dx = \int \frac{\cos \theta (\cos \theta \, d\theta)}{\sin^4 \theta}$$

$$= \int \cot^2 \theta \csc^2 \theta \, d\theta$$

$$= -\frac{1}{3} \cot^3 \theta + C$$

$$= -\frac{(1 - x^2)^{3/2}}{3x^3} + C$$

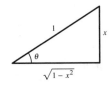

**31.** Same substitution as in Exercise 30

$$x = \frac{3}{2} \tan \theta, \quad dx = \frac{3}{2} \sec^2 \theta \, d\theta$$

$$\int \frac{1}{x \sqrt{4x^2 + 9}} \, dx = \int \frac{(3/2) \sec^2 \theta \, d\theta}{(3/2) \tan \theta \, 3 \sec \theta}$$

$$= \frac{1}{3} \int \csc \theta \, d\theta$$

$$= -\frac{1}{3} \ln|\csc \theta + \cot \theta| + C$$

$$= -\frac{1}{3} \ln \left| \frac{\sqrt{4x^2 + 9} + 3}{2x} \right| + C$$

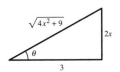

**33.** Let $u = x^2 + 3$, $du = 2x \, dx$.

$$\int \frac{-3x}{(x^2 + 3)^{3/2}} \, dx = -\frac{3}{2} \int (x^2 + 3)^{-3/2} (2x) \, dx$$

$$= -\frac{3}{2} \frac{(x^2 + 3)^{-1/2}}{(-1/2)} + C$$

$$= \frac{3}{\sqrt{x^2 + 3}} + C$$

**35.** Let $u = 1 + e^{2x}$, $du = 2e^{2x} \, dx$.

$$\int e^{2x} \sqrt{1 + e^{2x}} \, dx = \frac{1}{2} \int (1 + e^{2x})^{1/2} (2e^{2x}) \, dx$$

$$= \frac{1}{3} (1 + e^{2x})^{3/2} + C$$

**37.** Let $e^x = \sin \theta$, $e^x \, dx = \cos \theta \, d\theta$, $\sqrt{1 - e^{2x}} = \cos \theta$.

$$\int e^x \sqrt{1 - e^{2x}} \, dx = \int \cos^2 \theta \, d\theta$$

$$= \frac{1}{2} \int (1 + \cos 2\theta) \, d\theta$$

$$= \frac{1}{2} \left( \theta + \frac{\sin 2\theta}{2} \right)$$

$$= \frac{1}{2} (\theta + \sin \theta \cos \theta) + C$$

$$= \frac{1}{2} \left( \arcsin e^x + e^x \sqrt{1 - e^{2x}} \right) + C$$

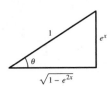

**39.** Let $x = \sqrt{2}\, \tan\theta$, $dx = \sqrt{2}\, \sec^2\theta\, d\theta$, $x^2 + 2 = 2\sec^2\theta$.

$$\int \frac{1}{4 + 4x^2 + x^4}\, dx = \int \frac{1}{\left(x^2 + 2\right)^2}\, dx = \int \frac{\sqrt{2}\, \sec^2\theta\, d\theta}{4\sec^4\theta}$$

$$= \frac{\sqrt{2}}{4}\int \cos^2\theta\, d\theta$$

$$= \frac{\sqrt{2}}{4}\left(\frac{1}{2}\right)\int (1 + \cos 2\theta)\, d\theta$$

$$= \frac{\sqrt{2}}{8}\left(\theta + \frac{1}{2}\sin 2\theta\right) + C$$

$$= \frac{\sqrt{2}}{8}\left(\theta + \sin\theta\cos\theta\right) + C$$

$$= \frac{\sqrt{2}}{8}\left(\arctan \frac{x}{\sqrt{2}} + \frac{x}{\sqrt{x^2 + 2}}\cdot\frac{\sqrt{2}}{\sqrt{x^2 + 2}}\right) = \frac{1}{4}\left(\frac{x}{x^2 + 2} + \frac{1}{\sqrt{2}}\arctan \frac{x}{\sqrt{2}}\right) + C$$

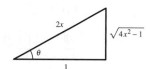

**41.** Use integration by parts. Because $x > \dfrac{1}{2}$,

$$u = \operatorname{arcsec} 2x \;\Rightarrow\; du = \frac{1}{x\sqrt{4x^2 - 1}}\, dx, \quad dv = dx \;\Rightarrow\; v = x$$

$$\int \operatorname{arcsec} 2x\, dx = x \operatorname{arcsec} 2x - \int \frac{1}{\sqrt{4x^2 - 1}}\, dx$$

$$2x = \sec\theta, \quad dx = \frac{1}{2}\sec\theta\tan\theta\, d\theta, \quad \sqrt{4x^2 - 1} = \tan\theta$$

$$\int \operatorname{arcsec} 2x\, dx = x \operatorname{arcsec} 2x - \int \frac{(1/2)\sec\theta\tan\theta\, d\theta}{\tan\theta} = x \operatorname{arcsec} 2x - \frac{1}{2}\int \sec\theta\, d\theta$$

$$= x \operatorname{arcsec} 2x - \frac{1}{2}\ln\left|\sec\theta + \tan\theta\right| + C = x \operatorname{arcsec} 2x - \frac{1}{2}\ln\left|2x + \sqrt{4x^2 - 1}\right| + C.$$

**43.** $\displaystyle\int \frac{1}{\sqrt{4x - x^2}}\, dx = \int \frac{1}{\sqrt{4 - (x - 2)^2}}\, dx = \arcsin\left(\frac{x - 2}{2}\right) + C$

**45.** $x^2 + 6x + 12 = x^2 + 6x + 9 + 3 = (x + 3)^2 + \left(\sqrt{3}\right)^2$

Let $x + 3 = \sqrt{3}\, \tan\theta$, $dx = \sqrt{3}\, \sec^2\theta\, d\theta$.

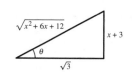

$$\sqrt{x^2 + 6x + 12} = \sqrt{(x + 3)^2 + \left(\sqrt{3}\right)^2} = \sqrt{3}\, \sec\theta$$

$$\int \frac{x}{\sqrt{x^2 + 6x + 12}}\, dx = \int \frac{\sqrt{3}\, \tan\theta - 3}{\sqrt{3}\, \sec\theta}\sqrt{3}\, \sec^2\theta\, d\theta$$

$$= \int \sqrt{3}\, \sec\theta\tan\theta\, d\theta - 3\int \sec\theta\, d\theta$$

$$= \sqrt{3}\, \sec\theta - 3\ln\left|\sec\theta + \tan\theta\right| + C$$

$$= \sqrt{3}\left(\frac{\sqrt{x^2 + 6x + 12}}{\sqrt{3}}\right) - 3\ln\left|\frac{\sqrt{x^2 + 6x + 12}}{\sqrt{3}} + \frac{x + 3}{\sqrt{3}}\right| + C$$

$$= \sqrt{x^2 + 6x + 12} - 3\ln\left|\sqrt{x^2 + 6x + 12} + (x + 3)\right| + C$$

**47.** Let $t = \sin\theta, \, dt = \cos\theta\, d\theta, 1 - t^2 = \cos^2\theta.$

(a) $\displaystyle\int \frac{t^2}{\left(1 - t^2\right)^{3/2}}\, dt = \int \frac{\sin^2\theta\cos\theta\, d\theta}{\cos^3\theta} = \int \tan^2\theta\, d\theta = \int\left(\sec^2\theta - 1\right) d\theta = \tan\theta - \theta + C = \frac{t}{\sqrt{1 - t^2}} - \arcsin t + C$

So, $\displaystyle\int_0^{\sqrt{3}/2} \frac{t^2}{\left(1 - t^2\right)^{3/2}}\, dt = \left[\frac{t}{\sqrt{1 - t^2}} - \arcsin t\right]_0^{\sqrt{3}/2} = \frac{\sqrt{3}/2}{\sqrt{1/4}} - \arcsin\frac{\sqrt{3}}{2} = \sqrt{3} - \frac{\pi}{3} \approx 0.685.$

(b) When $t = 0, \theta = 0.$ When $t = \sqrt{3}/2, \theta = \pi/3.$ So,

$\displaystyle\int_0^{\sqrt{3}/2} \frac{t^2}{\left(1 - t^2\right)^{3/2}}\, dt = \left[\tan\theta - \theta\right]_0^{\pi/3} = \sqrt{3} - \frac{\pi}{3} \approx 0.685.$

**49.** (a) Let $x = 3\tan\theta, \, dx = 3\sec^2\theta\, d\theta, \sqrt{x^2 + 9} = 3\sec\theta.$

$\displaystyle\int \frac{x^3}{\sqrt{x^2 + 9}}\, dx = \int \frac{\left(27\tan^3\theta\right)\left(3\sec^2\theta\, d\theta\right)}{3\sec\theta}$

$\displaystyle\qquad = 27\int\left(\sec^2\theta - 1\right)\sec\theta\tan\theta\, d\theta$

$\displaystyle\qquad = 27\left[\frac{1}{3}\sec^3\theta - \sec\theta\right] + C = 9\left[\sec^3\theta - 3\sec\theta\right] + C$

$\displaystyle\qquad = 9\left[\left(\frac{\sqrt{x^2 + 9}}{3}\right)^3 - 3\left(\frac{\sqrt{x^2 + 9}}{3}\right)\right] + C = \frac{1}{3}\left(x^2 + 9\right)^{3/2} - 9\sqrt{x^2 + 9} + C$

So, $\displaystyle\int_0^3 \frac{x^3}{\sqrt{x^2 + 9}}\, dx = \left[\frac{1}{3}\left(x^2 + 9\right)^{3/2} - 9\sqrt{x^2 + 9}\right]_0^3$

$\displaystyle\qquad = \left(\frac{1}{3}\left(54\sqrt{2}\right) - 27\sqrt{2}\right) - (9 - 27) = 18 - 9\sqrt{2} = 9\left(2 - \sqrt{2}\right) \approx 5.272.$

(b) When $x = 0, \theta = 0.$ When $x = 3, \theta = \pi/4.$ So,

$\displaystyle\int_0^3 \frac{x^3}{\sqrt{x^2 + 9}}\, dx = 9\left[\sec^3\theta - 3\sec\theta\right]_0^{\pi/4} = 9\left(2\sqrt{2} - 3\sqrt{2}\right) - 9(1 - 3) = 9\left(2 - \sqrt{2}\right) \approx 5.272.$

**51.** (a) Let $x = 3 \sec \theta$, $dx = 3 \sec \theta \tan \theta \, d\theta$, $\sqrt{x^2 - 9} = 3 \tan \theta$.

$$\int \frac{x^2}{\sqrt{x^2 - 9}} \, dx = \int \frac{9 \sec^2 \theta}{3 \tan \theta} \, 3 \sec \theta \tan \theta \, d\theta$$

$$= 9 \int \sec^3 \theta \, d\theta$$

$$= 9 \left( \frac{1}{2} \sec \theta \tan \theta + \frac{1}{2} \int \sec \theta \, d\theta \right) \quad \text{(8.3 Exercise 102 or Example 5, Section 8.2)}$$

$$= \frac{9}{2} \left( \sec \theta \tan \theta + \ln \left| \sec \theta + \tan \theta \right| \right)$$

$$= \frac{9}{2} \left( \frac{x}{3} \cdot \frac{\sqrt{x^2 - 9}}{3} + \ln \left| \frac{x}{3} + \frac{\sqrt{x^2 - 9}}{3} \right| \right)$$

So,

$$\int_4^6 \frac{x^2}{\sqrt{x^2 - 9}} \, dx = \frac{9}{2} \left[ \frac{x \sqrt{x^2 - 9}}{9} + \ln \left| \frac{x}{3} + \frac{\sqrt{x^2 - 9}}{3} \right| \right]_4^6$$

$$= \frac{9}{2} \left[ \left( \frac{6 \sqrt{27}}{9} + \ln \left| 2 + \frac{\sqrt{27}}{3} \right| \right) - \left( \frac{4 \sqrt{7}}{9} + \ln \left| \frac{4}{3} + \frac{\sqrt{7}}{3} \right| \right) \right]$$

$$= 9 \sqrt{3} - 2 \sqrt{7} + \frac{9}{2} \left[ \ln \left( \frac{6 + \sqrt{27}}{3} \right) - \ln \left( \frac{4 + \sqrt{7}}{3} \right) \right]$$

$$= 9 \sqrt{3} - 2 \sqrt{7} + \frac{9}{2} \ln \left( \frac{6 + 3\sqrt{3}}{4 + \sqrt{7}} \right) \approx 12.644$$

(b) When $x = 4$, $\theta = \operatorname{arcsec} \left( \frac{4}{3} \right)$. When $x = 6$, $\theta = \operatorname{arcsec}(2) = \frac{\pi}{3}$.

$$\int_4^6 \frac{x^2}{\sqrt{x^2 - 9}} \, dx = \frac{9}{2} \left[ \sec \theta \tan \theta + \ln \left| \sec \theta + \tan \theta \right| \right]_{\operatorname{arcsec}(4/3)}^{\pi/3}$$

$$= \frac{9}{2} \left( 2 \cdot \sqrt{3} + \ln \left| 2 + \sqrt{3} \right| \right) - \frac{9}{2} \left( \frac{4}{3} \frac{\sqrt{7}}{3} + \ln \left| \frac{4}{3} + \frac{\sqrt{7}}{3} \right| \right)$$

$$= 9 \sqrt{3} - 2 \sqrt{7} + \frac{9}{2} \ln \left( \frac{6 + 3\sqrt{3}}{4 + \sqrt{7}} \right) \approx 12.644$$

**53.** $x\dfrac{dy}{dx} = \sqrt{x^2 - 9}, \; x \geq 3, \; y(3) = 1$

$$y = \int \dfrac{\sqrt{x^2 - 9}}{x} \, dx$$

Let $x = 3, \sec \theta, \; dx = 3 \sec \theta \tan \theta \, d\theta, \; \sqrt{x^2 - 9} = 3 \tan \theta$.

$$y = \int \dfrac{3 \tan \theta}{3 \sec \theta} 3 \sec \theta \tan \theta \, d\theta = 3 \int \tan^2 \theta \, d\theta$$

$$= 3 \int \left(\sec^2 \theta - 1\right) d\theta = 3 \left[\tan \theta - \theta\right] + C$$

$$= 3 \left[\dfrac{\sqrt{x^2 - 9}}{3} - \arctan\left(\dfrac{\sqrt{x^2 - 9}}{3}\right)\right] + C$$

$$= \sqrt{x^2 - 9} - 3 \arctan\left(\dfrac{\sqrt{x^2 - 9}}{3}\right) + C$$

$y(3) = 1: \; 1 = 0 - 3(0) + C \Rightarrow C = 1$

$$y = \sqrt{x^2 - 9} - 3 \arctan\left(\dfrac{\sqrt{x^2 - 9}}{3}\right) + 1$$

**55.** $\displaystyle \int \dfrac{x^2}{\sqrt{x^2 + 10x + 9}} \, dx = \dfrac{1}{2}\sqrt{x^2 + 10x + 9}(x - 15) + 33 \ln\left|(x + 5) + \sqrt{x^2 + 10x + 9}\right| + C$

**57.** $\displaystyle \int \dfrac{x^2}{\sqrt{x^2 - 1}} \, dx = \dfrac{1}{2}\left(x\sqrt{x^2 - 1} + \ln\left|x + \sqrt{x^2 - 1}\right|\right) + C$

**59.** (a) Let $u = a \sin \theta, \; \sqrt{a^2 - u^2} = a \cos \theta,$ where $-\pi/2 \leq \theta \leq \pi/2$.

   (b) Let $u = a \tan \theta, \; \sqrt{a^2 + u^2} = a \sec \theta,$ where $-\pi/2 < \theta < \pi/2$.

   (c) Let $u = a \sec \theta, \; \sqrt{u^2 - a^2} = \tan \theta$ if $u > a$ and $\sqrt{u^2 - a^2} = -\tan \theta$
   if $u < -a,$ where $0 \leq \theta < \pi/2$ or $\pi/2 < \theta \leq \pi$.

**61.** Trigonometric substitution: $x = \sec \theta$

**63.** True

$$\int \dfrac{dx}{\sqrt{1 - x^2}} = \int \dfrac{\cos \theta \, d\theta}{\cos \theta} = \int d\theta$$

**65.** False

$$\int_0^{\sqrt{3}} \dfrac{dx}{\left(\sqrt{1 + x^2}\right)^3} = \int_0^{\pi/3} \dfrac{\sec^2 \theta \, d\theta}{\sec^3 \theta} = \int_0^{\pi/3} \cos \theta \, d\theta$$

**67.** $A = 4 \displaystyle \int_0^a \dfrac{b}{a}\sqrt{a^2 - x^2} \, dx$

$$= \dfrac{4b}{a} \int_0^a \sqrt{a^2 - x^2} \, dx$$

$$= \left[\dfrac{4b}{a}\left(\dfrac{1}{2}\right)\left(a^2 \arcsin\dfrac{x}{a} + x\sqrt{a^2 - x^2}\right)\right]_0^a$$

$$= \dfrac{2b}{a}\left(a^2\left(\dfrac{\pi}{2}\right)\right) = \pi ab$$

**Note:** See Theorem 8.2 for $\displaystyle \int \sqrt{a^2 - x^2} \, dx$.

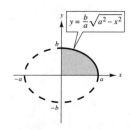

**69.** (a) $x^2 + (y - k)^2 = 25$

Radius of circle $= 5$

$k^2 = 5^2 + 5^2 = 50$

$k = 5\sqrt{2}$

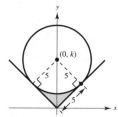

(b) Area $= \text{square} - \dfrac{1}{4}(\text{circle}) = 25 - \dfrac{1}{4}\pi(5)^2 = 25\left(1 - \dfrac{\pi}{4}\right)$

(c) Area $= r^2 - \dfrac{1}{4}\pi r^2 = r^2\left(1 - \dfrac{\pi}{4}\right)$

**71.** Let $x - 3 = \sin\theta,\ dx = \cos\theta\,d\theta,\ \sqrt{1 - (x - 3)^2} = \cos\theta$.

**Shell Method:**

$V = 4\pi \displaystyle\int_2^4 x\sqrt{1 - (x - 3)^2}\,dx$

$= 4\pi \displaystyle\int_{-\pi/2}^{\pi/2} (3 + \sin\theta)\cos^2\theta\,d\theta$

$= 4\pi\left[\dfrac{3}{2}\displaystyle\int_{-\pi/2}^{\pi/2}(1 + \cos 2\theta)\,d\theta + \displaystyle\int_{-\pi/2}^{\pi/2}\cos^2\theta\sin\theta\,d\theta\right]$

$= 4\pi\left[\dfrac{3}{2}\left(\theta + \dfrac{1}{2}\sin 2\theta\right) - \dfrac{1}{3}\cos^3\theta\right]_{-\pi/2}^{\pi/2} = 6\pi^2$

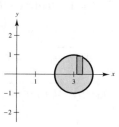

**73.** $y = \ln x,\ y' = \dfrac{1}{x},\ 1 + (y')^2 = 1 + \dfrac{1}{x^2} = \dfrac{x^2 + 1}{x^2}$

Let $x = \tan\theta,\ dx = \sec^2\theta\,d\theta,\ \sqrt{x^2 + 1} = \sec\theta$.

$s = \displaystyle\int_1^5 \sqrt{\dfrac{x^2 + 1}{x^2}}\,dx = \displaystyle\int_1^5 \dfrac{\sqrt{x^2 + 1}}{x}\,dx$

$= \displaystyle\int_a^b \dfrac{\sec\theta}{\tan\theta}\sec^2\theta\,d\theta = \displaystyle\int_a^b \dfrac{\sec\theta}{\tan\theta}(1 + \tan^2\theta)\,d\theta$

$= \displaystyle\int_a^b (\csc\theta + \sec\theta\tan\theta)\,d\theta = \left[-\ln|\csc\theta + \cot\theta| + \sec\theta\right]_a^b$

$= \left[-\ln\left|\dfrac{\sqrt{x^2 + 1}}{x} + \dfrac{1}{x}\right| + \sqrt{x^2 + 1}\right]_1^5$

$= \left[-\ln\left(\dfrac{\sqrt{26} + 1}{5}\right) + \sqrt{26}\right] - \left[-\ln(\sqrt{2} + 1) + \sqrt{2}\right]$

$= \ln\left[\dfrac{5(\sqrt{2} + 1)}{\sqrt{26} + 1}\right] + \sqrt{26} - \sqrt{2} \approx 4.367$ or $\ln\left[\dfrac{\sqrt{26} - 1}{5(\sqrt{2} - 1)}\right] + \sqrt{26} - \sqrt{2}$

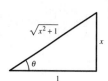

**75.** Length of one arch of sine curve: $y = \sin x$, $y' = \cos x$

$$L_1 = \int_0^\pi \sqrt{1 + \cos^2 x} \; dx$$

Length of one arch of cosine curve: $y = \cos x$, $y' = -\sin x$

$$L_2 = \int_{-\pi/2}^{\pi/2} \sqrt{1 + \sin^2 x} \; dx$$

$$= \int_{-\pi/2}^{\pi/2} \sqrt{1 + \cos^2\left(x - \frac{\pi}{2}\right)} \; dx, \quad u = x - \frac{\pi}{2}, du = dx$$

$$= \int_{-\pi}^{0} \sqrt{1 + \cos^2 u} \; du$$

$$= \int_{0}^{\pi} \sqrt{1 + \cos^2 u} \; du = L_1$$

**77.** (a)

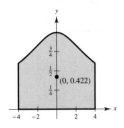

(b) $y = 0$ for $x = 200$    (range)

(c) $y = x - 0.005x^2$, $y' = 1 - 0.01x$, $1 + (y')^2 = 1 + (1 - 0.01x)^2$

Let $u = 1 - 0.01x$, $du = -0.01 \, dx$, $a = 1$.    (See Theorem 8.2.)

$$s = \int_0^{200} \sqrt{1 + (1 - 0.01x)^2} \; dx = -100 \int_0^{200} \sqrt{(1 - 0.01x)^2 + 1} \, (-0.01) \, dx$$

$$= -50 \left[ (1 - 0.01x)\sqrt{(1 - 0.01x)^2 + 1} + \ln\left| (1 - 0.01x) + \sqrt{(1 - 0.01x)^2 + 1} \right| \right]_0^{200}$$

$$= -50\left[ \left( -\sqrt{2} + \ln\left| -1 + \sqrt{2} \right| \right) - \left( \sqrt{2} + \ln\left| 1 + \sqrt{2} \right| \right) \right] = 100\sqrt{2} + 50 \ln\left( \frac{\sqrt{2} + 1}{\sqrt{2} - 1} \right) \approx 229.559$$

**79.** Let $x = 3 \tan\theta$, $dx = 3 \sec^2\theta \; d\theta$, $\sqrt{x^2 + 9} = 3 \sec\theta$.

$$A = 2\int_0^4 \frac{3}{\sqrt{x^2 + 9}} \; dx = 6\int_0^4 \frac{dx}{\sqrt{x^2 + 9}} = 6\int_a^b \frac{3\sec^2\theta \; d\theta}{3 \sec\theta}$$

$$= 6\int_a^b \sec\theta \; d\theta = \left[ 6 \ln\left| \sec\theta + \tan\theta \right| \right]_a^b = \left[ 6 \ln\left| \frac{\sqrt{x^2 + 9} + x}{3} \right| \right]_0^4 = 6 \ln 3$$

$\bar{x} = 0 \,(\text{by symmetry})$

$$\bar{y} = \frac{1}{2}\left( \frac{1}{A} \right)\int_{-4}^{4} \left( \frac{3}{\sqrt{x^2 + 9}} \right)^2 dx = \frac{9}{12 \ln 3} \int_{-4}^{4} \frac{1}{x^2 + 9} \; dx = \frac{3}{4 \ln 3}\left[ \frac{1}{3} \arctan\frac{x}{3} \right]_{-4}^{4} = \frac{2}{4 \ln 3} \arctan\frac{4}{3} \approx 0.422$$

$$(\bar{x}, \bar{y}) = \left( 0, \frac{1}{2 \ln 3} \arctan\frac{4}{3} \right) \approx (0, 0.422)$$

**81.** $y = x^2,\ y' = 2x,\ 1 + (y')^2 = 1 + 4x^2$

$2x = \tan\theta,\ dx = \dfrac{1}{2}\sec^2\theta\,d\theta,\ \sqrt{1 + 4x^2} = \sec\theta$

(For $\displaystyle\int \sec^5\theta\,d\theta$ and $\displaystyle\int\sec^3\theta\,d\theta$, see Exercise 102 in Section 8.3.)

$$S = 2\pi\int_0^{\sqrt{2}} x^2\sqrt{1 + 4x^2}\,dx = 2\pi\int_a^b\left(\frac{\tan\theta}{2}\right)^2(\sec\theta)\left(\frac{1}{2}\sec^2\theta\right)d\theta$$

$$= \frac{\pi}{4}\int_a^b\sec^3\theta\,\tan^2\theta\,d\theta = \frac{\pi}{4}\left[\int_a^b\sec^5\theta\,d\theta - \int_a^b\sec^3\theta\,d\theta\right]$$

$$= \frac{\pi}{4}\left\{\frac{1}{4}\left[\sec^3\theta\,\tan\theta + \frac{3}{2}(\sec\theta\,\tan\theta + \ln|\sec\theta + \tan\theta|)\right] - \frac{1}{2}(\sec\theta\,\tan\theta + \ln|\sec\theta + \tan\theta|)\right\}\Bigg]_a^b$$

$$= \frac{\pi}{4}\left[\frac{1}{4}\left[(1 + 4x^2)^{3/2}(2x)\right] - \frac{1}{8}\left[(1 + 4x^2)^{1/2}(2x) + \ln\left|\sqrt{1 + 4x^2} + 2x\right|\right]\right]_0^{\sqrt{2}}$$

$$= \frac{\pi}{4}\left[\frac{54\sqrt{2}}{4} - \frac{6\sqrt{2}}{8} - \frac{1}{8}\ln\!\left(3 + 2\sqrt{2}\right)\right]$$

$$= \frac{\pi}{4}\left(\frac{51\sqrt{2}}{4} - \frac{\ln\!\left(3 + 2\sqrt{2}\right)}{8}\right) = \frac{\pi}{32}\left[102\sqrt{2} - \ln\!\left(3 + 2\sqrt{2}\right)\right] \approx 13.989$$

**83.** (a) Area of representative rectangle: $2\sqrt{1 - y^2}\,\Delta y$

Force: $2(62.4)(3 - y)\sqrt{1 - y^2}\,\Delta y$

$$F = 124.8\int_{-1}^{1}(3 - y)\sqrt{1 - y^2}\,dy$$

$$= 124.8\left[3\int_{-1}^{1}\sqrt{1 - y^2}\,dy - \int_{-1}^{1}y\sqrt{1 - y^2}\,dy\right]$$

$$= 124.8\left[\frac{3}{2}\left(\arcsin y + y\sqrt{1 - y^2}\right) + \frac{1}{2}\left(\frac{2}{3}\right)\left(1 - y^2\right)^{3/2}\right]_{-1}^{1} = (62.4)3\left[\arcsin 1 - \arcsin(-1)\right] = 187.2\pi \text{ lb}$$

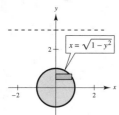

(b) $F = 124.8\displaystyle\int_{-1}^{1}(d - y)\sqrt{1 - y^2}\,dy = 124.8d\int_{-1}^{1}\sqrt{1 - y^2}\,dy - 124.8\int_{-1}^{1}y\sqrt{1 - y^2}\,dy$

$$= 124.8\left(\frac{d}{2}\right)\left[\arcsin y + y\sqrt{1 - y^2}\right]_{-1}^{1} - 124.8(0) = 62.4\pi d \text{ lb}$$

**85.** Let $u = a \sin \theta$, $du = a \cos \theta \, d\theta$, $\sqrt{a^2 - u^2} = a \cos \theta$.

$$\int \sqrt{a^2 - u^2} \, du = \int a^2 \cos^2 \theta \, d\theta = a^2 \int \frac{1 + \cos 2\theta}{2} \, d\theta$$

$$= \frac{a^2}{2}\left(\theta + \frac{1}{2}\sin 2\theta\right) + C = \frac{a^2}{2}(\theta + \sin \theta \cos \theta) + C$$

$$= \frac{a^2}{2}\left[\arcsin \frac{u}{a} + \left(\frac{u}{a}\right)\left(\frac{\sqrt{a^2 + u^2}}{a}\right)\right] + C = \frac{1}{2}\left(a^2 \arcsin \frac{u}{a} + u\sqrt{a^2 - u^2}\right) + C$$

Let $u = a \sec \theta$, $du = a \sec \theta \tan \theta \, d\theta$, $\sqrt{u^2 - a^2} = a \tan \theta$.

$$\int \sqrt{u^2 - a^2} \, du = \int a \tan \theta(a \sec \theta \tan \theta) \, d\theta = a^2 \int \tan^2 \theta \sec \theta \, d\theta$$

$$= a^2 \int (\sec^2 \theta - 1) \sec \theta \, d\theta = a^2 \int (\sec^3 \theta - \sec \theta) \, d\theta$$

$$= a^2\left[\frac{1}{2}\sec \theta \tan \theta + \frac{1}{2}\int \sec \theta \, d\theta\right] - a^2 \int \sec \theta \, d\theta = a^2\left[\frac{1}{2}\sec \theta \tan \theta - \frac{1}{2}\ln|\sec \theta + \tan \theta|\right]$$

$$= \frac{a^2}{2}\left[\frac{u}{a} \cdot \frac{\sqrt{u^2 - a^2}}{a} - \ln\left|\frac{u}{a} + \frac{\sqrt{u^2 - a^2}}{a}\right|\right] + C_1 = \frac{1}{2}\left[u\sqrt{u^2 - a^2} - a^2 \ln\left|u + \sqrt{u^2 - a^2}\right|\right] + C$$

Let $u = a \tan \theta$, $du = a \sec^2 \theta \, d\theta$, $\sqrt{u^2 + a^2} = a \sec \theta$.

$$\int \sqrt{u^2 + a^2} \, du = \int (a \sec \theta)(a \sec^2 \theta) \, d\theta$$

$$= a^2 \int \sec^3 \theta \, d\theta = a^2\left[\frac{1}{2}\sec \theta \tan \theta + \frac{1}{2}\ln|\sec \theta + \tan \theta|\right] + C_1$$

$$= \frac{a^2}{2}\left[\frac{\sqrt{u^2 + a^2}}{a} \cdot \frac{u}{a} + \ln\left|\frac{\sqrt{u^2 + a^2}}{a} + \frac{u}{a}\right|\right] + C_1 = \frac{1}{2}\left[u\sqrt{u^2 + a^2} + a^2 \ln\left|u + \sqrt{u^2 + a^2}\right|\right] + C$$

**87.** Large circle: $x^2 + y^2 = 25$

$$y = \sqrt{25 - x^2}, \quad \text{upper half}$$

From the right triangle, the center of the small circle is $(0, 4)$.

$$x^2 + (y - 4)^2 = 9$$

$$y = 4 + \sqrt{9 - x^2}, \quad \text{upper half}$$

$$A = 2\int_0^3 \left[\left(4 + \sqrt{9 - x^2}\right) - \sqrt{25 - x^2}\right] dx$$

$$= 2\left[4x + \frac{1}{2}\left[9 \arcsin\left(\frac{x}{3}\right) + x\sqrt{9 - x^2}\right] - \frac{1}{2}\left[25 \arcsin\left(\frac{x}{5}\right) + x\sqrt{25 - x^2}\right]\right]_0^3$$

$$= 2\left[12 + \frac{9}{2}\arcsin(1) - \frac{25}{2}\arcsin\frac{3}{5} - 6\right]$$

$$= 12 + \frac{9\pi}{2} - 25 \arcsin\frac{3}{5} \approx 10.050$$

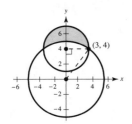

**89.** Let $I = \int_0^1 \dfrac{\ln(x + 1)}{x^2 + 1}\, dx$

Let $x = \dfrac{1 - u}{1 + u}, \quad dx = \dfrac{-2}{(1 + u)^2}\, du$

$x + 1 = \dfrac{2}{1 + u}, \quad x^2 + 1 = \dfrac{2 + 2u^2}{(1 + u)^2}$

$I = \int_1^0 \dfrac{\ln\left(\dfrac{2}{1 + u}\right)}{\left(\dfrac{2 + 2u^2}{(1 + u)^2}\right)}\left(\dfrac{-2}{(1 + u)^2}\right) du$

$= \int_1^0 \dfrac{-\ln\left(\dfrac{2}{1 + u}\right)}{1 + u^2}\, du = \int_0^1 \dfrac{\ln\left(\dfrac{2}{1 + u}\right)}{1 + u^2}\, du = \int_0^1 \dfrac{\ln 2}{1 + u^2}\, du - \int_0^1 \dfrac{\ln(1 + u)}{1 + u^2}\, du = (\ln 2)\left[\arctan u\right]_0^1 - I$

$\Rightarrow 2I = \ln 2\left(\dfrac{\pi}{4}\right)$

$I = \dfrac{\pi}{8}\ln 2 \approx 0.272198$

# Section 8.5   Partial Fractions

**1.** $\dfrac{4}{x^2 - 8x} = \dfrac{4}{x(x - 8)} = \dfrac{A}{x} + \dfrac{B}{x - 8}$

**3.** $\dfrac{2x - 3}{x^3 + 10x} = \dfrac{2x - 3}{x(x^2 + 10)} = \dfrac{A}{x} + \dfrac{Bx + C}{x^2 + 10}$

**5.** $\dfrac{x - 9}{x^2 - 6x} = \dfrac{x - 9}{x(x - 6)} = \dfrac{A}{x} + \dfrac{B}{x - 6}$

**7.** $\dfrac{1}{x^2 - 9} = \dfrac{1}{(x - 3)(x + 3)} = \dfrac{A}{x + 3} + \dfrac{B}{x - 3}$

$1 = A(x - 3) + B(x + 3)$

When $x = 3, \quad 1 = 6B \Rightarrow B = \dfrac{1}{6}$.

When $x = -3, \quad 1 = -6A \Rightarrow A = -\dfrac{1}{6}$.

$\int \dfrac{1}{x^2 - 9}\, dx = -\dfrac{1}{6}\int \dfrac{1}{x + 3}\, dx + \dfrac{1}{6}\int \dfrac{1}{x - 3}\, dx$

$= -\dfrac{1}{6}\ln|x + 3| + \dfrac{1}{6}\ln|x - 3| + C$

$= \dfrac{1}{6}\ln\left|\dfrac{x - 3}{x + 3}\right| + C$

**9.** $\dfrac{5}{x^2 + 3x - 4} = \dfrac{5}{(x + 4)(x - 1)} = \dfrac{A}{x + 4} + \dfrac{B}{x - 1}$

$5 = A(x - 1) + B(x + 4)$

When $x = 1, \quad 5 = 5B \Rightarrow B = 1$.

When $x = -4, \quad 5 = -5A \Rightarrow A = -1$.

$\int \dfrac{5}{x^2 + 3x - 4}\, dx = \int \dfrac{-1}{x + 4}\, dx + \int \dfrac{1}{x - 1}\, dx$

$= -\ln|x + 4| + \ln|x - 1| + C$

$= \ln\left|\dfrac{x - 1}{x + 4}\right| + C$

**11.** $\dfrac{5 - x}{2x^2 + x - 1} = \dfrac{5 - x}{(2x - 1)(x + 1)} = \dfrac{A}{2x - 1} + \dfrac{B}{x + 1}$

$5 - x = A(x + 1) + B(2x - 1)$

When $x = \dfrac{1}{2}, \dfrac{9}{2} = \dfrac{3}{2}A \Rightarrow A = 3$.

When $x = -1, 6 = -3B \Rightarrow B = -2$.

$\int \dfrac{5 - x}{2x^2 + x - 1}\, dx = 3\int \dfrac{1}{2x - 1}\, dx - 2\int \dfrac{1}{x + 1}\, dx$

$= \dfrac{3}{2}\ln|2x - 1| - 2\ln|x + 1| + C$

**13.** $\dfrac{x^2 + 12x + 12}{x(x + 2)(x - 2)} = \dfrac{A}{x} + \dfrac{B}{x + 2} + \dfrac{C}{x - 2}$

$\quad x^2 + 12x + 12 = A(x + 2)(x - 2) + Bx(x - 2) + Cx(x + 2)$

When $x = 0, 12 = -4A \Rightarrow A = -3$.

When $x = -2, -8 = 8B \Rightarrow B = -1$.

When $x = 2, 40 = 8C \Rightarrow C = 5$.

$\displaystyle\int \dfrac{x^2 + 12x + 12}{x^3 - 4x}\, dx = 5\int \dfrac{1}{x - 2}\, dx - \int \dfrac{1}{x + 2}\, dx - 3\int \dfrac{1}{x}\, dx = 5\ln|x - 2| - \ln|x + 2| - 3\ln|x| + C$

**15.** $\dfrac{2x^3 - 4x^2 - 15x + 5}{x^2 - 2x - 8} = 2x + \dfrac{x + 5}{(x - 4)(x + 2)} = 2x + \dfrac{A}{x - 4} + \dfrac{B}{x + 2}$

$\quad\quad\quad x + 5 = A(x + 2) + B(x - 4)$

When $x = 4, 9 = 6A \Rightarrow A = \dfrac{3}{2}$.

When $x = -2, 3 = -6B \Rightarrow B = -\dfrac{1}{2}$.

$\displaystyle\int \dfrac{2x^3 - 4x^2 - 15x + 5}{x^2 - 2x - 8}\, dx = \int\left(2x + \dfrac{3/2}{x - 4} - \dfrac{1/2}{x + 2}\right) dx = x^2 + \dfrac{3}{2}\ln|x - 4| - \dfrac{1}{2}\ln|x + 2| + C$

**17.** $\dfrac{4x^2 + 2x - 1}{x^2(x + 1)} = \dfrac{A}{x} + \dfrac{B}{x^2} + \dfrac{C}{x + 1}$

$\quad 4x^2 + 2x - 1 = Ax(x + 1) + B(x + 1) + Cx^2$

When $x = 0, B = -1$.

When $x = -1, C = 1$.

When $x = 1, A = 3$.

$\displaystyle\int \dfrac{4x^2 + 2x - 1}{x^3 + x^2}\, dx = \int\left(\dfrac{3}{x} - \dfrac{1}{x^2} + \dfrac{1}{x + 1}\right) dx$

$\quad\quad\quad\quad\quad\quad\quad\quad = 3\ln|x| + \dfrac{1}{x} + \ln|x + 1| + C$

$\quad\quad\quad\quad\quad\quad\quad\quad = \dfrac{1}{x} + \ln|x^4 + x^3| + C$

**19.** $\dfrac{x^2 + 3x - 4}{x^3 - 4x^2 + 4x} = \dfrac{x^2 + 3x - 4}{x(x - 2)^2} = \dfrac{A}{x} + \dfrac{B}{(x - 2)} + \dfrac{C}{(x - 2)^2}$

$\quad x^2 + 3x - 4 = A(x - 2)^2 + Bx(x - 2) + Cx$

When $x = 0, -4 = 4A \Rightarrow A = -1$.

When $x = 2, 6 = 2C \Rightarrow C = 3$.

When $x = 1, 0 = -1 - B + 3 \Rightarrow B = 2$.

$\displaystyle\int \dfrac{x^2 + 3x - 4}{x^3 - 4x^2 + 4x}\, dx = \int \dfrac{-1}{x}\, dx + \int \dfrac{2}{(x - 2)}\, dx + \int \dfrac{3}{(x - 2)^2}\, dx = -\ln|x| + 2\ln|x - 2| - \dfrac{3}{(x - 2)} + C$

**21.** $\dfrac{x^2 - 1}{x(x^2 + 1)} = \dfrac{A}{x} + \dfrac{Bx + C}{x^2 + 1}$

$\qquad x^2 - 1 = A(x^2 + 1) + (Bx + C)x$

When $x = 0$, $A = -1$.

When $x = 1$, $0 = -2 + B + C$.

When $x = -1$, $0 = -2 + B - C$.

Solving these equations you have $A = -1$, $B = 2$, $C = 0$.

$\displaystyle\int \frac{x^2 - 1}{x^3 + x}\,dx = -\int \frac{1}{x}\,dx + \int \frac{2x}{x^2 + 1}\,dx = -\ln|x| + \ln|x^2 + 1| + C = \ln\left|\frac{x^2 + 1}{x}\right| + C$

**23.** $\dfrac{x^2}{x^4 - 2x^2 - 8} = \dfrac{A}{x - 2} + \dfrac{B}{x + 2} + \dfrac{Cx + D}{x^2 + 2}$

$\qquad\qquad x^2 = A(x + 2)(x^2 + 2) + B(x - 2)(x^2 + 2) + (Cx + D)(x + 2)(x - 2)$

When $x = 2$, $4 = 24A$.

When $x = -2$, $4 = -24B$.

When $x = 0$, $0 = 4A - 4B - 4D$.

When $x = 1$, $1 = 9A - 3B - 3C - 3D$.

Solving these equations you have $A = \dfrac{1}{6}$, $B = -\dfrac{1}{6}$, $C = 0$, $D = \dfrac{1}{3}$.

$\displaystyle\int \frac{x^2}{x^4 - 2x^2 - 8}\,dx = \frac{1}{6}\left(\int \frac{1}{x - 2}\,dx - \int \frac{1}{x + 2}\,dx + 2\int \frac{1}{x^2 + 2}\,dx\right) = \frac{1}{6}\left(\ln\left|\frac{x - 2}{x + 2}\right| + \sqrt{2}\arctan\frac{x}{\sqrt{2}}\right) + C$

**25.** $\dfrac{x}{(2x - 1)(2x + 1)(4x^2 + 1)} = \dfrac{A}{2x - 1} + \dfrac{B}{2x + 1} + \dfrac{Cx + D}{4x^2 + 1}$

$\qquad\qquad x = A(2x + 1)(4x^2 + 1) + B(2x - 1)(4x^2 + 1) + (Cx + D)(2x - 1)(2x + 1)$

When $x = \dfrac{1}{2}$, $\dfrac{1}{2} = 4A$.

When $x = -\dfrac{1}{2}$, $-\dfrac{1}{2} = -4B$.

When $x = 0$, $0 = A - B - D$.

When $x = 1$, $1 = 15A + 5B + 3C + 3D$.

Solving these equations you have $A = \dfrac{1}{8}$, $B = \dfrac{1}{8}$, $C = -\dfrac{1}{2}$, $D = 0$.

$\displaystyle\int \frac{x}{16x^4 - 1}\,dx = \frac{1}{8}\left(\int \frac{1}{2x - 1}\,dx + \int \frac{1}{2x + 1}\,dx - 4\int \frac{x}{4x^2 + 1}\,dx\right) = \frac{1}{16}\ln\left|\frac{4x^2 - 1}{4x^2 + 1}\right| + C$

**27.** $\dfrac{x^2 + 5}{(x + 1)(x^2 - 2x + 3)} = \dfrac{A}{x + 1} + \dfrac{Bx + C}{x^2 - 2x + 3}$

$\qquad\qquad x^2 + 5 = A(x^2 - 2x + 3) + (Bx + C)(x + 1)$

$\qquad\qquad\qquad = (A + B)x^2 + (-2A + B + C)x + (3A + C)$

When $x = -1$, $A = 1$.

By equating coefficients of like terms, you have $A + B = 1$, $-2A + B + C = 0$, $3A + C = 5$.

Solving these equations you have $A = 1$, $B = 0$, $C = 2$.

$\displaystyle\int \frac{x^2 + 5}{x^3 - x^2 + x + 3}\,dx = \int \frac{1}{x + 1}\,dx + 2\int \frac{1}{(x - 1)^2 + 2}\,dx = \ln|x + 1| + \sqrt{2}\arctan\left(\frac{x - 1}{\sqrt{2}}\right) + C$

**29.** $\dfrac{3}{4x^2 + 5x + 1} = \dfrac{3}{(4x + 1)(x + 1)} = \dfrac{A}{4x + 1} + \dfrac{B}{x + 1}$

$$3 = A(x + 1) + B(4x + 1)$$

When $x = -1$, $3 = -3B \Rightarrow B = -1$.

When $-\dfrac{1}{4}$, $3 = \dfrac{3}{4}A \Rightarrow A = 4$.

$\displaystyle\int_0^2 \dfrac{3}{4x^2 + 5x + 1}\,dx = \int_0^2 \dfrac{4}{4x + 1}\,dx + \int_0^2 \dfrac{-1}{x + 1}\,dx$

$\qquad = \Big[\ln|4x + 1| - \ln|x + 1|\Big]_0^2$

$\qquad = \ln 9 - \ln 3$

$\qquad = 2\ln 3 - \ln 3 = \ln 3$

**31.** $\dfrac{x + 1}{x(x^2 + 1)} = \dfrac{A}{x} + \dfrac{Bx + C}{x^2 + 1}$

$$x + 1 = A(x^2 + 1) + (Bx + C)x$$

When $x = 0$, $A = 1$.

When $x = 1$, $2 = 2A + B + C$.

When $x = -1$, $0 = 2A + B - C$.

Solving these equations we have

$A = 1$, $B = -1$, $C = 1$.

$\displaystyle\int_1^2 \dfrac{x + 1}{x(x^2 + 1)}\,dx = \int_1^2 \dfrac{1}{x}\,dx - \int_1^2 \dfrac{x}{x^2 + 1}\,dx + \int_1^2 \dfrac{1}{x^2 + 1}\,dx$

$\qquad = \Big[\ln|x| - \dfrac{1}{2}\ln(x^2 + 1) + \arctan x\Big]_1^2$

$\qquad = \dfrac{1}{2}\ln\dfrac{8}{5} - \dfrac{\pi}{4} + \arctan 2$

$\qquad \approx 0.557$

**37.** $\displaystyle\int \dfrac{2x^2 - 2x + 3}{x^3 - x^2 - x - 2}\,dx = \ln|x - 2| + \dfrac{1}{2}\ln|x^2 + x + 1| - \sqrt{3}\arctan\left(\dfrac{2x + 1}{\sqrt{3}}\right) + C$

$(3, 10): 0 + \dfrac{1}{2}\ln 13 - \sqrt{3}\arctan\dfrac{7}{\sqrt{3}} + C = 10 \Rightarrow C = 10 - \dfrac{1}{2}\ln 13 + \sqrt{3}\arctan\dfrac{7}{\sqrt{3}}$

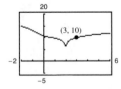

**39.** $\displaystyle\int \dfrac{1}{x^2 - 25}\,dx = \dfrac{1}{10}\ln\left|\dfrac{x - 5}{x + 5}\right| + C$

$(7, 2): \dfrac{1}{10}\ln\left|\dfrac{2}{12}\right| + C = 2$

$\qquad\qquad C = 2 - \dfrac{1}{10}\ln\left(\dfrac{1}{6}\right) = 2 + \dfrac{1}{10}\ln 6$

$y = \dfrac{1}{10}\ln\left|\dfrac{x - 5}{x + 5}\right| + 2 + \dfrac{1}{10}\ln 6$

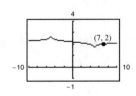

**33.** $\displaystyle\int \dfrac{5x}{x^2 - 10x + 25}\,dx = 5\ln|x - 5| - \dfrac{5x}{x - 5} + C$

$(6, 0): 5\ln(1) - \dfrac{30}{1} + C = 0 \Rightarrow C = 30$

$y = 5\ln|x - 5| - \dfrac{5x}{x - 5} + 30$

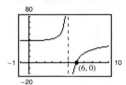

**35.** $\displaystyle\int \dfrac{x^2 + x + 2}{(x^2 + 2)^2}\,dx = \dfrac{\sqrt{2}}{2}\arctan\dfrac{x}{\sqrt{2}} - \dfrac{1}{2(x^2 + 2)} + C$

$(0, 1): 0 - \dfrac{1}{4} + C = 1 \Rightarrow C = \dfrac{5}{4}$

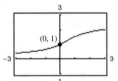

**41.** Let $u = \cos x\; du = -\sin x\, dx$.

$$\frac{1}{u(u-1)} = \frac{A}{u} + \frac{B}{u-1}$$

$$1 = A(u-1) + Bu$$

When $u = 0$, $A = -1$.

When $u = 1$, $B = 1$.

$$\int \frac{\sin x}{\cos x(\cos x - 1)}\, dx = -\int \frac{1}{u(u-1)}\, du$$

$$= \int \frac{1}{u}\, du - \int \frac{1}{u-1}\, du$$

$$= \ln|u| - \ln|u-1| + C$$

$$= \ln\left|\frac{u}{u-1}\right| + C$$

$$= \ln\left|\frac{\cos x}{\cos x - 1}\right| + C$$

**43.** Let $u = \sin x$, $du = \cos x\, dx$

$$\frac{1}{u + u^2} = \frac{1}{u(1+u)} = \frac{A}{u} + \frac{B}{u+1}$$

$$1 = A(u+1) + Bu$$

When $u = 0$, $1 = A$.

When $u = -1$, $1 = -B \Rightarrow B = -1$.

$$\int \frac{\cos x}{\sin x + \sin^2 x}\, dx = \int \frac{1}{u + u^2}\, du$$

$$= \int \frac{1}{u}\, du - \int \frac{1}{u+1}\, du$$

$$= \ln|u| - \ln|u+1| + C$$

$$= \ln\left|\frac{\sin x}{\sin x + 1}\right| + C$$

**45.** Let $u = \tan x$, $du = \sec^2 x\, dx$.

$$\frac{1}{u^2 + 5u + 6} = \frac{1}{(u+3)(u+2)} = \frac{A}{u+3} + \frac{B}{u+2}$$

$$1 = A(u+2) + B(u+3)$$

When $u = -2$, $1 = B$.

When $u = -3$, $1 = -A \Rightarrow A = -1$.

$$\int \frac{\sec^2 x}{\tan^2 x + 5\tan x + 6}\, dx = \int \frac{1}{u^2 + 5u + 6}\, du$$

$$= \int \frac{-1}{u+3}\, du + \int \frac{1}{u+2}\, du$$

$$= -\ln|u+3| + \ln|u+2| + C$$

$$= \ln\left|\frac{\tan x + 2}{\tan x + 3}\right| + C$$

**47.** Let $u = e^x$, $du = e^x\, dx$.

$$\frac{1}{(u-1)(u+4)} = \frac{A}{u-1} + \frac{B}{u+4}$$

$$1 = A(u+4) + B(u-1)$$

When $u = 1$, $A = \frac{1}{5}$.

When $u = -4$, $B = -\frac{1}{5}$.

$$\int \frac{e^x}{(e^x - 1)(e^x + 4)}\, dx = \int \frac{1}{(u-1)(u+4)}\, du$$

$$= \frac{1}{5}\left(\int \frac{1}{u-1}\, du - \int \frac{1}{u+4}\, du\right)$$

$$= \frac{1}{5}\ln\left|\frac{u-1}{u+4}\right| + C$$

$$= \frac{1}{5}\ln\left|\frac{e^x - 1}{e^x + 4}\right| + C$$

**49.** Let $u = \sqrt{x}$, $u^2 = x$, $2u\, du = dx$.

$$\int \frac{\sqrt{x}}{x-4}\, dx = \int \frac{u(2u)du}{u^2 - 4} = \int \left(\frac{2u^2 - 8}{u^2 - 4} + \frac{8}{u^2 - 4}\right) du = \int \left(2 + \frac{8}{u^2 - 4}\right) du$$

$$\frac{8}{u^2 - 4} = \frac{8}{(u-2)(u+2)} = \frac{A}{u-2} + \frac{B}{u+2}$$

$$8 = A(u+2) + B(u-2)$$

When $u = -2$, $8 = -4B \Rightarrow B = -2$.

When $u = 2$, $8 = 4A \Rightarrow A = 2$.

$$\int \left(2 + \frac{8}{u^2 - 4}\right) du = 2u + \int \left(\frac{2}{u-2} - \frac{2}{u+2}\right) du$$

$$= 2u + 2\ln|u-2| - 2\ln|u+2| + C$$

$$= 2\sqrt{x} + 2\ln\left|\frac{\sqrt{x} - 2}{\sqrt{x} + 2}\right| + C$$

**51.** $\dfrac{1}{x(a + bx)} = \dfrac{A}{x} + \dfrac{B}{a + bx}$

$$1 = A(a + bx) + Bx$$

When $x = 0, 1 = aA \Rightarrow A = 1/a$.

When $x = -a/b, 1 = -(a/b)B \Rightarrow B = -b/a$.

$$\int \dfrac{1}{x(a + bx)}\, dx = \dfrac{1}{a} \int \left( \dfrac{1}{x} - \dfrac{b}{a + bx} \right) dx$$

$$= \dfrac{1}{a} \left( \ln|x| - \ln|a + bx| \right) + C$$

$$= \dfrac{1}{a} \ln \left| \dfrac{x}{a + bx} \right| + C$$

**53.** $\dfrac{x}{(a + bx)^2} = \dfrac{A}{a + bx} + \dfrac{B}{(a + bx)^2}$

$$x = A(a + bx) + B$$

When $x = -a/b, B = -a/b$.

When $x = 0, 0 = aA + B \Rightarrow A = 1/b$.

$$\int \dfrac{x}{(a + bx)^2}\, dx = \int \left( \dfrac{1/b}{a + bx} + \dfrac{-a/b}{(a + bx)^2} \right) dx$$

$$= \dfrac{1}{b} \int \dfrac{1}{a + bx}\, dx - \dfrac{a}{b} \int \dfrac{1}{(a + bx)^2}\, dx$$

$$= \dfrac{1}{b^2} \ln|a + bx| + \dfrac{a}{b^2} \left( \dfrac{1}{a + bx} \right) + C$$

$$= \dfrac{1}{b^2} \left( \dfrac{a}{a + bx} + \ln|a + bx| \right) + C$$

**55.** $\dfrac{dy}{dx} = \dfrac{6}{4 - x^2}, y(0) = 3$

$$y = \dfrac{3}{2} \ln \left| \dfrac{2 + x}{2 - x} \right| + 3$$

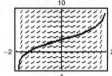

**57.** Dividing $x^3$ by $x - 5$

**59.** $A = \displaystyle\int_0^1 \dfrac{12}{x^2 + 5x + 6}\, dx$

$$\dfrac{12}{x^2 + 5x + 6} = \dfrac{12}{(x + 2)(x + 3)} = \dfrac{A}{x + 2} + \dfrac{B}{x + 3}$$

$$12 = A(x + 3) + B(x + 2)$$

Let $x = -3: 12 = B(-1) \Rightarrow B = -12$

Let $x = -2: 12 = A(1) \Rightarrow A = 12$

$$A = \int_0^1 \left( \dfrac{12}{x + 2} - \dfrac{12}{x + 3} \right) dx$$

$$= \Big[ 12 \ln|x + 2| - 12 \ln|x + 3| \Big]_0^1$$

$$= 12(\ln 3 - \ln 4 - \ln 2 + \ln 3)$$

$$= 12 \ln \left( \dfrac{9}{8} \right) \approx 1.4134$$

**61.** $A = 2 \displaystyle\int_0^3 \left( 1 - \dfrac{7}{16 - x^2} \right) dx = 2 \int_0^3 dx - 14 \int_0^3 \dfrac{1}{16 - x^2}\, dx$

$$= \left[ 2x - \dfrac{14}{8} \ln \left| \dfrac{4 + x}{4 - x} \right| \right]_0^3 \qquad \text{(From Exercise 52)}$$

$$= 6 - \dfrac{7}{4} \ln 7$$

$$\approx 2.595$$

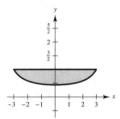

**63.** Average cost $= \dfrac{1}{80 - 75} \displaystyle\int_{75}^{80} \dfrac{124p}{(10 + p)(100 - p)}\, dp$

$$= \dfrac{1}{5} \int_{75}^{80} \left( \dfrac{-124}{(10 + p)11} + \dfrac{1240}{(100 - p)11} \right) dp$$

$$= \dfrac{1}{5} \left[ \dfrac{-124}{11} \ln(10 + p) - \dfrac{1240}{11} \ln(100 - p) \right]_{75}^{80}$$

$$\approx \dfrac{1}{5}(24.51) = 4.9$$

Approximately \$490,000

**65.** $V = \pi \int_0^3 \left(\dfrac{2x}{x^2+1}\right)^2 dx = 4\pi \int_0^3 \dfrac{x^2}{\left(x^2+1\right)^2} dx$

$\qquad\qquad = 4\pi \int_0^3 \left(\dfrac{1}{x^2+1} - \dfrac{1}{\left(x^2+1\right)^2}\right) dx \qquad\qquad \text{(partial fractions)}$

$\qquad\qquad = 4\pi \left[\arctan x - \dfrac{1}{2}\left(\arctan x + \dfrac{x}{x^2+1}\right)\right]_0^3 \qquad \text{(trigonometric substitution)}$

$\qquad\qquad = 2\pi \left[\arctan x - \dfrac{x}{x^2+1}\right]_0^3 = 2\pi\left(\arctan 3 - \dfrac{3}{10}\right) \approx 5.963$

$A = \int_0^3 \dfrac{2x}{x^2+1} dx = \left[\ln\left(x^2+1\right)\right]_0^3 = \ln 10$

$\bar{x} = \dfrac{1}{A}\int_0^3 \dfrac{2x^2}{x^2+1} dx = \dfrac{1}{\ln 10}\int_0^3 \left(2 - \dfrac{2}{x^2+1}\right) dx = \dfrac{1}{\ln 10}\left[2x - 2\arctan x\right]_0^3 = \dfrac{2}{\ln 10}(3 - \arctan 3) \approx 1.521$

$\bar{y} = \dfrac{1}{A}\left(\dfrac{1}{2}\right)\int_0^3 \left(\dfrac{2x}{x^2+1}\right)^2 dx = \dfrac{2}{\ln 10}\int_0^3 \dfrac{x^2}{\left(x^2+1\right)^2} dx$

$\qquad = \dfrac{2}{\ln 10}\int_0^3 \left(\dfrac{1}{x^2+1} - \dfrac{1}{\left(x^2+1\right)^2}\right) dx \qquad\qquad \text{(partial fractions)}$

$\qquad = \dfrac{2}{\ln 10}\left[\arctan x - \dfrac{1}{2}\left(\arctan x + \dfrac{x}{x^2+1}\right)\right]_0^3 \qquad \text{(trigonometric substitution)}$

$\qquad = \dfrac{2}{\ln 10}\left[\dfrac{1}{2}\arctan x - \dfrac{x}{2\left(x^2+1\right)}\right]_0^3 = \dfrac{1}{\ln 10}\left[\arctan x - \dfrac{x}{x^2+1}\right]_0^3 = \dfrac{1}{\ln 10}\left(\arctan 3 - \dfrac{3}{10}\right) \approx 0.412$

$\left(\bar{x}, \bar{y}\right) \approx \left(1.521, 0.412\right)$

**67.** $\qquad\qquad \dfrac{1}{(x+1)(n-x)} = \dfrac{A}{x+1} + \dfrac{B}{n-x}, \ A = B = \dfrac{1}{n+1}$

$\qquad \dfrac{1}{n+1}\int\left(\dfrac{1}{x+1} + \dfrac{1}{n-x}\right) dx = kt + C$

$\qquad\qquad \dfrac{1}{n+1}\ln\left|\dfrac{x+1}{n-x}\right| = kt + C$

When $t = 0$, $x = 0$, $C = \dfrac{1}{n+1}\ln\dfrac{1}{n}$.

$\qquad\qquad \dfrac{1}{n+1}\ln\left|\dfrac{x+1}{n-x}\right| = kt + \dfrac{1}{n+1}\ln\dfrac{1}{n}$

$\qquad \dfrac{1}{n+1}\left[\ln\left|\dfrac{x+1}{n-x}\right| - \ln\dfrac{1}{n}\right] = kt$

$\qquad\qquad \ln\dfrac{nx+n}{n-x} = (n+1)kt$

$\qquad\qquad\quad \dfrac{nx+n}{n-x} = e^{(n+1)kt}$

$\qquad\qquad\qquad x = \dfrac{n\left[e^{(n+1)kt} - 1\right]}{n + e^{(n+1)kt}} \qquad \textbf{Note: } \lim_{t\to\infty} x = n$

**69.** $\dfrac{x}{1 + x^4} = \dfrac{Ax + B}{x^2 + \sqrt{2}x + 1} + \dfrac{Cx + D}{x^2 - \sqrt{2}x + 1}$

$\quad\quad x = (Ax + B)(x^2 - \sqrt{2}x + 1) + (Cx + D)(x^2 + \sqrt{2}x + 1)$

$\quad\quad\quad = (A + C)x^3 + (B + D - \sqrt{2}A + \sqrt{2}C)x^2 + (A + C - \sqrt{2}B + \sqrt{2}D)x + (B + D)$

$0 = A + C \Rightarrow C = -A$

$0 = B + D - \sqrt{2}A + \sqrt{2}C \quad\quad -2\sqrt{2}A = 0 \Rightarrow A = 0 \text{ and } C = 0$

$1 = A + C - \sqrt{2}B + \sqrt{2}D \quad\quad -2\sqrt{2}B = 1 \Rightarrow B = -\dfrac{\sqrt{2}}{4} \text{ and } D = \dfrac{\sqrt{2}}{4}$

$0 = B + D \Rightarrow D = -B$

So,

$\displaystyle\int_0^1 \frac{x}{1 + x^4}\, dx = \int_0^1 \left( \frac{-\sqrt{2}/4}{x^2 + \sqrt{2}x + 1} + \frac{\sqrt{2}/4}{x^2 - \sqrt{2}x + 1} \right) dx$

$\quad\quad\quad = \dfrac{\sqrt{2}}{4} \displaystyle\int_0^1 \left[ \frac{-1}{\left[ x + \left( \sqrt{2}/2 \right) \right]^2 + (1/2)} + \frac{1}{\left[ x - \left( \sqrt{2}/2 \right) \right]^2 + (1/2)} \right] dx$

$\quad\quad\quad = \dfrac{\sqrt{2}}{4} \cdot \dfrac{1}{1/\sqrt{2}} \left[ -\arctan\left( \dfrac{x + \left( \sqrt{2}/2 \right)}{1/\sqrt{2}} \right) + \arctan\left( \dfrac{x - \left( \sqrt{2}/2 \right)}{1/\sqrt{2}} \right) \right]_0^1$

$\quad\quad\quad = \dfrac{1}{2} \left[ -\arctan\left( \sqrt{2}x + 1 \right) + \arctan\left( \sqrt{2}x - 1 \right) \right]_0^1$

$\quad\quad\quad = \dfrac{1}{2} \left[ \left( -\arctan\left( \sqrt{2} + 1 \right) + \arctan\left( \sqrt{2} - 1 \right) \right) - \left( -\arctan 1 + \arctan(-1) \right) \right]$

$\quad\quad\quad = \dfrac{1}{2} \left[ \arctan\left( \sqrt{2} - 1 \right) - \arctan\left( \sqrt{2} + 1 \right) + \dfrac{\pi}{4} + \dfrac{\pi}{4} \right].$

Because $\arctan x - \arctan y = \arctan\left[ (x - y)/(1 + xy) \right]$, you have:

$\displaystyle\int_0^1 \frac{x}{1 + x^4}\, dx = \dfrac{1}{2} \left[ \arctan\left( \frac{\left( \sqrt{2} - 1 \right) - \left( \sqrt{2} + 1 \right)}{1 + \left( \sqrt{2} - 1 \right)\left( \sqrt{2} + 1 \right)} \right) + \dfrac{\pi}{2} \right] = \dfrac{1}{2} \left[ \arctan\left( \dfrac{-2}{2} \right) + \dfrac{\pi}{2} \right] = \dfrac{1}{2} \left( -\dfrac{\pi}{4} + \dfrac{\pi}{2} \right) = \dfrac{\pi}{8}$

## Section 8.6 Integration by Tables and Other Integration Techniques

**1.** By Formula 6: $(a = 5, b = 1)$

$\quad \displaystyle\int \frac{x^2}{5 + x}\, dx = \left[ -\frac{x}{2}(10 - x) + 25 \ln|5 + x| \right] + C$

**3.** By Formula 26: $\displaystyle\int e^x \sqrt{1 + e^{2x}}\, dx = \frac{1}{2} \left[ e^x \sqrt{e^{2x} + 1} + \ln\left( e^x + \sqrt{e^{2x} + 1} \right) \right] + C$

$\quad\quad u = e^x,\, du = e^x\, dx$

**5.** By Formula 44: $\displaystyle\int \frac{1}{x^2 \sqrt{1 - x^2}}\, dx = -\frac{\sqrt{1 - x^2}}{x} + C$

**7.** By Formulas 51 and 49:

$$\int \cos^4 3x \, dx = \frac{1}{3} \int \cos^4 3x \, (3) \, dx$$

$$= \frac{1}{3} \left[ \frac{\cos^3 3x \sin 3x}{4} + \frac{3}{4} \int \cos^2 3x \, dx \right]$$

$$= \frac{1}{12} \cos^3 3x \sin 3x + \frac{1}{4} \cdot \frac{1}{3} \int \cos^2 3x \, (3) \, dx$$

$$= \frac{1}{12} \cos^3 3x \sin 3x + \frac{1}{12} \cdot \frac{1}{2} (3x + \sin 3x \cos 3x) + C$$

$$= \frac{1}{24} \left( 2 \cos^3 3x \sin 3x + 3x + \sin 3x \cos 3x \right) + C$$

**9.** By Formula 57: $\int \dfrac{1}{\sqrt{x} \left( 1 - \cos \sqrt{x} \right)} \, dx = 2 \int \dfrac{1}{1 - \cos \sqrt{x}} \left( \dfrac{1}{2\sqrt{x}} \right) dx = -2 \left( \cot \sqrt{x} + \csc \sqrt{x} \right) + C$

$$u = \sqrt{x}, \, du = \frac{1}{2\sqrt{x}} \, dx$$

**11.** By Formula 84:

$$\int \frac{1}{1 + e^{2x}} \, dx = x - \frac{1}{2} \ln\left( 1 + e^{2x} \right) + C$$

**13.** By Formula 89: $(n = 7)$

$$\int x^7 \ln x \, dx = \frac{x^8}{64} [-1 + 8 \ln x] + C = \frac{1}{64} x^8 (8 \ln x - 1) + C$$

**15.** (a) Let $u = 3x, \, x = \dfrac{u}{3}, \, du = 3 \, dx$.

$$\int x^2 e^{3x} \, dx = \int \left( \frac{u}{3} \right)^2 e^u \frac{1}{3} du = \frac{1}{27} \int u^2 e^u \, du$$

By Formulas 83 and 82:

$$\int x^2 e^{3x} \, dx = \frac{1}{27} \left[ u^2 e^u - 2 \int u e^u \, du \right]$$

$$= \frac{1}{27} \left[ u^2 e^u - 2 \left( (u - 1) e^u \right) \right] + C$$

$$= \frac{1}{27} e^{3x} \left( 9x^2 - 6x + 2 \right) + C$$

(b) Integration by parts: $u = x^2, \, du = 2x \, dx,$

$$dv = e^{3x} \, dx, v = \frac{1}{3} e^{3x}$$

$$\int x^2 e^{3x} \, dx = x^2 \frac{1}{3} e^{3x} - \int \frac{2}{3} x e^{3x} \, dx$$

Parts again: $u = x, \, du = dx, \, dv = e^{3x}, \, v = \dfrac{1}{3} e^{3x}$

$$\int x^2 e^{3x} \, dx = \frac{1}{3} x^2 e^{3x} - \frac{2}{3} \left[ \frac{x}{3} e^{3x} - \int \frac{1}{3} e^{3x} \, dx \right]$$

$$= \frac{1}{3} x^2 e^{3x} - \frac{2}{9} x e^{3x} + \frac{2}{27} e^{3x} + C$$

$$= \frac{1}{27} e^{3x} \left[ 9x^2 - 6x + 2 \right] + C$$

**17.** (a) By Formula 12: $(a = b = 1, u = x)$

$$\int \frac{1}{x^2 (x + 1)} \, dx = \frac{-1}{1} \left( \frac{1}{x} + \frac{1}{1} \ln \left| \frac{x}{1 + x} \right| \right) + C$$

$$= \frac{-1}{x} - \ln \left| \frac{x}{1 + x} \right| + C$$

$$= \frac{-1}{x} + \ln \left| \frac{x + 1}{x} \right| + C$$

(b) Partial fractions:

$$\frac{1}{x^2 (x + 1)} = \frac{A}{x} + \frac{B}{x^2} + \frac{C}{x + 1}$$

$$1 = Ax(x + 1) + B(x + 1) + Cx^2$$

$$x = 0: 1 = B$$

$$x = -1: 1 = C$$

$$x = 1: 1 = 2A + 2 + 1 \Rightarrow A = -1$$

$$\int \frac{1}{x^2 (x + 1)} \, dx = \int \left[ \frac{-1}{x} + \frac{1}{x^2} + \frac{1}{x + 1} \right] dx$$

$$= -\ln|x| - \frac{1}{x} + \ln|x + 1| + C$$

$$= -\frac{1}{x} - \ln \left| \frac{x}{x + 1} \right| + C$$

**19.** By Formula 80:

$$\int x \, \text{arccsc}\left(x^2 + 1\right) dx = \frac{1}{2} \int \text{arccsc}\left(x^2 + 1\right)(2x) \, dx$$

$$= \frac{1}{2}\left[\left(x^2 + 1\right)\text{arccsc}\left(x^2 + 1\right) + \ln\left|x^2 + 1 + \sqrt{\left(x^2 + 1\right)^2 - 1}\right|\right] + C$$

$$= \frac{1}{2}\left(x^2 + 1\right)\text{arccsc}\left(x^2 + 1\right) + \frac{1}{2}\ln\left(x^2 + 1 + \sqrt{x^4 + 2x^2}\right) + C$$

**21.** By Formula 35: $\displaystyle\int\frac{1}{x^2\sqrt{x^2 - 4}} \, dx = \frac{\sqrt{x^2 - 4}}{4x} + C$

**23.** By Formula 4: $(a = 2, b = -5)$

$$\int\frac{4x}{\left(2 - 5x\right)^2} \, dx = 4\left[\frac{1}{25}\left(\frac{2}{2 - 5x} + \ln|2 - 5x|\right)\right] + C$$

$$= \frac{4}{25}\left(\frac{2}{2 - 5x} + \ln|2 - 5x|\right) + C$$

**25.** By Formula 76:

$$\int e^x \arccos e^x \, dx = e^x \arccos e^x - \sqrt{1 - e^{2x}} + C$$

$u = e^x, \, du = e^x \, dx$

**27.** By Formula 73:

$$\int\frac{x}{1 - \sec x^2} \, dx = \frac{1}{2}\int\frac{2x}{1 - \sec x^2} \, dx$$

$$= \frac{1}{2}\left(x^2 + \cot x^2 + \csc x^2\right) + C$$

**29.** By Formula 14: $\displaystyle\int\frac{\cos\theta}{3 + 2\sin\theta + \sin^2\theta} \, d\theta = \frac{\sqrt{2}}{2}\arctan\left(\frac{1 + \sin\theta}{\sqrt{2}}\right) + C$   $\left(b^2 = 4 < 12 = 4ac\right)$

$u = \sin\theta, \, du = \cos\theta \, d\theta$

**31.** By Formula 35: $\displaystyle\int\frac{1}{x^2\sqrt{2 + 9x^2}} \, dx = 3\int\frac{3}{\left(3x\right)^2\sqrt{\left(\sqrt{2}\right)^2 + \left(3x\right)^2}} \, dx = -\frac{3\sqrt{2 + 9x^2}}{6x} + C = -\frac{\sqrt{2 + 9x^2}}{2x} + C$

**33.** By Formula 3: $\displaystyle\int\frac{\ln x}{x(3 + 2\ln x)} \, dx = \frac{1}{4}\left(2\ln|x| - 3\ln|3 + 2\ln|x||\right) + C$

$u = \ln x, \, du = \frac{1}{x} \, dx$

**35.** By Formulas 1, 25, and 33: $\displaystyle\int\frac{x}{\left(x^2 - 6x + 10\right)^2} \, dx = \frac{1}{2}\int\frac{2x - 6 + 6}{\left(x^2 - 6x + 10\right)^2} \, dx$

$$= \frac{1}{2}\int\left(x^2 - 6x + 10\right)^{-2}(2x - 6) \, dx + 3\int\frac{1}{\left[\left(x - 3\right)^2 + 1\right]^2} \, dx$$

$$= -\frac{1}{2\left(x^2 - 6x + 10\right)} + \frac{3}{2}\left[\frac{x - 3}{x^2 - 6x + 10} + \arctan(x - 3)\right] + C$$

$$= \frac{3x - 10}{2\left(x^2 - 6x + 10\right)} + \frac{3}{2}\arctan(x - 3) + C$$

**37.** By Formula 31: $\int \dfrac{x}{\sqrt{x^4 - 6x^2 + 5}}\,dx = \dfrac{1}{2}\int \dfrac{2x}{\sqrt{\left(x^2 - 3\right)^2 - 4}}\,dx = \dfrac{1}{2}\ln\left| x^2 - 3 + \sqrt{x^4 - 6x^2 + 5}\right| + C$

$u = x^2 - 3,\, du = 2x\,dx$

**39.** $\int \dfrac{x^3}{\sqrt{4 - x^2}}\,dx = \int \dfrac{8\sin^3 \theta\left(2\cos\theta\,d\theta\right)}{2\cos\theta}$

$\qquad = 8\int\left(1 - \cos^2 \theta\right)\sin\theta\,d\theta$

$\qquad = 8\int\left[\sin\theta - \cos^2 \theta\left(\sin\theta\right)\right]d\theta$

$\qquad = -8\cos\theta + \dfrac{8\cos^3 \theta}{3} + C$

$\qquad = -8\dfrac{\sqrt{4 - x^2}}{2} + \dfrac{8}{3}\left(\dfrac{\sqrt{4 - x^2}}{2}\right)^3 + C$

$\qquad = \sqrt{4 - x^2}\left[-4 + \dfrac{1}{3}\left(4 - x^2\right)\right] + C = \dfrac{-\sqrt{4 - x^2}}{3}\left(x^2 + 8\right) + C$

$x = 2\sin\theta,\, dx = 2\cos\theta\,d\theta,\, \sqrt{4 - x^2} = 2\cos\theta$

**41.** By Formula 8:

$\int \dfrac{e^{3x}}{\left(1 + e^x\right)^3}\,dx = \int \dfrac{\left(e^x\right)^2}{\left(1 + e^x\right)^3}\left(e^x\right)dx = \dfrac{2}{1 + e^x} - \dfrac{1}{2\left(1 + e^x\right)^2} + \ln\left|1 + e^x\right| + C$

$u = e^x,\, du = e^x\,dx$

**43.** By Formula 81:

$\int_0^1 xe^{x^2}\,dx = \left[\tfrac{1}{2}e^{x^2}\right]_0^1 = \tfrac{1}{2}(e - 1) \approx 0.8591$

**45.** By Formula 89: $(n = 4)$

$\int_1^2 x^4 \ln x\,dx = \dfrac{x^5}{25}\left[-1 + 5\ln x\right]_1^2 = \dfrac{32}{25}\left[-1 + 5\ln 2\right] - \dfrac{1}{25}\left[-1 + 0\right] = -\dfrac{31}{25} + \dfrac{32}{5}\ln 2 \approx 3.1961$

**47.** By Formula 23, and letting $u = \sin x$:

$\int_{-\pi/2}^{\pi/2} \dfrac{\cos x}{1 + \sin^2 x}\,dx = \left[\arctan\left(\sin x\right)\right]_{-\pi/2}^{\pi/2} = \arctan(1) - \arctan(-1) = \dfrac{\pi}{2}$

**49.** By Formulas 54 and 55:

$\int t^3 \cos t\,dt = t^3 \sin t - 3\int t^2 \sin t\,dt$

$\qquad = t^3 \sin t - 3\left(-t^2 \cos t + 2\int t \cos t\,dt\right)$

$\qquad = t^3 \sin t + 3t^2 \cos t - 6\left(t \sin t - \int \sin t\,dt\right)$

$\qquad = t^3 \sin t + 3t^2 \cos t - 6t \sin t - 6\cos t + C$

So,

$\int_0^{\pi/2} t^3 \cos t\,dt = \left[t^3 \sin t + 3t^2 \cos t - 6t \sin t - 6\cos t\right]_0^{\pi/2}$

$\qquad = \left(\dfrac{\pi^3}{8} - 3\pi\right) + 6 = \dfrac{\pi^3}{8} + 6 - 3\pi \approx 0.4510$

**51.** $\dfrac{u^2}{(a+bu)^2} = \dfrac{1}{b^2} - \dfrac{(2a/b)u + (a^2/b^2)}{(a+bu)^2} = \dfrac{1}{b^2} + \dfrac{A}{a+bu} + \dfrac{B}{(a+bu)^2}$

$-\dfrac{2a}{b}u - \dfrac{a^2}{b^2} = A(a+bu) + B = (aA+B) + bAu$

Equating the coefficients of like terms you have $aA + B = -a^2/b^2$ and $bA = -2a/b$. Solving these equations you have $A = -2a/b^2$ and $B = a^2/b^2$.

$\displaystyle\int \dfrac{u^2}{(a+bu)^2}\,du = \dfrac{1}{b^2}\int du - \dfrac{2a}{b^2}\left(\dfrac{1}{b}\right)\int\dfrac{1}{a+bu}\,b\,du + \dfrac{a^2}{b^2}\left(\dfrac{1}{b}\right)\int\dfrac{1}{(a+bu)^2}\,b\,du = \dfrac{1}{b^2}u - \dfrac{2a}{b^3}\ln|a+bu| - \dfrac{a^2}{b^3}\left(\dfrac{1}{a+bu}\right) + C$

$\hspace{3cm} = \dfrac{1}{b^3}\left(bu - \dfrac{a^2}{a+bu} - 2a\ln|a+bu|\right) + C$

**53.** When you have $u^2 + a^2$:

$\hspace{2cm} u = a\tan\theta$

$\hspace{2cm} du = a\sec^2\theta\,d\theta$

$\hspace{2cm} u^2 + a^2 = a^2\sec^2\theta$

$\displaystyle\int\dfrac{1}{(u^2+a^2)^{3/2}}\,du = \int\dfrac{a\sec^2\theta\,d\theta}{a^3\sec^3\theta} = \dfrac{1}{a^2}\int\cos\theta\,d\theta = \dfrac{1}{a^2}\sin\theta + C = \dfrac{u}{a^2\sqrt{u^2+a^2}} + C$

When you have $u^2 - a^2$:

$\hspace{2cm} u = a\sec\theta$

$\hspace{2cm} du = a\sec\theta\tan\theta\,d\theta$

$\hspace{2cm} u^2 - a^2 = a^2\tan^2\theta$

$\displaystyle\int\dfrac{1}{(u^2-a^2)^{3/2}}\,du = \int\dfrac{a\sec\theta\tan\theta\,d\theta}{a^3\tan^3\theta} = \dfrac{1}{a^2}\int\dfrac{\cos\theta}{\sin^2\theta}\,d\theta = \dfrac{1}{a^2}\int\csc\theta\cot\theta\,d\theta = -\dfrac{1}{a^2}\csc\theta + C = \dfrac{-u}{a^2\sqrt{u^2-a^2}} + C$

**55.** $\displaystyle\int(\arctan u)\,du = u\arctan u - \dfrac{1}{2}\int\dfrac{2u}{1+u^2}\,du$

$\hspace{2.5cm} = u\arctan u - \dfrac{1}{2}\ln(1+u^2) + C$

$\hspace{2.5cm} = u\arctan u - \ln\sqrt{1+u^2} + C$

$\hspace{1.5cm} w = \arctan u,\ dv = du,\ dw = \dfrac{du}{1+u^2},\ v = u$

**57.** $\displaystyle\int\dfrac{1}{x^{3/2}\sqrt{1-x}}\,dx = \dfrac{-2\sqrt{1-x}}{\sqrt{x}} + C$

$\left(\dfrac{1}{2},5\right)\colon\ \dfrac{-2\sqrt{1/2}}{\sqrt{1/2}} + C = 5 \Rightarrow C = 7$

$\hspace{1.5cm} y = \dfrac{-2\sqrt{1-x}}{\sqrt{x}} + 7$

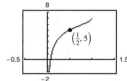

**59.** $\displaystyle\int \frac{1}{\left(x^2 - 6x + 10\right)^2}\, dx = \frac{1}{2}\left[\arctan(x - 3) + \frac{x - 3}{x^2 - 6x + 10}\right] + C$

$(3, 0)\colon \frac{1}{2}\left[0 + \frac{0}{10}\right] + C = 0 \Rightarrow C = 0$

$y = \frac{1}{2}\left[\arctan(x - 3) + \frac{x - 3}{x^2 - 6x + 10}\right]$

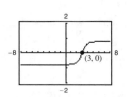

**61.** $\displaystyle\int \frac{1}{\sin \theta \tan \theta}\, d\theta = -\csc \theta + C$

$\left(\frac{\pi}{4}, 2\right)\colon -\frac{2}{\sqrt{2}} + C = 2 \Rightarrow C = 2 + \sqrt{2}$

$y = -\csc \theta + 2 + \sqrt{2}$

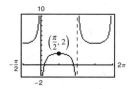

**63.** $\displaystyle\int \frac{1}{2 - 3 \sin \theta}\, d\theta = \int \left[\dfrac{\dfrac{2\, du}{1 + u^2}}{2 - 3\left(\dfrac{2u}{1 + u^2}\right)}\right],\ u = \tan \frac{\theta}{2}$

$= \displaystyle\int \frac{2}{2\left(1 + u^2\right) - 6u}\, du$

$= \displaystyle\int \frac{1}{u^2 - 3u + 1}\, du$

$= \displaystyle\int \frac{1}{\left(u - \dfrac{3}{2}\right)^2 - \dfrac{5}{4}}\, du$

$= \dfrac{1}{\sqrt{5}} \ln \left|\dfrac{\left(u - \dfrac{3}{2}\right) - \dfrac{\sqrt{5}}{2}}{\left(u - \dfrac{3}{2}\right) + \dfrac{\sqrt{5}}{2}}\right| + C$

$= \dfrac{1}{\sqrt{5}} \ln \left|\dfrac{2u - 3 - \sqrt{5}}{2u - 3 + \sqrt{5}}\right| + C$

$= \dfrac{1}{\sqrt{5}} \ln \left|\dfrac{2 \tan\left(\dfrac{\theta}{2}\right) - 3 - \sqrt{5}}{2 \tan\left(\dfrac{\theta}{2}\right) - 3 + \sqrt{5}}\right| + C$

**65.** $\displaystyle\int_0^{\pi/2} \frac{1}{1 + \sin \theta + \cos \theta}\, d\theta = \int_0^1 \left[\dfrac{\dfrac{2\, du}{1 + u^2}}{1 + \dfrac{2u}{1 + u^2} + \dfrac{1 - u^2}{1 + u^2}}\right]$

$= \displaystyle\int_0^1 \frac{1}{1 + u}\, du$

$= \Big[\ln|1 + u|\Big]_0^1$

$= \ln 2$

$u = \tan \dfrac{\theta}{2}$

**67.** $\displaystyle\int \frac{\sin \theta}{3 - 2 \cos \theta}\, d\theta = \frac{1}{2}\int \frac{2 \sin \theta}{3 - 2 \cos \theta}\, d\theta$

$= \dfrac{1}{2} \ln|u| + C$

$= \dfrac{1}{2} \ln(3 - 2 \cos \theta) + C$

$u = 3 - 2 \cos \theta,\ du = 2 \sin \theta\, d\theta$

**69.** $\displaystyle\int \frac{\sin \sqrt{\theta}}{\sqrt{\theta}}\, d\theta = 2 \int \sin \sqrt{\theta}\left(\frac{1}{2\sqrt{\theta}}\right) d\theta$

$= -2 \cos \sqrt{\theta} + C$

$u = \sqrt{\theta},\ du = \dfrac{1}{2\sqrt{\theta}}\, d\theta$

**71.** By Formula 21: $(a = 3, b = 1)$

$A = \displaystyle\int_0^6 \frac{x}{\sqrt{x + 3}}\, dx = \left[\frac{-2(6 - x)}{3}\sqrt{x + 3}\right]_0^6$

$= 4\sqrt{3} \approx 6.928$ square units

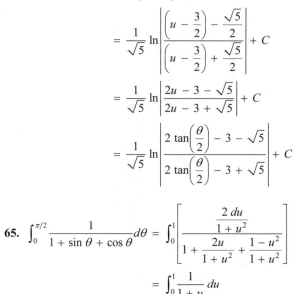

**73. (a)** $n = 1$: $u = \ln x$, $du = \dfrac{1}{x}\,dx$, $dv = x\,dx$, $v = \dfrac{x^2}{2}$

$$\int x \ln x\,dx = \frac{x^2}{2}\ln x - \int \frac{x^2}{2}\frac{1}{x}\,dx = \frac{x^2}{2}\ln x - \frac{x^2}{4} + C$$

$n = 2$: $u = \ln x$, $du = \dfrac{1}{x}\,dx$, $dv = x^2\,dx$, $v = \dfrac{x^3}{3}$

$$\int x^2 \ln x\,dx = \frac{x^3}{3}\ln x - \int \frac{x^3}{3}\frac{1}{x}\,dx = \frac{x^3}{3}\ln x - \frac{x^3}{9} + C$$

$n = 3$: $u = \ln x$, $du = \dfrac{1}{x}\,dx$, $dv = x^3\,dx$, $v = \dfrac{x^4}{4}$

$$\int x^3 \ln x\,dx = \frac{x^4}{4}\ln x - \int \frac{x^4}{4}\frac{1}{x}\,dx = \frac{x^4}{4}\ln x - \frac{x^4}{16} + C$$

**(b)** $\displaystyle \int x^n \ln x\,dx = \frac{x^{n+1}}{n+1}\ln x - \frac{x^{n+1}}{(n+1)^2} + C$

**75.** False. You might need to convert your integral using substitution or algebra.

**77.**

$$V = 2\pi \int_0^4 x\left(x\sqrt{16 - x^2}\right) dx$$

$$= 2\pi \int_0^4 x^2\sqrt{16 - x^2}\,dx$$

By Formula 38: $(a = 4)$

$$V = 2\pi\left[\frac{1}{8}\left(x(2x^2 - 16)\sqrt{16 - x^2} + 256 \arcsin\left(\frac{x}{4}\right)\right)\right]_0^4$$

$$= 2\pi\left[32\left(\frac{\pi}{2}\right)\right] = 32\pi^2$$

**79.** $W = \displaystyle\int_0^5 2000xe^{-x}\,dx$

$$= -2000 \int_0^5 -xe^{-x}\,dx$$

$$= 2000 \int_0^5 (-x)e^{-x}(-1)\,dx$$

$$= 2000\left[(-x)e^{-x} - e^{-x}\right]_0^5$$

$$= 2000\left(-\frac{6}{e^5} + 1\right)$$

$$\approx 1919.145 \text{ ft-lb}$$

**81. (a)** $V = 20(2)\displaystyle\int_0^3 \frac{2}{\sqrt{1 + y^2}}\,dy$

$$= \left[80 \ln\left|y + \sqrt{1 + y^2}\right|\right]_0^3$$

$$= 80 \ln\left(3 + \sqrt{10}\right)$$

$$\approx 145.5 \text{ ft}^3$$

$W = 148\left(80 \ln\left(3 + \sqrt{10}\right)\right)$

$$= 11{,}840 \ln\left(3 + \sqrt{10}\right)$$

$$\approx 21{,}530.4 \text{ lb}$$

**(b)** By symmetry, $\bar{x} = 0$.

$$M = \rho(2)\int_0^3 \frac{2}{\sqrt{1 + y^2}}\,dy$$

$$= \left[4\rho \ln\left|y + \sqrt{1 + y^2}\right|\right]_0^3$$

$$= 4\rho \ln\left(3 + \sqrt{10}\right)$$

$$M_x = 2\rho \int_0^3 \frac{2y}{\sqrt{1 + y^2}}\,dy$$

$$= \left[4\rho\sqrt{1 + y^2}\right]_0^3$$

$$= 4\rho\left(\sqrt{10} - 1\right)$$

$$\bar{y} = \frac{M_x}{M} = \frac{4\rho\left(\sqrt{10} - 1\right)}{4\rho \ln\left(3 + \sqrt{10}\right)} \approx 1.19$$

Centroid: $(\bar{x}, \bar{y}) \approx (0, 1.19)$

**83.** (a) $\displaystyle\int_0^4 \frac{k}{2+3x}\,dx = 10$

$$k = \frac{10}{\displaystyle\int_0^4 \frac{1}{2+3x}\,dx} = \frac{10}{\frac{1}{3}\ln 7} \approx \frac{10}{0.6486} = 15.417\left(=\frac{30}{\ln 7}\right)$$

(b) $\displaystyle\int_0^4 \frac{15.417}{2+3x}\,dx$

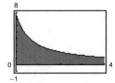

**85.** Let $I = \displaystyle\int_0^{\pi/2} \frac{dx}{1+\left(\tan x\right)^{\sqrt{2}}}$.

For $x = \dfrac{\pi}{2} - u,\ dx = -du$, and

$$I = \int_{\pi/2}^0 \frac{-du}{1+\left(\tan\left(\pi/2 - u\right)\right)^{\sqrt{2}}} = \int_0^{\pi/2} \frac{du}{1+\left(\cot u\right)^{\sqrt{2}}} = \int_0^{\pi/2} \frac{\left(\tan u\right)^{\sqrt{2}}}{\left(\tan u\right)^{\sqrt{2}}+1}\,du.$$

$$2I = \int_0^{\pi/2} \frac{dx}{1+\left(\tan x\right)^{\sqrt{2}}} + \int_0^{\pi/2} \frac{\left(\tan x\right)^{\sqrt{2}}}{\left(\tan x\right)^{\sqrt{2}}+1}\,dx = \int_0^{\pi/2} dx = \frac{\pi}{2}$$

So, $I = \dfrac{\pi}{4}$.

## Section 8.7.   Integration Forms and L'Hôpital's Rule

**1.** $\displaystyle\lim_{x\to 0} \frac{\sin 4x}{\sin 3x} \approx 1.3333 \ \left(\text{exact: } \frac{4}{3}\right)$

| $x$ | $-0.1$ | $-0.01$ | $-0.001$ | $0.001$ | $0.01$ | $0.1$ |
|---|---|---|---|---|---|---|
| $f(x)$ | 1.3177 | 1.3332 | 1.3333 | 1.3333 | 1.3332 | 1.3177 |

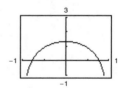

**3.** $\displaystyle\lim_{x\to\infty} x^5 e^{-x/100} \approx 0$

| $x$ | 1 | 10 | $10^2$ | $10^3$ | $10^4$ | $10^5$ |
|---|---|---|---|---|---|---|
| $f(x)$ | 0.9900 | 90,484 | $3.7\times 10^9$ | $4.5\times 10^{10}$ | 0 | 0 |

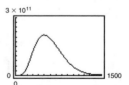

**5.** (a) $\displaystyle\lim_{x\to 4} \frac{3(x-4)}{x^2-16} = \lim_{x\to 4} \frac{3(x-4)}{(x-4)(x+4)} = \lim_{x\to 4} \frac{3}{x+4} = \frac{3}{8}$

(b) $\displaystyle\lim_{x\to 4} \frac{3(x-4)}{x^2-16} = \lim_{x\to 4} \frac{d/dx\left[3(x-4)\right]}{d/dx\left[x^2-16\right]} = \lim_{x\to 4} \frac{3}{2x} = \frac{3}{8}$

**7.** (a) $\displaystyle\lim_{x\to 6} \frac{\sqrt{x+10}-4}{x-6} = \lim_{x\to 6} \frac{\sqrt{x+10}-4}{x-6} \cdot \frac{\sqrt{x+10}+4}{\sqrt{x+10}+4} = \lim_{x\to 6} \frac{(x+10)-16}{(x-6)\left(\sqrt{x+10}+4\right)} = \lim_{x\to 6} \frac{1}{\sqrt{x+10}+4} = \frac{1}{8}$

(b) $\displaystyle\lim_{x\to 6} \frac{\sqrt{x+10}-4}{x-6} = \lim_{x\to 6} \frac{d/dx\left[\sqrt{x+10}-4\right]}{d/dx\left[x-6\right]} = \lim_{x\to 6} \frac{\frac{1}{2}(x+10)^{-1/2}}{1} = 1/8$

**9. (a)** $\lim\limits_{x\to\infty} \dfrac{5x^2 - 3x + 1}{3x^2 - 5} = \lim\limits_{x\to\infty} \dfrac{5 - (3/x) + (1/x^2)}{3 - (5/x^2)} = \dfrac{5}{3}$

**(b)** $\lim\limits_{x\to\infty} \dfrac{5x^2 - 3x + 1}{3x^2 - 5} = \lim\limits_{x\to\infty} \dfrac{(d/dx)\left[5x^2 - 3x + 1\right]}{(d/dx)\left[3x^2 - 5\right]} = \lim\limits_{x\to\infty} \dfrac{10x - 3}{6x} = \lim\limits_{x\to\infty} \dfrac{(d/dx)[10x - 3]}{(d/dx)[6x]} = \lim\limits_{x\to\infty} \dfrac{10}{6} = \dfrac{5}{3}$

**11.** $\lim\limits_{x\to 3} \dfrac{x^2 - 2x - 3}{x - 3} = \lim\limits_{x\to 3} \dfrac{2x - 2}{1} = 4$

**13.** $\lim\limits_{x\to 0} \dfrac{\sqrt{25 - x^2} - 5}{x} = \lim\limits_{x\to 0} \dfrac{\frac{1}{2}(25 - x^2)^{-1/2}(-2x)}{1}$

$= \lim\limits_{x\to 0} \dfrac{-x}{\sqrt{25 - x^2}} = 0$

**15.** $\lim\limits_{x\to 0} \dfrac{e^x - (1 - x)}{x} = \lim\limits_{x\to 0} \dfrac{e^x + 1}{1} = 2$

**17.** $\lim\limits_{x\to 0^+} \dfrac{e^x - (1 + x)}{x^3} = \lim\limits_{x\to 0^+} \dfrac{e^x - 1}{3x^2} = \lim\limits_{x\to 0^+} \dfrac{e^x}{6x} = \infty$

**19.** $\lim\limits_{x\to 1} \dfrac{x^{11} - 1}{x^4 - 1} = \lim\limits_{x\to 1} \dfrac{11x^{10}}{4x^3} = \dfrac{11}{4}$

**21.** $\lim\limits_{x\to 0} \dfrac{\sin 3x}{\sin 5x} = \lim\limits_{x\to 0} \dfrac{3\cos 3x}{5\cos 5x} = \dfrac{3}{5}$

**23.** $\lim\limits_{x\to 0} \dfrac{\arcsin x}{x} = \lim\limits_{x\to 0} \dfrac{1/\sqrt{1 - x^2}}{1} = 1$

**25.** $\lim\limits_{x\to\infty} \dfrac{5x^2 + 3x - 1}{4x^2 + 5} = \lim\limits_{x\to\infty} \dfrac{10x + 3}{8x} = \lim\limits_{x\to\infty} \dfrac{10}{8} = \dfrac{5}{4}$

**27.** $\lim\limits_{x\to\infty} \dfrac{x^2 + 4x + 7}{x - 6} = \lim\limits_{x\to\infty} \dfrac{2x + 4}{1} = \infty$

**29.** $\lim\limits_{x\to\infty} \dfrac{x^3}{e^{x/2}} = \lim\limits_{x\to\infty} \dfrac{3x^2}{(1/2)e^{x/2}}$

$= \lim\limits_{x\to\infty} \dfrac{6x}{(1/4)e^{x/2}} = \lim\limits_{x\to\infty} \dfrac{6}{(1/8)e^{x/2}} = 0$

**31.** $\lim\limits_{x\to\infty} \dfrac{x}{\sqrt{x^2 + 1}} = \lim\limits_{x\to\infty} \dfrac{1}{\sqrt{1 + (1/x^2)}} = 1$

**Note:** L'Hôpital's Rule does not work on this limit. See Exercise 89.

**33.** $\lim\limits_{x\to\infty} \dfrac{\cos x}{x} = 0$ by Squeeze Theorem

$\left(\dfrac{\cos x}{x} \le \dfrac{1}{x}, \text{ for } x > 0\right)$

**35.** $\lim\limits_{x\to\infty} \dfrac{\ln x}{x^2} = \lim\limits_{x\to\infty} \dfrac{1/x}{2x} = \lim\limits_{x\to\infty} \dfrac{1}{2x^2} = 0$

**37.** $\lim\limits_{x\to\infty} \dfrac{e^x}{x^4} = \lim\limits_{x\to\infty} \dfrac{e^x}{4x^3}$

$= \lim\limits_{x\to\infty} \dfrac{e^x}{12x^2}$

$= \lim\limits_{x\to\infty} \dfrac{e^x}{24x}$

$= \lim\limits_{x\to\infty} \dfrac{e^x}{24} = \infty$

**39.** $\lim\limits_{x\to 0} \dfrac{\sin 5x}{\tan 9x} = \lim\limits_{x\to 0} \dfrac{5\cos 5x}{9\sec^2 9x} = \dfrac{5}{9}$

**41.** $\lim\limits_{x\to 0} \dfrac{\arctan x}{\sin x} = \lim\limits_{x\to 0} \dfrac{1/(1 + x^2)}{\cos x} = 1$

**43.** $\lim\limits_{x\to\infty} \dfrac{\int_1^x \ln\left(e^{4t-1}\right) dt}{x}$

$= \lim\limits_{x\to\infty} \dfrac{\int_1^x (4t - 1)\, dt}{x}$

$= \lim\limits_{x\to\infty} \dfrac{4x - 1}{1} = \infty$

**45. (a)** $\lim\limits_{x\to\infty} x \ln x$, not indeterminate

**(b)** $\lim\limits_{x\to\infty} x \ln x = (\infty)(\infty) = \infty$

**(c)**

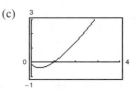

**47. (a)** $\lim\limits_{x\to\infty}\left(x \sin \dfrac{1}{x}\right) = (\infty)(0)$

**(b)** $\lim\limits_{x\to\infty} x \sin \dfrac{1}{x} = \lim\limits_{x\to\infty} \dfrac{\sin(1/x)}{1/x}$

$= \lim\limits_{x\to\infty} \dfrac{(-1/x^2)\cos(1/x)}{-1/x^2}$

$= \lim\limits_{x\to\infty} \cos\left(\dfrac{1}{x}\right) = 1$

**(c)**

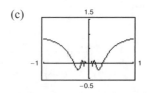

**49.** (a)  $\lim\limits_{x \to 0^+} x^{1/x} = 0^\infty = 0$, not indeterminate

(See Exercise 116).

(b) Let   $y = x^{1/x}$

$$\ln y = \ln x^{1/x} = \frac{1}{x} \ln x.$$

Because $x \to 0^+$, $\dfrac{1}{x} \ln x \to (\infty)(-\infty) = -\infty$. So,

$\ln y \to -\infty \Rightarrow y \to 0^+$.

Therefore, $\lim\limits_{x \to 0^+} x^{1/x} = 0$.

(c)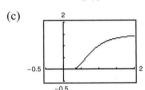

**51.** (a)  $\lim\limits_{x \to \infty} x^{1/x} = \infty^0$

(b) Let $y = \lim\limits_{x \to \infty} x^{1/x}$.

$$\ln y = \lim\limits_{x \to \infty} \frac{\ln x}{x} = \lim\limits_{x \to \infty} \left( \frac{1/x}{1} \right) = 0$$

So, $\ln y = 0 \Rightarrow y = e^0 = 1$. Therefore,

$\lim\limits_{x \to \infty} x^{1/x} = 1$.

(c)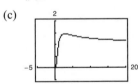

**53.** (a)  $\lim\limits_{x \to 0^+} (1 + x)^{1/x} = 1^\infty$

(b) Let $y = \lim\limits_{x \to 0^+} (1 + x)^{1/x}$.

$$\ln y = \lim\limits_{x \to 0^+} \frac{\ln(1 + x)}{x}$$

$$= \lim\limits_{x \to 0^+} \left( \frac{1/(1 + x)}{1} \right) = 1$$

So, $\ln y = 1 \Rightarrow y = e^1 = e$.

Therefore, $\lim\limits_{x \to 0^+} (1 + x)^{1/x} = e$.

(c)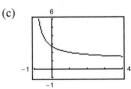

**55.** (a)  $\lim\limits_{x \to 0^+} \left[ 3(x)^{x/2} \right] = 0^0$

(b) Let $y = \lim\limits_{x \to 0^+} 3(x)^{x/2}$.

$$\ln y = \lim\limits_{x \to 0^+} \left[ \ln 3 + \frac{x}{2} \ln x \right]$$

$$= \lim\limits_{x \to 0^+} \left[ \ln 3 + \frac{\ln x}{2/x} \right]$$

$$= \lim\limits_{x \to 0^+} \ln 3 + \lim\limits_{x \to 0^+} \frac{1/x}{-2/x^2}$$

$$= \lim\limits_{x \to 0^+} \ln 3 - \lim\limits_{x \to 0^+} \frac{x}{2}$$

$$= \ln 3$$

So, $\lim\limits_{x \to 0^+} 3(x)^{x/2} = 3$.

(c)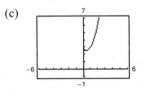

**57.** (a)  $\lim\limits_{x \to 1^+} (\ln x)^{x-1} = 0^0$

(b) Let $y = (\ln x)^{x-1}$.

$$\ln y = \ln \left[ (\ln x)^{x-1} \right] = (x - 1) \ln(\ln x)$$

$$= \frac{\ln(\ln x)}{(x - 1)^{-1}}$$

$$\lim\limits_{x \to 1^+} \ln y = \lim\limits_{x \to 1^+} \frac{\ln(\ln x)}{(x - 1)^{-1}}$$

$$= \lim\limits_{x \to 1^+} \frac{1/(x \ln x)}{-(x - 1)^{-2}}$$

$$= \lim\limits_{x \to 1^+} \frac{-(x - 1)^2}{x \ln x}$$

$$= \lim\limits_{x \to 1^+} \frac{-2(x - 1)}{1 + \ln x} = 0$$

Because $\lim\limits_{x \to 1^+} \ln y = 0$, $\lim\limits_{x \to 1^+} y = 1$.

(c)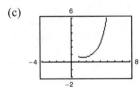

**59.** (a) $\displaystyle\lim_{x\to2^+}\left(\frac{8}{x^2-4}-\frac{x}{x-2}\right)=\infty-\infty$

(b) $\displaystyle\lim_{x\to2^+}\left(\frac{8}{x^2-4}-\frac{x}{x-2}\right)=\lim_{x\to2^+}\frac{8-x(x+2)}{x^2-4}=\lim_{x\to2^+}\frac{(2-x)(4+x)}{(x+2)(x-2)}=\lim_{x\to2^+}\frac{-(x+4)}{x+2}=\frac{-3}{2}$

(c)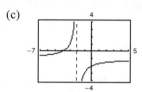

**61.** (a) $\displaystyle\lim_{x\to1^+}\left(\frac{3}{\ln x}-\frac{2}{x-1}\right)=\infty-\infty$

(b) $\displaystyle\lim_{x\to1^+}\left(\frac{3}{\ln x}-\frac{2}{x-1}\right)=\lim_{x\to1^+}\frac{3x-3-2\ln x}{(x-1)\ln x}=\lim_{x\to1^+}\frac{3-(2/x)}{\left[(x-1)/x\right]+\ln x}=\infty$

(c)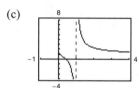

**63.** (a)

(b) $\displaystyle\lim_{x\to3}\frac{x-3}{\ln(2x-5)}=\lim_{x\to3}\frac{1}{2/(2x-5)}=\lim_{x\to3}\frac{2x-5}{2}=\frac{1}{2}$

**65.** (a)

(b) $\displaystyle\lim_{x\to\infty}\left(\sqrt{x^2+5x+2}-x\right)=\lim_{x\to\infty}\left(\sqrt{x^2+5x+2}-x\right)\frac{\left(\sqrt{x^2+5x+2}+x\right)}{\left(\sqrt{x^2+5x+2}+x\right)}$

$=\displaystyle\lim_{x\to\infty}\frac{\left(x^2+5x+2\right)-x^2}{\sqrt{x^2+5x+2}+x}$

$=\displaystyle\lim_{x\to\infty}\frac{5x+2}{\sqrt{x^2+5x+2}+x}=\lim_{x\to\infty}\frac{5+(2/x)}{\sqrt{1+(5/x)+(2/x^2)}+1}=\frac{5}{2}$

**67.** $\dfrac{0}{0},\dfrac{\infty}{\infty},0\cdot\infty,1^\infty,0^0,\infty-\infty,\infty^0$

**69.** (a) Let $f(x)=x^2-25$ and $g(x)=x-5$.

(b) Let $f(x)=(x-5)^2$ and $g(x)=x^2-25$.

(c) Let $f(x)=x^2-25$ and $g(x)=(x-5)^3$.

(Answers will vary.)

**71.**

| $x$ | 10 | $10^2$ | $10^4$ | $10^6$ | $10^8$ | $10^{10}$ |
|---|---|---|---|---|---|---|
| $\dfrac{(\ln x)^4}{x}$ | 2.811 | 4.498 | 0.720 | 0.036 | 0.001 | 0.000 |

**73.** $\lim\limits_{x \to \infty} \dfrac{x^2}{e^{5x}} = \lim\limits_{x \to \infty} \dfrac{2x}{5e^{5x}} = \lim\limits_{x \to \infty} \dfrac{2}{25e^{5x}} = 0$

**75.** $\lim\limits_{x \to \infty} \dfrac{(\ln x)^3}{x} = \lim\limits_{x \to \infty} \dfrac{3(\ln x)^2(1/x)}{1}$

$\qquad = \lim\limits_{x \to \infty} \dfrac{3(\ln x)^2}{x}$

$\qquad = \lim\limits_{x \to \infty} \dfrac{6(\ln x)(1/x)}{1}$

$\qquad = \lim\limits_{x \to \infty} \dfrac{6(\ln x)}{x} = \lim\limits_{x \to \infty} \dfrac{6}{x} = 0$

**77.** $\lim\limits_{x \to \infty} \dfrac{(\ln x)^n}{x^m} = \lim\limits_{x \to \infty} \dfrac{n(\ln x)^{n-1}/x}{mx^{m-1}}$

$\qquad = \lim\limits_{x \to \infty} \dfrac{n(\ln x)^{n-1}}{mx^m}$

$\qquad = \lim\limits_{x \to \infty} \dfrac{n(n-1)(\ln x)^{n-2}}{m^2 x^m}$

$\qquad = \cdots = \lim\limits_{x \to \infty} \dfrac{n!}{m^n x^m} = 0$

**79.** $y = x^{1/x}, \ x > 0$

Horizontal asymptote: $y = 1$ (See Exercise 51.)

$\ln y = \dfrac{1}{x} \ln x$

$\dfrac{1}{y}\dfrac{dy}{dx} = \dfrac{1}{x}\left(\dfrac{1}{x}\right) + (\ln x)\left(-\dfrac{1}{x^2}\right)$

$\dfrac{dy}{dx} = x^{1/x}\left(\dfrac{1}{x^2}\right)(1 - \ln x) = x^{(1/x)-2}(1 - \ln x) = 0$

Critical number: $\qquad x = e$

Intervals: $\qquad (0, e) \qquad (e, \infty)$

Sign of $dy/dx$: $\qquad + \qquad\qquad -$

$y = f(x)$: $\quad$ Increasing $\qquad$ Decreasing

Relative maximum: $\quad \left(e, e^{1/e}\right)$

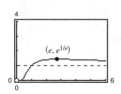

**81.** $y = 2xe^{-x}$

$\lim\limits_{x \to \infty} \dfrac{2x}{e^x} = \lim\limits_{x \to \infty} \dfrac{2}{e^x} = 0$

Horizontal asymptote: $y = 0$

$\dfrac{dy}{dx} = 2x\left(-e^{-x}\right) + 2e^{-x}$

$\qquad = 2e^{-x}(1 - x) = 0$

Critical number: $x = 1$

Intervals: $(-\infty, 1) \ \ (1, \infty)$

Sign of $dy/dx$: $+ -$

$y = f(x)$: Increasing $\ $ Decreasing

Relative maximum: $\left(1, \dfrac{2}{e}\right)$

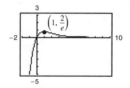

**83.** $\lim\limits_{x \to 2}\left(3x^2 + 4x + 1\right) = \infty$

Limit is not of the form $\dfrac{0}{0}$ or $\dfrac{\infty}{\infty}$.

L'Hôpital's Rule does not apply.

**85.** $\lim\limits_{x \to \infty} \dfrac{\sin \pi x - 1}{x} = 0 \qquad$ (Numerator is bounded)

Limit is not of the form $0/0$ or $\infty/\infty$.

L'Hôpital's Rule does not apply.

**87.** $\lim\limits_{x \to \infty} \dfrac{e^{-x}}{1 + e^{-x}} = \dfrac{0}{1 + 0} = 0$

Limit is not of the form $0/0$ or $\infty/\infty$.

L'Hôpital's Rule does not apply.

**89.** (a) Applying L'Hôpital's Rule twice results in the original limit, so L'Hôpital's Rule fails:

$$\lim_{x \to \infty} \frac{x}{\sqrt{x^2 + 1}} = \lim_{x \to \infty} \frac{1}{x/\sqrt{x^2 + 1}}$$

$$= \lim_{x \to \infty} \frac{\sqrt{x^2 + 1}}{x}$$

$$= \lim_{x \to \infty} \frac{x/\sqrt{x^2 + 1}}{1}$$

$$= \lim_{x \to \infty} \frac{x}{\sqrt{x^2 + 1}}$$

(b) $$\lim_{x \to \infty} \frac{x}{\sqrt{x^2 + 1}} = \lim_{x \to \infty} \frac{x/x}{\sqrt{x^2 + 1}/x}$$

$$= \lim_{x \to \infty} \frac{1}{\sqrt{1 + 1/x^2}} = \frac{1}{\sqrt{1 + 0}} = 1$$

(c)

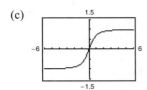

**91.** $f(x) = \sin(3x)$, $g(x) = \sin(4x)$

$f'(x) = 3\cos(3x)$, $g'(x) = 4\cos(4x)$

$$y_1 = \frac{f(x)}{g(x)} = \frac{\sin 3x}{\sin 4x},$$

$$y_2 = \frac{f'(x)}{g'(x)} = \frac{3\cos 3x}{4\cos 4x}$$

As $x \to 0$, $y_1 \to 0.75$ and $y_2 \to 0.75$

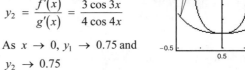

By L'Hôpital's Rule,

$$\lim_{x \to 0} \frac{\sin 3x}{\sin 4x} = \lim_{x \to 0} \frac{3\cos 3x}{4\cos 4x} = \frac{3}{4}$$

**93.** $$\lim_{k \to 0} \frac{32\left(1 - e^{-kt} + \dfrac{v_0 k e^{-kt}}{32}\right)}{k} = \lim_{k \to 0} \frac{32(1 - e^{-kt})}{k} + \lim_{k \to 0}\left(v_0 e^{-kt}\right) = \lim_{k \to 0} \frac{32(0 + t e^{-kt})}{1} + \lim_{k \to 0}\left(\frac{v_0}{e^{kt}}\right) = 32t + v_0$$

**95.** Let $N$ be a fixed value for $n$. Then

$$\lim_{x \to \infty} \frac{x^{N-1}}{e^x} = \lim_{x \to \infty} \frac{(N-1)x^{N-2}}{e^x} = \lim_{x \to \infty} \frac{(N-1)(N-2)x^{N-3}}{e^x} = \cdots = \lim_{x \to \infty}\left[\frac{(N-1)!}{e^x}\right] = 0. \quad \text{(See Exercise 78.)}$$

**97.** $f(x) = x^3$, $g(x) = x^2 + 1$, $[0, 1]$

$$\frac{f(b) - f(a)}{g(b) - g(a)} = \frac{f'(c)}{g'(c)}$$

$$\frac{f(1) - f(0)}{g(1) - g(0)} = \frac{3c^2}{2c}$$

$$\frac{1}{1} = \frac{3c}{2}$$

$$c = \frac{2}{3}$$

**99.** $f(x) = \sin x$, $g(x) = \cos x$, $\left[0, \dfrac{\pi}{2}\right]$

$$\frac{f(\pi/2) - f(0)}{g(\pi/2) - g(0)} = \frac{f'(c)}{g'(c)}$$

$$\frac{1}{-1} = \frac{\cos c}{-\sin c}$$

$$-1 = -\cot c$$

$$c = \frac{\pi}{4}$$

**101.** False. L'Hôpital's Rule does not apply because

$$\lim_{x \to 0}\left(x^2 + x + 1\right) \neq 0.$$

$$\lim_{x \to 0^+} \frac{x^2 + x + 1}{x} = \lim_{x \to 0^+}\left(x + 1 + \frac{1}{x}\right) = 1 + \infty = \infty$$

**103.** True

**105.** Area of triangle: $\dfrac{1}{2}(2x)(1 - \cos x) = x - x \cos x$

Shaded area: Area of rectangle − Area under curve

$$2x(1 - \cos x) - 2\int_0^x (1 - \cos t)dt = 2x(1 - \cos x) - 2\big[t - \sin t\big]_0^x$$
$$= 2x(1 - \cos x) - 2(x - \sin x)$$
$$= 2 \sin x - 2x \cos x$$

Ratio: $\displaystyle\lim_{x \to 0} \dfrac{x - x \cos x}{2 \sin x - 2x \cos x} = \lim_{x \to 0} \dfrac{1 + x \sin x - \cos x}{2 \cos x + 2x \sin x - 2 \cos x}$

$$= \lim_{x \to 0} \dfrac{1 + x \sin x - \cos x}{2x \sin x}$$
$$= \lim_{x \to 0} \dfrac{x \cos x + \sin x + \sin x}{2x \cos x + 2 \sin x}$$
$$= \lim_{x \to 0} \dfrac{x \cos x + 2 \sin x}{2x \cos x + 2 \sin x} \cdot \dfrac{1/\cos x}{1/\cos x} = \lim_{x \to 0} \dfrac{x + 2 \tan x}{2x + 2 \tan x} = \lim_{x \to 0} \dfrac{1 + 2 \sec^2 x}{2 + 2 \sec^2 x} = \dfrac{3}{4}$$

**107.** $\displaystyle\lim_{x \to 0} \dfrac{4x - 2 \sin 2x}{2x^3} = \lim_{x \to 0} \dfrac{4 - 4 \cos 2x}{6x^2} = \lim_{x \to 0} \dfrac{8 \sin 2x}{12x} = \lim_{x \to 0} \dfrac{16 \cos 2x}{12} = \dfrac{16}{12} = \dfrac{4}{3}$

Let $c = \dfrac{4}{3}$.

**109.** $\displaystyle\lim_{x \to 0} \dfrac{a - \cos bx}{x^2} = 2$

Near $x = 0$, $\cos bx \approx 1$ and $x^2 \approx 0 \Rightarrow a = 1$.

Using L'Hôpital's Rule,

$$\lim_{x \to 0} \dfrac{1 - \cos bx}{x^2} = \lim_{x \to 0} \dfrac{b \sin bx}{2x} = \lim_{x \to 0} \dfrac{b^2 \cos bx}{2} = 2.$$

So, $b^2 = 4$ and $b = \pm 2$.

Answer: $a = 1, b = \pm 2$

**111.** (a) $\displaystyle\lim_{h \to 0} \dfrac{f(x + h) - f(x - h)}{2h} = \lim_{h \to 0} \dfrac{f'(x + h)(1) - f'(x - h)(-1)}{2} = \lim_{h \to 0}\left[\dfrac{f'(x + h) + f'(x - h)}{2}\right] = \dfrac{f'(x) + f'(x)}{2} = f'(x)$

(b)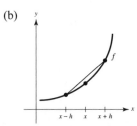

Graphically, the slope of the line joining $(x - h, f(x - h))$ and $(x + h, f(x + h))$ is approximately $f'(x)$. And, as

$h \to 0$, $\displaystyle\lim_{h \to 0} \dfrac{f(x + h) - f(x - h)}{2h} = f'(x)$.

**113.** $g(x) = \begin{cases} e^{-1/x^2}, & x \neq 0 \\ 0, & x = 0 \end{cases}$

$$g'(0) = \lim_{x \to 0} \frac{g(x) - g(0)}{x - 0} = \lim_{x \to 0} \frac{e^{-1/x^2}}{x}$$

Let $y = \dfrac{e^{-1/x^2}}{x}$, then $\ln y = \ln\left(\dfrac{e^{-1/x^2}}{x}\right) = -\dfrac{1}{x^2} - \ln x = \dfrac{-1 - x^2 \ln x}{x^2}$. Because

$$\lim_{x \to 0} x^2 \ln x = \lim_{x \to 0} \frac{\ln x}{1/x^2} = \lim_{x \to 0} \frac{1/x}{-2/x^3} = \lim_{x \to 0} \left(-\frac{x^2}{2}\right) = 0$$

you have $\displaystyle\lim_{x \to 0} \left(\dfrac{-1 - x^2 \ln x}{x^2}\right) = -\infty$. So, $\displaystyle\lim_{x \to 0} y = e^{-\infty} = 0 \Rightarrow g'(0) = 0$.

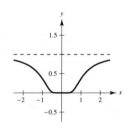

**Note:** The graph appears to support this conclusion—the tangent line is horizontal at $(0, 0)$.

**115.** (a)   $\displaystyle\lim_{x \to 0^+} (-x \ln x)$ is the form $0 \cdot \infty$.

(b)   $\displaystyle\lim_{x \to 0^+} \frac{-\ln x}{1/x} = \lim_{x \to 0^+} \frac{-1/x}{-1/x^2} = \lim_{x \to 0^+} (x) = 0$

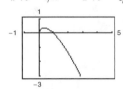

**117.** $\displaystyle\lim_{x \to a} f(x)^{g(x)}$

$$y = f(x)^{g(x)}$$

$$\ln y = g(x) \ln f(x)$$

$$\lim_{x \to a} g(x) \ln f(x) = (-\infty)(-\infty) = \infty$$

As $x \to a$, $\ln y \Rightarrow \infty$, and therefore $y = \infty$. So, $\displaystyle\lim_{x \to a} f(x)^{g(x)} = \infty$.

**119.** (a)   $\displaystyle\lim_{x \to 0^+} x^{(\ln 2)/(1 + \ln x)}$ is of form $0^0$.

Let $y = x^{(\ln 2)/(1 + \ln x)}$

$$\ln y = \frac{\ln 2}{1 + \ln x} \ln x$$

$$\lim_{x \to 0^+} \ln y = \frac{\ln 2(1/x)}{1/x} = \ln 2.$$

So, $\displaystyle\lim_{x \to 0^+} x^{(\ln 2)/(1 + \ln x)} = 2.$

(b)   $\displaystyle\lim_{x \to \infty} x^{(\ln 2)/(1 + \ln x)}$ is of form $\infty^0$.

Let $y = x^{(\ln 2)/(1 + \ln x)}$

$$\ln y = \frac{\ln 2}{1 + \ln x} \ln x$$

$$\lim_{x \to \infty} \ln y = \frac{\ln 2(1/x)}{1/x} = \ln 2.$$

So, $\displaystyle\lim_{x \to \infty} x^{(\ln 2)/(1 + \ln x)} = 2.$

(c) $\lim_{x \to 0}(x + 1)^{(\ln 2)/(x)}$ is of form $1^{\infty}$.

Let $y = (x + 1)^{(\ln 2)/(x)}$

$\ln y = \dfrac{\ln 2}{x} \ln(x + 1)$

$\lim_{x \to 0} \ln y = \lim_{x \to 0} \dfrac{(\ln 2)1/(x + 1)}{1} = \ln 2.$

So, $\lim_{x \to 0}(x + 1)^{(\ln 2)/(x)} = 2.$

**121.** (a) $h(x) = \dfrac{x + \sin x}{x}$

$\lim_{x \to \infty} h(x) = 1$

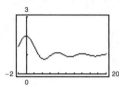

(b) $h(x) = \dfrac{x + \sin x}{x} = \dfrac{x}{x} + \dfrac{\sin x}{x} = 1 + \dfrac{\sin x}{x}, x > 0$

So, $\lim_{x \to \infty} h(x) = \lim_{x \to \infty}\left[1 + \dfrac{\sin x}{x}\right] = 1 + 0 = 1.$

(c) No. $h(x)$ is not an indeterminate form.

**123.** Let $f(x) = \left[\dfrac{1}{x} \cdot \dfrac{a^x - 1}{a - 1}\right]^{1/x}.$

For $a > 1$ and $x > 0$,

$\ln f(x) = \dfrac{1}{x}\left[\ln \dfrac{1}{x} + \ln(a^x - 1) - \ln(a - 1)\right] = -\dfrac{\ln x}{x} + \dfrac{\ln(a^x - 1)}{x} - \dfrac{\ln(a - 1)}{x}.$

As $x \to \infty$, $\dfrac{\ln x}{x} \to 0$, $\dfrac{\ln(a - 1)}{x} \to 0$, and $\dfrac{\ln(a^x - 1)}{x} = \dfrac{\ln\left[(1 - a^{-x})a^x\right]}{x} = \dfrac{\ln(1 - a^{-x})}{x} + \ln a \to \ln a.$

So, $\ln f(x) \to \ln a.$

For $0 < a < 1$ and $x > 0$,

$\ln f(x) = \dfrac{-\ln x}{x} + \dfrac{\ln(1 - a^x)}{x} - \dfrac{\ln(1 - a)}{x} \to 0$ as $x \to \infty.$

Combining these results, $\lim_{x \to \infty} f(x) = \begin{cases} a & \text{if} \quad a > 1 \\ 1 & \text{if} \quad 0 < a < 1\end{cases}.$

# Section 8.8   Improper Integrals

**1.** $\displaystyle\int_0^1 \dfrac{dx}{5x - 3}$ is improper because $5x - 3 = 0$ when

$x = \dfrac{3}{5}$, and $0 \le \dfrac{3}{5} \le 1.$

**3.** $\displaystyle\int_0^1 \dfrac{2x - 5}{x^2 - 5x + 6}\, dx = \int_0^1 \dfrac{2x - 5}{(x - 2)(x - 3)}\, dx$ is not

improper because

$\dfrac{2x - 5}{(x - 2)(x - 3)}$ is continuous on $[0, 1].$

**5.** $\int_0^2 e^{-x}\, dx$ is not improper because $f(x) = e^{-x}$ is continuous on $[0, 2]$.

**7.** $\int_{-\infty}^{\infty} \dfrac{\sin x}{4 + x^2}\, dx$ is improper because the limits of integration are $-\infty$ and $\infty$.

**9.** Infinite discontinuity at $x = 0$.

$$\int_0^4 \frac{1}{\sqrt{x}}\, dx = \lim_{b \to 0^+} \int_b^4 \frac{1}{\sqrt{x}}\, dx$$
$$= \lim_{b \to 0^+} \left[ 2\sqrt{x} \right]_b^4$$
$$= \lim_{b \to 0^+} \left( 4 - 2\sqrt{b} \right) = 4$$

Converges

**11.** Infinite discontinuity at $x = 1$.

$$\int_0^2 \frac{1}{(x - 1)^2}\, dx = \int_0^1 \frac{1}{(x - 1)^2}\, dx + \int_1^2 \frac{1}{(x - 1)^2}\, dx$$
$$= \lim_{b \to 1^-} \int_0^b \frac{1}{(x - 1)^2}\, dx + \lim_{c \to 1^+} \int_c^2 \frac{1}{(x - 1)^2}\, dx$$
$$= \lim_{b \to 1^-} \left[ -\frac{1}{x - 1} \right]_0^b + \lim_{c \to 1^+} \left[ -\frac{1}{x - 1} \right]_c^2$$
$$= (\infty - 1) + (-1 + \infty)$$

Diverges

**13.** Infinite limit of integration.

$$\int_0^{\infty} e^{-4x}\, dx = \lim_{b \to \infty} \int_0^b e^{-4x}\, dx$$
$$= \lim_{b \to \infty} \left[ -\tfrac{1}{4} e^{-4x} \right]_0^b$$
$$= \lim_{b \to \infty} \left[ -\tfrac{1}{4} e^{-4b} + \tfrac{1}{4} \right] = \tfrac{1}{4}$$

Converges

**15.** $\int_{-1}^1 \dfrac{1}{x^2}\, dx \neq -2$

because the integrand is not defined at $x = 0$. The integral diverges.

**17.** $\int_0^{\infty} e^{-x}\, dx \neq 0$. You need to evaluate the limit.

$$\lim_{b \to \infty} \int_0^b e^{-x}\, dx = \lim_{b \to \infty} \left[ -e^{-x} \right]_0^b$$
$$= \lim_{b \to \infty} \left[ -e^{-b} + 1 \right] = 1$$

**19.** $\int_1^{\infty} \dfrac{1}{x^3}\, dx = \lim_{b \to \infty} \int_1^b x^{-3}\, dx$

$$= \lim_{b \to \infty} \left[ \frac{x^{-2}}{-2} \right]_1^b$$
$$= \lim_{b \to \infty} \left[ \frac{-1}{2b^2} + \frac{1}{2} \right] = \frac{1}{2}$$

**21.** $\int_1^{\infty} \dfrac{3}{\sqrt[3]{x}}\, dx = \lim_{b \to \infty} \int_1^b 3x^{-1/3}\, dx$

$$= \lim_{b \to \infty} \left[ \frac{9}{2} x^{2/3} \right]_1^b = \infty$$

Diverges

**23.** $\int_{-\infty}^0 xe^{-4x}\, dx = \lim_{b \to -\infty} \int_b^0 xe^{-4x}\, dx$

$$= \lim_{b \to -\infty} \left[ \left( \frac{-x}{4} - \frac{1}{16} \right) e^{-4x} \right]_b^0 \qquad \text{(Integration by parts)}$$
$$= \lim_{b \to -\infty} \left[ -\frac{1}{16} + \frac{b}{4} + \frac{1}{16} e^{-4b} \right] = -\infty$$

Diverges

**25.** $\int_0^{\infty} x^2 e^{-x}\, dx = \lim_{b \to \infty} \int_0^b x^2 e^{-x}\, dx$

$$= \lim_{b \to \infty} \left[ -e^{-x} \left( x^2 + 2x + 2 \right) \right]_0^b$$
$$= \lim_{b \to \infty} \left( -\frac{b^2 + 2b + 2}{e^b} + 2 \right) = 2$$

Because $\lim_{b \to \infty} \left( -\dfrac{b^2 + 2b + 2}{e^b} \right) = 0$ by L'Hôpital's Rule.

**27.** $\displaystyle\int_0^\infty e^{-x}\cos x\,dx = \lim_{b\to\infty}\tfrac{1}{2}\Big[e^{-x}(-\cos x+\sin x)\Big]_0^b$

$\qquad\qquad = \tfrac{1}{2}\big[0-(-1)\big] = \tfrac{1}{2}$

**29.** $\displaystyle\int_4^\infty \frac{1}{x(\ln x)^3}\,dx = \lim_{b\to\infty}\int_4^b (\ln x)^{-3}\frac{1}{x}\,dx$

$\qquad\qquad = \lim_{b\to\infty}\left[-\frac{1}{2}(\ln x)^{-2}\right]_4^b$

$\qquad\qquad = \lim_{b\to\infty}\left[-\frac{1}{2}(\ln b)^{-2} + \frac{1}{2}(\ln 4)^{-2}\right]$

$\qquad\qquad = \frac{1}{2}\frac{1}{(2\ln 2)^2} = \frac{1}{2(\ln 4)^2}$

**31.** $\displaystyle\int_{-\infty}^\infty \frac{4}{16+x^2}\,dx = \int_{-\infty}^0 \frac{4}{16+x^2}\,dx + \int_0^\infty \frac{4}{16+x^2}\,dx$

$\qquad\qquad = \lim_{b\to-\infty}\int_b^0 \frac{4}{16+x^2}\,dx + \lim_{c\to\infty}\int_0^c \frac{4}{16+x^2}\,dx$

$\qquad\qquad = \lim_{b\to-\infty}\left[\arctan\left(\frac{x}{4}\right)\right]_b^0 + \lim_{c\to\infty}\left[\arctan\left(\frac{x}{4}\right)\right]_0^c$

$\qquad\qquad = \lim_{b\to-\infty}\left[0 - \arctan\left(\frac{b}{4}\right)\right] + \lim_{c\to\infty}\left[\arctan\left(\frac{c}{4}\right) - 0\right]$

$\qquad\qquad = -\left(-\frac{\pi}{2}\right) + \frac{\pi}{2} = \pi$

**33.** $\displaystyle\int_0^\infty \frac{1}{e^x+e^{-x}}\,dx = \lim_{b\to\infty}\int_0^b \frac{e^x}{1+e^{2x}}\,dx$

$\qquad\qquad = \lim_{b\to\infty}\Big[\arctan(e^x)\Big]_0^b$

$\qquad\qquad = \frac{\pi}{2} - \frac{\pi}{4} = \frac{\pi}{4}$

**35.** $\displaystyle\int_0^\infty \cos \pi x\,dx = \lim_{b\to\infty}\left[\frac{1}{\pi}\sin \pi x\right]_0^b$

Diverges because $\sin \pi b$ does not approach a limit as $b\to\infty$.

**37.** $\displaystyle\int_0^1 \frac{1}{x^2}\,dx = \lim_{b\to0^+}\left[\frac{-1}{x}\right]_b^1 = \lim_{b\to0^+}\left(-1+\frac{1}{b}\right) = -1+\infty$

Diverges

**39.** $\displaystyle\int_0^8 \frac{1}{\sqrt[3]{8-x}}\,dx = \lim_{b\to8^-}\int_0^b \frac{1}{\sqrt[3]{8-x}}\,dx$

$\qquad\qquad = \lim_{b\to8^-}\left[\frac{-3}{2}(8-x)^{2/3}\right]_0^b = 6$

**41.** $\displaystyle\int_0^1 x\ln x\,dx = \lim_{b\to0^+}\left[\frac{x^2}{2}\ln|x| - \frac{x^2}{4}\right]_b^1$

$\qquad\qquad = \lim_{b\to0^+}\left(\frac{-1}{4} - \frac{b^2\ln b}{2} + \frac{b^2}{4}\right) = \frac{-1}{4}$

because $\displaystyle\lim_{b\to0^+}\left(b^2\ln b\right) = 0$ by L'Hôpital's Rule.

**43.** $\displaystyle\int_0^{\pi/2} \tan\theta\,d\theta = \lim_{b\to(\pi/2)^-}\Big[\ln|\sec\theta|\Big]_0^b = \infty$

Diverges

**45.** $\displaystyle\int_2^4 \frac{2}{x\sqrt{x^2-4}}\,dx = \lim_{b\to2^+}\int_b^4 \frac{2}{x\sqrt{x^2-4}}\,dx$

$\qquad\qquad = \lim_{b\to2^+}\left[\operatorname{arcsec}\left|\frac{x}{2}\right|\right]_b^4$

$\qquad\qquad = \lim_{b\to2^+}\left(\operatorname{arcsec}2 - \operatorname{arcsec}\left(\frac{b}{2}\right)\right)$

$\qquad\qquad = \frac{\pi}{3} - 0 = \frac{\pi}{3}$

**47.** $\displaystyle\int_2^4 \frac{1}{\sqrt{x^2-4}} = \lim_{b\to2^+}\left[\ln\left|x+\sqrt{x^2-4}\right|\right]_b^4$

$\qquad\qquad = \ln\left(4+2\sqrt{3}\right) - \ln 2$

$\qquad\qquad = \ln\left(2+\sqrt{3}\right) \approx 1.317$

**49.** $\int_0^2 \dfrac{1}{\sqrt[3]{x-1}} \, dx = \int_0^1 \dfrac{1}{\sqrt[3]{x-1}} \, dx + \int_1^2 \dfrac{1}{\sqrt[3]{x-1}} \, dx = \lim_{b\to 1^-} \left[\dfrac{3}{2}(x-1)^{2/3}\right]_0^b + \lim_{c\to 1^+} \left[\dfrac{3}{2}(x-1)^{2/3}\right]_c^2 = \dfrac{-3}{2} + \dfrac{3}{2} = 0$

**51.** $\int_3^\infty \dfrac{1}{x\sqrt{x^2-9}} \, dx = \lim_{b\to 3^+} \int_b^5 \dfrac{1}{x\sqrt{x^2-9}} \, dx + \lim_{c\to\infty} \int_5^\infty \dfrac{1}{x\sqrt{x^2-9}} \, dx$

$= \lim_{b\to 3^+} \left[\dfrac{1}{3}\operatorname{arcsec}\dfrac{x}{3}\right]_b^5 + \lim_{c\to\infty} \left[\dfrac{1}{3}\operatorname{arcsec}\left(\dfrac{x}{3}\right)\right]_5^\infty$

$= \lim_{b\to 3^+} \left[\dfrac{1}{3}\operatorname{arcsec}\left(\dfrac{5}{3}\right) - \dfrac{1}{3}\operatorname{arcsec}\left(\dfrac{b}{3}\right)\right] + \lim_{c\to\infty} \left[\dfrac{1}{3}\operatorname{arcsec}\left(\dfrac{c}{3}\right) - \dfrac{1}{3}\operatorname{arcsec}\left(\dfrac{5}{3}\right)\right] = -0 + \dfrac{1}{3}\left(\dfrac{\pi}{2}\right) = \dfrac{\pi}{6}$

**53.** $\int_0^\infty \dfrac{4}{\sqrt{x}(x+6)} \, dx = \int_0^1 \dfrac{4}{\sqrt{x}(x+6)} \, dx + \int_1^\infty \dfrac{4}{\sqrt{x}(x+6)} \, dx$

Let $u = \sqrt{x}, u^2 = x, 2u \, du = dx$.

$\int \dfrac{4}{\sqrt{x}(x+6)} \, dx = \int \dfrac{4(2u \, du)}{u(u^2+6)} = 8\int \dfrac{du}{u^2+6} = \dfrac{8}{\sqrt{6}}\arctan\left(\dfrac{u}{\sqrt{6}}\right) + C = \dfrac{8}{\sqrt{6}}\arctan\left(\dfrac{\sqrt{x}}{\sqrt{6}}\right) + C$

So, $\int_0^\infty \dfrac{4}{\sqrt{x}(x+6)} \, dx = \lim_{b\to 0^+} \left[\dfrac{8}{\sqrt{6}}\arctan\left(\dfrac{\sqrt{x}}{\sqrt{6}}\right)\right]_b^1 + \lim_{c\to\infty} \left[\dfrac{8}{\sqrt{6}}\arctan\left(\dfrac{\sqrt{x}}{\sqrt{6}}\right)\right]_1^c$

$= \left[\dfrac{8}{\sqrt{6}}\arctan\left(\dfrac{1}{\sqrt{6}}\right) - \dfrac{8}{\sqrt{6}}(0)\right] + \left[\dfrac{8}{\sqrt{6}}\dfrac{\pi}{2} - \dfrac{8}{\sqrt{6}}\arctan\left(\dfrac{1}{\sqrt{6}}\right)\right] = \dfrac{8\pi}{2\sqrt{6}} = \dfrac{2\pi\sqrt{6}}{3}.$

**55.** If $p = 1$, $\int_1^\infty \dfrac{1}{x} \, dx = \lim_{b\to\infty} \int_1^b \dfrac{1}{x} \, dx = \lim_{b\to\infty}\left[\ln x\right]_1^b$

$= \lim_{b\to\infty}(\ln b) = \infty.$

Diverges. For $p \neq 1$,

$\int_1^\infty \dfrac{1}{x^p} \, dx = \lim_{b\to\infty}\left[\dfrac{x^{1-p}}{1-p}\right]_1^b = \lim_{b\to\infty}\left(\dfrac{b^{1-p}}{1-p} - \dfrac{1}{1-p}\right).$

This converges to $\dfrac{1}{p-1}$ if $1 - p < 0$ or $p > 1$.

**57.** For $n = 1$:

$\int_0^\infty xe^{-x} \, dx = \lim_{b\to\infty} \int_0^b xe^{-x} \, dx$

$= \lim_{b\to\infty}\left[-e^{-x}x - e^{-x}\right]_0^b \qquad \left(\text{Parts: } u = x, dv = e^{-x} \, dx\right)$

$= \lim_{b\to\infty}\left(-e^{-b}b - e^{-b} + 1\right)$

$= \lim_{b\to\infty}\left(\dfrac{-b}{e^b} - \dfrac{1}{e^b} + 1\right) = 1 \quad \text{(L'Hôpital's Rule)}$

Assume that $\int_0^\infty x^n e^{-x} \, dx$ converges. Then for $n + 1$ you have

$\int x^{n+1} e^{-x} \, dx = -x^{n+1}e^{-x} + (n+1)\int x^n e^{-x} \, dx$

by parts $\left(u = x^{n+1}, du = (n+1)x^n \, dx, dv = e^{-x} \, dx, v = -e^{-x}\right).$

So,

$\int_0^\infty x^{n+1} e^{-x} \, dx = \lim_{b\to\infty}\left[-x^{n+1}e^{-x}\right]_0^b + (n+1)\int_0^\infty x^n e^{-x} \, dx = 0 + (n+1)\int_0^\infty x^n e^{-x} \, dx$, which converges.

**59.** $\int_0^1 \dfrac{1}{x^5}\,dx$ diverges by Exercise 56. $(p = 5)$

**61.** $\int_1^\infty \dfrac{1}{x^5}\,dx$ converges by Exercise 55. $(p = 5)$

**63.** Because $\dfrac{1}{x^2 + 5} \le \dfrac{1}{x^2}$ on $[1, \infty)$ and

$\int_1^\infty \dfrac{1}{x^2}\,dx$ converges by Exercise 55,

$\int_1^\infty \dfrac{1}{x^2 + 5}\,dx$ converges.

**65.** Because $\dfrac{1}{\sqrt[3]{x(x-1)}} \ge \dfrac{1}{\sqrt[3]{x^2}}$ on $[2, \infty)$ and

$\int_2^\infty \dfrac{1}{\sqrt[3]{x^2}}\,dx$ diverges by Exercise 55,

$\int_2^\infty \dfrac{1}{\sqrt[3]{x(x-1)}}\,dx$ diverges.

**67.** Because $\dfrac{2 + e^{-x}}{x} \ge \dfrac{2}{x}$ on $[1, \infty)$ and $\int_1^\infty \dfrac{2}{x}\,dx$ diverges

by Exercise 55, $\int_1^\infty \dfrac{2 + e^{-x}}{x}\,dx$ diverges.

**69.** $\int_1^\infty \dfrac{2}{x^2}\,dx$ converges, and $\dfrac{1 - \sin x}{x^2} \le \dfrac{2}{x^2}$ on $[1, \infty)$, so

$\int_1^\infty \dfrac{1 - \sin x}{x^2}\,dx$ converges.

**71.** Answers will vary. *Sample answer:*

An integral with infinite integration limits or an integral
with an infinite discontinuity at or between the
integration limits

**73.** $\int_{-1}^1 \dfrac{1}{x^3}\,dx = \int_{-1}^0 \dfrac{1}{x^3}\,dx + \int_0^1 \dfrac{1}{x^3}\,dx$

These two integrals diverge by Exercise 56.

**75.** $A = \int_{-\infty}^1 e^x\,dx$

$= \lim_{b \to -\infty} \int_b^1 e^x\,dx$

$= \lim_{b \to -\infty} \left[ e^x \right]_b^1$

$= \lim_{b \to -\infty} \left( e - e^b \right) = e$

**77.** $A = \int_{-\infty}^\infty \dfrac{1}{x^2 + 1}\,dx$

$= \lim_{b \to -\infty} \int_b^0 \dfrac{1}{x^2 + 1}\,dx + \lim_{b \to \infty} \int_0^b \dfrac{1}{x^2 + 1}\,dx$

$= \lim_{b \to -\infty} \left[ \arctan(x) \right]_b^0 + \lim_{b \to \infty} \left[ \arctan(x) \right]_0^b$

$= \lim_{b \to -\infty} \left[ 0 - \arctan(b) \right] + \lim_{b \to \infty} \left[ \arctan(b) - 0 \right]$

$= -\left( -\dfrac{\pi}{2} \right) + \dfrac{\pi}{2} = \pi$

**79. (a)** $A = \int_0^\infty e^{-x}\,dx$

$= \lim_{b \to \infty} \left[ -e^{-x} \right]_0^b = 0 - (-1) = 1$

**(b) Disk:**

$V = \pi \int_0^\infty \left( e^{-x} \right)^2\,dx$

$= \lim_{b \to \infty} \pi \left[ -\dfrac{1}{2} e^{-2x} \right]_0^b = \dfrac{\pi}{2}$

**(c) Shell:**

$V = 2\pi \int_0^\infty x e^{-x}\,dx$

$= \lim_{b \to \infty} 2\pi \left[ -e^{-x}(x + 1) \right]_0^b = 2\pi$

**81.**   $x^{2/3} + y^{2/3} = 4$

$\dfrac{2}{3} x^{-1/3} + \dfrac{2}{3} y^{-1/3} y' = 0$

$y' = \dfrac{-y^{1/3}}{x^{1/3}}$

$\sqrt{1 + (y')^2} = \sqrt{1 + \dfrac{y^{2/3}}{x^{2/3}}} = \sqrt{\dfrac{x^{2/3} + y^{2/3}}{x^{2/3}}} = \sqrt{\dfrac{4}{x^{2/3}}} = \dfrac{2}{x^{1/3}}, \quad (x > 0)$

$s = 4 \int_0^8 \dfrac{2}{x^{1/3}}\,dx = \lim_{b \to 0^+} \left[ 8 \cdot \dfrac{3}{2} x^{2/3} \right]_b^8 = 48$

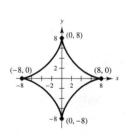

**83.** $(x - 2)^2 + y^2 = 1$

$2(x - 2) + 2yy' = 0$

$y' = \dfrac{-(x - 2)}{y}$

$\sqrt{1 + (y')^2} = \sqrt{1 + \left[(x - 2)^2 / y^2\right]} = \dfrac{1}{y}$ (Assume $y > 0$.)

$S = 4\pi \displaystyle\int_1^3 \dfrac{x}{y}\, dx = 4\pi \int_1^3 \dfrac{x}{\sqrt{1 - (x - 2)^2}}\, dx = 4\pi \int_1^3 \left[\dfrac{x - 2}{\sqrt{1 - (x - 2)^2}} + \dfrac{2}{\sqrt{1 - (x - 2)^2}}\right] dx$

$\quad = \displaystyle\lim_{\substack{a \to 1^+ \\ b \to 3^-}} 4\pi \left[-\sqrt{1 - (x - 2)^2} + 2\arcsin(x - 2)\right]_a^b = 4\pi\left[0 + 2\arcsin(1) - 2\arcsin(-1)\right] = 8\pi^2$

**85.** (a) $F(x) = \dfrac{K}{x^2},\ 5 = \dfrac{K}{(4000)^2},\ K = 80{,}000{,}000$

$\quad W = \displaystyle\int_{4000}^{\infty} \dfrac{80{,}000{,}000}{x^2}\, dx = \lim_{b \to \infty}\left[\dfrac{-80{,}000{,}000}{x}\right]_{4000}^{b} = 20{,}000 \text{ mi-ton}$

(b) $\dfrac{W}{2} = 10{,}000 = \left[\dfrac{-80{,}000{,}000}{x}\right]_{4000}^{b} = \dfrac{-80{,}000{,}000}{b} + 20{,}000$

$\dfrac{80{,}000{,}000}{b} = 10{,}000$

$b = 8000$

Therefore, 4000 miles above the earth's surface.

**87.** (a) $\displaystyle\int_{-\infty}^{\infty} \tfrac{1}{7}e^{-t/7}\, dt = \int_0^{\infty} \tfrac{1}{7}e^{-t/7}\, dt = \lim_{b \to \infty}\left[-e^{-t/7}\right]_0^b = 1$

(b) $\displaystyle\int_0^4 \tfrac{1}{7}e^{-t/7}\, dt = \left[-e^{-t/7}\right]_0^4 = -e^{-4/7} + 1$

$\quad \approx 0.4353 = 43.53\%$

(c) $\displaystyle\int_0^{\infty} t\left[\tfrac{1}{7}e^{-t/7}\right] dt = \lim_{b \to \infty}\left[-te^{-t/7} - 7e^{-t/7}\right]_0^b$

$\quad = 0 + 7 = 7$

**89.** (a) $C = 650{,}000 + \displaystyle\int_0^5 25{,}000\, e^{-0.06t}\, dt = 650{,}000 - \left[\dfrac{25{,}000}{0.06}e^{-0.06t}\right]_0^5 \approx \$757{,}992.41$

(b) $C = 650{,}000 + \displaystyle\int_0^{10} 25{,}000e^{-0.06t}\, dt \approx \$837{,}995.15$

(c) $C = 650{,}000 + \displaystyle\int_0^{\infty} 25{,}000e^{-0.06t}\, dt = 650{,}000 - \lim_{b \to \infty}\left[\dfrac{25{,}000}{0.06}e^{-0.06t}\right]_0^b \approx \$1{,}066{,}666.67$

**91.** Let $K = \dfrac{2\pi NI\, r}{k}$. Then

$$P = K \int_c^\infty \frac{1}{\left(r^2 + x^2\right)^{3/2}}\, dx.$$

Let $x = r \tan\theta$, $dx = r\sec^2\theta\, d\theta$, $\sqrt{r^2 + x^2} = r\sec\theta$.

$$\int \frac{1}{\left(r^2 + x^2\right)^{3/2}}\, dx = \int \frac{r\sec^2\theta\, d\theta}{r^3\sec^3\theta} = \frac{1}{r^2}\int \cos\theta\, d\theta = \frac{1}{r^2}\sin\theta + C = \frac{1}{r^2}\frac{x}{\sqrt{r^2 + x^2}} + C$$

So,

$$P = K\frac{1}{r^2}\lim_{b\to\infty}\left[\frac{x}{\sqrt{r^2 + x^2}}\right]_c^b = \frac{K}{r^2}\left[1 - \frac{c}{\sqrt{r^2 + c^2}}\right] = \frac{K\left(\sqrt{r^2 + c^2} - c\right)}{r^2\sqrt{r^2 + c^2}} = \frac{2\pi NI\left(\sqrt{r^2 + c^2} - c\right)}{kr\sqrt{r^2 + c^2}}.$$

**93.** False. $f(x) = 1/(x + 1)$ is continuous on

$[0, \infty)$, $\displaystyle\lim_{x\to\infty} 1/(x + 1) = 0$, but

$$\int_0^\infty \frac{1}{x + 1}\, dx = \lim_{b\to\infty}\left[\ln|x + 1|\right]_0^b = \infty.$$

Diverges

**95.** True

**97.** (a) $\displaystyle\int_{-\infty}^\infty \sin x\, dx = \int_{-\infty}^0 \sin x\, dx + \int_0^\infty \sin x\, dx$

$$= \lim_{b\to-\infty}\int_b^0 \sin x\, dx + \lim_{c\to\infty}\int_0^c \sin x\, dx = \lim_{b\to-\infty}\left[-\cos x\right]_b^0 + \lim_{c\to\infty}\left[-\cos x\right]_0^c$$

Because $\displaystyle\lim_{b\to-\infty}\left[-\cos b\right]$ diverges, as does $\displaystyle\lim_{c\to\infty}\left[-\cos c\right]$,

$\displaystyle\int_{-\infty}^\infty \sin x\, dx$ diverges.

(b) $\displaystyle\lim_{a\to\infty}\int_{-a}^a \sin x\, dx = \lim_{a\to\infty}\left[-\cos x\right]_{-a}^a = \lim_{a\to\infty}\left[-\cos(a) + \cos(-a)\right] = 0$

(c) The definition of $\displaystyle\int_{-\infty}^\infty f(x)\, dx$ is not

$$\lim_{a\to\infty}\int_{-a}^a f(x)\, dx.$$

**99.** (a) $\displaystyle\int_1^\infty \frac{1}{x}\, dx = \lim_{b\to\infty}\left[\ln|x|\right]_1^b = \infty$

$\displaystyle\int_1^\infty \frac{1}{x^2}\, dx = \lim_{b\to\infty}\left[-\frac{1}{x}\right]_1^b = 1$

$\displaystyle\int_1^\infty \frac{1}{x^n}\, dx$ will converge if $n > 1$ and will diverge if $n \le 1$.

(b) It would appear to converge.

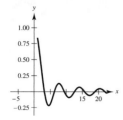

(c) Let $dv = \sin x \, dx \Rightarrow v = -\cos x$

$$u = \frac{1}{x} \Rightarrow du = -\frac{1}{x^2} \, dx.$$

$$\int_1^\infty \frac{\sin x}{x} \, dx = \lim_{b \to 0} \left[ -\frac{\cos x}{x} \right]_1^b - \int_1^\infty \frac{\cos x}{x^2} \, dx = \cos 1 - \int_1^\infty \frac{\cos x}{x^2} \, dx$$

Converges

**101.** $\Gamma(n) = \int_0^\infty x^{n-1} e^{-x} \, dx$

(a) $\Gamma(1) = \int_0^\infty e^{-x} \, dx = \lim_{b \to \infty} \left[ -e^{-x} \right]_0^b = 1$

$\Gamma(2) = \int_0^\infty x e^{-x} \, dx = \lim_{b \to \infty} \left[ -e^{-x}(x+1) \right]_0^b = 1$

$\Gamma(3) = \int_0^\infty x^2 e^{-x} \, dx = \lim_{b \to \infty} \left[ -x^2 e^{-x} - 2x e^{-x} - 2e^{-x} \right]_0^b = 2$

(b) $\Gamma(n+1) = \int_0^\infty x^n e^{-x} \, dx = \lim_{b \to \infty} \left[ -x^n e^{-x} \right]_0^b + \lim_{b \to \infty} n \int_0^b x^{n-1} e^{-x} \, dx = 0 + n\Gamma(n) \qquad (u = x^n, \, dv = e^{-x} \, dx)$

(c) $\Gamma(n) = (n-1)!$

**103.** $f(t) = 1$

$$F(s) = \int_0^\infty e^{-st} \, dt = \lim_{b \to \infty} \left[ -\frac{1}{s} e^{-st} \right]_0^b = \frac{1}{s}, \, s > 0$$

**105.** $f(t) = t^2$

$$F(s) = \int_0^\infty t^2 e^{-st} \, dt = \lim_{b \to \infty} \left[ \frac{1}{s^3} \left( -s^2 t^2 - 2st - 2 \right) e^{-st} \right]_0^b = \frac{2}{s^3}, \, s > 0$$

**107.** $f(t) = \cos at$

$$F(s) = \int_0^\infty e^{-st} \cos at \, dt$$

$$= \lim_{b \to \infty} \left[ \frac{e^{-st}}{s^2 + a^2} (-s \cos at + a \sin at) \right]_0^b = 0 + \frac{s}{s^2 + a^2} = \frac{s}{s^2 + a^2}, \, s > 0$$

**109.** $f(t) = \cosh at$

$$F(s) = \int_0^\infty e^{-st} \cosh at \, dt = \int_0^\infty e^{-st} \left( \frac{e^{at} + e^{-at}}{2} \right) dt = \frac{1}{2} \int_0^\infty \left[ e^{t(-s+a)} + e^{t(-s-a)} \right] dt$$

$$= \lim_{b \to \infty} \frac{1}{2} \left[ \frac{1}{(-s+a)} e^{t(-s+a)} + \frac{1}{(-s-a)} e^{t(-s-a)} \right]_0^b = 0 - \frac{1}{2} \left[ \frac{1}{(-s+a)} + \frac{1}{(-s-a)} \right]$$

$$= \frac{-1}{2} \left[ \frac{1}{(-s+a)} + \frac{1}{(-s-a)} \right] = \frac{s}{s^2 - a^2}, \, s > |a|$$

**111.** (a)  $f(x) = \dfrac{1}{3\sqrt{2\pi}} e^{-(x-70)^2/18}$

$\displaystyle\int_{50}^{90} f(x)\, dx \approx 1.0$

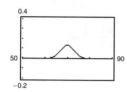

(b)  $P(72 \le x < \infty) \approx 0.2525$

(c)  $0.5 - P(70 \le x \le 72) \approx 0.5 - 0.2475 = 0.2525$

These are the same answers because of symmetry,  $P(70 \le x < \infty) = 0.5$  and

$0.5 = P(70 \le x < \infty) = P(70 \le x \le 72) + P(72 \le x < \infty).$

**113.**  $\displaystyle\int_0^{\infty}\left(\dfrac{1}{\sqrt{x^2+1}} - \dfrac{c}{x+1}\right) dx = \lim_{b\to\infty}\int_0^b\left(\dfrac{1}{\sqrt{x^2+1}} - \dfrac{c}{x+1}\right) dx$

$\qquad\qquad = \lim_{b\to\infty}\left[\ln\left|x + \sqrt{x^2+1}\right| - c\ln|x+1|\right]_0^b$

$\qquad\qquad = \lim_{b\to\infty}\left[\ln\left(b + \sqrt{b^2+1}\right) - \ln(b+1)^c\right] = \lim_{b\to\infty}\ln\left[\dfrac{b + \sqrt{b^2+1}}{(b+1)^c}\right]$

This limit exists for  $c = 1$ , and you have  $\displaystyle\lim_{b\to\infty}\ln\left[\dfrac{b + \sqrt{b^2+1}}{(b+1)}\right] = \ln 2.$

**115.**  $f(x) = \begin{cases} x\ln x, & 0 < x \le 2 \\ 0, & x = 0 \end{cases}$

$V = \pi\displaystyle\int_0^2 (x\ln x)^2\, dx$

Let  $u = \ln x,\, e^u = x,\, e^u\, du = dx.$

$V = \pi\displaystyle\int_{-\infty}^{\ln 2} e^{2u}u^2(e^u\, du)$

$\quad = \pi\displaystyle\int_{-\infty}^{\ln 2} e^{3u}u^2\, du$

$\quad = \lim_{b\to-\infty}\left[\pi\left[\dfrac{u^2}{3} - \dfrac{2u}{9} + \dfrac{2}{27}\right]e^{3u}\right]_b^{\ln 2}$

$\quad = 8\pi\left[\dfrac{(\ln 2)^2}{3} - \dfrac{2\ln 2}{9} + \dfrac{2}{27}\right] \approx 2.0155$

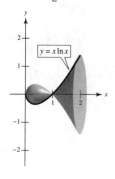

$y = x\ln x$

**117.**  $u = \sqrt{x},\, u^2 = x,\, 2u\, du = dx$

$\displaystyle\int_0^1 \dfrac{\sin x}{\sqrt{x}}\, dx = \int_0^1 \dfrac{\sin(u^2)}{u}(2u\, du) = \int_0^1 2\sin(u^2)\, du$

Trapezoidal Rule  $(n = 5)$ : 0.6278

**119.** (a)

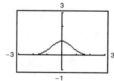

(b)  Let  $y = e^{-x^2},\, 0 \le x < \infty.$

$\ln y = -x^2$

$x = \sqrt{-\ln y}$  for  $0 < y \le 1$

The area bounded by  $y = e^{-x^2},\, x = 0$  and  $y = 0$  is

$\displaystyle\int_0^{\infty} e^{-x^2}\, dx = \int_0^1 \sqrt{-\ln y}\, dy, \qquad \left(= \dfrac{\sqrt{\pi}}{2}\right).$

# Review Exercises for Chapter 8

**1.** $\int x\sqrt{x^2 - 36}\, dx = \frac{1}{2}\int (x^2 - 36)^{1/2}(2x)\, dx$

$\qquad = \frac{1}{2}\frac{(x^2 - 36)^{3/2}}{3/2} + C$

$\qquad = \frac{1}{3}(x^3 - 36)^{3/2} + C$

**3.** $\int \frac{x}{x^2 - 49}\, dx = \frac{1}{2}\int \frac{2x}{x^2 - 49}\, dx = \frac{1}{2}\ln|x^2 - 49| + C$

**5.** Let $u = \ln(2x)$, $du = \frac{1}{x}\, dx$.

$\int_1^e \frac{\ln(2x)}{x}\, dx = \int_{\ln 2}^{1+\ln 2} u\, du$

$\qquad = \frac{u^2}{2}\Big]_{\ln 2}^{1+\ln 2}$

$\qquad = \frac{1}{2}\Big[1 + 2\ln 2 + (\ln 2)^2 - (\ln 2)^2\Big]$

$\qquad = \frac{1}{2} + \ln 2 \approx 1.1931$

**7.** $\int \frac{100}{\sqrt{100 - x^2}}\, dx = 100\arcsin\left(\frac{x}{10}\right) + C$

**9.** $\int xe^{3x}\, dx = \frac{x}{3}e^{3x} - \int \frac{1}{3}e^{3x}\, dx$

$\qquad = \frac{x}{3}e^{3x} - \frac{1}{9}e^{3x} + C$

$\qquad = \frac{1}{9}e^{3x}(3x - 1) + C$

$dv = e^{3x}\, dx \;\Rightarrow\; v = \frac{1}{3}e^{3x}$

$u = x \qquad \Rightarrow\; du = dx$

**11.** $\int e^{2x} \sin 3x\, dx = -\frac{1}{3}e^{2x}\cos 3x + \frac{2}{3}\int e^{2x}\cos 3x\, dx$

$\qquad = -\frac{1}{3}e^{2x}\cos 3x + \frac{2}{3}\left(\frac{1}{3}e^{2x}\sin 3x - \frac{2}{3}\int e^{2x}\sin 3x\, dx\right)$

$\frac{13}{9}\int e^{2x}\sin 3x\, dx = -\frac{1}{3}e^{2x}\cos 3x + \frac{2}{9}e^{2x}\sin 3x$

$\int e^{2x}\sin 3x\, dx = \frac{e^{2x}}{13}(2\sin 3x - 3\cos 3x) + C$

(1) $dv = \sin 3x\, dx \;\Rightarrow\; v = -\frac{1}{3}\cos 3x$

$\quad u = e^{2x} \qquad \Rightarrow\; du = 2e^{2x}\, dx$

(2) $dv = \cos 3x\, dx \;\Rightarrow\; v = \frac{1}{3}\sin 3x$

$\quad u = e^{2x} \qquad \Rightarrow\; du = 2e^{2x}\, dx$

**13.** $\int x\sqrt{x - 1}\, dx = \frac{2}{3}x(x - 1)^{3/2} - \int \frac{2}{3}(x - 1)^{3/2}\, dx$

$\qquad = \frac{2}{3}x(x - 1)^{3/2} - \frac{4}{15}(x - 1)^{5/2} + C$

$\qquad = \frac{2}{15}(x - 1)^{3/2}(5x - 2(x - 1)) + C$

$\qquad = \frac{2}{15}(x - 1)^{3/2}(3x + 2) + C$

$dv = (x - 1)^{1/2}\, dx \Rightarrow v = \frac{2}{3}(x - 1)^{3/2}$

$h = x \Rightarrow du = dx$

**15.** $\int x^2 \sin 2x\, dx = -\frac{1}{2}x^2\cos 2x + \int x\cos 2x\, dx$

$\qquad = -\frac{1}{2}x^2\cos 2x + \frac{1}{2}x\sin 2x - \frac{1}{2}\int \sin 2x\, dx$

$\qquad = -\frac{1}{2}x^2\cos 2x + \frac{x}{2}\sin 2x + \frac{1}{4}\cos 2x + C$

(1) $dv = \sin 2x\, dx \;\Rightarrow\; v = -\frac{1}{2}\cos 2x$

$\quad u = x^2 \qquad \Rightarrow\; du = 2x\, dx$

(2) $dv = \cos 2x\, dx \;\Rightarrow\; v = \frac{1}{2}\sin 2x$

$\quad u = x \qquad \Rightarrow\; du = dx$

**17.** $\displaystyle \int x \arcsin 2x \, dx = \frac{x^2}{2} \arcsin 2x - \int \frac{x^2}{\sqrt{1 - 4x^2}} \, dx$

$$= \frac{x^2}{2} \arcsin 2x - \frac{1}{8} \int \frac{2(2x)^2}{\sqrt{1 - (2x)^2}} \, dx$$

$$= \frac{x^2}{2} \arcsin 2x - \frac{1}{8}\left(\frac{1}{2}\right)\left[-(2x)\sqrt{1 - 4x^2} + \arcsin 2x\right] + C \text{ (by Formula 43 of Integration Tables)}$$

$$= \frac{1}{16}\left[(8x^2 - 1)\arcsin 2x + 2x\sqrt{1 - 4x^2}\right] + C$$

$\qquad dv = x \, dx \qquad \Rightarrow \qquad v = \dfrac{x^2}{2}$

$\qquad u = \arcsin 2x \quad \Rightarrow \quad du = \dfrac{2}{\sqrt{1 - 4x^2}} \, dx$

**19.** $\displaystyle \int \cos^3(\pi x - 1) \, dx = \int \left[1 - \sin^2(\pi x - 1)\right]\cos(\pi x - 1) \, dx$

$$= \frac{1}{\pi}\left[\sin(\pi x - 1) - \frac{1}{3}\sin^3(\pi x - 1)\right] + C$$

$$= \frac{1}{3\pi}\sin(\pi x - 1)\left[3 - \sin^2(\pi x - 1)\right] + C$$

$$= \frac{1}{3\pi}\sin(\pi x - 1)\left[3 - \left(1 - \cos^2(\pi x - 1)\right)\right] + C$$

$$= \frac{1}{3\pi}\sin(\pi x - 1)\left[2 + \cos^2(\pi x - 1)\right] + C$$

**21.** $\displaystyle \int \sec^4\left(\frac{x}{2}\right) dx = \int \left[\tan^2\left(\frac{x}{2}\right) + 1\right]\sec^2\left(\frac{x}{2}\right) dx$

$$= \int \tan^2\left(\frac{x}{2}\right)\sec^2\left(\frac{x}{2}\right) dx + \int \sec^2\left(\frac{x}{2}\right) dx$$

$$= \frac{2}{3}\tan^3\left(\frac{x}{2}\right) + 2\tan\left(\frac{x}{2}\right) + C = \frac{2}{3}\left[\tan^3\left(\frac{x}{2}\right) + 3\tan\left(\frac{x}{2}\right)\right] + C$$

**23.** $\displaystyle \int \frac{1}{1 - \sin \theta} \, d\theta = \int \frac{1}{1 - \sin \theta} \cdot \frac{1 + \sin \theta}{1 + \sin \theta} \, d\theta = \int \frac{1 + \sin \theta}{\cos^2 \theta} \, d\theta = \int \left(\sec^2 \theta + \sec \theta \tan \theta\right) d\theta = \tan \theta + \sec \theta + C$

**25.** $\displaystyle A = \int_{\pi/4}^{3\pi/4} \sin^4 x \, dx.$ Using the Table of Integrals,

$$\int \sin^4 x \, dx = -\frac{\sin^3 x \cos x}{4} + \frac{3}{4}\int \sin^2 x \, dx = \frac{-\sin^3 x \cos x}{4} + \frac{3}{4}\left[\frac{1}{2}(x - \sin x \cos x)\right] + C$$

$$\int_{\pi/4}^{3\pi/4} \sin^4 x \, dx = \left[\frac{-\sin^3 x \cos x}{4} + \frac{3}{8}x - \frac{3}{8}\sin x \cos x\right]_{\pi/4}^{3\pi/4} = \left(\frac{1}{16} + \frac{9\pi}{32} + \frac{3}{16}\right) - \left(\frac{-1}{16} + \frac{3\pi}{32} - \frac{3}{16}\right) = \frac{3\pi}{16} + \frac{1}{2} \approx 1.0890$$

**27.** $\displaystyle \int \frac{-12}{x^2\sqrt{4 - x^2}} \, dx = \int \frac{-24 \cos \theta \, d\theta}{(4 \sin^2 \theta)(2 \cos \theta)} = -3\int \csc^2 \theta \, d\theta = 3 \cot \theta + C = \frac{3\sqrt{4 - x^2}}{x} + C$

$\qquad x = 2 \sin \theta, \ dx = 2 \cos \theta \, d\theta, \ \sqrt{4 - x^2} = 2 \cos \theta$

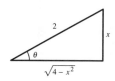

**29.**    $x = 2 \tan \theta$

$dx = 2 \sec^2 \theta \, d\theta$

$4 + x^2 = 4 \sec^2 \theta$

$$\int \frac{x^3}{\sqrt{4+x^2}} \, dx = \int \frac{8 \tan^3 \theta}{2 \sec \theta} \, 2 \sec^2 \theta \, d\theta$$

$$= 8 \int \tan^3 \theta \sec \theta \, d\theta$$

$$= 8 \int (\sec^2 \theta - 1) \tan \theta \sec \theta \, d\theta$$

$$= 8 \left[ \frac{\sec^3 \theta}{3} - \sec \theta \right] + C$$

$$= 8 \left[ \frac{(x^2+4)^{3/2}}{24} - \frac{\sqrt{x^2+4}}{2} \right] + C$$

$$= \sqrt{x^2+4} \left[ \frac{1}{3}(x^2+4) - 4 \right] + C$$

$$= \frac{1}{3} x^2 \sqrt{x^2+4} - \frac{8}{3} \sqrt{x^2+4} + C = \frac{1}{3}(x^2+4)^{1/2}(x^2-8) + C$$

**31.**  $\displaystyle\int_{-2}^{0} \sqrt{4-x^2} \, dx = \frac{1}{2} \left[ 4 \arcsin\left(\frac{x}{2}\right) + x\sqrt{4-x^2} \right]_{-2}^{0} = \frac{1}{2}[0 - 4\arcsin(-1)] = \frac{1}{2}\left[ -4\left(\frac{-\pi}{2}\right) \right] = \pi$

Note: The integral represents the area of a quarter circle of radius 2:  $A = \frac{1}{4}(\pi 2^2) = \pi$.

**33.** (a) Let  $x = 2 \tan \theta$,  $dx = 2 \sec^2 \theta \, d\theta$.

$$\int \frac{x^3}{\sqrt{4+x^2}} \, dx = \int \frac{8 \tan^3 \theta}{2 \sec \theta} \, 2 \sec^2 \theta \, d\theta$$

$$= 8 \int \tan^3 \theta \sec \theta \, d\theta$$

$$= 8 \int \frac{\sin^3 \theta}{\cos^4 \theta} \, d\theta$$

$$= 8 \int (1 - \cos^2 \theta) \cos^{-4} \theta \sin \theta \, d\theta$$

$$= 8 \int (\cos^{-4} \theta - \cos^{-2} \theta) \sin \theta \, d\theta$$

$$= 8 \left[ \frac{\cos^{-3} \theta}{3} - \frac{\cos^{-1} \theta}{-1} \right] + C$$

$$= \frac{8}{3} \sec \theta (\sec^2 \theta - 3) + C$$

$$= \frac{8}{3} \frac{\sqrt{4+x^2}}{2} \left( \frac{4+x^2}{4} - 3 \right) + C$$

$$= \frac{1}{3} \sqrt{4+x^2} (x^2 - 8) + C$$

(b) $\displaystyle\int \frac{x^3}{\sqrt{4 + x^2}}\,dx = \int \frac{x^2}{\sqrt{4 + x^2}}\,x\,dx$

$$= \int \frac{(u^2 - 4)u\,du}{u}$$

$$= \int (u^2 - 4)\,du$$

$$= \frac{1}{3}u^3 - 4u + C$$

$$= \frac{u}{3}(u^2 - 12) + C$$

$$= \frac{\sqrt{4 + x^2}}{3}(x^2 - 8) + C$$

$u^2 = 4 + x^2,\ 2u\,du = 2x\,dx$

(c) $\displaystyle\int \frac{x^3}{\sqrt{4 + x^2}}\,dx = x^2\sqrt{4 + x^2} - \int 2x\sqrt{4 + x^2}\,dx$

$$= x^2\sqrt{4 + x^2} - \frac{2}{3}(4 + x^2)^{3/2} + C = \frac{\sqrt{4 + x^2}}{3}(x^2 - 8) + C$$

$dv = \dfrac{x}{\sqrt{4 + x^2}}\,dx \quad\Rightarrow\quad v = \sqrt{4 + x^2}$

$u = x^2 \qquad\qquad\Rightarrow\quad du = 2x\,dx$

**35.** $\dfrac{x - 39}{x^2 - x - 12} = \dfrac{x - 39}{(x - 4)(x + 3)} = \dfrac{A}{x - 4} + \dfrac{B}{x + 3}$

$$x - 39 = A(x + 3) + B(x - 4)$$

Let $x = -3$: $-42 = -7B \Rightarrow B = 6$

Let $x = 4$: $-35 = 7A \Rightarrow A = -5$

$\displaystyle\int \frac{x - 39}{x^2 - x - 12}\,dx = \int \frac{-5}{x - 4}\,dx + \int \frac{6}{x + 3}\,dx$

$$= -5 \ln|x - 4| + 6 \ln|x + 3| + C$$

**37.** $\dfrac{x^2 + 2x}{(x - 1)(x^2 + 1)} = \dfrac{A}{x - 1} + \dfrac{Bx + C}{x^2 + 1}$

$$x^2 + 2x = A(x^2 + 1) + (Bx + C)(x - 1)$$

Let $x = 1$: $3 = 2A \Rightarrow A = \dfrac{3}{2}$   Let $x = 0$: $0 = A - C \Rightarrow C = \dfrac{3}{2}$   Let $x = 2$: $8 = 5A + 2B + C \Rightarrow B = -\dfrac{1}{2}$

$\displaystyle\int \frac{x^2 + 2x}{x^3 - x^2 + x - 1}\,dx = \frac{3}{2}\int \frac{1}{x - 1}\,dx - \frac{1}{2}\int \frac{x - 3}{x^2 + 1}\,dx$

$$= \frac{3}{2}\int \frac{1}{x - 1}\,dx - \frac{1}{4}\int \frac{2x}{x^2 + 1}\,dx + \frac{3}{2}\int \frac{1}{x^2 + 1}\,dx$$

$$= \frac{3}{2}\ln|x - 1| - \frac{1}{4}\ln|x^2 + 1| + \frac{3}{2}\arctan x + C$$

$$= \frac{1}{4}\Big[6 \ln|x - 1| - \ln(x^2 + 1) + 6 \arctan x\Big] + C$$

**39.** $\dfrac{x^2}{x^2 + 5x - 24} = 1 - \dfrac{5x - 24}{x^2 + 5x - 24} = 1 - \dfrac{5x - 24}{(x + 8)(x - 3)}$

$\dfrac{5x - 24}{(x + 8)(x - 3)} = \dfrac{A}{x + 8} + \dfrac{B}{x - 3}$

$5x - 24 = A(x - 3) + B(x + 8)$

Let $x = 3$: $-9 = 11B \Rightarrow B = -9/11$

Let $x = -8$: $-64 = -11A \Rightarrow A = 64/11$

$\displaystyle\int \dfrac{x^2}{x^2 + 5x - 24}\, dx = \int \left[1 - \dfrac{64/11}{x + 8} + \dfrac{9/11}{x - 3}\right] dx = x - \dfrac{64}{11}\ln|x + 8| + \dfrac{9}{11}\ln|x - 3| + C$

**41.** Using Formula 4: $(a = 4, b = 5)$

$\displaystyle\int \dfrac{x}{(4 + 5x)^2}\, dx = \dfrac{1}{25}\left(\dfrac{4}{4 + 5x} + \ln|4 + 5x|\right) + C$

**43.** Let $u = x^2$, $du = 2x\, dx$.

$\displaystyle\int_0^{\sqrt{\pi/2}} \dfrac{x}{1 + \sin x^2}\, dx = \dfrac{1}{2}\int_0^{\pi/4} \dfrac{1}{1 + \sin u}\, du$

$\qquad = \dfrac{1}{2}\Big[\tan u - \sec u\Big]_0^{\pi/4}$

$\qquad = \dfrac{1}{2}\Big[(1 - \sqrt{2}) - (0 - 1)\Big]$

$\qquad = 1 - \dfrac{\sqrt{2}}{2}$

**45.** $\displaystyle\int \dfrac{x}{x^2 + 4x + 8}\, dx = \dfrac{1}{2}\left[\ln|x^2 + 4x + 8| - 4\int \dfrac{1}{x^2 + 4x + 8}\, dx\right]$ (Formula 15)

$\qquad = \dfrac{1}{2}\left[\ln|x^2 + 4x + 8|\right] - 2\left[\dfrac{2}{\sqrt{32 - 16}} \arctan\left(\dfrac{2x + 4}{\sqrt{32 - 16}}\right)\right] + C$ (Formula 14)

$\qquad = \dfrac{1}{2}\ln|x^2 + 4x + 8| - \arctan\left(1 + \dfrac{x}{2}\right) + C$

**47.** $\displaystyle\int \dfrac{1}{\sin \pi x \cos \pi x}\, dx = \dfrac{1}{\pi}\int \dfrac{1}{\sin \pi x \cos \pi x}\,(\pi)\, dx \quad (u = \pi x)$

$\qquad = \dfrac{1}{\pi}\ln|\tan \pi x| + C$ (Formula 58)

**49.** $dv = dx \qquad \Rightarrow \quad v = x$

$u = (\ln x)^n \quad \Rightarrow \quad du = n(\ln x)^{n-1}\dfrac{1}{x}\, dx$

$\displaystyle\int (\ln x)^n\, dx = x(\ln x)^n - n\int (\ln x)^{n-1}\, dx$

**51.** $\displaystyle\int \theta \sin \theta \cos \theta\, d\theta = \dfrac{1}{2}\int \theta \sin 2\theta\, d\theta$

$\qquad = -\dfrac{1}{4}\theta \cos 2\theta + \dfrac{1}{4}\int \cos 2\theta\, d\theta = -\dfrac{1}{4}\theta \cos 2\theta + \dfrac{1}{8}\sin 2\theta + C = \dfrac{1}{8}(\sin 2\theta - 2\theta \cos 2\theta) + C$

$dv = \sin 2\theta\, d\theta \quad \Rightarrow \quad v = -\dfrac{1}{2}\cos 2\theta$

$u = \theta \qquad\qquad \Rightarrow \quad du = d\theta$

**53.** $\displaystyle \int \frac{x^{1/4}}{1 + x^{1/2}}\, dx = 4\int \frac{u(u^3)}{1 + u^2}\, du$

$$= 4\int \left( u^2 - 1 + \frac{1}{u^2 + 1} \right) du$$

$$= 4\left( \frac{1}{3}u^3 - u + \arctan u \right) + C$$

$$= \frac{4}{3}\left[ x^{3/4} - 3x^{1/4} + 3\arctan(x^{1/4}) \right] + C$$

$u = \sqrt[4]{x},\, x = u^4,\, dx = 4u^3\, du$

**55.** $\displaystyle \int \sqrt{1 + \cos x}\, dx = \int \frac{\sqrt{1 + \cos x}}{1} \cdot \frac{\sqrt{1 - \cos x}}{\sqrt{1 - \cos x}}\, dx$

$$= \int \frac{\sin x}{\sqrt{1 - \cos x}}\, dx$$

$$= \int (1 - \cos x)^{-1/2}(\sin x)\, dx$$

$$= 2\sqrt{1 - \cos x} + C$$

$u = 1 - \cos x,\, du = \sin x\, dx$

**57.** $\displaystyle \int \cos x \ln(\sin x)\, dx = \sin x \ln(\sin x) - \int \cos x\, dx = \sin x \ln(\sin x) - \sin x + C$

$dv = \cos x\, dx \;\Rightarrow\; v = \sin x$

$u = \ln(\sin x) \;\Rightarrow\; du = \dfrac{\cos x}{\sin x}\, dx$

**59.** $\displaystyle y = \int \frac{25}{x^2 - 25}\, dx = 25\left(\frac{1}{10}\right) \ln\left| \frac{x - 5}{x + 5} \right| + C$

$$= \frac{5}{2} \ln\left| \frac{x - 5}{x + 5} \right| + C$$

(Formula 24)

**61.** $\displaystyle y = \int \ln(x^2 + x)\, dx = x \ln\left| x^2 + x \right| - \int \frac{2x^2 + x}{x^2 + x}\, dx$

$$= x \ln\left| x^2 + x \right| - \int \frac{2x + 1}{x + 1}\, dx$$

$$= x \ln\left| x^2 + x \right| - \int 2\, dx + \int \frac{1}{x + 1}\, dx = x \ln\left| x^2 + x \right| - 2x + \ln\left| x + 1 \right| + C$$

$dv = dx \qquad\qquad \Rightarrow\; v = x$

$u = \ln(x^2 + x) \;\Rightarrow\; du = \dfrac{2x + 1}{x^2 + x}\, dx$

**63.** $\displaystyle \int_2^{\sqrt{5}} x(x^2 - 4)^{3/2}\, dx = \left[ \frac{1}{5}(x^2 - 4)^{5/2} \right]_2^{\sqrt{5}} = \frac{1}{5}$

**65.** $\displaystyle \int_1^4 \frac{\ln x}{x}\, dx = \left[ \frac{1}{2}(\ln x)^2 \right]_1^4 = \frac{1}{2}(\ln 4)^2 \approx 0.961$

**67.** $\displaystyle \int_0^\pi x \sin x\, dx = \left[ -x \cos x + \sin x \right]_0^\pi = \pi$

**69.** $\displaystyle A = \int_0^4 x\sqrt{4 - x}\, dx = \int_2^0 (4 - u^2)\, u(-2u)\, du$

$$= \int_2^0 2(u^4 - 4u^2)\, du$$

$$= \left[ 2\left( \frac{u^5}{5} - \frac{4u^3}{3} \right) \right]_2^0 = \frac{128}{15}$$

$u = \sqrt{4 - x},\, x = 4 - u^2,\, dx = -2u\, du$

**71.** By symmetry, $\bar{x} = 0$, $A = \dfrac{1}{2}\pi$.

$$\bar{y} = \frac{2}{\pi}\left(\frac{1}{2}\right)\int_{-1}^{1}\left(\sqrt{1-x^2}\right)^2 dx = \frac{1}{\pi}\left[x - \frac{1}{3}x^3\right]_{-1}^{1} = \frac{4}{3\pi}$$

$$(\bar{x}, \bar{y}) = \left(0, \frac{4}{3\pi}\right)$$

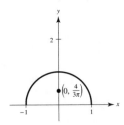

**73.** $s = \displaystyle\int_{0}^{\pi}\sqrt{1 + \cos^2 x}\, dx \approx 3.82$

**75.** $\displaystyle\lim_{x\to 1}\left[\frac{(\ln x)^2}{x-1}\right] = \lim_{x\to 1}\left[\frac{2(1/x)\ln x}{1}\right] = 0$

**77.** $\displaystyle\lim_{x\to\infty}\frac{e^{2x}}{x^2} = \lim_{x\to\infty}\frac{2e^{2x}}{2x} = \lim_{x\to\infty}\frac{4e^{2x}}{2} = \infty$

**79.** $y = \displaystyle\lim_{x\to\infty}(\ln x)^{2/x}$

$$\ln y = \lim_{x\to\infty}\frac{2\ln(\ln x)}{x} = \lim_{x\to\infty}\left[\frac{2/(x\ln x)}{1}\right] = 0$$

Because $\ln y = 0$, $y = 1$.

**81.** $\displaystyle\lim_{n\to\infty}1000\left(1 + \frac{0.09}{n}\right)^n = 1000\lim_{n\to\infty}\left(1 + \frac{0.09}{n}\right)^n$

Let $y = \displaystyle\lim_{n\to\infty}\left(1 + \frac{0.09}{n}\right)^n$.

$$\ln y = \lim_{n\to\infty}n\ln\left(1 + \frac{0.09}{n}\right) = \lim_{n\to\infty}\frac{\ln\left(1 + \frac{0.09}{n}\right)}{\frac{1}{n}} = \lim_{n\to\infty}\left(\frac{\frac{-0.09/n^2}{1 + (0.09/n)}}{-\frac{1}{n^2}}\right) = \lim_{n\to\infty}\frac{0.09}{1 + \left(\frac{0.09}{n}\right)} = 0.09$$

So, $\ln y = 0.09 \Rightarrow y = e^{0.09}$ and $\displaystyle\lim_{n\to\infty}1000\left(1 + \frac{0.09}{n}\right)^n = 1000e^{0.09} \approx 1094.17$.

**83.** $\displaystyle\int_{0}^{16}\frac{1}{\sqrt[4]{x}}\, dx = \lim_{b\to 0^+}\left[\frac{4}{3}x^{3/4}\right]_{b}^{16} = \frac{32}{3}$

**85.** $\displaystyle\int_{1}^{\infty}x^2\ln x\, dx = \lim_{b\to\infty}\left[\frac{x^3}{9}(-1 + 3\ln x)\right]_{1}^{b} = \infty$

Diverges

**87.** Let $u = \ln x$, $du = \dfrac{1}{x}\, dx$, $dv = x^{-2}\, dx$, $v = -x^{-1}$.

$$\int\frac{\ln x}{x^2}\, dx = \frac{-\ln x}{x} + \int\frac{1}{x^2}\, dx = \frac{-\ln x}{x} - \frac{1}{x} + C$$

$$\int_{1}^{\infty}\frac{\ln x}{x^2}\, dx = \lim_{b\to\infty}\left[\frac{-\ln x}{x} - \frac{1}{x}\right]_{1}^{b} = \lim_{b\to\infty}\left(\frac{-\ln b}{b} - \frac{1}{b}\right) - (-1) = 0 + 1 = 1$$

**89.** $\displaystyle\int_{2}^{\infty}\frac{1}{x\sqrt{x^2-4}}\, dx = \int_{2}^{3}\frac{1}{x\sqrt{x^2-4}}\, dx + \int_{3}^{\infty}\frac{1}{x\sqrt{x^2-4}}\, dx$

$$= \lim_{b\to 2^+}\left[\frac{1}{2}\operatorname{arcsec}\left(\frac{x}{2}\right)\right]_{b}^{3} + \lim_{c\to\infty}\left[\frac{1}{2}\operatorname{arcsec}\left(\frac{x}{2}\right)\right]_{3}^{c}$$

$$= \frac{1}{2}\operatorname{arcsec}\left(\frac{3}{2}\right) - \frac{1}{2}(0) + \frac{1}{2}\left(\frac{\pi}{2}\right) - \frac{1}{2}\operatorname{arcsec}\left(\frac{3}{2}\right) = \frac{\pi}{4}$$

**91.** $\displaystyle\int_{0}^{t_0}500{,}000e^{-0.05t}\, dt = \left[\frac{500{,}000}{-0.05}e^{-0.05t}\right]_{0}^{t_0} = \frac{-500{,}000}{0.05}\left(e^{-0.05t_0} - 1\right) = 10{,}000{,}000\left(1 - e^{-0.05t_0}\right)$

(a) $t_0 = 20$: \$6,321,205.59

(b) $t_0 \to \infty$: \$10,000,000

**93.** (a) $P(13 \leq x < \infty) = \dfrac{1}{0.95\sqrt{2\pi}} \displaystyle\int_{13}^{\infty} e^{-(x-12.9)^2/2(0.95)^2} \, dx \approx 0.4581$

(b) $P(15 \leq x < \infty) = \dfrac{1}{0.95\sqrt{2\pi}} \displaystyle\int_{15}^{\infty} e^{-(x-12.9)^2/2(0.95)^2} \, dx \approx 0.0135$

## Problem Solving for Chapter 8

**1.** (a) $\displaystyle\int_{-1}^{1} (1 - x^2) \, dx = \left[ x - \dfrac{x^3}{3} \right]_{-1}^{1} = 2\left( 1 - \dfrac{1}{3} \right) = \dfrac{4}{3}$

$\displaystyle\int_{-1}^{1} (1 - x^2)^2 \, dx = \int_{-1}^{1} (1 - 2x^2 + x^4) \, dx = \left[ x - \dfrac{2x^3}{3} + \dfrac{x^5}{5} \right]_{-1}^{1} = 2\left( 1 - \dfrac{2}{3} + \dfrac{1}{5} \right) = \dfrac{16}{15}$

(b) Let $x = \sin u, \, dx = \cos u \, du, \, 1 - x^2 = 1 - \sin^2 u = \cos^2 u$.

$\displaystyle\int_{-1}^{1} (1 - x^2)^n \, dx = \int_{-\pi/2}^{\pi/2} (\cos^2 u)^n \cos u \, du$

$\qquad = \displaystyle\int_{-\pi/2}^{\pi/2} \cos^{2n+1} u \, du$

$\qquad = 2\left[ \dfrac{2}{3} \cdot \dfrac{4}{5} \cdot \dfrac{6}{7} \cdots \dfrac{(2n)}{(2n+1)} \right]$   (Wallis's Formula)

$\qquad = 2\left[ \dfrac{2^2 \cdot 4^2 \cdot 6^2 \cdots (2n)^2}{2 \cdot 3 \cdot 4 \cdot 5 \cdots (2n)(2n+1)} \right] = \dfrac{2(2^{2n})(n!)^2}{(2n+1)!} = \dfrac{2^{2n+1}(n!)^2}{(2n+1)!}$

**3.**

$\displaystyle\lim_{x \to \infty} \left( \dfrac{x + c}{x - c} \right)^x = 9$

$\displaystyle\lim_{x \to \infty} x \ln\left( \dfrac{x + c}{x - c} \right) = \ln 9$

$\displaystyle\lim_{x \to \infty} \dfrac{\ln(x + c) - \ln(x - c)}{1/x} = \ln 9$

$\displaystyle\lim_{x \to \infty} \dfrac{\dfrac{1}{x + c} - \dfrac{1}{x - c}}{-\dfrac{1}{x^2}} = \ln 9$

$\displaystyle\lim_{x \to \infty} \dfrac{-2c}{(x + c)(x - c)}(-x^2) = \ln 9$

$\displaystyle\lim_{x \to \infty} \left( \dfrac{2cx^2}{x^2 - c^2} \right) = \ln 9$

$2c = \ln 9$

$2c = 2 \ln 3$

$c = \ln 3$

**5.** $\sin \theta = \dfrac{PB}{OP} = PB, \cos \theta = OB$

$AQ = \overparen{AP} = \theta$

$BR = OR + OB = OR + \cos \theta$

The triangles $\triangle AQR$ and $\triangle BPR$ are similar:

$\dfrac{AR}{AQ} = \dfrac{BR}{BP} \Rightarrow \dfrac{OR + 1}{\theta} = \dfrac{OR + \cos \theta}{\sin \theta}$

$\sin \theta(OR) + \sin \theta = (OR)\theta + \theta \cos \theta$

$OR = \dfrac{\theta \cos \theta - \sin \theta}{\sin \theta - \theta}$

$\displaystyle\lim_{\theta \to 0^+} OR = \lim_{\theta \to 0^+} \dfrac{\theta \cos \theta - \sin \theta}{\sin \theta - \theta}$

$\qquad = \displaystyle\lim_{\theta \to 0^+} \dfrac{-\theta \sin \theta + \cos \theta - \cos \theta}{\cos \theta - 1}$

$\qquad = \displaystyle\lim_{\theta \to 0^+} \dfrac{-\theta \sin \theta}{\cos \theta - 1}$

$\qquad = \displaystyle\lim_{\theta \to 0^+} \dfrac{-\sin \theta - \theta \cos \theta}{-\sin \theta}$

$\qquad = \displaystyle\lim_{\theta \to 0^+} \dfrac{\cos \theta + \cos \theta - \theta \sin \theta}{\cos \theta} = 2$

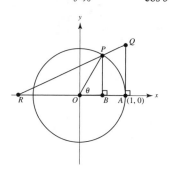

**7. (a)**

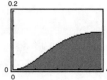

Area $\approx 0.2986$

**(b)** Let $x = 3 \tan \theta$, $dx = 3 \sec^2 \theta \, d\theta$, $x^2 + 9 = 9 \sec^2 \theta$.

$$\int \frac{x^2}{(x^2 + 9)^{3/2}} \, dx = \int \frac{9 \tan^2 \theta}{(9 \sec^2 \theta)^{3/2}} (3 \sec^2 \theta \, d\theta)$$

$$= \int \frac{\tan^2 \theta}{\sec \theta} \, d\theta$$

$$= \int \frac{\sin^2 \theta}{\cos \theta} \, d\theta$$

$$= \int \frac{1 - \cos^2 \theta}{\cos \theta} \, d\theta$$

$$= \ln|\sec \theta + \tan \theta| - \sin \theta + C$$

$$\text{Area} = \int_0^4 \frac{x^2}{(x^2 + 9)^{3/2}} \, dx = \Big[\ln|\sec \theta + \tan \theta| - \sin \theta\Big]_0^{\tan^{-1}(4/3)}$$

$$= \left[\ln\left(\frac{\sqrt{x^2 + 9}}{3} + \frac{x}{3}\right) - \frac{x}{\sqrt{x^2 + 9}}\right]_0^4$$

$$= \ln\left(\frac{5}{3} + \frac{4}{3}\right) - \frac{4}{5} = \ln 3 - \frac{4}{5}$$

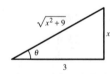

**(c)** $x = 3 \sinh u$, $dx = 3 \cosh u \, du$, $x^2 + 9 = 9 \sinh^2 u + 9 = 9 \cosh^2 u$

$$A = \int_0^4 \frac{x^2}{(x^2 + 9)^{3/2}} \, dx = \int_0^{\sinh^{-1}(4/3)} \frac{9 \sinh^2 u}{(9 \cosh^2 u)^{3/2}} (3 \cosh u \, du) = \int_0^{\sinh^{-1}(4/3)} \tanh^2 u \, du$$

$$= \int_0^{\sinh^{-1}(4/3)} (1 - \text{sech}^2 u) \, du = \Big[u - \tanh u\Big]_0^{\sinh^{-1}(4/3)}$$

$$= \sinh^{-1}\left(\frac{4}{3}\right) - \tanh\left(\sinh^{-1}\left(\frac{4}{3}\right)\right) = \ln\left(\frac{4}{3} + \sqrt{\frac{16}{9} + 1}\right) - \tanh\left[\ln\left(\frac{4}{3} + \sqrt{\frac{16}{9} + 1}\right)\right]$$

$$= \ln\left(\frac{4}{3} + \frac{5}{3}\right) - \tanh\left(\ln\left(\frac{4}{3} + \frac{5}{3}\right)\right) = \ln 3 - \tanh(\ln 3)$$

$$= \ln 3 - \frac{3 - (1/3)}{3 + (1/3)} = \ln 3 - \frac{4}{5}$$

**9.** $y = \ln(1 - x^2)$, $y' = \dfrac{-2x}{1 - x^2}$

$$1 + (y')^2 = 1 + \dfrac{4x^2}{(1 - x^2)^2}$$

$$= \dfrac{1 - 2x^2 + x^4 + 4x^2}{(1 - x^2)^2}$$

$$= \left(\dfrac{1 + x^2}{1 - x^2}\right)^2$$

Arc length $= \displaystyle\int_0^{1/2} \sqrt{1 + (y')^2}\, dx$

$$= \int_0^{1/2} \left(\dfrac{1 + x^2}{1 - x^2}\right) dx$$

$$= \int_0^{1/2} \left(-1 + \dfrac{2}{1 - x^2}\right) dx$$

$$= \int_0^{1/2} \left(-1 + \dfrac{1}{x + 1} + \dfrac{1}{1 - x}\right) dx$$

$$= \left[-x + \ln(1 + x) - \ln(1 - x)\right]_0^{1/2}$$

$$= \left(-\dfrac{1}{2} + \ln\dfrac{3}{2} - \ln\dfrac{1}{2}\right)$$

$$= -\dfrac{1}{2} + \ln 3 - \ln 2 + \ln 2$$

$$= \ln 3 - \dfrac{1}{2} \approx 0.5986$$

**11.** Consider $\displaystyle\int \dfrac{1}{\ln x}\, dx$.

Let $u = \ln x$, $du = \dfrac{1}{x}\, dx$, $x = e^u$. Then

$$\int \dfrac{1}{\ln x}\, dx = \int \dfrac{1}{u}e^u\, du = \int \dfrac{e^u}{u}\, du.$$

If $\displaystyle\int \dfrac{1}{\ln x}\, dx$ were elementary, then $\displaystyle\int \dfrac{e^u}{u}\, du$ would be too, which is false.

So, $\displaystyle\int \dfrac{1}{\ln x}\, dx$ is not elementary.

**13.** $x^4 + 1 = (x^2 + ax + b)(x^2 + cx + d)$

$$= x^4 + (a + c)x^3 + (ac + b + d)x^2 + (ad + bc)x + bd$$

$a = -c$, $b = d = 1$, $a = \sqrt{2}$

$$x^4 + 1 = (x^2 + \sqrt{2}x + 1)(x^2 - \sqrt{2}x + 1)$$

$$\int_0^1 \dfrac{1}{x^4 + 1}\, dx = \int_0^1 \dfrac{Ax + B}{x^2 + \sqrt{2}x + 1}\, dx + \int_0^1 \dfrac{Cx + D}{x^2 - \sqrt{2}x + 1}\, dx$$

$$= \int_0^1 \dfrac{\dfrac{1}{2} + \dfrac{\sqrt{2}}{4}x}{x^2 + \sqrt{2}x + 1}\, dx - \int_0^1 \dfrac{-\dfrac{1}{2} + \dfrac{\sqrt{2}}{4}x}{x^2 + \sqrt{2}x + 1}\, dx$$

$$= \dfrac{\sqrt{2}}{4}\left[\arctan(\sqrt{2}x + 1) + \arctan(\sqrt{2}x - 1)\right]_0^1 + \dfrac{\sqrt{2}}{8}\left[\ln(x^2 + \sqrt{2}x + 1) - \ln(x^2 - \sqrt{2}x + 1)\right]_0^1$$

$$= \dfrac{\sqrt{2}}{4}\left[\arctan(\sqrt{2} + 1) + \arctan(\sqrt{2} - 1)\right] + \dfrac{\sqrt{2}}{8}\left[\ln(2 + \sqrt{2}) - \ln(2 - \sqrt{2})\right] - \dfrac{\sqrt{2}}{4}\left[\dfrac{\pi}{4} - \dfrac{\pi}{4}\right] - \dfrac{\sqrt{2}}{8}[0]$$

$$\approx 0.5554 + 0.3116$$

$$\approx 0.8670$$

**15.** Using a graphing utility,

(a) $\displaystyle\lim_{x \to 0^+} \left( \cot x + \frac{1}{x} \right) = \infty$

(b) $\displaystyle\lim_{x \to 0^+} \left( \cot x - \frac{1}{x} \right) = 0$

(c) $\displaystyle\lim_{x \to 0^+} \left( \cot x + \frac{1}{x} \right)\left( \cot x - \frac{1}{x} \right) \approx -\frac{2}{3}$.

Analytically,

(a) $\displaystyle\lim_{x \to 0^+} \left( \cot x + \frac{1}{x} \right) = \infty + \infty = \infty$

(b) $\displaystyle\lim_{x \to 0^+} \left( \cot x - \frac{1}{x} \right) = \lim_{x \to 0^+} \frac{x \cot x - 1}{x} = \lim_{x \to 0^+} \frac{x \cos x - \sin x}{x \sin x}$

$$= \lim_{x \to 0^+} \frac{\cos x - x \sin x - \cos x}{\sin x + x \cos x}$$

$$= \lim_{x \to 0^+} \frac{-x \sin x}{\sin x + x \cos x}$$

$$= \lim_{x \to 0^+} \frac{-\sin x - x \cos x}{\cos x + \cos x - x \sin x} = 0.$$

(c) $\displaystyle \left( \cot x + \frac{1}{x} \right)\left( \cot x - \frac{1}{x} \right) = \cot^2 x - \frac{1}{x^2}$

$$= \frac{x^2 \cot^2 x - 1}{x^2}$$

$$\lim_{x \to 0^+} \frac{x^2 \cot^2 x - 1}{x^2} = \lim_{x \to 0^+} \frac{2x \cot^2 x - 2x^2 \cot x \csc^2 x}{2x}$$

$$= \lim_{x \to 0^+} \frac{\cot^2 x - x \cot x \csc^2 x}{1}$$

$$= \lim_{x \to 0^+} \frac{\cos^2 x \sin x - x \cos x}{\sin^3 x}$$

$$= \lim_{x \to 0^+} \frac{\left( 1 - \sin^2 x \right)\sin x - x \cos x}{\sin^3 x}$$

$$= \lim_{x \to 0^+} \frac{\sin x - x \cos x}{\sin^3 x} - 1$$

Now, $\displaystyle\lim_{x \to 0^+} \frac{\sin x - x \cos x}{\sin^3 x} = \lim_{x \to 0^+} \frac{\cos x - \cos x + x \sin x}{3 \sin^2 x \cos x}$

$$= \lim_{x \to 0^+} \frac{x}{3 \sin x \cdot \cos x}$$

$$= \lim_{x \to 0^+} \left( \frac{x}{\sin x} \right)\frac{1}{3 \cos x} = \frac{1}{3}.$$

So, $\displaystyle\lim_{x \to 0^+} \left( \cot x + \frac{1}{x} \right)\left( \cot x - \frac{1}{x} \right) = \frac{1}{3} - 1 = -\frac{2}{3}.$

The form $0 \cdot \infty$ is indeterminant.

**17.** $\dfrac{x^3 - 3x^2 + 1}{x^4 - 13x^2 + 12x} = \dfrac{P_1}{x} + \dfrac{P_2}{x - 1} + \dfrac{P_3}{x + 4} + \dfrac{P_4}{x - 3} \Rightarrow c_1 = 0,\ c_2 = 1,\ c_3 = -4,\ c_4 = 3$

$N(x) = x^3 - 3x^2 + 1$

$D'(x) = 4x^3 - 26x + 12$

$P_1 = \dfrac{N(0)}{D'(0)} = \dfrac{1}{12}$

$P_2 = \dfrac{N(1)}{D'(1)} = \dfrac{-1}{-10} = \dfrac{1}{10}$

$P_3 = \dfrac{N(-4)}{D'(-4)} = \dfrac{-111}{-140} = \dfrac{111}{140}$

$P_4 = \dfrac{N(3)}{D'(3)} = \dfrac{1}{42}$

So, $\dfrac{x^3 - 3x^2 + 1}{x^4 - 13x^2 + 12x} = \dfrac{1/12}{x} + \dfrac{1/10}{x - 1} + \dfrac{111/140}{x + 4} + \dfrac{1/42}{x - 3}.$

**19.** By parts,

$$\int_a^b f(x)g''(x)\,dx = \left[f(x)g'(x)\right]_a^b - \int_a^b f'(x)g'(x)\,dx \qquad \left[u = f(x),\, dv = g''(x)\,dx\right]$$

$$= -\int_a^b f'(x)g'(x)\,dx$$

$$= \left[-f'(x)g(x)\right]_a^b + \int_a^b g(x)f''(x)\,dx \qquad \left[u = f'(x),\, dv = g'(x)\,dx\right] = \int_a^b f''(x)g(x)\,dx.$$

**21.** $\displaystyle\int_2^\infty \left[\dfrac{1}{x^5} + \dfrac{1}{x^{10}} + \dfrac{1}{x^{15}}\right] dx < \int_2^\infty \dfrac{1}{x^5 - 1}\,dx < \int_2^\infty \left[\dfrac{1}{x^5} + \dfrac{1}{x^{10}} + \dfrac{2}{x^{15}}\right] dx$

$$\lim_{b\to\infty}\left[-\dfrac{1}{4x^4} - \dfrac{1}{9x^9} - \dfrac{1}{14x^{14}}\right]_2^b < \int_2^\infty \dfrac{1}{x^5 - 1}\,dx < \lim_{b\to\infty}\left[-\dfrac{1}{4x^4} - \dfrac{1}{9x^9} - \dfrac{1}{7x^{14}}\right]_2^b$$

$$0.015846 < \int_2^\infty \dfrac{2}{x^5 - 1}\,dx < 0.015851$$

**22.** $\dfrac{1}{2}V = \displaystyle\int_0^{\arcsin(c)} \pi\left(c - \sin x\right)^2 dx + \int_{\arcsin(c)}^{\pi/2} \pi\left(\sin x - c\right)^2 dx = \dfrac{2c^2\pi - 8c + \pi}{4}\pi = f(c)$

$f'(c) = \dfrac{4c\pi - 8}{4}\pi = 0 \Rightarrow c = \dfrac{2}{\pi}$

For $c = 0,\ \dfrac{1}{2}V = \dfrac{\pi^2}{4} \approx 2.4674$

For $c = 1,\ \dfrac{1}{2}V = \dfrac{\pi}{4}(3\pi - 8) \approx 1.1190$

For $c = \dfrac{2}{\pi},\ \dfrac{1}{2}V = \dfrac{\pi^2 - 8}{4} \approx 0.4674$

(a) Maximum: $c = 0$

(b) Minimum: $c = \dfrac{2}{\pi}$

# C H A P T E R   9
## Infinite Series

# C H A P T E R   9
## Infinite Series

## Section 9.1   Sequences

**1.** $a_n = 3^n$

$a_1 = 3^1 = 3$

$a_2 = 3^2 = 9$

$a_3 = 3^3 = 27$

$a_4 = 3^4 = 81$

$a_5 = 3^5 = 243$

**3.** $a_n = \left(-\frac{1}{4}\right)^n$

$a_1 = -\frac{1}{4}$

$a_2 = \left(-\frac{1}{4}\right)^2 = \frac{1}{16}$

$a_3 = \left(-\frac{1}{4}\right)^3 = -\frac{1}{64}$

$a_4 = \left(-\frac{1}{4}\right)^4 = \frac{1}{256}$

$a_5 = \left(-\frac{1}{4}\right)^5 = -\frac{1}{1024}$

**5.** $a_n = \sin \frac{n\pi}{2}$

$a_1 = \sin \frac{\pi}{2} = 1$

$a_2 = \sin \pi = 0$

$a_3 = \sin \frac{3\pi}{2} = -1$

$a_4 = \sin 2\pi = 0$

$a_5 = \sin \frac{5\pi}{2} = 1$

**7.** $a_n = \frac{(-1)^{n(n+1)/2}}{n^2}$

$a_1 = \frac{(-1)^1}{1^2} = -1$

$a_2 = \frac{(-1)^3}{2^2} = -\frac{1}{4}$

$a_3 = \frac{(-1)^6}{3^2} = \frac{1}{9}$

$a_4 = \frac{(-1)^{10}}{4^2} = \frac{1}{16}$

$a_5 = \frac{(-1)^{15}}{5^2} = -\frac{1}{25}$

**9.** $a_n = 5 - \frac{1}{n} + \frac{1}{n^2}$

$a_1 = 5 - 1 + 1 = 5$

$a_2 = 5 - \frac{1}{2} + \frac{1}{4} = \frac{19}{4}$

$a_3 = 5 - \frac{1}{3} + \frac{1}{9} = \frac{43}{9}$

$a_4 = 5 - \frac{1}{4} + \frac{1}{16} = \frac{77}{16}$

$a_5 = 5 - \frac{1}{5} + \frac{1}{25} = \frac{121}{25}$

**11.** $a_1 = 3, \ a_{k+1} = 2(a_k - 1)$

$a_2 = 2(a_1 - 1)$

$\quad = 2(3 - 1) = 4$

$a_3 = 2(a_2 - 1)$

$\quad = 2(4 - 1) = 6$

$a_4 = 2(a_3 - 1)$

$\quad = 2(6 - 1) = 10$

$a_5 = 2(a_4 - 1)$

$\quad = 2(10 - 1) = 18$

**13.** $a_1 = 32, \ a_{k+1} = \frac{1}{2}a_k$

$a_2 = \frac{1}{2}a_1 = \frac{1}{2}(32) = 16$

$a_3 = \frac{1}{2}a_2 = \frac{1}{2}(16) = 8$

$a_4 = \frac{1}{2}a_3 = \frac{1}{2}(8) = 4$

$a_5 = \frac{1}{2}a_4 = \frac{1}{2}(4) = 2$

**15.** $a_n = \frac{10}{n+1}, \ a_1 = \frac{10}{1+1} = 5, \ a_2 = \frac{10}{3}$

Matches (c)

**17.** $a_n = (-1)^n, \ a_1 = -1, \ a_2 = 1, \ a_3 = -1, \ldots$

Matches (d)

**19.** $-2, 0, \frac{2}{3}, 1, \ldots$ matches (b)

$a_n = 2 - \frac{4}{n}$

$a_1 = 2 - \frac{4}{1} = -2, \ a_2 = 2 - \frac{4}{2} = 0, \ a_3 = 2 - \frac{4}{3} = \frac{2}{3}, \ldots$

**21.** $\frac{2}{3}, \frac{4}{3}, 2, \frac{8}{3}, \ldots$ matches (a)

$$a_n = \frac{2}{3}n$$

$$a_1 = \frac{2}{3}, a_2 = \frac{4}{3}, a_3 = \frac{6}{3} = 2, \ldots$$

**23.** $a_n = 3n - 1$

$$a_5 = 3(5) - 1 = 14$$

$$a_6 = 3(6) - 1 = 17$$

Add 3 to preceding term.

**25.** $a_{n+1} = 2a_n, a_1 = 5$

$$a_5 = 2(40) = 80$$

$$a_6 = 2(80) = 160$$

Multiply the preceding term by 2.

**27.** $a_n = \dfrac{3}{(-2)^{n-1}}$

$$a_5 = \frac{3}{(-2)^4} = \frac{3}{16}$$

$$a_6 = \frac{3}{(-2)^5} = -\frac{3}{32}$$

Multiply the preceding term by $-\dfrac{1}{2}$.

**29.** $\dfrac{11!}{8!} = \dfrac{11(10)(9)8!}{8!} = 11(10)(9) = 990$

**31.** $\dfrac{(n+1)!}{n!} = \dfrac{n!(n+1)}{n!} = n+1$

**33.** $\dfrac{(2n-1)!}{(2n+1)!} = \dfrac{(2n-1)!}{(2n-1)!(2n)(2n+1)} = \dfrac{1}{2n(2n+1)}$

**35.** $\displaystyle\lim_{n\to\infty} \frac{5n^2}{n^2+2} = 5$

**37.** $\displaystyle\lim_{n\to\infty} \frac{2n}{\sqrt{n^2+1}} = \lim_{n\to\infty} \frac{2}{\sqrt{1+(1/n^2)}} = \frac{2}{1} = 2$

**39.** $\displaystyle\lim_{n\to\infty} \sin\left(\frac{1}{n}\right) = 0$

**41.**

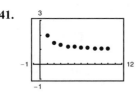

The graph seems to indicate that the sequence converges to 1. Analytically,

$$\lim_{n\to\infty} a_n = \lim_{n\to\infty} \frac{n+1}{n} = \lim_{x\to\infty} \frac{x+1}{x} = \lim_{x\to\infty} 1 = 1.$$

**43.**

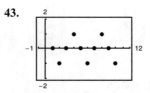

The graph seems to indicate that the sequence diverges. Analytically, the sequence is

$$\{a_n\} = \{0, -1, 0, 1, 0, -1, \ldots\}.$$

So, $\displaystyle\lim_{n\to\infty} a_n$ does not exist.

**45.** $\displaystyle\lim_{n\to\infty} \left[(0.3)^n - 1\right] = 0 - 1 = -1$, converges

**47.** $\displaystyle\lim_{n\to\infty} \frac{5}{n+2} = 0$, converges

**49.** $\displaystyle\lim_{n\to\infty} (-1)^n\left(\frac{n}{n+1}\right)$

does not exist (oscillates between $-1$ and 1), diverges.

**51.** $\displaystyle\lim_{n\to\infty} \frac{3n^2 - n + 4}{2n^2 + 1} = \frac{3}{2}$, converges

**53.** $a_n = \dfrac{1 \cdot 3 \cdot 5 \cdots (2n-1)}{(2n)^n}$

$$= \frac{1}{2n} \cdot \frac{3}{2n} \cdot \frac{5}{2n} \cdots \frac{2n-1}{2n} < \frac{1}{2n}$$

So, $\displaystyle\lim_{n\to\infty} a_n = 0$, converges.

**55.** $\displaystyle\lim_{n\to\infty} \frac{1 + (-1)^n}{n} = 0$, converges

**57.** $\displaystyle\lim_{n\to\infty} \frac{\ln(n^3)}{2n} = \lim_{n\to\infty} \frac{3}{2} \frac{\ln(n)}{n}$

$$= \lim_{n\to\infty} \frac{3}{2}\left(\frac{1}{n}\right) = 0, \text{ converges}$$

(L'Hôpital's Rule)

**59.** $\displaystyle\lim_{n\to\infty} \left(\frac{3}{4}\right)^n = 0$, converges

**61.** $\displaystyle\lim_{n\to\infty} \frac{(n+1)!}{n!} = \lim_{n\to\infty} (n+1) = \infty$, diverges

**63.** $\displaystyle\lim_{n\to\infty} \left(\frac{n-1}{n} - \frac{n}{n-1}\right) = \lim_{n\to\infty} \frac{(n-1)^2 - n^2}{n(n-1)}$

$$= \lim_{n\to\infty} \frac{1 - 2n}{n^2 - n} = 0, \text{ converges}$$

**65.** $\lim\limits_{n\to\infty} \dfrac{n^p}{e^n} = 0$, converges

$(p > 0, n \geq 2)$

**67.** $\lim\limits_{n\to\infty} 2^{1/n} = 2^0 = 1$, converges

**69.** $a_n = \left(1 + \dfrac{k}{n}\right)^n$

$\lim\limits_{n\to\infty} \left(1 + \dfrac{k}{n}\right)^n = \lim\limits_{u\to 0} \left[(1 + u)^{1/u}\right]^k = e^k$

where $u = k/n$, converges

**71.** $\lim\limits_{n\to\infty} \dfrac{\sin n}{n} = \lim\limits_{n\to\infty} (\sin n)\dfrac{1}{n} = 0$,

converges (because $(\sin n)$ is bounded)

**73.** $a_n = 3n - 2$

**75.** $a_n = n^2 - 2$

**77.** $a_n = \dfrac{n + 1}{n + 2}$

**79.** $a_n = 1 + \dfrac{1}{n} = \dfrac{n + 1}{n}$

**81.** $a_n = \dfrac{n}{(n + 1)(n + 2)}$

**83.** $a_n = \dfrac{(-1)^{n-1}}{1 \cdot 3 \cdot 5 \cdots (2n - 1)}$

$= \dfrac{(-1)^{n-1} 2^n n!}{(2n)!}$

**85.** $a_n = (2n)!, n = 1, 2, 3, \ldots$

**87.** $a_n = 4 - \dfrac{1}{n} < 4 - \dfrac{1}{n + 1} = a_{n+1}$,

Monotonic; $\left|a_n\right| < 4$, bounded

**89.** $\dfrac{n}{2^{n+2}} \overset{?}{\geq} \dfrac{n + 1}{2^{(n+1)+2}}$

$2^{n+3} n \overset{?}{\geq} 2^{n+2}(n + 1)$

$2n \overset{?}{\geq} n + 1$

$n \geq 1$

So,    $n \geq 1$

$2n \geq n + 1$

$2^{n+3} n \geq 2^{n+2}(n + 1)$

$\dfrac{n}{2^{n+2}} \geq \dfrac{n + 1}{2^{(n+1)+2}}$

$a_n \geq a_{n+1}.$

Monotonic; $\left|a_n\right| \leq \dfrac{1}{8}$, bounded

**91.** $a_n = (-1)^n\left(\dfrac{1}{n}\right)$

$a_1 = -1$

$a_2 = \dfrac{1}{2}$

$a_3 = -\dfrac{1}{3}$

Not monotonic; $\left|a_n\right| \leq 1$, bounded

**93.** $a_n = \left(\dfrac{2}{3}\right)^n > \left(\dfrac{2}{3}\right)^{n+1} = a_{n+1}$

Monotonic; $\left|a_n\right| \leq \frac{2}{3}$, bounded

**95.** $a_n = \sin\left(\dfrac{n\pi}{6}\right)$

$a_1 = 0.500$

$a_2 = 0.8660$

$a_3 = 1.000$

$a_4 = 0.8660$

Not monotonic; $\left|a_n\right| \leq 1$, bounded

**97.** $a_n = \dfrac{\cos n}{n}$

$a_1 = 0.5403$

$a_2 = -0.2081$

$a_3 = -0.3230$

$a_4 = -0.1634$

Not monotonic; $\left|a_n\right| \leq 1$, bounded

**99.** (a) $a_n = 5 + \dfrac{1}{n}$

$\left| 5 + \dfrac{1}{n} \right| \le 6 \Rightarrow \{a_n\}$, bounded

$a_n = 5 + \dfrac{1}{n} > 5 + \dfrac{1}{n+1}$

$\quad = a_{n+1} \Rightarrow \{a_n\}$, monotonic

Therefore, $\{a_n\}$ converges.

(b)

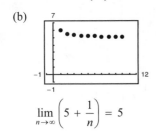

$\lim\limits_{n \to \infty} \left( 5 + \dfrac{1}{n} \right) = 5$

**101.** (a) $a_n = \dfrac{1}{3}\left( 1 - \dfrac{1}{3^n} \right)$

$\left| \dfrac{1}{3}\left( 1 - \dfrac{1}{3^n} \right) \right| < \dfrac{1}{3} \Rightarrow \{a_n\}$, bounded

$a_n = \dfrac{1}{3}\left( 1 - \dfrac{1}{3^n} \right) < \dfrac{1}{3}\left( 1 - \dfrac{1}{3^{n+1}} \right)$

$\quad = a_{n+1} \Rightarrow \{a_n\}$, monotonic

Therefore, $\{a_n\}$ converges.

(b)

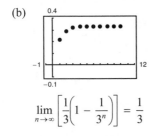

$\lim\limits_{n \to \infty} \left[ \dfrac{1}{3}\left( 1 - \dfrac{1}{3^n} \right) \right] = \dfrac{1}{3}$

**103.** $\{a_n\}$ has a limit because it is a bounded, monotonic sequence. The limit is less than or equal to 4, and greater than or equal to 2.

$2 \le \lim\limits_{n \to \infty} a_n \le 4$

**105.** $A_n = P\left( 1 + \dfrac{r}{12} \right)^n$

(a) Because $P > 0$ and $\left( 1 + \dfrac{r}{12} \right) > 1$, the sequence diverges. $\lim\limits_{n \to \infty} A_n = \infty$

(b) $P = 10{,}000$, $r = 0.055$, $A_n = 10{,}000\left( 1 + \dfrac{0.055}{12} \right)^n$

$A_0 = 10{,}000$
$A_1 = 10{,}045.83$
$A_2 = 10{,}091.88$
$A_3 = 10{,}138.13$
$A_4 = 10{,}184.60$
$A_5 = 10{,}231.28$
$A_6 = 10{,}278.17$
$A_7 = 10{,}325.28$
$A_8 = 10{,}372.60$
$A_9 = 10{,}420.14$
$A_{10} = 10{,}467.90$

**107.** No, it is not possible. See the "Definition of the Limit of a sequence". The number $L$ is unique.

**109.** The graph on the left represents a sequence with alternating signs because the terms alternate from being above the $x$-axis to being below the $x$-axis.

**111.** (a) $A_n = (0.8)^n\, 4{,}500{,}000{,}000$

(b) $A_1 = \$3{,}600{,}000{,}000$
$A_2 = \$2{,}880{,}000{,}000$
$A_3 = \$2{,}304{,}000{,}000$
$A_4 = \$1{,}843{,}200{,}000$

(c) $\lim\limits_{n \to \infty} A_n = \lim\limits_{n \to \infty} (0.8)^n (4.5) = 0$, converges

**113.** (a) $a_n = -5.364n^2 + 608.04n + 4998.3$

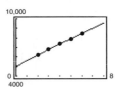

(b) For 2012, $n = 12$: $a_{12} \approx \$11{,}522.4$ billion

**115.** $a_n = \dfrac{10^n}{n!}$

(a) $a_9 = a_{10} = \dfrac{10^9}{9!} = \dfrac{1{,}000{,}000{,}000}{362{,}880} = \dfrac{1{,}562{,}500}{567}$

(b) Decreasing

(c) Factorials increase more rapidly than exponentials.

**117.** $a_n = \sqrt[n]{n} = n^{1/n}$

$a_1 = 1^{1/1} = 1$

$a_2 = \sqrt{2} \approx 1.4142$

$a_3 = \sqrt[3]{3} \approx 1.4422$

$a_4 = \sqrt[4]{4} \approx 1.4142$

$a_5 = \sqrt[5]{5} \approx 1.3797$

$a_6 = \sqrt[6]{6} \approx 1.3480$

Let $y = \lim_{n \to \infty} n^{1/n}$.

$\ln y = \lim_{n \to \infty} \left( \frac{1}{n} \ln n \right) = \lim_{n \to \infty} \frac{\ln n}{n} = \lim_{n \to \infty} \frac{1/n}{n} = 0$

Because $\ln y = 0$, you have $y = e^0 = 1$. Therefore, $\lim_{n \to \infty} \sqrt[n]{n} = 1$.

**119.** True

**121.** True

**123.** True

**125.** $a_{n+2} = a_n + a_{n+1}$

(a)
| | |
|---|---|
| $a_1 = 1$ | $a_7 = 8 + 5 = 13$ |
| $a_2 = 1$ | $a_8 = 13 + 8 = 21$ |
| $a_3 = 1 + 1 = 2$ | $a_9 = 21 + 13 = 34$ |
| $a_4 = 2 + 1 = 3$ | $a_{10} = 34 + 21 = 55$ |
| $a_5 = 3 + 2 = 5$ | $a_{11} = 55 + 34 = 89$ |
| $a_6 = 5 + 3 = 8$ | $a_{12} = 89 + 55 = 144$ |

(b) $b_n = \dfrac{a_{n+1}}{a_n}, n \geq 1$

| | |
|---|---|
| $b_1 = \dfrac{1}{1} = 1$ | $b_6 = \dfrac{13}{8} = 1.625$ |
| $b_2 = \dfrac{2}{1} = 2$ | $b_7 = \dfrac{21}{13} \approx 1.6154$ |
| $b_3 = \dfrac{3}{2} = 1.5$ | $b_8 = \dfrac{34}{21} \approx 1.6190$ |
| $b_4 = \dfrac{5}{3} \approx 1.6667$ | $b_9 = \dfrac{55}{34} \approx 1.6176$ |
| $b_5 = \dfrac{8}{5} = 1.6$ | $b_{10} = \dfrac{89}{55} \approx 1.6182$ |

(c) $1 + \dfrac{1}{b_{n-1}} = 1 + \dfrac{1}{a_n / a_{n-1}}$

$= 1 + \dfrac{a_{n-1}}{a_n} = \dfrac{a_n + a_{n-1}}{a_n} = \dfrac{a_{n+1}}{a_n} = b_n$

(d) If $\lim_{n \to \infty} b_n = \rho$, then $\lim_{n \to \infty} \left( 1 + \dfrac{1}{b_{n-1}} \right) = \rho$.

Because $\lim_{n \to \infty} b_n = \lim_{n \to \infty} b_{n-1}$, you have

$1 + (1/\rho) = \rho$.

$\rho + 1 = \rho^2$

$0 = \rho^2 - \rho - 1$

$\rho = \dfrac{1 \pm \sqrt{1 + 4}}{2} = \dfrac{1 \pm \sqrt{5}}{2}$

Because $a_n$, and therefore $b_n$, is positive,

$\rho = \dfrac{1 + \sqrt{5}}{2} \approx 1.6180$.

**127.** (a) $a_1 = \sqrt{2} \approx 1.4142$

$a_2 = \sqrt{2 + \sqrt{2}} \approx 1.8478$

$a_3 = \sqrt{2 + \sqrt{2 + \sqrt{2}}} \approx 1.9616$

$a_4 = \sqrt{2 + \sqrt{2 + \sqrt{2 + \sqrt{2}}}} \approx 1.9904$

$a_5 = \sqrt{2 + \sqrt{2 + \sqrt{2 + \sqrt{2 + \sqrt{2}}}}} \approx 1.9976$

(b) $a_n = \sqrt{2 + a_{n-1}}, \qquad n \geq 2, a_1 = \sqrt{2}$

(c) First use mathematical induction to show that $a_n \le 2$; clearly $a_1 \le 2$. So assume $a_k \le 2$. Then

$$a_k + 2 \le 4$$
$$\sqrt{a_k + 2} \le 2$$
$$a_{k+1} \le 2.$$

Now show that $\{a_n\}$ is an increasing sequence. Because $a_n \ge 0$ and $a_n \le 2$,

$$(a_n - 2)(a_n + 1) \le 0$$
$$a_n^2 - a_n - 2 \le 0$$
$$a_n^2 \le a_n + 2$$
$$a_n \le \sqrt{a_n + 2}$$
$$a_n \le a_{n+1}.$$

Because $\{a_n\}$ is a bounding increasing sequence, it converges to some number $L$, by Theorem 9.5.

$$\lim_{n \to \infty} a_n = L \Rightarrow \sqrt{2 + L} = L \Rightarrow 2 + L = L^2 \Rightarrow L^2 - L - 2 = 0$$
$$\Rightarrow (L - 2)(L + 1) = 0 \Rightarrow L = 2 \quad (L \ne -1)$$

**129.** (a) Use mathematical induction to show that $a_n \le \dfrac{1 + \sqrt{1 + 4k}}{2}$.

[Note that if $k = 2$, and $a_n \le 3$, and if $k = 6$, then $a_n \le 3$.] Clearly, $a_1 = \sqrt{k} \le \dfrac{\sqrt{1 + 4k}}{2} \le \dfrac{1 + \sqrt{1 + 4k}}{2}$.

Before proceeding to the induction step, note that

$$2 + 2\sqrt{1 + 4k} + 4k = 2 + 2\sqrt{1 + 4k} + 4k$$
$$\frac{1 + \sqrt{1 + 4k}}{2} + k = \frac{1 + 2\sqrt{1 + 4k} + 1 + 4k}{4}$$
$$\frac{1 + \sqrt{1 + 4k}}{2} + k = \left[\frac{1 + \sqrt{1 + 4k}}{2}\right]^2$$
$$\sqrt{\frac{1 + \sqrt{1 + 4k}}{2} + k} = \frac{1 + \sqrt{1 + 4k}}{2}.$$

So assume $a_n \le \dfrac{1 + \sqrt{1 + 4k}}{2}$. Then

$$a_n + k \le \frac{1 + \sqrt{1 + 4k}}{2} + k$$
$$\sqrt{a_n + k} \le \sqrt{\frac{1 + \sqrt{1 + 4k}}{2} + k}$$
$$a_{n+1} \le \frac{1 + \sqrt{1 + 4k}}{2}.$$

$\{a_n\}$ is increasing because

$$\left(a_n - \frac{1 + \sqrt{1 + 4k}}{2}\right)\left(a_n - \frac{1 - \sqrt{1 + 4k}}{2}\right) \le 0$$
$$a_n^2 - a_n - k \le 0$$
$$a_n^2 \le a_n + k$$
$$a_n \le \sqrt{a_n + k}$$
$$a_n \le a_{n+1}.$$

(b) Because $\{a_n\}$ is bounded and increasing, it has a limit $L$.

(c) $\lim\limits_{n \to \infty} a_n = L$ implies that

$$L = \sqrt{k + L} \;\Rightarrow\; L^2 = k + L$$
$$\Rightarrow\; L^2 - L - k = 0$$
$$\Rightarrow\; L = \frac{1 \pm \sqrt{1 + 4k}}{2}.$$

Because $L > 0$, $L = \dfrac{1 + \sqrt{1 + 4k}}{2}$.

**131.** (a) $f(x) = \sin x,\; a_n = n \sin\dfrac{1}{n}$

$f'(x) = \cos x,\; f'(0) = 1$

$$\lim\limits_{n \to \infty} a_n = \lim\limits_{n \to \infty} n \sin\frac{1}{n} = \lim\limits_{n \to \infty} \frac{\sin(1/n)}{(1/n)} = 1 = f'(0)$$

(b) $f'(0) = \lim\limits_{h \to 0^+} \dfrac{f(0 + h) - f(0)}{h} = \lim\limits_{h \to 0^+} \dfrac{f(h)}{h} = \lim\limits_{n \to \infty} \dfrac{f(1/n)}{(1/n)} = \lim\limits_{n \to \infty} nf\left(\dfrac{1}{n}\right) = \lim\limits_{n \to \infty} a_n$

**133.** (a)

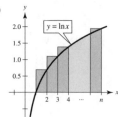

$$\int_1^n \ln x \, dx < \ln 2 + \ln 3 + \cdots + \ln n = \ln(1 \cdot 2 \cdot 3 \cdots n) = \ln(n!)$$

(b)

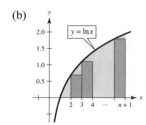

$$\int_1^{n+1} \ln x \, dx > \ln 2 + \ln 3 + \cdots + \ln n = \ln(n!)$$

(c) $\displaystyle\int \ln x \, dx = x \ln x - x + C$

$$\int_1^n \ln x \, dx = n \ln n - n + 1 = \ln n^n - n + 1$$

From part (a): $\ln n^n - n + 1 < \ln(n!)$

$$e^{\ln n^n - n + 1} < n!$$

$$\frac{n^n}{e^{n-1}} < n!$$

$$\int_1^{n+1} \ln x \, dx = (n + 1)\ln(n + 1) - (n + 1) + 1 = \ln(n + 1)^{n+1} - n$$

From part (b): $\ln(n + 1)^{n+1} - n > \ln(n!)$

$$e^{\ln(n+1)^{n+1} - n} > n!$$

$$\frac{(n + 1)^{n+1}}{e^n} > n!$$

(d) $\dfrac{n^n}{e^{n-1}} < n! < \dfrac{(n+1)^{n+1}}{e^n}$

$\dfrac{n}{e^{1-(1/n)}} < \sqrt[n]{n!} < \dfrac{(n+1)^{(n+1)/n}}{e}$

$\dfrac{1}{e^{1-(1/n)}} < \dfrac{\sqrt[n]{n!}}{n} < \dfrac{(n+1)^{1+(1/n)}}{ne}$

$\displaystyle \lim_{n\to\infty} \dfrac{1}{e^{1-(1/n)}} = \dfrac{1}{e}$

$\displaystyle \lim_{n\to\infty} \dfrac{(n+1)^{1+(1/n)}}{ne} = \lim_{n\to\infty} \dfrac{(n+1)}{n}\dfrac{(n+1)^{1/n}}{e} = (1)\dfrac{1}{e} = \dfrac{1}{e}$

By the Squeeze Theorem, $\displaystyle \lim_{n\to\infty} \dfrac{\sqrt[n]{n!}}{n} = \dfrac{1}{e}$.

(e) $n = 20$: $\dfrac{\sqrt[20]{20!}}{20} \approx 0.4152$

$n = 50$: $\dfrac{\sqrt[50]{50!}}{50} \approx 0.3897$

$n = 100$: $\dfrac{\sqrt[100]{100!}}{100} \approx 0.3799$

$\dfrac{1}{e} \approx 0.3679$

**135.** For a given $\varepsilon > 0$, you must find $M > 0$ such that

$$|a_n - L| = \left|\dfrac{1}{n^3}\right| < \varepsilon$$

whenever $n > M$. That is,

$$n^3 > \dfrac{1}{\varepsilon} \text{ or } n > \left(\dfrac{1}{\varepsilon}\right)^{1/3}.$$

So, let $\varepsilon > 0$ be given. Let $M$ be an integer satisfying $M > (1/\varepsilon)^{1/3}$. For $n > M$, you have

$$n > \left(\dfrac{1}{\varepsilon}\right)^{1/3}$$

$$n^3 > \dfrac{1}{\varepsilon}$$

$$\varepsilon > \dfrac{1}{n^3} \Rightarrow \left|\dfrac{1}{n^3} - 0\right| < \varepsilon.$$

So, $\displaystyle \lim_{n\to\infty} \dfrac{1}{n^3} = 0$.

**137.** Answers will vary. *Sample answer*:

$\{a_n\} = \{(-1)^n\} = \{-1, 1, -1, 1, \ldots\}$ diverges

$\{a_{2n}\} = \{(-1)^{2n}\} = \{1, 1, 1, 1, \ldots\}$ converges

**139.** If $\{a_n\}$ is bounded, monotonic and nonincreasing, then $a_1 \geq a_2 \geq a_3 \geq \cdots \geq a_n \geq \cdots$. Then

$-a_1 \leq -a_2 \leq -a_3 \leq \cdots \leq -a_n \leq \cdots$ is a bounded, monotonic, nondecreasing sequence which converges by the first half of the theorem. Because $\{-a_n\}$ converges, then so does $\{a_n\}$.

**141.**  $T_n = n! + 2^n$

Use mathematical induction to verify the formula.

$T_0 = 1 + 1 = 2$

$T_1 = 1 + 2 = 3$

$T_2 = 2 + 4 = 6$

Assume $T_k = k! + 2^k$. Then

$T_{k+1} = (k + 1 + 4)T_k - 4(k + 1)T_{k-1} + \left(4(k + 1) - 8\right)T_{k-2}$

$= (k + 5)\left[k! + 2^k\right] - 4(k + 1)\left((k - 1)! + 2^{k-1}\right) + (4k - 4)\left((k - 2)! + 2^{k-2}\right)$

$= \left[(k + 5)(k)(k - 1) - 4(k + 1)(k - 1) + 4(k - 1)\right](k - 2)! + \left[(k + 5)4 - 8(k + 1) + 4(k - 1)\right]2^{k-2}$

$= \left[k^2 + 5k - 4k - 4 + 4\right](k - 1)! + 8 \cdot 2^{k-2}$

$= (k + 1)! + 2^{k+1}.$

By mathematical induction, the formula is valid for all $n$.

# Section 9.2    Series and Convergence

**1.**  $S_1 = 1$

$S_2 = 1 + \frac{1}{4} = 1.2500$

$S_3 = 1 + \frac{1}{4} + \frac{1}{9} \approx 1.3611$

$S_4 = 1 + \frac{1}{4} + \frac{1}{9} + \frac{1}{16} \approx 1.4236$

$S_5 = 1 + \frac{1}{4} + \frac{1}{9} + \frac{1}{16} + \frac{1}{25} \approx 1.4636$

**3.**  $S_1 = 3$

$S_2 = 3 - \frac{9}{2} = -1.5$

$S_3 = 3 - \frac{9}{2} + \frac{27}{4} = 5.25$

$S_4 = 3 - \frac{9}{2} + \frac{27}{4} - \frac{81}{8} = -4.875$

$S_5 = 3 - \frac{9}{2} + \frac{27}{4} - \frac{81}{8} + \frac{243}{16} = 10.3125$

**5.**  $S_1 = 3$

$S_2 = 3 + \frac{3}{2} = 4.5$

$S_3 = 3 + \frac{3}{2} + \frac{3}{4} = 5.250$

$S_4 = 3 + \frac{3}{2} + \frac{3}{4} + \frac{3}{8} = 5.625$

$S_5 = 3 + \frac{3}{2} + \frac{3}{4} + \frac{3}{8} + \frac{3}{16} = 5.8125$

**7.**  $a_n = \dfrac{n + 1}{n}$

$\{a_n\} = \left\{\dfrac{2}{1}, \dfrac{3}{2}, \dfrac{4}{3}, \ldots\right\}$ converges to 1

$\displaystyle\sum_{n=1}^{\infty} a_n = \frac{2}{1} + \frac{3}{2} + \frac{4}{3} + \cdots$ diverges

**9.**  $\displaystyle\sum_{n=0}^{\infty} \left(\frac{7}{6}\right)^n$

Geometric series

$r = \frac{7}{6} > 1$

Diverges by Theorem 9.6

**11.**  $\displaystyle\sum_{n=0}^{\infty} 1000(1.055)^n$

Geometric series

$r = 1.055 > 1$

Diverges by Theorem 9.6

**13.**  $\displaystyle\sum_{n=1}^{\infty} \frac{n}{n + 1}$

$\displaystyle\lim_{n \to \infty} \frac{n}{n + 1} = 1 \neq 0$

Diverges by Theorem 9.9

**15.**  $\displaystyle\sum_{n=1}^{\infty} \frac{n^2}{n^2 + 1}$

$\displaystyle\lim_{n \to \infty} \frac{n^2}{n^2 + 1} = 1 \neq 0$

Diverges by Theorem 9.9

**17.**  $\displaystyle\sum_{n=1}^{\infty} \frac{2^n + 1}{2^{n+1}}$

$\displaystyle\lim_{n \to \infty} \frac{2^n + 1}{2^{n+1}} = \lim_{n \to \infty} \frac{1 + 2^{-n}}{2} = \frac{1}{2} \neq 0$

Diverges by Theorem 9.9

**19.** $\sum_{n=0}^{\infty} \frac{9}{4}\left(\frac{1}{4}\right)^n = \frac{9}{4}\left[1 + \frac{1}{4} + \frac{1}{16} + \cdots\right]$

$S_0 = \frac{9}{4}, S_1 = \frac{9}{4} \cdot \frac{5}{4} = \frac{45}{16}, S_2 = \frac{9}{4} \cdot \frac{21}{16} \approx 2.95, \ldots$

Matches graph (c). Analytically, the series is geometric:

$\sum_{n=0}^{\infty} \left(\frac{9}{4}\right)\left(\frac{1}{4}\right)^n = \frac{9/4}{1-1/4} = \frac{9/4}{3/4} = 3$

**21.** $\sum_{n=0}^{\infty} \frac{15}{4}\left(-\frac{1}{4}\right)^n = \frac{15}{4}\left[1 - \frac{1}{4} + \frac{1}{16} - \cdots\right]$

$S_0 = \frac{15}{4}, S_1 = \frac{45}{16}, S_2 \approx 3.05, \ldots$

Matches graph (a). Analytically, the series is geometric:

$\sum_{n=0}^{\infty} \frac{15}{4}\left(-\frac{1}{4}\right)^n = \frac{15/4}{1-(-1/4)} = \frac{15/4}{5/4} = 3$

**23.** $\sum_{n=0}^{\infty} \frac{17}{3}\left(-\frac{1}{2}\right)^n = \frac{17}{3}\left[1 - \frac{1}{2} + \frac{1}{4} - \cdots\right]$

$S_0 = \frac{17}{3}, S_1 = \frac{17}{6}, \ldots$

Matches (f ). The series is geometric:

$\sum_{n=0}^{\infty} \frac{17}{3}\left(-\frac{1}{2}\right)^n = \frac{17}{3} \frac{1}{1-(-1/2)} = \frac{17}{3}\frac{2}{3} = \frac{34}{9}$

**25.** $\sum_{n=0}^{\infty} \left(\frac{5}{6}\right)^n$

Geometric series with $r = \frac{5}{6} < 1$

Converges by Theorem 9.6

**27.** $\sum_{n=0}^{\infty} (0.9)^n$

Geometric series with $r = 0.9 < 1$

Converges by Theorem 9.6

**29.** $\sum_{n=1}^{\infty} \frac{1}{n(n+1)} = \sum_{n=1}^{\infty} \left(\frac{1}{n} - \frac{1}{n+1}\right) = \left(1 - \frac{1}{2}\right) + \left(\frac{1}{2} - \frac{1}{3}\right) + \left(\frac{1}{3} - \frac{1}{4}\right) + \left(\frac{1}{4} - \frac{1}{5}\right) + \cdots, \quad S_n = 1 - \frac{1}{n+1}$

$\sum_{n=1}^{\infty} \frac{1}{n(n+1)} = \lim_{n\to\infty} S_n = \lim_{n\to\infty} \left(1 - \frac{1}{n+1}\right) = 1$

**31. (a)** $\sum_{n=1}^{\infty} \frac{6}{n(n+3)} = 2\sum_{n=1}^{\infty} \left(\frac{1}{n} - \frac{1}{n+3}\right) = 2\left[\left(1 - \frac{1}{4}\right) + \left(\frac{1}{2} - \frac{1}{5}\right) + \left(\frac{1}{3} - \frac{1}{6}\right) + \left(\frac{1}{4} - \frac{1}{7}\right) + \cdots\right]$

$\left(S_n = 2\left[1 + \frac{1}{2} + \frac{1}{3} - \left(\frac{1}{n+1} + \frac{1}{n+2} + \frac{1}{n+3}\right)\right]\right) = 2\left(1 + \frac{1}{2} + \frac{1}{3}\right) = \frac{11}{3} \approx 3.667$

**(b)**

| $n$ | 5 | 10 | 20 | 50 | 100 |
|---|---|---|---|---|---|
| $S_n$ | 2.7976 | 3.1643 | 3.3936 | 3.5513 | 3.6078 |

**(c)**

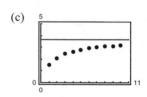

**(d)** The terms of the series decrease in magnitude slowly. So, the sequence of partial sums approaches the sum slowly.

**33. (a)** $\sum_{n=1}^{\infty} 2(0.9)^{n-1} = \sum_{n=0}^{\infty} 2(0.9)^n = \frac{2}{1-0.9} = 20$

**(b)**

| $n$ | 5 | 10 | 20 | 50 | 100 |
|---|---|---|---|---|---|
| $S_n$ | 8.1902 | 13.0264 | 17.5685 | 19.8969 | 19.9995 |

**(c)**

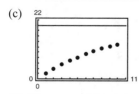

**(d)** The terms of the series decrease in magnitude slowly. So, the sequence of partial sums approaches the sum slowly.

**35.** (a) $\displaystyle\sum_{n=1}^{\infty} 10(0.25)^{n-1} = \frac{10}{1 - 0.25} = \frac{40}{3} \approx 13.3333$

(b)

| $n$ | 5 | 10 | 20 | 50 | 100 |
|---|---|---|---|---|---|
| $S_n$ | 13.3203 | 13.3333 | 13.3333 | 13.3333 | 13.3333 |

(c)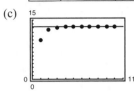

(d) The terms of the series decrease in magnitude rapidly. So, the sequence of partial sums approaches the sum rapidly.

**37.** $\displaystyle\sum_{n=0}^{\infty}\left(\frac{1}{2}\right)^{n} = \frac{1}{1 - (1/2)} = 2$

**39.** $\displaystyle\sum_{n=0}^{\infty}\left(-\frac{1}{3}\right)^{n} = \frac{1}{1 - (1/3)} = \frac{3}{4}$

**41.** $\displaystyle\sum_{n=2}^{\infty} \frac{1}{n^2 - 1} = \sum_{n=2}^{\infty}\left(\frac{1/2}{n-1} - \frac{1/2}{n+1}\right) = \frac{1}{2}\sum_{n=2}^{\infty}\left(\frac{1}{n-1} - \frac{1}{n+1}\right)$

$S_n = \frac{1}{2}\left[\left(1 - \frac{1}{3}\right) + \left(\frac{1}{2} - \frac{1}{4}\right) + \left(\frac{1}{3} - \frac{1}{5}\right) + \cdots + \left(\frac{1}{n-2} - \frac{1}{n}\right) + \left(\frac{1}{n-1} - \frac{1}{n+1}\right)\right] = \frac{1}{2}\left(1 + \frac{1}{2} - \frac{1}{n} - \frac{1}{n+1}\right)$

$\displaystyle\sum_{n=2}^{\infty} \frac{1}{n^2 - 1} = \lim_{n\to\infty} S_n = \lim_{n\to\infty} \frac{1}{2}\left(1 + \frac{1}{2} - \frac{1}{n} - \frac{1}{n+1}\right) = \frac{3}{4}$

**43.** $\displaystyle\sum_{n=1}^{\infty} \frac{8}{(n+1)(n+2)} = 8\sum_{n=1}^{\infty}\left(\frac{1}{n+1} - \frac{1}{n+2}\right)$

$S_n = 8\left[\left(\frac{1}{2} - \frac{1}{3}\right) + \left(\frac{1}{3} - \frac{1}{4}\right) + \left(\frac{1}{4} - \frac{1}{5}\right) + \cdots + \left(\frac{1}{n+1} - \frac{1}{n+2}\right)\right] = 8\left(\frac{1}{2} - \frac{1}{n+2}\right)$

$\displaystyle\sum_{n=1}^{\infty} \frac{8}{(n+1)(n+2)} = \lim_{n\to\infty} S_n = \lim_{n\to\infty} 8\left(\frac{1}{2} - \frac{1}{n+2}\right) = 4$

**45.** $\displaystyle\sum_{n=0}^{\infty}\left(\frac{1}{10}\right)^{n} = \frac{1}{1 - (1/10)} = \frac{10}{9}$

**47.** $\displaystyle\sum_{n=0}^{\infty} 3\left(-\frac{1}{3}\right)^{n} = \frac{3}{1 - (-1/3)} = \frac{9}{4}$

**49.** $\displaystyle\sum_{n=0}^{\infty}\left(\frac{1}{2^n} - \frac{1}{3^n}\right) = \sum_{n=0}^{\infty}\left(\frac{1}{2}\right)^{n} - \sum_{n=0}^{\infty}\left(\frac{1}{3}\right)^{n}$

$\displaystyle = \frac{1}{1 - (1/2)} - \frac{1}{1 - (1/3)} = 2 - \frac{3}{2} = \frac{1}{2}$

**51.** Note that $\sin(1) \approx 0.8415 < 1$. The series $\displaystyle\sum_{n=1}^{\infty}\left[\sin(1)\right]^{n}$ is geometric with $r = \sin(1) < 1$. So,

$\displaystyle\sum_{n=1}^{\infty}\left[\sin(1)\right]^{n} = \sin(1)\sum_{n=0}^{\infty}\left[\sin(1)\right]^{n} = \frac{\sin(1)}{1 - \sin(1)} \approx 5.3080.$

**53.** (a) $0.\overline{4} = \sum\limits_{n=0}^{\infty} \dfrac{4}{10}\left(\dfrac{1}{10}\right)^{n}$

(b) Geometric series with $a = \dfrac{4}{10}$ and $r = \dfrac{1}{10}$

$$S = \dfrac{a}{1-r} = \dfrac{4/10}{1-(1/10)} = \dfrac{4}{9}$$

**55.** (a) $0.\overline{81} = \sum\limits_{n=0}^{\infty} \dfrac{81}{100}\left(\dfrac{1}{100}\right)^{n}$

(b) Geometric series with $a = \dfrac{81}{100}$ and $r = \dfrac{1}{100}$

$$S = \dfrac{a}{1-r} = \dfrac{81/100}{1-(1/100)} = \dfrac{81}{99} = \dfrac{9}{11}$$

**57.** (a) $0.0\overline{75} = \sum\limits_{n=0}^{\infty} \dfrac{3}{40}\left(\dfrac{1}{100}\right)^{n}$

(b) Geometric series with $a = \dfrac{3}{40}$ and $r = \dfrac{1}{100}$

$$S = \dfrac{a}{1-r} = \dfrac{3/40}{99/100} = \dfrac{5}{66}$$

**59.** $\sum\limits_{n=0}^{\infty} (1.075)^{n}$

Geometric series with $r = 1.075$

Diverges by Theorem 9.6

**61.** $\sum\limits_{n=1}^{\infty} \dfrac{n+10}{10n+1}$

$$\lim_{n\to\infty} \dfrac{n+10}{10n+1} = \dfrac{1}{10} \neq 0$$

Diverges by Theorem 9.9

**63.** $\sum\limits_{n=1}^{\infty} \left(\dfrac{1}{n} - \dfrac{1}{n+2}\right)$

$$S_n = \left(1 - \dfrac{1}{3}\right) + \left(\dfrac{1}{2} - \dfrac{1}{4}\right) + \left(\dfrac{1}{3} - \dfrac{1}{5}\right) + \cdots + \left(\dfrac{1}{n-1} - \dfrac{1}{n+1}\right) + \left(\dfrac{1}{n} - \dfrac{1}{n+2}\right) = 1 + \dfrac{1}{2} - \dfrac{1}{n+1} - \dfrac{1}{n+2}$$

$$\sum\limits_{n=1}^{\infty} \left(\dfrac{1}{n} - \dfrac{1}{n+2}\right) = \lim_{n\to\infty} S_n = \lim_{n\to\infty}\left(1 + \dfrac{1}{2} - \dfrac{1}{n+1} - \dfrac{1}{n+2}\right) = \dfrac{3}{2}, \text{ converges}$$

**65.** $\sum\limits_{n=1}^{\infty} \dfrac{1}{n(n+3)} = \dfrac{1}{3}\sum\limits_{n=1}^{\infty}\left(\dfrac{1}{n} - \dfrac{1}{n+3}\right)$

$$S_n = \dfrac{1}{3}\left[\left(\dfrac{1}{1} - \dfrac{1}{4}\right) + \left(\dfrac{1}{2} - \dfrac{1}{5}\right) + \left(\dfrac{1}{3} - \dfrac{1}{6}\right) + \left(\dfrac{1}{4} - \dfrac{1}{7}\right) + \cdots + \left(\dfrac{1}{n-2} - \dfrac{1}{n+1}\right) - \left(\dfrac{1}{n-1} - \dfrac{1}{n+2}\right) - \left(\dfrac{1}{n} - \dfrac{1}{n+3}\right)\right]$$

$$= \dfrac{1}{3}\left(1 + \dfrac{1}{2} + \dfrac{1}{3} - \dfrac{1}{n+1} - \dfrac{1}{n+2} - \dfrac{1}{n+3}\right)$$

$$\sum\limits_{n=1}^{\infty} \dfrac{1}{n(n+3)} = \lim_{n\to\infty} S_n = \lim_{n\to\infty} \dfrac{1}{3}\left(1 + \dfrac{1}{2} + \dfrac{1}{3} - \dfrac{1}{n+1} - \dfrac{1}{n+2} - \dfrac{1}{n+3}\right) = \dfrac{1}{3}\left(\dfrac{11}{6}\right) = \dfrac{11}{8}, \text{ converges}$$

**67.** $\sum\limits_{n=1}^{\infty} \dfrac{3n-1}{2n+1}$

$$\lim_{n\to\infty} \dfrac{3n-1}{2n+1} = \dfrac{3}{2} \neq 0$$

Diverges by Theorem 9.9

**69.** $\sum\limits_{n=0}^{\infty} \dfrac{4}{2^n} = 4\sum\limits_{n=0}^{\infty}\left(\dfrac{1}{2}\right)^{n}$

Geometric series with $r = \dfrac{1}{2}$

Converges by Theorem 9.6

**71.** Because $n > \ln(n)$, the terms $a_n = \dfrac{n}{\ln(n)}$ do not approach 0 as $n \to \infty$. So, the series $\sum\limits_{n=2}^{\infty} \dfrac{n}{\ln(n)}$ diverges.

**73.** For $k \neq 0$,

$$\lim_{n\to\infty}\left(1 + \dfrac{k}{n}\right)^{n} = \lim_{n\to\infty}\left[\left(1 + \dfrac{k}{n}\right)^{n/k}\right]^{k}$$

$$= e^{k} \neq 0.$$

For $k = 0, \lim_{n\to\infty} (1+0)^{n} = 1 \neq 0.$

So, $\sum\limits_{n=1}^{\infty}\left[1 + \dfrac{k}{n}\right]^{n}$ diverges.

**75.** $\lim\limits_{n \to \infty} \arctan n = \dfrac{\pi}{2} \neq 0$

So, $\sum\limits_{n=1}^{\infty} \arctan n$ diverges.

**77.** See definitions on page 608.

**79.** The series given by

$$\sum_{n=0}^{\infty} ar^n = a + ar + ar^2 + \cdots + ar^n + \cdots, \; a \neq 0$$

is a geometric series with ratio $r$. When $0 < |r| < 1$, the series converges to $a/(1-r)$. The series diverges if $|r| \geq 1$.

**81.** (a) $\sum\limits_{n=1}^{\infty} a_n = a_1 + a_2 + a_3 + \cdots$

(b) $\sum\limits_{k=1}^{\infty} a_k = a_1 + a_2 + a_3 + \cdots$

These are the same. The third series is different, unless $a_1 = a_2 = \cdots = a$ is constant.

(c) $\sum\limits_{n=1}^{\infty} a_k = a_k + a_k + \cdots$

**83.** $\sum\limits_{n=1}^{\infty} \dfrac{x^n}{2^n} = \sum\limits_{n=1}^{\infty} \left(\dfrac{x}{2}\right)^n = \dfrac{x}{2}\sum\limits_{n=0}^{\infty} \left(\dfrac{x}{2}\right)^n$

Geometric series: converges for $\left|\dfrac{x}{2}\right| < 1$ or $|x| < 2$

$f(x) = \dfrac{x}{2}\sum\limits_{n=0}^{\infty} \left(\dfrac{x}{2}\right)^n$

$\qquad = \dfrac{x}{2}\dfrac{1}{1-(x/2)} = \dfrac{x}{2}\dfrac{2}{2-x} = \dfrac{x}{2-x}, \quad |x| < 2$

**85.** $\sum\limits_{n=1}^{\infty} (x-1)^n = (x-1)\sum\limits_{n=0}^{\infty} (x-1)^n$

Geometric series: converges for
$|x-1| < 1 \Rightarrow 0 < x < 2$

$f(x) = (x-1)\sum\limits_{n=0}^{\infty} (x-1)^n$

$\qquad = (x-1)\dfrac{1}{1-(x-1)} = \dfrac{x-1}{2-x}, \quad 0 < x < 2$

**87.** $\sum\limits_{n=0}^{\infty} (-1)^n x^n = \sum\limits_{n=0}^{\infty} (-x)^n$

Geometric series: converges for
$|-x| < 1 \Rightarrow |x| < 1 \Rightarrow -1 < x < 1$

$f(x) = \sum\limits_{n=0}^{\infty} (-x)^n = \dfrac{1}{1+x}, \quad -1 < x < 1$

**89.** $\sum\limits_{n=0}^{\infty} \left(\dfrac{1}{x}\right)^n$

Geometric series: converges if $\left|\dfrac{1}{x}\right| < 1$

$\Rightarrow |x| > 1 \Rightarrow x < -1$ or $x > 1$

$f(x) = \sum\limits_{n=0}^{\infty} \left(\dfrac{1}{x}\right)^n = \dfrac{1}{1-(1/x)} = \dfrac{x}{x-1}, \; x > 1$ or $x < -1$

**91.** $\sum\limits_{n=2}^{\infty} (1+c)^{-n} = \sum\limits_{n=0}^{\infty} (1+c)^{-n-2} = \dfrac{1}{(1+c)^2}\sum\limits_{n=0}^{\infty} \left(\dfrac{1}{1+c}\right)^n$

For convergence, $\left|\dfrac{1}{1+c}\right| < 1 \Rightarrow 1 < |1+c|$

$\Rightarrow c > 0$ or $c < -2$

$2 = \dfrac{1}{(1+c)^2}\sum\limits_{n=0}^{\infty} \left(\dfrac{1}{1+c}\right)^n$

$2(1+c)^2 = \dfrac{1}{1-\left(\dfrac{1}{1+c}\right)} = \dfrac{1+c}{c}$

$2(1+c) = \dfrac{1}{c}$

$2c^2 + 2c - 1 = 0$

$c = \dfrac{-2 + \sqrt{4+8}}{4} = \dfrac{-1 \pm \sqrt{3}}{2}$

Because $-2 < \dfrac{-1-\sqrt{3}}{2} < 0$, the answer is

$c = \dfrac{-1+\sqrt{3}}{2}$

**93.** Neither statement is true. The formula

$$\dfrac{1}{1-x} = 1 + x + x^2 + \cdots$$

holds for $-1 < x < 1$.

**95.** (a) $x$ is the common ratio.

(b) $1 + x + x^2 + \cdots = \sum\limits_{n=0}^{\infty} x^n = \dfrac{1}{1-x}, \quad |x| < 1$

(c) $y_1 = \dfrac{1}{1-x}$

$y_2 = S_3 = 1 + x + x^2$

$y_3 = S_5 = 1 + x + x^2 + x^3 + x^4$

Answers will vary.

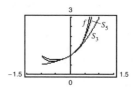

**97.** $f(x) = 3\left[\dfrac{1 - 0.5^x}{1 - 0.5}\right]$

Horizontal asymptote: $y = 6$

$\displaystyle\sum_{n=0}^{\infty} 3\left(\frac{1}{2}\right)^n$

$S = \dfrac{3}{1 - (1/2)} = 6$

The horizontal asymptote is the sum of the series.

$f(n)$ is the $n$th partial sum.

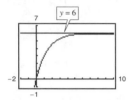

**99.** $\dfrac{1}{n(n + 1)} < 0.0001$

$10{,}000 < n^2 + n$

$0 < n^2 + n - 10{,}000$

$n = \dfrac{-1 \pm \sqrt{1^2 - 4(1)(-10{,}000)}}{2}$

Choosing the positive value for $n$ you have
$n \approx 99.5012$. The first *term* that is less than 0.0001 is
$n = 100$.

$\left(\dfrac{1}{8}\right)^n < 0.0001$

$10{,}000 < 8^n$

This inequality is true when $n = 5$. This series
converges at a faster rate.

**101.** $\displaystyle\sum_{i=0}^{n-1} 8000(0.95)^i = \dfrac{8000\left[1 - 0.95^n\right]}{1 - 0.95}$

$= 160{,}000\left[1 - 0.95^n\right], \quad n > 0$

**103.** $\displaystyle\sum_{i=0}^{\infty} 200(0.75)^i = 800$ million dollars

**105.** $D_1 = 16$

$D_2 = \underbrace{0.81(16)}_{up} + \underbrace{0.81(16)}_{down} = 32(0.81)$

$D_3 = 16(0.81)^2 + 16(0.81)^2 = 32(0.81)^2$

$\vdots$

$D = 16 + 32(0.81) + 32(0.81)^2 + \cdots$

$= -16 + \displaystyle\sum_{n=0}^{\infty} 32(0.81)^n = -16 + \dfrac{32}{1 - 0.81}$

$\approx 152.42$ feet

**107.** $P(n) = \dfrac{1}{2}\left(\dfrac{1}{2}\right)^n$

$P(2) = \dfrac{1}{2}\left(\dfrac{1}{2}\right)^2 = \dfrac{1}{8}$

$\displaystyle\sum_{n=0}^{\infty} \dfrac{1}{2}\left(\dfrac{1}{2}\right)^n = \dfrac{1/2}{1 - (1/2)} = 1$

**109.** (a) $\displaystyle\sum_{n=1}^{\infty} \left(\dfrac{1}{2}\right)^n = \sum_{n=0}^{\infty} \dfrac{1}{2}\left(\dfrac{1}{2}\right)^n = \dfrac{1}{2}\dfrac{1}{\left(1 - (1/2)\right)} = 1$

(b) No, the series is not geometric.

(c) $\displaystyle\sum_{n=1}^{\infty} n\left(\dfrac{1}{2}\right)^n = 2$

**111.** (a) $64 + 32 + 16 + 8 + 4 + 2 = 126$ in.$^2$

(b) $\displaystyle\sum_{n=0}^{\infty} 64\left(\dfrac{1}{2}\right)^n = \dfrac{64}{1 - (1/2)} = 128$ in.$^2$

**Note:** This is one-half of the area of the original square

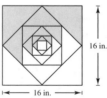

**113.** $\displaystyle\sum_{n=1}^{20} 100{,}000\left(\frac{1}{1.06}\right)^n = \dfrac{100{,}000}{1.06}\sum_{i=0}^{19}\left(\dfrac{1}{1.06}\right)^i = \dfrac{100{,}000}{1.06}\left[\dfrac{1 - 1.06^{-20}}{1 - 1.06^{-1}}\right] \quad \left(n = 20, r = 1.06^{-1}\right) \approx \$1{,}146{,}992.12$

The \$2,000,000 sweepstakes has a present value of \$1,146,992.12. After accruing interest over the 20-year period, it attains its
full value.

**115.** $w = \displaystyle\sum_{i=0}^{n-1} 0.01(2)^i = \dfrac{0.01\left(1 - 2^n\right)}{1 - 2} = 0.01\left(2^n - 1\right)$

(a) When $n = 29$: $w = \$5{,}368{,}709.11$

(b) When $n = 30$: $w = \$10{,}737{,}418.23$

(c) When $n = 31$: $w = \$21{,}474{,}836.47$

**117.** $P = 45, \quad r = 0.03, \quad t = 20$

   (a) $A = 45\left(\dfrac{12}{0.03}\right)\left[\left(1 + \dfrac{0.03}{12}\right)^{12(20)} - 1\right] \approx \$14{,}773.59$

   (b) $A = \dfrac{45\left(e^{0.03(20)} - 1\right)}{e^{0.03/12} - 1} \approx \$14{,}779.65$

**119.** $P = 100, \quad r = 0.04, \quad t = 35$

   (a) $A = 100\left(\dfrac{12}{0.04}\right)\left[\left(1 + \dfrac{0.04}{12}\right)^{12(35)} - 1\right] \approx \$91{,}373.09$

   (b) $A = \dfrac{100\left(e^{0.04(35)} - 1\right)}{e^{0.04/12} - 1} \approx \$91{,}503.32$

**121.** $T = 50{,}000 + 50{,}000(1.04) + \cdots + 50{,}000(1.04)^{39}$

$$= \sum_{n=0}^{39} 50{,}000(1.04)^{n}$$

$$= 50{,}000\left[\dfrac{1 - 1.04^{40}}{1 - 1.04}\right] \approx \$4{,}751{,}275.79$$

**123.** False. $\displaystyle\lim_{n\to\infty} \dfrac{1}{n} = 0,$ but $\displaystyle\sum_{n=1}^{\infty} \dfrac{1}{n}$ diverges.

**125.** False; $\displaystyle\sum_{n=1}^{\infty} ar^{n} = \left(\dfrac{a}{1 - r}\right) - a$

The formula requires that the geometric series begins with $n = 0$.

**127.** True

$$0.74999\ldots = 0.74 + \dfrac{9}{10^{3}} + \dfrac{9}{10^{4}} + \cdots$$

$$= 0.74 + \dfrac{9}{10^{3}} \sum_{n=0}^{\infty} \left(\dfrac{1}{10}\right)^{n}$$

$$= 0.74 + \dfrac{9}{10^{3}} \cdot \dfrac{1}{1 - (1/10)}$$

$$= 0.74 + \dfrac{9}{10^{3}} \cdot \dfrac{10}{9}$$

$$= 0.74 + \dfrac{1}{100} = 0.75$$

**129.** By letting $S_{0} = 0,$ you have

$$a_{n} = \sum_{k=1}^{n} a_{k} - \sum_{k=1}^{n-1} a_{k} = S_{n} - S_{n-1}. \text{ So,}$$

$$\sum_{n=1}^{\infty} a_{n} = \sum_{n=1}^{\infty} (S_{n} - S_{n-1})$$

$$= \sum_{n=1}^{\infty} (S_{n} - S_{n-1} + c - c)$$

$$= \sum_{n=1}^{\infty} \big[(c - S_{n-1}) - (c - S_{n})\big].$$

**131.** Let $\displaystyle\sum a_{n} = \sum_{n=0}^{\infty} 1$ and $\displaystyle\sum b_{n} = \sum_{n=0}^{\infty} (-1).$

Both are divergent series.

$$\sum(a_{n} + b_{n}) = \sum_{n=0}^{\infty} \big[1 + (-1)\big] = \sum_{n=0}^{\infty} [1 - 1] = 0$$

**133.** Suppose, on the contrary, that $\sum c a_{n}$ converges. Because $c \neq 0,$

$$\sum\left(\dfrac{1}{c}\right) c a_{n} = \sum a_{n}$$

converges. This is a contradiction since $\sum c a_{n}$ diverged. So, $\sum c a_{n}$ diverges.

**135.** (a) $\dfrac{1}{a_{n+1}a_{n+2}} - \dfrac{1}{a_{n+2}a_{n+3}} = \dfrac{a_{n+3} - a_{n+1}}{a_{n+1}a_{n+2}a_{n+3}} = \dfrac{a_{n+2}}{a_{n+1}a_{n+2}a_{n+3}} = \dfrac{1}{a_{n+1}a_{n+3}}$

   (b) $S_{n} = \displaystyle\sum_{k=0}^{n} \dfrac{1}{a_{k+1}a_{k+3}}$

$$= \sum_{k=0}^{n} \left[\dfrac{1}{a_{k+1}a_{k+2}} - \dfrac{1}{a_{k+2}a_{k+3}}\right]$$

$$= \left[\dfrac{1}{a_{1}a_{2}} - \dfrac{1}{a_{2}a_{3}}\right] + \left[\dfrac{1}{a_{2}a_{3}} - \dfrac{1}{a_{3}a_{4}}\right] + \cdots + \left[\dfrac{1}{a_{n+1}a_{n+2}} - \dfrac{1}{a_{n+2}a_{n+3}}\right] = \dfrac{1}{a_{1}a_{2}} - \dfrac{1}{a_{n+2}a_{n+3}} = 1 - \dfrac{1}{a_{n+2}a_{n+3}}$$

$$\sum_{n=0}^{\infty} \dfrac{1}{a_{n+1}a_{n+3}} = \lim_{n\to\infty} S_{n} = \lim_{n\to\infty}\left[1 - \dfrac{1}{a_{n+2}a_{n+3}}\right] = 1$$

**137.** $\dfrac{1}{r} + \dfrac{1}{r^2} + \dfrac{1}{r^3} + \cdots = \displaystyle\sum_{n=0}^{\infty} \dfrac{1}{r}\left(\dfrac{1}{r}\right)^n = \dfrac{1/r}{1 - (1/r)} = \dfrac{1}{r-1}$    $\left(\text{since} \left|\dfrac{1}{r}\right| < 1\right)$

This is a geometric series which converges if

$\left|\dfrac{1}{r}\right| < 1 \Leftrightarrow |r| > 1.$

**139.** (a)

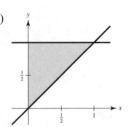  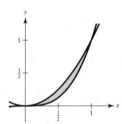

(b)    $\displaystyle\int_0^1 (1 - x)\, dx = \left[x - \dfrac{x^2}{2}\right]_0^1 = \dfrac{1}{2}$

$\displaystyle\int_0^1 (x - x^2)\, dx = \left[\dfrac{x^2}{2} - \dfrac{x^3}{3}\right]_0^1 = \dfrac{1}{2} - \dfrac{1}{3} = \dfrac{1}{6}$

$\displaystyle\int_0^1 (x^2 - x^3)\, dx = \left[\dfrac{x^3}{3} - \dfrac{x^4}{4}\right]_0^1 = \dfrac{1}{3} - \dfrac{1}{4} = \dfrac{1}{12}$

(c)    $a_n = \displaystyle\int_0^1 (x^{n-1} - x^n)\, dx = \left[\dfrac{x^n}{n} - \dfrac{x^{n+1}}{n+1}\right]_0^1 = \dfrac{1}{n} - \dfrac{1}{n+1}$

$\displaystyle\sum_{n=1}^{\infty} a_n = \left(1 - \dfrac{1}{2}\right) + \left(\dfrac{1}{2} - \dfrac{1}{3}\right) + \cdots = 1$

Note that the square has area $= 1$.

**141.** Let $H$ represent the half-life of the drug. If a patient receives $n$ equal doses of $P$ units each of this drug, administered at equal time interval of length $t$, the total amount of the drug in the patient's system at the time the last dose is administered is given by $T_n = P + Pe^{kt} + Pe^{2kt} + \cdots + Pe^{(n-1)kt}$ where $k = -(\ln 2)/H$. One time interval *after* the last dose is administered is given by $T_{n+1} = Pe^{kt} + Pe^{2kt} + Pe^{3kt} + \cdots + Pe^{nkt}$. Two time intervals *after* the last dose is administered is given by $T_{n+2} = Pe^{2kt} + Pe^{3kt} + Pe^{4kt} + \cdots + Pe^{(n+1)kt}$ and so on. Because $k < 0$, $T_{n+s} \to 0$ as $s \to \infty$, where $s$ is an integer.

**143.** $f(1) = 0,\ f(2) = 1,\ f(3) = 2,\ f(4) = 4, \ldots$

In general: $f(n) = \begin{cases} n^2/4, & n \text{ even} \\ (n^2 - 1)/4, & n \text{ odd.} \end{cases}$

(See below for a proof of this.)

$x + y$ and $x - y$ are either both odd or both even. If both even, then

$f(x + y) - f(x - y) = \dfrac{(x + y)^2}{4} - \dfrac{(x - y)^2}{4} = xy.$

If both odd,

$f(x + y) - f(x - y) = \dfrac{(x + y)^2 - 1}{4} - \dfrac{(x - y)^2 - 1}{4} = xy.$

Proof by induction that the formula for $f(n)$ is correct. It is true for $n = 1$. Assume that the formula is valid for $k$. If $k$ is even, then $f(k) = k^2/4$ and

$f(k + 1) = f(k) + \dfrac{k}{2} = \dfrac{k^2}{4} + \dfrac{k}{2} = \dfrac{k^2 + 2k}{4} = \dfrac{(k + 1)^2 - 1}{4}.$

The argument is similar if $k$ is odd.

# Section 9.3   The Integral Test and *p*-Series

**1.** $\sum_{n=1}^{\infty} \dfrac{1}{n+3}$

Let $f(x) = \dfrac{1}{x+3}$.

$f$ is positive, continuous, and decreasing for $x \geq 1$.

$\int_1^{\infty} \dfrac{1}{x+3} \, dx = \left[ \ln(x+3) \right]_1^{\infty} = \infty$

Diverges by Theorem 9.10

**3.** $\sum_{n=1}^{\infty} \dfrac{1}{2^n}$

Let $f(x) = \dfrac{1}{2^x}$.

$f$ is positive, continuous, and decreasing for $x \geq 1$.

$\int_1^{\infty} \dfrac{1}{2^x} \, dx = \left[ \dfrac{-1}{(\ln 2) \, 2^x} \right]_1^{\infty} = \dfrac{1}{2 \ln 2}$

Converges by Theorem 9.10.

**5.** $\sum_{n=1}^{\infty} e^{-n}$

Let $f(x) = e^{-x}$.

$f$ is positive, continuous, and decreasing for $x \geq 1$.

$\int_1^{\infty} e^{-x} \, dx = \left[ -e^{-x} \right]_1^{\infty} = \dfrac{1}{e}$

Converges by Theorem 9.10

**7.** $\sum_{n=1}^{\infty} \dfrac{1}{n^2+1}$

Let $f(x) = \dfrac{1}{x^2+1}$.

$f$ is positive, continuous, and decreasing for $x \geq 1$.

$\int_1^{\infty} \dfrac{1}{x^2+1} \, dx = \left[ \arctan x \right]_1^{\infty} = \dfrac{\pi}{4}$

Converges by Theorem 9.10

**9.** $\sum_{n=1}^{\infty} \dfrac{\ln(n+1)}{n+1}$

Let $f(x) = \dfrac{\ln(x+1)}{x+1}$.

$f$ is positive, continuous, and decreasing for

$x \geq 2$ because $f'(x) = \dfrac{1 - \ln(x+1)}{(x+1)^2} < 0$ for $x \geq 2$.

$\int_1^{\infty} \dfrac{\ln(x+1)}{x+1} \, dx = \left[ \dfrac{[\ln(x+1)]^2}{2} \right]_1^{\infty} = \infty$

Diverges by Theorem 9.10

**11.** $\sum_{n=1}^{\infty} \dfrac{1}{\sqrt{n}\left(\sqrt{n}+1\right)}$

Let $f(x) = \dfrac{1}{\sqrt{x}\left(\sqrt{x}+1\right)}$,

$f'(x) = -\dfrac{1 + 2\sqrt{x}}{2x^{3/2}\left(\sqrt{x}+1\right)^2} < 0.$

$f$ is positive, continuous, and decreasing for $x \geq 1$.

$\int_1^{\infty} \dfrac{1}{\sqrt{x}\left(\sqrt{x}+1\right)} \, dx = \left[ 2\ln\left(\sqrt{x}+1\right) \right]_1^{\infty} = \infty,$ diverges

So, the series diverges by Theorem 9.10.

**13.** $\sum_{n=1}^{\infty} \dfrac{1}{\sqrt{n+2}}$

Let $f(x) = \dfrac{1}{\sqrt{x+2}}$, $f'(x) = \dfrac{-1}{2(x+2)^{3/2}} < 0$

$f$ is positive, continuous, and decreasing for $x \geq 1$.

$\int_1^{\infty} \dfrac{1}{(x+2)^{1/2}} \, dx = \left[ 2\sqrt{x+2} \right]_1^{\infty} = \infty,$ diverges.

So, the series diverges by Theorem 9.10.

**15.** $\sum_{n=1}^{\infty} \dfrac{\ln n}{n^2}$

Let $f(x) = \dfrac{\ln x}{x^2}$, $f'(x) = \dfrac{1 - 2\ln x}{x^3}$.

$f$ is positive, continuous, and decreasing for

$x > e^{1/2} \approx 1.6$.

$\int_1^{\infty} \dfrac{\ln x}{x^2} \, dx = \left[ \dfrac{-(\ln x + 1)}{x} \right]_1^{\infty} = 1,$ converges

So, the series converges by Theorem 9.10.

**17.** $\sum_{n=1}^{\infty} \dfrac{\arctan n}{n^2+1}$

Let $f(x) = \dfrac{\arctan x}{x^2+1}$,

$f'(x) = \dfrac{1 - 2x \arctan x}{(x^2+1)^2} < 0$ for $x \geq 1$.

$f$ is positive, continuous, and decreasing for $x \geq 1$.

$\int_1^{\infty} \dfrac{\arctan x}{x^2+1} \, dx = \left[ \dfrac{(\arctan x)^2}{2} \right]_1^{\infty} = \dfrac{3\pi^2}{32},$ converges

So, the series converges by Theorem 9.10.

**19.** $\displaystyle\sum_{n=1}^{\infty} \frac{1}{(2n + 3)^3}$

Let $f(x) = (2x + 3)^{-3}$, $f'(x) = \dfrac{-6}{(2x + 3)^4} < 0$

$f$ is positive, continuous, and decreasing for $x \geq 1$.

$\displaystyle\int_1^{\infty} (2x + 3)^{-3}\, dx = \left[ \dfrac{-1}{4(2x + 3)^2} \right]_1^{\infty} = \dfrac{1}{100}$, converges.

So, $\displaystyle\sum_{n=1}^{\infty} \frac{1}{(2n + 3)^3}$ converges by Theorem 9.10.

**21.** $\displaystyle\sum_{n=1}^{\infty} \frac{4n}{2n^2 + 1}$

Let $f(x) = \dfrac{4x}{2x^2 + 1}$, $f'(x) = \dfrac{-4(2x^2 - 1)}{(2x^2 + 1)^2} < 0$

for $x \geq 1$

$f$ is positive, continuous, and decreasing for $x \geq 1$.

$\displaystyle\int_1^{\infty} \frac{4x}{2x^2 + 1}\, dx = \left[ \ln(2x^2 + 1) \right]_1^{\infty} = \infty$, diverges.

So, $\displaystyle\sum_{n=1}^{\infty} \frac{4n}{2n^2 + 1}$ diverges by Theorem 9.10.

**23.** $\displaystyle\sum_{n=1}^{\infty} \frac{n}{(4n + 5)^{3/2}}$

Let $f(x) = \dfrac{x}{(4x + 5)^{3/2}}$, $f'(x) = \dfrac{-(2x - 5)}{(4x + 5)^{5/2}} < 0$ for

$x \geq 3$

$f$ is positive, continuous, and decreasing for $x \geq 3$.

$\displaystyle\int_1^{\infty} \frac{x}{(4x + 5)^{3/2}} = \left[ \dfrac{2x + 5}{4\sqrt{4x + 5}} \right]_1^{\infty} = \infty$

(Integration by parts: $u = x$, $dv = (4x + 5)^{-3/2}\, dx$ )

So, $\displaystyle\sum_{n=1}^{\infty} \frac{n}{(4n + 5)^{3/2}}$ diverges by Theorem 9.10.

**25.** $\displaystyle\sum_{n=1}^{\infty} \frac{n^{k-1}}{n^k + c}$

Let $f(x) = \dfrac{x^{k-1}}{x^k + c}$.

$f$ is positive, continuous, and decreasing for

$x > \sqrt[k]{c(k - 1)}$ because

$f'(x) = \dfrac{x^{k-2}\left[c(k - 1) - x^k\right]}{(x^k + c)^2} < 0$

for $x > \sqrt[k]{c(k - 1)}$.

$\displaystyle\int_1^{\infty} \frac{x^{k-1}}{x^k + c}\, dx = \left[ \dfrac{1}{k} \ln(x^k + c) \right]_1^{\infty} = \infty$

Diverges by Theorem 9.10

**27.** Let $f(x) = \dfrac{(-1)^x}{x}$, $f(n) = a_n$.

The function $f$ is not positive for $x \geq 1$.

**29.** Let $f(x) = \dfrac{2 + \sin x}{x}$, $f(n) = a_n$.

The function $f$ is not decreasing for $x \geq 1$.

**31.** $\displaystyle\sum_{n=1}^{\infty} \frac{1}{n^3}$

Let $f(x) = \dfrac{1}{x^3}$.

$f$ is positive, continuous, and decreasing for $x \geq 1$.

$\displaystyle\int_1^{\infty} \frac{1}{x^3}\, dx = \left[ -\dfrac{1}{2x^2} \right]_1^{\infty} = \dfrac{1}{2}$

Converges by Theorem 9.10

**33.** Let $f(x) = \dfrac{1}{x^{1/4}}$, $f'(x) = \dfrac{-1}{4x^{5/4}} < 0$ for $x \geq 1$

$f$ is positive, continuous, and decreasing for $x \geq 1$

$\displaystyle\int_1^{\infty} \frac{1}{x^{1/4}}\, dx = \left[ \dfrac{4x^{3/4}}{3} \right]_1^{\infty} = \infty$, diverges

So, the series diverges by Theorem 9.10.

**35.** $\displaystyle\sum_{n=1}^{\infty} \frac{1}{\sqrt[5]{n}} = \sum_{n=1}^{\infty} \frac{1}{n^{1/5}}$

Divergent $p$-series with $p = \dfrac{1}{5} < 1$

**37.** $\displaystyle\sum_{n=1}^{\infty} \frac{1}{n^{1/2}}$

Divergent $p$-series with $p = \dfrac{1}{2} < 1$

**39.** $\displaystyle\sum_{n=1}^{\infty} \frac{1}{n^{3/2}}$

Convergent $p$-series with $p = \dfrac{3}{2} > 1$

**41.** $\displaystyle\sum_{n=1}^{\infty} \frac{1}{n^{1.04}}$

Convergent $p$-series with $p = 1.04 > 1$

**43.** $\displaystyle\sum_{n=1}^{\infty} \frac{2}{n^{3/4}}$

$S_1 = 2$

$S_2 \approx 3.1892$

$S_3 \approx 4.0666$

$S_4 \approx 4.7740$

Matches (c), diverges

**45.** $\displaystyle\sum_{n=1}^{\infty} \frac{2}{\sqrt{n^{\pi}}} = \sum_{n=1}^{\infty} \frac{2}{n^{\pi/2}}$

$S_1 = 2$

$S_2 \approx 2.6732$

$S_3 \approx 3.0293$

$S_4 \approx 3.2560$

Matches (b), converges

**Note:** The partial sums for 45 and 47 are very similar because $\pi/2 \approx 3/2$.

**47.** $\displaystyle\sum_{n=1}^{\infty} \frac{2}{n\sqrt{n}} = \sum_{n=1}^{\infty} \frac{2}{n^{3/2}}$

$S_1 = 2$

$S_2 \approx 2.7071$

$S_3 \approx 3.0920$

$S_4 \approx 3.3420$

Matches (d), converges

**Note:** The partial sums for 45 and 47 are very similar because $\pi/2 \approx 3/2$.

**49.** (a)

| $n$ | 5 | 10 | 20 | 50 | 100 |
|-----|-----|-----|-----|-----|-----|
| $S_n$ | 3.7488 | 3.75 | 3.75 | 3.75 | 3.75 |

The partial sums approach the sum 3.75 very rapidly.

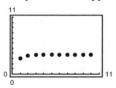

(b)

| $n$ | 5 | 10 | 20 | 50 | 100 |
|-----|-----|-----|-----|-----|-----|
| $S_n$ | 1.4636 | 1.5498 | 1.5962 | 1.6251 | 1.635 |

The partial sums approach the sum $\pi^2/6 \approx 1.6449$ slower than the series in part (a).

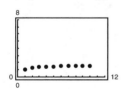

**51.** Let $f$ be positive, continuous, and decreasing for $x \geq 1$ and $a_n = f(n)$. Then,

$$\sum_{n=1}^{\infty} a_n \quad \text{and} \quad \int_1^{\infty} f(x)\, dx$$

either both converge or both diverge (Theorem 9.10). See Example 1, page 620.

**53.** Your friend is not correct. The series

$$\sum_{n=10,000}^{\infty} \frac{1}{n} = \frac{1}{10,000} + \frac{1}{10,001} + \cdots$$

is the harmonic series, starting with the 10,000[th] term, and therefore diverges.

**55.** $\sum\limits_{n=1}^{6} a_n \geq \int_{1}^{7} f(x)\,dx \geq \sum\limits_{n=2}^{7} a_n$

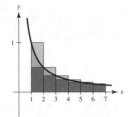

**57.** $\sum\limits_{n=2}^{\infty} \dfrac{1}{n(\ln n)^p}$

If $p = 1$, then the series diverges by the Integral Test. If $p \neq 1$,

$$\int_{2}^{\infty} \frac{1}{x(\ln x)^p}\,dx = \int_{2}^{\infty} (\ln x)^{-p}\frac{1}{x}\,dx = \left[\frac{(\ln x)^{-p+1}}{-p+1}\right]_{2}^{\infty}.$$

Converges for $-p + 1 < 0$ or $p > 1$

**59.** $\sum\limits_{n=1}^{\infty} \dfrac{n}{\left(1 + n^2\right)^p}$

If $p = 1$, $\sum\limits_{n=1}^{\infty} \dfrac{n}{1 + n^2}$ diverges (see Example 1). Let

$$f(x) = \frac{x}{\left(1 + x^2\right)^p}, \quad p \neq 1$$

$$f'(x) = \frac{1 - (2p - 1)x^2}{\left(1 + x^2\right)^{p+1}}.$$

For a fixed $p > 0$, $p \neq 1$, $f'(x)$ is eventually negative. $f$ is positive, continuous, and eventually decreasing.

$$\int_{1}^{\infty} \frac{x}{\left(1 + x^2\right)^p}\,dx = \left[\frac{1}{\left(x^2 + 1\right)^{p-1}(2 - 2p)}\right]_{1}^{\infty}$$

For $p > 1$, this integral converges. For $0 < p < 1$, it diverges.

**61.** $\sum\limits_{n=1}^{\infty} \dfrac{1}{p^n} = \sum\limits_{n=1}^{\infty} \left(\dfrac{1}{p}\right)^n$, Geometric series.

Converges for $\left|\dfrac{1}{p}\right| < 1 \Rightarrow p > 1$

**63.** $\sum\limits_{n=2}^{\infty} \dfrac{1}{n(\ln n)^p}$ converges for $p > 1$.

So, $\sum\limits_{n=2}^{\infty} \dfrac{1}{n \ln n}$, diverges.

**65.** $\sum\limits_{n=2}^{\infty} \dfrac{1}{n(\ln n)^p}$ converges for $p > 1$.

So, $\sum\limits_{n=2}^{\infty} \dfrac{1}{n(\ln n)^2}$, converges.

**67.**

$$S_N = \sum\limits_{n=1}^{N} a_n = a_1 + a_2 + \cdots + a_N$$

$$R_N = S - S_N = \sum\limits_{n=N+1}^{\infty} a_n > 0$$

$$R_N = S - S_N = \sum\limits_{n=N+1}^{\infty} a_n = a_{N+1} + a_{N+2} + \cdots$$

$$\leq \int_{N}^{\infty} f(x)\,dx$$

So, $0 \leq R_n \leq \int_{N}^{\infty} f(x)\,dx$

**69.** $S_6 = 1 + \dfrac{1}{2^4} + \dfrac{1}{3^4} + \dfrac{1}{4^4} + \dfrac{1}{5^4} + \dfrac{1}{6^4} \approx 1.0811$

$$R_6 \leq \int_{6}^{\infty} \frac{1}{x^4}\,dx = \left[-\frac{1}{3x^3}\right]_{6}^{\infty} \approx 0.0015$$

$$1.0811 \leq \sum\limits_{n=1}^{\infty} \frac{1}{n^4} \leq 1.0811 + 0.0015 = 1.0826$$

**71.** $S_{10} = \dfrac{1}{2} + \dfrac{1}{5} + \dfrac{1}{10} + \dfrac{1}{17} + \dfrac{1}{26} + \dfrac{1}{37} + \dfrac{1}{50} + \dfrac{1}{65} + \dfrac{1}{82} + \dfrac{1}{101} \approx 0.9818$

$R_{10} \le \displaystyle\int_{10}^{\infty} \dfrac{1}{x^2 + 1}\, dx = \left[\arctan x\right]_{10}^{\infty} = \dfrac{\pi}{2} - \arctan 10 \approx 0.0997$

$0.9818 \le \displaystyle\sum_{n=1}^{\infty} \dfrac{1}{n^2 + 1} \le 0.9818 + 0.0997 = 1.0815$

**73.** $S_4 = \dfrac{1}{e} + \dfrac{2}{e^4} + \dfrac{3}{e^9} + \dfrac{4}{e^{16}} \approx 0.4049$

$R_4 \le \displaystyle\int_{4}^{\infty} x e^{-x^2}\, dx = \left[-\dfrac{1}{2} e^{-x^2}\right]_{4}^{\infty} = \dfrac{e^{-16}}{2} \approx 5.6 \times 10^{-8}$

$0.4049 \le \displaystyle\sum_{n=1}^{\infty} n e^{-n^2} \le 0.4049 + 5.6 \times 10^{-8}$

**75.** $0 \le R_N \le \displaystyle\int_{N}^{\infty} \dfrac{1}{x^4}\, dx = \left[-\dfrac{1}{3x^3}\right]_{N}^{\infty} = \dfrac{1}{3N^3} < 0.001$

$\dfrac{1}{N^3} < 0.003$

$N^3 > 333.33$

$N > 6.93$

$N \ge 7$

**77.** $R_N \le \displaystyle\int_{N}^{\infty} e^{-5x}\, dx = \left[-\dfrac{1}{5} e^{-5x}\right]_{N}^{\infty} = \dfrac{e^{-5N}}{5} < 0.001$

$\dfrac{1}{e^{5N}} < 0.005$

$e^{5N} > 200$

$5N > \ln 200$

$N > \dfrac{\ln 200}{5}$

$N > 1.0597$

$N \ge 2$

**79.** $R_N \le \displaystyle\int_{N}^{\infty} \dfrac{1}{x^2 + 1}\, dx = \left[\arctan x\right]_{N}^{\infty}$

$\qquad = \dfrac{\pi}{2} - \arctan N < 0.001$

$-\arctan N < 0.001 - \dfrac{\pi}{2}$

$\arctan N > \dfrac{\pi}{2} - 0.001$

$N > \tan\left(\dfrac{\pi}{2} - 0.001\right)$

$N \ge 1000$

**81. (a)** $\displaystyle\sum_{n=2}^{\infty} \dfrac{1}{n^{1.1}}$. This is a convergent p-series with $p = 1.1 > 1$. $\displaystyle\sum_{n=2}^{\infty} \dfrac{1}{n \ln n}$ is a divergent series. Use the Integral Test.

$f(x) = \dfrac{1}{x \ln x}$ is positive, continuous, and decreasing for $x \ge 2$.

$\displaystyle\int_{2}^{\infty} \dfrac{1}{x \ln x}\, dx = \left[\ln|\ln x|\right]_{2}^{\infty} = \infty$

**(b)** $\displaystyle\sum_{n=2}^{6} \dfrac{1}{n^{1.1}} = \dfrac{1}{2^{1.1}} + \dfrac{1}{3^{1.1}} + \dfrac{1}{4^{1.1}} + \dfrac{1}{5^{1.1}} + \dfrac{1}{6^{1.1}} \approx 0.4665 + 0.2987 + 0.2176 + 0.1703 + 0.1393$

$\displaystyle\sum_{n=2}^{6} \dfrac{1}{n \ln n} = \dfrac{1}{2 \ln 2} + \dfrac{1}{3 \ln 3} + \dfrac{1}{4 \ln 4} + \dfrac{1}{5 \ln 5} + \dfrac{1}{6 \ln 6} \approx 0.7213 + 0.3034 + 0.1803 + 0.1243 + 0.0930$

For $n \ge 4$, the terms of the convergent series **seem** to be larger than those of the divergent series.

**(c)** $\dfrac{1}{n^{1.1}} < \dfrac{1}{n \ln n}$

$n \ln n < n^{1.1}$

$\ln n < n^{0.1}$

This inequality holds when $n \ge 3.5 \times 10^{15}$. Or, $n > e^{40}$. Then $\ln e^{40} = 40 < \left(e^{40}\right)^{0.1} = e^4 \approx 55$.

**83.** (a) Let $f(x) = 1/x$. $f$ is positive, continuous, and decreasing on $[1, \infty)$.

$$S_n - 1 \le \int_1^n \frac{1}{x}\, dx$$

$$S_n - 1 \le \ln n$$

So, $S_n \le 1 + \ln n$. Similarly,

$$S_n \ge \int_1^{n+1} \frac{1}{x}\, dx = \ln(n + 1).$$

So, $\ln(n + 1) \le S_n \le 1 + \ln n$.

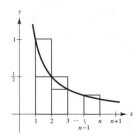

(b) Because $\ln(n + 1) \le S_n \le 1 + \ln n$, you have $\ln(n + 1) - \ln n \le S_n - \ln n \le 1$. Also, because $\ln x$ is an increasing function, $\ln(n + 1) - \ln n > 0$ for $n \ge 1$. So, $0 \le S_n - \ln n \le 1$ and the sequence $\{a_n\}$ is bounded.

(c) $a_n - a_{n+1} = [S_n - \ln n] - [S_{n+1} - \ln(n + 1)] = \int_n^{n+1} \frac{1}{x}\, dx - \frac{1}{n+1} \ge 0$

So, $a_n \ge a_{n+1}$ and the sequence is decreasing.

(d) Because the sequence is bounded and monotonic, it converges to a limit, $\gamma$.

(e) $a_{100} = S_{100} - \ln 100 \approx 0.5822$ (Actually $\gamma \approx 0.577216$.)

**85.** $\displaystyle\sum_{n=2}^{\infty} x^{\ln n}$

(a) $x = 1$: $\displaystyle\sum_{n=2}^{\infty} 1^{\ln n} = \sum_{n=2}^{\infty} 1$, diverges

(b) $x = \dfrac{1}{e}$: $\displaystyle\sum_{n=2}^{\infty} \left(\frac{1}{e}\right)^{\ln n} = \sum_{n=2}^{\infty} e^{-\ln n} = \sum_{n=2}^{\infty} \frac{1}{n}$, diverges

(c) Let $x$ be given, $x > 0$. Put $x = e^{-p} \Leftrightarrow \ln x = -p$.

$$\sum_{n=2}^{\infty} x^{\ln n} = \sum_{n=2}^{\infty} e^{-p \ln n} = \sum_{n=2}^{\infty} n^{-p} = \sum_{n=2}^{\infty} \frac{1}{n^p}$$

This series converges for $p > 1 \Rightarrow x < \dfrac{1}{e}$.

**87.** Let $f(x) = \dfrac{1}{3x - 2}$, $f'(x) = \dfrac{-3}{(3x - 2)^2} < 0$ for $x \ge 1$

$f$ is positive, continuous, and decreasing for $x \ge 1$.

$$\int_1^{\infty} \frac{1}{3x - 2}\, dx = \left[\frac{1}{3} \ln|3x - 2|\right]_1^{\infty} = \infty$$

So, the series $\displaystyle\sum_{n=1}^{\infty} \frac{1}{3n - 2}$

diverges by Theorem 9.10.

**89.** $\displaystyle\sum_{n=1}^{\infty} \frac{1}{n\sqrt[4]{n}} = \sum_{n=1}^{\infty} \frac{1}{n^{5/4}}$

*p*-series with $p = \dfrac{5}{4}$

Converges by Theorem 9.11

**91.** $\displaystyle\sum_{n=0}^{\infty} \left(\frac{2}{3}\right)^n$

Geometric series with $r = \dfrac{2}{3}$

Converges by Theorem 9.6

**93.** $\displaystyle\sum_{n=1}^{\infty} \frac{n}{\sqrt{n^2+1}}$

$\displaystyle\lim_{n\to\infty} \frac{n}{\sqrt{n^2+1}} = \lim_{n\to\infty} \frac{1}{\sqrt{1+(1/n^2)}} = 1 \neq 0$

Diverges by Theorem 9.9

**95.** $\displaystyle\sum_{n=1}^{\infty} \left(1+\frac{1}{n}\right)^n$

$\displaystyle\lim_{n\to\infty}\left(1+\frac{1}{n}\right)^n = e \neq 0$

Fails *n*th-Term Test

Diverges by Theorem 9.9

**97.** $\displaystyle\sum_{n=2}^{\infty} \frac{1}{n(\ln n)^3}$

Let $f(x) = \dfrac{1}{x(\ln x)^3}$.

*f* is positive, continuous, and decreasing for $x \geq 2$.

$\displaystyle\int_2^{\infty} \frac{1}{x(\ln x)^3}\,dx = \int_2^{\infty} (\ln x)^{-3}\frac{1}{x}\,dx$

$\displaystyle = \left[\frac{(\ln x)^{-2}}{-2}\right]_2^{\infty}$

$\displaystyle = \left[-\frac{1}{2(\ln x)^2}\right]_2^{\infty} = \frac{1}{2(\ln 2)^2}$

Converges by Theorem 9.10. See Exercise 57.

## Section 9.4   Comparisons of Series

**1.** (a) $\displaystyle\sum_{n=1}^{\infty} \frac{6}{n^{3/2}} = \frac{6}{1} + \frac{6}{2^{3/2}} + \cdots;\ S_1 = 6$

$\displaystyle\sum_{n=1}^{\infty} \frac{6}{n^{3/2}+3} = \frac{6}{4} + \frac{6}{2^{3/2}+3} + \cdots;\ S_1 = \frac{3}{2}$

$\displaystyle\sum_{n=1}^{\infty} \frac{6}{n\sqrt{n^2+0.5}} = \frac{6}{1\sqrt{1.5}} + \frac{6}{2\sqrt{4.5}} + \cdots;\ S_1 = \frac{6}{\sqrt{1.5}} \approx 4.9$

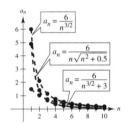

(b) The first series is a *p*-series. It converges $\left(p = \dfrac{3}{2} > 1\right)$.

(c) The magnitude of the terms of the other two series are less than the corresponding terms at the convergent *p*-series. So, the other two series converge.

(d) The smaller the magnitude of the terms, the smaller the magnitude of the terms of the sequence of partial sums.

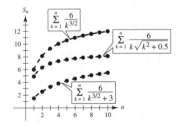

**3.** $0 < \dfrac{1}{n^2 + 1} < \dfrac{1}{n^2}$

Therefore,

$$\sum_{n=1}^{\infty} \frac{1}{n^2 + 1}$$

converges by comparison with the convergent $p$-series

$$\sum_{n=1}^{\infty} \frac{1}{n^2}.$$

**5.** $\dfrac{1}{2n - 1} > \dfrac{1}{2n} > 0$ for $n \geq 1$

Therefore,

$$\sum_{n=1}^{\infty} \frac{1}{2n - 1}$$

diverges by comparison with the divergent $p$-series

$$\frac{1}{2}\sum_{n=1}^{\infty} \frac{1}{n}.$$

**7.** $\dfrac{1}{4^n + 1} < \dfrac{1}{4^n}$

Therefore,

$$\sum_{n=0}^{\infty} \frac{1}{4^n + 1}$$

converges by comparison with the convergent geometric

series $\displaystyle\sum_{n=0}^{\infty} \frac{1}{4^n}.$

**9.** For $n \geq 3, \dfrac{\ln n}{n + 1} > \dfrac{1}{n + 1} > 0.$

Therefore,

$$\sum_{n=1}^{\infty} \frac{\ln n}{n + 1}$$

diverges by comparison with the divergent series

$$\sum_{n=1}^{\infty} \frac{1}{n + 1}.$$

**Note:** $\displaystyle\sum_{n=1}^{\infty} \frac{1}{n + 1}$ diverges by the Integral Test.

**11.** For $n > 3, \dfrac{1}{n^2} > \dfrac{1}{n!} > 0.$

Therefore,

$$\sum_{n=0}^{\infty} \frac{1}{n!}$$

converges by comparison with the convergent $p$-series

$$\sum_{n=1}^{\infty} \frac{1}{n^2}.$$

**13.** $0 < \dfrac{1}{e^{n^2}} \leq \dfrac{1}{e^n}$

Therefore,

$$\sum_{n=0}^{\infty} \frac{1}{e^{n^2}}$$

converges by comparison with the convergent geometric

series $\displaystyle\sum_{n=0}^{\infty} \left(\frac{1}{e}\right)^n.$

**15.** $\displaystyle\lim_{n \to \infty} \dfrac{n/(n^2 + 1)}{1/n} = \lim_{n \to \infty} \dfrac{n^2}{n^2 + 1} = 1$

Therefore,

$$\sum_{n=1}^{\infty} \frac{n}{n^2 + 1}$$

diverges by a limit comparison with the divergent $p$-

series $\displaystyle\sum_{n=1}^{\infty} \frac{1}{n}.$

**17.** $\displaystyle\lim_{n \to \infty} \dfrac{1/\sqrt{n^2 + 1}}{1/n} = \lim_{n \to \infty} \dfrac{n}{\sqrt{n^2 + 1}} = 1$

Therefore,

$$\sum_{n=0}^{\infty} \frac{1}{\sqrt{n^2 + 1}}$$

diverges by a limit comparison with the divergent $p$-

series $\displaystyle\sum_{n=1}^{\infty} \frac{1}{n}.$

**19.** $\displaystyle\lim_{n \to \infty} \dfrac{\dfrac{2n^2 - 1}{3n^5 + 2n + 1}}{1/n^3} = \lim_{n \to \infty} \dfrac{2n^5 - n^3}{3n^5 + 2n + 1} = \dfrac{2}{3}$

Therefore,

$$\sum_{n=1}^{\infty} \frac{2n^2 - 1}{3n^5 + 2n + 1}$$

converges by a limit comparison with the convergent

$p$-series $\displaystyle\sum_{n=1}^{\infty} \frac{1}{n^3}.$

**21.** $\displaystyle\lim_{n \to \infty} \dfrac{(n + 3)/n(n^2 + 4)}{1/n^2} = \lim_{n \to \infty} \dfrac{(n + 3)n^2}{n(n^2 + 4)} = 1$

Therefore,

$$\sum_{n=1}^{\infty} \frac{n + 3}{n(n^2 + 4)}$$

converges by a limit comparison with the convergent

$p$-series $\displaystyle\sum_{n=1}^{\infty} \frac{1}{n^2}.$

**23.** $\lim\limits_{n\to\infty} \dfrac{1/\left(n\sqrt{n^2+1}\right)}{1/n^2} = \lim\limits_{n\to\infty} \dfrac{n^2}{n\sqrt{n^2+1}} = 1$

Therefore,

$$\sum_{n=1}^{\infty} \frac{1}{n\sqrt{n^2+1}}$$

converges by a limit comparison with the convergent

*p*-series $\sum\limits_{n=1}^{\infty} \dfrac{1}{n^2}$.

**25.** $\lim\limits_{n\to\infty} \dfrac{\left(n^{k-1}\right)/\left(n^k+1\right)}{1/n} = \lim\limits_{n\to\infty} \dfrac{n^k}{n^k+1} = 1$

Therefore,

$$\sum_{n=1}^{\infty} \frac{n^{k-1}}{n^k+1}$$

diverges by a limit comparison with the divergent *p*-series

$$\sum_{n=1}^{\infty} \frac{1}{n}.$$

**27.** $\lim\limits_{n\to\infty} \dfrac{\sin(1/n)}{1/n} = \lim\limits_{n\to\infty} \dfrac{\left(-1/n^2\right)\cos(1/n)}{-1/n^2}$

$$= \lim_{n\to\infty} \cos\!\left(\frac{1}{n}\right) = 1$$

Therefore,

$$\sum_{n=1}^{\infty} \sin\!\left(\frac{1}{n}\right)$$

diverges by a limit comparison with the divergent *p*-series

$$\sum_{n=1}^{\infty} \frac{1}{n}.$$

**29.** $\displaystyle\sum_{n=1}^{\infty} \frac{\sqrt[3]{n}}{n} = \sum_{n=1}^{\infty} \frac{1}{n^{2/3}}$

Diverges;

*p*-series with $p = \dfrac{2}{3}$

**31.** $\displaystyle\sum_{n=1}^{\infty} \frac{1}{5^n+1}$

Converges;

Direct comparison with convergent geometric series

$$\sum_{n=1}^{\infty} \left(\frac{1}{5}\right)^n$$

**33.** $\displaystyle\sum_{n=1}^{\infty} \frac{2n}{3n-2}$

Diverges; $n^{\text{th}}$-Term Test

$$\lim_{n\to\infty} \frac{2n}{3n-2} = \frac{2}{3} \neq 0$$

**35.** $\displaystyle\sum_{n=1}^{\infty} \frac{n}{\left(n^2+1\right)^2}$

Converges; Integral Test

**37.** $\lim\limits_{n\to\infty} \dfrac{a_n}{1/n} = \lim\limits_{n\to\infty} na_n.$ By given conditions $\lim\limits_{n\to\infty} na_n$ is finite and nonzero. Therefore,

$$\sum_{n=1}^{\infty} a_n$$

diverges by a limit comparison with the *p*-series

$$\sum_{n=1}^{\infty} \frac{1}{n}.$$

**39.** $\dfrac{1}{2} + \dfrac{2}{5} + \dfrac{3}{10} + \dfrac{4}{17} + \dfrac{5}{26} + \cdots = \displaystyle\sum_{n=1}^{\infty} \frac{n}{n^2+1},$

which diverges because the degree of the numerator is only one less than the degree of the denominator.

**41.** $\displaystyle\sum_{n=1}^{\infty} \frac{1}{n^3 + 1}$

converges because the degree of the numerator is three less than the degree of the denominator.

**43.** $\displaystyle\lim_{n\to\infty} n\left(\frac{n^3}{5n^4 + 3}\right) = \lim_{n\to\infty} \frac{n^4}{5n^4 + 3} = \frac{1}{5} \neq 0$

Therefore, $\displaystyle\sum_{n=1}^{\infty} \frac{n^3}{5n^4 + 3}$ diverges.

**45.** $\displaystyle\frac{1}{200} + \frac{1}{400} + \frac{1}{600} + \cdots = \sum_{n=1}^{\infty} \frac{1}{200n}$

diverges, (harmonic)

**47.** $\displaystyle\frac{1}{201} + \frac{1}{204} + \frac{1}{209} + \frac{1}{216} = \sum_{n=1}^{\infty} \frac{1}{200 + n^2}$

converges

**49.** Some series diverge or converge very slowly. You cannot decide convergence or divergence of a series by comparing the first few terms.

**51.** See Theorem 9.13, page 628. One example is

$\displaystyle\sum_{n=2}^{\infty} \frac{1}{\sqrt{n} - 1}$ diverges because $\displaystyle\lim_{n\to\infty} \frac{1/\sqrt{n} - 1}{1/\sqrt{n}} = 1$ and $\displaystyle\sum_{n=2}^{\infty} \frac{1}{\sqrt{n}}$ diverges (*p*-series).

**53.** (a) $\displaystyle\sum_{n=1}^{\infty} \frac{1}{(2n-1)^2} = \sum_{n=1}^{\infty} \frac{1}{4n^2 - 4n + 1}$

converges because the degree of the numerator is two less than the degree of the denominator. (See Exercise 38.)

(b)

| $n$ | 5 | 10 | 20 | 50 | 100 |
|-----|------|------|------|------|------|
| $S_n$ | 1.1839 | 1.2087 | 1.2212 | 1.2287 | 1.2312 |

(c) $\displaystyle\sum_{n=3}^{\infty} \frac{1}{(2n-1)^2} = \frac{\pi^2}{8} - S_2 \approx 0.1226$

(d) $\displaystyle\sum_{n=10}^{\infty} \frac{1}{(2n-1)^2} = \frac{\pi^2}{8} - S_9 \approx 0.0277$

**55.** False. Let $a_n = \dfrac{1}{n^3}$ and $b_n = \dfrac{1}{n^2}$. $0 < a_n \leq b_n$ and both

$\displaystyle\sum_{n=1}^{\infty} \frac{1}{n^3}$ and $\displaystyle\sum_{n=1}^{\infty} \frac{1}{n^2}$ converge.

**57.** True

**59.** True

**61.** Because $\displaystyle\sum_{n=1}^{\infty} b_n$ converges, $\displaystyle\lim_{n\to\infty} b_n = 0$. There exists $N$

such that $b_n < 1$ for $n > N$. So, $a_n b_n < a_n$ for

$n > N$ and $\displaystyle\sum_{n=1}^{\infty} a_n b_n$ converges by comparison to the

convergent series $\displaystyle\sum_{i=1}^{\infty} a_n$.

**63.** $\displaystyle\sum \frac{1}{n^2}$ and $\displaystyle\sum \frac{1}{n^3}$ both converge, and therefore, so does

$\displaystyle\sum \left(\frac{1}{n^2}\right)\left(\frac{1}{n^3}\right) = \sum \frac{1}{n^5}.$

**65.** Suppose $\displaystyle\lim_{n\to\infty} \frac{a_n}{b_n} = 0$ and $\Sigma b_n$ converges.

From the definition of limit of a sequence, there exists $M > 0$ such that

$$\left| \frac{a_n}{b_n} - 0 \right| < 1$$

whenever $n > M$. So, $a_n < b_n$ for $n > M$. From the Comparison Test, $\Sigma a_n$ converges.

**67.** (a) Let $\displaystyle\sum a_n = \sum \frac{1}{(n+1)^3}$, and $\displaystyle\sum b_n = \sum \frac{1}{n^2}$,

converges.

$$\lim_{n\to\infty} \frac{a_n}{b_n} = \lim_{n\to\infty} \frac{1/\left[(n+1)^3\right]}{1/(n^2)} = \lim_{n\to\infty} \frac{n^2}{(n+1)^3} = 0$$

By Exercise 65, $\displaystyle\sum_{n=1}^{\infty} \frac{1}{(n+1)^3}$ converges.

(b) Let $\displaystyle\sum a_n = \sum \frac{1}{\sqrt{n}\pi^n}$, and $\displaystyle\sum b_n = \sum \frac{1}{\pi^n}$,

converges.

$$\lim_{n\to\infty} \frac{a_n}{b_n} = \lim_{n\to\infty} \frac{1/\left(\sqrt{n}\pi^n\right)}{1/(\pi^n)} = \lim_{n\to\infty} \frac{1}{\sqrt{n}} = 0$$

By Exercise 65, $\displaystyle\sum_{n=1}^{\infty} \frac{1}{\sqrt{n}\pi^n}$ converges.

**69.** Because $\displaystyle\lim_{n\to\infty} a_n = 0$, the terms of $\Sigma \sin(a_n)$ are positive

for sufficiently large $n$. Because

$$\lim_{n\to\infty} \frac{\sin(a_n)}{a_n} = 1 \text{ and } \sum a_n$$

converges, so does $\Sigma \sin(a_n)$.

**71.** First note that $f(x) = \ln x - x^{1/4} = 0$ when

$x \approx 5503.66$. That is,

$\ln n < n^{1/4}$ for $n > 5504$

which implies that

$$\frac{\ln n}{n^{3/2}} < \frac{1}{n^{5/4}} \text{ for } n > 5504.$$

Because $\displaystyle\sum_{n=1}^{\infty} \frac{1}{n^{5/4}}$ is a convergent $p$-series,

$$\sum_{n=1}^{\infty} \frac{\ln n}{n^{3/2}}$$

converges by direct comparison.

**73.** Consider two cases:

If $a_n \geq \dfrac{1}{2^{n+1}}$, then $a_n^{1/(n+1)} \geq \left(\dfrac{1}{2^{n+1}}\right)^{1/(n+1)} = \dfrac{1}{2}$, and

$$a_n^{n/(n+1)} = \frac{a_n}{a_n^{1/(n+1)}} \leq 2a_n.$$

If $a_n \leq \dfrac{1}{2^{n+1}}$, then $a_n^{n/(n+1)} \leq \left(\dfrac{1}{2^{n+1}}\right)^{n/(n+1)} = \dfrac{1}{2^n}$, and

combining, $a_n^{n/(n+1)} \leq 2a_n + \dfrac{1}{2^n}$.

Because $\displaystyle\sum_{n=1}^{\infty} \left(2a_n + \frac{1}{2^n}\right)$ converges, so does $\displaystyle\sum_{n=1}^{\infty} a_n^{n/(n+1)}$

by the Comparison Test.

# Section 9.5   Alternating Series

**1.** $\displaystyle\sum_{n=1}^{\infty} \frac{6}{n^2}$

$S_1 = 6$

$S_2 = 7.5$

$S_3 \approx 8.1667$

Matches (d).

**3.** $\displaystyle\sum_{n=1}^{\infty} \frac{3}{n!}$

$S_1 = 3$

$S_2 = 4.5$

$S_3 = 5.0$

Matches (a).

**5.** $\displaystyle\sum_{n=1}^{\infty} \frac{10}{n2^n}$

$S_1 = 5$

$S_2 = 6.25$

Matches (e).

**7.** $\displaystyle\sum_{n=1}^{\infty} \frac{(-1)^{n-1}}{2n-1} = \frac{\pi}{4} \approx 0.7854$

(a)

| $n$ | 1 | 2 | 3 | 4 | 5 | 6 | 7 | 8 | 9 | 10 |
|-----|---|--------|--------|--------|--------|--------|--------|--------|--------|--------|
| $S_n$ | 1 | 0.6667 | 0.8667 | 0.7238 | 0.8349 | 0.7440 | 0.8209 | 0.7543 | 0.8131 | 0.7605 |

(b)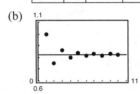

(c) The points alternate sides of the horizontal line $y = \dfrac{\pi}{4}$ that represents the sum of the series.

   The distance between successive points and the line decreases.

(d) The distance in part (c) is always less than the magnitude of the next term of the series.

**9.** $\displaystyle\sum_{n=1}^{\infty} \dfrac{(-1)^{n-1}}{n^2} = \dfrac{\pi^2}{12} \approx 0.8225$

(a)

| $n$ | 1 | 2 | 3 | 4 | 5 | 6 | 7 | 8 | 9 | 10 |
|---|---|---|---|---|---|---|---|---|---|---|
| $S_n$ | 1 | 0.75 | 0.8611 | 0.7986 | 0.8386 | 0.8108 | 0.8312 | 0.8156 | 0.8280 | 0.8180 |

(b) 1.1

   0.6  0  11

(c) The points alternate sides of the horizontal line $y = \dfrac{\pi^2}{12}$ that represents the sum of the series.

   The distance between successive points and the line decreases.

(d) The distance in part (c) is always less than the magnitude of the next term in the series.

**11.** $\displaystyle\sum_{n=1}^{\infty} \dfrac{(-1)^{n+1}}{n+1}$

$a_{n+1} = \dfrac{1}{n+2} < \dfrac{1}{n+1} = a_n$

$\displaystyle\lim_{n\to\infty} a_n = \lim_{n\to\infty} \dfrac{1}{n+1} = 0$

Converges by Theorem 9.14

**13.** $\displaystyle\sum_{n=1}^{\infty} \dfrac{(-1)^n}{3^n}$

$a_{n+1} = \dfrac{1}{3^{n+1}} < \dfrac{1}{3^n} = a_n$

$\displaystyle\lim_{n\to 0} \dfrac{1}{3^n} = 0$

Converges by Theorem 9.14

(**Note:** $\displaystyle\sum_{n=1}^{\infty} \left(\dfrac{-1}{3}\right)^n$ is a convergent geometric series)

**15.** $\displaystyle\sum_{n=1}^{\infty} \dfrac{(-1)^n (5n-1)}{4n+1}$

$\displaystyle\lim_{n\to\infty} \dfrac{5n-1}{4n+1} = \dfrac{5}{4}$

Diverges by $n$th-Term test

**17.** $\displaystyle\sum_{n=1}^{\infty} \dfrac{(-1)^{n+1}}{3n+2}$

$a_{n+1} = \dfrac{1}{3(n+1)+2} < \dfrac{1}{3n+2} = a_n$

$\displaystyle\lim_{n\to\infty} \dfrac{1}{3n+2} = 0$

Converges by Theorem 9.14

**19.** $\displaystyle\sum_{n=1}^{\infty} \dfrac{(-1)^n n^2}{n^2+5}$

$\displaystyle\lim_{n\to\infty} \dfrac{n^2}{n^2+5} = 1$

Diverges by $n$th-Term test

**21.** $\displaystyle\sum_{n=1}^{\infty} \dfrac{(-1)^n}{\sqrt{n}}$

$a_{n+1} = \dfrac{1}{\sqrt{n+1}} < \dfrac{1}{\sqrt{n}} = a_n$

$\displaystyle\lim_{n\to\infty} \dfrac{1}{\sqrt{n}} = 0$

Converges by Theorem 9.14

**23.** $\displaystyle\sum_{n=1}^{\infty} \frac{(-1)^{n+1}(n+1)}{\ln(n+1)}$

$$\lim_{n\to\infty} \frac{n+1}{\ln(n+1)} = \lim_{n\to\infty} \frac{1}{1/(n+1)} = \lim_{n\to\infty} (n+1) = \infty$$

Diverges by the *n*th-Term Test

**25.** $\displaystyle\sum_{n=1}^{\infty} \sin\left[\frac{(2n-1)\pi}{2}\right] = \sum_{n=1}^{\infty}(-1)^{n+1}$

Diverges by the *n*th-Term Test

**27.** $\displaystyle\sum_{n=1}^{\infty} \cos n\pi = \sum_{n=1}^{\infty}(-1)^n$

Diverges by the *n*th-Term Test

**33.** $\displaystyle\sum_{n=1}^{\infty} \frac{(-1)^{n+1} n!}{1\cdot 3\cdot 5\cdots(2n-1)}$

$$a_{n+1} = \frac{(n+1)!}{1\cdot 3\cdot 5\cdots(2n-1)(2n+1)} = \frac{n!}{1\cdot 3\cdot 5\cdots(2n-1)} \cdot \frac{n+1}{2n+1} = a_n\left(\frac{n+1}{2n+1}\right) < a_n$$

$$\lim_{n\to\infty} a_n = \lim_{n\to\infty} \frac{n!}{1\cdot 3\cdot 5\cdots(2n-1)} = \lim_{n\to\infty}\frac{1\cdot 2\cdot 3\cdots n}{1\cdot 3\cdot 5\cdots(2n-1)} = \lim_{n\to\infty} 2\left[\frac{3}{3}\cdot\frac{4}{5}\cdot\frac{5}{7}\cdots\frac{n}{2n-3}\right]\cdot\frac{1}{2n-1} = 0$$

Converges by Theorem 9.14

**35.** $\displaystyle\sum_{n=1}^{\infty} \frac{(-1)^{n+1}(2)}{e^n - e^{-n}} = \sum_{n=1}^{\infty}\frac{(-1)^{n+1}(2e^n)}{e^{2n}-1}$

Let $f(x) = \dfrac{2e^x}{e^{2x}-1}$. Then

$$f'(x) = \frac{-2e^x(e^{2x}+1)}{(e^{2x}-1)^2} < 0.$$

So, $f(x)$ is decreasing. Therefore, $a_{n+1} < a_n$, and

$$\lim_{n\to\infty}\frac{2e^n}{e^{2n}-1} = \lim_{n\to\infty}\frac{2e^n}{2e^{2n}} = \lim_{n\to\infty}\frac{1}{e^n} = 0.$$

The series converges by Theorem 9.14.

**37.** $S_6 = \displaystyle\sum_{n=0}^{5}\frac{2(-1)^n}{n!} \approx 0.7333$

$0.7333 - 0.002778 \le S \le 0.7333 + 0.002778$

$|R_6| = |S - S_6| \le a_7 = \dfrac{2}{6!} = 0.002778$

$0.7305 \le S \le 0.7361$

**29.** $\displaystyle\sum_{n=0}^{\infty}\frac{(-1)^n}{n!}$

$$a_{n+1} = \frac{1}{(n+1)!} < \frac{1}{n!} = a_n$$

$$\lim_{n\to\infty}\frac{1}{n!} = 0$$

Converges by Theorem 9.14

**31.** $\displaystyle\sum_{n=1}^{\infty}\frac{(-1)^{n+1}\sqrt{n}}{n+2}$

$$a_{n+1} = \frac{\sqrt{n+1}}{(n+1)+2} < \frac{\sqrt{n}}{n+2} \text{ for } n \ge 2$$

$$\lim_{n\to\infty}\frac{\sqrt{n}}{n+2} = 0$$

Converges by Theorem 9.14

**39.** $S_6 = \displaystyle\sum_{n=1}^{6}\frac{3(-1)^{n+1}}{n^2} = 2.4325$

$2.4325 - 0.0612 \le S \le 2.4325 + 0.0612$

$|R_6| = |S - S_6| \le a_7 = \dfrac{3}{49} \approx 0.0612$

$2.3713 \le S \le 2.4937$

**41.** $\displaystyle\sum_{n=0}^{\infty}\frac{(-1)^n}{n!}$

(a) By Theorem 9.15,

$$|R_N| \le a_{N+1} = \frac{1}{(N+1)!} < 0.001.$$

This inequality is valid when $N = 6$.

(b) Approximate the series by

$$\sum_{n=0}^{6}\frac{(-1)^n}{n!} = 1 - 1 + \frac{1}{2} - \frac{1}{6} + \frac{1}{24} - \frac{1}{120} + \frac{1}{720}$$

$$\approx 0.368.$$

(7 terms. Note that the sum begins with $n = 0$.)

**43.** $\displaystyle\sum_{n=0}^{\infty} \frac{(-1)^n}{(2n+1)!}$

(a) By Theorem 9.15,

$$|R_N| \le a_{N+1} = \frac{1}{[2(N+1)+1]!} < 0.001.$$

This inequality is valid when $N = 2$.

(b) Approximate the series by

$$\sum_{n=0}^{2} \frac{(-1)^n}{(2n+1)!} = 1 - \frac{1}{6} + \frac{1}{120} \approx 0.842.$$

(3 terms. Note that the sum begins with $n = 0$.)

**45.** $\displaystyle\sum_{n=1}^{\infty} \frac{(-1)^{n+1}}{n!}$

(a) By Theorem 9.15,

$$|R_N| \le a_{N+1} = \frac{1}{N+1} < 0.001.$$

This inequality is valid when $N = 1000$.

(b) Approximate the series by

$$\sum_{n=1}^{1000} \frac{(-1)^{n+1}}{n} = 1 - \frac{1}{2} + \frac{1}{3} - \frac{1}{4} + \cdots - \frac{1}{1000}$$

$$\approx 0.693.$$

(1000 terms)

**47.** $\displaystyle\sum_{n=1}^{\infty} \frac{(-1)^{n+1}}{n^3}$

By Theorem 9.15,

$$|R_N| \le a_{N+1} = \frac{1}{(N+1)^3} < 0.001$$

$$\Rightarrow (N+1)^3 > 1000 \Rightarrow N + 1 > 10.$$

Use 10 terms.

**49.** $\displaystyle\sum_{n=1}^{\infty} \frac{(-1)^{n+1}}{2n^3 - 1}$

By Theorem 9.15,

$$|R_N| \le a_{N+1} = \frac{1}{2(N+1)^3 - 1} < 0.001.$$

This inequality is valid when $N = 7$. Use 7 terms.

**51.** $\displaystyle\sum_{n=1}^{\infty} \frac{(-1)^n}{2^n}$

$\displaystyle\sum_{n=1}^{\infty} \frac{1}{2^n}$ is a convergent geometric series.

Therefore, $\displaystyle\sum_{n=1}^{\infty} \frac{(-1)^n}{2^n}$ converges absolutely.

**53.** $\displaystyle\sum_{n=1}^{\infty} \frac{(-1)^n}{n!}$

$$\frac{1}{n!} < \frac{1}{n^2} \text{ for } n \ge 4$$

and $\displaystyle\sum_{n=1}^{\infty} \frac{1}{n^2}$ is a convergent $p$-series.

So, $\displaystyle\sum_{n=1}^{\infty} \frac{1}{n!}$ converges, and

$\displaystyle\sum_{n=1}^{\infty} \frac{(-1)^n}{n!}$ converges absolutely.

**55.** $\displaystyle\sum_{n=1}^{\infty} \frac{(-1)^{n+1}}{(n+3)^2}$

$\displaystyle\sum_{n=1}^{\infty} \frac{1}{(n+3)^2}$ converges by a limit comparison to the

$p$-series $\displaystyle\sum_{n=1}^{\infty} \frac{1}{n^2}$. Therefore, $\displaystyle\sum_{n=1}^{\infty} \frac{(-1)^{n+1}}{(n+3)^2}$ converges

absolutely.

**57.** $\displaystyle\sum_{n=1}^{\infty} \frac{(-1)^{n+1}}{\sqrt{n}}$

The given series converges by the Alternating Series Test, but does not converge absolutely because

$$\sum_{n=1}^{\infty} \frac{1}{\sqrt{n}}$$

is a divergent $p$-series. Therefore, the series converges conditionally.

**59.** $\displaystyle\sum_{n=1}^{\infty} \frac{(-1)^{n+1} n^2}{(n+1)^2}$

$$\lim_{n \to \infty} \frac{n^2}{(n+1)^2} = 1$$

Therefore, the series diverges by the $n$th-Term Test.

**61.** $\displaystyle\sum_{n=2}^{\infty} \frac{(-1)^n}{n \ln n}$

The series converges by the Alternating Series Test.

Let $f(x) = \dfrac{1}{x \ln x}$.

$$\int_2^{\infty} \frac{1}{x \ln x} \, dx = \left[ \ln(\ln x) \right]_2^{\infty} = \infty$$

By the Integral Test, $\displaystyle\sum_{n=2}^{\infty} \frac{1}{n \ln n}$ diverges.

So, the series $\displaystyle\sum_{n=2}^{\infty} \frac{(-1)^n}{n \ln n}$ converges conditionally.

**63.** $\displaystyle\sum_{n=2}^{\infty} \frac{(-1)^n n}{n^3 - 5}$

$\displaystyle\sum_{n=2}^{\infty} \frac{n}{n^3 - 5}$ converges by a limit comparison to the

*p*-series $\displaystyle\sum_{n=2}^{\infty} \frac{1}{n^2}$. Therefore, the given series converges

absolutely.

**65.** $\displaystyle\sum_{n=0}^{\infty} \frac{(-1)^n}{(2n+1)!}$

$\displaystyle\sum_{n=0}^{\infty} \frac{1}{(2n+1)!}$

is convergent by comparison to the convergent geometric series

$\displaystyle\sum_{n=0}^{\infty} \left(\frac{1}{2}\right)^n$

because

$\dfrac{1}{(2n+1)!} < \dfrac{1}{2^n}$ for $n > 0$.

Therefore, the given series converges absolutely.

**67.** $\displaystyle\sum_{n=0}^{\infty} \frac{\cos n\pi}{n+1} = \sum_{n=0}^{\infty} \frac{(-1)^n}{n+1}$

The given series converges by the Alternating Series Test, but

$\displaystyle\sum_{n=0}^{\infty} \frac{\left|\cos n\pi\right|}{n+1} = \sum_{n=0}^{\infty} \frac{1}{n+1}$

diverges by a limit comparison to the divergent harmonic series,

$\displaystyle\sum_{n=1}^{\infty} \frac{1}{n}$.

$\displaystyle\lim_{n\to\infty} \frac{\left|\cos n\pi\right|/(n+1)}{1/n} = 1$, therefore, the series

converges conditionally.

**69.** $\displaystyle\sum_{n=1}^{\infty} \frac{\cos n\pi}{n^2} = \sum_{n=1}^{\infty} \frac{(-1)^n}{n^2}$

$\displaystyle\sum_{n=1}^{\infty} \frac{1}{n^2}$ is a convergent *p*-series.

Therefore, the given series converges absolutely.

**71.** An alternating series is a series whose terms alternate in sign.

**73.** $\left|S - S_N\right| = \left|R_N\right| \le a_{N+1}$   (Theorem 9.15)

**75.** (b). The partial sums alternate above and below the horizontal line representing the sum.

**77.** True. $S_{100} = -1 + \frac{1}{2} - \frac{1}{3} + \cdots + \frac{1}{100}$

Because the next term $-\frac{1}{101}$ is negative, $S_{100}$ is an overestimate of the sum.

**79.** $\displaystyle\sum_{n=1}^{\infty} (-1)^n \frac{1}{n^p}$

If $p = 0$, then $\displaystyle\sum_{n=1}^{\infty} (-1)^n$ diverges.

If $p < 0$, then $\displaystyle\sum_{n=1}^{\infty} (-1)^n n^{-p}$ diverges.

If $p > 0$, then $\displaystyle\lim_{n\to\infty} \frac{1}{n^p} = 0$ and

$a_{n+1} = \dfrac{1}{(n+1)^p} < \dfrac{1}{n^p} = a_n$.

Therefore, the series converges for $p > 0$.

**81.** Because

$\displaystyle\sum_{n=1}^{\infty} \left|a_n\right|$

converges you have $\displaystyle\lim_{n\to\infty} \left|a_n\right| = 0$. So, there must exist

an $N > 0$ such that $\left|a_N\right| < 1$ for all $n > N$ and it

follows that $a_n^2 \le \left|a_n\right|$ for all $n > N$. So, by the

Comparison Test,

$\displaystyle\sum_{n=1}^{\infty} a_n^2$

converges. Let $a_n = 1/n$ to see that the converse is false.

**83.** $\displaystyle\sum_{n=1}^{\infty} \frac{1}{n^2}$ converges, and so does $\displaystyle\sum_{n=1}^{\infty} \frac{1}{n^4}$.

**85.** (a) No, the series does not satisfy $a_{n+1} \le a_n$ for all $n$.

For example, $\dfrac{1}{9} < \dfrac{1}{8}$.

(b) Yes, the series converges.

$S_{2n} = \dfrac{1}{2} - \dfrac{1}{3} + \cdots + \dfrac{1}{2^n} - \dfrac{1}{3^n}$

$= \left(\dfrac{1}{2} + \cdots + \dfrac{1}{2^n}\right) - \left(\dfrac{1}{3} + \cdots + \dfrac{1}{3^n}\right)$

$= \left(1 + \dfrac{1}{2} + \cdots + \dfrac{1}{2^n}\right) - \left(1 + \dfrac{1}{3} + \cdots + \dfrac{1}{3^n}\right)$

As $n \to \infty$,

$S_{2n} \to \dfrac{1}{1 - (1/2)} - \dfrac{1}{1 - (1/3)} = 2 - \dfrac{3}{2} = \dfrac{1}{2}$.

**87.** $\displaystyle\sum_{n=1}^{\infty} \frac{10}{n^{3/2}} = 10\sum_{n=1}^{\infty} \frac{1}{n^{3/2}}$,

convergent $p$-series

**89.** Diverges by $n$th-Term Test

$\displaystyle\lim_{n\to\infty} a_n = \infty$

**91.** Convergent geometric series

$\left(r = \frac{7}{8} < 1\right)$

**99.** $s = 1 - \dfrac{1}{2} + \dfrac{1}{3} - \dfrac{1}{4} + \cdots$

$S = 1 + \dfrac{1}{3} - \dfrac{1}{2} + \dfrac{1}{5} + \dfrac{1}{7} - \dfrac{1}{4} + \dfrac{1}{9} + \dfrac{1}{11} - \dfrac{1}{6} + \cdots$

(i) $s_{4n} = 1 - \dfrac{1}{2} + \dfrac{1}{3} - \dfrac{1}{4} + \dfrac{1}{5} - \dfrac{1}{6} + \cdots + \dfrac{1}{4n-1} - \dfrac{1}{4n}$

$\dfrac{1}{2}s_{2n} = \dfrac{1}{2} - \dfrac{1}{4} + \dfrac{1}{6} - \dfrac{1}{8} + \dfrac{1}{10} - \cdots + \dfrac{1}{4n-2} - \dfrac{1}{4n}$

Adding: $s_{4n} + \dfrac{1}{2}s_{2n} = 1 + \dfrac{1}{3} - \dfrac{1}{2} + \dfrac{1}{5} + \dfrac{1}{7} - \dfrac{1}{4} + \cdots + \dfrac{1}{4n-3} + \dfrac{1}{4n-1} - \dfrac{1}{2n} = s_{3n}$

(ii) $\displaystyle\lim_{n\to\infty} s_n = s$   (In fact, $s = \ln 2$.)

$s \neq 0$ because $s > \dfrac{1}{2}$.

$S = \displaystyle\lim_{n\to\infty} S_{3n} = s_{4n} + \dfrac{1}{2}s_{2n} = s + \dfrac{1}{2}s = \dfrac{3}{2}s$

So, $S \neq s$.

**93.** Convergent geometric series $\left(r = 1/\sqrt{e}\right)$ or Integral Test

**95.** Converges (absolutely) by Alternating Series Test

**97.** The first term of the series is zero, not one. You cannot regroup series terms arbitrarily.

## Section 9.6   The Ratio and Root Tests

**1.** $\dfrac{(n+1)!}{(n+2)!} = \dfrac{(n+1)(n)(n-1)(n-2)!}{(n-2)!} = (n+1)(n)(n-1)$

**3.** Use the Principle of Mathematical Induction. When $k = 1$, the formula is valid because $1 = \dfrac{(2(1))!}{2^1 \cdot 1!}$. Assume that

$1 \cdot 3 \cdot 5 \cdots (2n-1) = \dfrac{(2n)!}{2^n \, n!}$

and show that

$1 \cdot 3 \cdot 5 \cdots (2n-1)(2n+1) = \dfrac{(2n+2)!}{2^{n+1}(n+1)!}$.

To do this, note that:

$1 \cdot 3 \cdot 5 \cdots (2n-1)(2n+1) = \left[1 \cdot 3 \cdot 5 \cdots (2n-1)\right](2n+1)$

$= \dfrac{(2n)!}{2^n \, n!} \cdot (2n+1)$   (Induction hypothesis)

$= \dfrac{(2n)!(2n+1)}{2^n \, n!} \cdot \dfrac{(2n+2)}{2(n+1)} = \dfrac{(2n)!(2n+1)(2n+2)}{2^{n+1} \, n!(n+1)} = \dfrac{(2n+2)!}{2^{n+1}(n+1)!}$

The formula is valid for all $n \geq 1$.

**5.** $\displaystyle\sum_{n=1}^{\infty} n\left(\frac{3}{4}\right)^n = 1\left(\frac{3}{4}\right) + 2\left(\frac{9}{16}\right) + \cdots$

$S_1 = \frac{3}{4}, S_2 \approx 1.875$

Matches (d).

**7.** $\displaystyle\sum_{n=1}^{\infty} \frac{(-3)^{n+1}}{n!} = 9 - \frac{3^3}{2} + \cdots$

$S_1 = 9$

Matches ( f ).

**9.** $\displaystyle\sum_{n=1}^{\infty} \left(\frac{4n}{5n-3}\right)^n = \frac{4}{2} + \left(\frac{8}{7}\right)^2 + \cdots$

$S_1 = 2, S_2 = 3.31$

Matches (a).

**11.** (a) Ratio Test: $\displaystyle\lim_{n\to\infty}\left|\frac{a_{n+1}}{a_n}\right| = \lim_{n\to\infty}\frac{(n+1)^2(5/8)^{n+1}}{n^2(5/8)^n} = \lim_{n\to\infty}\left(\frac{n+1}{n}\right)^2\frac{5}{8} = \frac{5}{8} < 1.$ Converges

(b)

| $n$ | 5 | 10 | 15 | 20 | 25 |
|-----|-----|-----|-----|-----|-----|
| $S_n$ | 9.2104 | 16.7598 | 18.8016 | 19.1878 | 19.2491 |

(c)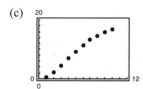

(d) The sum is approximately 19.26.

(e) The more rapidly the terms of the series approach 0, the more rapidly the sequence of the partial sums approaches the sum of the series.

**13.** $\displaystyle\sum_{n=1}^{\infty} \frac{1}{2^n}$

$\displaystyle\lim_{n\to\infty}\left|\frac{a_{n+1}}{a_n}\right| = \lim_{n\to\infty}\left|\frac{1/2^{n+1}}{1/2^n}\right| = \lim_{n\to\infty}\frac{2^n}{2^{n+1}} = \frac{1}{2} < 1$

Therefore, the series converges by the Ratio Test.

**15.** $\displaystyle\sum_{n=0}^{\infty} \frac{n!}{3^n}$

$\displaystyle\lim_{n\to\infty}\left|\frac{a_{n+1}}{a_n}\right| = \lim_{n\to\infty}\left|\frac{(n+1)!}{3^{n+1}} \cdot \frac{3^n}{n!}\right| = \lim_{n\to\infty}\frac{n+1}{3} = \infty$

Therefore, by the Ratio Test, the series diverges.

**17.** $\displaystyle\sum_{n=1}^{\infty} n\left(\frac{6}{5}\right)^n$

$\displaystyle\lim_{n\to\infty}\left|\frac{a_{n+1}}{a_n}\right| = \lim_{n\to\infty}\frac{(n+1)(6/5)^{n+1}}{n(6/5)^n}$

$\displaystyle = \lim_{n\to\infty}\frac{n+1}{n}\left(\frac{6}{5}\right) = \frac{6}{5} > 1$

Therefore, the series diverges by the Ratio Test.

**19.** $\displaystyle\sum_{n=1}^{\infty}\frac{n}{4^n}$

$$\lim_{n\to\infty}\left|\frac{a_{n+1}}{a_n}\right|=\lim_{n\to\infty}\frac{(n+1)/4^{n+1}}{n/4^n}=\lim_{n\to\infty}\frac{n+1}{4n}=1/4<1$$

Therefore, the series converges by the Ratio Test.

**21.** $\displaystyle\sum_{n=1}^{\infty}\frac{4^n}{n^2}$

$$\lim_{n\to\infty}\left|\frac{a_{n+1}}{a_n}\right|=\lim_{n\to\infty}\frac{4^{n+1}/(n+1)^2}{4^n/n^2}$$

$$=\lim_{n\to\infty}\frac{n^2}{(n+1)^2}(4)=4>1$$

Therefore, the series diverges by the Ratio Test.

**23.** $\displaystyle\sum_{n=0}^{\infty}\frac{(-1)^n 2^n}{n!}$

$$\lim_{n\to\infty}\left|\frac{a_{n+1}}{a_n}\right|=\lim_{n\to\infty}\left|\frac{2^{n+1}}{(n+1)!}\cdot\frac{n!}{2^n}\right|$$

$$=\lim_{n\to\infty}\frac{2}{n+1}=0$$

Therefore, by the Ratio Test, the series converges.

**29.** $\displaystyle\sum_{n=0}^{\infty}\frac{6^n}{(n+1)^n}$

$$\lim_{n\to\infty}\left|\frac{a_{n+1}}{a_n}\right|=\lim_{n\to\infty}\frac{6^{n+1}/(n+2)^{n+1}}{6^n/(n+1)^n}=\lim_{n\to\infty}\frac{6}{n+2}\left(\frac{n+1}{n+2}\right)^n=0\left(\frac{1}{e}\right)=0.$$

To find $\displaystyle\lim_{n\to\infty}\left(\frac{n+1}{n+2}\right)^n$: Let $y=\left(\frac{n+1}{n+2}\right)^n$

$$\ln y=n\ln\left(\frac{n+1}{n+2}\right)=\frac{\ln(n+1)-\ln(n+2)}{1/n}$$

$$\lim_{n\to\infty}\left[\ln y\right]=\lim_{n\to\infty}\left[\frac{1/(n+1)-1/(n+2)}{-1/n^2}\right]=\lim_{n\to\infty}\left[\frac{-n^2\left[(n+2)-(n+1)\right]}{(n+1)(n+2)}\right]=-1$$

by L'Hôpital's Rule. So, $y\to\dfrac{1}{e}$.

Therefore, the series converges by the Ratio Test.

**31.** $\displaystyle\sum_{n=0}^{\infty}\frac{5^n}{2^n+1}$

$$\lim_{n\to\infty}\left|\frac{a_{n+1}}{a_n}\right|=\lim_{n\to\infty}\frac{5^{n+1}/(2^{n+1}+1)}{5^n/(2^n+1)}=\lim_{n\to\infty}\frac{5(2^n+1)}{(2^{n+1}+1)}=\lim_{n\to\infty}\frac{5(1+1/2^n)}{2+1/2^n}=\frac{5}{2}>1$$

Therefore, the series diverges by the Ratio Test.

**25.** $\displaystyle\sum_{n=1}^{\infty}\frac{n!}{n3^n}$

$$\lim_{n\to\infty}\left|\frac{a_{n+1}}{a_n}\right|=\lim_{n\to\infty}\left|\frac{(n+1)!}{(n+1)3^{n+1}}\cdot\frac{n3^n}{n!}\right|=\lim_{n\to\infty}\frac{n}{3}=\infty$$

Therefore, by the Ratio Test, the series diverges.

**27.** $\displaystyle\sum_{n=0}^{\infty}\frac{e^n}{n!}$

$$\lim_{n\to\infty}\left|\frac{a_{n+1}}{a_n}\right|=\lim_{n\to\infty}\frac{e^{n+1}/(n+1)!}{e^n/n!}$$

$$=\lim_{n\to\infty}e\left(\frac{n!}{(n+1)!}\right)=\lim_{n\to\infty}\frac{e}{n+1}=0$$

Therefore, the series converges by the Ratio Test.

**33.** $\displaystyle\sum_{n=0}^{\infty} \frac{(-1)^{n+1} n!}{1 \cdot 3 \cdot 5 \cdots (2n+1)}$

$$\lim_{n \to \infty} \left| \frac{a_{n+1}}{a_n} \right| = \lim_{n \to \infty} \left| \frac{(n+1)!}{1 \cdot 3 \cdot 5 \cdots (2n+1)(2n+3)} \cdot \frac{1 \cdot 3 \cdot 5 \cdots (2n+1)}{n!} \right| = \lim_{n \to \infty} \frac{n+1}{2n+3} = \frac{1}{2}$$

Therefore, by the Ratio Test, the series converges.

**Note:** The first few terms of this series are $-1 + \dfrac{1}{1 \cdot 3} - \dfrac{2!}{1 \cdot 3 \cdot 5} + \dfrac{3!}{1 \cdot 3 \cdot 5 \cdot 7} - \cdots$.

**35.** $\displaystyle\sum_{n=1}^{\infty} \frac{1}{5^n}$

$$\lim_{n \to \infty} \sqrt[n]{|a_n|} = \lim_{n \to \infty} \left[ \frac{1}{5^n} \right]^{1/n} = \frac{1}{5} < 1$$

Therefore, by the Root Test, the series converges.

**37.** $\displaystyle\sum_{n=1}^{\infty} \left( \frac{n}{2n+1} \right)^n$

$$\lim_{n \to \infty} \sqrt[n]{|a_n|} = \lim_{n \to \infty} \sqrt[n]{\left( \frac{n}{2n+1} \right)^n} = \lim_{n \to \infty} \frac{n}{2n+1} = \frac{1}{2}$$

Therefore, by the Root Test, the series converges.

**39.** $\displaystyle\sum_{n=2}^{\infty} \left( \frac{2n+1}{n-1} \right)^n$

$$\lim_{n \to \infty} \sqrt[n]{|a_n|} = \lim_{n \to \infty} \sqrt[n]{\left( \frac{2n+1}{n-1} \right)^n} = \lim_{n \to \infty} \left( \frac{2n+1}{n-1} \right) = 2$$

Therefore, by the Root Test, the series diverges.

**41.** $\displaystyle\sum_{n=2}^{\infty} \frac{(-1)^n}{(\ln n)^n}$

$$\lim_{n \to \infty} \sqrt[n]{|a_n|} = \lim_{n \to \infty} \sqrt[n]{\left| \frac{(-1)^n}{(\ln n)^n} \right|} = \lim_{n \to \infty} \frac{1}{|\ln n|} = 0$$

Therefore, by the Root Test, the series converges.

**43.** $\displaystyle\sum_{n=1}^{\infty} \left( 2\sqrt[n]{n} + 1 \right)^n$

$$\lim_{n \to \infty} \sqrt[n]{|a_n|} = \lim_{n \to \infty} \sqrt[n]{\left( 2\sqrt[n]{n} + 1 \right)^n} = \lim_{n \to \infty} \left( 2\sqrt[n]{n} + 1 \right)$$

To find $\displaystyle\lim_{n \to \infty} \sqrt[n]{n}$, let $y = \lim_{n \to \infty} \sqrt[n]{n}$. Then

$$\ln y = \lim_{n \to \infty} \left( \ln \sqrt[n]{n} \right)$$

$$= \lim_{n \to \infty} \frac{1}{n} \ln n = \lim_{n \to \infty} \frac{\ln n}{n} = \lim_{n \to \infty} \frac{1/n}{1} = 0.$$

So, $\ln y = 0$, so $y = e^0 = 1$ and

$$\lim_{n \to \infty} \left( 2\sqrt[n]{n} + 1 \right) = 2(1) + 1 = 3.$$

Therefore, by the Root Test, the series diverges.

**45.** $\displaystyle\sum_{n=1}^{\infty} \frac{n}{3^n}$

$$\lim_{n \to \infty} \sqrt[n]{|a_n|} = \lim_{n \to \infty} \left( \frac{n}{3^n} \right)^{1/n} = \lim_{n \to \infty} \frac{n^{1/n}}{3} = \frac{1}{3}$$

Therefore, the series converges by the Root Test.

**Note:** You can use L'Hôpital's Rule to show $\displaystyle\lim_{n \to \infty} n^{1/n} = 1$:

Let $y = n^{1/n}$, $\ln y = \dfrac{1}{n} \ln n = \dfrac{\ln n}{n}$

$$\lim_{n \to \infty} \frac{\ln n}{n} = \lim_{n \to \infty} \frac{1/n}{1} = 0 \Rightarrow y \to 1$$

**47.** $\displaystyle\sum_{n=1}^{\infty} \left( \frac{1}{n} - \frac{1}{n^2} \right)^n$

$$\lim_{n \to \infty} \sqrt[n]{|a_n|} = \lim_{n \to \infty} \sqrt[n]{\left( \frac{1}{n} - \frac{1}{n^2} \right)^n}$$

$$= \lim_{n \to \infty} \left( \frac{1}{n} - \frac{1}{n^2} \right) = 0 - 0 = 0 < 1$$

Therefore, by the Root Test, the series converges.

**49.** $\displaystyle\sum_{n=2}^{\infty} \frac{n}{(\ln n)^n}$

$$\lim_{n \to \infty} \sqrt[n]{|a_n|} = \lim_{n \to \infty} \sqrt[n]{\frac{n}{(\ln n)^n}} = \lim_{n \to \infty} \frac{n^{1/n}}{\ln n} = 0$$

Therefore, by the Root Test, the series converges.

**51.** $\displaystyle\sum_{n=1}^{\infty} \frac{(-1)^{n+1} 5}{n}$

$$a_{n+1} = \frac{5}{n+1} < \frac{5}{n} = a_n$$

$$\lim_{n \to \infty} \frac{5}{n} = 0$$

Therefore, by the Alternating Series Test, the series converges (conditional convergence).

**53.** $\displaystyle\sum_{n=1}^{\infty} \frac{3}{n\sqrt{n}} = 3 \sum_{n=1}^{\infty} \frac{1}{n^{3/2}}$

This is a convergent *p*-series.

**55.** $\displaystyle\sum_{n=1}^{\infty} \frac{5n}{2n-1}$

$$\lim_{n\to\infty} \frac{5n}{2n-1} = \frac{5}{2}$$

Therefore, the series diverges by the $n$th-Term Test

**57.** $\displaystyle\sum_{n=1}^{\infty} \frac{(-1)^n 3^{n-2}}{2^n} = \sum_{n=1}^{\infty} \frac{(-1)^n 3^n 3^{-2}}{2^n} = \sum_{n=1}^{\infty} \frac{1}{9}\left(-\frac{3}{2}\right)^n$

Because $|r| = \dfrac{3}{2} > 1$, this is a divergent geometric series.

**59.** $\displaystyle\sum_{n=1}^{\infty} \frac{10n+3}{n2^n}$

$$\lim_{n\to\infty} \frac{(10n+3)/n2^n}{1/2^n} = \lim_{n\to\infty} \frac{10n+3}{n} = 10$$

Therefore, the series converges by a Limit Comparison Test with the geometric series

$$\sum_{n=0}^{\infty} \left(\frac{1}{2}\right)^n.$$

**61.** $\left|\dfrac{\cos n}{3^n}\right| \le \dfrac{1}{3^n}$

Therefore the series $\displaystyle\sum_{n=1}^{\infty} \left|\frac{\cos n}{3^n}\right|$ converges

by Direct comparison with the convergent geometric

series $\displaystyle\sum_{n=1}^{\infty} \frac{1}{3^n}$. So, $\displaystyle\sum \frac{\cos n}{3^n}$ converges.

**63.** $\displaystyle\sum_{n=1}^{\infty} \frac{n7^n}{n!}$

$$\lim_{n\to\infty} \left|\frac{a_{n+1}}{a_n}\right| = \lim_{n\to\infty} \left|\frac{(n+1)7^{n+1}}{(n+1)!} \cdot \frac{n!}{n7^n}\right| = \lim_{n\to\infty} \frac{7}{n} = 0$$

Therefore, by the Ratio Test, the series converges.

**65.** $\displaystyle\sum_{n=1}^{\infty} \frac{(-1)^n 3^{n-1}}{n!}$

$$\lim_{n\to\infty} \left|\frac{a_{n+1}}{a_n}\right| = \lim_{n\to\infty} \left|\frac{3^n}{(n+1)!} \cdot \frac{n!}{3^{n-1}}\right| = \lim_{n\to\infty} \frac{3}{n+1} = 0$$

Therefore, by the Ratio Test, the series converges. (Absolutely)

**67.** $\displaystyle\sum_{n=1}^{\infty} \frac{(-3)^n}{3 \cdot 5 \cdot 7 \cdots (2n+1)}$

$$\lim_{n\to\infty} \left|\frac{a_{n+1}}{a_n}\right| = \lim_{n\to\infty} \left|\frac{(-3)^{n+1}}{3 \cdot 5 \cdot 7 \cdots (2n+1)(2n+3)} \cdot \frac{3 \cdot 5 \cdot 7 \cdots (2n+1)}{(-3)^n}\right| = \lim_{n\to\infty} \frac{3}{2n+3} = 0$$

Therefore, by the Ratio Test, the series converges.

**69.** (a) and (c) are the same.

$$\sum_{n=1}^{\infty} \frac{n5^n}{n!} = \sum_{n=0}^{\infty} \frac{(n+1)5^{n+1}}{(n+1)!} = 5 + \frac{(2)(5)^2}{2!} + \frac{(3)(5)^3}{3!} + \frac{(4)(5)^4}{4!} + \cdots$$

**71.** (a) and (b) are the same.

$$\sum_{n=0}^{\infty} \frac{(-1)^n}{(2n+1)!} = \sum_{n=1}^{\infty} \frac{(-1)^{n-1}}{(2n-1)!} = 1 - \frac{1}{3!} + \frac{1}{5!} - \cdots$$

**73.** Replace $n$ with $n+1$.

$$\sum_{n=1}^{\infty} \frac{n}{7^n} = \sum_{n=0}^{\infty} \frac{n+1}{7^{n+1}}$$

**75.** (a) Because

$$\frac{3^{10}}{2^{10}10!} \approx 1.59 \times 10^{-5},$$

use 9 terms.

(b) $\displaystyle\sum_{k=1}^{9} \frac{(-3)^k}{2^k k!} \approx -0.7769$

**77.** $\displaystyle\lim_{n\to\infty} \left|\frac{a_{n+1}}{a_n}\right| = \lim_{n\to\infty} \left|\frac{(4n-1)/(3n+2)a_n}{a_n}\right|$

$$= \lim_{n\to\infty} \frac{4n-1}{3n+2} = \frac{4}{3} > 1$$

The series diverges by the Ratio Test.

**79.** $\displaystyle\lim_{n\to\infty} \left|\frac{a_{n+1}}{a_n}\right| = \lim_{n\to\infty} \left|\frac{(\sin n + 1)/(\sqrt{n})a_n}{a_n}\right|$

$$= \lim_{n\to\infty} \frac{\sin n + 1}{\sqrt{n}} = 0 < 1$$

The series converges by the Ratio Test.

**81.** $\displaystyle\lim_{n\to\infty} \left|\frac{a_{n+1}}{a_n}\right| = \lim_{n\to\infty} \left|\frac{(1+(1)/(n))a_n}{a_n}\right| = \lim_{n\to\infty} \left(1+\frac{1}{n}\right) = 1$

The Ratio Test is inconclusive.

But, $\displaystyle\lim_{n\to\infty} a_n \ne 0$, so the series diverges.

**83.** $\displaystyle \lim_{n \to \infty} \left| \frac{a_{n+1}}{a_n} \right| = \lim_{n \to \infty} \left| \frac{\dfrac{1 \cdot 2 \cdots n(n+1)}{1 \cdot 3 \cdots (2n-1)(2n+1)}}{\dfrac{1 \cdot 2 \cdots n}{1 \cdot 3 \cdots (2n-1)}} \right|$

$\displaystyle = \lim_{n \to \infty} \frac{n+1}{2n+1} = \frac{1}{2} < 1$

The series converges by the Ratio Test.

**85.** $\displaystyle \sum_{n=3}^{\infty} \frac{1}{(\ln n)^n}$

$\displaystyle \lim_{n \to \infty} \sqrt[n]{|a_n|} = \lim_{n \to \infty} \sqrt[n]{\frac{1}{(\ln n)^n}} = \lim_{n \to \infty} \frac{1}{\ln n} = 0$

Therefore, by the Root Test, the series converges.

**87.** $\displaystyle \lim_{n \to \infty} \left| \frac{a_{n+1}}{a_n} \right| = \lim_{n \to \infty} \left| \frac{2(x/3)^{n+1}}{2(x/3)^n} \right| = \lim_{n \to \infty} \left| \frac{x}{3} \right| = \left| \frac{x}{3} \right|$

For the series to converge: $\left| \dfrac{x}{3} \right| < 1 \Rightarrow -3 < x < 3$.

For $x = 3$, the series diverges.

For $x = -3$, the series diverges.

**89.** $\displaystyle \lim_{n \to \infty} \left| \frac{a_{n+1}}{a_n} \right| = \lim_{n \to \infty} \left| \frac{(x+1)^{n+1}/(n+1)}{x^n/n} \right|$

$\displaystyle = \lim_{n \to \infty} \left| \frac{n}{n+1}(x+1) \right| = |x+1|$

For the series to converge,

$|x+1| < 1 \Rightarrow -1 < x+1 < 1$

$\Rightarrow -2 < x < 0.$

For $x = 0$, $\displaystyle \sum_{n=1}^{\infty} \frac{(-1)^n}{n}$, converges.

For $x = -2$, $\displaystyle \sum_{n=1}^{\infty} \frac{(-1)^n(-1)^n}{n} = \sum_{n=1}^{\infty} \frac{1}{n}$, diverges.

**91.** $\displaystyle \lim_{n \to \infty} \left| \frac{a_{n+1}}{a_n} \right| = \lim_{n \to \infty} \frac{(n+1)! \left| \dfrac{x}{2} \right|^{n+1}}{n! \left| \dfrac{x}{2} \right|^n}$

$\displaystyle = \lim_{n \to \infty} (n+1) \left| \frac{x}{2} \right| = \infty$

The series converges only at $x = 0$.

**107.** $\displaystyle \sum_{n=1}^{\infty} \frac{(n!)^2}{(xn)!}$, $x$ positive integer

(a) $x = 1$: $\displaystyle \sum \frac{(n!)^2}{n!} = \sum n!$, diverges

**93.** See Theorem 9.17, page 641.

**95.** No. Let $a_n = \dfrac{1}{n + 10,000}$.

The series $\displaystyle \sum_{n=1}^{\infty} \frac{1}{n + 10,000}$ diverges.

**97.** The series converges absolutely. See Theorem 9.17.

**99.** Assume that

$\displaystyle \lim_{n \to \infty} |a_{n+1}/a_n| = L > 1$ or that $\displaystyle \lim_{n \to \infty} |a_{n+1}/a_n| = \infty$.

Then there exists $N > 0$ such that $|a_{n+1}/a_n| > 1$ for all $n > N$. Therefore,

$|a_{n+1}| > |a_n|, n > N \Rightarrow \displaystyle \lim_{n \to \infty} a_n \neq 0 \Rightarrow \sum a_n$ diverges.

**101.** $\displaystyle \sum_{n=1}^{\infty} \frac{1}{n^{3/2}}$

$\displaystyle \lim_{n \to \infty} \left| \frac{a_{n+1}}{a_n} \right| = \lim_{n \to \infty} \left| \frac{1}{(n+1)^{3/2}} \cdot \frac{n^{3/2}}{1} \right| = \lim_{n \to \infty} \left( \frac{n}{n+1} \right)^{3/2} = 1$

**103.** $\displaystyle \sum_{n=1}^{\infty} \frac{1}{n^4}$

$\displaystyle \lim_{n \to \infty} \left| \frac{a_{n+1}}{a_n} \right| = \lim_{n \to \infty} \left| \frac{1}{(n+1)^4} \cdot \frac{n^4}{1} \right| = \lim_{n \to \infty} \left( \frac{n}{n+1} \right)^4 = 1$

**105.** $\displaystyle \sum_{n=1}^{\infty} \frac{1}{n^p}$, $p$-series

$\displaystyle \lim_{n \to \infty} \sqrt[n]{|a_n|} = \lim_{n \to \infty} \sqrt[n]{\frac{1}{n^p}} = \lim_{n \to \infty} \frac{1}{n^{p/n}} = 1$

So, the Root Test is inconclusive.

**Note:** $\displaystyle \lim_{n \to \infty} n^{p/n} = 1$ because if $y = n^{p/n}$, then

$\ln y = \dfrac{p}{n} \ln n$ and $\dfrac{p}{n} \ln n \to 0$ as $n \to \infty$.

So $y \to 1$ as $n \to \infty$.

(b) $x = 2$: $\sum \dfrac{(n!)^2}{(2n)!}$ converges by the Ratio Test:

$$\lim_{n \to \infty} \dfrac{[(n+1)!]^2}{(2n+2)!} \bigg/ \dfrac{(n!)^2}{(2n)!} = \lim_{n \to \infty} \dfrac{(n+1)^2}{(2n+2)(2n+1)}$$

$$= \dfrac{1}{4} < 1$$

(c) $x = 3$: $\sum \dfrac{(n!)^2}{(3n)!}$ converges by the Ratio Test:

$$\lim_{n \to \infty} \dfrac{[(n+1)!]^2}{(3n+3)!} \bigg/ \dfrac{(n!)^2}{(3n)!} = \lim_{n \to \infty} \dfrac{(n+1)^2}{(3n+3)(3n+2)(3n+1)} = 0 < 1$$

(d) Use the Ratio Test:

$$\lim_{n \to \infty} \dfrac{[(n+1)!]^2}{[x(n+1)]!} \bigg/ \dfrac{(n!)^2}{(xn)!} = \lim_{n \to \infty} (n+1)^2 \dfrac{(xn)!}{(xn+x)!}$$

The cases $x = 1, 2, 3$ were solved above. For $x > 3$, the limit is 0. So, the series converges for all integers $x \geq 2$.

**109.** The differentiation test states that if

$$\sum_{n=1}^{\infty} U_n$$

is an infinite series with real terms and $f(x)$ is a real function such that $f(1/n) = U_n$ for all positive integers $n$ and $d^2 f / dx^2$ exists at $x = 0$, then

$$\sum_{n=1}^{\infty} U_n$$

converges absolutely if $f(0) = f'(0) = 0$ and diverges otherwise. Below are some examples.

| Convergent Series | Divergent Series |
|---|---|
| $\sum \dfrac{1}{n^3}, \ f(x) = x^3$ | $\sum \dfrac{1}{n}, \ f(x) = x$ |
| $\sum \left(1 - \cos \dfrac{1}{n}\right), \ f(x) = 1 - \cos x$ | $\sum \sin \dfrac{1}{n}, \ f(x) = \sin x$ |

**111.** First prove Abel's Summation Theorem:

If the partial sums of $\sum a_n$ are bounded and if $\{b_n\}$ decreases to zero, then $\sum a_n b_n$ converges.

Let $S_k = \displaystyle\sum_{i=1}^{k} a_i$. Let $M$ be a bound for $\{|S_k|\}$.

$$a_1 b_1 + a_2 b_2 + \cdots + a_n b_n = S_1 b_1 + (S_2 - S_1)b_2 + \cdots + (S_n - S_{n-1})b_n$$

$$= S_1(b_1 - b_2) + S_2(b_2 - b_3) + \cdots + S_{n-1}(b_{n-1} - b_n) + S_n b_n$$

$$= \sum_{i=1}^{n-1} S_i(b_i - b_{i+1}) + S_n b_n$$

The series $\displaystyle\sum_{i=1}^{\infty} S_i(b_i - b_{i+1})$ is absolutely convergent because $|S_i(b_i - b_{i+1})| \leq M(b_i - b_{i+1})$ and $\displaystyle\sum_{i=1}^{\infty}(b_i - b_{i+1})$ converges to $b_1$.

Also, $\displaystyle\lim_{n \to \infty} S_n b_n = 0$ because $\{S_n\}$ bounded and $b_n \to 0$. Thus, $\displaystyle\sum_{n=1}^{\infty} a_n b_n = \lim_{n \to \infty} \sum_{i=1}^{n} a_i b_i$ converges.

Now let $b_n = \dfrac{1}{n}$ to finish the problem.

# Section 9.7   Taylor Polynomials and Approximations

**1.** $y = -\frac{1}{2}x^2 + 1$

Parabola

Matches (d)

**3.** $y = e^{-1/2}[(x + 1) + 1]$

Linear

Matches (a)

**5.** $f(x) = \dfrac{8}{\sqrt{x}} = 8x^{-1/2}$   $\qquad f(4) = 4$

$f'(x) = -4x^{-3/2}$   $\qquad f'(4) = -\dfrac{1}{2}$

$P_1(x) = f(4) + f'(4)(x - 4)$

$\qquad\quad = 4 - \dfrac{1}{2}(x - 4) = -\dfrac{1}{2}x + 6$

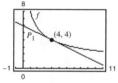

$P_1$ is the first degree Taylor polynomial for $f$ at 4.

**9.** $f(x) = \dfrac{4}{\sqrt{x}} = 4x^{-1/2}$   $\qquad f(1) = 4$

$f'(x) = -2x^{-3/2}$   $\qquad\quad f'(1) = -2$

$f''(x) = 3x^{-5/2}$   $\qquad\quad f''(1) = 3$

$P_2 = f(1) + f'(1)(x - 1) + \dfrac{f''(1)}{2}(x - 1)^2$

$\quad = 4 - 2(x - 1) + \dfrac{3}{2}(x - 1)^2$

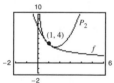

| $x$ | 0 | 0.8 | 0.9 | 1.0 | 1.1 | 1.2 | 2 |
|---|---|---|---|---|---|---|---|
| $f(x)$ | Error | 4.4721 | 4.2164 | 4.0 | 3.8139 | 3.6515 | 2.8284 |
| $P_2(x)$ | 7.5 | 4.46 | 4.215 | 4.0 | 3.815 | 3.66 | 3.5 |

**7.** $f(x) = \sec x$   $\qquad f\left(\dfrac{\pi}{4}\right) = \sqrt{2}$

$f'(x) = \sec x \tan x$   $\qquad f'\left(\dfrac{\pi}{4}\right) = \sqrt{2}$

$P_1(x) = f\left(\dfrac{\pi}{4}\right) + f'\left(\dfrac{\pi}{4}\right)\left(x - \dfrac{\pi}{4}\right)$

$P_1(x) = \sqrt{2} + \sqrt{2}\left(x - \dfrac{\pi}{4}\right)$

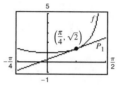

$P_1$ is called the first degree Taylor polynomial for $f$ at $\dfrac{\pi}{4}$.

**11.** $f(x) = \cos x$

$P_2(x) = 1 - \frac{1}{2}x^2$

$P_4(x) = 1 - \frac{1}{2}x^2 + \frac{1}{24}x^4$

$P_6(x) = 1 - \frac{1}{2}x^2 + \frac{1}{24}x^4 - \frac{1}{720}x^6$

(a)

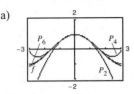

(b) $f'(x) = -\sin x$  $\qquad P_2'(x) = -x$

$f''(x) = -\cos x$  $\qquad P_2''(x) = -1$

$f''(0) = P_2''(0) = -1$

$f'''(x) = \sin x$  $\qquad P_4'''(x) = x$

$f^{(4)}(x) = \cos x$  $\qquad P_4^{(4)}(x) = 1$

$f^{(4)}(0) = 1 = P_4^{(4)}(0)$

$f^{(5)}(x) = -\sin x$  $\qquad P_6^{(5)}(x) = -x$

$f^{(6)}(x) = -\cos x$  $\qquad P^{(6)}(x) = -1$

$f^{(6)}(0) = -1 = P_6^{(6)}(0)$

(c) In general, $f^{(n)}(0) = P_n^{(n)}(0)$ for all $n$.

**13.** $f(x) = e^{3x}$  $\qquad f(0) = 1$

$f'(x) = 3e^{3x}$  $\qquad f'(0) = 3$

$f''(x) = 9e^{3x}$  $\qquad f''(0) = 9$

$f'''(x) = 27e^{3x}$  $\qquad f'''(0) = 27$

$f^{(4)}(x) = 81e^{3x}$  $\qquad f^{(4)}(0) = 81$

$$P_4(x) = f(0) + f'(0)x + \frac{f''(0)}{2!}x^2 + \frac{f'''(0)}{3!}x^3 + \frac{f^{(4)}(0)}{4!}x^4$$

$$= 1 + 3x + \frac{9}{2}x^2 + \frac{9}{2}x^3 + \frac{27}{8}x^4$$

**15.** $f(x) = e^{-x/2}$  $\qquad f(0) = 1$

$f'(x) = -\frac{1}{2}e^{-x/2}$  $\qquad f'(0) = -\frac{1}{2}$

$f''(x) = \frac{1}{4}e^{-x/2}$  $\qquad f''(0) = \frac{1}{4}$

$f'''(x) = -\frac{1}{8}e^{-x/2}$  $\qquad f'''(0) = -\frac{1}{8}$

$f^{(4)}(x) = \frac{1}{16}e^{-x/2}$  $\qquad f^{(4)}(0) = \frac{1}{16}$

$$P_4(x) = f(0) + f'(0)x + \frac{f''(0)}{2!}x^2 + \frac{f'''(0)}{3!}x^3 + \frac{f^{(4)}(0)}{4!}x^4$$

$$= 1 - \frac{1}{2}x + \frac{1}{8}x^2 - \frac{1}{48}x^3 + \frac{1}{384}x^4$$

**17.**   $f(x) = \sin x$         $f(0) = 0$
   $f'(x) = \cos x$         $f'(0) = 1$
   $f''(x) = -\sin x$         $f''(0) = 0$
   $f'''(x) = -\cos x$         $f'''(0) = -1$
   $f^{(4)}(x) = \sin x$         $f^{(4)}(0) = 0$
   $f^{(5)}(x) = \cos x$         $f^{(5)}(0) = 1$

$$P_5(x) = 0 + (1)x + \frac{0}{2!}x^2 + \frac{-1}{3!}x^3 + \frac{0}{4!}x^4 + \frac{1}{5!}x^5$$

$$= x - \frac{1}{6}x^3 + \frac{1}{120}x^5$$

**19.**   $f(x) = xe^x$         $f(0) = 0$
   $f'(x) = xe^x + e^x$         $f'(0) = 1$
   $f''(x) = xe^x + 2e^x$         $f''(0) = 2$
   $f'''(x) = xe^x + 3e^x$         $f'''(0) = 3$
   $f^{(4)}(x) = xe^x + 4e^x$         $f^{(4)}(0) = 4$

$$P_4(x) = 0 + x + \frac{2}{2!}x^2 + \frac{3}{3!}x^3 + \frac{4}{4!}x^4$$

$$= x + x^2 + \frac{1}{2}x^3 + \frac{1}{6}x^4$$

**21.**   $f(x) = \dfrac{1}{x+1} = (x+1)^{-1}$         $f(0) = 1$
   $f'(x) = -(x+1)^{-2}$         $f'(0) = -1$
   $f''(x) = 2(x+1)^{-3}$         $f''(0) = 2$
   $f'''(x) = -6(x+1)^{-4}$         $f'''(0) = -6$
   $f^{(4)}(x) = 24(x+1)^{-5}$         $f^{(4)}(0) = 24$
   $f^{(5)}(x) = -120(x+1)^{-6}$         $f^{(5)}(0) = -120$

$$P_5(x) = 1 - x + \frac{2x^2}{2!} - \frac{6x^3}{3!} + \frac{24x^4}{4!} - \frac{120x^5}{5!}$$

$$= 1 - x + x^2 - x^3 + x^4 - x^5$$

**23.**   $f(x) = \sec x$         $f(0) = 1$
   $f'(x) = \sec x \tan x$         $f'(0) = 0$
   $f''(x) = \sec^3 x + \sec x \tan^2 x$         $f''(0) = 1$

$$P_2(x) = 1 + 0x + \frac{1}{2!}x^2 = 1 + \frac{1}{2}x^2$$

**25.**   $f(x) = \dfrac{2}{x} = 2x^{-1}$         $f(1) = 2$
   $f'(x) = -2x^{-2}$         $f'(1) = -2$
   $f''(x) = 4x^{-3}$         $f''(1) = 4$
   $f'''(x) = -12x^{-4}$         $f'''(1) = -12$

$$P_3(x) = 2 - 2(x-1) + \frac{4}{2!}(x-1)^2 - \frac{12}{3!}(x-1)^3$$

$$= 2 - 2(x-1) + 2(x-1)^2 - 2(x-1)^3$$

**27.**   $f(x) = \sqrt{x} = x^{1/2}$         $f(4) = 2$

$$f'(x) = \frac{1}{2}x^{-1/2} \qquad f'(4) = \frac{1}{4}$$

$$f''(x) = -\frac{1}{4}x^{-3/2} \qquad f''(4) = -\frac{1}{32}$$

$$f'''(x) = \frac{3}{8}x^{-5/2} \qquad f'''(4) = \frac{3}{256}$$

$$P_3(x) = 2 + \frac{1}{4}(x-4) - \frac{1/32}{2!}(x-4)^2 + \frac{3/256}{3!}(x-4)^3$$

$$= 2 + \frac{1}{4}(x-4) - \frac{1}{64}(x-4)^2 + \frac{1}{512}(x-4)^3$$

**29.**   $f(x) = \ln x$         $f(2) = \ln 2$

$$f'(x) = \frac{1}{x} = x^{-1} \qquad f'(2) = 1/2$$

$$f''(x) = -x^{-2} \qquad f''(2) = -1/4$$

$$f'''(x) = 2x^{-3} \qquad f'''(2) = 1/4$$

$$f^{(4)}(x) = -6x^{-4} \qquad f^{(4)}(2) = -3/8$$

$$P_4(x) = \ln 2 + \frac{1}{2}(x-2) - \frac{1/4}{2!}(x-2)^2 + \frac{1/4}{3!}(x-2)^3 - \frac{3/8}{4!}(x-2)^4$$

$$= \ln 2 + \frac{1}{2}(x-2) - \frac{1}{8}(x-2)^2 + \frac{1}{24}(x-2)^3 - \frac{1}{64}(x-2)^4$$

**31.** (a) $P_3(x) = \pi x + \dfrac{\pi^3}{3}x^3$

(b) $Q_3(x) = 1 + 2\pi\left(x - \dfrac{1}{4}\right) + 2\pi^2\left(x - \dfrac{1}{4}\right)^2 + \dfrac{8}{3}\pi^3\left(x - \dfrac{1}{4}\right)^3$

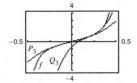

**33.** $f(x) = \sin x$

$P_1(x) = x$

$P_3(x) = x - \dfrac{1}{6}x^3$

$P_5(x) = x - \dfrac{1}{6}x^3 + \dfrac{1}{120}x^5$

(a)

| $x$ | 0.00 | 0.25 | 0.50 | 0.75 | 1.00 |
|---|---|---|---|---|---|
| $\sin x$ | 0.0000 | 0.2474 | 0.4794 | 0.6816 | 0.8415 |
| $P_1(x)$ | 0.0000 | 0.2500 | 0.5000 | 0.7500 | 1.0000 |
| $P_3(x)$ | 0.0000 | 0.2474 | 0.4792 | 0.6797 | 0.8333 |
| $P_5(x)$ | 0.0000 | 0.2474 | 0.4794 | 0.6817 | 0.8417 |

(b)

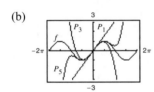

(c) As the distance increases, the accuracy decreases.

**35.** $f(x) = \arcsin x$

(a) $P_3(x) = x + \dfrac{x^3}{6}$

(b)

| $x$ | −0.75 | −0.50 | −0.25 | 0 | 0.25 | 0.50 | 0.75 |
|---|---|---|---|---|---|---|---|
| $f(x)$ | −0.848 | −0.524 | −0.253 | 0 | 0.253 | 0.524 | 0.848 |
| $P_3(x)$ | −0.820 | −0.521 | −0.253 | 0 | 0.253 | 0.521 | 0.820 |

(c)

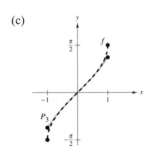

**37.** $f(x) = \cos x$

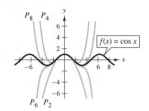

**39.** $f(x) = \ln(x^2 + 1)$

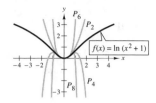

**41.** $f(x) = e^{3x} \approx 1 + 3x + \frac{9}{2}x^2 + \frac{9}{2}x^3 + \frac{27}{8}x^4$

$f\left(\frac{1}{2}\right) \approx 4.3984$

**43.** $f(x) = \ln x \approx \ln(2) + \frac{1}{2}(x - 2) - \frac{1}{8}(x - 2)^2 + \frac{1}{24}(x - 2)^3 - \frac{1}{64}(x - 2)^4$

$f(2.1) \approx 0.7419$

**45.** $f(x) = \cos x;\ f^{(5)}(x) = -\sin x \Rightarrow$ Max on $[0, 0.3]$ is 1.

$R_4(x) \le \frac{1}{5!}(0.3)^5 = 2.025 \times 10^{-5}$

Note: you could use $R_5(x)$: $f^{(6)}(x) = -\cos x$, max on $[0, 0.3]$ is 1.

$R_5(x) \le \frac{1}{6!}(0.3)^6 = 1.0125 \times 10^{-6}$

Exact error: $0.000001 = 1.0 \times 10^{-6}$

**47.** $f(x) = \arcsin x;\ f^{(4)}(x) = \frac{x(6x^2 + 9)}{(1 - x^2)^{7/2}} \Rightarrow$ Max on $[0, 0.4]$ is $f^{(4)}(0.4) \approx 7.3340$.

$R_3(x) \le \frac{7.3340}{4!}(0.4)^4 \approx 0.00782 = 7.82 \times 10^{-3}$. The exact error is $8.5 \times 10^{-4}$. [Note: You could use $R_4$.]

**49.** $g(x) = \sin x$

$\left| g^{(n+1)}(x) \right| \le 1$ for all $x$.

$R_n(x) \le \frac{1}{(n + 1)!}(0.3)^{n+1} < 0.001$

By trial and error, $n = 3$.

**51.** $f(x) = e^x$

$f^{(n+1)}(x) = e^x$

Max on $[0, 0.6]$ is $e^{0.6} \approx 1.8221$.

$R_n \le \frac{1.8221}{(n + 1)!}(0.6)^{n+1} < 0.001$

By trial and error, $n = 5$.

**53.** $f(x) = \ln(x + 1)$

$f^{(n+1)}(x) = \frac{(-1)^n n!}{(x + 1)^{n+1}} \Rightarrow$ Max on $[0, 0.5]$ is $n!$.

$R_n \le \frac{n!}{(n + 1)!}(0.5)^{n+1} = \frac{(0.5)^{n+1}}{n + 1} < 0.0001$

By trial and error, $n = 9$. (See Example 9.) Using 9 terms, $\ln(1.5) \approx 0.4055$.

**55.** $f(x) = e^{-\pi x},\ f(1.3)$

$f'(x) = (-\pi)e^{-\pi x}$

$f^{(n+1)}(x) = (-\pi)^{n+1}e^{-\pi x} \le \left| (-\pi)^{n+1} \right|$ on $[0, 1.3]$

$\left| R_n \right| \le \frac{(\pi)^{n+1}}{(n + 1)!}(1.3)^{n+1} < 0.0001$

By trial and error, $n = 16$. Using 16 terms, $e^{-\pi(1.3)} \approx 0.01684$.

**57.** $f(x) = e^x \approx 1 + x + \dfrac{x^2}{2} + \dfrac{x^3}{6}, \; x < 0$

$R_3(x) = \dfrac{e^z}{4!}x^4 < 0.001$

$e^z x^4 < 0.024$

$\left| xe^{z/4} \right| < 0.3936$

$|x| < \dfrac{0.3936}{e^{z/4}} < 0.3936, \; z < 0$

$-0.3936 < x < 0$

**59.** $f(x) = \cos x \approx 1 - \dfrac{x^2}{2!} + \dfrac{x^4}{4!}$, fifth degree polynomial

$\left| f^{(n+1)}(x) \right| \le 1$ for all $x$ and all $n$.

$\left| R_5(x) \right| \le \dfrac{1}{6!}|x|^6 < 0.001$

$|x|^6 < 0.72$

$|x| < 0.9467$

$-0.9467 < x < 0.9467$

**Note:** Use a graphing utility to graph
$y = \cos x - \left( 1 - x^2/2 + x^4/24 \right)$ in the viewing
window $[-0.9467, 0.9467] \times [-0.001, 0.001]$ to verify
the answer.

**61.** The graph of the approximating polynomial $P$ and the
elementary function $f$ both pass through the point
$(c, f(c))$ and the slopes of $P$ and $f$ agree at
$(c, f(c))$. Depending on the degree of $P$, the $n$th
derivatives of $P$ and $f$ agree at $(c, f(c))$.

**63.** See definition on page 652.

**65.** As the degree of the polynomial increases, the graph of
the Taylor polynomial becomes a better and better
approximation of the function within the interval of
convergence. Therefore, the accuracy is increased.

**67.** (a) $f(x) = e^x$

$P_4(x) = 1 + x + \dfrac{1}{2}x^2 + \dfrac{1}{6}x^3 + \dfrac{1}{24}x^4$

$g(x) = xe^x$

$Q_5(x) = x + x^2 + \dfrac{1}{2}x^3 + \dfrac{1}{6}x^4 + \dfrac{1}{24}x^5$

$Q_5(x) = x\, P_4(x)$

(b) $f(x) = \sin x$

$P_5(x) = x - \dfrac{x^3}{3!} + \dfrac{x^5}{5!}$

$g(x) = x \sin x$

$Q_6(x) = x\, P_5(x) = x^2 - \dfrac{x^4}{3!} + \dfrac{x^6}{5!}$

(c) $g(x) = \dfrac{\sin x}{x} = \dfrac{1}{x}P_5(x) = 1 - \dfrac{x^2}{3!} + \dfrac{x^4}{5!}$

**69.** (a) $Q_2(x) = -1 + \dfrac{\pi^2(x+2)^2}{32}$

(b) $R_2(x) = -1 + \dfrac{\pi^2(x-6)^2}{32}$

(c) No. The polynomial will be linear. Horizontal
translations of the result in part (a) are possible only
at $x = -2 + 8n$ (where $n$ is an integer) because the
period of $f$ is 8.

**71.** Let $f$ be an even function and $P_n$ be the $n$th Maclaurin
polynomial for $f$. Because $f$ is even, $f'$ is odd, $f''$ is
even, $f'''$ is odd, etc. All of the odd derivatives of $f$ are
odd and so, all of the odd powers of $x$ will have
coefficients of zero. $P_n$ will only have terms with even
powers of $x$.

**73.** As you move away from $x = c$, the Taylor Polynomial
becomes less and less accurate.

## Section 9.8 Power Series

**1.** Centered at 0

**3.** Centered at 2

**5.** $\displaystyle\sum_{n=0}^{\infty} (-1)^n \dfrac{x^n}{n+1}$

$L = \lim_{n\to\infty}\left| \dfrac{u_{n+1}}{u_n} \right| = \lim_{n\to\infty}\left| \dfrac{(-1)^{n+1}x^{n+1}}{n+2} \cdot \dfrac{n+1}{(-1)^n x^n} \right|$

$= \lim_{n\to\infty}\left| \dfrac{n+1}{n+2} \right||x| = |x|$

$|x| < 1 \Rightarrow R = 1$

**7.** $\displaystyle\sum_{n=1}^{\infty} \dfrac{(4x)^n}{n^2}$

$L = \lim_{n\to\infty}\left| \dfrac{u_{n+1}}{u_n} \right|$

$= \lim_{n\to\infty}\left| \dfrac{(4x)^{n+1}/(n+1)^2}{(4x)^n/n^2} \right| = \lim_{n\to\infty}\left| \dfrac{n^2}{(n+1)^2}(4x) \right| = 4|x|$

$4|x| < 1 \Rightarrow R = \dfrac{1}{4}$

**9.** $\displaystyle\sum_{n=0}^{\infty} \frac{x^{2n}}{(2n)!}$

$$L = \lim_{n\to\infty} \left| \frac{u_{n+1}}{u_n} \right|$$

$$= \lim_{n\to\infty} \left| \frac{x^{(2n+2)}/(2n+2)!}{x^{2n}/(2n)!} \right|$$

$$= \lim_{n\to\infty} \left| \frac{x^2}{(2n+2)(2n+1)} \right| = 0$$

So, the series converges for all $x \Rightarrow R = \infty$.

**11.** $\displaystyle\sum_{n=0}^{\infty} \left(\frac{x}{4}\right)^n$

Because the series is geometric, it converges only if

$$\left| \frac{x}{4} \right| < 1, \text{ or } -4 < x < 4.$$

**13.** $\displaystyle\sum_{n=1}^{\infty} \frac{(-1)^n x^n}{n}$

$$\lim_{n\to\infty} \left| \frac{u_{n+1}}{u_n} \right| = \lim_{n\to\infty} \left| \frac{(-1)^{n+1} x^{n+1}}{n+1} \cdot \frac{n}{(-1)^n x^n} \right|$$

$$= \lim_{n\to\infty} \left| \frac{nx}{n+1} \right| = |x|$$

Interval: $-1 < x < 1$

When $x = 1$, the alternating series $\displaystyle\sum_{n=1}^{\infty} \frac{(-1)^n}{n}$ converges.

When $x = -1$, the $p$-series $\displaystyle\sum_{n=1}^{\infty} \frac{1}{n}$ diverges.

Therefore, the interval of convergence is $(-1, 1]$.

**15.** $\displaystyle\sum_{n=0}^{\infty} \frac{x^{5n}}{n!}$

$$\lim_{n\to\infty} \left| \frac{u_{n+1}}{u_n} \right| = \lim_{n\to\infty} \left| \frac{x^{5(n+1)}/(n+1)!}{5^n/n!} \right| = \lim_{n\to\infty} \left| \frac{x^5}{n+1} \right| = 0$$

The series converges for all $x$. The interval of convergence is $(-\infty, \infty)$.

**17.** $\displaystyle\sum_{n=0}^{\infty} (2n)! \left(\frac{x}{3}\right)^n$

$$\lim_{n\to\infty} \left| \frac{u_{n+1}}{u_n} \right| = \lim_{n\to\infty} \left| \frac{(2n+2)!(x/3)^{n+1}}{(2n)!(x/3)^n} \right|$$

$$= \left| \frac{(2n+2)(2n+1)x}{3} \right| = \infty$$

The series converges only for $x = 0$.

**19.** $\displaystyle\sum_{n=1}^{\infty} \frac{(-1)^{n+1} x^n}{4^n}$

Because the series is geometric, it converges only if $|x/4| < 1$ or $-4 < x < 4$.

**21.** $\displaystyle\sum_{n=1}^{\infty} \frac{(-1)^{n+1} (x-4)^n}{n9^n}$

$$\lim_{n\to\infty} \left| \frac{u_{n+1}}{u_n} \right| = \lim_{n\to\infty} \left| \frac{(-1)^{n+2}(x-4)^{n+1}/((n+1)9^{n+1})}{(-1)^n(x-4)^n/(n9^n)} \right|$$

$$= \lim_{n\to\infty} \left| \frac{n}{n+1} \frac{(x-4)}{9} \right| = \frac{1}{9}|x-4|$$

$R = 9$

Interval: $-5 < x < 13$

When $x = 13$, $\displaystyle\sum_{n=1}^{\infty} \frac{(-1)^{n+1} 9^n}{n9^n} = \sum_{n=1}^{\infty} \frac{(-1)^{n+1}}{n}$ converges.

When $x = -5$, $\displaystyle\sum_{n=1}^{\infty} \frac{(-1)^{n+1}(-9)^n}{n9^n} = \sum_{n=1}^{\infty} \frac{-1}{n}$ diverges.

Therefore, the interval of convergence is $(-5, 13]$.

**23.** $\displaystyle\sum_{n=0}^{\infty} \frac{(-1)^{n+1} (x-1)^{n+1}}{n+1}$

$$\lim_{n\to\infty} \left| \frac{u_{n+1}}{u_n} \right| = \lim_{n\to\infty} \left| \frac{(-1)^{n+2}(x-1)^{n+2}}{n+2} \cdot \frac{n+1}{(-1)^{n+1}(x-1)^{n+1}} \right| = \lim_{n\to\infty} \left| \frac{(n+1)(x-1)}{n+2} \right| = |x-1|$$

$R = 1$

Center: $x = 1$

Interval: $-1 < x - 1 < 1$ or $0 < x < 2$

When $x = 0$, the series $\displaystyle\sum_{n=0}^{\infty} \frac{1}{n+1}$ diverges by the integral test.

When $x = 2$, the alternating series $\displaystyle\sum_{n=0}^{\infty} \frac{(-1)^{n+1}}{n+1}$ converges.

Therefore, the interval of convergence is $(0, 2]$.

**25.** $\displaystyle\sum_{n=1}^{\infty}\left(\frac{x-3}{3}\right)^{n-1}$ is geometric. It converges if

$$\left|\frac{x-3}{3}\right| < 1 \Rightarrow |x-3| < 3 \Rightarrow 0 < x < 6.$$

Therefore, the interval of convergence is $(0, 6)$.

**27.** $\displaystyle\sum_{n=1}^{\infty}\frac{n}{n+1}(-2x)^{n-1}$

$$\lim_{n\to\infty}\left|\frac{u_{n+1}}{u_n}\right| = \lim_{n\to\infty}\left|\frac{(n+1)(-2x)^n}{n+2}\cdot\frac{n+1}{n(-2x)^{n-1}}\right|$$

$$= \lim_{n\to\infty}\left|\frac{(-2x)(n+1)^2}{n(n+2)}\right| = 2|x|$$

$$R = \frac{1}{2}$$

Interval: $-\dfrac{1}{2} < x < \dfrac{1}{2}$

When $x = -\dfrac{1}{2}$, the series $\displaystyle\sum_{n=1}^{\infty}\frac{n}{n+1}$ diverges by the $n$th Term Test.

When $x = \dfrac{1}{2}$, the alternating series

$$\sum_{n=1}^{\infty}\frac{(-1)^{n-1}n}{n+1} \text{ diverges.}$$

Therefore, the interval of convergence is $\left(-\dfrac{1}{2}, \dfrac{1}{2}\right)$.

**29.** $\displaystyle\sum_{n=0}^{\infty}\frac{x^{3n+1}}{(3n+1)!}$

$$\lim_{n\to\infty}\left|\frac{u_{n+1}}{u_n}\right| = \lim_{n\to\infty}\left|\frac{x^{3n+4}/(3n+4)!}{x^{3n+1}/(3n+1)!}\right|$$

$$= \lim_{n\to\infty}\left|\frac{x^3}{(3n+4)(3n+3)(3n+2)}\right| = 0$$

Therefore, the interval of convergence is $(-\infty, \infty)$.

**31.** $\displaystyle\sum_{n=1}^{\infty}\frac{2\cdot 3\cdot 4\cdots(n+1)x^n}{n!} = \sum_{n=1}^{\infty}(n+1)x^n$

$$\lim_{n\to\infty}\left|\frac{a_{n+1}}{a_n}\right| = \lim_{n\to\infty}\left|\frac{(n+2)x^{n+1}}{(n+1)x^n}\right| = \lim_{n\to\infty}\left|\frac{n+2}{n+1}x\right| = |x|$$

Converges if $|x| < 1 \Rightarrow -1 < x < 1$.

At $x = \pm 1$, diverges.

Therefore the interval of convergence is $(-1, 1)$.

**33.** $\displaystyle\sum_{n=1}^{\infty}\frac{(-1)^{n+1}3\cdot 7\cdot 11\cdots(4n-1)(x-3)^n}{4^n}$

$$\lim_{n\to\infty}\left|\frac{u_{n+1}}{u_n}\right| = \lim_{n\to\infty}\left|\frac{(-1)^{n+2}\cdot 3\cdot 7\cdot 11\cdots(4n-1)(4n+3)(x-3)^{n+1}}{4^{n+1}}\cdot\frac{4^n}{(-1)^{n+1}\cdot 3\cdot 7\cdot 11\cdots(4n-1)(x-3)^n}\right|$$

$$= \lim_{n\to\infty}\left|\frac{(4n+3)(x-3)}{4}\right| = \infty$$

$R = 0$

Center: $x = 3$

Therefore, the series converges only for $x = 3$.

**35.** $\displaystyle\sum_{n=1}^{\infty} \frac{(x-c)^{n-1}}{c^{n-1}}$

$$\lim_{n\to\infty}\left|\frac{u_{n+1}}{u_n}\right| = \lim_{n\to\infty}\left|\frac{(x-c)^n}{c^n} \cdot \frac{c^{n-1}}{(x-c)^{n-1}}\right| = \frac{1}{c}|x-c|$$

$R = c$

Center: $x = c$

Interval: $-c < x - c < c$ or $0 < x < 2c$

When $x = 0$, the series $\displaystyle\sum_{n=1}^{\infty} (-1)^{n-1}$ diverges.

When $x = 2c$, the series $\displaystyle\sum_{n=1}^{\infty} 1$ diverges.

Therefore, the interval of convergence is $(0, 2c)$.

**37.** $\displaystyle\sum_{n=0}^{\infty} \left(\frac{x}{k}\right)^n$

Because the series is geometric, it converges only if $\left|x/k\right| < 1$ or $-k < x < k$.

Therefore, the interval of convergence is $(-k, k)$.

**39.** $\displaystyle\sum_{n=1}^{\infty} \frac{k(k+1)\cdots(k+n-1)x^n}{n!}$

$$\lim_{n\to\infty}\left|\frac{u_{n+1}}{u_n}\right| = \lim_{n\to\infty}\left|\frac{k(k+1)\cdots(k+n-1)(k+n)x^{n+1}}{(n+1)!} \cdot \frac{n!}{k(k+1)\cdots(k+n-1)x^n}\right| = \lim_{n\to\infty}\left|\frac{(k+n)x}{n+1}\right| = |x|$$

$R = 1$

When $x = \pm 1$, the series diverges and the interval of convergence is $(-1, 1)$.

$$\left[\frac{k(k+1)\cdots(k+n-1)}{1\cdot 2\cdots n} \geq 1\right]$$

**41.** $\displaystyle\sum_{n=0}^{\infty} \frac{x^n}{n!} = 1 + \frac{x}{1} + \frac{x^2}{2} + \cdots = \sum_{n=1}^{\infty} \frac{x^{n-1}}{(n-1)!}$

**43.** $\displaystyle\sum_{n=0}^{\infty} \frac{x^{2n+1}}{(2n+1)!} = \sum_{n=1}^{\infty} \frac{x^{2n-1}}{(2n-1)!}$

Replace $n$ with $n - 1$.

**45.** (a) $f(x) = \displaystyle\sum_{n=0}^{\infty} \left(\frac{x}{3}\right)^n,\ (-3, 3)$    (Geometric)

(b) $f'(x) = \displaystyle\sum_{n=1}^{\infty} \frac{n}{3}\left(\frac{x}{3}\right)^{n-1},\ (-3, 3)$

(c) $f''(x) = \displaystyle\sum_{n=2}^{\infty} \frac{n(n-1)}{9}\left(\frac{x}{3}\right)^{n-2},\ (-3, 3)$

(d) $\displaystyle\int f(x)\, dx = \sum_{n=0}^{\infty} \frac{3}{n+1}\left(\frac{x}{3}\right)^{n+1},\ [-3, 3)$

$$\left[\sum \frac{3}{n+1}\left(\frac{-3}{3}\right)^{n+1} = \sum \frac{(-1)^{n+1}3}{n+1},\text{ converges}\right]$$

**47.** (a) $f(x) = \displaystyle\sum_{n=0}^{\infty} \frac{(-1)^{n+1}(x-1)^{n+1}}{n+1},\ (0, 2]$

(b) $f'(x) = \displaystyle\sum_{n=0}^{\infty} (-1)^{n+1}(x-1)^n,\ (0, 2)$

(c) $f''(x) = \displaystyle\sum_{n=1}^{\infty} (-1)^{n+1}n(x-1)^{n-1},\ (0, 2)$

(d) $\displaystyle\int f(x)\, dx = \sum_{n=1}^{\infty} \frac{(-1)^{n+1}(x-1)^{n+2}}{(n+1)(n+2)},\ [0, 2]$

**49.** $g(1) = \displaystyle\sum_{n=0}^{\infty} \left(\tfrac{1}{3}\right)^n = 1 + \tfrac{1}{3} + \tfrac{1}{9} + \cdots$

$S_1 = 1,\ S_2 = 1.33.$ Matches (c).

**51.** $g(3.1) = \displaystyle\sum_{n=0}^{\infty} \left(\tfrac{3.1}{3}\right)^n$ diverges. Matches (b).

**53.** $g\left(\frac{1}{8}\right) = \displaystyle\sum_{n=0}^{\infty} \left[2\left(\frac{1}{8}\right)\right]^n = \sum_{n=0}^{\infty} \left(\frac{1}{4}\right)^n = 1 + \frac{1}{4} + \frac{1}{16} + \cdots,$

converges

$S_1 = 1, S_2 = 1.25, S_3 = 1.3125$ Matches (b).

**55.** $g\left(\frac{9}{16}\right) = \displaystyle\sum_{n=0}^{\infty} \left[2\left(\frac{9}{16}\right)\right]^n = \sum_{n=0}^{\infty} \left(\frac{9}{8}\right)^n,$ diverges

$S_1 = 1, S_2 = \frac{17}{8}$ Matches (d).

**57.** A series of the form

$$\sum_{n=0}^{\infty} a_n(x - c)^n = a_0 + a_1(x - c) + a_2(x - c)^2 + \cdots$$

$$+ a_n(x - c)^n + \cdots$$

is called a power series centered at $c$, where $c$ is constant.

**63. (a)** $f(x) = \displaystyle\sum_{n=0}^{\infty} \frac{x^{2n+1}}{(2n + 1)!}$

$\displaystyle\lim_{n\to\infty} \left|\frac{u_{n+1}}{u_n}\right| = \lim_{n\to\infty} \left|\frac{x^{2n+3}}{(2n + 3)!} \cdot \frac{(2n + 1)!}{x^{2n+1}}\right| = \lim_{n\to\infty} \left|\frac{x^2}{(2n + 2)(2n + 3)}\right| = 0$

Therefore, the interval of convergence is $(-\infty\ \infty)$.

$g(x) = \displaystyle\sum_{n=0}^{\infty} \frac{(-1)^n x^{2n}}{(2n)!}$

$\displaystyle\lim_{n\to\infty} \left|\frac{u_{n+1}}{u_n}\right| = \lim_{n\to\infty} \left|\frac{(-1)^{n+1} x^{2n+2}}{(2n + 2)!} \cdot \frac{(2n)!}{(-1)^n x^{2n}}\right| = \lim_{n\to\infty} \frac{1}{2n + 2} = 0$

Therefore, the interval of convergence is $(-\infty\ \infty)$.

**(b)** $f'(x) = \displaystyle\sum_{n=0}^{\infty} \frac{(-1)^n x^{2n}}{(2n)!} = g(x)$

**(c)** $g'(x) = \displaystyle\sum_{n=1}^{\infty} \frac{(-1)^n x^{2n-1}}{(2n - 1)!} = \sum_{n=0}^{\infty} \frac{(-1)^{n+1} x^{2n+1}}{(2n + 1)!} = -\sum_{n=0}^{\infty} \frac{(-1)^n x^{2n+1}}{(2n + 1)!} = -f(x)$

**(d)** $f(x) = \sin x$ and

$g(x) = \cos x$

**65.** $y = \displaystyle\sum_{n=0}^{\infty} \frac{(-1)^n x^{2n+1}}{(2n + 1)!} = \sum_{n=1}^{\infty} \frac{(-1)^{n-1} x^{2n-1}}{(2n - 1)!}$

$y' = \displaystyle\sum_{n=0}^{\infty} \frac{(-1)^n (2n + 1) x^{2n}}{(2n + 1)!} = \sum_{n=0}^{\infty} \frac{(-1)^n x^{2n}}{(2n)!}$

$y'' = \displaystyle\sum_{n=1}^{\infty} \frac{(-1)^n (2n) x^{2n-1}}{(2n)!} = \sum_{n=1}^{\infty} \frac{(-1) x^{2n-1}}{(2n - 1)!}$

$y'' + y = \displaystyle\sum_{n=1}^{\infty} \frac{(-1)^n x^{2n-1}}{(2n - 1)!} + \sum_{n=1}^{\infty} \frac{(-1)^{n-1} x^{2n-1}}{(2n - 1)!} = 0$

**59.** A single point, an interval, or the entire real line.

**61.** Answers will vary.

$\displaystyle\sum_{n=1}^{\infty} \frac{x^n}{n}$ converges for $-1 \le x < 1$. At $x = -1$, the

convergence is conditional because $\sum \frac{1}{n}$ diverges.

$\displaystyle\sum_{n=1}^{\infty} \frac{x^n}{n^2}$ converges for $-1 \le x \le 1$. At $x = \pm 1$, the

convergence is absolute.

**67.** $y = \displaystyle\sum_{n=0}^{\infty} \frac{x^{2n+1}}{(2n+1)!} = \sum_{n=1}^{\infty} \frac{x^{2n-1}}{(2n-1)!}$

$y' = \displaystyle\sum_{n=0}^{\infty} \frac{(2n+1)x^{2n}}{(2n+1)!} = \sum_{n=0}^{\infty} \frac{x^{2n}}{(2n)!}$

$y'' = \displaystyle\sum_{n=1}^{\infty} \frac{(2n)x^{2n-1}}{(2n)!} = \sum_{n=1}^{\infty} \frac{x^{2n-1}}{(2n-1)!} = y$

$y'' - y = 0$

**69.** $y = \displaystyle\sum_{n=0}^{\infty} \frac{x^{2n}}{2^n n!} \qquad y' = \sum_{n=1}^{\infty} \frac{2nx^{2n-1}}{2^n n!} \qquad y'' = \sum_{n=1}^{\infty} \frac{2n(2n-1)x^{2n-2}}{2^n n!}$

$y'' - xy' - y = \displaystyle\sum_{n=1}^{\infty} \frac{2n(2n-1)x^{2n-2}}{2^n n!} - \sum_{n=1}^{\infty} \frac{2nx^{2n}}{2^n n!} - \sum_{n=0}^{\infty} \frac{x^{2n}}{2^n n!}$

$\qquad = \displaystyle\sum_{n=1}^{\infty} \frac{2n(2n-1)x^{2n-2}}{2^n n!} - \sum_{n=0}^{\infty} \frac{(2n+1)x^{2n}}{2^n n!}$

$\qquad = \displaystyle\sum_{n=0}^{\infty} \left[ \frac{(2n+2)(2n+1)x^{2n}}{2^{n+1}(n+1)!} - \frac{(2n+1)x^{2n}}{2^n n!} \cdot \frac{2(n+1)}{2(n+1)} \right] = \sum_{n=0}^{\infty} \frac{2(n+1)x^{2n}\left[(2n+1)-(2n+1)\right]}{2^{n+1}(n+1)!} = 0$

**71.** $J_0(x) = \displaystyle\sum_{k=0}^{\infty} \frac{(-1)^k x^{2k}}{2^{2k}(k!)^2}$

(a) $\displaystyle\lim_{k\to\infty} \left| \frac{u_{k+1}}{u_k} \right| = \lim_{k\to\infty} \left| \frac{(-1)^{k+1}x^{2k+2}}{2^{2k+2}\left[(k+1)!\right]^2} \cdot \frac{2^{2k}(k!)^2}{(-1)^k x^{2k}} \right| = \lim_{k\to\infty} \left| \frac{(-1)x^2}{2^2(k+1)^2} \right| = 0$

Therefore, the interval of convergence is $-\infty < x < \infty$.

(b) $\qquad J_0 = \displaystyle\sum_{k=0}^{\infty} (-1)^k \frac{x^{2k}}{4^k(k!)^2}$

$\qquad J_0' = \displaystyle\sum_{k=1}^{\infty} (-1)^k \frac{2kx^{2k-1}}{4^k(k!)^2} = \sum_{k=0}^{\infty} (-1)^{k+1} \frac{(2k+2)x^{2k+1}}{4^{k+1}\left[(k+1)!\right]^2}$

$\qquad J_0'' = \displaystyle\sum_{k=1}^{\infty} (-1)^k \frac{2k(2k-1)x^{2k-2}}{4^k(k!)^2} = \sum_{k=0}^{\infty} (-1)^{k+1} \frac{(2k+2)(2k+1)x^{2k}}{4^{k+1}\left[(k+1)!\right]^2}$

$x^2 J_0'' + x J_0' + x^2 J_0 = \displaystyle\sum_{k=0}^{\infty} (-1)^{k+1} \frac{2(2k+1)x^{2k+2}}{4^{k+1}(k+1)!k!} + \sum_{k=0}^{\infty} (-1)^{k+1} \frac{2x^{2k+2}}{4^{k+1}(k+1)!k!} + \sum_{k=0}^{\infty} (-1)^k \frac{x^{2k+2}}{4^k(k!)^2}$

$\qquad = \displaystyle\sum_{k=0}^{\infty} \frac{(-1)^k x^{2k+2}}{4^k(k!)^2} \left[ (-1)\frac{2(2k+1)}{4(k+1)} + (-1)\frac{2}{4(k+1)} + 1 \right]$

$\qquad = \displaystyle\sum_{k=0}^{\infty} \frac{(-1)^k x^{2k+2}}{4^k(k!)^2} \left[ \frac{-4k-2}{4k+4} - \frac{2}{4k+4} + \frac{4k+4}{4k+4} \right] = 0$

(c) $P_6(x) = 1 - \dfrac{x^2}{4} + \dfrac{x^4}{64} - \dfrac{x^6}{2304}$

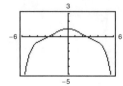

(d) $\displaystyle\int_0^1 J_0 \, dx = \int_0^1 \sum_{k=0}^{\infty} \frac{(-1)^k x^{2k}}{4^k(k!)^2} \, dx = \left[ \sum_{k=0}^{\infty} \frac{(-1)^k x^{2k+1}}{4^k(k!)^2(2k+1)} \right]_0^1 = \sum_{k=0}^{\infty} \frac{(-1)^k}{4^k(k!)^2(2k+1)} = 1 - \frac{1}{12} + \frac{1}{320} \approx 0.92$

(integral is approximately 0.9197304101)

**73.** $f(x) = \displaystyle\sum_{n=0}^{\infty} (-1)^n \frac{x^{2n}}{(2n)!} = \cos x$

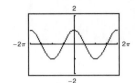

**75.** $f(x) = \displaystyle\sum_{n=0}^{\infty} (-1)^n x^n = \sum_{n=0}^{\infty} (-x)^n$    Geometric

$$= \frac{1}{1 - (-x)} = \frac{1}{1 + x} \text{ for } -1 < x < 1$$

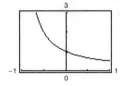

**77.** $\displaystyle\sum_{n=0}^{\infty} \left(\frac{x}{4}\right)^n, \quad (-4, 4)$

(a) $\displaystyle\sum_{n=0}^{\infty} \left(\frac{(5/2)}{4}\right)^n = \sum_{n=0}^{\infty} \left(\frac{5}{8}\right)^n = \frac{1}{1 - 5/8} = \frac{8}{3}$

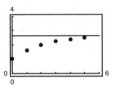

(b) $\displaystyle\sum_{n=0}^{\infty} \left(\frac{(-5/2)}{4}\right)^n = \sum_{n=0}^{\infty} \left(-\frac{5}{8}\right)^n = \frac{1}{1 + 5/8} = \frac{8}{13}$

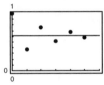

(c) The alternating series converges more rapidly. The partial sums of the series of positive terms approaches the sum from below. The partial sums of the alternating series alternate sides of the horizontal line representing the sum.

(d)

| $M$ | 10 | 100 | 1000 | 10,000 |
|---|---|---|---|---|
| $N$ | 5 | 14 | 24 | 35 |

**79.** False;

$$\sum_{n=1}^{\infty} \frac{(-1)^n x^n}{n2^n}$$

converges for $x = 2$ but diverges for $x = -2$.

**81.** True; the radius of convergence is $R = 1$ for both series.

**83.** $\displaystyle\lim_{n\to\infty} \left|\frac{a_{n+1}}{a_n}\right| = \lim_{n\to\infty} \left|\frac{(n + 1 + p)!}{(n + 1)!(n + 1 + q)!} x^{n+1} \bigg/ \frac{(n + p)!}{n!(n + q)!} x^n\right| = \lim_{n\to\infty} \left|\frac{(n + 1 + p)x}{(n + 1)(n + 1 + q)}\right| = 0$

So, the series converges for all $x$: $R = \infty$.

**85. (a)** $f(x) = \sum_{n=0}^{\infty} c_n x^n, \; c_{n+3} = c_n$

$$= c_0 + c_1 x + c_2 x^2 + c_0 x^3 + c_1 x^4 + c_2 x^5 + c_0 x^6 + \cdots$$

$$S_{3n} = c_0\left(1 + x^3 + \cdots + x^{3n}\right) + c_1 x\left(1 + x^3 + \cdots + x^{3n}\right) + c_2 x^2\left(1 + x^3 + \cdots + x^{3n}\right)$$

$$\lim_{n \to \infty} S_{3n} = c_0 \sum_{n=0}^{\infty} x^{3n} + c_1 x \sum_{n=0}^{\infty} x^{3n} + c_2 x^2 \sum_{n=0}^{\infty} x^{3n}$$

Each series is geometric, $R = 1$, and the interval of convergence is $(-1, 1)$.

**(b)** For $|x| < 1$, $f(x) = c_0 \dfrac{1}{1 - x^3} + c_1 x \dfrac{1}{1 - x^3} + c_2 x^2 \dfrac{1}{1 - x^3} = \dfrac{c_0 + c_1 x + c_2 x^2}{1 - x^3}$.

**87.** At $x = x_0 + R$, $\displaystyle\sum_{n=0}^{\infty} c_n(x - x_0)^n = \sum_{n=0}^{\infty} c_n R^n$, diverges.

At $x = x_0 - R$, $\displaystyle\sum_{n=0}^{\infty} c_n(x - x_0)^n = \sum_{n=0}^{\infty} c_n(-R)^n$, converges.

Furthermore, at $x = x_0 - R$,

$$\sum_{n=0}^{\infty} \left| c_n(x - x_0)^n \right| = \sum_{n=0}^{\infty} C_n R^n, \text{ diverges.}$$

So, the series converges conditionally at $x_0 - R$.

## Section 9.9   Representation of Functions by Power Series

**1. (a)** $\dfrac{1}{4 - x} = \dfrac{1/4}{1 - (x/4)}$

$$= \dfrac{a}{1 - r} = \sum_{n=0}^{\infty} \left(\dfrac{1}{4}\right)\left(\dfrac{x}{4}\right)^n = \sum_{n=0}^{\infty} \dfrac{x^n}{4^{n+1}}$$

This series converges on $(-4, 4)$.

**(b)**
$$\begin{array}{r}
\dfrac{1}{4} + \dfrac{x}{16} + \dfrac{x^2}{64} + \cdots \\
4 - x \overline{) \; 1 \phantom{xxxxxxx}} \\
1 - \dfrac{x}{4} \phantom{xxxxx} \\
\hline
\dfrac{x}{4} \phantom{xxxxxx} \\
\dfrac{x}{4} - \dfrac{x^2}{16} \phantom{xx} \\
\hline
\dfrac{x^2}{16} \phantom{xx} \\
\dfrac{x^2}{16} - \dfrac{x^3}{64} \\
\hline
\vdots
\end{array}$$

**3. (a)** $\dfrac{3}{4 + x} = \dfrac{3/4}{1 - (-x/4)} = \dfrac{a}{1 - r}$

$$= \sum_{n=0}^{\infty} \dfrac{3}{4}\left(-\dfrac{x}{4}\right)^n = \sum_{n=0}^{\infty} \dfrac{3(-1)^n x^n}{4^{n+1}}$$

This series converges on $(-4, 4)$.

**(b)**
$$\begin{array}{r}
\dfrac{3}{4} - \dfrac{3x}{16} + \dfrac{3x^2}{64} - \cdots \\
4 + x \overline{) \; 3 \phantom{xxxxxxx}} \\
3 + \dfrac{3}{4}x \phantom{xxxx} \\
\hline
-\dfrac{3}{4}x \phantom{xxxxx} \\
-\dfrac{3}{4}x - \dfrac{3x^2}{16} \phantom{x} \\
\hline
\dfrac{3x^2}{16} \phantom{x} \\
\dfrac{3x^2}{16} + \dfrac{3x^3}{64} \\
\hline
\vdots
\end{array}$$

**5.** $\dfrac{1}{3 - x} = \dfrac{1}{2 - (x - 1)} = \dfrac{1/2}{1 - \left(\dfrac{x - 1}{2}\right)} = \dfrac{a}{1 - r}$

$$= \sum_{n=0}^{\infty} \dfrac{1}{2}\left(\dfrac{x - 1}{2}\right)^n = \sum_{n=0}^{\infty} \dfrac{(x - 1)^n}{2^{n+1}}$$

Interval of convergence: $\left|\dfrac{x - 1}{2}\right| < 1 \Rightarrow (-1, 3)$

**7.** $\dfrac{1}{1 - 3x} = \dfrac{a}{1 - r} = \displaystyle\sum_{n=0}^{\infty} (3x)^n$

Interval of convergence: $|3x| < 1 \Rightarrow \left(\dfrac{1}{3}, \dfrac{1}{3}\right)$

**9.** $\dfrac{5}{2x - 3} = \dfrac{5}{-9 + 2(x + 3)} = \dfrac{-5/9}{1 - \dfrac{2}{9}(x + 3)} = \dfrac{a}{1 - r}$

$= -\dfrac{5}{9}\displaystyle\sum_{n=0}^{\infty} \left(\dfrac{2}{9}(x + 3)\right)^n, \ \left|\dfrac{2}{9}(x + 3)\right| < 1$

$= -5\displaystyle\sum_{n=0}^{\infty} \dfrac{2^n}{9^{n+1}}(x + 3)^n$

Interval of convergence: $\left|\dfrac{2}{9}(x + 3)\right| < 1 \Rightarrow \left(-\dfrac{15}{2}, \dfrac{3}{2}\right)$

**11.** $\dfrac{2}{2x + 3} = \dfrac{2/3}{1 + \dfrac{2}{3}x} = \dfrac{2/3}{1 - \left(-\dfrac{2x}{3}\right)} = \dfrac{a}{1 - r}$

$= \dfrac{2}{3}\displaystyle\sum_{n=0}^{\infty} \left(-\dfrac{2x}{3}\right)^n$

$= \displaystyle\sum_{n=0}^{\infty} \dfrac{(-1)^n 2^{n+1}}{3^{n+1}}x^n$

Interval of convergence: $\left|\dfrac{-2x}{3}\right| < 1 \Rightarrow \left(-\dfrac{3}{2}, \dfrac{3}{2}\right)$

**13.** $\dfrac{4x}{x^2 + 2x - 3} = \dfrac{3}{x + 3} + \dfrac{1}{x - 1}$

$= \dfrac{1}{1 - (-x/3)} + \dfrac{-1}{1 - x}$

$= \displaystyle\sum_{n=0}^{\infty} \left(-\dfrac{x}{3}\right)^n - \displaystyle\sum_{n=0}^{\infty} x^n = \displaystyle\sum_{n=0}^{\infty} \left[\dfrac{1}{(-3)^n} - 1\right]x^n$

Interval of convergence: $\left|\dfrac{x}{3}\right| < 1$ and $|x| < 1 \Rightarrow (-1, 1)$

**15.** $\dfrac{2}{1 - x^2} = \dfrac{1}{1 - x} + \dfrac{1}{1 + x}$

$= \displaystyle\sum_{n=0}^{\infty} \left(1 + (-1)^n\right)x^n = 2\displaystyle\sum_{n=0}^{\infty} x^{2n}$

Interval of convergence: $|x^2| < 1$ or $(1, 1)$ because

$\displaystyle\lim_{n \to \infty} \left|\dfrac{u_{n+1}}{u_n}\right| = \lim_{n \to \infty} \left|\dfrac{2x^{2n+2}}{2x^{2n}}\right| = |x^2|$

**17.** $\dfrac{1}{1 + x} = \displaystyle\sum_{n=0}^{\infty} (-1)^n x^n$

$\dfrac{1}{1 - x} = \displaystyle\sum_{n=0}^{\infty} (-1)^n(-x)^n = \displaystyle\sum_{n=0}^{\infty} (-1)^{2n} x^n = \displaystyle\sum_{n=0}^{\infty} x^n$

$h(x) = \dfrac{-2}{x^2 - 1} = \dfrac{1}{1 + x} + \dfrac{1}{1 - x} = \displaystyle\sum_{n=0}^{\infty} (-1)^n x^n + \displaystyle\sum_{n=0}^{\infty} x^n = \displaystyle\sum_{n=0}^{\infty} \left[(-1)^n + 1\right]x^n$

$= 2 + 0x + 2x^2 + 0x^3 + 2x^4 + 0x^5 + 2x^6 + \cdots = 2\displaystyle\sum_{n=0}^{\infty} x^{2n}, \ (-1, 1) \ \text{(See Exercise 15.)}$

**19.** By taking the first derivative, you have $\dfrac{d}{dx}\left[\dfrac{1}{x + 1}\right] = \dfrac{-1}{(x + 1)^2}$. Therefore,

$\dfrac{-1}{(x + 1)^2} = \dfrac{d}{dx}\left[\displaystyle\sum_{n=0}^{\infty} (-1)^n x^n\right] = \displaystyle\sum_{n=1}^{\infty} (-1)^n n x^{n-1} = \displaystyle\sum_{n=0}^{\infty} (-1)^{n+1}(n + 1)x^n, \ (-1, 1).$

**21.** By integrating, you have $\displaystyle\int \dfrac{1}{x + 1}\,dx = \ln(x + 1)$. Therefore,

$\ln(x + 1) = \displaystyle\int\left[\displaystyle\sum_{n=0}^{\infty} (-1)^n x^n\right]dx = C + \displaystyle\sum_{n=0}^{\infty} \dfrac{(-1)^n x^{n+1}}{n + 1}, \ -1 < x \le 1.$

To solve for $C$, let $x = 0$ and conclude that $C = 0$. Therefore,

$\ln(x + 1) = \displaystyle\sum_{n=0}^{\infty} \dfrac{(-1)^n x^{n+1}}{n + 1}, \ (-1, 1].$

**23.** $\dfrac{1}{x^2 + 1} = \displaystyle\sum_{n=0}^{\infty} (-1)^n \left(x^2\right)^n = \sum_{n=0}^{\infty} (-1)^n x^{2n},\ (-1,\ 1)$

**25.** Because, $\dfrac{1}{x + 1} = \displaystyle\sum_{n=0}^{\infty} (-1)^n x^n$, you have $\dfrac{1}{4x^2 + 1} = \displaystyle\sum_{n=0}^{\infty} (-1)^n \left(4x^2\right)^n = \sum_{n=0}^{\infty} (-1)^n 4^n x^{2n} = \sum_{n=0}^{\infty} (-1)^n (2x)^{2n},\ \left(-\dfrac{1}{2},\ \dfrac{1}{2}\right).$

**27.** $x - \dfrac{x^2}{2} \le \ln(x + 1) \le x - \dfrac{x^2}{2} + \dfrac{x^3}{3}$

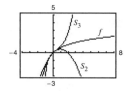

| $x$ | 0.0 | 0.2 | 0.4 | 0.6 | 0.8 | 1.0 |
|---|---|---|---|---|---|---|
| $S_2 = x - \dfrac{x^2}{2}$ | 0.000 | 0.180 | 0.320 | 0.420 | 0.480 | 0.500 |
| $\ln(x + 1)$ | 0.000 | 0.182 | 0.336 | 0.470 | 0.588 | 0.693 |
| $S_3 = x - \dfrac{x^2}{2} + \dfrac{x^3}{3}$ | 0.000 | 0.183 | 0.341 | 0.492 | 0.651 | 0.833 |

**29.** $\displaystyle\sum_{n=1}^{\infty} \dfrac{(-1)^{n+1}(x - 1)^n}{n} = \dfrac{(x - 1)}{1} - \dfrac{(x - 1)^2}{2} + \dfrac{(x - 1)^3}{3} - \cdots$

(a)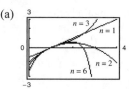

(b) From Example 4,

$$\sum_{n=1}^{\infty} \frac{(-1)^{n+1}(x - 1)^n}{n} = \sum_{n=0}^{\infty} \frac{(-1)^n (x - 1)^{n+1}}{n + 1} = \ln x,\ 0 < x \le 2,\ R = 1.$$

(c) $x = 0.5$:

$$\sum_{n=1}^{\infty} \frac{(-1)^{n+1}(-1/2)^n}{n} = \sum_{n=1}^{\infty} \frac{-(1/2)^n}{n} \approx -0.693147$$

(d) This is an approximation of $\ln\!\left(\dfrac{1}{2}\right)$. The error is approximately 0. [The error is less than the first omitted term,

$1/\left(51 \cdot 2^{51}\right) \approx 8.7 \times 10^{-18}.$ ]

**31.** $g(x) = x$ line

Matches (c)

**33.** $g(x) = x - \dfrac{x^3}{3} + \dfrac{x^5}{5}$

Matches (a)

**In Exercises 35 and 37,  arctan  $x = \displaystyle\sum_{n=0}^{\infty} (-1)^n \frac{x^{2n+1}}{2n+1}$.**

**35.**  $\arctan \dfrac{1}{4} = \displaystyle\sum_{n=0}^{\infty} (-1)^n \frac{(1/4)^{2n+1}}{2n+1} = \sum_{n=0}^{\infty} \frac{(-1)^n}{(2n+1)4^{2n+1}} = \frac{1}{4} - \frac{1}{192} + \frac{1}{5120} + \cdots$

Because  $\dfrac{1}{5120} < 0.001$, you can approximate the series by its first two terms:  $\arctan \dfrac{1}{4} \approx \dfrac{1}{4} - \dfrac{1}{192} \approx 0.245$.

**37.**    $\dfrac{\arctan x^2}{x} = \dfrac{1}{x}\displaystyle\sum_{n=0}^{\infty} (-1)^n \frac{\left(x^2\right)^{2n+1}}{2n+1} = \sum_{n=0}^{\infty} (-1)^2 \frac{x^{4n+1}}{2n+1}$

$\displaystyle\int \frac{\arctan x^2}{x}\, dx = \sum_{n=0}^{\infty} (-1)^n \frac{x^{4n+2}}{(4n+2)(2n+1)} + C \;(\text{Note: } C = 0)$

$\displaystyle\int_0^{1/2} \frac{\arctan x^2}{x}\, dx = \sum_{n=0}^{\infty} (-1)^n \frac{1}{(4n+2)(2n+1)2^{4n+2}} = \frac{1}{8} - \frac{1}{1152} + \cdots$

Because  $\dfrac{1}{1152} < 0.001$, you can approximate the series by its first term:  $\displaystyle\int_0^{1/2} \frac{\arctan x^2}{x}\, dx \approx 0.125$.

**In Exercises 39 and 41, use  $\dfrac{1}{1-x} = \displaystyle\sum_{n=0}^{\infty} x^n, |x| < 1$.**

**39.**  $\dfrac{1}{(1-x)^2} = \dfrac{d}{dx}\!\left[\dfrac{1}{1-x}\right] = \dfrac{d}{dx}\!\left[\displaystyle\sum_{n=0}^{\infty} x^n\right] = \sum_{n=1}^{\infty} nx^{n-1}, \; |x| < 1$

**41.**  $\dfrac{1+x}{(1-x)^2} = \dfrac{1}{(1-x)^2} + \dfrac{x}{(1-x)^2}$

$= \displaystyle\sum_{n=1}^{\infty} n\left(x^{n-1} + x^n\right), \quad |x| < 1$

$= \displaystyle\sum_{n=0}^{\infty} (2n+1)x^n, \quad |x| < 1$

**43.**  $P(n) = \left(\dfrac{1}{2}\right)^n$

$E(n) = \displaystyle\sum_{n=1}^{\infty} nP(n) = \sum_{n=1}^{\infty} n\left(\frac{1}{2}\right)^n = \frac{1}{2}\sum_{n=1}^{\infty} n\left(\frac{1}{2}\right)^{n-1}$

$= \dfrac{1}{2}\dfrac{1}{\left[1 - (1/2)\right]^2} = 2$

Because the probability of obtaining a head on a single toss is  $\dfrac{1}{2}$, it is expected that, on average, a head will be obtained in two tosses.

**45.**  Because  $\dfrac{1}{1+x} = \dfrac{1}{1-(-x)}$, substitute  $(-x)$  into the geometric series.

**47.**  Because  $\dfrac{1}{1+x} = 5\!\left(\dfrac{1}{1-(-x)}\right)$, substitute  $(-x)$  into the geometric series and then multiply the series by 5.

**49.**  Let  $\arctan x + \arctan y = \theta.$ Then,

$\tan(\arctan x + \arctan y) = \tan \theta$

$\dfrac{\tan(\arctan x) + \tan(\arctan y)}{1 - \tan(\arctan x)\tan(\arctan y)} = \tan \theta$

$\dfrac{x+y}{1-xy} = \tan \theta$

$\arctan\!\left(\dfrac{x+y}{1-xy}\right) = \theta.$

Therefore,

$\arctan x + \arctan y = \arctan\!\left(\dfrac{x+y}{1-xy}\right)$ for  $xy \neq 1$.

**51.** (a)   $2 \arctan \dfrac{1}{2} = \arctan \dfrac{1}{2} + \arctan \dfrac{1}{2} = \arctan \left[ \dfrac{\frac{1}{2} + \frac{1}{2}}{1 - (1/2)^2} \right] = \arctan \dfrac{4}{3}$

$2 \arctan \dfrac{1}{2} - \arctan \dfrac{1}{7} = \arctan \dfrac{4}{3} + \arctan \left( -\dfrac{1}{7} \right) = \arctan \left[ \dfrac{(4/3) - (1/7)}{1 + (4/3)(1/7)} \right] = \arctan \dfrac{25}{25} = \arctan 1 = \dfrac{\pi}{4}$

(b)   $\pi = 8 \arctan \dfrac{1}{2} - 4 \arctan \dfrac{1}{7} \approx 8 \left[ \dfrac{1}{2} - \dfrac{(0.5)^3}{3} + \dfrac{(0.5)^5}{5} - \dfrac{(0.5)^7}{7} \right] - 4 \left[ \dfrac{1}{7} - \dfrac{(1/7)^3}{3} + \dfrac{(1/7)^5}{5} - \dfrac{(1/7)^7}{7} \right] \approx 3.14$

**53.** From Exercise 21, you have

$\ln(x + 1) = \displaystyle\sum_{n=0}^{\infty} \dfrac{(-1)^n x^{n+1}}{n + 1} = \sum_{n=1}^{\infty} \dfrac{(-1)^{n-1} x^n}{n} = \sum_{n=1}^{\infty} \dfrac{(-1)^{n+1} x^n}{n}.$

So, $\displaystyle\sum_{n=1}^{\infty} (-1)^{n+1} \dfrac{1}{2^n n} = \sum_{n=1}^{\infty} \dfrac{(-1)^{n+1}(1/2)^n}{n} = \ln \left( \dfrac{1}{2} + 1 \right) = \ln \dfrac{3}{2} \approx 0.4055.$

**55.** From Exercise 53, you have

$\displaystyle\sum_{n=1}^{\infty} (-1)^{n+1} \dfrac{2^n}{5^n n} = \sum_{n=1}^{\infty} \dfrac{(-1)^{n+1}(2/5)^n}{n}$

$= \ln \left( \dfrac{2}{5} + 1 \right) = \ln \dfrac{7}{5} \approx 0.3365.$

**57.** From Exercise 56, you have

$\displaystyle\sum_{n=0}^{\infty} (-1)^n \dfrac{1}{2^{2n+1}(2n + 1)} = \sum_{n=0}^{\infty} (-1)^n \dfrac{(1/2)^{2n+1}}{2n + 1}$

$= \arctan \dfrac{1}{2} \approx 0.4636.$

**59.** $f(x) = \arctan x$ is an odd function (symmetric to the origin).

**61.** The series in Exercise 56 converges to its sum at a slower rate because its terms approach 0 at a much slower rate.

**63.** Because the first series is the derivative of the second series, the second series converges for $|x + 1| < 4$ (and perhaps at the endpoints, $x = 3$ and $x = -5$.)

**65.** $\displaystyle\sum_{n=0}^{\infty} \dfrac{(-1)^n}{3^n (2n + 1)}$

From Example 5 you have $\arctan x = \displaystyle\sum_{n=0}^{\infty} (-1)^n \dfrac{x^{2n+1}}{2n + 1}.$

$\displaystyle\sum_{n=0}^{\infty} \dfrac{(-1)^n}{3^n(2n + 1)} = \sum_{n=0}^{\infty} \dfrac{(-1)^n}{\left(\sqrt{3}\right)^{2n}(2n + 1)} \dfrac{\sqrt{3}}{\sqrt{3}}$

$= \sqrt{3} \displaystyle\sum_{n=0}^{\infty} \dfrac{(-1)^n \left(1/\sqrt{3}\right)^{2n+1}}{2n + 1}$

$= \sqrt{3} \arctan \left( \dfrac{1}{\sqrt{3}} \right)$

$= \sqrt{3} \left( \dfrac{\pi}{6} \right) \approx 0.9068997$

**67.** Using a graphing utility, you obtain the following partial sums for the left hand side. Note that $1/\pi \approx 0.3183098862.$

$n = 0: S_0 \approx 0.3183098784$

$n = 1: S_1 \approx 0.3183098862$

## Section 9.10   Taylor and Maclaurin Series

**1.** For $c = 0$, you have:

$f(x) = e^{2x}$

$f^{(n)}(x) = 2^n e^{2x} \Rightarrow f^{(n)}(0) = 2^n$

$e^{2x} = 1 + 2x + \dfrac{4x^2}{2!} + \dfrac{8x^3}{3!} + \dfrac{16x^4}{4!} + \cdots = \displaystyle\sum_{n=0}^{\infty} \dfrac{(2x)^n}{n!}.$

**3.** For $c = \pi/4$, you have:

$$f(x) = \cos(x) \qquad f\left(\frac{\pi}{4}\right) = \frac{\sqrt{2}}{2}$$

$$f'(x) = -\sin(x) \qquad f'\left(\frac{\pi}{4}\right) = -\frac{\sqrt{2}}{2}$$

$$f''(x) = -\cos(x) \qquad f''\left(\frac{\pi}{4}\right) = -\frac{\sqrt{2}}{2}$$

$$f'''(x) = \sin(x) \qquad f'''\left(\frac{\pi}{4}\right) = \frac{\sqrt{2}}{2}$$

$$f^{(4)}(x) = \cos(x) \qquad f^{(4)}\left(\frac{\pi}{4}\right) = \frac{\sqrt{2}}{2}$$

and so on. Therefore, you have:

$$\cos x = \sum_{n=0}^{\infty} \frac{f^{(n)}(\pi/4)\left[x - (\pi/4)\right]^n}{n!}$$

$$= \frac{\sqrt{2}}{2}\left[1 - \left(x - \frac{\pi}{4}\right) - \frac{\left[x - (\pi/4)\right]^2}{2!} + \frac{\left[x - (\pi/4)\right]^3}{3!} + \frac{\left[x - (\pi/4)\right]^4}{4!} - \cdots\right] = \frac{\sqrt{2}}{2}\sum_{n=0}^{\infty} \frac{(-1)^{n(n+1)/2}\left[x - (\pi/4)\right]^n}{n!}.$$

[**Note:** $(-1)^{n(n+1)/2} = 1, -1, -1, 1, 1, -1, -1, 1, 1, \ldots$]

**5.** For $c = 1$, you have

$$f(x) = \frac{1}{x} = x^{-1} \qquad f(1) = 1$$

$$f'(x) = -x^{-2} \qquad f'(1) = -1$$

$$f''(x) = 2x^{-3} \qquad f''(1) = 2$$

$$f'''(x) = -6x^{-4} \qquad f'''(1) = -6$$

and so on. Therefore, you have

$$\frac{1}{x} = \sum_{n=0}^{\infty} \frac{f^{(n)}(1)(x-1)^n}{n!} = 1 - (x-1) + \frac{2(x-1)^2}{2!} - \frac{6(x-1)^3}{3!} + \cdots = 1 - (x-1) + (x-1)^2 - (x-1)^3 + \cdots = \sum_{n=0}^{\infty} (-1)^n(x-1)^n$$

**7.** For $c = 1$, you have,

$$f(x) = \ln x \qquad f(1) = 0$$

$$f'(x) = \frac{1}{x} \qquad f'(1) = 1$$

$$f''(x) = -\frac{1}{x^2} \qquad f''(1) = -1$$

$$f'''(x) = \frac{2}{x^3} \qquad f'''(1) = 2$$

$$f^{(4)}(x) = -\frac{6}{x^4} \qquad f^{(4)}(1) = -6$$

$$f^{(5)}(x) = \frac{24}{x^5} \qquad f^{(5)}(1) = 24$$

and so on. Therefore, you have:

$$\ln x = \sum_{n=0}^{\infty} \frac{f^{(n)}(1)(x-1)^n}{n!}$$

$$= 0 + (x-1) - \frac{(x-1)^2}{2!} + \frac{2(x-1)^3}{3!} - \frac{6(x-1)^4}{4!} + \frac{24(x-1)^5}{5!} - \cdots$$

$$= (x-1) - \frac{(x-1)^2}{2} + \frac{(x-1)^3}{3} - \frac{(x-1)^4}{4} + \frac{(x-1)^5}{5} - \cdots = \sum_{n=0}^{\infty} (-1)^n \frac{(x-1)^{n+1}}{n+1}.$$

**9.** For $c = 0$, you have

$$f(x) = \sin 3x \qquad\qquad f(0) = 0$$
$$f'(x) = 3 \cos 3x \qquad\qquad f'(0) = 3$$
$$f''(x) = -9 \sin 3x \qquad\qquad f''(0) = 0$$
$$f'''(x) = -27 \cos 3x \qquad\qquad f'''(0) = -27$$
$$f^{(4)}(x) = 81 \sin 3x \qquad\qquad f^{(4)}(0) = 0$$

and so on. Therefore you have

$$\sin 3x = \sum_{n=0}^{\infty} \frac{f^{(n)}(0)x^n}{n!} = 0 + 3x + 0 - \frac{27x^3}{3!} + 0 + \cdots = \sum_{n=0}^{\infty} \frac{(-1)^n (3x)^{2n+1}}{(2n+1)!}$$

**11.** For $c = 0$, you have:

$$f(x) = \sec(x) \qquad\qquad\qquad\qquad f(0) = 1$$
$$f'(x) = \sec(x)\tan(x) \qquad\qquad\qquad\qquad f'(0) = 0$$
$$f''(x) = \sec^3(x) + \sec(x)\tan^2(x) \qquad\qquad\qquad f''(0) = 1$$
$$f'''(x) = 5\sec^3(x)\tan(x) + \sec(x)\tan^3(x) \qquad\qquad f'''(0) = 0$$
$$f^{(4)}(x) = 5\sec^5(x) + 18\sec^3(x)\tan^2(x) + \sec(x)\tan^4(x) \qquad f^{(4)}(0) = 5$$

$$\sec(x) = \sum_{n=0}^{\infty} \frac{f^{(n)}(0)x^n}{n!} = 1 + \frac{x^2}{2!} + \frac{5x^4}{4!} + \cdots.$$

**13.** The Maclaurin series for $f(x) = \cos x$ is $\displaystyle\sum_{n=0}^{\infty} \frac{(-1)^n x^{2n}}{(2n)!}$.

Because $f^{(n+1)}(x) = \pm\sin x$ or $\pm\cos x$, you have $\left| f^{(n+1)}(z) \right| \le 1$ for all $z$. So by Taylor's Theorem,

$$0 \le \left| R_n(x) \right| = \left| \frac{f^{(n+1)}(z)}{(n+1)!}x^{n+1} \right| \le \frac{|x|^{n+1}}{(n+1)!}.$$

Because $\displaystyle\lim_{n\to\infty} \frac{|x|^{n+1}}{(n+1)!} = 0$, it follows that $R_n(x) \to 0$ as $n \to \infty$. So, the Maclaurin series for $\cos x$ converges to $\cos x$ for all $x$.

**15.** The Maclaurin series for $f(x) = \sinh x$ is $\displaystyle\sum_{n=0}^{\infty} \frac{x^{2n+1}}{(2n+1)!}$.

$f^{(n+1)}(x) = \sinh x$ (or $\cosh x$). For fixed $x$,

$$0 \le \left| R_n(x) \right| = \left| \frac{f^{(n+1)}(z)}{(n+1)!}x^{n+1} \right| = \left| \frac{\sinh(z)}{(n+1)!}x^{n+1} \right| \to 0 \text{ as } n \to \infty.$$

(The argument is the same if $f^{(n+1)}(x) = \cosh x$ ). So, the Maclaurin series for $\sinh x$ converges to $\sinh x$ for all $x$.

**17.** Because $(1 + x)^{-k} = 1 - kx + \dfrac{k(k+1)x^2}{2!} - \dfrac{k(k+1)(k+2)x^3}{3!} + \cdots$, you have

$$(1+x)^{-2} = 1 - 2x + \frac{2(3)x^2}{2!} - \frac{2(3)(4)x^3}{3!} + \frac{2(3)(4)(5)x^4}{4!} - \cdots = 1 - 2x + 3x^2 - 4x^3 + 5x^4 - \cdots$$

$$= \sum_{n=0}^{\infty} (-1)^n (n+1)x^n.$$

**19.** Because $(1 + x)^{-k} = 1 - kx + \dfrac{k(k + 1)x^2}{2!} - \dfrac{k(k + 1)(k + 2)x^3}{3!} + \cdots$, you have

$$\left[1 + (-x)\right]^{-1/2} = 1 + \left(\dfrac{1}{2}\right)x + \dfrac{(1/2)(3/2)x^2}{2!} + \dfrac{(1/2)(3/2)(5/2)x^3}{3!} + \cdots$$

$$= 1 + \dfrac{x}{2} + \dfrac{(1)(3)x^2}{2^2\,2!} + \dfrac{(1)(3)(5)x^3}{2^3\,3!} + \cdots$$

$$= 1 + \sum_{n=1}^{\infty} \dfrac{1 \cdot 3 \cdot 5 \cdots (2n - 1)x^n}{2^n\, n!}.$$

**21.** $\dfrac{1}{\sqrt{4 + x^2}} = \left(\dfrac{1}{2}\right)\left[1 + \left(\dfrac{x}{2}\right)^2\right]^{-1/2}$ and because $(1 + x)^{-1/2} = 1 + \sum_{n=1}^{\infty} \dfrac{(-1)^n 1 \cdot 3 \cdot 5 \cdots (2n - 1)x^n}{2^n\, n!}$, you have

$$\dfrac{1}{\sqrt{4 + x^2}} = \dfrac{1}{2}\left[1 + \sum_{n=1}^{\infty} \dfrac{(-1)^n 1 \cdot 3 \cdot 5 \cdots (2n - 1)(x/2)^{2n}}{2^n\, n!}\right] = \dfrac{1}{2} + \sum_{n=1}^{\infty} \dfrac{(-1)^n 1 \cdot 3 \cdot 5 \cdots (2n - 1)x^{2n}}{2^{3n+1}\, n!}.$$

**23.** $\sqrt{1 + x} = (1 + x)^{1/2}, \qquad k = 1/2$

$$\sqrt{1 + x} = 1 + \dfrac{1}{2}x + \dfrac{1/2(-1/2)}{2!}x^2 + \dfrac{1/2(-1/2)(-3/2)}{3!}x^3 + \cdots = 1 + \dfrac{1}{2}x + \sum_{n=2}^{\infty} (-1)^{n+1} \dfrac{1 \cdot 3 \cdot 5 \cdots (2n - 3)}{2^n\, n!}x^n$$

**25.** Because $(1 + x)^{1/2} = 1 + \dfrac{x}{2} + \sum_{n=2}^{\infty} \dfrac{(-1)^{n+1} 1 \cdot 3 \cdot 5 \cdots (2n - 3)x^n}{2^n\, n!}$

you have $(1 + x^2)^{1/2} = 1 + \dfrac{x^2}{2} + \sum_{n=2}^{\infty} \dfrac{(-1)^{n+1} 1 \cdot 3 \cdot 5 \cdots (2n - 3)x^{2n}}{2^n\, n!}.$

**27.** $e^x = \sum_{n=0}^{\infty} \dfrac{x^n}{n!} = 1 + x + \dfrac{x^2}{2!} + \dfrac{x^3}{3!} + \dfrac{x^4}{4!} + \dfrac{x^5}{5!} + \cdots$

$$e^{x^2/2} = \sum_{n=0}^{\infty} \dfrac{(x^2/2)^n}{n!} = \sum_{n=0}^{\infty} \dfrac{x^{2n}}{2^n\, n!} = 1 + \dfrac{x^2}{2} + \dfrac{x^4}{2^2\, 2!} + \dfrac{x^6}{2^3\, 3!} + \dfrac{x^8}{2^4\, 4!} + \cdots$$

**29.** $\ln x = \sum_{n=1}^{\infty} (-1)^{n-1} \dfrac{(x - 1)^n}{n}, \qquad 0 < x \le 2$

$$\ln (x + 1) = \sum_{n=1}^{\infty} \dfrac{(-1)^{n-1}x^n}{n}, \quad -1 < x \le 1$$

**31.** $\sin x = \sum_{n=0}^{\infty} \dfrac{(-1)^n x^{2n+1}}{(2n + 1)!}$

$$\sin 3x = \sum_{n=0}^{\infty} \dfrac{(-1)^n (3x)^{2n+1}}{(2n + 1)!}$$

**33.** $\cos x = \sum_{n=0}^{\infty} \dfrac{(-1)^n x^{2n}}{(2n)!} = 1 - \dfrac{x^2}{2!} + \dfrac{x^4}{4!} - \dfrac{x^6}{6!} + \cdots$

$$\cos 4x = \sum_{n=0}^{\infty} \dfrac{(-1)^n (4x)^{2n}}{(2n)!} = \sum_{n=0}^{\infty} \dfrac{(-1)^n 4^{2n} x^{2n}}{(2n)!}$$

$$= 1 - \dfrac{16x^2}{2!} + \dfrac{256x^4}{4!} - \cdots$$

**35.** $\cos x = \sum_{n=0}^{\infty} \dfrac{(-1)^n x^{2n}}{(2n)!} = 1 - \dfrac{x^2}{2!} + \dfrac{x^4}{4!} - \cdots$

$$\cos x^{3/2} = \sum_{n=0}^{\infty} \dfrac{(-1)^n (x^{3/2})^{2n}}{(2n)!}$$

$$= \sum_{n=0}^{\infty} \dfrac{(-1)^n x^{3n}}{(2n)!} = 1 - \dfrac{x^3}{2!} + \dfrac{x^6}{4!} - \cdots$$

**37.** $e^x = 1 + x + \dfrac{x^2}{2!} + \dfrac{x^3}{3!} + \dfrac{x^4}{4!} + \dfrac{x^5}{5!} + \cdots$

$$e^{-x} = 1 - x + \dfrac{x^2}{2!} - \dfrac{x^3}{3!} + \dfrac{x^4}{4!} - \dfrac{x^5}{5!} + \cdots$$

$$e^x - e^{-x} = 2x + \dfrac{2x^3}{3!} + \dfrac{2x^5}{5!} + \dfrac{2x^7}{7!} + \cdots$$

$$\sinh(x) = \dfrac{1}{2}(e^x - e^{-x})$$

$$= x + \dfrac{x^3}{3!} + \dfrac{x^5}{5!} + \dfrac{x^7}{7!} + \cdots = \sum_{n=0}^{\infty} \dfrac{x^{2n+1}}{(2n + 1)!}$$

**39.** $\cos^2(x) = \dfrac{1}{2}\Big[1 + \cos(2x)\Big]$

$$= \frac{1}{2}\left[1 + 1 - \frac{(2x)^2}{2!} + \frac{(2x)^4}{4!} - \frac{(2x)^6}{6!} - \cdots\right] = \frac{1}{2}\left[1 + \sum_{n=0}^{\infty}\frac{(-1)^n(2x)^{2n}}{(2n)!}\right]$$

**41.** $x\sin x = x\left(x - \dfrac{x^3}{3!} + \dfrac{x^5}{5!} - \cdots\right) = x^2 - \dfrac{x^4}{3!} + \dfrac{x^6}{5!} - \cdots = \displaystyle\sum_{n=0}^{\infty}\frac{(-1)^n x^{2n+2}}{(2n+1)!}$

**43.** $\dfrac{\sin x}{x} = \dfrac{x - (x^3/3!) + (x^5/5!) - \cdots}{x} = 1 - \dfrac{x^2}{2!} + \dfrac{x^4}{4!} - \cdots = \displaystyle\sum_{n=0}^{\infty}\frac{(-1)^n x^{2n}}{(2n+1)!},\ x \neq 0$

$$= 1,\ x = 0$$

**45.**  $e^{ix} = 1 + ix + \dfrac{(ix)^2}{2!} + \dfrac{(ix)^3}{3!} + \dfrac{(ix)^4}{4!} + \cdots = 1 + ix - \dfrac{x^2}{2!} - \dfrac{ix^3}{3!} + \dfrac{x^4}{4!} + \dfrac{ix^5}{5!} - \dfrac{x^6}{6!} - \cdots$

$e^{-ix} = 1 - ix + \dfrac{(-ix)^2}{2!} + \dfrac{(-ix)^3}{3!} + \dfrac{(-ix)^4}{4!} + \cdots = 1 - ix - \dfrac{x^2}{2!} + \dfrac{ix^3}{3!} + \dfrac{x^4}{4!} - \dfrac{ix^5}{5!} - \dfrac{x^6}{6!} + \cdots$

$e^{ix} - e^{-ix} = 2ix - \dfrac{2ix^3}{3!} + \dfrac{2ix^5}{5!} - \dfrac{2ix^7}{7!} + \cdots$

$\dfrac{e^{ix} - e^{-ix}}{2i} = x - \dfrac{x^3}{3!} + \dfrac{x^5}{5!} - \dfrac{x^7}{7!} + \cdots = \displaystyle\sum_{n=0}^{\infty}\frac{(-1)^n x^{2n+1}}{(2n+1)!} = \sin(x)$

**47.**  $f(x) = e^x \sin x$

$$= \left(1 + x + \frac{x^2}{2} + \frac{x^3}{6} + \frac{x^4}{24} + \cdots\right)\left(x - \frac{x^3}{6} + \frac{x^5}{120} - \cdots\right)$$

$$= x + x^2 + \left(\frac{x^3}{2} - \frac{x^3}{6}\right) + \left(\frac{x^4}{6} - \frac{x^4}{6}\right) + \left(\frac{x^5}{120} - \frac{x^5}{12} + \frac{x^5}{24}\right) + \cdots$$

$$= x + x^2 + \frac{x^3}{3} - \frac{x^5}{30} + \cdots$$

$$P_5(x) = x + x^2 + \frac{x^3}{3} - \frac{x^5}{30}$$

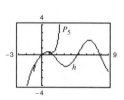

**49.**  $h(x) = \cos x \ln(1 + x)$

$$= \left(1 - \frac{x^2}{2} + \frac{x^4}{24} + \cdots\right)\left(x - \frac{x^2}{2} + \frac{x^3}{3} - \frac{x^4}{4} + \frac{x^5}{5} - \cdots\right)$$

$$= x - \frac{x^2}{2} + \left(\frac{x^3}{3} - \frac{x^3}{2}\right) + \left(\frac{x^4}{4} - \frac{x^4}{4}\right) + \left(\frac{x^5}{5} - \frac{x^5}{6} + \frac{x^5}{24}\right) + \cdots$$

$$= x - \frac{x^2}{2} - \frac{x^3}{6} + \frac{3x^5}{40} + \cdots$$

$$P_5(x) = x - \frac{x^2}{2} - \frac{x^3}{6} + \frac{3x^5}{40}$$

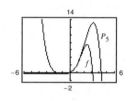

**51.** $g(x) = \dfrac{\sin x}{1 + x}$. Divide the series for $\sin x$ by $(1 + x)$.

$$1 + x \overline{\smash{\big)}\ x + 0x^2 - \dfrac{x^3}{6} + 0x^4 + \dfrac{x^5}{120} + \cdots}$$

with quotient $x - x^2 + \dfrac{5x^3}{6} - \dfrac{5x^4}{6} +$

$\underline{x + x^2}$

$-x^2 - \dfrac{x^3}{6}$

$\underline{-x^2 - x^3}$

$\dfrac{5x^3}{6} + 0x^4$

$\underline{\dfrac{5x^3}{6} + \dfrac{5x^4}{6}}$

$-\dfrac{5x^4}{6} + \dfrac{x^5}{120}$

$\underline{-\dfrac{5x^4}{6} - \dfrac{5x^5}{6}}$

$\vdots$

$g(x) = x - x^2 + \dfrac{5x^3}{6} - \dfrac{5x^4}{6} + \cdots$

$P_4(x) = x - x^2 + \dfrac{5x^3}{6} - \dfrac{5x^4}{6}$

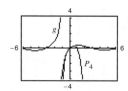

**53.** $y = x^2 - \dfrac{x^4}{3!} = x\left(x - \dfrac{x^3}{3!}\right)$

$f(x) = x \sin x$

Matches (c)

**55.** $y = x + x^2 + \dfrac{x^3}{2!} = x\left(1 + x + \dfrac{x^2}{2!}\right)$

$f(x) = xe^x$

Matches (a)

**57.** $\displaystyle\int_0^x \left(e^{-t^2} - 1\right) dt = \int_0^x \left[\left(\sum_{n=0}^{\infty} \dfrac{(-1)^n t^{2n}}{n!}\right) - 1\right] dt$

$\displaystyle = \int_0^x \left[\sum_{n=0}^{\infty} \dfrac{(-1)^{n+1} t^{2n+2}}{(n+1)!}\right] dt$

$\displaystyle = \left[\sum_{n=0}^{\infty} \dfrac{(-1)^{n+1} t^{2n+3}}{(2n+3)(n+1)!}\right]_0^x$

$\displaystyle = \sum_{n=0}^{\infty} \dfrac{(-1)^{n+1} x^{2n+3}}{(2n+3)(n+1)!}$

**59.** Because $\ln x = \displaystyle\sum_{n=0}^{\infty} \dfrac{(-1)^n (x-1)^{n+1}}{n+1} = (x-1) - \dfrac{(x-1)^2}{2} + \dfrac{(x-1)^3}{3} - \dfrac{(x-1)^4}{4} + \cdots, \quad (0 < x \le 2)$

you have $\ln 2 = 1 - \dfrac{1}{2} + \dfrac{1}{3} - \dfrac{1}{4} + \cdots = \displaystyle\sum_{n=1}^{\infty} (-1)^{n+1} \dfrac{1}{n} \approx 0.6931.$ (10,001 terms)

**61.** Because $e^x = \displaystyle\sum_{n=0}^{\infty} \dfrac{x^n}{n!} = 1 + x + \dfrac{x^2}{2!} + \dfrac{x^3}{3!} + \cdots,$

you have $e^2 = 1 + 2 + \dfrac{2^2}{2!} + \dfrac{2^3}{3!} + \cdots = \displaystyle\sum_{n=0}^{\infty} \dfrac{2^n}{n!} \approx 7.3891.$ (12 terms)

**63.** Because

$$\cos x = \sum_{n=0}^{\infty} \dfrac{(-1)^n x^{2n}}{(2n)!} = 1 - \dfrac{x^2}{2!} + \dfrac{x^4}{4!} - \dfrac{x^6}{6!} + \dfrac{x^8}{8!} - \cdots$$

$$1 - \cos x = \dfrac{x^2}{2!} - \dfrac{x^4}{4!} + \dfrac{x^6}{6!} - \dfrac{x^8}{8!} + \cdots = \sum_{n=0}^{\infty} \dfrac{(-1)^n x^{2n+2}}{(2n+2)!}$$

$$\dfrac{1 - \cos}{x} = \dfrac{x}{2!} - \dfrac{x^3}{4!} + \dfrac{x^5}{6!} - \dfrac{x^7}{8!} + \cdots = \sum_{n=0}^{\infty} \dfrac{(-1)^n x^{2n+1}}{(2n+2)!}$$

you have $\displaystyle\lim_{x \to 0} \dfrac{1 - \cos x}{x} = \lim_{x \to 0} \sum_{n=0}^{\infty} \dfrac{(-1) x^{2n+1}}{(2n+2)!} = 0.$

**65.** Because $e^x = 1 + x + \dfrac{x^2}{2!} + \dfrac{x^3}{3!} + \cdots$

$e^x - 1 = x + \dfrac{x^2}{2!} + \dfrac{x^3}{3!} + \cdots \displaystyle\sum_{n=0}^{\infty} \dfrac{x^{n+1}}{(n+1)!}$

and $\dfrac{e^x - 1}{x} = 1 + \dfrac{x}{2!} + \dfrac{x^2}{3!} + \cdots \displaystyle\sum_{n=0}^{\infty} \dfrac{x^n}{(n+1)!}$

you have $\displaystyle\lim_{x \to 0} \dfrac{e^x - 1}{x} = \lim_{x \to 0} \sum \dfrac{x^n}{(n+1)!} = 1.$

**67.** $\displaystyle\int_0^1 e^{-x^3}\, dx = \int_0^1 \left[ \sum_{n=0}^{\infty} \dfrac{\left(-x^3\right)^n}{n!} \right] dx$

$= \displaystyle\int_0^1 \left[ \sum_{n=0}^{\infty} \dfrac{(-1)^n x^{3n}}{n!} \right] dx$

$= \left[ \displaystyle\sum_{n=0}^{\infty} \dfrac{(-1)^n x^{3n+1}}{(3n+1)n!} \right]_0^1$

$= \displaystyle\sum_{n=0}^{\infty} \dfrac{(-1)^n}{(3n+1)n!}$

$= 1 - \dfrac{1}{4} + \dfrac{1}{14} - \cdots + (-1)^n \dfrac{1}{(3n+1)n!} + \cdots$

Because $\dfrac{1}{[3(6)+1]6!} < 0.0001$, you need 6 terms.

$\displaystyle\int_0^1 e^{-x^2}\, dx \approx \sum_{n=0}^{5} \dfrac{(-1)^n}{(3n+1)n!} \approx 0.8075$

**69.** $\displaystyle\int_0^1 \dfrac{\sin x}{x}\, dx = \int_0^1 \left[ \sum_{n=0}^{\infty} \dfrac{(-1)^n x^{2n}}{(2n+1)!} \right] dx = \left[ \sum_{n=0}^{\infty} \dfrac{(-1)^n x^{2n+1}}{(2n+1)(2n+1)!} \right]_0^1 = \sum_{n=0}^{\infty} \dfrac{(-1)^n}{(2n+1)(2n+1)!}$

Because $1/(7 \cdot 7!) < 0.0001$, you need three terms:

$\displaystyle\int_0^1 \dfrac{\sin x}{x}\, dx = 1 - \dfrac{1}{3 \cdot 3!} + \dfrac{1}{5 \cdot 5!} - \cdots \approx 0.9461.$     (using three nonzero terms)

**Note:** You are using $\displaystyle\lim_{x \to 0^+} \dfrac{\sin x}{x} = 1.$

**71.** $\displaystyle\int_0^{1/2} \dfrac{\arctan x}{x}\, dx = \int_0^{1/2} \left( 1 - \dfrac{x^2}{3} + \dfrac{x^4}{5} - \dfrac{x^6}{7} + \cdots \right) dx = \left[ x - \dfrac{x^3}{3^2} + \dfrac{x^5}{5^2} - \dfrac{x^7}{7^2} + \cdots \right]_0^{1/2}$

Because $1/\left(9^2 2^9\right) < 0.0001$, you have

$\displaystyle\int_0^{1/2} \dfrac{\arctan x}{x}\, dx \approx \left( \dfrac{1}{2} - \dfrac{1}{3^2 2^3} + \dfrac{1}{5^2 2^5} - \dfrac{1}{7^2 2^7} + \dfrac{1}{9^2 2^9} \right) \approx 0.4872.$

**Note:** You are using $\displaystyle\lim_{x \to 0^+} \dfrac{\arctan x}{x} = 1.$

**73.** $\displaystyle\int_{0.1}^{0.3} \sqrt{1 + x^3}\, dx = \int_{0.1}^{0.3} \left( 1 + \dfrac{x^3}{2} - \dfrac{x^6}{8} + \dfrac{x^9}{16} - \dfrac{5x^{12}}{128} + \cdots \right) dx = \left[ x + \dfrac{x^4}{8} - \dfrac{x^7}{56} + \dfrac{x^{10}}{160} - \dfrac{5x^{13}}{1664} + \cdots \right]_{0.1}^{0.3}$

Because $\dfrac{1}{56}\left(0.3^7 - 0.1^7\right) < 0.0001$, you need two terms.

$\displaystyle\int_{0.1}^{0.3} \sqrt{1 + x^3}\, dx = \left[ (0.3 - 0.1) + \dfrac{1}{8}\left(0.3^4 - 0.1^4\right) \right] \approx 0.201.$

**75.** $\displaystyle\int_0^{\pi/2} \sqrt{x}\, \cos x\, dx = \int_0^{\pi/2} \left[ \sum_{n=0}^{\infty} \dfrac{(-1)^n x^{(4n+1)/2}}{(2n)!} \right] dx = \left[ \sum_{n=0}^{\infty} \dfrac{(-1)^n x^{(4n+3)/2}}{\left( \dfrac{4n+3}{2} \right)(2n)!} \right]_0^{\pi/2} = \left[ \sum_{n=0}^{\infty} \dfrac{(-1)^n 2 x^{(4n+3)/2}}{(4n+3)(2n)!} \right]_0^{\pi/2}$

Because $2(\pi/2)^{23/2} / (23 \cdot 10!) < 0.0001$, you need five terms.

$\displaystyle\int_0^1 \sqrt{x}\, \cos x\, dx = 2\left[ \dfrac{(\pi/2)^{3/2}}{3} - \dfrac{(\pi/2)^{7/2}}{14} + \dfrac{(\pi/2)^{11/2}}{264} - \dfrac{(\pi/2)^{15/2}}{10{,}800} + \dfrac{(\pi/2)^{19/2}}{766{,}080} \right] \approx 0.7040.$

**77.** From Exercise 27, you have

$$\frac{1}{\sqrt{2\pi}}\int_0^1 e^{-x^2/2}\, dx = \frac{1}{\sqrt{2\pi}}\int_0^1 \sum_{n=0}^\infty \frac{(-1)^n x^{2n}}{2^n n!}\, dx = \frac{1}{\sqrt{2\pi}}\left[\sum_{n=0}^\infty \frac{(-1)^n x^{2n+1}}{2^n n!(2n+1)}\right]_0^1 = \frac{1}{\sqrt{2\pi}}\sum_{n=0}^\infty \frac{(-1)^n}{2^n n!(2n+1)}$$

$$\approx \frac{1}{\sqrt{2\pi}}\left(1 - \frac{1}{2\cdot 1\cdot 3} + \frac{1}{2^2\cdot 2!\cdot 5} - \frac{1}{2^3\cdot 3!\cdot 7}\right) \approx 0.3412.$$

**79.** $f(x) = x\cos 2x = \displaystyle\sum_{n=0}^\infty \frac{(-1)^n 4^n x^{2n+1}}{(2n)!}$

$P_5(x) = x - 2x^3 + \dfrac{2x^5}{3}$

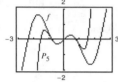

The polynomial is a reasonable approximation on the interval $\left[-\dfrac{3}{4}, \dfrac{3}{4}\right]$.

**81.** $f(x) = \sqrt{x}\ln x,\ c = 1$

$P_5(x) = (x - 1) - \dfrac{(x-1)^3}{24} + \dfrac{(x-1)^4}{24} - \dfrac{71(x-1)^5}{1920}$

The polynomial is a reasonable approximation on the interval $\left[\dfrac{1}{4}, 2\right]$.

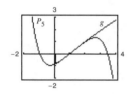

**83.** See Guidelines, page 682.

**85.** The binomial series is $(1 + x)^k = 1 + kx + \dfrac{k(k-1)}{2!}x^2 + \dfrac{k(k-1)(k-2)}{3!}x^3 + \cdots.$ The radius of convergence is $R = 1$.

**87.** $y = \left(\tan\theta - \dfrac{g}{kv_0\cos\theta}\right)x - \dfrac{g}{k^2}\ln\left(1 - \dfrac{kx}{v_0\cos\theta}\right)$

$$= (\tan\theta)x - \frac{gx}{kv_0\cos\theta} - \frac{g}{k^2}\left[-\frac{kx}{v_0\cos\theta} - \frac{1}{2}\left(\frac{kx}{v_0\cos\theta}\right)^2 - \frac{1}{3}\left(\frac{kx}{v_0\cos\theta}\right)^3 - \frac{1}{4}\left(\frac{kx}{v_0\cos\theta}\right)^4 - \cdots\right]$$

$$= (\tan\theta)x - \frac{gx}{kv_0\cos\theta} + \frac{gx}{kv_0\cos\theta} + \frac{gx^2}{2v_0^2\cos^2\theta} + \frac{gkx^3}{3v_0^3\cos^3\theta} + \frac{gk^2x^4}{4v_0^4\cos^4\theta} + \cdots$$

$$= (\tan\theta)x + \frac{gx^2}{2v_0^2\cos^2\theta} + \frac{kgx^3}{3v_0^3\cos^3\theta} + \frac{k^2gx^4}{4v_0^4\cos^4\theta} + \cdots$$

**89.** $f(x) = \begin{cases} e^{-1/x^2}, & x \neq 0 \\ 0, & x = 0 \end{cases}$

(a)

(b) $f'(0) = \lim_{x \to 0} \dfrac{f(x) - f(0)}{x - 0} = \lim_{x \to 0} \dfrac{e^{-1/x^2} - 0}{x}$

Let $y = \lim_{x \to 0} \dfrac{e^{-1/x^2}}{x}$. Then

$\ln y = \lim_{x \to 0} \ln\left(\dfrac{e^{-1/x^2}}{x}\right) = \lim_{x \to 0^+}\left[-\dfrac{1}{x^2} - \ln x\right] = \lim_{x \to 0^+}\left[\dfrac{-1 - x^2 \ln x}{x^2}\right] = -\infty.$

So, $y = e^{-\infty} = 0$ and you have $f'(0) = 0.$

(c) $\displaystyle\sum_{n=0}^{\infty} \dfrac{f^{(n)}(0)}{n!}x^n = f(0) + \dfrac{f'(0)x}{1!} + \dfrac{f''(0)x^2}{2!} + \cdots = 0 \neq f(x)$ This series converges to $f$ at $x = 0$ only.

(d) The curves are nearly identical for $0 < x < 1$. Hence, the integrals nearly agree on that interval.

**91.** By the Ratio Test: $\displaystyle\lim_{n \to \infty}\left|\dfrac{x^{n+1}}{(n+1)!} \cdot \dfrac{n!}{x^n}\right| = \lim_{n \to \infty}\dfrac{|x|}{n+1} = 0$ which shows that $\displaystyle\sum_{n=0}^{\infty} \dfrac{x^n}{n!}$ converges for all $x$.

**93.** $\dbinom{5}{3} = \dfrac{5 \cdot 4 \cdot 3}{3!} = \dfrac{60}{6} = 10$

**95.** $\dbinom{0.5}{4} = \dfrac{(0.5)(-0.5)(-1.5)(-2.5)}{4!} = -0.0390625 = -\dfrac{5}{128}$

**97.** $(1 + x)^k = \displaystyle\sum_{n=0}^{\infty} \dbinom{k}{n}x^n$

Example: $(1 + x)^2 = \displaystyle\sum_{n=0}^{\infty} \dbinom{2}{n}x^n = 1 + 2x + x^2$

**99.** $g(x) = \dfrac{x}{1 - x - x^2} = a_0 + a_1 x + a_2 x^2 + \cdots$

$x = (1 - x - x^2)(a_0 + a_1 x + a_2 x^2 + \cdots)$

$x = a_0 + (a_1 - a_0)x + (a_2 - a_1 - a_0)x^2 + (a_3 - a_2 - a_1)x^3 + \cdots$

Equating coefficients,

$a_0 = 0$

$a_1 - a_0 = 1 \Rightarrow a_1 = 1$

$a_2 - a_1 - a_0 = 0 \Rightarrow a_2 = 1$

$a_3 - a_2 - a_1 = 0 \Rightarrow a_3 = 2$

$a_4 = a_3 + a_2 = 3$, etc.

In general, $a_n = a_{n-1} + a_{n-2}$. The coefficients are the Fibonacci numbers.

# Review Exercises for Chapter 9

**1.** $a_n = \dfrac{1}{n! + 1}$

**3.** $a_n = 4 + \dfrac{2}{n}$: $6, 5, 4.67, \ldots$

Matches (a)

**5.** $a_n = 10(0.3)^{n-1}$: $10, 3, \ldots$

Matches (d)

**7.** $a_n = \dfrac{5n + 2}{n}$

The sequence seems to converge to 5.

$$\lim_{n \to \infty} a_n = \lim_{n \to \infty} \frac{5n + 2}{n}$$

$$= \lim_{n \to \infty} \left( 5 + \frac{2}{n} \right) = 5$$

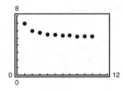

**9.** $\displaystyle\lim_{n \to \infty} \left[ \left( \tfrac{7}{8} \right)^n + 3 \right] = 3$

Converges

**11.** $\displaystyle\lim_{n \to \infty} \frac{n^3 + 1}{n^2} = \infty$

Diverges

**13.** $\displaystyle\lim_{n \to \infty} \frac{n}{n^2 + 1} = 0$

Converges

**23.** (a)

| $n$ | 5 | 10 | 15 | 20 | 25 |
|---|---|---|---|---|---|
| $S_n$ | 0.4597 | 0.4597 | 0.4597 | 0.4597 | 0.4597 |

The series converges by the Alternating Series Test.

(b)

**15.** $\displaystyle\lim_{n \to \infty} \left( \sqrt{n + 1} - \sqrt{n} \right) = \lim_{n \to \infty} \left( \sqrt{n + 1} - \sqrt{n} \right) \frac{\sqrt{n + 1} + \sqrt{n}}{\sqrt{n + 1} + \sqrt{n}}$

$$= \lim_{n \to \infty} \frac{1}{\sqrt{n + 1} + \sqrt{n}} = 0$$

Converges

**17.** $\displaystyle\lim_{n \to \infty} \frac{\sin \sqrt{n}}{\sqrt{n}} = 0$

Converges

**19.** (a) $A_n = 8000 \left( 1 + \dfrac{0.05}{4} \right)^n$, $\quad n = 1, 2, 3, \ldots$

$A_1 = 8000 \left( 1 + \dfrac{0.05}{4} \right)^1 = \$8100.00$

$A_2 = \$8201.25$

$A_3 = \$8303.77$

$A_4 = \$8407.56$

$A_5 = \$8512.66$

$A_6 = \$8619.07$

$A_7 = \$8726.80$

$A_8 = \$8835.89$

(b) $A_{40} = \$13{,}148.96$

**21.** (a)

| $n$ | 5 | 10 | 15 | 20 | 25 |
|---|---|---|---|---|---|
| $S_n$ | 13.2 | 113.3 | 873.8 | 6648.5 | 50,500.3 |

The series diverges $\left(\text{geometric } r = \tfrac{3}{2} > 1 \right)$.

(b)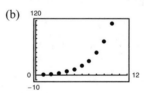

**25.** $\sum\limits_{n=0}^{\infty}\left(\dfrac{2}{3}\right)^{n}$ Geometric series with $a = 1$ and $r = \dfrac{2}{3}$.

$$S = \frac{a}{1-r} = \frac{1}{1-(2/3)} = \frac{1}{1/3} = 3$$

**27.** $\sum\limits_{n=1}^{\infty}\left[(0.6)^{n}+(0.8)^{n}\right] = \sum\limits_{n=0}^{\infty}0.6(0.6)^{n} + \sum\limits_{n=0}^{\infty}0.8(0.8)^{n} = (0.6)\dfrac{1}{1-0.6} + (0.8)\dfrac{1}{1-0.8} = \dfrac{6}{10}\cdot\dfrac{10}{4} + \dfrac{8}{10}\cdot\dfrac{10}{2} = \dfrac{11}{2} = 5.5$

**29.** (a) $0.\overline{09} = 0.09 + 0.0009 + 0.000009 + \cdots = 0.09(1 + 0.01 + 0.0001 + \cdots) = \sum\limits_{n=0}^{\infty}(0.09)(0.01)^{n}$

    (b) $0.\overline{09} = \dfrac{0.09}{1-0.01} = \dfrac{1}{11}$

**31.** Diverges. Geometric series with $a = 1$ and $|r| = 1.67 > 1$.

**33.** Diverges. $n$th-Term Test. $\lim\limits_{n\to\infty} a_{n} \neq 0$.

**35.** $D_{1} = 8$

    $D_{2} = 0.7(8) + 0.7(8) = 16(0.7)$

    $\vdots$

    $D = 8 + 16(0.7) + 16(0.7)^{2} + \cdots + 16(0.7)^{n} + \cdots$

    $= -8 + \sum\limits_{n=0}^{\infty}16(0.7)^{n} = -8 + \dfrac{16}{1-0.7} = 45\dfrac{1}{3}$ meters

**37.** (See Exercise 116 in Section 9.2)

    $A = \dfrac{P(e^{rt}-1)}{e^{r/12}-1} = \dfrac{300(e^{0.06(2)}-1)}{e^{0.06/12}-1} \approx \$7630.70$

**39.** $\displaystyle\int_{1}^{\infty}x^{-4}\ln(x)\,dx = \lim\limits_{b\to\infty}\left[-\dfrac{\ln x}{3x^{3}} - \dfrac{1}{9x^{3}}\right]_{1}^{b} = 0 + \dfrac{1}{9} = \dfrac{1}{9}$

    By the Integral Test, the series converges.

**41.** $\sum\limits_{n=1}^{\infty}\left(\dfrac{1}{n^{2}} - \dfrac{1}{n}\right) = \sum\limits_{n=1}^{\infty}\dfrac{1}{n^{2}} - \sum\limits_{n=1}^{\infty}\dfrac{1}{n}$

    Because the second series is a divergent $p$-series while the first series is a convergent $p$-series, the difference diverges.

**43.** $\sum\limits_{n=1}^{\infty}\dfrac{6}{5n-1}$

    $\lim\limits_{n\to\infty}\left[\dfrac{6/(5n-1)}{1/n}\right] = \dfrac{6}{5}$

    By a limit comparison test with the divergent harmonic series $\sum\limits_{n=1}^{\infty}\dfrac{1}{n}$, the series diverges.

**45.** $\sum\limits_{n=1}^{\infty}\dfrac{1}{\sqrt{n^{3}+2n}}$

    $\lim\limits_{n\to\infty}\dfrac{1/\sqrt{n^{3}+2n}}{1/(n^{3/2})} = \lim\limits_{n\to\infty}\dfrac{n^{3/2}}{\sqrt{n^{3}+2n}} = 1$

    By a limit comparison test with the convergent $p$-series $\sum\limits_{n=1}^{\infty}\dfrac{1}{n^{3/2}}$, the series converges.

**47.** $\sum\limits_{n=1}^{\infty}\dfrac{1\cdot3\cdot5\cdots(2n-1)}{2\cdot4\cdot6\cdots(2n)}$

    $a_{n} = \dfrac{1\cdot3\cdot5\cdots(2n-1)}{2\cdot4\cdot6\cdots(2n)} = \left(\dfrac{3}{2}\cdot\dfrac{5}{4}\cdots\dfrac{2n-1}{2n-2}\right)\dfrac{1}{2n} > \dfrac{1}{2n}$

    Because $\sum\limits_{n=1}^{\infty}\dfrac{1}{2n} = \dfrac{1}{2}\sum\limits_{n=1}^{\infty}\dfrac{1}{n}$ diverges (harmonic series), so does the original series.

**49.** $\sum\limits_{n=1}^{\infty}\dfrac{(-1)^{n}}{n^{5}}$ converges by the Alternating Series Test.

    $\lim\limits_{n\to\infty}\dfrac{1}{n^{5}} = 0$ and $a_{n+1} = \dfrac{1}{(n+1)^{5}} < \dfrac{1}{n^{5}} = a_{n}$.

**51.** $\sum\limits_{n=2}^{\infty}\dfrac{(-1)^{n}-n}{n^{2}-3}$ converges by the Alternating Series Test.

    $\lim\limits_{n\to\infty}\dfrac{n}{n^{2}-3} = 0$ and if $f(x) = \dfrac{n}{n^{2}-3}$,

    $f'(x) = \dfrac{-(n^{2}+3)}{(n^{2}-3)^{2}} < 0 \Rightarrow$ terms are decreasing. So,

    $a_{n+1} < a_{n}$.

**53.** Diverges by the $n$th-Term Test.

    $\lim\limits_{n\to\infty}\dfrac{n}{n-3} = 1 \neq 0$

**55.** $\lim\limits_{n\to\infty} \sqrt[n]{\left(\dfrac{3n-1}{2n+5}\right)^n} = \lim\limits_{n\to\infty} \left(\dfrac{3n-1}{2n+5}\right) = \dfrac{3}{2} > 1$

Diverges by Root Test.

**57.** $\displaystyle\sum_{n=1}^{\infty} \dfrac{n}{e^{n^2}}$

$\begin{aligned}
\lim\limits_{n\to\infty} \left| \dfrac{a_{n+1}}{a_n} \right| &= \lim\limits_{n\to\infty} \left| \dfrac{n+1}{e^{(n+1)^2}} \cdot \dfrac{e^{n^2}}{n} \right| \\
&= \lim\limits_{n\to\infty} \left| \dfrac{e^{n^2}(n+1)}{e^{n^2+2n+1}n} \right| \\
&= \lim\limits_{n\to\infty} \left( \dfrac{1}{e^{2n+1}} \right)\left( \dfrac{n+1}{n} \right) \\
&= (0)(1) = 0 < 1
\end{aligned}$

By the Ratio Test, the series converges.

**61.** (a)  Ratio Test: $\lim\limits_{n\to\infty} \left| \dfrac{a_{n+1}}{a_n} \right| = \lim\limits_{n\to\infty} \dfrac{(n+1)(3/5)^{n+1}}{n(3/5)^n} = \lim\limits_{n\to\infty} \left( \dfrac{n+1}{n} \right)\left( \dfrac{3}{5} \right) = \dfrac{3}{5} < 1$, Converges

(b)

| $n$ | 5 | 10 | 15 | 20 | 25 |
|-----|-----|-----|-----|-----|-----|
| $S_n$ | 2.8752 | 3.6366 | 3.7377 | 3.7488 | 3.7499 |

(c)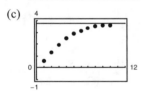

(d)  The sum is approximately 3.75.

**63.** (a)  $\displaystyle\int_N^\infty \dfrac{1}{x^2}\,dx = \left[ -\dfrac{1}{x} \right]_N^\infty = \dfrac{1}{N}$

| $N$ | 5 | 10 | 20 | 30 | 40 |
|-----|-----|-----|-----|-----|-----|
| $\displaystyle\sum_{n=1}^{N} \dfrac{1}{n^2}$ | 1.4636 | 1.5498 | 1.5962 | 1.6122 | 1.6202 |
| $\displaystyle\int_N^\infty \dfrac{1}{x^2}\,dx$ | 0.2000 | 0.1000 | 0.0500 | 0.0333 | 0.0250 |

(b)  $\displaystyle\int_N^\infty \dfrac{1}{x^5}\,dx = \left[ -\dfrac{1}{4x^4} \right]_N^\infty = \dfrac{1}{4N^4}$

| $N$ | 5 | 10 | 20 | 30 | 40 |
|-----|-----|-----|-----|-----|-----|
| $\displaystyle\sum_{n=1}^{N} \dfrac{1}{n^5}$ | 1.0367 | 1.0369 | 1.0369 | 1.0369 | 1.0369 |
| $\displaystyle\int_N^\infty \dfrac{1}{x^5}\,dx$ | 0.0004 | 0.0000 | 0.0000 | 0.0000 | 0.0000 |

The series in part (b) converges more rapidly. The integral values represent the remainders of the partial sums.

**59.** $\displaystyle\sum_{n=1}^{\infty} \dfrac{2^n}{n^3}$

$\lim\limits_{n\to\infty} \left| \dfrac{a_{n+1}}{a_n} \right| = \lim\limits_{n\to\infty} \left| \dfrac{2^{n+1}}{(n+1)^3} \cdot \dfrac{n^3}{2^n} \right| = \lim\limits_{n\to\infty} \dfrac{2n^3}{(n+1)^3} = 2$

By the Ratio Test, the series diverges.

**65.** $f(x) = e^{-3x}$      $f(0) = 1$

$f'(x) = -3e^{-3x}$      $f'(0) = -3$

$f''(x) = 9e^{-3x}$      $f''(0) = 9$

$f'''(x) = -27e^{-3x}$    $f'''(0) = -27$

$$P_3(x) = 1 - 3x + \frac{9x^2}{2!} - \frac{27x^3}{3!} = 1 - 3x + \frac{9}{2}x^2 - \frac{9}{2}x^3$$

**67.** Because $\dfrac{(95\pi)^9}{180^9 \cdot 9!} < 0.001$, use four terms.

$$\sin 95° = \sin\left(\frac{95\pi}{180}\right) \approx \frac{95\pi}{180} - \frac{(95\pi)^3}{180^3 3!} + \frac{(95\pi)^5}{180^5 5!} - \frac{(95\pi)^7}{180^7 7!} \approx 0.99594$$

**69.** Because $\dfrac{(0.75)^{15}}{15} < 0.001$, use 14 terms.

$$\ln(1.75) \approx (0.75) - \frac{(0.75)^2}{2} + \frac{(0.75)^3}{3} - \frac{(0.75)^4}{4} + \frac{(0.75)^5}{5} - \frac{(0.75)^6}{6} + \cdots - \frac{(0.75)^{14}}{14} \approx 0.559062$$

**71.** $f(x) = \cos x, \quad c = 0$

$$R_n(x) = \frac{f^{(n+1)}(z)}{(n+1)!}x^{n+1}$$

$$\left| f^{(n+1)}(z) \right| \le 1 \Rightarrow R_n(x) \le \frac{x^{n+1}}{(n+1)!}$$

(a) $R_n(x) \le \dfrac{(0.5)^{n+1}}{(n+1)!} < 0.001$

    This inequality is true for $n = 4$.

(b) $R_n(x) \le \dfrac{(1)^{n+1}}{(n+1)!} < 0.001$

    This inequality is true for $n = 6$.

(c) $R_n(x) \le \dfrac{(0.5)^{n+1}}{(n+1)!} < 0.0001$

    This inequality is true for $n = 5$.

(d) $R_n(x) \le \dfrac{2^{n+1}}{(n+1)!} < 0.0001$

    This inequality is true for $n = 10$.

**73.** $\displaystyle\sum_{n=0}^{\infty} \left(\frac{x}{10}\right)^n$

Geometric series which converges only if $|x/10| < 1$ or $-10 < x < 10$.

**75.** $\displaystyle\sum_{n=0}^{\infty} \frac{(-1)^n (x-2)^n}{(n+1)^2}$

$$\lim_{n\to\infty}\left|\frac{u_{n+1}}{u_n}\right| = \lim_{n\to\infty}\left|\frac{(-1)^{n+1}(x-2)^{n+1}}{(n+2)^2} \cdot \frac{(n+1)^2}{(-1)^n(x-2)^n}\right|$$

$$= |x - 2|$$

$R = 1$

Center: 2

Because the series converges when $x = 1$ and when $x = 3$, the interval of convergence is $[1, 3]$.

**77.** $\displaystyle\sum_{n=0}^{\infty} n!(x-2)^n$

$$\lim_{n\to\infty}\left|\frac{u_{n+1}}{u_n}\right| = \lim_{n\to\infty}\left|\frac{(n+1)!(x-2)^{n+1}}{n!(x-2)^n}\right| = \infty$$

which implies that the series converges only at the center $x = 2$.

**79.**
$$y = \sum_{n=0}^{\infty} (-1)^n \frac{x^{2n}}{4^n (n!)^2}$$

$$y' = \sum_{n=1}^{\infty} \frac{(-1)^n (2n) x^{2n-1}}{4^n (n!)^2} = \sum_{n=0}^{\infty} \frac{(-1)^{n+1} (2n + 2) x^{2n+1}}{4^{n+1} [(n + 1)!]^2}$$

$$y'' = \sum_{n=0}^{\infty} \frac{(-1)^{n+1} (2n + 2)(2n + 1) x^{2n}}{4^{n+1} [(n + 1)!]^2}$$

$$x^2 y'' + xy' + x^2 y = \sum_{n=0}^{\infty} \frac{(-1)^{n+1} (2n + 2)(2n + 1) x^{2n+2}}{4^{n+1} [(n + 1)!]^2} + \sum_{n=0}^{\infty} \frac{(-1)^{n+1} (2n + 2) x^{2n+2}}{4^{n+1} [(n + 1)!]^2} + \sum_{n=0}^{\infty} (-1)^n \frac{x^{2n+2}}{4^n (n!)^2}$$

$$= \sum_{n=0}^{\infty} \left[ (-1)^{n+1} \frac{(2n + 2)(2n + 1)}{4^{n+1} [(n + 1)!]^2} + \frac{(-1)^{n+1} (2n + 2)}{4^{n+1} [(n + 1)!]^2} + \frac{(-1)^n}{4^n (n!)^2} \right] x^{2n+2}$$

$$= \sum_{n=0}^{\infty} \left[ \frac{(-1)^{n+1} (2n + 2)(2n + 1 + 1)}{4^{n+1} [(n + 1)!]^2} + (-1)^n \frac{1}{4^n (n!)^2} \right] x^{2n+2}$$

$$= \sum_{n=0}^{\infty} \left[ \frac{(-1)^{n+1} 4(n + 1)^2}{4^{n+1} [(n + 1)!]^2} + (-1)^n \frac{1}{4^n (n!)^2} \right] x^{2n+2}$$

$$= \sum_{n=0}^{\infty} \left[ \frac{(-1)^{n+1} 1}{4^n (n!)^2} + (-1)^n \frac{1}{4^n (n!)^2} \right] x^{2n+2} = 0$$

**81.** $\dfrac{2}{3 - x} = \dfrac{2/3}{1 - (x/3)} = \dfrac{a}{1 - r}$

$$\sum_{n=0}^{\infty} \frac{2}{3} \left( \frac{x}{3} \right)^n = \sum_{n=0}^{\infty} \frac{2x^n}{3^{n+1}}$$

**83.** $g(x) = \dfrac{2}{3 - x}$. Power series $\displaystyle\sum_{n=0}^{\infty} \frac{2}{3} \left( \frac{x}{3} \right)^n$

Derivative: $\displaystyle\sum_{n=1}^{\infty} \frac{2}{3} n \left( \frac{x}{3} \right)^{n-1} \left( \frac{1}{3} \right) = \sum_{n=1}^{\infty} \frac{2}{9} n \left( \frac{x}{3} \right)^{n-1} = \sum_{n=0}^{\infty} \frac{2}{9} (n + 1) \left( \frac{x}{3} \right)^n$

**85.** $1 + \dfrac{2}{3} x + \dfrac{4}{9} x^2 + \dfrac{8}{27} x^3 + \cdots = \displaystyle\sum_{n=0}^{\infty} \left( \frac{2x}{3} \right)^n = \dfrac{1}{1 - (2x/3)} = \dfrac{3}{3 - 2x}, \quad \left( -\dfrac{3}{2}, \dfrac{3}{2} \right)$

**87.** $f(x) = \sin x$

$f'(x) = \cos x$

$f''(x) = -\sin x$

$f'''(x) = -\cos x, \cdots$

$\sin(x) = \displaystyle\sum_{n=0}^{\infty} \frac{f^{(n)}(x) [x - (3\pi/4)]^n}{n!}$

$= \dfrac{\sqrt{2}}{2} - \dfrac{\sqrt{2}}{2} \left( x - \dfrac{3\pi}{4} \right) - \dfrac{\sqrt{2}}{2 \cdot 2!} \left( x - \dfrac{3\pi}{4} \right)^2 + \cdots = \dfrac{\sqrt{2}}{2} \displaystyle\sum_{n=0}^{\infty} \frac{(-1)^{n(n+1)/2} [x - (3\pi/4)]^n}{n!}$

**89.** $3^x = \left(e^{\ln(3)}\right)^x = e^{x\ln(3)}$ and because $e^x = \displaystyle\sum_{n=0}^{\infty} \frac{x^n}{n!}$, you have

$$3^x = \sum_{n=0}^{\infty} \frac{(x\ln 3)^n}{n!} = 1 + x\ln 3 + \frac{x^2[\ln 3]^2}{2!} + \frac{x^3[\ln 3]^3}{3!} + \frac{x^4[\ln 3]^4}{4!} + \cdots.$$

**91.** $f(x) = \dfrac{1}{x}$

$f'(x) = -\dfrac{1}{x^2}$

$f''(x) = \dfrac{2}{x^3}$

$f'''(x) = -\dfrac{6}{x^4}, \cdots$

$$\frac{1}{x} = \sum_{n=0}^{\infty} \frac{f^{(n)}(-1)(x+1)^n}{n!} = \sum_{n=0}^{\infty} \frac{-n!(x+1)^n}{n!} = -\sum_{n=0}^{\infty}(x+1)^n, \quad -2 < x < 0$$

**93.** $(1+x)^k = 1 + kx + \dfrac{k(k-1)x^2}{2!} + \dfrac{k(k-1)(k-2)x^3}{3!} + \cdots$

$(1+x)^{1/5} = 1 + \dfrac{x}{5} + \dfrac{(1/5)(-4/5)x^2}{2!} + \dfrac{1/5(-4/5)(-9/5)x^3}{3!} + \cdots$

$= 1 + \dfrac{1}{5}x - \dfrac{1\cdot 4x^2}{5^2 2!} + \dfrac{1\cdot 4\cdot 9x^3}{5^3 3!} - \cdots = 1 + \dfrac{x}{5} + \displaystyle\sum_{n=2}^{\infty} \dfrac{(-1)^{n+1}4\cdot 9\cdot 14\cdots(5n-6)x^n}{5^n n!} = 1 + \dfrac{x}{5} - \dfrac{2}{25}x^2 + \dfrac{6}{125}x^3 - \cdots$

**95.** $\ln x = \displaystyle\sum_{n=1}^{\infty}(-1)^{n+1}\frac{(x-1)^n}{n}, \quad 0 < x \le 2$

$\ln\left(\dfrac{5}{4}\right) = \displaystyle\sum_{n=1}^{\infty}(-1)^{n+1}\left(\frac{(5/4)-1}{n}\right)^n = \sum_{n=1}^{\infty}(-1)^{n+1}\frac{1}{4^n n} \approx 0.2231$

**97.** $e^x = \displaystyle\sum_{n=0}^{\infty}\frac{x^n}{n!}, \quad -\infty < x < \infty$

$e^{1/2} = \displaystyle\sum_{n=0}^{\infty}\frac{(1/2)^n}{n!} = \sum_{n=0}^{\infty}\frac{1}{2^n n!} \approx 1.6487$

**99.** $\cos x = \displaystyle\sum_{n=0}^{\infty}(-1)^n\frac{x^{2n}}{(2n)!}, \quad -\infty < x < \infty$

$\cos\left(\dfrac{2}{3}\right) = \displaystyle\sum_{n=0}^{\infty}(-1)^n\frac{2^{2n}}{3^{2n}(2n)!} \approx 0.7859$

**101.** The series in Exercise 45 converges very slowly because the terms approach 0 at a slow rate.

**103.** (a)  $f(x) = e^{2x}$    $f(0) = 1$

$f'(x) = 2e^{2x}$    $f'(0) = 2$

$f''(x) = 4e^{2x}$    $f''(0) = 4$

$f'''(x) = 8e^{2x}$    $f'''(0) = 8$

$P(x) = 1 + 2x + \dfrac{4x^2}{2!} + \dfrac{8x^3}{3!} = 1 + 2x + 2x^2 + \dfrac{4}{3}x^3$

(b)  $e^x = \displaystyle\sum_{n=0}^{\infty} \dfrac{x^n}{n!}, e^{2x} = \sum_{n=0}^{\infty} \dfrac{(2x)^n}{n!}$

$P(x) = 1 + 2x + 2x^2 + \dfrac{4}{3}x^3$

(c)  $e^x \cdot e^x = \left(1 + x + \dfrac{x^2}{2!} + \cdots\right)\left(1 + x + \dfrac{x^2}{2!} + \cdots\right)$

$P(x) = 1 + 2x + 2x^2 + \dfrac{4}{3}x^3$

**105.**  $\sin t = \displaystyle\sum_{n=0}^{\infty} \dfrac{(-1)^n t^{2n+1}}{(2n+1)!}$

$\dfrac{\sin t}{t} = \displaystyle\sum_{n=0}^{\infty} \dfrac{(-1)^n t^{2n}}{(2n+1)!}$

$\displaystyle\int_0^x \dfrac{\sin t}{t}\, dt = \left[\sum_{n=0}^{\infty} \dfrac{(-1)^n t^{2n+1}}{(2n+1)(2n+1)!}\right]_0^x = \sum_{n=0}^{\infty} \dfrac{(-1)^n x^{2n+1}}{(2n+1)(2n+1)!}$

**107.**  $\dfrac{1}{1+t} = \displaystyle\sum_{n=0}^{\infty} (-1)^n t^n$

$\ln(1+t) = \displaystyle\int \dfrac{1}{1+t}\, dt = \sum_{n=0}^{\infty} \dfrac{(-1)^n t^{n+1}}{n+1}$

$\dfrac{\ln(t+1)}{t} = \displaystyle\sum_{n=0}^{\infty} \dfrac{(-1)^n t^n}{n+1}$

$\displaystyle\int_0^x \dfrac{\ln(t+1)}{t}\, dt = \left[\sum_{n=0}^{\infty} \dfrac{(-1)^n t^{n+1}}{(n+1)^2}\right]_0^x = \sum_{n=0}^{\infty} \dfrac{(-1)^n x^{n+1}}{(n+1)^2}$

**109.**  $\arctan x = x - \dfrac{x^3}{3} + \dfrac{x^5}{5} - \dfrac{x^7}{7} + \dfrac{x^9}{9} - \cdots$

$\dfrac{\arctan x}{\sqrt{x}} = \sqrt{x} - \dfrac{x^{5/2}}{3} + \dfrac{x^{9/2}}{5} - \dfrac{x^{13/2}}{7} + \dfrac{x^{17/2}}{9} - \cdots$

$\displaystyle\lim_{x \to 0^+} \dfrac{\arctan x}{\sqrt{x}} = 0$

By L'Hôpital's Rule,  $\displaystyle\lim_{x \to 0^+} \dfrac{\arctan x}{\sqrt{x}} = \lim_{x \to 0^+} \dfrac{\left(\dfrac{1}{1+x^2}\right)}{\left(\dfrac{1}{2\sqrt{x}}\right)} = \lim_{x \to 0^+} \dfrac{2\sqrt{x}}{1+x^2} = 0.$

# Problem Solving for Chapter 9

**1. (a)** $1\left(\dfrac{1}{3}\right) + 2\left(\dfrac{1}{9}\right) + 4\left(\dfrac{1}{27}\right) + \cdots = \displaystyle\sum_{n=0}^{\infty} \dfrac{1}{3}\left(\dfrac{2}{3}\right)^n = \dfrac{1/3}{1 - (2/3)} = 1$

**(b)** $0, \dfrac{1}{3}, \dfrac{2}{3}, 1$, etc.

**(c)** $\displaystyle\lim_{n\to\infty} C_n = 1 - \sum_{n=0}^{\infty} \dfrac{1}{3}\left(\dfrac{2}{3}\right)^n = 1 - 1 = 0$

**3.** Let $S = \displaystyle\sum_{n=1}^{\infty} \dfrac{1}{(2n-1)^2} = \dfrac{1}{1^2} + \dfrac{1}{3^2} + \dfrac{1}{5^2} + \cdots$.

Then $\dfrac{\pi^2}{6} = \dfrac{1}{1^2} + \dfrac{1}{2^2} + \dfrac{1}{3^2} + \dfrac{1}{4^2} + \cdots = S + \dfrac{1}{2^2} + \dfrac{1}{4^2} + \cdots = S + \dfrac{1}{2^2}\left[1 + \dfrac{1}{2^2} + \dfrac{1}{3^2} + \cdots\right] = S + \dfrac{1}{2^2}\left(\dfrac{\pi^2}{6}\right)$.

So, $S = \dfrac{\pi^2}{6} - \dfrac{1}{4}\dfrac{\pi^2}{6} = \dfrac{\pi^2}{6}\left(\dfrac{3}{4}\right) = \dfrac{\pi^2}{8}$.

**5. (a)** Position the three blocks as indicated in the figure.

The bottom block extends $1/6$ over the edge of the table, the middle block extends $1/4$ over the edge of the bottom block, and the top block extends $1/2$ over the edge of the middle block.

The centers of gravity are located at

bottom block: $\dfrac{1}{6} - \dfrac{1}{2} = -\dfrac{1}{3}$

middle block: $\dfrac{1}{6} + \dfrac{1}{4} - \dfrac{1}{2} = -\dfrac{1}{12}$

top block: $\dfrac{1}{6} + \dfrac{1}{4} + \dfrac{1}{2} - \dfrac{1}{2} = \dfrac{5}{12}$.

The center of gravity of the top 2 blocks is

$\left(-\dfrac{1}{12} + \dfrac{5}{12}\right)\Big/ 2 = \dfrac{1}{6}$, which lies over the bottom block. The center of gravity of the 3 blocks is

$\left(-\dfrac{1}{3} - \dfrac{1}{12} + \dfrac{5}{12}\right)\Big/ 3 = 0$ which lies over the table. So, the far edge of the top block lies $\dfrac{1}{6} + \dfrac{1}{4} + \dfrac{1}{2} = \dfrac{11}{12}$ beyond the edge of the table.

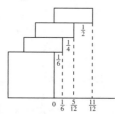

**(b)** Yes. If there are $n$ blocks, then the edge of the top block lies $\displaystyle\sum_{i=1}^{n} \dfrac{1}{2i}$ from the edge of the table. Using 4 blocks,

$\displaystyle\sum_{i=1}^{4} \dfrac{1}{2i} = \dfrac{1}{2} + \dfrac{1}{4} + \dfrac{1}{6} + \dfrac{1}{8} = \dfrac{25}{24}$

which shows that the top block extends beyond the table.

**(c)** The blocks can extend any distance beyond the table because the series diverges:

$\displaystyle\sum_{i=1}^{\infty} \dfrac{1}{2i} = \dfrac{1}{2}\sum_{i=1}^{\infty} \dfrac{1}{i} = \infty$.

**7.** $a - \dfrac{b}{2} + \dfrac{a}{3} - \dfrac{b}{4} + \cdots = \displaystyle\sum_{n=1}^{\infty} \dfrac{(-1)^{n+1}(a+b) + (a-b)}{2n}$

If $a = b$, $\displaystyle\sum_{n=1}^{\infty} \dfrac{(-1)^{n+1}(2a)}{2n} = a\displaystyle\sum_{n=1}^{\infty} \dfrac{(-1)^{n+1}}{n}$ converges conditionally.

If $a \neq b$, $\displaystyle\sum_{n=1}^{\infty} \dfrac{(-1)^{n+1}(a+b)}{2n} + \displaystyle\sum_{n=1}^{\infty} \dfrac{a-b}{2n}$ diverges.

No values of $a$ and $b$ give absolute convergence. $a = b$ implies conditional convergence.

**9.**
$$e^x = 1 + x + \dfrac{x^2}{2!} + \cdots$$

$$e^{x^2} = 1 + x^2 + \dfrac{x^4}{2!} + \cdots + \dfrac{x^{12}}{6!} + \cdots$$

$$\dfrac{f^{(12)}(0)}{12!} = \dfrac{1}{6!} \Rightarrow f^{(12)}(0) = \dfrac{12!}{6!} = 665{,}280$$

**11.** (a) If $p = 1$, $\displaystyle\int_2^{\infty} \dfrac{1}{x \ln x}\, dx = \left[\ln \ln x\right]_2^{\infty}$ diverges.

If
$$p > 1, \int_2^{\infty} \dfrac{1}{x(\ln x)^p}\, dx = \lim_{b \to \infty}\left[\dfrac{(\ln b)^{1-p}}{1-p} - \dfrac{(\ln 2)^{1-p}}{1-p}\right]$$
converges.

If $p < 1$, diverges.

(b) $\displaystyle\sum_{n=4}^{\infty} \dfrac{1}{n\ln(n^2)} = \dfrac{1}{2}\displaystyle\sum_{n=4}^{\infty} \dfrac{1}{n\ln n}$ diverges by part (a).

**13.** Let $b_n = a_n r^n$.

$$(b_n)^{1/n} = (a_n r^n)^{1/n} = a_n^{1/n} \cdot r \to Lr \text{ as } n \to \infty.$$

$$Lr < \dfrac{1}{r}r = 1.$$

By the Root Test, $\displaystyle\sum b_n$ converges $\Rightarrow \displaystyle\sum a_n r^n$ converges.

**15.** (a) $\dfrac{1}{0.99} = \dfrac{1}{1 - 0.01} = \displaystyle\sum_{n=0}^{\infty} (0.01)^n = 1 + 0.01 + (0.01)^2 + \cdots = 1.010101\cdots$

(b) $\dfrac{1}{0.98} = \dfrac{1}{1 - 0.02} = \displaystyle\sum_{n=0}^{\infty} (0.02)^n$

$$= 1 + 0.02 + (0.02)^2 + \cdots = 1 + 0.02 + 0.0004 + \cdots = 1.0204081632\cdots$$

**17.** (a) Height $= 2\left[1 + \dfrac{1}{\sqrt{2}} + \dfrac{1}{\sqrt{3}} + \cdots\right]$

$$= 2\displaystyle\sum_{n=1}^{\infty} \dfrac{1}{n^{1/2}} = \infty \left(p\text{-series}, p = \dfrac{1}{2} < 1\right)$$

(b) $S = 4\pi\left[1 + \dfrac{1}{2} + \dfrac{1}{3} + \cdots\right] = 4\pi\displaystyle\sum_{n=1}^{\infty} \dfrac{1}{n} = \infty$

(c) $W = \dfrac{4}{3}\pi\left[1 + \dfrac{1}{2^{3/2}} + \dfrac{1}{3^{3/2}} + \cdots\right] = \dfrac{4}{3}\pi\displaystyle\sum_{n=1}^{\infty} \dfrac{1}{n^{3/2}}$ converges.

# CHAPTER 10
# Conics, Parametric Equations, and Polar Coordinates

# CHAPTER 10
# Conics, Parametric Equations, and Polar Coordinates

## Section 10.1  Conics and Calculus

**1.** $y^2 = 4x$  Parabola

Vertex: $(0, 0)$

$p = 1 > 0$

Opens to the right

Matches (h)

**3.** $(x + 4)^2 = -2(y - 2)$  Parabola

Vertex: $(-4, 2)$

Opens downward

Matches (e)

**5.** $\dfrac{x^2}{4} + \dfrac{y^2}{9} = 1$  Ellipse

Center: $(0, 0)$

Vertices: $(0, \pm 3)$

Matches (f)

**7.** $\dfrac{y^2}{16} - \dfrac{x^2}{1} = 1$  Hyperbola

Vertices: $(0, \pm 4)$

Matches (c)

**9.** $y^2 = -8x = 4(-2)x$

Vertex: $(0, 0)$

Focus: $(-2, 0)$

Directrix: $x = 2$

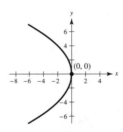

**11.** $(x + 5) + (y - 3)^2 = 0$

$(y - 3)^2 = -(x + 5) = 4\left(-\dfrac{1}{4}\right)(x + 5)$

Vertex: $(-5, 3)$

Focus: $\left(-\dfrac{21}{4}, 3\right)$

Directrix: $x = -\dfrac{19}{4}$

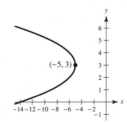

**13.** $y^2 - 4y - 4x = 0$

$y^2 - 4y + 4 = 4x + 4$

$(y - 2)^2 = 4(1)(x + 1)$

Vertex: $(-1, 2)$

Focus: $(0, 2)$

Directrix: $x = -2$

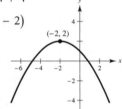

**15.** $x^2 + 4x + 4y - 4 = 0$

$x^2 + 4x + 4 = -4y + 4 + 4$

$(x + 2)^2 = 4(-1)(y - 2)$

Vertex: $(-2, 2)$

Focus: $(-2, 1)$

Directrix: $y = 3$

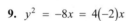

**17.** $y^2 + x + y = 0$

$y^2 + y + \dfrac{1}{4} = -x + \dfrac{1}{4}$

$\left(y + \dfrac{1}{2}\right)^2 = 4\left(-\dfrac{1}{4}\right)\left(x - \dfrac{1}{4}\right)$

Vertex: $\left(\dfrac{1}{4}, -\dfrac{1}{2}\right)$

Focus: $\left(0, -\dfrac{1}{2}\right)$

Directrix: $x = \dfrac{1}{2}$

$y_1 = -\dfrac{1}{2} + \sqrt{\dfrac{1}{4} - x}$

$y_2 = -\dfrac{1}{2} - \sqrt{\dfrac{1}{4} - x}$

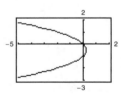

**19.** $y^2 - 4x - 4 = 0$

$y^2 = 4x + 4$

$= 4(1)(x + 1)$

Vertex: $(-1, 0)$

Focus: $(0, 0)$

Directrix: $x = -2$

$y_1 = 2\sqrt{x + 1}$

$y_2 = -2\sqrt{x + 1}$

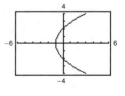

**21.** $(y - 4)^2 = 4(-2)(x - 5)$

$\qquad y^2 - 8y + 16 = -8x + 40$

$\qquad y^2 - 8y + 8x - 24 = 0$

**23.** $\qquad (x - 0)^2 = 4(8)(y - 5)$

$\qquad\qquad\quad x^2 = 4(8)(y - 5)$

$\qquad x^2 - 32y + 160 = 0$

**25.** $\qquad\qquad y = 4 - x^2$

$\qquad x^2 + y - 4 = 0$

**27.** Because the axis of the parabola is vertical, the form of the equation is $y = ax^2 + bx + c$. Now, substituting the values of the given coordinates into this equation, you obtain

$\qquad 3 = c, 4 = 9a + 3b + c, 11 = 16a + 4b + c.$

Solving this system, you have $a = \frac{5}{3}, b = -\frac{14}{3}, c = 3.$

So,

$\qquad y = \frac{5}{3}x^2 - \frac{14}{3}x + 3$ or $5x^2 - 14x - 3y + 9 = 0.$

**29.** $16x^2 + y^2 = 16$

$\qquad x^2 + \dfrac{y^2}{16} = 1$

$\qquad a^2 = 16, b^2 = 1, c^2 = 16 - 1 = 15$

Center: $(0, 0)$

Foci: $\left(0, \pm\sqrt{15}\right)$

Vertices: $(0, \pm4)$

$e = \dfrac{c}{a} = \dfrac{\sqrt{15}}{4}$

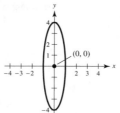

**31.** $\dfrac{(x - 3)^2}{16} + \dfrac{(y - 1)^2}{25} = 1$

$\qquad a^2 = 25, b^2 = 16, c^2 = 25 - 16 = 9$

Center: $(3, 1)$

Foci: $(3, 1 + 3) = (3, 4), (3, 1 - 3) = (3, -2)$

Vertices: $(3, 6), (3, -4)$

$e = \dfrac{c}{a} = \dfrac{3}{5}$

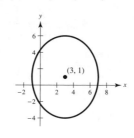

**33.** $\qquad 9x^2 + 4y^2 + 36x - 24y + 36 = 0$

$\qquad 9(x^2 + 4x + 4) + 4(y^2 - 6y + 9) = -36 + 36 + 36$

$\qquad\qquad\qquad\qquad\qquad\qquad\qquad = 36$

$\qquad\qquad \dfrac{(x + 2)^2}{4} + \dfrac{(y - 3)^2}{9} = 1$

$\qquad a^2 = 9, b^2 = 4, c^2 = 5$

Center: $(-2, 3)$

Foci: $\left(-2, 3 \pm \sqrt{5}\right)$

Vertices: $(-2, 6), (-2, 0)$

$e = \dfrac{\sqrt{5}}{3}$

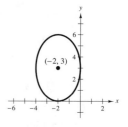

**35.** $\qquad 12x^2 + 20y^2 - 12x + 40y - 37 = 0$

$\qquad 12\left(x^2 - x + \dfrac{1}{4}\right) + 20(y^2 + 2y + 1) = 37 + 3 + 20 = 60$

$\qquad\qquad \dfrac{[x - (1/2)]^2}{5} + \dfrac{(y + 1)^2}{3} = 1$

$\qquad a^2 = 5, b^2 = 3, c^2 = 2$

Center: $\left(\dfrac{1}{2}, -1\right)$

Foci: $\left(\dfrac{1}{2} \pm \sqrt{2}, -1\right)$

Vertices: $\left(\dfrac{1}{2} \pm \sqrt{5}, -1\right)$

Solve for $y$:

$\qquad 20(y^2 + 2y + 1) = -12x^2 + 12x + 37 + 20$

$\qquad\qquad\qquad (y + 1)^2 = \dfrac{57 + 12x - 12x^2}{20}$

$\qquad\qquad\qquad\qquad y = -1 \pm \sqrt{\dfrac{57 + 12x - 12x^2}{20}}$

(Graph each of these separately.)

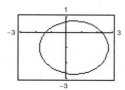

**37.**  $x^2 + 2y^2 - 3x + 4y + 0.25 = 0$

$$\left(x^2 - 3x + \frac{9}{4}\right) + 2\left(y^2 + 2y + 1\right) = -\frac{1}{4} + \frac{9}{4} + 2 = 4$$

$$\frac{\left[x - (3/2)\right]^2}{4} + \frac{(y + 1)^2}{2} = 1$$

$a^2 = 4, b^2 = 2, c^2 = 2$

Center: $\left(\dfrac{3}{2}, -1\right)$

Foci: $\left(\dfrac{3}{2} \pm \sqrt{2}, -1\right)$

Vertices: $\left(-\dfrac{1}{2}, -1\right), \left(\dfrac{7}{2}, -1\right)$

Solve for $y$: $2\left(y^2 + 2y + 1\right) = -x^2 + 3x - \dfrac{1}{4} + 2$

$$(y + 1)^2 = \frac{1}{2}\left(\frac{7}{4} + 3x - x^2\right)$$

$$y = -1 \pm \sqrt{\frac{7 + 12x - 4x^2}{8}}$$

(Graph each of these separately.)

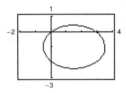

**39.** Center: $(0, 0)$

Focus: $(5, 0)$

Vertex: $(6, 0)$

Horizontal major axis

$a = 6, c = 5 \Rightarrow b = \sqrt{a^2 - c^2} = \sqrt{11}$

$\dfrac{x^2}{36} + \dfrac{y^2}{11} = 1$

**41.** Vertices: $(3, 1), (3, 9)$

Minor axis length: 6

Vertical major axis

Center: $(3, 5)$

$a = 4, b = 3$

$\dfrac{(x - 3)^2}{9} + \dfrac{(y - 5)^2}{16} = 1$

**43.** Center: $(0, 0)$

Horizontal major axis

Points on ellipse: $(3, 1), (4, 0)$

Because the major axis is horizontal,

$$\left(\frac{x^2}{a^2}\right) + \left(\frac{y^2}{b^2}\right) = 1.$$

Substituting the values of the coordinates of the given points into this equation, you have

$$\left(\frac{9}{a^2}\right) + \left(\frac{1}{b^2}\right) = 1, \text{ and } \frac{16}{a^2} = 1.$$

The solution to this system is $a^2 = 16, b^2 = \dfrac{16}{7}$.

So,

$$\frac{x^2}{16} + \frac{y^2}{16/7} = 1, \frac{x^2}{16} + \frac{7y^2}{16} = 1.$$

**45.** $y^2 - \dfrac{x^2}{9} = 1$

$a = 1, b = 3, c = \sqrt{10}$

Center: $(0, 0)$

Vertices: $(0, \pm 1)$

Foci: $\left(0, \pm\sqrt{10}\right)$

Asymptotes: $y = \pm\dfrac{a}{b}x = \pm\dfrac{1}{3}x$

**47.** $\dfrac{(x - 1)^2}{4} - \dfrac{(y + 2)^2}{1} = 1$

$a = 2, b = 1, c = \sqrt{5}$

Center: $(1, -2)$

Vertices: $(-1, -2), (3, -2)$

Foci: $\left(1 \pm \sqrt{5}, -2\right)$

Asymptotes: $y = -2 \pm \dfrac{1}{2}(x - 1)$

**49.**  $9x^2 - y^2 - 36x - 6y + 18 = 0$

$9\left(x^2 - 4x + 4\right) - \left(y^2 + 6y + 9\right) = -18 + 36 - 9$

$$\frac{(x - 2)^2}{1} - \frac{(y + 3)^2}{9} = 1$$

$a = 1, b = 3, c = \sqrt{10}$

Center: $(2, -3)$

Vertices: $(1, -3), (3, -3)$

Foci: $\left(2 \pm \sqrt{10}, -3\right)$

Asymptotes: $y = -3 \pm 3(x - 2)$

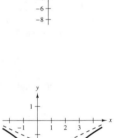

**51.**    $x^2 - 9y^2 + 2x - 54y - 80 = 0$

$$\left(x^2 + 2x + 1\right) - 9\left(y^2 + 6y + 9\right) = 80 + 1 - 81 = 0$$

$$(x + 1)^2 - 9(y + 3)^2 = 0$$

$$y + 3 = \pm\frac{1}{3}(x + 1)$$

$$y = 3 \pm \frac{1}{3}(x + 1)$$

Degenerate hyperbola is two lines intersecting at $(-1, -3)$.

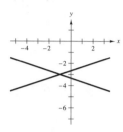

**53.**    $9y^2 - x^2 + 2x + 54y + 62 = 0$

$$9\left(y^2 + 6y + 9\right) - \left(x^2 - 2x + 1\right) = -62 - 1 + 81 = 18$$

$$\frac{(y + 3)^2}{2} - \frac{(x - 1)^2}{18} = 1$$

$a = \sqrt{2}, b = 3\sqrt{2}, c = 2\sqrt{5}$

Center: $(1, -3)$

Vertices: $\left(1, -3 \pm \sqrt{2}\right)$

Foci: $\left(1, -3 \pm 2\sqrt{5}\right)$

Asymptotes: $y = \dfrac{1}{3}x - \dfrac{1}{3} - 3$

$$y = -\frac{1}{3}x + \frac{1}{3} - 3$$

Solve for $y$:

$$9\left(y^2 + 6y + 9\right) = x^2 - 2x - 62 + 81$$

$$(y + 3)^2 = \frac{x^2 - 2x + 19}{9}$$

$$y = -3 \pm \frac{1}{3}\sqrt{x^2 - 2x + 19}$$

(Graph each curve separately.)

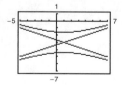

**55.**    $3x^2 - 2y^2 - 6x - 12y - 27 = 0$

$$3\left(x^2 - 2x + 1\right) - 2\left(y^2 + 6y + 9\right) = 27 + 3 - 18 = 12$$

$$\frac{(x - 1)^2}{4} - \frac{(y + 3)^2}{6} = 1$$

$a = 2, b = \sqrt{6}, c = \sqrt{10}$

Center: $(1, -3)$

Vertices: $(-1, -3), (3, -3)$

Foci: $\left(1 \pm \sqrt{10}, -3\right)$

Asymptotes: $y = \dfrac{\sqrt{6}x}{2} - \dfrac{\sqrt{6}}{2} - 3$

$$y = -\frac{\sqrt{6}x}{2} + \frac{\sqrt{6}}{2} - 3$$

Solve for $y$:

$$2\left(y^2 + 6y + 9\right) = 3x^2 - 6x - 27 + 18$$

$$(y + 3)^2 = \frac{3x^2 - 6x - 9}{2}$$

$$y = -3 \pm \sqrt{\frac{3\left(x^2 - 2x - 3\right)}{2}}$$

(Graph each curve separately.)

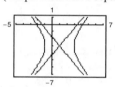

**57.** Vertices: $(\pm 1, 0)$

Asymptotes: $y = \pm 5x$

Horizontal transverse axis

Center: $(0, 0)$

$a = 1, \dfrac{b}{a} = 5 \Rightarrow b = 5$

$$\frac{x^2}{1} - \frac{y^2}{25} = 1$$

**59.** Vertices: $(2, \pm 3)$

Point on graph: $(0, 5)$

Vertical transverse axis

Center: $(2, 0)$

$a = 3$

So, the equation is of the form

$$\frac{y^2}{9} - \frac{(x - 2)^2}{b^2} = 1.$$

Substituting the coordinates of the point $(0, 5)$, you have

$$\frac{25}{9} - \frac{4}{b^2} = 1 \quad \text{or} \quad b^2 = \frac{9}{4}.$$

So, the equation is $\dfrac{y^2}{9} - \dfrac{(x - 2)^2}{9/4} = 1.$

**61.** Center: $(0, 0)$

Vertex: $(0, 2)$

Focus: $(0, 4)$

Vertical transverse axis

$a = 2, c = 4, b^2 = c^2 - a^2 = 12$

So, $\dfrac{y^2}{4} - \dfrac{x^2}{12} = 1$.

**63.** Vertices: $(0, 2), (6, 2)$

Asymptotes: $y = \dfrac{2}{3}x, y = 4 - \dfrac{2}{3}x$

Horizontal transverse axis

Center: $(3, 2)$

$a = 3$

Slopes of asymptotes: $\pm\dfrac{b}{a} = \pm\dfrac{2}{3}$

So, $b = 2$. Therefore,

$$\dfrac{(x - 3)^2}{9} - \dfrac{(y - 2)^2}{4} = 1.$$

**65. (a)** $\dfrac{x^2}{9} - y^2 = 1, \dfrac{2x}{9} - 2yy' = 0, \dfrac{x}{9y} = y'$

At $x = 6$: $y = \pm\sqrt{3}, y' = \dfrac{\pm 6}{9\sqrt{3}} = \dfrac{\pm 2\sqrt{3}}{9}$

At $\left(6, \sqrt{3}\right)$: $y - \sqrt{3} = \dfrac{2\sqrt{3}}{9}(x - 6)$ or $2x - 3\sqrt{3}y - 3 = 0$

At $\left(6, -\sqrt{3}\right)$: $y + \sqrt{3} = \dfrac{-2\sqrt{3}}{9}(x - 6)$ or $2x + 3\sqrt{3}y - 3 = 0$

**(b)** From part (a) you know that the slopes of the normal lines must be $\mp 9/\left(2\sqrt{3}\right)$.

At $\left(6, \sqrt{3}\right)$: $y - \sqrt{3} = -\dfrac{9}{2\sqrt{3}}(x - 6)$ or $9x + 2\sqrt{3}y - 60 = 0$

At $\left(6, -\sqrt{3}\right)$: $y + \sqrt{3} = \dfrac{9}{2\sqrt{3}}(x - 6)$ or $9x - 2\sqrt{3}y - 60 = 0$

**67.** $x^2 + 4y^2 - 6x + 16y + 21 = 0$

$\left(x^2 - 6x + 9\right) + 4\left(y^2 + 4y + 4\right) = -21 + 9 + 16$

$(x - 3)^2 + 4(y + 2)^2 = 4$

Ellipse

**69.** $y^2 - 8y - 8x = 0$

$y^2 - 8y + 16 = 8x + 16$

$(y - 4)^2 = 4(2)(x + 2)$

Parabola

**71.** $4x^2 + 4y^2 - 16y + 15 = 0$

$4x^2 + 4\left(y^2 - 4y + 4\right) = -15 + 16$

$4x^2 + 4(y - 2)^2 = 1$

Circle (Ellipse)

**73.** $9x^2 + 9y^2 - 36x + 6y + 34 = 0$

$9\left(x^2 - 4x + 4\right) + 9\left(y^2 + \dfrac{2}{3}y + \dfrac{1}{9}\right) = -34 + 36 + 1$

$9(x - 2)^2 + 9\left(y + \dfrac{1}{3}\right)^2 = 3$

Circle (Ellipse)

**75.** $3(x - 1)^2 = 6 + 2(y + 1)^2$

$3(x - 1)^2 - 2(y + 1)^2 = 6$

$\dfrac{(x - 1)^2}{2} - \dfrac{(y + 1)^2}{3} = 1$

Hyperbola

**77.** (a) A parabola is the set of all points $(x, y)$ that are

equidistant from a fixed line (directrix) and a fixed point (focus) not on the line.

(b) For directrix $y = k - p : (x - h)^2 = 4p(y - k)$

For directrix $x = h - p : (y - k)^2 = 4p(x - h)$

(c) If $P$ is a point on a parabola, then the tangent line to the parabola at $P$ makes equal angles with the line
passing through $P$ and the focus, and with the line passing through $P$ parallel to the axis of the parabola.

**79.** (a) A hyperbola is the set of all points $(x, y)$ for which

the absolute value of the difference between the distances from two distinct fixed points (foci) is constant.

(b) Transverse axis is horizontal: $\dfrac{(x - h)^2}{a^2} - \dfrac{(y - k)^2}{b^2} = 1$

Transverse axis is vertical: $\dfrac{(y - k)^2}{a^2} - \dfrac{(x - h)^2}{b^2} = 1$

(c) Transverse axis is horizontal: $y = k + (b/a)(x - h)$ and $y = k - (b/a)(x - h)$

Transverse axis is vertical: $y = k + (a/b)(x - h)$ and $y = k - (a/b)(x - h)$

**81.** Assume that the vertex is at the origin.

$x^2 = 4py$

$(3)^2 = 4p(1)$

$\dfrac{9}{4} = p$

The pipe is located $\dfrac{9}{4}$ meters from the vertex.

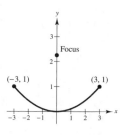

**83.** $y = ax^2$

$y' = 2ax$

The equation of the tangent line is

$y - ax_0^2 = 2ax_0(x - x_0)$

or $y = 2ax_0 x - ax_0^2$.

Let $y = 0$. Then: $-ax_0^2 = 2ax_0 x - 2ax_0^2$

$\qquad\qquad\qquad ax_0^2 = 2ax_0 x$

$\qquad\qquad\qquad\quad x = \dfrac{x_0}{2}$

So, $\left(\dfrac{x_0}{2}, 0\right)$ is the $x$-intercept.

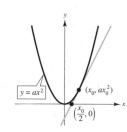

**85. (a)** Consider the parabola $x^2 = 4py$. Let $m_0$ be the slope of the one tangent line at $(x_1, y_1)$ and so,

$-\dfrac{1}{m_0}$ is the slope of the second at $(x_2, y_2)$. Differentiating, $2x = 4py'$ or $y' = \dfrac{x}{2p}$, and you have:

$$m_0 = \frac{1}{2p}x_1 \quad \text{or} \quad x_1 = 2pm_0$$

$$\frac{-1}{m_0} = \frac{1}{2p}x_2 \quad \text{or} \quad x_2 = \frac{-2p}{m_0}.$$

Substituting these values of $x$ into the equation $x^2 = 4py$, we have the coordinates of the points of tangency $\left(2pm_0,\ pm_0{}^2\right)$ and $\left(-2p/m_0,\ p/m_0{}^2\right)$ and the equations of the tangent lines are

$$\left(y - pm_0{}^2\right) = m_0(x - 2pm_0) \quad \text{and} \quad \left(y - \frac{p}{m_0{}^2}\right) = \frac{-1}{m_0}\left(x + \frac{2p}{m_0}\right).$$

The point of intersection of these lines is $\left(\dfrac{p\left(m_0{}^2 - 1\right)}{m_0},\ -p\right)$ and is on the directrix, $y = -p$.

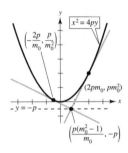

**(b)** $x^2 - 4x - 4y + 8 = 0$

$$(x - 2)^2 = 4(y - 1)$$

Vertex: $(2, 1)$

$$2x - 4 - 4\frac{dy}{dx} = 0$$

$$\frac{dy}{dx} = \frac{1}{2}x - 1$$

At $(-2, 5)$, $\dfrac{dy}{dx} = -2$. At $\left(3, \dfrac{5}{4}\right)$, $\dfrac{dy}{dx} = \dfrac{1}{2}$.

Tangent line at $(-2, 5)$: $y - 5 = -2(x + 2) \Rightarrow 2x + y - 1 = 0$.

Tangent line at $\left(3, \dfrac{5}{4}\right)$: $y - \dfrac{5}{4} = \dfrac{1}{2}(x - 3) \Rightarrow 2x - 4y - 1 = 0$.

Because $m_1 m_2 = (-2)\left(\dfrac{1}{2}\right) = -1$, the lines are perpendicular.

Point of intersection: $-2x + 1 = \dfrac{1}{2}x - \dfrac{1}{4}$

$$-\frac{5}{2}x = -\frac{5}{4}$$

$$x = \frac{1}{2}$$

$$y = 0$$

Directrix: $y = 0$ and the point of intersection $\left(\dfrac{1}{2}, 0\right)$ lies on this line.

**87.** $y = x - x^2$

$$\frac{dy}{dx} = 1 - 2x$$

At the point of tangency $(x_1, y_1)$ on the mountain, $m = 1 - 2x_1$. Also, $m = \dfrac{y_1 - 1}{x_1 + 1}$.

$$\frac{y_1 - 1}{x_1 + 1} = 1 - 2x_1$$

$$\left(x_1 - x_1^2\right) - 1 = \left(1 - 2x_1\right)\left(x_1 + 1\right)$$

$$-x_1^2 + x_1 - 1 = -2x_1^2 - x_1 + 1$$

$$x_1^2 + 2x_1 - 2 = 0$$

$$x_1 = \frac{-2 \pm \sqrt{2^2 - 4(1)(-2)}}{2(1)} = \frac{-2 \pm 2\sqrt{3}}{2} = -1 \pm \sqrt{3}$$

Choosing the positive value for $x_1$, we have $x_1 = -1 + \sqrt{3}$.

$$m = 1 - 2\left(-1 + \sqrt{3}\right) = 3 - 2\sqrt{3}$$

$$m = \frac{0 - 1}{x_0 + 1} = -\frac{1}{x_0 + 1}$$

So, $-\dfrac{1}{x_0 + 1} = 3 - 2\sqrt{3}$

$$\frac{-1}{3 - 2\sqrt{3}} = x_0 + 1$$

$$\frac{3 + 2\sqrt{3}}{3} - 1 = x_0$$

$$\frac{2\sqrt{3}}{3} = x_0.$$

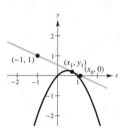

The closest the receiver can be to the hill is $\left(2\sqrt{3}/3\right) - 1 \approx 0.155$.

**89.** Parabola

Vertex: $(0, 4)$

$$x^2 = 4p(y - 4)$$

$$4^2 = 4p(0 - 4)$$

$$p = -1$$

$$x^2 = -4(y - 4)$$

$$y = 4 - \frac{x^2}{4}$$

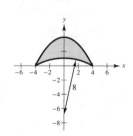

Circle

Center: $(0, k)$

Radius: 8

$$x^2 + (y - k)^2 = 64$$

$$4^2 + (0 - k)^2 = 64$$

$$k^2 = 48$$

$$k = -4\sqrt{3} \qquad \text{(Center is on the negative } y\text{-axis.)}$$

$$x^2 + \left(y + 4\sqrt{3}\right)^2 = 64$$

$$y = -4\sqrt{3} \pm \sqrt{64 - x^2}$$

Because the $y$-value is positive when $x = 0$, we have $y = -4\sqrt{3} + \sqrt{64 - x^2}$.

$$A = 2\int_0^4 \left[\left(4 - \frac{x^2}{4}\right) - \left(-4\sqrt{3} + \sqrt{64 - x^2}\right)\right] dx = 2\left[4x - \frac{x^3}{12} + 4\sqrt{3}x - \frac{1}{2}\left(x\sqrt{64 - x^2} + 64\arcsin\frac{x}{8}\right)\right]_0^4$$

$$= 2\left(16 - \frac{64}{12} + 16\sqrt{3} - 2\sqrt{48} - 32\arcsin\frac{1}{2}\right) = \frac{16\left(4 + 3\sqrt{3} - 2\pi\right)}{3} \approx 15.536 \text{ square feet}$$

**91.** (a) Assume that $y = ax^2$.

$$20 = a(60)^2 \Rightarrow a = \frac{2}{360} = \frac{1}{180} \Rightarrow y = \frac{1}{180}x^2$$

(b) $f(x) = \frac{1}{180}x^2$, $f'(x) = \frac{1}{90}x$

$$S = 2\int_0^{60} \sqrt{1 + \left(\frac{1}{90}x\right)^2}\, dx = \frac{2}{90}\int_0^{60} \sqrt{90^2 + x^2}\, dx$$

$$= \frac{2}{90}\frac{1}{2}\left[x\sqrt{90^2 + x^2} + 90^2 \ln\left|x + \sqrt{90^2 + x^2}\right|\right]_0^{60} \quad \text{(Formula 26)}$$

$$= \frac{1}{90}\left[60\sqrt{11,700} + 90^2 \ln\left(60 + \sqrt{11,700}\right) - 90^2 \ln 90\right]$$

$$= \frac{1}{90}\left[1800\sqrt{13} + 90^2 \ln\left(60 + 30\sqrt{13}\right) - 90^2 \ln 90\right]$$

$$= 20\sqrt{13} + 90 \ln\left(\frac{60 + 30\sqrt{13}}{90}\right)$$

$$= 10\left[2\sqrt{13} + 9 \ln\left(\frac{2 + \sqrt{13}}{3}\right)\right] \approx 128.4 \text{ m}$$

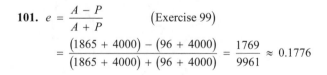

**93.** $x^2 = 4py$, $p = \frac{1}{4}, \frac{1}{2}, 1, \frac{3}{2}, 2$

As $p$ increases, the graph becomes wider.

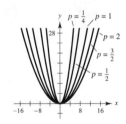

**95.** (a) At the vertices you notice that the string is horizontal and has a length of $2a$.

(b) The thumbtacks are located at the foci and the length of string is the constant sum of the distances from the foci.

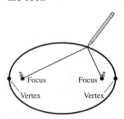

**97.**

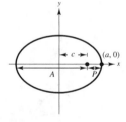

**99.**

$$e = \frac{c}{a}$$

$$A + P = 2a$$

$$a = \frac{A + P}{2}$$

$$c = a - P = \frac{A + P}{2} - P = \frac{A - P}{2}$$

$$e = \frac{c}{a} = \frac{(A - P)/2}{(A + P)/2} = \frac{A - P}{A + P}$$

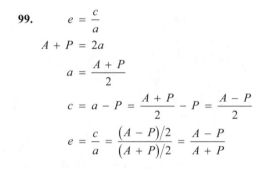

**101.** $e = \dfrac{A - P}{A + P}$ \quad (Exercise 99)

$$= \frac{(1865 + 4000) - (96 + 4000)}{(1865 + 4000) + (96 + 4000)} = \frac{1769}{9961} \approx 0.1776$$

**103.** $e = \dfrac{A - P}{A + P} = \dfrac{35.29 - 0.59}{35.29 + 0.59} = 0.9671$

**105.**  $\dfrac{x^2}{10^2} + \dfrac{y^2}{5^2} = 1$

$\dfrac{2x}{10^2} + \dfrac{2yy'}{5^2} = 0$

$y' = \dfrac{-5^2 x}{10^2 y} = \dfrac{-x}{4y}$

At $(-8, 3)$: $y' = \dfrac{8}{12} = \dfrac{2}{3}$

The equation of the tangent line is $y - 3 = \dfrac{2}{3}(x + 8)$. It

will cross the $y$-axis when $x = 0$ and

$y = \dfrac{2}{3}(8) + 3 = \dfrac{25}{3}$.

**107.**  $16x^2 + 9y^2 + 96x + 36y + 36 = 0$

$32x + 18yy' + 96 + 36y' = 0$

$y'(18y + 36) = -(32x + 96)$

$y' = \dfrac{-(32x + 96)}{18y + 36}$

$y' = 0$ when $x = -3$. $y'$ is undefined when $y = -2$.

At $x = -3$, $y = 2$ or $-6$.

Endpoints of major axis: $(-3, 2), (-3, -6)$

At $y = -2$, $x = 0$ or $-6$.

Endpoints of minor axis: $(0, -2), (-6, -2)$

**Note:** Equation of ellipse is $\dfrac{(x + 3)^2}{9} + \dfrac{(y + 2)^2}{16} = 1$

**109. (a)**  $A = 4\displaystyle\int_0^2 \dfrac{1}{2}\sqrt{4 - x^2}\, dx = \left[ x\sqrt{4 - x^2} + 4\arcsin\left(\dfrac{x}{2}\right) \right]_0^2 = 2\pi$   $\left[ \text{or, } A = \pi ab = \pi(2)(1) = 2\pi \right]$

**(b) Disk:**  $V = 2\pi\displaystyle\int_0^2 \dfrac{1}{4}(4 - x^2)\, dx = \dfrac{1}{2}\pi\left[ 4x - \dfrac{1}{3}x^3 \right]_0^2 = \dfrac{8\pi}{3}$

$y = \dfrac{1}{2}\sqrt{4 - x^2}$

$y' = \dfrac{-x}{2\sqrt{4 - x^2}}$

$\sqrt{1 + (y')^2} = \sqrt{1 + \dfrac{x^2}{16 - 4x^2}} = \sqrt{\dfrac{16 - 3x^2}{4y}}$

$S = 2(2\pi)\displaystyle\int_0^2 y\left( \dfrac{\sqrt{16 - 3x^2}}{4y} \right) dx = \pi\displaystyle\int_0^2 \sqrt{16 - 3x^2}\, dx$

$= \dfrac{\pi}{2\sqrt{3}}\left[ \sqrt{3}x\sqrt{16 - 3x^2} + 16\arcsin\left( \dfrac{\sqrt{3}x}{4} \right) \right]_0^2 = \dfrac{2\pi}{9}(9 + 4\sqrt{3}\pi) \approx 21.48$

**(c) Shell:**  $V = 2\pi\displaystyle\int_0^2 x\sqrt{4 - x^2}\, dx = -\pi\displaystyle\int_0^2 -2x(4 - x^2)^{1/2}\, dx = -\dfrac{2\pi}{3}\left[ (4 - x^2)^{3/2} \right]_0^2 = \dfrac{16\pi}{3}$

$x = 2\sqrt{1 - y^2}$

$x' = \dfrac{-2y}{\sqrt{1 - y^2}}$

$\sqrt{1 + (x')^2} = \sqrt{1 + \dfrac{4y^2}{1 - y^2}} = \dfrac{\sqrt{1 + 3y^2}}{\sqrt{1 - y^2}}$

$S = 2(2\pi)\displaystyle\int_0^1 2\sqrt{1 - y^2}\, \dfrac{\sqrt{1 + 3y^2}}{\sqrt{1 - y^2}}\, dy = 8\pi\displaystyle\int_0^1 \sqrt{1 + 3y^2}\, dy$

$= \dfrac{8\pi}{2\sqrt{3}}\left[ \sqrt{3}y\sqrt{1 + 3y^2} + \ln\left| \sqrt{3}y + \sqrt{1 + 3y^2} \right| \right]_0^1$

$= \dfrac{4\pi}{3}\left| 6 + \sqrt{3}\ln(2 + \sqrt{3}) \right| \approx 34.69$

**111.** From Example 5,

$$C = 4a \int_0^{\pi/2} \sqrt{1 - e^2 \sin^2 \theta} \, d\theta$$

For $\dfrac{x^2}{25} + \dfrac{y^2}{49} = 1$, you have

$$a = 7, b = 5, c = \sqrt{49 - 25} = 2\sqrt{6}, e = \frac{c}{a} = \frac{2\sqrt{6}}{7}.$$

$$C = 4(7) \int_0^{\pi/2} \sqrt{1 - \frac{24}{49} \sin^2 \theta} \, d\theta \approx 28(1.3558) \approx 37.96$$

**113.** Area circle $= \pi r^2 = 100\pi$

Area ellipse $= \pi ab = \pi a(10)$

$$2(100\pi) = 10\pi a \Rightarrow a = 20$$

So, the length of the major axis is $2a = 40$.

**115.** The transverse axis is horizontal since $(2, 2)$ and $(10, 2)$ are the foci (see definition of hyperbola).

Center: $(6, 2)$

$$c = 4, 2a = 6, b^2 = c^2 - a^2 = 7$$

So, the equation is $\dfrac{(x - 6)^2}{9} - \dfrac{(y - 2)^2}{7} = 1$.

**117.** $2a = 10 \Rightarrow a = 5$

$c = 6 \Rightarrow b = \sqrt{11}$

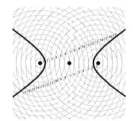

**119.** Time for sound of bullet hitting target to reach $(x, y)$: $\dfrac{2c}{v_m} + \dfrac{\sqrt{(x - c)^2 + y^2}}{v_s}$

Time for sound of rifle to reach $(x, y)$: $\dfrac{\sqrt{(x + c)^2 + y^2}}{v_s}$

Because the times are the same, you have: $\dfrac{2c}{v_m} + \dfrac{\sqrt{(x - c)^2 + y^2}}{v_s} = \dfrac{\sqrt{(x + c)^2 + y^2}}{v_s}$

$$\frac{4c^2}{v_m^2} + \frac{4c}{v_m v_s} \sqrt{(x - c)^2 + y^2} + \frac{(x - c)^2 + y^2}{v_s^2} = \frac{(x + c)^2 + y^2}{v_s^2}$$

$$\sqrt{(x - c)^2 + y^2} = \frac{v_m^2 x - v_s^2 c}{v_s v_m}$$

$$\left(1 - \frac{v_m^2}{v_s^2}\right) x^2 + y^2 = \left(\frac{v_s^2}{v_m^2} - 1\right) c^2$$

$$\frac{x^2}{c^2 v_s^2 / v_m^2} - \frac{y^2}{c^2 (v_m^2 - v_s^2) / v_m^2} = 1$$

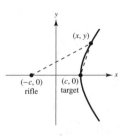

**121.** The point $(x, y)$ lies on the line between $(0, 10)$ and $(10, 0)$. So, $y = 10 - x$. The point also lies on the hyperbola $(x^2/36) - (y^2/64) = 1$. Using substitution, you have:

$$\frac{x^2}{36} - \frac{(10 - x)^2}{64} = 1$$

$$16x^2 - 9(10 - x)^2 = 576$$

$$7x^2 + 180x - 1476 = 0$$

$$x = \frac{-180 \pm \sqrt{180^2 - 4(7)(-1476)}}{2(7)} = \frac{-180 \pm 192\sqrt{2}}{14} = \frac{-90 \pm 96\sqrt{2}}{7}$$

Choosing the positive value for $x$ we have:

$$x = \frac{-90 + 96\sqrt{2}}{7} \approx 6.538 \text{ and}$$

$$y = \frac{160 - 96\sqrt{2}}{7} \approx 3.462$$

**123.**
$$\frac{x^2}{a^2} + \frac{2y^2}{b^2} = 1 \Rightarrow \frac{2y^2}{b^2} = 1 - \frac{x^2}{a^2}. \text{ Let } c^2 = a^2 - b^2.$$

$$\frac{x^2}{a^2 - b^2} - \frac{2y^2}{b^2} = 1 \Rightarrow \frac{2y^2}{b^2} = \frac{x^2}{a^2 - b^2} - 1$$

$$1 - \frac{x^2}{a^2} = \frac{x^2}{a^2 - b^2} - 1 \Rightarrow 2 = x^2\left(\frac{1}{a^2} + \frac{1}{a^2 - b^2}\right)$$

$$x^2 = \frac{2a^2(a^2 - b^2)}{2a^2 - b^2} \Rightarrow x = \pm\frac{\sqrt{2}a\sqrt{a^2 - b^2}}{\sqrt{2a^2 - b^2}} = \pm\frac{\sqrt{2}ac}{\sqrt{2a^2 - b^2}}$$

$$\frac{2y^2}{b^2} = 1 - \frac{1}{a^2}\left(\frac{2a^2c^2}{2a^2 - b^2}\right) \Rightarrow \frac{2y^2}{b^2} = \frac{b^2}{2a^2 - b^2}$$

$$y^2 = \frac{b^4}{2(2a^2 - b^2)} \Rightarrow y = \pm\frac{b^2}{\sqrt{2}\sqrt{2a^2 - b^2}}$$

There are four points of intersection: $\left(\dfrac{\sqrt{2}ac}{\sqrt{2a^2 - b^2}}, \pm\dfrac{b^2}{\sqrt{2}\sqrt{2a^2 - b^2}}\right)$, $\left(-\dfrac{\sqrt{2}ac}{\sqrt{2a^2 - b^2}}, \pm\dfrac{b^2}{\sqrt{2}\sqrt{2a^2 - b^2}}\right)$

$$\frac{x^2}{a^2} + \frac{2y^2}{b^2} = 1 \Rightarrow \frac{2x}{a^2} + \frac{4yy'}{b^2} = 0 \Rightarrow y_e' = -\frac{b^2x}{2a^2y}$$

$$\frac{x^2}{a^2 - b^2} - \frac{2y^2}{b^2} = 1 \Rightarrow \frac{2x}{c^2} - \frac{4yy'}{b^2} = 0 \Rightarrow y_h' = \frac{b^2x}{2c^2y}$$

At $\left(\dfrac{\sqrt{2}ac}{\sqrt{2a^2 - b^2}}, \dfrac{b^2}{\sqrt{2}\sqrt{2a^2 - b^2}}\right)$, the slopes of the tangent lines are:

$$y_e' = \frac{-b^2\left(\dfrac{\sqrt{2}ac}{\sqrt{2a^2 - b^2}}\right)}{2a^2\left(\dfrac{b^2}{\sqrt{2}\sqrt{2a^2 - b^2}}\right)} = -\frac{c}{a} \quad \text{and} \quad y_h' = \frac{b^2\left(\dfrac{\sqrt{2}ac}{\sqrt{2a^2 - b^2}}\right)}{2c^2\left(\dfrac{b^2}{\sqrt{2}\sqrt{2a^2 - b^2}}\right)} = \frac{a}{c}.$$

Because the slopes are negative reciprocals, the tangent lines are perpendicular. Similarly, the curves are perpendicular at the other three points of intersection.

**125.** False. See the definition of a parabola.

**127.** True

**129.** True

**131.** Let $\dfrac{x^2}{a^2} + \dfrac{y^2}{b^2} = 1$ be the equation of the ellipse with $a > b > 0$. Let $(\pm c, 0)$ be the foci,

$c^2 = a^2 - b^2$. Let $(u, v)$ be a point on the tangent line at $P(x, y)$, as indicated in the figure.

$$x^2 b^2 + y^2 a^2 = a^2 b^2$$

$$2xb^2 + 2yy'a^2 = 0$$

$$y' = -\dfrac{b^2 x}{a^2 y} \quad \text{Slope at } P(x, y)$$

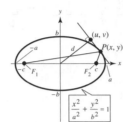

Now, $\dfrac{y - v}{x - u} = -\dfrac{b^2 x}{a^2 y}$

$$y^2 a^2 - a^2 vy = -b^2 x^2 + b^2 xu$$

$$y^2 a^2 + x^2 b^2 = a^2 vy + b^2 ux$$

$$a^2 b^2 = a^2 vy + b^2 ux$$

Because there is a right angle at $(u, v)$,

$$\dfrac{v}{u} = \dfrac{a^2 y}{b^2 x}$$

$$vb^2 x = a^2 uy.$$

You have two equations:

$$a^2 vy + b^2 ux = a^2 b^2$$

$$a^2 uy - b^2 vx = 0.$$

Multiplying the first by $v$ and the second by $u$, and adding,

$$a^2 v^2 y + a^2 u^2 y = a^2 b^2 v$$

$$y[u^2 + v^2] = b^2 v$$

$$yd^2 = b^2 v$$

$$v = \dfrac{yd^2}{b^2}.$$

Similarly, $u = \dfrac{xd^2}{a^2}.$

From the figure, $u = d \cos \theta$ and $v = d \sin \theta$. So, $\cos \theta = \dfrac{xd}{a^2}$ and $\sin \theta = \dfrac{yd}{b^2}.$

$$\cos^2 \theta + \sin^2 \theta = \dfrac{x^2 d^2}{a^4} + \dfrac{y^2 d^2}{b^4} = 1$$

$$x^2 b^4 d^2 + y^2 a^4 d^2 = a^4 b^4$$

$$d^2 = \dfrac{a^4 b^4}{x^2 b^4 + y^2 a^4}$$

Let $r_1 = PF_1$ and $r_2 = PF_2$, $r_1 + r_2 = 2a$.

$$r_1 r_2 = \dfrac{1}{2}\left[(r_1 + r_2)^2 - r_1^2 - r_2^2\right] = \dfrac{1}{2}\left[4a^2 - (x + c)^2 - y^2 - (x - c)^2 - y^2\right] = 2a^2 - x^2 - y^2 - c^2 = a^2 + b^2 - x^2 - y^2$$

Finally, $d^2 r_1 r_2 = \dfrac{a^4 b^4}{x^2 b^4 + y^2 a^4} \cdot \left[a^2 + b^2 - x^2 - y^2\right]$

$$= \dfrac{a^4 b^4}{b^2(b^2 x^2) + a^2(a^2 y^2)} \cdot \left[a^2 + b^2 - x^2 - y^2\right]$$

$$= \dfrac{a^4 b^4}{b^2(a^2 b^2 - a^2 y^2) + a^2(a^2 b^2 - b^2 x^2)} \cdot \left[a^2 + b^2 - x^2 - y^2\right]$$

$$= \dfrac{a^4 b^4}{a^2 b^2\left[a^2 + b^2 - x^2 - y^2\right]} \cdot \left[a^2 + b^2 - x^2 - y^2\right] = a^2 b^2, \text{ a constant!}$$

## Section 10.2  Plane Curves and Parametric Equations

**1.** $x = \sqrt{t}, y = 3 - t$

(a)

| $t$ | 0 | 1 | 2 | 3 | 4 |
|-----|---|---|---|---|---|
| $x$ | 0 | 1 | $\sqrt{2}$ | $\sqrt{3}$ | 2 |
| $y$ | 3 | 2 | 1 | 0 | -1 |

(b), (c)

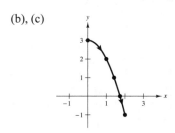

(d) $x^2 = t, y = 3 - x^2, x \ge 0$

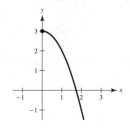

**3.** $x = 2t - 3$

$y = 3t + 1$

$t = \dfrac{x + 3}{2}$

$y = 3\left(\dfrac{x + 3}{2}\right) + 1 = \dfrac{3}{2}x + \dfrac{11}{2}$

$3x - 2y + 11 = 0$

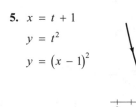

**5.** $x = t + 1$

$y = t^2$

$y = (x - 1)^2$

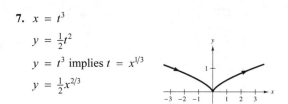

**7.** $x = t^3$

$y = \dfrac{1}{2}t^2$

$y = t^3$ implies $t = x^{1/3}$

$y = \dfrac{1}{2}x^{2/3}$

**9.** $x = \sqrt{t}$

$y = t - 5$

$x^2 = t$

$y = x^2 - 5, x \ge 0$

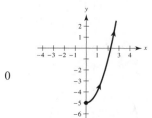

**11.** $x = t - 3$

$y = \dfrac{t}{t - 3}$

$t = x + 3$

$y = \dfrac{x + 3}{(x + 3) - 3}$

$= 1 + \dfrac{3}{x} = \dfrac{x + 3}{x}$

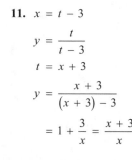

**13.** $x = 2t$

$y = |t - 2|$

$y = \left|\dfrac{x}{2} - 2\right| = \dfrac{|x - 4|}{2}$

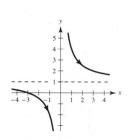

**15.** $x = e^t, x > 0$

$y = e^{3t} + 1$

$y = x^3 + 1, x > 0$

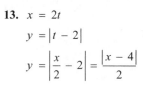
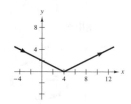

**17.** $x = \sec \theta$

$y = \cos \theta$

$0 \le \theta < \dfrac{\pi}{2}, \dfrac{\pi}{2} < \theta \le \pi$

$xy = 1$

$y = \dfrac{1}{x}$

$|x| \ge 1, |y| \le 1$

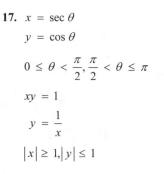
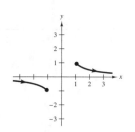

**19.** $x = 8 \cos \theta$

$y = 8 \sin \theta$

$x^2 + y^2 = 64 \cos^2 \theta + 64 \sin^2 \theta = 64(1) = 64$

Circle

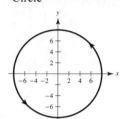

**21.** $x = 6 \sin 2\theta$

$y = 4 \cos 2\theta$

$\left(\dfrac{x}{6}\right)^2 + \left(\dfrac{y}{4}\right)^2 = \sin^2 2\theta + \cos^2 2\theta = 1$

$\dfrac{x^2}{36} + \dfrac{y^2}{16} = 1$

Ellipse

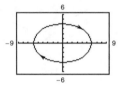

**23.**    $x = 4 + 2 \cos \theta$

$y = -1 + \sin \theta$

$\dfrac{(x-4)^2}{4} = \cos^2 \theta$

$\dfrac{(y+1)^2}{1} = \sin^2 \theta$

$\dfrac{(x-4)^2}{4} + \dfrac{(y+1)^2}{1} = 1$

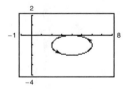

**25.** $x = -3 + 4 \cos \theta$

$y = 2 + 5 \sin \theta$

$x + 3 = 4 \cos \theta$

$y - 2 = 5 \sin \theta$

$\left(\dfrac{x+3}{4}\right)^2 + \left(\dfrac{y-2}{5}\right)^2 = \cos^2 \theta + \sin^2 \theta = 1$

$\dfrac{(x+3)^2}{16} + \dfrac{(y-2)^2}{25} = 1$

Ellipse

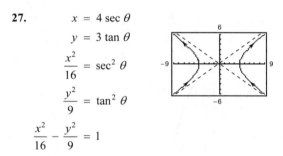

**27.**       $x = 4 \sec \theta$

$y = 3 \tan \theta$

$\dfrac{x^2}{16} = \sec^2 \theta$

$\dfrac{y^2}{9} = \tan^2 \theta$

$\dfrac{x^2}{16} - \dfrac{y^2}{9} = 1$

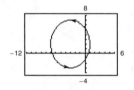

**29.** $x = t^3$

$y = 3 \ln t$

$y = 3 \ln \sqrt[3]{x} = \ln x$

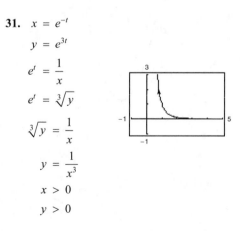

**31.** $x = e^{-t}$

$y = e^{3t}$

$e^t = \dfrac{1}{x}$

$e^t = \sqrt[3]{y}$

$\sqrt[3]{y} = \dfrac{1}{x}$

$y = \dfrac{1}{x^3}$

$x > 0$

$y > 0$

**33.** By eliminating the parameters in (a) – (d), you get
$y = 2x + 1$. They differ from each other in orientation
and in restricted domains. These curves are all smooth
except for (b).

(a)  $x = t, y = 2t + 1$

(b)  $x = \cos \theta \qquad y = 2 \cos \theta + 1$

$-1 \le x \le 1 \qquad -1 \le y \le 3$

$\dfrac{dx}{d\theta} = \dfrac{dy}{d\theta} = 0$ when $\theta = 0, \pm\pi, \pm 2\pi, \ldots$.

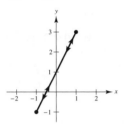

(c)  $x = e^{-t} \qquad y = 2e^{-t} + 1$

$\quad\;\; x > 0 \qquad\;\;\; y > 1$

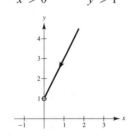

(d)  $x = e^{t} \qquad y = 2e^{t} + 1$

$\quad\;\; x > 0 \qquad\; y > 1$

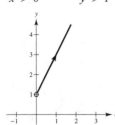

**35.** The curves are identical on $0 < \theta < \pi$. They are both
smooth. They represent $y = 2(1 - x^2)$ for $-1 \le x \le 1$.
The orientation is from right to left in part (a) and in part (b)

**37.** (a)

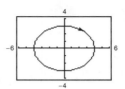

(b) The orientation of the second curve is reversed.

(c) The orientation will be reversed.

(d) Answers will vary. For example,

$\quad x = 2 \sec t \qquad x = 2 \sec (-t)$

$\quad y = 5 \sin t \qquad y = 5 \sin(-t)$

have the same graphs, but their orientations are
reversed.

**39.** $\qquad x = x_1 + t(x_2 - x_1)$

$\qquad\qquad y = y_1 + t(y_2 - y_1)$

$\qquad \dfrac{x - x_1}{x_2 - x_1} = t$

$\qquad\quad y = y_1 + \left(\dfrac{x - x_1}{x_2 - x_1}\right)(y_2 - y_1)$

$\qquad\quad y - y_1 = \dfrac{y_2 - y_1}{x_2 - x_1}(x - x_1)$

$\qquad\quad y - y_1 = m(x - x_1)$

**41.** $\qquad\qquad\qquad x = h + a \cos \theta$

$\qquad\qquad\qquad\qquad y = k + b \sin \theta$

$\qquad\qquad \dfrac{x - h}{a} = \cos \theta$

$\qquad\qquad \dfrac{y - k}{b} = \sin \theta$

$\qquad \dfrac{(x - h)^2}{a^2} + \dfrac{(y - k)^2}{b^2} = 1$

**43.** From Exercise 39 you have

$x = 4t$

$y = -7t$

Solution not unique

**45.** From Exercise 40 you have

$x = 3 + 2 \cos \theta$

$y = 1 + 2 \sin \theta$

Solution not unique

**47.** From Exercise 41 you have

$a = 10, c = 8 \Rightarrow b = 6$

$x = 10 \cos \theta$

$y = 6 \sin \theta$

Center: $(0, 0)$

Solution not unique

**49.** From Exercise 42 you have

$a = 4, c = 5 \Rightarrow b = 3$

$x = 4 \sec \theta$

$y = 3 \tan \theta.$

Center: $(0, 0)$

Solution not unique

**51.** $y = 6x - 5$

Examples:

$x = t, y = 6t - 5$

$x = t + 1, y = 6t + 1$

**53.** $y = x^3$

Example

$x = t, \qquad y = t^3$

$x = \sqrt[3]{t}, \qquad y = t$

$x = \tan t, \qquad y = \tan^3 t$

**55.** $y = 2x - 5$

At $(3, 1), t = 0$: $\qquad x = 3 - t$

$y = 2(3 - t) - 5 = -2t + 1$

or, $x = t + 3$

$y = 2t + 1$

**57.** $y = x^2$

$t = 4$ at $(4, 16)$: $\qquad x = t$

$y = t^2$

**59.** $x = 2(\theta - \sin \theta)$

$y = 2(1 - \cos \theta)$

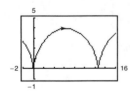

Not smooth at $\theta = 2n\pi$

**61.** $x = \theta - \frac{3}{2} \sin \theta$

$y = 1 - \frac{3}{2} \cos \theta$

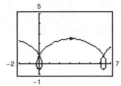

Smooth Everywhere

**63.** $x = 3 \cos^3 \theta$

$y = 3 \sin^3 \theta$

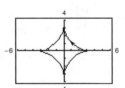

Not smooth at $(x, y) = (\pm 3, 0)$ and $(0, \pm 3)$, or

$\theta = \frac{1}{2} n\pi.$

**65.** $x = 2 \cot \theta$

$y = 2 \sin^2 \theta$

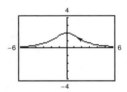

Smooth everywhere

**67.** Each point $(x, y)$ in the plane is determined by the plane curve $x = f(t), y = g(t)$. For each $t$, plot $(x, y)$. As $t$ increases, the curve is traced out in a specific direction called the orientation of the curve.

**69.** When the circle has rolled $\theta$ radians, you know that the center is at $(a\theta, a)$.

$\sin \theta = \sin(180° - \theta) = \dfrac{|AC|}{b} = \dfrac{|BD|}{b}$ or $|BD| = b \sin \theta$

$\cos \theta = -\cos(180° - \theta) = \dfrac{|AP|}{-b}$ or $|AP| = -b \cos \theta$

So, $x = a\theta - b \sin \theta$ and $y = a - b \cos \theta.$

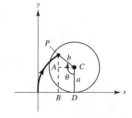

**71.** False

$$x = t^2 \Rightarrow x \geq 0$$

$$y = t^2 \Rightarrow y \geq 0$$

The graph of the parametric equations is only a portion of the line $y = x$ when $x \geq 0$.

**73.** True. $y = \cos x$

**75.** (a)  $100 \text{ mi/hr} = \dfrac{(100)(5280)}{3600} = \dfrac{440}{3} \text{ ft/sec}$

$$x = (v_0 \cos \theta)t = \left(\frac{440}{3} \cos \theta\right)t$$

$$y = h + (v_0 \sin \theta)t - 16t^2 = 3 + \left(\frac{440}{3} \sin \theta\right)t - 16t^2$$

(b)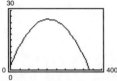

It is not a home run—when $x = 400$, $y < 10$.

(c)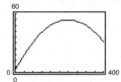

Yes, it's a home run when $x = 400$, $y > 10$.

(d) You need to find the angle $\theta$ (and time $t$) such that

$$x = \left(\frac{440}{3} \cos \theta\right)t = 400$$

$$y = 3 + \left(\frac{440}{3} \sin \theta\right)t - 16t^2 = 10.$$

From the first equation $t = 1200/440 \cos \theta$. Substituting into the second equation,

$$10 = 3 + \left(\frac{440}{3} \sin \theta\right)\left(\frac{1200}{440 \cos \theta}\right) - 16\left(\frac{1200}{440 \cos \theta}\right)^2$$

$$7 = 400 \tan \theta - 16\left(\frac{120}{44}\right)^2 \sec^2 \theta = 400 \tan \theta - 16\left(\frac{120}{44}\right)^2 (\tan^2 \theta + 1).$$

You now solve the quadratic for $\tan \theta$:

$$16\left(\frac{120}{44}\right)^2 \tan^2 \theta - 400 \tan \theta + 7 + 16\left(\frac{120}{44}\right)^2 = 0.$$

$\tan \theta \approx 0.35185 \Rightarrow \theta \approx 19.4°$

# Section 10.3   Parametric Equations and Calculus

**1.** $\dfrac{dy}{dx} = \dfrac{dy/dt}{dx/dt} = \dfrac{-6}{2t} = -\dfrac{3}{t}$

**3.** $\dfrac{dy}{dx} = \dfrac{dy/d\theta}{dx/d\theta} = \dfrac{-2 \cos \theta \sin \theta}{2 \sin \theta \cos \theta} = -1$

$\left[\text{Note: } x + y = 1 \Rightarrow y = 1 - x \text{ and } \dfrac{dy}{d\theta} = -1\right]$

**5.** $x = 4t, y = 3t - 2$

$$\frac{dy}{dx} = \frac{dy/dt}{dx/dt} = \frac{3}{4}$$

$$\frac{d^2y}{dx^2} = 0$$

At $t = 3$, slope is $\frac{3}{4}$. (Line)

Neither concave upward nor downward.

**7.** $x = t + 1, y = t^2 + 3t$

$$\frac{dy}{dx} = \frac{2t + 3}{1} = 1 \text{ when } t = -1.$$

$$\frac{d^2y}{dx^2} = 2 \text{ concave upward}$$

**9.** $x = 4 \cos \theta, y = 4 \sin \theta$

$$\frac{dy}{dx} = \frac{dy/d\theta}{dx/d\theta} = \frac{4 \cos \theta}{-4 \sin \theta} = \frac{-\cos \theta}{\sin \theta} = -\cot \theta$$

$$\frac{d^2y}{dx^2} = \frac{\frac{d}{d\theta}[-\cot \theta]}{dx/d\theta} = \frac{\csc^2 \theta}{-4 \sin \theta} = \frac{-1}{4 \sin^3 \theta} = -\frac{1}{4} \csc^3 \theta$$

At $\theta = \frac{\pi}{4}, \frac{dy}{dx} = -1.$

$$\frac{d^2y}{dx^2} = \frac{-1}{4\left(\sqrt{2}/2\right)^3} = \frac{-\sqrt{2}}{2}$$

concave downward

**11.** $x = 2 + \sec \theta, y = 1 + 2 \tan \theta$

$$\frac{dy}{dx} = \frac{2 \sec^2 \theta}{\sec \theta \tan \theta}$$

$$= \frac{2 \sec \theta}{\tan \theta} = 2 \csc \theta = 4 \text{ when } \theta = \frac{\pi}{6}.$$

$$\frac{d^2y}{dx^2} = \frac{\frac{d\left[\frac{dy}{dx}\right]}{dx}}{\frac{dx}{d\theta}} = \frac{-2 \csc \theta \cot \theta}{\sec \theta \tan \theta}$$

$$= -2 \cot^3 \theta = -6\sqrt{3} \text{ when } \theta = \frac{\pi}{6}.$$

concave downward

**13.** $x = \cos^3 \theta, y = \sin^3 \theta$

$$\frac{dy}{dx} = \frac{3 \sin^2 \theta \cos \theta}{-3 \cos^2 \theta \sin \theta} = -\tan \theta = -1 \text{ when } \theta = \frac{\pi}{4}.$$

$$\frac{d^2y}{dx^2} = \frac{-\sec^2 \theta}{-3 \cos^2 \theta \sin \theta} = \frac{1}{3 \cos^4 \theta \sin \theta}$$

$$= \frac{\sec^4 \theta \csc \theta}{3} = \frac{4\sqrt{2}}{3} \text{ when } \theta = \frac{\pi}{4}.$$

concave upward

**15.** $x = 2 \cot \theta, y = 2 \sin^2 \theta$

$$\frac{dy}{dx} = \frac{4 \sin \theta \cos \theta}{-2 \csc^2 \theta} = -2 \sin^3 \theta \cos \theta$$

At $\left(-\frac{2}{\sqrt{3}}, \frac{3}{2}\right)$, $\theta = \frac{2\pi}{3}$, and $\frac{dy}{dx} = \frac{3\sqrt{3}}{8}.$

Tangent line: $y - \frac{3}{2} = \frac{3\sqrt{3}}{8}\left(x + \frac{2}{\sqrt{3}}\right)$

$$3\sqrt{3}x - 8y + 18 = 0$$

At $(0, 2)$, $\theta = \frac{\pi}{2}$, and $\frac{dy}{dx} = 0.$

Tangent line: $y - 2 = 0$

At $\left(2\sqrt{3}, \frac{1}{2}\right)$, $\theta = \frac{\pi}{6}$, and $\frac{dy}{dx} = -\frac{\sqrt{3}}{8}.$

Tangent line: $y - \frac{1}{2} = -\frac{\sqrt{3}}{8}\left(x - 2\sqrt{3}\right)$

$$\sqrt{3}x + 8y - 10 = 0$$

**17.** $x = t^2 - 4$

$y = t^2 - 2t$

$$\frac{dy}{dx} = \frac{dy/dt}{dx/dt} = \frac{2t - 2}{2t}$$

At $(0, 0)$, $t = 2, \frac{dy}{dx} = \frac{1}{2}.$

Tangent line: $y = \frac{1}{2}x$

$$2y - x = 0$$

At $(-3, -1)$, $t = 1, \frac{dy}{dx} = 0.$

Tangent line: $y = -1$

$$y + 1 = 0$$

At $(-3, 3)$, $t = -1, \frac{dy}{dx} = 2.$

Tangent line: $y - 3 = 2(x + 3)$

$$2x - y + 9 = 0$$

**19.** $x = 6t, \ y = t^2 + 4, \ t = 1$

(a), (d)

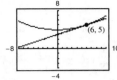

(b)  At $t = 1, \ (x, y) = (6, 5)$

$$\frac{dx}{dt} = 6, \frac{dy}{dt} = 2, \frac{dy}{dx} = \frac{1}{3}.$$

(c)  $y - 5 = \dfrac{1}{3}(x - 6)$

$$y = \frac{1}{3}x + 3$$

**21.** $x = t^2 - t + 2, \ y = t^3 - 3t, \ t = -1$

(a), (d)

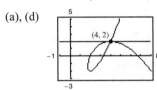

(b)  At $t = -1, \ (x, y) = (4, 2),$ and

$$\frac{dx}{dt} = -3, \frac{dy}{dt} = 0, \frac{dy}{dx} = 0.$$

(c)  $\dfrac{dy}{dx} = 0.$ At $(4, 2), \ y - 2 = 0(x - 4)$

$$y = 2.$$

**23.** $x = 2 \sin 2t, \ y = 3 \sin t$ crosses itself at the origin,
$(x, y) = (0, 0).$

At this point, $t = 0$ or $t = \pi.$

$$\frac{dy}{dx} = \frac{3 \cos t}{4 \cos 2t}$$

At $t = 0: \dfrac{dy}{dx} = \dfrac{3}{4}$ and $y = \dfrac{3}{4}x.$    Tangent Line

At $t = \pi, \dfrac{dy}{dx} = -\dfrac{3}{4}$ and $y = -\dfrac{3}{4}x.$    Tangent Line

**25.** $x = t^2 - t, \ y = t^3 - 3t - 1$ crosses itself at the
point $(x, y) = (2, 1).$

At this point, $t = -1$ or $t = 2.$

$$\frac{dy}{dx} = \frac{3t^2 - 3}{2t - 1}$$

At $t = -1, \dfrac{dy}{dx} = 0$ and $y = 1.$    Tangent Line

At $t = 2, \dfrac{dy}{dt} = \dfrac{9}{3} = 3$ and $y - 1 = 3(x - 2)$ or

$$y = 3x - 5.$$

Tangent Line

**27.** $x = \cos \theta + \theta \sin \theta, \ y = \sin \theta - \theta \cos \theta$

Horizontal tangents: $\dfrac{dy}{d\theta} = \theta \sin \theta = 0$ when

$\theta = \pm \pi, \pm 2\pi, \pm 3\pi, \ldots$

Points: $(-1, [2n - 1]\pi), (1, 2n\pi)$ where $n$ is an integer.

Points shown: $(1, 0), (-1, \pi), (1, -2\pi)$

Vertical tangents: $\dfrac{dx}{d\theta} = \theta \cos \theta = 0$ when

$\theta = \pm \dfrac{\pi}{2}, \pm \dfrac{3\pi}{2}, \pm \dfrac{5\pi}{2}, \ldots$

**Note:** $\theta = 0$ corresponds to the cusp at $(x, y) = (1, 0).$

$$\frac{dy}{dx} = \frac{\theta \sin \theta}{\theta \cos \theta} = \tan \theta = 0 \text{ at } \theta = 0$$

Points: $\left( \dfrac{(-1)^{n+1}(2n - 1)\pi}{2}, (-1)^{n+1} \right)$

Points shown: $\left( \dfrac{\pi}{2}, 1 \right), \left( -\dfrac{3\pi}{2}, -1 \right), \left( \dfrac{5\pi}{2}, 1 \right)$

**29.** $x = 4 - t, \ y = t^2$

Horizontal tangents: $\dfrac{dy}{dt} = 2t = 0$ when $t = 0.$

Point: $(4, 0)$

Vertical tangents: $\dfrac{dx}{dt} = -1 \neq 0$  None

**31.** $x = t + 4, \ y = t^3 - 3t$

Horizontal tangents:

$$\frac{dy}{dt} = 3t^2 - 3 = 3(t - 1)(t + 1) = 0 \Rightarrow t = \pm 1$$

Points: $(5, -2), (3, 2)$

Vertical tangents: $\dfrac{dx}{dt} = 1 \neq 0$  None

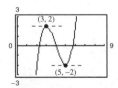

**33.** $x = 3\cos\theta$, $y = 3\sin\theta$

Horizontal tangents: $\dfrac{dy}{d\theta} = 3\cos\theta = 0$ when

$\theta = \dfrac{\pi}{2}, \dfrac{3\pi}{2}$.

Points: $(0, 3), (0, -3)$

Vertical tangents: $\dfrac{dx}{d\theta} = -3\sin\theta = 0$ when $\theta = 0, \pi$.

Points: $(3, 0), (-3, 0)$

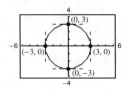

**35.** $x = 5 + 3\cos\theta$, $y = -2 + \sin\theta$

Horizontal tangents: $\dfrac{dy}{dt} = \cos\theta = 0 \Rightarrow \theta = \dfrac{\pi}{2}, \dfrac{3\pi}{2}$

Points: $(5, -1), (5, -3)$

Vertical tangents: $\dfrac{dx}{dt} = -3\sin\theta = 0 \Rightarrow \theta = 0, \pi$

Points: $(8, -2), (2, -2)$

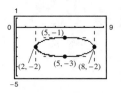

**37.** $x = \sec\theta$, $y = \tan\theta$

Horizontal tangents: $\dfrac{dy}{d\theta} = \sec^2\theta \neq 0$; None

Vertical tangents: $\dfrac{dx}{d\theta} = \sec\theta\tan\theta = 0$ when

$x = 0, \pi$.

Points: $(1, 0), (-1, 0)$

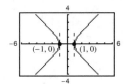

**39.** $x = 3t^2$, $y = t^3 - t$

$$\frac{dy}{dx} = \frac{dy/dt}{dx/dt} = \frac{3t^2 - 1}{6t} = \frac{t}{2} - \frac{1}{6t}$$

$$\frac{d^2y}{dx^2} = \frac{\dfrac{d}{dt}\left[\dfrac{t}{2} - \dfrac{1}{6t}\right]}{dx/dt} = \frac{\dfrac{1}{2} + \dfrac{1}{6t^2}}{6t} = \frac{6t^2 + 2}{36t^3}$$

Concave upward for $t > 0$

Concave downward for $t < 0$

**41.** $x = 2t + \ln t$, $y = 2t - \ln t$, $t > 0$

$$\frac{dy}{dx} = \frac{2 - (1/t)}{2 + (1/t)} = \frac{2t - 1}{2t + 1}$$

$$\frac{d^2y}{dx^2} = \left[\frac{(2t + 1)2 - (2t - 1)2}{(2t + 1)^2}\right]\bigg/\left(2 + \frac{1}{t}\right)$$

$$= \frac{4}{(2t + 1)^2} \cdot \frac{t}{2t + 1} = \frac{4t}{(2t + 1)^3}$$

Because $t > 0$, $\dfrac{d^2y}{dx^2} > 0$

Concave upward for $t > 0$

**43.** $x = \sin t$, $y = \cos t$, $0 < t < \pi$

$$\frac{dy}{dx} = -\frac{\sin t}{\cos t} = -\tan t$$

$$\frac{d^2y}{dx^2} = -\frac{\sec^2 t}{\cos t} = -\frac{1}{\cos^3 t}$$

Concave upward on $\pi/2 < t < \pi$

Concave downward on $0 < t < \pi/2$

**45.** $x = 3t - t^2$, $y = 2t^{3/2}$, $1 \leq t \leq 3$

$$\frac{dx}{dt} = 3 - 2t, \quad \frac{dy}{dt} = 3t^{1/2}$$

$$s = \int_a^b \sqrt{\left(\frac{dx}{dt}\right)^2 + \left(\frac{dy}{dt}\right)^2}\, dt$$

$$= \int_1^3 \sqrt{(3 - 2t)^2 + 9t}\, dt = \int_1^3 \sqrt{4t^2 - 3t + 9}\, dt$$

**47.** $x = e^t + 2$, $y = 2t + 1$, $-2 \leq t \leq 2$

$$\frac{dx}{dt} = e^t, \quad \frac{dy}{dt} = 2$$

$$s = \int_a^b \sqrt{\left(\frac{dx}{dt}\right)^2 + \left(\frac{dy}{dt}\right)^2}\, dt$$

$$= \int_{-2}^2 \sqrt{e^{2t} + 4}\, dt$$

**49.** $x = 3t + 5, y = 7 - 2t, -1 \le t \le 3$

$$\frac{dx}{dt} = 3, \frac{dy}{dt} = -2$$

$$s = \int_a^b \sqrt{\left(\frac{dx}{dt}\right)^2 + \left(\frac{dy}{dt}\right)^2}\, dt$$

$$= \int_{-1}^3 \sqrt{9 + 4}\, dt$$

$$\left[\sqrt{13}\, t\right]_{-1}^3 = 4\sqrt{13} \approx 14.422$$

**51.** $x = 6t^2, y = 2t^3, 1 \le t \le 4$

$$\frac{dx}{dt} = 12t, \frac{dy}{dt} = 6t^2$$

$$s = \int_a^b \sqrt{\left(\frac{dx}{dt}\right)^2 + \left(\frac{dy}{dt}\right)^2}\, dt$$

$$= \int_1^4 \sqrt{144t^2 + 36t^4}\, dt$$

$$= \int_1^4 6t\sqrt{4 + t^2}\, dt$$

$$= \left[2\left(4 + t^2\right)^{3/2}\right]_1^4$$

$$= 2\left(20^{3/2} - 5^{3/2}\right)$$

$$= 70\sqrt{5} \approx 156.525$$

**53.** $x = e^{-t} \cos t, y = e^{-t} \sin t, 0 \le t \le \dfrac{\pi}{2}$

$$\frac{dx}{dt} = -e^{-t}(\sin t + \cos t), \frac{dy}{dt} = e^{-t}(\cos t - \sin t)$$

$$s = \int_0^{\pi/2} \sqrt{\left(\frac{dx}{dt}\right)^2 + \left(\frac{dy}{dt}\right)^2}\, dt$$

$$= \int_0^{\pi/2} \sqrt{2e^{-2t}}\, dt = -\sqrt{2}\int_0^{\pi/2} e^{-t}(-1)\, dt$$

$$= \left[-\sqrt{2}e^{-t}\right]_0^{\pi/2}$$

$$= \sqrt{2}\left(1 - e^{-\pi/2}\right) \approx 1.12$$

**55.** $x = \sqrt{t}, y = 3t - 1, \dfrac{dx}{dt} = \dfrac{1}{2\sqrt{t}}, \dfrac{dy}{dt} = 3$

$$s = \int_0^1 \sqrt{\frac{1}{4t} + 9}\, dt = \frac{1}{2}\int_0^1 \frac{\sqrt{1 + 36t}}{\sqrt{t}}\, dt$$

$$= \frac{1}{6}\int_0^6 \sqrt{1 + u^2}\, du$$

$$= \frac{1}{12}\left[\ln\left(\sqrt{1 + u^2} + u\right) + u\sqrt{1 + u^2}\right]_0^6$$

$$= \frac{1}{12}\left[\ln\left(\sqrt{37} + 6\right) + 6\sqrt{37}\right] \approx 3.249$$

$$u = 6\sqrt{t}, du = \frac{3}{\sqrt{t}}\, dt$$

**57.** $x = a\cos^3\theta, y = a\sin^3\theta, \dfrac{dx}{d\theta} = -3a\cos^2\theta\sin\theta,$

$$\frac{dy}{d\theta} = 3a\sin^2\theta\cos\theta$$

$$s = 4\int_0^{\pi/2} \sqrt{9a^2\cos^4\theta\sin^2\theta + 9a^2\sin^4\theta\cos^2\theta}\, d\theta$$

$$= 12a\int_0^{\pi/2} \sin\theta\cos\theta\sqrt{\cos^2\theta + \sin^2\theta}\, d\theta$$

$$= 6a\int_0^{\pi/2} \sin 2\theta\, d\theta = \left[-3a\cos 2\theta\right]_0^{\pi/2} = 6a$$

**59.** $x = a(\theta - \sin\theta), y = a(1 - \cos\theta),$

$$\frac{dx}{d\theta} = a(1 - \cos\theta), \frac{dy}{d\theta} = a\sin\theta$$

$$s = 2\int_0^\pi \sqrt{a^2(1 - \cos\theta)^2 + a^2\sin^2\theta}\, d\theta$$

$$= 2\sqrt{2}a\int_0^\pi \sqrt{1 - \cos\theta}\, d\theta$$

$$= 2\sqrt{2}a\int_0^\pi \frac{\sin\theta}{\sqrt{1 + \cos\theta}}\, d\theta$$

$$= \left[-4\sqrt{2}a\sqrt{1 + \cos\theta}\right]_0^\pi = 8a$$

**61.** $x = (90\cos 30°)t, y = (90\sin 30°)t - 16t^2$

(a)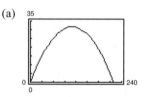

(b) Range: 219.2 ft, $\left(t = \dfrac{45}{16}\right)$

(c) $\dfrac{dx}{dt} = 90\cos 30°, \dfrac{dy}{dt} = 90\sin 30° - 32t$

$$y = 0 \text{ for } t = \frac{45}{16}.$$

$$s = \int_0^{45/16} \sqrt{(90\cos 30°)^2 + (90\sin 30° - 32t)^2}\, dt$$

$$\approx 230.8 \text{ ft}$$

**63.** $x = \dfrac{4t}{1+t^3},\ y = \dfrac{4t^2}{1+t^3}$

(a) $x^3 + y^3 = 4xy$

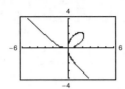

(b) $\dfrac{dy}{dt} = \dfrac{\left(1+t^3\right)(8t) - 4t^2\left(3t^2\right)}{\left(1+t^3\right)^2}$

$= \dfrac{4t\left(2-t^3\right)}{\left(1+t^3\right)^2} = 0$ when $t = 0$ or $t = \sqrt[3]{2}$.

Points: $(0,0),\ \left(\dfrac{4\sqrt[3]{2}}{3},\ \dfrac{4\sqrt[3]{4}}{3}\right) \approx (1.6799,\ 2.1165)$

(c) $s = 2\displaystyle\int_0^1 \sqrt{\left[\dfrac{4\left(1-2t^3\right)}{\left(1+t^3\right)^2}\right]^2 + \left[\dfrac{4t\left(2-t^3\right)}{\left(1+t^3\right)^2}\right]^2}\ dt$

$= 2\displaystyle\int_0^1 \sqrt{\dfrac{16}{\left(1+t^3\right)^4}\left[t^8 + 4t^6 - 4t^5 - 4t^3 + 4t^2 + 1\right]}\ dt$

$= 8\displaystyle\int_0^1 \dfrac{\sqrt{t^8 + 4t^6 - 4t^5 - 4t^3 + 4t^2 + 1}}{\left(1+t^3\right)^2}\ dt \approx 6.557$

**65.** (a) $x = t - \sin t$
$y = 1 - \cos t$
$0 \le t \le 2\pi$

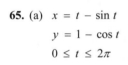

$x = 2t - \sin(2t)$
$y = 1 - \cos(2t)$
$0 \le t \le \pi$

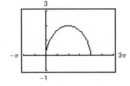

(b) The average speed of the particle on the second path is twice the average speed of a particle on the first path.

(c) $x = \frac{1}{2}t - \sin\left(\frac{1}{2}t\right)$

$y = 1 - \cos\left(\frac{1}{2}t\right)$

The time required for the particle to traverse the same path is $t = 4\pi$.

**67.** $x = 3t,\ \dfrac{dx}{dt} = 3$

$y = t + 2,\ \dfrac{dy}{dt} = 1$

$S = 2\pi \displaystyle\int_0^4 (t+2)\sqrt{3^2 + 1^2}\ dt$

$= 2\pi\sqrt{10}\left[\dfrac{t^2}{2} + 2t\right]_0^4$

$= 2\pi\sqrt{10}[8 + 8] = 32\sqrt{10}\,\pi \approx 317.9068$

**69.** $x = \cos^2\theta,\ \dfrac{dx}{d\theta} = -2\cos\theta\sin\theta$

$y = \cos\theta,\ \dfrac{dy}{d\theta} = -\sin\theta$

$S = 2\pi \displaystyle\int_0^{\pi/2} \cos\theta\sqrt{4\cos^2\theta\sin^2\theta + \sin^2\theta}\ d\theta$

$= 2\pi \displaystyle\int_0^{\pi/2} \cos\theta\sin\theta\sqrt{4\cos^2\theta + 1}\ d\theta$

$= \dfrac{\left(5\sqrt{5} - 1\right)\pi}{6}$

$\approx 5.3304$

**71.** $x = 2t,\ \dfrac{dx}{dt} = 2$

$y = 3t,\ \dfrac{dy}{dt} = 3$

(a) $S = 2\pi \displaystyle\int_0^3 3t\sqrt{4 + 9}\ dt$

$= 6\sqrt{13}\pi\left[\dfrac{t^2}{2}\right]_0^3 = 6\sqrt{13}\pi\left(\dfrac{9}{2}\right) = 27\sqrt{13}\pi$

(b) $S = 2\pi \displaystyle\int_0^3 2t\sqrt{4 + 9}\ dt$

$= 4\sqrt{13}\pi\left[\dfrac{t^2}{2}\right]_0^3 = 4\sqrt{13}\pi\left(\dfrac{9}{2}\right) = 18\sqrt{13}\pi$

**73.** $x = 5\cos\theta,\ \dfrac{dx}{d\theta} = -5\sin\theta$

$y = 5\sin\theta,\ \dfrac{dy}{d\theta} = 5\cos\theta$

$S = 2\pi \displaystyle\int_0^{\pi/2} 5\cos\theta\sqrt{25\sin^2\theta + 25\cos^2\theta}\ d\theta$

$= 10\pi \displaystyle\int_0^{\pi/2} 5\cos\theta\ d\theta$

$= 50\pi\left[\sin\theta\right]_0^{\pi/2} = 50\pi$

[**Note:** This is the surface area of a hemisphere of radius 5]

**75.** $x = a \cos^3 \theta$, $y = a \sin^3 \theta$, $\dfrac{dx}{d\theta} = -3a \cos^2 \theta \sin \theta$, $\dfrac{dy}{d\theta} = 3a \sin^2 \theta \cos \theta$

$$S = 4\pi \int_0^{\pi/2} a \sin^3 \theta \sqrt{9a^2 \cos^4 \theta \sin^2 \theta + 9a^2 \sin^4 \theta \cos^2 \theta} \, d\theta$$

$$= 12a^2 \pi \int_0^{\pi/2} \sin^4 \theta \cos \theta \, d\theta = \frac{12\pi a^2}{5} \Big[ \sin^5 \theta \Big]_0^{\pi/2} = \frac{12}{5} \pi a^2$$

**77.** $\dfrac{dy}{dx} = \dfrac{dy/dt}{dx/dt}$

See Theorem 10.7.

**79.** $x = t$, $y = 6t - 5 \Rightarrow \dfrac{dy}{dx} = \dfrac{6}{1} = 6$

**81.** (a) $S = 2\pi \int_a^b g(t) \sqrt{\left(\dfrac{dx}{dt}\right)^2 + \left(\dfrac{dy}{dt}\right)^2} \, dt$

(b) $S = 2\pi \int_a^b f(t) \sqrt{\left(\dfrac{dx}{dt}\right)^2 + \left(\dfrac{dy}{dt}\right)^2} \, dt$

**83.** Let $y$ be a continuous function of $x$ on $a \le x \le b$.
Suppose that $x = f(t)$, $y = g(t)$, and $f(t_1) = a$,
$f(t_2) = b$. Then using integration by substitution,
$dx = f'(t) \, dt$ and

$$\int_a^b y \, dx = \int_{t_1}^{t_2} g(t) f'(t) \, dt.$$

**85.** $x = 2 \sin^2 \theta$

$y = 2 \sin^2 \theta \tan \theta$

$\dfrac{dx}{d\theta} = 4 \sin \theta \cos \theta$

$A = \displaystyle\int_0^{\pi/2} 2 \sin^2 \theta \tan \theta (4 \sin \theta \cos \theta) \, d\theta$

$= 8 \displaystyle\int_0^{\pi/2} \sin^4 \theta \, d\theta$

$= 8 \left[ \dfrac{-\sin^3 \theta \cos \theta}{4} - \dfrac{3}{8} \sin \theta \cos \theta + \dfrac{3}{8} \theta \right]_0^{\pi/2} = \dfrac{3\pi}{2}$

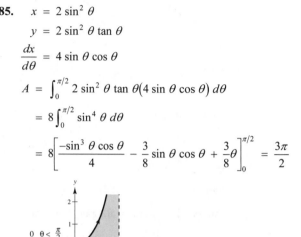

**87.** $\pi ab$ is area of ellipse (d).

**89.** $6\pi a^2$ is area of cardioid (f).

**91.** $\dfrac{8}{3} ab$ is area of hourglass (a).

**93.** $x = \sqrt{t}$, $y = 4 - t$, $0 < t < 4$

$$A = \int_0^2 y \, dx = \int_0^4 (4 - t) \frac{1}{2\sqrt{t}} \, dt = \frac{1}{2} \int_0^4 \left(4t^{-1/2} - t^{1/2}\right) dt = \left[ \frac{1}{2} \left( 8\sqrt{t} - \frac{2}{3} t \sqrt{t} \right) \right]_0^4 = \frac{16}{3}$$

$$\bar{x} = \frac{1}{A} \int_0^2 yx \, dx = \frac{3}{16} \int_0^4 (4 - t) \sqrt{t} \left( \frac{1}{2\sqrt{t}} \right) dt = \frac{3}{32} \int_0^4 (4 - t) \, dt = \left[ \frac{3}{32} \left( 4t - \frac{t^2}{2} \right) \right]_0^4 = \frac{3}{4}$$

$$\bar{y} = \frac{1}{A} \int_0^2 \frac{y^2}{2} \, dx = \frac{3}{32} \int_0^4 (4 - t)^2 \frac{1}{2\sqrt{t}} \, dt = \frac{3}{64} \int_0^4 \left( 16t^{-1/2} - 8t^{1/2} + t^{3/2} \right) dt = \frac{3}{64} \left[ 32\sqrt{t} - \frac{16}{3} t\sqrt{t} + \frac{2}{5} t^2 \sqrt{t} \right]_0^4 = \frac{8}{5}$$

$$(\bar{x}, \bar{y}) = \left( \frac{3}{4}, \frac{8}{5} \right)$$

**95.** $x = 6 \cos \theta$, $y = 6 \sin \theta$, $\dfrac{dx}{d\theta} = -6 \sin \theta \, d\theta$

$$V = 2\pi \int_{\pi/2}^0 (6 \sin \theta)^2 (-6 \sin \theta) \, d\theta = -432\pi \int_{\pi/2}^0 \sin^3 \theta \, d\theta$$

$$= -432\pi \int_{\pi/2}^0 \left( 1 - \cos^2 \theta \right) \sin \theta \, d\theta = -432\pi \left[ -\cos \theta + \frac{\cos^3 \theta}{3} \right]_{\pi/2}^0 = -432\pi \left( -1 + \frac{1}{3} \right) = 288\pi$$

Note: Volume of sphere is $\dfrac{4}{3} \pi (6^3) = 288\pi$.

**97.** $x = a(\theta - \sin \theta), \; y = a(1 - \cos \theta)$

(a) $\dfrac{dy}{d\theta} = a \sin \theta, \; \dfrac{dx}{d\theta} = a(1 - \cos \theta)$

$\dfrac{dy}{dx} = \dfrac{a \sin \theta}{a(1 - \cos \theta)} = \dfrac{\sin \theta}{1 - \cos \theta}$

$\dfrac{d^2y}{dx^2} = \left[ \dfrac{(1 - \cos \theta) \cos \theta - \sin \theta(\sin \theta)}{(1 - \cos \theta)^2} \right] \Big/ \left[ a(1 - \cos \theta) \right] = \dfrac{\cos \theta - 1}{a(1 - \cos \theta)^3} = \dfrac{-1}{a(\cos \theta - 1)^2}$

(b) At $\theta = \dfrac{\pi}{6}, \; x = a\left( \dfrac{\pi}{6} - \dfrac{1}{2} \right), \; y = a\left( 1 - \dfrac{\sqrt{3}}{2} \right), \; \dfrac{dy}{dx} = \dfrac{1/2}{1 - \sqrt{3}/2} = 2 + \sqrt{3}.$

Tangent line: $y - a\left( 1 - \dfrac{\sqrt{3}}{2} \right) = \left( 2 + \sqrt{3} \right)\left( x - a\left( \dfrac{\pi}{6} - \dfrac{1}{2} \right) \right)$

(c) $\dfrac{dy}{dx} = \dfrac{\sin \theta}{1 - \cos \theta} = 0 \Rightarrow \sin \theta = 0, \; 1 - \cos \theta \neq 0$

Points of horizontal tangency: $(x, y) = \left( a(2n + 1)\pi, 2a \right)$

(d) Concave downward on all open $\theta$-intervals:

$\dots, (-2\pi, 0), (0, 2\pi), (2\pi, 4\pi), \dots$

(e) $s = \displaystyle\int_0^{2\pi} \sqrt{a^2 \sin^2 \theta + a^2(1 - \cos \theta)^2} \; d\theta$

$= a \displaystyle\int_0^{2\pi} \sqrt{2 - 2 \cos \theta} \; d\theta = a \displaystyle\int_0^{2\pi} \sqrt{4 \sin^2 \dfrac{\theta}{2}} \; d\theta = 2a \displaystyle\int_0^{2\pi} \sin \dfrac{\theta}{2} \; d\theta = \left[ -4a \cos \left( \dfrac{\theta}{2} \right) \right]_0^{2\pi} = 8a$

**99.** $x = t + u = r \cos \theta + r\theta \sin \theta$

$\quad = r(\cos \theta + \theta \sin \theta)$

$y = v - w = r \sin \theta - r\theta \cos \theta$

$\quad = r(\sin \theta - \theta \cos \theta)$

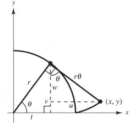

**101.** (a)

(b) $x = \dfrac{1 - t^2}{1 + t^2}, \; y = \dfrac{2t}{1 + t^2}, \; -20 \le t \le 20$

The graph (for $-\infty < t < \infty$) is the circle $x^2 + y^2 = 1$, except the point $(-1, 0)$.

Verify:

$x^2 + y^2 = \left( \dfrac{1 - t^2}{1 + t^2} \right)^2 + \left( \dfrac{2t}{1 + t^2} \right)^2 = \dfrac{1 - 2t^2 + t^4 + 4t^2}{\left( 1 + t^2 \right)^2} = \dfrac{\left( 1 + t^2 \right)^2}{\left( 1 + t^2 \right)^2} = 1$

(c) As $t$ increases from $-20$ to $0$, the speed increases, and as $t$ increases from $0$ to $20$, the speed decreases.

**103.** False. $\dfrac{d^2y}{dx^2} = \dfrac{\dfrac{d}{dt}\left[\dfrac{g'(t)}{f'(t)}\right]}{f'(t)} = \dfrac{f'(t)g''(t) - g'(t)f''(t)}{\left[f'(t)\right]^3}$

**105.** $A = \pi(2)^2 - \pi\left(\dfrac{1}{2}\right)^2 = \left(4 - \dfrac{1}{4}\right)\pi = \dfrac{15}{4}\pi$

$L = \dfrac{\dfrac{15}{4}\pi}{0.001} \approx 11780.97$ in. $\approx 981.7$ ft

# Section 10.4   Polar Coordinates and Polar Graphs

**1.** $\left(8, \dfrac{\pi}{2}\right)$

$x = 8\cos\dfrac{\pi}{2} = 0$

$y = 8\sin\dfrac{\pi}{2} = 8$

$(x, y) = (0, 8)$

**3.** $\left(-4, -\dfrac{3\pi}{4}\right)$

$x = -4\cos\left(\dfrac{-3\pi}{4}\right) = -4\left(-\dfrac{\sqrt{2}}{2}\right) = 2\sqrt{2}$

$y = -4\sin\left(\dfrac{-3\pi}{4}\right) = -4\left(-\dfrac{\sqrt{2}}{2}\right) = 2\sqrt{2}$

$(x, y) = \left(2\sqrt{2}, 2\sqrt{2}\right)$

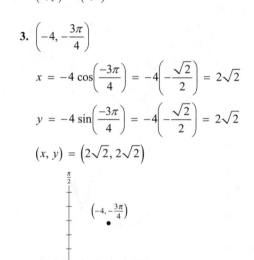

**5.** $\left(\sqrt{2}, 2.36\right)$

$x = \sqrt{2}\cos(2.36) \approx -1.004$

$y = \sqrt{2}\sin(2.36) \approx 0.996$

$(x, y) = (-1.004, 0.996)$

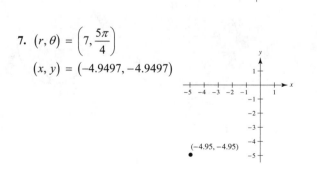

**7.** $(r, \theta) = \left(7, \dfrac{5\pi}{4}\right)$

$(x, y) = (-4.9497, -4.9497)$

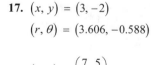
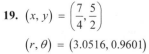

**9.** $(r, \theta) = (-4.5, 3.5)$

$(x, y) = (4.2141, 1.5785)$

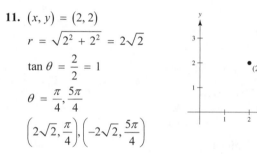

**11.** $(x, y) = (2, 2)$

$r = \sqrt{2^2 + 2^2} = 2\sqrt{2}$

$\tan\theta = \dfrac{2}{2} = 1$

$\theta = \dfrac{\pi}{4}, \dfrac{5\pi}{4}$

$\left(2\sqrt{2}, \dfrac{\pi}{4}\right), \left(-2\sqrt{2}, \dfrac{5\pi}{4}\right)$

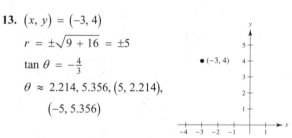

**13.** $(x, y) = (-3, 4)$

$r = \pm\sqrt{9 + 16} = \pm 5$

$\tan\theta = -\dfrac{4}{3}$

$\theta \approx 2.214, 5.356, (5, 2.214),$

$(-5, 5.356)$

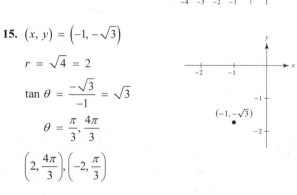

**15.** $(x, y) = \left(-1, -\sqrt{3}\right)$

$r = \sqrt{4} = 2$

$\tan\theta = \dfrac{-\sqrt{3}}{-1} = \sqrt{3}$

$\theta = \dfrac{\pi}{3}, \dfrac{4\pi}{3}$

$\left(2, \dfrac{4\pi}{3}\right), \left(-2, \dfrac{\pi}{3}\right)$

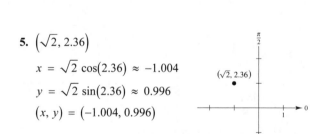

**17.** $(x, y) = (3, -2)$

$(r, \theta) = (3.606, -0.588)$

**19.** $(x, y) = \left(\dfrac{7}{4}, \dfrac{5}{2}\right)$

$(r, \theta) = (3.0516, 0.9601)$

**21.** (a) $(x, y) = (4, 3.5)$

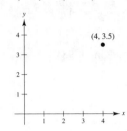

(b) $(r, \theta) = (4, 3.5)$

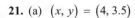

**23.** $r = 2 \sin \theta$ circle

Matches (c)

**25.** $r = 3(1 + \cos \theta)$

Cardioid

Matches (a)

**27.** $x^2 + y^2 = 9$

$r^2 = 9$

$r = 3$

Circle

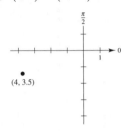

**29.** $x^2 + y^2 = a^2$

$r = a$

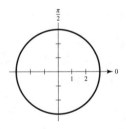

**31.** $y = 8$

$r \sin \theta = 8$

$r = 8 \csc \theta$

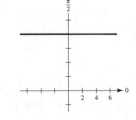

**33.**

$$3x - y + 2 = 0$$
$$3r \cos \theta - r \sin \theta + 2 = 0$$
$$r(3 \cos \theta - \sin \theta) = -2$$
$$r = \frac{-2}{3 \cos \theta - \sin \theta}$$

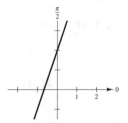

**35.**

$$y^2 = 9x$$
$$r^2 \sin^2 \theta = 9r \cos \theta$$
$$r = \frac{9 \cos \theta}{\sin^2 \theta}$$
$$r = 9 \csc^2 \theta \cos \theta$$

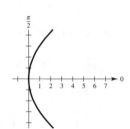

**37.** $r = 4$

$r^2 = 16$

$x^2 + y^2 = 16$

Circle

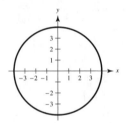

**39.** $r = 3 \sin \theta$

$r^2 = 3r \sin \theta$

$x^2 + y^2 = 3y$

$x^2 + \left(y^2 - 3y + \frac{9}{4}\right) = \frac{9}{4}$

$x^2 + \left(y - \frac{3}{2}\right)^2 = \frac{9}{4}$

Circle

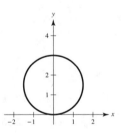

**41.**

$$r = \theta$$
$$\tan r = \tan \theta$$
$$\tan \sqrt{x^2 + y^2} = \frac{y}{x}$$
$$\sqrt{x^2 + y^2} = \arctan \frac{y}{x}$$

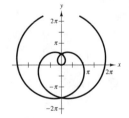

**43.**   $r = 3 \sec \theta$
$r \cos \theta = 3$
$x = 3$
$x - 3 = 0$

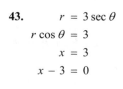

**45.**   $r = \sec \theta \tan \theta$
$r \cos \theta = \tan \theta$
$x = \dfrac{y}{x}$
$y = x^2$

Parabola

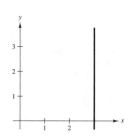

**47.** $r = 2 - 5 \cos \theta$
$0 \le \theta < 2\pi$

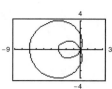

**49.** $r = 2 + \sin \theta$
$0 \le \theta < 2\pi$

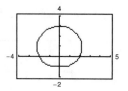

**51.** $r = \dfrac{2}{1 + \cos \theta}$

Traced out once on $-\pi < \theta < \pi$

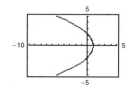

**53.** $r = 2 \cos\left(\dfrac{3\theta}{2}\right)$
$0 \le \theta < 4\pi$

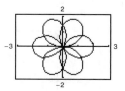

**55.** $r^2 = 4 \sin 2\theta$
$r_1 = 2\sqrt{\sin 2\theta}$
$r_2 = -2\sqrt{\sin 2\theta}$
$0 \le \theta < \dfrac{\pi}{2}$

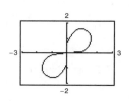

**57.**
$$r = 2(h \cos \theta + k \sin \theta)$$
$$r^2 = 2r(h \cos \theta + k \sin \theta)$$
$$r^2 = 2\big[h(r \cos \theta) + k(r \sin \theta)\big]$$
$$x^2 + y^2 = 2(hx + ky)$$
$$x^2 + y^2 - 2hx - 2ky = 0$$
$$\left(x^2 - 2hx + h^2\right) + \left(y^2 - 2ky + k^2\right) = 0 + h^2 + k^2$$
$$(x - h)^2 + (y - k)^2 = h^2 + k^2$$

Radius: $\sqrt{h^2 + k^2}$

Center: $(h, k)$

**59.** $\left(1, \dfrac{5\pi}{6}\right), \left(4, \dfrac{\pi}{3}\right)$

$$d = \sqrt{1^2 + 4^2 - 2(1)(4) \cos\left(\dfrac{5\pi}{6} - \dfrac{\pi}{3}\right)}$$
$$= \sqrt{17 - 8 \cos \dfrac{\pi}{2}} = \sqrt{17}$$

**61.** $(2, 0.5), (7, 1.2)$

$$d = \sqrt{2^2 + 7^2 - 2(2)(7) \cos(0.5 - 1.2)}$$
$$= \sqrt{53 - 28 \cos(-0.7)} \approx 5.6$$

**63.** $r = 2 + 3 \sin \theta$

$$\dfrac{dy}{dx} = \dfrac{3 \cos \theta \sin \theta + \cos \theta(2 + 3 \sin \theta)}{3 \cos \theta \cos \theta - \sin \theta(2 + 3 \sin \theta)}$$
$$= \dfrac{2 \cos \theta(3 \sin \theta + 1)}{3 \cos 2\theta - 2 \sin \theta} = \dfrac{2 \cos \theta(3 \sin \theta + 1)}{6 \cos^2 \theta - 2 \sin \theta - 3}$$

At $\left(5, \dfrac{\pi}{2}\right)$, $\dfrac{dy}{dx} = 0$.

At $(2, \pi)$, $\dfrac{dy}{dx} = -\dfrac{2}{3}$.

At $\left(-1, \dfrac{3\pi}{2}\right)$, $\dfrac{dy}{dx} = 0$.

**65. (a), (b)** $r = 3(1 - \cos\theta)$

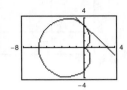

$$(r, \theta) = \left(3, \frac{\pi}{2}\right) \Rightarrow (x, y) = (0, 3)$$

Tangent line: $y - 3 = -1(x - 0)$

$$y = -x + 3$$

**(c)** At $\theta = \frac{\pi}{2}, \frac{dy}{dx} = -1.0.$

**67. (a), (b)** $r = 3\sin\theta$

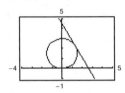

$$(r, \theta) = \left(\frac{3\sqrt{3}}{2}, \frac{\pi}{3}\right) \Rightarrow (x, y) = \left(\frac{3\sqrt{3}}{4}, \frac{9}{4}\right)$$

Tangent line: $y - \frac{9}{4} = -\sqrt{3}\left(x - \frac{3\sqrt{3}}{4}\right)$

$$y = -\sqrt{3}x + \frac{9}{2}$$

**(c)** At $\theta = \frac{\pi}{3}, \frac{dy}{dx} = -\sqrt{3} \approx -1.732.$

**69.**   $r = 1 - \sin\theta$

$$\frac{dy}{d\theta} = (1 - \sin\theta)\cos\theta - \cos\theta\sin\theta$$

$$= \cos\theta(1 - 2\sin\theta) = 0$$

$$\cos\theta = 0 \text{ or } \sin\theta = \frac{1}{2} \Rightarrow \theta = \frac{\pi}{2}, \frac{3\pi}{2}, \frac{\pi}{6}, \frac{5\pi}{6}$$

Horizontal tangents: $\left(2, \frac{3\pi}{2}\right), \left(\frac{1}{2}, \frac{\pi}{6}\right), \left(\frac{1}{2}, \frac{5\pi}{6}\right)$

$$\frac{dx}{d\theta} = (-1 + \sin\theta)\sin\theta - \cos\theta\cos\theta$$

$$= -\sin\theta + \sin^2\theta + \sin^2\theta - 1$$

$$= 2\sin^2\theta - \sin\theta - 1$$

$$= (2\sin\theta + 1)(\sin\theta - 1) = 0$$

$$\sin\theta = 1 \text{ or } \sin\theta = -\frac{1}{2} \Rightarrow \theta = \frac{\pi}{2}, \frac{7\pi}{6}, \frac{11\pi}{6}$$

Vertical tangents: $\left(\frac{3}{2}, \frac{7\pi}{6}\right), \left(\frac{3}{2}, \frac{11\pi}{6}\right)$

**71.**   $r = 2\csc\theta + 3$

$$\frac{dy}{d\theta} = (2\csc\theta + 3)\cos\theta + (-2\csc\theta\cot\theta)\sin\theta$$

$$= 3\cos\theta = 0$$

$$\theta = \frac{\pi}{2}, \frac{3\pi}{2}$$

Horizontal: $\left(5, \frac{\pi}{2}\right), \left(1, \frac{3\pi}{2}\right)$

**73.** $r = 4\sin\theta\cos^2\theta$

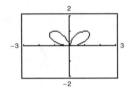

Horizontal tangents:

$$(r, \theta) = (0, 0), (1.4142, 0.7854), (1.4142, 2.3562)$$

**75.** $r = 2\csc\theta + 5$

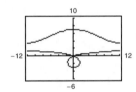

Horizontal tangents: $(r, \theta) = \left(7, \frac{\pi}{2}\right), \left(3, \frac{3\pi}{2}\right)$

**77.**   $r = 5\sin\theta$

$$r^2 = 5r\sin\theta$$

$$x^2 + y^2 = 5y$$

$$x^2 + \left(y^2 - 5y + \frac{25}{4}\right) = \frac{25}{4}$$

$$x^2 + \left(y - \frac{5}{2}\right)^2 = \frac{25}{4}$$

Circle: center: $\left(0, \frac{5}{2}\right)$, radius: $\frac{5}{2}$

Tangent at pole: $\theta = 0$

Note: $f(\theta) = r = 5\sin\theta$

$$f(0) = 0, f'(0) \neq 0$$

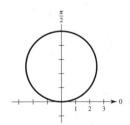

**79.** $r = 2(1 - \sin\theta)$

Cardioid

Symmetric to $y$-axis, $\theta = \dfrac{\pi}{2}$

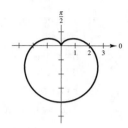

**81.** $r = 4\cos 3\theta$

Rose curve with three petals.

Tangents at pole: $(r = 0, r' \neq 0)$:

$\theta = \dfrac{\pi}{6}, \dfrac{\pi}{2}, \dfrac{5\pi}{6}$

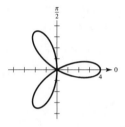

**83.** $r = 3\sin 2\theta$

Rose curve with four petals

Symmetric to the polar axis, $\theta = \dfrac{\pi}{2}$, and pole

Relative extrema: $\left(\pm 3, \dfrac{\pi}{4}\right), \left(\pm 3, \dfrac{5\pi}{4}\right)$

Tangents at the pole: $\theta = 0, \dfrac{\pi}{2}$

$\left(\theta = \pi, \dfrac{3\pi}{2} \text{ give the same tangents.}\right)$

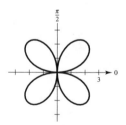

**85.** $r = 8$

Circle radius 8

$x^2 + y^2 = 64$

**87.** $r = 4(1 + \cos\theta)$

Cardioid

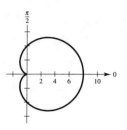

**89.** $r = 3 - 2\cos\theta$

Limaçon

Symmetric to polar axis

| $\theta$ | 0 | $\dfrac{\pi}{3}$ | $\dfrac{\pi}{2}$ | $\dfrac{2\pi}{3}$ | $\pi$ |
|---|---|---|---|---|---|
| $r$ | 1 | 2 | 3 | 4 | 5 |

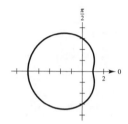

**91.**  $r = 3\csc\theta$

$r\sin\theta = 3$

$y = 3$

Horizontal line

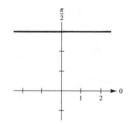

**93.** $r = 2\theta$

Spiral of Archimedes

Symmetric to $\theta = \dfrac{\pi}{2}$

| $\theta$ | 0 | $\dfrac{\pi}{4}$ | $\dfrac{\pi}{2}$ | $\dfrac{3\pi}{4}$ | $\pi$ | $\dfrac{5\pi}{4}$ | $\dfrac{3\pi}{2}$ |
|---|---|---|---|---|---|---|---|
| $r$ | 0 | $\dfrac{\pi}{2}$ | $\pi$ | $\dfrac{3\pi}{2}$ | $2\pi$ | $\dfrac{5\pi}{2}$ | $3\pi$ |

Tangent at the pole: $\theta = 0$

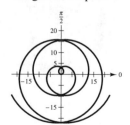

**95.** $r^2 = 4\cos(2\theta)$

$r = 2\sqrt{\cos 2\theta}, \quad 0 \le \theta \le 2\pi$

Lemniscate

Symmetric to the polar axis, $\theta = \dfrac{\pi}{2}$, and pole

Relative extrema: $(\pm 2, 0)$

| $\theta$ | 0 | $\dfrac{\pi}{6}$ | $\dfrac{\pi}{4}$ |
|---|---|---|---|
| $r$ | $\pm 2$ | $\pm\sqrt{2}$ | 0 |

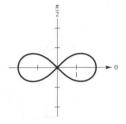

Tangents at the pole:

$\theta = \dfrac{\pi}{4}, \dfrac{3\pi}{4}$

**97.** Because

$r = 2 - \sec\theta = 2 - \dfrac{1}{\cos\theta},$

the graph has polar axis symmetry and the tangents at the pole are

$\theta = \dfrac{\pi}{3}, -\dfrac{\pi}{3}.$

Furthermore,

$r \Rightarrow -\infty$ as $\theta \Rightarrow \dfrac{\pi}{2^-}$

$r \Rightarrow \infty$ as $\theta \Rightarrow -\dfrac{\pi}{2^+}.$

Also,

$r = 2 - \dfrac{1}{\cos\theta}$

$= 2 - \dfrac{r}{r\cos\theta} = 2 - \dfrac{r}{x}$

$rx = 2x - r$

$r = \dfrac{2x}{1 + x}.$

So, $r \Rightarrow \pm\infty$ as $x \Rightarrow -1.$

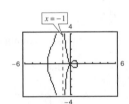

**99.** $r = \dfrac{2}{\theta}$

Hyperbolic spiral

$r \Rightarrow \infty$ as $\theta \Rightarrow 0$

$r = \dfrac{2}{\theta} \Rightarrow \theta = \dfrac{2}{r} = \dfrac{2\sin\theta}{r\sin\theta} = \dfrac{2\sin\theta}{y}$

$y = \dfrac{2\sin\theta}{\theta}$

$\lim_{\theta\to 0} \dfrac{2\sin\theta}{\theta} = \lim_{\theta\to 0} \dfrac{2\cos\theta}{1}$

$= 2$

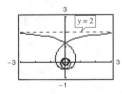

**101.** The rectangular coordinate system consists of all points of the form $(x, y)$ where $x$ is the directed distance from the $y$-axis to the point, and $y$ is the directed distance from the $x$-axis to the point.

Every point has a unique representation.

The polar coordinate system uses $(r, \theta)$ to designate the location of a point.

$r$ is the directed distance to the origin and $\theta$ is the angle the point makes with the positive $x$-axis, measured counterclockwise.

Points do not have a unique polar representation.

**103.** Slope of tangent line to graph of $r = f(\theta)$ at $(r, \theta)$ is

$\dfrac{dy}{dx} = \dfrac{f(\theta)\cos\theta + f'(\theta)\sin\theta}{-f(\theta)\sin\theta + f'(\theta)\cos\theta}.$

If $f(\alpha) = 0$ and $f'(\alpha) \ne 0$, then $\theta = \alpha$ is tangent at the pole.

**105.** $r = 4\sin\theta$

(a) $0 \le \theta \le \dfrac{\pi}{2}$

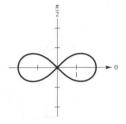

(b) $\dfrac{\pi}{2} \le \theta \le \pi$

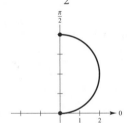

(c) $-\dfrac{\pi}{2} \le \theta \le \dfrac{\pi}{2}$

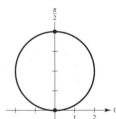

**107.** Let the curve $r = f(\theta)$ be rotated by $\phi$ to form the curve $r = g(\theta)$. If $(r_1, \theta_1)$ is a point on $r = f(\theta)$, then $(r_1, \theta_1 + \phi)$ is on $r = g(\theta)$. That is,

$$g(\theta_1 + \phi) = r_1 = f(\theta_1).$$

Letting $\theta = \theta_1 + \phi$, or $\theta_1 = \theta - \phi$, you see that

$$g(\theta) = g(\theta_1 + \phi) = f(\theta_1) = f(\theta - \phi).$$

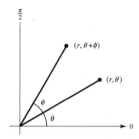

**109.** $r = 2 - \sin \theta$

(a) $r = 2 - \sin\left(\theta - \dfrac{\pi}{4}\right) = 2 - \dfrac{\sqrt{2}}{2}(\sin \theta - \cos \theta)$

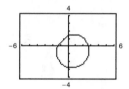

(b) $r = 2 - \sin\left(\theta - \dfrac{\pi}{2}\right) = 2 - (-\cos \theta) = 2 + \cos \theta$

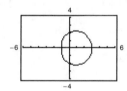

(c) $r = 2 - \sin(\theta - \pi) = 2 - (-\sin \theta) = 2 + \sin \theta$

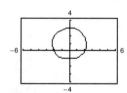

(d) $r = 2 - \sin\left(\theta - \dfrac{3\pi}{2}\right) = 2 - \cos \theta$

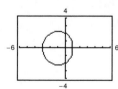

**111.** (a) $r = 1 - \sin \theta$

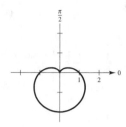

(b) $r = 1 - \sin\left(\theta - \dfrac{\pi}{4}\right)$

Rotate the graph of

$r = 1 - \sin \theta$

through the angle $\pi/4$.

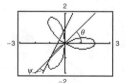

**113.** $\tan \psi = \dfrac{r}{dr/d\theta} = \dfrac{2(1 - \cos \theta)}{2 \sin \theta}$

At $\theta = \pi$, $\tan \psi$ is undefined $\Rightarrow \psi = \dfrac{\pi}{2}$.

**115.** $r = 2 \cos 3\theta$

$\tan \psi = \dfrac{r}{dr/d\theta} = \dfrac{2 \cos 3\theta}{-6 \sin 3\theta} = -\dfrac{1}{3} \cot 3\theta$

At $\theta = \dfrac{\pi}{4}$, $\tan \psi = -\dfrac{1}{3} \cot\left(\dfrac{3\pi}{4}\right) = \dfrac{1}{3}$.

$\psi = \arctan\left(\dfrac{1}{3}\right) \approx 18.4°$

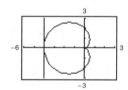

**117.** $r = \dfrac{6}{1 - \cos\theta} = 6(1 - \cos\theta)^{-1} \Rightarrow \dfrac{dr}{d\theta} = \dfrac{6\sin\theta}{(1 - \cos\theta)^2}$

$\tan\psi = \dfrac{r}{\dfrac{dr}{d\theta}} = \dfrac{\dfrac{6}{(1 - \cos\theta)}}{\dfrac{-6\sin\theta}{(1 - \cos\theta)^2}} = \dfrac{1 - \cos\theta}{-\sin\theta}$

At $\theta = \dfrac{2\pi}{3}$, $\tan\psi = \dfrac{1 - \left(-\dfrac{1}{2}\right)}{-\dfrac{\sqrt{3}}{2}} = -\sqrt{3}.$

$\psi = \dfrac{\pi}{3}, (60°)$

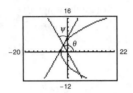

**119.** True

**121.** True

# Section 10.5   Area and Arc Length in Polar Coordinates

**1.** $A = \dfrac{1}{2}\int_\alpha^\beta \left[f(\theta)\right]^2 d\theta$

$= \dfrac{1}{2}\int_0^{\pi/2} \left[4\sin\theta\right]^2 d\theta = 8\int_0^{\pi/2}\sin^2\theta\, d\theta$

**3.** $A = \dfrac{1}{2}\int_\alpha^\beta \left[f(\theta)\right]^2 d\theta = \dfrac{1}{2}\int_{\pi/2}^{3\pi/2}\left[3 - 2\sin\theta\right]^2 d\theta$

**5.** $A = \dfrac{1}{2}\int_0^\pi \left[6\sin\theta\right]^2 d\theta$

$= 18\int_0^\pi \dfrac{1 - \cos 2\theta}{2}\, d\theta = 9\left[\theta - \dfrac{\sin 2\theta}{2}\right]_0^\pi = 9\pi$

Note: $r = 6\sin\theta$ is circle of radius 3, $0 \le \theta \le \pi$.

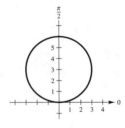

**7.** $A = 2\left[\dfrac{1}{2}\int_0^{\pi/6}(2\cos 3\theta)^2 d\theta\right] = 2\left[\theta + \dfrac{1}{6}\sin 6\theta\right]_0^{\pi/6} = \dfrac{\pi}{3}$

**9.** $A = \dfrac{1}{2}\int_0^{\pi/2}\left[\sin 2\theta\right]^2 d\theta = \dfrac{1}{2}\int_0^{\pi/2}\dfrac{1 - \cos 4\theta}{2}\, d\theta$

$= \dfrac{1}{4}\left[\theta - \dfrac{\sin 4\theta}{4}\right]_0^{\pi/2} = \dfrac{1}{4}\left[\dfrac{\pi}{2}\right] = \dfrac{\pi}{8}$

**11.** $A = 2\left[\dfrac{1}{2}\int_{-\pi/2}^{\pi/2}(1 - \sin\theta)^2 d\theta\right]$

$= \left[\dfrac{3}{2}\theta + 2\cos\theta - \dfrac{1}{4}\sin 2\theta\right]_{-\pi/2}^{\pi/2} = \dfrac{3\pi}{2}$

**13.** $A = \dfrac{1}{2}\int_0^{2\pi}\left[5 + 2\sin\theta\right]^2 d\theta$

$= \dfrac{1}{2}\int_0^{2\pi}\left[25 + 20\sin\theta + 4\sin^2\theta\right] d\theta$

$= \dfrac{1}{2}\int_0^{2\pi}\left[25 + 20\sin\theta + 2(1 - \cos 2\theta)\right] d\theta$

$= \dfrac{1}{2}\left[27\theta - 20\cos\theta - \sin 2\theta\right]_0^{2\pi}$

$= \dfrac{1}{2}\left[27(2\pi)\right] = 27\pi$

**15.** On the interval $-\dfrac{\pi}{4} \le \theta \le 0$, $r = 2\sqrt{\cos 2\theta}$ traces out one-half of one leaf of the lemniscate. So,

$A = 4\dfrac{1}{2}\int_{-\pi/4}^0 4\cos 2\theta\, d\theta$

$= 8\left[\dfrac{\sin 2\theta}{2}\right]_{-\pi/4}^0 = 8\left[\dfrac{1}{2}\right] = 4.$

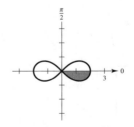

**17.** $A = \left[2\dfrac{1}{2}\displaystyle\int_{2\pi/3}^{\pi}\left(1 + 2\cos\theta\right)^2 d\theta\right]$

$= \left[3\theta + 4\sin\theta + \sin 2\theta\right]_{2\pi/3}^{\pi} = \dfrac{2\pi - 3\sqrt{3}}{2}$

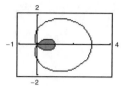

**19.** The inner loop of $r = 1 + 2\sin\theta$ is traced out on the

interval $\dfrac{7\pi}{6} \le \theta \le \dfrac{11\pi}{6}$. So,

$A = \dfrac{1}{2}\displaystyle\int_{7\pi/6}^{11\pi/6}\left[1 + 2\sin\theta\right]^2 d\theta$

$= \dfrac{1}{2}\displaystyle\int_{7\pi/6}^{11\pi/6}\left[1 + 4\sin\theta + 4\sin^2\theta\right] d\theta$

$= \dfrac{1}{2}\displaystyle\int_{7\pi/6}^{11\pi/6}\left[1 + 4\sin\theta + 2(1 - \cos 2\theta)\right] d\theta$

$= \dfrac{1}{2}\left[3\theta - 4\cos\theta - \sin 2\theta\right]_{7\pi/6}^{11\pi/6}$

$= \dfrac{1}{2}\left[\left(\dfrac{11\pi}{2} - 2\sqrt{3} + \dfrac{\sqrt{3}}{2}\right) - \left(\dfrac{7\pi}{2} + 2\sqrt{3} - \dfrac{\sqrt{3}}{2}\right)\right]$

$= \dfrac{1}{2}\left[2\pi - 3\sqrt{3}\right].$

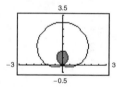

**21.** The area inside the outer loop is

$2\left[\dfrac{1}{2}\displaystyle\int_{0}^{2\pi/3}\left(1 + 2\cos\theta\right)^2 d\theta\right] = \left[3\theta + 4\sin\theta + \sin 2\theta\right]_{0}^{2\pi/3}$

$= \dfrac{4\pi + 3\sqrt{3}}{2}.$

From the result of Exercise 17, the area between the loops is

$A = \left(\dfrac{4\pi + 3\sqrt{3}}{2}\right) - \left(\dfrac{2\pi - 3\sqrt{3}}{2}\right) = \pi + 3\sqrt{3}.$

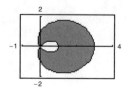

**23.** The area inside the outer loop is

$A = 2 \cdot \dfrac{1}{2}\displaystyle\int_{5\pi/6}^{3\pi/2}\left[3 - 6\sin\theta\right]^2 d\theta$

$= \displaystyle\int_{5\pi/6}^{3\pi/2}\left[9 - 36\sin\theta + 36\sin^2\theta\right] d\theta$

$= \displaystyle\int_{5\pi/6}^{3\pi/2}\left[9 - 36\sin\theta + 18(1 - \cos 2\theta)\right] d\theta$

$= \left[27\theta + 36\cos\theta - 9\sin 2\theta\right]_{5\pi/6}^{3\pi/2}$

$= \left[\dfrac{81\pi}{2} - \left(\dfrac{45\pi}{2} - 18\sqrt{3} + \dfrac{9\sqrt{3}}{2}\right)\right]$

$= 18\pi + \dfrac{27\sqrt{3}}{2}.$

The area inside the inner loop is

$A = 2 \cdot \dfrac{1}{2}\displaystyle\int_{\pi/6}^{\pi/2}\left[3 - 6\sin\theta\right]^2 d\theta$

$= \left[27\theta + 36\cos\theta - 9\sin 2\theta\right]_{\pi/6}^{\pi/2}$

$= \left[\dfrac{27\pi}{2} - \left(\dfrac{9\pi}{2} + 18\sqrt{3} - \dfrac{9\sqrt{3}}{2}\right)\right]$

$= 9\pi - \dfrac{27\sqrt{3}}{2}.$

Finally, the area between the loops is

$\left[18\pi + \dfrac{27\sqrt{3}}{2}\right] - \left[9\pi - \dfrac{27\sqrt{3}}{2}\right] = 9\pi + 27\sqrt{3}.$

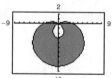

**25.** $r = 1 + \cos\theta$

$r = 1 - \cos\theta$

Solving simultaneously,

$1 + \cos\theta = 1 - \cos\theta$

$2\cos\theta = 0$

$\theta = \dfrac{\pi}{2}, \dfrac{3\pi}{2}.$

Replacing $r$ by $-r$ and $\theta$ by $\theta + \pi$ in the first equation and solving, $-1 + \cos\theta = 1 - \cos\theta$, $\cos\theta = 1$,

$\theta = 0$. Both curves pass through the pole, $(0, \pi)$, and

$(0, 0)$, respectively.

Points of intersection: $\left(1, \dfrac{\pi}{2}\right), \left(1, \dfrac{3\pi}{2}\right), (0, 0)$

**27.** $r = 1 + \cos \theta$

$r = 1 - \sin \theta$

Solving simultaneously,

$1 + \cos \theta = 1 - \sin \theta$

$\cos \theta = -\sin \theta$

$\tan \theta = -1$

$\theta = \dfrac{3\pi}{4}, \dfrac{7\pi}{4}.$

Replacing $r$ by $-r$ and $\theta$ by $\theta + \pi$ in the first equation and solving, $-1 + \cos \theta = 1 - \sin \theta$,

$\sin \theta + \cos \theta = 2$, which has no solution. Both curves pass through the pole, $(0, \pi)$, and $(0, \pi/2)$, respectively.

Points of intersection:

$\left(\dfrac{2 - \sqrt{2}}{2}, \dfrac{3\pi}{4}\right), \left(\dfrac{2 + \sqrt{2}}{2}, \dfrac{7\pi}{4}\right), (0, 0)$

**29.** $r = 4 - 5 \sin \theta$

$r = 3 \sin \theta$

Solving simultaneously,

$4 - 5 \sin \theta = 3 \sin \theta$

$\sin \theta = \dfrac{1}{2}$

$\theta = \dfrac{\pi}{6}, \dfrac{5\pi}{6}.$

Both curves pass through the pole, $(0, \arcsin 4/5)$, and $(0, 0)$, respectively.

Points of intersection: $\left(\dfrac{3}{2}, \dfrac{\pi}{6}\right), \left(\dfrac{3}{2}, \dfrac{5\pi}{6}\right), (0, 0)$

**31.** $r = \dfrac{\theta}{2}$

$r = 2$

Solving simultaneously, you have

$\theta/2 = 2, \theta = 4.$

Points of intersection:

$(2, 4), (-2, -4)$

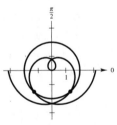

**33.** $r = 2 \sin 2\theta$

$r = 1$

$r = 2 \sin 2\theta$ is a 4-leaved rose curve. The circle $r = 1$ intersects at 8 points. For the petal in the first quadrant,

$2 \sin 2\theta = 1$

$2\theta = \dfrac{\pi}{6}, \dfrac{5\pi}{6}$

$\theta = \dfrac{\pi}{12}, \dfrac{5\pi}{12}$

The points of intersection for one petal are $\left(1, \dfrac{\pi}{12}\right),$

$\left(1, \dfrac{5\pi}{12}\right).$ By symmetry, the other points are

$\left(1, \dfrac{7\pi}{12}\right), \left(1, \dfrac{11\pi}{12}\right), \left(1, \dfrac{13\pi}{12}\right), \left(1, \dfrac{17\pi}{12}\right),$

$\left(1, \dfrac{19\pi}{12}\right), \left(1, \dfrac{23\pi}{12}\right).$

**35.** $r = 2 + 3 \cos \theta$

$r = \dfrac{\sec \theta}{2}$

The graph of $r = 2 + 3 \cos \theta$ is a limaçon with an inner loop $(b > a)$ and is symmetric to the polar axis. The graph of $r = (\sec \theta)/2$ is the vertical line $x = 1/2$. So, there are four points of intersection. Solving simultaneously,

$2 + 3 \cos \theta = \dfrac{\sec \theta}{2}$

$6 \cos^2 \theta + 4 \cos \theta - 1 = 0$

$\cos \theta = \dfrac{-2 \pm \sqrt{10}}{6}$

$\theta = \arccos\left(\dfrac{-2 + \sqrt{10}}{6}\right) \approx 1.376$

$\theta = \arccos\left(\dfrac{-2 - \sqrt{10}}{6}\right) \approx 2.6068.$

Points of intersection:

$(-0.581, \pm 2.607), (2.581, \pm 1.376)$

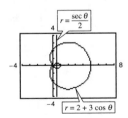

**37.** $r = \cos\theta$

$r = 2 - 3\sin\theta$

Points of intersection:

$(0, 0), (0.935, 0.363), (0.535, -1.006)$

The graphs reach the pole at different times ($\theta$ values).

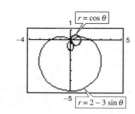

**39.** From Exercise 25, the points of intersection for one petal are $(2, \pi/12)$ and $(2, 5\pi/12)$. The area within one petal is

$$A = \frac{1}{2}\int_0^{\pi/12} (4\sin 2\theta)^2\, d\theta + \frac{1}{2}\int_{\pi/12}^{5\pi/12} (2)^2\, d\theta + \frac{1}{2}\int_{5\pi/12}^{\pi/2} (4\sin 2\theta)^2\, d\theta$$

$$= 16\int_0^{\pi/12} \sin^2(2\theta)\, d\theta + 2\int_{\pi/12}^{5\pi/12} d\theta \;\left(\text{by symmetry of the petal}\right)$$

$$= 8\left[\theta - \frac{1}{4}\sin 4\theta\right]_0^{\pi/12} + \left[2\theta\right]_{\pi/12}^{5\pi/12} = \frac{4\pi}{3} - \sqrt{3}.$$

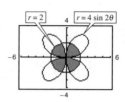

Total area $= 4\left(\dfrac{4\pi}{3} - \sqrt{3}\right) = \dfrac{16\pi}{3} - 4\sqrt{3} = \dfrac{4}{3}\left(4\pi - 3\sqrt{3}\right)$

**41.** $A = 4\left[\dfrac{1}{2}\displaystyle\int_0^{\pi/2} (3 - 2\sin\theta)^2\, d\theta\right]$

$$= 2\left[11\theta + 12\cos\theta - \sin(2\theta)\right]_0^{\pi/2} = 11\pi - 24$$

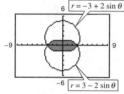

**43.** $A = 2\left[\dfrac{1}{2}\displaystyle\int_0^{\pi/6} (4\sin\theta)^2\, d\theta + \dfrac{1}{2}\displaystyle\int_{\pi/6}^{\pi/2} (2)^2\, d\theta\right]$

$$= 16\left[\frac{1}{2}\theta - \frac{1}{4}\sin(2\theta)\right]_0^{\pi/6} + \left[4\theta\right]_{\pi/6}^{\pi/2}$$

$$= \frac{8\pi}{3} - 2\sqrt{3} = \frac{2}{3}\left(4\pi - 3\sqrt{3}\right)$$

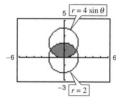

**45.** $r = 2\cos\theta = 1 \Rightarrow \theta = \pi/3$

$$A = 2\cdot\frac{1}{2}\int_0^{\pi/3}\left([2\cos\theta]^2 - 1\right) d\theta$$

$$= \int_0^{\pi/3}\left[2(1 + \cos 2\theta) - 1\right] d\theta$$

$$= \left[\theta + \sin 2\theta\right]_0^{\pi/3}$$

$$= \frac{\pi}{3} + \frac{\sqrt{3}}{2}$$

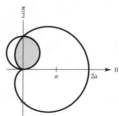

**47.** $A = 2\left[\dfrac{1}{2}\displaystyle\int_0^{\pi} \left[a(1 + \cos\theta)\right]^2 d\theta\right] - \dfrac{a^2\pi}{4}$

$$= a^2\left[\frac{3}{2}\theta + 2\sin\theta + \frac{\sin 2\theta}{4}\right]_0^{\pi} - \frac{a^2\pi}{4}$$

$$= \frac{3a^2\pi}{2} - \frac{a^2\pi}{4} = \frac{5a^2\pi}{4}$$

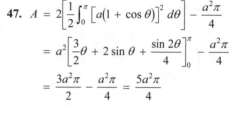

**49.** $A = \dfrac{\pi a^2}{8} + \dfrac{1}{2}\displaystyle\int_{\pi/2}^{\pi}\left[a(1 + \cos\theta)\right]^2 d\theta$

$$= \frac{\pi a^2}{8} + \frac{a^2}{2}\int_{\pi/2}^{\pi}\left(\frac{3}{2} + 2\cos\theta + \frac{\cos 2\theta}{2}\right) d\theta$$

$$= \frac{\pi a^2}{8} + \frac{a^2}{2}\left[\frac{3}{2}\theta + 2\sin\theta + \frac{\sin 2\theta}{4}\right]_{\pi/2}^{\pi}$$

$$= \frac{\pi a^2}{8} + \frac{a^2}{2}\left[\frac{3\pi}{2} - \frac{3\pi}{4} - 2\right] = \frac{a^2}{2}[\pi - 2]$$

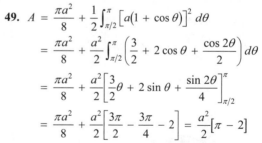

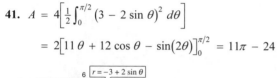

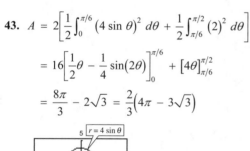

**51.** (a)  $r = a \cos^2 \theta$

$$r^3 = ar^2 \cos^2 \theta$$

$$\left(x^2 + y^2\right)^{3/2} = ax^2$$

(b)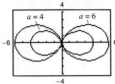

(c)  $A = 4\left(\dfrac{1}{2}\right)\displaystyle\int_0^{\pi/2}\left[\left(6\cos^2\theta\right)^2 - \left(4\cos^2\theta\right)^2\right]d\theta = 40\int_0^{\pi/2}\cos^4\theta\,d\theta$

$$= 10\int_0^{\pi/2}\left(1+\cos 2\theta\right)^2 d\theta = 10\int_0^{\pi/2}\left(1 + 2\cos 2\theta + \frac{1-\cos 4\theta}{2}\right)d\theta$$

$$= 10\left[\frac{3}{2}\theta + \sin 2\theta + \frac{1}{8}\sin 4\theta\right]_0^{\pi/2} = \frac{15\pi}{2}$$

**53.** $r = a\cos(n\theta)$

For $n = 1$:

$r = a\cos\theta$

$A = \pi\left(\dfrac{a}{2}\right)^2 = \dfrac{\pi a^2}{4}$

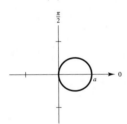

For $n = 2$:

$r = a\cos 2\theta$

$A = 8\left(\dfrac{1}{2}\right)\displaystyle\int_0^{\pi/4}\left(a\cos 2\theta\right)^2 d\theta = \dfrac{\pi a^2}{2}$

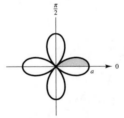

For $n = 3$:

$r = a\cos 3\theta$

$A = 6\left(\dfrac{1}{2}\right)\displaystyle\int_0^{\pi/6}\left(a\cos 3\theta\right)^2 d\theta = \dfrac{\pi a^2}{4}$

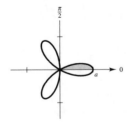

For $n = 4$:

$r = a\cos 4\theta$

$A = 16\left(\dfrac{1}{2}\right)\displaystyle\int_0^{\pi/8}\left(a\cos 4\theta\right)^2 d\theta = \dfrac{\pi a^2}{2}$

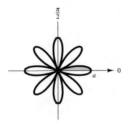

In general, the area of the region enclosed by $r = a\cos(n\theta)$ for $n = 1, 2, 3, \ldots$ is $\left(\pi a^2\right)\big/4$ if $n$ is odd and is $\left(\pi a^2\right)\big/2$ if $n$ is even.

**55.** $r = 8, r^1 = 0$

$$s = \int_0^{2\pi} \sqrt{8^2 + 0^2}\, d\theta = 8\theta \Big]_0^{2\pi} = 16\pi$$

(circumference of circle of radius 8)

**57.** $r = 4 \sin \theta$

$r' = 4 \cos \theta$

$$s = \int_0^{\pi} \sqrt{(4 \sin \theta)^2 + (4 \cos \theta)^2}\, d\theta$$

$$= \int_0^{\pi} 4\, d\theta = [4\theta]_0^{\pi} = 4\pi$$

(circumference of circle of radius 2)

**59.** $r = 1 + \sin \theta$

$r' = \cos \theta$

$$s = 2\int_{\pi/2}^{3\pi/2} \sqrt{(1 + \sin \theta)^2 + (\cos \theta)^2}\, d\theta$$

$$= 2\sqrt{2}\int_{\pi/2}^{3\pi/2} \sqrt{1 + \sin \theta}\, d\theta$$

$$= 2\sqrt{2}\int_{\pi/2}^{3\pi/2} \frac{-\cos \theta}{\sqrt{1 - \sin \theta}}\, d\theta$$

$$= \left[4\sqrt{2}\sqrt{1 - \sin \theta}\right]_{\pi/2}^{3\pi/2}$$

$$= 4\sqrt{2}\left(\sqrt{2} - 0\right) = 8$$

**61.** $r = 2\theta, 0 \le \theta \le \dfrac{\pi}{2}$

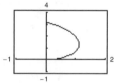

Length $\approx 4.16$

**63.** $r = \dfrac{1}{\theta}, \pi \le \theta \le 2\pi$

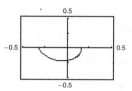

Length $\approx 0.71$

**65.** $r = \sin(3 \cos \theta), 0 \le \theta \le \pi$

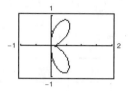

Length $\approx 4.39$

**67.** $r = 6 \cos \theta$

$r' = -6 \sin \theta$

$$S = 2\pi \int_0^{\pi/2} 6 \cos \theta \sin \theta \sqrt{36 \cos^2 \theta + 36 \sin^2 \theta}\, d\theta$$

$$= 72\pi \int_0^{\pi/2} \sin \theta \cos \theta\, d\theta$$

$$= \left[36\pi \sin^2 \theta\right]_0^{\pi/2}$$

$$= 36\pi$$

**69.** $r = e^{a\theta}$

$r' = ae^{a\theta}$

$$S = 2\pi \int_0^{\pi/2} e^{a\theta} \cos \theta \sqrt{\left(e^{a\theta}\right)^2 + \left(ae^{a\theta}\right)^2}\, d\theta$$

$$= 2\pi\sqrt{1 + a^2} \int_0^{\pi/2} e^{2a\theta} \cos \theta\, d\theta$$

$$= 2\pi\sqrt{1 + a^2} \left[\frac{e^{2a\theta}}{4a^2 + 1}(2a \cos \theta + \sin \theta)\right]_0^{\pi/2}$$

$$= \frac{2\pi\sqrt{1 + a^2}}{4a^2 + 1}\left(e^{\pi a} - 2a\right)$$

**71.** $r = 4 \cos 2\theta$

$r' = -8 \sin 2\theta$

$$S = 2\pi \int_0^{\pi/4} 4 \cos 2\theta \sin \theta \sqrt{16 \cos^2 2\theta + 64 \sin^2 \theta\ 2\theta}\, d\theta = 32\pi \int_0^{\pi/4} \cos 2\theta \sin \theta \sqrt{\cos^2 2\theta + 4 \sin^2 2\theta}\, d\theta \approx 21.87$$

**73.** You will only find simultaneous points of intersection. There may be intersection points that do not occur with the same coordinates in the two graphs.

**75.** (a) $S = 2\pi \int_{\alpha}^{\beta} f(\theta) \sin \theta \sqrt{f(\theta)^2 + f'(\theta)^2}\, d\theta$

   (b) $S = 2\pi \int_{\alpha}^{\beta} f(\theta) \cos \theta \sqrt{f(\theta)^2 + f'(\theta)^2}\, d\theta$

**77.** Revolve $r = 2$ about the line $r = 5 \sec \theta$.

$$f(\theta) = 2, \; f'(\theta) = 0$$

$$S = 2\pi \int_0^{2\pi} (5 - 2\cos\theta)\sqrt{2^2 + 0^2} \; d\theta$$

$$= 4\pi \int_0^{2\pi} (5 - 2\cos\theta) \; d\theta$$

$$= 4\pi \left[ 5\theta - 2\sin\theta \right]_0^{2\pi} = 40\pi^2$$

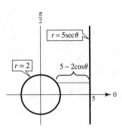

**79.** $r = 8\cos\theta, \; 0 \le \theta \le \pi$

(a) $A = \dfrac{1}{2}\displaystyle\int_0^\pi r^2 \, d\theta = \dfrac{1}{2}\int_0^\pi 64\cos^2\theta \, d\theta = 32\int_0^\pi \dfrac{1 + \cos 2\theta}{2} \, d\theta = 16\left[\theta + \dfrac{\sin 2\theta}{2}\right]_0^\pi = 16\pi$

$\left(\text{Area circle} = \pi r^2 = \pi 4^2 = 16\pi\right)$

(b)

| $\theta$ | 0.2 | 0.4 | 0.6 | 0.8 | 1.0 | 1.2 | 1.4 |
|---|---|---|---|---|---|---|---|
| $A$ | 6.32 | 12.14 | 17.06 | 20.80 | 23.27 | 24.60 | 25.08 |

(c), (d) For $\dfrac{1}{4}$ of area $(4\pi \approx 12.57)$: 0.42

For $\dfrac{1}{2}$ of area $(8\pi \approx 25.13)$: $1.57\left(\dfrac{\pi}{2}\right)$

For $\dfrac{3}{4}$ of area $(12\pi \approx 37.70)$: 2.73

(e) No, it does not depend on the radius.

**81.**
$$r = a\sin\theta + b\cos\theta$$
$$r^2 = ar\sin\theta + br\cos\theta$$
$$x^2 + y^2 = ay + bx$$
$$x^2 + y^2 - bx - ay = 0 \text{ represents a circle.}$$

**83.** (a) $r = \theta, \; \theta \ge 0$

As $a$ increases, the spiral opens more rapidly. If $\theta < 0$, the spiral is reflected about the $y$-axis.

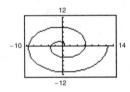

(b) $r = a\theta, \; \theta \ge 0$, crosses the polar axis for $\theta = n\pi, \; n$ and integer. To see this

$r = a\theta \Rightarrow r\sin\theta = y = a\theta\sin\theta = 0$

for $\theta = n\pi$. The points are $(r, \theta) = (an\pi, n\pi), \; n = 1, 2, 3, \ldots$.

(c) $f(\theta) = \theta, \; f'(\theta) = 1$

$$s = \int_0^{2\pi} \sqrt{\theta^2 + 1} \, d\theta = \dfrac{1}{2}\left[\ln\left(\sqrt{x^2 + 1} + x\right) + x\sqrt{x^2 + 1}\right]_0^{2\pi}$$

$$= \dfrac{1}{2}\ln\left(\sqrt{4\pi^2 + 1} + 2\pi\right) + \pi\sqrt{4\pi^2 + 1} \approx 21.2563$$

(d) $A = \dfrac{1}{2}\displaystyle\int_\alpha^\beta r^2 \, dr = \dfrac{1}{2}\int_0^{2\pi} \theta^2 \, d\theta = \left[\dfrac{\theta^3}{6}\right]_0^{2\pi} = \dfrac{4}{3}\pi^3$

**85.** The smaller circle has equation $r = a \cos \theta$. The area of the shaded lune is:

$$A = 2\left(\frac{1}{2}\right)\int_0^{\pi/4}\left[(a\cos\theta)^2 - 1\right]d\theta$$

$$= \int_0^{\pi/4}\left[\frac{a^2}{2}(1 + \cos 2\theta) - 1\right]d\theta$$

$$= \left[\frac{a^2}{2}\left(\theta + \frac{\sin 2\theta}{2}\right) - \theta\right]_0^{\pi/4} = \frac{a^2}{2}\left(\frac{\pi}{4} + \frac{1}{2}\right) - \frac{\pi}{4}$$

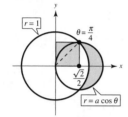

This equals the area of the square, $\left(\dfrac{\sqrt{2}}{2}\right)^2 = \dfrac{1}{2}$.

$$\frac{a^2}{2}\left(\frac{\pi}{4} + \frac{1}{2}\right) - \frac{\pi}{4} = \frac{1}{2}$$

$$\pi a^2 + 2a^2 - 2\pi - 4 = 0$$

$$a^2 = \frac{4 + 2\pi}{2 + \pi} = 2$$

$$a = \sqrt{2}$$

Smaller circle: $r = \sqrt{2}\cos\theta$

**87.** False. $f(\theta) = 1$ and $g(\theta) = -1$ have the same graphs.

**89.** In parametric form,

$$s = \int_a^b \sqrt{\left(\frac{dx}{dt}\right)^2 + \left(\frac{dy}{dt}\right)^2}\, dt.$$

Using $\theta$ instead of $t$, you have
$x = r\cos\theta = f(\theta)\cos\theta$ and
$y = r\sin\theta = f(\theta)\sin\theta$. So,

$$\frac{dx}{d\theta} = f'(\theta)\cos\theta - f(\theta)\sin\theta \text{ and}$$

$$\frac{dy}{d\theta} = f'(\theta)\sin\theta + f(\theta)\cos\theta.$$

It follows that

$$\left(\frac{dx}{d\theta}\right)^2 + \left(\frac{dy}{d\theta}\right)^2 = \left[f(\theta)\right]^2 + \left[f'(\theta)\right]^2.$$

So, $s = \displaystyle\int_\alpha^\beta \sqrt{\left[f(\theta)\right]^2 + \left[f'(\theta)\right]^2}\, d\theta.$

## Section 10.6   Polar Equations of Conics and Kepler's Laws

**1.** $r = \dfrac{2e}{1 + e\cos\theta}$

(a) $e = 1, r = \dfrac{2}{1 + \cos\theta}$,  parabola

(b) $e = 0.5$,

$\quad r = \dfrac{1}{1 + 0.5\cos\theta} = \dfrac{2}{2 + \cos\theta}$, ellipse

(c) $e = 1.5$,

$\quad r = \dfrac{3}{1 + 1.5\cos\theta} = \dfrac{6}{2 + 3\cos\theta}$, hyperbola

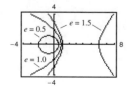

**3.** $r = \dfrac{2e}{1 - e\sin\theta}$

(a) $e = 1, r = \dfrac{2}{1 - \sin\theta}$, parabola

(b) $e = 0.5$,

$\quad r = \dfrac{1}{1 - 0.5\sin\theta} = \dfrac{2}{2 - \sin\theta}$, ellipse

(c) $e = 1.5$,

$\quad r = \dfrac{3}{1 - 1.5\sin\theta} = \dfrac{6}{2 - 3\sin\theta}$, hyperbola

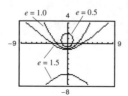

**5.** $r = \dfrac{4}{1 + e \sin \theta}$

(a) The conic is an ellipse. As $e \to 1^-$, the ellipse becomes more elliptical, and as $e \to 0^+$, it becomes more circular.

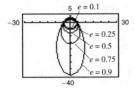

(b) The conic is a parabola.

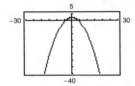

(c) The conic is a hyperbola. As $e \to 1^+$, the hyperbola opens more slowly, and as $e \to \infty$, it opens more rapidly.

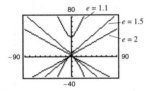

**7.** Parabola; Matches (c)

**9.** Hyperbola; Matches (a)

**11.** Ellipse; Matches (b)

**13.** $r = \dfrac{1}{1 - \cos \theta}$

Parabola because $e = 1, d = 1$.

Distance from pole to directrix: $|d| = 1$

Directrix: $x = -d = -1$

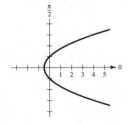

**15.** $r = \dfrac{-4}{1 - \sin \theta} = \dfrac{1(-4)}{1 - \sin \theta}$

$e = 1, d = -4$    Parabola

Distance from pole to directrix: $|d| = 4$

Directrix: $y = 4$

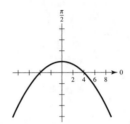

**17.** $r = \dfrac{6}{2 + \cos \theta} = \dfrac{3}{1 + (1/2) \cos \theta}$

Ellipse because $e = \dfrac{1}{2}; d = 6$

Directrix: $x = 6$

Distance from pole to directrix: $|d| = 6$

Vertices: $(r, \theta) = (2, 0), (6, \pi)$

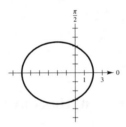

**19.** $r(2 + \sin \theta) = 4$

$r = \dfrac{4}{2 + \sin \theta} = \dfrac{2}{1 + (1/2) \sin \theta}$

Ellipse because $e = 1/2; d = 4$

Directrix: $y = 4$

Distance from pole to directrix: $|d| = 4$

Vertices: $(r, \theta) = (4/3, \pi/2), (4, 3\pi/2)$

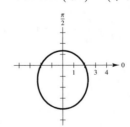

**21.** $r = \dfrac{5}{-1 + 2\cos\theta} = \dfrac{-5}{1 - 2\cos\theta}$

Hyperbola because $e = 2 > 1$; $d = -5/2$

Directrix: $x = 5/2$

Distance from pole to directrix: $|d| = 5/2$

Vertices: $(r, \theta) = (5, 0), (-5/3, \pi)$

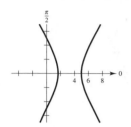

**23.** $r = \dfrac{3}{2 + 6\sin\theta} = \dfrac{3/2}{1 + 3\sin\theta}$

Hyperbola because $e = 3 > 0$; $d = 1/2$

Directrix: $y = 1/2$

Distance from pole to directrix: $|d| = 1/2$

Vertices: $(r, \theta) = (3/8, \pi/2), (-3/4, 3\pi/2)$

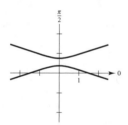

**25.** $r = \dfrac{300}{-12 + 6\sin\theta} = \dfrac{-25}{1 - \frac{1}{2}\sin\theta} = \dfrac{\frac{1}{2}(-50)}{1 - \frac{1}{2}\sin\theta}$

$e = \dfrac{1}{2}, d = -50$,  Ellipse

Distance from pole to directrix: $|d| = 50$

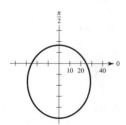

**27.** $r = \dfrac{3}{-4 + 2\sin\theta} = \dfrac{-\frac{3}{4}}{1 - \frac{1}{2}\sin\theta}$

$e = \dfrac{1}{2}$, Ellipse

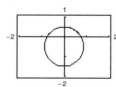

**29.** $r = \dfrac{-10}{1 - \cos\theta}$

$e = 1$,  Parabola

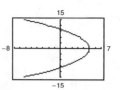

**31.** $r = \dfrac{-4}{1 - \sin\left(\theta - \dfrac{\pi}{4}\right)}$

Rotate the graph of

$r = \dfrac{-4}{1 - \sin\theta}$

$\dfrac{\pi}{4}$ radian counterclockwise.

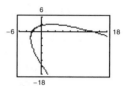

**33.** $r = \dfrac{6}{2 + \cos\left(\theta + \dfrac{\pi}{6}\right)}$

Rotate the graph of

$r = \dfrac{6}{2 + \cos\theta}$

$\dfrac{\pi}{6}$ radian clockwise.

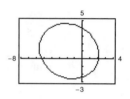

**35.** Change $\theta$ to $\theta + \dfrac{\pi}{6}$

$r = \dfrac{8}{8 + 5\cos\left(\theta + \dfrac{\pi}{6}\right)}$

**37.** Parabola

$e = 1$

$x = -3 \Rightarrow d = 3$

$r = \dfrac{ed}{1 - e\cos\theta} = \dfrac{3}{1 - \cos\theta}$

**39.** Ellipse

$e = \dfrac{1}{2}, y = 1, d = 1$

$r = \dfrac{ed}{1 + e\sin\theta} = \dfrac{1/2}{1 + (1/2)\sin\theta} = \dfrac{1}{2 + \sin\theta}$

**41.** Hyperbola

$e = 2, x = 1, d = 1$

$r = \dfrac{ed}{1 + e\cos\theta} = \dfrac{2}{1 + 2\cos\theta}$

**43.** Parabola

Vertex: $\left(1, -\dfrac{\pi}{2}\right)$

$e = 1, \, d = 2, \, r = \dfrac{2}{1 - \sin\theta}$

**45.** Ellipse

Vertices: $(2, 0), (8, \pi)$

$e = \dfrac{3}{5}, \, d = \dfrac{16}{3}$

$r = \dfrac{ed}{1 + e\cos\theta} = \dfrac{16/5}{1 + (3/5)\cos\theta} = \dfrac{16}{5 + 3\cos\theta}$

**47.** Hyperbola

Vertices: $\left(1, \dfrac{3\pi}{2}\right), \left(9, \dfrac{3\pi}{2}\right)$

$e = \dfrac{5}{4}, \, d = \dfrac{9}{5}$

$r = \dfrac{ed}{1 - e\sin\theta} = \dfrac{9/4}{1 - (5/4)\sin\theta} = \dfrac{9}{4 - 5\sin\theta}$

**49.** Ellipse, $e = \dfrac{1}{2}$,

Directrix: $r = 4\sec\theta \Rightarrow x = r\cos\theta = 4$

$r = \dfrac{ed}{1 + e\cos\theta} = \dfrac{\left(\dfrac{1}{2}\right)4}{1 + \dfrac{1}{2}\cos\theta} = \dfrac{4}{2 + \cos\theta}$

**51.** Ellipse if $0 < e < 1$, parabola if $e = 1$, hyperbola if $e > 1$.

**53.** If the foci are fixed and $e \to 0$, then $d \to \infty$. To see this, compare the ellipses

$r = \dfrac{1/2}{1 + (1/2)\cos\theta}, \, e = 1/2, \, d = 1$

$r = \dfrac{5/16}{1 + (1/4)\cos\theta}, \, e = 1/4, \, d = 5/4.$

**55.**

$$\dfrac{x^2}{a^2} + \dfrac{y^2}{b^2} = 1$$

$$x^2 b^2 + y^2 a^2 = a^2 b^2$$

$$b^2 r^2 \cos^2\theta + a^2 r^2 \sin^2\theta = a^2 b^2$$

$$r^2\left[b^2\cos^2\theta + a^2\left(1 - \cos^2\theta\right)\right] = a^2 b^2$$

$$r^2\left[a^2 + \cos^2\theta\left(b^2 - a^2\right)\right] = a^2 b^2$$

$$r^2 = \dfrac{a^2 b^2}{a^2 + \left(b^2 - a^2\right)\cos^2\theta} = \dfrac{a^2 b^2}{a^2 - c^2\cos^2\theta}$$

$$= \dfrac{b^2}{1 - (c/a)^2\cos^2\theta} = \dfrac{b^2}{1 - e^2\cos^2\theta}$$

**57.** $a = 5, \, c = 4, \, e = \dfrac{4}{5}, \, b = 3$

$r^2 = \dfrac{9}{1 - (16/25)\cos^2\theta}$

**59.** $a = 3, \, b = 4, \, c = 5, \, e = \dfrac{5}{3}$

$r^2 = \dfrac{-16}{1 - (25/9)\cos^2\theta}$

**61.** $A = 2\left[\dfrac{1}{2}\displaystyle\int_0^\pi \left(\dfrac{3}{2 - \cos\theta}\right)^2 d\theta\right]$

$= 9\displaystyle\int_0^\pi \dfrac{1}{(2 - \cos\theta)^2}\, d\theta \approx 10.88$

**63.** $A = 2\left[\dfrac{1}{2}\displaystyle\int_{-\pi/2}^{\pi/2} \left(\dfrac{2}{3 - 2\sin\theta}\right)^2 d\theta\right]$

$= 4\displaystyle\int_{-\pi/2}^{\pi/2} \dfrac{1}{(3 - 2\sin\theta)^2}\, d\theta \approx 3.37$

**65.** Vertices: $(123{,}000 + 4000, 0) = (127{,}000, 0)$

$(119 + 4000, \pi) = (4119, \pi)$

$a = \dfrac{127{,}000 + 4119}{2} = 65{,}559.5$

$c = 65{,}559.5 - 4119 = 61{,}440.5$

$e = \dfrac{c}{a} = \dfrac{122{,}881}{131{,}119} \approx 0.93717$

$r = \dfrac{ed}{1 - e\cos\theta}$

$\theta = 0: r = \dfrac{ed}{1 - e}, \quad \theta = \pi: r = \dfrac{ed}{1 + e}$

$2a = 2(65{,}559.5) = \dfrac{ed}{1 - e} + \dfrac{ed}{1 + e}$

$131{,}119 = d\left(\dfrac{e}{1 - e} + \dfrac{e}{1 + e}\right) = d\left(\dfrac{2e}{1 - e^2}\right)$

$d = \dfrac{131{,}119\left(1 - e^2\right)}{2e} \approx 8514.1397$

$r = \dfrac{7979.21}{1 - 0.93717\cos\theta} = \dfrac{1{,}046{,}226{,}000}{131{,}119 - 122{,}881\cos\theta}$

When $\theta = 60° = \dfrac{\pi}{3}, \, r \approx 15{,}015.$

Distance between earth and the satellite is $r - 4000 \approx 11{,}015$ miles.

**67.** $a = 1.496 \times 10^8$, $e = 0.0167$

$$r = \frac{(1 - e^2)a}{1 - e\cos\theta} = \frac{149,558,278.1}{1 - 0.0167\cos\theta}$$

Perihelion distance: $a(1 - e) \approx 147,101,680$ km

Aphelion distance: $a(1 + e) \approx 152,098,320$ km

**69.** $a = 4.498 \times 10^9$, $e = 0.0086$

$$r = \frac{(1 - e^2)a}{1 - e\cos\theta} = \frac{4,497,667,328}{1 - 0.0086\cos\theta}$$

Perihelion distance: $a(1 - e) \approx 4,459,317,200$ km

Aphelion distance: $a(1 + e) \approx 4,536,682,800$ km

**71.** $r = \dfrac{4.498 \times 10^9}{1 - 0.0086\cos\theta}$

(a) $A = \dfrac{1}{2}\displaystyle\int_0^{\pi/9} r^2\,d\theta \approx 3.591 \times 10^{18}$ km$^2$

$$165\left[\frac{\frac{1}{2}\int_0^{\pi/2} r^2\,d\theta}{\frac{1}{2}\int_0^{2\pi} r^2\,d\theta}\right] \approx 9.322 \text{ yrs}$$

(b) $\dfrac{1}{2}\displaystyle\int_\pi^\alpha r^2\,d\theta = 3.591 \times 10^{18}$

By trial and error, $\alpha \approx \pi + 0.361$

$0.361 > \pi/9 \approx 0.349$ because the rays in part (a) are longer than those in part (b)

(c) For part (a),

$$s = \int_0^{\pi/9} \sqrt{r^2 + (dr/d\theta)^2} \approx 1.583 \times 10^9 \text{ km}$$

Average per year $= \dfrac{1.583 \times 10^9}{9.322}$

$\approx 1.698 \times 10^8$ km/yr

For part (b),

$$s = \int_\pi^{\pi + 0.361} \sqrt{r^2 + (dr/d\theta)^2}\,d\theta$$

$\approx 1.610 \times 10^9$ km

Average per year $= \dfrac{1.610 \times 10^9}{9.322}$

$\approx 1.727 \times 10^8$ km/yr

**73.** $r_1 = a + c$, $r_0 = a - c$, $r_1 - r_0 = 2c$, $r_1 + r_0 = 2a$

$$e = \frac{c}{a} = \frac{r_1 - r_0}{r_1 + r_0}$$

$$\frac{1 + e}{1 - e} = \frac{1 + \dfrac{c}{a}}{1 - \dfrac{c}{a}} = \frac{a + c}{a - c} = \frac{r_1}{r_0}$$

**75.** $r_1 = \dfrac{ed}{1 + \sin\theta}$ and $r_2 = \dfrac{ed}{1 - \sin\theta}$

Points of intersection: $(ed, 0), (ed, \pi)$

$$r_1: \frac{dy}{dx} = \frac{\left(\dfrac{ed}{1 + \sin\theta}\right)(\cos\theta) + \left(\dfrac{-ed\cos\theta}{(1 + \sin\theta)^2}\right)(\sin\theta)}{\left(\dfrac{-ed}{1 + \sin\theta}\right)(\sin\theta) + \left(\dfrac{-ed\cos\theta}{(1 + \sin\theta)^2}\right)(\cos\theta)}$$

At $(ed, 0)$, $\dfrac{dy}{dx} = -1$. At $(ed, \pi)$, $\dfrac{dy}{dx} = 1$.

$$r_2: \frac{dy}{dx} = \frac{\left(\dfrac{ed}{1 - \sin\theta}\right)(\cos\theta) + \left(\dfrac{ed\cos\theta}{(1 - \sin\theta)^2}\right)(\sin\theta)}{\left(\dfrac{-ed}{1 - \sin\theta}\right)(\sin\theta) + \left(\dfrac{ed\cos\theta}{(1 - \sin\theta)^2}\right)(\cos\theta)}$$

At $(ed, 0)$, $\dfrac{dy}{dx} = 1$. At $(ed, \pi)$, $\dfrac{dy}{dx} = -1$.

So, at $(ed, 0)$ you have $m_1 m_2 = (-1)(1) = -1$, and at $(ed, \pi)$ you have $m_1 m_2 = 1(-1) = -1$. The curves intersect at right angles.

# Review Exercises for Chapter 10

**1.** $4x^2 + y^2 = 4$

Ellipse

Vertex: $(1, 0)$.

Matches (e)

**3.** $y^2 = -4x$

Parabola opening to left.

Matches (b)

**5.** $x^2 + 4y^2 = 4$

Ellipse

Vertex: $(0, 1)$

Matches (a)

**7.** $16x^2 + 16y^2 - 16x + 24y - 3 = 0$

$\left(x^2 - x + \frac{1}{4}\right) + \left(y^2 + \frac{3}{2}y + \frac{9}{16}\right) = \frac{3}{16} + \frac{1}{4} + \frac{9}{16}$

$\left(x - \frac{1}{2}\right)^2 + \left(y + \frac{3}{4}\right)^2 = 1$

Circle

Center: $\left(\frac{1}{2}, -\frac{3}{4}\right)$

Radius: 1

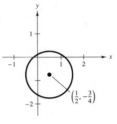

**9.** $3x^2 - 2y^2 + 24x + 12y + 24 = 0$

$3\left(x^2 + 8x + 16\right) - 2\left(y^2 - 6y + 9\right) = -24 + 48 - 18$

$\frac{(x + 4)^2}{2} - \frac{(y - 3)^2}{3} = 1$

Hyperbola

Center: $(-4, 3)$

Vertices: $\left(-4 \pm \sqrt{2}, 3\right)$

Asymptotes:

$y = 3 \pm \sqrt{\frac{3}{2}}(x + 4)$

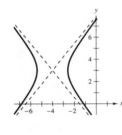

**11.** $3x^2 + 2y^2 - 12x + 12y + 29 = 0$

$3\left(x^2 - 4x + 4\right) + 2\left(y^2 + 6y + 9\right) = -29 + 12 + 18$

$\frac{(x - 2)^2}{1/3} + \frac{(y + 3)^2}{1/2} = 1$

Ellipse

Center: $(2, -3)$

Vertices: $\left(2, -3 \pm \frac{\sqrt{2}}{2}\right)$

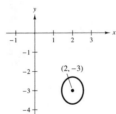

**13.** Vertex: $(0, 2)$

Directrix: $x = -3$

Parabola opens to the right.

$p = 3$

$(y - 2)^2 = 4(3)(x - 0)$

$y^2 - 4y - 12x + 4 = 0$

**15.** Vertices: $(-5, 0), (7, 0)$

Foci: $(-3, 0), (5, 0)$

Horizontal major axis

$a = 6, c = 4, b = \sqrt{36 - 16} = 2\sqrt{5}$

Center: $(1, 0)$

$\frac{(x - 1)^2}{36} + \frac{y^2}{20} = 1$

**17.** Vertices: $(\pm 7, 0)$

Foci: $(\pm 9, 0)$

Horizontal transverse axis

Center $(0, 0)$

$a = 7, c = 9, b = \sqrt{81 - 49} = \sqrt{32}$

$\frac{x^2}{49} - \frac{y^2}{32} = 1$

**19.** $\frac{x^2}{9} + \frac{y^2}{4} = 1, a = 3, b = 2, c = \sqrt{5}, e = \frac{\sqrt{5}}{3}$

By Example 5 of Section 10.1,

$C = 12 \int_0^{\pi/2} \sqrt{1 - \left(\frac{5}{9}\right) \sin^2 \theta} \; d\theta \approx 15.87.$

**21.** $y = x - 2$ has a slope of 1. The perpendicular slope is $-1$.

$y = x^2 - 2x + 2$

$\frac{dy}{dx} = 2x - 2 = -1$ when $x = \frac{1}{2}$ and $y = \frac{5}{4}$.

Perpendicular line: $\quad y - \frac{5}{4} = -1\left(x - \frac{1}{2}\right)$

$4x + 4y - 7 = 0$

**23.** $y = \frac{1}{200}x^2$

(a) $x^2 = 200y$

$x^2 = 4(50)y$

Focus: $(0, 50)$

(b) $\quad y = \frac{1}{200}x^2$

$y' = \frac{1}{100}x$

$\sqrt{1 + (y')^2} = \sqrt{1 + \frac{x^2}{10,000}}$

$S = 2\pi \int_0^{100} x\sqrt{1 + \frac{x^2}{10,000}} \; dx \approx 38,294.49$

**25.** $x = 1 + 8t, \ y = 3 - 4t$

$t = \dfrac{x-1}{8} \Rightarrow y = 3 - 4\left(\dfrac{x-1}{8}\right) = \dfrac{7}{2} - \dfrac{x}{2}$

$x + 2y - 7 = 0, \quad$ Line

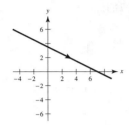

**27.** $x = e^t - 1, \ y = e^{3t}$

$e^t = x + 1 \Rightarrow y = (x+1)^3, \ x > -1$

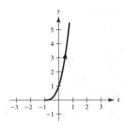

**29.** $x = 6 \cos \theta, \ y = 6 \sin \theta$

$\left(\dfrac{x}{6}\right)^2 + \left(\dfrac{y}{6}\right)^2 = 1$

$x^2 + y^2 = 36$

Circle

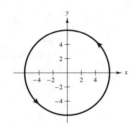

**31.** $x = 2 + \sec \theta, \ y = 3 + \tan \theta$

$(x-2)^2 = \sec^2 \theta = 1 + \tan^2 \theta = 1 + (y-3)^2$

$(x-2)^2 - (y-3)^2 = 1$

Hyperbola

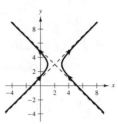

**33.** $x = 3 + \big(3 - (-2)\big)t = 3 + 5t$

$y = 2 + (2-6)t = 2 - 4t$

(other answers possible)

**35.** $\dfrac{(x+3)^2}{16} + \dfrac{(y-4)^2}{9} = 1$

Let $\dfrac{(x+3)^2}{16} = \cos^2 \theta$ and $\dfrac{(y-4)^2}{9} = \sin^2 \theta.$

Then $x = -3 + 4 \cos \theta$ and $y = 4 + 3 \sin \theta.$

**37.** $x = \cos 3\theta + 5 \cos \theta$

$y = \sin 3\theta + 5 \sin \theta$

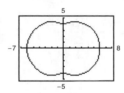

**39.** $x = 2 + 5t, \ y = 1 - 4t$

(a) $\dfrac{dy}{dx} = -\dfrac{4}{5}$

　　No horizontal tangent

(b) $t = \dfrac{x-2}{5} \Rightarrow y = 1 - 4\left(\dfrac{x-2}{5}\right) = 1 - \dfrac{4}{5}x + \dfrac{8}{5} = -\dfrac{4}{5}x + \dfrac{13}{5}$

$4x + 5y - 13 = 0$

(c)

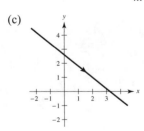

**41.** $x = \dfrac{1}{t}$

$y = 2t + 3$

(a) $\dfrac{dy}{dx} = \dfrac{2}{-1/t^2} = -2t^2$

No horizontal tangents, $(t \neq 0)$

(b) $t = \dfrac{1}{x}$

$y = \dfrac{2}{x} + 3$

(c)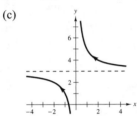

**43.** $x = \dfrac{1}{2t + 1}$

$y = \dfrac{1}{t^2 - 2t}$

(a) $\dfrac{dy}{dx} = \dfrac{\dfrac{-(2t - 2)}{\left(t^2 - 2t\right)^2}}{\dfrac{-2}{(2t + 1)^2}} = \dfrac{(t - 1)(2t + 1)^2}{t^2(t - 2)^2} = 0$

when $t = 1$.

Point of horizontal tangency: $\left(\dfrac{1}{3}, -1\right)$

(b) $2t + 1 = \dfrac{1}{x} \Rightarrow t = \dfrac{1}{2}\left(\dfrac{1}{x} - 1\right)$

$y = \dfrac{1}{\dfrac{1}{2}\left(\dfrac{1 - x}{x}\right)\left[\dfrac{1}{2}\left(\dfrac{1 - x}{x}\right) - 2\right]}$

$= \dfrac{4x^2}{(1 - x)^2 - 4x(1 - x)} = \dfrac{4x^2}{(5x - 1)(x - 1)}, (x \neq 0)$

(c)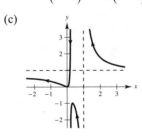

**45.** $x = 5 + \cos\theta, \; y = 3 + 4\sin\theta$

(a) $\dfrac{dy}{dx} = \dfrac{4\cos\theta}{-\sin\theta} = -4\cot\theta$

Horizontal tangents:

$\theta = \dfrac{\pi}{2}, \dfrac{3\pi}{2} \Rightarrow (5, 7), (5, -1)$

(b) $(x - 5)^2 + \left(\dfrac{(y - 3)}{4}\right)^2 = 1$

$(x - 5)^2 + \dfrac{(y - 3)^2}{16} = 1,$ Ellipse

(c)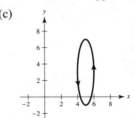

**47.** $x = \cos^3\theta$

$y = 4\sin^3\theta$

(a) $\dfrac{dy}{dx} = \dfrac{12\sin^2\theta\cos\theta}{3\cos^2\theta(-\sin\theta)} = \dfrac{-4\sin\theta}{\cos\theta} = -4\tan\theta = 0$

when $\theta = 0, \pi$.

But, $\dfrac{dy}{dt} = \dfrac{dx}{dt} = 0$ at $\theta = 0, \pi$. So no points of horizontal tangency.

(b) $x^{2/3} + \left(\dfrac{y}{4}\right)^{2/3} = 1$

(c)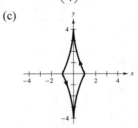

**49.** $x = 5 - t, \; y = 2t^2$

$\dfrac{dx}{dt} = -1, \dfrac{dy}{dt} = 4t$

Horizontal tangent at $t = 0$: $(5, 0)$

No vertical tangents

**51.** $x = 2 + 2 \sin \theta, \ y = 1 + \cos \theta$

$$\frac{dx}{d\theta} = 2 \cos \theta, \ \frac{dy}{d\theta} = -\sin \theta$$

$$\frac{dy}{d\theta} = 0 \text{ for } \theta = 0, \pi, 2\pi, \dots$$

Horizontal tangents: $(x, y) = (2, 2), (2, 0)$

$$\frac{dx}{d\theta} = 0 \text{ for } \theta = \frac{\pi}{2}, \frac{3\pi}{2}, \dots$$

Vertical tangents: $(x, y) = (4, 1), (0, 1)$

**53.** $x = \cot \theta$

$y = \sin 2\theta = 2 \sin \theta \cos \theta$

(a), (c)

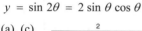
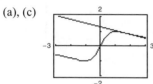

(b) At $\theta = \frac{\pi}{6}, \frac{dx}{d\theta} = -4, \frac{dy}{d\theta} = 1, \text{ and } \frac{dy}{dx} = -\frac{1}{4}$.

**55.** $x = r(\cos \theta + \theta \sin \theta)$

$y = r(\sin \theta - \theta \cos \theta)$

$$\frac{dx}{d\theta} = r\theta \cos \theta$$

$$\frac{dy}{d\theta} = r\theta \sin \theta$$

$$s = r \int_0^\pi \sqrt{\theta^2 \cos^2 \theta + \theta^2 \sin^2 \theta} \ d\theta$$

$$= r \int_0^\pi \theta \ d\theta = \frac{r}{2} \left[ \theta^2 \right]_0^\pi = \frac{1}{2}\pi^2 r$$

**57.** $x = t, \ y = 3t, \ 0 \le t \le 2$

$$\frac{dx}{dt} = 1, \frac{dy}{dt} = -3, \sqrt{\left(\frac{dx}{dt}\right)^2 + \left(\frac{dy}{dt}\right)^2} = \sqrt{1 + 9} = \sqrt{10}$$

(a) $S = 2\pi \int_0^2 3t\sqrt{10} \ dt = 6\sqrt{10} \ \pi \left[\frac{t^2}{2}\right]_0^2 = 12\sqrt{10} \ \pi$

$\approx 119.215$

(b) $S = 2\pi \int_0^2 \sqrt{10} \ dt = 2\pi \left[\sqrt{10} t\right]_0^2 = 4\pi\sqrt{10}$

$\approx 39.738$

**59.** $x = 3 \sin \theta, \ y = 2 \cos \theta$

$$A = \int_a^b y \ dx = \int_{-\pi/2}^{\pi/2} 2 \cos \theta (3 \cos \theta) \ d\theta$$

$$= 6 \int_{-\pi/2}^{\pi/2} \frac{1 + \cos 2\theta}{2} \ d\theta$$

$$= 3 \left[ \theta + \frac{\sin 2\theta}{2} \right]_{-\pi/2}^{\pi/2}$$

$$= 3 \left[ \frac{\pi}{2} + \frac{\pi}{2} \right] = 3\pi$$

**61.** $(r, \theta) = \left( 5, \frac{3\pi}{2} \right)$

$$x = r \cos \theta = 5 \cos \frac{3\pi}{2} = 0$$

$$y = r \sin \theta = 5 \sin \frac{3\pi}{2} = -5$$

$$(x, y) = (0, -5)$$

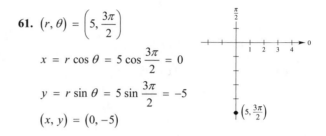

**63.** $(r, \theta) = \left( \sqrt{3}, 1.56 \right)$

$$(x, y) = \left( \sqrt{3} \cos(1.56), \sqrt{3} \sin(1.56) \right)$$

$$\approx (0.0187, 1.7319)$$

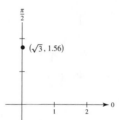

**65.** $(x, y) = (4, -4)$

$$r = \sqrt{4^2 + (-4)^2} = 4\sqrt{2}$$

$$\theta = \frac{7\pi}{4}$$

$$(r, \theta) = \left( 4\sqrt{2}, \frac{7\pi}{4} \right), \left( -4\sqrt{2}, \frac{3\pi}{4} \right)$$

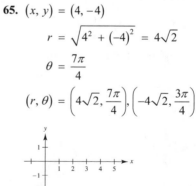

**67.** $(x, y) = (-1, 3)$

$$r = \sqrt{(-1)^2 + 3^2} = \sqrt{10}$$

$$\theta = \arctan(-3) \approx 1.89(108.43°)$$

$$(r, \theta) = \left(\sqrt{10}, 1.89\right), \left(-\sqrt{10}, 5.03\right)$$

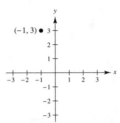

**69.**

$$r = 3\cos\theta$$

$$r^2 = 3r\cos\theta$$

$$x^2 + y^2 = 3x$$

$$x^2 + y^2 - 3x = 0$$

**71.**

$$r = -2(1 + \cos\theta)$$

$$r^2 = -2r(1 + \cos\theta)$$

$$x^2 + y^2 = -2\left(\pm\sqrt{x^2 + y^2}\right) - 2x$$

$$\left(x^2 + y^2 + 2x\right)^2 = 4\left(x^2 + y^2\right)$$

**73.**

$$r^2 = \cos 2\theta = \cos^2\theta - \sin^2\theta$$

$$r^4 = r^2\cos^2\theta - r^2\sin^2\theta$$

$$\left(x^2 + y^2\right)^2 = x^2 - y^2$$

**75.**

$$r = 4\cos 2\theta \sec\theta$$

$$= 4\left(2\cos^2\theta - 1\right)\left(\frac{1}{\cos\theta}\right)$$

$$r\cos\theta = 8\cos^2\theta - 4$$

$$x = 8\left(\frac{x^2}{x^2 + y^2}\right) - 4$$

$$x^3 + xy^2 = 4x^2 - 4y^2$$

$$y^2 = x^2\left(\frac{4 - x}{4 + x}\right)$$

**77.** $\left(x^2 + y^2\right)^2 = ax^2 y$

$$r^4 = a\left(r^2\cos^2\theta\right)(r\sin\theta)$$

$$r = a\cos^2\theta\sin\theta$$

**79.** $x^2 + y^2 = a^2\left(\arctan\dfrac{y}{x}\right)^2$

$$r^2 = a^2\theta^2$$

**81.** $r = 6$, Circle radius 6

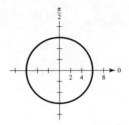

**83.** $r = -\sec\theta = \dfrac{-1}{\cos\theta}$

$$r\cos\theta = -1, x = -1$$

Vertical line

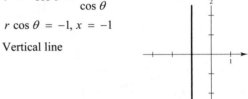

**85.** $r = -2(1 + \cos\theta)$

Cardioid

Symmetric to polar axis

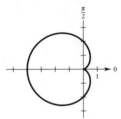

| $\theta$ | 0 | $\dfrac{\pi}{3}$ | $\dfrac{\pi}{2}$ | $\dfrac{2\pi}{3}$ | $\pi$ |
|---|---|---|---|---|---|
| $r$ | $-4$ | $-3$ | $-2$ | $-1$ | 0 |

**87.** $r = 4 - 3\cos\theta$

Limaçon

Symmetric to polar axis

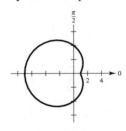

| $\theta$ | 0 | $\dfrac{\pi}{3}$ | $\dfrac{\pi}{2}$ | $\dfrac{2\pi}{3}$ | $\pi$ |
|---|---|---|---|---|---|
| $r$ | 1 | $\dfrac{5}{2}$ | 4 | $\dfrac{11}{2}$ | 7 |

**89.** $r = -3\cos 2\theta$

Rose curve with four petals

Symmetric to polar axis, $\theta = \dfrac{\pi}{2}$, and pole

Relative extrema: $(-3, 0), \left(3, \dfrac{\pi}{2}\right), (-3, \pi), \left(3, \dfrac{3\pi}{2}\right)$

Tangents at the pole: $\theta = \dfrac{\pi}{4}, \dfrac{3\pi}{4}$

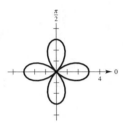

**91.** $r^2 = 4\sin^2 2\theta$

$r = \pm 2\sin(2\theta)$

Rose curve with four petals

Symmetric to the polar axis, $\theta = \dfrac{\pi}{2}$, and pole

Relative extrema: $\left(\pm 2, \dfrac{\pi}{4}\right), \left(\pm 2, \dfrac{3\pi}{4}\right)$

Tangents at the pole: $\theta = 0, \dfrac{\pi}{2}$

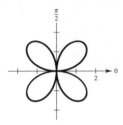

**93.** $r = \dfrac{3}{\cos\theta - (\pi/4)}$

Graph of $r = 3\sec\theta$ rotated through an angle of $\pi/4$

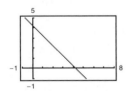

**95.** $r = 4\cos 2\theta \sec\theta$

Strophoid

Symmetric to the polar axis

$r \Rightarrow \infty$ as $\theta \Rightarrow \dfrac{\pi^-}{2}$

$r \Rightarrow \infty$ as $\theta \Rightarrow \dfrac{-\pi^+}{2}$

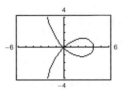

**97.** $r = 1 - 2\cos\theta$

(a) The graph has polar symmetry and the tangents at the pole are $\theta = \dfrac{\pi}{3}, -\dfrac{\pi}{3}$.

(b) $\dfrac{dy}{dx} = \dfrac{2\sin^2\theta + (1 - 2\cos\theta)\cos\theta}{2\sin\theta\cos\theta - (1 - 2\cos\theta)\sin\theta}$

Horizontal tangents: $-4\cos^2\theta + \cos\theta + 2 = 0$,

$\cos\theta = \dfrac{-1 \pm \sqrt{1 + 32}}{-8} = \dfrac{1 \pm \sqrt{33}}{8}$

When

$\cos\theta = \dfrac{1 \pm \sqrt{33}}{8}, r = 1 - 2\left(\dfrac{1 + \sqrt{33}}{8}\right) = \dfrac{3 \mp \sqrt{33}}{4}$,

$\left[\dfrac{3 - \sqrt{33}}{4}, \arccos\left(\dfrac{1 + \sqrt{33}}{8}\right)\right] \approx (-0.686, 0.568)$

$\left[\dfrac{3 - \sqrt{33}}{4}, -\arccos\left(\dfrac{1 + \sqrt{33}}{8}\right)\right] \approx (-0.686, -0.568)$

$\left[\dfrac{3 + \sqrt{33}}{4}, \arccos\left(\dfrac{1 - \sqrt{33}}{8}\right)\right] \approx (2.186, 2.206)$

$\left[\dfrac{3 + \sqrt{33}}{4}, -\arccos\left(\dfrac{1 - \sqrt{33}}{8}\right)\right] \approx (2.186, -2.206)$.

Vertical tangents:

$\sin\theta(4\cos\theta - 1) = 0, \sin\theta = 0, \cos\theta = \dfrac{1}{4}$,

$\theta = 0, \pi, \theta = \pm\arccos\left(\dfrac{1}{4}\right), (-1, 0), (3, \pi)$

$\left(\dfrac{1}{2}, \pm\arccos\dfrac{1}{4}\right) \approx (0.5, \pm 1.318)$

(c)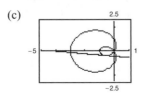

**99.** $r = 1 + \cos\theta, r = 1 - \cos\theta$

The points $(1, \pi/2)$ and $(1, 3\pi/2)$ are the two points of intersection (other than the pole). The slope of the graph of $r = 1 + \cos\theta$ is

$$m_1 = \frac{dy}{dx}$$

$$= \frac{r'\sin\theta + r\cos\theta}{r'\cos\theta - r\sin\theta}$$

$$= \frac{-\sin^2\theta + \cos\theta(1 + \cos\theta)}{-\sin\theta\cos\theta - \sin\theta(1 + \cos\theta)}.$$

At $(1, \pi/2)$, $m_1 = -1/-1 = 1$ and at $(1, 3\pi/2)$, $m_1 = -1/1 = -1$. The slope of the graph of $r = 1 - \cos\theta$ is

$$m_2 = \frac{dy}{dx} = \frac{\sin^2\theta + \cos\theta(1 - \cos\theta)}{\sin\theta\cos\theta - \sin\theta(1 - \cos\theta)}.$$

At $(1, \pi/2)$, $m_2 = 1/-1 = -1$ and at $(1, 3\pi/2)$, $m_2 = 1/1 = 1$. In both cases, $m_1 = -1/m_2$ and you conclude that the graphs are orthogonal at $(1, \pi/2)$ and $(1, 3\pi/2)$.

**101.** $A = 2 \cdot \frac{1}{2}\int_0^{\pi/10} [3\cos 5\theta]^2 \, d\theta$

$$= \int_0^{\pi/10} 9\left(\frac{1 + \cos(10\theta)}{2}\right) d\theta$$

$$= \frac{9}{2}\left[\theta + \frac{\sin(10\theta)}{2}\right]_0^{\pi/10} = \frac{9}{2}\left[\frac{\pi}{10}\right] = \frac{9\pi}{20}$$

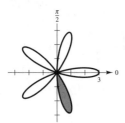

**103.** $r = 2 + \cos\theta$

$$A = 2\left[\frac{1}{2}\int_0^{\pi} (2 + \cos\theta)^2 \, d\theta\right] \approx 14.14, \left(\frac{9\pi}{2}\right)$$

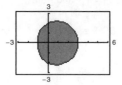

**105.** $r^2 = 4\sin 2\theta$

$$A = 2\left[\frac{1}{2}\int_0^{\pi/2} 4\sin 2\theta \, d\theta\right] = 4$$

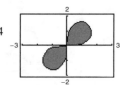

**107.** $r = 1 - \cos\theta$

$r = 1 + \sin\theta$

The cardioids intersect at 3 points:

$1 - \cos\theta = 1 + \sin\theta$

$$\tan\theta = -1 \Rightarrow \theta = \frac{3\pi}{4}, \frac{7\pi}{4}.$$

$$\left(1 + \frac{\sqrt{2}}{2}, \frac{3\pi}{4}\right), \left(1 - \frac{\sqrt{2}}{2}, \frac{7\pi}{4}\right), (0, 0)$$

**109.** $r = \sin\theta\cos^2\theta$

$$A = 2\left[\frac{1}{2}\int_0^{\pi/2} (\sin\theta\cos^2\theta)^2 \, d\theta\right] \approx 0.10, \left(\frac{\pi}{32}\right)$$

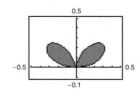

**111.** $r = 3, r^2 = 18\sin 2\theta$

$9 = r^2 = 18\sin 2\theta$

$$\sin 2\theta = \frac{1}{2}$$

$$\theta = \frac{\pi}{12}$$

$$A = 2\left[\frac{1}{2}\int_0^{\pi/12} 18\sin 2\theta \, d\theta + \frac{1}{2}\int_{\pi/12}^{5\pi/12} 9 \, d\theta + \frac{1}{2}\int_{5\pi/12}^{\pi/2} 18\sin 2\theta \, d\theta\right] \approx 1.2058 + 9.4248 + 1.2058 \approx 11.84$$

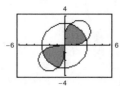

**113.** $r = a(1 - \cos\theta), 0 \le \theta \le \pi$

$$\frac{dr}{d\theta} = a\sin\theta$$

$$s = \int_0^{\pi} \sqrt{a^2(1 - \cos\theta)^2 + a^2\sin^2\theta}\, d\theta$$

$$= \sqrt{2}a \int_0^{\pi} \sqrt{1 - \cos\theta}\, d\theta$$

$$= \sqrt{2}a \int_0^{\pi} \frac{\sin\theta}{\sqrt{1 + \cos\theta}}\, d\theta$$

$$= -2\sqrt{2}a\left[(1 + \cos\theta)^{1/2}\right]_0^{\pi} = 4a$$

**115.** $f(\theta) = 1 + 4\cos\theta$

$$f'(\theta) = -4\sin\theta$$

$$\sqrt{f(\theta)^2 + f'(\theta)^2} = \sqrt{(1 + 4\cos\theta)^2 + (-4\sin\theta)^2}$$

$$= \sqrt{17 + 8\cos\theta}$$

$$S = 2\pi \int_0^{\pi/2} (1 + 4\cos\theta)\sin\theta\sqrt{17 + 8\cos\theta}\, d\theta$$

$$= \frac{34\pi\sqrt{17}}{5} \approx 88.08$$

**117.** $r = \dfrac{6}{1 - \sin\theta}$

$e = 1,$

Parabola

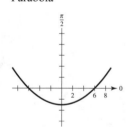

**119.** $r = \dfrac{6}{3 + 2\cos\theta} = \dfrac{2}{1 + (2/3)\cos\theta}, e = \dfrac{2}{3}$

Ellipse

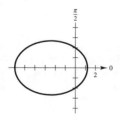

**121.** $r = \dfrac{4}{2 - 3\sin\theta} = \dfrac{2}{1 - (3/2)\sin\theta}, e = \dfrac{3}{2}$

Hyperbola

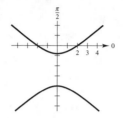

**123.** Circle

Center: $\left(5, \dfrac{\pi}{2}\right) = (0, 5)$ in rectangular coordinates

Solution point: $(0, 0)$

$$x^2 + (y - 5)^5 = 25$$

$$x^2 + y^2 - 10y = 0$$

$$r^2 - 10r\sin\theta = 0$$

$$r = 10\sin\theta$$

**125.** Parabola

Vertex: $(2, \pi)$

Focus: $(0, 0)$

$e = 1, d = 4$

$$r = \frac{4}{1 - \cos\theta}$$

**127.** Ellipse

Vertices: $(5, 0), (1, \pi)$

Focus: $(0, 0)$

$$a = 3, c = 2, e = \frac{2}{3}, d = \frac{5}{2}$$

$$r = \frac{\left(\dfrac{2}{3}\right)\left(\dfrac{5}{2}\right)}{1 - \left(\dfrac{2}{3}\right)\cos\theta} = \frac{5}{3 - 2\cos\theta}$$

# Problem Solving for Chapter 10

**1.** (a)

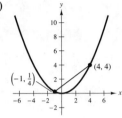

(b) $x^2 = 4y$

$2x = 4y'$

$y' = \dfrac{1}{2}x$

$y - 4 = 2(x - 4) \implies y = 2x - 4$  Tangent line at $(4, 4)$

$y - \dfrac{1}{4} = -\dfrac{1}{2}(x + 1) \implies y = -\dfrac{1}{2}x - \dfrac{1}{4}$  Tangent line at $\left(-1, \dfrac{1}{4}\right)$

Tangent lines have slopes of $2$ and $-\dfrac{1}{2} \implies$ perpendicular.

(c)  Intersection:

$2x - 4 = -\dfrac{1}{2}x - \dfrac{1}{4}$

$8x - 16 = -2x - 1$

$10x = 15$

$x = \dfrac{3}{2} \implies \left(\dfrac{3}{2}, -1\right)$

Point of intersection, $\left(\dfrac{3}{2}, -1\right)$, is on directrix $y = -1$.

**3.** Consider $x^2 = 4py$ with focus $F = (0, p)$.

Let $P(a, b)$ be point on parabola.

$2x = 4py' \implies y' = \dfrac{x}{2p}$

$y - b = \dfrac{a}{2p}(x - a)$   Tangent line at $P$

For $x = 0$, $y = b + \dfrac{a}{2p}(-a) = b - \dfrac{a^2}{2p} = b - \dfrac{4pb}{2p} = -b$.

So, $Q = (0, -b)$.

$\triangle FQP$ is isosceles because

$|FQ| = p + b$

$|FP| = \sqrt{(a - 0)^2 + (b - p)^2} = \sqrt{a^2 + b^2 - 2bp + p^2} = \sqrt{4pb + b^2 - 2bp + p^2} = \sqrt{(b + p)^2} = b + p.$

So, $\angle FQP = \angle BPA = \angle FPQ$.

**5. (a)** In $\triangle OCB$, $\cos\theta = \dfrac{2a}{OB} \Rightarrow OB = 2a \cdot \sec\theta$.

In $\triangle OAC$, $\cos\theta = \dfrac{OA}{2a} \Rightarrow OA = 2a \cdot \cos\theta$.

$$r = OP = AB = OB - OA = 2a(\sec\theta - \cos\theta)$$
$$= 2a\left(\frac{1}{\cos\theta} - \cos\theta\right)$$
$$= 2a \cdot \frac{\sin^2\theta}{\cos\theta}$$
$$= 2a \cdot \tan\theta\sin\theta$$

**(b)** $x = r\cos\theta = (2a\tan\theta\sin\theta)\cos\theta = 2a\sin^2\theta$

$y = r\sin\theta = (2a\tan\theta\sin\theta)\sin\theta = 2a\tan\theta \cdot \sin^2\theta, -\dfrac{\pi}{2} < \theta < \dfrac{\pi}{2}$

Let $t = \tan\theta, -\infty < t < \infty$.

Then $\sin^2\theta = \dfrac{t^2}{1 + t^2}$ and $x = 2a\dfrac{t^2}{1 + t^2}, y = 2a\dfrac{t^3}{1 + t^2}$.

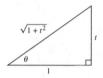

**(c)** $r = 2a\tan\theta\sin\theta$

$r\cos\theta = 2a\sin^2\theta$

$r^3\cos\theta = 2a\,r^2\sin^2\theta$

$(x^2 + y^2)x = 2ay^2$

$$y^2 = \frac{x^3}{(2a - x)}$$

**7. (a)** $y^2 = \dfrac{t^2(1 - t^2)^2}{(1 + t^2)^2}, x^2 = \dfrac{(1 - t^2)^2}{(1 + t^2)^2}$

$$\frac{1 - x}{1 + x} = \frac{1 - \left(\dfrac{1 - t^2}{1 + t^2}\right)}{1 + \left(\dfrac{1 - t^2}{1 + t^2}\right)} = \frac{2t^2}{2} = t^2$$

So, $y^2 = x^2\left(\dfrac{1 - x}{1 + x}\right)$.

**(b)** $r^2\sin^2\theta = r^2\cos^2\theta\left(\dfrac{1 - r\cos\theta}{1 + r\cos\theta}\right)$

$\sin^2\theta(1 + r\cos\theta) = \cos^2\theta(1 - r\cos\theta)$

$r\cos\theta\sin^2\theta + \sin^2\theta = \cos^2\theta - r\cos^3\theta$

$r\cos\theta(\sin^2\theta + \cos^2\theta) = \cos^2\theta - \sin^2\theta$

$r\cos\theta = \cos 2\theta$

$r = \cos 2\theta \cdot \sec\theta$

(c)

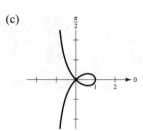

(d) $r(\theta) = 0$ for $\theta = \dfrac{\pi}{4}, \dfrac{3\pi}{4}$.

So, $y = x$ and $y = -x$ are tangent lines to curve at the origin.

(e) $y'(t) = \dfrac{\left(1 + t^2\right)\left(1 - 3t^2\right) - \left(t - t^3\right)(2t)}{\left(1 + t^2\right)^2} = \dfrac{1 - 4t^2 - t^4}{\left(1 + t^2\right)^2} = 0$

$t^4 + 4t^2 - 1 = 0 \Rightarrow t^2 = -2 \pm \sqrt{5} \Rightarrow x = \dfrac{1 - \left(-2 \pm \sqrt{5}\right)}{1 + \left(-2 \pm \sqrt{5}\right)} = \dfrac{3 \mp \sqrt{5}}{-1 \pm \sqrt{5}} = \dfrac{3 - \sqrt{5}}{-1 + \sqrt{5}} = \dfrac{\sqrt{5} - 1}{2}$

$\left(\dfrac{\sqrt{5} - 1}{2}, \pm \dfrac{\sqrt{5} - 1}{2}\sqrt{-2 + \sqrt{5}}\right)$

9. (a)

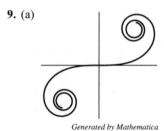

*Generated by Mathematica*

(b) $(-x, -y) = \left(-\displaystyle\int_0^t \cos\dfrac{\pi u^2}{2}\,du, \ -\displaystyle\int_0^t \sin\dfrac{\pi u^2}{2}\,du\right)$ is

on the curve whenever $(x, y)$ is on the curve.

(c) $x'(t) = \cos\dfrac{\pi t^2}{2}, \ y'(t) = \sin\dfrac{\pi t^2}{2}$,

$x'(t)^2 + y'(t)^2 = 1$

So, $s = \displaystyle\int_0^a dt = a$.

On $[-\pi, \pi]$, $s = 2\pi$.

11. $r = \dfrac{ab}{a \sin\theta + b \cos\theta}, \quad 0 \le \theta \le \dfrac{\pi}{2}$

$r(a \sin\theta + b \cos\theta) = ab$

$ay + bx = ab$

$\dfrac{y}{b} + \dfrac{x}{a} = 1$

Line segment

Area $= \dfrac{1}{2}ab$

13. Let $(r, \theta)$ be on the graph.

$\sqrt{r^2 + 1 + 2r\cos\theta}\sqrt{r^2 + 1 - 2r\cos\theta} = 1$

$\left(r^2 + 1\right)^2 - 4r^2\cos^2\theta = 1$

$r^4 + 2r^2 + 1 - 4r^2\cos^2\theta = 1$

$r^2\left(r^2 - 4\cos^2\theta + 2\right) = 0$

$r^2 = 4\cos^2\theta - 2$

$r^2 = 2\left(2\cos^2\theta - 1\right)$

$r^2 = 2\cos 2\theta$

**15. (a)** The first plane makes an angle of 70° with the positive *x*-axis, and is 150 miles from P:

$$x_1 = \cos 70°(150 - 375t)$$

$$y_1 = \sin 70°(150 - 375t)$$

Similarly for the second plane,

$$x_2 = \cos 135°(190 - 450t)$$

$$= \cos 45°(-190 + 450t)$$

$$y_2 = \sin 135°(190 - 450t)$$

$$= \sin 45°(190 - 450t).$$

**(b)** $d = \sqrt{(x_2 - x_1)^2 + (y_2 - y_1)^2}$

$$= \left[\left[\cos 45°(-190 + 450t) - \cos 70°(150 - 375t)\right]^2 + \left[\sin 45°(190 - 450t) - \sin 70°(150 - 375t)\right]^2\right]^{1/2}$$

**(c)**

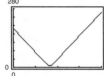

The minimum distance is 7.59 miles when $t = 0.4145$; Yes.

**17.**

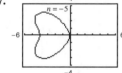 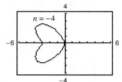 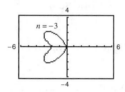 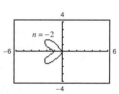

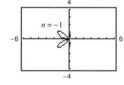 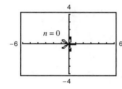 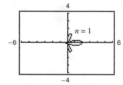 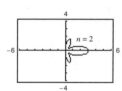

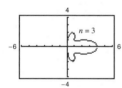 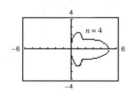 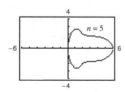

$n = 1, 2, 3, 4, 5$ produce "bells"; $n = -1, -2, -3, -4, -5$ produce "hearts".

# C H A P T E R  1 1
# Vectors and the Geometry of Space

# C H A P T E R  1 1
## Vectors and Geometry of Space

### Section 11.1   Vectors in the Plane

**1.** (a)  $\mathbf{v} = \langle 5 - 1, 4 - 2 \rangle = \langle 4, 2 \rangle$

(b)

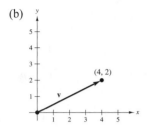

**3.** (a)  $\mathbf{v} = \langle -4 - 2, -3 - (-3) \rangle = \langle -6, 0 \rangle$

(b)

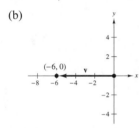

**5.**  $\mathbf{u} = \langle 5 - 3, 6 - 2 \rangle = \langle 2, 4 \rangle$

$\mathbf{v} = \langle 3 - 1, 8 - 4 \rangle = \langle 2, 4 \rangle$

$\mathbf{u} = \mathbf{v}$

**7.**  $\mathbf{u} = \langle 6 - 0, -2 - 3 \rangle = \langle 6, -5 \rangle$

$\mathbf{v} = \langle 9 - 3, 5 - 10 \rangle = \langle 6, -5 \rangle$

$\mathbf{u} = \mathbf{v}$

**9.** (b)  $\mathbf{v} = \langle 5 - 2, 5 - 0 \rangle = \langle 3, 5 \rangle$

(c)  $\mathbf{v} = 3\mathbf{i} + 5\mathbf{j}$

(a), (d)

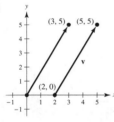

**11.** (b)  $\mathbf{v} = \langle 6 - 8, -1 - 3 \rangle = \langle -2, -4 \rangle$

(c)  $\mathbf{v} = -2\mathbf{i} - 4\mathbf{j}$

(a), (d)

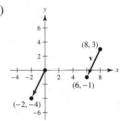

**13.** (b)  $\mathbf{v} = \langle 6 - 6, 6 - 2 \rangle = \langle 0, 4 \rangle$

(c)  $\mathbf{v} = 4\mathbf{j}$

(a) and (d).

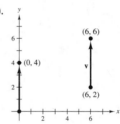

**15.** (b)  $\mathbf{v} = \left\langle \frac{1}{2} - \frac{3}{2}, 3 - \frac{4}{3} \right\rangle = \left\langle -1, \frac{5}{3} \right\rangle$

(c)  $\mathbf{v} = -\mathbf{i} + \frac{5}{3}\mathbf{j}$

(a) and (d)

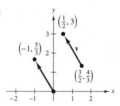

**17.** (a) $2\mathbf{v} = 2\langle 3, 5\rangle = \langle 6, 10\rangle$

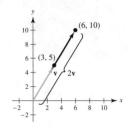

(b) $-3\mathbf{v} = \langle -9, -15\rangle$

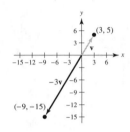

(c) $\frac{7}{2}\mathbf{v} = \left\langle \frac{21}{2}, \frac{35}{2}\right\rangle$

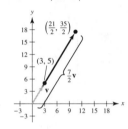

(d) $\frac{2}{3}\mathbf{v} = \left\langle 2, \frac{10}{3}\right\rangle$

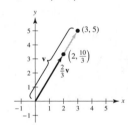

**19.**

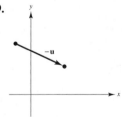

**21.**

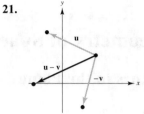

**23.** (a) $\frac{2}{3}\mathbf{u} = \frac{2}{3}\langle 4, 9\rangle = \left\langle \frac{8}{3}, 6\right\rangle$

(b) $\mathbf{v} - \mathbf{u} = \langle 2, -5\rangle - \langle 4, 9\rangle = \langle -2, -14\rangle$

(c) $2\mathbf{u} + 5\mathbf{v} = 2\langle 4, 9\rangle + 5\langle 2, -5\rangle = \langle 18, -7\rangle$

**25.** $\mathbf{v} = \frac{3}{2}(2\mathbf{i} - \mathbf{j}) = 3\mathbf{i} - \frac{3}{2}\mathbf{j} = \left\langle 3, -\frac{3}{2}\right\rangle$

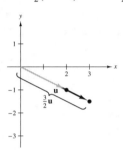

**27.** $\mathbf{v} = (2\mathbf{i} - \mathbf{j}) + 2(\mathbf{i} + 2\mathbf{j}) = 4\mathbf{i} + 3\mathbf{j} = \langle 4, 3\rangle$

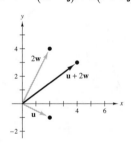

**29.** $u_1 - 4 = -1$       $u_1 = 3$

    $u_2 - 2 = 3$       $u_2 = 5$

                               $Q = (3, 5)$

**31.** $\|\mathbf{v}\| = \sqrt{0 + 7^2} = 7$

**33.** $\|\mathbf{v}\| = \sqrt{4^2 + 3^2} = 5$

**35.** $\|\mathbf{v}\| = \sqrt{6^2 + (-5)^2} = \sqrt{61}$

**37.** $\mathbf{v} = \langle 3, 12 \rangle$

$\|\mathbf{v}\| = \sqrt{3^2 + 12^2} = \sqrt{153}$

$\mathbf{u} = \dfrac{\mathbf{v}}{\|\mathbf{v}\|} = \dfrac{\langle 3, 12 \rangle}{\sqrt{153}} = \left\langle \dfrac{3}{\sqrt{153}}, \dfrac{12}{\sqrt{153}} \right\rangle$

$\qquad\qquad = \left\langle \dfrac{\sqrt{17}}{17}, \dfrac{4\sqrt{17}}{17} \right\rangle$ unit vector

**39.** $\mathbf{v} = \left\langle \dfrac{3}{2}, \dfrac{5}{2} \right\rangle$

$\|\mathbf{v}\| = \sqrt{\left(\dfrac{3}{2}\right)^2 + \left(\dfrac{5}{2}\right)^2} = \dfrac{\sqrt{34}}{2}$

$\mathbf{u} = \dfrac{\mathbf{v}}{\|\mathbf{v}\|} = \dfrac{\left\langle \left(\dfrac{3}{2}\right), \left(\dfrac{5}{2}\right) \right\rangle}{\dfrac{\sqrt{34}}{2}} = \left\langle \dfrac{3}{\sqrt{34}}, \dfrac{5}{\sqrt{34}} \right\rangle$

$\qquad\qquad = \left\langle \dfrac{3\sqrt{34}}{34}, \dfrac{5\sqrt{34}}{34} \right\rangle$ unit vector

**41.** $\mathbf{u} = \langle 1, -1 \rangle, \mathbf{v} = \langle -1, 2 \rangle$

(a) $\|\mathbf{u}\| = \sqrt{1 + 1} = \sqrt{2}$

(b) $\|\mathbf{v}\| = \sqrt{1 + 4} = \sqrt{5}$

(c) $\mathbf{u} + \mathbf{v} = \langle 0, 1 \rangle$

$\|\mathbf{u} + \mathbf{v}\| = \sqrt{0 + 1} = 1$

(d) $\dfrac{\mathbf{u}}{\|\mathbf{u}\|} = \dfrac{1}{\sqrt{2}} \langle 1, -1 \rangle$

$\left\| \dfrac{\mathbf{u}}{\|\mathbf{u}\|} \right\| = 1$

(e) $\dfrac{\mathbf{v}}{\|\mathbf{v}\|} = \dfrac{1}{\sqrt{5}} \langle -1, 2 \rangle$

$\left\| \dfrac{\mathbf{v}}{\|\mathbf{v}\|} \right\| = 1$

(f) $\dfrac{\mathbf{u} + \mathbf{v}}{\|\mathbf{u} + \mathbf{v}\|} = \langle 0, 1 \rangle$

$\left\| \dfrac{\mathbf{u} + \mathbf{v}}{\|\mathbf{u} + \mathbf{v}\|} \right\| = 1$

**43.** $\mathbf{u} = \left\langle 1, \dfrac{1}{2} \right\rangle, \mathbf{v} = \langle 2, 3 \rangle$

(a) $\|\mathbf{u}\| = \sqrt{1 + \dfrac{1}{4}} = \dfrac{\sqrt{5}}{2}$

(b) $\|\mathbf{v}\| = \sqrt{4 + 9} = \sqrt{13}$

(c) $\mathbf{u} + \mathbf{v} = \left\langle 3, \dfrac{7}{2} \right\rangle$

$\|\mathbf{u} + \mathbf{v}\| = \sqrt{9 + \dfrac{49}{4}} = \dfrac{\sqrt{85}}{2}$

(d) $\dfrac{\mathbf{u}}{\|\mathbf{u}\|} = \dfrac{2}{\sqrt{5}} \left\langle 1, \dfrac{1}{2} \right\rangle$

$\left\| \dfrac{\mathbf{u}}{\|\mathbf{u}\|} \right\| = 1$

(e) $\dfrac{\mathbf{v}}{\|\mathbf{v}\|} = \dfrac{1}{\sqrt{13}} \langle 2, 3 \rangle$

$\left\| \dfrac{\mathbf{v}}{\|\mathbf{v}\|} \right\| = 1$

(f) $\dfrac{\mathbf{u} + \mathbf{v}}{\|\mathbf{u} + \mathbf{v}\|} = \dfrac{2}{\sqrt{85}} \left\langle 3, \dfrac{7}{2} \right\rangle$

$\left\| \dfrac{\mathbf{u} + \mathbf{v}}{\|\mathbf{u} + \mathbf{v}\|} \right\| = 1$

**45.**
$\mathbf{u} = \langle 2, 1 \rangle$

$\|\mathbf{u}\| = \sqrt{5} \approx 2.236$

$\mathbf{v} = \langle 5, 4 \rangle$

$\|\mathbf{v}\| = \sqrt{41} \approx 6.403$

$\mathbf{u} + \mathbf{v} = \langle 7, 5 \rangle$

$\|\mathbf{u} + \mathbf{v}\| = \sqrt{74} \approx 8.602$

$\|\mathbf{u} + \mathbf{v}\| \le \|\mathbf{u}\| + \|\mathbf{v}\|$

$\sqrt{74} \le \sqrt{5} + \sqrt{41}$

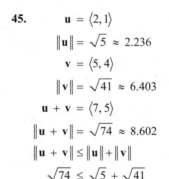
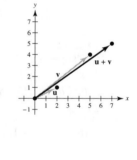

**47.** $\dfrac{\mathbf{u}}{\|\mathbf{u}\|} = \dfrac{1}{3} \langle 0, 3 \rangle = \langle 0, 1 \rangle$

$6 \left( \dfrac{\mathbf{u}}{\|\mathbf{u}\|} \right) = 6 \langle 0, 1 \rangle = \langle 0, 6 \rangle$

$\mathbf{v} = \langle 0, 6 \rangle$

**49.** $\dfrac{\mathbf{u}}{\|\mathbf{u}\|} = \dfrac{1}{\sqrt{5}} \langle -1, 2 \rangle = \left\langle -\dfrac{1}{\sqrt{5}}, \dfrac{2}{\sqrt{5}} \right\rangle$

$5 \left( \dfrac{\mathbf{u}}{\|\mathbf{u}\|} \right) = 5 \left\langle -\dfrac{1}{\sqrt{5}}, \dfrac{2}{\sqrt{5}} \right\rangle = \langle -\sqrt{5}, 2\sqrt{5} \rangle$

$\mathbf{v} = \langle -\sqrt{5}, 2\sqrt{5} \rangle$

**51.** $\mathbf{v} = 3[(\cos 0°)\mathbf{i} + (\sin 0°)\mathbf{j}] = 3\mathbf{i} = \langle 3, 0 \rangle$

**53.** $\mathbf{v} = 2[(\cos 150°)\mathbf{i} + (\sin 150°)\mathbf{j}]$

$\quad = -\sqrt{3}\mathbf{i} + \mathbf{j} = \langle -\sqrt{3}, 1 \rangle$

**55.** $\quad \mathbf{u} = (\cos 0°)\mathbf{i} + (\sin 0°)\mathbf{j} = \mathbf{i}$

$\quad \mathbf{v} = 3(\cos 45°)\mathbf{i} + 3(\sin 45°)\mathbf{j} = \dfrac{3\sqrt{2}}{2}\mathbf{i} + \dfrac{3\sqrt{2}}{2}\mathbf{j}$

$\mathbf{u} + \mathbf{v} = \left( \dfrac{2 + 3\sqrt{2}}{2} \right)\mathbf{i} + \dfrac{3\sqrt{2}}{2}\mathbf{j} = \left\langle \dfrac{2 + 3\sqrt{2}}{2}, \dfrac{3\sqrt{2}}{2} \right\rangle$

**57.** $\quad \mathbf{u} = 2(\cos 4)\mathbf{i} + 2(\sin 4)\mathbf{j}$

$\quad \mathbf{v} = (\cos 2)\mathbf{i} + (\sin 2)\mathbf{j}$

$\mathbf{u} + \mathbf{v} = (2\cos 4 + \cos 2)\mathbf{i} + (2\sin 4 + \sin 2)\mathbf{j}$

$\quad = \langle 2\cos 4 + \cos 2, 2\sin 4 + \sin 2 \rangle$

**59.** Answers will vary. *Sample answer:* A scalar is a real number such as 2. A vector is represented by a directed line segment. A vector has both magnitude and direction. For example $\langle \sqrt{3}, 1 \rangle$ has direction $\dfrac{\pi}{6}$ and a magnitude of 2.

**61.** (a) Vector. The velocity has both magnitude and direction.

(b) Scalar. The price is a number.

**For Exercises 63–67,**

$\mathbf{au} + \mathbf{bw} = a(\mathbf{i} + 2\mathbf{j}) + b(\mathbf{i} - \mathbf{j}) = (a + b)\mathbf{i} + (2a - b)\mathbf{j}.$

**63.** $\mathbf{v} = 2\mathbf{i} + \mathbf{j}$. So, $a + b = 2, 2a - b = 1$. Solving simultaneously, you have $a = 1, b = 1$.

**65.** $\mathbf{v} = 3\mathbf{i}$. So, $a + b = 3, 2a - b = 0$. Solving simultaneously, you have $a = 1, b = 2$.

**67.** $\mathbf{v} = \mathbf{i} + \mathbf{j}$. So, $a + b = 1, 2a - b = 1$. Solving simultaneously, you have $a = \frac{2}{3}, b = \frac{1}{3}$.

**69.** $f(x) = x^2, f'(x) = 2x, f'(3) = 6$

(a) $m = 6$. Let $\mathbf{w} = \langle 1, 6 \rangle, \|\mathbf{w}\| = \sqrt{37}$, then $\pm\dfrac{\mathbf{w}}{\|\mathbf{w}\|} = \pm\dfrac{1}{\sqrt{37}}\langle 1, 6 \rangle$.

(b) $m = -\frac{1}{6}$. Let $\mathbf{w} = \langle -6, 1 \rangle, \|\mathbf{w}\| = \sqrt{37}$, then $\pm\dfrac{\mathbf{w}}{\|\mathbf{w}\|} = \pm\dfrac{1}{\sqrt{37}}\langle -6, 1 \rangle$.

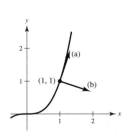

**71.** $f(x) = x^3, f'(x) = 3x^2 = 3$ at $x = 1$.

(a) $m = 3$. Let $\mathbf{w} = \langle 1, 3 \rangle, \|\mathbf{w}\| = \sqrt{10}$, then $\dfrac{\mathbf{w}}{\|\mathbf{w}\|} = \pm\dfrac{1}{\sqrt{10}}\langle 1, 3 \rangle$.

(b) $m = -\dfrac{1}{3}$. Let $\mathbf{w} = \langle 3, -1 \rangle, \|\mathbf{w}\| = \sqrt{10}$, then $\dfrac{\mathbf{w}}{\|\mathbf{w}\|} = \pm\dfrac{1}{\sqrt{10}}\langle 3, -1 \rangle$.

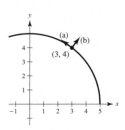

**73.** $f(x) = \sqrt{25 - x^2}$

$f'(x) = \dfrac{-x}{\sqrt{25 - x^2}} = \dfrac{-3}{4}$ at $x = 3$.

(a) $m = -\dfrac{3}{4}$. Let $\mathbf{w} = \langle -4, 3 \rangle, \|\mathbf{w}\| = 5$, then $\dfrac{\mathbf{w}}{\|\mathbf{w}\|} = \pm\dfrac{1}{5}\langle -4, 3 \rangle$.

(b) $m = \dfrac{4}{3}$. Let $\mathbf{w} = \langle 3, 4 \rangle, \|\mathbf{w}\| = 5$, then $\dfrac{\mathbf{w}}{\|\mathbf{w}\|} = \pm\dfrac{1}{5}\langle 3, 4 \rangle$.

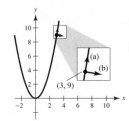

**75.**   $u = \dfrac{\sqrt{2}}{2}\mathbf{i} + \dfrac{\sqrt{2}}{2}\mathbf{j}$

$u + v = \sqrt{2}\mathbf{j}$

$v = (u + v) - u = -\dfrac{\sqrt{2}}{2}\mathbf{i} + \dfrac{\sqrt{2}}{2}\mathbf{j} = \left\langle -\dfrac{\sqrt{2}}{2}, \dfrac{\sqrt{2}}{2} \right\rangle$

**77.** (a)–(c) Programs will vary.

(d)  Magnitude $\approx 63.5$

Direction $\approx -8.26°$

**79.**  $\|F_1\| = 2, \theta_{F_1} = 33°$

$\|F_2\| = 3, \theta_{F_2} = -125°$

$\|F_3\| = 2.5, \theta_{F_3} = 110°$

$\|R\| = \|F_1 + F_2 + F_3\| \approx 1.33$

$\theta_R = \theta_{F_1 + F_2 + F_3} \approx 132.5°$

**81.**  $F_1 + F_2 = (500\cos 30°\mathbf{i} + 500\sin 30°\mathbf{j}) + (200\cos(-45°)\mathbf{i} + 200\sin(-45°)\mathbf{j}) = (250\sqrt{3} + 100\sqrt{2})\mathbf{i} + (250 - 100\sqrt{2})\mathbf{j}$

$\|F_1 + F_2\| = \sqrt{(250\sqrt{3} + 100\sqrt{2})^2 + (250 - 100\sqrt{2})^2} \approx 584.6 \text{ lb}$

$\tan\theta = \dfrac{250 - 100\sqrt{2}}{250\sqrt{3} + 100\sqrt{2}} \Rightarrow \theta \approx 10.7°$

**83.**  $F_1 + F_2 + F_3 = (75\cos 30°\mathbf{i} + 75\sin 30°\mathbf{j}) + (100\cos 45°\mathbf{i} + 100\sin 45°\mathbf{j}) + (125\cos 120°\mathbf{i} + 125\sin 120°\mathbf{j})$

$= \left(\tfrac{75}{2}\sqrt{3} + 50\sqrt{2} - \tfrac{125}{2}\right)\mathbf{i} + \left(\tfrac{75}{2} + 50\sqrt{2} + \tfrac{125}{2}\sqrt{3}\right)\mathbf{j}$

$\|R\| = \|F_1 + F_2 + F_3\| \approx 228.5 \text{ lb}$

$\theta_R = \theta_{F_1 + F_2 + F_3} \approx 71.3°$

**85.** (a)  The forces act along the same direction. $\theta = 0°$.

(b)  The forces cancel out each other. $\theta = 180°$.

(c)  No, the magnitude of the resultant can not be greater than the sum.

**87.**  $(-4, -1), (6, 5), (10, 3)$

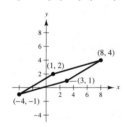

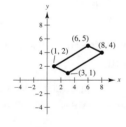

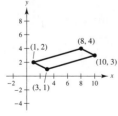

**89.**  $u = \overrightarrow{CB} = \|u\|(\cos 30°\mathbf{i} + \sin 30°\mathbf{j})$

$v = \overrightarrow{CA} = \|v\|(\cos 130°\mathbf{i} + \sin 130°\mathbf{j})$

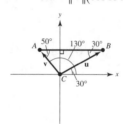

Vertical components: $\|u\|\sin 30° + \|v\|\sin 130° = 3000$

Horizontal components: $\|u\|\cos 30° + \|v\|\cos 130° = 0$

Solving this system, you obtain

$\|u\| \approx 1958.1 \text{ pounds}$

$\|v\| \approx 2638.2 \text{ pounds}$

**91.**  Horizontal component $= \|v\|\cos\theta$

$= 1200\cos 6° \approx 1193.43 \text{ ft/sec}$

Vertical component $= \|v\|\sin\theta$

$= 1200\sin 6° \approx 125.43 \text{ ft/sec}$

**93.** $\mathbf{u} = 900(\cos 148°\,\mathbf{i} + \sin 148°\,\mathbf{j})$

$\mathbf{v} = 100(\cos 45°\,\mathbf{i} + \sin 45°\,\mathbf{j})$

$\mathbf{u} + \mathbf{v} = (900 \cos 148° + 100 \cos 45°)\mathbf{i} + (900 \sin 148° + 100 \sin 45°)\mathbf{j}$

$\approx -692.53\mathbf{i} + 547.64\mathbf{j}$

$\theta \approx \arctan\left(\dfrac{547.64}{-692.53}\right) \approx -38.34°; \; 38.34°$ North of West

$\|\mathbf{u} + \mathbf{v}\| \approx \sqrt{(-692.53)^2 + (547.64)^2} \approx 882.9$ km/h

**95.** True

**97.** True

**99.** False

$\|a\mathbf{i} + b\mathbf{j}\| = \sqrt{2}\,|a|$

**101.** $\|\mathbf{u}\| = \sqrt{\cos^2\theta + \sin^2\theta} = 1,$

$\|\mathbf{v}\| = \sqrt{\sin^2\theta + \cos^2\theta} = 1$

**103.** Let $\mathbf{u}$ and $\mathbf{v}$ be the vectors that determine the parallelogram, as indicated in the figure. The two diagonals are $\mathbf{u} + \mathbf{v}$ and $\mathbf{v} - \mathbf{u}$. So,

$\mathbf{r} = x(\mathbf{u} + \mathbf{v}), \mathbf{s} = 4(\mathbf{v} - \mathbf{u})$. But,

$\mathbf{u} = \mathbf{r} - \mathbf{s}$

$= x(\mathbf{u} + \mathbf{v}) - y(\mathbf{v} - \mathbf{u}) = (x + y)\mathbf{u} + (x - y)\mathbf{v}.$

So, $x + y = 1$ and $x - y = 0$. Solving you have

$x = y = \frac{1}{2}.$

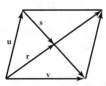

**105.** The set is a circle of radius 5, centered at the origin.

$\|\mathbf{u}\| = \|\langle x, y\rangle\| = \sqrt{x^2 + y^2} = 5 \Rightarrow x^2 + y^2 = 25$

# Section 11.2 Space Coordinates and Vectors in Space

**1.** $A(2, 3, 4)$

$B(-1, -2, 2)$

**3.**

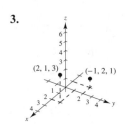

**5.**

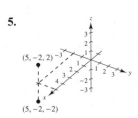

**7.** $x = -3, y = 4, z = 5$: $(-3, 4, 5)$

**9.** $y = z = 0, x = 12$: $(12, 0, 0)$

**11.** The $z$-coordinate is 0.

**13.** The point is 6 units above the $xy$-plane.

**15.** The point is on the plane parallel to the $yz$-plane that passes through $x = -3$.

**17.** The point is to the left of the $xz$-plane.

**19.** The point is on or between the planes $y = 3$ and $y = -3$.

**21.** The point $(x, y, z)$ is 3 units below the $xy$-plane, and below either quadrant I or III.

**23.** The point could be above the $xy$-plane and so above quadrants II or IV, or below the $xy$-plane, and so below quadrants I or III.

**25.** $d = \sqrt{(-4 - 0)^2 + (2 - 0)^2 + (7 - 0)^2}$

$= \sqrt{16 + 4 + 49} = \sqrt{69}$

**27.** $d = \sqrt{(6 - 1)^2 + (-2 - (-2))^2 + (-2 - 4)^2}$

$= \sqrt{25 + 0 + 36} = \sqrt{61}$

**29.** $A(0, 0, 4), B(2, 6, 7), C(6, 4, -8)$

$\|AB\| = \sqrt{2^2 + 6^2 + 3^2} = \sqrt{49} = 7$

$\|AC\| = \sqrt{6^2 + 4^2 + (-12)^2} = \sqrt{196} = 14$

$\|BC\| = \sqrt{4^2 + (-2)^2 + (-15)^2} = \sqrt{245} = 7\sqrt{5}$

$\|BC\|^2 = 245 = 49 + 196 = \|AB\|^2 + \|AC\|^2$

Right triangle

**31.** $A(-1, 0, -2), B(-1, 5, 2), C(-3, -1, 1)$

$\|AB\| = \sqrt{0 + 25 + 16} = \sqrt{41}$

$\|AC\| = \sqrt{4 + 1 + 9} = \sqrt{14}$

$\|BC\| = \sqrt{4 + 36 + 1} = \sqrt{41}$

Because $\|AB\| = \|BC\|$, the triangle is isosceles.

**41.**
$$x^2 + y^2 + z^2 - 2x + 6y + 8z + 1 = 0$$
$$\left(x^2 - 2x + 1\right) + \left(y^2 + 6y + 9\right) + \left(z^2 + 8z + 16\right) = -1 + 1 + 9 + 16$$
$$(x - 1)^2 + (y + 3)^2 + (z + 4)^2 = 25$$

Center: $(1, -3, -4)$

Radius: 5

**43.**
$$9x^2 + 9y^2 + 9z^2 - 6x + 18y + 1 = 0$$
$$x^2 + y^2 + z^2 - \tfrac{2}{3}x + 2y + \tfrac{1}{9} = 0$$
$$\left(x^2 - \tfrac{2}{3}x + \tfrac{1}{9}\right) + \left(y^2 + 2y + 1\right) + z^2 = -\tfrac{1}{9} + \tfrac{1}{9} + 1$$
$$\left(x - \tfrac{1}{3}\right)^2 + (y + 1)^2 + (z - 0)^2 = 1$$

Center: $\left(\tfrac{1}{3}, -1, 0\right)$

Radius: 1

**45.** $x^2 + y^2 + z^2 \leq 36$

Solid sphere of radius 6 centered at origin.

**47.**
$$x^2 + y^2 + z^2 < 4x - 6y + 8z - 13$$
$$\left(x^2 - 4x + 4\right) + \left(y^2 + 6y + 9\right) + \left(z^2 - 8z + 16\right) < 4 + 9 + 16 - 13$$
$$(x - 2)^2 + (y + 3)^2 + (z - 4)^2 < 16$$

Interior of sphere of radius 4 centered at $(2, -3, 4)$.

**49.** (a) $\mathbf{v} = \langle 2 - 4, 4 - 2, 3 - 1 \rangle = \langle -2, 2, 2 \rangle$

(b) $\mathbf{v} = -2\mathbf{i} + 2\mathbf{j} + 2\mathbf{k}$

(c)

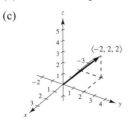

**33.** The $z$-coordinate is changed by 5 units:

$(0, 0, 9), (2, 6, 12), (6, 4, -3)$

**35.** $\left(\dfrac{5 + (-2)}{2}, \dfrac{-9 + 3}{2}, \dfrac{7 + 3}{2}\right) = \left(\dfrac{3}{2}, -3, 5\right)$

**37.** Center: $(0, 2, 5)$

Radius: 2

$(x - 0)^2 + (y - 2)^2 + (z - 5)^2 = 4$

**39.** Center: $\dfrac{(2, 0, 0) + (0, 6, 0)}{2} = (1, 3, 0)$

Radius: $\sqrt{10}$

$(x - 1)^2 + (y - 3)^2 + (z - 0)^2 = 10$

**51.** (a) $\mathbf{v} = \langle 0 - 3, 3 - 3, 3 - 0 \rangle = \langle -3, 0, 3 \rangle$

(b) $\mathbf{v} = -3\mathbf{i} + 3\mathbf{k}$

(c)

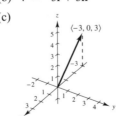

**53.** $\langle 4 - 3, 1 - 2, 6 - 0 \rangle = \langle 1, -1, 6 \rangle$

$\|\langle 1, -1, 6 \rangle\| = \sqrt{1 + 1 + 36} = \sqrt{38}$

Unit vector: $\dfrac{\langle 1, -1, 6 \rangle}{\sqrt{38}} = \left\langle \dfrac{1}{\sqrt{38}}, \dfrac{-1}{\sqrt{38}}, \dfrac{6}{\sqrt{38}} \right\rangle$

**55.** $\langle -5 - (-4), 3 - 3, 0 - 1 \rangle = \langle -1, 0, -1 \rangle$

$\|\langle -1, 0, -1 \rangle\| = \sqrt{1 + 1} = \sqrt{2}$

Unit vector: $\dfrac{\langle -1, 0, -1 \rangle}{\sqrt{2}} = \left\langle \dfrac{-1}{\sqrt{2}}, 0, \dfrac{-1}{\sqrt{2}} \right\rangle$

**57.** (b) $\mathbf{v} = \langle 3 - (-1), 3 - 2, 4 - 3 \rangle = \langle 4, 1, 1 \rangle$

(c) $\mathbf{v} = 4\mathbf{i} + \mathbf{j} + \mathbf{k}$

(a), (d)

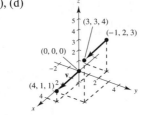

**59.** $(q_1, q_2, q_3) - (0, 6, 2) = (3, -5, 6)$

$Q = (3, 1, 8)$

**61.** (a) $2\mathbf{v} = \langle 2, 4, 4 \rangle$

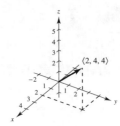

(b) $-\mathbf{v} = \langle -1, -2, -2 \rangle$

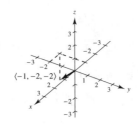

(c) $\frac{3}{2}\mathbf{v} = \left\langle \frac{3}{2}, 3, 3 \right\rangle$

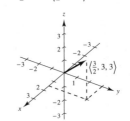

(d) $0\mathbf{v} = \langle 0, 0, 0 \rangle$

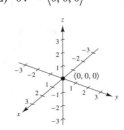

**63.** $\mathbf{z} = \mathbf{u} - \mathbf{v} = \langle 1, 2, 3 \rangle - \langle 2, 2, -1 \rangle = \langle -1, 0, 4 \rangle$

**65.** $\mathbf{z} = 2\mathbf{u} + 4\mathbf{v} - \mathbf{w} = \langle 2, 4, 6 \rangle + \langle 8, 8, -4 \rangle - \langle 4, 0, -4 \rangle = \langle 6, 12, 6 \rangle$

**67.** $2\mathbf{z} - 3\mathbf{u} = 2\langle z_1, z_2, z_3 \rangle - 3\langle 1, 2, 3 \rangle = \langle 4, 0, -4 \rangle$

$2z_1 - 3 = 4 \Rightarrow z_1 = \frac{7}{2}$

$2z_2 - 6 = 0 \Rightarrow z_2 = 3$

$2z_3 - 9 = -4 \Rightarrow z_3 = \frac{5}{2}$

$\mathbf{z} = \left\langle \frac{7}{2}, 3, \frac{5}{2} \right\rangle$

**69.** (a) and (b) are parallel because $\langle -6, -4, 10 \rangle = -2\langle 3, 2, -5 \rangle$ and $\left\langle 2, \frac{4}{3}, -\frac{10}{3} \right\rangle = \frac{2}{3}\langle 3, 2, -5 \rangle$.

**71.** $\mathbf{z} = -3\mathbf{i} + 4\mathbf{j} + 2\mathbf{k}$

(a) is parallel because $-6\mathbf{i} + 8\mathbf{j} + 4\mathbf{k} = 2\mathbf{z}$.

**73.** $P(0, -2, -5), Q(3, 4, 4), R(2, 2, 1)$

$$\overline{PQ} = \langle 3, 6, 9 \rangle$$

$$\overline{PR} = \langle 2, 4, 6 \rangle$$

$$\langle 3, 6, 9 \rangle = \tfrac{3}{2}\langle 2, 4, 6 \rangle$$

So, $\overline{PQ}$ and $\overline{PR}$ are parallel, the points are collinear.

**75.** $P(1, 2, 4), Q(2, 5, 0), R(0, 1, 5)$

$$\overline{PQ} = \langle 1, 3, -4 \rangle$$

$$\overline{PR} = \langle -1, -1, 1 \rangle$$

Because $\overline{PQ}$ and $\overline{PR}$ are not parallel, the points are not collinear.

**77.** $A(2, 9, 1), B(3, 11, 4), C(0, 10, 2), D(1, 12, 5)$

$$\overline{AB} = \langle 1, 2, 3 \rangle$$

$$\overline{CD} = \langle 1, 2, 3 \rangle$$

$$\overline{AC} = \langle -2, 1, 1 \rangle$$

$$\overline{BD} = \langle -2, 1, 1 \rangle$$

Because $\overline{AB} = \overline{CD}$ and $\overline{AC} = \overline{BD}$, the given points form the vertices of a parallelogram.

**79.** $\mathbf{v} = \langle 0, 0, 0 \rangle$

$$\|\mathbf{v}\| = 0$$

**81.** $\mathbf{v} = 3\mathbf{j} - 5\mathbf{k} = \langle 0, 3, -5 \rangle$

$$\|\mathbf{v}\| = \sqrt{0 + 9 + 25} = \sqrt{34}$$

**83.** $\mathbf{v} = \mathbf{i} - 2\mathbf{j} - 3\mathbf{k} = \langle 1, -2, -3 \rangle$

$$\|\mathbf{v}\| = \sqrt{1 + 4 + 9} = \sqrt{14}$$

**85.** $\mathbf{v} = \langle 2, -1, 2 \rangle$

$$\|\mathbf{v}\| = \sqrt{4 + 1 + 4} = 3$$

(a) $\dfrac{\mathbf{v}}{\|\mathbf{v}\|} = \dfrac{1}{3}\langle 2, -1, 2 \rangle$

(b) $-\dfrac{\mathbf{v}}{\|\mathbf{v}\|} = -\dfrac{1}{3}\langle 2, -1, 2 \rangle$

**87.** $\mathbf{v} = \langle 3, 2, -5 \rangle$

$$\|\mathbf{v}\| = \sqrt{9 + 4 + 25} = \sqrt{38}$$

(a) $\dfrac{\mathbf{v}}{\|\mathbf{v}\|} = \dfrac{1}{\sqrt{38}}\langle 3, 2, -5 \rangle$

(b) $-\dfrac{\mathbf{v}}{\|\mathbf{v}\|} = -\dfrac{1}{\sqrt{38}}\langle 3, 2, -5 \rangle$

**89.** (a)–(d) Programs will vary.

(e) $\mathbf{u} + \mathbf{v} = \langle 4, 7.5, -2 \rangle$

$$\|\mathbf{u} + \mathbf{v}\| \approx 8.732$$

$$\|\mathbf{u}\| \approx 5.099$$

$$\|\mathbf{v}\| \approx 9.019$$

**91.** $\|c\mathbf{v}\| = \|c(2\mathbf{i} + 2\mathbf{j} - \mathbf{k})\| = \sqrt{4c^2 + 4c^2 + c^2} = 7$

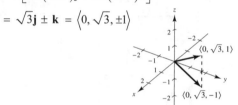

$$\sqrt{9c^2} = 7$$

$$9c^2 = 49$$

$$c = \pm\tfrac{7}{3}$$

**93.** $\mathbf{v} = 10\dfrac{\mathbf{u}}{\|\mathbf{u}\|} = 10\dfrac{\langle 0, 3, 3 \rangle}{3\sqrt{2}}$

$$= 10\left\langle 0, \dfrac{1}{\sqrt{2}}, \dfrac{1}{\sqrt{2}} \right\rangle = \left\langle 0, \dfrac{10}{\sqrt{2}}, \dfrac{10}{\sqrt{2}} \right\rangle$$

**95.** $\mathbf{v} = \dfrac{3}{2}\dfrac{\mathbf{u}}{\|\mathbf{u}\|} = \dfrac{3}{2}\dfrac{\langle 2, -2, 1 \rangle}{3} = \dfrac{3}{2}\left\langle \dfrac{2}{3}, \dfrac{-2}{3}, \dfrac{1}{3} \right\rangle = \left\langle 1, -1, \dfrac{1}{2} \right\rangle$

**97.** $\mathbf{v} = 2\left[\cos(\pm 30°)\mathbf{j} + \sin(\pm 30°)\mathbf{k}\right]$

$$= \sqrt{3}\mathbf{j} \pm \mathbf{k} = \langle 0, \sqrt{3}, \pm 1 \rangle$$

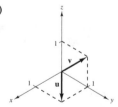

**99.**

$$\mathbf{v} = \langle -3, -6, 3 \rangle$$

$$\tfrac{2}{3}\mathbf{v} = \langle -2, -4, 2 \rangle$$

$$(4, 3, 0) + (-2, -4, 2) = (2, -1, 2)$$

**101.** (a)

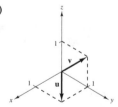

(b) $\mathbf{w} = a\mathbf{u} + b\mathbf{v} = a\mathbf{i} + (a + b)\mathbf{j} + b\mathbf{k} = 0$

$$a = 0, a + b = 0, b = 0$$

So, $a$ and $b$ are both zero.

(c) $a\mathbf{i} + (a + b)\mathbf{j} + b\mathbf{k} = \mathbf{i} + 2\mathbf{j} + \mathbf{k}$

$$a = 1, a + b = 2, b = 1$$

$$\mathbf{w} = \mathbf{u} + \mathbf{v}$$

(d) $a\mathbf{i} + (a + b)\mathbf{j} + b\mathbf{k} = \mathbf{i} + 2\mathbf{j} + 3\mathbf{k}$

$$a = 1, a + b = 2, b = 3$$

Not possible

**103.** $x_0$ is directed distance to *yz*-plane.

$y_0$ is directed distance to *xz*-plane.

$z_0$ is directed distance to *xy*-plane.

**105.** $(x - x_0)^2 + (y - y_0)^2 + (z - z_0)^2 = r^2$

**107.**

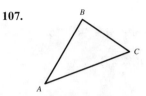

$$\overrightarrow{AB} + \overrightarrow{BC} = \overrightarrow{AC}$$

So, $\overrightarrow{AB} + \overrightarrow{BC} + \overrightarrow{CA} = \overrightarrow{AC} + \overrightarrow{CA} = \mathbf{0}$

**109.** (a) The height of the right triangle is $h = \sqrt{L^2 - 18^2}$.

The vector $\overrightarrow{PQ}$ is given by

$$\overrightarrow{PQ} = \langle 0, -18, h \rangle.$$

The tension vector **T** in each wire is

$$\mathbf{T} = c\langle 0, -18, h \rangle \text{ where } ch = \frac{24}{3} = 8.$$

So, $\mathbf{T} = \dfrac{8}{h}\langle 0, -18, h \rangle$ and

$$T = \|\mathbf{T}\| = \frac{8}{h}\sqrt{18^2 + h^2} = \frac{8}{\sqrt{L^2 - 18^2}}\sqrt{18^2 + \left(L^2 - 18^2\right)} = \frac{8L}{\sqrt{L^2 - 18^2}}, \ L > 18.$$

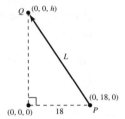

(b)

| $L$ | 20 | 25 | 30 | 35 | 40 | 45 | 50 |
|---|---|---|---|---|---|---|---|
| $T$ | 18.4 | 11.5 | 10 | 9.3 | 9.0 | 8.7 | 8.6 |

(c)

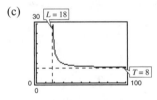

$x = 18$ is a vertical asymptote and $y = 8$ is a horizontal asymptote.

(d) $\displaystyle\lim_{L \to 18^+} \frac{8L}{\sqrt{L^2 - 18^2}} = \infty$

$$\lim_{L \to \infty} \frac{8L}{\sqrt{L^2 - 18^2}} = \lim_{L \to \infty} \frac{8}{\sqrt{1 - (18/L)^2}} = 8$$

(e) From the table, $T = 10$ implies $L = 30$ inches.

**111.** Let $\alpha$ be the angle between **v** and the coordinate axes.

$$\mathbf{v} = (\cos\alpha)\mathbf{i} + (\cos\alpha)\mathbf{j} + (\cos\alpha)\mathbf{k}$$

$$\|\mathbf{v}\| = \sqrt{3}\cos\alpha = 1$$

$$\cos\alpha = \frac{1}{\sqrt{3}} = \frac{\sqrt{3}}{3}$$

$$\mathbf{v} = \frac{\sqrt{3}}{3}(\mathbf{i} + \mathbf{j} + \mathbf{k}) = \frac{\sqrt{3}}{3}\langle 1, 1, 1 \rangle$$

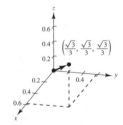

**113.** $\overrightarrow{AB} = \langle 0, 70, 115 \rangle$, $\mathbf{F_1} = C_1\langle 0, 70, 115 \rangle$

$\overrightarrow{AC} = \langle -60, 0, 115 \rangle$, $\mathbf{F_2} = C_2\langle -60, 0, 115 \rangle$

$\overrightarrow{AD} = \langle 45, -65, 115 \rangle$, $\mathbf{F_3} = C_3\langle 45, -65, 115 \rangle$

$\mathbf{F} = \mathbf{F_1} + \mathbf{F_2} + \mathbf{F_3} = \langle 0, 0, 500 \rangle$

So:       $-60C_2 + 45C_3 = 0$

$\quad\quad 70C_1 \quad\quad\quad - 65C_3 = 0$

$\quad 115(C_1 + C_2 + C_3) = 500$

Solving this system yields $C_1 = \frac{104}{69}$, $C_2 = \frac{28}{23}$, and $C_3 = \frac{112}{69}$. So:

$\|\mathbf{F_1}\| \approx 202.919\,\text{N}$

$\|\mathbf{F_2}\| \approx 157.909\,\text{N}$

$\|\mathbf{F_3}\| \approx 226.521\,\text{N}$

**115.** $d(AP) = 2d(BP)$

$$\sqrt{x^2 + (y+1)^2 + (z-1)^2} = 2\sqrt{(x-1)^2 + (y-2)^2 + z^2}$$

$$x^2 + y^2 + z^2 + 2y - 2z + 2 = 4(x^2 + y^2 + z^2 - 2x - 4y + 5)$$

$$0 = 3x^2 + 3y^2 + 3z^2 - 8x - 18y + 2z + 18$$

$$-6 + \frac{16}{9} + 9 + \frac{1}{9} = \left(x^2 - \frac{8}{3}x + \frac{16}{9}\right) + (y^2 - 6y + 9) + \left(z^2 + \frac{2}{3}z + \frac{1}{9}\right)$$

$$\frac{44}{9} = \left(x - \frac{4}{3}\right)^2 + (y-3)^2 + \left(z + \frac{1}{3}\right)^2$$

Sphere; center: $\left(\frac{4}{3}, 3, -\frac{1}{3}\right)$, radius: $\frac{2\sqrt{11}}{3}$

# Section 11.3    The Dot Product of Two Vectors

**1.** $\mathbf{u} = \langle 3, 4 \rangle$, $\mathbf{v} = \langle -1, 5 \rangle$

(a) $\mathbf{u} \cdot \mathbf{v} = 3(-1) + 4(5) = 17$

(b) $\mathbf{u} \cdot \mathbf{u} = 3(3) + 4(4) = 25$

(c) $\|\mathbf{u}\|^2 = 3^2 + 4^2 = 25$

(d) $(\mathbf{u} \cdot \mathbf{v})\mathbf{v} = 17\langle -1, 5 \rangle = \langle -17, 85 \rangle$

(e) $\mathbf{u} \cdot (2\mathbf{v}) = 2(\mathbf{u} \cdot \mathbf{v}) = 2(17) = 34$

**3.** $\mathbf{u} = \langle 6, -4 \rangle$, $\mathbf{v} = \langle -3, 2 \rangle$

(a) $\mathbf{u} \cdot \mathbf{v} = 6(-3) + (-4)(2) = -26$

(b) $\mathbf{u} \cdot \mathbf{u} = 6(6) + (-4)(-4) = 52$

(c) $\|\mathbf{u}\|^2 = 6^2 + (-4)^2 = 52$

(d) $(\mathbf{u} \cdot \mathbf{v})\mathbf{v} = -26\langle -3, 2 \rangle = \langle 78, -52 \rangle$

(e) $\mathbf{u} \cdot (2\mathbf{v}) = 2(\mathbf{u} \cdot \mathbf{v}) = 2(-26) = -52$

**5.** $\mathbf{u} = \langle 2, -3, 4 \rangle$, $\mathbf{v} = \langle 0, 6, 5 \rangle$

(a) $\mathbf{u} \cdot \mathbf{v} = 2(0) + (-3)(6) + (4)(5) = 2$

(b) $\mathbf{u} \cdot \mathbf{u} = 2(2) + (-3)(-3) + 4(4) = 29$

(c) $\|\mathbf{u}\|^2 = 2^2 + (-3)^2 + 4^2 = 29$

(d) $(\mathbf{u} \cdot \mathbf{v})\mathbf{v} = 2\langle 0, 6, 5 \rangle = \langle 0, 12, 10 \rangle$

(e) $\mathbf{u} \cdot (2\mathbf{v}) = 2(\mathbf{u} \cdot \mathbf{v}) = 2(2) = 4$

**7.** $\mathbf{u} = 2\mathbf{i} - \mathbf{j} + \mathbf{k}$, $\mathbf{v} = \mathbf{i} - \mathbf{k}$

(a) $\mathbf{u} \cdot \mathbf{v} = 2(1) + (-1)(0) + 1(-1) = 1$

(b) $\mathbf{u} \cdot \mathbf{u} = 2(2) + (-1)(-1) + (1)(1) = 6$

(c) $\|\mathbf{u}\|^2 = 2^2 + (-1)^2 + 1^2 = 6$

(d) $(\mathbf{u} \cdot \mathbf{v})\mathbf{v} = \mathbf{v} = \mathbf{i} - \mathbf{k}$

(e) $\mathbf{u} \cdot (2\mathbf{v}) = 2(\mathbf{u} \cdot \mathbf{v}) = 2$

**9.** $\dfrac{\mathbf{u} \cdot \mathbf{v}}{\|\mathbf{u}\|\,\|\mathbf{v}\|} = \cos\theta$

$\mathbf{u} \cdot \mathbf{v} = (8)(5)\cos\dfrac{\pi}{3} = 20$

**11.** $\mathbf{u} = \langle 1, 1\rangle,\ \mathbf{v} = \langle 2, -2\rangle$

$\cos\theta = \dfrac{\mathbf{u}\cdot\mathbf{v}}{\|\mathbf{u}\|\,\|\mathbf{v}\|} = \dfrac{0}{\sqrt{2}\sqrt{8}} = 0$

$\theta = \dfrac{\pi}{2}$

**13.** $\mathbf{u} = 3\mathbf{i} + \mathbf{j},\ \mathbf{v} = -2\mathbf{i} + 4\mathbf{j}$

$\cos\theta = \dfrac{\mathbf{u}\cdot\mathbf{v}}{\|\mathbf{u}\|\,\|\mathbf{v}\|} = \dfrac{-2}{\sqrt{10}\sqrt{20}} = \dfrac{-1}{5\sqrt{2}}$

$\theta = \arccos\left(-\dfrac{1}{5\sqrt{2}}\right) \approx 98.1°$

**15.** $\mathbf{u} = \langle 1, 1, 1\rangle,\ \mathbf{v} = \langle 2, 1, -1\rangle$

$\cos\theta = \dfrac{\mathbf{u}\cdot\mathbf{v}}{\|\mathbf{u}\|\,\|\mathbf{v}\|} = \dfrac{2}{\sqrt{3}\sqrt{6}} = \dfrac{\sqrt{2}}{3}$

$\theta = \arccos\dfrac{\sqrt{2}}{3} \approx 61.9°$

**17.** $\mathbf{u} = 3\mathbf{i} + 4\mathbf{j},\ \mathbf{v} = 2\mathbf{j} + 3\mathbf{k}$

$\cos\theta = \dfrac{\mathbf{u}\cdot\mathbf{v}}{\|\mathbf{u}\|\,\|\mathbf{v}\|} = \dfrac{-8}{5\sqrt{13}} = \dfrac{-8\sqrt{13}}{65}$

$\theta = \arccos\left(-\dfrac{8\sqrt{13}}{65}\right) \approx 116.3°$

**19.** $\mathbf{u} = \langle 4, 0\rangle,\ \mathbf{v} = \langle 1, 1\rangle$

$\mathbf{u} \neq c\mathbf{v} \Rightarrow$ not parallel

$\mathbf{u} \cdot \mathbf{v} = 4 \neq 0 \Rightarrow$ not orthogonal

Neither

**21.** $\mathbf{u} = \langle 4, 3\rangle,\ \mathbf{v} = \left\langle \frac{1}{2}, -\frac{2}{3}\right\rangle$

$\mathbf{u} \neq c\mathbf{v} \Rightarrow$ not parallel

$\mathbf{u} \cdot \mathbf{v} = 0 \Rightarrow$ orthogonal

**23.** $\mathbf{u} = \mathbf{j} + 6\mathbf{k},\ \mathbf{v} = \mathbf{i} - 2\mathbf{j} - \mathbf{k}$

$\mathbf{u} \neq c\mathbf{v} \Rightarrow$ not parallel

$\mathbf{u} \cdot \mathbf{v} = -8 \neq 0 \Rightarrow$ not orthogonal

Neither

**25.** $\mathbf{u} = \langle 2, -3, 1\rangle,\ \mathbf{v} = \langle -1, -1, -1\rangle$

$\mathbf{u} \neq c\mathbf{v} \Rightarrow$ not parallel

$\mathbf{u} \cdot \mathbf{v} = 0 \Rightarrow$ orthogonal

**27.** The vector $\langle 1, 2, 0\rangle$ joining $(1, 2, 0)$ and $(0, 0, 0)$ is perpendicular to the vector $\langle -2, 1, 0\rangle$ joining $(-2, 1, 0)$ and $(0, 0, 0)$: $\langle 1, 2, 0\rangle \cdot \langle -2, 1, 0\rangle = 0$

The triangle has a right angle, so it is a right triangle.

**29.** $A(2, 0, 1),\ B(0, 1, 2),\ C\left(-\frac{1}{2}, \frac{3}{2}, 0\right)$

$\overrightarrow{AB} = \langle -2, 1, 1\rangle \qquad \overrightarrow{BA} = \langle 2, -1, -1\rangle$

$\overrightarrow{AC} = \left\langle -\frac{5}{2}, \frac{3}{2}, -1\right\rangle \qquad \overrightarrow{CA} = \left\langle \frac{5}{2}, -\frac{3}{2}, 1\right\rangle$

$\overrightarrow{BC} = \left\langle -\frac{1}{2}, \frac{1}{2}, -2\right\rangle \qquad \overrightarrow{CB} = \left\langle \frac{1}{2}, -\frac{1}{2}, 2\right\rangle$

$\overrightarrow{AB} \cdot \overrightarrow{AC} = 5 + \frac{3}{2} - 1 > 0$

$\overrightarrow{BA} \cdot \overrightarrow{BC} = -1 - \frac{1}{2} + 2 > 0$

$\overrightarrow{CA} \cdot \overrightarrow{CB} = \frac{5}{4} + \frac{3}{4} + 2 > 0$

The triangle has three acute angles, so it is an acute triangle.

**31.** $\mathbf{u} = \mathbf{i} + 2\mathbf{j} + 2\mathbf{k},\ \|\mathbf{u}\| = 3$

$\cos\alpha = \frac{1}{3}$

$\cos\beta = \frac{2}{3}$

$\cos\gamma = \frac{2}{3}$

$\cos^2\alpha + \cos^2\beta + \cos^2\gamma = \frac{1}{9} + \frac{4}{9} + \frac{4}{9} = 1$

**33.** $\mathbf{u} = \langle 0, 6, -4\rangle,\ \|\mathbf{u}\| = \sqrt{52} = 2\sqrt{13}$

$\cos\alpha = 0$

$\cos\beta = \dfrac{3}{\sqrt{13}}$

$\cos\gamma = -\dfrac{2}{\sqrt{13}}$

$\cos^2\alpha + \cos^2\beta + \cos^2\gamma = 0 + \dfrac{9}{13} + \dfrac{4}{13} = 1$

**35.** $\mathbf{u} = \langle 3, 2, -2\rangle\ \|\mathbf{u}\| = \sqrt{17}$

$\cos\alpha = \dfrac{3}{\sqrt{17}} \Rightarrow \alpha \approx 0.7560 \text{ or } 43.3°$

$\cos\beta = \dfrac{2}{\sqrt{17}} \Rightarrow \beta \approx 1.0644 \text{ or } 61.0°$

$\cos\gamma = \dfrac{-2}{\sqrt{17}} \Rightarrow \gamma \approx 2.0772 \text{ or } 119.0°$

**37.** $\mathbf{u} = \langle -1, 5, 2 \rangle$ $\|\mathbf{u}\| = \sqrt{30}$

$$\cos \alpha = \frac{-1}{\sqrt{30}} \Rightarrow \alpha \approx 1.7544 \text{ or } 100.5°$$

$$\cos \beta = \frac{5}{\sqrt{30}} \Rightarrow \beta \approx 0.4205 \text{ or } 24.1°$$

$$\cos \gamma = \frac{2}{\sqrt{30}} \Rightarrow \gamma \approx 1.1970 \text{ or } 68.6°$$

**39.** $\mathbf{F}_1$: $C_1 = \dfrac{50}{\|\mathbf{F}_1\|} \approx 4.3193$

$\mathbf{F}_2$: $C_2 = \dfrac{80}{\|\mathbf{F}_2\|} \approx 5.4183$

$\mathbf{F} = \mathbf{F}_1 + \mathbf{F}_2$

$\approx 4.3193 \langle 10, 5, 3 \rangle + 5.4183 \langle 12, 7, -5 \rangle$

$= \langle 108.2126, 59.5246, -14.1336 \rangle$

$\|\mathbf{F}\| \approx 124.310 \text{ lb}$

$$\cos \alpha \approx \frac{108.2126}{\|\mathbf{F}\|} \Rightarrow \alpha \approx 29.48°$$

$$\cos \beta \approx \frac{59.5246}{\|\mathbf{F}\|} \Rightarrow \beta \approx 61.39°$$

$$\cos \gamma \approx \frac{-14.1336}{\|\mathbf{F}\|} \Rightarrow \gamma \approx 96.53°$$

**41.** $\overrightarrow{OA} = \langle 0, 10, 10 \rangle$

$$\cos \alpha = \frac{0}{\sqrt{0^2 + 10^2 + 10^2}} = 0 \Rightarrow \alpha = 90°$$

$$\cos \beta = \cos \gamma = \frac{10}{\sqrt{0^2 + 10^2 + 10^2}}$$

$$= \frac{1}{\sqrt{2}} \Rightarrow \beta = \gamma = 45°$$

**43.** $\mathbf{u} = \langle 6, 7 \rangle$, $\mathbf{v} = \langle 1, 4 \rangle$

(a) $\mathbf{w}_1 = \text{proj}_{\mathbf{v}} \mathbf{u} = \left( \dfrac{\mathbf{u} \cdot \mathbf{v}}{\|\mathbf{v}\|^2} \right) \mathbf{v}$

$= \dfrac{6(1) + 7(4)}{1^2 + 4^2} \langle 1, 4 \rangle$

$= \dfrac{34}{17} \langle 1, 4 \rangle = \langle 2, 8 \rangle$

(b) $\mathbf{w}_2 = \mathbf{u} - \mathbf{w}_1 = \langle 6, 7 \rangle - \langle 2, 8 \rangle = \langle 4, -1 \rangle$

**45.** $\mathbf{u} = 2\mathbf{i} + 3\mathbf{j} = \langle 2, 3 \rangle$, $\mathbf{v} = 5\mathbf{i} + \mathbf{j} = \langle 5, 1 \rangle$

(a) $\mathbf{w}_1 = \text{proj}_{\mathbf{v}} \mathbf{u} = \left( \dfrac{\mathbf{u} \cdot \mathbf{v}}{\|\mathbf{v}\|^2} \right) \mathbf{v}$

$= \dfrac{2(5) + 3(1)}{5^2 + 1} \langle 5, 1 \rangle$

$= \dfrac{13}{26} \langle 5, 1 \rangle = \left\langle \dfrac{5}{2}, \dfrac{1}{2} \right\rangle$

(b) $\mathbf{w}_2 = \mathbf{u} - \mathbf{w}_1 = \langle 2, 3 \rangle - \left\langle \dfrac{5}{2}, \dfrac{1}{2} \right\rangle = \left\langle -\dfrac{1}{2}, \dfrac{5}{2} \right\rangle$

**47.** $\mathbf{u} = \langle 0, 3, 3 \rangle$, $\mathbf{v} = \langle -1, 1, 1 \rangle$

(a) $\mathbf{w}_1 = \text{proj}_{\mathbf{v}} \mathbf{u} = \left( \dfrac{\mathbf{u} \cdot \mathbf{v}}{\|\mathbf{v}\|^2} \right) \mathbf{v}$

$= \dfrac{0(-1) + 3(1) + 3(1)}{1 + 1 + 1} \langle -1, 1, 1 \rangle$

$= \dfrac{6}{3} \langle -1, 1, 1 \rangle = \langle -2, 2, 2 \rangle$

(b) $\mathbf{w}_2 = \mathbf{u} - \mathbf{w}_1 = \langle 0, 3, 3 \rangle - \langle -2, 2, 2 \rangle = \langle 2, 1, 1 \rangle$

**49.** $\mathbf{u} = 2\mathbf{i} + \mathbf{j} + 2\mathbf{k} = \langle 2, 1, 2 \rangle$

$\mathbf{v} = 3\mathbf{j} + 4\mathbf{k} = \langle 0, 3, 4 \rangle$

(a) $\mathbf{w}_1 = \text{proj}_{\mathbf{v}} \mathbf{u} = \left( \dfrac{\mathbf{u} \cdot \mathbf{v}}{\|\mathbf{v}\|^2} \right) \mathbf{v}$

$= \dfrac{2(0) + 1(3) + 2(4)}{3^2 + 4^2} \langle 0, 3, 4 \rangle$

$= \dfrac{11}{25} \langle 0, 3, 4 \rangle = \left\langle 0, \dfrac{33}{25}, \dfrac{44}{25} \right\rangle$

(b) $\mathbf{w}_2 = \mathbf{u} - \mathbf{w}_1 = \langle 2, 1, 2 \rangle - \left\langle 0, \dfrac{33}{25}, \dfrac{44}{25} \right\rangle = \left\langle 2, -\dfrac{8}{25}, \dfrac{6}{25} \right\rangle$

**51.** $\mathbf{u} \cdot \mathbf{v} = \langle u_1, u_2, u_3 \rangle \cdot \langle v_1, v_2, v_3 \rangle = u_1 v_1 + u_2 v_2 + u_3 v_3$

**53.** (a) and (b) are defined. (c) and (d) are not defined because it is not possible to find the dot product of a scalar and a vector or to add a scalar to a vector.

**55.** See figure 11.29, page 787.

**57.** Yes, $\left\| \dfrac{\mathbf{u} \cdot \mathbf{v}}{\|\mathbf{v}\|^2} \mathbf{v} \right\| = \left\| \dfrac{\mathbf{v} \cdot \mathbf{u}}{\|\mathbf{u}\|^2} \mathbf{u} \right\|$

$|\mathbf{u} \cdot \mathbf{v}| \dfrac{\|\mathbf{v}\|}{\|\mathbf{v}\|^2} = |\mathbf{v} \cdot \mathbf{u}| \dfrac{\|\mathbf{u}\|}{\|\mathbf{u}\|^2}$

$\dfrac{1}{\|\mathbf{v}\|} = \dfrac{1}{\|\mathbf{u}\|}$

$\|\mathbf{u}\| = \|\mathbf{v}\|$

**59.** $\mathbf{u} = \langle 3240, 1450, 2235 \rangle$

$\mathbf{v} = \langle 1.35, 2.65, 1.85 \rangle$

$\mathbf{u} \cdot \mathbf{v} = 3240(1.35) + 1450(2.65) + 2235(1.85)$

$\qquad = \$12,351.25$

This represents the total amount that the restaurant earned on its three products.

**61.** (a)–(c) Programs will vary.

**63.** Programs will vary.

**65.** Because $\mathbf{u}$ and $\mathbf{v}$ are parallel, $\text{proj}_\mathbf{v}\mathbf{u} = \mathbf{u}$

**67.** Answers will vary. *Sample answer:*

$\mathbf{u} = -\dfrac{1}{4}\mathbf{i} + \dfrac{3}{2}\mathbf{j}$. Want $\mathbf{u} \cdot \mathbf{v} = 0$.

$\mathbf{v} = 12\mathbf{i} + 2\mathbf{j}$ and $-\mathbf{v} = -12\mathbf{i} - 2\mathbf{j}$ are orthogonal to $\mathbf{u}$.

**69.** Answers will vary. *Sample answer:*

$\mathbf{u} = \langle 3, 1, -2 \rangle$. Want $\mathbf{u} \cdot \mathbf{v} = 0$.

$\mathbf{v} = \langle 0, 2, 1 \rangle$ and $-\mathbf{v} = \langle 0, -2, -1 \rangle$ are orthogonal to $\mathbf{u}$.

**71.** (a) Gravitational Force $\mathbf{F} = -48,000\mathbf{j}$

$\mathbf{v} = \cos 10°\mathbf{i} + \sin 10°\mathbf{j}$

$\mathbf{w}_1 = \dfrac{\mathbf{F} \cdot \mathbf{v}}{\|\mathbf{v}\|^2}\mathbf{v} = (\mathbf{F} \cdot \mathbf{v})\mathbf{v}$

$\qquad = (-48,000)(\sin 10°)\mathbf{v}$

$\qquad \approx -8335.1(\cos 10°\mathbf{i} + \sin 10°\mathbf{j})$

$\|\mathbf{w}_1\| \approx 8335.1 \text{ lb}$

(b) $\mathbf{w}_2 = \mathbf{F} - \mathbf{w}_1$

$\qquad = -48,000\mathbf{j} + 8335.1(\cos 10°\mathbf{i} + \sin 10°\mathbf{j})$

$\qquad = 8208.5\mathbf{i} - 46,552.6\mathbf{j}$

$\|\mathbf{w}_2\| \approx 47,270.8 \text{ lb}$

**73.** $\mathbf{F} = 85\left(\dfrac{1}{2}\mathbf{i} + \dfrac{\sqrt{3}}{2}\mathbf{j}\right)$

$\mathbf{v} = 10\mathbf{i}$

$W = \mathbf{F} \cdot \mathbf{v} = 425 \text{ ft-lb}$

**75.** $\mathbf{F} = 1600(\cos 25° \, \mathbf{i} + \sin 25° \, \mathbf{j})$

$\mathbf{v} = 2000\mathbf{i}$

$W = \mathbf{F} \cdot \mathbf{v} = 1600(2000)\cos 25°$

$\qquad \approx 2,900,184.9 \text{ Newton meters (Joules)}$

$\qquad \approx 2900.2 \quad \text{km-N}$

**77.** False.

For example, let $\mathbf{u} = \langle 1, 1 \rangle$, $\mathbf{v} = \langle 2, 3 \rangle$ and

$\mathbf{w} = \langle 1, 4 \rangle$. Then $\mathbf{u} \cdot \mathbf{v} = 2 + 3 = 5$ and

$\mathbf{u} \cdot \mathbf{w} = 1 + 4 = 5$.

**79.** Let $s = $ length of a side.

$\mathbf{v} = \langle s, s, s \rangle$

$\|\mathbf{v}\| = s\sqrt{3}$

$\cos \alpha = \cos \beta = \cos \gamma = \dfrac{s}{s\sqrt{3}} = \dfrac{1}{\sqrt{3}}$

$\alpha = \beta = \gamma = \arccos\left(\dfrac{1}{\sqrt{3}}\right) \approx 54.7°$

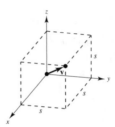

**81.** (a) The graphs $y_1 = x^2$ and $y_2 = x^{1/3}$ intersect at $(0, 0)$ and $(1, 1)$.

(b) $y_1' = 2x$ and $y_2' = \dfrac{1}{3x^{2/3}}$.

At $(0, 0), \pm \langle 1, 0 \rangle$ is tangent to $y_1$ and $\pm\langle 0, 1 \rangle$ is tangent to $y_2$.

At $(1, 1), y_1' = 2$ and $y_2' = \dfrac{1}{3}$.

$\pm\dfrac{1}{\sqrt{5}}\langle 1, 2 \rangle$ is tangent to $y_1$, $\pm\dfrac{1}{\sqrt{10}}\langle 3, 1 \rangle$ is tangent to $y_2$.

(c) At $(0, 0)$, the vectors are perpendicular $(90°)$.

At $(1, 1)$,

$\cos \theta = \dfrac{\dfrac{1}{\sqrt{5}}\langle 1, 2 \rangle \cdot \dfrac{1}{\sqrt{10}}\langle 3, 1 \rangle}{(1)(1)} = \dfrac{5}{\sqrt{50}} = \dfrac{1}{\sqrt{2}}$

$\theta = 45°$

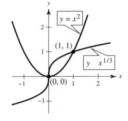

**83.** (a) The graphs of $y_1 = 1 - x^2$ and $y_2 = x^2 - 1$ intersect at $(1, 0)$ and $(-1, 0)$.

(b) $y_1' = -2x$ and $y_2' = 2x$.

At $(1, 0)$, $y_1' = -2$ and $y_2' = 2$. $\pm\dfrac{1}{\sqrt{5}}\langle 1, -2\rangle$ is tangent to $y_1$, $\pm\dfrac{1}{\sqrt{5}}\langle 1, 2\rangle$ is tangent to $y_2$.

At $(-1, 0)$, $y_1' = 2$ and $y_2' = -2$. $\pm\dfrac{1}{\sqrt{5}}\langle 1, 2\rangle$ is tangent to $y_1$, $\pm\dfrac{1}{\sqrt{5}}\langle 1, -2\rangle$ is tangent to $y_2$.

(c) At $(1, 0)$, $\cos\theta = \dfrac{1}{\sqrt{5}}\langle 1, -2\rangle \cdot \dfrac{-1}{\sqrt{5}}\langle 1, -2\rangle = \dfrac{3}{5}$.

$\theta \approx 0.9273$ or $53.13°$

By symmetry, the angle is the same at $(-1, 0)$.

**85.** In a rhombus, $\|\mathbf{u}\| = \|\mathbf{v}\|$. The diagonals are $\mathbf{u} + \mathbf{v}$ and $\mathbf{u} - \mathbf{v}$.

$(\mathbf{u} + \mathbf{v}) \cdot (\mathbf{u} - \mathbf{v}) = (\mathbf{u} + \mathbf{v}) \cdot \mathbf{u} - (\mathbf{u} + \mathbf{v}) \cdot \mathbf{v}$

$\qquad = \mathbf{u} \cdot \mathbf{u} + \mathbf{v} \cdot \mathbf{u} - \mathbf{u} \cdot \mathbf{v} - \mathbf{v} \cdot \mathbf{v}$

$\qquad = \|\mathbf{u}\|^2 - \|\mathbf{v}\|^2 = 0$

So, the diagonals are orthogonal.

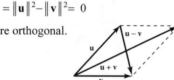

**87.** (a)

(b) Length of each edge: $\sqrt{k^2 + k^2 + 0^2} = k\sqrt{2}$

(c) $\cos\theta = \dfrac{k^2}{(k\sqrt{2})(k\sqrt{2})} = \dfrac{1}{2}$

$\theta = \arccos\left(\dfrac{1}{2}\right) = 60°$

(d) $\vec{r_1} = \langle k, k, 0\rangle - \left\langle \dfrac{k}{2}, \dfrac{k}{2}, \dfrac{k}{2}\right\rangle = \left\langle \dfrac{k}{2}, \dfrac{k}{2}, -\dfrac{k}{2}\right\rangle$

$\vec{r_2} = \langle 0, 0, 0\rangle - \left\langle \dfrac{k}{2}, \dfrac{k}{2}, \dfrac{k}{2}\right\rangle = \left\langle -\dfrac{k}{2}, -\dfrac{k}{2}, -\dfrac{k}{2}\right\rangle$

$\cos\theta = \dfrac{-\dfrac{k^2}{4}}{\left(\dfrac{k}{2}\right)^2 \cdot 3} = -\dfrac{1}{3}$

$\theta = 109.5°$

**89.** $\|\mathbf{u} - \mathbf{v}\|^2 = (\mathbf{u} - \mathbf{v}) \cdot (\mathbf{u} - \mathbf{v})$

$\qquad = (\mathbf{u} - \mathbf{v}) \cdot \mathbf{u} - (\mathbf{u} - \mathbf{v}) \cdot \mathbf{v}$

$\qquad = \mathbf{u} \cdot \mathbf{u} - \mathbf{v} \cdot \mathbf{u} - \mathbf{u} \cdot \mathbf{v} + \mathbf{v} \cdot \mathbf{v}$

$\qquad = \|\mathbf{u}\|^2 - \mathbf{u} \cdot \mathbf{v} - \mathbf{u} \cdot \mathbf{v} + \|\mathbf{v}\|^2$

$\qquad = \|\mathbf{u}\|^2 + \|\mathbf{v}\|^2 - 2\mathbf{u} \cdot \mathbf{v}$

**91.** $\|\mathbf{u} + \mathbf{v}\|^2 = (\mathbf{u} + \mathbf{v}) \cdot (\mathbf{u} + \mathbf{v})$

$\qquad = (\mathbf{u} + \mathbf{v}) \cdot \mathbf{u} + (\mathbf{u} + \mathbf{v}) \cdot \mathbf{v}$

$\qquad = \mathbf{u} \cdot \mathbf{u} + \mathbf{v} \cdot \mathbf{u} + \mathbf{u} \cdot \mathbf{v} + \mathbf{v} \cdot \mathbf{v}$

$\qquad = \|\mathbf{u}\|^2 + 2\mathbf{u} \cdot \mathbf{v} + \|\mathbf{v}\|^2$

$\qquad \leq \|\mathbf{u}\|^2 + 2\|\mathbf{u}\|\|\mathbf{v}\| + \|\mathbf{v}\|^2 \leq (\|\mathbf{u}\| + \|\mathbf{v}\|)^2$

So, $\|\mathbf{u} + \mathbf{v}\| \leq \|\mathbf{u}\| + \|\mathbf{v}\|$.

## Section 11.4   The Cross Product of Two Vectors in Space

**1.** $\mathbf{j} \times \mathbf{i} = \begin{vmatrix} \mathbf{i} & \mathbf{j} & \mathbf{k} \\ 0 & 1 & 0 \\ 1 & 0 & 0 \end{vmatrix} = -\mathbf{k}$

**3.** $\mathbf{j} \times \mathbf{k} = \begin{vmatrix} \mathbf{i} & \mathbf{j} & \mathbf{k} \\ 0 & 1 & 0 \\ 0 & 0 & 1 \end{vmatrix} = \mathbf{i}$

**5.** $\mathbf{i} \times \mathbf{k} = \begin{vmatrix} \mathbf{i} & \mathbf{j} & \mathbf{k} \\ 1 & 0 & 0 \\ 0 & 0 & 1 \end{vmatrix} = -\mathbf{j}$

**7.** (a) $\mathbf{u} \times \mathbf{v} = \begin{vmatrix} \mathbf{i} & \mathbf{j} & \mathbf{k} \\ -2 & 4 & 0 \\ 3 & 2 & 5 \end{vmatrix} = 20\mathbf{i} + 10\mathbf{j} - 16\mathbf{k}$

(b) $\mathbf{v} \times \mathbf{u} = -(\mathbf{u} \times \mathbf{v}) = -20\mathbf{i} - 10\mathbf{j} + 16\mathbf{k}$

(c) $\mathbf{v} \times \mathbf{v} = \mathbf{0}$

**9.** (a) $\mathbf{u} \times \mathbf{v} = \begin{vmatrix} \mathbf{i} & \mathbf{j} & \mathbf{k} \\ 7 & 3 & 2 \\ 1 & -1 & 5 \end{vmatrix} = 17\mathbf{i} - 33\mathbf{j} - 10\mathbf{k}$

(b) $\mathbf{v} \times \mathbf{u} = -(\mathbf{u} \times \mathbf{v}) = -17\mathbf{i} + 33\mathbf{j} + 10\mathbf{k}$

(c) $\mathbf{v} \times \mathbf{v} = \mathbf{0}$

**11.** $\mathbf{u} = \langle 12, -3, 0 \rangle$, $\mathbf{v} = \langle -2, 5, 0 \rangle$

$\mathbf{u} \times \mathbf{v} = \begin{vmatrix} \mathbf{i} & \mathbf{j} & \mathbf{k} \\ 12 & -3 & 0 \\ -2 & 5 & 0 \end{vmatrix} = 54\mathbf{k} = \langle 0, 0, 54 \rangle$

$\mathbf{u} \cdot (\mathbf{u} \times \mathbf{v}) = 12(0) + (-3)(0) + 0(54)$
$= 0 \Rightarrow \mathbf{u} \perp \mathbf{u} \times \mathbf{v}$

$\mathbf{v} \cdot (\mathbf{u} \times \mathbf{v}) = -2(0) + 5(0) + 0(54)$
$= 0 \Rightarrow \mathbf{v} \perp \mathbf{u} \times \mathbf{v}$

**13.** $\mathbf{u} = \langle 2, -3, 1 \rangle$, $\mathbf{v} = \langle 1, -2, 1 \rangle$

$\mathbf{u} \times \mathbf{v} = \begin{vmatrix} \mathbf{i} & \mathbf{j} & \mathbf{k} \\ 2 & -3 & 1 \\ 1 & -2 & 1 \end{vmatrix} = -\mathbf{i} - \mathbf{j} - \mathbf{k} = \langle -1, -1, -1 \rangle$

$\mathbf{u} \cdot (\mathbf{u} \times \mathbf{v}) = 2(-1) + (-3)(-1) + (1)(-1)$
$= 0 \Rightarrow \mathbf{u} \perp \mathbf{u} \times \mathbf{v}$

$\mathbf{v} \cdot (\mathbf{u} \times \mathbf{v}) = 1(-1) + (-2)(-1) + (1)(-1)$
$= 0 \Rightarrow \mathbf{v} \perp \mathbf{u} \times \mathbf{v}$

**15.** $\mathbf{u} = \mathbf{i} + \mathbf{j} + \mathbf{k}$, $\mathbf{v} = 2\mathbf{i} + \mathbf{j} - \mathbf{k}$

$\mathbf{u} \times \mathbf{v} = \begin{vmatrix} \mathbf{i} & \mathbf{j} & \mathbf{k} \\ 1 & 1 & 1 \\ 2 & 1 & -1 \end{vmatrix} = -2\mathbf{i} + 3\mathbf{j} - \mathbf{k} = \langle -2, 3, -1 \rangle$

$\mathbf{u} \cdot (\mathbf{u} \times \mathbf{v}) = 1(-2) + 1(3) + 1(-1)$
$= 0 \Rightarrow \mathbf{u} \perp \mathbf{u} \times \mathbf{v}$

$\mathbf{v} \cdot (\mathbf{u} \times \mathbf{v}) = 2(-2) + 1(3) + (-1)(-1)$
$= 0 \Rightarrow \mathbf{v} \perp \mathbf{u} \times \mathbf{v}$

$(-\mathbf{v}) \times \mathbf{u} = -(\mathbf{v} \times \mathbf{u}) = \mathbf{u} \times \mathbf{v}$

**17.**

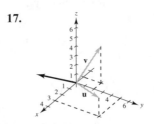

**19.**

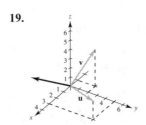

**21.** $\mathbf{u} = \langle 4, -3.5, 7 \rangle$, $\mathbf{v} = \langle 2.5, 9, 3 \rangle$

$\mathbf{u} \times \mathbf{v} = \langle -73.5, 5.5, 44.75 \rangle$

$\dfrac{\mathbf{u} \times \mathbf{v}}{\|\mathbf{u} \times \mathbf{v}\|} = \left\langle -\dfrac{2.94}{\sqrt{11.8961}}, \dfrac{0.22}{\sqrt{11.8961}}, \dfrac{1.79}{\sqrt{11.8961}} \right\rangle$

**23.** $\mathbf{u} = \langle -3, 2, -5 \rangle$, $\mathbf{v} = \langle 0.4, -0.8, 0.2 \rangle$

$\mathbf{u} \times \mathbf{v} = \langle -3.6, -1.4, 1.6 \rangle$

$\dfrac{\mathbf{u} \times \mathbf{v}}{\|\mathbf{u} \times \mathbf{v}\|} = \left\langle -\dfrac{1.8}{\sqrt{4.37}}, -\dfrac{0.7}{\sqrt{4.37}}, \dfrac{0.8}{\sqrt{4.37}} \right\rangle$

**25.** Programs will vary.

**27.** $\mathbf{u} = \mathbf{j}$
$\mathbf{v} = \mathbf{j} + \mathbf{k}$

$\mathbf{u} \times \mathbf{v} = \begin{vmatrix} \mathbf{i} & \mathbf{j} & \mathbf{k} \\ 0 & 1 & 0 \\ 0 & 1 & 1 \end{vmatrix} = \mathbf{i}$

$A = \|\mathbf{u} \times \mathbf{v}\| = \|\mathbf{i}\| = 1$

**29.**
$$\mathbf{u} = \langle 3, 2, -1 \rangle$$
$$\mathbf{v} = \langle 1, 2, 3 \rangle$$
$$\mathbf{u} \times \mathbf{v} = \begin{vmatrix} \mathbf{i} & \mathbf{j} & \mathbf{k} \\ 3 & 2 & -1 \\ 1 & 2 & 3 \end{vmatrix} = \langle 8, -10, 4 \rangle$$
$$A = \|\mathbf{u} \times \mathbf{v}\| = \|\langle 8, -10, 4 \rangle\| = \sqrt{180} = 6\sqrt{5}$$

**31.** $A(0, 3, 2), B(1, 5, 5), C(6, 9, 5), D(5, 7, 2)$

$$\overline{AB} = \langle 1, 2, 3 \rangle$$
$$\overline{DC} = \langle 1, 2, 3 \rangle$$
$$\overline{BC} = \langle 5, 4, 0 \rangle$$
$$\overline{AD} = \langle 5, 4, 0 \rangle$$

Because $\overline{AB} = \overline{DC}$ and $\overline{BC} = \overline{AD}$, the figure $ABCD$ is a parallelogram.

$\overline{AB}$ and $\overline{AD}$ are adjacent sides

$$\overline{AB} \times \overline{AD} = \begin{vmatrix} \mathbf{i} & \mathbf{j} & \mathbf{k} \\ 1 & 2 & 3 \\ 5 & 4 & 0 \end{vmatrix} = \langle -12, 15, -6 \rangle$$
$$A = \|\overline{AB} \times \overline{AD}\| = \sqrt{144 + 225 + 36} = 9\sqrt{5}$$

**33.** $A(0, 0, 0), B(1, 0, 3), C(-3, 2, 0)$

$$\overline{AB} = \langle 1, 0, 3 \rangle, \ \overline{AC} = \langle -3, 2, 0 \rangle$$
$$\overline{AB} \times \overline{AC} = \begin{vmatrix} \mathbf{i} & \mathbf{j} & \mathbf{k} \\ 1 & 0 & 3 \\ -3 & 2 & 0 \end{vmatrix} = \langle -6, -9, 2 \rangle$$
$$A = \tfrac{1}{2}\|\overline{AB} \times \overline{AC}\| = \tfrac{1}{2}\sqrt{36 + 81 + 4} = \tfrac{11}{2}$$

**35.** $A(2, -7, 3), B(-1, 5, 8), C(4, 6, -1)$

$$\overline{AB} = \langle -3, 12, 5 \rangle, \ \overline{AC} = \langle 2, 13, -4 \rangle$$
$$\overline{AB} \times \overline{AC} = \begin{vmatrix} \mathbf{i} & \mathbf{j} & \mathbf{k} \\ -3 & 12 & 5 \\ 2 & 13 & -4 \end{vmatrix} = \langle -113, -2, -63 \rangle$$
$$A = \tfrac{1}{2}\|\overline{AB} \times \overline{AC}\| = \tfrac{1}{2}\sqrt{16,742}$$

**37.** $\mathbf{F} = -20\mathbf{k}$

$$\overline{PQ} = \tfrac{1}{2}(\cos 40°\mathbf{j} + \sin 40°\mathbf{k})$$
$$\overline{PQ} \times \mathbf{F} = \begin{vmatrix} \mathbf{i} & \mathbf{j} & \mathbf{k} \\ 0 & \cos 40°/2 & \sin 40°/2 \\ 0 & 0 & -20 \end{vmatrix} = -10 \cos 40°\mathbf{i}$$
$$\|\overline{PQ} \times \mathbf{F}\| = 10 \cos 40° \approx 7.66 \text{ ft-lb}$$

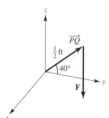

**39.** (a) Place the wrench in the $xy$-plane, as indicated in the figure.

The angle from $\overline{AB}$ to $\mathbf{F}$ is $30° + 180° + \theta = 210° + \theta$

$\|\overline{OA}\| = 18$ inches $= 1.5$ feet

$$\overline{OA} = 1.5[\cos(30°)\mathbf{i} + \sin(30°)\mathbf{j}] = \frac{3\sqrt{3}}{4}\mathbf{i} + \frac{3}{4}\mathbf{j}$$
$$\mathbf{F} = 56[\cos(210° + \theta)\mathbf{i} + \sin(210° + \theta)\mathbf{j}]$$

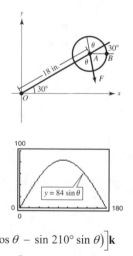

$$\overline{OA} \times \mathbf{F} = \begin{vmatrix} \mathbf{i} & \mathbf{j} & \mathbf{k} \\ \dfrac{3\sqrt{3}}{4} & \dfrac{3}{4} & 0 \\ 56 \cos(210° + \theta) & 56 \sin(210° + \theta) & 0 \end{vmatrix}$$

$$= \left[42\sqrt{3} \sin(210° + \theta) - 42 \cos(210° + \theta)\right]\mathbf{k}$$

$$= \left[42\sqrt{3}(\sin 210° \cos \theta + \cos 210° \sin \theta) - 42(\cos 210° \cos \theta - \sin 210° \sin \theta)\right]\mathbf{k}$$

$$= \left[42\sqrt{3}\left(-\frac{1}{2}\cos \theta - \frac{\sqrt{3}}{2}\sin \theta\right) - 42\left(-\frac{\sqrt{3}}{2}\cos \theta + \frac{1}{2}\sin \theta\right)\right]\mathbf{k} = (-84 \sin \theta)\mathbf{k}$$

$$\|\overline{OA} \times \mathbf{F}\| = 84 \sin \theta, \ 0 \le \theta \le 180°$$

(b) When $\theta = 45°$, $\left\|\overrightarrow{OA} \times \mathbf{F}\right\| = 84 \dfrac{\sqrt{2}}{2} = 42\sqrt{2} \approx 59.40$

(c) Let $T = 84 \sin \theta$

$\dfrac{dT}{d\theta} = 84 \cos \theta = 0$ when $\theta = 90°$.

This is reasonable. When $\theta = 90°$, the force is perpendicular to the wrench.

**41.** $\mathbf{u} \cdot (\mathbf{v} \times \mathbf{w}) = \begin{vmatrix} 1 & 0 & 0 \\ 0 & 1 & 0 \\ 0 & 0 & 1 \end{vmatrix} = 1$

**43.** $\mathbf{u} \cdot (\mathbf{v} \times \mathbf{w}) = \begin{vmatrix} 2 & 0 & 1 \\ 0 & 3 & 0 \\ 0 & 0 & 1 \end{vmatrix} = 6$

**45.** $\mathbf{u} \cdot (\mathbf{v} \times \mathbf{w}) = \begin{vmatrix} 1 & 1 & 0 \\ 0 & 1 & 1 \\ 1 & 0 & 1 \end{vmatrix} = 2$

$V = \left| \mathbf{u} \cdot (\mathbf{v} \times \mathbf{w}) \right| = 2$

**47.** $\mathbf{u} = \langle 3, 0, 0 \rangle$

$\mathbf{v} = \langle 0, 5, 1 \rangle$

$\mathbf{w} = \langle 2, 0, 5 \rangle$

$\mathbf{u} \cdot (\mathbf{v} \times \mathbf{w}) = \begin{vmatrix} 3 & 0 & 0 \\ 0 & 5 & 1 \\ 2 & 0 & 5 \end{vmatrix} = 75$

$V = \left| \mathbf{u} \cdot (\mathbf{v} \times \mathbf{w}) \right| = 75$

**49.** $\mathbf{u} \times \mathbf{v} = \mathbf{0} \Rightarrow \mathbf{u}$ and $\mathbf{v}$ are parallel.

$\mathbf{u} \cdot \mathbf{v} = 0 \Rightarrow \mathbf{u}$ and $\mathbf{v}$ are orthogonal.

So, $\mathbf{u}$ or $\mathbf{v}$ (or both) is the zero vector.

**51.** $\mathbf{u} \times \mathbf{v} = \langle u_1, u_2, u_3 \rangle \cdot \langle v_1, v_2, v_3 \rangle$

$= (u_2 v_3 - u_3 v_2)\mathbf{i} - (u_1 v_3 - u_3 v_1)\mathbf{j} + (u_1 v_2 - u_2 v_1)\mathbf{k}$

**53.** The magnitude of the cross product will increase by a factor of 4.

**55.** False. If the vectors are ordered pairs, then the cross product does not exist.

**57.** False. Let $\mathbf{u} = \langle 1, 0, 0 \rangle$, $\mathbf{v} = \langle 1, 0, 0 \rangle$, $\mathbf{w} = \langle -1, 0, 0 \rangle$.

Then, $\mathbf{u} \times \mathbf{v} = \mathbf{u} \times \mathbf{w} = \mathbf{0}$, but $\mathbf{v} \neq \mathbf{w}$.

**59.** $\mathbf{u} = \langle u_1, u_2, u_3 \rangle$, $\mathbf{v} = \langle v_1, v_2, v_3 \rangle$, $\mathbf{w} = \langle w_1, w_2, w_3 \rangle$

$\mathbf{u} \times (\mathbf{v} + \mathbf{w}) = \begin{vmatrix} \mathbf{i} & \mathbf{j} & \mathbf{k} \\ u_1 & u_2 & u_3 \\ v_1 + w_1 & v_2 + w_2 & v_3 + w_3 \end{vmatrix}$

$= \big[u_2(v_3 + w_3) - u_3(v_2 + w_2)\big]\mathbf{i} - \big[u_1(v_3 + w_3) - u_3(v_1 + w_1)\big]\mathbf{j} + \big[u_1(v_2 + w_2) - u_2(v_1 + w_1)\big]\mathbf{k}$

$= (u_2 v_3 - u_3 v_2)\mathbf{i} - (u_1 v_3 - u_3 v_1)\mathbf{j} + (u_1 v_2 - u_2 v_1)\mathbf{k} + (u_2 w_3 - u_3 w_2)\mathbf{i} - (u_1 w_3 - u_3 w_1)\mathbf{j} + (u_1 w_2 - u_2 w_1)\mathbf{k}$

$= (\mathbf{u} \times \mathbf{v}) + (\mathbf{u} \times \mathbf{w})$

**61.** $\mathbf{u} = \langle u_1, u_2, u_3 \rangle$

$\mathbf{u} \times \mathbf{u} = \begin{vmatrix} \mathbf{i} & \mathbf{j} & \mathbf{k} \\ u_1 & u_2 & u_3 \\ u_1 & u_2 & u_3 \end{vmatrix} = (u_2 u_3 - u_3 u_2)\mathbf{i} - (u_1 u_3 - u_3 u_1)\mathbf{j} + (u_1 u_2 - u_2 u_1)\mathbf{k} = \mathbf{0}$

**63.** $\mathbf{u} \times \mathbf{v} = (u_2 v_3 - u_3 v_2)\mathbf{i} - (u_1 v_3 - u_3 v_1)\mathbf{j} + (u_1 v_2 - u_2 v_1)\mathbf{k}$

$(\mathbf{u} \times \mathbf{v}) \cdot \mathbf{u} = (u_2 v_3 - u_3 v_2)u_1 + (u_3 v_1 - u_1 v_3)u_2 + (u_1 v_2 - u_2 v_1)u_3 = 0$

$(\mathbf{u} \times \mathbf{v}) \cdot \mathbf{v} = (u_2 v_3 - u_3 v_2)v_1 + (u_3 v_1 - u_1 v_3)v_2 + (u_1 v_2 - u_2 v_1)v_3 = 0$

So, $\mathbf{u} \times \mathbf{v} \perp \mathbf{u}$ and $\mathbf{u} \times \mathbf{v} \perp \mathbf{v}$.

**65.** $\|\mathbf{u} \times \mathbf{v}\| = \|\mathbf{u}\|\|\mathbf{v}\| \sin \theta$

If $\mathbf{u}$ and $\mathbf{v}$ are orthogonal, $\theta = \pi/2$ and $\sin \theta = 1$. So, $\|\mathbf{u} \times \mathbf{v}\| = \|\mathbf{u}\|\|\mathbf{v}\|$.

# Section 11.5  Lines and Planes in Space

**1.** $x = 1 + 3t, y = 2 - t, z = 2 + 5t$

(a)

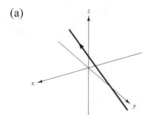

(b) When $t = 0, P = (1, 2, 2)$. When $t = 3, Q = (10, -1, 17)$.

$\overrightarrow{PQ} = \langle 9, -3, 15 \rangle$

The components of the vector and the coefficients of $t$ are proportional because the line is parallel to $\overrightarrow{PQ}$.

(c) $y = 0$ when $t = 2$. So, $x = 7$ and $z = 12$.

Point: $(7, 0, 12)$

$x = 0$ when $t = -\frac{1}{3}$. So, $y = \frac{7}{3}$ and $z = \frac{1}{3}$. Point: $\left(0, \frac{7}{3}, \frac{1}{3}\right)$

$z = 0$ when $t = -\frac{2}{5}$. So, $x = -\frac{1}{5}$ and $y = \frac{12}{5}$. Point: $\left(-\frac{1}{5}, \frac{12}{5}, 0\right)$

**3.** $x = -2 + t, y = 3t, z = 4 + t$

(a) $(0, 6, 6)$: For $x = 0 = -2 + t$, you have
$t = 2$. Then $y = 3(2) = 6$ and
$z = 4 + 2 = 6$. Yes, $(0, 6, 6)$ lies on the line.

(b) $(2, 3, 5)$: For $x = 2 = -2 + t$, you have
$t = 4$. Then $y = 3(4) = 12 \neq 3$. No, $(2, 3, 5)$ does
not lie on the line.

**5.** Point: $(0, 0, 0)$

Direction vector: $\langle 3, 1, 5 \rangle$

Direction numbers: 3, 1, 5

(a) Parametric: $x = 3t, y = t, z = 5t$

(b) Symmetric: $\dfrac{x}{3} = y = \dfrac{z}{5}$

**7.** Point: $(-2, 0, 3)$

Direction vector: $\mathbf{v} = \langle 2, 4, -2 \rangle$

Direction numbers: 2, 4, -2

(a) Parametric: $x = -2 + 2t, y = 4t, z = 3 - 2t$

(b) Symmetric: $\dfrac{x + 2}{2} = \dfrac{y}{4} = \dfrac{z - 3}{-2}$

**9.** Point: $(1, 0, 1)$

Direction vector: $\mathbf{v} = 3\mathbf{i} - 2\mathbf{j} + \mathbf{k}$

Direction numbers: 3, -2, 1

(a) Parametric: $x = 1 + 3t, y = -2t, z = 1 + t$

(b) Symmetric: $\dfrac{x - 1}{3} = \dfrac{y}{-2} = \dfrac{z - 1}{1}$

**11.** Points: $(5, -3, -2), \left(-\dfrac{2}{3}, \dfrac{2}{3}, 1\right)$

Direction vector: $\mathbf{v} = \dfrac{17}{3}\mathbf{i} - \dfrac{11}{3}\mathbf{j} - 3\mathbf{k}$

Direction numbers: 17, -11, -9

(a) Parametric:
$x = 5 + 17t, y = -3 - 11t, z = -2 - 9t$

(b) Symmetric: $\dfrac{x - 5}{17} = \dfrac{y + 3}{-11} = \dfrac{z + 2}{-9}$

**13.** Points: $(7, -2, 6), (-3, 0, 6)$

Direction vector: $\langle -10, 2, 0 \rangle$

Direction numbers: $-10, 2, 0$

(a) Parametric: $x = 7 - 10t, \ y = -2 + 2t, \ z = 6$

(b) Symmetric: Not possible because the direction number for $z$ is 0. But, you could describe the line as $\dfrac{x - 7}{10} = \dfrac{y + 2}{-2}, \ z = 6.$

**15.** Point: $(2, 3, 4)$

Direction vector: $\mathbf{v} = \mathbf{k}$

Direction numbers: $0, 0, 1$

Parametric: $x = 2, \ y = 3, \ z = 4 + t$

**17.** Point: $(2, 3, 4)$

Direction vector: $\mathbf{v} = 3\mathbf{i} + 2\mathbf{j} - \mathbf{k}$

Direction numbers: $3, 2, -1$

Parametric: $x = 2 + 3t, \ y = 3 + 2t, \ z = 4 - t$

**19.** Point: $(5, -3, -4)$

Direction vector: $\mathbf{v} = \langle 2, -1, 3 \rangle$

Direction numbers: $2, -1, 3$

Parametric: $x = 5 + 2t, \ y = -3 - t, \ z = -4 + 3t$

**21.** Point: $(2, 1, 2)$

Direction vector: $\langle -1, 1, 1 \rangle$

Direction numbers: $-1, 1, 1$

Parametric: $x = 2 - t, \ y = 1 + t, \ z = 2 + t$

**23.** Let $t = 0$: $P = (3, -1, -2)$ (other answers possible)

$\mathbf{v} = \langle -1, 2, 0 \rangle$ (any nonzero multiple of $\mathbf{v}$ is correct)

**25.** Let each quantity equal 0:

$P = (7, -6, -2)$ (other answers possible)

$\mathbf{v} = \langle 4, 2, 1 \rangle$ (any nonzero multiple of $\mathbf{v}$ is correct)

**27.** $L_1: \mathbf{v} = \langle -3, 2, 4 \rangle \quad (6, -2, 5)$ on line

$L_2: \mathbf{v} = \langle 6, -4, -8 \rangle \quad (6, -2, 5)$ on line

$L_3: \mathbf{v} = \langle -6, 4, 8 \rangle \quad (6, -2, 5)$ not online

$L_4: \mathbf{v} = \langle 6, 4, -6 \rangle \quad$ not parallel to $L_1, L_2,$ nor $L_3$

$L_1$ and $L_2$ are identical. $L_1 = L_2$ and is parallel to $L_3$.

**29.** $L_1: \mathbf{v} = \langle 4, -2, 3 \rangle \quad (8, -5, -9)$ on line

$L_2: \mathbf{v} = \langle 2, 1, 5 \rangle$

$L_3: \mathbf{v} = \langle -8, 4, -6 \rangle \quad (8, -5, -9)$ on line

$L_4: \mathbf{v} = \langle -2, 1, 1.5 \rangle$

$L_1$ and $L_3$ are identical.

**31.** At the point of intersection, the coordinates for one line equal the corresponding coordinates for the other line. So,

(i) $4t + 2 = 2s + 2$, (ii) $3 = 2s + 3$, and

(iii) $-t + 1 = s + 1$.

From (ii), you find that $s = 0$ and consequently, from (iii), $t = 0$. Letting $s = t = 0$, you see that equation (i) is satisfied and so the two lines intersect. Substituting zero for $s$ or for $t$, you obtain the point $(2, 3, 1)$.

$\mathbf{u} = 4\mathbf{i} - \mathbf{k} \qquad$ (First line)

$\mathbf{v} = 2\mathbf{i} + 2\mathbf{j} + \mathbf{k} \quad$ (Second line)

$\cos \theta = \dfrac{|\mathbf{u} \cdot \mathbf{v}|}{\|\mathbf{u}\| \|\mathbf{v}\|} = \dfrac{8 - 1}{\sqrt{17}\sqrt{9}} = \dfrac{7}{3\sqrt{17}} = \dfrac{7\sqrt{17}}{51}$

**33.** Writing the equations of the lines in parametric form you have

$x = 3t \qquad y = 2 - t \qquad z = -1 + t$

$x = 1 + 4s \qquad y = -2 + s \qquad z = -3 - 3s.$

For the coordinates to be equal, $3t = 1 + 4s$ and $2 - t = -2 + s$. Solving this system yields $t = \frac{17}{7}$ and $s = \frac{11}{7}$. When using these values for $s$ and $t$, the $z$ coordinates are not equal. The lines do not intersect.

**35.** $x = 2t + 3 \qquad x = -2s + 7$

$y = 5t - 2 \qquad y = s + 8$

$z = -t + 1 \qquad z = 2s - 1$

Point of intersection: $(7, 8, -1)$

**Note:** $t = 2$ and $s = 0$

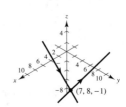

**37.** $4x - 3y - 6z = 6$

(a) $P(0, 0, -1), Q(0, -2, 0), R(3, 4, -1)$

$\overrightarrow{PQ} = \langle 0, -2, 1 \rangle, \ \overrightarrow{PR} = \langle 3, 4, 0 \rangle$

(b) $\overrightarrow{PQ} \times \overrightarrow{PR} = \begin{vmatrix} \mathbf{i} & \mathbf{j} & \mathbf{k} \\ 0 & -2 & 1 \\ 3 & 4 & 0 \end{vmatrix} = \langle -4, 3, 6 \rangle$

The components of the cross product are proportional to the coefficients of the variables in the equation. The cross product is parallel to the normal vector.

**39.** $x + 2y - 4z - 1 = 0$

(a) $(-7, 2, -1)$: $(-7) + 2(2) - 4(-1) - 1 = 0$

Point is in plane

(b) $(5, 2, 2)$: $5 + 2(2) - 4(2) - 1 = 0$

Point is in plane

**41.** Point: $(1, 3, -7)$

Normal vector: $\mathbf{n} = \mathbf{j} = \langle 0, 1, 0 \rangle$

$0(x - 1) + 1(y - 3) + 0(z - (-7)) = 0$

$y - 3 = 0$

**43.** Point: $(3, 2, 2)$

Normal vector: $\mathbf{n} = 2\mathbf{i} + 3\mathbf{j} - \mathbf{k}$

$2(x - 3) + 3(y - 2) - 1(z - 2) = 0$

$2x + 3y - z = 10$

**45.** Point: $(-1, 4, 0)$

Normal vector: $\mathbf{v} = \langle 2, -1, -2 \rangle$

$2(x + 1) - 1(y - 4) - 2(z - 0) = 0$

$2x - y - 2z + 6 = 0$

**47.** Let $\mathbf{u}$ be the vector from $(0, 0, 0)$ to

$(2, 0, 3)$: $\mathbf{u} = \langle 2, 0, 3 \rangle$

Let $\mathbf{u}$ be the vector from $(0, 0, 0)$ to

$(-3, -1, 5)$: $\mathbf{v} = \langle -3, -1, 5 \rangle$

Normal vectors: $\mathbf{u} \times \mathbf{v} = \begin{vmatrix} \mathbf{i} & \mathbf{j} & \mathbf{k} \\ 2 & 0 & 3 \\ -3 & -1 & 5 \end{vmatrix} = \langle 3, -19, -2 \rangle$

$3(x - 0) - 19(y - 0) - 2(z - 0) = 0$

$3x - 19y - 2z = 0$

**49.** Let $\mathbf{u}$ be the vector from $(1, 2, 3)$ to

$(3, 2, 1)$: $\mathbf{u} = 2\mathbf{i} - 2\mathbf{k}$

Let $\mathbf{v}$ be the vector from $(1, 2, 3)$ to

$(-1, -2, 2)$: $\mathbf{v} = -2\mathbf{i} - 4\mathbf{j} - \mathbf{k}$

Normal vector:

$\left(\tfrac{1}{2}\mathbf{u}\right) \times (-\mathbf{v}) = \begin{vmatrix} \mathbf{i} & \mathbf{j} & \mathbf{k} \\ 1 & 0 & -1 \\ 2 & 4 & 1 \end{vmatrix} = 4\mathbf{i} - 3\mathbf{j} + 4\mathbf{k}$

$4(x - 1) - 3(y - 2) + 4(z - 3) = 0$

$4x - 3y + 4z = 10$

**51.** $(1, 2, 3)$, Normal vector: $\mathbf{v} = \mathbf{k}$, $1(z - 3) = 0$, $z = 3$

**53.** The direction vectors for the lines are

$\mathbf{u} = -2\mathbf{i} + \mathbf{j} + \mathbf{k}$, $\mathbf{v} = -3\mathbf{i} + 4\mathbf{j} - \mathbf{k}$.

Normal vector: $\mathbf{u} \times \mathbf{v} = \begin{vmatrix} \mathbf{i} & \mathbf{j} & \mathbf{k} \\ -2 & 1 & 1 \\ -3 & 4 & -1 \end{vmatrix} = -5(\mathbf{i} + \mathbf{j} + \mathbf{k})$

Point of intersection of the lines: $(-1, 5, 1)$

$(x + 1) + (y - 5) + (z - 1) = 0$

$x + y + z = 5$

**55.** Let $\mathbf{v}$ be the vector from $(-1, 1, -1)$ to

$(2, 2, 1)$: $\mathbf{v} = 3\mathbf{i} + \mathbf{j} + 2\mathbf{k}$

Let $\mathbf{n}$ be a vector normal to the plane

$2x - 3y + z = 3$: $\mathbf{n} = 2\mathbf{i} - 3\mathbf{j} + \mathbf{k}$

Because $\mathbf{v}$ and $\mathbf{n}$ both lie in the plane $P$, the normal vector to $P$ is

$\mathbf{v} \times \mathbf{n} = \begin{vmatrix} \mathbf{i} & \mathbf{j} & \mathbf{k} \\ 3 & 1 & 2 \\ 2 & -3 & 1 \end{vmatrix} = 7\mathbf{i} - \mathbf{j} - 11\mathbf{k}$

$7(x - 2) + 1(y - 2) - 11(z - 1) = 0$

$7x + y - 11z = 5$

**57.** Let $\mathbf{u} = \mathbf{i}$ and let $\mathbf{v}$ be the vector from $(1, -2, -1)$ to

$(2, 5, 6)$: $\mathbf{v} = \mathbf{i} + 7\mathbf{j} + 7\mathbf{k}$

Because $\mathbf{u}$ and $\mathbf{v}$ both lie in the plane $P$, the normal vector to $P$ is:

$\mathbf{u} \times \mathbf{v} = \begin{vmatrix} \mathbf{i} & \mathbf{j} & \mathbf{k} \\ 1 & 0 & 0 \\ 1 & 7 & 7 \end{vmatrix} = -7\mathbf{j} + 7\mathbf{k} = -7(\mathbf{j} - \mathbf{k})$

$\left[y - (-2)\right] - \left[z - (-1)\right] = 0$

$y - z = -1$

**59.** *xy*-plane: Let $z = 0$.

Then $0 = 4 - t \Rightarrow t = 4 \Rightarrow x = 1 - 2(4) = -7$ and

$\quad y = -2 + 3(4) = 10$. Intersection: $(-7, 10, 0)$

*xz*-plane: Let $y = 0$.

Then $0 = -2 + 3t \Rightarrow t = \frac{2}{3} \Rightarrow x = 1 - 2\left(\frac{2}{3}\right) = -\frac{1}{3}$ and

$\quad z = -4 + \frac{2}{3} = -\frac{10}{3}$. Intersection: $\left(-\frac{1}{3}, 0, -\frac{10}{3}\right)$

*yz*-plane: Let $x = 0$.

Then $0 = 1 - 2t \Rightarrow t = \frac{1}{2} \Rightarrow y = -2 + 3\left(\frac{1}{2}\right) = -\frac{1}{2}$ and

$\quad z = -4 + \frac{1}{2} = -\frac{7}{2}$. Intersection: $\left(0, -\frac{1}{2}, -\frac{7}{2}\right)$

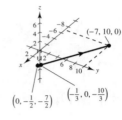

**61.** Let $(x, y, z)$ be equidistant from $(2, 2, 0)$ and $(0, 2, 2)$.

$$\sqrt{(x - 2)^2 + (y - 2)^2 + (z - 0)^2} = \sqrt{(x - 0)^2 + (y - 2)^2 + (z - 2)^2}$$

$$x^2 - 4x + 4 + y^2 - 4y + 4 + z^2 = x^2 + y^2 - 4y + 4 + z^2 - 4z + 4$$

$$-4x + 8 = -4z + 8$$

$$x - z = 0 \quad \text{Plane}$$

**63.** Let $(x, y, z)$ be equidistant from $(-3, 1, 2)$ and $(6, -2, 4)$.

$$\sqrt{(x + 3)^2 + (y - 1)^2 + (z - 2)^2} = \sqrt{(x - 6)^2 + (y + 2)^2 + (z - 4)^2}$$

$$x^2 + 6x + 9 + y^2 - 2y + 1 + z^2 - 4z + 4 = x^2 - 12x + 36 + y^2 + 4y + 4 + z^2 - 8z + 16$$

$$6x - 2y - 4z + 14 = -12x + 4y - 8z + 56$$

$$18x - 6y + 4z - 42 = 0$$

$$9x - 3y + 2z - 21 = 0 \quad \text{Plane}$$

**65.** The normal vectors to the planes are

$\mathbf{n}_1 = \langle 5, -3, 1 \rangle$, $\mathbf{n}_2 = \langle 1, 4, 7 \rangle$, $\cos \theta = \dfrac{|\mathbf{n}_1 \cdot \mathbf{n}_2|}{\|\mathbf{n}_1\| \|\mathbf{n}_2\|} = 0$.

So, $\theta = \pi/2$ and the planes are orthogonal.

**67.** The normal vectors to the planes are

$\mathbf{n}_1 = \mathbf{i} - 3\mathbf{j} + 6\mathbf{k}$, $\mathbf{n}_2 = 5\mathbf{i} + \mathbf{j} - \mathbf{k}$,

$\cos \theta = \dfrac{|\mathbf{n}_1 \cdot \mathbf{n}_2|}{\|\mathbf{n}_1\| \|\mathbf{n}_2\|} = \dfrac{|5 - 3 - 6|}{\sqrt{46}\sqrt{27}} = \dfrac{4\sqrt{138}}{414} = \dfrac{2\sqrt{138}}{207}$.

So, $\theta = \arccos\left(\dfrac{2\sqrt{138}}{207}\right) \approx 83.5°$.

**69.** The normal vectors to the planes are $\mathbf{n}_1 = \langle 1, -5, -1 \rangle$ and

$\mathbf{n}_2 = \langle 5, -25, -5 \rangle$. Because $\mathbf{n}_2 = 5\mathbf{n}_1$, the planes are parallel, but not equal.

**71.** $4x + 2y + 6z = 12$

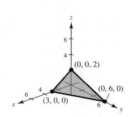

**73.** $2x - y + 3z = 4$

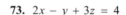
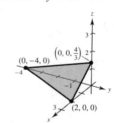

**75.** $x + z = 6$

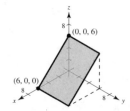

**77.** $x = 5$

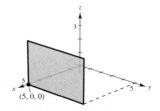

**79.** $2x + y - z = 6$

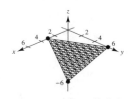

*Generated by Maple*

**81.** $-5x + 4y - 6z + 8 = 0$

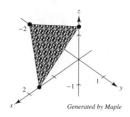

*Generated by Maple*

**83.** $P_1: \mathbf{n} = \langle 15, -6, 24 \rangle$      $(0, -1, -1)$ not on plane

$P_2: \mathbf{n} = \langle -5, 2, -8 \rangle$      $(0, -1, -1)$ on plane

$P_3: \mathbf{n} = \langle 6, -4, 4 \rangle$

$P_4: \mathbf{n} = \langle 3, -2, -2 \rangle$

Planes $P_1$ and $P_2$ are parallel.

**85.** $P_1: \mathbf{n} = \langle 3, -2, 5 \rangle$      $(1, -1, 1)$ on plane

$P_2: \mathbf{n} = \langle -6, 4, -10 \rangle$      $(1, -1, 1)$ not on plane

$P_3: \mathbf{n} = \langle -3, 2, 5 \rangle$

$P_4: \mathbf{n} = \langle 75, -50, 125 \rangle$      $(1, -1, 1)$ on plane

$P_1$ and $P_4$ are identical.

$P_1 = P_4$ and is parallel to $P_2$.

**87.** Each plane passes through the points

$(c, 0, 0), (0, c, 0),$ and $(0, 0, c)$.

**89.** If $c = 0, z = 0$ is $xy$-plane.

If $c \neq 0, cy + z = 0 \Rightarrow y = \dfrac{-1}{c}z$ is a plane parallel to

$x$-axis and passing through the points $(0, 0, 0)$ and

$(0, 1, -c)$.

**91.** (a) $\mathbf{n}_1 = 3\mathbf{i} + 2\mathbf{j} - \mathbf{k}$ and $\mathbf{n}_2 = \mathbf{i} - 4\mathbf{j} + 2\mathbf{k}$

$$\cos \theta = \frac{|\mathbf{n}_1 \cdot \mathbf{n}_2|}{\|\mathbf{n}_1\| \|\mathbf{n}_2\|} = \frac{|-7|}{\sqrt{14}\sqrt{21}} = \frac{\sqrt{6}}{6}$$

$$\Rightarrow \theta \approx 1.1503 \approx 65.91°$$

(b) The direction vector for the line is

$$\mathbf{n}_2 \times \mathbf{n}_1 = \begin{vmatrix} \mathbf{i} & \mathbf{j} & \mathbf{k} \\ 1 & -4 & 2 \\ 3 & 2 & -1 \end{vmatrix} = 7(\mathbf{j} + 2\mathbf{k}).$$

Find a point of intersection of the planes.

$$6x + 4y - 2z = 14$$
$$x - 4y + 2z = 0$$
$$7x \qquad\qquad = 14$$
$$x = 2$$

Substituting 2 for $x$ in the second equation, you have $-4y + 2z = -2$ or $z = 2y - 1$. Letting $y = 1$, a point of intersection is $(2, 1, 1)$.

$$x = 2, y = 1 + t, z = 1 + 2t$$

**93.** Writing the equation of the line in parametric form and substituting into the equation of the plane you have:

$$x = \frac{1}{2} + t, y = \frac{-3}{2} - t, z = -1 + 2t$$

$$2\left(\frac{1}{2} + t\right) - 2\left(\frac{-3}{2} - t\right) + (-1 + 2t) = 12, t = \frac{3}{2}$$

Substituting $t = 3/2$ into the parametric equations for the line you have the point of intersection $(2, -3, 2)$. The line does not lie in the plane.

**95.** Writing the equation of the line in parametric form and substituting into the equation of the plane you have:

$$x = 1 + 3t, y = -1 - 2t, z = 3 + t$$

$$2(1 + 3t) + 3(-1 - 2t) = 10, -1 = 10, \text{ contradiction}$$

So, the line does not intersect the plane.

**97.** Point: $Q(0, 0, 0)$

Plane: $2x + 3y + z - 12 = 0$

Normal to plane: $\mathbf{n} = \langle 2, 3, 1 \rangle$

Point in plane: $P(6, 0, 0)$

Vector $\overline{PQ} = \langle -6, 0, 0 \rangle$

$$D = \frac{\left| \overline{PQ} \cdot \mathbf{n} \right|}{\|\mathbf{n}\|} = \frac{|-12|}{\sqrt{14}} = \frac{6\sqrt{14}}{7}$$

**99.** Point: $Q(2, 8, 4)$

Plane: $2x + y + z = 5$

Normal to plane: $\mathbf{n} = \langle 2, 1, 1 \rangle$

Point in plane: $P\langle 0, 0, 5 \rangle$

Vector: $\overline{PQ} = \langle 2, 8, -1 \rangle$

$$D = \frac{\left| \overline{PQ} \cdot \mathbf{n} \right|}{\|\mathbf{n}\|} = \frac{11}{\sqrt{6}} = \frac{11\sqrt{6}}{6}$$

**101.** The normal vectors to the planes are $\mathbf{n}_1 = \langle 1, -3, 4 \rangle$ and $\mathbf{n}_2 = \langle 1, -3, 4 \rangle$. Because $\mathbf{n}_1 = \mathbf{n}_2$, the planes are parallel. Choose a point in each plane.

$P(10, 0, 0)$ is a point in $x - 3y + 4z = 10$.

$Q(6, 0, 0)$ is a point in $x - 3y + 4z = 6$.

$$\overline{PQ} = \langle -4, 0, 0 \rangle, \; D = \frac{\left| \overline{PQ} \cdot \mathbf{n}_1 \right|}{\|\mathbf{n}_1\|} = \frac{4}{\sqrt{26}} = \frac{2\sqrt{26}}{13}$$

**103.** The normal vectors to the planes are $\mathbf{n}_1 = \langle -3, 6, 7 \rangle$ and $\mathbf{n}_2 = \langle 6, -12, -14 \rangle$. Because $\mathbf{n}_2 = -2\mathbf{n}_1$, the planes are parallel. Choose a point in each plane.

$P(0, -1, 1)$ is a point in $-3x + 6y + 7z = 1$.

$Q\left( \frac{25}{6}, 0, 0 \right)$ is a point in $6x - 12y - 14z = 25$.

$$\overline{PQ} = \left\langle \frac{25}{6}, 1, -1 \right\rangle$$

$$D = \frac{\left| \overline{PQ} \cdot \mathbf{n}_1 \right|}{\|\mathbf{n}_1\|} = \frac{|-27/2|}{\sqrt{94}} = \frac{27}{2\sqrt{94}} = \frac{27\sqrt{94}}{188}$$

**105.** $\mathbf{u} = \langle 4, 0, -1 \rangle$ is the direction vector for the line.

$Q(1, 5, -2)$ is the given point, and $P(-2, 3, 1)$ is on the line.

$$\overline{PQ} = \langle 3, 2, -3 \rangle$$

$$\overline{PQ} \times \mathbf{u} = \begin{vmatrix} \mathbf{i} & \mathbf{j} & \mathbf{k} \\ 3 & 2 & -3 \\ 4 & 0 & -1 \end{vmatrix} = \langle -2, -9, -8 \rangle$$

$$D = \frac{\left\| \overline{PQ} \times \mathbf{u} \right\|}{\|\mathbf{u}\|} = \frac{\sqrt{149}}{\sqrt{17}} = \frac{\sqrt{2533}}{17}$$

**107.** $\mathbf{u} = \langle -1, 1, -2 \rangle$ is the direction vector for the line.

$Q(-2, 1, 3)$ is the given point, and $P(1, 2, 0)$ is on the line (let $t = 0$ in the parametric equations for the line).

$$\overline{PQ} = \langle -3, -1, 3 \rangle$$

$$\overline{PQ} \times \mathbf{u} = \begin{vmatrix} \mathbf{i} & \mathbf{j} & \mathbf{k} \\ -3 & -1 & 3 \\ -1 & 1 & -2 \end{vmatrix} = \langle -1, -9, -4 \rangle$$

$$D = \frac{\left\| \overline{PQ} \times \mathbf{u} \right\|}{\|\mathbf{u}\|} = \frac{\sqrt{1 + 81 + 16}}{\sqrt{1 + 1 + 4}} = \frac{\sqrt{98}}{6} = \frac{7}{\sqrt{3}} = \frac{7\sqrt{3}}{3}$$

**109.** The direction vector for $L_1$ is $\mathbf{v}_1 = \langle -1, 2, 1 \rangle$.

The direction vector for $L_2$ is $\mathbf{v}_2 = \langle 3, -6, -3 \rangle$.

Because $\mathbf{v}_2 = -3\mathbf{v}_1$, the lines are parallel.

Let $Q(2, 3, 4)$ to be a point on $L_1$ and $P(0, 1, 4)$ a point on $L_2$. $\overline{PQ} = \langle 2, 0, 0 \rangle$.

$\mathbf{u} = \mathbf{v}_2$ is the direction vector for $L_2$.

$$\overline{PQ} \times \mathbf{v}_2 = \begin{vmatrix} \mathbf{i} & \mathbf{j} & \mathbf{k} \\ 2 & 2 & 0 \\ 3 & -6 & -3 \end{vmatrix} = \langle -6, 6, -18 \rangle$$

$$D = \frac{\left\| \overline{PQ} \times \mathbf{v}_2 \right\|}{\|\mathbf{v}_2\|}$$

$$= \frac{\sqrt{36 + 36 + 324}}{\sqrt{9 + 36 + 9}} = \sqrt{\frac{396}{54}} = \sqrt{\frac{22}{3}} = \frac{\sqrt{66}}{3}$$

**111.** The parametric equations of a line $L$ parallel to $\mathbf{v} = \langle a, b, c, \rangle$ and passing through the point $P(x_1, y_1, z_1)$ are

$x = x_1 + at, \; y = y_1 + bt, \; z = z_1 + ct$.

The symmetric equations are

$$\frac{x - x_1}{a} = \frac{y - y_1}{b} = \frac{z - z_1}{c}.$$

**113.** Simultaneously solve the two linear equations representing the planes and substitute the values back into one of the original equations. Then choose a value for $t$ and form the corresponding parametric equations for the line of intersection.

**115.** (a) The planes are parallel if their normal vectors are parallel:

$$\langle a_1, b_1, c_1 \rangle = t \langle a_2, b_2, c_2 \rangle, \ t \neq 0$$

(b) The planes are perpendicular if their normal vectors are perpendicular:

$$\langle a_1, b_1, c_1 \rangle \cdot \langle a_2, b_2, c_2 \rangle = 0$$

**117.** An equation for the plane is

$$\frac{x}{a} + \frac{y}{b} + \frac{z}{c} = 1 \Rightarrow bcx + acy + abz = abc$$

For example, letting $y = z = 0$, the $x$-intercept is $(a, 0, 0)$.

**119.** Sphere

$$(x - 3)^2 + (y + 2)^2 + (z - 5)^2 = 16$$

**121.** $0.92x - 1.03y + z = 0.02 \Rightarrow z = 0.02 - 0.92x + 1.03y$

(a)

| Year | 1999 | 2000 | 2001 | 2002 | 2003 | 2004 | 2005 |
|---|---|---|---|---|---|---|---|
| $x$ | 1.4 | 1.4 | 1.4 | 1.6 | 1.6 | 1.7 | 1.7 |
| $y$ | 7.3 | 7.1 | 7.0 | 7.0 | 6.9 | 6.9 | 6.9 |
| $z$ | 6.2 | 6.1 | 5.9 | 5.8 | 5.6 | 5.5 | 5.6 |
| Model $z$ | 6.25 | 6.05 | 5.94 | 5.76 | 5.66 | 5.56 | 5.56 |

The approximations are close to the actual values.

(b) According to the model, if $x$ and $z$ decrease, then so will $y$. (Answers will vary.)

**123.** $L_1$: $x_1 = 6 + t$; $y_1 = 8 - t$, $z_1 = 3 + t$

$L_2$: $x_2 = 1 + t$, $y_2 = 2 + t$, $z_2 = 2t$

(a) At $t = 0$, the first insect is at $P_1(6, 8, 3)$ and the second insect is at $P_2(1, 2, 0)$.

$$\text{Distance} = \sqrt{(6 - 1)^2 + (8 - 2)^2 + (3 - 0)^2} = \sqrt{70} \approx 8.37 \text{ inches}$$

(b) $\text{Distance} = \sqrt{(x_1 - x_2)^2 + (y_1 - y_2)^2 + (z_1 - z_2)^2} = \sqrt{5^2 + (6 - 2t)^2 + (3 - t)^2} = \sqrt{5t^2 - 30t + 70},\ 0 \leq t \leq 10$

(c) The distance is never zero.

(d) Using a graphing utility, the minimum distance is 5 inches when $t = 3$ minutes.

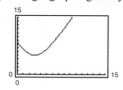

**125.** The direction vector $\mathbf{v}$ of the line is the normal to the plane, $\mathbf{v} = \langle 3, -1, 4 \rangle$.

The parametric equations of the line are $x = 5 + 3t$, $y = 4 - t$, $z = -3 + 4t$.

To find the point of intersection, solve for $t$ in the following equation:

$$3(5 + 3t) - (4 - t) + 4(-3 + 4t) = 7$$
$$26t = 8$$
$$t = \frac{4}{13}$$

Point of intersection:

$$\left( 5 + 3\left(\tfrac{4}{13}\right), 4 - \tfrac{4}{13}, -3 + 4\left(\tfrac{4}{13}\right) \right) = \left( \tfrac{77}{13}, \tfrac{48}{13}, -\tfrac{23}{13} \right)$$

**127.** The direction vector of the line $L$ through $(1, -3, 1)$ and $(3, -4, 2)$ is $\mathbf{v} = \langle 2, -1, 1 \rangle$.

The parametric equations for $L$ are $x = 1 + 2t$, $y = -3 - t$, $z = 1 + t$.

Substituting these equations into the equation of the plane gives

$$(1 + 2t) - (-3 - t) + (1 + t) = 2$$
$$4t = -3$$
$$t = -\tfrac{3}{4}.$$

Point of intersection:

$$\left(1 + 2\left(-\tfrac{3}{4}\right), -3 + \tfrac{3}{4}, 1 - \tfrac{3}{4}\right) = \left(-\tfrac{1}{2}, -\tfrac{9}{4}, \tfrac{1}{4}\right)$$

**129.** True

**131.** True

**133.** False. Planes $7x + y - 11z = 5$ and $5x + 2y - 4z = 1$ are both perpendicular to plane $2x - 3y + z = 3$, but are not parallel.

# Section 11.6   Surfaces in Space

**1.** Ellipsoid

Matches graph (c)

**3.** Hyperboloid of one sheet

Matches graph (f)

**5.** Elliptic paraboloid

Matches graph (d)

**7.** $y = 5$

Plane is parallel to the $xz$-plane.

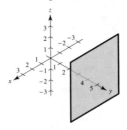

**9.** $y^2 + z^2 = 9$

The $x$-coordinate is missing so you have a right circular cylinder with rulings parallel to the $x$-axis. The generating curve is a circle.

**11.** $y = x^2$

The $z$-coordinate is missing so you have a parabolic cylinder with rulings parallel to the $z$-axis. The generating curve is a parabola.

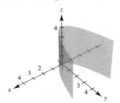

**13.** $4x^2 + y^2 = 4$

$$\frac{x^2}{1} + \frac{y^2}{4} = 1$$

The $z$-coordinate is missing so you have an elliptic cylinder with rulings parallel to the $z$-axis. The generating curve is an ellipse.

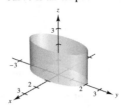

**15.** $z = \sin y$

The $x$-coordinate is missing so you have a cylindrical surface with rulings parallel to the $x$-axis. The generating curve is the sine curve.

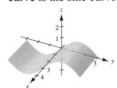

**17.** $z = x^2 + y^2$

  (a) You are viewing the paraboloid from the *x*-axis:
$(20, 0, 0)$

  (b) You are viewing the paraboloid from above, but not on the *z*-axis: $(10, 10, 20)$

  (c) You are viewing the paraboloid from the *z*-axis: $(0, 0, 20)$

  (d) You are viewing the paraboloid from the *y*-axis: $(0, 20, 0)$

**19.** $\dfrac{x^2}{1} + \dfrac{y^2}{4} + \dfrac{z^2}{1} = 1$

Ellipsoid

*xy*-trace: $\dfrac{x^2}{1} + \dfrac{y^2}{4} = 1$ ellipse

*xz*-trace: $x^2 + z^2 = 1$  circle

*yz*-trace: $\dfrac{y^2}{4} + \dfrac{z^2}{1} = 1$ ellipse

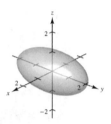

**21.** $16x^2 - y^2 + 16z^2 = 4$

$4x^2 - \dfrac{y^2}{4} + 4z^2 = 1$

Hyperboloid of one sheet

*xy*-trace: $4x^2 - \dfrac{y^2}{4} = 1$  hyperbola

*xz*-trace: $4(x^2 + z^2) = 1$  circle

*yz*-trace: $\dfrac{-y^2}{4} + 4z^2 = 1$ hyperbola

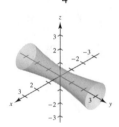

**23.** $4x^2 - y^2 - z^2 = 1$

Hyperboloid of two sheets

*xy*-trace: $4x^2 - y^2 = 1$   hyperbola

*yz*-trace: none

*xz*-trace: $4x^2 - z^2 = 1$   hyperbola

**25.** $x^2 - y + z^2 = 0$

Elliptic paraboloid

*xy*-trace: $y = x^2$

*xz*-trace: $x^2 + z^2 = 0$,
    point $(0, 0, 0)$

*yz*-trace: $y = z^2$

$y = 1$: $x^2 + z^2 = 1$

**27.** $x^2 - y^2 + z = 0$

Hyperbolic paraboloid

*xy*-trace: $y = \pm x$

*xz*-trace: $z = -x^2$

*yz*-trace: $z = y^2$

$y = \pm 1$: $z = 1 - x^2$

**29.** $z^2 = x^2 + \dfrac{y^2}{9}$

Elliptic cone

*xy*-trace: point $(0, 0, 0)$

*xz*-trace: $z = \pm x$

*yz*-trace: $z = \pm \dfrac{y}{3}$

When $z = \pm 1$, $x^2 + \dfrac{y^2}{9} = 1$ ellipse

**31.**    $16x^2 + 9y^2 + 16z^2 - 32x - 36y + 36 = 0$

$16(x^2 - 2x + 1) + 9(y^2 - 4y + 4) + 16z^2 = -36 + 16 + 36$

$16(x - 1)^2 + 9(y - 2)^2 + 16z^2 = 16$

$\dfrac{(x - 1)^2}{1} + \dfrac{(y - 2)^2}{16/9} + \dfrac{z^2}{1} = 1$

Ellipsoid with center $(1, 2, 0)$.

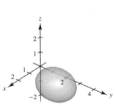

**33.** $z = 2\cos x$

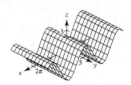

**35.** $z^2 = x^2 + 7.5y^2$

$z = \pm\sqrt{x^2 + 7.5y^2}$

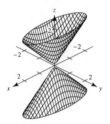

**37.** $x^2 + y^2 = \left(\dfrac{2}{z}\right)^2$

$y = \pm\sqrt{\dfrac{4}{z^2} - x^2}$

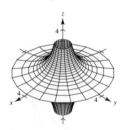

**39.** $z = 10 - \sqrt{|xy|}$

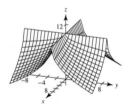

**41.** $6x^2 - 4y^2 + 6z^2 = -36$

$6z^2 = 4y^2 - 6x^2 - 36$

$3z^2 = 2y^2 - 3x^2 - 18$

$z = \pm\dfrac{1}{\sqrt{3}}\sqrt{2y^2 - 3x^2 - 18}$

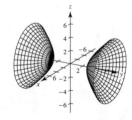

**43.**
$$z = 2\sqrt{x^2 + y^2}$$
$$z = 2$$
$$2\sqrt{x^2 + y^2} = 2$$
$$x^2 + y^2 = 1$$

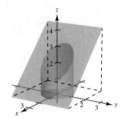

**45.** $x^2 + y^2 = 1$

$x + z = 2$

$z = 0$

**47.** $x^2 + z^2 = \left[r(y)\right]^2$ and $z = r(y) = \pm2\sqrt{y}$; so,

$x^2 + z^2 = 4y$.

**49.** $x^2 + y^2 = \left[r(z)\right]^2$ and $y = r(z) = \dfrac{z}{2}$; so,

$x^2 + y^2 = \dfrac{z^2}{4}$, $4x^2 + 4y^2 = z^2$.

**51.** $y^2 + z^2 = \left[r(x)\right]^2$ and $y = r(x) = \dfrac{2}{x}$; so,

$y^2 + z^2 = \left(\dfrac{2}{x}\right)^2$, $y^2 + z^2 = \dfrac{4}{x^2}$.

**53.** $x^2 + y^2 - 2z = 0$

$x^2 + y^2 = \left(\sqrt{2z}\right)^2$

Equation of generating curve: $y = \sqrt{2z}$ or $x = \sqrt{2z}$

**55.** Let $C$ be a curve in a plane and let $L$ be a line not in a parallel plane. The set of all lines parallel to $L$ and intersecting $C$ is called a cylinder. $C$ is called the generating curve of the cylinder, and the parallel lines are called rulings.

**57.** See pages 814 and 815.

**59.** $V = 2\pi \int_0^4 x(4x - x^2)\, dx$

$= 2\pi \left[ \dfrac{4x^3}{3} - \dfrac{x^4}{4} \right]_0^4 = \dfrac{218\pi}{3}$

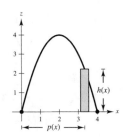

**61.** $z = \dfrac{x^2}{2} + \dfrac{y^2}{4}$

(a) When $z = 2$ we have $2 = \dfrac{x^2}{2} + \dfrac{y^2}{4}$, or

$1 = \dfrac{x^2}{4} + \dfrac{y^2}{8}$

Major axis: $2\sqrt{8} = 4\sqrt{2}$

Minor axis: $2\sqrt{4} = 4$

$c^2 = a^2 - b^2, c^2 = 4, c = 2$

Foci: $(0, \pm 2, 2)$

(b) When $z = 8$ we have $8 = \dfrac{x^2}{2} + \dfrac{y^2}{4}$, or

$1 = \dfrac{x^2}{16} + \dfrac{y^2}{32}$.

Major axis: $2\sqrt{32} = 8\sqrt{2}$

Minor axis: $2\sqrt{16} = 8$

$c^2 = 32 - 16 = 16, c = 4$

Foci: $(0, \pm 4, 8)$

**63.** If $(x, y, z)$ is on the surface, then

$(y + 2)^2 = x^2 + (y - 2)^2 + z^2$

$y^2 + 4y + 4 = x^2 + y^2 - 4y + 4 + z^2$

$x^2 + z^2 = 8y$

Elliptic paraboloid

Traces parallel to $xz$-plane are circles.

**65.** $\dfrac{x^2}{3963^2} + \dfrac{y^2}{3963^2} + \dfrac{z^2}{3950^2} = 1$

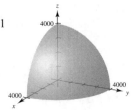

**67.** $z = \dfrac{y^2}{b^2} - \dfrac{x^2}{a^2}, z = bx + ay$

$bx + ay = \dfrac{y^2}{b^2} - \dfrac{x^2}{a^2}$

$\dfrac{1}{a^2}\left( x^2 + a^2bx + \dfrac{a^4b^2}{4} \right) = \dfrac{1}{b^2}\left( y^2 - ab^2y + \dfrac{a^2b^4}{4} \right)$

$\dfrac{\left( x + \dfrac{a^2b}{2} \right)^2}{a^2} = \dfrac{\left( y - \dfrac{ab^2}{2} \right)^2}{b^2}$

$y = \pm\dfrac{b}{a}\left( x + \dfrac{a^2b}{2} \right) + \dfrac{ab^2}{2}$

Letting $x = at$, you obtain the two intersecting lines
$x = at, \ y = -bt, \ z = 0$ and $x = at,$
$y = bt + ab^2, \ z = 2abt + a^2b^2$.

**69.** True. A sphere is a special case of an ellipsoid (centered at origin, for example)

$\dfrac{x^2}{a^2} + \dfrac{y^2}{b^2} + \dfrac{z^2}{c^2} = 1$

having $a = b = c$.

**71.** False. The trace $x = 2$ of the ellipsoid

$\dfrac{x^2}{4} + \dfrac{y^2}{9} + z^2 = 1$ is the point $(2, 0, 0)$.

**73.** The Klein bottle *does not* have both an "inside" and an "outside." It is formed by inserting the small open end through the side of the bottle and making it contiguous with the top of the bottle.

## Section 11.7    Cylindrical and Spherical Coordinates

**1.** $(-7, 0, 5)$, cylindrical

$x = r\cos\theta = -7\cos 0 = -7$

$y = r\sin\theta = -7\sin 0 = 0$

$z = 5$

$(-7, 0, 5)$, rectangular

**3.** $\left( 3, \dfrac{\pi}{4}, 1 \right)$, cylindrical

$x = 3\cos\dfrac{\pi}{4} = \dfrac{3\sqrt{2}}{2}$

$y = 3\sin\dfrac{\pi}{4} = \dfrac{3\sqrt{2}}{2}$

$z = 1$

$\left( \dfrac{3\sqrt{2}}{2}, \dfrac{3\sqrt{2}}{2}, 1 \right)$, rectangular

**5.** $\left(4, \dfrac{7\pi}{6}, 3\right)$, cylindrical

$$x = 4 \cos \frac{7\pi}{6} = -2\sqrt{3}$$

$$y = 4 \sin \frac{7\pi}{6} = -2$$

$$z = 3$$

$\left(-2\sqrt{3}, -2, 3\right)$, rectangular

**7.** $(0, 5, 1)$, rectangular

$$r = \sqrt{(0)^2 + (5)^2} = 5$$

$$\theta = \arctan \frac{5}{0} = \frac{\pi}{2}$$

$$z = 1$$

$\left(5, \dfrac{\pi}{2}, 1\right)$, cylindrical

**9.** $(2, -2, -4)$, rectangular

$$r = \sqrt{2^2 + (-2)^2} = 2\sqrt{2}$$

$$\theta = \arctan(-1) = -\frac{\pi}{4}$$

$$z = -4$$

$\left(2\sqrt{2}, -\dfrac{\pi}{4}, -4\right)$, cylindrical

**11.** $\left(1, \sqrt{3}, 4\right)$, rectangular

$$r = \sqrt{1^2 + \left(\sqrt{3}\right)^2} = 2$$

$$\theta = \arctan\sqrt{3} = \frac{\pi}{3}$$

$$z = 4$$

$\left(2, \dfrac{\pi}{3}, 4\right)$, cylindrical

**13.** $z = 4$ is the equation in cylindrical coordinates. (plane)

**15.** $x^2 + y^2 + z^2 = 17$, rectangular equation

$r^2 + z^2 = 17$, cylindrical equation

**17.** $y = x^2$, rectangular equation

$$r \sin \theta = (r \cos \theta)^2$$

$$\sin \theta = r \cos^2 \theta$$

$$r = \sec \theta \cdot \tan \theta, \text{ cylindrical equation}$$

**19.** $y^2 = 10 - z^2$, rectangular equation

$$(r \sin \theta)^2 = 10 - z^2$$

$r^2 \sin^2 \theta + z^2 = 10$, cylindrical equation

**21.** $r = 3$

$$\sqrt{x^2 + y^2} = 3$$

$$x^2 + y^2 = 9$$

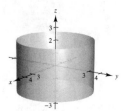

**23.** $\theta = \dfrac{\pi}{6}$

$$\tan \frac{\pi}{6} = \frac{y}{x}$$

$$\frac{1}{\sqrt{3}} = \frac{y}{x}$$

$$x = \sqrt{3}y$$

$$x - \sqrt{3}y = 0$$

**25.** $r^2 + z^2 = 5$

$$x^2 + y^2 + z^2 = 5$$

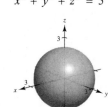

**27.** $r = 2 \sin \theta$

$$r^2 = 2r \sin \theta$$

$$x^2 + y^2 = 2y$$

$$x^2 + y^2 - 2y = 0$$

$$x^2 + (y - 1)^2 = 1$$

**29.** $(4, 0, 0)$, rectangular

$$\rho = \sqrt{4^2 + 0^2 + 0^2} = 4$$

$$\tan \theta = \frac{y}{x} = 0 \Rightarrow \theta = 0$$

$$\phi = \arccos 0 = \frac{\pi}{2}$$

$\left(4, 0, \dfrac{\pi}{2}\right)$, spherical

**31.** $\left(-2, 2\sqrt{3}, 4\right)$, rectangular

$$\rho = \sqrt{(-2)^2 + \left(2\sqrt{3}\right)^2 + 4^2} = 4\sqrt{2}$$

$$\tan \theta = \frac{y}{x} = \frac{2\sqrt{3}}{-2} = -\sqrt{3}$$

$$\theta = \frac{2\pi}{3}$$

$$\phi = \arccos\frac{1}{\sqrt{2}} = \frac{\pi}{4}$$

$\left(4\sqrt{2}, \frac{2\pi}{3}, \frac{\pi}{4}\right)$, spherical

**33.** $\left(\sqrt{3}, 1, 2\sqrt{3}\right)$, rectangular

$$\rho = \sqrt{3 + 1 + 12} = 4$$

$$\tan \theta = \frac{y}{x} = \frac{1}{\sqrt{3}}$$

$$\theta = \frac{\pi}{6}$$

$$\phi = \arccos\frac{\sqrt{3}}{2} = \frac{\pi}{6}$$

$\left(4, \frac{\pi}{6}, \frac{\pi}{6}\right)$, spherical

**35.** $\left(4, \frac{\pi}{6}, \frac{\pi}{4}\right)$, spherical

$$x = 4 \sin\frac{\pi}{4} \cos\frac{\pi}{6} = \sqrt{6}$$

$$y = 4 \sin\frac{\pi}{4} \sin\frac{\pi}{6} = \sqrt{2}$$

$$z = 4 \cos\frac{\pi}{4} = 2\sqrt{2}$$

$\left(\sqrt{6}, \sqrt{2}, 2\sqrt{2}\right)$, rectangular

**37.** $\left(12, -\frac{\pi}{4}, 0\right)$, spherical

$$x = 12 \sin 0 \cos\left(-\frac{\pi}{4}\right) = 0$$

$$y = 12 \sin 0 \sin\left(-\frac{\pi}{4}\right) = 0$$

$$z = 12 \cos 0 = 12$$

$\left(0, 0, 12\right)$, rectangular

**39.** $\left(5, \frac{\pi}{4}, \frac{3\pi}{4}\right)$, spherical

$$x = 5 \sin\frac{3\pi}{4} \cos\frac{\pi}{4} = \frac{5}{2}$$

$$y = 5 \sin\frac{3\pi}{4} \sin\frac{\pi}{4} = \frac{5}{2}$$

$$z = 5 \cos\frac{3\pi}{4} = -\frac{5\sqrt{2}}{2}$$

$\left(\frac{5}{2}, \frac{5}{2}, -\frac{5\sqrt{2}}{2}\right)$, rectangular

**41.** $y = 2$, rectangular equation

$$\rho \sin \phi \sin \theta = 2$$

$$\rho = 2 \csc \phi \csc \theta, \text{ spherical equation}$$

**43.** $x^2 + y^2 + z^2 = 49$, rectangular equation

$$\rho^2 = 49$$

$$\rho = 7, \quad \text{spherical equation}$$

**45.** $x^2 + y^2 = 16$, rectangular equation

$$\rho^2 \sin^2 \phi \sin^2 \theta + \rho^2 \sin^2 \phi \cos^2 \theta = 16$$

$$\rho^2 \sin^2 \phi\left(\sin^2 \theta + \cos^2 \theta\right) = 16$$

$$\rho^2 \sin^2 \phi = 16$$

$$\rho \sin \phi = 4$$

$$\rho = 4 \csc \phi, \text{ spherical equation}$$

**47.**

$$x^2 + y^2 = 2z^2, \text{ rectangular equation}$$

$$\rho^2 \sin^2 \phi \cos^2 \theta + \rho^2 \sin^2 \phi \sin^2 \theta = 2\rho^2 \cos^2 \phi$$

$$\rho^2 \sin^2 \phi\left[\cos^2 \theta + \sin^2 \theta\right] = 2\rho^2 \cos^2 \phi$$

$$\rho^2 \sin^2 \phi = 2\rho^2 \cos^2 \theta$$

$$\frac{\sin^2 \phi}{\cos^2 \phi} = 2$$

$$\tan^2 \phi = 2$$

$$\tan \phi = \pm\sqrt{2}, \text{ spherical equation}$$

**49.** $\rho = 5$

$x^2 + y^2 + z^2 = 25$

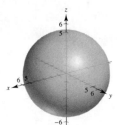

**51.** $\phi = \dfrac{\pi}{6}$

$$\cos \phi = \frac{z}{\sqrt{x^2 + y^2 + z^2}}$$

$$\frac{\sqrt{3}}{2} = \frac{z}{\sqrt{x^2 + y^2 + z^2}}$$

$$\frac{3}{4} = \frac{z^2}{x^2 + y^2 + z^2}$$

$$3x^2 + 3y^2 - z^2 = 0, z \ge 0$$

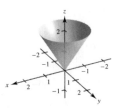

**53.** $\rho = 4 \cos \phi$

$$\sqrt{x^2 + y^2 + z^2} = \frac{4z}{\sqrt{x^2 + y^2 + z^2}}$$

$$x^2 + y^2 + z^2 - 4z = 0$$

$$x^2 + y^2 + (z - 2)^2 = 4, z \ge 0$$

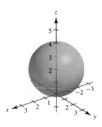

**55.** $\rho = \csc \phi$

$$\rho \sin \phi = 1$$

$$\sqrt{x^2 + y^2} = 1$$

$$x^2 + y^2 = 1$$

**57.** $\left(4, \dfrac{\pi}{4}, 0\right)$, cylindrical

$$\rho = \sqrt{4^2 + 0^2} = 4$$

$$\theta = \frac{\pi}{4}$$

$$\phi = \arccos 0 = \frac{\pi}{2}$$

$$\left(4, \frac{\pi}{4}, \frac{\pi}{2}\right), \text{ spherical}$$

**59.** $\left(4, \dfrac{\pi}{2}, 4\right)$, cylindrical

$$\rho = \sqrt{4^2 + 4^2} = 4\sqrt{2}$$

$$\theta = \frac{\pi}{2}$$

$$\phi = \arccos\left(\frac{4}{4\sqrt{2}}\right) = \frac{\pi}{4}$$

$$\left(4\sqrt{2}, \frac{\pi}{2}, \frac{\pi}{4}\right), \text{ spherical}$$

**61.** $\left(4, -\dfrac{\pi}{6} -\dfrac{\pi}{6}, 6\right)$, cylindrical

$$\rho = \sqrt{4^2 + 6^2} = 2\sqrt{13}$$

$$\theta = -\frac{\pi}{6}$$

$$\phi = \arccos \frac{3}{\sqrt{13}}$$

$$\left(2\sqrt{13}, -\frac{\pi}{6}, \arccos \frac{3}{\sqrt{13}}\right), \text{ spherical}$$

**63.** $(12, \pi, 5)$, cylindrical

$$\rho = \sqrt{12^2 + 5^2} = 13$$

$$\theta = \pi$$

$$\phi = \arccos \frac{5}{13}$$

$$\left(13, \pi, \arccos \frac{5}{13}\right), \text{ spherical}$$

**65.** $\left(10, \dfrac{\pi}{6}, \dfrac{\pi}{2}\right)$, spherical

$$r = 10 \sin \frac{\pi}{2} = 10$$

$$\theta = \frac{\pi}{6}$$

$$z = 10 \cos \frac{\pi}{2} = 0$$

$$\left(10, \frac{\pi}{6}, 0\right), \text{ cylindrical}$$

**67.** $\left(36, \pi, \dfrac{\pi}{2}\right)$, spherical

$r = \rho \sin \phi = 36 \sin \dfrac{\pi}{2} = 36$

$\theta = \pi$

$z = \rho \cos \phi = 36 \cos \dfrac{\pi}{2} = 0$

$(36, \pi, 0)$, cylindrical

**69.** $\left(6, -\dfrac{\pi}{6}, \dfrac{\pi}{3}\right)$, spherical

$r = 6 \sin \dfrac{\pi}{3} = 3\sqrt{3}$

$\theta = -\dfrac{\pi}{6}$

$z = 6 \cos \dfrac{\pi}{3} = 3$

$\left(3\sqrt{3}, -\dfrac{\pi}{6}, 3\right)$, cylindrical

**71.** $\left(8, \dfrac{7\pi}{6}, \dfrac{\pi}{6}\right)$, spherical

$r = 8 \sin \dfrac{\pi}{6} = 4$

$\theta = \dfrac{7\pi}{6}$

$z = 8 \cos \dfrac{\pi}{6} = \dfrac{8\sqrt{3}}{2}$

$\left(4, \dfrac{7\pi}{6}, 4\sqrt{3}\right)$, cylindrical

| *Rectangular* | *Cylindrical* | *Spherical* |
|---|---|---|
| **73.** $(4, 6, 3)$ | $(7.211, 0.983, 3)$ | $(7.810, 0.983, 1.177)$ |
| **75.** $(4.698, 1.710, 8)$ | $\left(5, \dfrac{\pi}{9}, 8\right)$ | $(9.434, 0.349, 0.559)$ |
| **77.** $(-7.071, 12.247, 14.142)$ | $(14.142, 2.094, 14.142)$ | $\left(20, \dfrac{2\pi}{3}, \dfrac{\pi}{4}\right)$ |
| **79.** $(3, -2, 2)$ | $(3.606, -0.588, 2)$ | $(4.123, -0.588, 1.064)$ |
| **81.** $\left(\dfrac{5}{2}, \dfrac{4}{3}, -\dfrac{3}{2}\right)$ | $(2.833, 0.490, -1.5)$ | $(3.206, 0.490, 2.058)$ |
| **83.** $(-3.536, 3.536, -5)$ | $\left(5, \dfrac{3\pi}{4}, -5\right)$ | $(7.071, 2.356, 2.356)$ |
| **85.** $(2.804, -2.095, 6)$ | $(-3.5, 2.5, 6)$ | $(6.946, 5.642, 0.528)$ |

$\Big[$**Note:** Use the cylindrical coordinates $(3.5, 5.642, 6)\Big]$

| | | |
|---|---|---|
| **87.** $(-1.837, 1.837, 1.5)$ | $(2.598, 2.356, 1.5)$ | $\left(3, \dfrac{3\pi}{4}, \dfrac{\pi}{3}\right)$ |

**89.** $r = 5$

Cylinder

Matches graph (d)

**91.** $\rho = 5$

Sphere

Matches graph (c)

**93.** $r^2 = z,\ x^2 + y^2 = z$

Paraboloid

Matches graph (f )

**95.** Rectangular to cylindrical: $r^2 = x^2 + y^2$

$$\tan \theta = \frac{y}{x}$$

$$z = z$$

Cylindrical to rectangular: $x = r \cos \theta$

$$y = r \sin \theta$$

$$z = z$$

**97.** Rectangular to spherical: $\rho^2 = x^2 + y^2 + z^2$

$$\tan \theta = \frac{y}{x}$$

$$\phi = \arccos\left(\frac{z}{\sqrt{x^2 + y^2 + z^2}}\right)$$

Spherical to rectangular: $x = \rho \sin \phi \cos \theta$

$$y = \rho \sin \phi \sin \theta$$

$$z = \rho \cos \phi$$

**99.** $x^2 + y^2 + z^2 = 25$

(a) $r^2 + z^2 = 25$

(b) $\rho^2 = 25 \Rightarrow \rho = 5$

**101.** $x^2 + y^2 + z^2 - 2z = 0$

(a) $r^2 + z^2 - 2z = 0 \Rightarrow r^2 + (z - 1)^2 = 1$

(b) $\rho^2 - 2\rho \cos \phi = 0$

$\rho(\rho - 2 \cos \phi) = 0$

$$\rho = 2 \cos \phi$$

**103.** $x^2 + y^2 = 4y$

(a) $r^2 = 4r \sin \theta, r = 4 \sin \theta$

(b) $\rho^2 \sin^2 \phi = 4\rho \sin \phi \sin \theta$

$\rho \sin \phi(\rho \sin \phi - 4 \sin \theta) = 0$

$$\rho = \frac{4 \sin \theta}{\sin \phi}$$

$$\rho = 4 \sin \theta \csc \phi$$

**105.** $x^2 - y^2 = 9$

(a) $r^2 \cos^2 \theta - r^2 \sin^2 \theta = 9$

$$r^2 = \frac{9}{\cos^2 \theta - \sin^2 \theta}$$

(b) $\rho^2 \sin^2 \phi \cos^2 \theta - \rho^2 \sin^2 \phi \sin^2 \theta = 9$

$$\rho^2 \sin^2 \phi = \frac{9}{\cos^2 \theta - \sin^2 \theta}$$

$$\rho^2 = \frac{9 \csc^2 \phi}{\cos^2 \theta - \sin^2 \theta}$$

**107.** $0 \le \theta \le \dfrac{\pi}{2}$

$0 \le r \le 2$

$0 \le z \le 4$

**109.** $0 \le \theta \le 2\pi$

$0 \le r \le a$

$r \le z \le a$

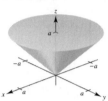

**111.** $0 \le \theta \le 2\pi$

$0 \le \phi \le \dfrac{\pi}{6}$

$0 \le \rho \le a \sec \phi$

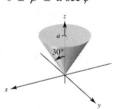

**113.** $0 \le \theta \le \dfrac{\pi}{2}$

$0 \le \phi \le \dfrac{\pi}{2}$

$0 \le \rho \le 2$

**115.** Rectangular

$$0 \le x \le 10$$
$$0 \le y \le 10$$
$$0 \le z \le 10$$

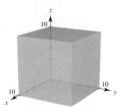

**117.** Spherical

$$4 \le \rho \le 6$$

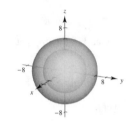

**119.** Cylindrical coordinates:

$$r^2 + z^2 \le 9,$$
$$r \le 3 \cos \theta, 0 \le \theta \le \pi$$

**121.** False. $r = z \Rightarrow x^2 + y^2 = z^2$ is a cone.

**123.** False. $(r, \theta, z) = (0, 0, 1)$ and $(r, \theta, z) = (0, \pi, 1)$
represent the same point $(x, y, z) = (0, 0, 1)$.

**125.** $z = \sin \theta, r = 1$

$$z = \sin \theta = \frac{y}{r} = \frac{y}{1} = y$$

The curve of intersection is the ellipse formed by the
intersection of the plane $z = y$ and the cylinder $r = 1$.

# Review Exercises for Chapter 11

**1.** $P = (1, 2), Q = (4, 1), R = (5, 4)$

  (a) $\mathbf{u} = \overline{PQ} = \langle 4 - 1, 1 - 2 \rangle = \langle 3, -1 \rangle$

      $\mathbf{v} = \overline{PR} = \langle 5 - 1, 4 - 2 \rangle = \langle 4, 2 \rangle$

  (b) $\mathbf{u} = 3\mathbf{i} - \mathbf{j}$

  (c) $\|\mathbf{v}\| = \sqrt{4^2 + 2^2} = \sqrt{20} = 2\sqrt{5}$

  (d) $2\mathbf{u} + \mathbf{v} = 2\langle 3, -1 \rangle + \langle 4, 2 \rangle = \langle 10, 0 \rangle$

**3.** $\mathbf{v} = \|\mathbf{v}\|(\cos \theta \, \mathbf{i} + \sin \theta \, \mathbf{j})$

    $= 8(\cos 60° \, \mathbf{i} + \sin 60° \, \mathbf{j})$

    $= 8\left(\frac{1}{2}\mathbf{i} + \frac{\sqrt{3}}{2}\mathbf{j}\right) = 4\mathbf{i} + 4\sqrt{3}\,\mathbf{j} = \langle 4, 4\sqrt{3} \rangle$

**5.** $z = 0, y = 4, x = -5: (-5, 4, 0)$

**7.** Looking down from the positive $x$-axis towards the
$yz$-plane, the point is either in the first quadrant
$(y > 0, z > 0)$ or in the third quadrant $(y < 0, z < 0)$.
The $x$-coordinate can be any number.

**9.** $(x - 3)^2 + (y + 2)^2 + (z - 6)^2 = \left(\frac{15}{2}\right)^2$

**11.** $(x^2 - 4x + 4) + (y^2 - 6y + 9) + z^2 = -4 + 4 + 9$

    $(x - 2)^2 + (y - 3)^3 + z^2 = 9$

Center: $(2, 3, 0)$

Radius: 3

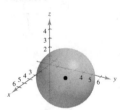

**13.** (a), (d)

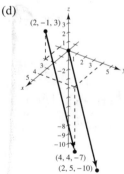

  (b) $\mathbf{v} = \langle 4 - 2, 4 - (-1), -7 - 3 \rangle = \langle 2, 5, -10 \rangle$

  (c) $\mathbf{v} = 2\mathbf{i} + 5\mathbf{j} - 10\mathbf{k}$

**15.** $\mathbf{v} = \langle -1 - 3, 6 - 4, 9 + 1 \rangle = \langle -4, 2, 10 \rangle$

$\mathbf{w} = \langle 5 - 3, 3 - 4, -6 + 1 \rangle = \langle 2, -1, -5 \rangle$

Because $-2\mathbf{w} = \mathbf{v}$, the points lie in a straight line.

**17.** Unit vector: $\dfrac{\mathbf{u}}{\|\mathbf{u}\|} = \left\langle \dfrac{2, 3, 5}{\sqrt{38}} \right\rangle = \left\langle \dfrac{2}{\sqrt{38}}, \dfrac{3}{\sqrt{38}}, \dfrac{5}{\sqrt{38}} \right\rangle$

**19.** $P = \langle 5, 0, 0 \rangle, Q = \langle 4, 4, 0 \rangle, R = \langle 2, 0, 6 \rangle$

(a) $\mathbf{u} = \overrightarrow{PQ} = \langle -1, 4, 0 \rangle$

$\mathbf{v} = \overrightarrow{PR} = \langle -3, 0, 6 \rangle$

(b) $\mathbf{u} \cdot \mathbf{v} = (-1)(-3) + 4(0) + 0(6) = 3$

(c) $\mathbf{v} \cdot \mathbf{v} = 9 + 36 = 45$

**21.** $\mathbf{u} = \langle 7, -2, 3 \rangle, \mathbf{v} = \langle -1, 4, 5 \rangle$

Because $\mathbf{u} \cdot \mathbf{v} = 0$, the vectors are orthogonal.

**23.** $\mathbf{u} = 5\left( \cos \dfrac{3\pi}{4} \mathbf{i} + \sin \dfrac{3\pi}{4} \mathbf{j} \right) = \dfrac{5\sqrt{2}}{2}[-\mathbf{i} + \mathbf{j}]$

$\mathbf{v} = 2\left( \cos \dfrac{2\pi}{3} \mathbf{i} + \sin \dfrac{2\pi}{3} \mathbf{j} \right) = -\mathbf{i} + \sqrt{3} \mathbf{j}$

$\mathbf{u} \cdot \mathbf{v} = \dfrac{5\sqrt{2}}{2}\left( 1 + \sqrt{3} \right)$

$\|\mathbf{u}\| = 5 \qquad \|\mathbf{v}\| = 2$

$\cos \theta = \dfrac{|\mathbf{u} \cdot \mathbf{v}|}{\|\mathbf{u}\|\|\mathbf{v}\|} = \dfrac{\left( 5\sqrt{2}/2 \right)\left( 1 + \sqrt{3} \right)}{5(2)} = \dfrac{\sqrt{2} + \sqrt{6}}{4}$

$\theta = \arccos \dfrac{\sqrt{2} + \sqrt{6}}{4} = 15° \left[ \text{or, } \dfrac{3\pi}{4} \cdot \dfrac{2\pi}{3} = \dfrac{\pi}{12} \text{ or } 15° \right]$

**25.** $\mathbf{u} = \langle 10, -5, 15 \rangle, \mathbf{v} = \langle -2, 1, -3 \rangle$

$\mathbf{u} = -5\mathbf{v} \Rightarrow \mathbf{u}$ is parallel to $\mathbf{v}$ and in the opposite direction.

$\theta = \pi$

**27.** There are many correct answers.

For example: $\mathbf{v} = \pm\langle 6, -5, 0 \rangle$.

In Exercises 29–37, $\mathbf{u} = \langle 3, -2, 1 \rangle, \mathbf{v} = \langle 2, -4, -3 \rangle, \mathbf{w} = \langle -1, 2, 2 \rangle$.

**29.** $\mathbf{u} \cdot \mathbf{u} = 3(3) + (-2)(-2) + (1)(1) = 14 = \left( \sqrt{14} \right)^2 = \|\mathbf{u}\|^2$

**31.** $\text{proj}_{\mathbf{u}} \mathbf{w} = \left( \dfrac{\mathbf{u} \cdot \mathbf{w}}{\|\mathbf{u}\|^2} \right) \mathbf{u} = -\dfrac{5}{14}\langle 3, -2, 1 \rangle$

$= \left\langle -\dfrac{15}{14}, \dfrac{10}{14}, -\dfrac{5}{14} \right\rangle = \left\langle -\dfrac{15}{14}, \dfrac{5}{7}, -\dfrac{5}{14} \right\rangle$

**33.** $\mathbf{n} = \mathbf{v} \times \mathbf{w} = \begin{bmatrix} \mathbf{i} & \mathbf{j} & \mathbf{k} \\ 2 & -4 & -3 \\ -1 & 2 & 2 \end{bmatrix} = -2\mathbf{i} - \mathbf{j}$

$\|\mathbf{n}\| = \sqrt{5}$

$\dfrac{\mathbf{n}}{\|\mathbf{n}\|} = \dfrac{1}{\sqrt{5}}(-2\mathbf{i} - \mathbf{j}), \text{ unit vector or } \dfrac{1}{\sqrt{5}}(2\mathbf{i} + \mathbf{j})$

**35.** $V = \left| \mathbf{u} \cdot (\mathbf{v} \times \mathbf{w}) \right| = \left| \langle 3, -2, 1 \rangle \cdot \langle -2, -1, 0 \rangle \right| = |-4| = 4$

**37.** Area parallelogram $= \|\mathbf{u} \times \mathbf{v}\| = \|\langle 10, 11, -8 \rangle\| = \sqrt{10^2 + 11^2 + (-8)^2}$ (See Exercises 34, 36)

$= \sqrt{285}$

**39.** $\mathbf{F} = c(\cos 20°\mathbf{j} + \sin 20°\mathbf{k})$

$\overline{PQ} = 2\mathbf{k}$

$\overline{PQ} \times \mathbf{F} = \begin{vmatrix} \mathbf{i} & \mathbf{j} & \mathbf{k} \\ 0 & 0 & 2 \\ 0 & c\cos 20° & c\sin 20° \end{vmatrix} = -2c\cos 20°\mathbf{i}$

$200 = \left\|\overline{PQ} \times \mathbf{F}\right\| = 2c\cos 20°$

$c = \dfrac{100}{\cos 20°}$

$\mathbf{F} = \dfrac{100}{\cos 20°}(\cos 20°\mathbf{j} + \sin 20°\mathbf{k}) = 100(\mathbf{j} + \tan 20°\mathbf{k})$

$\|\mathbf{F}\| = 100\sqrt{1 + \tan^2 20°} = 100\sec 20° \approx 106.4 \text{ lb}$

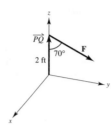

**41.** $\mathbf{v} = \langle 9 - 3, 11 - 0, 6 - 2\rangle = \langle 6, 11, 4\rangle$

(a) Parametric equations:
$x = 3 + 6t,\ y = 11t,\ z = 2 + 4t$

(b) Symmetric equations: $\dfrac{x - 3}{6} = \dfrac{y}{11} = \dfrac{z - 2}{4}$

**43.** $\mathbf{v} = \mathbf{j}$

(a) $x = 1,\ y = 2 + t,\ z = 3$

(b) None

(c)

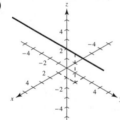

**45.** $3x - 3y - 7z = -4,\ x - y + 2z = 3$

Solving simultaneously, you have $z = 1$. Substituting $z = 1$ into the second equation, you have $y = x - 1$. Substituting for $x$ in this equation you obtain two points on the line of intersection, $(0, -1, 1),\ (1, 0, 1)$. The direction vector of the line of intersection is $\mathbf{v} = \mathbf{i} + \mathbf{j}$.

(a) $x = t,\ y = -1 + t,\ z = 1$

(b) $x = y + 1,\ z = 1$

(c)

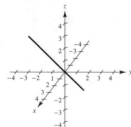

**47.** $P = (-3, -4, 2),\ Q = (-3, 4, 1),\ R = (1, 1, -2)$

$\overline{PQ} = \langle 0, 8, -1\rangle,\ \overline{PR} = [4, 5, -4]$

$\mathbf{n} = \overline{PQ} \times \overline{PR} = \begin{vmatrix} \mathbf{i} & \mathbf{j} & \mathbf{k} \\ 0 & 8 & -1 \\ 4 & 5 & -4 \end{vmatrix} = -27\mathbf{i} - 4\mathbf{j} - 32\mathbf{k}$

$-27(x + 3) - 4(y + 4) - 32(z - 2) = 0$

$27x + 4y + 32z = -33$

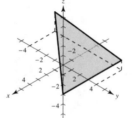

**49.** The two lines are parallel as they have the same direction numbers, $-2, 1, 1$. Therefore, a vector parallel to the plane is $\mathbf{v} = -2\mathbf{i} + \mathbf{j} + \mathbf{k}$. A point on the first line is $(1, 0, -1)$ and a point on the second line is $(-1, 1, 2)$. The vector $\mathbf{u} = 2\mathbf{i} - \mathbf{j} - 3\mathbf{k}$ connecting these two points is also parallel to the plane. Therefore, a normal to the plane is

$\mathbf{v} \times \mathbf{u} = \begin{vmatrix} \mathbf{i} & \mathbf{j} & \mathbf{k} \\ -2 & 1 & 1 \\ 2 & -1 & -3 \end{vmatrix}$

$= -2\mathbf{i} - 4\mathbf{j} = -2(\mathbf{i} + 2\mathbf{j}).$

Equation of the plane: $(x - 1) + 2y = 0$

$x + 2y = 1$

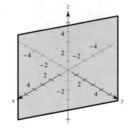

**51.** $Q(1, 0, 2)$ point

$2x - 3y + 6z = 6$

A point $P$ on the plane is $(3, 0, 0)$.

$\overrightarrow{PQ} = \langle -2, 0, 2 \rangle$

$\mathbf{n} = \langle 2, -3, 6 \rangle$ normal to plane

$D = \dfrac{|\overrightarrow{PQ} \cdot \mathbf{n}|}{\|\mathbf{n}\|} = \dfrac{8}{7}$

**53.** The normal vectors to the planes are the same,

$\mathbf{n} = \langle 5, -3, 1 \rangle.$

Choose a point in the first plane $P(0, 0, 2)$. Choose a point in the second plane, $Q(0, 0, -3)$.

$\overrightarrow{PQ} = \langle 0, 0, -5 \rangle$

$D = \dfrac{|\overrightarrow{PQ} \cdot \mathbf{n}|}{\|\mathbf{n}\|} = \dfrac{|-5|}{\sqrt{35}} = \dfrac{5}{\sqrt{35}} = \dfrac{\sqrt{35}}{7}$

**55.** $x + 2y + 3z = 6$

Plane

Intercepts: $(6, 0, 0), (0, 3, 0), (0, 0, 2),$

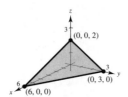

**57.** $y = \dfrac{1}{2}z$

Plane with rulings parallel to the $x$-axis.

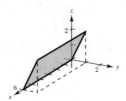

**59.** $\dfrac{x^2}{16} + \dfrac{y^2}{9} + z^2 = 1$

Ellipsoid

$xy$-trace: $\dfrac{x^2}{16} + \dfrac{y^2}{9} = 1$

$xz$-trace: $\dfrac{x^2}{16} + z^2 = 1$

$yz$-trace: $\dfrac{y^2}{9} + z^2 = 1$

**61.** $\dfrac{x^2}{16} - \dfrac{y^2}{9} + z^2 = -1$

$\dfrac{y^2}{9} - \dfrac{x^2}{16} - z^2 = 1$

Hyperboloid of two sheets

$xy$-trace: $\dfrac{y^2}{9} - \dfrac{x^2}{16} = 1$

$xz$-trace: None

$yz$-trace: $\dfrac{y^2}{9} - z^2 = 1$

**63.** $x^2 + z^2 = 4.$

Cylinder of radius 2 about $y$-axis

**65.** Let $y = r(x) = 2\sqrt{x}$ and revolve the curve about the $x$-axis.

**67.** $z^2 = 2y$ revolved about $y$-axis

$z = \pm\sqrt{2y}$

$x^2 + z^2 = [r(y)]^2 = 2y$

$x^2 + z^2 = 2y$

**69.** $\left(-2\sqrt{2}, 2\sqrt{2}, 2\right)$, rectangular

(a) $r = \sqrt{\left(-2\sqrt{2}\right)^2 + \left(2\sqrt{2}\right)^2} = 4,$

$\theta = \arctan(-1) = \dfrac{3\pi}{4}, z = 2,$

$\left(4, \dfrac{3\pi}{4}, 2\right),$ cylindrical

(b) $\rho = \sqrt{\left(-2\sqrt{2}\right)^2 + \left(2\sqrt{2}\right)^2 + (2)^2} = 2\sqrt{5},$

$\theta = \dfrac{3\pi}{4}, \phi = \arccos\dfrac{2}{2\sqrt{5}} = \arccos\dfrac{1}{\sqrt{5}},$

$\left(2\sqrt{5}, \dfrac{3\pi}{4}, \arccos\dfrac{\sqrt{5}}{5}\right),$ spherical

**71.** $\left(100, -\dfrac{\pi}{6}, 50\right)$, cylindrical

$$\rho = \sqrt{100^2 + 50^2} = 50\sqrt{5}$$

$$\theta = -\dfrac{\pi}{6}$$

$$\phi = \arccos\left(\dfrac{50}{50\sqrt{5}}\right) = \arccos\left(\dfrac{1}{\sqrt{5}}\right) \approx 63.4° \text{ or } 1.107$$

$$\left(50\sqrt{5}, -\dfrac{\pi}{6}, 63.4°\right), \text{ sperical or } \left(50\sqrt{5}, -\dfrac{\pi}{6}, 1.1071\right)$$

**73.** $\left(25, -\dfrac{\pi}{4}, \dfrac{3\pi}{4}\right)$, spherical

$$r^2 = \left(25\sin\left(\dfrac{3\pi}{4}\right)\right)^2 \Rightarrow r = \dfrac{25\sqrt{2}}{2}$$

$$\theta = -\dfrac{\pi}{4}$$

$$z = \rho\cos\phi = 25\cos\dfrac{3\pi}{4} = -\dfrac{25\sqrt{2}}{2}$$

$$\left(\dfrac{25\sqrt{2}}{2}, -\dfrac{\pi}{4}, -\dfrac{25\sqrt{2}}{2}\right), \text{ cylindrical}$$

**75.** $x^2 - y^2 = 2z$

  (a) Cylindrical:

$$r^2\cos^2\theta - r^2\sin^2\theta = 2z \Rightarrow r^2\cos 2\theta = 2z$$

  (b) Spherical:

$$\rho^2\sin^2\phi\cos^2\theta - \rho^2\sin^2\phi\sin^2\theta = 2\rho\cos\phi$$

$$\rho\sin^2\phi\cos 2\theta - 2\cos\phi = 0$$

$$\rho = 2\sec 2\theta\cos\phi\csc^2\phi$$

# Problem Solving for Chapter 11

**1.**

$$\mathbf{a} + \mathbf{b} + \mathbf{c} = \mathbf{0}$$

$$\mathbf{b} \times (\mathbf{a} + \mathbf{b} + \mathbf{c}) = \mathbf{0}$$

$$(\mathbf{b} \times \mathbf{a}) + (\mathbf{b} \times \mathbf{c}) = \mathbf{0}$$

$$\|\mathbf{a} \times \mathbf{b}\| = \|\mathbf{b} \times \mathbf{c}\|$$

$$\|\mathbf{b} \times \mathbf{c}\| = \|\mathbf{b}\|\|\mathbf{c}\|\sin A$$

$$\|\mathbf{a} \times \mathbf{b}\| = \|\mathbf{a}\|\|\mathbf{b}\|\sin C$$

Then, $\dfrac{\sin A}{\|\mathbf{a}\|} = \dfrac{\|\mathbf{b} \times \mathbf{c}\|}{\|\mathbf{a}\|\|\mathbf{b}\|\|\mathbf{c}\|} = \dfrac{\|\mathbf{a} \times \mathbf{b}\|}{\|\mathbf{a}\|\|\mathbf{b}\|\|\mathbf{c}\|} = \dfrac{\sin C}{\|\mathbf{c}\|}.$

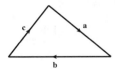

The other case, $\dfrac{\sin A}{\|\mathbf{a}\|} = \dfrac{\sin B}{\|\mathbf{b}\|}$ is similar.

**77.** $r = 5\cos\theta$, cylindrical equation

$$r^2 = 5r\cos\theta$$

$$x^2 + y^2 = 5x$$

$$x^2 - 5x + \dfrac{25}{4} + y^2 = \dfrac{25}{4}$$

$$\left(x - \dfrac{5}{2}\right)^2 + y^2 = \left(\dfrac{5}{2}\right)^2, \text{ rectangular equation}$$

**79.** $\theta = \dfrac{\pi}{4}$, spherical coordinates

$$\tan\theta = \tan\dfrac{\pi}{4} = 1$$

$$\dfrac{y}{x} = 1$$

$$y = x, x \geq 0, \text{ rectangular coordinates, half-plane}$$

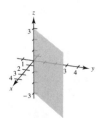

**3.** Label the figure as indicated.

From the figure, you see that

$$\overline{SP} = \dfrac{1}{2}\mathbf{a} - \dfrac{1}{2}\mathbf{b} = \overline{RQ} \text{ and } \overline{SR} = \dfrac{1}{2}\mathbf{a} + \dfrac{1}{2}\mathbf{b} = \overline{PQ}.$$

Because $\overline{SP} = \overline{RQ}$ and $\overline{SR} = \overline{PQ}$,

*PSRQ* is a parallelogram.

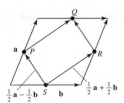

**5. (a)** $\mathbf{u} = \langle 0, 1, 1 \rangle$ is the direction vector of the line determined by $P_1$ and $P_2$.

$$D = \frac{\|\overrightarrow{P_1Q} \times \mathbf{u}\|}{\|\mathbf{u}\|}$$

$$= \frac{\|\langle 2, 0, -1 \rangle \times \langle 0, 1, 1 \rangle\|}{\sqrt{2}}$$

$$= \frac{\|\langle 1, -2, 2 \rangle\|}{\sqrt{2}} = \frac{3}{\sqrt{2}} = \frac{3\sqrt{2}}{2}$$

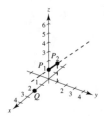

**(b)** The shortest distance to the line **segment** is $\|\overrightarrow{P_1Q}\| = \|\langle 2, 0, -1 \rangle\| = \sqrt{5}$.

**7. (a)** $V = \pi \int_0^1 \left(\sqrt{z}\right)^2 dz = \left[\pi \frac{z^2}{2}\right]_0^1 = \frac{1}{2}\pi$

Note: $\frac{1}{2}$ (base)(altitude) $= \frac{1}{2}\pi(1) = \frac{1}{2}\pi$

**(b)** $\dfrac{x^2}{a^2} + \dfrac{y^2}{b^2} = z$: (slice at $z = c$)

$$\frac{x^2}{\left(\sqrt{c}a\right)^2} + \frac{y^2}{\left(\sqrt{c}b\right)^2} = 1$$

At $z = c$, figure is ellipse of area

$$\pi\left(\sqrt{c}a\right)\left(\sqrt{c}b\right) = \pi abc.$$

$$V = \int_0^k \pi abc \cdot dc = \left[\frac{\pi abc^2}{2}\right]_0^k = \frac{\pi abk^2}{2}$$

**(c)** $V = \dfrac{1}{2}(\pi abk)k = \dfrac{1}{2}$ (area of base)(height)

**9. (a)** $\rho = 2 \sin \phi$

Torus

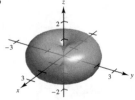

**(b)** $\rho = 2 \cos \phi$

Sphere

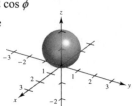

**11.** From Exercise 66, Section 11.4,

$$(\mathbf{u} \times \mathbf{v}) \times (\mathbf{w} \times \mathbf{z}) = \left[(\mathbf{u} \times \mathbf{v}) \cdot \mathbf{z}\right]\mathbf{w} - \left[(\mathbf{u} \times \mathbf{v}) \cdot \mathbf{w}\right]\mathbf{z}.$$

**13. (a)** $\mathbf{u} = \|\mathbf{u}\|(\cos 0\, \mathbf{i} + \sin 0\, \mathbf{j}) = \|\mathbf{u}\|\mathbf{i}$

Downward force $\mathbf{w} = -\mathbf{j}$

$$\mathbf{T} = \|\mathbf{T}\|(\cos(90° + \theta)\mathbf{i} + \sin(90° + \theta)\mathbf{j})$$

$$= \|\mathbf{T}\|(-\sin\theta\, \mathbf{i} + \cos\theta\, \mathbf{j})$$

$$\mathbf{0} = \mathbf{u} + \mathbf{w} + \mathbf{T} = \|\mathbf{u}\|\mathbf{i} - \mathbf{j} + \|\mathbf{T}\|(-\sin\theta\, \mathbf{i} + \cos\theta\, \mathbf{j})$$

$$\|\mathbf{u}\| = \sin\theta\|\mathbf{T}\|$$

$$1 = \cos\theta\|\mathbf{T}\|$$

If $\theta = 30°, \|\mathbf{u}\| = (1/2)\|\mathbf{T}\|$ and $1 = \left(\sqrt{3}/2\right)\|\mathbf{T}\| \Rightarrow \|\mathbf{T}\| = \dfrac{2}{\sqrt{3}} \approx 1.1547$ lb and $\|\mathbf{u}\| = \dfrac{1}{2}\left(\dfrac{2}{\sqrt{3}}\right) \approx 0.5774$ lb

**(b)** From part (a), $\|\mathbf{u}\| = \tan\theta$ and $\|\mathbf{T}\| = \sec\theta$.

Domain: $0 \le \theta \le 90°$

(c)

| $\theta$ | 0° | 10° | 20° | 30° | 40° | 50° | 60° |
|---|---|---|---|---|---|---|---|
| **T** | 1 | 1.0154 | 1.0642 | 1.1547 | 1.3054 | 1.5557 | 2 |
| $\|\mathbf{u}\|$ | 0 | 0.1763 | 0.3640 | 0.5774 | 0.8391 | 1.1918 | 1.7321 |

(d)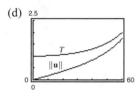

(e) Both are increasing functions.

(f) $\lim\limits_{\theta \to \pi/2^-} T = \infty$ and $\lim\limits_{\theta \to \pi/2^-} \|\mathbf{u}\| = \infty$.

Yes. As $\theta$ increases, both $T$ and $\|\mathbf{u}\|$ increase.

**15.** Let $\theta = \alpha - \beta$, the angle between **u** and **v.** Then

$$\sin(\alpha - \beta) = \frac{\|\mathbf{u} \times \mathbf{v}\|}{\|\mathbf{u}\|\|\mathbf{v}\|} = \frac{\|\mathbf{v} \times \mathbf{u}\|}{\|\mathbf{u}\|\|\mathbf{v}\|}.$$

For $\mathbf{u} = \langle \cos\alpha, \sin\alpha, 0 \rangle$ and $\mathbf{v} = \langle \cos\beta, \sin\beta, 0 \rangle$, $\|\mathbf{u}\| = \|\mathbf{v}\| = 1$ and

$$\mathbf{v} \times \mathbf{u} = \begin{vmatrix} \mathbf{i} & \mathbf{j} & \mathbf{k} \\ \cos\beta & \sin\beta & 0 \\ \cos\alpha & \sin\alpha & 0 \end{vmatrix} = (\sin\alpha\cos\beta - \cos\alpha\sin\beta)\mathbf{k}.$$

So, $\sin(\alpha - \beta) = \|\mathbf{v} \times \mathbf{u}\| = \sin\alpha\cos\beta - \cos\alpha\sin\beta$.

**17.** From Theorem 11.13 and Theorem 11.7 (6) you have

$$D = \frac{\left|\overrightarrow{PQ} \cdot \mathbf{n}\right|}{\|\mathbf{n}\|}$$

$$= \frac{\left|\mathbf{w} \cdot (\mathbf{u} \times \mathbf{v})\right|}{\|\mathbf{u} \times \mathbf{v}\|} = \frac{\left|(\mathbf{u} \times \mathbf{v}) \cdot \mathbf{w}\right|}{\|\mathbf{u} \times \mathbf{v}\|} = \frac{\left|\mathbf{u} \cdot (\mathbf{v} \times \mathbf{w})\right|}{\|\mathbf{u} \times \mathbf{v}\|}.$$

**19.** $x^2 + y^2 = 1$      cylinder

$z = 2y$      plane

Introduce a coordinate system in the plane $z = 2y$.

The new $u$-axis is the original $x$-axis.

The new $v$-axis is the line $z = 2y, x = 0$.

Then the intersection of the cylinder and plane satisfies the equation of an ellipse:

$$x^2 + y^2 = 1$$

$$x^2 + \left(\frac{z}{2}\right)^2 = 1$$

$$x^2 + \frac{z^2}{4} = 1 \qquad \text{ellipse}$$

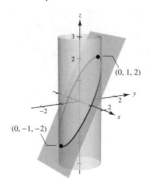